Dictionary of Microbiology

Dictionary of Microbiology

PAUL SINGLETON B.Sc., M.Sc.
DIANA SAINSBURY B.Sc.

A Wiley–Interscience Publication

JOHN WILEY & SONS
Chichester · New York · Brisbane · Toronto

Reprinted August 1980

Library of Congress Cataloging in Publication Data:

Singleton, Paul.
Dictionary of microbiology.

'A Wiley Interscience publication.'
1. Microbiology—Dictionaries. I. Sainsbury, Diana, joint author. II. Title.
QR9.S56 576'.03 78–4532

ISBN 0 471 99658 0

Text set in 9 pt Photon Times, printed and bound in Great Britain at The Pitman Press, Bath

Preface

In preparing the Dictionary of Microbiology we have attempted to produce a compact source of readily-available and up-to-date information for undergraduates and postgraduates in microbiology and for those studying—or otherwise engaged in—any of the wide range of subjects and disciplines for which some knowledge of microbiology is necessary. The Dictionary deals with terms, concepts, techniques, tests, and other topics, and covers over one thousand microbial taxa; the entries—which range from short definitions to descriptions and concise reviews—relate to microbiology (pure and applied), biochemistry, immunology, and genetics, and to the microbiological aspects of allied subjects such as medicine, veterinary science, and plant pathology.

We would like to acknowledge the cooperation of Mr T. F. Lawrence of Baird & Tatlock (London) Ltd. in connection with the line drawing of the Series 225 autoclave; we are also indebted to Carl Zeiss, Oberkochen, West Germany, for the line drawing of the light microscope, and to N. V. Philips' Gloeilampenfabrieken, Eindhoven, Netherlands, for the line drawing of the electron microscope.

Paul Singleton
Diana Sainsbury
London, January 1978.

Notes for the User

Alphabetization. Entries are arranged in alphabetical order on a 'nothing before something' principle—i.e. a space (between words, or between a letter and a word) precedes a letter. For example:

blast disease	**h mutants**	**R factors**
blast transformation	**HAA**	**R mutant**
Blastocladiales	**hadacidin**	**rabies**

Subscripts, superscripts, and numbers affect the order of entries *only* if there is no other difference between the entry headings; apostrophes are ignored for the purposes of alphabetization. For example:

C3 convertase	**ascolichen**
C_{55} lipid carrier	**Ascoli's thermoprecipitin test**
C_{27} organisms	**ascolocular**

When a hyphen connects two complete words, or connects a letter and a complete or incomplete word, the hyphen is regarded as a space. (In many such cases the hyphen is omitted by some authors.) If a hyphenated word can be written alternatively as a single, non-hyphenated word it is regarded as such for the purposes of alphabetization. For example:

C substances	**F′ donor**	**red tide**	**copper**
C-type particles	**F-duction**	**red-water fever**	**co-precipitation**
cabbage, diseases of	**F factor**	**redox potential**	**Coprinaceae**

When a Greek letter forms a *significant* part of an entry heading it is counted as a word; however, if it is a relatively minor qualification it is ignored for the purposes of alphabetization. For example:

gametothallus	**omasum**	**polyhedroses**	**prophylaxis**
***gamma* globulins**	***omega*-oxidation**	**poly-β-hydroxybutyrate**	**β-propiolactone**
***gamma* rays**	**omnivorous**	**poly(I:C)**	***Propionibacterium***
gamone	**oncogene**	**polykaryocytosis**	**propionic acid fermentation**

Certain entry headings defy any logical system of alphabetization. An example of such a heading is **T2H test**; this heading has been arbitrarily regarded as 'T H test' and hence:

T antigen
T-even phages
T2H test
T-independent antigens

Cross references. (These are indicated by SMALL CAPITAL letters.) In order to effect maximum economy of space, information given in any particular entry is seldom repeated elsewhere. In some cases a complete understanding of an entry is dependent on a knowledge of information given in other entries—which are indicated by cross references; in such cases the cross reference(s) frequently form an integral part of a sentence within the text, or may be introduced by 'see ....'. In other cases cross references may be used to link one topic with another or to extend the scope of a given topic; in such cases a cross reference is often placed within brackets and preceded by 'see also' or 'cf.'.

Certain entry headings are followed simply by (see CROSS REFERENCE); this is *not* intended to indicate that synonomy necessarily exists between the term which forms the entry heading and that to which the reader is referred—such referral signifies only that the meaning of the term is given under the heading indicated.

Numbered definitions. In a number of cases a term is used with different meanings by different authors—or it may have different meanings in different contexts; for such terms the various definitions are indicated by (1), (2), (3) etc.—though *all* the definitions which can be ascribed to the term are not necessarily included. The position in which any given definition appears in a list of definitions is *not* intended to reflect in any way the appropriateness or frequency of usage of that definition of the term.

Taxonomy. In general, the characteristics described for a given taxon apply to all subordinate taxa; thus, e.g. the distinguishing characteristics of a given family apply to all the genera within that family—though such characteristics are usually *not* repeated in the separate entries for each genus.

The taxonomic scheme used for the bacteria is based on that given in the 8th edition of *Bergey's Manual of Determinative Bacteriology* (1974). For the fungi the taxonomic scheme is based largely on 'A General Purpose Classification of the Fungi' proposed by G. C. Ainsworth in *Bibliography of Systematic Mycology* part 1 (1966) pages 1–4; the scheme adopted in the dictionary incorporates some of the suggestions given in volumes IVa and IVb of *The Fungi* (editors: G. C. Ainsworth, F. K. Sparrow, and A. S. Sussman)—both volumes published in 1973 by Academic Press. Protozoological classification is based partly on 'A Revised Classification of the Phylum Protozoa' by B. M. Honigberg *et al.* (*J. Protozool.* **11** (1) 7–20, 1964) and partly on more recent literature. A detailed classification of the algae was considered to be beyond the scope of this work; the taxa given are those currently accepted by a number of authors. In general, the animal viruses have been grouped according to the scheme given in *The Biology of Animal Viruses* by F. Fenner *et al.* (Academic Press, 2nd edition, 1974).

A

Å Ångström unit. A unit of length: 10^{-10} metres.

a_w (see WATER ACTIVITY)

A-protein (1) (of the TOBACCO MOSAIC VIRUS) Experimentally produced oligomers of protein coat subunits. (2) (of BACTERIOPHAGE Qβ and other RNA phages) A protein *maturation factor* present in the phage coat; it is distinct from other coat proteins. (cf. PROTEIN A)

A site (of a RIBOSOME) (see PROTEIN SYNTHESIS)

A-type particle (*virol.*) (see B-TYPE PARTICLE)

AAA pathway (see AMINOADIPIC ACID PATHWAY)

AAV *Adenovirus*-associated virus (see PARVOVIRUS).

ab (*immunol.*) Antibody.

abequose (3,6-dideoxy-D-galactose) A sugar which occurs e.g. in the O-specific chains of LIPOPOLYSACCHARIDES in certain serotypes of *Salmonella*. Abequose, first isolated from *S. abortus-equi*, contributes to the specificity of somatic (O) antigen 4 in group B salmonellae.

aberrations, chromosomal (see CHROMOSOME ABERRATION)

abhymenial (*mycol.*) Refers to the surface of a fruiting body which is opposite that of the hymenium, *or* may refer to any surface of a fruiting body *other than* that of the hymenium.

abiogenesis (spontaneous generation) The spontaneous formation of living organisms from non-living material; apart from its application to the evolutionary origin of life, the doctrine of abiogenesis has long since been abandoned.

ablastins (*immunol.*) Antibodies which specifically inhibit microbial reproduction by combining with homologous cell-surface antigens.

abomasum (see RUMEN)

abortive transduction (see TRANSDUCTION)

abortus Bang reaction (abortus Bang ringprobe) The MILK RING TEST.

ABR Abbreviation for *abortus Bang reaction* or *abortus Bang ringprobe*—both of which refer to the MILK RING TEST.

Absidia A genus of fungi of the ZYGOMYCETES.

absorption (*immunol.*) The removal or effective removal of particular antibodies (or antigens) from a given sample (e.g. serum) by the addition of homologous antigens (or antibodies) to that sample; the resulting antigen–antibody complex may or may not be removed from the sample.

acaricide Any chemical agent which kills members of the order Acarina (mites and ticks).

accessory pigments (in PHOTOSYNTHESIS) Pigments which harvest light energy and transfer it to the *reaction centres*. They include CAROTENOIDS, PHYCOBILIPROTEINS and certain CHLOROPHYLLS.

AcCoA Acetyl-coenzyme A (see COENZYME A).

acellular (see CELLULAR)

acellular slime moulds (Myxomycetes, Myxomycetales, 'true' slime moulds) A group of eucaryotic organisms in which the stages of the life cycle include a uninucleate amoeboid cell (the *myxamoeba*) and/or a uninucleate biflagellate (heterokont) *swarm cell*, and a motile multinucleate body, the *plasmodium*. (cf. CELLULAR SLIME MOULDS) The acellular slime moulds—which are usually classified among the fungi—occur in soil and on decomposing vegetation; they feed on bacteria and on small particles of organic detritus.

The life cycle includes a vegetative (feeding) phase in which the plasmodium—a shapeless mass of multinucleate protoplasm (sometimes brightly coloured)—increases in size to several centimetres or to a metre or more; the plasmodial nuclei are diploid, and the protoplasm of the plasmodium exhibits active streaming. Under certain environmental conditions (e.g. starvation *and* darkness) the plasmodium may give rise to a dry, brittle or pliable thick-walled mass, a *sclerotium*, which is a dormant, resistant form; in e.g. *Physarum polycephalum* the sclerotium is divided, internally, into a number of multinucleate sections ('spherules'). Normally, in the presence of light, a starved plasmodium gives rise to one or more *fruiting bodies*—the form of which varies widely among species of acellular slime moulds. In *Physarum polycephalum* the fruiting body consists of a short, cell-free stalk surmounted by a lobed *sporangium*; the sporangium consists of an outer sac, the *peridium*, which contains a number of thick-walled spores—among which ramifies a system of (cellulosic and/or chitinous?) threads, the *capillitium*. Although many species form stalked fruiting bodies, some (e.g. *Lycogalia* spp.) form non-stalked sporangia (*aethelia—sing.*: *aethelium*). *Ceratiomyxa* spp. differ from all other slime moulds in that e.g. the spores are borne *externally* on the fruiting body, i.e. the spores are not enveloped by a peridium. Spore germination gives rise either to a uninucleate myxamoeba or to a uninucleate, biflagellate *swarm cell*—depending on species and environment; for example, germination on a moist substratum may give rise to a myxamoeba, while germination in water may give rise to a swarm cell. In either case, the product of germination is a haploid cell; there is some evidence that meiosis occurs within the pre-formed spore. Myxamoebae may subsequently become swarm cells, and *vice versa*. The life cycle is completed when a pair of myxamoebae, or swarm cells, fuse to form a (diploid) zygote; flagella, if present, are subsequently lost. The zygote grows and develops to form the multinucleate, diploid plasmodium. Genera include *Ceratiomyxa*, *Didymium*, *Fuligo*, *Hemitrichia*, *Lycogala*, *Physarum*, and *Stemonitis*.

acentric (of a chromosome) Having no centromere.

acephaline (*protozool.*) Refers to those members of the Eugregarinida which form aseptate trophozoites.

acervulus An asexual reproductive structure formed by certain plant-pathogenic fungi (see MELANCONIALES); an acervulus is essentially a mat of plectenchymatous tissue—or of plectenchyma and plant tissue—which supports a mass of densely-packed conidiophores. Acervuli commonly develop sub-cuticularly or sub-epidermally in the plant host; when mature, the mass of conidiophores forces an opening in the overlaying layer of host tissue—thus permitting dispersal of the conidia.

Acetabularia A genus of tropical or subtropical marine algae of the division CHLOROPHYTA. The vegetative organism consists of a single cell which is differentiated to form a vertical *stalk* (axis) (up to several centimetres in height) and a branching rhizoid-like structure which anchors the organism to the substratum; the single nucleus is located in one branch of the rhizoid. The cell wall, which is frequently calcified, contains mannan as a major component. As the stalk of the alga grows, whorls of sterile hair-like appendages develop around the distal end; subsequently these appendages are shed leaving rings of scars around the stalk. The nucleus increases in size by up to twenty times but does not divide until 'gametangia' begin to develop; at this point the nucleus undergoes many mitotic divisions to form numerous diploid nuclei. The gametangia occur as a whorl of elongated sac-like structures which (depending on species) may or may not be joined to form a characteristic cup-shaped structure (the *cap*) at the distal end of the stalk. The nuclei migrate from the rhizoid to the gametangia by cytoplasmic streaming; within each gametangium a large number of (diploid) *cysts* are formed. These cysts, which may have calcified walls, are liberated into the sea. Meiosis occurs within the cysts and numerous biflagellate, haploid isogametes are liberated; pairs of gametes fuse to form zygotes which subsequently germinate to form new individuals.

Acetobacter A genus of gram-negative, obligately aerobic, chemoorganotrophic, ovoid or rod-shaped, peritrichously flagellate (or non-motile) non-sporing bacteria of uncertain taxonomic affiliation. Species occur e.g. in a variety of fermented products; typically they oxidize ethanol to acetic acid, and acetic acid to carbon dioxide ('overoxidation'). (cf. ACETOMONAS) Optimum growth temperature: about 30 °C; optimum pH about 6. (see also CARR'S MEDIUM) In a recent classification, *A. aceti*, *A. orleanensis*, *A. liquefaciens*, and the cellulose-producing *A. xylinum* have been regarded as subspecies of one of three species, *A. aceti*. (see also VINEGAR)

acetoin (acetylmethylcarbinol) 2-Keto-3-hydroxybutane. Acetoin is formed e.g. during the BUTANEDIOL FERMENTATION (see also VOGES–PROSKAUER TEST), and in the metabolism of citrate by *Leuconostoc citrovorum* (= *L. cremoris*) (see also BUTTER).

Acetomonas (syn: *Gluconobacter*) A genus of gram-negative, obligately aerobic, chemoorganotrophic, ovoid or rod-shaped, polarly flagellate (or non-motile) non-sporing bacteria of the family Pseudomonadaceae. Species occur e.g. in a variety of fermented products. *Acetomonas* oxidizes ethanol to acetic acid (cf. ACETOBACTER) and glucose to gluconate. Optimum growth temperature: about 25 °C; optimum pH about 5.5. (see also CARR'S MEDIUM)

acetone–butanol fermentation (solvent fermentation) A type of fermentation carried out under certain conditions by certain saccharolytic *Clostridium* species—e.g. *C. acetobutylicum*. Initially, glucose is fermented as in the BUTYRIC ACID FERMENTATION—acetic acid, butyric acid, CO_2 and H_2 being formed. As the acidic products accumulate and the pH of the medium drops to about 4–4.5, acetone and *n*-butanol (butyl alcohol) replace the acids as major end products (see Appendix III(d)). This fermentation has been used as an important industrial source of acetone and butanol, but has now been superceded by alternative processes.

***N*-acetyl-D-glucosamine** (*N*-acetyl-(2-amino-2-deoxy-D-glucose)) An amino sugar which occurs in a wide range of structural polysaccharides—see e.g. CHITIN; HYALURONIC ACID; LIPOPOLYSACCHARIDES; PEPTIDOGLYCAN (q.v. for formula); TEICHOIC ACIDS.

acetylmethylcarbinol (see ACETOIN)

***N*-acetylmuramic acid** The 3-*O*-D-lactyl derivative of *N*-ACETYL-D-GLUCOSAMINE: a constituent of the polysaccharide backbone of PEPTIDOGLYCAN.

***N*-acetylmuramidase** (see LYSOZYME)

***N*-acetylneuraminic acid** (see NEURAMINIC ACIDS)

achlorophyllous (of an organism) Lacking chlorophyll.

Achlya A genus of fungi of the OOMYCETES.

Acholeplasmataceae A family of the Mycoplasmatales, class MOLLICUTES. Species of *Acholeplasma*, the sole genus, differ from members of the MYCOPLASMATACEAE in that they have no sterol growth requirements. *A. laidlawii* (formerly *Mycoplasma laidlawii*) is a pigmented saprophytic or parasitic organism.

achromatic objective (achromat) An objective lens (see MICROSCOPY) in which chromatic aberration has been corrected for *two* colours and spherical aberration for *one* colour. (cf. APOCHROMATIC OBJECTIVE)

Achromobacter anitratum (see ACINETOBACTER)

Achromobacter fischeri Obsolete name for *Vibrio fischeri*.

achromycin TETRACYCLINE. (see also TETRACYCLINES)

acicular Needle-shaped.

acid-fast organisms Those organisms (e.g. *Mycobacterium* spp.) which, once effectively stained by certain procedures (e.g. ZIEHL–NEELSEN'S STAIN), cannot be decolorized by mineral acids or mixtures of acid and ethanol. While some organisms are strongly acid-fast, others tend to be decolorized on prolonged treatment with acid-alcohol. (see also AURAMINE–RHODAMINE STAIN)

acid-fast stain (see ZIEHL–NEELSEN'S STAIN and AURAMINE–RHODAMINE STAIN)

Acidaminococcus A genus of bacteria of the family VEILLONELLACEAE.

acidosis (in ruminants) (see RUMEN)

Acineta A genus of protozoans of the SUCTORIA.

Acinetobacter A genus of gram-negative, aerobic, chemoorganotrophic, asporogenous, aflagellate bacteria of the family NEISSERIACEAE; the organisms occur as saprophytes in soil and water, and may be isolated from various animals, including man—in whom they may behave as opportunist pathogens. The cells are rod-shaped (approximately 2 × 1.5 microns) during exponential growth but tend to be coccoid in the stationary phase. The organisms are oxidase-negative and catalase-positive; they are insensitive to penicillin. Intracellular granules of poly-β-hydroxybutyrate are not formed. Metabolism is oxidative (respiratory); some strains attack certain carbohydrates with acid-production. (Acid-producing strains of *A. calcoaceticus*—the type species—have appeared in the literature as e.g. *Herellea vaginicola*, *Achromobacter anitratum*, and *Acinetobacter anitratum*. Strains of *A. calcoaceticus* which do not form acid include those formerly known as *Acinetobacter lwoffi* and *Mima polymorpha*.)

Aconta A proposed taxon which includes only members of the RHODOPHYTA (red algae). (cf. CONTOPHORA)

acquired immunity (*immunol.*) (1) *Specific* IMMUNITY acquired through exposure to an ANTIGEN. (2) NON-SPECIFIC IMMUNITY such as the condition in which INTERFERONS are synthesized following effective contact with viruses or other inducers. (3) Used also (loosely) to refer to PASSIVE IMMUNITY.

Acrasiales (see CELLULAR SLIME MOULDS)

Acrasida (see CELLULAR SLIME MOULDS)

acrasin (see DICTYOSTELIUM)

Acrasina (see CELLULAR SLIME MOULDS)

Acrasiomycetes (see CELLULAR SLIME MOULDS)

Acrasis A genus of the CELLULAR SLIME MOULDS.

acridine dyes (as antimicrobial agents) Many DYES of the *acridine* series—which include ACRIDINE ORANGE, proflavine (3,6-diaminoacridine) and ACRIFLAVINE—are bacteriostatic or bactericidal in low concentrations; such dyes appear to act by inhibiting the synthesis of nucleic acids (both DNA and RNA)—possibly by binding to the microbial DNA. A wide range of bacteria are susceptible—including gram-positive and gram-negative species. Some acridines are active against certain protozoa (e.g. *Plasmodium* spp.) and inhibit the normal replication of certain viruses.

acridine orange (*synonyms*: euchrysine; basic orange) A basic dye and FLUOROCHROME. (see also ACRIDINES (as MUTAGENS) and ACRIDINE DYES (as antimicrobial agents)) This dye can be used e.g. to demonstrate the nucleoid bodies in bacteria: in cells stained with acridine orange and viewed by fluorescence microscopy the DNA fluoresces *green* while the RNA fluoresces *orange-red*. (This differential reaction is apparently due to the fact that naturally occuring bacterial DNA is *double* stranded while the RNA is *single* stranded.)

Sub-lethal concentrations of the dye eliminate the F FACTOR in growing populations of F^+ bacterial donor cells.

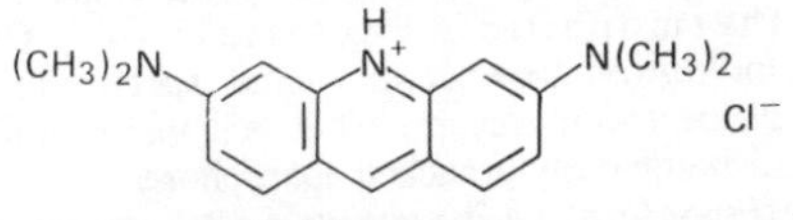

Acridine orange

acridines (as MUTAGENS) Among *bacteriophages* acridine dyes are effective mutagenic agents; they cause PHASE-SHIFT MUTATIONS. A prerequisite for an acridine-induced mutation appears to be a breakage in one strand of a double-stranded nucleic acid—such as occurs during DNA replication or recombination. The mechanism of acridine-induced mutagenesis is not understood; it is usually assumed to result from the intercalation of dye molecules with the base residues of the nucleic acid.

Acridines are generally not effective mutagens in bacteria; however, compounds in which an acridine nucleus is linked to an alkylating side chain (*ICR compounds*) induce in bacteria both insertions/deletions and point mutations.

acriflavine 3,6-Diamino-10-methylacridinium chloride: a basic yellow acridine dye and FLUOROCHROME. Acriflavine (which is soluble in water and in ethanol) has been used e.g. for the detection of smooth and rough strains of *Brucella*, for the induction of petite colonies of the yeast *Saccharomyces cerevisiae*, and as an antiseptic. (see also ACRIDINE DYES (as antimicrobial agents) and ACRIDINES (as mutagens))

acro- Prefix meaning tip, terminal, or outermost part.

acronematic Refers to a smooth *eucaryotic* FLAGELLUM which tapers to a fine point at the distal end.

acropetal (*mycol.*) (*adjective*) One of the modes in which a chain of spores may develop; in acropetal development the earliest-formed spores occupy positions in the chain nearest to the spore-bearing structure (e.g. conidiophore) while spores formed later occupy positions in the terminal or distal parts of the chain. (cf. BASIPETAL)

Actidione (*proprietary name*) (see CYCLOHEXIMIDE)

actin (see MACROTETRALIDES)

actino- A prefix signifying a ray, or rays.

Actinobacillus A genus of gram-negative, facultatively anaerobic, non-sporing bacteria of uncertain taxonomic affiliation; the cells are non-motile, pleomorphic bacilli, coccobacilli or cocci, and are usually less than 1.5 microns (maximum dimension). *A. lignieresii* is sometimes isolated from granulomatous lesions in cattle and sheep, and is the causal agent of 'wooden tongue'. (*A. muris* is now classified as *Streptobacillus moniliformis*, and *A. mallei* as *Pseudomonas mallei*.)

Actinobifida A genus of gram-positive, aerobic, thermophilic bacteria of the MICROMONOSPORACEAE; some species form *pigmented* mycelium, while all species form dichotomously-branched sporophores.

Actinobolina (*Actinobolus*) A genus of protozoans of the RHABDOPHORINA.

Actinomyces A genus of gram-positive bacteria of the ACTINOMYCETACEAE; the organisms—rods or branched filaments—are anaerobes with varying degrees of aerotolerance. Some species can be pathogenic—e.g. *A. bovis*, the causal agent of LUMPY JAW.

Actinomycetaceae A family of non-sporing, non-motile, non-acid fast, gram-positive bacteria of the ACTINOMYCETALES; species form readily-fragmenting filaments but do not form mycelium. The family includes aerobic, facultatively anaerobic, and anaerobic species. Genera include ACTINOMYCES, BIFIDOBACTERIUM and *Rothia*.

Actinomycetales (the actinomycetes*) An order of bacteria in which species form branching filaments and/or true mycelium; species are found e.g. in the soil—thermophilic species occurring in decaying vegetation and compost. The organisms differ from fungi e.g. in that they are procaryotic, have bacterial-type cell walls, and are inhibited by antibacterial agents; hyphae (when formed) are thin—commonly 1 micron or less in diameter, maximum about 2 microns—and flagella (when formed) are of the simple bacterial type. The organisms are gram-positive or gram-variable, the former sometimes becoming gram-negative in ageing cultures; some species are ACID FAST. Members of the Actinomycetales are chemo-organotrophs; substances attacked by various species include agar, cellulose, chitin, paraffins, and rubber. The majority are saprophytes, but some species are pathogenic for man, animals or plants; diseases incited by various species include JOHNE'S DISEASE, LUMPY JAW, TUBERCULOSIS, and common scab of POTATO. Most species of the Actinomycetales are aerobic, and some produce spores. Some species (Frankiaceae) form nitrogen fixing associations with higher plants (see ROOT NODULES). Many synthesize important ANTIBIOTICS, e.g. ERYTHROMYCIN, STREPTOMYCIN.

Eight families are distinguished: ACTINOMYCETACEAE, ACTINOPLANACEAE, DERMATOPHILACEAE, Frankiaceae, MICROMONOSPORACEAE, MYCOBACTERIACEAE, NOCARDIACEAE, and STREPTOMYCETACEAE. The chemical nature of the cell wall is a taxonomically-important feature among members of the Actinomycetales. Thus, four cell wall types have been distinguished: Type I species contain LL-2,6-diaminopimelic acid (LL-DAP) and glycine, Type II contain *meso*-DAP and glycine, Type III contain *meso*-DAP, and Type IV contain *meso*-DAP, arabinose and galactose. *The term 'actinomycetes' is often used in a sense which excludes members of the Mycobacteriaceae.

actinomycetes (see ACTINOMYCETALES)

actinomycins A group of ANTIBIOTICS produced by *Streptomyces* spp; each contains a substituted phenoxazone ring system (a red *chromophore* group) linked to two cyclic peptides—each incorporating a lactone bond. The principal member of the group is *actinomycin D* (= actinomycin C_1) which has two identical pentapeptide lactones. The actinomycins are extremely toxic to many types of cell—although certain organisms (e.g. some gram-negative bacilli) exhibit resistance; such resistance is presumably due to the impermeability of these organisms to the drug. Actinomycins also exhibit antitumour activity. The actinomycins inhibit DNA-directed RNA synthesis (and thus, eventually and indirectly, PROTEIN SYNTHESIS). Specifically they bind to *double-stranded* helical DNA—thus probably preventing the movement of RNA polymerase along the DNA template, i.e. chain *elongation* is prevented. (DNA replication may also be inhibited by considerably higher concentrations.) Actinomycin D is a useful biochemical tool in the study of nucleic acid synthesis.

actinomycosis Any disease, of man or other animals, in which the causal agent is a member of the ACTINOMYCETALES (excluding the genus *Mycobacterium*); examples include LUMPY JAW (*Actinomyces bovis*) and some cases of MADURA FOOT (*Nocardia, Streptomyces*).

Actinophrys sol (see HELIOZOANS)

Actinoplanaceae A family of mycelium-forming, non-acid fast bacteria of the ACTINOMYCETALES. The organisms are gram-positive, obligately-aerobic chemoorganotrophs which occur e.g. in humus-rich soils; spores—which are motile or non-motile, according to species—are formed in closed sacs (*sporangia*) at the tips of branched or unbranched aerial hyphae. The size and shape of the sporangia (and spores) depend on species; each sporangium may contain several to many hundreds of spores.

Actinosphaerium (see HELIOZOANS)

actins (see MACROTETRALIDES)

activated sludge process (see SEWAGE TREATMENT)

activation (of bacterial spores) (see SPORES (bacterial))

active bud (see LIPOMYCES)

active immunity *Specific* IMMUNITY afforded by the body's own immunological defence mechanisms following exposure to antigen. (cf. PASSIVE IMMUNITY)

active immunization (see IMMUNIZATION)

active transport (see TRANSPORT)

acute (*adjective*) (of diseases) Refers to any disease which has a rapid onset and which persists for a relatively short period of time (e.g. days, weeks) terminating either in recovery or death. Examples of acute diseases include CHOLERA and bovine ANTHRAX. The term is also used to denote an exceptionally severe or painful condition. (cf. CHRONIC)

acute-phase serum Serum derived from blood taken from a patient in the acute phase of a disease.

Acytostelium A genus of the CELLULAR SLIME MOULDS.

α-adamantanamine (amantadine) A polycyclic compound which inhibits certain VIRUSES in tissue culture (e.g. some influenza viruses). Adamantanamine is believed to act at some stage during the penetration of the host cell by the virus; it appears not to prevent adsorption of the virus to the cell surface. Adamantanamine hydrochloride has been used clinically in the prophylaxis and treatment of influenza caused by the A2 strain of INFLUENZA VIRUS—A/Singapore/1/57(H2N2); it is more effective as a prophylactic than as a therapeutic agent. (see also ANTIVIRAL AGENTS)

Adansonian taxonomy A method of classifying organisms proposed by Michel Adanson in the 18th Century. Relationships between organisms are defined by the *number* of characteristics which each organism has in common with those organisms with which it is compared—the same degree of importance being attached to all characteristics used in the comparison. (This contrasts with other systems of classification in which certain characteristics are considered to be of greater significance than others.) The modern application of this concept has been termed NUMERICAL TAXONOMY.

adaptation (*microbiol.*) Change(s) in an organism, or population of organisms, through which the organism(s) become more suited to the prevailing environmental conditions. (a) *Genetic adaptation* involves mutation and selection: in a given population, those organisms which are genetically more suited to the existing environment thrive and become numerically dominant. This results from the selection of preexisting mutants which arise, spontaneously, in any population of viable organisms. (b) *Non-genetic* (*phenotypic*) *adaptation* occurs independently of genotype. An organism which adapts non-genetically undergoes a change in metabolic activity, e.g. by enzyme induction or repression—see OPERON. (see also DIAUXIE)

Adeleina (Adeleorina) (*protozool.*) A sub-order of protozoans of the EUCOCCIDIA; species are distinguished from those of the other sub-orders by the occurrence of SYZYGY.

adenosine-5′-monophosphate (AMP) (biosynthesis of) (see Appendix V(a))

adenosine-5′-triphosphate (see ATP)

Adenovirus A genus of VIRUSES whose hosts include a variety of vertebrates—including man (and other primates), bovines, rodents, and fowl. The icosahedral *Adenovirus* virion is 70–80 nm in diameter; the capsid consists of 252 capsomeres and encloses a genome of linear double-stranded DNA. Each PENTON carries a projecting protein *fibre* which terminates in a distal knob. (Two such fibres, per penton, occur in the CELO virus—an avian strain.) The *Adenovirus* virion is non-enveloped and is ether-resistant. Human strains of *Adenovirus* share a common antigen. Many adenoviruses agglutinate the erythrocytes of the rat and/or rhesus monkey. Adenoviruses are assembled within the *nucleus* of a host cell; they can be cultivated readily in certain tissue cultures—e.g. those prepared from human amnion, human embryonic kidney, HeLa, HEp-2 (for human strains) and chick embryo kidney (for avian strains). Cytopathic effects (CPE) in tissue cultures typically include rounding of cells and the clumping of cells to form grape-like clusters. In man, adenoviruses are associated with e.g. some instances of acute respiratory tract infection and conjunctival disease, and with latent infections of the lymphoid tissues; some human strains of *Adenovirus*—and the CELO virus—can induce tumours in newborn rodents.

adjuvant (*immunol.*) Any substance which, when administered with (or before) an antigen, heightens, or affects *qualitatively*, the immune response to that antigen. Adjuvants are commonly administered with the object of increasing the *immunogenicity* of an antigen in order to stimulate a higher rate of antibody formation (or a more vigorous response in cell-mediated immunity) in respect of that antigen. Examples of adjuvants include aluminium hydroxide gel, certain alums, and water-in-oil emulsions. (see also FREUND'S ADJUVANT)

adnate (*mycol.*) In an agaric: refers to a gill the lower edge of which does not appreciably curve upwards or downwards as it approaches the stipe (to which it is attached).

adnexed (*mycol.*) In an agaric: refers to a gill the lower edge of which curves upwards as it approaches the stipe (to which it is narrowly attached).

adonitol Synonym of *ribitol*, a reduction product of ribose.

adoptive immunity Synonym of PASSIVE IMMUNITY.

adoral zone of membranelles (see AZM)

adsorption (*immunol.*) *Non-specific* adherence of substances (in solution or suspension) to cells or other forms of particulate matter. (cf. ABSORPTION) (see also BOYDEN PROCEDURE)

aecidiospores (aeciospores) (see RUSTS (1))

aecidium (*pl.* aecidia) An *aecium*—see RUSTS (1)

aeciospores (see RUSTS (1))

aecium (see RUSTS (1))

Aedes aegypti A species of mosquito (order Diptera, suborder Orthorrhapha, family Culicidae); one of the vectors of YELLOW FEVER. *A. simpsoni* and *Haemagogus* spp are among the other (mosquito) vectors of this disease.

aequihymeniiferous (aequihymenial) (see GILLS)

aerial mycelium (*bacteriol.*) In certain actinomycetes: that part of the mycelium which projects above the level of the medium (cf. SUBSTRATE MYCELIUM).

Aerobacter An obsolete bacterial genus; many of the organisms previously referred to as *Aerobacter* spp. are currently classified as species of *Klebsiella* or *Enterobacter*.

aerobe Any organism which grows in the *presence* of air or oxygen; a *strict* aerobe grows *only* under such conditions. (see also FACULTATIVE; cf. ANAEROBE)

aerobic (*adjective*) (1) (of an environment) Refers to an environment in which the partial pressure of oxygen is similar to that which occurs under normal atmospheric conditions (cf. ANAEROBIC and MICROAEROPHILIC). (2) (of organisms) See: AEROBE (see also FACULTATIVE).

Aerococcus A genus of the STREPTOCOCCACEAE.

aerogenic (*adjective*) Refers to those organisms which form gas (as well as other metabolic by-products) from particular substrates.

Aeromonas A genus of gram-negative, oxidase-positive, catalase-positive bacteria of the VIBRIONACEAE; the cells may be round-ended rods (approx. 1–4 × 0.5–1 microns) or short filaments. A single polar flagellum is usually present. All species are insensitive to O/129 (cf. *Vibrio*), all liquefy gelatin, and none produce acid from inositol (cf. *Plesiomonas*). Some strains form acid from glucose. Most strains fail to grow in media containing 7.5% NaCl. Growth occurs on common bacteriological media (e.g. nutrient agar); optimal growth temperature : 20–30 °C.

aerotaxis A particular form of CHEMOTAXIS in which a (motile) organism migrates along a concentration gradient to a location where the concentration of oxygen is optimal.

aerotolerant (*adjective*) Refers to an ANAEROBE which can (i) survive, but not grow, in the presence of oxygen (i.e. an aerotolerant OBLIGATE anaerobe), or (ii) grow, at sub-optimal rates, under aerobic conditions. (cf. MICROAEROPHILIC)

aeruginocins (pyocins) BACTERIOCINS produced by *Pseudomonas aeruginosa*.

aesculin The 6-β-D-glucosyl derivative of 6,7-dihydroxycoumarin. Aesculin is hydrolysed e.g. by certain streptococci (e.g. many strains of group D) to yield 6,7-dihydroxycoumarin—which gives a brown coloration with soluble ferric salts. (see also AESCULIN BROTH)

aesculin broth and agar *Broth*: PEPTONE WATER containing 0.1% AESCULIN and 0.05% ferric citrate sterilized at 115 °C/10 minutes. *Agar*: the broth gelled by 2% agar.

aethelium (see ACELLULAR SLIME MOULDS)

aetiology (*American*: etiology) The study of *causation*; the aetiological agent of an infectious disease is that organism which incites or causes the disease.

affinity (*immunol.*) The strength with which an

antibody binds to a simple, monovalent hapten or to one determinant of an antigen; factors which effect the strength of binding include the size (area) of the combining site, the closeness of fit, and the nature and strength of the intermolecular forces (e.g. van der Waals' forces). (cf. AVIDITY)

affinity labelling A technique used to identify the precise location of the COMBINING SITE of an antibody. In one form of affinity labelling the antibody is allowed to combine with an homologous HAPTEN (the 'affinity label') which, on subsequent irradiation, forms a covalent bond with the antibody; the point of attachment of the hapten to the antibody (i.e. the combining site) is then determined by chemical analysis following separation of the heavy and light chains. (In this particular method the (aromatic) hapten carries an azide side-chain which, on irradiation with u.v. light, forms the highly reactive *nitrene* group which readily combines (non-specifically) with any of a wide variety of organic groups.)

aflatoxins A range of low-M.Wt. compounds (M.Wt. in the region of 350) produced by strains of *Aspergillus flavus* and *A. parasiticus*; the aflatoxin molecule contains a coumarin-bifuran nucleus. Aflatoxins are toxic and carcinogenic for a range of animals. Aflatoxin-contaminated foodstuffs (particularly cereals and peanuts) have caused outbreaks of food-intoxication among cattle, pigs, etc. In England, in the 1960s, a high incidence of fatal aflatoxin poisoning among turkey poults (*turkey X disease*) was traced to a contaminated groundnut feed; symptoms of turkey X disease include lack of appetite and general weakness, and necrotic liver lesions are found *post mortem*.

Aflatoxins are soluble in organic solvents; some are fluorescent. (see also FOOD POISONING (fungal))

ag (*immunol.*) Antigen.

agar A complex sulphated galactan which is widely used (in *gel* form) as a base for many kinds of solid and semi-solid microbiological MEDIUM. Agar contains two main components: *agarose* and *agaropectin*. Agarose consists of linear chains of alternating residues of 1,3-linked β-D-galactose and 1,4-linked 3,6-anhydro-L-galactose. The structure of agaropectin has not been determined; it may consist of a mixture of polysaccharides and contains D-galactose, 3,6-anhydro-L-galactose, and mono-esterified sulphuric acid.

An agar gel is a translucent or transparent jelly-like substance formed when a mixture of agar and water is heated to above 100 °C and subsequently cooled; gelling occurs at about 45 °C. Agar media usually contain 1.5% to 2% w/v agar in water—though *semi-solid* agar contains 0.5% or less. *Stiff* agars contain e.g. 8% w/v agar. (The above concentrations refer to Japanese agars; when using New Zealand agar the above quantities should be halved to give gels of similar strengths.) Sterile gels are prepared by autoclaving (see AUTOCLAVE) a suspension of agar in water; the addition of nutrients and other substances, prior or subsequent to autoclaving, enables the gel to support the growth of particular types of microorganism (see e.g. NUTRIENT AGAR). Various agar-based media are used e.g. for the selective culture of particular organisms and for the display of particular cultural features (e.g. colony form) which assist in the identification of microorganisms. *Blood agar* is made by mixing 5–10% (v/v) citrated or defibrinated blood with molten agar (at about 50 °C) and allowing it to set. Culturing on blood agar allows the detection of bacterial *haemolysin* production; since a given microbial haemolysin may be active only against the erythrocytes of particular species (e.g. horse, sheep) the choice of blood is important. (see also CHOCOLATE AGAR) *Ion-free agar* is a refined form of agar used e.g. in *immunoelectrophoresis*.

Agar is produced by certain species of the Rhodophyta (red algae)—e.g. *Gelidium*. (The term 'agar' derives from the Malay 'agar-agar' which refers to certain edible seaweeds.) The ability to degrade agar is confined to a few organisms—including certain strains of *Streptomyces coelicolor*, certain marine vibrios, and marine species of *Cytophaga*.

agar gel diffusion (see GEL DIFFUSION)

agar plate (see PLATE)

agaric (1) Any fungus of the order AGARICALES. (2) Any *gill-bearing* member of the Agaricales—particularly those species which are mushroom-shaped (as opposed to species which are non-stipitate or which have a laterally situated stipe).

Agaricaceae A family of fungi of the order AGARICALES. Constituent species form fruiting bodies in which the gills are free or almost free from the stipe, in which an annulus is typically present, and in which the basidiospores (which *lack* an apical germ pore) are dark chocolate brown or of a similar *dark* coloration; the stipe separates cleanly from the cap. A UNIVERSAL VEIL is not formed. The genera include e.g. AGARICUS.

Agaricales An order of fungi of the sub-class HOLOBASIDIOMYCETIDAE. Constituent species form well-developed fruiting bodies which are typically gymnocarpic* and in which the hymenium occurs on the surfaces of GILLS (lamellae), lines the tubules of porous tissue (which is *easily* detached from the remainder of the pileus—cf. APHYLLOPHORALES), or (exceptionally) covers a smooth hymenophore; the holobasidia bear small, inconspicuous sterigmata. The basidiocarp is typically 'mushroom-shaped'—i.e. consists of a discoid, convex, conical, or umbonate pileus, with radially arranged gills on the underside, supported on a thin or stout stipe; the stipe is typically central but may be eccentric or lateral, or may be absent—according to species. The fruiting body is fleshy in the great majority of species; species which form tough or leathery fruiting bodies do not form porous hymenophores. Members of the Agaricales are typically saprophytic fungi which occur principally on soil and wood. The families of the order include e.g. AGARICACEAE, AMANITACEAE, BOLETACEAE, COPRINACEAE, HYGROPHORACEAE, Lepiotaceae, Paxillaceae, RUSSULACEAE, Strophariaceae, TRICHOLOMATACEAE. *The order now includes a number of species whose fruiting bodies are essentially angiocarpic or 'semiangiocarpic'; during at least certain stages, the fruiting bodies of these organisms—*Secotium* and other 'secotioid gasteromycetes'—develop in a manner analogous to that of the agarics. Thus, e.g. the early stages of fruiting body development in *Secotium agaricoides* correspond closely with those in *Agaricus*; however, in the mature fruiting body of *S. agaricoides* only remnants of the gill tissue may remain, and the statismospores are released into the unopened peridium—which subsequently disintegrates.

Agaricus A genus of fungi of the family AGARICACEAE; the organisms include a number of edible species (see MUSHROOM)—though some species are reported to be poisonous. The spores are dark chocolate brown; the gills are pinkish when immature.

agaropectin A component of AGAR.

agarophyte (agarphyte) Any AGAR-producing seaweed.

agarose A component of AGAR.

agglutination (*immunol.*) The formation of visible clumps of *particulate* antigens (e.g. bacteria, red blood cells) following the linkage of the antigen particles by specific antibodies (or other factors which have agglutinating properties—e.g. LECTINS). (cf. PRECIPITATION; see also HAEMAGGLUTINATION and PASSIVE AGGLUTINATION) Agglutinated particles may form a mat or reticulum over a relatively large area of the bottom of the test tube in which agglutination has occurred; by contrast, non-agglutinated particles form a smaller, dense button in the control tube.

agglutinin (*immunol.*) Any substance, usually an ANTIBODY, which brings about the AGGLUTINATION of *particulate* antigens, e.g. bacteria. (see also LECTINS)

agglutinogen (*immunol.*) The antigen which stimulates the production of a particular agglutinin.

aggressins Diffusible substances or cellular components produced by certain pathogenic microorganisms; aggressins promote the invasiveness of a pathogen. An aggressin may e.g. protect a pathogen from phagocytosis, or it may enable the pathogen to resist intracellular digestion within a phagocyte.

aglycon The non-sugar portion of a glycoside.

Agonomycetes (*Mycelia Sterilia*) A class of fungi within the sub-division DEUTEROMYCOTINA. Constituent species were originally distinguished from those of other classes in that they appeared not to produce reproductive structures of any kind (either sexual or asexual); however, since the inception of the *Mycelia Sterilia*, a number of species have been found to have an ascomycete or basidiomycete sexual stage. Constituent species include a number of plant pathogens—e.g. species of RHIZOCTONIA and SCLEROTIUM.

agranulocyte Any white blood cell which has a non-granular cytoplasm, e.g. a *lymphocyte*.

Agrobacterium A genus of gram-negative, aerobic, non-sporing, rod-shaped bacteria of the family Rhizobiaceae. Cells have either several (peritrichous) flagella or a single (commonly non-polar) flagellum. *Agrobacterium* species (all chemoorganotrophs) are found in the soil; most strains can initiate plant GALLS. (The type species, *A. tumefaciens*, causes CROWN GALL.) Growth on nutrient media is non-pigmented; in the presence of carbohydrates it is often highly mucoid. The temperature optimum is about 25 °C and the pH range wide. *Agrobacterium* spp do not fix nitrogen (cf. *Rhizobium*).

AHG (*immunol.*) Anti-human globulin—an antibody against human globulin.

air bladders, vesicles (see PNEUMATOCYSTS)

Ajellomyces dermatitidis (see BLASTOMYCES DERMATITIDIS)

akinete (1) In blue-green algae: a thick-walled spore, commonly larger than a vegetative cell, which is rich in food reserves; akinetes are often formed adjacent to heterocysts. An akinete is appreciably resistant to desiccation; it germinates to form a *hormogonium*. (2) (*mycol.*) A non-motile spore or resting cell.

alamethicin A cyclic polypeptide which shows ANTIBIOTIC activity against gram-positive bacteria, apparently by acting as an IONOPHORE.

alanine (biosynthesis of) (see Appendix IV(b))

alastrim (amaas) (see SMALLPOX)

alazopeptin (see 6-DIAZO-5-OXO-L-NORLEUCINE)

albamycin Synonym of NOVOBIOCIN.

Albert's stain A stain used for the demonstration of METACHROMATIC GRANULES; Albert's stain is used e.g. as an aid to the identification of

Corynebacterium diphtheriae. Two solutions are used: (a) Albert's stain: toluidine blue (0.15 g) and MALACHITE GREEN (0.2 g) are dissolved in 95% ethanol (2 ml); this is added to 1% glacial acetic acid (100 ml) and the whole filtered after standing for 24 hours. (b) LUGOL'S IODINE. The smear is stained with Albert's stain for 3–5 minutes, washed in tap water, and blotted dry; Lugol's iodine is applied and left for 1 minute, and the smear is subsequently washed and blotted dry. Granules stain darkly (or black); cytoplasm stains pale green.

albomycin (see SIDEROMYCINS)

Albugo A genus of fungi within the class OOMYCETES; *Albugo* spp. are obligate parasites and pathogens of a range of higher plants—e.g. cabbage, horse-radish, mustard. *A. candida* (economically the most important species) forms a well-developed, branched, aseptate mycelium within the tissues of the host; the (*inter*cellular) mycelium abstracts nutrients from the host by means of spherical haustoria. Asexual reproduction involves the formation of chains of multinucleate sporangia which develop, basipetally, on short club-shaped sporangiophores beneath the epidermis of the host plant; the compacted masses of sporangia appear as white 'blisters' on the surface of the host plant. When mature, the sporangia form a loose powdery mass which, on rupture of the host's epidermis, is dispersed by wind and rain. On germination, each sporangium produces a number of zoospores. Sexual reproduction is initiated when parts of the intercellular thallus differentiate to form an oogonium and an adjacent antheridium; subsequently, a fertilization tube is established between the two gametangia, and plasmogamy and karyogamy precede the development of a thick-walled oospore. Meiosis occurs within the oospore. On germination an oospore liberates a number of biflagellate zoospores.

albumen The white of an egg. (cf. ALBUMINS)

albumins A class of proteins. (cf. ALBUMEN)

Alcaligenes (*Alkaligenes*) A genus of gram-negative, aerobic, non-fermentative, chemoorganotrophic*, non-sporing bacteria of uncertain taxonomic position; *Alcaligenes* spp. are common in the mammalian intestine and in soil and aquatic environments. The cells are rod-shaped or coccoid, maximum size about 1.2 × 2.5 microns; they are sparsely flagellated. Growth occurs e.g. on nutrient agar at pH7, 37 °C. *Alcaligenes* spp. may oxidize glucose but do not form acid in glucose peptone water; they are oxidase positive. GC ratio: 58–70%. **A. eutrophus* can grow chemoautotrophically when provided with hydrogen, oxygen, and carbon dioxide.

alcoholic fermentation (ethanol fermentation) A type of fermentation carried out by a range of fungi—e.g. yeasts (including *Saccharomyces*) and species of *Aspergillus*, *Fusarium*, and *Mucor*. In the alcoholic fermentation glucose is converted to ethanol and carbon dioxide; small amounts of other compounds (e.g. glycerol) may be formed. Initially, glucose is converted to pyruvate *via* the EMBDEN–MEYERHOF–PARNAS PATHWAY. Subsequently pyruvate is decarboxylated to acetaldehyde in a reaction catalysed by pyruvate decarboxylase and thiamine pyrophosphate; the acetaldehyde is reduced to ethanol by a NAD-dependent alcohol dehydrogenase—resulting in the reoxidation of the $NADH_2$ formed during the EMP pathway. Alcoholic fermentation by *Saccharomyces* spp. has been used in the commercial production of ethanol. (cf. ZYMOMONAS; see also BREWING; CIDER; WINE-MAKING; GLYCEROL, MICROBIAL PRODUCTION OF)

alcohols (as antimicrobial agents) Under appropriate conditions, alcohols may be rapidly bactericidal, fungicidal, or virucidal (to certain viruses); they are virtually without effect on spores. The mechanism of antimicrobial activity may involve the denaturation of structural or enzymic proteins and/or a solubilizing effect on the lipids of the bacterial or fungal CELL MEMBRANE or on those of the ENVELOPE of certain viruses (e.g. influenza viruses). The antimicrobial activity of alcohols increases with molecular weight and with chain length up to about C8 or C10; above this, insolubility becomes an important factor. Activity decreases in the order: primary → *iso*-primary → secondary → tertiary. Penetration tends to be poor and alcohols are unsuitable for use in the presence of organic matter. *Ethanol* exerts maximum antimicrobial activity at 60–95% (by volume) in an ethanol–water mixture. *Isopropyl alcohol* ($CH_3.CHOH.CH_3$) is less volatile and somewhat more effective than ethanol; it is used primarily as a skin antiseptic. *Phenoxyethanol* (Phenoxetol) and *benzyl alcohol* are used as preservatives in pharmaceutical preparations. The activity of benzyl alcohol is improved by halogenation: 2,4-dichlorobenzyl alcohol is used as a skin antiseptic. *Ethylene glycol*, *propylene glycol*, and *trimethylene glycol* (dihydric alcohols) have been used, in aerosol form, for the disinfection of air; a relative humidity of some 60% is required. *Glycerol* (a trihydric alcohol) is bacteriostatic at concentrations above 50%; it has been used as a preservative in vaccines. (see also DISINFECTANTS and STERILIZATION)

Alectoria A genus of fruticose LICHENS in which the phycobiont is *Trebouxia*; the thallus may be attached to the substratum by a holdfast. Apothecia are rarely formed. Species include e.g.:

A. fuscescens. The thallus is brown to black, glossy, filamentous, branching, pendulous or entangled, and may reach 20 centimetres in length; small greenish-white soralia are generally present. *A. fuscescens* occurs on trees, fences, and occasionally on rocks.

Aleuria A genus of fungi of the order PEZIZALES; the organisms form sessile or stipitate, discoid or cup-shaped, minute or conspicuous apothecia in which the hymenium may be red, orange, or yellow.

aleuriospore (*mycol.*) A type of spore which is liberated from the sporophore by breakage of the latter close to its junction with the spore.

alexin (1) (*immunol.*) A historical term for COMPLEMENT. (2) (*plant pathol.*) (see PHYTOALEXINS)

algae An heterogenous group of *eucaryotic*, photosynthetic, unicellular and multicellular organisms. (cf. BLUE-GREEN ALGAE and PROCHLOROPHYTA) Algae resemble higher plants e.g. in that oxygen is evolved during PHOTOSYNTHESIS (cf. RHODOSPIRILLALES) and in that algal photosynthetic pigments invariably include CHLOROPHYLL *a*; the multicellular algae differ from vascular plants e.g. in that they lack the specialized conducting systems characteristic of the latter—though phloem (but not xylem) occurs in some of the larger seaweeds. Algae differ from bryophytes (mosses, liverworts) e.g. in that the algal reproductive structure, when formed, lacks the peripheral envelope of sterile cells found in analogous structures in every bryophyte—though see Charophyta, below. Some organisms have characteristics intermediate between those of algae and PROTOZOA. The aquatic algae occur in fresh, brackish, and/or marine waters—according to species—where they are of major ecological importance as primary producers (see also PLANKTON); terrestrial algae occur e.g. on damp soil, on ice and snow (the 'cryoflora'), on rocks, tree-trunks etc., and as the phycobionts of LICHENS. Apart from their ecological significance, algae are useful to man e.g. as sources of AGAR, ALGINIC ACID, CARRAGEENAN, FUNORAN, and FURCELLARAN; iodine (as iodide) has been extracted from the ash of burnt seaweed. (see also DIATOMACEOUS EARTH and MICROORGANISMS AS FOOD) A few algae are pathogenic—see e.g. RED RUST.

Algae are classified on the basis of their pigments, type of food reserve material, type and arrangement of flagella, CHLOROPLAST ultrastructure, and CELL WALL composition. The algal divisions are BACILLARIOPHYTA, CHAROPHYTA, CHLOROPHYTA, CHRYSOPHYTA, Cryptophyta, DINOPHYTA, EUGLENOPHYTA, HAPTOPHYTA, PHAEOPHYTA, Prasinophyta, RHODOPHYTA and XANTHOPHYTA. (Some authors divide the algae into a number of *classes* which correspond with the above divisions—e.g. Bacillariophyceae, Charophyceae, etc.)

According to species algae range in size from unicellular organisms (several microns in diameter) to the largest seaweeds (some 50 or more metres in length). Unicellular species occur in most of the algal divisions; none occurs in the Phaeophyta. In certain unicellular species (e.g. the amoeboid alga *Rhizochrysis*) the thallus (i.e. algal body) is naked—i.e. it lacks a cell wall; in many other unicellular algae (e.g. *Chlamydomonas*, *Chlorella*)—and in all multicellular algae—a cell wall is present (see CELL WALL (algal) for details). The multicellular algae exhibit great diversity of form. Examples of common morphological types include (a) the branched or unbranched *filament*—i.e. one or more chains of cells which form a thread-like or ribbon-like thallus (e.g. *Cladophora*, *Spirogyra*); (b) the *parenchymatous* thallus—i.e. a sheet of cells one or more cells thick (e.g. *Ulva*), or a tubular arrangement of cells (e.g. *Enteromorpha*); (c) the SIPHONACEOUS thallus. Some multicellular algae exhibit considerable differentiation; thus, e.g. in certain seaweeds (e.g. *Laminaria* spp.) the thallus includes root-like, stem-like, and leaf-like structures (the *holdfast*, *stipe*, and *lamina* (*blade*) respectively) and a conducting system, the *phloem*. (see also PNEUMATOCYST) (Some differentiation also occurs e.g. in the *unicellular* alga ACETABULARIA.) In the multicellular algae meristematic tissue may occur in apical, intercalary, and/or diffuse regions—according to species and to the stage of development. Algae may be non-motile (e.g. *Chorella*, *Scenedesmus*, members of the Rhodophyta) or motile (e.g. *Chlamydomonas*, *Euglena*, *Ochromonas*, *Volvox*); according to species, flagellate vegetative cells (and gametes) may have one, two or more flagella of equal or unequal length. (see also *diatoms* under MOTILITY) Some algae exhibit colonial organization—see COENOBIUM, PALMELLOID STAGE, and TETRASPORAL COLONY.

Although normally photosynthetic a number of algae can grow, in the dark, on certain organic substrates e.g. glucose and/or acetate—i.e. they behave as facultative chemoorganotrophs; such algae include species of *Chlamydomonas*, *Chlorella*, *Nitzschia*, and *Scenedesmus*. (Facultative chemoorganotrophy could be important to algae under conditions of low light intensity.) Chemoorganotrophy has not been demonstrated e.g. in members of the Phaeophyta. Under certain conditions some algae (e.g. *Scenedesmus* spp, *Ulva* spp.) carry out PHOTOREDUCTION.

Asexual reproduction. In some filamentous and colonial algae (e.g. *Spirogyra*, *Synura*) asexual reproduction occurs by fragmentation—a process referred to by some authors as *vegetative* reproduction. Some of the many modes of asexual reproduction are given in the entries CHLAMYDOMONAS, CHLORELLA, SCENEDESMUS, VAUCHERIA, and VOLVOX.

Sexual reproduction. For some examples of modes of sexual reproduction see ACETABULARIA, CHLAMYDOMONAS, SPIROGYRA, and VOLVOX. In members of the Charophyta sexual reproduction is strictly oogamous; according to species a given thallus may contain the gametangia of one or both sexes. This group is characterized by a complex multicellular male gametangium which is enclosed by an envelope of sterile cells (*shield cells*); such a structure is atypical of the algae, and it has been argued by some authors that members of the Charophyta should be excluded from the algae on this basis. AUTOGAMY can occur in certain diatoms. Sexual reproduction has not been observed in members of the Euglenophyta or e.g. in *Chlorella* and *Pleurococcus* or certain genera of the Bacillariophyta.

Many algae exhibit an ALTERNATION OF GENERATIONS; such algae may produce sporophytes and gametophytes which are morphologically similar (e.g. *Ectocarpus*, *Ulva*) or morphologically dissimilar (e.g. *Laminaria*). The vegetative cells of certain algae (e.g. *Chlamydomonas*, *Chlorella*, and many other green algae) are haploid; the *siphonaceous* green algae and e.g. all diatoms have diploid vegetative cells.

algicides Chemical agents which (selectively) kill algae. Algicides are used e.g. to inhibit algal growth in swimming pools (e.g. copper sulphate, DICHLONE) and in reservoirs (see BLOOM) and as components of the anti-fouling paints used on ships' hulls (e.g. THIURAM DISULPHIDE and tributyltin oxide, $(C_4H_9)_3Sn.O.Sn(C_4H_9)_3$).

alginic acid A complex polysaccharide which contains β-(1 → 4)-linked D-mannuronic acid residues and α-(1 → 4)-linked L-guluronic acid residues. Alginates are associated with the cell wall in members of the Phaeophyta. Alginic acid is insoluble in water, but certain alginates (e.g. sodium alginate) are readily soluble; gels are formed in the presence of calcium ions. Alginates have a wide range of industrial uses, e.g. in cosmetics, pharmaceuticals, and as emulsifiers or thickeners in foodstuffs. Certain bacteria (e.g. *Nocardia* spp.) can degrade alginates. (see also SWABS)

aliphatic hydrocarbons (microbial utilization of) A wide range of microorganisms can use aliphatic hydrocarbons as substrates for growth. Straight-chain paraffins (*n*-alkanes) appear to be the most widely used, but branched-chain paraffins (alkyl-alkanes) and unsaturated hydrocarbons (alkenes) can also be used by some organisms; in general, long-chain hydrocarbons (e.g. C_{10}) are more readily utilized than are short-chain hydrocarbons. (see also METHANE (microbial utilization of)) Microorganisms which can degrade aliphatic hydrocarbons include e.g. certain species of the bacteria *Actinomyces*, *Corynebacterium*, *Mycobacterium*, *Nocardia*, *Pseudomonas*, of the yeasts *Candida*, *Debaromyces*, *Hansenula*, *Pichia*, and *Torulopsis*, and of the filamentous fungi *Aspergillus*, *Botrytis*, *Cladosporium*, *Fusarium*, *Helminthosporium*, and *Penicillium*. Hydrocarbon metabolism is a strictly aerobic process and appears to occur most commonly by *omega-oxidation*: a terminal methyl group of the hydrocarbon is oxidized—in a reaction involving a mono-oxygenase—to form a primary alcohol. (In *Pseudomonas* spp, for example, the *omega*-oxidation enzyme system is complex and involves a mono-oxygenase (hydroxylase), the iron–sulphur protein *rubredoxin*, and an FAD-containing $NADH_2$-rubredoxin oxidoreductase.) The alcohol is oxidized to the corresponding fatty acid which can undergo degradation by *beta*-oxidation. Occasionally *both* ends of a hydrocarbon substrate are oxidized—resulting in the formation of a dicarboxylic acid; subterminal oxidation may also occur. Unsaturated hydrocarbons may be metabolised in the same way(s) as saturated hydrocarbons, but may also be oxidized at the double bond; this results in the formation of a diol—sometimes (at least) with the prior formation of an epoxide.

Hydrocarbon-utilizing organisms can cause spoilage of a number of hydrocarbon-containing materials; thus, e.g. microbial growth in jet aircraft fuels results in the formation of sludge and may lead to the corrosion of fuel tanks etc. Growth of such organisms tends to occur at a hydrocarbon–water interface. The commercial exploitation of hydrocarbon-utilizing microorganisms is being attempted e.g. in the production of *single-cell protein* (see MICROORGANISMS AS FOOD). Ethane-utilizing organisms have been used in oil prospecting; ethane associated with oil deposits may diffuse to the surface and allow localized growth of these organisms in the soil.

Alkalescens-Dispar group Non-motile members of the ENTEROBACTERIACEAE which ferment glucose anaerogenically, and in which lactose fermentation is delayed or absent; the organisms are otherwise similar to the typical form of *Escherichia coli*, and they are now regarded as variants of *E. coli*.

Alkaligenes Synonym of *Alcaligenes*.

alkanes (microbial utilization of) (see ALIPHATIC HYDROCARBONS (microbial utilization of))

alkenes (microbial utilization of) (see ALIPHATIC HYDROCARBONS (microbial utilization of))

alkylating agents (as antimicrobial agents and MUTAGENS) Alkylating agents react with nucleophilic groups (e.g. sulphhydryl, amino, carboxyl groups)—an alkyl group becoming covalently linked to the nucleophile. (*Bifunctional* alkylating agents have two groups per molecule capable of such reaction.) Alkylating agents substitute nucleophilic groups in proteins and nucleic acids and thus have inhibitory and/or mutagenic effects on microorganisms.

Volatile alkylating agents (e.g. ETHYLENE OXIDE, β-PROPIOLACTONE) may be used for gaseous STERILIZATION of enclosed spaces, or of objects not conveniently sterilized by other methods. (see also FORMALDEHYDE)

Effects of alkylating agents on DNA. Alkylation of DNA (e.g. by ethylmethanesulphonate: EMS, or ethylethanesulphonate: EES) generally occurs primarily at the N7 position of guanine; lower levels of alkylation may also occur at many other positions on the bases—and possibly on the phosphodiester groups. Alkylation of guanine and adenine can render the sugar-base bond labile so that the base may be eliminated (*depurination*); this may be followed by rupture of the DNA backbone. Bifunctional alkylating agents (e.g. NITROGEN MUSTARDS) may form inter- or intra-strand cross-links in DNA—the effects of which may be lethal to the cell. (see also MITOMYCINS) The mechanism of mutagenesis following alkylation is unknown; N7-alkylguanine appears to pair normally (i.e. as guanine), while depurination now appears unlikely as a direct cause of mutation. It is thought that mutagenesis may result from mispairing (during DNA replication) induced by the alkylation of certain positions in certain bases.

alkyldimethylbenzylammonium chloride (see QUATERNARY AMMONIUM COMPOUNDS)

allele (allelomorph) Any of one or more alternative forms of a given GENE. Both (or all) alleles of a given gene are concerned with the same trait or characteristic; however, the product or function coded for by a particular allele differs, qualitatively and/or quantitatively, from that coded for by other alleles of the same gene. Three or more alleles of a given gene constitute an *allelomorphic series*. In a diploid cell or organism the members of an *allelic pair* (i.e. two alleles of a given gene) occupy corresponding positions (loci) on a pair of homologous chromosomes; if these alleles are genetically identical the cell or organism is said to be *homozygous*—if genetically different, *heterozygous*—with respect to the particular gene. A *wild type allele* is one which codes for a particular phenotypic characteristic found in the WILD TYPE STRAIN of a given organism. (see also DOMINANT GENE)

allelomorph A synonym of ALLELE.

allergen (*immunol.*) An antigen (or autocoupling hapten—e.g. certain drugs) capable of producing a state of HYPERSENSITIVITY* or of provoking a reaction in a *sensitized* (hypersensitive) subject. *Commonly refers to *immediate* rather than delayed hypersensitivity. (see also REAGINIC ANTIBODY)

allergy (1) Synonym of HYPERSENSITIVITY. (2) Sometimes used as a synonym of IMMUNITY.

allergy of infection (see DELAYED HYPERSENSITIVITY)

allogeneic (allogenic) (*adjective*; of tissue transplants etc.) Derived from another individual of the same species but of different genotype.

allolactose D-Galactopyranosyl-β-(1→6)-D-glucose; the natural inducer of the *lac* OPERON.

Allomyces A genus of fungi of the order BLASTOCLADIALES; species occur in moist soils, muds, and aquatic environments. The vegetative form of the organisms is a branched, coenocytic mycelium which is anchored to the substratum by means of a system of branching rhizoids; the hyphae contain incomplete (partial) septa. Chitin has been detected in the cell wall in a number of species. Some species exhibit an ALTERNATION OF GENERATIONS. In e.g. *A. macrogynus* the SPOROTHALLUS gives rise to thick-walled resistant sporangia (meiosporangia) and thin-walled mitosporangia, while each branch of the GAMETOTHALLUS gives rise to an orange-coloured terminal male gametangium and a colourless sub-terminal female gametangium. Certain species—placed in the sub-genus *Brachyallomyces* (*Brachy-Allomyces*)—do not exhibit sexual phenomena.

allotypes (*immunol.*) The antigenically-different forms of a given type of plasma protein which occur in different individuals of the same species.

almond, diseases of These include: (a) *Leaf curl*—see PEACH, DISEASES OF. (b) CROWN GALL.

Alphavirus A genus of mosquito-borne ARBOVIRUSES of the family TOGAVIRIDAE; members of the genus exhibit cross-reactivity in haemagglutination inhibition tests and are distinguished from viruses of the genus FLAVOVIRUS in that e.g. the latter are inactivated by trypsin. Certain alphaviruses are the causal agents of encephalitis in the horse and in other mammals.

ALS Antilymphocyte serum (*q.v.*).

Alternaria A genus of fungi within the order

MONILIALES; species include a number of plant pathogens—e.g. the causal agents of *early blight* of potato and *black rot* of carrot. The vegetative form of the organisms is a septate mycelium. The conidia are produced (typically in chains) by acropetal development on conidiophores which closely resemble the somatic hyphae; the dark-coloured, pyriform conidia usually have both transverse and longitudinal septa.

alternaria rot Some species of *Alternaria* cause firm, dark ROTS in a variety of plant hosts. A phytotoxic agent, ALTERNARIC ACID, is produced by *A. solani*; this agent may be an important factor in pathogenesis.

alternaric acid A complex compound containing a diketotetrahydropyran group linked to a long-chain carboxylic acid; it is produced by *Alternaria solani.* Alternaric acid retards or inhibits germination of the spores of certain fungi, and brings about wilting and necrosis in the tissues of higher plants. (see also MYCOTOXINS)

alternate host In the life cycle of *heteroecious* RUSTS: the (plant) host in which the *telial* stage does *not* occur (cf. PRIMARY HOST). More loosely, either host of an heteroecious rust.

alternation of generations (*microbiol.*) In the life cycles of certain organisms: the alternate formation of mature haploid individuals and mature diploid individuals; an alternation of generations is exhibited e.g. by certain species of the fungus *Allomyces*, by some members of the FORAMINIFERIDA, and by many algae (e.g. *Laminaria*).

An individual which occurs during the diploid phase (a *sporophyte*, *sporothallus*, or *agamont*) subsequently undergoes meiosis to give rise to haploid *meiospores*; each meiospore develops to form a mature haploid individual (a *gametophyte, gametothallus*, or *gamont*). Subsequently, male and female gametes are produced—either by the same gametophyte (as e.g. in *Allomyces arbuscula*) or by different gametophytes (as e.g. in *Laminaria*). Following fusion between a male and a female gamete, the (diploid) zygote develops to form a mature diploid individual (sporophyte).

amaas (see SMALLPOX)

amanin (see AMATOXINS)

Amanita A genus of fungi of the family AMANITACEAE (order AGARICALES); species occur frequently in deciduous and/or coniferous woodlands—some species being commonly associated with particular type(s) of tree. The pileus of the fruiting body may be e.g. flat, slightly convex, dome-shaped, or umbonate, according to species; similarly, the colour of the cap may be e.g. white, yellowish, or any of a variety of shades of brown or red-brown, and *A. muscaria* forms a bright red cap to which often adhere a scatter of white scales (remnants of the universal veil). The gills (and spores) of *Amanita* spp are generally white or off-white. A number of species are highly poisonous; these include e.g. *A. phalloides*, *A. virosa*, *A. pantherina*, and *A. muscaria.* (see also AMATOXINS and PHALLOTOXINS)

Amanitaceae A family of fungi of the order AGARICALES; constituent species are distinguished from those of other families in that e.g. the tramae of the gills exhibit *bilateral* organization (see BILATERAL TRAMA), the gills are free from the stipe, and the fruiting body generally has both a UNIVERSAL VEIL and a PARTIAL VEIL. The family contains two genera: *Amanita* (in which a *volva* is formed), and *Limacella* (in which a volva is not formed).

amanitins (α-, β-, γ-, and ε-) (see AMATOXINS; see also FOOD POISONING (fungal))

amantadine (see α-ADAMANTANAMINE)

amanullin A *non*-toxic substance found in the fungus *Amanita phalloides.* (see also AMATOXINS)

amastigote (leishmanial form) A form assumed by some species of the Trypanosomatidae during a particular stage of development. The amastigote is typically a rounded cell; it lacks a free (external) flagellum. (see TRYPANOSOMA) Amastigotes ('Leishman–Donovan bodies') are produced by *Leishmania* spp in a vertebrate host. Species of *Crithidia* and *Leptomonas* produce amastigotes in invertebrate hosts.

amatoxins A group of toxic cyclic peptides present in the fungus *Amanita phalloides* and in certain other species of *Amanita*, e.g. *A. verna.* Amatoxins are more toxic than the PHALLOTOXINS which occur in the same fungi; clinical effects are produced 8–24 hours after ingestion. Initial symptoms include severe vomiting and diarrhoea; degenerative changes occur in cells of the liver and kidneys, and death may follow within a few days of toxin intake. In man, small quantities of toxin (e.g. 5–10 mg) may be lethal. The amatoxins include α*-, β-, γ-, and ε-amanitin, and amanin. A *non*-toxic substance of similar chemical composition (amanullin) also occurs in *A. phalloides.*
*α-Amanitin inhibits RNA synthesis in eucaryotic cells. (see also MYCOTOXINS and FOOD POISONING (fungal))

amber codon The codon UAG (in mRNA) which codes for polypeptide chain termination—see PROTEIN SYNTHESIS and GENETIC CODE. (cf. OCHRE CODON, UMBER CODON)

amber mutation A NONSENSE MUTATION which gives rise to the AMBER CODON.

amber suppressor (see SUPPRESSOR MUTATION)

amboceptor (*immunol.*) *Current usage*: antibody to the surface antigens of erythrocytes (red blood cells); by combining with homologous amboceptor erythrocytes become *sensitized* and thus able to function as an indicator system for the detection of free COMPLEMENT. (see also HAEMOLYTIC SYSTEM and COMPLEMENT-FIXATION TEST)

ambrosia fungi Those species of fungi which grow in the tunnels made by wood-boring *ambrosia beetles* (pinhole borers) in felled timber and logs. Ambrosia fungi include deuteromycetes (e.g. *Ambrosiella*) and ascomycetes (e.g. *Endomycopsis*) while the (scolytoid) ambrosia beetles include species of *Platypus* and *Xyleborus*; a particular species of beetle is usually associated with a particular species of fungus. The beetles derive their nutrients mainly or solely from the fungal growth within their tunnels; thus, wood plays a minor or negligible part in their nutrition, and the fungus—sustained by insect wastes—does not cause significant damage (decay) to the infected wood. The fungus-beetle relationship is a stable one; the beetle carries fungal spores in specially adapted parts of its body (*mycetangia*)—thus ensuring the effective inoculation of new tunnels. (Ambrosia—'food of the Gods'—was the name given, in the 19th Century, to the (then) unknown source of food which lined the tunnels of the pinhole borers; when subsequently identified as fungal growth, the fungi found in such locations were termed *ambrosia fungi*.)

amerosporae (see SACCARDIAN SYSTEM)

aminoadipic acid pathway (AAA pathway) (for lysine biosynthesis) A pathway (see Appendix IV(e) (ii)) which occurs in the higher fungi and in certain algae (e.g. *Euglena*). (cf. DIAMINOPIMELIC ACID PATHWAY)

***p*-aminobenzoic acid** (PAB; PABA) A constituent of FOLIC ACID; PAB is synthesized from chorismic acid (an intermediate in the biosynthesis of aromatic amino acids—see Appendix IV(f)). Certain organisms—e.g. species of *Clostridium* and *Lactobacillus*—require PAB as a growth factor. (see also SULPHONAMIDES)

aminoglycoside antibiotics Antibiotics which contain an aminosugar, an amino- or guanido-substituted inositol ring, and one or more residues of other sugars. They include FRAMYCETIN, GENTAMICIN, KANAMYCIN, NEOMYCIN, PAROMOMYCIN, and STREPTOMYCIN; SPECTINOMYCIN is sometimes included. All (except spectinomycin) are bactericidal; they are broad-spectrum antibiotics but toxicity limits their clinical usage. Aminoglycoside antibiotics bind to the 30S subunit of 70S RIBOSOMES and inhibit PROTEIN SYNTHESIS; their effectiveness is reduced under anaerobic conditions, and they are ineffective against obligate anaerobes.

Mutations giving one-step high-level resistance to aminoglycoside antibiotics are uncommon—but see STREPTOMYCIN. Other mutations may produce lower levels of resistance, e.g. by decreasing the permeability of the cell envelope to particular antibiotics. Resistance may also be conferred by a PLASMID; thus, e.g. plasmid-mediated resistance to both kanamycin and neomycin has been demonstrated in *Staphylococcus*. Such plasmids may code for enzymes which structurally alter the antibiotic and thus nullify its activity; examples include streptomycin phosphotransferase, kanamycin acetyltransferase, and gentamicin adenylate synthetase.

aminoquinolines (as antimalarial agents) 4-Aminoquinolines (e.g. *chloroquine*) are active against the intraerythrocyte stages of PLASMODIUM, the malarial parasite; they are not active against those parasites which have become localized in the liver. These drugs appear to act by binding to the plasmodial DNA and thus inhibiting nucleic acid synthesis. 8-Aminoquinolines (e.g. *primaquine*) are effective against exoerythrocyte stages of the parasite; their mode of action is not clear—though it has been suggested that they may affect the mitochondria of *Plasmodium*. (see also QUININE and MALARIA)

***para*-aminosalicylic acid** (see PAS)

amitosis In the cells of certain eucaryotes: a form of nuclear division in which the chromosomes do not become visible (i.e. do not 'condense') and in which a spindle is not formed (cf. MITOSIS); the nuclear membrane persists throughout division, and the nucleus divides by constriction. Amitosis occurs e.g. in the macronuclei of ciliates and in certain fungi.

amixis In some *haploid* organisms (e.g. certain fungi): a quasi-sexual process in which karyogamy and meiosis do not occur. Thus, e.g. the eight (haploid) ascospores formed by *Podospora arizonensis* are reported to be derived by *mitotic* divisions of each nucleus of the dikaryon in the ascus mother cell.

ammonia assimilation Ammonia can be used as the sole source of nitrogen by a range of microorganisms. It may be incorporated e.g. in the reactions α-ketoglutarate $\rightarrow$ glutamate, glutamate $\rightarrow$ glutamine. (Glutamate, for example, functions as a donor of amino groups in a wide range of reactions.)

ammonification The release of ammonia from nitrogenous organic matter by microbial action; a number of reactions may be involved, e.g. deamination of amino acids. Ammonia does not normally accumulate under aerobic conditions—being readily assimilated by a wide range of organisms; it may accumulate under polluted or anaerobic conditions. (see also NITROGEN CYCLE and SAPROBITY SYSTEM)

Amoeba A genus of free-living protozoans within the order AMOEBIDA (superclass SARCODINA); species occur in soil and water. The organisms are naked, uninucleate, amoeboid cells (see MOTILITY) which feed mainly on small flagellates, ciliates, and bacteria; they reproduce asexually by binary fission. A common species, *A. proteus*, may reach 500 microns or more in the elongated form.

amoebiasis (1) *Amoebic dysentery.* An infectious disease of man and other animals; the causal agent is *Entamoeba histolytica.* (see also ENTAMOEBA) The clinical spectrum of disease ranges from intermittent diarrhoea to severe (sometimes fatal) dysentery; a carrier state is recognized. For disease to occur the CYSTS of *E. histolytica* must be ingested. Following excystment the organism localizes in the large intestine and attacks the mucosa—producing non-inflammatory lesions; the stools subsequently consist almost entirely of blood and mucus which contain large numbers of *trophozoites.* The pathogen may cause liver abscesses and/or may be disseminated to the lungs, brain etc. In chronic infections an equilibrium is apparently set up between host and parasite; the stools of such patients often contain cysts. EMETINE and METRONIDAZOLE have been used as chemotherapeutic agents. *Laboratory diagnosis* involves the demonstration of trophozoites and/or cysts in stools or lesions. (2) *Meningoencephalitis*: see NAEGLERIA.

amoebicides and amoebistats Chemical agents which (respectively) kill, or inhibit the growth and reproduction of, pathogenic or other amoebae. They include *glycobiarsol* (see ARSENICALS), FUMAGILLIN, EMETINE, and *Enterovioform* and *Diodoquin* (see IODINE). In earlier times certain dyes, e.g. MALACHITE GREEN, were used as antiamoebal agents. (see also ANTIPROTOZOAL AGENTS)

Amoebida An order of protozoans within the superclass SARCODINA. The organisms are naked, amoeboid cells in which locomotion is typically effected by *lobose* pseudopodia (see MOTILITY); some species are uninucleate while others are binucleate or multinucleate. Many species are free-living inhabitants of the soil and/or fresh or salt water, but some are parasitic or pathogenic in animals—including man. The order includes carnivorous, herbivorous, and omnivorous species. According to species, the organisms may grow aerobically, microaerophilically, or anaerobically—typically with some tolerance of non-optimal conditions. Reproduction occurs asexually, commonly by binary fission or PLASMOTOMY. CYST formation is common. Genera include AMOEBA, ENDAMOEBA, ENTAMOEBA, NAEGLERIA, and PELOMYXA. (see also DIENTAMOEBA FRAGILIS, IODAMOEBA BÜTSCHLII, and AMOEBICIDES AND AMOEBISTATS)

amoeboid movement (see MOTILITY)

Amorphoderma A genus of ascomycetous fungi whose precise taxonomic position is undecided. The organisms form ascocarps in which the peridium is a dark, non-cellular, amorphous membrane. Imperfect stages are classified in the *form*-genus *Cladosporium.*

Amorphotheca resina The ascigerous stage of *Cladosporium resinae.*

amphibiotic Refers to the ability of a given microorganism to behave either as a symbiont or as a parasite in respect of a given host. (see also OPPORTUNIST PATHOGENS)

amphibolic pathway (*biochem.*) A metabolic pathway which has both catabolic and anabolic functions—e.g. the TRICARBOXYLIC ACID CYCLE.

Amphidinium A genus of algae of the DINOPHYTA.

amphiesma (see THECA)

amphigenous (*mycol.*) Refers to an hymenium which is not restricted to one side or one surface of the fruiting body.

amphithecium (*lichenol.*) A synonym of THALLINE MARGIN.

amphitrichous (*bacteriol.*) Refers to the arrangement of a cell's flagella: *two* flagella, one located at each pole of the cell. (cf. MONOTRICHOUS; PERITRICHOUS)

amphitrophic (*adjective*) Refers to the ability of an organism to live photosynthetically in the light and chemotrophically in the dark.

amphotericin B (see POLYENE ANTIBIOTICS)

ampicillin (see PENICILLINS)

amygdaliform (*adjective*) Shaped like an almond.

amygdalin A compound which occurs in bitter almonds. It consists of the disaccharide gentiobiose linked to mandelonitrile. Each molecule of amygdalin yields, on acid hydrolysis, one molecule of benzaldehyde, two of glucose, and one of hydrogen cyanide. Amygdalin is metabolized by some strains of *Lactobacillus.*

α-amylases (α-(1→4)-D-glucan glucanhydrolases) Enzymes which effect random cleavage of α-(1 → 4) glucosidic linkages (e.g. in STARCH and GLYCOGEN) but which do not cleave α-(1→6) linkages. Thus, e.g. amylose is initially broken down to DEXTRINS, then to a mixture of maltose and glucose; amylopectin and glycogen are broken down to a mixture of *limit dextrins*, maltose, and glucose. α-Amylases are widely distributed among microorganisms.

β-amylases (α-(1 → 4)-D-glucan maltohydrolases) Enzymes which act on linear α-(1 → 4)-linked glucans—cleaving alternate bonds, from the non-reducing end of the chain, to form *maltose.* When e.g. amylopectin or glycogen is the substrate, the action of a β-amylase is halted at the

α-(1 → 6) branch points; in such cases the products are maltose and *limit dextrins*. β-Amylases are common in plants but are rare in bacteria.

amyloid (*mycol.*) Refers to those structures which stain blue, or darkly, in iodine-containing solutions—e.g. Meltzer's solution.

amylopectin (see STARCH)

amyloplast A starch-storing PLASTID.

amylose (see STARCH)

amytal A barbiturate which acts as a RESPIRATORY INHIBITOR. The main site of action is believed to occur between $NADH_2$-dehydrogenase and ubiquinone in the mitochondrial ELECTRON TRANSPORT CHAIN. Many bacterial systems are partially or completely insensitive to amytal.

Anabaena A genus of filamentous BLUE-GREEN ALGAE; species occur in freshwater and marine environments, and *A. azollae* forms a symbiotic relationship with the water fern *Azolla*. *Anabaena* species form AKINETES and HETEROCYSTS and carry out NITROGEN FIXATION. At least some species are motile. (see also GAS VACUOLE)

anabolism (*biochem.*) In a cell: those reactions in which cell components, products etc. are built up from organic and/or inorganic precursors. Anabolism requires an input of energy which may be derived from e.g. CATABOLISM. (see also ANAPLEROTIC SEQUENCES)

anaerobe Any organism which grows in the *absence* of air or oxygen (see also ANAEROBIOSIS); such an organism may be an OBLIGATE or FACULTATIVE anaerobe. (see also AEROTOLERANT) Most obligate anaerobes are bacteria; they include e.g. species of BACTEROIDES, BUTYRIVIBRIO, CLOSTRIDIUM, DESULPHOVIBRIO, and VEILLONELLA. No *obligately* anaerobic fungi are known. Obligately anaerobic protozoans occur e.g. in the RUMEN and in the intestinal tract of termites.

The energy-yielding metabolism of an obligate anaerobe may be *fermentative* (see FERMENTATION), *respiratory* (oxidative) (see ANAEROBIC RESPIRATION), or *photosynthetic* (see PHOTOSYNTHESIS).

Culture of obligate anaerobes. (a) *Surface culture.* A *strict* anaerobe requires an oxygen-free gaseous phase above the surface of the medium (see ANAEROBIC JAR) and an absence of dissolved oxygen in the medium; even under these conditions, however, many anaerobes fail to grow unless the medium has been *pre-reduced*, i.e. *poised* at or below a particular REDOX POTENTIAL. Poising agents include cysteine hydrochloride, ascorbic acid, and thioglycollate; the E_h of pre-reduced media is usually −150 mV to −350 mV at pH 7. The precise medium-E_h which supports the growth of a given anaerobe depends e.g. on (i) the size of the inoculum—ongoing growth tending to lower the E_h of the surrounding medium; (ii) the identity of the poising agent. The necessity for particular medium-E_h values has yet to be explained; some authors suggest e.g. that an unacceptably high medium-E_h (leading to a high intracellular E_h) may prevent reduction in redox couple(s) which are normally freely reversible within the growing organism. Apart from incubation, the handling of a strict anaerobe (e.g. transfer, inoculation) must be carried out in an oxygen-free environment—e.g. in a stream of (oxygen-free) nitrogen or carbon dioxide. *Aerotolerant* anaerobes do not require sophisticated methods of handling and generally grow on media which have not been pre-reduced. (b) *Culture within a medium.* See POUR PLATE, ROLL TUBE TECHNIQUE and SHAKE CULTURE. (see also BREWER'S THIOGLYCOLLATE MEDIUM, ROBERTSON'S COOKED MEAT MEDIUM)

Oxygen-sensitivity of anaerobes. Some anaerobes are killed by a brief exposure to molecular (gaseous) oxygen, while the growth of others may be stopped or merely retarded. The lethal or inhibitory effect of oxygen on anaerobes is the subject of a number of hypotheses. One hypothesis suggests that, owing to the absence of CATALASE, anaerobes are unable to cope with the hydrogen peroxide formed on exposure to oxygen; this hypothesis cannot have universal application since catalase is present in some oxygen-sensitive anaerobes. Other hypotheses suggest that oxygen may e.g. (i) raise the E_h of the medium (and cell interior) to unacceptably high levels, and/or (ii) bring about the reversible or irreversible oxidation of vital cellular components. Recently, oxygen-sensitivity has been associated with the lack of—or possession of sub-effective amounts of—the enzyme SUPEROXIDE DISMUTASE. This enzyme detoxifies the potentially harmful free radicals formed in all procaryotic and eucaryotic cells in the presence of oxygen. Superoxide dismutase appears to be present in all aerobic organisms and (often in smaller amounts) in aerotolerant anaerobes; it has been found to be absent from all strict, highly oxygen-sensitive anaerobes.

anaerobic (*adjective*) (1) (of an environment) Refers to the absence of oxygen—see ANAEROBIOSIS. (2) (of organisms) See ANAEROBE. (cf. AEROBIC; FACULTATIVE)

anaerobic jar (McIntosh and Fildes' anaerobic jar) A container used for the incubation of materials (e.g. inoculated media) in the absence of oxygen or, in general, under gaseous conditions other than those of the atmosphere. Essentially it consists of a strong cylindrical metal chamber with a flat, circular, gas-tight lid. Media etc. are placed in the jar, and the lid is secured by means of a screw clamp. The jar is then evacuated, by a suction pump, *via* one of two screw-controlled needle valves in the lid; this valve is then closed. A rubber bladder filled with e.g. hydrogen is attached to the other valve which is then opened to permit the entry of the gas under atmospheric pressure; this valve is then closed. The cycle of evacuation and re-filling may be carried out several times in succession. Attached to the inside of the lid is a gauze envelope which contains a catalyst—e.g. palladium-coated pellets of alumina—to promote chemical combination between hydrogen and the last traces of oxygen. The anaerobic jar often incorporates a side-arm to which is attached a small, thick glass vessel which communicates with the interior of the jar and which contains an indicator of anaerobiosis.

anaerobic respiration In certain bacteria: the use of inorganic electron acceptor(s) *other than oxygen* as terminal electron acceptors for energy-yielding oxidative metabolism (see RESPIRATION); the use of such electron acceptors may be either *facultative* (as in NITRATE RESPIRATION) or *obligate* (as in DISSIMILATORY SULPHATE REDUCTION). Members of the METHANOBACTERIACEAE use carbon dioxide as electron acceptor—methane being produced; certain species can also use carbon monoxide. (cf. FERMENTATION)

anaerobiosis Refers to the complete absence, or (loosely) the paucity, of oxygen in a given environment; such an environment may be suitable for the growth of ANAEROBES. In nature, anaerobic conditions occur e.g. in the bottom muds of ponds and rivers, and in the intestinal tracts of animals (see e.g. RUMEN); such conditions may be simulated e.g. by the use of an ANAEROBIC JAR. In aerobic habitats, anaerobic micro-environments may be created by the utilization of locally available oxygen by aerobic, MICROAEROPHILIC or FACULTATIVEly anaerobic organisms.

anaerogenic Refers to an organism which does not produce gas from a given substrate.

anamnaestic response (anamnestic response) A heightened immunological response (in persons or animals) to the second or subsequent administration of a particular antigen given some time after the initial administration. Thus, e.g. following further contact with the antigen, the concentration of specific antibody in the plasma increases at a much greater rate than would occur in subjects receiving their first administration of the antigen. (see also BOOSTER)

anaphase The third stage in MITOSIS and analogous stages in MEIOSIS.

anaphylactic shock A manifestation of IMMEDIATE HYPERSENSITIVITY which results from the union of allergen (antigen) with homologous *reaginic* (IgE) antibodies which are bound to MAST CELLS; the degranulation of mast cells (a complement-*independent* process) involves the release of HISTAMINE and other physiologically active agents. In man, the symptoms include difficulty in breathing and cyanosis; in severe cases death may occur within minutes or hours. In the guinea-pig death from asphyxiation is a common sequel to anaphylactic shock.

anaphylatoxins Biologically active fragments of certain components of COMPLEMENT; they are liberated during the fixation of complement. Anaphylatoxins (e.g. C3a) cause degranulation of MAST CELLS.

anaphylaxis (see IMMEDIATE HYPERSENSITIVITY)

Anaplasma A genus of gram-negative bacteria of the ANAPLASMATACEAE; species are parasitic within the erythrocytes of *ruminants*. Within the erythrocyte, the organism multiplies (by binary fission) to form an *inclusion body*—i.e. a group of coccoid bacterial cells (each less than 0.5 microns in diameter) enclosed by a membranous envelope; the inclusion body typically occupies either a marginal or central location within the erythrocyte. *Anaplasma* infection may give rise to symptoms of severe anaemia; anaplasmosis (transmitted by arthropod vectors) occurs in tropical and sub-tropical regions. *Anaplasma* was once thought to be a protozoan of the class Piroplasmea.

Anaplasmataceae A family of gram-negative bacteria of the RICKETTSIALES; species are parasitic within, or in association with, the erythrocytes of certain vertebrates—in which they may or may not give rise to disease. The organisms have not been grown *in vitro*. ANAPLASMA is one of five genera.

anaplerotic sequences (*biochem.*) Sequences of reactions which serve to replenish 'pools' of intermediates which have been depleted as a result of biosynthesis—see e.g. TRICARBOXYLIC ACID CYCLE.

Anaptychia A genus of fruticose or foliose LICHENS in which the phycobiont is *Trebouxia*. The apothecia are lecanorine. Species include e.g.:

A. fusca. The thallus is dark green when wet and brown when dry—the narrow (3 millimetres or less), branching, radial lobes forming a rosette; dark-coloured rhizinae occur on the lower surface. Apothecia may be abundant and have dark brown to black discs. *A. fusca* occurs on rocks on the seashore down to high tide mark, and occasionally on inland rocks.

A. ciliaris. The thallus consists of a tangle of narrow, flattened branches 1–3 millimetres wide; the upper surface is light grey to brownish-grey while the lower surface is pale to white and lacks rhizinae. The thallus is attached to the substratum by long marginal cilia. The apothecia have brownish-black discs with a blue-grey pruina. *A. ciliaris* occurs on rocks, bark, soil, etc.

anatoxin Synonymous with TOXOID.

Ancistrocoma A genus of protozoans of the THIGMOTRICHIDA.

Andrade's indicator A pH indicator often incorporated in PEPTONE-WATER SUGARS. *Preparation*: To 100 ml of a 0.5% solution of *acid fuchsin* in distilled water is added 15 ml of 4% sodium hydroxide. The solution is subsequently kept at room temperature for 1 day—or until the colour changes to a pale (almost colourless) yellow; the addition of a further 1 or 2 ml of alkali (no more) may be required if this has not been achieved after 24 hours. The indicator thus prepared may be autoclaved before or after its incorporation in the medium; 1 ml of indicator is added to 90 ml of medium. Andrade's indicator becomes red at approximately pH5.5.

androphage Any BACTERIOPHAGE which infects only *donor* strains of bacteria. (see CONJUGATION) Androphages include the filamentous DNA phages f1 and fd and the minute RNA phages f2 and Qβ; f1 and fd adsorb to the *tip* of a sex pilus (see SEX PILI) while Qβ and f2 adsorb to the *side* of a sex pilus.

anemochoric (of spores etc.) Dispersed by wind.

anergy (*immunol.*) The failure of an immunologically PRIMED person or animal to exhibit a HYPERSENSITIVITY reaction (on administration of antigen) when another similarly primed subject would normally exhibit such a response.

aneuploid Refers to any (eucaryotic) nucleus or cell which exhibits ANEUPLOIDY. (cf. EUPLOID)

aneuploidy (anorthoploidy) The condition in which the nucleus of a (eucaryotic) cell contains one or more complete *chromosomes* in excess of, or less than, the normal haploid, diploid, or polyploid number characteristic of cells of the species. (see also CHROMOSOME ABERRATION)

aneurin (vitamin B_1) (see THIAMINE)

angiocarpic development (angiocarpous development) (*mycol.*) In a fruiting body: a form of growth in which the spores develop and reach maturity within a closed chamber or cavity in the fruiting body. Fungi whose fruiting bodies develop angiocarpically include e.g. species of *Amanita* and *Scleroderma*. (cf. GYMNOCARPIC DEVELOPMENT)

Ångström unit (Å) A unit of length: 10^{-10} metres.

aniline dyes Dyes derived from aniline (aminobenzene, $C_6H_5NH_2$)—see e.g. TRIPHENYLMETHANE DYES.

animalcules (Animaliculae) An archaic term for microscopic organisms.

anisogamy Fertilization in which the gametes differ e.g. in shape, size, and/or behaviour.

anlage (*pl.* anlagen) (*ciliate protozool.*) Certain structures, particularly the macronucleus, during the early stages of their development.

annulus (*mycol.*) (see PARTIAL VEIL)

anonymous mycobacteria (see ATYPICAL MYCOBACTERIA)

Anopheles A genus of mosquito (order Diptera, suborder Orthorrhapha, family Culicidae, tribe Anophelini). *Anopheles* spp. are vectors of MALARIA in man and other animals.

ansamycins A term sometimes used to refer to a group of ANTIBIOTICS which includes the RIFAMYCINS and the STREPTOVARICINS.

antagonism (in antibiotic action) When two or more ANTIBIOTICS are acting together: the production of inhibitory effects (on a given organism) which are less than the additive effects of those antibiotics acting independently. (see e.g. LINCOMYCIN)

anterior station (*parasitol.*) The anterior part of the alimentary canal, or the salivary glands, of an arthropod vector. Pathogens which become infective in the anterior station of the vector can be transmitted to the vertebrate host by the bite of the vector, i.e. infection is *inoculative*.

antheridiol (see HORMONAL CONTROL (of sexual processes in fungi))

antheridium (*mycol.*) A male gametangium.

anthracnose (*plant pathol.*) (1) In e.g. beans, cucumbers, watermelons, etc.: a disease characterized by discrete, dark, necrotic lesions on the leaves and/or fruits; causal agents: members of the Melanconiales. In anthracnose of beans (*Phaseolus* spp), caused by *Colletotrichum lindemuthianum*, brown or black lesions—sometimes bearing masses of pink conidia—are formed on the leaves, pods and seeds. (2) The term may refer to diseases with symptoms similar to those described above, but which are caused by agents other than members of the Melanconiales.

anthrax ('splenic fever') An infectious disease of animals, including man, cattle, sheep, and pigs; the causal agent is *Bacillus anthracis* (see BACILLUS). In man, infection may occur *via* cuts etc. giving rise to cutaneous anthrax (*anthrax boil* or *malignant pustule*); spores may be breathed in, leading to pulmonary anthrax (*woolsorters' disease*), or they may be ingested, resulting in intestinal anthrax. Any form of anthrax may be fatal (with septicaemia and/or haemorrhages) if allowed to develop unchecked. A toxin formed by *B. anthracis* produces tissue necrosis and promotes oedema-formation. Among cattle etc. infection commonly leads to a rapidly fatal, septicaemic intestinal anthrax. Anthrax is a NOTIFIABLE DISEASE.

antibacterial agents Agents which kill, or inhibit

the growth of, bacteria—see e.g. ANTIBIOTICS, ANTISEPTICS, DISINFECTANTS.

antibiosis The antagonism of one organism towards another—e.g. by the production of ANTIBIOTICS.

antibiotic (1) In the original sense: any microbial product which, in low concentrations (of the order of micrograms/millilitre), is capable of inhibiting or killing (susceptible) microorganisms. The term was first used by Waksman in the 1940s; it distinguished the (then) newly-discovered drug penicillin from the (synthetic) sulphonamides which had been developed in the 1930s. Difficulties in terminology became apparent when the penicillins (and other natural antimicrobial products) were subjected to chemical modification—or synthesized *de novo* in the laboratory. (2) Currently, the term antibiotic is often extended to include drugs such as the SULPHONAMIDES, NALIDIXIC ACID, and the semi-synthetic PENICILLINS.

Many antibiotics are CHEMOTHERAPEUTIC AGENTS but others are too highly toxic to be of any clinical value; certain antibiotics (e.g. COLICINS) are used for non-therapeutic purposes (see e.g. COLICIN TYPING). Most of the antibiotics used as chemotherapeutic agents are active against bacterial pathogens; some (e.g. GRISEOFULVIN) are antifungal, others (e.g. FUMAGILLIN, SURAMIN) are antiprotozoal, while a few (e.g. the RIFAMYCINS) have limited antiviral activity. (See accompanying table; see also separate entries for further details of individual antibiotics.)

Some antibiotics are microbistatic while others are microbicidal; certain antibiotics may be microbistatic or microbicidal depending on conditions—e.g. the concentration of the antibiotic. (see also MIC) Some antibiotics have BROAD-SPECTRUM activity, while others are more selective. (Microorganisms may possess or acquire resistance to a given antibiotic: see ANTIBIOTICS, RESISTANCE TO.) Two or more antibiotics, used together, may exhibit SYNERGISM, ANTAGONISM, or merely an additive effect. The activities of a given antibiotic *in vivo* and *in vitro* do not necessarily correspond. Thus, e.g. an antibiotic which is active *in vitro*

ANTIBIOTICS: GENERALIZED FEATURES*

ANTIBIOTIC	SOURCE: SPECIES OF	SUSCEPTIBLE ORGANISMS: SPECIES OF	SITE/PROCESS AFFECTED
Azaserine	*Streptomyces*	(see entry)	glutamine metabolism
Bacitracin	*Bacillus*	Bacteria (Gram+ and –)	cell wall synthesis
Cephalosporins	*Cephalosporium*	Bacteria (Gram+ and –)	cell wall synthesis
Chloramphenicol	synthetic	Bacteria (Gram+ and –)	protein synthesis
Cycloheximide	*Streptomyces*	Fungi; other eucaryotes	protein synthesis
D-Cycloserine	*Streptomyces*	Bacteria (Gram+ and –)	cell wall synthesis
Enniatins	*Fusarium*	Bacteria (Gram+)	cell membrane
Erythromycin	*Streptomyces*	Bacteria (mainly Gram+)	protein synthesis
Fumagillin	*Aspergillus*	Amoebae; viruses	unknown
Fusidic acid	*Fusidium*	Bacteria (mainly Gram+)	protein synthesis
Gentamicin	*Micromonospora*	Bacteria (Gram+ and –)	protein synthesis
Griseofulvin	*Penicillium*	Fungi	DNA replication (?)
Isoniazid	synthetic	*Mycobacterium*	unknown
Kanamycin	*Streptomyces*	Bacteria (Gram+ and –)	protein synthesis
Lincomycin	*Streptomyces*	Bacteria (mainly Gram+)	protein synthesis
Nalidixic acid	synthetic	Bacteria (mainly Gram–)	DNA replication
Neomycin	*Streptomyces*	Bacteria (Gram+ and –)	protein synthesis
Nitrofurans	synthetic	Bacteria, fungi, protozoa	unknown
Novobiocin	*Streptomyces*	Bacteria (mainly Gram+)	DNA replication
Oleandomycin	*Streptomyces*	Bacteria (mainly Gram+)	protein synthesis
Penicillins	*Penicillium*	Bacteria (mainly Gram+)	cell wall synthesis
Polyenes	*Streptomyces*	Fungi, protozoa	cell membrane
Polymyxins	*Bacillus*	Bacteria (mainly Gram–)	cell membrane
Polyoxins	*Streptomyces*	Fungi	chitin synthesis (?)
Psicofuranine	*Streptomyces*	(see entry)	GMP biosynthesis
Puromycin	*Streptomyces*	Bacteria, protozoa	protein synthesis
Rifamycins	*Streptomyces*	Bacteria (mainly Gram+)	RNA biosynthesis
Ristocetin	*Nocardia*	Bacteria (Gram+)	cell wall synthesis
Streptomycin	*Streptomyces*	Bacteria (Gram+ and –)	protein synthesis
Sulphonamides	synthetic	Bacteria (Gram+ and –)	folic acid synthesis
Sulphones	synthetic	Bacteria (Gram+ and –)	folic acid synthesis
Tetracyclines	*Streptomyces*	Bacteria (Gram+ and –)	protein synthesis
Usnic acid	Lichen	Bacteria (Gram+); fungi	unknown
Vancomycin	Actinomycetes	Bacteria (Gram+)	cell wall synthesis

* See also individual entries for further details.

may, *in vivo*, become bound (e.g. to plasma proteins) so that only a fraction of the administered dose remains available for antimicrobial action. (see also e.g. *p-N*-succinylsulphathiazole under **SULPHONAMIDES**) In some cases an antibiotic (e.g. penicillin) may provoke an immunological **HYPERSENSITIVITY** reaction in certain patients.

Certain antibiotics (e.g. **NISIN**, **TETRACYCLINES**) have been used for the preservation of foodstuffs; this practice is generally discouraged since it tends to promote the development of antibiotic-resistant microbial strains.

antibiotics, resistance to Microorganisms may display an *intrinsic* resistance to the inhibitory or lethal effects of particular antibiotics. Such resistance may depend e.g. on the absence, or inaccessibility, of those structural and/or functional features against which the antibiotic is effective. Thus, e.g. *Mycoplasma* spp (which lack cell walls) are insensitive (resistant) to e.g. the **PENICILLINS**, while the resistance of some microorganisms to certain antibiotics results from the *impermeability* of the cells to those antibiotics.

Species which are normally inhibited or killed by a given antibiotic may give rise to strains which are resistant to that antibiotic. Resistance to a particular antibiotic is often associated with resistance to other—chemically-related—antibiotics; such *cross-resistance* occurs e.g. among the **SULPHONAMIDES** and among the **TETRACYCLINES**. (see however **PENICILLINS**) Some organisms are not only resistant to certain antibiotics—their growth actually depends on the presence of such antibiotics: see **STREPTOMYCIN**.

Mechanisms of acquired resistance. (a) *Mutation.* The cellular feature (e.g. enzyme) against which an antibiotic is effective may be so modified, as a result of a **MUTATION**, that it continues to function in the presence of otherwise inhibitory concentrations of that antibiotic. (see e.g. **STREPTOMYCIN** and **SULPHONAMIDES**) A single (resistant) mutant in a population of sensitive cells may thus give rise to a dominant clone in the presence of the particular antibiotic. Resistance acquired as a result of a single mutation is commonly operative against a single antibiotic or a group of cross-resistant antibiotics. (b) *Genetic transfer.* In bacteria, *chromosomal* genes governing resistance to an antibiotic may be transferred from a resistant to a sensitive cell by **CONJUGATION**, **TRANSDUCTION**, or **TRANSFORMATION**. Resistance genes may also be **PLASMID**-borne (see **R FACTORS**); plasmid-borne resistance may be transferred by conjugation or by transduction. The transfer of an R factor to a sensitive cell may confer on that cell resistance to one antibiotic, or to two or more unrelated antibiotics (i.e. *multiple drug resistance*). An R factor may code for the synthesis of (constitutive or inducible) *enzymes* which inactivate specific antibiotics (see e.g. **PENICILLINASE**), or may bring about changes in cell permeability to certain antibiotics.

Sensitivity tests. Following the isolation and identification of a pathogen, it is often necessary to carry out *in vitro* tests to determine the susceptibility of that pathogen to each of a range of appropriate antibiotics. The results of such tests assist the clinician in the selection of optimally active agents for chemotherapy. In one common form of sensitivity test a **FLOOD PLATE** is initially prepared from a diluted broth culture of the pathogen and several small absorbent paper discs—each impregnated with a different antibiotic—are placed on the surface of the agar. During subsequent incubation the antibiotics diffuse out into the surrounding agar, and zones of growth-inhibition occur around those discs which contain antibiotics to which the organism is sensitive. In an alternative form of the test, solutions of antibiotics are placed, individually, into a number of wells cut into the agar. (Strains of penicillin-*resistant Staphylococcus aureus* commonly form zones of growth-inhibition around discs containing penicillin. These zones differ from those formed by sensitive organisms in that their outer edges are bordered by heavy growth containing large, discrete colonies; zones formed by sensitive organisms exhibit a gradual or abrupt transition between regions of growth and non-growth. The no-growth zones formed by resistant strains of *S. aureus* are produced because **PENICILLINASE**-production in such strains is *inducible*; thus, cells close to the penicillin disc, which are exposed immediately to high concentrations of the antibiotic, are killed before adequate amounts of penicillinase can be synthesized. Cells at the periphery of the zone are exposed to gradually increasing concentrations of the outwardly-diffusing antibiotic; such cells can synthesize adequate amounts of penicillinase and thus give rise to full-sized colonies—using the nutrients forfeited by the inactivated cells in their immediate vicinity.)

In all such tests the standardization of procedure is important. Commonly used is a standardized medium, a *sensitivity agar*; this is a solid medium which contains no constituents which excessively enhance or reduce the inhibitory or lethal effects of the particular antibiotics being tested. (For example, the use of an agar containing blood or serum tends to reduce the inhibitory activity of those antibiotics which show a high degree of protein-binding.) The pH of the test medium is also important—as is the inoculum size and period of incubation. (see also TAMBOUR)

antibodies (see ANTIBODY)

antibody (ab) A glycoprotein molecule produced in the body in direct response to the introduction of an ANTIGEN or autocoupling HAPTEN. (cf. NATURAL ANTIBODIES; see also ANTIBODY-FORMATION) Antibodies exhibit the properties of IMMUNOGLOBULINS (for structures see IGG and IGM). An antibody is capable of combining specifically, non-covalently and reversibly with the antigen which elicited its formation; specificity is not necessarily absolute: see CROSS-REACTING ANTIBODIES. Combination takes place between specific COMBINING SITES on the antibody (see also VALENCY) and corresponding DETERMINANTS on the antigen; antigen and antibody must be present in OPTIMAL PROPORTIONS for maximum precipitation to occur. The union of antigen with certain types of antibody can initiate COMPLEMENT–FIXATION. (see COMPLEMENT-FIXING ANTIBODIES) Antigen–antibody combination *in vivo* may confer certain benefits on the host animal—e.g. when the antigenic stimulus derives from an invading pathogen: see e.g. ABLASTINS, OPSONINS; nevertheless, such combination may give rise to harmful effects: see HYPERSENSITIVITY.

Antibodies may be qualitatively investigated *in vitro* by e.g. GEL DIFFUSION and IMMUNOFLUORESCENCE techniques, and quantitatively investigated e.g. by AGGLUTINATION and PRECIPITIN TESTS; tests to detect and quantify specific antibodies are important in the diagnosis of a number of diseases (e.g. SYPHILIS). (see also e.g. AGGLUTININ, BLOCKING ANTIBODIES, CELL-BOUND ANTIBODIES, CYTOPHILIC ANTIBODY, INCOMPLETE ANTIBODIES, and REAGINIC ANTIBODY.)

antibody-absorption (see ABSORPTION)

antibody-excess In a PRECIPITIN TEST, a condition under which precipitation of antigen–antibody complex may not be detected. A visible precipitate is not formed under conditions of gross antibody excess since, according to the LATTICE HYPOTHESIS, the antigenic determinants of every molecule of antigen have been satisfied—leaving no sites free to form the links of larger (more readily precipitable) complexes. In *agglutination* tests, excess antibody may give rise to a PROZONE for analogous reasons. (see also OPTIMAL PROPORTIONS)

antibody-formation One form of IMMUNE RESPONSE. On initial contact with a given ANTIGEN a small CLONE of *specifically-reactive* B LYMPHOCYTES proliferates mitotically and undergoes TRANSFORMATION; cells of the enlarged clone either differentiate to become PLASMA CELLS (which liberate homologous antibody) or become *memory cells*—which have been 'primed' to react rapidly (in terms of antibody-production) on the second or subsequent contact with specific antigen. The formation of plasma cells and memory cells on initial contact with antigen constitutes the *primary response*. On subsequent contact with the *same* antigen the enlarged clone expands further and promptly gives rise to increased amounts of specific antibody; this is the *secondary* or *anamnaestic response*.

In the antibody-forming response to many antigens and haptens, B lymphocytes require the active cooperation of *T* lymphocytes and MACROPHAGES; the corresponding antigens are termed *T-dependent antigens*. The initial response to many T-dependent antigens is the early formation of low titres of IgM antibodies which persist for short periods; the secondary response typically consists of the production of both IgM and IgG antibodies—the latter being greatly in excess of the former. *T-independent antigens*—those to which the B lymphocytes can respond unaided—include bacterial LIPOPOLYSACCHARIDES, pneumococcal capsular polysaccharides and polymerized FLAGELLIN; antibodies formed against these antigens tend to be mainly or exclusively of the IgM type.

When produced in response to invasion by pathogenic microorganisms (or their products—e.g. toxins) antibodies commonly have a protective function—cf. IMMEDIATE HYPERSENSITIVITY. (see also ANTIGENIC COMPETITION and IMMUNOSUPPRESSION)

anticoagulant Any agent which inhibits the normal clotting (coagulation) of blood, e.g. EDTA, heparin, the sodium salts of oxalic and citric acids.

anticodon (see GENETIC CODE)

anticomplementary (*immunol.*) A term applied to any substance or system, *other than* the specific combination of antibody with antigen, which brings about COMPLEMENT-FIXATION. Such substances or systems, if present in the materials of a COMPLEMENT-FIXATION TEST, could give rise to a *false positive* reaction, i.e. by fixing complement they would simulate the effects of a specific antibody–antigen union.

antifungal agents (see FUNGICIDES AND FUNGISTATS)

antigen (ag) Any agent which initiates ANTIBODY-FORMATION* and/or induces a state of active immunological HYPERSENSITIVITY**. (cf. HAPTEN; see also IMMUNOGEN) The DETERMINANT of an antigen combines *specifically* with the *combining site* of the antibody which it elicits; more than one determinant may be present on a given antigen—microbial antigens often consist of a pattern of repeated antigenic determinants. (see also ANTIGENIC VARIATION) (Some microorganisms are characterized and classified on the basis of their surface antigens:

see e.g. KAUFFMANN–WHITE CLASSIFICATION SCHEME.)

Antigenic substances include proteins and polysaccharides; in general, large molecules are more antigenic than smaller ones. An antigen may be *particulate* (e.g. a bacterial cell) or *soluble* (e.g. a microbial toxin or extract). Antigens are commonly more effective (i.e. more *antigenic*) when administered with an ADJUVANT. Minute quantities of antigen are commonly sufficient to elicit antibody-formation; the actual amount required depends e.g. on the route of administration (oral, parenteral, etc.) and on whether or not an adjuvant is used. In some cases a few micrograms of antigen may stimulate a significant response. (see also ANTIGENIC COMPETITION and ALLERGEN, FORSSMAN ANTIGEN, H ANTIGENS, HETEROPHILE ANTIGENS, K ANTIGEN, NEOANTIGEN, O ANTIGENS, REITER ANTIGENS, SOMATIC ANTIGEN, VI ANTIGENS) *An individual does not normally produce antibodies against its own antigens—see however AUTOANTIGEN. **An antigen may also give rise to IMMUNOLOGICAL TOLERANCE.

antigen deletion In microorganisms, the loss of particular antigenic determinants due (for example) to (a) a mutation, or (b) the loss of an episome.

antigen excess In a PRECIPITIN TEST, a condition under which precipitation of antigen–antibody complex may not occur. Under conditions of gross antigen excess precipitate is not formed since, according to the LATTICE HYPOTHESIS, the combining sites of every molecule of antibody have been satisfied—leaving no sites free to form the links of larger (insoluble) complexes. (see also OPTIMAL PROPORTIONS)

antigen gain In microorganisms, the development of new antigenic determinants—due e.g. to: (a) A mutation. (b) The acquisition of a PLASMID. In bacterial CONJUGATION the SEX FACTOR specifies certain *donor-specific* antigens. (c) BACTERIOPHAGE CONVERSION—e.g. certain cell wall antigens in *Salmonella* spp are present only when the cells are lysogenized by a particular phage. (see also ANTIGENIC VARIATION)

antigenic competition Inhibition of the IMMUNE RESPONSE to a given antigen (or DETERMINANT) due to the administration of another, unrelated (i.e. non-cross-reacting) antigen (or determinant); antigenic competition may be regarded as a form of *natural* IMMUNOSUPPRESSION.

In *sequential competition* (one form of *intermolecular competition*) the immune response to a given antigen is inhibited as a result of the *previous* administration of another antigen. In a different form of intermolecular competition the immune response to an antigen is inhibited as a result of the *simultaneous* administration of the second antigen; by carefully adjusting ('balancing') the relative amounts of the antigens, antigenic competition may be reduced or abolished. (Such balancing is of importance e.g. in the preparation of TRIPLE VACCINE.)

In *intramolecular competition* the response to an antigenic determinant is inhibited by the presence of another determinant on the same molecule. For example, with IgG as antigen, good titres of anti-*Fc* antibodies are produced but relatively few anti-*Fab* antibodies are formed—fewer than would be formed against isolated Fab fragments. (*Fc* might be said to be the *immunodominant* portion of the IgG molecule.)

The mechanisms of antigenic competition are incompletely understood.

antigenic determinant (*immunol.*) (see DETERMINANT)

antigenic drifting (in viruses) Genetically-determined changes which occur, from time to time, in the specificity of viral antigens. (see also DRIFT (genetic)) Antigenic drifting occurs, for example, in the viral envelope haemagglutinins and neuraminidases of the influenza viruses; such drifting is an important factor in the epidemiology of influenza.

antigenic formula (of bacteria) A symbolic notation which indicates the identity of the known somatic and (where applicable) capsular and/or flagellar antigens. See also KAUFFMANN–WHITE CLASSIFICATION SCHEME.

antigenic variation (1) Periodic changes in the antigenic structure of the wild-type strain of a given microorganism. (2) Differences in the antigenic structure of individual organisms within a population. See: ANTIGEN GAIN, ANTIGEN DELETION, ANTIGENIC DRIFTING, PHASE VARIATION, SMOOTH–ROUGH VARIATION.

antigenicity (1) (*immunogenicity*) The capacity of a substance to function as an ANTIGEN, i.e. to elicit an IMMUNE RESPONSE. (2) The capacity of an antigen to unite with its homologous antibody.

antiglobulin Antibody formed against the antigenic DETERMINANTS of a serum globulin—usually an IMMUNOGLOBULIN. Antiglobulin–globulin complexes may be formed without involving the specific antibody *combining sites* (if any) of the globulin.

antiglobulin consumption test (*immunol.*) Any CONSUMPTION TEST which detects the combination of an IMMUNOGLOBULIN with an antigen by the subsequent consumption of added ANTIGLOBULIN. For example, the *in vitro* union of autoantibodies with homologous cell-surface antigens can be detected by washing the cells free from uncombined antibody and subsequently incubating them with antiglobulin; if antibody–antigen combination has occurred the antiglobulin will be taken up (consumed) by the antibody to form a cell–antibody–antiglobulin complex. The quantity of antiglobulin consumed can be determined in various ways, e.g. by titrating that which is *not* consumed against globulin-coated erythrocytes.

antilymphocyte serum (ALS) An antiserum raised in one species against the LYMPHOCYTES of another species. ALS is used in IMMUNOSUPPRESSION. Antilymphocyte globulin (ALG) is the globulin fraction of ALS.

antimetabolite A substance which competitively inhibits the utilization, by an organism, of an exogenous or endogenous metabolite. Sulphanilamide (for example) acts as an antimetabolite in respect of PABA.

antimony (and compounds) (as antimicrobial agents) (see HEAVY METALS)

antimutator gene (see MUTATOR GENE)

antimycin A An antibiotic which functions as a RESPIRATORY INHIBITOR in mitochondria, but generally not in bacteria. The main site of inhibition appears to be associated with *site II* of the ELECTRON TRANSPORT CHAIN.

antiplectic metachrony (see METACHRONAL WAVES)

antiport (counter-transport) Refers to the *linked* TRANSPORT of two solutes across a membrane *in opposite directions*. (cf. SYMPORT and UNIPORT)

antiprotozoal agents These include (agents followed by susceptible protozoans): EMETINE (*Entamoeba histolytica*); ETHIDIUM BROMIDE (*Trypanosoma spp*); METRONIDAZOLE (*E. histolytica*, *Giardia lamblia*, *Trichomonas vaginalis*); PYRIMETHAMINE (*Plasmodium* spp, *Toxoplasma*); SURAMIN (*Trypanosoma* spp). (see also AMOEBICIDES AND AMOEBISTATS and STERILIZATION)

antisepsis The promotion of ASEPSIS either by DISINFECTION or STERILIZATION—according to the sense in which *asepsis* is used. (see also ASEPTIC TECHNIQUE) *Antisepsis* most commonly refers to the disinfection of particular human or animal tissues (usually the skin) with non-toxic and non-injurious chemical agents known as ANTISEPTICS; in the United States this process is also known as *degerming*.

antiseptic paint A term sometimes used to refer to the ANTIBODY-containing secretions which bathe the mucous surfaces of the body.

antiseptics Chemical agents used in ANTISEPSIS; the term generally refers to those substances which are used to treat human and animal tissues (usually the skin) with the object of killing or inactivating those microorganisms which are capable of causing an infection. (see also DISINFECTANTS) Antiseptics include: ACRIFLAVINE, BISPHENOLS, CHLORHEXIDINE, HALOGENS, HEAVY METALS, QUARTERNARY AMMONIUM COMPOUNDS, and SALICYLIC ACID AND DERIVATIVES.

antiserum Serum containing the antibodies to a particular antigen or antigens.

antistreptolysin-O test (ASOT; ASLT) A test used to detect and quantify serum antibodies to STREPTOLYSIN-O. The serum is initially inactivated (see INACTIVATION). Serial dilutions of the serum are prepared, and to each dilution is added a fixed volume of a solution of streptolysin-O. The whole is incubated at 4 °C for several hours. A fixed volume of an erythrocyte suspension is then added to each dilution, the whole being re-incubated at 37 °C for 30 minutes. A high titre of serum antibodies is indicated by an *absence* of haemolysis in all tubes up to the higher dilutions of antiserum—or in all dilutions. The absence of haemolysis, in a given tube, indicates that the streptolysin-O has been fully neutralized by serum antibodies. Other forms of the test (e.g. a LATEX PARTICLE TEST) are also used.

antitoxin Antibody to a toxin; the term also refers to an antiserum which contains a particular antitoxin.

antiviral agents Compared with e.g. antibacterial or antifungal agents, the number of antiviral agents which are effective *in vivo* is relatively small. The majority of antiviral agents are obtained by chemical synthesis. Some have been used clinically (either for treatment or prophylaxis) while others are unsuitable as chemotherapeutic agents due e.g. to toxicity or to the rapid emergence of resistant mutants; some of the agents which are unsuitable for systemic use may be administered topically. Antiviral agents include: α-ADAMANTANAMINE, BENZIMIDAZOLE DERIVATIVES, CYTARABINE, GUANIDINE, 5-IODO-2′-DEOXYURIDINE, ISATIN-β-THIOSEMICARBAZONES, and RIFAMYCINS. (see also INTERFERONS) Agents which may inactivate viruses *in vitro* include: FORMALDEHYDE, GLUTARALDEHYDE, β-PROPIOLACTONE. Enveloped viruses can be inactivated by lipid solvents and lipases. (see also STERILIZATION)

Aphanocapsa A genus of unicellular BLUE-GREEN ALGAE.

Aphanomyces A genus of fungi of the OOMYCETES.

aphthous fever Synonym of FOOT AND MOUTH DISEASE.

Aphthovirus (see RHINOVIRUS)

Aphyllophorales An order of fungi of the subclass HOLOBASIDIOMYCETIDAE. Constituent species form well-developed, gymnocarpous fruiting bodies in which the hymenium may occur in flat or wrinkled sheets, may line tubules (which are *firmly* attached to the pileus—cf. AGARICALES), or may cover the surfaces of a layer of tooth-like processes; the holobasidia bear small, inconspicuous sterigmata. According to species the basidiocarp may be e.g. crust-like, sessile, or stipitate, and may be corky, leathery, or woody (cf. AGARICALES); the hymenium may develop on one side or surface of the fruiting body or it may be amphigenous. Members of the Aphyllophorales are typically saprophytic fungi which occur principally on soil and wood. The families of the order include e.g. CONIOPHORACEAE, Fistulinaceae, GANODERMATACEAE, HYDNACEAE, POLYPORACEAE, and Stereaceae.

apical (*mycol.*) Terminal: at the tip or distal part of a structure. *Antonym*: *basal.*

apical granule (*mycol.*) (see SPITZENKÖRPER)

Apistonema A genus of algae of the HAPTOPHYTA.

aplanogamete A non-motile gamete.

aplanospore A non-motile spore.

apochlorotic (*adjective*) Refers to algae which have become 'colourless' (achlorophyllous) either through loss of chloroplasts or through loss of chlorophyll; in the latter case a colourless plastid (*leucoplast*) is formed. (see also BLEACHING)

apochromatic objective (apochromat) An objective lens (see MICROSCOPY) in which chromatic aberration has been corrected for *three* colours and spherical aberration for *two* colours. (cf. ACHROMATIC OBJECTIVE)

apoenzyme (see COFACTOR)

apomict Any organism which undergoes APOMIXIS.

apomixis A type of asexual reproduction considered to be a deviant form of the sexual process in that meiosis and syngamy do not occur; an apomict may develop morphologically differentiated regions—i.e. the outward appearances of sexuality may be exhibited. Progeny produced apomictically are genetically identical with their parent cells. (cf. AMIXIS)

apo-repressor (see OPERON)

Aporpium A genus of fungi of the order TREMELLALES.

Apostomatida An order of protozoans of the subclass HOLOTRICHIA (subphylum CILIOPHORA). The *apostomes* are mainly or exclusively marine organisms whose life cycles are intimately associated with crustacean hosts. *Mature* apostomes have a spirally arranged somatic ciliature, and typically possess an organelle (?), the *rosette*—function unknown, near the non-apical cytostome. The mature form (*trophont*) is the actively feeding and growing stage. The trophont commonly encysts (enters the *tomont* stage), undergoes repeated divisions, and liberates (from the cyst) a number of small, motile individuals (*tomites*) in which the (longitudinal) ciliature does not exhibit the conspicuous spiral arrangement evident in the trophont. The tomites subsequently encyst (*phoront* stage) at an appropriate site on a suitable host (e.g. the gills of certain species of crab). The phoront develops into a young trophont. Genera include *Foettingeria*, *Gymnodinioides*, *Polyspira*.

apothecioid pseudothecia (see ASCOSTROMA)

apothecium (*mycol.*, *lichenol.*) Typically: a dish-shaped or cup-shaped ASCOCARP—the inner surface of which is lined with a palisade-like layer of asci and paraphyses (collectively referred to as the *hymenium*); apothecia are the characteristic sexual fruiting bodies of fungi of the class DISCOMYCETES. The fleshy (sterile) body of the apothecium, the *excipulum*, may be borne upon a stalk or may be in direct contact with the substratum; the thin layer of mycelial mesh between the hymenium and the excipulum is termed the *subhymenium* or *hypothecium*. Some apothecia are not cup-like; thus, certain members of the PEZIZALES form hollow, ridged and pitted ovoid structures which bear asci on their outer surfaces (see MORCHELLA) or saddle-shaped structures which bear asci on their outer (upper) surfaces (*Helvella* spp). Apothecia may be barely visible to the naked eye or may attain diameters of 10 or more centimetres, according to species; those of some species are brightly coloured.

apple, diseases of These include: (a) *Scab*. Apple scab is important in temperate, humid regions. The causative fungus, *Venturia inaequalis*, may enter the host plant *via* wounds or by means of appressoria. (see also VENTURIA) Small dark spots appear initially on the undersides of the leaves but later develop on both surfaces. On the fruit: small, dark, velvety spots enlarge to form brown corky lesions. (b) FIRE BLIGHT. (see also PLANT PATHOGENS AND DISEASES)

appressed Of e.g. an organism: lying close to, or flattened against, the substratum.

appressorium (*mycol.*) A specialized, flattened region of a hypha or germ tube produced by certain plant-pathogenic fungi (e.g. *Puccinia graminis*, *Venturia inaequalis*); an appressorium facilitates penetration, by the fungus, of the outer layer(s) of the host organism. The appressorium adheres closely to the surface of the host and gives rise to a small *infection peg* (an outgrowth of fungal tissue) which penetrates the superficial layer(s) of the host.

APS Adenosine-5′-phosphosulphate (also called adenosine-5′-sulphatophosphate *or* adenylylsulphate). (see also ASSIMILATORY SULPHATE REDUCTION and Appendix V(c))

arabinose An aldopentose found e.g. in plants (see HEMICELLULOSES) and in the cell walls of certain bacteria e.g. *Nocardia* spp. It is metabolized by a number of bacteria including many strains of *Escherichia coli* and some strains of *Bacillus*, *Lactobacillus*, *Salmonella*, and *Streptococcus*. Arabinose is heat-labile: solutions should be sterilized by filtration.

arachnoid Resembling a spider's web.

arboviruses (arthropod-borne viruses) Those VIRUSES which can multiply in both arthropod and vertebrate hosts. Arboviruses occur in various taxa—e.g. the genus ORBIVIRUS, the family TOGAVIRIDAE, and the heterogenous assembly of viruses known as the Bunyamwera Supergroup.

arbuscule (arbuscle) The branched, rhizoidal terminus of an intracellular hypha found in certain types of endotrophic MYCORRHIZA. Arbuscules, and vesicles (swollen intercellular or intracellular hyphal endings), are characteristically formed by certain lower fungi—e.g. species of *Endogone* and *Pythium*; the association of such fungi with the plant root is referred to as an *arbuscular-vesicular* mycorrhiza. Such mycorrhizas are common in angiosperms.

Arcella A genus of protozoans: see ARCELLINIDA.

Arcellinida (*synonym*: Testacida) An order of freeliving, testate (shelled), amoeboid protozoans within the superclass SARCODINA; the organisms are found predominantly or exclusively in fresh water and/or damp soil and vegetation—their food consisting mainly of small algae (e.g. diatoms). The tests of the arcellinids are constructed from secreted organic materials which, in some species, are reinforced either with grains of sand (e.g. in *Difflugia*) or with silicaceous plates which are pre-formed internally (e.g. in *Euglypha*); the translucent or transparent test of *Arcella* lacks reinforcement. The arcellinid test has a single aperture through which the pseudopodia are extruded; pseudopodia may be lobose (e.g. in *Arcella*, *Difflugia*) or filopodial (e.g. in *Euglypha*). According to species the organisms have one or more nuclei. Reproduction occurs asexually by binary fission—a new test being formed by one of the daughter cells. *Arcella* spp are commonly 50–150 microns (maximum dimension), *Difflugia* spp 200–250 microns, and *Euglypha* spp 125–150 microns.

archicarp (*mycol.*) (1) The cell, or cells, from which a fruiting body develops. (2) The entire female gametangium, or that part of it which receives the male nucleus or nuclei.

arcuate Curved (like a bow); arched.

arenaceous (of the test or shell of a protozoan) Incorporating, or bearing a surface covering of, fine grains of sand.

Arenavirus A genus of VIRUSES whose hosts include rodents and other vertebrates; arenaviruses include the causal agents of e.g. LASSA FEVER and lymphocytic choriomeningitis (LCM) in man. The pleomorphic, enveloped (ether-sensitive) *Arenavirus* virion (80–300 nm, maximum dimension) contains a fragmented genome of single-stranded RNA; within each virion there is a number of electron-dense granules (each 20–30 nm) which appear to have the properties of the host cell ribosomes. The arenaviruses share a group-specific antigen; LCM and Lassa viruses are more closely related (antigenically) to each other than to the other members of the genus. Arenaviruses multiply within the *cytoplasm* of the host cell and mature by budding through the cytoplasmic membrane.

areolate Refers e.g. to a crustose lichen thallus which is divided up into small, scale-like areas (*areolae*) which are separated from each other by cracks or depressions in the thallus.

argentophilic Staining well with silver stains.

arginine (biosynthesis of) (see Appendix IV(a))

arginine dihydrolase An enzyme system present e.g. in species of *Pseudomonas*, *Streptococcus*. It consists of (i) *Arginine deiminase* which catalyses a hydrolytic conversion of arginine to citrulline and ammonia. (ii) *Ornithine carbamoyltransferase* which catalyses a phosphorolytic cleavage of citrulline—forming carbamoyl phosphate and ornithine. (iii) *Carbamate kinase* which transfers, to ADP, the phosphate group of carbamoyl phosphate; ATP is thus formed, the carbamic acid dissociating to CO_2 and NH_3.

argyrol (see *silver* under HEAVY METALS)

argyrophilic Staining well with silver stains.

Arhynchodina A sub-order of protozoans of the THIGMOTRICHIDA.

Arizona (see SALMONELLA)

Armillaria A genus of fungi of the family TRICHOLOMATACEAE. *A. mellea* (the 'honey fungus') grows saprophytically on many types of wood, and parasitically on a range of types of tree and shrub; the organism also grows in symbiotic association with the roots of certain orchids (see MYCORRHIZA). The fruiting body of *A.mellea* typically consists of a convex or flattish, yellow-brown, fleshy pileus (5–15 centimetres in diameter) which, when young, bears on its upper surface brownish or olivaceous scales; the pileus is supported on a central stipe which usually bears a conspicuous annulus. The white to flesh-coloured gills are generally adnate to decurrent, and the smooth-walled, ellipsoidal, hyaline, non-amyloid basidiospores (each approximately 8–9 x 5–6 microns) appear pale cream-coloured *en masse*. The fruiting bodies typically occur in clusters, and the organism commonly gives rise to tough, black rhizomorphs ('boot laces'). (see also TIMBER, DISEASES OF)

arsenicals (as antimicrobial agents) Many aromatic, arsenic-containing compounds ('arsenicals') have antimicrobial activity, and some have been used as therapeutic agents;

however, owing to their toxicity, arsenicals have now been largely superceded by ANTIBIOTICS. The antimicrobial action of arsenicals appears to involve their reaction with thiol (–SH) groups within cells—thus leading to the inhibition of many enzymes; reactions involving the coenzyme LIPOIC ACID are particularly sensitive since arsenic bridges the two thiol groups in this coenzyme. It is believed that pentavalent arsenic (in an arsenical) must be reduced, *in vivo*, to trivalent arsenic before it can function as an antimicrobial agent.

Atoxyl (*p*-aminophenylarsonic acid, $NH_2.C_6H_4.AsO_3HNa$) was the first arsenical to be used therapeutically—being used in the treatment of SLEEPING SICKNESS. Certain arsenicals are still used in the treatment of the later stages of certain forms of trypanosomiasis; these include *tryparsamide* (*p-N*-phenylglycineamidoarsonic acid) and *melarsoprol* (a derivative of benzenearsenous acid). (Arsenicals are not active against *Trypanosoma cruzi*.) Resistance to the arsenicals develops fairly readily in trypanosomes which have been subjected to sublethal concentrations; this is thought to be due to the decreased uptake of the drug by the organisms. *Glycobiarsol* (bismuth *N*-glycolyl-*p*-arsanilate) has been used in the treatment of *amoebic* DYSENTERY. *Salvarsan* (3,3′-diamino-4,4′-dihydroxyarsenobenzene) used to be a standard therapeutic agent for the treatment of SYPHILIS but has now been superceded by antibiotics.

The selective action of arsenicals is thought to be due to differences in the permeability of different types of cell to these compounds.

artefact (artifact) Any feature which does not occur in a cell or organism under natural conditions but which may be observed in that cell or organism (or part thereof) during experimentation. Artefacts are due to the disturbance introduced by the process of experimentation or observation; they may arise, for example, as a result of FIXATION.

Arthrobacter A genus of gram-positive*, obligately aerobic, chemoorganotrophic, asporogenous soil bacteria of uncertain taxonomic affinity; the organisms exhibit a growth cycle of the type: coccoid cells → (pleomorphic) rods → coccoid cells—the latter form predominating in old cultures. All species have a non-fermentative, non-cellulolytic metabolism, and most reduce nitrates; growth occurs optimally at about 25 °C. Type species: *A. globiformis*; GC% range for the genus: about 60–70. *Only parts of the rod-shaped forms may exhibit gram-positivity.

Arthroderma A genus of fungi of the order EUROTIALES; the conidial (imperfect) stages of at least some species are members of the *form*-genus TRICHOPHYTON.

arthrospore (1) (*mycol.*) Synonym of OIDIUM. (2) (*bacteriol.*) The smaller of the two types of coccoid cells formed by *Arthrobacter*.

Arthus reaction A form of type 3 reaction (see IMMEDIATE HYPERSENSITIVITY) exhibited some hours after the administration of a precipitating antigen to an individual whose plasma contains a high titre of the homologous antibody; characteristically, a severe localized inflammatory reaction occurs, within 3–12 hours, at the site of the subcutaneous injection. The *insoluble* antigen–antibody complexes precipitate intravascularly at the site of the injection; this may cause blockage of capillaries by aggregations of platelets. COMPLEMENT-FIXATION causes local tissue damage (e.g. by REACTIVE LYSIS) and, owing to the release of chemotactic factors, promotes infiltration by neutrophil polymorphs. The region becomes erythematous and oedematous and (locally) haemorrhagic and necrotic. (see also SERUM SICKNESS)

artifact (see ARTEFACT)

arylsulphatase test (see MYCOBACTERIUM)

ascarylose (3,6-dideoxy-L-mannose) A sugar found e.g. in the cell wall LIPOPOLYSACCHARIDE of certain strains of *Yersinia pseudotuberculosis*. (Ascarylose was first isolated from the eggs of *Ascaris* worms.)

Aschaffenburg–Mullen phosphatase test (see PHOSPHATASE TEST (for milk); see also MILK (microbiological aspects))

ascigerous (*adjective*) (1) Refers to the stage in the life cycle of an ascomycete in which *asci* are formed (see ASCUS): the perfect (i.e. sexual) stage. (2) Refers to those species which form asci. (3) Bearing or giving rise to asci.

ascitic fluid (*med.*) Fluid which collects in the peritoneal cavity during certain pathological conditions.

ascocarp (*mycol.*) Any of the various types of fruiting body formed during the sexual (ascigerous) stage in the life cycles of the ascomycetes; the asci (see ASCUS) develop within or at the surface of the ascocarp. (Members of the class Hemiascomycetes do not form ascocarps.) The main forms of ascocarp are the APOTHECIUM, ASCOSTROMA, CLEISTOTHECIUM, and PERITHECIUM. In addition to asci an ascocarp may bear or contain certain sterile structures: see PARAPHYSES, PERIPHYSES, and PSEUDOPARAPHYSES. In some (*ascohymenial*) species ascocarp development appears to *follow* the sexual stimulus—i.e. the fruiting body begins to develop only after plasmogamy has occurred; in these species the ascus-forming system is initiated in or on undifferentiated somatic mycelium and the wall (peridium) of the ascocarp develops subsequently. In other (*ascolocular*) species (e.g. members of the class LOCULOASCOMYCETES) the fruiting body begins to develop (in the form of a stroma) before plasmogamy has taken place; in these species the ascus-forming system is initiated *within* the preformed stroma or stromatal initial (see also ASCOSTROMA).

Ascochyta A genus of fungi within the order SPHAEROPSIDALES; species include e.g. the causal agent of *leaf spot* of pea (see PEA, DISEASES OF; see also *pisatin* under PHYTOALEXINS). *Ascochyta* spp form brown, elongated, transversely-septate, appendage-free conidia which are characteristically rounded at one end and tapered at the other end; the spores develop singly on conidiophores which are borne in brown, thick-walled, non-setose, unilocular, ostiolate pycnidia that are immersed in the substratum.

ascogenous hyphae (see ASCUS)

ascogone A synonym of ASCOGONIUM.

ascogonium (archicarp; ascogone) (*mycol.*) In ascomycetes: the female gametangium. An ascogonium may be single-celled or multicellular, and either simple or complex in form; it may or may not bear a TRICHOGYNE, according to species.

ascohymenial (see ASCOCARP)

Ascoideaceae A family of fungi of the ENDOMYCETALES.

ascolichen A lichen in which the mycobiont is an ascomycete.

Ascoli's thermoprecipitin test (Ascoli's test) A serological PRECIPITIN TEST used to detect the soluble antigens of *Bacillus anthracis* (the causative agent of ANTHRAX) in various animal products (e.g. hides). The material is extracted with boiling saline (or other solvent) and the extract is layered over a known positive antiserum in a RING TEST.

ascolocular (see ASCOCARP)

ascoma (*pl.* ascomata) A synonym of ASCOCARP.

ascomycetes Fungi of the sub-division ASCOMYCOTINA. (In some taxonomic schemes these fungi form the *class* Ascomycetes.)

Ascomycotina (the 'ascomycetes') A sub-division of FUNGI of the division EUMYCOTA; constituent species are distinguished from those of other sub-divisions in that their sexually derived (or, exceptionally, parthenogenically derived) spores (ascospores) are formed within *asci* (see ASCUS). The classes of ascomycetes are HEMIASCOMYCETES, PLECTOMYCETES, PYRENOMYCETES, DISCOMYCETES, LABOULBENIOMYCETES, and LOCULOASCOMYCETES.

ascorbic acid (vitamin C) L-*threo*-2,3,4,5,6-Pentahydroxy-2-hexenoic acid-4-lactone (oxidized form: dehydroascorbic acid); it is sometimes used as a reducing agent in culture media for anaerobic microorganisms. Some authors state that ascorbic acid is a growth requirement of certain parasitic flagellate protozoans; however, this view is not universally accepted. Ascorbic acid is synthesized e.g. by certain algae and by strains of *Aspergillus niger*.

ascospores Uninucleate, binucleate, or multinucleate spores formed within an ASCUS; ascospores may be non-septate, transversely-septate, or muriform, and may be any of a variety of shapes, sizes, and colours—according to species. Ascospores typically germinate to form one or more germ tubes; those of yeasts characteristically give rise to budding cells.

ascospores, discharge of In many species of ascomycete the mature ascospores are ejected forcibly or explosively from their asci; the precise mechanism involved is not fully understood. In species which form unitunicate asci the ascospores may be expelled *via* a pore, slit, or operculum in the apex of the ascus, or may be scattered on the explosive disintegration of the ascus wall (according to species); the mature ascus seems to absorb water and become turgid—subsequently bursting or ejecting the ascospores through the pre-formed apical discharge apparatus. In some cases this behaviour has been attributed to an increase in osmotic pressure within the ascus as a result of the rapid enzymic breakdown of a polysaccharide (e.g. glycogen); in other cases it has been postulated that ascus swelling may result from the uptake of water by a hydrophilic gel or mucilage within the ascus. Different mechanism(s) may be involved in different species.

Asci which are formed in closed ascocarps (cleistothecia) appear to swell up, when mature, and to thus rupture the cleistothecial wall; in some species the protruding (or totally freed) asci release their ascospores by violent disintegration. In *Myriangium* spp (in which the asci are borne in uniascal locules within an ascostroma) ascospore discharge is necessarily preceded by the disintegration of the stromatic tissue—which occurs as a result of weathering; once exposed to the environment the asci absorb water, swell, and forcibly discharge their spores *via* the ascal apex. Certain ascomycetes (e.g. *Sordaria fimicola*) produce their asci in an hymenial layer within a *short*-necked perithecium; when mature, each ascus in turn elongates, discharges its spores *via* the ostiolar pore, and then retracts. Asci formed in *long*-necked perithecia may, at maturity, become detached from the hymenium and subsequently blown through the ostiole as a result of pressure within the perithecial cavity; however, in species of *Ceratocystis* (which also form long-necked perithecia) the asci are evanescent, and the mature *ascospores* are slowly extruded in a slimy mass through the ostiole. Discomycetes often release their ascospores synchronously in a visible spore cloud; such a phenomenon is known as 'puffing'. The discharge of *bitunicate* asci initially involves the rupture of the outer ascus wall and the subsequent elongation and enlargement of the inner ascal chamber.

In e.g. *Tuber* spp the ascospores are not discharged violently; in such hypogean species spore dispersal is mediated by burrowing and digging animals.

Ascospore discharge may be strongly influenced by environmental conditions (e.g. light, humidity, etc.) and a distinct periodicity of spore discharge (a so-called endogenous rhythm) has been detected in a number of species. In fungi which form deeply concave (cupulate) apothecia, the asci are commonly positively phototropic; asci which line the near vertical walls of the apothecium thus direct their spores outwards (into the environment) rather than towards the opposite wall of the ascocarp. In some fungi, e.g. *Daldinia concentrica*, ascospore discharge occurs mainly or exclusively in the dark.

ascostroma (*mycol.*) An ASCOCARP which consists of a pseudoparenchymatous or prosenchymatous STROMA within which one or more cavities (*locules*) subsequently develop—each locule containing one or more asci; ascostromata are formed e.g. by members of the class LOCULOASCOMYCETES. This type of ascocarp is said to be formed by *ascolocular* development—see ASCOCARP.

Each locule in an ascostroma consists of an *unwalled* CENTRUM—i.e. the asci (with or without pseudoparaphyses, according to species) are bounded only by the stromatic tissue of the locule; this contrasts with the ascocarp structure of the typical stromatic pyrenomycetes in which a distinct peridium is present. A uniloculate ascostroma (i.e. a stroma containing a single locule) may contain an hymenial layer of asci and may thus resemble a perithecium or (infrequently, but e.g. in *Rhytidhysterium*) an apothecium; such structures are referred to, respectively, as a *perithecioid pseudothecium* (= *pseudoperithecium*) and an *apothecioid pseudothecium*. An ascostroma may develop on the surface of, or within, a substratum; ascostromata which develop within a substratum may or may not be erumpent. In some species (e.g. *Myriangium* spp) the asci may develop within uniascal locules which are scattered irregularly in the stromal tissue; such an ascostroma is termed a *conceptacle*.

ascus (*pl.* asci) A microscopic sac-like structure in which are formed the sexually derived (or, exceptionally, the parthenogenically derived) spores (ASCOSPORES) of fungi of the subdivision ASCOMYCOTINA. Asci are commonly elongate, usually clavate or cylindrical, but may be spherical or otherwise; each ascus typically contains eight ascospores—but some species regularly form asci which each contain less than eight (often four) or more than eight (sometimes several hundred) ascospores. Asci may be formed in or on an ASCOCARP, and may be formed singly, in groups, or in a closely-packed layer called an *hymenium*; according to species, the asci of an hymenium may or may not be interspersed with certain sterile structures: see PARAPHYSES and PSEUDOPARAPHYSES. An ascus may be UNITUNICATE or BITUNICATE (an important taxonomic feature), and may or may not possess an apical pore or an OPERCULUM through which the ascospores may be discharged. (see also ASCOSPORES, DISCHARGE OF) According to species, an ascus may be evanescent or persistent. Ascospores are liberated from an evanescent ascus by the dissolution of the ascus wall; ascospores formed within persistent asci are commonly discharged forcibly or explosively.

Ascus formation is usually preceded by plasmogamy. Plasmogamy may occur e.g. by GAMETANGIAL CONTACT, SPERMATIZATION, or SOMATOGAMY, according to species; in the ascogenous yeasts plasmogamy occurs between somatic cells and may occur between ascospores. Some ascomycetes are HOMOTHALLIC; other species undergo sexual processes only with compatible mating partners—see COMPATIBILITY (in fungal sexuality). Following plasmogamy the course of ascus formation depends on species.

Ascus formation in ascogenous yeasts. After plasmogamy, karyogamy commonly follows without delay. The zygote may undergo meiosis immediately (as e.g. in *Schizosaccharomyces octosporus*) so that the ascus develops directly from the zygote; alternatively, meiosis may be delayed for one or more mitotic divisions (e.g. the diploid budding phase in *Saccharomyces cerevisiae*) in which case an ascus develops from one of the diploid descendents of the zygote. Following meiotic division the four (haploid) nuclei may develop directly into ascospores (as e.g. in *S. cerevisiae*) or there may be a subsequent mitotic division with the eventual formation of eight ascospores (as e.g. in *S.octosporus*). The ascospores develop by *free cell formation*—see below. Some yeasts form less than four, or more than eight, ascospores per ascus. Some ascogenous yeasts are heterothallic.

Ascus formation in other ascomycetes. The

details of the ascigerous stage vary from species to species; the following is a *generalized* account. After plasmogamy, karyogamy is typically delayed. One or more hyphae (the *ascogenous hyphae*) arise from the ascogonium, and the (haploid) male and female nuclei migrate into these hyphae. The tip of each hypha then curves to form a crook, or *crozier*, such that two nuclei (presumably one from each gametangium) occur in the curved upper portion of the crozier. These nuclei then undergo mitosis, simultaneously, with their mitotic spindles arranged parallel to the long axis of the ascogenous hypha; the subsequent formation of a septum creates a terminal cell (the *ascus mother cell*) which contains one daughter nucleus from each of the two original nuclei. Karyogamy now occurs. The zygote undergoes meiosis and one or more mitotic divisions to produce the number of haploid nuclei (ascospore initials) characteristic of the species. Ascospores develop by a process known as *free cell formation*: a wall develops around each haploid nucleus so as to include that portion of cytoplasm immediately adjacent to it; the epiplasm may provide nourishment for the developing ascospores. A number of croziers (and hence asci) may develop from the branches of a single ascogenous hypha.

ascus mother cell (see ASCUS)

asepsis (1) The state in which potentially harmful microorganisms are absent from particular human or animal tissues, from materials, or from an environment; in this sense *asepsis* does not necessarily involve sterility (see STERILE). (2) A state of sterility. (see also STERILIZATION, ANTISEPSIS and ASEPTIC TECHNIQUE)

aseptate (*mycol.*) Lacking *septa*—see SEPTUM.

aseptic meningitis (see MENINGITIS)

aseptic technique Precautionary measures taken in microbiological work and clinical practice to prevent the contamination of cultures, sterile media, etc.—and/or the infection of persons, animals, or plants—by extraneous microorganisms. The working surfaces of instruments (forceps, bacteriological LOOPS etc.), and the rims of bottles and other vessels used for dispensing sterile materials, are rendered STERILE by FLAMING. No non-sterile object is allowed to come into contact with the protected material. The risk of contamination may be further reduced by treating the surfaces of benches etc. with antimicrobial solutions (DISINFECTANTS) and/or with e.g. ultraviolet radiation (see STERILIZATION); the ambient atmosphere may be filtered to remove the spores of bacteria and fungi, and other particulate matter. The inoculation of e.g. a medium is carried out in such a way that exposure to the atmosphere is reduced to a minimum—or is eliminated. Use is often made of pre-sterilized, disposable instruments and materials—e.g. syringes, petri-dishes, etc.

ASOT (ASLT) Antistreptolysin-O test (*q.v.*).

asparagine (biosynthesis of) (see Appendix IV(d))

aspartate (biosynthesis of) (see Appendix IV(d))

aspergillic acid A pyrazine derivative synthesized by *Aspergillus flavus*; it has antibiotic properties.

aspergilloma (*pl.* aspergillomata) (see ASPERGILLOSIS)

aspergillosis Any disease in which the causal agent is a species of ASPERGILLUS, often *A. fumigatus* or *A. niger*.

In man, aspergilloses are uncommon—infections being associated mainly with the lungs or ears. Pulmonary infection may lead e.g. to pneumonitis and may subsequently give rise to systemic involvement; alternatively, the pathogen may form fungal masses (*aspergillomata*) in pre-formed lung cavities (e.g. in recovered tuberculosis patients) without necessarily invading the tissues. A positive laboratory diagnosis generally requires both cultural and microscopical (or histological) evidence of infection, but *repeated* isolations of *Aspergillus* (e.g. from sputa) provide presumptive evidence of infection. In some types of aspergillosis specific antibodies may be demonstrated by gel diffusion techniques.

Among animals aspergillosis is particularly important in fowl, and economically important losses may occur in domestic flocks; respiratory symptoms are the most common, but lesions may also be formed in the intestine, liver, ears, and many other parts of the body. (see also AFLATOXIN)

Aspergillus A genus of fungi within the order MONILIALES; species include saprophytes and opportunist pathogens—the latter including *A. fumigatus* and *A.niger*: see ASPERGILLOSIS. The vegetative form of the organisms is a septate mycelium. Conidiophores arise directly from the somatic hyphae; a cell which branches to form a conidiophore is termed a *foot cell*. The conidiophore consists of an erect hypha which has a more or less spherical swelling (the 'vesicle') at its distal (free) end; in some species the vesicle is partly or completely covered by a confluent layer of phialids ('primary sterigmata') while in most species it is covered by a confluent layer of sterigmata ('primary sterigmata' or 'metulae') and a confluent, superimposed (outer) layer of phialids ('secondary sterigmata'). A chain of spherical, pigmented, aseptate conidia is produced basipetally from each of the phialids; according to species the conidia may be e.g. black, green, or yellow—although spore colour also depends on the content of particular trace elements in the medium. (*A. niger* may form yellow conidia—instead of black—if the copper content of the medium is low.) The vesicle—together with its sterigmata and/or phialids and associated chains of conidia—is termed a *conidial head.*

A parasexual cycle operates in some species (e.g. *A.nidulans*)—see PARASEXUALITY IN FUNGI.

A number of species of *Aspergillus* are known to have ascigerous stages; thus, e.g. *A. glaucus* and related species are imperfect stages of *Eurotium*, *A. nidulans* and related species are imperfect stages of *Emericella*, and *A. fumigatus* and related species are imperfect stages of *Sartorya*. Of those species of *Aspergillus* with known ascigerous stages some are known to be homothallic—while others are reported to be heterothallic. Species include:

A. fumigatus: the most important of the causal agents of aspergillosis in man. In *A. fumigatus* the vesicle often appears dome-shaped (rather than spherical) owing to the way in which the distal end of the conidiophore gradually widens to form the vesicle; the vesicle bears a confluent layer of phialids ('primary sterigmata') over the distal half/three-quarters of the surface of the vesicle. The more proximally situated phialids are longer than the others, and are inclined such that their free ends tend to be parallel with the more distal phialids; as a result, the chains of grey-green spherical spores may form a parallel-sided columnar spore mass which may reach one millimetre in length. *A. fumigatus* is reported to grow at temperatures up to 50 °C.

A. niger. In *A. niger* the spherical vesicle bears a confluent surface layer of radially orientated sterigmata and a superimposed layer of phialids; the entire surface of the vesicle is covered by spore-bearing structures. Conidia are black and spherical.

A. phialiseptus. A species which is similar, morphologically, to *A. fumigatus* but which differs e.g. in that the phialids are septate and are longer than those in *A. fumigatus*.

Aspidisca A genus of ciliate protozoans of the order Hypotrichida, subclass SPIROTRICHIA.

asporogenous (*adjective*) Refers to an organism which does not form spores.

assimilatory nitrate reduction Many bacteria, some fungi, and most algae and plants can use nitrate as the sole source of nitrogen. Nitrate is reduced to ammonia *via* nitrite, an unknown intermediate (possibly hyponitrite, HON=NOH), and hydroxylamine. Nitrate is reduced to nitrite by *nitrate reductase*. In most bacteria, and in fungi, nitrite is reduced to ammonia in a number of enzymic steps; in plants, algae and (apparently) *Clostridium pasteurianum*, nitrite is reduced to ammonia by a ferredoxin-linked nitrite reductase. Ammonia is assimilated by the reaction:

$$\alpha\text{-ketoglutarate} + NH_3 \xrightarrow[\text{dehydrogenase}]{\text{glutamate}} \text{glutamate.}$$

Assimilatory nitrate reduction is typically an aerobic process; it may be inhibited in the presence of ammonia or reduced nitrogenous organic compounds. ATP is not formed, and the enzymes involved appear to differ from those operative in NITRATE RESPIRATION.

assimilatory sulphate reduction Sulphate can be used as the sole source of sulphur by most microorganisms; the assimilatory pathway in bacteria and fungi can be outlined thus:

$$SO_4^{2-} + ATP \xrightarrow{\text{sulphate adenylyltransferase}} APS + PPi$$

$$APS + ATP \xrightarrow{\text{APS kinase}} PAPS + ADP$$

$$PAPS + NADPH_2 \xrightarrow{\text{PAPS reductase (thiol, FAD)}} SO_3^{2-} + NADP + PAP$$

$$SO_3^{2-} + 3NADPH_2 \xrightarrow{\text{sulphite reductase}} S^{2-} + 3NADP + 3H_2O$$

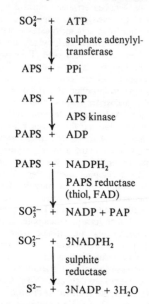

APS = adenosine-5′-phosphosulphate (see Appendix V(c)); PPi = pyrophosphate; PAPS = 3′-phosphoadenosine-5′-phosphosulphate; PAP = adenosine-3′,5′-diphosphate; SO_3^{2-} occurs as 'bound sulphite'. The sulphide thus formed is assimilated during the synthesis of cysteine from serine (see Appendix IV(c)). (cf. DISSIMILATORY SULPHATE REDUCTION; see also SULPHUR CYCLE)

Astasia A genus of ALGAE of the division EUGLENOPHYTA.

Asterionella A genus of free-floating pennate diatoms (division BACILLARIOPHYTA) which occur in freshwater and marine environments; species are particularly common in lakes and reservoirs during spring. The individual frustule is elongated and rectangular with slightly flared ends; commonly, seven or eight frustules form a star-shaped cluster in which each corner of one end of a given frustule is joined to one corner of another frustule.

Astomatida An order of protozoans of the subclass HOLOTRICHIA (subphylum CILIOPHORA); the absence of a cytostome is a feature common to all species. The *astomes* occur principally as intestinal parasites of annelid worms; they are often large, and some possess endoskeletal structures and well-developed holdfast organelles. Typically, the body ciliature is uniform. Asexual reproduction may occur by binary fission or budding, and in some genera, e.g. *Intoshellina*, chain-like colonies may be produced; in this, and some other genera, the cells often contain many contractile vacuoles.

astomatous (*adjective*) Refers to any structure which does not have a pore, opening, or mouth—e.g. a *cleistothecium*: a fruiting body in which the peridium (wall) forms a *closed* vessel which encloses the asci/ascospores.

AT types (of bacteria) (see BASE RATIO)

Athiorhodaceae (purple non-sulphur bacteria) (see RHODOSPIRILLACEAE)

athlete's foot (*tinea pedis*) (see RINGWORM)

atmungsferment Obsolete name for CYTOCHROME aa_3.

atomic mass unit (AMU) A unit of 'atomic weight' (i.e. relative atomic mass); the unit was proposed in 1960 by the International Union of Pure and Applied Physics (IUPAP) and the International Union of Pure and Applied Chemistry (IUPAC). The AMU is defined as one-twelfth of the mass of a neutral carbon-12 atom—approximately 1.66×10^{-24} gram.

atopy An inherited tendency to develop immunological HYPERSENSITIVITY conditions.

atoxyl (see ARSENICALS)

ATP Adenosine-5′-triphosphate. (See Appendix V(c) for structure.) ATP functions in cells as a carrier of energy; free energy yielded on ATP hydrolysis (e.g. to ADP—adenosine-5′-diphosphate) is used for a variety of energy-requiring processes—including biosynthesis, active TRANSPORT, and mechanical work (e.g. flagellar movements). ADP may be phosphorylated to ATP e.g. by OXIDATIVE PHOSPHORYLATION or by SUBSTRATE LEVEL PHOSPHORYLATION. In some reactions other nucleotides (particularly GTP—guanosine-5′-triphosphate) may function in an analogous manner.

attenuated vaccine Any VACCINE containing *live* organisms whose virulence for a given host has been suitably reduced or abolished.

attenuation (1) Any procedure in which the pathogenicity of a given organism (for a particular host) is reduced or abolished; attenuation is often achieved by SERIAL PASSAGE and may be carried out in order to prepare a VACCINE—e.g. the Sabin anti-poliomyelitis vaccine. (see also CAPRINIZED and LAPINIZED VACCINES) (2) (*mycol.*) The gradual narrowing of a fungal structure, e.g. the *stipe* of a basidiocarp. (3) For another meaning of *attenuation* see SPIRITS.

atypical mycobacteria (anonymous mycobacteria) Species of *Mycobacterium* which can cause disease in man but which (unlike the human and bovine tubercle bacilli) have little or no pathogenicity for the guinea-pig. Atypical mycobacteria may cause a disease resembling mild pulmonary tuberculosis and/or other types of disease; species include *M. fortuitum*, *M. intracellulare*, *M. kansasii*, and *M. xenopi*.

aufwuchs (periphyton community) Organisms which colonize and form a coating on submerged objects (stones, plants, etc.) in aquatic environments. Such organisms include certain algae and various sessilinids.

auramine O (canary yellow) A yellow dye and FLUOROCHROME. Auramine O is a substituted diphenylmethane; its uses include the fluorescent staining of acid-fast organisms. The dye fluoresces yellow.

auramine–rhodamine stain A fluorescent staining procedure used to detect *acid-fast* organisms—particularly cells of *Mycobacterium tuberculosis* in smears of sputum. The filtered stain, which contains glycerol and phenol in addition to the dyes auramine and rhodamine, is applied to a heat-fixed smear for 15–30 minutes (37 °C). Decolorization is carried out with acid-alcohol (5–10 minutes) and the slide rinsed thoroughly in water. The smear is then 'counterstained' with a 0.5% aqueous solution of potassium permanganate for approximately 2 minutes, rinsed in water, and dried. The object of counterstaining is to render non-fluorescent the various tissue debris etc. which may be present in the smear and which often exhibit non-specific fluorescence; a prolonged period of counter-staining tends to mask the fluorescence of any acid-fast organisms on subsequent examination of the smear by fluorescence microscopy.

aureomycin CHLORTETRACYCLINE. (see also TETRACYCLINES)

Auricularia A genus of fungi of the AURICULARIALES.

Auriculariales An order of fungi of the sub-class PHRAGMOBASIDIOMYCETIDAE; constituent species typically form elongated, *transversely* septate basidia in gymnocarpous fruiting bodies which may be e.g. resupinate or sessile and which are often gelatinous or waxy. The basidiocarp may be e.g. cup-shaped or ear-shaped. Some species are saprophytes; others are parasitic e.g. on mosses and on other fungi. The genera include *Auricularia*.

Australia antigen (see HEPATITIS B ANTIGEN)

auto-agglutination (saline-agglutination) Spontaneous agglutination which may occur when bacterial cells (or other particulate antigens) are suspended in saline.

autoantibody An ANTIBODY produced by an individual against one of its own antigens.

autoantigen An ANTIGEN which is normally present in the tissues of a given individual. In *autoimmune disease* the body develops an immunological reaction against such antigens.

autochthonous (1) Indigenous; an epithet sometimes used when referring to the common microflora of the body, or to those species of soil microflora which tend to remain constant (in abundance) despite fluctuations in the quantity of fermentable organic matter in the soil. (2) (*immunol.*) Self-donated—used e.g. of skin grafts etc. in which the donor is also the recipient.

autoclave An apparatus in which objects or materials may be sterilized by air-free *saturated* steam (under pressure) at temperatures in excess of 100 °C. (see also STERILIZATION) In order to achieve steam temperatures in the sterilizing range (e.g. 115 °C, 121 °C, 134 °C) water must be heated in a closed system, e.g. a boiler; steam at these temperatures is therefore necessarily under pressure—although the pressure itself plays little or no part in the sterilization process which depends on the combined effects of temperature and moisture. The simplest autoclaves resemble the domestic pressure cooker both in principle and appearance. The load is placed on a rack in the chamber and the lid clamped securely in position with the valve open; water in the bottom of the chamber is vaporized by external heating, and the steam is allowed to issue from the valve until all the air has been driven from the chamber. The valve is then closed. On further heating water continues to vaporize within the chamber, and the pressure and temperature of the steam rise to a level dictated by the setting of the steam exhaust valve; this valve then opens to maintain the required level of steam pressure and temperature. For a given steam temperature the time required for sterilization depends to a large extent on the nature of the load—the rate of steam penetration and the heat capacity of the load being important considerations. Sterilizing temperature/time combinations used with these or with larger autoclaves are commonly 115 °C(10lbs/square inch)/35 minutes, 121 °C(15lbs)/15–20 minutes, or 134 °C(30lbs)/4 minutes. These times may be varied according to the nature of the load; corresponding exposure times are somewhat shorter with vacuum autoclaves (see later).

Large autoclaves, used in hospitals and in industry, often use steam piped direct from a boiler, and frequently incorporate automatic controls for timing etc. The correct usage of an autoclave requires that the air be completely purged from the chamber and that all the free space within the chamber be filled with saturated steam at the appropriate temperature and pressure. (see also STEAM TRAP) In most autoclaves steam enters near the top of the chamber and the air is purged by downward displacement; thus, pockets of air may become trapped at the bottom of deep empty vessels—particularly when such vessels have small openings or long narrow necks, e.g. bottles, flasks etc. At such locations saturated steam may become locally superheated (see later) with consequent loss of sterilizing ability. (*Clean* flasks—and other glassware—are usually sterilized in a hot-air oven.) In general, if air is present in an operating autoclave the temperature corresponding to a given pressure will be lower than that attained in the complete absence of air. In some modern autoclaves air is initially purged from the chamber by connecting it, briefly, to a vacuum line; steam subsequently introduced into the chamber is then able to penetrate more effectively and more rapidly into flasks and between the fibres of surgical dressings etc. so that the overall sterilizing time is reduced.

Steam quality is important. In a small autoclave the steam is necessarily saturated since it remains in contact with the water in the chamber. However, when steam is piped from a boiler to a large autoclave, a change in steam quality may occur between the boiler and the chamber of the autoclave—particularly at a pressure-reduction valve. Steam entering the chamber should be either *dry saturated* or *wet saturated* steam. Dry saturated steam is steam in equilibrium with water at the same temperature; wet saturated steam is steam permeated by water in the liquid phase. *Saturated* steam is used in an autoclave for two reasons: (a) It effects optimum heat transfer; when condensing on objects at slightly lower temperatures such steam readily delivers up an appreciable quantity of latent heat. (b) Effective moist-heat sterilization requires the presence of free water (for reasons see STERILIZATION). *Superheated steam* is an *unsaturated vapour.* It is formed when the pressure (but not the temperature) of dry saturated steam is reduced, or when the temperature of dry saturated steam is raised in the absence of free water. Superheated steam tends to behave as a gas and requires temperatures higher than those of saturated steam for effective sterilization (see dry-heat sterilization under STERILIZATION).

At the end of the sterilization cycle the steam inlet valve is closed (or closes automatically) and the steam within the chamber allowed to

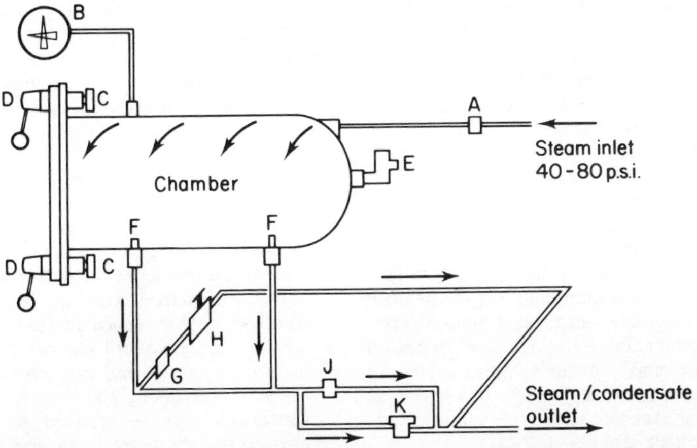

AUTOCLAVE. Simplified, diagrammatic representation of a Series 225 steam–mains autoclave. (Courtesy of Baird and Tatlock (London) Ltd) **A.** Steam inlet valve. The valve is opened/closed electromagnetically—its operation being governed by the pressure controller/indicator. (Control by temperature—*via* the thermocouple—is an available option.) **B.** Pressure controller/indicator. **C.** Door microswitch—a safety device; electrical power is connected to the autoclave only when the microswitch has been actuated, i.e., when the door has been fully closed. Additionally, this device automatically opens the steam exit valve if any attempt is made to open the door while the chamber is under pressure. **D.** Door closure bolts which operate microswitches. **E.** Safety valve set to open at 40 p.s.i. **F.** Filters which prevent blockage of steam discharge lines and steam trap. **G.** Thermocouple pocket: temperature control/indicator/recording option. **H.** Manually-operated bypass valve—can be operated in case of electrical power failure. (Under such conditions valve **J** could not be opened.) The valve contains a built-in bleed which permits a small but continual flow of steam from the chamber to the exterior. **J.** Electrically-operated steam exit valve—automatic/manual control. **K.** Steam trap.

Operational sequence. 1. Insert load. 2. Secure door. 3. Set controls for required pressure/temperature and sterilizing time. 4. Switch on power. Steam enters the chamber through **A**, and the downwardly-displaced air is purged, *via* filters **F**, through (a) the steam trap **K** and (b) the permanent bleed **H**; **J** is closed, and, when *steam* leaves the chamber, **K** closes. **A** closes when the chamber pressure reaches the selected value. Steam continues to leave the chamber *via* the permanent bleed **H**—permitting continual temperature monitoring by the thermocouple **G**; since **G** is located at the lowest (and hence coolest) part of the system, it measures the 'worst case' temperature—the chamber temperature always being higher than that registered by **G**. Sterilizing conditions within the chamber are maintained (against heat losses and against the continual steam bleed) by periodically admitting a pulse of steam via **A**. At the end of the sterilization cycle steam is discharged from the chamber *via* **J**; when atmospheric pressure has been re-established in the chamber the door can be opened and the load withdrawn.

Automatic timing of the sterilization cycle commences when the chamber pressure (or, optionally, temperature—measured by **G**) reaches the desired value.

escape (*via* the steam exit valve) until the chamber steam-pressure is equal to that of the atmosphere—when the chamber can be safely opened. (NB. When bottles of liquid have been autoclaved the chamber steam pressure should be reduced *slowly*; this prevents boiling and the consequent loss of contents and/or damage to the (glass) bottles. Screw-caps on bottles should always be loosened prior to autoclaving.) In some types of autoclave a partial vacuum is allowed to form in the chamber as the steam condenses; this is advantageous since it tends to dry the surfaces of objects and materials within the chamber. When 'breaking the vacuum' i.e. allowing the ingress of air to the chamber, various types of filter etc. are used in the air intake line—to prevent the entry of contaminated air.

Some materials cannot be effectively sterilized by autoclaving; examples include vaseline (petroleum jelly), liquid paraffin, and similar substances which do not permit adequate penetration of steam. Other materials may be unsuitable owing to heat lability.

Regular monitoring of autoclave performance is important, and monitoring devices should be used to check the temperatures reached at a number of different locations within the chamber. Monitoring devices may be physical (e.g. thermocouples), chemical (e.g. BROWNE'S TUBES) or bacteriological (e.g. spores of a thermoduric organism such as *Bacillus stearothermophilus*). Biological monitoring involves testing the autoclaved spores for viability by attempting to culture them; clearly, such a process is directly related to the aim of sterilization, but—in constrast to chemical indicators—involves a built-in delay, i.e. the incubation time. Different types of temperature and pressure-sensitive tapes are available commercially; such tapes change colour after exposure to appropriate sterilizing conditions. (see also STERILIZER)

autocoupling (*adjective*) (of a *hapten*) Capable of spontaneously forming an *immunogenic* complex with a tissue component when injected into the body.

autoecious (*adjective*) (*mycol.*) Refers to those parasitic fungi which complete their life cycles on or within a single host species. (see also HETEROECISM)

autogamy A series of nuclear events which occur, within a single cell, in certain protozoans (including ciliates, flagellates, and sarcodines) and in some diatoms; these events include meiotic division of the organism's nucleus, or nuclei, and the subsequent formation of a zygote, or *syncaryon*. In *Paramecium aurelia* each of the two micronuclei undergoes meiosis; of the eight haploid *pronuclei* produced all except one disintegrate. The surviving pronucleus divides, mitotically, to form a pair of gametic nuclei—which subsequently fuse to form the (homozygous) syncaryon. The zygotic nucleus then undergoes two mitotic divisions; two of the resulting (diploid) nuclei become new micronuclei while the other two give rise to a pair of macronuclei—the original macronucleus having disintegrated during the preceding events. When the cell next undergoes binary fission, one macronucleus passes to each daughter cell; both micronuclei divide, mitotically, and a pair of micronuclei passes to each daughter—thus re-establishing the normal nuclear constitution. Since only the original complement of genes is involved, the potential for genetic change in autogamy is less than that inherent in sexual events such as CONJUGATION; nevertheless, new combinations of alleles may be formed.

In some organisms autogamy occurs spontaneously; in others it may be induced by starvation, ageing, radiation, etc.

autogenous vaccine Any vaccine prepared by the culture and inactivation of pathogens isolated from a particular patient and subsequently used to inoculate that patient.

autoimmune disease A condition in which the body develops an immunological reaction against its own tissues. (see also AUTOANTIGEN)

autologous (*immunol.*) Derived from the subject's own tissues.

autolysins (autolytic enzymes) A range of endogenous enzymes which are involved in the breakdown of certain structural components of the cell during particular phases of cellular growth and development, and/or which are involved in AUTOPHAGY or in AUTOLYSIS. Autolysin activity appears to be necessary e.g. for the extension of cell wall PEPTIDOGLYCAN*, septum-formation, and cell separation during bacterial GROWTH; the active, wall-bound autolysins may be derived from inactive precursors—the latter being activated e.g. by proteolytic enzymes. (see also BILE SOLUBILITY) Mutant bacteria with defective autolysin activity may exhibit traits such as (abnormal) chain formation and resistance to antibiotics to which they were formerly sensitive. In fungi, hyphal growth appears to require the activity of wall-lytic enzymes —coupled with coordinated wall synthesis. Autolytic enzymes also appear to be involved e.g. in the germination of bacterial and fungal spores. *Autolysins active against peptidoglycan include *N*-acetylglucosaminidase and *N*-acetylmuramyl-L-alanine amidase.

autolysis Breakdown of the components of a cell or tissue by *endogenous* enzymes—usually following the death of the cell or tissue. (see also BILE SOLUBILITY)

autophagy The digestion, by a cell, of some of its own internal structures; autophagy may occur e.g. during periods of starvation—apparently as a means of obtaining vital materials. The structures are initially enclosed within a membranous sac—similar to a FOOD VACUOLE; subsequently the sac coalesces with a LYSOSOME and enzymatic degradation occurs.

autoradiography (radioautography) A process in which photography is used to locate and quantify radio-actively labelled substances in e.g. cells or in liberated cell contents. With a judicious choice of material, autoradiography enables the researcher to obtain information on e.g. the sites of biosynthesis, and the turnover rates, of particular cell components. Initially, radio-active precursors are taken up by the living cell; these are incorporated into cellular structures etc. as would be the corresponding non-radio-active precursors. If the cell is part of a tissue, the tissue is fixed (see FIXATION) and is prepared as a thin section on a glass slide. The section is coated with a thin layer of photographic emulsion (in the dark room) and left, in the dark, for a number of hours. Each radio-active precursor in the section acts as a point source of radiation, and causes chemical changes in that part of the emulsion immediately above it. The emulsion is subsequently developed and fixed, as in conventional photographic processing, and the whole examined under the microscope. Regions of the emulsion directly above the point sources of radiation are indicated by (opaque, black) grains of silver which have been formed from silver iodide (in the emulsion) as a result of the incident radiation. Since these grains are superimposed on the tissue section, the cellular location of the radio-active components can readily be determined. (The tissue section may be stained, prior to application of the emulsion, although certain substances, e.g. the salts of heavy metals, should not be used, either in the fixative or stain, since such substances may inhibit the response of the emulsion to incident radiation from the section.) The brief description above refers to the *coating technique*—the earliest form of autoradiography developed in the 1940s; other modes of autoradiography have since been devised. Essential features of autoradiography are (a) the use of very thin sections of tissue and thin layers of emulsion, and (b) the use of low-energy sources of radiation, e.g. tritium. (Low-energy radiators—such as tritium—tend to affect the emulsion *immediately above* the point source, i.e. only radiation emitted at, or near, right-angles to the section produces a response in the emulsion; radiation emitted at other angles passes through the section for a greater distance and is absorbed before reaching the emulsion. With high-energy emitters, each point source gives rise to a wide-angled cone of effective radiation; the silver grains in the final preparation do not, therefore, indicate with such precision the location of the point sources.) Specific radio-active precursors are used for studying particular cell components; thus, tritium-labelled thymidine is used for studying DNA metabolism.

autospore (*algol.*) A non-motile spore, formed asexually, which resembles the parent cell morphologically; such spores are produced e.g. by *Chlorella* spp.

autotroph (lithotroph) (1) Any organism for whom environmental carbon dioxide is the major (or sole) source of carbon—thus including those microorganisms which can, or must, use small amounts of organic compounds—e.g. trace amounts of vitamins. (cf. HETEROTROPH) (2) In a more restrictive sense: any organism whose growth and reproduction are entirely independent of external sources of organic compounds—the carbon requirement (for example) being completely satisfied by environmental carbon dioxide.

An autotroph which lives PHOTOTROPHically is referred to as a *photoautotroph* or *photolithotroph*; one which lives CHEMOTROPHically is referred to as a *chemoautotroph* or *chemolithotroph*. Photoautotrophism is characteristic of the algae and of some procaryotic organisms—e.g. some photosynthetic bacteria. Chemoautotrophic organisms probably occur only among certain groups of bacteria (e.g. *Thiobacillus* spp).

Some organisms are *obligate* autotrophs; others (e.g. *Pseudomonas facilis*) are facultative autotrophs.

auxanographic technique A procedure used for identifying the growth factor(s) required by a suspected AUXOTROPH. The surface of a solid MINIMAL MEDIUM is first inoculated with the suspected auxotroph and is then dried. Separate, discrete inoculations are then made with (a) a mixture of amino acids, (b) a mixture of vitamins, and (c) one or more possible requirements for nucleic acid synthesis (e.g. adenine). Following incubation, growth at the site of one of the inoculations indicates the type of growth factor required. If, for example, growth occurs at the site of the mixed amino acid inoculation, the analysis is continued by repeating the procedure with separate, discrete inoculations of individual amino acids until the specific requirement(s) is/are identified. (see also AUXOTROPHIC MUTANTS, ISOLATION OF)

auxins A group of substances which function as plant growth regulators; they promote stem elongation and, in conjunction with GIBBERELLINS and/or CYTOKININS, play an important role in many plant processes. Auxins probably occur universally in higher plants—being produced mainly at stem apices and in leaves; they are also formed by certain algae, bacteria, and fungi. In many plants the main auxin appears to be 3-indole-acetic acid (IAA—also known as 'auxin' or 'heteroauxin'). Indole-3-acetonitrile (IAN) and certain other indole derivatives show auxin activity—although (at least) some of these are probably converted by the plant to IAA. Some non-indole compounds also have auxin activity.

Changes in auxin levels may contribute to the symptoms of certain plant diseases. *Hyperauxiny* (abnormally high auxin levels in plants) may be due to auxin production by the pathogen or by the host under stimulation from the pathogen. (see e.g. CROWN GALL) Hyperauxiny may also be caused by inhibition of IAA-*oxidase*, a normal plant enzyme believed to be important in the regulation of IAA concentration; compounds which inhibit IAA-oxidase may be synthesized by the pathogen or by the diseased host (see e.g.

SCOPOLETIN). Hyperauxiny occurs in many diseases, e.g. in wilt of tomato caused by *Verticillium alboatrum*. The relationship between hyperauxiny and disease development is not clear. In some diseases, e.g. coffee blight caused by *Omphalia flavida*, the pathogen produces enzymes which destroy plant auxins. (see also MYCORRHIZA and ROOT NODULES)

auxospore (*algol.*) The zygote of a *diatom*; some authors use the term to refer only to the naked zygote, while others extend the meaning to include the stage at which new cell walls have been synthesized.

auxotroph A microorganism which lacks the ability to synthesize one or more essential growth factors; an auxotroph may arise by the occurrence of one or more mutations in a PROTOTROPH. (A medium used to culture a given auxotroph must contain those factor(s) which the organism cannot synthesize; see also MINIMIAL MEDIUM and COMPLETE MEDIUM.) Auxotrophy results from a cell's inability to produce functional enzyme(s) which catalyse particular stage(s) in the synthesis of essential growth factor(s); such genetic block(s) may involve (a) the complete absence of enzyme; (b) the presence of normal enzyme in sub-normal amounts; (c) the presence of an abnormal enzyme—which is devoid of or has a lowered enzymic activity. (see also SYNTROPHISM TESTS)

Apparent auxotrophy may be manifested under certain conditions—e.g. when transport mechanisms are defective (see CRYPTIC MUTANT).

auxotrophic mutants, isolation of Since AUXOTROPHs have nutritional requirements *in excess* of those of the corresponding wild type strains, they cannot be isolated from mixed auxotroph–prototroph populations by common culture methods. In the *limited enrichment* method, a dilute suspension of mutagenized cells is inoculated onto a MINIMAL MEDIUM which has been enriched with limiting amounts of nutrients. (The dilution is chosen such that isolated colonies are obtained following incubation.) A colony of auxotrophic cells quickly exhausts the nutrient supply at its location: thus, colony size is restricted; a colony of prototrophic cells, not restricted by nutrient supply, attains a greater size. Thus, small colonies may be presumed to be those of auxotrophs. In the *delayed enrichment* method, the mutagenized preparation of cells is first inoculated onto a minimal medium. A small quantity of molten minimal agar is then layered onto the surface of the inoculated medium and allowed to set. The plate is then incubated. Colonies are formed only by prototrophs, and the positions occupied by these colonies are recorded. Finally, complete medium is layered onto the surface of the plate and allowed to set; the plate is then re-incubated. As nutrients diffuse into the minimal agar the auxotrophic cells begin to grow and form colonies.

Currently, auxotrophic mutants are commonly isolated by a REPLICA PLATING process. A plate of *complete medium* is inoculated with the mutagenized preparation of cells and is then incubated. From this *master plate* replica plates are prepared on *minimal agar*. By comparing the positions of colonies on the master plate with those on the replica plates, presumptive auxotrophs can be identified by their absence from the latter. A given auxotroph can then be obtained in pure culture and its particular growth requirements determined. (see AUXANOGRAPHIC TECHNIQUE)

Another method is based on the differing effects of PENICILLIN on growing and non-growing cells of certain (penicillin-sensitive) species. If a well-washed mixture of auxotrophic and prototrophic cells is suspended in *minimal medium* with an appropriate concentration of penicillin, the prototrophs are lysed while the auxotrophs—being unable to grow—remain viable. The suspension can then be washed and plated on a complete medium to recover the auxotrophs. In this method it is essential that the mutagenized cells be grown in a complete medium for several cell-division cycles *prior to penicillin treatment*; this permits the development of auxotrophy in mutant cells. (This procedure is essential since the newly-mutant cell—as opposed to its progeny—contains the full complement of enzymes etc. found in the prototroph; only after several cell divisions do the progeny cells exhibit a truly auxotrophic phenotype.) Additionally, only low concentrations of mutagenized cells should be treated with penicillin since the contents of the lysed cells may be used as nutrients by the auxotrophs—thereby rendering the latter susceptible to lysis by penicillin.

avenacin A complex, strongly antifungal fluorescent glycoside present in the roots of the oat (*Avena*). Avenacin is believed to be responsible for the resistance of the oat to may strains of *Gaeumannomyces graminis* (formerly *Ophiobolus graminis*)—the causal agent of take-all disease of CEREALS.

aversion zone (*mycol.*) A zone of growth inhibition which separates two fungal colonies; such zones (of the order of one or two millimetres) are formed e.g. when strains of *Phycomyces blakesleeanus* of similar mating type (i.e. + and +, or − and −) are cultured on the same plate.

avianized vaccine Any vaccine containing microorganisms (usually viruses) whose virulence for a given host has been suitably attenuated by a period of adaptation in live chicks and/or serial passage through chick embryos (eggs). Such attentuated organisms may or may not be inactivated prior to use in a vaccine. The FLURY VIRUS is an avianized strain.

avidin A protein present in the white of raw hens' eggs; it binds BIOTIN, thus rendering it unavailable as a cofactor.

avidity (*immunol.*) A term used to refer to the stability of antibody–antigen complexes; the degree of stability in any given case depends not only on the AFFINITY of the individual determinant-combining site unions but also e.g on the number of satisfied valencies of the antibodies and antigens.

axenic (*adjective*) Refers to the condition in which only one type (i.e. species or strain) of organism is present—e.g. a pure CULTURE.

axial fibrils (axial filaments) (in spirochaetes) (see SPIROCHAETALES)

axial filament (1) (in bacterial sporulation) See SPORES (bacterial). (2) (axial fibril) See SPIROCHAETALES.

axistyle (in *Leptospira*) (see SPIROCHAETALES)

axoneme (flagellar) (see FLAGELLUM (eucaryotic))

axopodia Contractile, rod-like *pseudopodia* (see MOTILITY) which emanate, radially, from the (spherical) cells of heliozoans and some species of radiolarian. Each axopodium has a (longitudinal) core consisting of an orderly array of MICROTUBULES.

axostyle An endo-skeletal structure found in some protozoans, e.g. *Trichomonas*—in which it resembles a longitudinal bar which extends to both poles of the cell and projects a short distance beyond the posterior end.

azaserine ($COOH.CHNH_2.CH_2.O.CO.CH{=}N^+{=}N^-$) An ANTIBIOTIC produced by *Streptomyces* sp; it inhibits a range of microorganisms but, being toxic, is of minimal use clinically. Azaserine is an analogue of glutamine and blocks a number of reactions in which glutamine acts as an NH_2 donor; in particular it inhibits the formation of 5′-phosphoribosyl-*N*-formylglycinamidine from 5′-phosphoribosyl-*N*-formylglycinamide during the biosynthesis of purine nucleotides—thus effectively blocking nucleic acid synthesis. (see Appendix V(a); see also 6-DIAZO-5-OXO-L-NORLEUCINE and HADACIDIN)

azide (N_3^-) (as a respiratory inhibitor) Azide combines with, and prevents the reduction of, the oxidized a_3 component of cytochrome aa_3 (= cytochrome *c* oxidase—see CYTOCHROMES)—thus preventing the operation of the ELECTRON TRANSPORT CHAIN. Azide appears also to inhibit OXIDATIVE PHOSPHORYLATION in a manner similar to that of OLIGOMYCIN. (see also SODIUM AZIDE)

AZM (adoral zone of membranelles) (*ciliate protozool.*) A number of MEMBRANELLES arranged in a definite (and taxonomically important) pattern; such an arrangement is found in certain holotrichs (e.g. *Tetrahymena*) and in all spirotrichs (e.g. *Euplotes*). The AZM is primarily concerned with feeding although, in some cases, it may be involved in locomotion.

azoferredoxin (see NITROGENASE)

Azomonas A genus of gram-negative (or gram-variable), motile, pleomorphic bacteria (family AZOTOBACTERACEAE) which occur in the soil and in water. *Azomonas* spp do not form cysts or spores; they are catalase positive and have GC ratios of 53–59%. The following species were formerly classified in the genus AZOTOBACTER: *A. agilis*, *A. insignis*, and *A. macrocytogenes*.

Azotobacter A genus of gram-negative, coccoid, ovoid or rod-shaped non-sporing bacteria (family AZOTOBACTERACEAE) found in soil and water. The cells (long axis commonly between 2 and 7 microns) may be peritrichously flagellate or non-motile. CYSTS are formed. Mucoid growth is produced on nitrogen-free carbohydrate-containing media, and some species form a water-soluble fluorescent pigment. (*A. chroococcum* usually develops a water-insoluble brown or black pigment.) Growth conditions: temperature optimum about 25 °C, pH optimum about 7–7.5 (poor acid tolerance). Most species can metabolize sodium benzoate; gaseous or bound nitrogen can be used. (For NITROGEN FIXATION either molybdenum or vanadium must be supplied.) *Azotobacter* spp are catalase positive; their GC ratios are 63–66%. Stimulation of plant growth by *Azotobacter* spp is believed not to be related to the organisms' ability to fix nitrogen. (see also AZOMONAS)

Azotobacteraceae A family of gram-negative (or gram-variable) obligately aerobic, chemoorganotrophic, rod-shaped, coccoid or pleomorphic bacteria found in the soil, the phyllosphere, and in water. Cells are either non-motile or motile with peritrichous or polar flagella. Some species form cysts; none form spores. All species are capable of NITROGEN FIXATION under free-living conditions. Constituent genera: *Azomonas*, *Azotobacter*, *Beijerinckia*, and *Derxia*.

azotoflavin (see FLAVODOXINS)

azurin (1) See BLUE PROTEIN. (2) A solution of copper sulphate and ammonium hydroxide used as an agricultural antifungal agent.

azygospores (*mycol.*) *Parthenogenically*-derived structures which are similar to zygospores; they are formed e.g. by *Entomophthora muscae*.

B

B lymphocyte (see LYMPHOCYTE)

B-type particles (*virol.*) One of two types of particle (A and B) observed (by electron microscopy) in mouse mammary tumour tissue; the B-type particle, subsequently identified as the causal agent of such tumours, is now referred to as the mouse (or murine) mammary tumour virus (MMTV)—a LEUKOVIRUS. B-type particles are spheroidal, enveloped viruses, approximately 80 nm in diameter, which differ from C-TYPE PARTICLES in that e.g the genome is situated eccentrically in the virion; they are believed to develop from A-type particles (observed only in intracellular locations) when the latter become enveloped at the cytoplasmic membrane or intracytoplasmic membranes of the host cell.

B virus (see HERPESVIRUS)

Babes–Ernst granules (see METACHROMATIC GRANULES)

Babesia A genus of protozoans of the class PIROPLASMEA. Species of *Babesia* are parasitic in the *erythrocytes* of vertebrates (cf. THEILERIA). Some species are important pathogens; thus, the tick-borne disease of cattle, RED-WATER FEVER, may be caused by *B. bigemina*, *B. bovis*, or *B. divergens*, and infections (some fatal) have occurred in man. Within an erythrocyte, the organism is a uninucleate, pyriform or spherical cell of size (long axis) between 1 and 6 microns; two or four daughter cells are formed by fission, and these infect fresh erythrocytes when the host cell ruptures. The development of *Babesia* in the vector appears to be incompletely known; in *B. bigemina* (at least) the forms infective for the vertebrate are believed to develop in cells of the tick's salivary glands. Organisms ingested by the *adult* vector can be transmitted transovarially to its offspring—and thence to a vertebrate host; the protozoans are believed to multiply schizogonously in the tick's gut epithelium and to produce motile merozoites (*vermicules*) which pass into the eggs.

Vertebrates which are normally resistant to *Babesia* appear to become (more) susceptible following removal of the spleen.

Bacillaceae A family of SPORE-forming bacteria comprising the genera BACILLUS, CLOSTRIDIUM, DESULPHOTOMACULUM, SPOROLACTOBACILLUS, and SPOROSARCINA.

Bacillaria A genus of pennate diatoms of the BACILLARIOPHYTA.

Bacillariophyta (diatoms) A division of the ALGAE; species of the Bacillariophyta include freshwater, marine, and terrestrial organisms. Diatoms are typically unicellular organisms although some species form colonies and others (e.g. *Melosira* spp) form sheathed filaments; the CELL WALL (*frustule*) of a diatom is a silicaceous structure which consists of two parts (*valves*) that fit together like the two parts of a petri dish—or like a box with an overlapping lid. Individual cells exhibit *pennate* (bilateral) or *centric* (radial) symmetry. Certain species (apparently confined to pennate diatoms with a *raphe* system) exhibit MOTILITY. All species contain CHLOROPHYLLS *a* and *c*, β-carotene, and a number of xanthophylls (including *fucoxanthin*) (see CAROTENOIDS); intracellular storage products include CHRYSOLAMINARIN. Sexual processes (isogamy, anisogamy, or oogamy) occur in certain species; gametes each have one *tinsel* flagellum. (see also AUXOSPORE) AUTOGAMY occurs in some species. The vegetative cells of all diatoms are diploid. Genera include ASTERIONELLA, *Bacillaria, Cylindrotheca, Fragilaria,* MELOSIRA, *Navicula, Nitzschia, Stauroneis,* and *Tabellaria.*

bacille Calmette–Guérin (see BCG)

bacillus The term for any *rod-shaped* bacterial cell. The important morphological details of a bacillus include: length and width; longitudinal axis—straight or curved; the ends of the cell—rounded or truncate; occurrence of cells—singly, in groups, pairs, chains etc; presence or absence of a CAPSULE; presence or absence of appendages—e.g. FLAGELLA, FIMBRIAE; position of spore (if formed)—central, terminal, subterminal; characteristics of spore—thin-walled or thick-walled, wider (or otherwise) than the cell. (cf. COCCUS)

Bacillus A genus of SPORE-forming, chemoorganotrophic, facultatively anaerobic (or strictly aerobic), rod-shaped bacteria of the BACILLACEAE. The organisms are *typically* gram-positive, motile (flagellate), and CATALASE-positive. *Bacillus* spp are widespread as saprophytes in soil and water and some species are pathogens of animals; diseases caused by *Bacillus* spp include ANTHRAX, FOULBROOD of the honey bee, and *milky disease* of the Japanese beetle (see BIOLOGICAL CONTROL).

Cells are within the approximate range: 0.5–2 x 2–6 microns. Some strains form a CAPSULE; some are pigmented. L FORMS occur. Growth generally occurs on simple media (e.g. NUTRIENT AGAR); on blood agar some strains exhibit haemolysis. Metabolism may be respiratory or facultatively fermentative; some strains carry out NITRATE RESPIRATION, and many strains of *B. polymyxa* and *B. macerans* carry out NITROGEN FIXATION. Typically, acid—but not gas—is produced from glucose. Storage products include POLY-β-HYDROXYBUTYRATE. Antibiotics produced by *Bacillus* spp include CIRCULINS, POLYMYXINS, and TYROCIDINS. (see also SURFACTIN) Some strains produce BACTERIOCINS.

Vegetative cells may or may not be killed by pasteurization. The *spores* are highly resistant to common disinfectants and to temperatures below 100 °C; see STERILIZATION for methods of inactivation.

TRANSDUCTION and TRANSFORMATION, but not conjugation, occur among *Bacillus* spp.

B. subtilis. The type species. The cells are motile. Acid only is produced from glucose, arabinose and mannitol; most strains form α-AMYLASES. VP TEST positive. Nitrates are reduced. Lecithinases are not produced. Growth is poor/absent under anaerobiosis.

B. anthracis. The causal agent of anthrax. *Non*-motile cells, approximately 1–1.5 × 3–6 microns, which are typically square-ended and often in long chains. Virulent strains form an anti-phagocytic γ-linked poly-D-glutamic acid capsule; a complex exotoxin is produced. Growth occurs on nutrient agar; capsule-formation is enhanced by increased CO_2 and by the use of serum-enriched media containing bicarbonate. Biochemically similar to *B. cereus*. Weak lecithinase activity is exhibited. Haemolysis is weak or absent in cultures on *sheep* blood agar. *B. anthracis* is susceptible to the γ phage (*B. cereus* is not susceptible) and reacts with a near-specific fluorescent antibody conjugate. On agar containing benzylPENICILLIN (0.5 unit/ml agar) *B. anthracis* gives rise to chains of enlarged, spherical cells which, when examined *in situ*, exhibit a 'string of pearls' effect.

B. cereus. Motile or non-motile. Acid (no gas) formed in anaerobic glucose broth; no acid from arabinose or mannitol. VP-positive. Nitrates are reduced. Starch is hydrolysed. LECITHINASES are produced.

B. circulans. Similar to *B. polymyxa* but ferments carbohydrates anaerogenically. VP-negative. Often forms MOTILE COLONIES.

B. larvae. The causal agent of *American* foulbrood.

B. licheniformis. Morphologically and biochemically similar to *B. subtilis* but growth occurs anaerobically; acid (and sometimes gas) formed in anaerobic glucose broth, and acid formed from arabinose and mannitol.

B. macerans. Similar to *B. polymyxa* but VP-negative, and does not attack casein. Produces SCHARDINGER DEXTRINS from starch or glycogen.

B. megaterium. Commonly motile; a capsule is often formed. *Aerobic*. Acid is produced from glucose, and often from arabinose and mannitol. Starch is hydrolysed. Lecithinase not produced. VP-negative.

B. polymyxa. Usually motile. Grows well anaerobically. Acid and gas usually produced from glucose, arabinose and mannitol. Metabolically very active—starch, GELATIN, casein, and PECTINS are commonly attacked. VP-positive.

B. popilliae. Cells motile or non-motile; oval spore, formed sub-terminally or terminally, causing distension of the sporangium. Sporulation poor on artificial media. A pathogen of the Japanese beetle, *Popillia japonica*.

B. stearothermophilus. A thermophilic species which can grow at 65°C and above. The spores are highly heat resistant, and are sometimes used to monitor AUTOCLAVE performance. The organism sometimes causes spoilage of canned foods.

B. thuringiensis. Usually motile; morphology and biochemistry similar to *B. cereus*. During sporulation each cell produces an intracellular protein crystal which is situated near the endospore and which stains deeply with aniline dyes; the crystal typically appears bi-pyramidal under the electron microscope. The crystalline protein (= *δ-endotoxin*) causes intestinal paralysis in the larvae of most species of the Lepidoptera; it is used for the biological control of pests of cabbage and other brassicae. *B. thuringiensis* also produces exotoxins.

bacitracin A cyclic dodecapeptide ANTIBIOTIC produced by strains of *Bacillus subtilis* and *B. licheniformis*. It is bactericidal for many gram-positive and certain gram-negative species. Bacitracin inhibits the dephosphorylation of the C_{55}-pyrophosphate intermediate in PEPTIDOGLYCAN biosynthesis.

back cross (*genetics*) A cross between a HYBRID and one of its parents.

back focal plane (of a convex lens) The plane, perpendicular to the optical axis, which passes through the focal point of a lens on the side furthest from the source of light. (cf. FRONT FOCAL PLANE)

back mutation (reverse mutation) A MUTATION which reverses the effects of a FORWARD MUTATION by restoring the original nucleotide sequence. (cf. SUPPRESSOR MUTATION)

bacteraemia The condition in which viable bacteria are present in the blood. SEPTICAEMIA is one form of bacteraemia. (see also PYAEMIA)

bacteria (*sing*. bacterium) A group of diverse and ubiquitous PROCARYOTIC single-celled organisms; the photosynthetic bacteria (RHODOSPIRILLALES) differ from the blue-green algae and members of the Prochlorophyta in that they contain *bacterio*CHLOROPHYLLS and carry out PHOTOSYNTHESIS anaerobically. (see also PROTISTA)

General characteristics, structure, size. The basic shapes of bacteria are the COCCUS, BACILLUS and SPIRILLUM. (see also PLEOMORPHISM, INVOLUTION FORMS and L FORMS) Bacterial cells occur singly or in chains, clusters, PACKETS, pairs or PALISADES; some *actinomycetes* form a MYCELIUM. The PROTOPLAST (see also CELL MEMBRANE) contains the cytoplasm, genetic apparatus (see CHROMOSOME), enzymes, RIBOSOMES etc. normally associated with cellular structure and metabolism; CYTOCHROMES are common components, and granules or globules of the storage compounds GLYCOGEN, POLY-β-HYDROXYBUTYRATE, or volutin may also be present. (see also CYTOMEMBRANES, GAS VACUOLES, and MESOSOMES) Some bacteria form pigments (e.g. CAROTENOIDS, PRODIGIOSINS). Most species have a rigid CELL WALL containing PEPTIDOGLYCAN; exceptions include *Halobacterium* and *Mycoplasma* spp. (see also SPIROCHAETALES). CAPSULE-formation is characteristic of some bacteria. Bacterial appendages may include flagella (see FLAGELLUM), FIMBRIAE and/or SEX PILI. (see also PROSTHECA) Certain bacteria (e.g. *Bacillus* spp, *Clostridium* spp) form SPORES. (see also CYST) Bacteria usually carry a surface charge—see ZETA POTENTIAL. According to species, strain etc. a bacterium may exhibit MOTILITY. In size, bacteria may be less than 1 micron (e.g. *Chlamydia, Francisella*) to about 500 microns (*Spirochaeta*) but most species are approximately 1–10 microns, maximum dimension.

Antigens. Bacterial antigens include O ANTIGENS, H ANTIGENS, K ANTIGENS, and VI ANTIGENS. Antigenic characteristics are important in the identification and classification of certain groups of bacteria—see e.g. KAUFFMANN–WHITE CLASSIFICATION SCHEME. Bacterial antigens are not necessarily stable: see e.g. SMOOTH–ROUGH VARIATION. (see also BORRELIA)

Occurrence. Bacteria occur in soil, water, and air, and as symbionts, parasites or pathogens of man and other animals, plants, and other microorganisms. Saprophytic species are important in the cycles of matter—see e.g. NITROGEN CYCLE, SULPHUR CYCLE. Certain species form symbiotic relationships with higher plants (see ROOT NODULES), protozoans (see KAPPA PARTICLES), or with other organisms (see SYMBIOSIS). Diseases caused by bacteria include ANTHRAX, BRAXY, FIRE BLIGHT, PLAGUE, SYPHILIS, TETANUS, TUBERCULOSIS, and WEIL'S DISEASE (see also INSECT DISEASES and PLANT PATHOGENS AND DISEASES). Some parasitic/pathogenic species (e.g. *Anaplasma*, *Rickettsia*) grow intracellularly (i.e. within the host's cells). (see also OPPORTUNIST PATHOGENS)

Growth conditions. In the absence of inimical agents, bacteria usually grow on or in any MEDIUM which contains appropriate nutrients (see also TRANSPORT) and which has a suitable WATER ACTIVITY; a particular species grows only within a certain range of pH and temperature. (see also PSYCHROPHILE, THERMOPHILE) Obligate aerobes require the presence of oxygen or air—cf. ANAEROBES. *Halophilic* species require high salt concentrations, and photosynthetic bacteria require radiation of appropriate wavelength(s) for phototrophic growth. Some bacteria (e.g. *Treponema pallidum*) have not been grown outside living cells.

Nutrition, metabolism. Nutrients enter the bacterial cell in solution (see also TRANSPORT). Most bacteria are chemoHETEROTROPHS (chemoorganotrophs) while a few (e.g. *Nitrobacter*, *Thiobacillus*) are chemoAUTOTROPHS (chemolithotrophs). Radiant energy is used by photoorganotrophs (e.g. *Rhodospirillum*), by photolithotrophs (e.g. *Chlorobium*), and by *Halobacterium halobium*—see PURPLE MEMBRANE. Metabolism may be *fermentative* (see FERMENTATION) and/or respiratory (oxidative)—see RESPIRATION and ANAEROBIC RESPIRATION. (see also MIXOTROPHY and BIOLUMINESCENCE)

Growth, reproduction. See GROWTH (bacterial). Reproduction occurs *asexually*—commonly by transverse symmetrical BINARY FISSION; asymmetrical binary fission occurs in *Caulobacter* (in which there is morphological differentiation), and *ternary fission* occurs in *Pelodictyon. Nitrobacter winogradskyi* reproduces by budding. A developmental cycle (involving morphological differentiation) occurs in *Chlamydia*. Species of the MYXOBACTERALES form *fruiting bodies*.

Genetic transfer, mutations. Changes in the genetic constitution of a bacterium may occur through a MUTATION or through CONJUGATION, TRANSDUCTION, or TRANSFORMATION—e.g. by the acquisition of a PLASMID (e.g. COLICIN FACTOR, R FACTOR). (see also AUXOTROPH, BACTERIOPHAGE CONVERSION, REPLICA PLATING) The characteristics of a bacterial *population* may change as a result of e.g. PERIODIC SELECTION.

Methods of studying bacteria. Bacteria are usually manipulated by a LOOP or STRAIGHT WIRE—sometimes by a SWAB. (see also ASEPTIC TECHNIQUE, MICROMANIPULATION) Certain features of bacterial cells (e.g. morphology, structure, reactions to dyes, motility) are generally determined by MICROSCOPY. For *light* microscopy the organisms are usually prepared as a SMEAR on a SLIDE—or in a HANGING DROP arrangement. (see also FIXATION, STAINING, FREEZE-ETCHING) Antigenic structure is studied e.g. by serological techniques, e.g. AGGLUTINATION reactions. The metabolic activities of a given strain of bacterium are determined by carrying out biochemical and other tests (e.g. carbohydrate-utilization tests, tests for particular enzymes) on a *pure* CULTURE of that strain. CONTINUOUS-FLOW CULTURE and/or SYNCHRONOUS CULTURE are sometimes used for metabolic and other studies. (see also ENUMERATION OF MICROORGANISMS)

Identification of bacteria. Few bacteria are sufficiently distinctive to permit identification by simple microscopical examination; the appearance of the COLONY—which may vary e.g. with medium and growth conditions—may or may not assist in identification. In general an organism may be identified by making certain observations and tests on an axenic population (i.e. a pure culture) of that organism. Features commonly determined include gram-reaction (see GRAM STAIN), morphology, motility, the ability to form spores, ACID-FASTness, and a requirement for aerobic or anaerobic conditions; other (biochemical) tests include e.g. determination of the ability of the organism to utilize particular substrates (perhaps with the characteristic production of acidity and/or gas), to produce particular enzymes (e.g. CATALASE), or to reduce, or oxidize, particular components of the medium. A knowledge of the environment from which the organism was isolated may assist in identification.

Extracellular products. According to species, strain etc. these may include enzymes, HAEMOLYSINS, toxins, diffusible pigments, BACTERIOCINS, polysaccharides etc.

Preservation. A *viable* population of bacteria may be preserved e.g. by DESICCATION, FREEZING, FREEZE-DRYING or repeated subculture.

Destruction/inactivation and parasitization of bacteria. Bacteria may be destroyed/inactivated by STERILIZATION techniques, by ULTRASONICATION, and e.g. by DISINFECTANTS and ANTIBIOTICS; some species are lysed in the presence of specific antibodies and COMPLEMENT. (see also LYSOZYME) Some species are lysed by members of the Myxobacterales. Bacteria may be parasitized by BACTERIOPHAGE. (see also BDELLOVIBRIO)

Exploitation of bacteria by man. Bacteria are involved in the preparation of e.g. BUTTER, CHEESE, SAUERKRAUT, and YOGHURT, and in flax RETTING. Some species are important in SEWAGE TREATMENT. Certain bacteria (particularly *Streptomyces* spp) produce important ANTIBIOTICS (e.g. POLYMYXINS, STREPTOMYCIN, TETRACYCLINES). Certain species are used in BIOLOGICAL CONTROL.

Taxonomy. See TAXONOMY. Currently, the tendency is to abolish certain of the higher taxa which were formerly widely-accepted; thus, e.g. it is now recognized that there is no valid basis for the orders Eubacteriales and Pseudomonadales. In the current taxonomic hierarchy certain families are not allocated to orders, and certain genera are not allocated to families. Orders and families which continue to be recognized (with or without modification) *include*: ACTINOMYCETALES, BACILLACEAE, BACTEROIDACEAE, ENTEROBACTERIACEAE, MICROCOCCACEAE, MYCOPLASMATACEAE, MYXOBACTERALES, NEISSERIACEAE, NITROBACTERACEAE, PSEUDOMONADACEAE, RHODOSPIRILLALES, RICKETTSIALES, SPIROCHAETALES, and STREPTOCOCCACEAE. Recently, more interest has been shown in NUMERICAL TAXONOMY and in the use of fundamental characteristics (e.g. GC VALUES) for taxonomic purposes. (see also NOMENCLATURE)

bactericidal (*adjective*) Refers to any agent (chemical or physical) which is able to kill (at least) some types of (vegetative) bacteria; some agents can also irreversibly inactivate bacterial spores. (see also DISINFECTANTS and STERILANT)

bactericidin (1) Any antibody (either an induced or a NATURAL ANTIBODY) which, under appropriate conditions, is able to function as a bactericidal agent. (2) Non-specific bactericidal plasma factors such as PROPERDIN.

bacteriochlorophylls (see CHLOROPHYLLS)

bacteriocin (see BACTERIOCINS)

bacteriocinogenic factors PLASMIDS which occur in certain gram-positive and gram-negative bacteria and which may enable such bacteria to synthesize BACTERIOCINS. In general, only a small proportion of individuals in a population of cells which contain bacteriocinogenic factor actually produce bacteriocins; bacteriocin production can be induced by treating the cells with certain mutagens—e.g. ultraviolet light. The synthesis of bacteriocins appears to be a lethal event for those cells in which it occurs. Bacteriocinogenic factors may be transferred from one cell to another e.g. by CONJUGATION. (see also COLICIN FACTOR)

bacteriocins (bacteriocines) Substances which may be secreted by certain bacteria and which are lethal for other (often related) strains of bacteria; the ability to synthesize bacteriocins is conferred on a bacterium by a BACTERIOCINOGENIC FACTOR. Bacteriocins include the COLICINS (produced e.g. by *Escherichia coli*), pyocins (produced by *Pseudomonas aeruginosa*), megacins (produced by *Bacillus megaterium*), and vibriocins (produced by *Vibrio* spp); a bacteriocin is also produced by species of *Rhizobium*. Staphylococcins, produced by certain strains of *Staphylococcus aureus*, are lethal e.g. for *Corynebacterium diphtheriae*. Most bacteriocins appear to be proteins, and some (e.g. those produced by *Rhizobium*, and some produced by *Pseudomonas*) resemble the tail portion of a bacteriophage. Different bacteriocins may function in different ways: thus, e.g. a given bacteriocin may inhibit protein synthesis or may affect the cell's DNA; cell lysis does not normally occur. In some cases a single molecule or particle of bacteriocin may be lethal for a sensitive cell. At least some bacteriocins are inactivated by trypsin.

bacteriocuprein A copper and zinc-containing SUPEROXIDE DISMUTASE found in a strain of *Photobacterium*.

bacteriocytes (see MYCETOCYTES)

bacteriolysis The lysis (rupture) of bacteria. Bacteriolysis may be brought about e.g. *mechanically* (see e.g. ULTRASONICATION), *serologically* (see e.g. PFEIFFER PHENOMENON), or *enzymically*—e.g. some gram-positive bacteria are easily lysed by the action of LYSOZYME. Bacteriolysis also occurs as a result of the induction of a lysogenic bacteriophage.

bacteriophage (phage) Any virus (see VIRUSES) whose host is a bacterium. (see also CYANOPHAGE) Commonly, a given phage can infect only a particular species, or strain, or closely-related strains, of a given bacterium. A wide range of bacteria is susceptible to phage infection, and a given strain may be susceptible to a number of different phages. A *virulent* phage replicates within, and lyses, the host bacterium (for details see below); lysis occurs only in bacteria which are living at the time of infection. A *temperate* phage has the ability to form a stable (non-disruptive) relationship (called LYSOGENY) with a (living) bacterium; during the infection of a bacterial population by temperate phage some cells become lysogenized while others are lysed. Lysogenization may bring about phenotypic changes in the host cell—see BACTERIOPHAGE CONVERSION. Certain phages (the filamentous ANDROPHAGES) are unusual in that, following multiplication, the phage progeny pass out through the cell wall of the host—which subsequently remains viable. Phages can effect the intercellular transfer of bacterial genes—see TRANSDUCTION.

Size, Composition, and Morphology of Bacteriophages. The smallest phages (e.g. ϕX174, Qβ) are approximately 25 nm in diameter; the T-even phages of *Escherichia coli* (e.g. T2) are 200–250 nm in length, while the filamentous phages f1 and fd are approximately 800 × 5 nm. Phage virions can be examined by ELECTRON MICROSCOPY. Many phages appear to consist largely, or solely, of nucleic acid and protein—the nucleic acid being contained within an outer coat of protein (the *capsid*). (Other components have been detected in some phages; thus, e.g lipid has been found in the PM2 and ϕ6 phages of *Pseudomonas* spp, and ATP occurs in the tail sheath of the T-even phages of *E. coli*.) According to phage type, the nucleic acid may be single-stranded or double-stranded DNA or single-stranded RNA; within the virion the nucleic acid may be in the form of a loop (as e.g. in ϕX174) or may be linear (as e.g. in the T-even phages and phage *lambda*). Unusual bases occur in the nucleic acids of certain phages; thus, e.g. the coliphages T2, T4 and T6 contain 5-hydroxymethylcytosine (HMC) in place of cytosine, and phage SP8 (host: *Bacillus subtilis*) contains 5-hydroxymethyluracil in place of thymine. Phages may be filamentous (e.g. f1), polyhedral (e.g. ϕX174), or tailed (polyhedral with a long or short, contractile or non-contractile, straight tubular *tail*)—see e.g. BACTERIOPHAGE T2, BACTERIOPHAGE LAMBDA. In phage ϕX174 certain of the individual morphological units of the capsid (*capsomeres*) bear a single projecting fibre (spikelet).

Culture, Assay, Inactivation, Preservation. Phages can be cultured and assayed on a LAWN PLATE of susceptible bacteria (see PLAQUE, PLAQUE ASSAY); they may also be cultivated in a broth culture of susceptible bacteria—phage activity being indicated by a progressive decrease in turbidity due to cell lysis. A suspension of phage may be prepared from the lysate by the removal of bacterial débris by membrane FILTRATION or by CENTRIFUGATION. Phage may be *inactivated*—i.e. deprived of the ability to infect/replicate—by any of a variety of means which include: (a) Heat. Rapid inactivation occurs above 80 °C. (b) Desiccation. Most phages have a low degree of resistance to desiccation. (c) Antisera. Phage activity can be neutralized by a serum containing antibodies homologous to the surface antigens of the phage. (d) Ionizing radiations (see STERILIZATION). (e) Chemical agents. Many antibacterial agents are directly active against phage; these agents include PHENOLS, OXIDIZING AGENTS, and certain HEAVY METALS (particularly mercury and silver). FORMALDEHYDE is particularly active against phages with *single*-stranded nucleic acid. A suspension of infective phage may be maintained for months (or longer) at 4 °C or in the frozen state; the buffered (neutral) lysate should be filtered or centrifuged before storage in order to remove bacterial debris.

The lytic cycle. (i) *Adsorption of bacteriophage.* A phage adsorbs to its bacterial host following contact between the attachment site(s) of the phage and specific receptor site(s) on the cell; more than one phage can adsorb to a cell at one time. While many phages (e.g. the T phages, and phage *lambda* of *E. coli*) adsorb to sites on the bacterial cell wall, others (e.g. phage PBS1 of *Bacillus subtilis*) adsorb to the flagellum, and the donor-specific androphages adsorb only to particular regions of the sex pili. Certain mutations in a bacterial host can so alter the receptor sites that a given phage can no longer adsorb; however, a HOST-RANGE MUTANT of that phage may be able to adsorb to the altered receptor sites. Adsorption commonly takes place only in the presence of appropriate concentrations of certain cations—e.g. calcium, magnesium; for some phages the cation requirement is highly specific. In addition, some phages adsorb only in the presence of certain organic substances (sometimes called *adsorption cofactors*); thus, e.g. T4 adsorbs only in the presence of tryptophan—which is believed to be required for mobilization of the phage's tail fibres. If a large number of phage virions adsorb to a cell the cell may lyse immediately; this phenomenon ('lysis from without') may occur as a result of the weakening of the cell wall. (ii) *Penetration (passage of phage nucleic acid into the bacterial cytoplasm)*. During infection by the filamentous phages (e.g. fd) the *entire virion* appears to enter the cell; in most other phages only the nucleic acid enters the host cell. During the adsorption of a T-even phage (T2, T4 or T6) the end of the tubular tail is brought into close contact with the bacterial cell wall—the tail being perpendicular to the wall surface. Subsequently, the tail *sheath* (see BACTERIOPHAGE T2) contracts—an energy-dependent process which occurs at the expense of sheath-bound ATP; this causes the end of the tail *inner* tube (tail core) to pass through the bacterial cell wall in a syringe-like manner. Phage DNA is then injected into the host cell. The stimulus for sheath-contraction and the mechanism of DNA injection are both unknown. Following adsorption of the small RNA phages, part of the phage protein coat (the 'A-protein') appears to enter the cell along with the nucleic acid; again, the mechanism of penetration is unknown. (iii) *Intracellular events*. Following penetration, phage development can occur only if the injected nucleic acid is not degraded by the host's RESTRICTION endonucleases (see also HOST-CONTROLLED MODIFICATION). (iii)(a) *Transcription and translation*. (For a generalized account of these processes see entries RNA and PROTEIN SYNTHESIS.) In many *DNA phages* (e.g. T4, T7, *lambda*) transcription of the phage genes occurs in a definite time sequence—'early' and 'late' genes, respectively, being transcribed during the initial and subsequent parts of the infection cycle. Transcription may be carried out, throughout, by the unmodified bacterial RNA polymerase (as e.g. in ϕX174); alternatively, some genes may be transcribed by a new (phage-coded) polymerase, or by the phage-modified bacterial polymerase (as e.g. in T7 and T4 respectively). In some phages (e.g. T7) only one strand of the DNA duplex is transcribed, while in others (e.g. T4) parts of each strand are transcribed at different times during the infection cycle. In *RNA phages* the nucleic acid of the virion is equivalent to mRNA. Translation of phage mRNA employs the ribosomes (and, usually, the tRNAs) of the host cell—a feature common to all viruses. In some phages (e.g. T4) regulation of phage protein synthesis appears to occur largely at the level of transcription; in others (e.g.Qβ) control occurs at the level of translation. (For further details see entries BACTERIOPHAGE T4, BACTERIOPHAGE T7, and BACTERIOPHAGE Qβ.) (iii)(b) *Replication of the phage genome*. Some phages (e.g. ϕX174) rely entirely on the host bacterium for nucleic acid precursors, and are heavily dependent on host cell enzymes; others code for many of their essential enzymes—e.g. specific polymerases, and enzymes required for the synthesis of unusual bases. A generalized account of DNA synthesis is given in the entry DNA (*q.v.*). For details of some different modes of genome replication see entries BACTERIOPHAGE LAMBDA, BACTERIOPHAGE T7, BACTERIOPHAGE T4, BACTERIOPHAGE ϕX174, BACTERIOPHAGE Qβ. (iii)(c) *Maturation: the intracellular assembly of mature (infective) phage virions*. In the structurally simpler phages maturation appears to occur largely or exclusively by *self-assembly*—i.e. phage components aggregate spontaneously in an orderly fashion to produce mature virions. In more complex phages (e.g. the T-even phages of *E. coli*) maturation occurs partly by self-assembly but is also dependent on the activity of certain non-structural phage-coded proteins; thus, e.g. in phage T4 several non-structural gene products are required for the assembly and attachment of the tail fibres. The amount of DNA packed into the head of a large, tailed phage appears to be determined by head size; mutant phages with small or large heads have been found to contain amounts of DNA smaller or larger, respectively, than that found in phages with normal-sized heads. (iv) *Release of mature virions from the host cell*. The filamentous phages are extruded from host cells which remain viable after extrusion. The release of most other phages appears to depend on the activity of several phage-coded gene products; one of these products brings about certain changes in the cell membrane, while another (LYSOZYME) brings about cell lysis. (see also ONE-STEP GROWTH EXPERIMENT)

Exploitation of phage. Various types of phage have been used for the investigation of many fundamental problems in genetics and biochemistry, and phage TYPING has proved to be a useful technique in diagnostic bacteriology.

Importance of phage in industry. Certain manufacturing processes (e.g. antibiotic production, cheese-making) which depend on the activity of particular species or strains of bacteria can be inhibited by phage active against those bacteria.

bacteriophage conversion (phage conversion; lysogenic conversion) Following infection of a bacterium by a bacteriophage: the acquisition, by the bacterium, of one or more new characteristics as a result of the expression of particular *phage* genes (cf. TRANSDUCTION). Bacteriophage conversion is observed most commonly in *lysogenized* bacteria (see LYSOGENY)—such characteristics being exhibited by cells and their progeny for as long as the lysogenic state is maintained; in bacteria infected with virulent phage new characteristics have been observed during the brief period between infection and lysis. One example of bacteriophage conversion is toxin-production by *Corynebacterium diphtheriae*; toxinogenicity is dependent on the lysogenization of *C. diphtheriae* by a particular phage.

bacteriophage f1 A filamentous BACTERIOPHAGE, approximately 800 × 5 nm, which contains single-stranded DNA; it infects conjugal *donor* strains of e.g. *Escherichia coli.* (see ANDROPHAGE) Following adsorption, the entire virion appears to enter the cell. Progeny phages are subsequently extruded from the host cell—which remains viable.

bacteriophage f2 A phage whose characteristics are similar to those of BACTERIOPHAGE Qβ.

bacteriophage fd A phage whose characteristics are similar to those of BACTERIOPHAGE F1.

bacteriophage *lambda* (λ) A temperate BACTERIOPHAGE which contains linear, double-stranded DNA. Host: *Escherichia coli.* The virion consists of a polyhedral *head* (which contains the DNA) and a long, non-contractile, tubular *tail* which lacks a base plate (cf. BACTERIOPHAGE T2) but which terminates in a single *tail fibre*. The overall length of the virion is approximately 250 nm; the head is about 60 nm wide and the tail approximately 150 × 10 nm. Specific site(s) located at the distal end of the tail are involved in the adsorption of the phage to its host. Subsequent infection may be either lytic or lysogenic; see LYSOGENY for details of lysogenic infection. *Lytic infection.* Shortly after infection, the intracellular phage genome circularizes; circularization occurs by hybridization (and subsequent ligase action) between the single-stranded, mutually-complementary regions at either end of the genome. As with other large, complex phages, early and late phases of transcription can be distinguished; products of the 'early genes' include factors involved in the control of transcription and in the synthesis of phage DNA, while 'late gene' products include the head and tail proteins of the virion and the enzyme LYSOZYME (concerned with cell lysis and phage release). The transcription of 'late' genes is not coupled with the replication of phage DNA (cf. phage T4). Transcription on *each* of the strands occurs during different stages in the cycle of infection (cf. phage T7). Replication of the circularized λ genome appears to occur by the process described by the CAIRNS MODEL; a number of circular, double-stranded progeny genomes are produced. Subsequently, double-stranded concatenates appear to be formed from these genomes by the mechanism described by the ROLLING CIRCLE MODEL. Finally, specific endonucleases are believed to split the concatenates into individual genomes—each having the terminal single-stranded regions characteristic of mature (virion) DNA. Replication of the λ genome involves both host-specified and phage-specified factors.

bacteriophage neutralization test (see NEUTRALIZATION TEST)

bacteriophage φX174 A polyhedral, tailless, virulent BACTERIOPHAGE, diameter approximately 25 nm, which contains circular, single-stranded DNA. Host: *Escherichia coli.* During infection, the phage genome (*plus* strand) initially attaches to a site on the bacterial membrane where a complementary strand (*minus* strand) is synthesized by a *bacterial* DNA polymerase from host-derived precursors; the resulting circular, double-stranded molecule is referred to as the *replicative form* (RF). The (−) strand serves as a template for transcription—which appears to be carried out by the unmodified bacterial RNA polymerase. (Until recently it appeared that the number of phage-specified proteins exceeded that permitted by the size of the phage genome; this paradox seems to have been resolved by recent evidence which indicates that the nucleotide sequences of (at least) some genes partly coincide—i.e. the nucleotide sequence of one gene occurs within, or overlaps, the sequence of another gene.) The parental RF appears to replicate by the mechanism described by the ROLLING CIRCLE MODEL; both the unwinding and circular strands function as templates and a number of circular, double-stranded daughter RFs are formed. In subsequent replication of the RFs, synthesis on the (+) strand template is repressed and single-stranded (+) genomes are synthesized on the circular (–) strand template. Phage maturation appears to occur largely or solely by self-assembly.

bacteriophage Qβ A polyhedral, tailless BACTERIOPHAGE, diameter approximately 25 nm, which contains single-stranded RNA; it infects conjugal *donor* strains of e.g. *Escherichia coli.* (see also ANDROPHAGE) The phage RNA functions as mRNA (see entry RNA)—i.e. it is a *plus* strand. The regulation of phage protein synthesis occurs at the level of translation: it has been suggested that the mRNA undergoes certain conformational change(s) which expose particular ribosome binding sites. Phage-coded products include a coat protein, a maturation protein ('A-protein')—associated with the adsorption site of the virion, and an RNA replicase subunit; synthesis of the subunit begins shortly after infection and is later terminated by a repressor (the coat protein) which binds to the mRNA. The phage-coded replicase subunit binds to three host-specified protein factors to form a multicomponent RNA-dependent RNA polymerase (= RNA replicase, RNA synthetase); the host-specified proteins are the elongation factor components Ts and Tu and an interference factor—all of which are normally involved in bacterial PROTEIN SYNTHESIS. During replication of the phage genome a complementary (–) strand is synthesized on the parental genome (+ strand); subsequently, new genomes (+ strands) are synthesized on the minus strand(s). Maturation appears to occur largely or exclusively by self-assembly.

bacteriophage T2 A virulent, tailed BACTERIOPHAGE whose host is *Escherichia coli.* The *head* of the virion, which contains linear, double-stranded DNA, is a hollow, polyhedral, protein structure approximately 110 nm long and 80 nm wide. To the head is attached, *via* a disc-like *collar*, a composite *tail* which is approximately 100 nm long and 15 nm in diameter; the tail consists of two coaxial tubes: an inner *tail core*, and an outer, contractile *sheath*—with which is associated a number of molecules of ATP. Perpendicular to the distal end of the tail is a roughly hexagonal plate (*tail plate*, *base plate*, *end plate*)—each corner of which bears a *tail pin* (aligned parallel with the tail) and a long, jointed fibre (*tail fibre*). (Other T-even phages are similar to T2 in morphology and size.) T2 adsorbs to a host cell *via* specific sites at the free ends of the tail fibres and tail pins; the fibres adsorb first. (Other T-even phages adsorb in a similar fashion.) Penetration is described under BACTERIOPHAGE.

bacteriophage T4 A virulent, tailed BACTERIOPHAGE similar to BACTERIOPHAGE T2 in form and size; the virion contains linear, double-stranded DNA. Host: *Escherichia coli.* T4 adsorbs to its host by means of tail fibres and tail pins (see BACTERIOPHAGE T2); adsorption occurs only in the presence of appropriate cations and the adsorption cofactor tryptophan. Early and late transcription can be distinguished during the infection cycle; in general, early and late species of mRNA appear to be transcribed from different strands of the phage DNA. Initially, transcription is mediated by the unmodified host RNA polymerase, but later this enzyme undergoes phage-directed modification—e.g. 5′-adenylation of the *alpha* subunits of the core enzyme; only modified RNA polymerase can transcribe the late genes. (The modified RNA polymerase remains sensitive to RIFAMYCINS.) The products of the early genes include those required for viral DNA synthesis, for inhibiting host macromolecular synthesis, and for modifying the host's RNA polymerase. The products of late transcription include phage structural proteins and factors involved in cell lysis and phage release; in contrast to e.g. BACTERIOPHAGE LAMBDA, late transcription occurs only *during* the replication of phage DNA—suggesting, perhaps, that transcription of the late genes requires the existence of single-strand discontinuities in the phage DNA. Replication of the phage genome requires the activity of a number of phage-coded enzymes, and involves the formation of concatenates. Within a given, infected cell T4 produces a number of *circularly permuted*, terminally redundant genomes; such genomes are of the type: abcdefghijkab, bcdefghijkabc, cdefghijkabcd etc., where a, b, c etc. represent genes. Certain aspects of phage assembly (e.g. attachment of the tail fibres) appear to require non-structural phage-directed proteins.

bacteriophage T7 A virulent, polyhedral BACTERIOPHAGE with a short, non-contractile tail (cf. BACTERIOPHAGE T2); the virion contains linear, double-stranded, terminally-redundant DNA which, unlike that in phage λ, does not have single-stranded terminal regions. Host: *Escherichia coli.* As in other tailed phages, the adsorption site(s) of T7 are located at the distal end of the tail. During the infection cycle, bacterial RNA polymerase transcribes the 'early' mRNA from the phage nucleic acid; one of the products of the 'early' genes (gene 1) is a (RIFAMYCIN-resistant) phage-specific RNA polymerase which is necessary for transcription of the remaining ('late') phage genes. The transcription of 'late' genes is not coupled with the replication of phage DNA (cf. BACTERIOPHAGE T4); enzymes coded for by the 'late' genes include those required for phage

DNA synthesis, and those which inhibit the host's macromolecular synthesis and degrade the host's DNA. (The products of host DNA degradation are used in the synthesis of phage DNA.) Other 'late' gene products include enzymes concerned with the synthesis of phage protein structures and those involved in maturation. Only one strand of the T7 DNA is transcribed (cf. BACTERIOPHAGE T4). The linear, terminally redundant genome present in the virion remains linear throughout the infection cycle; DNA replication, which occurs bidirectionally, is initiated at a site near (but not at) one end of the genome. DNA synthesis appears to require only phage-specified enzymes—which include nucleases, a ligase, and a DNA polymerase. The replicated T7 DNA occurs in the form of *concatenates*—i.e. linear molecules, each consisting of several phage genomes covalently linked end-to-end. (Concatenate formation appears to occur by linkage between the (terminally redundant) newly synthesized T7 genomes—in each of which a short nucleotide sequence at the 3′-end of one strand is not paired with a complementary nucleotide sequence; concatenate formation could thus occur by hybridization between the single-stranded regions of two progeny genomes followed by ligase action.) Mature phage genomes could be formed from the concatenates by appropriate endonuclease action with synthesis in the terminally redundant regions. Maturation is believed to involve phage-coded non-structural protein factors.

bacteriophage typing (phage typing) (see TYPING)

bacteriorhodopsin (see PURPLE MEMBRANE)

bacteriostatic (*adjective*) Refers to any agent which inhibits the growth and (particularly) the reproduction of (at least) some types of (vegetative) bacteria.

bacterium (see BACTERIA)

Bacterium anitratum Obsolete name for an acid-forming strain of *Acinetobacter calcoaceticus*.

bacteroid (see BACTEROIDS)

Bacteroidaceae A family of gram-negative, anaerobic, asporogenous, non-motile or motile, chemoorganotrophic rod-shaped bacteria; species are parasitic or pathogenic in man and other animals. The genera include BACTEROIDES and FUSOBACTERIUM.

Bacteroides A genus of gram-negative anaerobic, asporogenous, non-motile or peritrichously flagellate bacteria of the family BACTEROIDACEAE; species occur in the intestinal tract of man and other animals, and may be the predominant organisms in the large intestine. Some species are found in the RUMEN. *Bacteroides* spp may behave as opportunist pathogens—e.g. following bowel surgery; they have been isolated from abscesses and other lesions.

The cells are pleomorphic, fusiform or non-fusiform bacilli, or coccobacilli; the size and shape of individual cells appear to be strongly influenced by the nature of the growth medium. Growth occurs on blood agar; the organisms are usually non-haemolytic. Some species attack a range of sugars, while others attack sugars weakly or not at all—but attack peptones. Fermentation products include mixtures of acetic, butyric, formic, isobutyric, isovaleric, lactic, and succinic acids.

Most strains are sensitive to erythromycin and resistant to neomycin.

The type species is *B. fragilis*; GC% range for the genus: 40–55.

B. fragilis. A non-motile species common in the human intestine. The organism attacks a range of carbohydrates producing acetic and succinic acids as the main fermentation products. On blood agar *B. fragilis* forms smooth, greyish, round, entire, low convex colonies. Optimum growth temperature: 37 °C; optimum pH: 7.

B. ruminicola and *B. succinogenes* occur in the rumen; *B. succinogenes* is known to be cellulolytic.

B. pneumosintes (formerly *Dialister pneumosintes*) is a minute bacillus reported to have a maximum *in vivo* length of 0.3 microns.

B. melaninogenicus produces a black pigment (a haem derivative); *B. niger* appears to form a dark melanin pigment.

bacteroids Bacterium-like cells. The term is used e.g. to refer to: (a) morphologically-differentiated cells of *Rhizobium* in ROOT NODULES; (b) individual cells of *Frankia* formed by mycelial fragmentation following an abortive invasion of a host organism.

bactoprenol A lipid-soluble, membrane-bound polyisoprenyl alcohol consisting of a linear chain of eleven isoprenoid units (*undecaprenol*):

$$CH_3.C(CH_3){=}CH.CH_2[CH_2.C(CH_3){=}CH.CH_2]_9.CH_2.C(CH_3){=}CH.CH_2OH$$

The bactoprenol monophosphate acts as a carrier in the synthesis of a number of bacterial polymers, e.g. PEPTIDOGLYCAN, LIPOPOLYSACCHARIDES; it presumably facilitates the transfer of lipophobic sugar nucleotides across the lipophilic CELL MEMBRANE. A similar carrier appears to be involved in the synthesis of TEICHOIC ACIDS and certain CAPSULE polymers.

baculoviruses Agents of GRANULOSES and *nuclear* POLYHEDROSES.

Baeomyces A genus of LICHENS in which there is a granular or squamulose *primary* (basal) thallus and short, erect *podetia*; each PODETIUM bears a convex apothecium—forming a mushroom-shaped structure. The phycobiont is *Coccomyxa*. Species include e.g.: *B. rufus*. The primary thallus is greenish-grey, granular and often minutely squamulose. Apothecia are 1–2 millimetres across and occur on short, more or less flattened, furrowed podetia; the disc is pink to reddish-brown. *B. rufus* occurs e.g. on peaty or sandy soils—mainly in hilly regions.

bagassosis A pulmonary disease associated with inhalation of the dust of sugar cane waste (bagasse); the actinomycete *Thermoactinomyces sacchari* has been implicated as a causal agent.

bakanae disease of rice (foolish seedling disease) A disease, caused by the fungus *Gibberella fujikuroi*, in which the early symptoms (e.g. stem elongation) are due to the effects of GIBBERELLINS formed by the fungus.

bakers' yeast (see YEAST)

baking Yeast (*Saccharomyces cerevisiae*) is used as a leavening agent in the preparation of bread and similar products. By fermenting sugars in the dough, the yeast produces large numbers of carbon dioxide bubbles which give rise to the characteristic honeycomb texture of the final product. Certain other products of fermentation—and the yeast itself—contribute to the flavour of the bread.

balanced growth (see GROWTH (bacterial))

balanced salt solution (BSS) (see HANKS' BSS)

Balantidium coli A species of parasitic ciliate protozoan within the order TRICHOSTOMATIDA; the causative agent of balantidial DYSENTERY in man. (*B. coli* is the only ciliate pathogenic for man.) The trophozoite is ovoid, usually about 65 × 50 microns, with spirally arranged longitudinal rows of short cilia. At the anterior end there is a deep invagination (the vestibule) which leads to the *cytostome*. The large macronucleus is generally visible only in stained preparations; it stains deeply with haematoxylin. Reproduction occurs by binary fission; conjugation also occurs. Trophozoites can be cultured *in vitro* in media which support the parasitic intestinal amoebae.* Encystment occurs in the gut; the spherical cyst (approx. 50–60 microns in diameter) contains a large macronucleus which can be easily seen on staining with a solution of methyl green in dilute acetic acid. (*see also ENTAMOEBA)

Ballerup–Bethesda group (see CITROBACTER)

ballistospores (*mycol.*) Spores which, at maturity, are forcibly projected from their points of attachment on the sporophore; ballistospores are formed by many basidiomycetes and by certain imperfect yeasts (e.g. *Sporobolomyces*). (Ballistospores are not formed e.g. by the gasteromycetes or by the smut fungi.) A number of hypotheses have been advanced to explain the mechanism of spore discharge but none of these appears to have gained universal support; it seems possible that different mechanisms may be operative in different species. Some authors believe that the mechanism may involve the formation of a gas bubble or a drop of liquid; the gas or liquid is supposed to accumulate beneath a membrane in the region where the spore is attached to the sterigma, and spore ejection is supposed to be achieved on the bursting of the membrane and the release of the gas or liquid. An alternative suggestion supposes that that part of the sterigma adjacent to the spore rounds off (as a result of turgor pressure in the basidium) and in so doing exerts a force on the spore.

Baltimore classification (of viruses) A classification based on the pathway: viral genome → mRNA synthesis. Six classes (groups) of viruses are distinguished: *Class I*. Viruses containing double-stranded DNA. In certain viruses of this class (e.g. phage T4) some mRNA is transcribed from each strand of the DNA duplex; in others (e.g. phage T7) mRNA is transcribed only from one strand of the duplex (the *minus* strand)—the other strand being the *plus* strand. (see PLUS AND MINUS STRANDS OF DNA AND RNA) *Class II*. Viruses containing single-stranded DNA. mRNA is transcribed from a complementary (*minus*) DNA strand which is synthesized on the genome (a *plus* strand)—see e.g. BACTERIOPHAGE ϕX174. *Class III*. Viruses containing double-stranded RNA—e.g. REOVIRUSES. mRNA is transcribed only from one of the strands of the duplex (the *minus* strand). *Class IV*. Viruses containing single-stranded RNA (*plus* strand); transcription (= genome replication in these viruses) involves the initial formation of an RNA duplex (a complementary, minus, strand is synthesized on the genome) and the subsequent synthesis of plus strands on a minus strand template. (see e.g. BACTERIOPHAGE Qβ) *Class V*. Viruses containing single-stranded RNA (*minus* strand); in at least some viruses of this class (which includes e.g. the influenza viruses) an RNA-dependent RNA polymerase is associated with the virion. *Class VI*. Single-stranded RNA viruses in which mRNA is synthesized on a DNA template produced from the genome by a REVERSE TRANSCRIPTASE.

Bangia A genus of red algae (RHODOPHYTA).

Bang's disease (see BRUCELLOSIS)

barley, diseases of (see CEREALS, DISEASES OF)

barley yellow dwarf virus A virus whose hosts include many members of the Gramineae—e.g. wheat, oats, barley; the polyhedral virion, approximately 25 nm in diameter, contains single-stranded RNA. On barley (*Hordeum vulgare*) typical symptoms of infection include obvious stunting and the development of an orange-yellow coloration in the leaves. The virus is commonly transmitted circulatively by aphids; mechanical transmission is unknown. (see also VIRUS DISEASES OF PLANTS)

barrier filter (stopping filter) An optical filter which transmits visible light but which arrests ultra-violet radiation. A barrier filter is commonly fitted in the eyepiece of a microscope used for fluorescence MICROSCOPY; its function is to prevent the entry of ultraviolet light into the eye. (Prolonged exposure of the eye to ultraviolet radiation may result in serious injury.)

Bartonella A genus of gram-negative bacteria of the family BARTONELLACEAE. The organisms appear to be pleomorphic rod-shaped or ellipsoidal cells which are flagellate in culture and which divide by binary fission; a cell wall is present. The sole species, *B. bacilliformis*, is parasitic in erythrocytes and in other tissue cells; this organism, which occurs mainly or exclusively in South America, is the causal agent of OROYA FEVER.

Bartonellaceae A family of gram-negative bacteria of the RICKETTSIALES; species are intracellular parasites or pathogens in the erythrocytes (and, in some cases, other cells) in a variety of animals, including man. The organisms can be grown in cell-free media. The genera are BARTONELLA and GRAHAMELLA.

basal body (of a bacterial flagellum) A structure which appears to attach the FLAGELLUM to the cell envelope. The proximal end of the flagellum appears to be curved and somewhat thickened and is referred to as the *hook*; the thickening is believed to indicate the presence of a sheath. The hook leads into the basal body which, like the hook, consists of protein. In *Escherichia coli* the basal body is believed to consist of four parallel ring-shaped structures arranged on a rod-shaped core; the rings may provide effective contact between the basal body and the various layers of the cell envelope. (Fewer rings appear to occur on the basal bodies in gram-positive cells—in which the cell envelope consists of fewer layers.)

basal body (of a eucaryotic flagellum or cilium) (Synonyms: basal granule, blepharoplast, kinetosome, mastigosome; the term KINETOPLAST is sometimes used incorrectly as a synonym.) A small cylindrical body, within the cytoplasm, which forms the base (proximal end) of every cilium (see CILIUM (1)) and *eucaryotic* FLAGELLUM. Structurally, the basal body resembles an extension of the flagellar *axoneme* (see FLAGELLUM)—with the principal exceptions that (a) the central pair of tubules is absent, and (b) the peripheral system consists of nine *triplets* of MICROTUBULES. Basal bodies appear to contain DNA. (The basal bodies of some flagellates can also act as CENTRIOLES during mitotic cell division.)

basal granule (*protozool.*) (see BASAL BODY)

basal medium (see NUTRIENT BROTH)

base composition (of microbial DNA) (see GC VALUES and BASE RATIO; see also DNA BASE COMPOSITION, DETERMINATION OF)

base ratio (dissymmetry ratio) (of microbial DNA) The ratio: $\frac{A + T}{G + C}$ in which A, T, G, and C represent the amounts of adenine, thymine, guanine, and cytosine, respectively, in a sample of DNA. (cf. GC VALUE) The base ratio of bacterial DNA ranges from below 0.4 (*Micrococcus* spp) to over 2.5 (*Clostridium* spp). Organisms which have a base ratio greater than unity are sometimes referred to as *AT types*; those having a base ratio lower than unity are described as *GC types*.

Basidiobolus A genus of fungi of the ZYGOMYCETES.

basidiocarp The fruiting body of a basidiomycete.

basidiole (1) An immature BASIDIUM at the *binucleate* stage—i.e. prior to the appearance of the sterigmata. (2) An aborted basidium. (3) A sterile cell which is similar in appearance to an immature basidium.

basidiolichen A lichen in which the mycobiont is a basidiomycete.

basidiomycetes Fungi of the sub-division BASIDIOMYCOTINA. (In some taxonomic schemes these fungi form the *class* Basidiomycetes.)

Basidiomycotina (the 'basidiomycetes') A sub-division of FUNGI of the division EUMYCOTA; constituent species are distinguished from those of other sub-divisions in that their sexually-derived spores (basidiospores) are formed on *basidia* (see BASIDIUM). The classes of basidiomycetes are TELIOMYCETES, HYMENOMYCETES, and GASTEROMYCETES.

basidiospores Sexually-derived spores formed on basidia (see BASIDIUM) by fungi of the sub-division BASIDIOMYCOTINA. According to species, basidiospores may be any of a variety of shapes and sizes, and they may be pigmented or non-pigmented; in most species the basidiospores develop on sterigmata (which form on the basidia), while in some species (e.g. the smut fungi) sterigmata are not formed—the spores arising directly from the basidium. Basidiospores are commonly uninucleate and haploid; however, some species produce basidiospores which are binucleate. In many basidiomycetes the basidiospores are forcibly discharged from the basidium—see BALLISTOSPORES and BASIDIOSPORES, DISCHARGE OF.

On germination, the basidiospores of most species form germ tubes; in certain lower basidiomycetes (e.g. many smut fungi) the basidiospores give rise to secondary spores (*germination by repetition*).

basidiospores, discharge of The basidiospores of many basidiomycetes are forcibly discharged from their positions on the basidia; such spores are referred to as BALLISTOSPORES. (Forcibly discharged spores are not formed e.g. by the gasteromycetes or by the smut fungi.)

In agarics with aequihymeniiferous GILLS the ballistospores are projected, more or less horizontally, for distances of some 0.1–0.2 mm before gravity initiates a vertical fall; the spores on a given basidium are commonly discharged in rapid succession—the time intervals between discharges being of the order of seconds/minutes. In these agarics the shape of the gills is such the spore dispersal is facilitated; thus, the tapered gills permit the dispersal of spores discharged from the most proximal parts of the gills even when the gills (or the stipe) are orientated several degrees from the vertical.

In *Coprinus* spp the inaequihymeniiferous gills are extremely thin, are rectangular in transverse cross section, and are situated close together on the pileus. In these fungi the spores are forcibly discharged from their sterigmata but the shape of the gills militates against spore dispersal; however, the presence of *cystidia* prevents contact between adjacent gills, and the process of autodigestion facilitates the dispersal of spores from basidia situated some distance from the free end of the gill. (Autodigestion involves the progressive liquefaction of the gill tissue; digestion begins at the free end of the gill and progresses towards the pileus as fresh crops of basidia become mature. Thus, spores discharged by a given basidium are required to fall for only a short distance between adjacent gill surfaces before becoming free of the fruiting body.)

In the gasteromycete *Sphaerobolus* the entire glebal mass is violently discharged; discharge involves the initial rupture of the multilayered peridium and the subsequent eversion of the inner peridial layer. The gleba is projected for a distance of several metres.

basidium (*pl.* basidia) A microscopic structure at the surface of which are formed (as extrusions) the sexually-derived spores (BASIDIOSPORES) in fungi of the sub-division BASIDIOMYCOTINA. In some species (class HYMENOMYCETES) the basidia occur as a layer (*hymenium*) at the surface of some part of the basidiocarp (fruiting body)—e.g. the gill surface of the common mushroom; the hymenium may or may not contain cystidia (see CYSTIDIUM; see also BASIDIOLE)—according to species. In many species the basidium is a simple cylindrical or clavate structure which bears two or four apically-situated basidiospores on short sterigmata; in other species the basidium may be apically bifurcate (each fork bearing a single terminal basidiospore—see DACRYMYCETALES) or may be divided, apically, into four finger-like processes (sterigmata)—each sterigma bearing a terminal basidiospore. In each of the above cases the basidium is a *holobasidium*—i.e. it is not divided by septa into segments; such basidia are formed by members of the sub-class HOLOBASIDIOMYCETIDAE. In other cases, a basidium may be segmented by a number of transverse or longitudinal septa; such a *phragmobasidium* is formed by members of the sub-class PHRAGMOBASIDIOMYCETIDAE and members of the class TELIOMYCETES.

Basidium-formation: a simple holobasidium. Typically, a basidium develops from a terminal, dikaryotic hyphal cell in the fertile tissue of a developing hymenium; the boundary between this cell and the parent hypha is often marked by a CLAMP CONNECTION. Within the developing basidium karyogamy and meiosis typically give rise to four (haploid) nuclei. Four small protuberances, the *sterigmata*, then form at the apical (free) end of the developing basidium, and the terminal part of each sterigma swells to form the basidiospore initial; one haploid nucleus migrates into each basidiospore initial and the spores are subsequently delimited.

Basidium-formation in the teliomycetes. In the rust and smut fungi the basidium develops from a germ tube which arises from a thick-walled spore, the *teliospore*; the diploid nucleus of the teliospore passes into the short germ tube (also called a *promycelium*) and there undergoes meiosis to form a number of haploid nuclei—each of which subsequently becomes the nucleus of a basidiospore. Typically, the rust fungi form four basidiospores per basidium—each spore being borne on a sterigma which develops laterally on the transversely-septate basidium. In many smut fungi the basidium forms an indefinite number of basidiospores as a result of repeated mitotic divisions of the haploid nuclei within the transversely-septate or (according to species) non-septate promycelium; the basidiospores are borne directly (laterally and apically) on the promycelium—i.e. sterigmata are not formed. (see also METABASIDIUM and PROBASIDIUM)

basipetal (*mycol.*) (*adjective*) One of the modes in which a chain of spores or sporangia may develop; in basipetal development the earliest-formed spores or sporangia occupy positions in the terminal or distal parts of the chain while spores formed later occupy positions nearer the spore-bearing structure. (cf. ACROPETAL)

batch culture (closed culture) The common, simple form of CULTURE in which a fixed volume of liquid medium is inoculated and incubated for an appropriate period of time. Cells grown under such conditions are exposed to a continually changing environment—due e.g. to the gradual accumulation of metabolic wastes (cf. CONTINUOUS-FLOW CULTURE).

Battey bacillus A strain of non-chromogenic ATYPICAL MYCOBACTERIA similar to *M. avium* but only weakly pathogenic for chickens. Strictly, the name refers to a strain of *M. intracellulare* but has been used to refer to any non-chromogenic atypical species and even to *M. avium*. (see also MYCOBACTERIUM)

BCG (bacille Calmette–Guérin) An attenuated strain of *Mycobacterium bovis* currently used in a vaccine against tuberculosis and leprosy.

Bdellovibrio A genus of gram-negative, chemoorganotrophic bacteria of uncertain taxonomic affinity; on primary isolation from the soil or water (the natural habitat) all strains are obligate parasites of other (living) bacteria e.g. *Escherichia*, *Pseudomonas*, *Spirillum*. Host specificity may be exhibited. *Bdellovibrio** attaches to the cell wall of the host organism; later, it passes through the wall into the periplasmic space where it grows and divides—the progeny eventually occupying much or all of the host cell. Free cells of parasitic strains are round-ended curved rods, 0.3–0.4 × 0.8–2 microns; each cell carries a long, sheathed flagellum. (Daughter cells formed within the host are initially non-flagellated.) Mutant (*host-independent*) strains of *Bdellovibrio* grow saprophytically; they are motile or non-motile organisms which may form yellow or orange pigments. The metabolism of host-independent strains is obligately oxidative (respiratory); sugars are not utilized. Optimum growth temperature: about 30 °C. *Bdellovibrio* may be isolated e.g. from soil infusions by membrane filtration—the *filtrate* being inoculated onto a lawn plate prepared from a sensitive strain of bacteria; after incubation, macroscopic plaques are formed by viable cells of *Bdellovibrio*. (**Bdello* derives from the Greek for 'leech'.)

beans, bacterial diseases of Bacterial diseases of *Phaseolus* species include: (a) *Common blight*. Typical symptoms include small translucent lesions on leaves, and brown or red spots on the pods; brown vascular discoloration may follow. Systemic infection may result from the growth of infected seeds. Causal agent: *Xanthomonas phaseoli*. (b) *Halo blight*. Similar to (a) but each leaf lesion is surrounded by a chlorotic zone. Causal agent: *Pseudomonas phaseolicola*. (see also ANTHRACNOSE)

beard lichens A common name for lichens of the genus USNEA.

Beauveria A genus of fungi within the order MONILIALES; species include *B. bassiana*, the causal agent of *muscardine* disease of the silkworm.

B. bassiana The vegetative form of the organism is a septate mycelium. Spherical conidia are borne on the branches of conidiophores which develop sympodially; the basal (proximal) asporogenic region of the conidiophore is swollen and flask-shaped.

Bedsonia Obsolete name for the genus *Chlamydia*.

beefsteak fungus The edible basidiomycete *Fistulina hepatica*. (see also BROWN OAK)

beers (see BREWING)

beet, diseases of These include: (a) *Common scab*. A disease of the root surface; causal agent: *Streptomyces scabies* (see also POTATO, BACTERIAL DISEASES OF). (b) *Black root*. A dry, black rot of the root in which deep fissures subsequently form; causal agent: *Rhizoctonia* sp. (c) *Leaf spot*. Typical symptoms include small round leaf spots with clearly-defined dark margins. Causal agent: *Cercospora beticola*.

beet yellows virus A single-stranded RNA virus which can affect a wide range of hosts; the virion is approximately 10 × 1000 nm. On sugar beet (*Beta vulgaris*) the symptoms include yellowing and brittleness of the older leaves—the chlorotic areas (which may become orange or red) spreading from leaf tips and margins; chlorotic areas subsequently become necrotic. (Under certain conditions necrosis may be the only symptom on leaves.) Cytoplasmic inclusion bodies may or may not be found. Transmission of the virus is commonly effected by aphids.

Beggiatoa A genus of gram-negative bacteria of the BEGGIATOACEAE; species occur in marine and freshwater habitats and are indicator organisms in the SAPROBITY SYSTEM. *Beggiatoa* spp are non-pigmented gliding filaments of width between 1 and 30 microns; intracellular sulphur granules are formed when growth occurs in environments rich in hydrogen sulphide, and at least some species form granules of poly-β-hydroxybutyrate. The organism becomes disseminated by fragmentation of the filaments.

Beggiatoaceae A family of bacteria of the CYTOPHAGALES. The organisms are aerobic or microaerophilic non-pigmented gliding *filaments* (chains of longitudinally-flattened cells) of width between 1 and 30 microns; they are either* chemoorganotrophs or *mixotrophs* (organisms which can use, concurrently, an inorganic electron donor and an organic source of carbon). Intracellular sulphur granules are formed in some species (e.g. *Beggiatoa* spp) when the organisms are grown in an environment rich in hydrogen sulphide. The genera include BEGGIATOA and *Thioploca*—the latter organism consisting of longitudinal bundles (*fascicles*) of filaments, each fascicle being bounded by a *sheath*. *Chemolithotrophic strains have been claimed.

Beijerinckia A genus of gram-negative, aerobic, catalase-positive bacteria of the family AZOTOBACTERACEAE; species occur in soil and in the phyllosphere. The organisms are coccoid, pyriform, or rod-shaped non-sporing cells—each 0.5–2 × 2–4.5 microns; some species (e.g. *B. mobilis*) are motile. Typically, each cell contains two polar bodies (poly-β-hydroxybutyrate). Some species form multicellular capsules—in which a single capsule encloses a small group of cells. Growth occurs within the pH range 3–10, and the optimum temperature is about 25 °C; highly mucoid growth occurs on nitrogen-free glucose-containing media. *Beijerinckia* spp have GC ratios within the approximate range 55–59%.

bell morels The fruiting bodies of VERPA spp.

Benlate Trade name for BENOMYL.

benomyl Methyl (1-*n*-butylcarbamoyl)-benzimidazole-2-carbamate. An important agricultural *systemic* antifungal agent which has ERADICANT and PROTECTANT activity against a wide range of plant diseases, including many *powdery* MILDEWS, APPLE scab, *Botrytis* infections, black spot of roses etc. As a seed dressing it protects against seed-borne diseases such as smuts and bunts of CEREALS. It is also used to prevent rots of stored fruit and vegetables. Benomyl is not effective against downy mildews, late blight of POTATO, or DAMPING-OFF. In aqueous solution, benomyl is readily hydrolysed to methyl benzimidazole-2-carbamate; this product is believed to be the fungitoxic agent. (see also BENZIMIDAZOLE DERIVATIVES (as antifungal agents))

benquinox (1,2-benzoquinone *N′*-benzoylhydrazone-4-oxime) A QUINONE derivative used as an agricultural antifungal agent; it is used mainly as a seed dressing.

benthic zone The mud, sand etc. at the bottom of a lake, sea, or other body of water. (cf. PELAGIC ZONE) The term *littoral*, as applied to the *marine* benthic zone, is used with different meanings by different authors. Many biologists regard the marine littoral zone as the 'intertidal region'—i.e. the zone uncovered at low tide and under water during high tide; some have defined the region as that zone below the average low tide mark—i.e. a zone uncovered only occasionally. The term is also used e.g. to refer to that region between high tide and a depth of 200 metres. In *freshwater* environments the littoral region of the benthic zone refers to those parts of the benthic zone which occur in shallower waters of unspecified depth.

benzalkonium chloride (see QUATERNARY AMMONIUM COMPOUNDS)

benzimidazole derivatives (as antifungal agents) An important group of agricultural *systemic* antifungal agents, e.g. BENOMYL, THIABENDAZOLE and FUBERIDAZOLE; these agents have a similar range of antifungal activity and exhibit a high degree of cross-resistance—thus their modes of action are assumed to be similar.

benzimidazole derivatives (as antiviral agents) A number of benzimidazole derivatives exhibit antiviral activity. At present, the most important of these is HBB (2-(α-hydroxybenzyl)-benzimidazole); HBB selectively inhibits the replication of a number of picornaviruses in tissue culture systems—apparently by inhibiting viral RNA synthesis, possibly at the initiation stage. At antiviral concentrations HBB shows no significant effects on essential cell processes; however, it is not used as a chemotherapeutic agent since resistant mutants rapidly emerge. (HBB-dependent mutants have also been isolated.) (see also GUANIDINE)

benzoic acid (C_6H_5.COOH) (as a food preservative) Benzoic acid and benzoates are bactericidal and fungicidal—particularly in acid solutions (approximately pH 4.5 or below). Benzoates are widely used e.g. in fruit juices, cordials, wines and sauces. The PARABENZOATES (*p*-hydroxybenzoates) find similar uses; their maximum effectiveness occurs at approximately pH 5 or below. (see also FOOD, PRESERVATION OF)

benzoquinones (see QUINONES (in electron transport))

Berkefeld candle (see FILTRATION)

Besnoitia A genus of protozoans of the suborder EIMERIINA. *Besnoitia* resembles TOXOPLASMA but appears not to have a sexual phase. Species are parasitic or pathogenic in domestic and other animals, though not in man; *B. besnoiti* causes a chronic disease of the skin and eye tissues in cattle.

Betacoccus Obsolete name for LEUCONOSTOC.

Bethesda–Ballerup group (see CITROBACTER)

BFP or BFPR (*immunol.*) Biological false positive reaction (*q.v.*).

BGAV Blue-green algal virus—see CYANOPHAGE.

BGG Bovine *gamma* globulin.

bifid (*bacteriol.*) A rod-shaped bacterium which is forked at one or both ends.

Bifidobacterium A genus of gram-positive bacteria of the ACTINOMYCETACEAE; the organisms—pleomorphic rods and *bifids*—include strict anaerobes and facultative aerobes. Species occur e.g. in the alimentary tract of man and animals; *B. bifidum* (formerly *Lactobacillus bifidus*) is found in the intestine of breast-fed infants. The products of glucose fermentation include acetic and lactic acids in the molar ratio 3:2 (see Appendix III(a)(iii)).

biguttulate (*mycol.*) Refers to a spore which contains two GUTTULES.

bijou A small glass screw-cap bottle, approximately 5–7 mls capacity, used e.g. for liquid and solid CULTURES.

bilabiate (of the asci of certain discomycetes) Refers to those asci in which spore discharge is preceded by the development of a vertical slit in the apex of the ascus.

bilateral trama (divergent trama) Refers to the arrangement of hyphae in the gill tramae in certain types of agaric—e.g. *Amanita* spp; in the bilateral arrangement the hyphae of a trama *diverge* from the central plane of the gill—i.e. the hyphae run downwards (i.e. towards the gill margin) and outwards (i.e. towards the face of the gill) from the central plane.

bile acids A class of steroid carboxylic acids which form amides with certain nitrogenous compounds—e.g. *cholic acid* (= 3,7,12-trihydroxycholanic acid) may react with glycine to form glycocholic acid, or with taurine ($NH_2(CH_2)_2SO_3H$) to form taurocholic acid; salts of these amides function as surfactants and are involved in the emulsification of lipids in the intestine. The salts of bile acids and of their amides are inhibitory to certain organisms and are used in certain selectively-inhibitory media, e.g. MACCONKEY'S AGAR. (see also BILE SOLUBILITY)

bile solubility (of pneumococci) The great majority of freshly-isolated strains of pneumococci undergo rapid lysis, under alkaline conditions, in the presence of bile, sodium deoxycholate, sodium taurocholate, or certain other substances. This reaction characterizes the pneumococcus and serves to distinguish it from most of the culturally-similar streptococci. In the *bile solubility test* a few drops of 10% sodium deoxycholate solution are added to a liquid culture (or saline suspension) of the test organism which is brought to pH 7.4–7.6; incubation is then carried out for 15 minutes at 37 °C. In a positive test the culture or suspension becomes less optically dense, i.e it clears; no clearing should occur in the control experiment. (Some workers prefer to use a saline suspension since significant amounts of protein may inhibit the test.) Deoxycholate stimulates the action of a pneumococcal autolytic enzyme, alanine-muramyl amidase, which ruptures the cell wall PEPTIDOGLYCAN. The replacement of choline by ethanolamine in the cell wall teichoic acid (by culture in a defined, synthetic medium) renders the pneumococcus resistant to autolysis.

biliproteins (see PHYCOBILIPROTEINS)

binapacryl (2,4-dinitro-6-*sec*-butylphenyl 3-methylcrotonate) (*plant pathol.*) An agricultural antifungal and acaricidal agent. It is particularly active against powdery MILDEW diseases, e.g. apple mildew. (see also DINITROPHENOLS and FUNGICIDES AND FUNGISTATS)

binary fission A process in which two cells—usually of similar size and shape—are formed by the division of one cell. In binary fission cytoplasmic division is preceded by nuclear division—which involves chromosome replication in procaryotes and e.g. MITOSIS in eucaryotes. If a cell divides *across* its longitudinal axis the process is termed *transverse binary fission*; if division occurs along the longitudinal axis it is termed *longitudinal binary fission*. Binary fission is a common mode of asexual reproduction in many types of microorganism.

Among bacteria, binary fission is typically of the transverse type and usually results in the formation of two morphologically-similar daughter cells; asymmetrical binary fission occurs e.g. in *Caulobacter*. Following binary fission the daughter cells may separate; alternatively, with subsequent progeny, they may form associations of cells such as chains, PACKETS, PALISADES etc. (see also GROWTH (bacterial), SEPTUM (bacterial) and TERNARY FISSION)

Among protozoans, binary fission in ciliates is typically HOMOTHETOGENIC, and in flagellates it is typically SYMMETROGENIC.

Among yeasts, binary fission occurs e.g. in *Schizosaccharomyces* spp; in many yeasts, e.g. *Saccharomyces*, reproduction occurs by BUDDING (cf. BUD-FISSION).

binding hyphae (see HYPHA)

binding proteins Soluble proteins which specifically and reversibly bind a variety of substances (including sugars, amino acids, inorganic ions and vitamins) but which appear to have no enzymic activity. Binding proteins are believed to occur in the periplasmic space of many gram-negative bacteria, and have also been found in certain yeasts. They are believed to play some role in the TRANSPORT of the substances bound, but the precise nature of this role remains uncertain; it is thought that they may act as sites of recognition for the substances to be transported. Osmotic shock has been used to release the binding proteins from gram-negative bacteria.

binomial (Latin binomial; binominal) (*noun*) A designation consisting of two names; thus, the name of a SPECIES consists of the name of the genus to which it belongs followed by the SPECIFIC EPITHET. (see also NOMENCLATURE)

bioassay (biological assay) The use of an organism (or other living system) for assay purposes. Thus, e.g. the pathogenicity of an entomopathogenic microorganism can be assayed by the use of susceptible insects, and microorganisms may be used for the measurement of low concentrations of certain vitamins, amino acids, etc. For the assay of e.g. *pantothenic acid* use is made of an organism (e.g. *Tetrahymena pyriformis*) for whom this

vitamin is an essential growth requirement. The organism is inoculated into a medium containing adequate concentrations of all growth requirements *except* pantothenic acid (which is present at a known, sub-optimal concentration) and incubated under standard conditions. (Care must be taken to avoid carrying over any pantothenic acid in the inoculum.) After incubation the amount of growth is determined—the amount of growth being proportional to the concentration of pantothenic acid in the medium.* A *standard curve* is constructed by plotting amount of growth (e.g. in mg dry weight) *versus* concentration of pantothenic acid under standard conditions of incubation; using the standard curve an unknown concentration of pantothenic acid can thus be quantified. In a similar manner *Euglena gracilis* may be used to assay vitamin B_{12}, and *Saccharomyces* sp to assay biotin; bacteria have been used for the bioassay of amino acids.
*The medium should not contain substances which affect the restriction on growth imposed by the limiting concentration of vitamin (or other substance being quantified); thus, e.g. if an organism can synthesize biotin from pimelic acid, the presence of the latter in the medium will permit growth which is not proportional to the concentration of biotin present.

bioautography A method for detecting small amounts of one or more substances (e.g. vitamins) in a complex mixture, such substance(s) being essential growth requirements for the test organism(s) used. (see also BIOASSAY) The components of the mixture are initially separated by chromatography. The presence of a particular component (in a given chromatographic band) is then detected by extracting the band, with an appropriate solvent, and adding the extract to a medium which lacks the relevant growth factor. The medium is inoculated with an appropriate test organism, and incubated. Growth indicates that the particular component was present in the original mixture.

biochemical oxygen demand (BOD) (see SEWAGE TREATMENT)

biocide (1) Synonym of STERILANT. (2) The term has also been used to refer to certain industrial PRESERVATIVES. (see also DISINFECTANT)

biocytin *ε-N*-Biotinyl-L-lysine (see BIOTIN)

biological assay (see BIOASSAY)

biological control The deliberate use (by man) of one species of organism to control or eliminate another. Examples involving microorganisms include: (a) Control of sawflies in conifer forests by the use of *nuclear polyhedrosis viruses*—see POLYHEDROSES. (b) Control of the soil pest *Popillia japonica* (Japanese beetle) by *Bacillus popilliae* or *B. lentimorbus*—which cause a fatal septicaemia (*milky disease* types A and B respectively) in these insects. (c) *Bacillus thuringiensis* (see BACILLUS) is pathogenic for a range of insects, particularly caterpillars (e.g. *Trichoplusia ni*, the cabbage looper); these bacteria are being developed as 'broad spectrum' microbial insecticides. (d) Control of rabbits by MYXOMATOSIS.

biological false positive reaction (BFPR) (*immunol.*) A positive result in the WASSERMANN REACTION (or other serological test for SYPHILIS) in the *absence* of infection by *Treponema pallidum*. Such reactions may be obtained from the sera of patients suffering from tuberculosis, malaria and other diseases—including autoimmune disease.

biological filter (see SEWAGE TREATMENT)

bioluminescence (*microbiol.*) The generation of light by certain microorganisms. Bioluminescence involves the production of an electronically excited entity following the oxidation of a reduced substrate (*luciferin*) by an enzyme (*luciferase*); the excited entity is returned to the (unexcited) ground state by the emission of a photon of light energy. (Luciferin and luciferase are general terms which are applied to any bioluminescent system; the nature of the components involved differs in different organisms.)

Bacterial bioluminescence. Most luminescent bacteria are marine species (see PHOTOBACTERIUM, VIBRIO; see also SYMBIOSIS). Light of wavelength 475–505 nm is emitted continuously. In *cell-free* systems light production requires oxygen, $NADH_2$, FMN, luciferase, and a long-chain aldehyde (C_7–C_{18}). A postulated mechanism involves the reduction of FMN by $NADH_2$ followed by the formation of a luciferase-$FMNH_2$ complex; in the presence of oxygen this complex forms an entity which, in the presence of the aldehyde, emits light. The nature of the emitter is unknown—$FMNH^+$ has been suggested. During the reaction the aldehyde becomes oxidized to the corresponding carboxylic acid; its function is unknown, and there is no evidence that it is required *in vivo*. Bacterial bioluminescence competes with the respiratory chain for reduced NAD; luminescence is rapidly enhanced following cyanide inhibition of respiration.

Fungal bioluminescence occurs in certain basidiomycetes—e.g. *Armillaria mellea* (mycelium and rhizomorphs), *Omphalia flavida* (mycelium), and *Pleurotus* spp (fruiting bodies); luminescence may be e.g. greenish-blue (wavelength about 530 nm). Luminescence is continuous but, in some species, exhibits a diurnal fluctuation in intensity; it requires $NADH_2$ (or $NADPH_2$) and oxygen, and appears to involve two enzymes.

Bioluminescence in dinoflagellates. Species of e.g. *Gonyaulax*, *Noctiluca* and *Pyrodinium* emit flashes of light (duration: approx. 0.1

second) on appropriate stimulation (e.g. acidification, agitation); bioluminescence in *G. polyedra* exhibits a CIRCADIAN RHYTHM. In *Noctiluca* luminescence is associated with subcellular organelles—the *microsources*. Two luminescent systems have been isolated from *Gonyaulax*: (a) a soluble, luciferin-luciferase system (which requires oxygen and a high salt concentration), and (b) a system involving subcellular, crystal-like structures—*scintillons*—which emit a flash of light on acidification in the presence of oxygen. Scintillons contain crystalline guanine; similar structures have been observed in non-luminescent dinoflagellates. *In vivo*, systems (a) and (b) may function independently or as a unified system; the existence of two systems *in vitro* may or may not be an artefact of isolation.

A number of radiolarians also exhibit bioluminescence.

The function of microbial bioluminescence, if any, is unknown. It has been suggested that it is the vestige of a system once used for eliminating low concentrations of oxygen from the environment.

Luminescent bacteria have been used in a highly sensitive method for detecting low concentrations of dissolved oxygen.

biomass The dry-weight, volume (or other *quantitative* estimation) of organisms (commonly microorganisms—sometimes specifically *photosynthetic* microorganisms) in a body of water (e.g. a lake) or in a synthetic medium. *Biomass* is also used to refer to the sum total of living organisms in an ecosystem.

biopterin 2-Amino-4-hydroxy-6-(1′,2′-dihydroxypropyl)-pteridine: a growth factor required e.g. by *Crithidia fasciculata* and *Leishmania tarentolae*; the requirement for biopterin can be partially or totally overcome by high concentrations of FOLIC ACID. Biopterin is believed to act as a coenzyme in certain hydroxylation reactions.

biotin (vitamin H; coenzyme R) A cofactor which is involved in transcarboxylation reactions and in (non-photosynthetic) carbon dioxide fixation reactions; biotin is bound, *via* its carboxyl group (see formula), to the ε-amino group of a lysine residue in the apoenzyme.

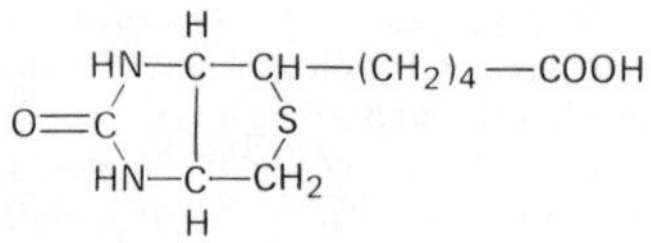

Biotin

Biotin is required as a growth factor e.g. by many fungi—particularly yeasts (e.g. *Saccharomyces cerevisiae*, most *Candida* spp), by several algae, and by species of the bacterial genera *Lactobacillus* and *Clostridium*. Many microorganisms can synthesize biotin; a precursor of biotin is believed to be *dethiobiotin* (= desthiobiotin)—in which the sulphur atom of biotin is absent. (see also VITAMINS and BIOASSAY)

biotope (1) In general, the environment occupied by an organism, or organisms. (2) (*med.* and *vet.*) The location of (particular) parasitic organisms within the body. (3) The spatial distribution of the biomass in a cross-section of a river, lake etc.

biotroph A parasite which derives nutrients from the *living* tissues of its hosts. (cf. PERTHOPHYTE)

biotype (*microbiol.*) A variant form of a given SPECIES, or SEROTYPE, distinguishable by biochemical (metabolic) or other means.

biplicity (in flagellation) (see FLAGELLUM (bacterial))

bipolar heterothallism (see HETEROTHALLISM)

bird's-nest fungi Certain members of the order NIDULARIALES—e.g. CYATHUS spp.

birefringence (double refraction) The phenomenon in which a beam of light is split into two parts on entering certain types of medium (*anisotropic* media); within the medium the two parts travel in different directions.

birth scar (see SCAR)

bisdithiocarbamates (see DITHIOCARBAMATE DERIVATIVES)

bismuth compounds (as antimicrobial agents) (see HEAVY METALS)

bisphenols (as antimicrobial agents) Bisphenols are compounds which contain two phenolic residues per molecule; these residues may be linked directly (carbon-carbon) or *via* an alkane group (*dihydroxydiphenylalkanes*) or sulphur (*dihydroxydiphenylsulphides*). In the most widely used of these compounds each of the two phenyl groups has a hydroxyl group in the *ortho* position (2-position), and is usually chlorinated—the degree of chlorination affecting the antimicrobial activity of the compound; compounds containing two *mono*chlorophenyl groups are more effective against fungi and gram-negative bacteria, while those containing two *tri*chlorophenyl groups are more active against gram-positive bacteria. The activity of the diphenylalkanes increases with the length of the alkane group, becoming maximum with a 4-carbon group. Bisphenols are believed to exert antimicrobial action by altering the permeability of the CELL MEMBRANE. Bisphenols are bacteriostatic at low concentrations and may be bactericidal at higher concentrations; they are slow-acting, so that prolonged exposure is necessary. Bisphenols are insoluble in water but form colloidal suspensions in the presence of surface-active agents, e.g. SOAPS, TWEEN; they are generally soluble in dilute alkali and organic solvents. Since bisphenols retain antimicrobial activity in the presence of soaps, and because they are relatively non-toxic, they are widely used as commercial antiseptics.

2-Hydroxydiphenyl (*ortho*-phenylphenol) is a skin antiseptic which inhibits the growth of bacteria and fungi. *Hexachlorophene* (= hexachlorophane; 2,2′-dihydroxy-3,5,6,3′,5′,6′-hexachlorodiphenylmethane) is widely used as a skin antiseptic—being particularly active against staphylococci, though generally inactive against gram-negative bacteria and without effect on spores. It is used in a number of toilet preparations (e.g. deodorants, antiseptic soaps such as *Cidal*) and in the preparations *pHisoHex* and *Ster-zac*. *Dichlorophene* (2,2′-dihydroxy-5,5′-dichlorodiphenylmethane) is active against fungi and is used as a preservative for textiles and paper. *Bithionol* (2,2′-dihydroxy-3,5,3′,5′-tetrachlorodiphenylsulphide) inhibits the growth of a range of fungi and gram-positive and gram-negative bacteria and finds uses e.g. in antiseptic soaps.

bithionol (see BISPHENOLS)

bitunicate (*mycol.*) (*adjective*) Refers to that type of ASCUS whose wall consists of a more or less rigid outer envelope and a relatively flexible inner envelope. During the process of spore release the inner envelope expands and is extruded from a split in the outer envelope; the enclosed ascospores are subsequently discharged from the inner envelope. (cf. UNITUNICATE)

biuret test A colour reaction used to detect the presence of peptides or proteins; when treated with a dilute, alkaline solution of copper sulphate, peptides and proteins give a violet/pink coloration.

bivalent (see MEIOSIS)

black death (see PLAGUE)

black mildews Fungi of the family PERISPORIACEAE

black rot, of crucifers (see CRUCIFERS, BACTERIAL DISEASES OF)

black scours (of calves, cattle) (see DYSENTERY (bacterial))

black-stem rust (see CEREALS, DISEASES OF)

blackfellows' bread The SCLEROTIUM of *Polyporus mylittae* which may attain many pounds in weight and which is used as a food by Australian aboriginals.

blackhead (histomoniasis, enterohepatitis) (in turkeys) An ulcerative and inflammatory disease of the caecum and liver. The causal agent is *Histomonas meleagridis*. Infected birds are listless, lack appetite, and pass yellowish faeces; death may occur rapidly in a high proportion of diseased birds. *Post mortem*, the caeca are seen to be swollen, and circular necrotic lesions may be seen on the liver. Cells of *H. meleagridis* may be observed in scrapings from caecal lesions. *H. meleagridis*, which does not form cysts, is transmitted transovarially by *Heterakis gallinae* (= *H. gallinarum*)—a nematode worm commonly found within the alimentary tract of the turkey; transmission is complete when the eggs of *H. gallinae* are ingested by a healthy bird and the pathogen is liberated within its intestinal tract. Since both pathogen and vector are common in chickens (in which histomoniasis seldom occurs) chickens and turkeys should not be reared together. *Furazolidone* (see NITROFURANS) has been used as a prophylactic.

blackleg (blackquarter) Endogenous GAS GANGRENE which affects the muscles of the quarters in cattle and other animals. The causal agent, generally *Clostridium chauvoei*, appears to reach the affected site *via* the intestine and bloodstream; bruising of the muscles appears to be a predisposing factor.

blackleg, of crucifers (see CRUCIFERS, FUNGAL DISEASES OF)

blackleg, of potato (see POTATO, BACTERIAL DISEASES OF)

bladders (algal) (see PNEUMATOCYSTS)

bladderwrack The seaweed *Fucus vesiculosus*.

blade (*algol.*) (see LAMINA)

blast disease (of rice) (see RICE, DISEASES OF)

blast transformation (see TRANSFORMATION, OF LYMPHOCYTES)

Blastocladiales An order of FUNGI within the class CHYTRIDIOMYCETES; the typical vegetative form of the organisms is a rhizoid-bearing hyphal or mycelial thallus, and sexual reproduction involves the fusion of motile isogametes or anisogametes (cf. CHYTRIDIALES, MONOBLEPHARIDALES). Most species live saprophytically in water or soil; the species of one genus, *Coelomomyces* (which form wall-less, coenocytic mycelia lacking rhizoids) are obligate parasites of certain insects. Constituent genera include ALLOMYCES and *Blastocladiella*.

Blastomyces dermatitidis A dimorphic fungus which is the causal agent of BLASTOMYCOSIS; *B. dermatitidis* occurs within the tissues in the form of yeast-like spherical or spheroid cells which are commonly 10–15 microns in diameter. In this unicellular form the organism reproduces by budding—a given cell producing one or more buds; each bud is attached by a wide neck to the parent cell. The yeast-like form of the organism can be cultured on blood agar at 37 °C. When cultured (e.g. on Sabouraud's agar) at room temperature, or 30 °C, the organism grows as a septate mycelium; the mycelial form produces spherical or spheroid conidia terminally or laterally on short, straight conidiophores. The ascigerous stage of *B. dermatitidis* is *Ajellomyces dermatitidis*. *A. dermatitidis*, which is heterothallic, produces eight-spored asci in spherical or spheroid cleistothecia.

Blastomycetes A class of fungi within the subdivision DEUTEROMYCOTINA; constituent species consist of imperfect yeasts and allied fungi. The class is divided into two families: the SPOROBOLOMYCETACEAE (species which form ballistospores) and the CRYPTOCOCCACEAE (species which do not form ballistospores).

blastomycosis (North American blastomycosis) A chronic mycosis in which lesions develop e.g. in the lungs, bones, and skin; the disease occurs principally in North America and in many parts of Africa. The causal agent is *Blastomyces dermatitidis*. (see also PARACOCCIDIOIDOMYCOSIS)

blastospore (*mycol.*) A spore formed by BUDDING.

bleaches (as antimicrobial agents) (see *hydrogen peroxide* under OXIDIZING AGENTS, and *hypochlorites* under CHLORINE)

bleaching (in photosynthetic organisms) Loss of CHLOROPHYLL due e.g. to growth in the absence of light.

blepharoplast (*protozool.*) (see BASAL BODY)

blewit (blewitt) The edible fruiting body of the basidiomycete *Tricholoma saevum* (*T. personatum*).

blight The name of each of a variety of unrelated plant diseases. Examples include: FIRE BLIGHT, *common* and *halo blights* of BEANS, *late* and *early blights* of POTATO etc. (see also SCAB, ROT, MILDEW etc)

blister blight (of tea) A disease of the tea plant (*Camellia sinensis*—also known as *Thea sinensis*) which occurs throughout the Asian tea-growing regions; the causal agent is the fungus *Exobasidium vexans*. Leaf symptoms include the formation of translucent (sometimes pinkish), circular, depressed (concave) lesions which commonly occur on the upper surface of the leaf; the lesions subsequently enlarge, and the convex underside of each lesion (on the lower surface of the leaf) becomes white and downy with spores. Only young leaves are susceptible to infection. Copper sprays are included in control measures. (see also RED RUST)

blocking antibodies (1) (incomplete antibodies; *non-agglutinating* antibodies) Antibodies which fail to agglutinate homologous antigens even though antigen–antibody combination may occur. Lack of agglutination may be due e.g. to an insufficiency of *effective* combining sites on the antibody molecules. By combining with the antigen, blocking antibodies prevent ('block') agglutination by homologous *agglutinating* antibodies. The combination of antigens with blocking antibodies may be detected e.g. by an ANTIGLOBULIN CONSUMPTION TEST. (2) IgG antibodies which combine with those antigens which would otherwise combine with reaginic (IgE) antibodies and initiate an immediate hypersensitivity reaction.

blood agar (see AGAR)

blood culture (*bacteriol.*) A procedure used for detecting and identifying bacteria in blood (see BACTERAEMIA and SEPTICAEMIA). Pathogenic bacteria commonly appear in the blood during the course of certain diseases, e.g. the early phase of TYPHOID FEVER. Since blood-borne bacteria may be present only in very small numbers, attempts are not generally made to culture them by direct plating methods; instead, isolation is attempted only after a sample of blood (in a suitable liquid medium) has been incubated to permit an increase in the numbers of any pathogenic bacteria which may be present. Blood (5–10ml) is added, aseptically—and immediately after removal from the patient—to some 50ml glucose broth (or other medium) in a specially designed blood culture bottle; the bottle is then incubated at 37 °C. Periodically, a suitable solid medium is inoculated from the bottle and incubated at 37 °C. The medium in the bottle usually contains an anticoagulant; additionally it may contain e.g. *p*-AMINOBENZOIC ACID to counteract the anti-microbial effect of any SULPHONAMIDE drugs present in the blood, or PENICILLINASE to inactivate any penicillin in the blood. The blood culture bottle is commonly a plain 100ml screw-cap glass bottle; a rubber disc or diaphragm, which covers the opening of the bottle, is held in place by the (metal) screw cap which contains a small central hole. The entire screw cap, and part of the neck of the bottle, is covered (and kept in a sterile condition) by a tight-fitting plastic film. After withdrawal of the blood a fresh, sterile needle is fitted to the syringe and the sample is injected into the culture bottle, *via* the rubber diaphragm, after removal of the protective plastic film. (see also CLOT CULTURE and CASTAÑEDA'S METHOD)

blood poisoning Synonym of SEPTICAEMIA.

bloom (water bloom) A visible (sometimes striking) abundance of plankton at the surface of a body of water (reservoir, lake, ocean etc). In reservoirs, blooms are often caused by certain members of the blue-green algae—particularly species which form GAS VACUOLES. Excessive algal growth in a reservoir inhibits filtration (see WATER SUPPLIES), and the subsequent decay of algal cells may cause the water to become tainted and unpalatable. Reservoir blooms are encouraged by thermal stratification; pumping, to discourage stratification, is sometimes used to prevent or control blooms. Control may involve the use of algicides, e.g. copper sulphate (a few parts per million)—although this agent sometimes encourages smaller (more troublesome) species by decimating larger (more susceptible) species. Phenanthraquinone and DICHLONE have been reported to be effective against species of bloom-forming blue-green algae.

blooming (of lenses) A procedure in which certain substances (e.g. magnesium fluoride) are used to form a thin layer or coating (of calculated thickness) on the surface of a lens. Blooming reduces the amount of incident light which is *reflected* from the surface of a lens; in the microscope (see MICROSCOPY) light reflected from e.g. the lower surface of an eyepiece lens forms a background of stray light rays which affect the sharpness of the final image. The thickness of the fluoride coating is such that *destructive interference* occurs between the rays reflected from the surface of the bloomed lens and those reflected at the fluoride-glass interface, thus eliminating the effects of reflection. In practice, blooming can eliminate the reflected light of one wavelength only (usually green or yellow); the thickness of the coating is one-quarter of the chosen wavelength.

blue-green algae (Cyanobacteria, Cyanochloronta, Cyanophyceae, Cyanophyta, Myxophyceae, Schizophyceae) An heterogenous group of PROCARYOTIC, photosynthetic organisms which contain CHLOROPHYLL *a*; they include unicellular, colonial, and filamentous species. Blue-green algae differ from the photosynthetic bacteria (RHODOSPIRILLALES) e.g. in that they evolve oxygen during PHOTOSYNTHESIS; they differ from ALGAE e.g. in that they lack chloroplasts (and other eucaryotic features) and in that they do not have algal-type cell walls. The aquatic blue-green algae occur in freshwater, brackish, or marine environments (see also PLANKTON) and may or may not be free-floating; some (e.g. *Anabaena* spp, *Microcystis* spp) are often the dominant species in water BLOOMS. Terrestrial species occur e.g. on moist soils, and in symbiotic or epiphytic associations with certain higher plants. Certain blue-green algae (e.g. *Nostoc* spp) form the phycobionts of some LICHENS. A number of species of blue-green algae carry out NITROGEN FIXATION, and their role in the NITROGEN CYCLE is believed to be a significant one. Organisms resembling blue-green algae occur among the oldest of fossil microorganisms. (see also PROCHLOROPHYTA and CYANOPHAGE)

In size, blue-green algae range from unicellular organisms of the order of one micron (e.g. *Synechococcus*) to TRICHOMES in which each cell may have a width greater than thirty microns (e.g. *Oscillatoria princeps*). Individual cells have a typical procaryotic organization, and the RIBOSOMES (in those species examined) resemble bacterial ribosomes (i.e. they have a sedimentation coefficient of 70S). The cell wall is similar, in structure and composition, to that of certain types of bacteria—e.g. it contains PEPTIDOGLYCAN; blue-green algae are thus susceptible to penicillins and to lysozyme. A feature common to many species of blue-green algae is the presence of a slimy mucilaginous outer envelope or sheath. The photosynthetic pigments occur in THYLAKOIDS which are typically arranged peripherally within the cell; pigments associated with the thylakoids are chlorophyll *a*, CAROTENOIDS, and PHYCOBILIPROTEINS. The relative proportions of these pigments in a given organism determine the colour of the organism—which, according to species and environmental conditions, may be e.g. blue-green, olive-green, yellow-green, red, purple, or black. Structures found in many (but not all) blue-green algae include GAS VACUOLES and/or HETEROCYSTS. No blue-green alga is flagellated; however, many species exhibit a gliding form of motility (mechanism unknown) and some species exhibit phototaxis.

Blue-green algae are characteristically photolithotrophic organisms, and many species appear to be obligate phototrophs—i.e. they grow only in the presence of light; the intracellular storage product is *cyanophycean* STARCH. A number of species (e.g. of *Synechococcus*, *Aphanocapsa*) have been found to be capable of photoheterotrophic growth—i.e. they can assimilate certain organic compounds (e.g. glucose) in a light-dependent process. Facultative chemoorganotrophy occurs in certain strains of *Aphanocapsa* and has been reported in the filamentous species *Plectonema boryanum*.

Sexual processes have not been detected among the blue-green algae. Unicellular species reproduce by binary fission; in some (e.g. *Chroococcus* spp, *Gloeocapsa* spp) the products of division consist of groups of two or more cells—each group of cells occurring within its own gelatinous envelope. A trichome grows by division of the intercalary cells; trichome fragmentation gives rise to disseminative units (see HORMOGONIUM and NECRIDIUM) and, in some species, vegetative cell(s) within the trichome may develop into thick-walled spores (*akinetes*).

Genera include *Aphanocapsa*, ANABAENA, *Chroococcus*, *Gloeocapsa*, LYNGBYA, MICROCYSTIS, NOSTOC, OSCILLATORIA, *Phormidium*, *Plectonema*, *Spirulina*, *Synechococcus*, and *Tolypothrix*.

blue moulds Common name for species of *Penicillium*.

blue proteins A group of copper-containing proteins. Blue proteins include: (a) The *azurin* found in *Pseudomonas aeruginosa.* Azurin is believed to play some part in respiration; reduced azurin is readily oxidized by cytochrome oxidase derived from *P. aeruginosa.* Similar copper-containing proteins have been found in *Bordetella pertussis* and *Alcaligenes* spp. (b) PLASTOCYANIN.

blue-stain (sap-stain) Discoloration in certain types of timber (e.g. sapwood of pine) caused by a fungal infection; blue-stain rarely occurs in *growing* trees. The bluish-grey 'stain' is due to the appearance of the fungal hyphae as seen through the cells of the wood. Any of a range of fungi can cause blue-stain, the commonest in temperate zones being species of the ascomycete *Ceratocystis*, e.g. *C. pilifera.* (In tropical regions *Diplodia* is an important staining fungus.) Blue-stain reduces the value of timber but does not significantly affect its mechanical strength—the CELLULOSE and LIGNIN components of the wood being little affected. Blue-stain can be prevented by the immediate treatment of felled timber with an anti-fungal agent, e.g. borax, pentachlorophenate. (see also TIMBER, STAINING OF and TIMBER, DISEASES OF)

bluetongue An acute viral disease of cattle, sheep and other ruminants, characterized by catarrhal inflammation of the buccal, nasal, and intestinal mucous membranes; infection may also result in abortion. The causal agent, an *orbivirus*, may be transmitted by the 'bite' of a biological vector (*Culicoides* spp) or (inoculatively) by any of a range of mechanical vectors; the incubation period is approximately 2 days to 2 weeks. Infection can be subclinical in some breeds of ruminant.

blusher, the Popular name for *Amanita rubescens.*

Boas–Oppler bacillus A slender, gram-positive rod, possibly a species of *Lactobacillus*, observed by Oppler (1895) in the stomach contents of patients suffering from cancer of the stomach.

BOD (see SEWAGE TREATMENT)

Bodo A genus of flagellate protozoans of the order KINETOPLASTIDA; species occur e.g. in sewage-polluted waters. The organisms are ovoid or elongated biflagellate cells, some 10–15 microns in length; both flagella arise anteriorly (near the cytostome)—one flagellum being directed posteriorly. *Bodo* spp thrive in anaerobic or microaerophilic conditions; they feed on bacteria.

Boivin antigen A synonym of O ANTIGENS.

Boletaceae A family of fungi of the order AGARICALES; constituent species form fruiting bodies in which the hymenophore varies from porous through semilamellate to lamellate (according to species). The genera include e.g. BOLETUS (hymenophore porous) and *Phylloporus* (hymenophore lamellate).

Boletus A genus of fungi of the family BOLETACEAE; constituent species form fleshy fruiting bodies in which the hymenophore is porous and in which the stipe is central. *Boletus* spp occur commonly in woodland areas, and many species appear to form mycorrhizal associations with trees; one species, *B. parasiticus*, is parasitic on *Scleroderma.* Some species of *Boletus* are edible—e.g. *B. edulis* (the cèpe).

booster (*immunol.*) A second or subsequent dose of a particular vaccine given with the object of eliciting a higher rate of antibody formation.

boot-laces A common name for the black rhizomorphs of *Armillaria mellea.*

Bordeaux mixture (*plant pathol.*) An anti-fungal preparation first used in Bordeaux in the late 19th Century for the treatment of downy mildew of vines; it has since been widely used for the treatment of a number of fungal plant diseases. Bordeaux mixture is prepared from a solution of copper sulphate and an aqueous suspension of calcium hydroxide; it appears to contain a mixture of basic copper sulphates ($CuSO_4 . xCu(OH)_2$). The freshly-prepared mixture is used in the form of a spray; the insoluble deposit thus formed on leaves, stems etc. slowly releases soluble copper (the effective agent) as a result of various influences, e.g. the action of plant or fungal secretions and/or reaction with atmospheric carbon dioxide. Bordeaux mixture is effective against many blights, mildews, rusts and leaf-spot diseases, but is now less frequently used owing to phytotoxicity. (see also BURGUNDY MIXTURE and CHESHUNT COMPOUND)

Bordet–Gengou medium A medium used for the primary isolation of *Bordetella* spp, especially *B. pertussis.* The agar-based medium contains glycerol and soluble starch and incorporates between 20% and 30% fresh horse or sheep blood. Some workers include peptone (1%) in the medium—although certain peptones are inhibitory to the growth of *Bordetella* spp. Certain antibiotics (e.g. PENICILLIN) may also be incorporated to suppress the growth of non-pathogenic organisms of the nasopharyngeal flora. (see also COUGH PLATE)

Bordetella A genus of gram-negative, aerobic, chemoorganotrophic bacteria of uncertain taxonomic affinity; *Bordetella* spp are parasites or pathogens of the respiratory tract in man and other animals. The cells are coccobacilli, about 0.3×0.5–1 micron. Metabolism is oxidative (respiratory); optimum growth temperature: 36 ± 1 °C. Three species are recognized:

B. pertussis. The causal agent of WHOOPING

COUGH; a non-motile, urease-negative species which may produce an indistinct zone of haemolysis on blood media. Primary isolation is usually carried out on BORDET-GENGOU MEDIUM. (*Bordetella* spp are inhibited by e.g. unsaturated fatty acids, sulphides.) Colonies (smooth, domed, 'metallic') develop after 2 or more days incubation. Subculture on non-enriched media generally results in a typical smooth-rough transition—with a loss of virulence and other characteristics.

B. parapertussis. A non-motile, urease-positive species which forms a brown, water-soluble pigment. Colonies develop within 2 days (cf. *B. pertussis*).

B. bronchiseptica occurs in the dog, and other animals, and infrequently causes a respiratory-tract infection in man; a motile (peritrichous) urease-positive species which can reduce nitrate.

Species-specific heat-labile somatic antigens(*) permit serological species differentiation. *B. pertussis* antigens: 1*,2,3,4,5,6,7; *B. parapertussis*: 7,8,9,10,14*; *B. bronchiseptica*: 7,8,9,10,11,12*.

boric acid (H_3BO_3) Boric acid is widely used as a mild antiseptic for delicate tissues, e.g. the eyes; it is also used as a bacteriostatic agent in urine samples. Boric acid has very weak antimicrobial properties. (see also ANTISEPTICS) Borax ($Na_2B_4O_7$)—sometimes in combination with boric acid—is used as a wood preservative.

Borrelia A genus of anaerobic, parasitic and pathogenic bacteria of the SPIROCHAETALES. The cells are commonly 5–20 microns in length, 0.5 microns or less in width; they stain well with aniline dyes. The details of metabolism are largely unknown. *B. recurrentis* and other species are the causal agents of RELAPSING FEVER; the organisms are believed to be capable of a considerable degree of *in vivo* ANTIGENIC VARIATION.

Bostrychia A genus of red algae (RHODOPHYTA).

Botrydium A genus of siphonaceous algae of the division XANTHOPHYTA; species occur on mud and damp soil. The mature organism consists of a green, pear-shaped vesicle (of the order of 1mm in diameter) anchored to the substratum by colourless branching rhizoids; the cellulosic wall of the vesicle is lined with a layer of protoplasm which contains chloroplasts and numerous nuclei and which surrounds a large central vacuole. Asexual reproduction may involve the formation of either zoospores or aplanospores. Sexual reproduction has been reported to occur by the fusion of motile isogametes. Cysts are formed under dry conditions.

Botrytis A genus of fungi within the order MONILIALES; species include a number of plant pathogens—e.g. *B. allii* (see ONION, DISEASES OF) and *B. cinerea* (see e.g. STRAWBERRY, DISEASES OF and NOBLE ROT). The vegetative form of the organisms is a septate mycelium. Dark-coloured conidia are formed as a layer over the surface of the terminal, globose swelling formed on each of the short branches of the conidiophore; individual conidia are ovoid and non-septate.

bottom yeast (see BREWING)

bottromycins A group of peptide ANTIBIOTICS which inhibit bacterial PROTEIN SYNTHESIS; they are produced by *Streptomyces* sp.

botulism (see FOOD POISONING (bacterial))

Boucherie process A process in which the sap of newly-felled trees is replaced by a preservative in order to prevent subsequent fungal decay of the timber.

Bouin's fluid A fixative used in protozoological and bacteriological work: a mixture of picric acid (75ml saturated, aqueous), formalin (25ml), and glacial acetic acid (5ml). (see also FIXATION)

boutonneuse (fièvre boutonneuse; Marseilles fever) In man, a mild, rickettsial *spotted fever* which occurs in the Mediterranean area; the causal agent, *Rickettsia conorii*, is transmitted by *tick* bite. Following an incubation period of 1–2 weeks, a characteristic lesion (the *tâche noire*) develops at the location of the bite; this lesion (up to 0.5cms diameter) consists of a black necrotic centre and a reddened margin. The regional lymph nodes are often affected. Fever, headache, muscular pains etc. are typically followed by the appearance of a maculopapular rash. *Laboratory diagnosis* may include the WEIL-FELIX TEST.

Boveria A genus of protozoans of the THIGMOTRICHIDA.

bovine typhus (see RINDERPEST)

Boyden procedure (*immunol.*) Treatment of erythrocytes with tannic acid (see TANNINS)—thus enabling their surfaces to effectively adsorb soluble proteins. Such *tanned cells* are used e.g. in PASSIVE AGGLUTINATION tests. (Chromium salts or e.g. *bis*-diazobenzidine have been used in place of tannic acid in the preparation of coated erythrocytes.)

BPL Abbreviation for β-PROPIOLACTONE.

Brachyallomyces (*Brachy–Allomyces*) (see ALLOMYCES)

Brachybasidiales An order of fungi of the subclass HOLOBASIDIOMYCETIDAE; constituent species are leaf endoparasites which do not form either a fruiting body or an hymenium on the surface of the parasitized plant (cf. EXOBASIDIALES).

bracket fungi (shelf fungi) Those fungi of the order Aphyllophorales whose fruiting bodies jut out like shelves or brackets from the surfaces of infected trees and rotting logs; one example is *Piptoporus betulinus*, a pathogen of the birch (*Betula*).

bradyzoites (of *Toxoplasma*) (*synonyms:* zoites; merozoites) Slowly-multiplying, uninucleate cells which occur within cysts—'tissue cysts'—in various regions of the host (brain, musculature etc.) in chronic or latent TOXOPLASMOSIS. The cyst wall is a definite structure (i.e. distinct from the plasmalemma of the host cell) which may persist for years; cysts may be 50–100 or more microns in diameter. (see also TACHYZOITES)

brand spores Smut spores—see SMUTS (1).

brandy (cognac) (see SPIRITS)

braxy An acute, frequently fatal disease of lambs. The causal agent, *Clostridium septicum*, apparently invades the tissues *via* the fourth stomach (abomasum) and gives rise to severe toxaemia, fever and collapse; death may occur within hours.

bread moulds The common (black) bread mould is *Rhizopus stolonifer* (*R. nigricans*). *Red* bread mould is *Neurospora sitophila*.

break-point chlorination (see WATER SUPPLIES)

breaking (*plant pathol.*) A form of *variegation* in the petals of flowers; breaking may result from viral infection. Flowers (particularly tulips) affected in this way are often prized for their attractive appearance. (see also MOSAIC)

Brewer's thioglycollate medium A liquid medium used for the culture of ANAEROBES. The medium (pH 7.0–7.2) contains (w/v): beef extract (0.1%), yeast extract (0.2%), peptone (0.5%), glucose (0.5%), sodium chloride (0.5%), sodium thioglycollate (0.11%), methylene blue (0.0002%), and agar (0.1–0.2%). Sodium thioglycollate acts as a *poising* agent, and methylene blue as an oxidation-reduction indicator—see ANAEROBE and REDOX POTENTIAL. The medium is particularly useful for testing the sterility of those products which contain mercurial preservatives—the antimicrobial activity of the latter being neutralized by the thioglycollate.

brewers' yeast (see YEAST)

brewing Beer and lager are prepared by the FERMENTATION of an aqueous extract of *malted* barley containing the essential oils and resins of the dried (female) flowers of the hop (*Humulus lupulus*). *The malting process.* Water-soaked barley is allowed to germinate; in this process, STARCH and certain of the grain proteins are attacked by endogenous AMYLASES and proteases. Germination is then halted by heating the grain to about 80 °C (i.e. *kilning*); this reduces, but does not eliminate, enzymic activity. The grain husks are then split (by *grinding*) to form *grist* (barley grains with exposed endosperms). *The brewing process* begins with *mashing*: the grist—in a mash tun (container)—is mixed with water and held at temperature(s) between 40°C and 65° C for a total period of 1–2 hours; starch is broken down to glucose, maltose and DEXTRINS, and proteins are degraded to amino acids and polypeptides. (Dextrins and polypeptides contribute to foam-retention—enabling the beer to form a good *head*.) The resulting liquor (*wort*) is separated from the spent grain and boiled together with hops. (Boiling helps to prevent microbial spoilage, and precipitates soluble proteins which would otherwise form cloudy suspensions, or precipitate, in the beer.) The wort is run into a fermentation vessel and seeded ('pitched') with a strain of the yeast *Saccharomyces*. In *top-fermentation* a top-fermenting* strain of *S. cerevisiae* ferments the wort at 15–20 °C; the yeast rises to the surface of the liquor—excess yeast being periodically removed. Fermentation is complete within a week. In *bottom-fermentation*, *S. carlsbergensis* or bottom-fermenting* strains of *S. cerevisiae* are used; fermentation is carried out at lower temperatures for longer periods of time—during which the yeast tends to sediment.

The fermentation process is appreciably exothermic and periodic cooling of the fermentation vessels is essential. The alcohol and carbon dioxide derive from the metabolism of glucose *via* the Embden–Meyerhof–Parnas pathway. Beers are filtered and pasteurized prior to canning or bottling.

Spoilage of beers. Off-flavours and/or ropiness may be caused by *Acetobacter* if the wort becomes sufficiently aerobic. *Sarcina sickness* (caused by *Pediococcus* sp) results in the formation of beer with the odour and flavour of diacetyl. Species of *Lactobacillus* and the yeasts *Candida* and *Pichia* also cause spoilage.

*In general, top yeasts hydrolyse RAFFINOSE incompletely (raffinose → melibiose + fructose) while bottom yeasts hydrolyse raffinose to galactose, glucose and fructose. (see also SPIRITS)

bright-field microscopy (see MICROSCOPY)

brights (*ciliate protozool.*) *B*-type KAPPA PARTICLES.

Brill–Zinsser disease (see TYPHUS FEVERS)

brilliant green (see TRIPHENYLMETHANE DYES)

brittleworts Certain species of the CHAROPHYTA.

broad spectrum antibiotics ANTIBIOTICS which are active against a wide range of bacteria—including both gram-positive and gram-negative species. The description 'broad spectrum' was originally applied to the TETRACYCLINES; other broad spectrum antibiotics include CHLORAMPHENICOL and the AMINOGLYCOSIDE ANTIBIOTICS.

bromatia (see FUNGUS GARDENS)

bromcresol green (bromocresol green) (PH INDICATOR) pH3.8 (yellow) to pH5.4 (green). pK_a 4.7.

bromcresol purple (bromocresol purple) (PH INDICATOR) pH5.2 (yellow) to pH6.8 (purple). pK_a 6.3.

bromine and compounds (as antimicrobial agents) Bromine is a strong oxidizing agent both in the vapour phase and in solution; its mode of antimicrobial activity appears to be similar to that of CHLORINE. Hypobromous acid and the hypobromites are strongly microbicidal. (see also IODINE and FLUORINE)

5-bromouracil (BU) (as a MUTAGEN) Bromouracil is an analogue of thymine (5-methyluracil). BU can exist in either of two tautomeric forms: the *keto* and *enol* forms—the former occurring with a much higher frequency than the latter. If present during DNA replication, keto-BU may be incorporated into the new DNA strand in place of thymine. While in the keto form, BU can base-pair with adenine, and can thus simulate thymine during subsequent rounds of replication; under such conditions, no mutation will occur. However, *enol*-BU pairs with guanine rather than with adenine; thus, if a BU residue in a template strand undergoes a keto→enol shift immediately prior to replication, the new DNA strand will contain a guanine residue opposite enol-BU. In the subsequent replication, guanine pairs (normally) with cytosine—hence an A-T to G-C transition will have occurred. Less frequently, the enol form of BU may be incorporated in the new DNA strand during replication by pairing with a template guanine residue. If this BU residue is in the keto form during the next round of replication, adenine will be incorporated into the new strand opposite the keto-BU—hence G-C to A-T transitions are also possible.

bromphenol blue (bromophenol blue) (PH INDICATOR) pH3.0 (yellow) to pH4.6 (blue). pK_a 4.0.

bromthymol blue (bromothymol blue) (PH INDICATOR) pH6.0 (yellow) to pH7.6 (blue). pK_a 7.0.

bronchitis Inflammation of the bronchi. Common causal agents include *Streptococcus pneumoniae*, *Haemophilus influenzae*, and certain viruses; in some cases tracheobronchitis may be due to infection by *Mycoplasma pneumoniae*.

broth A term used in bacteriology to refer to any of a variety of liquid media (see MEDIUM)—though commonly referring to NUTRIENT BROTH or any medium based on nutrient broth.

broth sugars ('sugars') Bacteriological media used to determine the range of carbohydrates metabolized by a given strain of bacterium. Except in that they are *broth*-based, broth sugars are similar to PEPTONE WATER SUGARS both in composition and mode of usage.

brown algae (see PHAEOPHYTA)

brown oak Timber from oak (*Quercus*) which has been infected by *Fistulina hepatica* (the *beefsteak fungus*). Such infection causes the heartwood of the tree to slowly become a rich brown colour. The colour appears to derive partly from pigment within the fungal hyphae and partly from extracellular fungal products. Brown oak is somewhat weaker than the normal wood but it is a valued medium for furniture making etc.

brown rot (of timber) (see TIMBER, DISEASES OF)

Browne's tubes Sealed glass tubes containing a fluid which changes from red to green when exposed to certain temperature/time combinations; they are used to indicate the nature of the conditions which have occurred, at various positions, within an operating AUTOCLAVE or hot-air oven. Browne's type I and type II tubes are used in autoclaves for indicating different temperature/time ranges, while the type III tube (160 °C/1 hour) is used in hot-air ovens. Each tube may be used only once.

Brownian movement *Random* movements which may be observed when small particles (of the order of 1 micron) are freely suspended in a fluid medium; the movements are due to bombardment of the particles by molecules of the medium. When examining an organism for MOTILITY (e.g. by the HANGING DROP METHOD) it is important to distinguish between true motility and Brownian movement; motile organisms may be seen to change their relative positions, while non-motile cells tend merely to oscillate about more or less fixed positions. (The hanging drop technique does not permit the development of micro-currents within the medium; such currents would tend to give a false impression of motility.)

Brown's tubes A set of sealed glass tubes containing aqueous suspensions of barium sulphate in increasing concentrations; the tubes thus exhibit a range of optical densities which vary from transparent, though translucent, to turbid. Brown's tubes are used for standardizing microbial suspensions; thus, e.g. the opacity of a given bacterial suspension may be matched (visually) with that of a particular tube, and the suspension described as a Brown's number 2 or a Brown's number 5 suspension etc. (Higher numbers correspond to greater opacities.) The unknown suspension should be examined in a tube of size and thickness equivalent to those containing the standard suspensions.

Brucella A genus of gram-negative, aerobic, chemoorganotrophic, catalase-positive bacteria of uncertain taxonomic affinity; *Brucella* spp are extracellular or intracellular parasites or pathogens in a range of mammalian hosts (see BRUCELLOSIS). The cells are non-motile bacilli or coccobacilli, about 0.5 × 0.6–1.5 microns; they stain with aniline dyes (see also KOSTER'S STAIN). Metabolism is oxidative (respiratory). Growth requirements generally include certain amino acids, biotin, nicotinic acid, and thiamine; 5–10% carbon dioxide may be essential—particularly in primary isolation. Primary culture may be carried out on serum-glucose agar, chocolate agar etc. Optimum growth temperature: 37 °C. The organisms are killed by PASTEURIZATION and are generally sensitive to streptomycin and tetracyclines. In aqueous acriflavine solution (1/1000 w/v) *smooth* strains are easily emulsifiable, *rough* strains agglutinate, and *mucoid* strains form stringy threads. Type species: *B. melitensis*; GC% range for the genus: 56–58.

B. abortus. Many strains require 5–10% CO_2. Colonies are small, round and translucent, blue-grey in obliquely-transmitted light. All strains are lysed by the *Tbilisi* phage at RTD—a characteristic which distinguishes *B. abortus* from all other *Brucella* spp.

B. melitensis, *B. suis*. CO_2 enrichment not required for growth.

B. abortus, *B. melitensis* and *B. suis* are each divided into a number of *biotypes* based e.g. on metabolic characteristics and sensitivity to the dyes thionin and basic fuchsin. Differentiation between biotypes is also achieved by the use of *monospecific* antisera prepared against each of the two main somatic antigens in the brucellae (the *A* and *M* antigens). Monospecific *A* antiserum can agglutinate certain biotypes of *B. abortus* and *B. melitensis*, and all biotypes of *B. suis*; monospecific *M* antiserum agglutinates certain biotypes of *B. abortus* and *B. melitensis*, and one biotype of *B. suis*. (Some biotypes in each species are agglutinated by the *A* or the *M* antiserum.) *B. abortus*, *B. melitensis* and *B. suis* produce characteristic and distinguishable colonies when grown on a medium containing diethylDITHIOCARBAMATE.

B. ovis, *B. canis*. Pathogens of the reproductive organs of sheep and dogs respectively; infections may lead to abortion. Only *rough* strains are known, and the organisms do not agglutinate with *A* or *M* antisera. *B. ovis* requires 5–10% CO_2 for growth.

brucellergen A nucleoprotein extract of the cells of *Brucella* spp used in a SKIN TEST for detecting DELAYED HYPERSENSITIVITY in respect of *Brucella* spp.

brucellin An extract of the cells of *Brucella* spp used either in a SKIN TEST (for detecting DELAYED HYPERSENSITIVITY in respect of *Brucella* spp) or as an agent for desensitizing a subject suffering from BRUCELLOSIS. (see also DESENSITIZATION)

brucellosis Any disease of man or animals in which the causal agent is a species of *Brucella*. In *man*, brucellosis (undulant fever, Malta fever) is an acute or chronic systemic disease caused by *B. melitensis*, *B. abortus* or *B. suis*. Infection occurs orally, *via* the conjunctivae, or *via* wounds; the incubation period may be days, weeks or months. Typically, brucellosis produces intermittent periods of fever. Lesions may be formed e.g. in the spleen, liver, bone marrow; endotoxins and/or hypersensitivity reactions may account for the tissue damage. *Laboratory diagnosis*: (a) Blood cultures are frequently positive during the first few weeks of the (acute) disease. (b) The culture of bone marrow or lymph node aspirates may be useful in chronic cases. (c) Serological assay for specific serum agglutinins. (NB. Agglutinins formed against cholera vibrios cross-react with those formed against *Brucella* spp.) Tetracyclines in conjunction with streptomycin have been used for treatment. *Among animals*, disease in cattle is commonly due to *B. abortus*, in pigs *B. suis*, and in goats *B. melitensis*—but host specificity is by no means absolute. *Cattle*. Contagious abortion of cattle (Bang's disease) is economically important in the U.S.A. and other countries. Infection may occur *via* the mouth or vagina or by wound contamination. In cows, the pathogen has a predilection for the mammary glands and for the foetus and placenta; the latter affinity is believed to be due to the presence, in the foetal and placental tissues, of *erythritol*—a substance which encourages development of the pathogen. Abortion and sterility are frequent sequelae of brucellosis. *Swine*. Lesions occur in the reproductive organs, lymph nodes and joints. The abortion rate is lower than that found among cattle. *Goats*. Regions affected include the reproductive organs of both sexes and the mammary glands. Abortion may occur. (see also MILK RING TEST)

Bryopsis A genus of green algae (CHLOROPHYTA).

BSA Bovine serum albumin.

BSS Balanced salt solution (see HANKS' BSS).

BU 5-BROMOURACIL.

bubo An inflamed and enlarged lymph node which may be suppurative. *Buboes* are formed in a number of diseases, including PLAGUE.

bubonic plague (see PLAGUE)

buccal cavity (*ciliate protozool.*) In some ciliates, a cavity or chamber which is open to the environment—either directly or *via* a VESTIBULUM—and at the base of which is situated the cell mouth, or CYTOSTOME. The walls of the buccal cavity are lined with specialized oral ciliature—which is often organized into complex membranelles (see AZM and UNDULATING MEMBRANE). Certain holotrichs lack a true buccal cavity—although a vestibulum is found in some of these species.

buccal overture (*ciliate protozool.*) The entrance to the BUCCAL CAVITY.

bud-fission A form of asexual reproduction carried out e.g. by *Saccharomycodes ludwigii* (see SACCHAROMYCODES).

bud scar (see SCAR)

budding (1) (*mycol.*) A form of vegetative (asexual) reproduction in which a daughter cell (or a spore) develops from a small outgrowth or protrusion of the parent cell; budding requires the synthesis of additional cell wall material and appears to involve the coordinated activities of autolytic and synthetic enzymes. During the budding process mitotic division occurs in the parent cell, and one of the post-division nuclei passes into the daughter cell. Budding occurs in a number of fungi—particularly in certain yeasts. In *multipolar* budding yeasts (e.g. *Saccharomyces*) a daughter cell may arise at any of a number of sites on the parent cell (see also SCAR). In *bipolar* budding yeasts (e.g. *Kloeckera*, *Saccharomycodes*) budding occurs only at the poles of the cell. In *monopolar* budding yeasts (e.g. *Pityrosporum*) repeated budding occurs from one region of the cell. If, following budding, daughter and parent cells fail to separate, a branched or unbranched chain of cells (a *sprout mycelium*) is eventually formed. (see also BINARY FISSION) (2) (*bacteriol.*) A type of asexual reproduction analogous to that described in (1) occurs in certain bacteria—e.g. *Nitrobacter winogradskyi*. (see also BINARY FISSION). (3) (*virol.*) See *assembly* under VIRUSES.

BUdR 5-Bromodeoxyuridine, the deoxyribonucleotide corresponding to 5-BROMOURACIL.

Buellia A genus of crustose or squamulose LICHENS in which the phycobiont is *Trebouxia*. Apothecia are lecideine with black discs. Species include e.g.:

B. canescens. The thallus is placodioid, white to grey, usually darker and areolate in the centre; white soralia are generally scattered over the surface. Apothecia are frequently absent. *B. canescens* occurs on rocks, walls and trees.

B. punctata. The thallus is crustose, grey, smooth to warty and areolate. Apothecia are generally numerous; they are initially plane and subsequently convex. *B. punctata* occurs on trees and occasionally on rocks.

buffy coat The thin layer of white cells which forms at the surface of the packed red cells during the centrifugation of unclotted blood.

bullate With bubble- or blister-like swellings.

Buller phenomenon (*mycol.*) A form of DIKARYOTIZATION in which a nucleus is donated by a dikaryotic cell or mycelium to a monokaryotic (haploid) cell or mycelium.

Bullera A genus of non-fermentative, unicellular fungi within the family SPOROBOLOMYCETACEAE; asexual reproduction occurs by budding, and sexual reproduction involves the formation of ballistospores—each of which is orientated *symmetrically* on its sterigma (cf. SPOROBOLOMYCES). *Bullera* spp do not form pseudomycelium or true mycelium. In culture the cells appear cream-coloured or yellow.

bunt (of wheat) (see CEREALS, DISEASES OF)

Burgundy mixture A mixture of copper sulphate and sodium carbonate; it is sometimes used (in aqueous solution) as an anti-fungal plant spray. (see also BORDEAUX MIXTURE and CHESHUNT COMPOUND)

Burkitt's lymphoma In man, a type of LYMPHOMA with which the Epstein–Barr virus has been associated as a probable causal agent; freshly-obtained specimens of tumour material appear not to contain the *virions* of EBV—although the (common) intracellular presence of virus-specific surface antigens in such material indicates the presence of the viral genome.

Burnet's clonal selection theory (see CLONAL SELECTION THEORY)

bursa of Fabricius In the chicken, a lymphoid organ in which B LYMPHOCYTE precursors undergo *lymphopoiesis* i.e. the (maturation) process following which they are able to produce immunoglobulins. The *bursa-equivalent* (if any) in mammals has not yet been identified.

burst size (in bacteriophage infection) The average number of phages released, per infected cell, following the lytic infection of a sensitive bacterial population. (see also ONE-STEP GROWTH EXPERIMENT)

butanediol fermentation (butylene glycol fermentation) A type of fermentation which is carried out by certain members of the Enterobacteriaceae (e.g. species of *Enterobacter*, *Erwinia*, *Klebsiella*, *Serratia*) and of the Vibrionaceae (e.g. some species of *Aeromonas*) (see Appendix III(c)). Glucose is fermented with the formation of 2,3-butanediol in addition to smaller amounts of those products formed in the MIXED ACID FERMENTATION (see Appendix III(b)); since only one molecule of $NADH_2$ is reoxidized for every two molecules of pyruvate converted to butanediol, the surplus reducing power (generated during the formation of pyruvate from glucose) is used to form ethanol from acetyl-CoA. Those (aerogenic) organisms which possess a FORMATE HYDROGEN LYASE system can decompose formate to CO_2 and H_2; CO_2 is also formed during the synthesis of butanediol.

The amount of acid produced during butanediol fermentation is generally insufficient to give a positive METHYL RED TEST. Butanediol producers give a positive VOGES-PROSKAUER TEST.

***n*-butanol fermentation** (see ACETONE-BUTANOL FERMENTATION)

butt The thickest (widest) part of an agar or gelatin SLOPE. When the butt of a freshly-prepared slope is stab-inoculated (by means of a STRAIGHT WIRE) the inoculum is introduced to a microaerophilic or anaerobic region of the medium.

butt rot A rot affecting the *base* of a tree trunk. (see also TIMBER, DISEASES OF)

butter Butter is manufactured from soured cream, i.e. cream which has been allowed to become acidic through the metabolic activities of its natural flora, commonly strains of *Streptococcus lactis* and/or *Strep. cremoris*; these organisms form lactic acid from the milk sugar, lactose. Following PASTEURIZATION, the soured cream is seeded with cultures of *Leuconostoc citrovorum*. These bacteria attack citric acid (present in milk in low concentrations) and produce acetylmethylcarbinol (*acetoin*) and dimethylglyoxal (*diacetyl*)—the latter being the main flavour component of butter. The soured cream is then *churned* until the fat globules coalesce, thus producing the consistency typical of butter. Salt—up to about 2%—may be added and worked into the butter; this reduces the possibility of spoilage since, although overall concentration is low, the salt dissolves only the aqueous phase (water droplets) in which it forms solutions of high concentration which inhibit certain of the lipolytic (spoilage) bacteria. As an alternative to the above mode of manufacture, the cream may be *initially* pasteurized and the processes of souring and flavour-production controlled by the addition of specific bacterial cultures.

The liquid which remains at the end of the churning process is called buttermilk.

buttermilk (see BUTTER)

button stage (of the fruiting bodies of agarics) That stage of growth which occurs before the gills (lamellae) are exposed to the environment and during which the immature pileus is more or less dome-shaped.

butyl alcohol fermentation (see ACETONE-BUTANOL FERMENTATION)

butylene glycol fermentation (see BUTANEDIOL FERMENTATION)

butyric acid fermentation A type of fermentation, carried out by certain saccharolytic CLOSTRIDIUM species (e.g. *C. butyricum*), in which glucose is fermented with the formation of acetic acid, butyric acid, CO_2 and H_2 (see Appendix III(d)). Pyruvate (formed from glucose *via* the EMBDEN–MEYERHOF–PARNAS PATHWAY) undergoes oxidative decarboxylation in a reaction involving FERREDOXIN and coenzyme A—acetyl-CoA being formed; electrons transferred to ferredoxin during the reaction may be transferred to hydrogen ions—in the presence of hydrogenase—to form hydrogen gas. Butyric acid formation utilizes $NADH_2$ formed during the EMP pathway; the formation of acetic acid generates ATP but does not require reducing power. The formation of a high proportion of acetate thus results in a surplus of reducing power which may be eliminated as hydrogen *via* ferredoxin and hydrogenase, i.e. $NADH_2$ may donate electrons to ferredoxin. Certain clostridia, e.g. *C. perfringens*, form lactate and ethanol in addition to the above products; lactate becomes a major product when these organisms are subjected to iron deficiency. In certain other species of *Clostridium*—e.g. *C. acetobutylicum*—accumulation of the acidic products causes a switch from the butyric acid fermentation to the ACETONE-BUTANOL FERMENTATION.

Butyrivibrio A genus of gram-negative, obligately anaerobic, asporogenous, motile, curved rod-shaped bacteria of uncertain taxonomic affinity; strains of *Butyrivibrio* are common in the RUMEN. The organisms are chemoorganotrophs; metabolism is fermentative—butyric acid usually being a major product of carbohydrate fermentation. Some strains are cellulolytic. Individual cells carry a single polar or sub-polar flagellum; cells may also occur in chains. All strains are temporarily included within the type species, *B. fibrisolvens*.

butyrous (*adjective*) Having the consistency of butter.

Byssochlamys A genus of fungi of the order EUROTIALES; species occur e.g. in the soil. Asexual reproduction involves the formation of chains of conidia—each chain arising from a phialid which develops on the mycelium. During the development of the ascocarp the ascogonium becomes coiled around the antheridium and subsequently gives rise to a number of ascogenous hyphae; the mature ascocarp consists of a group of asci surrounded by a few wisps of hyphae. *B. fulva* has been implicated in some instances of spoilage of canned and bottled foods; the organism produces extracellular pectinases, and its ascospores are reported to withstand temperatures of approximately 85 °C for 30 minutes—and higher temperatures for shorter periods of time.

byssoid Composed of slender fibrils.

C

C (also C′) (*immunol.*) The symbol for COMPLEMENT.

C3 convertase (see COMPLEMENT-FIXATION)

C_{55} lipid carrier A term which refers to BACTOPRENOL.

C_{27} organisms Bacteria of the genus PLESIOMONAS.

C-reactive protein A protein produced in the body during the course of certain inflammatory conditions, e.g. rheumatic fever. C-reactive protein reacts with a variety of agents—e.g. it is precipitated by cellular extracts of *Streptococcus pneumoniae* (the basis for the C-reactive protein diagnostic test) and by antigenic extracts of *Aspergillus* spp.

C-region (of an ANTIBODY) (see V-REGION)

C_H regions (*immunol.*) (see V-REGION)

C substances (of streptococci) A range of serologically distinct carbohydrates—only one of which may occur in a given strain of *Streptococcus*. C substances are used as a basis for the identification and classification of streptococci: see LANCEFIELD'S STREPTOCOCCAL GROUPING TEST. In the majority of *Lancefield groups* the C substance is a cell wall constituent. In Lancefield group *A* streptococci the C substance is a rhamnose polymer with a terminal residue of *N*-acetylglucosamine—the latter being immunodominant; in strains of the *A-variant* and *B* groups the main determinant appears to be rhamnose while in group *C* it is a terminal residue of *N*-acetylgalactosamine. The C substance of group *D* occurs on the cell membrane or (perhaps) in the periplasmic space; it has been identified as a glycerol TEICHOIC ACID. (see also M PROTEIN (streptococcal))

C-type particles (C-type viruses) Particles first observed in neoplastic mouse cells; C-type particles are now known to be oncogenic RNA viruses (subgenus A of LEUKOVIRUS) and to include the causal agents of a range of avian and mammalian leukaemias and sarcomas. The complex, spheroidal, enveloped C-type virion, approximately 80–130nm in diameter, contains a fragmented single-stranded RNA genome and an RNA-dependent DNA polymerase; unlike the B-TYPE PARTICLE, the C-type particle has a *centrally* situated genome and has no obvious intracellular precursor form—appearing to develop *during* the process of budding at the cytoplasmic membrane or intracytoplasmic membranes of the host cell.

cabbage, diseases of (see CRUCIFERS, DISEASES OF)

cabbage B virus (see CAULIFLOWER MOSAIC VIRUS)

caeomatoid (caeomoid) Refers to an aecium which lacks a peridium.

caffeic acid 3-(3′, 4′-dihydroxyphenyl)-propenoic acid; a substance produced by certain higher plants—sometimes as a precursor of LIGNIN. Caffeic acid appears to be implicated in the resistance exhibited by certain plants to certain diseases; thus, e.g. potatoes which are resistant to *late blight* have been found to accumulate caffeic acid on infection with *Phytophthora infestans*. (Caffeic acid has been shown to inhibit *P. infestans in vitro*.) (see also SPORES (FUNGAL), DORMANCY IN)

Cairns model (of DNA replication) A model that accounts for certain instances of replication of circular, double-stranded DNA—e.g replication of the bacterial chromosome and certain instances of PLASMID and BACTERIOPHAGE replication. Both strands are replicated—replication proceeding (generally in *both* directions) from a specific starting point; two *replication forks* (see entry DNA) thus move around the circle with the eventual formation of two new double-stranded circles—each consisting of one new and one parental strand. (cf. ROLLING CIRCLE MODEL)

calcium nutrient agar (CNA) NUTRIENT AGAR containing a calculated amount of calcium chloride. A lawn plate prepared on CNA is often used for the cultivation of bacteriophage.

calf pneumonia A non-specific disease, the causal agent being any of a range of viruses (e.g. adenoviruses) or species of *Chlamydia*. Poor hygiene and housing conditions are believed to predispose towards the disease.

Calicium A genus of crustose LICHENS in which the phycobiont is a green alga; species generally occur on the bark of trees. The thallus may be green, grey or white—according to species; the ascocarps are stalked *mazaedia* (see MAZAEDIUM).

Calicivirus A genus of VIRUSES of the family PICORNAVIRIDAE. Caliciviruses differ morphologically from the other picornaviruses; the *Calicivirus* virion (35–40nm in diameter) is inactivated at pH3. The genus includes the causal agent of vesicular exanthema of swine.

Calocera A genus of fungi of the order DACRYMYCETALES; species occur as saprophytes on wood. *C. cornea* forms an erect, unbranched, columnar, gelatinous fruiting body about one centimetre in height; the fructification is pale yellow and is pointed at the apex. The organism occurs on a variety of types of wood. *C. viscosa* forms erect, antler-like fruiting bodies on coniferous wood; the fructifications are commonly three to five centimetres high and are orange-coloured and gelatinous.

calomel (see *mercury* under HEAVY METALS)

Caloplaca A genus of crustose or squamulose LICHENS in which the phycobiont is *Trebouxia*. The thallus frequently contains the orange anthraquinone pigment *parietin*. Apothecia are formed; the ascospores are polarilocular or, very rarely, simple. Species include e.g.:
C. citrina. The thallus is greenish-yellow to orange-yellow and consists of granules which may be scattered or may form an areolate crust; the granules are sorediate or isidiate. The apothecia are 0.3–1.0 millimetres in diameter; each has a yellow-orange disc and, when young, a thalline margin. *C. citrina* occurs on a variety of substrata—e.g. trees, concrete, soil.
C. marina. The thallus is orange-yellow to orange and usually consists of scattered granules which do not bear isidia or soredia. The apothecia are 0.5–1.0 millimetres in diameter and have orange discs. *C. marina* occurs on rocks on the seashore at, or just above, high tide mark.

Calvatia A genus of fungi of the LYCOPERDALES.

Calvin cycle (reductive pentose cycle) The chief pathway for carbon dioxide fixation in photolithotrophs and chemolithotrophs—see PHOTOSYNTHESIS for details.

calyciform (*adjective*) Cup-shaped.

calymma (see RADIOLARIANS)

Calymmatobacterium A genus of gram-negative, facultatively anaerobic, pleomorphic bacilli of uncertain taxonomic affiliation; *C. granulomatis* is pathogenic for man—causing various granulomatous lesions.

CAM (*virol.*) The chorioallantoic membrane of an embryonated hen's egg.

cAMP 3′,5′-Cyclic adenosine monophosphate (see Appendix V(c) for structure).

CAMP test A procedure for detecting *Streptococcus agalactiae*—the causal agent of bovine mastitis. A suspected strain (isolated e.g. from milk) is inoculated in a fine streak onto blood agar (prepared from ox or sheep blood); another streak, at right angles to the first—and separated from it by a few millimetres, is made from a culture of a β-haemolytic strain of *Staphylococcus*. Following incubation (12 hours or more at 37 °C) a positive test is indicated by the appearance of *clear* haemolysis in the area which separates the two lines of bacterial growth.

If milk is moderately or heavily contaminated with *S. agalactiae* the milk itself may give a positive result if used to streak the test plate (in place of the suspected isolate).

Campanella A genus of ciliate protozoans of the sub-order Sessilina, subclass PERITRICHIA. *Campanella* is found in a variety of fresh and brackish waters. In general appearance the organism is not unlike *Vorticella* (*q.v.*); however, *Campanella* is larger and is a *colonial* organism—one zooid being carried on each branch of the *non*-contractile stalk.

Campylobacter A genus of gram-negative, microaerophilic or anaerobic, chemoorganotrophic bacteria of the SPIRILLACEAE; species occur in the mouth, alimentary tract and reproductive organs in man and other animals. The cells are non-pigmented, less than 1 micron in diameter and about 1–5 microns in length; a single flagellum is carried at one or at each pole of the cell—the organism moving with a corkscrew-like motion. *C. fetus* (formerly *Vibrio fetus*) may cause abortion in sheep and cattle.

Canada balsam A clear resin obtained from fir trees (*Abies* sp); the high boiling-point fraction of the resin, dissolved in xylene or benzene, is used as a MOUNTANT. (Canada balsam is not soluble in ethanol.) Permanent preparations are made by using the solution to cement a cover slip to that part of a slide bearing the (dehydrated and cleared) specimen or smear. Canada balsam may take a month or more to dry (harden) completely. The refractive index of the dried balsam is approximately 1.53. (see also DPX)

cancer The common term for any of a variety of types of malignant (i.e. invasive, progressive) NEOPLASM.

Candelariella A genus of crustose LICHENS in which the phycobiont is *Trebouxia*. The apothecia are lecanorine. Species include e.g.:
C. vitellina. The yellow thallus is granular and may be contiguous or scattered. The apothecia have dull-yellow discs which darken with age. *C. vitellina* occurs on rocks, walls, trees etc.—particularly where these are enriched with nitrogen (e.g. by bird droppings).

candicidin B (see POLYENE ANTIBIOTICS)

Candida (*syn*: *Monilia*) A genus of imperfect yeasts within the family CRYPTOCOCCACEAE; species have been isolated e.g. from the tissues and/or secretions of warm-blooded animals, including man. (see also CANDIDIASIS) The organisms occur typically in the form of spheroid, ovoid or elongated cells which, according to species and/or conditions, may form (or may be accompanied by) pseudomycelium and/or true mycelium; reproduction occurs by multipolar budding and by the formation of blastospores. Certain species form chlamydospores. Arthrospores are not formed. Orange or red pigments are not produced. Some strains are non-fermentative, but many ferment glucose and at least one other sugar. Species include:
C. albicans. Individual cells are usually 3–6 × 5–10 microns; pseudomycelium is com-

mon and true mycelium may be formed. Blastospores are formed, and thick-walled chlamydospores develop when the organism is cultured on CORNMEAL AGAR at 20 °C. *C. albicans* ferments glucose and maltose (forming acid and gas) and does not ferment lactose.

C. stellatoidea. A species which is similar to *C. albicans* both serologically and in fermentative activity; it differs from *C. albicans* e.g. in that it fails to assimilate sucrose and in that it rarely forms chlamydospores.

C. tropicalis. A species which differs from *C. albicans* and *C. stellatoidea* in that e.g. it regularly ferments sucrose with the formation of acid and gas. Chlamydospores are formed under certain conditions of culture.

candidiasis Any disease in which the causal agent is a species of the imperfect yeast *Candida*—usually *C. albicans*. (Other pathogenic species include *C. parapsilosis* and *C. tropicalis*.) In man the common candidiases include THRUSH, vaginitis, and bronchocandidiasis; systemic infections (including septicaemia) also occur. Infection is probably endogenous in most cases; pre-disposing factors may include the suppression of the normal bacterial flora—e.g. by antibiotic therapy. *Laboratory diagnosis* may include both direct microscopical examination and cultural examination of specimens. It is important to determine whether or not a given specimen contains *significantly* greater numbers of *Candida* cells than might be expected from corresponding specimens taken from healthy individuals; prompt examination is essential since delay may permit a (misleading) increase in cell numbers. *Chemotherapy* generally involves the topical or systemic administration of a POLYENE ANTIBIOTIC (e.g. amphotericin B, Nystatin).

candle The name applied to several types of microbiological filter (see FILTRATION).

candle snuff fungus A common name for the ascomycete *Xylaria hypoxylon*.

candling (*virol.*) Examination of the interior of an intact egg by transmitted light. The egg is placed over an aperture in a box which contains an electric lamp; when switched on, the lamp reveals certain internal details of the egg—e.g. the position of the embryo and blood vessels. Thus, e.g. candling facilitates the inoculation (by syringe) of a particular membrane or cavity within the egg—see VIRUSES, CULTURE OF. (Candling should be carried out rapidly in order to avoid overheating the egg; it is best performed in a darkened room.)

canning (appertization) A method used for the preservation of foodstuffs. Suitably-prepared foods are placed in metal containers which are then heated, exhausted, and hermetically sealed. Foods of neutral or near-neutral pH are then heated to sterilizing temperatures (e.g. 115 °C) for periods of time which depend on container size; acidic foods (e.g. fruits) are commonly processed at lower temperatures. The object of the canning process is to destroy vegetative microorganisms and microbial spores—particularly the heat-resistant bacterial spores—thus preventing microbial spoilage of the food; an essential requirement of the process is the destruction of spores of *Clostridium botulinum*—the causal agent of botulism. Spores of other bacteria (e.g. *Bacillus stearothermophilus*) which may survive the canning process may give rise e.g. to *flat sours*—see FOOD, MICROBIAL SPOILAGE OF.

CAP (see CATABOLITE REPRESSION)

capillitium (*mycol.*) A system of threads or filaments which ramify among the spores in the fruiting bodies of certain species of the acellular slime moulds and of members of the class Gasteromycetes.

caprinized vaccine A vaccine containing live organisms whose virulence (for a given host) has been suitably attenuated by serial passage through goats.

capsid (*virol.*) Of a virion: a protein coat or shell which may enclose a CORE or naked nucleic acid, or which may be complexed with the nucleic acid of the virion to form a NUCLEOCAPSID. The capsid may form the outermost layer of a virion—as e.g. in reoviruses and the tobacco mosaic virus—or it may be enclosed within a lipoprotein ENVELOPE—as e.g. in herpesviruses and rhabdoviruses. (see also CAPSOMERE)

capsomere (capsomer; morphological unit) (*virol.*) Any one of the identical (or similar) protein particles—distinguishable by electron microscopy—which form the CAPSID or the protein part of a NUCLEOCAPSID of a virus.

capsule (bacterial) A layer of material which lies external to the bacterial CELL WALL. A capsule may have either a clearly-defined outer limit or no definite external border; limiting membranes have not been detected. Capsules may be demonstrated by negative STAINING (e.g. with India ink or NIGROSIN)—when the capsule is revealed as a clear zone between the opaque medium and the cell wall; they may also be demonstrated by the use of antibodies homologous to the capsular antigens—see e.g. QUELLUNG PHENOMENON.

Capsules may be divided into three categories: (a) *Macrocapsules* or 'true' capsules. These are sufficiently thick to be easily visible (with negative staining) under the light microscope. (b) *Microcapsules*. These cannot be detected by light microscopy but their presence may be revealed by serological techniques; some bacteria which have lost the ability to form a macrocapsule may form a serologically-identical microcapsule. (c) *Slime layers*. These are diffuse secretions which may adhere loosely to the cell surface; such materials commonly diffuse into the medium when the organism is grown in liquid culture. The composition of capsules and slime layers varies with species; antigenically distinguishable capsules may be formed by different members of the same species, and serotypes of e.g. *Streptococcus pneumoniae* have been distinguished on this basis. Water is commonly the principal constituent of a capsule. The commonest organic constituents are polysaccharides (e.g. DEXTRANS) but capsules composed of polypeptides (or a mixture of polysaccharide and polypeptide) may occur—e.g. in species of *Bacillus*. Certain species of *Acetobacter* produce extracellular CELLULOSE. The nature of the binding between the capsule and the cell wall is largely unknown; ionic bonding may be involved, and there is some evidence for covalent linkage in certain cases—e.g. capsular material covalently bound to PEPTIDOGLYCAN has been isolated from *Bacillus anthracis*. Cells may be decapsulated by treatment with dilute acids or alkalis, with specific enzymes, or sometimes merely by shaking the (liquid) culture. Usually, capsules can be removed without loss of the cell's viability.

Among capsule-forming pathogens a high degree of correlation has been observed between capsulation and pathogenicity. Thus, enzymic removal of capsules from virulent type 3 pneumococci renders these organisms non-virulent for a number of host species. Organisms which have lost their capsules commonly appear to be more susceptible to PHAGOCYTOSIS. Virulence and capsulation also appear to be associated in *Bacillus anthracis*. (see also VI ANTIGEN and SMOOTH-ROUGH VARIATION)

Biosynthesis of capsules. Polysaccharide capsules are formed by transglycolation from sugar-nucleotide precursors. In at least some cases the subunits are transferred to the exterior of the cell *via* a BACTOPRENOL-type lipid in the cell membrane.

Capsule-forming bacteria include e.g.: *Acetobacter xylinum*, *Klebsiella pneumoniae*, *Leuconostoc mesenteroides*, *Streptococcus pneumoniae*, *Streptococcus salivarious*, *Yersinia pestis*.

capsule (*entomopathol.*) (see GRANULOSES)

capsule swelling reaction (see QUELLUNG PHENOMENON)

captafol (difolatan) *N*-(1,1,2,2-tetrachloroethylthio)-cyclohex-4-ene-1,2-dicarboximide. A relatively new agricultural antifungal agent which is closely related to CAPTAN. It is used as a protectant against e.g. *late blight* of POTATO and several other diseases.

captan *N*-(trichloromethylthio)-cyclohex-4-ene-1,2-dicarboximide. An important agricultural antifungal *protectant*; it is effective against a number of fungal plant pathogens and is widely used e.g. for the prevention of apple scab and for the protection of foliage from leaf-spot diseases, e.g. black spot of roses. It is also used as a component of seed dressings etc. The activity of captan has been attributed to the $-SCCl_3$ group which may interact with essential $-SH$ groups within the fungal cell. Captan is almost insoluble in water and is generally formulated as a *wettable powder*. (see also CAPTAFOL and FUNGICIDES)

carbamic acid derivatives (as antimicrobial agents) Derivatives of carbamic acid ($NH_2.CO.OH$) include UREA ($NH_2.CO.NH_2$), derivatives of thiocarbamic acid ($R_2N.CS.OH$) (see e.g. TOLNAFTATE), and an important group of antifungal agents, the DITHIOCARBAMATE DERIVATIVES ($R_2N.CS.SH$). (see also BENOMYL).

carbamide A synonym of UREA.

carbanilides (as antimicrobial agents) (see UREA)

carbenicillin (see PENICILLINS)

carbolfuchsin (Ziehl's carbolfuchsin) A red dye which is used e.g. in the GRAM STAIN (5% aqueous, as a COUNTERSTAIN) and in ZIEHL-NEELSEN'S STAIN (hot, concentrated). Concentrated carbolfuchsin is prepared by the addition of 10% *basic* FUCHSIN in absolute ethanol (10 ml) to 5% aqueous phenol (100 ml); after standing overnight the solution is filtered.

carbolic acid Phenol. (see also PHENOLS)

carbon disulphide (CS_2) (*plant pathol.*) Carbon disulphide (boiling point 46 °C) is occasionally used as a soil fumigant for the control of *Armillaria mellea* root infections.

carbon monoxide (as a RESPIRATORY INHIBITOR) Carbon monoxide combines with, and inhibits the oxidation of, the reduced a_3 component of eucaryotic cytochrome oxidase (see CYTOCHROMES); many bacterial terminal oxidases may be inhibited to a greater or lesser degree. Carbon monoxide-cytochrome complexes can be dissociated by strong illumination with light of wavelengths absorbed by the complex—with concomitant relief of inhibition.

Carbowax (*proprietary name*) Polyethylene glycol; a substance used e.g. for concentrating, by osmosis, serum and other aqueous solutions or suspensions. Serum etc. may be concentrated by first placing it in a U-shaped length of *Visking* tubing (made from transparent, semi-permeable material); the section of tubing which contains the serum is then immersed in a thick, syrupy aqueous solution of Carbowax.

Carboxide (*trade name*) (see ETHYLENE OXIDE)

carboxydismutase (see *Calvin cycle* under PHOTOSYNTHESIS)

carboxysomes (polyhedral bodies) Polyhedral, intracellular bodies (of unknown function) observed in species of THIOBACILLUS (e.g. *T. thioparus*, *T. neapolitanus*), in many members of the family NITROBACTERACEAE, in *Beggiatoa*, and in certain blue-green algae; it is unknown whether or not the structures observed in all these organisms are equivalent. When present, more than one carboxysome is usually observed in each cell. Carboxysomes isolated from *T. neapolitanus* each consist of a number of 10 nm particles enclosed within a nonunit-type membrane; these particles contain *ribulose-1,5-diphosphate carboxylase*—an enzyme involved in the autotrophic fixation of carbon dioxide. Whether or not carboxysomes contain other components is unknown.

Carchesium A genus of ciliate protozoans of the sub-order Sessilina, subclass PERITRICHIA. *Carchesium* is found in a variety of freshwater habitats (e.g. ponds) and some species (e.g. *C. polypinum*) are common in British SEWAGE TREATMENT plants (both activated-sludge and biological filter types). In general appearance, size, and other features, *Carchesium* is not unlike *Vorticella* (*q.v.*); however, *Carchesium* is a colonial peritrich, i.e. in the mature organism a number of zooids are carried on branched stalks. Each zooid, and its branch of the stalk, is capable of independent contraction.

carcinogen Any agent which promotes the formation of any type of NEOPLASM; carcinogens include certain types of virus and radiation and certain chemicals.

carcinoma A malignant tumour of epithelial cells.

cardiolipin Diphosphatidyl glycerol. Cardiolipin may be prepared by alcoholic extraction from beef heart; it is used to prepare an artificial antigen for use in certain serological tests for syphilis. (see also VDRL TEST and WASSERMANN REACTION)

caries (see DENTAL CARIES)

cariogenic (*adjective*) Refers to any factor, substance or organism which promotes DENTAL CARIES.

carnivorous (of protozoans) Refers to protozoans which feed on other protozoans.

Carnoy's fluid A fixative used in protozoological work. Different formulations are used: (a) Ethanol (absolute or 95%) and glacial acetic acid in the ratio 3:1. (b) Ethanol (absolute or 95%), chloroform, and glacial acetic acid in the ratio 6:3:1.

carotenes (see CAROTENOIDS)

carotenoids A class of pigments (commonly yellow or orange) widely distributed among microorganisms. They include: (a) *Carotenes*—hydrocarbons which commonly have the formula:

$$R'{-}(CH{=}CH{-}C(CH_3){=}CH)_2{-}CH{=}CH{-}(CH{=}C(CH_3){-}CH{=}CH)_2{-}R''$$

where R′ and R″ may be linear or cyclic; e.g. in *β*-carotene: R′ = R″ = 2,6,6-trimethyl-1-cyclohexenyl, and in *lycopene*: R′ = R″ = 2,6-dimethyl-1,5-heptadienyl. (b) *Xanthophylls*—oxygen-containing derivatives of carotenes, e.g. *lutein* (= 3,3′-dihydroxy-*α*-carotene).

A variety of carotenoids is found among photosynthetic microorganisms. *β*-Carotene appears to occur in most algal groups. Members of the Chlorophyta contain e.g. *lutein* and *neoxanthin*—carotenoids which also occur in higher plants. *Fucoxanthin* is characteristic of the Phaeophyta and Bacillariophyta, and *peridinin* of the dinoflagellates. The carotenoids of blue-green algae include glycoside derivatives (e.g. the rhamnose-containing *myxoxanthophyll*). Characteristic carotenoids occur in photosynthetic bacteria—e.g. *lycopene*, *γ-carotene*, *spirilloxanthins*. In photosynthetic organisms, carotenoids are believed to contribute to 'light-harvesting' by transferring light energy to CHLOROPHYLL (see PHOTOSYNTHESIS); they may also prevent photodegradation of the chlorophyll. (see also SINGLET OXYGEN)

Certain non-photosynthetic bacteria (e.g. *Xanthomonas* spp) and fungi (e.g. *Rhodotorula*, *Trichophyton violaceum*) contain carotenoid pigments. (see also SPOROPOLLENIN)

carpomycetes A term sometimes used to refer, collectively, to the ascomycetes and basidiomycetes.

carrageenan (carrageenin; carragheen) A group of sulphated galactans present in many members of the Rhodophyta, e.g. *Chondrus crispus* ('Irish moss') and *Gigartina* spp. *κ*-Carrageenan consists principally of a linear sequence of alternating residues: D-galactose-4-sulphate-*β*-(1 → 4)-3,6-anhydro-D-galactose-*α*-(1 → 3)-D-galactose-4-sulphate.... etc. Similarly, *λ*-carrageenan consists principally of D-galactose-2-sulphate-*β*-(1 → 4)-D-galactose-2,6-disulphate-*α*-(1 → 3)-D-galactose-2-sulphate...etc. Other types of carrageenan are known. Carrageenans are associated with the algal cell wall; they appear to prevent desiccation of the alga during periods of exposure to the atmosphere. Carrageenans can form gels and have many uses e.g. in the pharmaceutical, textile and food industries.

Carrel flask A squat cylindrical glass vessel which opens to the exterior, at the side, *via* a straight tubular neck inclined at an angle to the horizontal.

carrier An individual who harbours a particular pathogenic microorganism but who shows no clinical signs of the disease, and who is potentially able to transmit the pathogen to others. The carrier state is well established for certain diseases, e.g. diphtheria and salmonellosis, but apparently never occurs in others. (see also TYPHOID MARY)

carrier state (in pseudo-lysogeny) (see PSEUDOLYSOGENY)

carrots, diseases of These include: (a) *Bacterial soft rot.* The causal agent, *Erwinia carotovora*, survives in soil and in decaying vegetation; certain insects are believed to aid the spread of the disease. Infection, which is favoured by humid conditions, may occur *via* wounds; lesions consist of watery masses of disintegrated host tissue. Carrots may be infected in the ground or after harvesting. (b) *Black rot* caused by *Alternaria* sp. (c) *Rhizoctonia* root rot. (see also PLANT PATHOGENS AND DISEASES)

Carr's medium A medium used to distinguish between strains of *Acetobacter* and *Acetomonas*. The medium contains yeast extract (3%), agar (2%), BROMCRESOL GREEN indicator (0.002%), and ethanol (final concentration 2%). *Acetomonas* spp (which oxidize ethanol to acetic acid) produce colonies surrounded by a yellow halo. *Acetobacter* spp (which oxidize ethanol to acetic acid, and acetic acid to carbon dioxide) initially produce a similar reaction; however, on further incubation the acid is oxidized, and the indicator reverts to its original colour.

carvacrol (see PHENOLS)

caryonide (karyonide) In *Paramecium*: a clone in which all the cells have genetically similar macronuclei.

caryotype (see KARYOTYPE)

Castañeda's method (of blood culture) A form of BLOOD CULTURE similar to the conventional technique except in that the culture bottle contains an agar slant in addition to the liquid medium; the slant projects above the level of the liquid medium when the bottle is standing vertically. Inoculation of the liquid medium is carried out in the usual way and the bottle is incubated at 37 °C. After 24/48 hours the bottle is tilted so that the blood-containing medium washes over the exposed agar; the bottle is then incubated (vertically) and examined after an appropriate period of time for bacterial growth. The agar may be inoculated (by tilting the bottle) at intervals of one or two days. Castañeda's method eliminates the risk of contamination which attends subculturing in the conventional technique.

catabolism (*biochem.*) In a cell: those reactions which involve the enzymic degradation of organic compounds to simpler organic or inorganic compounds. Energy released during catabolism may be used—with varying degrees of efficiency—for energy-requiring cellular processes such as biosynthesis, the maintenance of the intracellular environment, etc. (cf. ANABOLISM and AMPHIBOLIC PATHWAY)

catabolite gene activator protein (see CATABOLITE REPRESSION)

catabolite repression (glucose effect; glucose repression) Repression of the transcription of genes coding for certain inducible enzyme systems by glucose or by certain other readily-utilizable carbon sources. Thus, e.g. in *Escherichia coli*, the *lac* and *ara* OPERONS—among others—are repressed ('turned off') in the presence of glucose, even in the presence of their respective inducers; the inducibility of such systems can be restored by the addition of 3′,5′-cyclic adenosine monophosphate (cAMP). A protein—the *catabolite gene activator protein** (CAP or CGA protein)—is required, together with cAMP, for the transcription of those operons which are subject to catabolite repression; the cAMP–CAP complex binds to the promoters of these operons. Glucose causes a reduction in the cellular levels of cAMP, and cAMP–CAP is released by the promoter so that transcription can no longer occur. Thus, if *E. coli* grows in a medium containing glucose *and* lactose, the *lac* operon remains repressed until all the glucose has been utilized; subsequently cAMP levels rise, cAMP–CAP binds to its site on the *lac* promoter, and the *lac* operon can be transcribed. (This biphasic response is referred to as DIAUXIE.)

Similar systems have been detected in other bacteria—and apparently in certain fungi. In e.g. *Pseudomonas* spp a number of inducible systems (e.g. those for the degradation of tryptophan and histidine) are repressed in the presence of succinate or certain other intermediates of the tricarboxylic acid cycle.
*Also called *cyclic AMP receptor protein*, CRP.

catalase (hydrogen peroxide : hydrogen peroxide oxidoreductase, EC 1.11.1.6) A protohaem-containing enzyme (see PORPHYRINS) which catalyses the reaction:

$$2H_2O_2 \longrightarrow 2H_2O + O_2$$

(Hydrogen peroxide, H_2O_2, is the highly toxic product of certain cellular processes—being produced e.g. by SUPEROXIDE DISMUTASE action and by the oxidation of reduced flavoproteins by oxygen.) Catalase is present, often in high concentrations, in the majority of aerobic microorganisms, but is absent from most obligate ANAEROBES. (see also CATALASE TEST)

catalase test A test used to determine whether or not a given bacterial strain produces CATALASE. One drop of hydrogen peroxide is placed on a bacterial colony (or added to an emulsion of bacteria); the presence of catalase (a positive test) is indicated by the appearance of bubbles of gas (oxygen)—either immediately or within a few seconds. (If blood-containing media are used care should be taken to prevent contact between hydrogen peroxide and the medium since erythrocytes, which contain catalase, can give a false-positive reaction.) Certain bacteria—e.g. strains of *Lactobacillus, Leuconostoc, Streptococcus*—can produce catalase only in the presence of haem; thus, if grown on haem-containing media such bacteria give a positive test—though they give a negative test if grown on haem-free media.

catenate (*adjective*) Formed in chains.

cationic detergents (as ANTISEPTICS and DISINFECTANTS) (see QUATERNARY AMMONIUM COMPOUNDS)

cattle plague (see RINDERPEST)

cattle tick fever (see RED-WATER FEVER)

Caulerpa A genus of green algae (CHLOROPHYTA).

cauliflower mosaic virus (cabbage B virus) A virus (see VIRUSES) whose hosts include a number of crucifers; the polyhedral virion, approximately 50 nm in diameter, contains double-stranded DNA. On cauliflower (*Brassica oleracea* var *botrytis*) typical symptoms of infection include vein-clearing and/or vein-banding; enations are occasionally produced. Rounded, virus-containing amorphous cytoplasmic inclusion bodies are formed. Stylet-borne transmission of the virus is effected by aphids; mechanical transmission may be effected with difficulty.

Caulobacter A genus of gram-negative, obligately aerobic, chemoorganotrophic bacteria of uncertain taxonomic affinity; species occur in soil and water. The vegetative, non-motile cell is rod-shaped or ovoid, about 0.5–1.0 × 1–2 microns, with a polar stalk (PROSTHECA); the stalk, which has an adhesive tip, may be almost as long as the cell itself—the length varying with the strain and with growth conditions. Binary fission gives rise to a stalked, non-motile daughter cell, and a motile daughter cell carrying a single polar flagellum; the latter cell (*swarmer*) must become stalked and non-flagellate before being able to undergo cell division. The type species: *C. vibrioides*; GC% for the genus: about 65 ± 2.

cavity slide (see SLIDE)

CDP Cytidine-5′-diphosphate (see NUCLEOTIDE).

cecidia (see GALLS)

cedar apples (see GYMNOSPORANGIUM)

celery, diseases of These include: (a) *Late blight* (*leaf spot*). A seed-borne fungal disease caused by *Septoria apiicola*. The leaves, petioles etc. bear small chlorotic lesions which later become necrotic and exhibit numerous black *pycnidia*. (b) *Yellows*. A disease in which chlorosis and retarded growth result from vascular infection by *Fusarium oxysporum*.

cell-bound antibodies Antibodies bound to the surfaces of cells either by their *combining sites* or (see CYTOPHILIC ANTIBODY) by other sites.

cell culture (see TISSUE CULTURE)

cell cycle (in mammalian cells) The sequence of phases in the growth-division cycle; these phases are:
mitosis (*M* phase) → gap phase 1 (G_1) → DNA synthesis (*S* phase) → gap phase 2 (G_2) → mitosis... The period between the commencement of G_1 and the end of G_2 corresponds to *interphase*, a period of cell growth. The *S* phase commonly occupies about $\frac{1}{3}$ to $\frac{1}{2}$ of the cycle.

cell envelope (bacterial) A term used to refer to the CELL MEMBRANE together with all structures external to it—including the CELL WALL and, where applicable, the CAPSULE.

cell line (tissue culture) Any population of cells—*other than* those of the PRIMARY CULTURE—prepared either by the first subculture or at any stage during the serial subculture of a primary culture. Such a population of cells is commonly heterogenous. Using e.g. MICROMANIPULATION techniques, a *cloned line* may be prepared, i.e. a population consisting of a single CLONE. (see also ESTABLISHED CELL LINE)

cell-mediated hypersensitivity (see CELL-MEDIATED IMMUNITY)

cell-mediated immunity (type 4 reaction; cell-mediated hypersensitivity) Antibody-*independent* HYPERSENSITIVITY. Cell-mediated immunity (CMI) involves the T LYMPHOCYTES and certain of their extracellular products; the reactions of CMI generally depend upon MACROPHAGE-lymphocyte co-operation.

On *initial* exposure to a given antigen the specifically reactive T lymphocytes form *memory cells* (see TRANSFORMATION); an individual in whom these events have occurred is said to be *primed* in respect of the given antigen. A second (or subsequent) exposure to the *same* antigen may bring about a manifestation of cell-mediated immunity—e.g. a DELAYED HYPERSENSITIVITY reaction or a tissue transplant rejection. These effects are due to a complex reaction involving the antigen, macrophages, and the specifically reactive memory cells—a reaction which results in the release of LYMPHOKINES. The mechanism of CMI is not fully understood.

cell membrane (bacterial) (*synonyms*: plasma membrane; cytoplasmic membrane; protoplast membrane; plasmalemma) The selectively-permeable membrane which forms the outer limit of the protoplast; in most bacteria it is bordered externally by the CELL WALL.

Composition of the membrane. As with other biological membranes, lipids are the characteristic structural components. The types and quantities of lipid depend on the organism, the conditions of growth, and the age of the culture; the proportion of lipids in the membrane is generally greater in the stationary phase than in the exponential phase. Membrane lipids consist mainly of phospholipids. Phosphatidylglycerols may be universally present in bacterial membranes, while phosphatidylethanolamine seems to be more common in the membranes of gram-negative species; phosphatidylcholine, sphingolipids and sterols are uncommon—however, see MYCOPLASMA. Glycolipids and diglycerides are common in small quantities. The relative proportions of saturated and unsaturated fatty acids in the membrane appear to depend on species and growth temperature. Thus, e.g. in some organisms, it has been observed that the proportion of unsaturated fatty acids appears to increase when growth occurs at lower temperatures, and conversely; this is believed to be a compensatory response—a higher proportion of unsaturated fatty acids enabling the fluidity and permeability of the membrane to be maintained at lower temperatures. Membrane fatty acids may be branched (apparently more common in gram-positive species) or straight-chained; some contain a cyclopropane ring. The membranes of certain bacteria (e.g. *Clostridium perfringens*, *Staphylococcus aureus*) appear to contain *lipoamino acids*—esters of phosphatidylglycerol and basic amino acids such as lysine, arginine or ornithine; the content of lipoamino acids appears to be influenced by the pH of the growth medium. The membranes of gram-positive bacteria may contain glycerol TEICHOIC ACIDS linked covalently to lipid as *lipoteichoic acid*.

Much of the cell membrane consists of protein; many of the protein constituents have proven roles in TRANSPORT and/or in catalysis—but none appears to fulfill a purely structural function.

Structure of the membrane. Under the electron microscope the cell membrane has a trilaminar structure: an electron-transparent layer sandwiched between two electron-dense layers, the structure having a thickness of some 75Å. The most widely-accepted theory of bacterial membrane structure is that based on the unit membrane concept. It is postulated that the membrane contains a double layer of phospholipid molecules orientated such that their hydrocarbon (fatty acid) side chains occur at right angles to the plane of the membrane; the hydrophilic groups of these molecules form the outer and inner limits of the bilayer, while the hydrophobic portions of the molecules form the lipophilic internal environment of the membrane. It is believed that the side chains do not form part of a rigid structure—they are currently believed to be in a *fluid* state. The membrane proteins are believed to lie within, partly within, and on the surfaces of the lipid bilayer—sometimes extending across the width of the membrane. Such a structure has been referred to as the *fluid mosaic model*. In many bacteria, the inner surface of the membrane is the site of a complex which contains 'ATPase'; this enzyme, which is involved in oxidative phosphorylation, appears to form the terminal spheres of 'stalked particles' similar to those which occur on the inner membrane of the MITOCHONDRION. (The protein which forms the stalk region is known as *nectin*.) The bacterial 'ATPase' (also referred to as BF_1) closely resembles the 'ATPases' of mitochondria and chloroplasts (F_1, CF_1 respectively).

Functions of the membrane. The membrane regulates the passage of various metabolites, waste products etc. into and out of the cell (see also TRANSPORT); it is also the site of biosynthesis of a number of exocellular substances, and the site of the ELECTRON TRANSPORT CHAIN. The lipid fraction of the membrane provides the permeability barrier of the cell and the lipophilic environment necessary for the activity of many membrane-bound enzymes. The latter may include: (a) enzymes concerned with the synthesis of polymers such as LIPOPOLYSACCHARIDES, PEPTIDOGLYCAN, TEICHOIC ACIDS, and CAPSULE polymers (see also BACTOPRENOL); (b) enzymes involved in the synthesis of complex lipids; (c) enzymes associated with electron transport and OXIDATIVE PHOSPHORYLATION.

Biosynthesis of the membrane. Currently, little information is available concerning membrane biosynthesis. Although fatty acids are synthesized by soluble (cytoplasmic) enzymes, the glycerol esters appear to be synthesized by membrane-bound enzymes—i.e. the phospholipids are apparently synthesized *in situ*.

Membrane differentiation. (see CYTOMEMBRANES and PURPLE MEMBRANE)

Antibiotics affecting the cell membrane. These include DEPSIPEPTIDE ANTIBIOTICS, GRAMICIDIN, POLYMYXINS, and TYROCIDINS; many DISINFECTANTS and ANTISEPTICS probably act by interfering with the integrity of the cell membrane.

cell membrane (fungal) (*synonyms*: plasma membrane, cytoplasmic membrane, protoplast membrane, plasmalemma) The selectively-permeable membrane which forms the outer limit of the protoplast; it is bordered externally by the CELL WALL. *Composition of the membrane.* The available data are based on the analyses of membranes of only a few fungi—in particular that of the yeast *Saccharomyces cerevisiae*. The membrane of *S. cerevisiae* appears to contain some 50% (dry weight) of protein, approximately 40% lipid, and small amounts of carbohydrate—particularly mannans (which may be precursors of cell wall constituents); the lipid fraction appears to consist mainly of glycerophospholipids (e.g. phosphatidylcholine, phosphatidylethanolamine) and sterols. *Structure of the membrane.* The structure of the membrane is believed to resemble the *fluid mosaic model* described under *bacterial* CELL MEMBRANE (*q.v.*)—with the additional presence of e.g. sterol molecules within the lipid bilayer. *Functions of the membrane.* The membrane regulates the transport of solutes (e.g. metabolites, waste products) into and out of the cell; this function probably involves some of the membrane proteins, and may also involve the ATPases which are associated with the membrane. The membrane also appears to be the site of biosynthesis and/or assembly of membrane/cell wall/capsular structural components and the location of enzymes associated with these processes. *Antibiotics which affect the fungal cell membrane* include the POLYENE ANTIBIOTICS.

cell sap Synonym of CYTOSOL.

cell wall (algal) A structure, external to the cell membrane, which forms the outermost boundary of an algal cell. (cf. the euglenoid *pellicle* (see EUGLENA) and the THECA of dinoflagellates) Relatively few algal cell walls have been examined in detail but those which have reveal a great variety of types of structure and composition. In the majority of unicellular algae, and in all multicellular algae, there is a rigid cell wall which commonly appears to consist of two or more distinct layers, and which sometimes bears an external mucilaginous coating. The cell wall typical of many species consists of one or more layers of CELLULOSE microfibrils—orientated randomly or in definite configurations—embedded in an amorphous matrix which (depending on species) may consist of, or include, materials such as PECTINS, ALGINIC ACID, FUCOIDIN. Cellulose microfibrils occur in the triple-layered cell walls of *Chlorella* and *Scenedesmus*, and e.g. in the walls of certain species of the Chlorophyta (e.g. *Chaetomorpha*, *Chlamydomonas*, *Cladophora*, *Pediastrum*, *Ulva*, *Valonia*), Xanthophyta (e.g. *Botrydium*, *Vaucheria*), Rhodophyta (e.g. *Griffithsia*), and Phaeophyta (e.g. *Fucus*, *Laminaria*); the proportion of cellulose in the cell wall differs from species to species. In some algae (e.g. the red alga *Porphyra*, and the siphonaceous green algae *Bryopsis*, *Caulerpa*, *Halimeda*, *Penicillus*, and *Udotea*) XYLANS replace cellulose as the main structural component of the cell wall. Mannan is the main structural polysaccharide in the cell walls of *Acetabularia* and *Codium*.

In diatoms (BACILLARIOPHYTA) there is a rigid silicaceous cell wall (*frustule*) external to the cell membrane; the frustule is composed of two parts (*valves*) which fit together in the manner of a box with an overlapping lid. During cell division two protoplasts, each surrounded by its own cell membrane, develop within the parent frustule; each protoplast then synthesizes one valve which, together with one of the parental valves, forms the complete frustule of the daughter diatom. Studies on the diatom *Melosira* indicate that wall development involves the deposition of silica within a membrane system (the *silicalemma*) which lies beneath the cell membrane (plasmalemma); subsequently each daughter develops a new region of cell membrane beneath the newly-formed silicaceous valve.

In some algae (e.g. certain members of the Haptophyta) the cell wall consists of a system of plates or scales of *calcified* organic material (*coccoliths*) external to the cell membrane. In at least some species coccoliths are known to be synthesized intracellularly and are derived from vesicles of the Golgi apparatus*; the organic skeleton of the coccolith becomes impregnated with calcium carbonate and is passed to the exterior of the cell. Coccoliths are usually oval in shape, and their external surfaces commonly bear various forms of ornamentation; in some species (e.g. *Cyclococcolithus leptoporus*) the coccoliths are imbricated. The *silicaceous* scales of members of the Chrysophyta (e.g. *Mallomonas*) appear to be formed and assembled in a manner analogous to that of the coccoliths. Some algae (e.g. *Chara*) typically bear a surface *encrustation* of calcium carbonate. In most calcified algae the calcium carbonate occurs in the form of calcite; in some it occurs in the form of aragonite.

A cell wall is absent in some unicellular algae (e.g. the green alga *Dunaliella*) and in the gametes and spores of a wide range of algae; the wall-less unicellular red alga *Porphyridium* is bounded by a gelatinous capsule. *In general, the synthesis and assembly of algal cell walls—even those of the 'conventional' layered type—are believed to involve the activity of the Golgi apparatus.

cell wall (bacterial) The outer layer of material—external to the CELL MEMBRANE—whose mechanical strength protects the cell from osmotic lysis and determines cellular morphology. (Osmotic pressures in gram-positive and gram-negative bacteria may reach approximately 25 and 10 atmospheres respectively.) Cell walls may account for some 10–50% of the dry weight of bacteria; isolated cell walls may be obtained by the differential centrifugation of disrupted cells. In most bacteria the main structural component of the cell wall is PEPTIDOGLYCAN.

The inner surface of the cell wall may be in contact with the cell membrane or may be separated from it by the *periplasmic space*. The outer surface of the cell wall may be in direct contact with the environment or it may be covered by a CAPSULE; in either case, the external boundary of the cell is important antigenically. Permeability differences between the cell walls of gram-positive and gram-negative bacteria are believed to account for the differential action of the GRAM STAIN.

Cell walls of gram-positive bacteria. No fine structure has been determined by electron microscopy. The walls consist of peptidoglycan (up to 80%) intimately associated with TEICHOIC ACIDS and/or other substances, e.g. proteins, TEICHURONIC ACIDS.

Cell walls of gram-negative bacteria. These appear multilayered by electron microscopy. Peptidoglycan forms the innermost layer, and the outer layers (sometimes collectively termed the 'outer membrane') consist of protein, lipoprotein, and LIPOPOLYSACCHARIDES. Owing to the relative inaccessibility of the peptidoglycan, gram-negative bacteria are usually much less susceptible than gram-positive species to PENICILLINS and functionally-similar antibiotics.

The cell walls of species of the HALOBACTERIACEAE contain no peptidoglycan, and are unique in that they appear to depend on electrostatic forces for the maintenance of their structural integrity; they disintegrate in solutions which contain low concentrations of salt. Cell walls are lacking altogether e.g. in *Mycoplasma* spp. Some bacteria give rise to wall-defective variants: see L FORMS.

Antibiotics which effect bacterial cell walls include BACITRACIN, CEPHALOSPORINS, CYCLOSERINE, PENICILLINS and VANCOMYCIN; see also LYSOZYME. (see also GROWTH (bacterial) and ZETA POTENTIAL)

cell wall (fungal) The rigid envelope, external to the CELL MEMBRANE, which determines the shape of an organism and which protects it from osmotic lysis. The fungal cell wall is composed largely of polysaccharides (typically 80–90% of the dry weight) which are complexed with enzymic and/or structural proteins and with lipids. The wall is generally a lamellate structure which typically contains both microfibrillar components (frequently of CHITIN, sometimes of—or containing—CELLULOSE), and amorphous (matrix) components (consisting chiefly of polysaccharides—particularly of glucans, mannans etc. intimately associated with proteins).

Fungi have been divided into a number of categories on the basis of the identities of the two principal cell wall polysaccharides; these categories appear to accord well with the major groupings produced by conventional taxonomic schemes—although, to date, the walls of only a few species have been examined. Thus, chitin and non-cellulosic glucans appear to be commonly associated with ascomycetes, basidiomycetes, *fungi imperfecti*, and chytrids; chitin and CHITOSAN occur in *Mucor* and other zygomycetes; cellulose and non-cellulosic glucans occur in *Saprolegnia* and other oomycetes; non-cellulosic glucans and mannans occur in the ascomycetous yeasts (e.g. *Saccharomyces*); chitin and mannans occur in the basidiomycetous yeasts (e.g. *Sporobolomyces*). Some fungi do not fit easily into categories predicted on the basis of their taxonomic affinities; thus, e.g. the ascomycetous fission yeast *Schizosaccharomyces* contains little or no mannan—and none of the *minor* component, chitin, which is found in *Saccharomyces* and related yeasts.

In some (dimorphic) fungi the composition of the cell wall varies according to growth conditions. Thus, e.g. at 25 °C *Paracoccidioides brasiliensis* grows in the mycelial form—the principal wall polysaccharides being β-glucans and galactomannans; at 37 °C the organism develops in the yeast (single cell) phase—with α-glucans forming the chief wall polysaccharides. Some species of *Mucor* exhibit a temperature-independent switch to the yeast phase when grown in the presence of increased carbon dioxide tensions; the cell walls of the yeast-phase cells contain levels of mannans significantly greater than those in the walls of the mycelial-phase organism.

cellar fungus A common name for the basidiomycete *Coniophora cerebella* (= *Coniophora puteana*) (see also WET ROT).

cellobiose D-Glucopyranosyl-β-(1 $\rightarrow$ 4)-D-glucopyranose; the disaccharide subunit of CELLULOSE.

cellular (*microbiol.*) (*adjective*) (1) Of (or belonging to) a cell. (2) Having the form and characteristics of a (single) cell. The term *unicellular* is commonly used with this meaning. (3) Having a body, thallus etc. composed of two or more cells. The term *multicellular* is commonly used with this meaning. Ambiguity sometimes arises from the use of the terms *acellular* and *non-cellular*. *Acellular* is sometimes used to refer to *single-celled* organisms (e.g. *Paramecium*) in which occur functionally-specialized regions. Thus, the term (supposedly) indicates that such organisms are equivalent, in organization, to a differentiated multicellular organism. *Non-cellular* has been used (a) as a synonym of acellular, and (b) to refer to organisms which are not cellular in any sense (e.g. viruses).

cellular immunity (1) An aspect of NON-SPECIFIC IMMUNITY: the activities of MACROPHAGES and other phagocytic cells. (2) May refer to CELL-MEDIATED IMMUNITY.

cellular slime moulds (Acrasiales, Acrasida, Acrasina, Acrasiomycetes) A group of eucaryotic organisms in which the stages of the life cycle include a uninucleate amoeboid cell (the *myxamoeba*) and a motile aggregate of myxamoebae (the *pseudoplasmodium*, *grex*, or *slug*). (cf. ACELLULAR SLIME MOULDS) The cellular slime moulds—which are usually classified among the fungi—occur in soil and on decomposing vegetation; they feed primarily on bacteria—though some species consume bacteria and/or yeasts.

The life cycle includes a vegetative (feeding) phase in which the individual myxamoebae grow and divide mitotically. When the food supply diminishes to a certain level the phase of differentiation commences: on a suitable moist, solid substratum the myxamoebae converge (chemotactically) to form one, several, or many migratory pseudoplasmodia; a pseudoplasmodium is a mass of cells (up to several millimetres, maximum dimension) which—in many species—moves in response to light, heat, and other types of stimulus. When the pseudoplasmodium stops moving it differentiates (in a process known as *culmination*) to form a fruiting body (*sorocarp*). During culmination certain cells of the pseudoplasmodium form one or more *stalks*; the remainder of the cells develop into thick-walled spores which form mass(es) at the distal (free) end(s) of the stalk(s). In most species the stalk of the sorocarp consists of a column of cells in a sheath of secreted cellulose; in *Acytostelium leptostomum* the stalk is a fine, cell-free cellulosic tube. In *Dictyostelium polycephalum* the sorocarp consists of a number of stalks which are fused together for most of their length—becoming free at their distal, spore-bearing ends; in *Polysphondylium* spp the long sorocarp stalk bears, at intervals, a group of whorled branchlets—each of which carries a terminal spore mass smaller than that found at the free end of the main stalk. The spores are highly resistant e.g. to desiccation, and may remain viable for extended periods (months, years); they are dispersed by wind and rain—dispersal being aided by the elevation of the spore masses at the tips of the stalks. The colour of the spores depends on species; e.g. they are yellow in *Dictyostelium discoideum*, purple in *Polysphondylium violaceum*. On germination each spore gives rise to a myxamoeba. Parasexual and/or sexual processes appear to occur in at least some strains of cellular slime mould. Genera include *Acrasis*, *Acytostelium*, DICTYOSTELIUM, and *Polysphondylium*.

cellulitis A localized, spreading inflammation of certain tissues (usually the skin) (cf. ERYSIPELAS); the causal agent is often *Streptococcus pyogenes*. *Anaerobic* cellulitis is commonly due to species of *Clostridium*.

cellulolytic Capable of degrading CELLULOSE.

Cellulomonas A genus of gram-positive*, aerobic or facultatively anaerobic, chemoorganotrophic, cellulolytic, pleomorphic rod-shaped soil bacteria of uncertain taxonomic affinity. Type species: *C. flavigena*; GC%: approximately 72. *Cells easily decolorized.

cellulosans (see HEMICELLULOSE)

cellulose A linear polysaccharide consisting of β-$(1 \rightarrow 4)$-linked D-glucopyranose residues. Cellulose occurs in the cell walls of plants, most ALGAE, and species of the fungal classes Oomycetes and Hyphochytridiomycetes; typically, microfibrils of cellulose are embedded in an amorphous matrix—see CELL WALL (algal) and CELL WALL (fungal). A cellulose sheath surrounds the stalk of the *sorocarp* in the slime mould *Dictyostelium discoideum*. Few procaryotes can synthesize cellulose; *Acetobacter xylinum* forms a cellulose PELLICLE, and *Sarcina ventriculi* forms a cellulose microCAPSULE.

Cellulose-decomposing (*cellulolytic*) organisms are ecologically important e.g. in the decay of plant litter. Cellulolytic bacteria include *Cellulomonas*, *Cytophaga*, many actinomycetes, and certain of the RUMEN microflora (e.g. *Bacteroides succinogenes*, *Ruminococcus flavefaciens*). Cellulolytic fungi include certain species of the Ascomycetes (e.g. *Chaetomium globosum*, *Stachybotrys atra*), fungi imperfecti, and Basidiomycetes. Certain plant-pathogenic fungi are cellulolytic—e.g. species of *Botrytis*, *Fusarium*, and *Pythium*. (see also TIMBER, DISEASES OF; PAPER (microbial spoilage of); *Trichonympha campanula* under SYMBIOSIS) *Cellulases* (i.e. enzymes which hydrolyse cellulose) show a range of specificities—e.g. some split the β-$(1 \rightarrow 4)$ glucosidic linkages at random to form oligosaccharides, while others remove CELLOBIOSE units from one end of the chain. Cellobiose is hydrolysed (to glucose) by another enzyme—*cellobiase*. (see also CONGO RED)

cellulytic Capable of bringing about the lysis of cells.

CELO virus (chick embryo lethal orphan virus) (see ADENOVIRUS)

central capsule (see RADIOLARIANS)

central dogma (*mol. biol.*) Originally, it was believed that the flow of genetic information could occur only in the direction DNA → RNA → protein. (see GENETIC CODE and PROTEIN SYNTHESIS) However, with the discovery of REVERSE TRANSCRIPTASE, the dogma must now be stated: DNA⇌RNA → protein.

centric diatoms Those diatoms which are radially symmetrical in valve view—e.g. *Melosira.* (cf. PENNATE DIATOMS).

centrifugation A method used for the separation of fine particulate matter from a liquid medium, or the separation of different types of particulate matter within a given suspension. The sample, in a suitable container, is placed in a metal holder (*bucket*) which is either fixed or hinged at the end of one arm of a spider-shaped or cruciform motor-driven metal *rotor*; the latter is rotated, at high speed, in a horizontal plane about its central axis. (In practice, one or more additional samples—or simulated samples—of equal weight must be carried by the rotor in order to achieve balance.) The apparatus described—suitably enclosed for safety—constitutes a simple centrifuge. Spinning the sample subjects it to a centrifugal field; the forces associated with this field tend to move the particulate matter in the sample in an outward (radial) direction, i.e. towards the bottom of the bucket. The strength of the centrifugal field (G) is given by:

$$G = \frac{4\pi^2 \,(\text{revs min}^{-1})^2 r}{3600} \text{ centimetres sec}^{-2}$$

where *revs* min^{-1} is the number of revolutions of the rotor, per minute, and r is the distance, in centimetres, between the bucket and the axis of rotation. The field strength is given in terms of *acceleration* (cms sec^{-2}); since acceleration is proportional to the force which produces it, the formula for G gives a measure of the force which acts on a body subjected to a centrifugal field. The strength of the field increases with (a) increased rate of rotation, and (b) increase in r (which, for a given centrifuge, is fixed). Since acceleration due to the force of *gravity* (g) is 980 cms sec^{-2}, G/980 gives the factor by which the strength of the centrifugal field is greater than that of the gravitational field; G/980 is termed the *relative centrifugal field*, or RCF.

The acceleration given by G is that which the field would give to a particle *in a vacuum*. Since particles subjected to centrifugation are suspended in a *liquid* medium, factors in addition to the applied force must be taken into account. These factors, which considerably influence the rates of movement of the particles, are the size, shape, and density of the particles, and the viscosity and density of the liquid. For example, the rate of movement of a *spherical* particle in a given centrifugal field is proportional to:

$$\frac{G r_p^2 \,(d_p - d)}{\text{viscosity of medium}}$$

where r_p is the particle's radius, d_p the particle's density, and d the density of the suspending medium. Thus, given a mixed population of spherical particles which differ e.g. in size and/or density, separation into homogenous populations may be achieved on the basis of their different sedimentation rates in a centrifugal field. If particle density is equal to that of the suspending medium ($d_p = d$) the rate of movement in the field is zero. If d_p is *less* than d, centrifugally accelerated *flotation* occurs; this has been used e.g. for the purification of GAS VACUOLES. *Density gradient centrifugation.* Initially, the sample container is partly filled with a solution of caesium chloride, sucrose, or other substance—the concentration of which varies—continuously or discontinuously—from the bottom to the top of the liquid column. Thus a density gradient is set up, the density being greatest at the bottom of the liquid column and lowest at the top. In one form of density gradient centrifugation (*rate zonal centrifugation*, *s zonal* or *gradient differential* centrifugation) the sample is carefully layered at the top of the density gradient column and the whole centrifuged until the various components form layers or bands at different levels within the column—at which time centrifugation is discontinued. The separation of components into bands is achieved on the basis of the different sedimentation rates of the components. In *isopycnic centrifugation* (*equal density* or *equilibrium zonal* centrifugation) a similar initial arrangement is set up. In this case, however, centrifugation is continued until equilibrium is reached, i.e. each particle has reached that part of the gradient which corresponds to its own density—its isopycnic or equal density position. Particles are thus separated into bands solely on the basis of their individual densities—as opposed to their sedimentation rates as in the rate zonal method. The use of a density gradient has an important practical advantage: it militates against the effects of convection—which can disrupt the liquid column in the absence of a gradient.

Ultracentrifugation. Ultracentrifugation is used e.g. for the sedimentation of macromolecules and for molecular weight determinations. The principles involved are similar to those described above. In the ultracentrifuge, the chamber which houses the rotor is refrigerated and evacuated. While the common laboratory centrifuge develops a maximum field of about 5000 g the maximum field obtainable with the ultracentrifuge is about 500,000 g. The ultracentrifuge incorporates various photographic and optical systems (e.g. the Schlieren system) so that the progress of sedimentation may be followed and recorded at all times. (see also SVEDBERG UNIT)

centriolar plaque (in *Saccharomyces*) (see MITOSIS)

centriole An intracellular organelle structurally similar or identical to the *eucaryotic* BASAL BODY; one or more centrioles (per cell) occur in a range of eucaryotic cells. Centrioles are important in the organization of the SPINDLE during MITOSIS and MEIOSIS.

Centrohelida (see HELIOZOANS)

centromere (primary constriction) The constricted region of a CHROMOSOME which is the main (or sole) site of attachment of sister CHROMATIDS. (cf. KINETOCHORE)

centroplast (see HELIOZOANS)

centrum (*mycol.*) The totality of fertile and sterile structures *or* the whole of the interior (structures plus space) within a CLEISTOTHECIUM, a PERITHECIUM, or the cavity (locule) of an ASCOSTROMA.

cèpe (cep) A common name for the edible fungus *Boletus edulis*.

Cepedea A genus of protozoans of the superclass OPALINATA.

Cephaleuros A genus of epiphytic and/or parasitic algae of the division CHLOROPHYTA. A species of *Cephaleuros* causes RED RUST disease of the tea plant.

cephaline (*protozool.*) Refers to those members of the Eugregarinida which form septate trophozoites.

cephalodium (*lichenol.*) A discrete group of BLUE-GREEN ALGAE found on or within the thallus in certain of those lichens whose phycobiont is a member of the Chlorophyta (green algae). *External* cephalodia generally occur (e.g. as warty protuberances) on the upper or lower surface of the thallus (depending on species); in certain lichens various species of blue-green algae may be found in the cephalodium, while in other lichens the cephalodium contains a specific blue-green alga. External cephalodia are formed e.g. by species of *Peltigera* and *Stereocaulon. Internal* cephalodia consist of colonies of blue-green algae (usually or always *Nostoc*) which lie in the medulla of the thallus and may be apparent as swellings on the surface of the thallus; they occur e.g. in species of *Lobaria* and *Nephroma* and in *Solorina saccata*. In most cases, cephalodia are formed from free-living blue-green algae which, on coming into contact with the lichen thallus, stimulate the growth of—and subsequently become enveloped by—hyphae of the mycobiont. Blue-green algae in the cephalodia carry out NITROGEN FIXATION—the fixed nitrogen being transferred to the mycobiont in the thallus.

cephaloridine (see CEPHALOSPORINS)

cephalosporins An heterogenous group of natural and semi-synthetic ANTIBIOTICS. Cephalosporins *C*, *N*, and *P* are obtained e.g. from cultures of the fungus *Cephalosporium acremonium*. *Cephalosporin N* (synonyms: synnematin *B*, penicillin *N*) possesses a PENICILLIN nucleus; *cephalosporin P* is a steroid structurally related to FUSIDIC ACID. From *cephalosporin C* (see formula) is prepared a range of (semi-synthetic) chemotherapeutic agents (e.g. *cephaloridine*, *cephalothin*) by appropriate replacement of one or both of the side chains X and Y (see formula).

The Cephalosporins

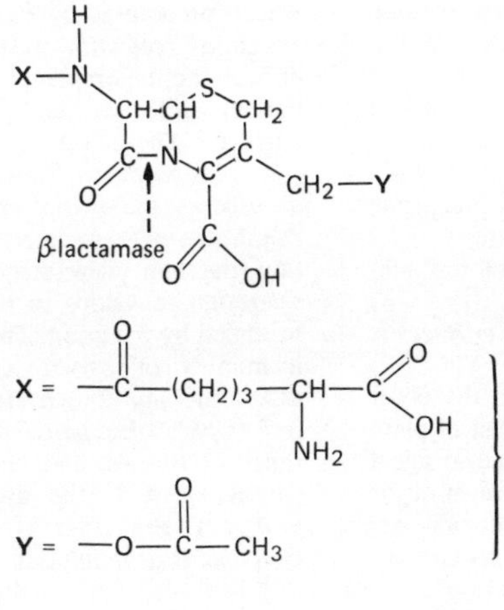

Cephalosporin C

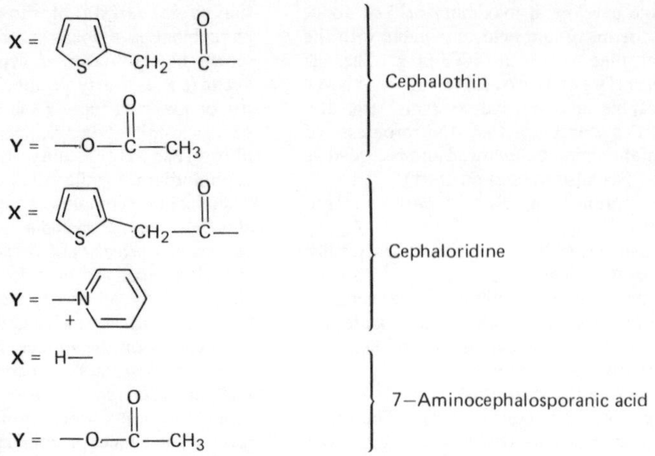

The semi-synthetic cephalosporins are active against a range of gram-positive and gram-negative bacteria; they appear to act by inhibiting the formation of cross-links in PEPTIDOGLYCAN.

Cephalosporium A genus of fungi within the order MONILIALES. The vegetative form of the organisms is a septate mycelium; conidia, which form gelatinous masses, are produced from phialids. *C. acremonium* is reported to form arthrospores in culture. (see also CEPHALOSPORINS)

cephalothin (see CEPHALOSPORINS)

Ceramium A genus of red algae (RHODOPHYTA).

Ceratiomyxa A genus of the ACELLULAR SLIME MOULDS.

Ceratium A genus of algae of the DINOPHYTA.

Ceratocystis (*syn*: *Ophiostoma*) A genus of fungi of the order SPHAERIALES, class PYRENOMYCETES; the genus includes a number of plant pathogens (e.g. the causal agents of certain wilt diseases and of DUTCH ELM DISEASE) and some species give rise to BLUE-STAIN of timber. *Ceratocystis* spp—which include homothallic and heterothallic organisms—form spherical, evanescent asci at various levels within globose, long-necked perithecia; according to species the ascospores may be e.g. crescentic or ovoid. *Ceratocystis* spp form various types of conidia, and some species (e.g. *C. ulmi*) form synnemata.

Ceratomyxa A genus of protozoa of the subphylum CNIDOSPORA.

Ceratostomella ulmi A synonym of *Ceratocystis ulmi*.

Cercospora A genus of fungi within the order MONILIALES; species include a number of plant pathogens—e.g. *C. beticola*, the causal agent of *leaf spot* of beet. The vegetative form of the organisms is a septate mycelium; conidia are elongated and multiseptate. (One species, *C. apii*, is reported to be the causal agent of a single case of human disease in Indonesia; the disease was characterized by the formation of ulcerated cutaneous and sub-cutaneous lesions, and hypertrophy of the affected regions.)

cereals, diseases of These include: (a) *Rusts*. (i) *Black stem rust* affects a range of cereals and other grasses and is of worldwide occurrence. The causal agent is the macrocyclic, heteroecious rust *Puccinia graminis* (see RUSTS (1) for details of life cycle); the pycnial and aecial stages of *P. graminis* develop on the barberry (*Berberis*). *Aeciospores* germinate on the surface of the cereal host and the fungus enters the leaf by means of appressoria. Masses of reddish-brown *uredospores* subsequently appear on the stems and leaves; the surface lesions later become black owing to the formation of *teliospores*. Elimination of the barberry may assist in disease control and inhibits the emergence of new strains of the pathogen. (ii) *Yellow rust* and *brown rust* (particularly of wheat) are caused by *Puccinia striiformis* and *Puccinia recondita* respectively. (b) *Ergot* affects the inflorescence of e.g. rye, wheat, barley, and oats, and is of worldwide occurrence. The causal agent is the ascomycete *Claviceps purpurea*. Subsequent to infection by *ascospores*, the ovaries exhibit a white mass of *conidiospores* associated with a sticky secretion (the 'honeydew' stage); later, a hard black *sclerotium* develops in the position normally occupied by the grain. The sclerotium contains a number of ERGOT ALKALOIDS. (see also CLAVICEPS and ERGOTISM) (c) *Take-all* affects turf (*Ophiobolus patch disease*) as well as a range of cereals, and occurs worldwide. The causal agent is the ascomycete *Gaeumannomyces graminis* (formerly known as *Ophiobolus graminis*). In wheat, the symptoms include retarded growth, the absence of tillers, and a dry rot (with associated surface mycelium) in the root, crown, and lower stem. (d) *Smuts*. (see also SMUTS (1) and (2)) (i) *Loose smut* (in wheat and barley). The causal agent, *Ustilago nuda*, infects the inflorescence and gives rise to black powdery spore masses which form in place of the grains. Stems which bear diseased ears are often conspicuously taller than those bearing healthy ears. (ii) *Covered smuts* of oats and barley are caused by *Ustilago kolleri* and *U. hordei* respectively. (iii) *Stinking smut* (*bunt*) is a seed-borne covered smut of wheat; the causal agent may be *Tilletia caries* or *T. foetida*. Symptoms may include retarded growth and the ears are thinner and have a blue-green coloration; grains are replaced by grey or black powdery spore masses which have an odour resembling that of decaying fish. (e) *Powdery mildew* may affect a range of cereals and other grasses, and is particularly important in wheat and barley; the causal agent is the ascomycete *Erysiphe graminis* (see also ERYSIPHE). The fungus penetrates the host by means of appressoria; white or pale grey powdery patches of mycelium are formed on the leaves and stems and may spread to cover much of the plant. Pinhead-sized black *cleistothecia* develop on the external mycelium. (f) *Leaf blotch* (of barley and rye) is caused by *Rhynchosporium secalis* and can be economically important in areas of high rainfall. Leaf symptoms may include grey or brown lesions with darker margins. (g) *Gibberella diseases*. Several species of the ascomycete *Gibberella* (*G. roseum*, *G. fujikuroi*) cause disease in maize, barley, wheat, rye, and rice. (see also RICE, DISEASES OF) (h) *Viral* pathogens of cereals include BARLEY YELLOW DWARF VIRUS and RICE DWARF VIRUS. (see also PLANT PATHOGENS AND DISEASES)

cetavlon (see QUATERNARY AMMONIUM COMPOUNDS)

Cetraria A genus of foliose or fruticose LICHENS in which the phycobiont is *Trebouxia*. The thallus is dorsiventral. Apothecia—when present—occur at or near the margins of the thallus and are lecanorine with reddish-brown to brown discs. Species include e.g.:

C. islandica (Iceland moss). The thallus is erect, strap-like and branching—approximately 2–7 centimetres in height; it is reddish near the base. The strap-like branches are chestnut-brown to dark brown on one surface while the other surface is paler and bears scattered white pseudocyphellae. The margin of the thallus bears a row of short black spines. *C. islandica* grows on the ground (e.g. on heather moors); it is used as a food in certain countries.

C. chlorophylla. The thallus is thin and leaf-like, deeply-cut, crinkled and sorediate along the margins; it is olive green when wet, olive-brown when dry. *C. chlorophylla* occurs mainly on trees and fences and occasionally on walls.

cetrimide (see QUATERNARY AMMONIUM COMPOUNDS)

cetyltrimethylammonium bromide (see QUATERNARY AMMONIUM COMPOUNDS)

CFT (see COMPLEMENT FIXATION TEST)

CGA (see CATABOLITE REPRESSION)

CH_{50} (see HD_{50})

Chaetocladium A genus of fungi of the ZYGOMYCETES.

Chaetomella A genus of fungi of the SPHAEROPSIDALES.

Chaetomium A genus of fungi of the order SPHAERIALES, class PYRENOMYCETES; species occur e.g. in the soil and on decomposing plant matter. The somatic mycelium occurs within the substratum; most species of *Chaetomium* do not produce conidia. *Chaetomium* spp (many of which are homothallic) form evanescent asci in perithecia which develop at the surface of the substratum; the surface of each perithecium bears a number of long hairs or bristles which may be curved, coiled and/or branched. The unicellular, commonly dark-coloured ascospores are discharged from the perithecium in a gelatinous mass. Many or all species are strongly cellulolytic (see CELLULOSE) and some are important in the spoilage of PAPER and other cellulose-containing materials.

Chagas' disease An infectious, systemic disease of man (mainly children), and domestic animals, caused by the intracellular pathogen, *Trypanosoma cruzi*; the disease occurs in Central and South America. *T. cruzi* is transmitted by blood-sucking triatomid bugs of the family *Reduviidae*; wild animals, e.g. armadillos, rodents, form a reservoir of infection. In man, the disease is contracted *contaminatively*—the infected faeces of bugs being rubbed into bites or other lesions. Chagas' disease may be acute or chronic, and may be fatal. Symptoms include anaemia and disorders of the heart, glands and nervous system. *Laboratory diagnosis*. *T. cruzi* may be observed in blood smears in acute cases; a CFT may be useful in chronic cases. (see also XENODIAGNOSIS)

challenge (*immunol.*) (1) Inoculation of an animal with a pathogen (or with other material) to test the efficacy of a previously-administered VACCINE. (2) To otherwise examine the state of IMMUNITY by exposure to a specific agent or agents.

chalybeate (of water) Containing iron.

Chamberland candle (see FILTRATION)

chancre The primary ulcerative lesion in SYPHILIS.

chancroid (*soft* chancre) A venereal ulcer affecting the generative organs or nearby regions; causal organism: *Haemophilus ducreyi.*

chanterelle The edible fruiting body of the basidiomycete *Cantharellus cibarius.*

Chaos chaos (see PELOMYXA)

Chara A genus of algae of the division CHAROPHYTA.

Charcot–Leyden crystals Microscopic crystals often present in the stools of patients suffering from AMOEBIASIS (or other ulcerative conditions of the intestine), and which may also occur in the sputum of patients with asthma or other chest conditions. The colourless crystals stain with iodine, and may be diamond-shaped or resemble short double-pointed needles.

Charophyta A division of the ALGAE which includes the brittleworts and stoneworts—organisms which occur in freshwater and brackish environments and which often bear a surface encrustation of calcium carbonate. Species of the Charophyta are macroscopic algae which exhibit whorled branching, have complex gametangia (see under *sexual reproduction* in ALGAE), and which are anchored to the substratum by rhizoids. (Some authors exclude members of the Charophyta from the algae on the basis of their complex gametangia.) All species contain CHLOROPHYLLS *a* and *b*, β-carotene, and several xanthophylls (see CAROTENOIDS); the intracellular storage product is STARCH. Asexual reproduction is unknown; sexual reproduction is oogamous in all species. Genera include *Chara* and *Nitella.*

checker colony (see DRAUGHTSMAN COLONY)

cheese A partially-fermented, coagulated product of milk. The characteristics of the various types of cheese reflect differences in e.g. the mode of manufacture, the nature of the milk used, the microorganisms employed. *Cheddar cheese.* Pasteurized whole milk is seeded with a *starter culture* of *Streptococcus lactis* (or *S. cremoris*) and incubated at about 30 °C. The streptococci form lactic acid from lactose—decreasing the pH. A calculated amount of *rennin* is added; this modifies the milk protein *casein* to *paracasein* which, with Ca^{++}, coagulates to form a clot, or *curd.* The fluid which remains (*whey*) is drained off and/or expressed from the curd. The curd is then cut into blocks which are waxed (or otherwise coated) and left to *ripen.* During ripening, many changes occur in the chemistry and texture of the cheese, some of which result from the activities of bacteria; the metabolic products of lactobacilli (particularly *Lactobacillus casei*) are considered to be important in the development of flavour. *Swiss* (*Emmentaler*) *cheeses. Propionibacterium* spp are included in the starter culture. During ripening, the propionibacteria ferment lactic acid to form propionic and acetic acids and carbon dioxide—the latter contributing to the formation of the holes (*eyes*) in the cheese. Propionic acid is responsible for the characteristic flavour of Swiss cheese. *Cottage cheese.* A soft, unripened cheese with a high moisture content; it is prepared from skim milk. The starter culture usually contains strains of *Streptococcus* and *Leuconostoc. Blue cheeses.* The milk or curd is seeded with the spores of certain fungi (e.g. *Penicillium roqueforti*)—the growth and metabolism of which determine the characteristic flavour and texture of the cheese. *Camembert cheese.* A soft cheese which has a relatively high moisture content; its flavour is due to the *surface* growth of e.g. yeasts, *Geotrichum* spp, *Penicillium camemberti.*

chemiosmotic hypothesis (see OXIDATIVE PHOSPHORYLATION)

chemoautotroph (see AUTOTROPH)

chemoheterotroph (see HETEROTROPH)

chemolithotroph (see AUTOTROPH)

chemoorganotroph (see HETEROTROPH)

chemostat (bactogen) A type of apparatus used for CONTINUOUS-FLOW CULTURE.

chemotaxis A particular type of TAXIS in which the stimulus is a chemical concentration gradient; a cell may respond to such a stimulus when it occupies a physiologically sub-optimal position within the gradient. Substances which *bacteria* tend to move (indirectly) *towards* are called *attractants*; these may be nutrients or non-metabolizable substances. Attractants are known to bind to receptor sites (*chemoreceptors*) on the cell surface; the chemoreceptors have narrow specificities—i.e. each binds a single attractant or a very limited range of attractants. The mechanism of chemotaxis is at present unknown. (see also AEROTAXIS and PHOTOTAXIS)

In *immunology* chemotaxis refers to the migration of granulocytes or macrophages in the direction of higher concentrations of certain agents called *cytotaxins.* Powerful cytotaxins are formed e.g. during the intermediate stages of COMPLEMENT-fixation; these agents—the fragments C3a, C5a, and the complex C567—possess a strong attraction for polymorphs. (see also LYMPHOKINES)

chemotherapeutic agent Any substance used as an antimicrobial agent for the treatment of disease in man or animals; chemotherapeutic agents include synthetic compounds (e.g. the SULPHONAMIDES) and compounds of microbial origin (including semi-synthetic derivatives of these). (Some authors use the term to refer only to synthetic agents—cf. ANTIBIOTICS; others stipulate that a chemotherapeutic agent must act *systemically* following injection or oral administration.)

chemotherapy The use of chemical agents for the treatment of disease.

chemotrophs Organisms whose energy is derived solely from endogenous chemical reactions (cf. PHOTOTROPHS). Such organisms may be chemoAUTOTROPHS or chemoHETEROTROPHS.

chemotropism (see TROPISM)

Cheshunt compound (*plant pathol.*) An antifungal preparation consisting of copper sulphate and ammonium carbonate. In aqueous solution Cheshunt compound is highly effective against DAMPING OFF. (see also HEAVY METALS)

chiasma (*pl.* chiasmata) A connection between a pair of homologous chromatids which occurs during the *diplotene* stage of MEIOSIS. Chiasmata are believed to be manifestations of CROSSING-OVER. (see also RECOMBINATION; INTERFERENCE; SYNAPTONEMAL COMPLEX)

chiasma interference (see INTERFERENCE)

chiastobasidial (*adjective*) Refers to the *transverse* arrangement of meiotic spindles in a developing BASIDIUM (cf. STICHOBASIDIAL).

Chick–Martin test A test devised by Chick and Martin (in 1908) to determine the PHENOL COEFFICIENT of a *phenolic* disinfectant. The test was designed to assess the efficacy of a disinfectant under conditions likely to be encountered during its practical usage (cf. RIDEAL–WALKER TEST). In the original test an evaluation was made of the efficacy of a disinfectant against a test organism in the presence of 3% sterilized faeces. (Many disinfectants are inactivated in the presence of organic matter.) A modern version of the test uses a yeast suspension in place of faeces. The disinfectant is serially diluted (in water) and each dilution is inoculated with a fixed volume of a suspension containing *Salmonella typhi* and yeast. 30 minutes after inoculation, each dilution is used to inoculate each of two tubes of broth; the broths are incubated at 37 °C for two days and subsequently examined for viable *S. typhi*. The same procedure is carried out using phenol in place of the disinfectant under test. For the *phenol* test, two dilutions are identified: (a) that dilution containing the highest concentration of phenol which permitted growth of *S. typhi* in *both* tubes of broth inoculated from it, and (b) that dilution containing the lowest concentration of phenol which permitted no viable *S. typhi* to be recovered from *either* of the tubes of broth inoculated from it. The *mean* of these two concentrations (P_{mean}) is calculated. For the *disinfectant* test, a mean figure (D_{mean}) is calculated in a similar manner. The phenol coefficient (*Chick–Martin coefficient*) is given by P_{mean} / D_{mean}

chickenpox (varicella) An acute, highly contagious disease which usually occurs in epidemic form among young children. The causal agent, a herpesvirus, invades *via* the oral or nasal route. An incubation period of 2–3 weeks is followed by fever and by the appearance of cutaneous lesions which rapidly become vesicular and subsequently crusty; successive crops of lesions are formed. The disease is normally self-limiting; complications, e.g. encephalitis, are uncommon. *Laboratory diagnosis* is rarely required; scrapings from the base of a vesicle may be stained and examined for typical varicella *giant cells* and cellular *inclusion bodies*. (see also SHINGLES)

childbed fever (see PUERPERAL FEVER)

Chilodochona A genus of marine protozoans of the order CHONOTRICHIDA.

Chilodonella A genus of protozoans of the suborder CYRTOPHORINA. *C. uncinata* is common in percolating filters (biological filters) in British SEWAGE TREATMENT plants.

Chilomastix A genus of flagellate protozoans of the order RETORTAMONADIDA. *Chilomastix mesnili*, a non-pathogenic intestinal parasite of man and other animals (e.g. the pig), is a pear-shaped organism—usually 10–15 microns in length—with three anteriorly-directed and one recurrent flagella arising from the blunt anterior end; a single nucleus is located anteriorly. The prominent cytostomal groove is visible both in the trophozoite and in the (stained) cyst—the latter thus resembling (somewhat) the cysts of *Giardia lamblia*.

Chinese letter arrangement Refers to the forms assumed by groups of cells of certain bacteria (e.g. *Corynebacterium* spp) as seen in microscopical preparations; to the occidental eye, at least, these forms resemble Chinese characters or 'letters'.

chiniofon (see 8-HYDROXYQUINOLINE)

chitin A polysaccharide consisting of β-(1 → 4)-linked *N*-acetyl-D-glucosamine residues; it is abundant in nature e.g. in arthropod exoskeletons and in fungal CELL WALLS. Chitin is attacked by a number of marine and soil bacteria—particularly by certain species of the Actinomycetales. (see also POLYOXINS)

chitosan (*syn*: mycosin). A polymer of β-(1 → 4)-linked D-glucosamine—i.e. non-acetylated CHITIN. Chitosan occurs in the cell walls of certain fungi, e.g. *Mucor*, *Zygorhynchus*.

Chlamydia (*synonyms*: *Bedsonia*; *Miyagawanella*; TRIC agents) A genus of gram-negative bacteria formally placed within the family Chlamydiaceae of the single-genus order Chlamydiales; species are obligate intracellular parasites or pathogens in a variety of vertebrates (including man and other mammals) and have been (infrequently) isolated from arthropods.

The *infectious* form of the organism (the *elementary body*) is a non-motile coccoid cell, less than 0.4 microns in diameter, which has a rigid, bacterial-type cell wall, and which stains well by the Giemsa or Giménez methods. Within the host cell the elementary body enlarges to form the (*non-infectious*) thin-walled, coccoid *initial body* (about 1–1.5 microns in diameter) which divides repeatedly (by binary fission) to produce a *microcolony* within the host cell's cytoplasm. The daughter cells subsequently differentiate to become the smaller, thick-walled infectious elementary bodies which, when the host cell ruptures, may infect fresh cells. This reproductive cycle does not occur extracellularly—although limited metabolism has been detected in cell-free systems. Reproduction is inhibited by tetracyclines and other antibiotics. In the laboratory the organisms can be cultured e.g. in the yolk sac membrane of the hen's egg. Type species: *C. trachomatis*; GC% for strains studied: 39–45.

C. trachomatis (formerly e.g. *Chlamydozoon trachomatis*). The causal agent of inclusion conjunctivitis, lymphogranuloma venereum, TRACHOMA and urethritis etc. in man. Microcolonies of *C. trachomatis* stain with iodine. Yolk sac culture is inhibited by sodium sulphadiazine (1mg/embryo).

C. psittaci (formerly e.g. *Chlamydozoon psittaci*). The causal agent of PSITTACOSIS, chlamydial abortion in sheep, enteritis in cattle etc. Microcolonies of *C. psittaci* do not stain with iodine, and yolk sac culture is not inhibited by sulphadiazine.

Chlamydomonas A genus of unicellular algae of the division CHLOROPHYTA; species occur in a wide range of freshwater environments and in soil. The uninucleate, haploid vegetative cell is pyriform, ovoid, or spherical, and is of the order of 25 microns in length/diameter; it has a cellulose-containing cell wall which may be surrounded by a layer of mucilage. Normally, two similar flagella project from the anterior end of the cell. The single, large CHLOROPLAST is generally cup-shaped but may be e.g. stellate or laminate—according to species; the chloroplast contains an EYESPOT and generally contains one or more pyrenoids. Two contractile vacuoles typically occur at the anterior end of the cell.

Asexual reproduction occurs by endodyogeny or endopolygeny: a cell comes to rest, the flagella are lost, and the protoplast undergoes one or more longitudinal divisions; the two, four, or eight daughters so formed develop cell walls and flagella and are released by gelatinization or rupture of the parental wall. (In some cases flagella do not develop and the cells remain enveloped in a gelatinous matrix for several subsequent generations, giving rise to a *palmelloid phase*. This stage is usually temporary and depends on the environmental conditions; under suitable conditions cells can develop flagella and escape from the matrix.) *Sexual reproduction* may be isogamous, anisogamous, or oogamous—according to species. In the isogamous process cells of appropriate mating type form pairs—becoming attached at their anterior ends; pairing is followed by plasmogamy and karyogamy. The zygote develops a thick wall and undergoes a period of dormancy; on germination, meiosis gives rise to four motile cells which are released by dissolution of the zygote wall.

chlamydospores Thick-walled, typically spherical or ovoid resting spores produced (asexually) by certain types of fungi from cells of the somatic hyphae. (*Chlamydospore* is also used to refer specifically to the *teliospore* of the smut fungi—see SMUTS (1).)

chloramines (as antimicrobial agents) (see CHLORINE)

chloramphenicol (chloromycetin) An ANTIBIOTIC formerly obtained from *Streptomyces venezuelae*, but now produced synthetically; it is bacteriostatic, and its spectrum of activity is similar to that of the TETRACYCLINES. Chloramphenicol inhibits PROTEIN SYNTHESIS by binding to the 50S subunit of 70S-type RIBOSOMES; it inhibits *peptidyl transferase* and thus prevents peptide bond formation. *Resistance to chloramphenicol.* Certain PLASMIDS code for *chloramphenicol transacetylase* which acetylates the antibiotic (at the expense of acetyl-CoA) to form 3-acetoxychloramphenicol—which has no antibiotic activity. In a mutant strain of a species of *Bacillus* the ribosomes fail to bind chloramphenicol.

Chloramphenicol antagonises those antibiotics which are active only against actively growing and dividing cells—e.g. the PENICILLINS.

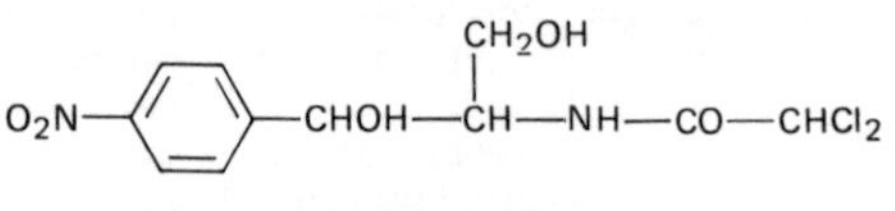

Chloramphenicol

chloranil (tetrachloro-*p*-benzoquinone) An agricultural antifungal agent used for the prevention of DAMPING-OFF, seed rots etc, and as a seed dressing (particularly for the seeds of legumes). Chloranil decomposes in the presence of light.

Chlorella A genus of non-motile, unicellular algae of the division CHLOROPHYTA; species occur in freshwater environments, on soil, tree-trunks etc., and in symbiotic association with other organisms (e.g. as zoochlorellae in *Sten-*

tor). The spherical, green (non-flagellate) cell may reach 10 microns in diameter; it has a cellulose-containing cell wall and contains a single haploid nucleus and a large (usually cup-shaped) chloroplast—which may or may not contain a pyrenoid. Reproduction is asexual: the protoplast of a cell divides to form two, four, or eight daughter protoplasts; these develop cell walls and are released on rupture of the parental wall. (The daughter cells resemble the parent cell and are referred to as *autospores*.) Sexual processes are unknown.

chlorguanide (paludrine; paludrin; proguanil) N^1-*p*-chlorophenyl-N^5-isopropyldiguanide. An antimalarial drug which, itself, has low activity but which is converted *in vivo* to a highly active cyclic form (a substituted dihydrotriazine). Chlorguanide is used mainly as a prophylactic; resistance to the drug develops readily. (see also PYRIMETHAMINE)

chlorhexidine (*proprietary name*:hibitane) 1,6-Di-(4′-chlorophenyldiguanido)-hexane. An ANTISEPTIC used e.g. in the treatment of burns and other wounds, and in urology and gynaecology. It is bactericidal for a wide range of gram-positive and gram-negative species, maximum activity occurring at high pH (e.g. pH 8); under acid conditions (e.g. pH 5) activity may be insignificant. It has been reported to be sporicidal at relatively high concentrations. Chlorhexidine is believed to function primarily by disrupting the bacterial CELL MEMBRANE. Chlorhexidine may be used in aqueous or alcoholic solution—sometimes in conjunction with a wetting agent such as *cetrimide*; since its water-solubility is low, a more soluble derivative of chlorhexidine may be used. Chlorhexidine remains active in the presence of organic matter and is relatively non-toxic.

chlorin Dihydroporphyrin. (see also PORPHYRINS and HAEM)

chlorine and compounds (as antimicrobial agents) Chlorine is an effective antimicrobial agent which is widely used e.g. in the disinfection of WATER SUPPLIES. Chlorine reacts with water to form *hypochlorous acid*, HOCl, ($Cl_2 + H_2O \rightleftharpoons HCl + HOCl$). Hypochlorous acid and its salts (*hypochlorites*) are effective microbicides—particularly at neutral or acid pH. The antimicrobial action of hypochlorites is believed to depend mainly on the liberation of nascent oxygen ($HOCl \rightarrow HCl + O$) which oxidizes cell constituents; in addition, chlorine may react directly with a variety of groups found in cell constituents—e.g. sulphhydryl groups in enzymes. Sodium hypochlorite is used in a range of household and hospital DISINFECTANTS and bleaches and is effective, in appropriate concentrations, against most microorganisms—including many types of spore and virus. Aerosols of hypochlorite solution have been used for the disinfection of air. *Chloramines* (which contain the group $\rangle$N-Cl) decompose slowly to release chlorine; thus, the antimicrobial activity of chloramines is much less rapid than that of the hypochlorites. *Monochloramine* (NH_2Cl) is formed by the reaction of hypochlorous acid with ammonia; such inorganic chloramines are used in the treatment of water supplies. Organic chloramines tend to be more stable and to release chlorine over longer periods of time; they are also less toxic and less irritating to animal tissues and are sometimes used for the disinfection of wounds. Examples of organic chloramines used as ANTISEPTICS and/or DISINFECTANTS include: *Chloramine T* (sodium *p*-toluenesulphonchloramide)—a white, crystalline, water-soluble solid; *Dichloramine T* (*p*-toluenesulphondichloramide)—water-insoluble and less rapidly effective than Chloramine T; *Halazone* (*p*-carboxy-*N*,*N*-dichlorobenzenesulphonamide)—used, in tablet form, for the disinfection of small quantities of drinking water.

The antimicrobial action of chlorine (and its compounds) is decreased by organic matter and by other compounds with which it can react. (see also IODINE; BROMINE; FLUORINE)

Chlorobacteriaceae (see CHLOROBIINEAE)

Chlorobiaceae A family of photosynthetic bacteria (suborder CHLOROBIINEAE) in which the *bacterio*CHLOROPHYLLS are located in CHLOROBIUM VESICLES; species occur in anaerobic sulphide-rich environments. The organisms are obligate anaerobic phototrophs which, according to species or strain, may be lithotrophic or mixotrophic*; some species require vitamin B_{12}. Sulphide or sulphur may be used as photosynthetic electron donor; when sulphur is deposited it accumulates *extra*cellularly. Most species are non-motile, but *Chloropseudomonas* spp are polarly flagellate. GAS VACUOLES are found in *Pelodictyon* and *Clathrochloris*. Type genus: CHLOROBIUM. *i.e. able to use organic carbon sources concurrently with inorganic electron donors.

Chlorobiineae (Chlorobacteriaceae; green sulphur bacteria) A suborder of the RHODOSPIRILLALES; species contain *bacterio*CHLOROPHYLLS *c* or *d* and relatively small amounts of bacteriochlorophyll *a*—these pigments occuring in CHLOROBIUM VESICLES. All species are placed in the single family CHLOROBIACEAE.

Chlorobium A genus of gram-negative photosynthetic bacteria of the CHLOROBIACEAE; species occur in anaerobic environments rich in sulphide. The cells are non-motile rods, cocci or vibrioids (according to species) which are 1 micron or less in length;

they divide by binary fission. Vitamin B_{12} is a growth requirement for some species. Nitrogen-fixation is unknown among *Chlorobium* spp.

chlorobium chlorophylls (see CHLOROPHYLLS)

chlorobium vesicles Elongated, membranous vesicles which underlie the cell membrane in members of the *green* photosynthetic bacteria (Chlorobiineae); the structure of the vesicular membrane differs from that of the cell membrane. (cf. CHROMATOPHORES) Associated with the vesicles are the bacterioCHLOROPHYLLS, CAROTENOIDS and electron carriers involved in PHOTOSYNTHESIS.

chlorogenic acid 1,3,4,5-tetrahydroxycyclohexane-1-carboxylic acid (quinic acid) esterified at the 3-hydroxy position with CAFFEIC ACID; an antifungal metabolite found in certain higher plants. Chlorogenic acid has been implicated in the resistance shown by certain plants to particular diseases, e.g. the resistance of certain strains of apple to scab (*Venturia inaequalis*) or of potato to common scab (*Streptomyces scabies*). Since chlorogenic acid tends to accumulate around damaged tissues at sites of infection it is sometimes regarded as a PHYTOALEXIN.

chloromycetin (see CHLORAMPHENICOL)

chloronitrobenzenes (as antifungal agents) These include a group of useful agricultural antifungal agents: DICLORAN, QUINTOZENE and TECNAZENE. (see also DINITROPHENOLS and FUNGICIDES AND FUNGISTATS)

chlorophycean starch (see STARCH)

chlorophylls A class of pigments involved in PHOTOSYNTHESIS; they may be regarded as derivatives of protoporphyrin IX (see PORPHYRINS) complexed with magnesium (see formula). Chlorophyll *a* occurs in all organisms which evolve oxygen during photosynthesis. Chlorophyll *b* occurs in most higher plants and in species of the Chlorophyta and Euglenophyta. Chlorophylls c_1 and c_2 occur in many algae e.g. diatoms and brown algae. Chlorophyll *d* occurs in some red algae. Chlorophyll *e* has been detected in two species of the Xanthophyta; it may be an artefact formed from another chlorophyll.

Bacteriochlorophyll a is found in members of the Rhodospirillineae and Chlorobiineae. Bacteriochlorophyll *b* occurs in *Rhodopseudomonas viridis* and *Thiocapsa pfennigii*. In addition to bacteriochlorophyll *a*, members of the Chlorobiineae contain a bacteriochlorophyll of the *c* series (= *chlorobium chlorophyll 660* series) *or* of the *d* series (= *chlorobium chlorophyll 650* series). Bacteriochlorophyll *e* has recently been identified as the major bacteriochlorophyll component of *Chlorobium phaeobacteroides* and *C. phaeovibrioides*; it occurs together with small amounts of bacteriochlorophyll *a*.

In the cell, chlorophylls occur in specialized organelles, e.g. see CHLOROPLAST, CHROMATOPHORE, THYLAKOID. Purified

Chlorophylls

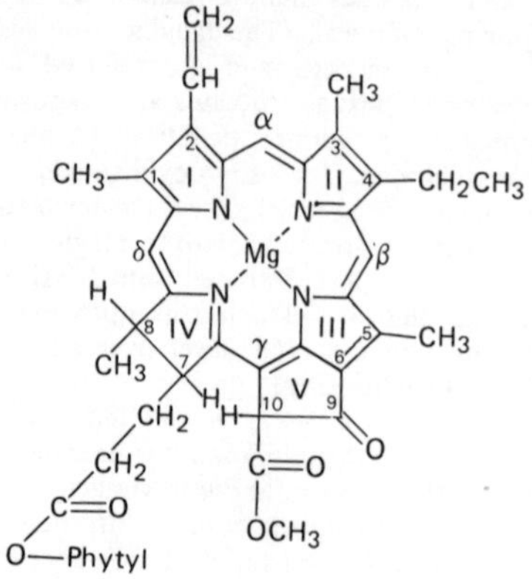

Chlorophyll a. (Phytyl = $C_{20}H_{39}$)

Other chlorophylls:

Chlorophyll	Differs from chlorophyll *a*: *Position*	*Substituent*
b	3	–CHO
c_1	7	–CH=CH.COOH
	7,8	double bond
c_2	7	–CH=CH.COOH
	7,8	double bond
	4	–CH=CH_2
d	2	–CHO
Bchl**a*	2	–CO.CH_3
	3,4	No double bond (H at C3 and C4)
Bchl *b*	2	–CO.CH_3
	4	=CH.CH_3
	3,4	No double bond (H at C3)
Bchls *c*	2	–CHOH.CH_3
	4	–C_2H_5, –C_3H_7, –C_4H_9(*iso*)
	5	–C_2H_5
	10	H replaces –CO.OCH_3
	Phytyl replaced by farnesyl ($C_{15}H_{25}$)	
	δ-methene	–CH_3
Bchls *d*	As for Bchls *c*, but δ-methene is unsubstituted.	
Bchls e	As for Bchls *c*, with 3	–CHO

* Bacteriochlorophyll.

chlorophylls are unstable in the presence of light and oxygen.

Chlorophyta (green algae) A division of the ALGAE. Species of the Chlorophyta occur in freshwater, marine, and terrestrial environments; of the terrestrial species, some are epiphytic, some form the phycobionts of lichens, and some occur as endosymbionts (zoochlorellae) in protozoans. The green algae include unicellular, coenobial, filamentous, siphonaceous, and parenchymatous organisms (e.g. *Chlorella*, *Scenedesmus*, *Spirogyra*, *Bryopsis*, and *Ulva* respectively); those vegetative cells (and gametes) which are motile characteristically have two or four apical isokont 'whiplash' flagella. All species contain CHLOROPHYLLS *a* and *b*, β-carotene (and α-carotene in some), and several xanthophylls (see CAROTENOIDS); the intracellular storage product is *chlorophycean* STARCH. Sexual processes (isogamous, anisogamous, or oogamous) occur in most species, and some species (e.g. *Ulva*) exhibit an ALTERNATION OF GENERATIONS; sexual reproduction is unknown in some species—e.g. *Chlorella* spp, *Pleurococcus* spp. In most groups of green algae the vegetative cells are haploid; all siphonaceous species appear to be diploid. Genera include: ACETABULARIA, *Bryopsis*, *Caulerpa*, CHLAMYDOMONAS, CHLORELLA, CLADOPHORA, *Halimeda*, *Penicillus*, SCENEDESMUS, SPIROGYRA, TREBOUXIA, *Udotea*, ULVA, and VOLVOX. (see also DESMIDS)

chloropicrin Trichloronitromethane, $Cl_3C(NO_3)$; a soil fumigant with fungicidal as well as herbicidal and insecticidal properties. Chloropicrin is injected into the soil.

chloroplast A semi-autonomous PLASTID—one or more of which occur (per cell) in *eucaryotic* photosynthetic organisms; the chloroplast is the site of PHOTOSYNTHESIS. *Algal* chloroplasts may be cup-shaped, discoid, lobate, spiral, stellate, or irregularly-shaped—according e.g. to species; they are commonly 2–20 microns in size.

Structure of the chloroplast. The chloroplast consists basically of a matrix (*stroma*) enclosed by a double membrane* (chloroplast envelope); within the stroma is a system of flattened, membranous sacs (the THYLAKOIDS) which may occur in one or more stacks (*grana*—singular: *granum*). (The arrangement of thylakoids in algal chloroplasts appears to have taxonomic significance.) The thylakoid membrane contains the CHLOROPHYLLS and electron carriers associated with the *light reaction* (see PHOTOSYNTHESIS); it appears to consist of, or to carry, a system of repeated particulate subunits which have been termed *quantasomes*. Associated with the *outer* surface of the thylakoid membrane are the chloroplast coupling factors ('ATPases') which are involved in photophosphorylation (cf. MITOCHONDRION). The *stroma* contains e.g. the enzymes of the Calvin cycle, granules of storage compounds (e.g. STARCH—see also PYRENOID), and the genophore(s). (see also EYESPOT and PLASTOGLOBULI)

Origin and semi-autonomy of the chloroplast. Chloroplasts are believed to arise either by the division of existing chloroplasts, or by the development of relatively-undifferentiated precursor organelles—the *proplastids*—which lack both the internal membrane system (thylakoids) and the specialized pigments and electron carriers; it has been suggested that thylakoids develop from the inner membrane of the proplastid envelope. Biochemical and genetic studies indicate that some chloroplast components are coded for by 'cytoplasmic' (chloroplast) genes, while the synthesis of others is directed by the nuclear genes; some chloroplast components may be synthesized in the cytoplasm while others may be synthesized *in situ*. Little is currently known of the function of chloroplast genes—or of the way in which the coordinated direction of nuclear and chloroplast genes may be involved in the synthesis of chloroplast components. Chloroplast DNA differs from nuclear DNA e.g. in buoyant density and in that it is not associated with histones; many copies of the 'genome'—which appears to be circular—may exist in a given chloroplast. The chloroplast is equipped with all the components necessary for PROTEIN SYNTHESIS, although chloroplast and cytoplasmic protein synthesis differ in certain respects—the former resembling more closely that which occurs in bacteria. Thus, e.g. chloroplast ribosomes resemble 70S bacterial ribosomes; in chloroplasts *N*-formylmethionine is the first amino acid to be incorporated in the polypeptide chain; and chloroplast protein synthesis is inhibited by certain antibiotics which inhibit bacterial protein synthesis—e.g. CHLORAMPHENICOL—to which cytoplasmic protein synthesis is insensitive. It has been suggested that endosymbiotic blue-green algae may have been the evolutionary antecedents of the present-day chloroplast. *A *triple* membrane has been detected in members of the Dinophyta and Euglenophyta.

Chloropseudomonas A genus of the CHLOROBIACEAE.

chloroquine An antimalarial AMINOQUINOLINE.

chlorosis (*plant pathol.*) The yellowing of (normally green) leaves and/or plant components. Chlorosis is symptomatic of many plant diseases of microbial causation, but may be due to iron deficiency, lack of light, etc.

chlorphenol red (chlorophenol red) (PH INDICATOR) pH 4.8 (yellow) to pH 6.4 (red). pK_a 6.0.

chlortetracycline (aureomycin) 7-Chlorotetracycline: an ANTIBIOTIC obtained from *Streptomyces aureofaciens* (see TETRACYCLINES).

Choanephora A genus of fungi of the ZYGOMYCETES.

choanomastigote A form assumed by *Crithidia* spp; typically, the choanomastigote is somewhat pear-shaped with a flagellum emerging from the base of a depression at the narrow (anterior) end of the cell.

chocolate agar Blood agar which has been heated (either in a molten or set condition) until the colour becomes chocolate-brown. (Temperatures of about 70–80 °C are satisfactory.) Chocolate agar is used for the isolation and culture of certain fastidious organisms, e.g. *Neisseria gonorrhoeae*, *Haemophilus* spp. (see also X FACTOR)

choke (*plant pathol.*) A disease which affects members of the Gramineae; the causal agent is the ascomycete *Epichloë typhina*—see EPICHLOË.

cholera (Asiatic cholera) In man, an acute, infectious disease caused by the *cholerae* or *el tor* biotype of *Vibrio cholerae*. Infection occurs orally; factors such as stomach acidity, and the number of organisms ingested, appear to determine the severity (or occurrence) of the disease. Pathogenesis is related to the production, by *V. cholerae*, of a protein ENTEROTOXIN. Symptoms include nausea, vomiting, abdominal cramps, and the frequent passage of *rice-water stools*—watery, mucus-containing stools which may be flecked with blood; dehydration occurs and there is a reduction in body electrolytes. Treatment involves intravenous replacement of fluid and electrolytes; certain antibiotics, e.g. TETRACYCLINES, have been used. *Laboratory diagnosis*: examination of stools by microscopy (including fluorescent-antibody methods) and cultural techniques.

cholera red test A test formerly used to distinguish between cholera vibrios and similar organisms (see VIBRIO). The unknown organism is grown in peptone water containing 0.1% KNO_3 or $NaNO_3$; to the culture is added one or two drops of concentrated sulphuric acid. The formation of a red nitrosoindole compound constitutes a *positive* reaction—given by any organism which forms both indole and nitrite under these conditions.

choleragen The cholera ENTEROTOXIN.

chondrioid Synonym of MESOSOME.

chondriome A composite intracellular structure comprising several organelles—one being the mitochondrion; in e.g. *Trypanosoma* spp the chondriome consists of a KINETOPLAST located within a diffuse mitochondrion.

chondriosome A synonym of MITOCHONDRION.

Chondromyces A genus of gram-negative bacteria of the MYXOBACTERALES; the vegetative cells are non-fusiform, about 1–1.4 × 3–14 microns. Fruiting bodies consist of *sporangia*—singly or in chains, clusters etc.—on branched or unbranched *stalks*; the *myxospores* are similar to vegetative cells. Microcysts are not formed.

Chondrus A genus of marine algae of the division RHODOPHYTA; species occur in intertidal and subtidal zones. The organism consists of dichotomously-branching fronds which are attached to rocks by a holdfast; the form tends to be variable. *Chondrus crispus* (Irish moss, carragheen moss) is used commercially as a source of CARRAGEENAN and—in some countries—as a food.

Chonotrichida An order of protozoans of the subclass HOLOTRICHIA (subphylum CILIOPHORA). *Chonotrichs* occur in freshwater and marine environments; they are sessile and are most commonly found attached to crustaceans. Individual cells are vase-shaped, i.e. the body, which is narrow at the point of attachment, broadens out before constricting and finally becoming flared and convoluted at the distal end. Somatic cilia are lacking in the mature cell, though cilia are present within the *peristome*—the funnel-shaped distal part of the cell which leads to the vestibulum and the cytostome. Reproduction occurs by budding, the bud developing into a ciliated, motile larva. Genera include *Chilodochona* (marine) and *Spirochona* (freshwater).

chorismate An intermediate formed in the biosynthesis of aromatic amino acids—see Appendix IV(f).

Christensen's agar (see UREASES)

Chromatiaceae (Thiorhodaceae; purple sulphur bacteria) A family of photosynthetic bacteria of the RHODOSPIRILLINEAE. Species occur in sulphide-rich aquatic and terrestrial environments (e.g. the SAPROPEL); they grow photolithotrophically under strict anaerobiosis* although many (or all) species are believed to be capable of photoassimilating various organic compounds. Elemental sulphur, *which can be oxidized to sulphate*, may be deposited intracellularly or, in *Ectothiorhodospira*, extracellularly. Genera include *Thiocapsa*, CHROMATIUM, and THIOSPIRILLUM. **Thiocapsa roseopersicina* can grow in the dark under microaerophilic conditions.

chromatid Either of the two fibrils formed from a (eucaryotic) CHROMOSOME when the latter replicates prior to MEIOSIS or MITOSIS. *Sister* chromatids are derived from the replication of a single chromosome. A chromatid is usually assumed to contain a single DNA duplex—although there is evidence to suggest that it may be multistranded.

chromatin The material which constitutes a CHROMOSOME.

chromatinic body (see NUCLEUS)

Chromatium A genus of gram-negative, non-sporing, non-motile or motile (lophotrichous) photosynthetic bacteria of the CHROMATIACEAE; species are found e.g. in contaminated pools, freshwater and marine muds. The cells are rod-shaped, ovoid or reniform, and may be large (e.g. 15 microns, long axis) or small (e.g. 2–5 microns, long axis); they divide by binary fission. Cells contain granules of elemental sulphur; storage materials include poly-β-hydroxybutyrate. The large-celled species often require vitamin B_{12}.

chromatoid bodies Nucleoprotein bodies (commonly rod-shaped) which occur within the CYSTS of certain amoebae (see ENTAMOEBA). Chromatoid bodies stain well with iron-haematoxylin; they do not stain with iodine. One, two or more bodies (per cyst) may be visible. The nucleoprotein tends to disperse in the cytoplasm as the cyst ages.

chromatophore (1) A system of vesicular, tubular or lamellate membranous structures which occur in members of the RHODOSPIRILLINEAE. (Some authors refer to these membranes as THYLAKOIDS—a term commonly used for the functionally-analogous membranes in algae, blue-green algae and higher plants.) The chromatophore membrane is continuous with the cell membrane; it contains bacterioCHLOROPHYLL, CAROTENOIDS, and the electron carriers involved in PHOTOSYNTHESIS. (see also CHLOROBIUM VESICLE) (2) One of a number of particles (some 500Å in diameter) obtained (experimentally) by the disruption of photosynthetic bacteria; the particles contain photosynthetic pigments and protein and can be used for *in vitro* photophosphorylation. (3) An algal CHLOROPLAST. (4) Synonym of CHROMOPLAST.

Chromobacterium A genus of gram-negative, aerobic or facultatively anaerobic, chemoorganotrophic, asporogenous bacteria of uncertain taxonomic affiliation; species occur in soil and water. The cells, which are about 0.5–1 × 1–5 microns, are polarly (or polarly and laterally) flagellate; they produce a violet, ethanol-soluble pigment, VIOLACEIN. Most strains are oxidase-positive. Type species: *C. violaceum*; GC% range for the genus: 63–72.

C. violaceum grows at 37 °C but not at 4 °C; it has been implicated in pyogenic and septicaemic infections in man.

C. lividum grows at 4 °C but not at 37 °C.

chromophore That part of a dye molecule which is responsible for the colour of the dye.

chromoplast (*syn*: chromatophore) Any pigmented PLASTID—e.g. an algal CHLOROPLAST or a CAROTENOID-rich plastid found in ripening fruit. (cf. LEUCOPLAST)

chromosome (see CHROMOSOMES)

chromosome aberration A change in the number or structure of chromosomes in a cell; chromosome aberration may lead to (heritable) changes in the characteristics of the organism. Alterations in the *number* of chromosomes may involve: (a) Changes in PLOIDY. (b) The loss or gain of one or more individual chromosomes—e.g. a *monosomic* cell has the diploid number of chromosomes *minus one*; a *trisomic* cell has the diploid number of chromosomes *plus one*. *Structural* changes in chromosomes include: (a) *Deletions* of segments from one or more chromosomes. (b) *Inversion* of one or more chromosomal segments. (In effect, a segment is excised, rotated through 180°, and re-inserted at the same site.) (c) *Translocation* of a chromosomal segment from one site to another on the same or a different chromosome. (d) *Duplication* of segments of one or more chromosomes.

chromosome mutation (1) Synonym of CHROMOSOME ABERRATION. (2) Any *structural* alteration in a chromosome.

chromosomes Structures which contain the nuclear DNA of a cell. (see also NUCLEUS and DNA) Chromosomes may be observed by light microscopy following appropriate staining—see e.g. FEULGEN REACTION. (a) *Bacterial chromosomes*. The chromosomes of several bacteria (e.g. *Escherichia coli*, *Bacillus subtilis*) have been shown to consist of a single, circular DNA duplex. In *E. coli* the DNA duplex is approximately 1.2 mm long; within the cell it is extensively folded to form the *nucleoid body*. On treatment with RNase the structure unfolds; a model has been proposed in which the DNA is folded into approximately 50 loops (each highly coiled) maintained by a central core of RNA. The chromosome appears to be attached to a MESOSOME; such attachment may be important in the segregation of daughter chromosomes during cell division. (see GROWTH (bacterial); see also CONJUGATION, TRANSDUCTION, TRANSFORMATION, LYSOGENY, and PLASMIDS) (b) *Eucaryotic chromosomes*. Eucaryotic cells generally have a characteristic number of chromosomes per cell (see HAPLOID, DIPLOID, POLYPLOID, ANEUPLOID). The early interphase chromosome is believed to contain a single DNA duplex which is frequently complexed with HISTONES. Non-basic proteins are associated with this nucleoprotein fibril; these proteins are quantitatively and qualitatively highly variable and may include enzymes—e.g. RNA polymerase, DNA polymerase. Chromosomes often contain small quantities of RNA—the significance of which is uncertain; it may be newly-transcribed RNA. The nucleoprotein fibrils must be extensively folded to form the chromosomes, although details of the manner and mechanism of folding are unknown. (see also MITOSIS and MEIOSIS)

Chromulina A genus of algae of the CHRYSOPHYTA.

chronic (*adjective*) (of diseases) Any disease which persists for a relatively long period of time (e.g. months, years) terminating either in recovery or death. Examples of such diseases include LEPROSY, LUMPY JAW and TUBERCULOSIS. (cf. ACUTE)

Chroococcus A genus of unicellular (colonial) BLUE-GREEN ALGAE.

Chrysochromulina A genus of algae of the HAPTOPHYTA.

chrysolaminarin (leucosin; chrysose) A reserve polysaccharide synthesized by members of the Chrysophyta and Bacillariophyta. Chrysolaminarin is a branched D-glucan with β-(1 → 3) and β-(1 → 6) links.

Chrysophyta (golden algae) A division of the ALGAE. Species of the Chrysophyta are predominantly freshwater algae; the division includes filamentous and colonial organisms but most species are unicellular. Motile species bear one or two apical flagella—a single *tinsel* FLAGELLUM, or one tinsel and one whiplash flagellum; in biflagellate species the flagella may be isokont or heterokont. Non-flagellate, amoeboid species occur. Some species bear silicaceous scales (see CELL WALL), and the ability to form endogenous silicaceous cysts is characteristic. All species contain CHLOROPHYLLS *a* and *c*, β-carotene, and fucoxanthin (see CAROTENOIDS); intracellular storage products include CHRYSOLAMINARIN. Sexual reproduction is isogamous in those species in which it occurs. Genera include *Chromulina*, *Mallomonas*, *Ochromonas*, and *Synura*.

chrysose (see CHRYSOLAMINARIN)

chytrid A member of the CHYTRIDIALES.

Chytridiales (the 'chytrids') An order of FUNGI within the class CHYTRIDIOMYCETES; constituent species do not form true hyphae (although some species form a 'rhizomycelium'—see NOWAKOWSKIELLA) and sexual reproduction (where known) may involve e.g. the fusion of motile isogametes or gametangial copulation (cf. BLASTOCLADIALES and MONOBLEPHARIDALES). The chytrids are typically aquatic saprophytes or parasites—the latter growing in or on e.g. algae and other fungi; of the terrestrial species some are parasitic on the underground and/or aerial parts of higher plants. (Certain chytrids cause important diseases of the potato and other cultivated plants.) The various species of chytrid—all *microfungi*—exhibit a considerable morphological range; constituent genera include (in the order 'primitive' to 'advanced') OLPIDIUM, SYNCHYTRIUM, RHIZOPHYDIUM, and NOWAKOWSKIELLA.

Chytridiomycetes A class of FUNGI of the subdivision MASTIGOMYCOTINA; constituent species are distinguished from those of other classes in that they give rise to zoospores each of which possesses a single posteriorly-directed whiplash flagellum. The chytridiomycetes are typically aquatic organisms—though some live in the soil, and a few are important pathogens of higher plants. According to species the form of the vegetative organism ranges from a simple, sac-like, holocarpic thallus to a well-developed, branched mycelium; chitin has been reported to occur in the cell wall in many species. Asexually-derived zoospores develop within closed, sac-like *sporangia*—see SPORANGIUM (*mycol.*); sexual reproduction typically involves the fusion of motile gametes—though oogamy occurs in the members of one order (Monoblepharidales) and in some of the species of another order (Chytridiales). The orders are BLASTOCLADIALES, CHYTRIDIALES, and MONOBLEPHARIDALES.

Cidal (*proprietary name*) An antiseptic soap containing *hexachlorophene* (see BISPHENOLS).

cider A beverage made by the fermentation of apple juice. *Cider apples* are characterized by a high content of TANNIN (contributing astringency) and malic acid (contributing sharpness). Fermentation may be carried out by the natural apple flora—which may include bacteria, yeasts (e.g. *Candida*, *Pichia*, *Kloeckera*, *Saccharomyces*) and other fungi (e.g. *Aspergillus*, *Penicillium*). *Kloeckera* spp begin fermentation, suppressing both *Candida* and *Pichia*; later, species of *Saccharomyces* predominate. In modern processes the natural flora may be suppressed (with sulphur dioxide) and the juice inoculated with cider yeasts (e.g. *Saccharomyces uvarum*); yeasts present on the milling and pressing equipment form an indirect inoculum.

cider sickness The tainting of cider due to the production of acetaldehyde by the bacterial contaminant *Zymomonas anaerobia*.

cidex (*proprietary name*) A disinfectant: 2% GLUTARALDEHYDE.

cilia (see CILIUM)

Ciliatea A class of ciliate protozoans which includes all species within the subphylum CILIOPHORA; it comprises the subclasses HOLOTRICHIA, PERITRICHIA, SUCTORIA and SPIROTRICHIA.

ciliates (*microbiol*) Members of the CILIOPHORA.

Ciliophora ('the ciliates') A subphylum of the PROTOZOA; the organisms are ciliated (see CILIUM) during at least some stage in their life cycles. (Cilia are not present in *mature* organisms of the Suctoria, and certain ciliate protozoans—the OPALINATA—are not classified with the Ciliophora.) Another feature typical of ciliates is the possession of two types of nucleus—the MICRONUCLEUS and the MACRONUCLEUS.

Many of the 6000 or so ciliate species are freeliving freshwater or marine organisms; some freshwater types are important in SEWAGE TREATMENT. Pathogenic species include *Balantidium coli*, the causal agent of an enteric disease in man, and *Ichthyophthirius multifiliis*, an ectoparasite of freshwater fish. Members of the Astomatida are parasitic in invertebrates, particularly annelids.

Ciliate species cover a wide morphological range, and cell sizes range from about 10 microns to several millimetres. The organisms generally have a substantial PELLICLE; in some, e.g. *Coleps*, the pellicular alveoli contain calcareous deposits which form a semi-rigid endoskeleton. Endoskeletal microtubular structures (see TRICHITES) occur in the holotrichs. The ciliature is often arranged in *kineties* which, on appropriate staining, exhibit the SILVER LINE SYSTEM. Other ciliate features include CONTRACTILE VACUOLES and TRICHOCYSTS; many ciliates form CYSTS. The ciliate is typically a colourless organism but some species (e.g. *Stentor*) take the colour of their algal endosymbionts (*zoochlorellae*). Most ciliates (not e.g. the Astomatida) possess a CYTOSTOME (cell mouth); in primitive species the cytostome is situated apically (polarly). Ciliates commonly feed holozoically—i.e. they ingest particulate food, e.g. algae, bacteria, or other protozoans. Astomes presumably absorb nutrients direct through the pellicle. (see also FOOD VACUOLES) Some species, e.g. *Paramecium*, *Tetrahymena*, are motile; others, e.g. *Carchesium*, *Vorticella*, are sessile.

Asexual reproduction occurs in all species—typically by transverse, perkinetal binary fission; multiple fission occurs in *Stephanopogon*. (see also CONJUGATION, SYNGEN, AUTOGAMY)

All members of the Ciliophora are included in the single class Ciliatea. Four subclasses are distinguished: HOLOTRICHIA (containing the most primitive ciliates), PERITRICHIA, SUCTORIA and SPIROTRICHIA.

cilium (*pl.* cilia) (1) (*protozool*) A thread-like appendage which arises from the cell in members of the Ciliophora and the Opalinata; cilia may occur in groups—often as specialized organelles—and/or as a surface layer covering a major or minor part of the cell. The structure of a cilium is similar to that of the *eucaryotic* FLAGELLUM, though cilia are usually shorter than flagella; each cilium terminates, proximally, in a BASAL BODY. Cilia are generally involved in either locomotion or feeding, sometimes both. The basic ciliary movement consists of the *effective* stroke—a rapid swing (in a relatively rigid state) through a wide angle (almost 180°)—followed by the slower *recovery* stroke—in which the cilium moves back in a relatively relaxed state to its original position; currents are thus generated in the ambient fluid. (see also METACHRONAL WAVES) The mechanism by which coordinated movement is achieved by numbers of cilia is unknown. Many varieties of ciliary movement are found among ciliated organisms, and some are able to change the direction of their ciliary beating. (see also AZM; MEMBRANELLE; UNDULATING MEMBRANE) (2) (*lichenol*) A vegetative appendage which structurally resembles a RHIZINA; cilia arise from the margins of the thallus or apothecium (or occasionally from the upper surface of the thallus) in certain LICHENS (e.g. *Anaptychia ciliaris*, *Physcia adscendens*). Cilia may be branched or unbranched, according to species.

cinchona alkaloids Antimalarial compounds obtained from the bark of the South American cinchona tree, e.g. QUININE, cinchonidine.

circadian rhythms Diurnal rhythmical changes which occur in an organism even when it is isolated from the natural diurnal fluctuations of the environment. Such innate rhythms—often referred to in terms such as the 'biological clock' etc.—have been observed in certain eucaryotic microorganisms; thus, for example, photosynthetic activity in the green alga *Acetabularia* has been shown to peak with 24-hour periodicity even when the organism is kept in continuous light. Circadian rhythms have been observed to continue for days, weeks, or longer, in other organisms isolated from their natural environments.

circularly-permuted genomes (see BACTERIOPHAGE T4)

circulative viruses (persistent viruses) (*plant pathol.*) Viruses which are transmitted, from plant to plant, *via* the haemocoel and salivary glands of the insect vector. Such transmission commonly involves a latent period during which the vector, having acquired virus from an infected plant, is unable to transmit the virus to a healthy plant. In this form of transmission—which is common in leafhoppers, and which occurs in some aphids—the ability of a virus-carrying insect to infect a healthy plant may be retained for days, weeks, or for an indefinite period. (cf. STYLET-BORNE VIRUSES; see also PROPAGATIVE VIRUSES)

circulins A group of polypeptide ANTIBIOTICS closely allied to the POLYMYXINS both in structure and activity. They are produced by *Bacillus circulans*.

cirrus (*pl. cirri*) (*ciliate protozool.*) A discrete group of *somatic* cilia which act as a unified locomotive organelle. Cirri are characteristic of the Hypotrichida (subclass Spirotrichia). (cf. MEMBRANELLE)

cis-trans test A procedure used to ascertain whether, in a particular haploid organism or cell, a given phenotypic characteristic is determined by a single gene or by a number of genes. The procedure comprises two separate tests: a *cis* test and a *trans* test; the *trans* test is a COMPLEMENTATION TEST, while the *cis* test provides a control. The terms *cis* and *trans* refer to the arrangement of mutations on two genomes being tested for complementation; in the *trans* arrangement one mutation occurs in each genome, while in the *cis* arrangement both mutations are located in one genome—the other genome being non-mutant (i.e. wild type). In both the *cis* and *trans* tests, two (haploid) genomes are brought together in the same organism or cell. In the *trans* test the two genomes each have one mutation, and both display the same mutant phenotype with respect to the characteristic being investigated; in the *cis* test one genome is doubly-mutant, the other wild type. (The doubly-mutant genome used in the *cis* test is constructed, by recombination, from two (haploid) cells: one cell has one of the two genomes present in the *trans* test, while the other cell has the other genome; thus, mutations in the *cis* arrangement occupy the same *loci* as those in the *trans* arrangement, but they occur on the same genome.)

The *cis-trans* test compares, qualitatively and/or quantitatively, the phenotype displayed in the *cis* test with that displayed in the *trans* test. If both tests display *identical* phenotypes the *cis-trans* test is *positive*; this indicates that the particular phenotypic characteristic being investigated is determined by more than one gene. If the two tests display different phenotypes—i.e. if the phenotype displayed in the *trans* test differs from the wild type more than does that displayed in the *cis* test—the test is *negative*; this indicates monogenic determination of the particular phenotypic characteristic being investigated.

cistron (1) A segment of genetic nucleic acid from which can be translated one specific polypeptide chain (see PROTEIN SYNTHESIS). The cistron has been defined in terms of COMPLEMENTATION: in a diploid cell (or merozygote etc.), either of two homologous regions of genetic nucleic acid in which two mutations in the *trans* configuration (see CIS-TRANS TEST) fail to exhibit *complete* complementation. (However, this definition does not take into account polar effects—see POLAR MUTATION.) Thus, the cistron can be described as the functional unit of genetic inheritance, i.e. a unit whose function is reduced, or lost completely, if it is not present *in its entirety*. (2) A synonym of GENE.

citrate test (citrate utilization test) A test used in the identification of certain bacteria, e.g. members of the *Enterobacteriaceae*. (see also IMVIC TESTS) The test determines the ability of an organism to use citrate as the sole source of carbon*. Media used for the test, e.g. *Koser's citrate medium*, contain, as essential constituents, citric acid or citrate, an ammonium salt (as a nitrogen source), sodium chloride and magnesium sulphate. A saline suspension of the test strain is prepared from growth on a solid medium—care being taken to avoid contaminating the saline with nutrient materials from the solid medium. The test medium is inoculated from the saline suspension by means of a STRAIGHT WIRE; such a small inoculum is used (a) to avoid or minimise carry-over of nutrients with the inoculum, and (b) so as not to produce *turbidity*—which is used as the criterion of growth following appropriate incubation. Incubation may be carried out at the optimum growth temperature (if known); a duplicate test may be put up at a non-optimal temperature. The medium is examined daily for evidence of growth, i.e. turbidity.

*For one mode of citrate utilization see TRICARBOXYLIC ACID CYCLE.

citric acid cycle (see TRICARBOXYLIC ACID CYCLE)

Citrobacter A genus of gram-negative, motile bacteria of the family ENTEROBACTERIACEAE; strains occur e.g. in sewage-polluted waters. The organisms give a positive CITRATE TEST and grow on deoxycholate-citrate agar and on media containing potassium cyanide; they give a postive METHYL RED TEST and a negative VOGES-PROSKAUER TEST. Acid and gas are formed from glucose and other sugars; lactose may or may not be fermented—or may be fermented slowly (by members of the 'Ballerup–Bethesda' group). Lysine decarboxylase is not formed. *Citrobacter* spp share O antigens with other members of the Enterobacteriaceae—including *Salmonella* spp; certain strains have a VI ANTIGEN.

citrovorum factor N^5-formyltetrahydroFOLIC ACID.

citrulline An amino acid formed e.g. as an intermediate in the biosynthesis of arginine—see Appendix IV(a).

Cladonia A large genus of LICHENS in which there is a granular or squamulose (often evanescent) *primary* (basal) thallus and erect, usually hollow *podetia* (see PODETIUM) which may bear apothecia. The phycobiont is *Trebouxia*. Species include e.g.:

C. chlorophaea. The primary thallus is grey-green and squamulose. Each podetium is of the order of 1 centimetre in height and is in the form of a funnel-shaped cup which expands from the base; the surface of the lower part is warty while that of the upper part bears granular soredia—particularly within the cup. Globular, brown, stalked or sessile apothecia may occur on the margin of the cup. *C. chlorophaea* occurs in damp situations among rocks, on rotting logs, soil etc.

C. digitata. The primary thallus is dominant and consists of large scale-like lobes; the upper surface is grey-green while the lower surface is white and sorediate. The podetia are of the order of 1 centimetre in height, sorediate, and either simple (columnar) or opening into broad, irregular cups. Stalked red apothecia may occur on the margins of the cups. *C. digitata* occurs e.g. on decaying logs and tree-stumps and occasionally on peaty soil.

C. rangiferina ('reindeer moss'). The primary thallus is evanescent. The podetia are white to grey, 4–8 centimetres in height, and extensively branched; the slender apices of the branches are curved—predominantly in one direction. There are no scyphi. The surface of the podetium has a loosely-felted appearance. The podetia may bear small, terminal, brown apothecia which become globular as they mature. *C. rangiferina* occurs e.g. on peat in mountainous regions.

C. uncialis. The primary thallus is rudimentary and evanescent. The podetia are yellowish-grey, up to 6 centimetres in height, sparsely-branched, and hollow with a swollen appearance; the tips are brown and are divided into 2–4 points. The surface of the podetium is smooth. The brown, apical apothecia are rarely formed. *C. uncialis* occurs commonly on heaths and moors.

Cladophora A genus of extensively-branched filamentous algae of the division CHLOROPHYTA; the filaments, which may reach several centimetres in length, are each composed of a chain of cylindrical, multinucleate cells. Species are widely distributed in freshwater and marine environments; they may be free-floating or anchored to the substratum by rhizoids. Asexual reproduction involves the formation of uninucleate, quadriflagellate zoospores. Sexual reproduction is isogamous—biflagellate gametes being formed. The organism shows an isomorphic ALTERNATION OF GENERATIONS.

Cladosporium A genus of fungi within the order MONILIALES; species include saprophytes and plant pathogens—e.g. *C. fulvum*, the causal agent of *leaf mould* of tomato. The vegetative form of the organisms is a septate mycelium; elongated, ellipsoidal, non-septate or uniseptate, dark-coloured conidia are produced in acropetally-developed chains on conidiophores which bear terminal clusters of sterigmata. The perfect stage of *C. resinae* (the CREOSOTE FUNGUS) is the ascomycete *Amorphotheca resina*.

clamp connections (clamps) (*mycol.*) In many basidiomycetes: hyphal structures which are formed during cell division in dikaryotic (heterokaryotic) hyphal cells (i.e. cells which each contain two nuclei of different mating type).

Prior to nuclear division a cell develops a short, angled, lateral branch—the incipient clamp connection—into which *one* of the nuclei migrates. Both nuclei then undergo mitotic division simultaneously (*conjugate division*)—the mitotic spindles being orientated so that one daughter nucleus from each of the two original nuclei subsequently occupies a position at one end of the dividing cell. The formation of a septum isolates the nucleus in the developing clamp, and a second septum delimits a binucleate (heterokaryotic) cell from an adjacent mononucleate cell. The incipient clamp now grows towards the mononucleate cell and fuses with it—thus forming the clamp connection. The outcome of the entire process is the formation of two adjacent binucleate (heterokaryotic) hyphal cells. The clamp connection persists as a hyphal bulge in the region of the septation; it is indicative of the dikaryophase—although many dikaryotic basidiomycetes do not have clamp connections. Cultures of *Coniophora cerebella* have been reported to form more than one clamp-like structure around an individual septation (whorled clamps); some of these structures bear short hyphal branches.

Medallions (medallion clamps) are formed when the terminal part only of the incipient clamp fuses with the parent hypha—a small space or eyelet remaining between the curved clamp and the parent hypha. Medallion clamps are formed by a number of wood-rotting fungi—e.g. species of *Lentinus*, *Lenzites*, *Poria*, and *Polyporus*.

Some mycologists believe that the basidiomycetes have evolved from the ascomycetes, and parallels have been drawn between clamp connections and the *croziers* formed during ASCUS development.

class A taxonomic group: see TAXONOMY and NOMENCLATURE.

Clathrulina A genus of protozoans: see HELIOZOANS.

clavate Club-shaped, i.e. thicker at one end.

Claviceps A genus of fungi of the order SPHAERIALES, class PYRENOMYCETES; species occur e.g. as parasites and pathogens of graminiferous hosts. (see also *ergot* under CEREALS, DISEASES OF)

C. purpurea is the causal agent of ergot diseases of cereals. The infection cycle begins when the ascospores of *C. purpurea* infect the ovaries of susceptible species; the ovarian tissue subsequently becomes covered by a conidiophore-bearing mycelial mat. The infection may be spread by insect-borne conidia. Later, the mycelial mat on a given ovary develops to form a hard, dark-purple or black *sclerotium* ('ergot') in the position normally occupied by the grain. The sclerotium—an elongated, often curved, body approximately one centimetre in length—is the overwintering stage. In the spring the sclerotium germinates and gives rise to one or more stalked, mushroom-shaped stromata—each of which is several millimetres in height. A number of perithecia develop within the surface layer of the stroma—their ostiolar pores level with the stromal surface. Each ascus contains eight long, thread-like ascospores.

clearing The process in which a dehydrated tissue section (or e.g. a preparation containing protozoans or algae) becomes permeated by a solvent (*clearing agent*) that is miscible with both the dehydrating agent (often ethanol) and the MOUNTANT; the clearing agent has a refractive index of such value that the specimen becomes optimally visible under the microscope. (Clearing is essential when the dehydrating agent and the mountant are not mutually miscible.) Clearing agents include terpineol, xylene, benzene.

cleistothecium (cleistocarp) A closed, hollow, typically spherical or spheroid ASCOCARP in which develop one or more *asci* (see ASCUS); at maturity the asci (or ascospores) are liberated on rupture of the cleistothecial wall. In some species the cleistothecial wall (*peridium*) is a more or less rigid pseudoparenchymatous structure; in other species the wall may be little more than an envelope of loosely-woven hyphae. The cleistothecia of e.g. *Erysiphe* spp are approximately 100–200 microns in diameter.

clindamycin 7-Chloro-7-deoxy-LINCOMYCIN.

Clitocybe A genus of fungi of the family TRICHOLOMATACEAE.

clonal selection theory A theory which attempts to account for the observed features of ANTIBODY-FORMATION. Essentially it supposes that each B LYMPHOCYTE is capable of responding to a *specific* antigen with which the body may or may not have had previous contact. The concept of *pre-existing specificity* is thus central to the theory. (cf. INSTRUCTIVE THEORIES OF ANTIBODY-FORMATION)

clone A population of cells derived *asexually* from a single cell. Such a population is often assumed to be genetically homogenous.

cloned line (see CELL LINE)

closed culture A closed-system culture: BATCH CULTURE.

Closterium A genus of algae—see DESMIDS.

Clostridium A genus of SPORE-forming, chemoorganotrophic, obligately anaerobic (or aerotolerant), pleomorphic rod-shaped bacteria of the family BACILLACEAE. (see also ANAEROBE) The organisms are *typically* gram-positive (often gram-negative in old cultures), motile (peritrichously flagellate) and catalase-negative. *Clostridium* spp are widespread in soil, muds etc., and in the intestinal tract of animals, including man. Diseases caused by *Clostridium* spp include botulism (see FOOD POISONING), GAS GANGRENE, TETANUS, BLACKLEG and BRAXY.

The cells are commonly 0.5–1.5 × 2–10 microns; CAPSULE-formation is uncommon. Many strains grow poorly on basal media (e.g. nutrient agar), and particular growth requirements (e.g. BIOTIN) are common. Metabolism is fermentative. Strains which digest cooked meat media, casein etc., and which grow well in nutrient broth (without carbohydrate supplement) are described as *proteolytic*. The growth of *saccharolytic* strains is improved by, or dependent on, the presence of a suitable carbohydrate; both acid and gas are usually formed when carbohydrates are fermented*. Nitrates are reduced by some species, but sulphates are not reduced (cf. DESULPHOTOMACULUM). Pathogenic species produce a range of toxins. On blood agar many strains exhibit haemolysis. SWARMING is common in some species. Pathogens of warm-blooded animals grow optimally at or about 37 °C.

Clostridium spp are commonly sensitive e.g. to PENICILLINS, certain MACROLIDE ANTIBIOTICS and TETRACYCLINES, but resistant to AMINOGLYCOSIDE ANTIBIOTICS (which are ineffective under anaerobic conditions). For methods of spore-inactivation see STERILIZATION.

Clostridium spp are identified e.g. by morphological characteristics, by biochemical and cultural reactions, by agglutination with specific antisera, and by determining—by means of NEUTRALIZATION TESTS—the identity of any toxins formed. Species include e.g.:

C. butyricum. The type species. The cells are

capsulated and motile. *C. butyricum* is saccharolytic, producing acid and gas from many sugars, e.g. glucose, maltose, lactose and sucrose. Gelatin is not hydrolysed, H_2S is not produced, and lecithinases are not formed. Non-pathogenic.

C. bifermentans. Motile, non-capsulated cells; an oval spore is formed subterminally. Saccharolytic and proteolytic. Gelatin is hydrolysed. Lecithinase (antigenically-similar to the α-toxin of *C. perfringens*) is produced. Indole-positive.

C. botulinum. The causal agent of *botulism*. Motile, non-capsulated cells; an oval spore is formed subterminally in a swollen sporangium. Strictly anaerobic. Saccharolytic or saccharolytic and proteolytic. Gelatin may or may not be hydrolysed. Each strain produces a protein neurotoxin; botulinum toxins are reported to be labile at 100 °C/10 minutes or 80 °C/45–60 minutes. The species is divided into a number of types (Types *A* to *G*) on the basis of antigenic differences between toxins; a given strain appears to produce only one form of toxin. The toxins of *C. botulinum* types *A*, *B*, *E* and *F* are the commonest causes of botulism in man; those of types *C* and *D* appear to cause botulism only in animals other than man. (*C. botulinum* type *C* produces toxin only when lysogenized by a particular phage.) Botulinum toxin in e.g. food may be detected by injecting into mice the filtrate of a saline suspension of the food; the identity of a particular toxin (or of the strain producing it) may be determined by finding out whether the specific *A*, *B*, *C*, *D*, *E*, *F* or *G* antitoxin can render the toxin innocuous to mice. (see also FOOD POISONING (bacterial) and FOOD, PRESERVATION OF)

C. novyi (= *C. oedematiens*). A causal agent of gas gangrene. Motile or non-motile cells; an oval spore is formed subterminally. Strictly anaerobic. Types *A*, *B*, *C* and *D* are distinguished by the particular range of toxins which they produce.

C. pasteurianum. Motile, non-capsulated cells; an oval spore is formed subterminally. Saccharolytic and proteolytic. NITROGEN FIXATION carried out. Nitrates are not reduced. H_2S is not produced.

C. perfringens (= *C. welchii*). A causal agent of e.g. gas gangrene, one form of FOOD POISONING, and lamb dysentery. The cells are non-motile, and many strains are capsulated. An oval spore is formed subterminally—but sporulation is uncommon. Strains may be aerotolerant. Predominantly saccharolytic; acid and gas are formed from a range of sugars. GELATIN is usually hydrolysed. Typically, a STORMY CLOT is produced in litmus milk. Five types (*A* to *E*) are distinguished by the range of toxins/extracellular enzymes which they produce; *all* strains form the α-toxin (a LECITHINASE). (see also NAGLER'S REACTION) Type *A* strains (food poisoning, gas gangrene) additionally form toxins which include (a) the θ-toxin (an oxygen-labile haemolysin active against the erythrocytes of e.g. sheep and goats); (b) the κ-toxin (a COLLAGENase); (c) the μ-toxin (a HYALURONIDASE). Type *B* strains (lamb dysentery) form the β, ε, and ι-toxins (all lethal, necrotizing).

C. sordellii. Similar to *C. bifermentans* but produces UREASES.

C. sporogenes. Motile, non-capsulated cells; an oval spore is formed subterminally in the swollen sporangium. Saccharolytic and proteolytic. Acid and gas are produced from glucose but not from lactose or sucrose. Gelatin hydrolysed. H_2S is produced. Indole is not produced. A violet coloration is given with VANILLIN reagent. *C. sporogenes* is similar to certain strains of *C. botulinum* but is atoxigenic.

C. tetani. The causal agent of tetanus. The cells are commonly motile; a spherical spore is formed terminally in the swollen sporangium. Strictly anaerobic. Swarming is common. Non-proteolytic and non-saccharolytic; mixed acids are produced from the fermentation of peptones. Gelatin is hydrolysed. A neurotoxin is produced.

C. welchii (see *C. perfringens*)

* see also BUTYRIC ACID FERMENTATION and ACETONE-BUTANOL FERMENTATION.

clot culture A form of BLOOD CULTURE; clotted blood, from which the serum has been removed, is liquefied (e.g. with STREPTOKINASE) and the resulting fluid used as the inoculum in a conventional blood culture procedure. Some workers claim that clot culture is more effective than conventional blood culture for the recovery of certain pathogens—particularly the salmonellae. Furthermore the serum may be used for serological tests, e.g. the WIDAL TEST. However, clot culture appears not to be widely used; one major disadvantage of the method is the risk of contamination during the processing stages which precede inoculation of the blood culture bottle. (see also CASTAÑEDA'S METHOD)

clotrimazole (*bis*-phenyl-(2-chlorophenyl)-1-imidazole methane) A drug which has a wide range of antifungal activity; it has been used in the treatment of e.g. aspergillosis and candidiasis.

clover, rot of A fungal disease of clover in which dark spots appear on the leaves and a soft rot affects the roots and lower stem; a white mycelium develops on the surface of the plant. The causal agent is an ascomycete.

clover wound tumour virus (see WOUND TUMOUR VIRUS)

cloxacillin (see PENICILLINS)

clubroot, of crucifers (see CRUCIFERS, FUNGAL DISEASES OF)

clumping factor (see COAGULASE)

cluster cup (see *stage I* under RUSTS (1))

Cnidospora A sub-phylum of parasitic PROTOZOA. The Cnidospora form complex spores which contain one or more extensile filaments (*polar filaments*) together with one or more cells (*sporoplasms*) which, on germination, become amoeboid invasive forms. The two classes are: (a) *Myxosporidea*. Myxosporideans are regarded as extracellular parasites of invertebrates and cold-blooded vertebrates—particularly fish. The spores are spherical, ovoid or irregularly-shaped, from five to several hundred microns in size; the spore wall consists of two or more *valves* (parts). Each polar filament is coiled within an intrasporal sac, the *polar capsule*. A generalized myxosporidean life cycle is as follows: Ingested spores germinate within the host's gut—the spore filament(s) extruding and appearing to attach to (or penetrate) the gut wall. The emergent amoeboid cells pass into (or through) the gut wall and subsequently localize in a suitable tissue, e.g. skin, bladder, muscle etc. Each cell gives rise to a syncytium which eventually encysts. Spores develop within the cyst, the syncytial nuclei initially forming *sporonts*. Each sporont undergoes several nuclear divisions to form a pair of multinucleate cells (*sporoblasts*) enclosed within a common coat, the whole being termed a *pansporoblast*. Each multinucleate cell develops into a multicellular unit which gives rise to a single spore—some cells forming the spore wall, others the polar capsule(s) and filament(s), while one or more form the sporoplasm(s). Spores are liberated when the cyst ruptures; spores in deep tissues (e.g. muscle) are presumably ingested by a new host following the death of the infected host. Myxosporidean genera include: *Ceratomyxa*, *Myxobolus*, and *Triactinomyxon* (a parasite of the *Tubifex* worm). (b) *Microsporidea*. Microsporideans are intracellular parasites of invertebrates and vertebrates—including many insects, fish, amphibians and (rarely) mammals. The spores are small (commonly less than 10 microns) and are of *unicellular* origin; they are usually ovoid or spherical, and typically have a *non*-composite wall. Each spore contains a single sporoplasm and a single, tubular, coiled polar filament which is not enclosed within a capsule; an intrasporal body, known as the *polaroplast*, is believed to be involved in the process of filament extrusion. The filament of an ingested spore is extruded within the host's gut; the sporoplasm passes through the filament and emerges as an amoeboid cell—which penetrates the gut wall and subsequently localizes in a suitable tissue. Cell division and spore formation occur intracellularly. The mode of spore release depends on the site of infection; some species, e.g. *Nosema bombycis*, may be transmitted transovarially by the infected host. Microsporidean genera (all within the single order Microsporida) include: NOSEMA, *Plistophora*, and *Thelohania* (a parasite of the mosquito and other insects).

CoA (see COENZYME A)

coagulase Any of a variety of serologically-distinguishable bacterial enzymes capable of coagulating citrated or oxalated *plasma*. Coagulase producers include *Yersinia pestis* and strains of *Staphylococcus aureus*; staphylocoagulases appear to have been investigated more intensively than the corresponding enzymes of other organisms. *Staphylocoagulases*. These comprise a number of antigenically-distinct *free coagulases* (enzymes released into the medium) and a *bound coagulase* (*clumping factor*) which is found in the (external) surface of the cell wall.

Free coagulase reacts with a plasma cofactor (the *coagulase reacting factor*, or CRF) to produce a thrombin-like entity—'coagulase-thrombin'—which promotes the change fibrinogen → fibrin, thus forming a coagulum or *clot*. Staphylocoagulases derived from strains pathogenic for humans exhibit maximal activity in human or rabbit plasma. The *in vitro* activity of free coagulase is inhibited by a variety of chemicals, particularly oxidizing agents, and by a number of antibiotics, including certain of the penicillins. Free coagulase may be detected by the *tube test* (see COAGULASE TEST).

Bound coagulase appears to act *directly* on fibrinogen, i.e. no auxiliary plasma factor appears to be necessary for coagulation; it is detected by the *slide test* (see COAGULASE TEST). Bound coagulase is entirely distinct from free coagulase—antibodies to one being unable to neutralize the activity of the other.

Staphylocoagulases may be regarded as AGGRESSINS: the fibrin barrier formed by coagulase-producing strains *in vivo* inhibits phagocytosis and thus promotes the establishment of the pathogen within the host's tissues. The property of coagulase production is commonly used to distinguish pathogenic cocci (*Staph. aureus*) from non-pathogenic strains (*Staph. albus*); this practice is questioned by some authors who point out that coagulase-negative staphylococci are not necessarily harmless—such organisms may form any of a variety of potent extracellular enzymes, e.g. HYALURONIDASES. Among staphylococci there is a high correlation between coagulase production and *enterotoxin* formation. (see also FOOD POISONING (bacterial) and PARA-COAGULATION)

coagulase test Any procedure used to determine whether or not a given strain of bacteria produces a COAGULASE. A coagulase test is used, for example, to distinguish *Staphylococcus aureus* (coagulase-positive) from *Staph. albus* (coagulase-negative).

The *slide test* detects *bound* coagulase (i.e. clumping factor). A loopful of citrated or oxalated human or rabbit plasma is stirred into a drop of thick bacterial suspension on a slide; the clumping of cells within 5 seconds indicates that the particular strain produces coagulase, i.e. is *coagulase-positive*. (It should be noted that some organisms agglutinate spontaneously when prepared as a saline suspension.) Organisms known to be coagulase-positive should be used as positive control strains, and coagulase-negative organisms should be used as negative controls.

The *tube-test* is commonly used to detect *free* coagulase; many different ways of performing the test have been recommended. Usually an equal (or smaller) volume of an 18–24 hour broth culture of the organism under test is added to a volume (e.g. 0.5 ml, 1 ml) of plasma in a small test-tube; the tube is then incubated (37 °C) and examined for clot-formation at one-hourly intervals—and following overnight incubation if no clot appears previously. Oxalated or citrated plasma may be used; plasma from diabetic patients is said to provide a more sensitive test. Many workers report that the anticoagulant *heparin* does not interfere with coagulase clotting—although at least one author has indicated that the clotting time (i.e. the time which elapses before a clot is formed) may be extended if heparinized plasma is used. It has also been reported that the preservative THIOMERSAL may inhibit the activity of free coagulase. Most authors agree that filtered plasma is not suitable for the coagulase test. Purified fibrinogen, or freeze-dried plasma, may be used in place of anticoagulant-treated plasma. As in the slide test, controls should be used. Strains which produce relatively large amounts of *fibrinolysin* may not form clots, or may bring about the lysis of any clot formed; such strains are more common among staphylococci isolated from human infections than among those isolated from animals. Before testing an organism for coagulase-production it is wise to ensure that the organism is unable to metabolise the anticoagulant; thus, citrate-utilizing organisms may cause clotting of plasma (usually on prolonged incubation) when citrate is used as anticoagulant. There is a high, though not total, correlation between positive results in the slide and tube tests.

CoASH (see COENZYME A)

cobalamin (see VITAMIN B_{12})

cobamide coenzymes (see VITAMIN B_{12})

cocarboxylase (see THIAMINE)

cocci (see COCCUS)

coccidia The common name for protozoans of the suborder EIMERIINA (order EUCOCCIDIA); many species are pathogens (see COCCIDIOSIS). The following is a generalized account of the life cycles of EIMERIA, ISOSPORA and related, monoxenous genera. The disseminative forms are resistant oocysts (see later)—each of which contains a number of infective forms (*sporozoites*). Oocysts ingested by the host break down in the intestine, liberating sporozoites (*excystation*). Sporozoites are banana-shaped, motile but non-flagellated cells, about 10–25 microns in length, which have organelles such as the CONOID, MICROPORE and RHOPTRIES. Sporozoites penetrate cells of the intestinal mucosa (or other tissues) and round off—becoming *trophozoites*; these grow and undergo SCHIZOGONY, becoming *schizonts*. Schizonts, and the host cells which contain them, subsequently rupture and release *merozoites*. Morphologically and structurally, these uninucleate cells are similar to sporozoites; sizes range from about 3 to 30 microns. Merozoites penetrate fresh host cells, become trophozoites, and schizogony is repeated. The number of schizogonous cycles is characteristic for a given parasite. Later generations of trophozoites develop into *gametocytes*. Some become large, uninucleate (female) *macrogametes*; others undergo multiple nuclear division and produce small, uninucleate, flagellated cells, the (male) *microgametes*, which escape from the ruptured host cells. Macrogametes are fertilized *in situ*; the encysted zygote (oocyst) is liberated from the host cell and voided with the faeces. Oocysts are spherical, ovoid, or otherwise; the form of the oocyst varies between species and may vary in a given species. The oocyst wall consists of one, two, or more layers (see also CYST) and may or may not include a MICROPYLE—according to species. Some oocyst characteristics: *E. tenella*: ovoid, 15–30 microns (long axis), micropyle absent; *E. necatrix*: ovoid, 15–25 microns, micropyle absent; *E. stiedae*: ovoid, 30–40 microns, micropyle present; *Isospora bigemina*: spherical or ovoid, 8–20 microns, micropyle absent. In the external* *aerobic* environment the oocyst undergoes SPOROGONY—in which the first nuclear division is *meiotic*—thus forming a number of (haploid) sporozoites; such an oocyst is said to be *sporulated*. In many species (e.g. *Eimeria*, *Isospora*) the sporozoites are formed within *sporocysts* which develop within the oocyst. A sporocyst is a spherical, ovoid, pyriform or elongated sac containing one or more sporozoites; those of some species possess a STIEDA BODY. The number of sporozoites per sporocyst and the number of sporocysts per oocyst are taxonomically-important features. Oocysts in fresh faeces are usually unsporulated; for identification purposes, oocysts are concentrated (by FLOTATION) and sporulated (usually 2–7 days, 10–30 °C) in a 2% solution of potassium dichromate—which inhibits bacterial growth and helps to maintain essential *aerobic* conditions. *The oocysts of some species (e.g. *I. bigemina*; *E. carpelli* and other eimerian parasites of fish) have been reported to sporulate within the host. (see also TOXOPLASMA)

Coccidia A subclass of protozoans of the Telosporea (subphylum SPOROZOA); the organisms parasitize mainly vertebrates, and the mature trophozoites are found *intra*cellularly (cf. GREGARINIA). SYZYGY occurs in the suborder Adeleina. The subclass is divided into the orders PROTOCOCCIDIA and EUCOCCIDIA. *NB*. 'Coccidia' is also used to refer specifically to members of the EIMERIINA: see adjacent entry COCCIDIA.

Coccidioides A genus of fungi within the order MONILIALES; the sole species, *C. immitis*, occurs as a saprophyte in certain soils and is the causal agent of COCCIDIOIDOMYCOSIS in man and other animals. In culture, *C. immitis* forms a grey mycelium which subsequently fragments to form barrel-shaped arthrospores. In man, *C. immitis* occurs within the tissues primarily as a multinucleate, spherical, thick-walled cell (*sporangium* or 'spherule') which, when mature, has a diameter of approximately 50 microns; at maturity the sporangial protoplast cleaves first into multinucleate masses (*protospores*) and subsequently into uninucleate sporangiospores (*endospores*). The endospores are liberated on rupture of the sporangial wall, and the reproductive cycle of the pathogen is repeated, within the tissues, by growth and nuclear division of the individual spores. Hyphae may be formed e.g. in the tissues bordering lung cavities.

coccidioidomycosis A disease of man and domestic animals in which the causal agent is *Coccidioides immitis*; infection usually occurs *via* the respiratory route. In man infection may be inapparent—or may give rise e.g. to an acute, self-limiting disease resembling influenza; the primary infection may lead to progressive, disseminated (often fatal) coccidioidomycosis—in which lesions may develop e.g. in the viscera, bones, or central nervous system. In cattle lesions have been reported to occur e.g. in the bronchial and mediastinal lymph nodes and in the lungs. *Laboratory diagnosis* may involve microscopical and cultural examination of sputum, pus etc. and/or serological tests. Certain antibiotics (e.g. amphotericin B) have been used to treat the disease.

coccidiosis Any disease caused by a member of the COCCIDIA; coccidioses occur infrequently in man but are not uncommon in domestic animals (except horses). Typically, coccidioses are diseases of the intestinal tract; symptoms often include bloody diarrhoea, debility and emaciation. Infection occurs by ingestion of *sporulated* oocysts*, and the severity of the disease is usually related to the number of oocysts ingested. Excystment occurs in the intestine: see COCCIDIA for details of subsequent asexual and sexual stages of development.

Eimeria. In *chickens*, *E. necatrix*, *E. tenella* and *E. brunetti* are among the most pathogenic species. After an incubation period of some 4–6 days the animals lose vitality and their stools become bloody; death may occur shortly after. Haemorrhages and necrosis occur in the small intestine and/or caeca. Recovery may be attended by long-term immunity. In *cattle*, *E. zürnii* (= *E. zuernii*) is one of several pathogenic species. Young animals are the most commonly affected; older (immune) animals may be cyst-shedding carriers. The pathogen develops in the mucosae of the large and/or small intestine, and haemorrhagic diarrhoea occurs in acute manifestations. In *rabbits*, *E. stiedae* produces a severe *hepatic* coccidiosis; the pathogen develops in the epithelium of the bile ducts. Loss of appetite, diarrhoea and enlargement of the liver occur before death. Other species (e.g. *E. irresidua*, *E. magna*) develop in the small intestine.

Species pathogenic for other animals include *E. scabra* (pigs), *E. anseris* (geese), *E. ahsata* (sheep) and *E. phasiani* (pheasant).

Isospora. In *man*, infections by *I. belli* and *I. hominis* are believed to be commonly asymptomatic, but *I. belli* may cause a mild intestinal illness. In the *cat* and *dog*, *I. bigemina* may cause an intestinal disease characterized by bloody diarrhoea and emaciation.

Toxoplasma. See TOXOPLASMOSIS.

In the coccidioses, immunity to re-infection may follow recovery, and may be specific for a particular pathogen; copro-antibodies may be involved. Diagnosis of coccidioses necessitates identification of *sporulated* oocysts: oocysts from fresh faeces are concentrated by FLOTATION and sporulated aerobically—commonly 2–7 days in a 2% solution of potassium dichromate. *See TOXOPLASMOSIS for an alternative mode.

coccobacillus (*pl.* coccobacilli) The term for any bacterial cell intermediate in form between a COCCUS and a BACILLUS.

coccolith (see CELL WALL (algal))

Coccolithus A genus of algae of the HAPTOPHYTA.

Coccomyxa A genus of algae of the division CHLOROPHYTA. Species occur e.g. as the phycobiont in certain LICHENS—e.g. species of BAEOMYCES, NEPHROMA, PELTIGERA, and SOLORINA.

coccus (*pl.* cocci) Any spherical or near-spherical bacterial cell. Cocci of different species may vary considerably in size, and may occur singly, in pairs, in regular groups of four or more, in chains, or in irregular clusters.

codon (see GENETIC CODE)

Coelomomyces A genus of obligately-parasitic fungi of the order BLASTOCLADIALES.

Coelomycetes A class of filamentous (mycelial) FUNGI within the sub-division DEUTEROMYCOTINA; constituent species are distinguished from those of the class HYPHOMYCETES in that their conidia develop within *cavities* which are—at least initially—enclosed by fungal tissue and/or the tissue of the (plant) host. The somatic form of the organisms is a septate mycelium; (asexual) reproductive structures include acervuli (see ACERVULUS), pycnidia (see PYCNIDIUM), and stromata (see STROMA). A number of species have an ascomycetous sexual stage. The constituent orders are MELANCONIALES and SPHAEROPSIDALES.

coenobium (*algol.*) In certain types of ALGAE: a colonial form which consists of a number of cells that are either contiguous (as e.g. in *Scenedesmus*) or associated within a common matrix (as e.g. in *Volvox*). A coenobium of a given species commonly contains a number of cells which is characteristic of that species.

coenocyte Among microorganisms: any *multinucleate* cell, structure, or organism formed by the division of an existing multinucleate entity, or formed when nuclear divisions are not accompanied by the formation of dividing walls or septa. Coenocytes occur e.g. among the *siphonaceous* algae (e.g. BOTRYDIUM); the mycelium of many species of fungi may be regarded as coenocytic. (see also SYNCYTIUM)

coenzyme (see COFACTOR)

coenzyme A (CoA) A coenzyme which is derived from PANTOTHENIC ACID (see Appendix V(c) for structure); coenzyme A functions as a carrier of acyl groups—with which it forms *thioesters* (CoA.S.CO.R). (Uncombined coenzyme A may thus be represented as CoASH.) Acyl-coenzyme A thioesters have a high free energy of hydrolysis and can thus be involved in SUBSTRATE LEVEL PHOSPHORYLATION.

coenzyme Q (see QUINONES (in electron transport))

coenzyme R (see BIOTIN)

cofactor (*biochem.*) Some enzymes are functional only when the protein portion (*apoenzyme*) is conjugated with a relatively small, non-protein organic molecule (the *cofactor*). The terms *coenzyme* and *prosthetic group* are frequently used as synonyms of cofactor; however, a distinction is sometimes made between those cofactors which dissociate readily from the apoenzyme (*coenzymes*) and those which are firmly bound (*prosthetic groups*). The complete, conjugated enzyme (apoenzyme + cofactor) is termed a *holoenzyme*. Metal ions, often needed for enzyme activity, are called *activators*.

Col factor (see COLICIN FACTOR)

colanic acid A capsular heteropolysaccharide produced under certain conditions by many strains of *Escherichia* and *Salmonella*. Colanic acid consists of a repeating trisaccharide unit (-glucose-fucose-fucose-)—each unit carrying a trisaccharide branch (galactose-glucuronic acid-galactose-) on the central fucose residue. Non-carbohydrate substituents (e.g. acetyl, pyruvyl) may be present—the nature and position of such groups varying with strain.

colchicine A compound which occurs naturally in the autumn crocus (*Colchicum autumnale*); it has important applications in experimental genetics—e.g. in the study of chromosomes (see also KARYOTYPE) and in the promotion of polyploidy in plants. Colchicine inhibits the polymerization of TUBULIN and encourages the de-polymerization of MICROTUBULES; it thus inhibits SPINDLE-formation and arrests MITOSIS at metaphase.

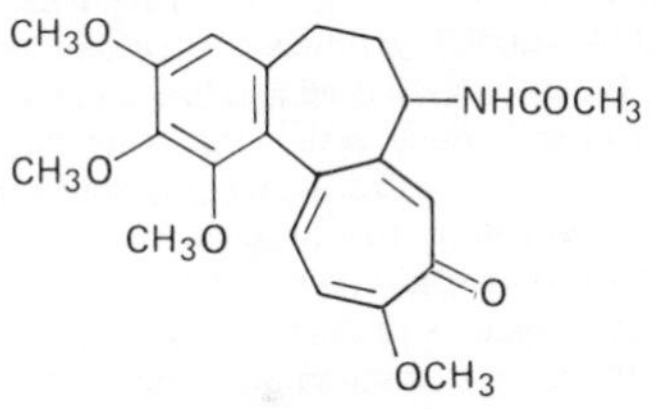

Colchicine

cold, common (see COMMON COLD)

cold agglutinin Any AGGLUTININ capable of maximum combination with homologous antigen at low temperatures (e.g. 4 °C) but which combines minimally, or not at all, at room temperature or 37 °C. Cold agglutinins often appear in the plasma of patients suffering from primary atypical pneumonia; such agglutinins bring about agglutination of human group O erythrocytes.

cold sore (herpes-simplex) A thin-walled vesicular lesion which may occur e.g. on the lips or conjunctivae; the causal agent is a HERPESVIRUS.

Coleps A genus of freshwater and marine

protozoans of the suborder **RHABDOPHORINA.** The cells are barrel-shaped, between 50 and 110 microns in length, with uniform somatic ciliation. The cytostome is apical, and there is a semi-rigid endoskeleton—formed by the deposition of calcium phosphocarbonate in the pellicular alveoli; due to the calcareous deposits the body appears to be covered by a regular series of platelets. A number of posteriorly-directed spines are found at the posterior end of the cell. Food consists of algae and other protozoans.

colicin (see **COLICINS**)

colicin factor (colicinogenic factor; Col factor) A **PLASMID**, the intracellular presence of which confers on certain members of the Enterobacteriaceae the ability to synthesize a **COLICIN**. The majority of Col factors are *transmissible* (see **SEX FACTOR**); most transmissible Col factors are F-like while some are I-like. Non-transmissible Col factors may be co-transferred with a superinfecting sex factor (e.g. an **F FACTOR** or the Col I factor). Col I factor can also promote the transfer of bacterial genes during **CONJUGATION**—although the mechanism is unknown; integration of Col I factor with the bacterial chromosome has not been demonstrated.

colicin typing A form of **TYPING** in which closely-related strains of a given bacterial species may be distinguished on the basis of the **COLICINS** they produce (*colicinogenic* strains) or the colicins to which they are sensitive; colicin typing has been used e.g. to distinguish between some fifteen types of *Shigella sonnei*.

colicins A class of protein **ANTIBIOTICS** (M.Wt. 40,000–80,000) produced by certain strains of the Enterobacteriaceae and bactericidal for certain other strains within the same family; the ability to synthesize a colicin is conferred on a bacterial cell by the intracellular presence of a **COLICIN FACTOR**—such a cell being said to be *colicinogenic*. A cell may carry the potential for synthesizing more than one type of colicin. In a colicinogenic population only a small proportion of cells actually synthesizes colicins—a synthesis which, in at least some cases, appears to be a lethal event for the cell; other cells which possess the same colicin factor are immune to the effects of the particular colicin.

Several types of colicin can be distinguished on the basis of diffusibility and host specificity; examples include colicins E, I, K, and V. (A colicin may also be designated according to the strain from which it was derived—e.g. colicin E-K30, obtained from *Escherichia coli* strain K30.) Colicins bind to specific receptor sites on the surface of a sensitive bacterium; binding may be followed by at least a partial penetration of the cell membrane. The mechanisms of the bactericidal action of colicins appear to be various and imperfectly understood. Colicins of the E1 type appear to interfere with membrane function, e.g. altering permeability and apparently dissipating the high-energy intermediate formed during respiration (see **OXIDATIVE PHOSPHORYLATION**); colicins of the E3 type inhibit protein synthesis by causing excision of part of the 16S ribosomal RNA. (see also **BACTERIOCINS**)

coliforms A name generally used to refer to gram-negative, lactose-fermenting, enteric bacilli, but which has also been used, more loosely, to refer e.g. to any gram-negative enteric bacilli. In water bacteriology the term 'coliforms' refers to gram-negative, oxidase-negative, asporogenous bacilli which can ferment lactose within 48 hours at 37 °C with the production of acid and gas, and which are capable of aerobic growth on bile-salts media.

coliphage Any **BACTERIOPHAGE** whose host is a strain of *Escherichia coli*.

colistin (see **POLYMYXINS**)

colitose (3,6-dideoxy-L-galactose) A sugar which occurs e.g. in the O-specific chains of **LIPOPOLYSACCHARIDES** in certain serotypes of *Salmonella* and in the lipopolysaccharide of *Escherichia coli* serotypes of the 0111 group. Colitose, first isolated from *E. coli*, contributes to the specificity of somatic (O) antigen 35 in the group O salmonellae.

collagen The main protein component of bone, cartilage, connective tissue. Collagen has a high glycine content and contains hydroxyproline. It is generally resistant to enzymic degradation, although some bacteria (e.g. *Clostridium perfringens, Cl. histolyticum*) produce specific *collagenases* which may enhance invasiveness in pathogenic species. Collagenolytic organisms may cause spoilage of leather.

Collema A genus of gelatinous, foliose to fruticose **LICHENS** in which the phycobiont is *Nostoc*; the thallus is dark-coloured, homoiomerous, and lacks a distinct cortex. The apothecia, when present, generally have dark reddish-brown discs. Species include e.g.:

C. crispum. The thallus is green-brown to black and consists of small, rounded, flattened lobes which may be imbricated or may form rosettes; scale-like isidia occur towards the centre of the thallus. *C. crispum* occurs on limestone rocks and walls.

C. auriculatum. The thallus is dark olive-green when wet, brown with fine striations when dry; the lobes are rounded, thick, and swollen when wet. The centre of the thallus usually bears dense, globular isidia. *C. auriculatum* occurs on limestone rocks and walls and sometimes on the ground or among mosses.

Colletotrichum A genus of fungi within the order **MELANCONIALES**; species include a number of important plant pathogens—e.g. the causal agents of onion *smudge* and *red-rot* of sugarcane (see also ANTHRACNOSE and ONION, DISEASES OF). *Colletotrichum* spp form elongated, aseptate conidia which lack appendages; the conidia are formed on conidiophores which develop in setose acervuli.

Collybia A genus of fungi of the TRICHOLOMATACEAE.

colominic acid A homopolysaccharide consisting of *N*-acetylneuraminic acid pyranose units apparently linked by (2 → 8)-ketosidic bonds; it occurs in certain strains of *Escherichia coli*. Colominic acid resists hydrolysis by most neuraminidases although it is completely hydrolysed by an enzyme from *Clostridium perfringens*. (see also NEURAMINIC ACIDS)

colon bacillus Synonym of *Escherichia coli*.

colony (*microbiol.*) Collectively, a number of individual cells or organisms of a given species which, during their development, have formed a discrete aggregate or group.

Bacterial colonies. On or within a solid MEDIUM (e.g. an agar plate): a bacterial colony commonly consists of a compact mass of individual cells which have usually resulted from the multiplication of a single cell at that location. (Certain bacteria—e.g. *Streptomyces* spp—form *mycelial* colonies.) In order to prepare a culture containing *discrete* colonies (i.e. colonies which do not overlap or coalesce) the cells in the inoculum must be adequately dispersed on or within the medium—e.g. by STREAKING or by the POUR PLATE method. The *size* of a bacterial colony depends on species, on the type of medium and growth conditions, and on the time during which growth (multiplication) has occurred; a colony may be microscopic or macroscopic. Features which tend to be more stable (in a given species of bacterium) include: (a) the *colour* of the colony under given conditions; (b) the *elevation* of the (surface) colony—e.g. flat, low convex, domed; (c) the *edge* or margin of the colony—e.g. entire (i.e. circular and unbroken) or crenate (i.e. scalloped); (d) the *texture* of the colony—e.g. butyrous, friable, mucoid; (e) the *surface appearance* of the colony—e.g. matt, glossy. When growth occurs on an appropriate type of blood agar the colonies of certain species are bordered by zones of HAEMOLYSIS. Colony characteristics are often used to aid in the identification of bacteria. (see also DAISY HEAD COLONY, DRAUGHTSMAN COLONY, ENUMERATION OF MICROORGANISMS, MOTILE COLONIES, and SMOOTH-ROUGH VARIATION)

Fungal colonies. Certain types of fungi (e.g. yeasts) form colonies which resemble those typical of many types of bacteria—i.e. round, convex, glossy or matt, butyrous or mucoid colonies. Other types of fungi commonly give rise to circular 'fluffy' colonies each of which consists of a mass of MYCELIUM that may or may not include reproductive structures. (see also PELLET and PETITE COLONIES)

Algal colonies. While bacterial and fungal colonies arise merely as a result of growth at a given location—and are of indeterminate size—the algal colony is a more or less stable form which, in some species, is the only form in which the organism occurs. (see e.g. COENOBIUM, cf. PALMELLOID PHASE; see also e.g. SCENEDESMUS, VOLVOX)

Protozoal colonies. In (the few) colonial protozoans the colony typically consists of attached individuals that form a group of variable size. Thus, e.g. in the colonial peritrich CARCHESIUM a (variable) number of zooids is carried on a single multi-branched stalk (cf. VORTICELLA). Chain-like colonies are formed by certain members of the Astomatida.

colostrum A secretion of the mammary gland which is produced prior to the commencement of lactation proper. In man the IMMUNOGLOBULIN content of colostrum is relatively low, but in some animals—e.g. ruminants (in which immunoglobulins do not pass the placenta)—large amounts are normally present. The immunoglobulins found in colostrum are mainly IgA and IgG.

colourless sulphur bacteria The group of obligately or facultatively chemolithotrophic, *non*-photosynthetic bacteria which obtain energy by the oxidation of sulphur and/or reduced sulphur compounds, e.g. THIOBACILLUS.

Colpidium A genus of freshwater protozoans of the TETRAHYMENINA; *Colpidium* species occur particularly in waters containing decomposing organic matter. *C. campylum* is an ovoid or reniform organism, some 50 to 80 microns in length; the somatic ciliature is uniform, and the buccal ciliature is similar to that of TETRAHYMENA. Each organism has one macronucleus and one micronucleus, and there is a single contractile vacuole. *C. colpoda* is similar but larger.

Colpoda (*Kolpoda*) A genus of protozoans of the order TRICHOSTOMATIDA. Species of *Colpoda* are found in various freshwater habitats and in the soil. *C. cucullus* is reniform, between 40 and 120 microns in length, with uniform somatic ciliation; the cytostome is lateral and is found at the base of a vestibulum. Food consists of algae, bacteria and small flagellates. Asexual reproduction occurs within a cyst—in which four or more individuals may be formed. *C. steini* is a smaller species, about 20 to 60 microns in length, with a conspicuous group of cilia—the 'beard'—projecting from the posterior part of the vestibulum.

columella (*mycol.*) An axial sterile body within a spore-bearing structure—e.g. a short extension of the sporangiophore into the cavity of a sporangium.

combining site (*immunol.*) That part of an ANTIBODY which combines with the *determinant* of the homologous antigen or hapten; its location has been determined e.g. by AFFINITY LABELLING. In the basic immunoglobulin molecule (see IGG) the antibody combining sites occur near the *N*-terminals of the LIGHT CHAINS and HEAVY CHAINS—amino acid residues of both types of chain being associated with each site; one combining site is found on each FAB FRAGMENT. (see also VALENCY and V-REGION)

commensal One of the two organisms which co-exist in a state of COMMENSALISM.

commensalism A term used with any of the following meanings: (1) A stable condition in which two species of organism live in close physical association—one of the organisms deriving benefit and the other organism appearing to be neither benefited nor harmed as a result of the association. (2) A stable condition in which two species of organism live in close physical association and in which each organism derives some benefit from the association (i.e. synonymous with SYMBIOSIS, sense (2)). (3) A stable condition in which two species of organism live in close physical association but in which neither benefit nor harm accrues to either organism as a result of the association.

comminutor (see SEWAGE TREATMENT)

common blight (of beans) (see BEANS, BACTERIAL DISEASES OF)

common cold In man, an acute self-limiting catarrhal inflammation of the mucous membranes of the nasopharyngeal region; the causal agent may be any of a number of viruses—usually a member of the genus RHINOVIRUS or CORONAVIRUS. Secondary (bacterial) infections (e.g. otitis, sinusitis) may occasionally occur.

common scab (of potato) (see POTATO, BACTERIAL DISEASES OF)

comparative test (*vet.*) A form of TUBERCULIN TEST, applied to cattle, in which two intradermal injections are given, in different regions of the skin, on the same occasion. One injection contains tuberculin derived from a strain of *Mycobacterium avium* while the other contains tuberculin from a strain of *Mycobacterium tuberculosis*. The relative intensities of the reaction(s) (if any) indicate the probable identity of the particular type of *Mycobacterium* with which the animal has been in contact.

compatibility (in fungal sexuality) In fungi, karyogamy may or may not occur between two given nuclei according to whether the nuclei are compatible or incompatible. In many species of fungi it appears that any nucleus is compatible (i.e can fuse) with any other nucleus of the same species, and such (self-fertile) fungi are said to exhibit HOMOTHALLISM; homothallic species occur in all the major groups of fungi, and homothallism appears to be the most common form of sexuality in fungi as a whole. (cf. HETEROTHALLISM) Bipolar (dimictic) heterothallism occurs among the lower fungi (e.g. species of *Mucor*, *Rhizopus*), the ascomycetes (e.g. species of *Neurospora*, *Saccharomyces*), and basidiomycetes (e.g. many species of the Uredinales and Ustilaginales). Multiallele (diaphoromictic) heterothallism appears to occur only in certain basidiomycetes (e.g. *Piptoporus betulinus*, *Schizophyllum commune*, *Ustilago maydis*, and species of *Coprinus* and *Fomes*). SECONDARY HOMOTHALLISM (homoheteromixis) occurs in certain hemiascomycetes and in some basidiomycetes. In general, the ways in which sexual reproduction may be subject to genetic control appear to increase in number and complexity from the lower fungi, through the ascomycetes, to the basidiomycetes.

compatible (*plant pathol.*) Used to describe a plant-pathogen relationship in which disease occurs.

competence (in bacterial transformation) (see TRANSFORMATION (bacterial))

complement (symbol C or C′; historical synonym: *alexin*) (*immunol.*) A group of proteins normally present in plasma and tissue fluids and in *freshly-isolated* serum; when activated (see later) the *complement system* is responsible for a number of important physiological activities.

Species-specific varieties of complement are produced. The components of *human* complement are designated C1 (comprising C1q, C1r and C1s), C2, C3....C9; the majority are β or γ-globulins. Components C1r, C2 and C6 are heat-labile; components C3 and C4 are destroyed by ammonia or hydrazine.

Prior to activation the components of complement exist as proenzymes and/or as forms which, on activation, combine with other components to produce physiologically-active complexes and/or fragments. The complement system may be activated e.g. by an antigen–antibody complex; activation is referred to as COMPLEMENT-FIXATION. The activated complement system is responsible e.g. for immunological CHEMOTAXIS and for IMMUNE ADHERENCE. The action of complement on *sensitized* erythrocytes commonly leads to IMMUNE HAEMOLYSIS; however, the complement of some species will not lyse the erythrocytes of certain other species—e.g. sheep erythrocytes cannot be lysed by horse complement.

Complement activity (in serum) is destroyed

if the sample is left at room temperature for a few days or is heated to 56 °C for 30 minutes. Activity is also abolished if calcium and magnesium ions are effectively withdrawn from the system e.g. by EDTA, oxalate or citrate.

complement-fixation The *activation* of COMPLEMENT. (a) *Fixation by the 'Classical pathway'*. Fixation is initiated on the formation of a complex between an antigen and a COMPLEMENT-FIXING ANTIBODY. C 1q binds to the Fc portion of the antibody; subsequently, and in a regular sequence, each of the remaining components becomes activated (functional) following the activation of the previous component. Components become activated in the sequence: 1 4 2 3 5 6 7 8 9. (The numbering of components preceded a knowledge of their activation sequence.) *Calcium ions* are required for the binding of C 1q, C 1r and C 1s to the antibody. Bound C 1 acts enzymically on C4 producing the modified component C4b which, in the presence of *magnesium ions*, binds to C2—forming the complex C42. C42 functions as a *C3 convertase* i.e. an enzyme which splits C 3 to form the fragment C 3a (an ANAPHYLATOXIN with chemotactic properties) and C 3b—which is important in IMMUNE ADHERENCE and PHAGOCYTOSIS. At this juncture, KAF (*q.v.*) may enzymically attack C 3b—exposing a reactive site; this site can combine with CONGLUTININ. If C 3b is not modified (by KAF) it acts (enzymically) on C5—splitting the latter into the chemotactic fragment C5a, and C5b. With C6, C5b forms the complex C56 which may or may not remain at the initial complement binding site; if it does not it may become involved in REACTIVE LYSIS. Either way, C56 complexes with C7, C567 binding strongly to membranes. Membrane-bound C567 binds C8 and C9, and the complex C56789 brings about irreversible damage to the membrane to which it is attached; such damage may be inflicted on erythrocytes (*immune haemolysis*), other cells—including bacteria, and even LIPOSOMES. (Since the sequence C 14235 involves a series of *enzymic* reactions, *amplification* occurs at each step, i.e. each molecule of an enzymic component initiates the formation of a *number* of molecules of the succeeding enzymic component.)

(b) *Fixation by the 'alternative pathway'*. Fixation from C3 to C9 (i.e. omitting the initial sequence C 142) is brought about by PROPERDIN in conjunction with any one of the following: ENDOTOXINS, ZYMOSAN, inulin, aggregates of human IgA, cobra venom factor, and certain other entities; *magnesium ions* are required. The mechanism of the alternative pathway is not clearly understood; a C3 convertase may be formed from the properdin complex. (Other enzymes which split C3 to C3a and C3b include *thrombin*.)

Classical and alternative fixation can be distinguished *in vitro* by the reagent ethylene-glycol - *bis*(β - aminoethylether) - *N,N,N′,N′* - tetraacetic acid (EGTA) which chelates calcium ions but does not chelate magnesium ions.

complement-fixation test (CFT) Any serological test in which COMPLEMENT-FIXATION is detected and quantified in order to indicate the presence or absence (or quantity) of a particular antigen, or a specific COMPLEMENT-FIXING ANTIBODY, in a given sample. Thus, e.g. a specific serum antibody may be quantified by adding to each of a number of serial dilutions of the *inactivated* serum fixed quantities of antigen and complement. Following incubation, each serum dilution is examined for the presence of free (i.e. unfixed) complement by the addition of a HAEMOLYTIC SYSTEM. In a *positive* test (i.e. one in which serum antibody is detected) the antibody *titre* may be quoted as that serum dilution in which a specified degree of haemolysis occurs. A dilution in which none of the complement has been fixed contains no detectable antibody; if haemolysis is observed in the *lowest* serum dilution the test is *negative*—i.e. the original serum contained no detectable antibody.

Control procedures are essential in order e.g. to preclude interpretations based on the effects of ANTICOMPLEMENTARY substances.

complement-fixing antibodies Antibodies (IgG, IgM) which bind ('fix') COMPLEMENT following combination with their homologous antigens.

complementation (genetic complementation) (a) *Intergenic* (*intercistronic*) *complementation*. In a diploid nucleus, heterokaryon (or, analogously, in a merozygote): the ability of two *non*-allelic wild type genes (one on each of two *homologous* chromosomes) to compensate for the functional deficiency of their homologous, mutant (recessive) alleles. Intergenic complementation permits the expression of the wild type phenotype—i.e. the mutant phenotype of each chromosome is suppressed. (b) *Intragenic* (*intracistronic, inter-allelic*) *complementation*. In a diploid nucleus, heterokaryon (or, analogously, in a merozygote): the joint expression of two mutant *homologous* alleles—the resultant phenotype resembling the wild type phenotype more closely than that expressed, individually, by either mutant allele. (Most pairs of mutant homologous alleles express a fully mutant phenotype—i.e. the absence of complementation is more common than its occurrence; whether or not complementation occurs depends on the loci of the mutations.). (see also COMPLEMENTATION TEST and CIS-TRANS TEST)

complementation test A procedure used e.g. to

determine whether, in a particular haploid genome, a given phenotypic characteristic is controlled by one gene or by more than one gene. Two haploid genomes (both associated with the same mutant phenotype) are brought together within the same organism or cell—e.g. by the formation of a heterokaryon (in fungi) or of a merozygote (in bacteria); the occurrence of complete COMPLEMENTATION—giving a wild type phenotype—indicates digenic (or polygenic) control of the phenotypic characteristic, while partial complementation (phenotype intermediate between mutant and wild type) indicates monogenic control. (If complete *or* partial complementation occurs the test is said to be *positive*.) The absence of complementation (a *negative* complementation test) indicates monogenic control.

Compared with the simple complementation test (above) the CIS-TRANS TEST is a more rigorous test of the type of genetic control exercised on a given phenotype characteristic.

complete medium (CM) A type of culture medium used e.g. in experimental genetics. The medium contains nutrients sufficient to support the growth of prototrophs and auxotrophs.

concanavalin A A LECTIN which, under appropriate conditions, can bring about transformation in lymphocytes.

concatenate (concatener, concatemer) Two or more viral genomes covalently linked end to end.

conceptacle (*mycol.*) (see ASCOSTROMA)

conchate (*mycol.*) Refers to a bracket-type fruiting body in which the upper surface is convex, and the lower, hymenial surface is concave or planar.

conditional lethal mutant A mutant organism (e.g. a bacterium or virus) in which the mutation does not significantly affect the phenotype under one set of (*permissive*) conditions, but which prevents growth/replication under another set of (*restrictive*, or *non-permissive*) conditions.

Temperature-sensitive mutants. The mutation (generally a MIS-SENSE MUTATION) alters an essential protein in such a way that it can function at certain temperatures, but is inactive at other (usually higher) temperatures. For example, some strains of *Escherichia coli* can grow at 37 °C but not at 42 °C; some phage T4 mutants exhibit near-normal virulence at 25 °C (permissive temperature), but are not virulent at 37 °C (restrictive temperature).

Host-dependent mutants (in bacteriophages). A bacteriophage containing an AMBER MUTATION in an essential gene can grow only in a *permissive host* which contains an (*intergenic*) amber SUPPRESSOR MUTATION.

confluent growth (of microorganisms) (see CULTURE)

conglutination (*immunol.*) (see CONGLUTININ)

conglutinin A protein normally present in *bovine* serum. Conglutinin binds to CONGLUTINOGEN and can agglutinate COMPLEMENT-fixing entities—such agglutination being termed *conglutination*. Conglutinin is distinct from IMMUNOCONGLUTININS.

conglutinogen A site on the bound C3b component of COMPLEMENT which becomes exposed by the activity of KAF (*q.v.*); this site binds to CONGLUTININ during *conglutination*.

conglutinogen activating factor (see KAF)

Congo red A water-soluble acid dye; Congo red is used e.g. for staining cellulose, as a vital stain, and as a pH indicator—pH 3.0 (blue) to pH 4.5 (red).

conidia (*sing.* conidium) (*mycol.*) Thin-walled, asexually-derived spores which are borne singly or in groups or clusters *on* specialized hyphae (CONIDIOPHORES); conidia are typically deciduous at maturity, and each germinates by the formation of one or more germ tubes. (cf. SPORANGIOSPORES, THALLOSPORES) Conidia may be uninucleate, binucleate, or multinucleate according to species. Conidium-formation is typical of the higher fungi—though conidia are formed by certain lower fungi*. The conidia of the various species differ greatly in shape, colour, and size; the conidiophores which bear them may develop randomly on the vegetative hyphae, may be organized into distinct structures (see ACERVULUS, COREMIUM), or may develop—often in an hymenium-like layer—within closed or partly-closed fructifications (see PYCNIDIUM).

Conidia may develop e.g. by (a) *blastogenous development* (i.e. BUDDING)—which occurs e.g. in species of *Candida* and *Taphrina*; (b) *phialogenous development* (e.g. in *Aspergillus* and *Penicillium*) in which conidia are formed in PHIALIDS; (c) *porogenous development*, in which the conidia ('porospores') are extruded from small pores in the walls of the conidiophores; (d) *murogenous development*, in which conidia are formed by the expansion and separation of the extremities of a conidiophore.

*Some species of lower fungi, e.g. *Cunninghamella* spp, form true conidia; however, deciduous, conidium-like sporangiola and sporangia (see SPORANGIOLUM and SPORANGIUM) are sometimes referred to as conidia.

conidial head (*mycol.*) The expanded, globular, terminal portion of the conidiophore in certain fungi—e.g. ASPERGILLUS.

conidiation The phenomenon in which conidia function as gametes.

conidiophore (*mycol.*) A differentiated or undifferentiated, branched or unbranched hypha which bears one or more asexually-derived spores (CONIDIA).

conidiospore A synonym of conidium—see CONIDIA.

conidium (see CONIDIA)

Coniophora A genus of fungi of the family CONIOPHORACEAE; species include *C. cerebella* (= *C. puteana*)—the causal agent of WET ROT of timber. *C. cerebella* forms a flattened, warty fruiting body which subsequently becomes greenish-brown as the spores develop; the ovoid spores are approximately 12 × 8 microns. The hyphae of *C. cerebella* are reported to exhibit whorled clamp connections in young cultures.

Coniophoraceae A family of fungi of the order APHYLLOPHORALES. Constituent species form smooth, pigmented spores in which the inner layer of the thick, bilayered spore wall is strongly cyanophilous; according to species, the hymenial surface may be smooth, wrinkled, warty, or (under certain circumstances) may consist of a layer of tooth-like processes. Cystidia are not formed. The fruiting body fails to develop a green coloration on application of 10% aqueous ferrous sulphate. The genera include CONIOPHORA and SERPULA.

conjugate (*immunol.*) (see CONJUGATION (*immunol.*))

conjugate division (*mycol.*) (see CLAMP CONNECTIONS)

conjugation (*algol.*) A sexual process which involves pairing of organisms and the subsequent fusion of gametes derived from their protoplasts; conjugation occurs e.g. in SPIROGYRA and in the DESMIDS.

conjugation (bacterial) A process in which genetic material is transferred from one bacterium (the *donor* or 'male') to another (the *recipient* or 'female') *via* a physical connection between the two cells. The following refers to conjugation in *Escherichia coli*.

The *donor* characteristic is conferred upon a cell by the intracellular presence of a SEX FACTOR, e.g. the F FACTOR (*fertility factor*). A cell which possesses an *extrachromosomal* F factor is designated F^+; such a cell normally contains one F factor per chromosome. (F factor replication is normally synchronous with that of the chromosome.) A *recipient* cell lacks the F factor and is designated F^- (but see PHENOCOPY). On mixing F^+ and F^- cells, donor-recipient pairs are formed; the cells of each pair are connected by the *sex pilus* (*sex fimbria*, *F-pilus*) of the donor (F^+) cell—see SEX PILI. During conjugation F^- cells become converted to F^+ cells—i.e. they gain an F factor; donor cells retain their donor characteristics. A widely-accepted model for this transfer of fertility is as follows. At a specific site in *one* strand of the circular, double-stranded F factor a break occurs. DNA synthesis is initiated at this site, using the unbroken, circular strand as template; the synthesis is *unidirectional* (cf. entry DNA) and results in the unwinding of the broken strand and its replacement with a new strand. The (unwound) single strand passes (5′-end leading) into the recipient cell—presumably *via* the sex pilus. The DNA polymerase is thought to be membrane-bound—in which case the circular template must rotate as synthesis proceeds; if the polymerase occurs at a site near the sex pilus, this rotation may propel the unwinding strand into the recipient cell. (cf. ROLLING CIRCLE MODEL) In the recipient cell, a complementary strand is synthesized on the donor strand and a circular, double-stranded F factor is formed; the recipient cell is thus converted to the F^+ state. F^+ cells confer fertility on F^- populations with high efficiency (often approaching 100%).

In an F^+ population, a second type of donor cell arises spontaneously at a rate of the order of 10^{-5}—i.e. 1 per 10^5 F^+ cells. Conjugation between this donor and an F^- recipient rarely confers fertility (the *donor* characteristic) on the recipient, but gives rise to a high frequency of RECOMBINATION in exconjugant recipients. These donors are designated *Hfr* (high-frequency recombination) donors. In Hfr donors the F factor has become integrated with the bacterial chromosome; the F factor thus loses its autonomy and behaves as part of the bacterial chromosome. In a given Hfr strain the F factor is integrated at a particular site, and in a particular orientation, within the chromosome; the site and orientation of the F factor are maintained during cell divisions. In other Hfr strains the F factor may occur at different chromosomal sites and/or may be present in converse orientation. (Integration of the F factor may occur at any of a number of sites where some homology exists between chromosome and plasmid.) During conjugation a break occurs in *one* strand (apparently always the *same* strand) of the (integrated) F factor. The 5′-end of the broken strand unwinds and enters the recipient in a manner analogous to that of F factor transfer in $F^+ \times F^-$ crosses (see above). However, because the F factor is integrated, the free (5′) end (the *origin*, O) is followed by the corresponding strand of the donor chromosome; thus donor genes adjacent to the F factor will be transferred in sequence in the $5' \rightarrow 3'$ order. (DNA synthesis in the donor replaces the transferred strand, so that the exconjugant Hfr donor remains viable.) The *position* of the integrated F factor within the chromosome determines the *order* in which bacterial genes are transferred to the recipient; an F factor in the same *position* but opposite *orientation* results in the transfer of genes in the reverse sequence. Since the initial strand breakage apparently occurs *within* the F factor, the complete F factor can be transmitted only when the entire donor strand enters the recipient. This occurs only rarely since strand breakage generally supervenes; cells which do receive the entire donor strand inherit the Hfr character. If donor strand breakage occurs, DNA synthesis (in the recipient) on the donor fragment template gives rise to a double-stranded fragment which can undergo recombination with an homologous region of the recipient's chromosome. (A recipient which contains such a donor fragment is called a MEROZYGOTE (partial zygote).) Recombination is generally assumed to occur by two or more CROSSING-OVER events. (see also INTERRUPTED MATING) Hfr donors may be responsible for at least part of the low level of recombination sometimes observed following $F^+ \times F^-$ crosses. Hfr cells revert to the F^+ state at rates similar to those of their formation—i.e. about 10^{-5}.

A third type of donor—the *F′ donor* (*F-prime* or *intermediate donor*)—occasionally arises, spontaneously, in a population of Hfr donors. An F′ donor contains an *F′ factor* (*F-genote, F-merogenote*) which arises during detachment of the F factor from the chromosome of an Hfr donor: during detachment the F factor carries with it adjacent region(s) of the donor chromosome. (As much as one-third of the bacterial genome may be carried by an F′ factor.) The nature of the genes carried by an F′ factor depends on the Hfr strain from which it was derived. The cell in which an F′ factor arises is referred to as a *primary* F′ donor; its chromosome suffers a *deletion* corresponding to those gene(s) which become incorporated in the F′ factor. During $F' \times F^-$ crosses the F′ donor transfers the F′ factor in a manner analogous to that of the transfer of the F factor in $F^+ \times F^-$ crosses; the recipient becomes a *secondary* F′ donor. Since the F′ factor carries bacterial gene(s), a secondary F′ donor will be *diploid* with respect to the gene(s) transferred; the transfer of bacterial genes in this manner is referred to as *sexduction* (*F-duction, F-mediated transduction*). (cf. *restricted* TRANSDUCTION) Recombination may occur between the recipient's chromosome and bacterial gene(s) carried by the F′ factor. The region of homology (the diploid region) facilitates the integration of an F′ factor with the chromosome of the secondary F′ donor. In a population of secondary F′ donors the F′ factor occurs in the integrated state in some 10% of the cells—the rates of integration and detachment being similar; hence, the population consists of some 90% F′ donors and 10% Hfr donors. Such a population thus displays characteristics *intermediate* between those of F^+ and Hfr donors.

During F′ factor formation, *part* of the F factor may remain integrated with the bacterial chromosome; such a chromosome contains a region of F factor homology which acts as a site of preferential attachment for an F factor which may subsequently infect the cell. Such regions of homology are termed *sex-factor affinity* (*sfa*) loci.

Conjugation occurs in a wide range of gram-negative bacteria—including species of *Azotobacter, Klebsiella, Proteus, Pseudomonas, Salmonella, Serratia, Shigella,* and *Vibrio.* Conjugation may occur between species of different genera—e.g. *Escherichia* × *Salmonella, Salmonella* × *Shigella*; *recombination* occurs only when there is sufficient homology between the donor fragment and the recipient chromosome.

conjugation (*ciliate protozool.*) A sexual process which occurs in many ciliate species. In *Paramecium aurelia* (see PARAMECIUM) conjugation takes place between two individuals of appropriate *mating type* (see SYNGEN). Cells pair with their ventral surfaces in contact, and cytoplasmic bridges are formed in the region of the gullet. In each conjugant the macronucleus begins to disintegrate, and both micronuclei undergo meiosis; thus eight haploid *pronuclei* are formed in each conjugant—of which seven disintegrate. The remaining pronucleus (sometimes called the *gonal* nucleus) divides mitotically to form a pair of gametic nuclei. One gametic nucleus from each conjugant then passes into its conjugal partner and fuses with the stationary nucleus, thus forming the zygote, or *syncaryon*. Subsequently, mitotic divisions lead to the formation of four diploid nuclei in each conjugant—which, by now, has separated from its partner. Two of the nuclei give rise to two new micronuclei while the other two form two new macronuclei. During the first binary fission after conjugation, the micronuclei—but not the macronuclei—undergo mitotic division; each daughter cell receives one macronucleus and two micronuclei—thus re-establishing the normal nuclear constitution of the species. Conjugation in *P. aurelia* usually takes between twelve and eighteen hours.

Conjugation in certain sessile ciliates, e.g. *Vorticella*, involves the formation of two morphologically dissimilar conjugants from a given individual; the larger of the two conjugants, the *macroconjugant*, remains attached to the stalk, while the smaller *microconjugant*—which bears special, locomotive cilia—swims in search of a macroconjugant. Following conjugal fusion, and nuclear exchange, the microconjugant either disintegrates or is absorbed by the macroconjugant.

Conjugation has not been recorded in certain ciliates—e.g. *Stephanopogon*. (see also AUTOGAMY and SELFING)

conjugation (*immunol.*) Covalent bonding between two or more different species of molecule; e.g. a dye may be conjugated with a protein—such as in FLUORESCEIN conjugated antiglobulin (referred to as a *conjugate*).

conjunctivitis Inflammation of the conjunctiva. Conjunctivitis may or may not be of microbial causation.

conk A colloquial name for the fructification of certain wood-rotting basidiomycetes—e.g. the hoof-like fruiting body of *Fomes igniarius*.

conoid A hollow cone of spirally-arranged fibrils, open at the apex, found at the extreme anterior end in coccidian sporozoites and merozoites; it may be as much as one twentieth of the length of the cell but is often smaller. The conoid may help the protozoan parasite to penetrate the cells of its host.

consolidation (of lung tissue, in pneumonia) The condition in which lung alveoli become densely packed with leucocytes and serous fluid.

constriction (in bacterial cell division) (see SEPTUM)

consumption test (*immunol.*) Any test in which an assessment is made of the amount of antibody (or antigen) removed (consumed) from a system in the form of an antigen–antibody complex. If the initial concentration of a reactant is known, the amount consumed may be determined by titrating that which remains unbound at the end of the reaction. (see also ANTIGLOBULIN CONSUMPTION TEST)

contagious abortion of cattle (see BRUCELLOSIS)

contagious disease A disease which is transmissible by physical contact between infected and healthy persons or animals. (see also INFECTIOUS DISEASE)

contagious pustular dermatitis In man, a localized, self-limiting disease involving the formation of vesicular lesion(s) on the hands, arms, or eyelids. In sheep (in which the disease is also called *orf*) vesicular lesions are formed on the lips, in the mouth, and elsewhere; the disease affects mainly young lambs. The causal agent is the *orf* virus (see POXVIRUS); infection occurs *via* abrasions etc.

contaminative infection Generally, the contamination of a pre-formed lesion by pathogens. Sometimes used to refer to infection by the ingestion of material containing pathogens. (cf. INOCULATIVE INFECTION; see also POSTERIOR STATION)

context (*mycol.*) The inner, structural tissues of a fruiting body—particularly those of the pileus in a member of the Hymenomycetes.

continuous cell line Synonym of ESTABLISHED CELL LINE.

continuous-flow culture (continuous culture; open culture) The culture of organisms (in a liquid medium) under *controlled* environmental conditions for extended periods of time. Thus, e.g. it is possible to achieve a close approach to steady-state or *balanced* GROWTH; this implies a constant growth rate, i.e. one *analogous* to the log phase of BATCH CULTURE. Additionally, accurately-controlled adjustments may be made, during growth, in one or more independent variables—level of substrate(s), temperature, pH etc. Essentially, the process involves a continual inflow of fresh medium to the culture (with which it is *well* mixed) and the simultaneous outflow of fluid from the culture vessel; the latter fluid contains organisms (and their metabolic products) and consists of a mixture of fresh and used medium. Flow may be achieved by pumps or by gravity-dependent systems; the flow rate (to and from the culture vessel) may be determined by automatic monitoring devices. Aerobic or anaerobic culture may be used. In one type of apparatus, the *chemostat*, the organism is cultured at a *sub-maximum* growth rate by controlling a growth-limiting substrate. In another type of apparatus, the *turbidostat*, the organism is maintained at its *maximum* growth rate—control of culture density (cells/unit volume) being achieved by a photosensitive device which monitors the opacity of the culture and makes appropriate adjustments in the flow rate. Difficulties in continuous culture include: (a) the inability to achieve perfect (100% efficient or instantaneous) mixing between the inflowing medium and the culture fluid, and (b) the emergence of mutants during the (extended) periods of culture. Nevertheless, in e.g. studies on microbial metabolism, data from continuous culture methods are generally more reproducible than those from batch culture methods.

Contophora A proposed taxon which includes all *eucaryotic* algae except the Rhodophyta (red algae).

contractile vacuoles (*protozool.*) Cytoplasmic organelles which occur (one or more per cell) in a variety of protozoans—including sarcodines, flagellates, and ciliates; they are generally believed to have an osmoregulatory function (i.e. the maintenance of a particular osmotic differential across the cell membrane). The contractile vacuole is a membrane-bounded region within the cytoplasm which—under normal environmental conditions—periodically fills with fluid and discharges the fluid to the exterior of the cell. Filling and discharge are known, respectively, as *diastole* and *systole*. In some species (e.g. *Paramecium*) discharge occurs always at a fixed location in the pellicle. During diastole the vacuole is fed either by small vesicles which appear to coalesce with the main vacuole—as in amoebae—or *via* a system of feeder tubules which ramify into the cytoplasm and which apparently drain areas remote from the vacuole; the latter system is characteristic of ciliates. The mechanism of vacuolar activity is not known; the systolic force may involve cytoplasmic turgor pressure, the vacuolar wall may be inherently contractile (at least in ciliates), or both of these factors may be involved. In general, the length of the diastole-systole cycle (seconds, minutes, or hours) appears to be temperature-dependent and to vary with species and with the osmolarity of the suspending medium. Vacuolar activity is higher in freshwater species than in marine species.

A similar organelle (the *pusule*) occurs in some dinoflagellates; the pusule is thought to be osmoregulatory, but does not undergo regular cycles of diastole and systole. Contractile vacuoles may occur together with the pusule in the same cell.

convalescent serum SERUM obtained from a patient in the convalescent stage of a disease. (In certain cases convalescent serum may be used for the treatment of other patients who are suffering from the same disease—see PASSIVE IMMUNITY.)

conversion (gene conversion) (see RECOMBINATION)

cooked meat medium (see ROBERTSON'S COOKED MEAT MEDIUM)

Coombs' reagent (*immunol.*). Serum containing antibodies to the antigenic DETERMINANTS of immunoglobulins, e.g. an antiserum taken from an animal (horse, sheep etc.) previously inoculated with human immunoglobulins. Such antisera are used, for example, in ANTIGLOBULIN CONSUMPTION TESTS.

Cooper–Helmstetter model (see GROWTH (bacterial))

copper (and compounds) (as antimicrobial agents) (see HEAVY METALS)

co-precipitation (*immunol.*) The precipitation of otherwise non-precipitable molecules or complexes etc. as part of, or enmeshed with, a normal immunological precipitate.

Coprinaceae A family of fungi of the order AGARICALES; the organisms live saprophytically on soil, wood, and dung. Constituent species form fruiting bodies in which the pileus bears a cutis that is typically cellular or hymeniform (i.e. composed of a palisade-like layer of cells); the basidiospores generally appear black or dark-coloured *en masse*, and each spore usually has an apical pore. The genera include e.g. COPRINUS and *Pseudocoprinus*.

Coprinus A genus of fungi of the family COPRINACEAE; the organisms occur e.g. on soil and dung. Species of *Coprinus* differ from those of other genera in the Coprinaceae in that they form fruiting bodies in which the hymenophore undergoes autodigestion at maturity (see also GILLS); typically, the fruiting bodies are fragile, and the spores *en masse* are black or dark-coloured. (see also INK CAP FUNGI)

copro-antibodies Antibodies found in the lumen of the intestine, particularly in the faeces. They appear to be produced locally, i.e. in the intestinal mucosa, and are believed to be of the IgA type.

coprogen (see SIDEROCHROMES)

coprophilous fungi Fungi which grow preferentially or exclusively on the dung of certain animals (particularly herbivores); such fungi may exhibit a special affinity for the dung of particular species of animal. The coprophilous fungi include species of *Coprinus*, *Pilobolus*, and *Sordaria*; other fungi (e.g. *Mucor* and other zygomycetes) may be found on dung—but these organisms are also found on a wide range of alternative substrates.

copy-choice model (of genetic RECOMBINATION) An early model which postulated that DNA replication occurs during recombination; the DNA polymerase was supposed to switch, periodically, from one DNA strand to another homologous strand—resulting in a new recombinant strand.

CoQ Coenzyme Q—see QUINONES (in electron transport).

coral spot fungus A common name for the ascomycete *Nectria cinnabarina*.

cord factor A glycolipid which may be extracted with petroleum ether (or other non-polar solvent) from the cell walls of certain strains of *Mycobacterium*; the cord factor is a MYCOLIC ACID 6,6′-diester of trehalose (6,6′-dimycolyltrehalose). The name *cord factor* derives from the fact that the substance can be isolated mainly from strains which characteristically grow in long cords when cultured in appropriate liquid media. The cord factor has been shown to be highly toxic for certain laboratory animals; its concentration in virulent strains of *Mycobacterium* appears to be greater than that in avirulent or attenuated strains. (see also WAX D)

Cordyceps A genus of fungi of the order SPHAERIALES.

core (1) (*bacteriol.*) The protoplast in a mature endospore—see SPORES (bacterial). (2) (*virol.*) In certain viruses: a complex consisting of nucleic acid (the viral genome) and (non-capsid) protein—the whole being enclosed by the CAPSID.

coremium An erect bundle or tuft of CONIDIOPHORES formed by certain fungi—e.g. *Penicillium expansum*. In the basal portion of the coremium the conidiophores may adhere to one another to form a plectenchymous rod. The term *synnema* is often used synonymously with coremium, although some authors reserve this term for structures in which there is an appreciable degree of plectenchyma formation. (see also SPORODOCHIUM)

co-repressor (see OPERON)

coriaceous Leathery in texture.

Coriolus A genus of fungi of the family POLYPORACEAE. Constituent species form non-stipitate, typically dimidiate, bracket-type fruiting bodies which are leathery, rubbery, or woody, and in which the whitish or pale-coloured context contains hyphae which exhibit trimitic organization. The upper surface of the fruiting body is often velvety; the hymenophore is porous (the pores being *non*-hexagonal) and the underside of the fructification is white or cream-coloured. Cystidia are not present in the hymenium. The elongated basidiospores are less than 8 microns in length. *C. versicolor* (also referred to as *Trametes versicolor*, *Polystictus versicolor* etc.) occurs e.g. on felled timber of many kinds. The upper surface of the fruiting body exhibits a number of bands of dark but contrasting colours concentric about the region of attachment of the fruiting body to the substratum; the fruiting bodies are often imbricated. The non-pigmented basidiospores are approximately 7 × 3 microns.

cornmeal agar A medium used in mycology e.g. to demonstrate chlamydospore formation in *Candida albicans*. Cornmeal (4 grams) is heated in distilled water (100 ml) for 1 hour at 60 °C; the filtrate is made up to 100 ml, agar (1.5 grams) is added, and the whole autoclaved for 15 minutes at 121 °C. While molten, the cornmeal agar is filtered through absorbent cotton wool, dispensed into tubes, and autoclaved for 15 minutes at 121 °C. (The addition of Tween-80 is reported to encourage chlamydospore formation.)

cornute (*mycol.*) (see ROESTELIOID)

Coronavirus A genus of VIRUSES whose hosts include man, rodents, avians and other vertebrates; coronaviruses are generally associated with diseases of the respiratory and enteric tracts—e.g. the common cold in man, avian infectious bronchitis, and gastroenteritis of the pig. The enveloped (ether-sensitive) *Coronavirus* virion (80–160 nm maximum dimension) contains a genome of single-stranded RNA; a characteristic feature of the virion is the presence of large closely-spaced club-shaped peplomers (orientated broad end outwards). Coronaviruses multiply in the *cytoplasm* of the host cell and mature by budding into cytoplasmic vacuoles.

correction collar An integral part of some microscope objectives by means of which it is possible to adjust the instrument to suit any of a range of coverslip thicknesses. (see also MICROSCOPY)

corticolous (corticicolous) Growing on (and/or in) the bark of trees. (cf. EPIPHLOEODAL and ENDOPHLOEODAL)

cortina (*mycol.*) The remnant of the PARTIAL VEIL (inner veil) which remains attached to the margin of the pileus in the mature fruiting bodies of certain species of agaric. The term *cortina* is sometimes used to refer to the partial veil itself.

Corynebacterium A genus of gram-positive, facultatively anaerobic, chemoorganotrophic, non-acid fast, asporogenous bacteria; the cells are pleomorphic, straight or curved (often coryneform) bacilli which are usually non-motile. Corynebacteria are regarded as relatives of the *actinomycetes*—certain *Corynebacterium* spp exhibiting features in common with *Mycobacterium* and *Nocardia*; these features include the presence of cell wall MYCOLIC ACIDS—the cell wall of *C. diphtheriae*, for example, containing an analogue of the CORD FACTOR. Type species: *C. diphtheriae*.

Corynebacterium spp fall into at least two groups: (a) Species parasitic in man and other animals (GC% range said to be about 58 ± 2). These include:

C. diphtheriae (Klebs-Löffler bacillus). Pleomorphic bacilli, 0.3–0.8 × 0.8–8 microns—the size varying according to (i) cultural type (see below) and (ii) conditions of culture. *Toxinogenic* strains of *C. diphtheriae* are the causal agents of DIPHTHERIA; the organism may be cultured from e.g. throat swabs of patients (or carriers)—primary culture usually being made on tellurite blood agar with subculture to LOEFFLER'S SERUM medium. The optimum growth temperature is 37 °C. All strains are catalase-positive, urease-negative, and produce acid (but no gas) from glucose and maltose. *C. diphtheriae* occurs in three main cultural types: *gravis*, *intermedius* and *mitis*; these may be distinguished on the basis of colonial morphology, cell morphology and biochemical activity. *Gravis*. Typically forms DAISY HEAD colonies; the bacilli tend to be short, and very few cells in a stained preparation (see ALBERT'S STAIN) exhibit METACHROMATIC GRANULES. Some strains are haemolytic. Typically, *gravis* strains produce acid from *starch* within 24–48 hours. *Intermedius*. 24-hour colonies on tellurite blood agar are grey or black and are less than 1 mm in diameter (cf. 24-hour colonies of *gravis* which are 2–3 mm in diameter); the cells may consist of a mixture of long and short bacilli in which metachromatic granules are seen more frequently than in *gravis*. No strain is haemolytic. *Mitis*. 24-hour colonies on blood tellurite agar are grey or black, 1–2 mm in diameter; the cells, each of which generally contains several metachromatic granules, tend to be long rods. Haemolytic strains are common.

All toxinogenic strains of *C. diphtheriae* are lysogenic—the prophage carrying the genetic determinants which specify toxin production. A strain may be tested for toxinogenicity e.g. by the ELEK PLATE method.

C. pseudotuberculosis (formerly *C. ovis*; the Preisz-Nocard bacillus). The causal agent of pyogenic infections in e.g. sheep and horses; the organism forms an exotoxin which differs from that of *C. diphtheriae*.

C. renale. A urease-positive species isolated e.g. from urinary tract infections in cattle.

C. xerosis. A non-pathogenic species which reduces nitrate and forms acid from maltose and sucrose.

C. equi. A pathogen of horses and other animals; the organism forms a pink pigment, reduces nitrate, and fails to form acid from the common test carbohydrates.

C. pyogenes. A species found (either alone or as part of a mixed flora) in pyogenic lesions in e.g. cattle and pigs. Most strains produce a soluble haemolysin and are catalase-negative. Similarities have been shown to exist between the cell walls of *C. pyogenes* and those of Lancefield group G streptococci; *C. pyogenes* is not universally accepted as a valid species of *Corynebacterium*.

(b) Plant-parasitic species (GC% range said to be about 60–80). These include a number of species which are motile and/or pigmented. Some cause disease in various grasses (*C. tritici*, *C. agropyri*) or lucerne (*C. insidiosum*), while *C. fascians* causes FASCIATION in a range of plants.

coryneform (1) Club-shaped. (2) The name applied to any gram-positive, asporogenous, pleomorphic, rod-shaped bacterium—particularly to *Corynebacterium* spp and morphologically-similar bacteria; in this sense DIPHTHEROID is synonymous.

Cosmarium A genus of algae—see DESMIDS.

costa A deeply-staining, curved (endoskeletal?) rod present in *Trichomonas* species; it corresponds, in position, with the proximal limit of the undulating membrane. (see also AXOSTYLE)

cotransduction The simultaneous transfer of more than one bacterial gene by TRANSDUCTION.

cotrimoxazole (see TRIMETHOPRIM)

cotton wool A fibrous, cellulosic material obtained from the seed-coats of the cotton plant (*Gossypium*); it is used e.g. for the preparation of SWABS. *Non-absorbent* cotton wool contains substances (e.g. unsaturated fatty acids) which are inhibitory to certain organisms (e.g. *Bordetella* spp); *absorbent* cotton wool has been treated with organic solvents, and bleached, so that these inhibitory substances have been largely or completely removed. Plugs of steam-sterilized non-absorbent cotton wool are often used as stoppers in tubes of bacteriological media.

cough plate A method by which *Bordetella pertussis* may be isolated from a patient suspected of suffering from WHOOPING COUGH. A plate of BORDET-GENGOU MEDIUM is exposed to droplet inoculation from the patient's cough and is subsequently incubated and examined bacteriologically. (In an alternative method use is made of a nasopharyngeal swab.)

Coulter counter An instrument used for counting the cells (or spores etc.) in a suspension. Essentially, it consists of two chambers which are separated by a thin partition of non-conducting material; the partition is perforated by a single hole of a size similar to that of the particles being counted. Each chamber contains an electrode. The sample is placed in one chamber and forced (under pressure) into the other; during this transfer the suspension forms a liquid bridge between the electrodes (*via* the hole). As each cell passes through the hole its presence, within the hole, alters the electrical conductivity of the path between the electrodes; this (momentary) change in conductivity is recorded (as a pulse) by an electronic counting circuit which is controlled by the two electrodes. The cell count may be indicated on a digital display. (see also ENUMERATION OF MICROORGANISMS)

Councilman body A group of *hyaline* necrotic cells in the liver; Councilman bodies are characteristic features of YELLOW FEVER.

counterstain (1) A *second* stain used on a microscopical preparation with the object of staining, in a contrasting way, those features of the preparation (e.g. cells, or parts of cells) which have not taken up the initial stain. (2) In the GRAM STAIN, a counterstain (often dilute carbolfuchsin) is used to stain those organisms, or tissues, which have been decolorized during the differentiation stage. (3) In the AURAMINE-RHODAMINE STAIN a solution of potassium permanganate (the 'counterstain') is used to reduce or eliminate any background (non-specific) fluorescence.

counting (of microorganisms) (see ENUMERATION OF MICROORGANISMS)

counting chamber An instrument used for determining the total cell count, viable cell count, or spore count etc. of a suspension of cells or

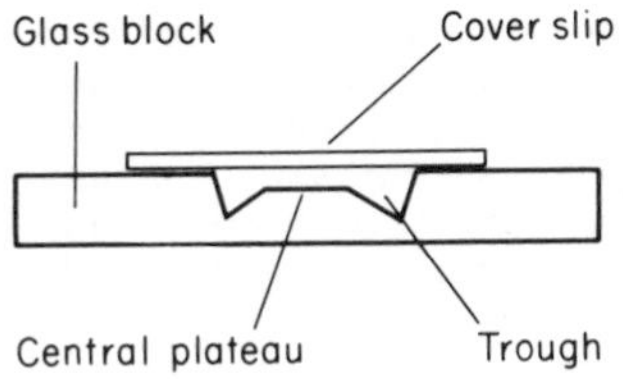

A COUNTING CHAMBER (Cross section)

spores. (see ENUMERATION OF MICROORGANISMS) The basic design of a chamber is shown in the diagram. A grid, etched on the central plateau of the glass block, typically consists of a square of side 1.0 mm divided into 400 squares, each 0.0025 mm^2; the distance between grid and cover-slip may be 0.1 mm (as in the Thoma chamber) or 0.02 mm (as in the Helber chamber). When correctly positioned, Newton's rings should be visible through those parts of the cover slip which are in contact with the glass shoulders—indicating close contact. The suspension is introduced (with a Pasteur pipette) into the space between cover-slip and central plateau; the troughs should remain empty. The cells are allowed to settle and are counted under the microscope; since the volume between grid and cover-slip is known, the count per unit volume can be calculated. For the determination of a *viable* cell count the cells are first suspended in a solution containing a vital stain—see VITAL STAINING.

cover-slip (see SLIDE)

covirus (multicomponent or multiparticle virus) A virus which exists as two or more entirely separate particles—all of which must concurrently infect the host organism for the *complete* replication cycle to occur. (cf. SATELLITE VIRUS) Coviruses include the cowpea mosaic, alfalfa mosaic and tobacco rattle viruses. The *tobacco rattle virus* consists of one long and one short rod—each rod consisting of protein-coated RNA. Alone, the *long* rod can replicate within a plant and cause lesions; however, under these conditions, only the *nucleic acid* of the long rod is replicated. This form of infection is not readily sap-transmissible. Alone, the *short* rod cannot replicate or cause lesions in a host plant. Infection with complete viruses (long *and* short rods) is readily sap-transmissible; the short rod codes for the coat protein of both long and short rods. (Tobacco rattle virus causes disease in tobacco, potato, tomato and other plants; transmission may occur either mechanically or *via* parasitic nematodes of the genus *Trichodorus*.)

cowpox In man, a mild, self-limiting disease involving the formation of vesicular lesion(s)—usually on the hands or arms—accompanied by local oedema and lymphangitis; healing and recovery are rapid. In the cow vesicular lesions are formed on the udder and teats. The causal agent is the *cowpox virus* (see POXVIRUS); infection occurs *via* abrasions etc.

Coxiella A genus of gram-negative bacteria of the RICKETTSIACEAE; the sole species, *C. burnetii*, is an obligate intracellular parasite in a range of vertebrate and arthropod hosts, and is the causal agent of Q FEVER in man. The (non-motile) cells are usually rod-shaped (about 0.2–0.6 microns in length) but coccoid forms (less than 0.5 microns in diameter) have been reported; a bacterial-type cell wall is present. The organisms stain well with the Giménez stain. Laboratory culture of *C. burnetii* is carried out e.g. in the yolk sac membrane of the hen's egg; incubation temperature: 35 °C. *C. burnetii* is appreciably more resistant than other rickettsiae to physical and chemical inhibitory agents.

coxsackievirus (see ENTEROVIRUS)

CPE (*virol.*) CYTOPATHIC EFFECTS.

Craigie's tube method A simple method for (a) separating motile from non-motile organisms, or (b) isolating alternative-phase strains of *Salmonella* (see PHASE VARIATION, IN SALMONELLA). Into a test-tube containing a *semi-solid* agar medium is placed a length of glass tubing of internal diameter about 3 mm; the upper end of the tubing projects above the level of the medium, and the lower end—being obliquely cut—permits free communication between the agar in the tubing and that in the remainder of the test-tube. If the medium *within the tubing* is inoculated with a mixture of motile and non-motile organisms, and the whole incubated, motile strains may migrate downwards and pass into the medium surrounding the tubing—from which they may subsequently be isolated. For the isolation of alternative-phase strains of *Salmonella* the medium is supplemented with antiserum to the flagella antigens of the existing phase; the medium within the tubing is inoculated and the whole incubated. Antiserum inhibits the motility of strains of the existing phase but is without effect on those of the alternative phase—which may be subsequently isolated from the medium outside the tubing.

Credé procedure (see GONORRHOEA)

crenate (of the edge of a colony etc.) Scalloped.

creosote A highly effective antifungal wood preservative widely used e.g. for the treatment of railway sleepers, telegraph poles etc. Creosote consists of a complex mixture of coal-tar distillation products obtained between 200 °C and 400 °C; these include e.g. phenols, cresols, quinoline, anthracene, naphthalene etc. (see *Lentinus lepideus* under TIMBER, DISEASES OF and CREOSOTE FUNGUS)

creosote fungus The deuteromycete *Cladosporium resinae* (= *Hormodendrum resinae*)—a fungus able to utilize creosote as a carbon source.

cresol red (PH INDICATOR) pH 7.2 (yellow) to pH 8.8 (red).

cresols (as antimicrobial agents) (see PHENOLS)

CRF COAGULASE reacting factor.

crista (1) (in *Cristispira*) The longitudinal ridge formed by the axial filament—see SPIROCHAETALES. (2) (see MITOCHONDRION)

Cristispira A genus of bacteria (order SPIROCHAETALES); species are parasitic in molluscs and in certain other aquatic animals. (see also CRISTA)

Crithidia (synonym: *Strigomonas*) A genus of flagellate protozoans of the sub-order TRYPANOSOMATINA; the organisms are gut parasites of invertebrates (chiefly insects). Amastigote and choanomastigote stages occur.

crithidial form (see EPIMASTIGOTE)

critical illumination (in MICROSCOPY) A form of illumination in which the source of light (e.g. the filament of a lamp) is focused (by the sub-stage condenser) to form an image in the plane of the object (specimen). Critical illumination tends to illuminate the specimen unevenly—particularly when the field of view is large. (cf. KÖHLER ILLUMINATION)

CRM (cross-reacting material) A protein, specified by a mutant gene, which is *serologically* related to (i.e. cross-reacts with) that specified by the wild-type gene—but which lacks e.g. the enzymic activity of the latter.

Cronartium A genus of fungi of the order UREDINALES. (see also RUSTS (1)) One species, *C. ribicola*, is the causal agent of 'blister rust' of the white pine.

cross-feeding (see SYNTROPHISM)

cross-reacting antibody (1) An ANTIBODY which is able to combine with a *non-homologous* antigen (i.e. an antigen which is not identical to the one which elicited the antibody). For such *cross-reaction* to occur the antigenic determinants of the homologous and non-homologous antigens must have a high degree of stereochemical similarity. Cross-reacting antibodies often have appreciably less affinity than the homologous antibody for a given antigen. (2) Antibodies raised against a particular microorganism may combine with organisms of a related species or, sometimes, with organisms which are quite unrelated. Such organisms may share a common antigenic determinant.

cross-reacting material (see CRM)

cross-resistance (see ANTIBIOTICS, RESISTANCE TO)

cross wall (bacterial) (see SEPTUM)

crossing-over (genetics) A process in which, in effect, a break occurs in each of two adjacent DNA strands and the exposed 5′ and 3′ ends of each strand unite with the exposed 3′ and 5′ ends, respectively, of the adjacent strand; for a possible mechanism see *reciprocal* RECOMBINATION. Crossing-over occurs e.g. between homologous chromatids during MEIOSIS and, occasionally, during mitosis—see PARASEXUALITY IN FUNGI; in bacteria, crossing-over is believed to occur in merozygotes formed by CONJUGATION, TRANSDUCTION and TRANSFORMATION, and during insertion and excision of episomes (e.g. the F FACTOR) and phage (see LYSOGENY). (*Crossing-over* is sometimes (incorrectly) used as a synonym of *recombination*.) (see also CHIASMA)

crown gall A callus-like tumour which may develop on the stems or roots of a range of plants (e.g. stone fruits, tomatoes, sunflowers) following infection (*via* a wound) by *Agrobacterium tumefaciens*. This organism appears to cause an hormonal imbalance in the host plant—possibly by stimulating the production of AUXINS by the plant; some strains of *A. tumefaciens* produce 3-indoleacetic acid. The pathogen appears to produce a factor known as the *tumour inducing principle* (TIP) which is thought to consist of DNA. TIP is believed to cause the host's cells to undergo uncontrolled division—even in the absence of the pathogen. (see also GALLS)

crozier (see ASCUS)

CRP (see CATABOLITE REPRESSION)

crucifers, bacterial diseases of An important example is *black rot* of cabbage, cauliflower, broccoli etc.; the causal agent is *Xanthomonas campestris*. Infection may be seed-borne, or the pathogen may enter the plant *via* stomata or wounds; symptoms include blackening of the veins and generalized chlorosis of the leaf. *Soft rot*, caused by *Erwinia* spp, may develop as a secondary disease.

crucifers, fungal diseases of These include: (a) *Clubroot*. A disease characterized by hyperplastic root swellings and stem GALLS; the upper plant may subsequently wilt or exhibit retarded growth. The causal agent, *Plasmodiophora brassicae*, is an intracellular parasite which stimulates the host cells to enlarge and divide. The fungus may invade young roots or gain entry *via* wounds. Diseased plants may suffer secondary invasion by soft rot bacteria. (b) *Blackleg* may affect e.g. turnips, cabbage, cauliflower etc.; the causal agent is *Phoma* sp. Lesions—which may bear pycnidia—appear on leaves and stems; gradual decay of the root system may occur.

crustose (crustaceous) (1) Forming or resembling a crust. (2) (of a lichen) Refers to a lichen whose entire thallus forms a crust which adheres strongly to the substratum (rock, bark etc.) and which lacks a lower cortex; attachment is effected by hyphae of the medulla. Crustose lichens include e.g. members of the genera CANDELARIELLA, ENTEROGRAPHA, OCHROLECHIA, and RHIZOCARPON. (cf. FOLIOSE and FRUTICOSE)

cryo- A combining form which indicates a connection with *cold*, e.g. *cryoprotectant* (see FREEZING).

cryptic mutant A cell which lacks one or more components of a membrane TRANSPORT system so that a particular substrate, or substrates, cannot enter the cell, and hence cannot be utilised—even though the cell may possess all the relevant metabolic enzymes. For example, the failure of a cell to produce *Enzyme I* (see TRANSPORT) as a result of a (pleiotropic) mutation would prevent that cell (a cryptic mutant) from taking up and metabolising a range of sugars.

cryptobiosis (see DORMANCY)

Cryptococcaceae A family of fungi within the class BLASTOMYCETES; saprophytic species have been isolated from e.g. fruit, the intestine of *Drosophila* (the fruit fly), from tree bark, and from mosses and lichens associated with trees—while parasitic and pathogenic species occur e.g. on the skin and within the tissues of man and other animals. The vegetative form of the organisms may be (according to species and/or conditions of growth) a single cell and/or a pseudomycelium or true mycelium; reproduction occurs by budding, and some species may form arthrospores—though none gives rise to ballistospores (cf. SPOROBOLOMYCETACEAE). In culture the organisms may appear cream, yellow, orange, pink, or red—according to species. Members of the family Cryptococcaceae appear to be imperfect stages of ascomycetes and basidiomycetes; genera include CANDIDA, CRYPTOCOCCUS, KLOECKERA, PITYROSPORUM, RHODOTORULA, *Torulopsis*, and TRICHOSPORON.

cryptococcosis (formerly *torulosis*) A disease of man and other animals; the causal agent is *Cryptococcus neoformans*. In man, infection probably occurs chiefly *via* the respiratory tract; *C. neoformans* is often present in the excrement of pigeons, and may remain viable for months in dried faeces. Pulmonary infection may be mild or inapparent, and lung lesions do not calcify on healing. In systemic cryptococcosis the liver, bones, skin etc. may be affected; cryptococcal meningitis is a common form of the disseminated disease. Amphotericin B has been used for treatment. Among domestic animals the disease may affect the upper respiratory tract, lungs, meninges or viscera; cryptococcal mastitis has been recorded in the cow. *Laboratory diagnosis* may include microscopical and cultural examination of sputum, pus, milk, CSF etc.

Cryptococcus A genus of fungi within the family CRYPTOCOCCACEAE; species have been isolated from e.g. various plants, seawater, the atmosphere, soil, the human respiratory tract, and from the milk of cows suffering from mastitis. The vegetative form of the organisms is a single spherical or ovoid cell which typically bears a polysaccharide capsule; some species may form a rudimentary pseudomycelium though none forms a true mycelium. Reproduction occurs by multipolar budding. The organisms are non-fermentative; they assimilate inositol. Species include:

C. neoformans. Individual cells are typically 3–10 microns, maximum dimension; pseudomycelium is not formed. *C. neoformans* does not assimilate nitrate and fails to grow in vitamin-free media; it is one of the few species of *Cryptococcus* capable of growth at 37 °C. The (recently-discovered) basidiomycetous stage of *Cryptococcus neoformans* has been designated *Filobasidiella neoformans*.

Cryptogamia (*bot.*) An obsolete taxonomic category in which were placed all those organisms which lack *obvious* reproductive structures (flowers, cones). The cryptogams included the fungi and algae (and other microorganisms) as well as the bryophytes (e.g. liverworts) and pteridophytes (e.g. ferns, horsetails etc).

Cryptophyta A division of the ALGAE.

crystal violet A deep-violet, non-fluorescent basic dye of the triphenylmethane group; it is used e.g. in the GRAM STAIN. Crystal violet is selectively bacteriostatic and, for this reason, was formerly used in media employed for the isolation of pathogenic mycobacteria. The dye is appreciably more soluble in ethanol than in water. (see also TRIPHENYLMETHANE DYES)

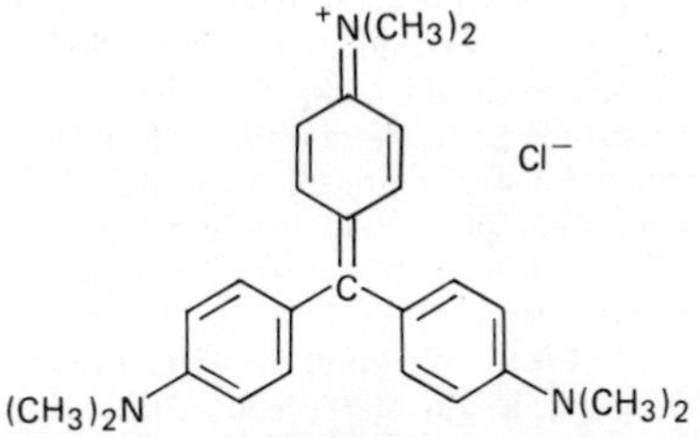

Crystal violet

CSF Cerebrospinal fluid.

C.S.S.D. Central sterile supply department. The department in a hospital (or other organization) responsible for supplying appropriate sterile materials.

CTAB (see QUATERNARY AMMONIUM COMPOUNDS)

Ctenostomatida (see Odontostomatida in SPIROTRICHIA)

CTP Cytidine-5′-triphosphate (see NUCLEOTIDE).

cudbear A purple dye obtained from the 'cudbear lichen' *Ochrolechia tartarea.*

cultivar A variety or strain of a cultivated species of plant.

culture (1) (*noun*) A growth of particular type(s) of microorganism on or within a solid MEDIUM, or in a liquid medium, formed as a result of the prior INOCULATION and INCUBATION of that medium. (cf. TISSUE CULTURE) Growth on a solid medium may be present as a continuous layer or film (*confluent growth*) or as discrete (individual) colonies (see COLONY)—depending e.g. on the method of inoculation (see e.g. LAWN PLATE and STREAKING). Cultures in liquid media may be turbid or clear, and may or may not have a PELLICLE.

Aerobic culture: a culture produced by the incubation of an inoculated medium in the presence of air. *Anaerobic culture*: a culture produced by the incubation of an inoculated medium in the absence of oxygen. (see also ANAEROBE) *Broth culture*: one in NUTRIENT BROTH or in a similar liquid medium. *Contaminated culture*: one which has been (unintentionally) exposed to non-STERILE conditions and which has become contaminated with extraneous organisms. *Mixed culture*: one containing two or more species or strains of organism. *Old culture*: one which has been incubated or stored for an excessive period of time and in which, as a consequence, degenerative changes in the organisms may be expected to have occurred (see e.g. INVOLUTION FORMS). *Plate culture* (a *plate*): one on a solid medium in a PETRI DISH. *Primary culture*: see PRIMARY CULTURE. *Pure culture*: one comprising organisms which are all of the same species or strain (see also AXENIC). *Shake culture*: see SHAKE CULTURE. *Slant culture* (a *slant* or SLOPE): one grown on a solid medium (usually agar) which has been allowed to set in a test-tube or bottle placed at an angle to the horizontal. *Stab culture*: one produced by deep inoculation of a solid medium (e.g. agar or gelatine) with a STRAIGHT WIRE; the wire is plunged vertically into the medium so that the INOCULUM (on the tip of the wire) is distributed along the length of the stab. *Subculture*: see SUBCULTURE. *Tissue culture*: see TISSUE CULTURE.

(2) (*verb*). To encourage the growth of particular type(s) of microorganism under controlled conditions. Essential requirements for growth are: suitable nutrients (and, sometimes, specific growth factors—e.g. particular VITAMINS); the absence of biological, chemical, and/or physical inhibitory factors; oxygen for obligate AEROBES and the absence of oxygen for obligate ANAEROBES; adequate light for obligately photosynthetic organisms; a suitable ionic environment; adequate water (see also WATER ACTIVITY); an optimum or suitable temperature. The incubation period (i.e. the time required for the development of detectable or significant amounts of growth) may be of the order of hours, days, or weeks—depending on the organism and growth conditions.

Cultures of bacteria or fungi are started by seeding, or inoculating, a medium with viable cells or SPORES from another culture, or by inoculating the medium with material expected to contain viable organisms or spores of particular type(s) of organism; the nature of the medium and incubating conditions required differ widely among the different types of microorganism. BACTERIOPHAGES are propagated within appropriate bacteria; see also VIRUSES, CULTURE OF. A living system must also be used for the culture of certain bacteria—e.g. *Treponema pallidum*, *Rickettsia* spp. (see also ASEPTIC TECHNIQUE; ENRICHMENT; LOOP; POUR PLATE)

Cunninghamella A genus of fungi within the class ZYGOMYCETES; *Cunninghamella* spp occur typically in the soil. Asexual reproduction involves the formation of deciduous conidia which develop as a single layer on the swollen distal ends of branched or unbranched conidiophores; some species (e.g. *C. elegans*) produce spiny (echinulate) conidia. Sporangia and sporangiola are not formed. Sexual reproduction involves gametangial fusion and zygospore formation.

cup fungi Common name for those *ascomycetes* which produce a cup-shaped fruiting body—particularly members of the order Pezizales.

cup lichens A common name for those species of CLADONIA whose podetia have conspicuous scyphi.

cupulate (1) Cup-shaped. (2) (aecidioid) (*mycol.*) Of *aecia* (see RUSTS (1)): cup-shaped with near-parallel sides—the edge of the ruptured peridium projecting very little beyond the level of the host's epidermis. Cupulate aecia are formed e.g. by *Puccinia* spp. (cf. ROESTELIOID)

curing (*bacteriol.*) (1) During the induction of a lysogen (see LYSOGENY): elimination of the phage by the inducing agent without loss of bacterial viability. (2) See PLASMID.

curling factor (see GRISEOFULVIN)

cutis (*mycol.*) A thin outer layer of tissue which invests the pileus and/or stipe in certain fungi.

Cutleria A genus of algae of the PHAEOPHYTA.

cyanellae Blue-green algal endosymbionts found e.g. in the flagellate *Cyanophora*.

cyanide (as a RESPIRATORY INHIBITOR) Cyanide binds to and inhibits component a_3 of eucaryotic cytochrome oxidase (see CYTOCHROMES); many bacterial cytochrome oxidases are also inhibited by cyanide.

cyanide test (KCN test) (*bacteriol.*) A test used in the identification of certain species of bacteria—particularly members of the family Enterobacteriaceae. The medium used (Møller's medium—or a modification of it) contains (%w/v): peptone or equivalent (0.3), sodium chloride (0.5), potassium dihydrogen phosphate (0.0225), disodium hydrogen phosphate (0.564), and potassium cyanide (0.0075). Following inoculation, the medium is incubated for 24 or 48 hours and examined for growth; growth constitutes a positive reaction. The rationale of the test is unknown; it is possible that different mechanisms of cyanide-insensitivity occur among those organisms which exhibit this property.

cyanobacteria (see BLUE-GREEN ALGAE)

Cyanochloronta (see BLUE-GREEN ALGAE)

cyanocobalamin (see VITAMIN B_{12})

cyanophage Any virus whose host is a blue-green alga; cyanophages have been isolated from both filamentous and unicellular BLUE-GREEN ALGAE (e.g. species of *Lyngbya*, *Nostoc*, *Plectonema*, and *Microcystis*). The cyanophages include 'tailed' viruses which are similar, morphologically, to certain types of BACTERIOPHAGE; all cyanophages appear to contain double-stranded DNA. Most of the known cyanophages are virulent—i.e. they lyse their host cells; the cells of a blue-green algal filament appear to be lysed at random—the death of intercalary cells resulting in the progressive fragmentation of the filament. Cyanophages which appear to 'lysogenize' their hosts have been reported. (see LYSOGENY) The nomenclature of cyanophages is currently based on the name(s) of the relevant host organism(s); thus, the first cyanophage to be discovered (hosts: *Lyngbya*, *Plectonema*, *Phormidium*) was designated LPP-1, and a second, antigenically-distinct, strain of this virus was designated LPP-2. Other cyanophages include SM-1 (hosts: *Synechococcus*, *Microcystis*), and N-1 (host: *Nostoc*).

cyanophilic (*mycol.*) Refers to those structures which have a strong affinity for certain blue dyes—e.g. LACTOPHENOL COTTON BLUE.

Cyanophyceae (see BLUE-GREEN ALGAE)

cyanophycean starch (see STARCH)

Cyanophyta (see BLUE-GREEN ALGAE)

cyanosomes PHYCOBILISOMES of the Cyanophyta.

Cyathus A genus of fungi of the order NIDULARIALES. In *Cyathus* spp ('bird's-nest fungi') the fruiting body is funnel-shaped. (In the immature fruiting body an *epiphragm* covers the opening of the funnel.) The wall of the funnel is the peridium of the fruiting body; attached to the inner surface of the peridium are a number of rounded, flattened *peridioles*—each peridiole consisting essentially of a sac (the *tunica*) which contains basidiospores. Each peridiole is attached to the peridium by a complex thread (the *funiculus* or *funicle*) which consists basically of three sections (joined end to end). Attached directly to the peridium is the *sheath* (a short cord of aggregated hyphae) which passes directly into the *middle piece*—a narrower extension of the sheath; the middle piece is attached to an elongated sac, the *purse*, which is itself attached to the peridiole. The purse contains a long thread of coiled hyphae (the *funicular cord*)—one end of which is firmly attached to the peridiole, while the other (unattached) end bears a mass of strongly adhesive hyphae, the *hapteron*. The peridioles are projected from the fruiting body when rain drops fall into the funnel-shaped peridium (sometimes referred to as a 'splash cup'); during the ejection of a peridiole the purse ruptures, and the peridiole flies into the air—trailing the funicular cord with its terminal hapteron. The hapteron adheres to nearby vegetation (e.g. a blade of grass).

cyclic AMP receptor protein (see CATABOLITE REPRESSION)

cyclical transmission, of pathogenic microorganisms (see VECTORS)

Cyclidium A genus of protozoans of the sub-order PLEURONEMATINA.

cyclitol antibiotics ANTIBIOTICS which contain a cyclic alcohol. Examples include the AMINOGLYCOSIDE ANTIBIOTICS (which contain a substituted inositol).

cycloheximide (Actidione; β-(2-(3,5-dimethyl-2-oxocyclohexyl)-2-hydroxyethyl)glutarimide) An ANTIBIOTIC produced by certain *Streptomyces* spp, e.g. *S. griseus*; it is a by-product in the manufacture of streptomycin. Cycloheximide is active against many fungi and other eucaryotes; bacteria are not affected. It apparently inhibits PROTEIN SYNTHESIS by preventing translocation—the primary site of action probably being the 60S subunit of 80S RIBOSOMES. 80S ribosomes from different sources may differ in sensitivity; thus, e.g. *Saccharomyces cerevisiae* is highly susceptible while *S. fragilis* is resistant. Only the yeast-like forms of certain dimorphic fungi, e.g. *Histoplasma capsulatum*, *Blastomyces dermatitidis*, are susceptible to cycloheximide. Certain stereoisomers, e.g. *isocycloheximide*, *naramycin B*, have been isolated from *Streptomyces* cultures; in general these, and synthetic cycloheximide derivatives (e.g. the thiosemicarbazone), are less active against fungi. *Streptovitacins*, produced by *S. griseus*, are monohydroxyl-substituted cycloheximides.

D-cycloserine (D-4-amino-3-isoxazolidone) An antibiotic which can be derived from cultures of *Streptomyces orchidaceus* or prepared synthetically. D-Cycloserine inhibits an early stage in the biosynthesis of the PEPTIDOGLYCAN component of bacterial cell walls. By acting as an analogue of D-alanine, this antibiotic competitively inhibits the two enzymes (*alanine racemase* and *D-alanyl-D-alanine synthetase*) involved in the synthesis of D-alanyl-D-alanine; incorporation of this dipeptide is normally the final stage in the synthesis of the PARK NUCLEOTIDE—a precursor of peptidoglycan. The antibiotic effect is enhanced owing to the accumulation of D-cycloserine by the cell. D-cycloserine is bacteriostatic; its spectrum of activity includes gram-positive and gram-negative organisms.

Cylindrocystis A genus of algae—see DESMIDS.

Cylindrotheca A genus of pennate diatoms of the BACILLARIOPHYTA.

cyphella (*lichenol.*) A round depression or pore in the lower surface of the thallus in foliose lichens of the genus *Sticta*; the lower cortex forms a protruding rim around the edge of the pore, while a layer of globose fungal cells lines the depression. Cyphellae can be observed macroscopically as small white pits in the lower surface of the thallus; they are believed to facilitate gaseous exchange in the thallus. (cf. PSEUDOCYPHELLA)

Cyrtophorina A sub-order of protozoans of the GYMNOSTOMATIDA (sub-phylum CILIOPHORA). In cyrtophorine gymnostomes the cell is often flattened dorso-ventrally, the cytostome being on the ventral side; the cytopharyngeal structure is generally more complex than that of the rhabdophorines. The cyrtophorines are herbivorous. Genera include e.g. *Chilodonella* and *Nassula*.

cyst The form assumed by certain microorganisms during specific stages in their life cycles or as a response to particular environmental conditions; during encystment an organism produces a thin or thick-walled membranous structure within which it becomes totally enclosed. In general, cyst-formation serves either a protective (and disseminative) or a reproductive function.

Cyst-formation is uncommon among bacteria. *Azotobacter* forms a thick-walled spherical or oval cyst which protects the organism from desiccation, and *microcysts* and/or *cysts* (*sporangia*) are produced by some species of the MYXOBACTERALES. Structures known as cysts (equivalent to sporangial initials) are formed by some members of the lower fungi (e.g. *Rhizophidium*); cyst-formation also occurs in diplanetic fungi (e.g. *Saprolegnia*). Certain types of free-living and parasitic protozoans commonly form cysts; in the free-living amoebae encystment permits survival (and dissemination) under dry conditions, while the cysts of *Entamoeba histolytica* and those of the COCCIDIA are important factors in the transmission of the organisms from one host to another. Cysts are formed by many ciliates (e.g. *Balantidium coli, Ichthyophthirius*) and some flagellates (e.g. *Giardia lamblia*).

Chemical composition of cyst walls. The cyst wall may consist of a single layer or of several layers of similar or different chemical composition. In most cases definitive information on the nature of the cyst wall is sparse. In the soil amoeba *Acanthamoeba* the cyst walls are thought to consist of an outer protein layer and an inner layer of cellulose. The walls of at least some coccidian oocysts are believed to contain a high proportion of protein together with some lipid and/or carbohydrate.

Staining. The cyst *wall* typically fails to take up the common dyes—though the cyst interior may be readily stained. The cysts of *Entamoeba histolytica* may be identified by staining the cyst interior with iodine; LUGOL'S IODINE ($\frac{1}{2}$–$\frac{1}{4}$ strength) is often used for staining wet mounts. The search for cysts in wet mounts of faeces can often be facilitated by the addition of one or two drops of eosin (1%): faecal debris takes up the stain while the cysts remain colourless. For methods of concentrating cysts (from faeces etc.) see FLOTATION and FORMOL-ETHER METHOD.

cysteine (biosynthesis of) (see Appendix IV(c))

cystidium (*mycol.*) A large, elongated, sterile cell which occurs among the basidia in the hymenia of certain basidiomycetes; cystidia vary greatly in size and form but usually project some way beyond the tips of the basidia. The function of cystidia is largely unknown; however, those which arise from the gill surfaces of certain species of *Coprinus* appear to function as 'spacers'—i.e. they maintain a small but significant distance between adjacent gills (lamellae), thus facilitating spore dispersal.

cystite A term formerly used for the larger of the two types of coccoid cells formed by *Arthrobacter*.

cytarabine (1-β-D-arabinofuranosyl-cytosine) A drug which inhibits the replication of certain DNA viruses; it also inhibits DNA synthesis in animal cells. Cytarabine has limited application in the treatment of e.g. severe herpes simplex infections.

cytidine-5′-triphosphate (biosynthesis of) (see Appendix V(b))

cytochromes A class of *haemoproteins*; the protein and HAEM of a cytochrome are linked *via* the 5th and 6th coordinate positions of the haem iron. A cytochrome effects *electron transfer* by the oxidation/reduction of its haem iron—see e.g. ELECTRON TRANSPORT CHAIN and PHOTOSYNTHESIS.

Cytochrome nomenclature. Cytochromes are classified by their haem group(s). Cytochromes *a* contain *haem a*—which carries one formyl group and one long hydrophobic side chain. Cytochromes *b* contain *protohaem* (see PORPHYRINS). In cytochromes *c*, *mesohaem* is bonded to the protein moiety by (covalent) thioether linkages as well as by the coordinate bonds. Cytochromes *d* contain an *iron-chlorin* (chlorin = dihydroporphyrin). When the protein moiety of a reduced cytochrome is replaced (experimentally) by two pyridine molecules, the resulting *dipyridine ferrohaemochrome* exhibits a characteristic absorption spectrum which may be used for identification. For each class of cytochromes the corresponding dipyridine ferrohaemochrome α-absorption bands fall within a narrow band of wavelengths: cytochromes *a* 580–590 nm; cyts *b* 556–558 nm; cyts *c* 549–551 nm; cyts *d* 600–620 nm.

Individual cytochromes within a given class may be distinguished by subscripts (e.g. c, c_1, c_2 etc.); alternatively, they may be identified by the position of the α band in the absorption spectrum of the (intact) *reduced cytochrome*—e.g. *b*-557 or b_{557}. Certain cytochromes have tended to retain their original names—even though such names do not reflect their current classification; thus e.g. cyt a_2 is now cyt *d*, cyt *f* is cyt b_6, and cyt *o* is a cytochrome of the *b* type.

Cytochromes containing more than one prosthetic group. Some cytochromes contain, in addition to the haem, a second prosthetic group; thus, e.g. cyt b_2 from certain yeasts contains FMN. The cytochrome *c* oxidase of eucaryotes (= cyt aa_3) is believed to contain two haem *a* groups linked to a single protein; it also contains two copper atoms and a lipid fraction. Cyt d_1c (containing iron-chlorin *and* mesohaem) functions as a nitrite reductase in *Pseudomonas aeruginosa*.

Cytochromes in eucaryotes. Mitochondria contain cyts aa_3, b_T, b_K, c_1 and *c*; many cells also contain b_5 and P-450 associated with the endoplasmic reticulum. Cyt P-450 (the cytochrome-carbon monoxide compound absorbs at 450 nm) is a *b*-type cytochrome known to be involved in a number of hydroxylation reactions e.g. of sterols, drugs. Among protozoans, many species appear to have systems similar to that of higher organisms, but others do show certain differences—e.g. some appear to contain cytochrome *o*. Some parasitic protozoans lack cytochromes—e.g. bloodstream forms of trypanosomes, intracellular stages of *Leishmania donovani*.

Cytochromes in bacteria. In bacteria the cytochromes are normally located in the CELL MEMBRANE and vary (quantitatively and qualitatively) from one species to another, and sometimes even in a single species under different cultural conditions; cytochromes appear to be absent from some facultative and obligate anaerobes. In bacteria a variety of cytochromes may act as terminal oxidases—including cyts a_1, a_3, *d* and *o*; more than one functional oxidase may be present in a given species (e.g. *Haemophilus parainfluenzae* contains cyts a_1, *d* and *o*). *Pseudomonas putida* contains a soluble P-450 which (in conjunction with an iron-sulphur protein and a FAD-containing reductase) effects hydroxylation of camphor and related compounds in a cyclic reaction sequence. A P-450 is also found in certain strains of *Rhizobium*. Some bacteria contain modified *c*-type cytochromes (*cytochromoids c* or cytochromes *c′*)—see e.g. RHP; these contain covalently-bound haem but differ chemically and spectroscopically from normal *c*-type cytochromes (e.g. RHP reacts with carbon monoxide, cyts *c* do not).

cytochromoid (see CYTOCHROMES)

cytocidal (*adjective*) Capable of killing cells.

cytohet (cytoplasmic HETEROZYGOTE) A cell in which occur two genetically-different copies of a given type of *cytoplasmic* DNA (e.g. mitochondrial DNA).

cytokinesis Those events—apart from nuclear division (e.g. MITOSIS, MEIOSIS)—which occur during the division of a eucaryotic cell into two or more progeny cells; such events include the apportionment of the cytoplasm (and cytoplasmic organelles) and the synthesis of new cell membrane/cell wall material.

cytokinins (phytokinins, kinins) Plant hormones (phytohormones) which, in conjunction with AUXINS, stimulate cell division and possibly differentiation; they are synthesized at the root apex and translocated *via* the xylem. The majority of cytokinins are derivatives of adenine—although similar activity is also shown by certain urea derivatives.

A substance which functions as a cytokinin, *kinetin* (6-furfuryl-aminopurine), was first isolated as a breakdown product of yeast and animal DNA; kinetin has not been found in plants. A number of compounds with cytokinin activity have been found in a variety of higher plants and algae. There is some evidence that cytokinins are involved in the production of symptoms in certain plant diseases—particularly where these include cell proliferation. For example, *Corynebacterium fascians* produces a cytokinin which may be responsible for the WITCHES' BROOM condition caused by this organism. Certain (nodule-forming) *Rhizobium* species have also been reported to synthesize cytokinins. (see also GIBBERELLINS)

cytomegaloviruses (see HERPESVIRUS)

cytomembranes (bacterial) Membranous systems or structures which occur within the bacterial cytoplasm; according to type, cytomembranes may or may not be continuous with the CELL MEMBRANE. Examples of cytomembranes include CHLOROBIUM VESICLES, CHROMATOPHORES, GAS VACUOLES, and MESOSOMES. (see also NITROBACTER and NITROSOCOCCUS)

cytomere (*protozool.*) A detached portion of a *schizont*.

cytopathic effects (CPE) (*virol.*) Any form of macroscopic or microscopic, localized or generalized degenerative change or abnormality in the cells of a monolayer TISSUE CULTURE (or membrane(s) of an embryonated egg) due to infection by viruses; whether or not CPE are produced in a virus-infected tissue culture (or egg membrane) depends mainly on the nature of the virus and of the infected cells. (Some viruses—e.g. the mouse leukaemia viruses—characteristically fail to give rise to CPE even when viral replication occurs.) According to the virus-cell system, CPE may include e.g. cell death (see also PLAQUE), cytoplasmic vacuolation (see FOAMY AGENTS), the formation of syncytia (e.g. by paramyxoviruses), the formation of discrete foci of cell proliferation—'microtumours' (see e.g. ROUS SARCOMA VIRUS). Certain forms of CPE are exploited in the assay of viruses; for assay purposes, conditions are arranged such that any given discrete lesion (or region of degenerative change) in a monolayer (or egg membrane) can be regarded as the result of infection by a single virion. (see also INCLUSION BODIES)

Cytophaga A genus of gram-negative nonsporing bacteria of the CYTOPHAGACEAE; species occur in the soil and in freshwater and marine environments. The cells are flexible rods, or unbranched filaments, some 0.2–0.7 × 2–30 microns; orange or yellow carotenoid pigments are common. All species are chemoorganotrophs, and the genus includes both strict aerobes and facultative anaerobes; one or more of the following may be attacked: AGAR, ALGINIC ACID, CELLULOSE, CHITIN.

Cytophagaceae A family of the CYTOPHAGALES; the organisms—which exhibit a gliding movement—are single, rod-shaped cells or non-sessile filaments. All species contain carotenoid pigments and appear yellow, orange, pink or red. Genera include CYTOPHAGA and *Flexibacter*. (see also BEGGIATOACEAE)

Cytophagales An order of GLIDING BACTERIA; constituent species do not form fruiting bodies (cf. MYXOBACTERALES). The organisms are single, rod-shaped cells, or filaments (chains of cells); all are gram-negative. Species occur in soil and in freshwater and marine habitats. Constituent families *include* CYTOPHAGACEAE, BEGGIATOACEAE and LEUCOTRICHACEAE.

cytopharynx (*ciliate protozool.*) A tube-like or pouch-like structure found, beneath the CYTOSTOME, in the cytoplasm; FOOD VACUOLES formed at the cytostome pass through the cytopharyngeal region.

cytophilic antibody An antibody which is capable of binding to a cell surface without the involvement of the specific antibody *combining sites*; such an antibody, when bound to a cell, possesses free combining sites which are able to bind antigen. (see also HOMOCYTOTROPIC ANTIBODY and HETEROCYTOTROPIC ANTIBODY)

cytoplasmic genes *Extranuclear* genes: genes located in certain eucaryotic organelles—e.g. mitochondria. (see also CYTOPLASMIC INHERITANCE)

cytoplasmic inheritance (Non-Mendelian inheritance, extrachromosomal inheritance) Inheritance governed by, or influenced by, CYTOPLASMIC GENES—which are not subject to the Mendelian laws of segregation and assortment. (see also MATERNAL INHERITANCE)

cytoplasmic membrane (see CELL MEMBRANE)

cytoproct Synonym of CYTOPYGE.

cytopyge In some protozoans, a permanent pore through which are voided non-digestible materials etc.

cytosegresome An intracellular, membrane-bounded vacuole within which a cell has enclosed some of its own constituents. Cytosegresomes are formed during autophagy. (cf. PHAGOCYTOSIS)

cytosol (*synonyms:* hyaloplasm; cell sap) The non-particulate, fluid portion of the cytoplasm.

cytosome Synonym of *cytoplasm*.

cytostome (*protozool.*) The cell mouth: a specific, modified region of the cell through which particulate food is ingested; at the cytostome the cell is bounded only by a simple unit-type membrane. A cytostome is found in the majority of *ciliates*—but is absent in the Astomatida and Suctoria. In some ciliates, e.g. the primitive *holotrichs*, the cytostome is situated at the level of the cell surface; in others, e.g. *Paramecium*, it is found at the base of a BUCCAL CAVITY. (see also VESTIBULUM) In ciliates, an apical (polar) cytostome is generally considered to be a primitive feature. Ciliature associated with the cytostomal region may be simple—or may occur in the form of complex membranelles (see AZM); oral ciliature is important taxonomically. Among other protozoans, a cytostome is absent in the sarcodines and members of the Opalinata, and in a number of flagellates, e.g. *Trypanosoma*. (see also FOOD VACUOLE)

cytotaxin (*immunol.*) Any substance which has the ability to promote CHEMOTAXIS.

D

D value (D_{10} value) (see DECIMAL REDUCTION TIME)

Dacrymyces A genus of fungi of the order DACRYMYCETALES; species are common as saprophytes on wood. *D. deliquescens* forms bright orange, gelatinous, globular, asexual fruiting bodies on decaying wood; the fructifications are each up to five millimetres in diameter, and each consists of a mass of radiating hyphae which, at the surface of the fruiting body, break up into oidia. The basidia subsequently develop at the surface of the same fructification—which concurrently becomes yellow. Each basidiospore forms three transverse septa prior to germination.

Dacrymycetales An order of fungi of the subclass HOLOBASIDIOMYCETIDAE; constituent species are distinguished from those of other orders e.g. in that they form the 'tuning fork' type of basidium. (The 'tuning fork' basidium is apically bifurcate (forked)—the orientation of the two forks (the *sterigmata*) being such that the whole basidium resembles a tuning fork.) The organisms form gelatinous or waxy fruiting bodies which may be e.g. pulvinate or erect and branched; the fruiting bodies are often brightly coloured—e.g. yellow, orange. Species occur typically as saprophytes on wood. The genera include e.g. DACRYMYCES and CALOCERA.

dactinomycin Synonym of ACTINOMYCIN D.

Daedalea A genus of lignicolous fungi of the family POLYPORACEAE; the organisms occur as perthophytes and saprophytes on various types of wood. *Daedalea* spp commonly form corky, hoof-like, sessile fruiting bodies in which the upper surface is brown or grey-brown and the hymenophore is labyrinthiform with thick dissepiments; *D. quercina* (which occurs particularly on oak—*Quercus* spp) may also give rise to imbricated fruiting bodies. The context exhibits trimitic organization. The spores of *D. quercina* are thin-walled and hyaline; they are essentially ellipsoidal, approximately 6 × 3 microns.

daisy head colony The form of colony typically produced by *Corynebacterium diphtheriae* (*gravis* type). After 48 hours (at 37 °C) on tellurite blood agar, the colony is 3–5 mm in diameter with a crenated edge, and is grey or black with a narrow translucent margin; radial striations occur around the periphery of the colony.

Daldinia A genus of fungi of the order SPHAERIALES, class PYRENOMYCETES; the sole species, *D. concentrica*, grows on dead or fallen trunks and branches of certain kinds of tree—particularly ash (*Fraxinus*). *D. concentrica* forms a hemispherical or sub-globose stroma on the surface of the substratum; when young, the stroma may bear a surface layer of light brown conidia borne on branched conidiophores. The *mature* stroma, some 5–10 centimetres across, is a hard, brittle, solid, dark brown or black structure which, in vertical cross-section, exhibits alternating light and dark bands which are concentric about the centre of the base of the stroma; perithecia develop within the superficial layer of the stroma with their ostiolar pores at the surface. The bulk of the stromatal tissue appears to act as a reservoir for water. Ascospores are discharged mainly or exclusively during the hours of darkness.

dalton A unit of atomic weight which has been defined by various authors in the following (mutually-incompatible) ways: (a) one-twelfth of the weight of a carbon-12 atom (i.e. synonymous with ATOMIC MASS UNIT); (b) one-sixteenth of the weight of an oxygen-16 atom; (c) the weight of a hydrogen-1 atom.

damping-off (*plant pathol.*) A disease in which seedlings weaken and collapse following microbial attack at soil level; infected plants may even fail to emerge from the ground. Common causative agents include species of the fungi *Pythium* and *Rhizoctonia.*

Dane particle (see HEPATITIS B ANTIGEN)

Danish agar (see FURCELLARAN)

Danysz' phenomenon (*immunol.*) When *equivalent* quantities of toxin and antitoxin are mixed, the toxicity of the resulting mixture varies according to the method of admixture; such variation constitutes *Danysz' phenomenon*. Thus, if an equivalent amount of toxin is added to antitoxin in one stage (i.e. all at once) the resulting mixture is non-toxic; if however the same quantity of toxin is added in two halves—the second half some thirty minutes after the first—the resulting mixture is toxic, i.e. it contains free (uncombined, non-neutralized) toxin. In the second method of admixture the first portion of toxin combines with *more than its equivalent* of antitoxin; consequently, insufficient free antitoxin is available to neutralize the second portion of toxin.

DAP pathway (see DIAMINOPIMELIC ACID PATHWAY)

dapsone (DDS) (see SULPHONES)

dark-field microscopy (see MICROSCOPY)

dark mildews Fungi of the family PERISPORIACEAE.

dark reaction (in photosynthesis) (see PHOTOSYNTHESIS)

dark repair (excision repair) (genetics) An intracellular, light-independent enzymic mechanism for the repair of DNA (*q.v.*) damaged e.g. by ULTRAVIOLET RADIATION, ALKYLATING AGENTS etc. (cf. PHOTOREACTIVATION) The mechanism is initiated by the distortion of the DNA duplex at the site of damage. In bacteria, the sequence of events appears to be: (a) the *excision* (by endonuclease and exonuclease) from the damaged DNA strand of an oligonucleotide containing the site of damage; (b) the *resynthesis* (by a polymerase) of the missing section of DNA by elongation from the exposed 3′-OH terminus of the original strand—the complementary strand acting as the template; (c) the *connection* (by a ligase) of the 3′-OH terminus of the newly-synthesized oligonucleotide to the 5′-phosphate terminus of the original strand.

Certain bacteriophages code for their own repair system, while others use that of their bacterial host.

Darling's disease (see HISTOPLASMOSIS)

DAT (*immunol.*) Differential agglutination test; any diagnostic or other test in which the result consists of two agglutinating titres: that of the test proper, and that of the *control*. (The titres may be quoted in full or expressed as a ratio.)

dazomet A cyclic DITHIOCARBAMATE which is used as a soil fumigant; in the soil it breaks down to form a volatile isothiocyanate. (see also METHAM SODIUM)

DCA (see DEOXYCHOLATE-CITRATE AGAR)

DCMU 3-(3,4-dichlorophenyl)-1,1-dimethylurea, an agent which blocks photosystem II in PHOTOSYNTHESIS.

DDS (dapsone) (see SULPHONES)

dead man's fingers The common name for fruiting bodies (stromata) of *Xylaria polymorpha*.

Dean and Webb titration A serological titration in which a constant volume of a given antiserum is added to each of a number of serial dilutions of antigen. The end point is indicated by the tube in which precipitation first occurs. In this tube antibody and antigen are present in OPTIMAL PROPORTIONS. (cf. RAMON TITRATION)

death cap fungus A common name for the basidiomycete *Amanita phalloides*. (see also AMATOXINS)

death phase (see GROWTH (bacterial))

death rate constant (see STERILIZATION)

Debaromyces A genus of fungi of the family SACCHAROMYCETACEAE.

débridement The surgical removal of necrotic and/or infected tissue.

decarboxylase test (*bacteriol.*) A test used to determine the ability of a given strain of bacteria to decarboxylate one or more of the amino acids arginine, lysine, and ornithine. In the test, three media—each incorporating one of the amino acids—are inoculated with the test organism and subsequently incubated. In a reaction requiring pyridoxal phosphate a given amino acid is cleaved, by a specific decarboxylase, to the corresponding amine—with resultant increase in alkalinity of the medium; in a *positive* test the final pH of the medium becomes such that an appropriate colour reaction occurs in the pH indicators included in the medium. (Arginine, lysine, and ornithine are decarboxylated to agmatine, cadaverine, and putrescine respectively.)

decimal reduction time (D value; D_{10} value) The time required, at a given temperature, to heat-inactivate (kill) 90% of a given population of cells or spores; the D value is usually quoted in minutes. (see also F VALUE)

declomycin DEMETHYLCHLORTETRACYCLINE. (see also TETRACYCLINES)

decoyinine (see PSICOFURANINE)

decurrent (*adjective*) (*mycol.*) In a fruiting body: refers to a type of gill (lamella) in which the lower edge of the gill is curved downwards near the stipe so that the gill is attached not only to the underside of the pileus but also to the surface of the upper portion of the stipe.

deep rough (see SMOOTH-ROUGH VARIATION (in bacteria))

defined medium (see MEDIUM)

degeneracy (genetics) (see GENETIC CODE)

degerm (see ANTISEPSIS)

degranulation (*immunol.*) (see MAST CELLS)

dehiscence (*mycol.*) The opening, at maturity, of a structure which is closed during its development.

dehydrogenase An OXIDOREDUCTASE which catalyses the removal of hydrogen atom(s) from a substrate—the hydrogen being donated to an acceptor *other than* molecular oxygen; e.g. in the dehydrogenation of L-lactate by 'lactate dehydrogenase' (= L-lactate:NAD oxidoreductase), NAD functions as the hydrogen acceptor. (cf. OXIDASE)

delayed hypersensitivity (historical synonym: *allergy of infection*) A form of CELL-MEDIATED IMMUNITY in which the causal antigen is often a component of the causal agent of a chronic infectious disease; delayed hypersensitivity reactions are the bases of a number of SKIN TESTS (e.g. the TUBERCULIN TEST) which may be used in the diagnosis of certain of these diseases. Following the subcutaneous injection of specific antigen into a *primed* subject, a local inflammatory reaction generally commences within 24 hours and reaches peak intensity within 48/72 hours. Typically, this consists of an erythematous, indurated lesion which forms at the site of the injection; the lesion contains a high density of macrophages—probably owing to the influence of the LYMPHOKINES *chemotactic factor for macrophages* (CFM) and *macrophage migration inhibition factor* (MIF). This reaction constitutes a *positive* skin test. In a severe reaction local tissue necrosis may occur.

The ability to give a delayed hypersensitivity reaction can be transferred from a *primed* subject to an individual who has had no previous contact with the given antigen; this can be achieved by the transfer of *lymphocytes*—but not by serum transfer.

Inability to express delayed hypersensitivity reactions has been recorded e.g. for certain patients suffering from chronic candidiasis.

deletion (genetics) A type of MUTATION in which a nucleotide, or two or more contiguous nucleotides, are lost from DNA (or, in certain viruses, RNA). If the number of nucleotides lost is not 3 or a multiple of 3 (i.e. not one or more complete codons) the deletion will be a PHASE-SHIFT MUTATION.

Delhi boil (see ORIENTAL SORE)

demethylchlortetracycline (declomycin) 6-Demethyl-7-chlorotetracycline. An ANTIBIOTIC obtained from a mutant strain of *Streptomyces aureofaciens*. (see TETRACYCLINES)

demicyclic (of rust fungi) Refers to those rusts which exhibit all the spore stages except stage II (the uredial stage). Demicyclic rusts include e.g. most species of *Gymnosporangium*.

Dendrosoma A genus of protozoans of the SUCTORIA.

dengue (dengue fever; breakbone fever) A disease of man characterized by fever, rash, and severe pains in joints and muscles; dengue occurs in the tropics and sub-tropics. The causal agent, a togavirus, is transmitted by mosquitoes. Dengue tends to occur in low-mortality epidemics. In the frequently-fatal *haemorrhagic* dengue, the severity of the disease has been attributed to the effects of re-infection in patients immunologically sensitized by a previous attack.

denitrification The formation of gaseous nitrogen (and/or certain oxides of nitrogen) from nitrate (or nitrite) by certain bacteria during NITRATE RESPIRATION. Denitrifying bacteria occur e.g. in the soil and in marine and freshwater environments; aquatic species can be useful in the elimination of nitrate from waste waters. Denitrification can occur only under anaerobic or microaerophilic conditions (e.g. in waterlogged soils) and at neutral to slightly alkaline pH. (see also NITROGEN CYCLE)

denitrifying bacteria Those species capable of DENITRIFICATION. They include: *Bacillus licheniformis*, *Paracoccus denitrificans* (formerly *Micrococcus denitrificans*), *Pseudomonas aeruginosa*, *Ps. stutzeri*, and *Thiobacillus denitrificans*.

Denitrobacillus licheniformis Synonym of *Bacillus licheniformis*. (see also DENITRIFYING BACTERIA)

density gradient centrifugation (see CENTRIFUGATION)

dental caries Tooth decay—a process promoted by dental PLAQUE. The extracellular acids and enzymes of the cariogenic bacteria (e.g. *Streptococcus mutans*) are major factors in dental decay.

deoxycholate A salt of the BILE ACID deoxycholic acid (= 3,12-dihydroxycholanic acid).

deoxycholate-citrate agar (DCA) A medium used e.g. for the primary isolation, from faeces, of certain pathogens—particularly *Salmonella* and *Shigella*; most strains of *Escherichia* and *Proteus* fail to grow on DCA. The medium generally includes (approximate w/v) meat extract and peptone (1%), lactose (1%), sodium citrate (1%), ferric citrate (0.1%), sodium deoxycholate (0.5%), neutral red (0.002%), and agar (1.5%); the pH of the medium is about 7.3. DCA should *not* be sterilized by autoclaving.

deoxyribonuclease (staphylococcal) (see DNASE)

deoxyribonuclease (streptococcal) (see STREPTODORNASE)

deoxyribonucleic acid (see DNA)

depside An ester formed by the condensation of two or more phenolic acids—e.g. COOH.ArO.CO.ArOH, a generalized *di*depside where Ar = an aromatic group. Depsides occur among the lichen acids and as constituents of TANNINS, e.g. digallic acid.

depsipeptide antibiotics A group of ANTIBIOTICS in each of which the molecule consists of amino acid residues alternating with hydroxy acid residues—the residues being linked by alternate peptide and ester bonds. The *enniatins* are a group of cyclic depsipeptides obtained from *Fusarium* spp; they are active against certain gram-positive bacteria, e.g. *Mycobacterium tuberculosis*. Enniatin *A* contains three residues each of D-α-hydroxyisovaleric acid and *N*-methyl-L-isoleucine; in enniatin *B*, *N*-methyl-L-isoleucine is replaced by *N*-methyl-L-valine. A closely related antibiotic, *valinomycin*, has a cyclic molecule containing 12 residues: (-D-valine—L-lactic acid—L-valine—D-α-hydroxyisovaleric acid-$)_3$. Depsipeptide antibiotics alter the permeability of the bacterial CELL MEMBRANE to potassium ions and, to a lesser extent, other cations—i.e. they function as *ionophores*; OXIDATIVE PHOSPHORYLATION is blocked as a secondary effect. Characteristics of the depsipeptides which affect their antibacterial activity include ring size, optical configuration, and the hydrophobicity of the amino acid side chains.

Dermatocarpon A genus of squamulose or foliose LICHENS whose phycobiont is *Hyalococcus* or *Myrmecia*—according to species.

Perithecia occur immersed in the thallus—their ostioles appearing as black dots on the thallus surface. *Dermatocarpon* species occur on soil or on rocks and are attached to the substratum by rhizinae or by a central holdfast—according to species; certain species play an important role in the stabilization of the soil surface in some arid semi-desert regions. Species include e.g.:

D. miniatum. The thallus is more or less discoid, of the order of 2 centimetres in diameter, with an approximately central point of attachment; the upper surface of the thallus is pale grey to brownish-grey while the lower surface is brown. *D. miniatum* generally occurs in colonies on rocks—particularly near lakes or the sea.

D. fluviatile. The extensively-lobed thallus is dark green when wet, brownish-black when dry. *D. fluviatile* occurs on rocks in, or adjacent to, unpolluted streams or lakes.

D. hepaticum. The squamulose thallus is pale brown to reddish-brown—tinged with green when wet; it occurs on soil and on rocks.

Dermatophilaceae A family of non-acid-fast bacteria of the ACTINOMYCETALES; the organisms form mycelium which, on division, produces numbers of flagellated coccoid forms. One species, *Dermatophilus congolensis*, is an epidermal pathogen of man and other animals.

dermatophytes A category of fungi which are characterized by their ability to metabolize KERATIN; all species are members of the order MONILIALES, and many are known to have perfect (sexual) stages within the (ascomycete) order Eurotiales. The dermatophytes occur in soil and as parasites and pathogens of man and other animals (see RINGWORM); parasitic and pathogenic species (see EPIDERMOPHYTON, MICROSPORUM, TRICHOPHYTON) are found in the hair and/or skin and nails of their animal hosts—only rarely causing disease of deeper tissues. The vegetative form of the organisms is a septate mycelium which, in many species, has a tendency to fragment with the formation of rows of arthrospores on and/or within (according to species) the shaft of a parasitized hair. Most species form both microconidia and macroconidia in culture; both types of conidia are reported to be aleuriospores. Media used for the isolation and culture of dermatophytes include SABOURAUD'S DEXTROSE AGAR. (see also GRISEOFULVIN; TOLNAFTATE; WOOD'S LAMP)

dermatophytosis (see RINGWORM)

Derxia A genus of gram-negative, non-sporing, rod-shaped bacteria (family AZOTOBACTERACEAE) found in tropical soils. The cells are polarly flagellate. *Derxia* species are catalase negative.

desensitization (hyposensitization) The regular administration of small amounts of antigen to a subject in whom that antigen usually provokes an IMMEDIATE HYPERSENSITIVITY reaction. The object is to stimulate the production of BLOCKING ANTIBODIES (2) and thus to reduce or abolish the condition of hypersensitivity in that subject. Desensitization has also been tried as a therapeutic process e.g. in brucellosis in which at least some of the tissue damage is thought to result from DELAYED HYPERSENSITIVITY reactions; this practice has not received universal approval.

desiccation The removal of water: a process used for the preservation of viable populations of certain types of vegetative microorganism. The cells are initially suspended in an appropriate medium—e.g. broth, serum, 5–10% nutrient gelatin. Drops of the suspension are then placed, singly, onto a sterile filter paper, or a piece of waxed paper, which is inserted into a glass ampoule. The ampoule is placed in a desiccator which contains phosphorus pentoxide and which is subsequently evacuated. The use of a vacuum is essential since it reduces the time for which cells are exposed to the deleterious effects of high solute concentrations. When the sample is fully desiccated the ampoule is sealed. This method of preservation is unsatisfactory for a number of aquatic organisms and for certain pathogenic bacteria—e.g. *Neisseria gonorrhoeae*—for all of which desiccation is a lethal process.

Desmidium A genus of algae—see DESMIDS.

desmids A group of freshwater algae within the division CHLOROPHYTA; the group includes the genera *Closterium*, *Cosmarium*, *Cylindrocystis*, *Desmidium*, *Staurastrum*, and *Staurodesmus*. The typical desmid is a bilaterally-symmetrical unicellular organism—though a few species are filamentous. Most desmids (including e.g. *Cosmarium*, *Staurastrum*) exhibit an obvious equatorial constriction (the *sinus*) which divides the cell into two 'semi-cells' joined by a narrow isthmus of cytoplasm; in the crescent-shaped desmid *Closterium* there is no sinus but the cell contents appear to be organized into two parts—a line of demarcation often being visible across the widest part of the cell. Some desmids exhibit motility. Sexual reproduction occurs by conjugation involving non-flagellate, amoeboid gametes which develop from the protoplasts.

desmodexy, rule of (*protozool.*) An expression of the observation that, in all known cases, each KINETODESMA in a ciliate organism lies on the (organism's) right-hand side of the corresponding row of basal bodies.

Desmothoracida (see HELIOZOANS)

desoxy- A prefix equivalent to 'deoxy'; thus, e.g. for 'desoxycholate' see DEOXYCHOLATE.

destroying angel A common name for the basidiomycete *Amanita virosa*.

Desulphotomaculum A genus of gram-negative, SPORE-forming, chemoorganotrophic, strictly anaerobic bacteria of the family BACILLACEAE; species occur e.g. in soil and in the RUMEN. The cells are rod-shaped with peritrichous flagella; maximum length: about 5–6 microns. Metabolism is respiratory (oxidative); the organisms carry out DISSIMILATORY SULPHATE REDUCTION using substrates such as lactate and pyruvate.

Desulphovibrio A genus of gram-negative, obligately anaerobic, asporogenous, polarly flagellate, curved or spirillar rod-shaped bacteria of uncertain taxonomic affinity; species occur in fresh and marine waters (particularly those which are polluted with organic matter), in muds, and in the soil. The organisms are chemoorganotrophs which obtain energy by anaerobic respiration using reducible sulphur compounds—see DISSIMILATORY SULPHATE REDUCTION. The type species, *D. desulphuricans*, carries a single polar flagellum and is approximately 4 × 0.5–1 microns.

determinant (antigenic determinant; determinant group) (*immunol.*) That region of the surface of an ANTIGEN with which the *homologous* antibody combines. The determinant is responsible for the *specificity* of the antibodies elicited by the antigen. The (*non*-covalent) union of antigen and antibody occurs between the determinant and the antibody *combining site*.

dethiobiotin (see BIOTIN)

Dettol (*trade name*) (see PHENOLS)

deuteromycetes Fungi of the DEUTEROMYCOTINA. (In some taxonomic schemes these fungi constitute the form-class Deuteromycetes.)

Deuteromycotina (deuteromycetes; *Fungi Imperfecti*) A category of FUNGI (of subdivision rank) within the division EUMYCOTINA; the deuteromycetes include fungi with no known perfect (sexual) stage, the *imperfect* (asexual, conidial) stages of a number of ascomycetes and basidiomycetes, and fungi which appear not to produce reproductive structures of any kind. The purpose of this non-phylogenetic classification is to facilitate the identification of those fungi which have a dominant asexual (conidial) phase or which have no sexual phase; accordingly the deuteromycetes are divided into a number of *form*-taxa (e.g. *form*-genus, *form*-species) on the basis of the characteristics of their asexual reproductive structures, conidia etc. (see also SACCARDIAN SYSTEM) The designations form-genus, form-species etc. signify that the organisms which constitute a given taxon are not necessarily related (and are often unrelated) in an evolutionary sense—evolutionary relationships usually being supposed to be made evident on the basis of the sexual (perfect) stages of fungi. (In most texts the prefix 'form'—although taxonomically correct—is omitted. Further, the name of the deuteromycete genus (i.e. form-genus) is often used to refer to an organism even after the sexual stage of that organism has been discovered and placed in a named ascomycete or basidiomycete genus.) Species with known sexual stages and species with no known sexual stage may occur together within a given deuteromycete genus. The deuteromycetes are divided into four classes: AGONOMYCETES, BLASTOMYCETES, COELOMYCETES, and HYPHOMYCETES.

deutomerite In members of the Eugregarinida, the outer (distal) part of a (developing) trophozoite which has yet to become detached from the host cell; it is separated (by a cross wall) from the PROTOMERITE. The nucleus of the trophozoite usually occurs in the deutomerite.

dexioplectic metachrony (see METACHRONAL WAVES)

dextrans A class of polysaccharides composed of glucose residues predominantly linked α-(1 → 6); branches are formed by occasional α-(1 → 4) and (less frequently) α-(1 → 3) linkages. The nature and extent of branching depends on the source of the dextran. Dextrans are relatively inert and can withstand autoclaving. They are not strongly antigenic and (partly for this reason) they are employed as plasma extenders in blood transfusion. (Dextran for this purpose is obtained from the bacterium *Leuconostoc mesenteroides*.) Dextrans which have undergone artificial cross-linking (commercially known as *Sephadex*) are used in gel filtration—a technique by which molecules of different sizes can be separated. Dextran is produced by a wide range of microorganisms, sometimes in copious amounts (see CAPSULES).

dextrins Products formed by the partial degradation of STARCH or GLYCOGEN by heat, by acid hydrolysis, or by enzymic action. *Limit dextrins* are those formed by enzymes which are unable to effect complete hydrolysis—e.g. β-AMYLASES. (see also SCHARDINGER DEXTRINS)

dextrose Dextrorotatory glucose.

diakinesis (see MEIOSIS)

Dialister pneumosintes (see BACTEROIDES)

diaminopimelic acid pathway (DAP pathway) (for lysine biosynthesis) A pathway (see Appendix IV(e)) which occurs in procaryotes, in certain lower fungi, and in most algae. (cf. AMINOADIPIC ACID PATHWAY)

diamonds, the (see SWINE ERYSIPELAS)

diaphorase A term which has been applied to any enzyme which can mediate in the oxidation of reduced NAD by an artificial electron acceptor.

diaphoromixis Bipolar or tetrapolar *multi*-allele HETEROTHALLISM.

diaspore A propagule, i.e. a disseminative unit.

diastatic (of organisms) Able to metabolize starch.

diastole (vacuolar) (see CONTRACTILE VACUOLES)

diatomaceous earth (diatomite; *kieselgühr*) A silicaceous material composed largely of fossil diatoms. Diatomite has been used in several types of microbiological filter (see FILTRATION) and is used in a number of industrial processes. Large deposits of diatomite are mined at Lompoc, California.

diatoms ALGAE of the division BACILLARIOPHYTA.

diauxie The phenomenon in which, given two carbon sources, an organism preferentially metabolizes one (completely) before commencing to metabolize the other; a lag phase commonly separates the two phases of growth. (see also CATABOLITE REPRESSION)

diauxy A synonym of DIAUXIE.

diazomycin A (see 6-DIAZO-5-OXO-L-NORLEUCINE)

6-diazo-5-oxo-L-norleucine (DON; $COOH.CHNH_2.CH_2.CH_2.CO.CH{=}N^+{=}N^-$) An ANTIBIOTIC produced by *Streptomyces* spp. DON exhibits antimicrobial activity in much the same way as does AZASERINE; it also exhibits antitumour activity. A number of DON derivatives are also produced by *Streptomyces* spp, e.g. *diazomycin A* (*N*-acetyl-DON) and *alazopeptin* (apparently consisting of one alanine and two DON residues); these also have antitumour activity. (see also HADACIDIN)

dicentric (of a chromosome) Having two centromeres.

dichlone (2,3-dichloro-1,4-naphthoQUINONE) An agricultural antifungal agent used as a seed dressing (e.g.for the seeds of legumes, cotton) and as a foliar spray. It is effective against a range of diseases, including apple scab, DAMPING-OFF etc. Dichlone is toxic to certain plants and to a number of algae.

dichloramine T (see CHLORINE)

dichlorophene (see BISPHENOLS)

dichotomous replication (of DNA) (see GROWTH (bacterial))

Dick test (*immunol.*) An *in vivo* test used to detect the presence of antibody (antitoxin) to the erythrogenic toxin of *Streptococcus pyogenes* in order to indicate the susceptibility or immunity of an individual to *scarlet fever*. The test procedure is similar to that employed in the SCHICK TEST; the erythrogenic toxin is used as an inoculum. A *positive* Dick test (indicative of *susceptibility* to the disease) consists of a local inflammatory response (at the site of the injection) which becomes evident within about 12 hours and reaches a peak within about 24 hours of the injection. In immune persons the toxin is neutralized by antitoxin—thus giving a *negative* Dick test.

dicloran (*plant pathol.*) 2,6-Dichloro-4-nitroaniline. An agricultural antifungal agent which is used against various *Botrytis* infections—particularly *B. cinerea* on lettuce. It is water-insoluble—hence very persistent—and is usually formulated as a dust. Dicloran has very low phytotoxicity. (see also CHLORONITROBENZENES)

dictyosome (1) A synonym of GOLGI APPARATUS. (2) Less frequently used to refer to one of the membranous saccules in the stack forming a Golgi apparatus.

dictyosporae (see SACCARDIAN SYSTEM)

Dictyostelium A genus of the CELLULAR SLIME MOULDS (*q.v.* for life cycle); species occur in soil and in decomposing vegetation—particularly in woodlands.

D. discoideum. The vegetative, haploid *myxamoeba* is generally 8–12 microns, maximum dimension. When food (bacteria) is abundant myxamoebae grow and divide mitotically by binary fission. Bacteria are ingested by phagocytosis; the myxamoebae are believed to be attracted to bacteria, chemotactically, as a result of the liberation, by the latter, of a particular compound—reported to be folic acid. (For wild type strains a bacterial diet is essential; however, certain laboratory-derived strains grow axenically—i.e. in a bacterium-free medium. Wild type isolates are generally cultured on SM AGAR.) In the absence of food, the chemotactic aggregation of myxamoebae is mediated by a product secreted by the myxamoebae: *acrasin* (= cyclic AMP, cAMP). The myxamoebae also secrete an *acrasinase* (cAMP phosphodiesterase) and an inhibitor of acrasinase; the aggregational event is governed by the relative proportions of these substances. The *pseudoplasmodium* is elongated and finger-like—commonly 1–4 mm, maximum dimension—and consists of a mass of cells enclosed by a (largely cellulosic) sheath; during migration, new sheath material is synthesized to replace that which remains behind as a slime trail in the track of the advancing pseudoplasmodium. The pseudoplasmodium exhibits tactic movements in response to light, heat, and/or humidity. Culmination results in the formation of a characteristic *sorocarp*: a single, unbranched stalk that tapers towards the free end—which carries a spheroidal or ovoid mass of yellow spores; that end of the stalk which is attached to the substratum is flared into a *basal disc*. The sorocarp is several millimetres in length; during its development, the stalk of the sorocarp becomes orientated such that the spore mass (*sorus*) it carries is the maximum distance from the substratum or from objects in the near vicinity. (Thus, e.g. a horizontally-orientated stalk develops when culmination occurs on a vertical substratum.) Orientation of the sorocarp is believed to be due to a tropism governed by an (unidentified) gas or vapour produced by the cells of the sorocarp. The time between aggregation and sorocarp formation appears to vary between eight and twenty-four hours (approximately). The ellipsoidal *spores* are each commonly 5–10 × 3–6 microns and each has a thick, cellulose-containing wall; germination normally occurs only in the presence of an adequate supply of amino acids. (Spores *in masses* fail to germinate owing to the presence of an inhibitor—reported to be *N,N*-dimethylguanosine—believed to be produced during culmination.) Parasexual and/or sexual processes are believed to occur in at least some strains.

Dictyuchus A genus of fungi of the OOMYCETES.

Didinium A genus of freshwater protozoans of the suborder RHABDOPHORINA. The cell is ovoid or barrel-shaped, between 80 and 200 microns in length, and is radially symmetrical; the flattened apical region has a prominent cone-shaped proboscis—at the apex of which is found the cytostome. Microtubular structures (see TRICHITES) are found in the cytopharyngeal region. The cell is circled with two bands of cilia—one equatorial and the other surrounding the base of the proboscis. The macronucleus is U-shaped. Food consists almost entirely of *Paramecium*; in the absence of food *Didinium* encysts.

Didymium A genus of the ACELLULAR SLIME MOULDS.

didymosporae (see SACCARDIAN SYSTEM)

Dienes phenomenon Mutual inhibition of growth which occurs when the swarms* of two different strains of *Proteus* meet on the culture plate. *see SWARMING.

Dientamoeba fragilis A species of parasitic amoeba within the order AMOEBIDA; it occurs in the large intestine of man but is not pathogenic. The trophozoite is some 5–15 microns in diameter and usually contains two nuclei—which do not exhibit peripheral chromatin (cf. ENTAMOEBA); there are four to eight karyosomes per nucleus. *D. fragilis* does not form cysts.

diethylpyrocarbonate (as a food preservative) A preservative which has been used in e.g. the USA for the preservation of fruit juices, beers, wines and other beverages; it is not a permitted preservative in the UK. (see also FOOD, PRESERVATION OF)

Dieudonné alkaline blood agar A medium used for the primary isolation of *Vibrio cholerae*. Defibrinated ox blood (15 ml) and 1N sodium hydroxide solution (15 ml) are mixed and steamed for 30 minutes on each of 3 successive days; the whole is well-mixed with sterile nutrient agar (70 ml of 3% agar at 56 °C) and the plates poured and allowed to stand for one or two days at room temperature. (see also MONSUR AGAR)

differential host (*plant pathol.*) A plant host which, on the basis of disease symptoms, serves to distinguish between various strains or races of a given pathogen.

Difflugia A genus of protozoans: see ARCELLINIDA.

difolatan Synonym of CAPTAFOL.

dikaryon (dicaryon) (*mycol.*) A binucleate cell. (*Dikaryon* is sometimes used to refer to a pair of associated *nuclei*.)

dikaryophase (dicaryophase) (*mycol.*) In the life cycles of ascomycetes and basidiomycetes: a phase which is characterized by the presence of heterokaryotic binucleate cells.

dikaryotization (*mycol.*) Any process which leads to the formation of a dikaryotic (binucleate) cell or a dikaryotic hypha; dikaryotization may occur e.g. by SOMATOGAMY or SPERMATIZATION, or by the fusion of basidiospores or of germ tubes from basidiospores of appropriate mating types.

Dileptus A protozoan genus of the RHABDOPHORINA.

dilution coefficient (of DISINFECTANTS) A number which expresses the relationship between *concentration* and *rate of disinfection* of a given disinfectant in respect of a given organism. For many disinfectants this relationship is approximated by:

$$tc^n = k$$

where t is the time required for 100% kill, c is the concentration of the disinfectant, n is the dilution coefficient, and k is a constant. Thus, if disinfectant X (n = 4) at unit concentration is diluted by a factor of 2, the expression c^n becomes $\frac{1}{2}^4$ i.e 1/16; then t must increase by a factor of 16 in order that the expression tc^n may remain numerically constant. Thus, disinfectants with high dilution coefficients (e.g. *phenol*—see PHENOLS) lose activity with dilution more rapidly than those with lower coefficients. (see also TEMPERATURE COEFFICIENT)

Dimastigamoeba gruberi A name formerly applied to *Naegleria gruberi*.

dimethirimol (*plant pathol.*) 5-*n*-Butyl-2-dimethylamino-4-hydroxy-6-methyl pyrimidine. An agricultural *systemic* antifungal agent which has been used for the treatment of powdery MILDEWS of cucumbers. Resistant strains of the fungus readily emerge. (see also ETHIRIMOL)

dimethyl sulphoxide (DMSO; $(CH_3)_2SO$) A non-ionized polar solvent used e.g. as a *cryoprotectant* in the preservation of microorganisms by

FREEZING. DMSO dissolves both hydrophilic and lipophilic substances.

dimethyldithiocarbamic acid (DMDC) $(CH_3)_2N.CS.SH$. A broad-spectrum antimicrobial agent; susceptible organisms include both fungi and bacteria. DMDC is widely used in agriculture as a broad-spectrum antifungal agent; it is used for the treatment of fruit and vegetables and as a constituent of seed and turf dressings etc. DMDC is used either as the ferric salt (*ferbam*: $((CH_3)_2N.CS.S)_3Fe$) or as the zinc salt (*ziram*: $((CH_3)_2N.CS.S)_2Zn$). Oxidation of DMDC gives THIRAM.

dimictic (see DIMIXIS)

dimidiate (*mycol.*) Refers to the type of fruiting body which is essentially semicircular in outline when viewed from the hymenial and/or abhymenial directions. (see also CONCHATE)

dimitic (see HYPHA)

dimixis That form of HETEROTHALLISM in which an organism has *two* types of compatible nucleus or gamete. Such (*dimictic*) fungi can be divided into two categories: (i) Fungi which form two types of thallus—one male, one female; compatibility between gametes is determined solely on a *sexual* basis without the restraint imposed by specific mating type loci—i.e. any male gamete can fuse with any female gamete. (This type of heterothallism is sometimes described as *morphological* heterothallism; heterothallism which involves specific mating type loci has been described as *physiological* heterothallism.) (ii) Fungi which exhibit bipolar 2-allele HETEROTHALLISM.

dimorphic fungi (1) Fungi which can adopt either of two morphologically-distinct vegetative states (according to environmental conditions). Thus e.g. *Candida albicans* commonly grows in a yeast-like form *in vitro* but may develop a mycelial form of growth within the tissues of a parasitized host. (2) Those fungi (e.g. *Saprolegnia* spp) which exhibit *diplanetism*. (3) Occasionally, dimorphic is used as a synonym of *dioecious*.

dinitrogen fixation (see NITROGEN FIXATION)

2,4-dinitrophenol (see UNCOUPLING AGENTS)

dinitrophenols (as antifungal agents) Dinitrophenols exhibit antifungal as well as insecticidal and herbicidal properties. As antifungal agents, the dinitrophenols are used mainly for the treatment of powdery MILDEWS. See BINAPACRYL, DINOBUTON and DINOCAP. (see also CHLORONITROBENZENES and FUNGICIDES AND FUNGISTATS)

dinobuton 2,4-Dinitro-6-*s*-butylphenyl isopropyl carbonate. An agricultural antifungal agent used against powdery MILDEWS. (see also DINITROPHENOLS)

dinocap (*plant pathol.*) An agricultural antifungal agent which is active against powdery MILDEWS in a wide range of plants. Dinocap is a mixture of three isomers of each of the compounds 2,4-dinitro-6-*s*-octylphenol and 2,6-dinitro-4-*s*-octylphenol; the octyl group may be 1-methylheptyl, 1-ethylhexyl or 1-propylpentyl. Dinocap is generally used as the crotonate. In its antifungal action dinocap behaves both as an ERADICANT and as a PROTECTANT. (see also DINITROPHENOLS and CAPTAN)

dinoflagellates Algae of the division DINOPHYTA.

Dinophyta (Pyrrhophyta; dinoflagellates) A division of the ALGAE; dinoflagellates are found typically in marine environments (see also PLANKTON, BIOLUMINESCENCE), though freshwater, brackish, and terrestrial species are known. Most dinoflagellates are free-living organisms, but some occur as endosymbionts in e.g. radiolarians, corals, and sea anemones. While the majority of dinoflagellates are unicellular, biflagellate organisms, non-motile unicellular and filamentous species also occur. A dinoflagellate is said to be armoured or unarmoured according to the nature of its THECA. All species contain CHLOROPHYLLS *a* and *c*, and β-carotene; the cellular content of xanthophylls (see CAROTENOIDS) includes *peridinin* (though fucoxanthin may replace peridinin in *Glenodinium foliaceum*). Intracellular storage products include STARCH. Sexual processes (isogamy, anisogamy) have been reported in some species. Genera include *Amphidinium*, *Ceratium*, *Gonyaulax*, *Gymnodinium*, *Gyrodinium*, *Noctiluca*, *Oxyrrhis*, *Peridinium*, *Prorocentrum*, and *Woloszynskia*. (see also RED TIDE)

diodoquin (see IODINE)

dioecious fungi Fungi in which the male and female reproductive structures occur in different individuals (cf. MONOECIOUS FUNGI).

diphasic serotypes, of *Salmonella* (see PHASE VARIATION, IN SALMONELLA)

diphenylthiourea (see UREA AND DERIVATIVES)

diphtheria (membranous croup) In man, an acute, infectious (sometimes fatal) disease caused by *toxinogenic* strains of *Corynebacterium diphtheriae*; the disease most commonly affects children. Infection occurs *via* the nose or mouth—e.g. in aerosols (droplet inhalation) or in contaminated milk; a carrier state is recognized. Typically, a tough, leathery *pseudomembrane*, which inhibits breathing, develops in the nasopharyngeal-laryngeal region; the membrane consists of fibrin, dead epithelial cells, inflammatory cells, erythrocytes and *Corynebacterium diphtheriae*. The potent EXOTOXIN may produce serious systemic effects—e.g. neurological, cardiac involvement. Treatment involves the administration of antitoxin and appropriate antibiotics (e.g. benzylpenicillin, cephalosporins). *Laboratory diagnosis*: *C. diphtheriae* may be cultured from swabs of the appropriate region. (see also CORYNEBACTERIUM and SCHICK TEST)

diphtheroid (1) Synonym of CORYNEFORM (sense 2). (2) *Corynebacterium* spp other than *C. diphtheriae*. (3) Any non-pathogenic species of *Corynebacterium*.

dipicolinic acid 2,6-Pyridinedicarboxylic acid; dipicolinic acid occurs within the protoplast in bacterial endospores—see SPORES (bacterial)—and has been reported to occur in certain fungi, e.g. *Penicillium citreo-viride*. In bacteria, dipicolinic acid synthesis occurs by a branch of the diaminopimelic acid pathway for lysine synthesis (see Appendix IV(e)).

diplanetism (*mycol.*) In certain species of oomycete (e.g. *Achlya*, *Saprolegnia*): a phenomenon in which two morphologically distinct forms of motile spore are produced during the asexual life cycle. In the first *swarm period* (i.e. period in which motile spores are prevalent) the zoospores (liberated from a SPORANGIUM) are pyriform and bear two flagella at the anterior end; these spores are said to be of the *primary type*. The primary spores encyst and subsequently germinate to form spores of the *secondary type*; these are reniform and bear two flagella which arise from the concavity. The secondary spores encyst and subsequently germinate to form the vegetative thallus (mycelium). In both types of zoospore, one flagellum is of the whiplash type and the other of the tinsel type (see FLAGELLUM (eucaryotic)). Species which produce both primary and secondary zoospores are said to be *diplanetic*.

Oomycetes which give rise to only one type of zoospore (and which have only one swarm period) are said to be *monoplanetic* and to exhibit *monoplanetism*; these include *Pythiopsis* and *Pythium*. Such species may form primary-type zoospores (e.g. *Phythiopsis*) or *secondary-type zoospores (e.g. Pythium)*.

diplococcus (1) A form commonly assumed by certain cocci (see COCCUS); following cell division cells remain together in pairs—each pair being a diplococcus. (2) *Diplococcus* is used by some authors (particularly in the United States) to designate a genus of gram-positive diplococci; the same organisms are included within the genus *Streptococcus* by other authors. Thus, e.g. *Diplococcus pneumoniae* = *Strep. pneumoniae*.

diploid (1) Refers to any (eucaryotic) nucleus or cell which has a PLOIDY of *two*. (cf. HAPLOID, POLYPLOID; see also HOMOLOGUE) (2) In bacteria: see MEROZYGOTE.

Diplomonadida An order of bilaterally symmetrical protozoans of the class ZOOMASTIGOPHOREA; each organism has two nuclei and eight flagella. Most species are parasitic; *Hexamita* spp occur e.g. in the intestinal tract in birds, mammals, and insects, and GIARDIA can be pathogenic in man.

diplont (1) (*noun*) An organism in whose life cycle only the gametes are haploid. (2) (*noun*) The diploid form, thallus etc. of an organism—e.g. a *sporophyte* (see ALTERNATION OF GENERATIONS). (3) (*adjective*) Refers to the diploid phase of an organism.

diplophase In the life cycles of those organisms which reproduce sexually: the diploid phase between KARYOGAMY and MEIOSIS. (cf. HAPLOPHASE)

diplornaviruses A name used by some authors to include viruses of the genera ORBIVIRUS and REOVIRUS.

diplotene stage (see MEIOSIS)

disc (*lichenol.*) The upper surface of the spore-producing region of an apothecium. (The term is used without regard to the shape of the apothecium.)

discocarp A synonym of APOTHECIUM.

Discomycetes A class of filamentous (mycelial) fungi within the sub-division ASCOMYCOTINA; in a typical species the fruiting body is an APOTHECIUM which bears an hymenial layer of unitunicate asci and paraphyses. (Members of the order Tuberales generally form their asci in closed, hypogean (i.e. subterranean) fruiting bodies; these organisms are usually included within the class since they are generally believed to be related to members of another discomycete order, Pezizales.) The main orders are HELOTIALES, OSTROPALES, PEZIZALES, and TUBERALES.

disinfectants Chemical agents used for DISINFECTION. Disinfectants for general use should be active against a range of common pathogenic microorganisms and should be microbicidal rather than microbistatic at the concentrations used. Many disinfectants are effective only against particular categories of organisms (e.g. gram-negative bacteria, certain fungi); even within such categories, species may be found which are resistant to—or minimally affected by—the particular disinfectant. Some disinfectants are non-injurious to human and animal tissues and may be used as ANTISEPTICS. Different disinfectants may require different conditions in order to be maximally effective; important factors include: (a) pH (e.g. for CHLORHEXIDINE and *hypochlorites*); (b) humidity (e.g. for FORMALDEHYDE and β-PROPIOLACTONE); (c) the presence or absence of organic matter (see e.g. CHLORINE); (d) the nature of the solvent used (see e.g. SOAPS). Certain disinfectants (e.g. QUATERNARY AMMONIUM COMPOUNDS) possess detergent properties which enable them to penetrate grease barriers. In general, disinfecting activity increases with rise in temperature and with increase in concentration of the disinfectant—see

TEMPERATURE COEFFICIENT and DILUTION COEFFICIENT. Disinfectants may be microbicidal or microbistatic according to concentration. (In low concentrations some disinfectants can be metabolised by certain organisms—e.g. certain *Pseudomonas* spp can metabolise phenol.) The action of some agents may be reversible—thus e.g. spores inactivated by mercuric ions (see HEAVY METALS) may germinate following exposure to hydrogen sulphide, the latter combining with and removing the mercuric ions. The majority of disinfectants are *antibacterial*; agents with *antifungal* activity include certain BISPHENOLS, some PHENOLS, SALICYLIC ACID AND DERIVATIVES (see also FUNGICIDES AND FUNGISTATS). Spores are resistant to many disinfectants but may be inactivated by STERILIZATION procedures. (See also ALCOHOLS; DYES; HALOGENS; OXIDIZING AGENTS; PINE-OIL DISINFECTANTS; PHENOL COEFFICIENT.)

disinfection The destruction, inactivation or removal of those microorganisms likely to cause infection or to give rise to other undesirable effects; disinfection does not necessarily involve STERILIZATION. In the most commonly used sense *disinfection* refers to the employment of chemical agents (DISINFECTANTS) for the treatment of inanimate objects; however, the use of heat, filtration etc. may also be properly referred to as disinfection processes. (cf. ANTISEPSIS) (see also PASTEURIZATION and TYNDALLIZATION)

disjunctor cells (*mycol.*) In a chain of spores: cells which alternate with the spores and which subsequently disintegrate—thus liberating the spores.

dissepiment (*mycol.*) The wall which occurs between the pores in the hymenophore of certain fungi.

dissimilatory nitrate reduction (see NITRATE RESPIRATION)

dissimilatory sulphate reduction The obligate use of sulphate as the terminal electron acceptor for (anaerobic) respiratory metabolism in the species of *Desulphovibrio* and *Desulphotomaculum*. (cf. NITRATE RESPIRATION) The first step in sulphate reduction requires energy for the 'activation' of sulphate—adenosine-5′-phosphosulphate (APS) and pyrophosphate being formed from sulphate and ATP. APS is subsequently reduced to sulphite and AMP, and sulphite is reduced to sulphide. Some of the sulphide can be assimilated to fulfil the cell's sulphur requirement, but most is eliminated as hydrogen sulphide—cf. ASSIMILATORY SULPHATE REDUCTION. Electrons are transferred to APS etc. from organic substrates or (in e.g. *Desulphovibrio desulphuricans*) from hydrogen. Electron transfer is accompanied by OXIDATIVE PHOSPHORYLATION; details of the electron transport chain are unknown.

Sulphate-reducing bacteria may cause spoilage of e.g. (imperfectly) canned foods, and may be responsible for the corrosion of certain metals. Plants growing in waterlogged soils may be damaged by H_2S produced by these bacteria. The reaction of metal ions with bacterially-formed sulphide is thought to be at least partly responsible for the formation of certain sulphide ores; in lakes, sulphur deposits may be formed by the oxidation of sulphide by photosynthetic bacteria. (see also SULPHUR CYCLE)

dissociation (*bacteriol.*) A term which has been used to refer to any transition of state within the context of SMOOTH-ROUGH VARIATION.

dissymmetry ratio (of microbial DNA) (see BASE RATIO)

distemper An acute, infectious disease which chiefly affects young animals of the *Canidae* (e.g. the dog). The causal agent (see PARAMYXOVIRUS) may enter the body in contaminated food or water, or may be inhaled. The incubation period appears to be 3–30 days. Symptoms may include fever, a nasal discharge, listlessness, and loss of appetite. When uncomplicated, the disease is rarely fatal. Secondary developments may include gastric or bowel disturbances, bronchopneumonia, or involvement of the central nervous system. Effective anti-distemper vaccines are available.

Distrene-80 (*trade name*) A high molecular weight polystyrene; it is a component of the mountant DPX.

distromatic (*algol.*) Refers to a parenchymatous thallus which is two cells thick.

dithanes See MANEB (dithane M-22), NABAM (dithane D-14) and ZINEB (dithane Z-78).

dithianon (2,3-dicyano-1,4-dihydro-1,4-dithia-anthraQUINONE) An agricultural antifungal agent used mainly for the control of apple scab and other diseases involving related fruits; it is inactive against apple mildew. (see also FUNGICIDES AND FUNGISTATS)

dithiocarbamate derivatives A group of compounds important in the control of fungal diseases of plants. There are three main classes: (a) the dithiocarbamates ($R_2N.CS.SH$) e.g. DIMETHYLDITHIOCARBAMIC ACID; (b) *thiuram disulphides*; (c) the metal salts of bisdithiocarbamates.

Many *dithiocarbamates* are active against bacteria as well as fungi; *diethyldithiocarbamate* is used in the differential diagnosis of *Brucella* spp. The antifungal activity of at least some dithiocarbamates (e.g. DAZOMET, METHAM SODIUM) appears to depend on their conversion to substances which are more highly fungitoxic—isothiocyanates in the case of *dazomet* and *metham sodium*.

Thiuram disulphides ($R_2N.CS.S.S.CS.NR_2$)

are obtained by mild oxidation of dithiocarbamates; the most important is **THIRAM** which is active against certain gram-positive bacteria as well as fungi. (see also **METIRAM**)

The *metal bisdithiocarbamates* are the most widely used of the dithiocarbamate derivatives in the range of agricultural antifungal agents. Included in this group are the ethylene bisdithiocarbamates* **MANCOZEB**, **MANEB**, **NABAM** and **ZINEB**; these are believed to be converted, under field conditions, to diisothiocyanate derivatives. (see also **CARBAMIC ACID DERIVATIVES**)

*XS.CS.NH.CH_2.CH_2.NH.CS.SX where X is a metal ion.

divergent (of a gill trama) (see **BILATERAL TRAMA**)

division A taxonomic group: see **TAXONOMY** and **NOMENCLATURE**. Although division is the preferred term in botanical taxonomy *phylum* is the common equivalent in zoological work.

DMDC (see **DIMETHYLDITHIOCARBAMIC ACID**)

DMSO (see **DIMETHYL SULPHOXIDE**)

DNA (deoxyribonucleic acid) A linear polymer of deoxyribo**NUCLEOTIDES** in which the deoxyribose residues are linked by 3′,5′-phosphodiester bridges. The nitrogenous base attached to each deoxyribose residue may be adenine, guanine, cytosine, thymine, or (infrequently) another base (e.g. 5-methylcytosine). (cf. **RNA**) DNA generally occurs as a *duplex* (but see **VIRUSES**) in which one polynucleotide strand is linked to a second (complementary) strand by hydrogen bonds formed between purine and pyrimidine bases; two hydrogen bonds can be formed between adenine and thymine, three between guanine and cytosine. These two *base-pairs* are approximately the same size and shape and can thus occur in any sequence. The two sugar-phosphate chains are *antiparallel*, i.e. a 3′....5′ chain runs alongside a 5′....3′ chain. The DNA duplex forms a right-handed helix (the *double helix*) in which one complete turn occurs every 10 bases; the two sugar-phosphate chains form the backbone of the helix and the (planar) base-pairs are stacked within. The helix is stabilized by hydrophobic interactions between the stacked bases, and by neutralization of the charges on the phosphate groups e.g. by divalent metal ions or polyamines (in bacteria) or by **HISTONES** (in eucaryotes). (see also **CHROMOSOMES**; **GC VALUES**; **DNA BASE COMPOSITION, DETERMINATION OF**; **PLUS AND MINUS STRANDS OF DNA AND RNA**)

DNA functions as genetic material in cells and some viruses—the sequence in which the bases occur constituting a **GENETIC CODE** which forms the basis of heredity. (see e.g. **GENE** and **PROTEIN SYNTHESIS**) In eucaryotes, the bulk of the DNA occurs in the **NUCLEUS**; DNA also occurs in the **MITOCHONDRION** and **CHLOROPLAST**. In bacteria, DNA occurs in the nucleoid body and as **PLASMIDS**. DNA may be detected *in situ* by the **FEULGEN REACTION**.

DNA synthesis. Since the genetic code is contained in the DNA base sequence, newly-synthesized DNA must contain an exact copy of this sequence; this is achieved by using the DNA of the original (parent) duplex as a pattern, or *template*. The correct sequence of bases is obtained by base-pairing between nucleoside-5′-triphosphates and template DNA, and the subsequent action of a DNA polymerase (pyrophosphate being eliminated). *Initiation* of DNA synthesis is not understood; it occurs at specific site(s) on the chromosome, the *origin(s)*. Bacteria appear to have one origin per chromosome, eucaryotes have many. Replication frequently proceeds in both directions from the origin—although it is unidirectional in mitochondrial DNA and in certain bacteriophages. (That length of DNA which is replicated from a given origin is termed a **REPLICON**.) For replication to proceed, some unwinding of the parent duplex must occur to reveal single-stranded templates; this may be assisted by 'unwinding proteins' which bind to, and stabilize, single-stranded DNA. (Unwinding imposes a strain on the parental helix which may be relieved by certain enzymes.) Several DNA polymerases have been isolated from *Escherichia coli*—of which *polymerase III* appears to be responsible for replication. DNA polymerases cannot *initiate* chain synthesis; they elongate *preformed* chains in the 5′ → 3′ direction—sequentially adding nucleotides to each successive free 3′-OH terminus. Short chains of RNA—synthesized on the DNA template by an RNA polymerase—appear to function as *primers*; DNA synthesis proceeds from the 3′-OH terminus of the RNA primer. Both strands of the parental DNA are replicated simultaneously so that, at the site of polymerase activity, a Y-shaped *replication fork* is formed—the stem of the Y constituting the parent duplex, and the arms the daughter duplexes. In unidirectional replication one replication fork is formed; when bi-directional, two forks move apart as synthesis proceeds. DNA synthesis generally appears to be *discontinuous*, i.e. a sequence of DNA fragments (*Okazaki fragments*) is synthesized along the template—each fragment being initiated by an RNA primer; the RNA sections are subsequently excised and replaced by DNA. (Overall synthesis in the 3′ → 5′ direction is achieved by sequential synthesis of fragments in the 5′ → 3′ direction.) Finally, phosphodiester bonds are formed between the Okazaki fragments by a ligase—the completed strand forming a duplex with its template strand. Each new duplex thus contains one new strand and one parental strand—a system of replication referred to as *semi-conservative replication*. (see also MODIFICATION and RESTRICTION)

DNA replication must be coordinated with cell division in order that daughter cells can each receive the full genetic complement. In *bacteria*, DNA synthesis is periodic in slow-growing cells, but continuous in fast-growing cells. The rate of polymerization—once initiated—is constant, but the overall rate of synthesis can be increased since additional rounds of synthesis may begin before previous rounds have been completed. (see GROWTH (bacterial)) The bacterial chromosome is attached to the CELL MEMBRANE at the origin and possibly also at the replication fork. The membrane may play a role in the initiation of replication and/or in the separation of the daughter chromosomes. (see also CAIRNS MODEL and ROLLING CIRCLE MODEL) In *mammalian* cells, nuclear DNA synthesis occurs during the *S* phase of the CELL CYCLE.

DNA can be damaged by a range of agents—e.g. ALKYLATING AGENTS, free radicals, ULTRAVIOLET RADIATION. (see also MUTAGENS) A number of cellular repair mechanisms can, to a greater or lesser degree, counteract such damage—see DARK REPAIR, PHOTOREACTIVATION, RECOMBINATION-REPAIR. Certain ANTIBIOTICS function by interfering with DNA synthesis, e.g. MITOMYCINS, NALIDIXIC ACID, NOVOBIOCIN.

DNA base composition, determination of The base composition of microbial DNA may be determined e.g. to evaluate the GC VALUE or the BASE RATIO of the DNA. Preliminary steps involve (a) cell lysis (by physical or physico-chemical methods) and (b) purification of the liberated DNA. The subsequent determination of base composition may be carried out by any of a variety of methods: e.g. (i) analysis of ultraviolet absorption spectra; (ii) hydrolysis of DNA followed by chemical or chromatographic estimation of the individual bases; (iii) determination of the *Tm* (*temperature mid-point*) value.

The Tm method. When the temperature is progressively increased above 75 °C, double-stranded DNA becomes progressively converted to single-stranded DNA as a result of the breakdown of interstrand hydrogen bonds. The absorption of ultraviolet radiation (at 260 nm) by two *single* strands of DNA is about 40% greater than that absorbed by the same strands combined in the form of a double helix; hence, with an increase in temperature as described above, DNA exhibits a progressive increase in the level of ultraviolet absorption. If the level of ultraviolet absorption is plotted against temperature a linear relationship will be found to exist over much of the denaturation curve—which passes from the absorption minimum (all double-stranded DNA) to the absorption maximum (all single-stranded DNA). The temperature corresponding to the mid-point between these minimum and maximum values is known as the *Tm*; under given conditions (e.g. of pH, nature of the solvent, ionic strength) the Tm is proportional to the %GC value of the DNA. Thus, using as a standard a DNA of known composition, it is possible to determine the %GC values of a range of DNAs which have different base compositions.

DNase (deoxyribonuclease) (staphylococcal) A thermostable extracellular deoxyribonuclease produced mainly by COAGULASE-*positive* strains of *Staphylococcus*; there is a high degree of correlation between the production of coagulase and the production of DNase. The staphylococcal DNase cleaves DNA to form 3′-nucleotides; calcium ions apparently function as activators, and optimal activity occurs at about pH 9.0. The enzyme is also able to cleave RNA. A test for DNase production is used to provide *presumptive* evidence of *Staphylococcus aureus*; the organism under test is cultured on a medium consisting of a suitable base (e.g. trypticase-soy agar) supplemented with DNA and a calcium salt. After growth has occurred the plate is flooded with dilute hydrochloric acid; this precipitates the DNA (causing opacity in the agar) but, if DNase is produced, the region of growth will be surrounded by a clear zone in which the DNA has been cleaved by the (diffusible) DNase.

DNase (deoxyribonuclease) (streptococcal) (see STREPTODORNASE)

Döderlein's bacillus An ill-defined species of *Lactobacillus* observed by Döderlein (1892) in human vaginal secretions. Considered by some to be *L. acidophilus*.

dodine (dodine acetate) *n*-Dodecylguanidine acetate. An agricultural antifungal agent used mainly for the treatment of apple scab—against which it has eradicant as well as protective action. Dodine acetate has the properties of a cationic surfactant.

dog lichen A common name for *Peltigera canina*.

doliform (*adjective*) Barrel-shaped.

dolipore septum (see SEPTUM (*mycol.*))

dominance (genetics) In diploid organisms or cells: the greater tendency of certain (*dominant*) genes to be expressed in the phenotype as compared with their corresponding, genetically-different (*recessive*) alleles. (see also DOMINANT GENE)

dominant gene That allele, in a pair of genetically-dissimilar (heterozygous) alleles, which is expressed in the phenotype; the other member of the allelic pair—which is not expressed—is termed the *recessive gene* (or recessive allele). The complete dominance of one allele over another occurs only in some cases; under such conditions the phenotype expressed by the heterozygote (Aa) is identical with that expressed by the homozygous dominant form (AA). In other cases partial or incomplete dominance may be exhibited; under these conditions the phenotype of the heterozygote Aa is intermediate between those of the homozygous dominant form (AA) and the homozygous recessive form (aa).

domiphen bromide (see QUATERNARY AMMONIUM COMPOUNDS)

DON (see 6-DIAZO-5-OXO-L-NORLEUCINE)

donor (*bacteriol.*) Any cell which contributes genetic information to another (*recipient*) cell—see CONJUGATION; TRANSDUCTION; TRANSFORMATION.

dormancy (*microbiol.*) Meanings ascribed to dormancy *include*: (1) (hypobiosis) The condition of an organism/spore that exhibits minimal physical and chemical change over an extended period of time; the extreme case of dormancy—in which *no* physical or chemical change is exhibited—is termed *cryptobiosis*. (2) The condition of a viable spore prior to *activation*—see SPORES (bacterial). (see also SPORES (FUNGAL), DORMANCY IN)

dorsiventral Having two unlike (upper and lower) sufaces.

Dothidea A genus of fungi of the LOCULOASCOMYCETES.

double diffusion (*immunol.*) (see GEL DIFFUSION)

double-dimension (see OUCHTERLONY TEST)

double helix, the (see DNA)

double recessive (genetics) An individual which is homozygous for a given recessive allele, or alleles.

double thymidine blockade (see SYNCHRONOUS CULTURE)

doubling dilutions (see SERIAL DILUTIONS)

Doulton filter (see FILTRATION)

dourine (mal du coit) A chronic, commonly fatal, sexually-transmitted disease of equines; the causal agent is *Trypanosoma equiperdum*. Symptoms include oedematous lesions of the genitalia and transient skin lesions (*plaques*) in various parts of the body.

downy mildews (see MILDEW)

doxycycline (vibramycin) 6-Deoxy-5-oxytetracycline. A semi-synthetic ANTIBIOTIC. (see TETRACYCLINES)

DPN Diphosphopyridine nucleotide—see NAD.

DPX (Distrene-plasticizer-xylol) A neutral, synthetic MOUNTANT consisting of Distrene and a plasticizer (tritolyl phosphate) dissolved in xylol. The refractive index of the dried mountant is approximately 1.53.

draughtsman colony A type of bacterial COLONY often seen e.g. in blood-agar cultures of *Streptococcus pneumoniae*—though certain strains of *S. pneumoniae* do not regularly form the draughtsman-type colony. The colony is round and steeply-elevated with a flat top—i.e. it resembles one of the pieces used in a game of draughts. (In the USA draughtsman colonies may be referred to as 'checker' colonies.) Very young colonies of *S. pneumoniae* are typically convex (domed); colonies with the flattened surface are said to result from the autolysis of constituent cells on continued incubation. Further incubation may result in the development of a concavity in the upper surface of the colony.

Dreyer's tube A small conically-based glass test-tube having a flared rim and an internal diameter of about 5 mm. Used in serology for clear observation of reactions involving precipitation or agglutination.

drift (genetic) In the genome of a species or strain: changes which are due not to selection pressures but to the persistence of *neutral mutations*, i.e. mutations which appear to be neither advantageous nor disadvantageous to the organism. (see also ANTIGENIC DRIFTING)

drop method (for counting bacteria) (see MILES AND MISRA'S METHOD)

drug resistance factor (see R FACTOR)

dry rot (of timber) Decay of structural and other timbers commonly caused (in Britain) by the cellulolytic fungus *Serpula lacrymans* (formerly *Merulius lacrymans*). Infected timber often exhibits longitudinal and cross-grained cracking and a surface growth of whitish mycelium containing yellow and/or lilac patches; the leathery fruiting bodies bear a mass of rust-coloured spores. (see also SERPULA) Only wood having a moisture content greater than about 20%* is attacked; however, water produced during the digestion of cellulose may enable the fungus to spread, across brick and stonework, to regions of drier wood—moisture-conducting rhizoids being formed for this purpose. Control involves removal of the decayed wood and the disinfection of remaining timbers by heat and/or by the application of fungicides; *S. lacrymans* is sensitive to many of the common wood preservatives. Corresponding infections in North America are often due to species of *Poria*. (see also WET ROT) *For explanation see TIMBER, DISEASES OF.

dryad's saddle A common name for the basidiomycete *Polyporus squamosus*. (cf. SADDLE FUNGUS and SADDLEBACK FUNGUS)

dulcitol (galactitol) The polyhydric alcohol corresponding to galactose. Dulcitol occurs, with SORBITOL, in certain algae e.g. *Iridaea* and *Bostrychia*.

Durham tube A transparent glass tube, commonly 2–4 centimetres long (internal diameter about 3 mm), one end of which is sealed. Durham tubes are used to demonstrate the production of gas during the growth of microorganisms in a liquid medium. A single, *inverted* Durham tube is placed in each tube of liquid medium during initial preparation; when the medium is sterilized (in an autoclave) all the air is driven from the Durham tube which, on becoming filled with the liquid medium, sinks to the bottom of the test-tube. Following inoculation and incubation of the medium, gas (if produced) collects in the Durham tube and remains trapped in the upper (sealed) end.

Dutch elm disease A disease which affects trees of the genus *Ulmus* (elm); the causal fungus is the ascomycete *Ceratocystis ulmi*. Symptoms range from the yellowing of leaves, or local defoliation, to the death of branches or the whole tree. The mechanism of disease development is unknown; the presence of the fungus within the xylem vessels leads to the formation of TYLOSES and there is some evidence that the fungus produces toxins. Commonly, the disease is transmitted (mechanically) by bark beetles of the genus *Scolytus*, but may be transmitted from diseased to adjacent healthy trees *via* the root system.

Duttonella A sub-genus of TRYPANOSOMA within the group SALIVARIA; the type-species, *T.* (*Duttonella*) *vivax*, is a pathogen of ruminants. The organisms are about 20 microns in length; they have a terminal (posterior) kinetoplast and a free flagellum.

dyes (as antimicrobial agents) A variety of chemically-unrelated dyes behave as effective antimicrobial agents. The specificity and the efficacy of such dyes may depend on their concentrations and on other factors. See: TRIPHENYLMETHANE DYES, FLUORESCENT DYES, and ACRIDINE DYES.

dyes (stains) (see STAINING)

dysentery (amoebic) (see AMOEBIASIS)

dysentery (bacterial) (a) An acute or chronic (recurrent) disease in man due to infection of the lower intestine by species of *Shigella*; *S. dysenteriae* serotype *1* causes the severest type of disease with the highest mortality rate. Infection occurs orally; incubation period: 1–6 days. Symptoms may include abdominal pain, fever, nausea, and diarrhoea containing blood and pus. The pathogen causes ulceration of the intestine but infrequently invades the bloodstream; it is uncommon in urine. *Laboratory diagnosis*: prompt culture from stool or rectal swab followed by serological or colicin typing. Treatment in severe cases may involve fluid replacement and the use of ampicillin, tetracyclines, chloramphenicol or other antibiotics. (b) Dysentery-like symptoms may be produced e.g. by *Salmonella* spp and—particularly in infants—by enteropathogenic strains of *Escherichia coli*; see also FOOD POISONING (bacterial). (c) Among animals, diarrhoea (*scours*, *scouring*) may be due to various causes. Calf scours (*white scours*) may be caused by certain serotypes of *E. coli* and may be serious or fatal; *Salmonella dublin* is another causal agent. *Black scours* (winter dysentery) of calves or cattle may be caused by strains of *Campylobacter*. In cattle, diarrhoea may be a symptom of JOHNE'S DISEASE. *Clostridium perfringens* is sometimes responsible for lamb dysentery.

dysentery (balantidial) In man, a severe or fatal dysentery caused by the ciliate *Balantidium coli*; the main source of infection appears to be the pig—in which *B. coli* usually lacks overt pathogenicity. On ingestion of the *cysts* of *B. coli*, vegetative forms (*trophozoites*) develop in the intestine and may cause extensive ulceration of the mucosa. Diiodohydroxyquin has been used for treatment. *Laboratory diagnosis* involves the identification of trophozoites or cysts in the faeces. (see also BALANTIDIUM COLI)

dysgonic (see EUGONIC)

E

E_h (see REDOX POTENTIAL)

E. coli An abbreviated form which *usually* signifies the bacterium *Escherichia coli*; the same abbreviation is also used for the amoeba *Entamoeba coli*.

Eagle's medium Any of a range of growth or maintenance media used in TISSUE CULTURE; such media consist basically of EARLE'S BSS (or HANKS' BSS) supplemented with a range of amino acids and vitamins together with serum and appropriate antibiotics.

ear, common microflora of The microflora commonly present in the human ear (external auditory canal) includes strains of *Staphylococcus*, diphtheroids, certain species of *Mycobacterium*, *Micrococcus roseus*, and various yeasts.

Earle's BSS One of a number of *balanced salt solutions* (see also HANKS' BSS). Earle's BSS contains (in grams/100 ml distilled water): NaCl (0.68), KCl (0.04), $CaCl_2$ (0.02), $MgSO_4.7H_2O$ (0.02), $NaH_2PO_4.H_2O$ (0.0125), $NaHCO_3$ (0.22), glucose (0.1) and phenol red (0.002). The pH is 7.6–7.8. Antibiotics may be added.

early blight, of potato (see POTATO, FUNGAL DISEASES OF)

earth balls A common name for the fructifications of the basidiomycete *Scleroderma* and related fungi.

earthstars The fruiting bodies of certain gasteromycetes, e.g. *Geastrum*; at maturity, the outer wall of the fructification ruptures and curls back to form a star-shaped fringe.

Eaton's agent *Mycoplasma pneumoniae.*

EB (*virol.*) Elementary body; a synonym of *virion.*

EB virus (Epstein–Barr virus) (see HERPESVIRUS)

EC Enzyme Commission—see ENZYMES.

echinulate Covered with pointed processes.

echovirus (enteric cytopathogenic human orphan virus) (see ENTEROVIRUS)

eclipse period (in bacteriophage infection) (see ONE-STEP GROWTH EXPERIMENT)

ectomycorrhiza (see MYCORRHIZA)

ectosymbiont (see SYMBIOSIS)

Ectothiorhodospira A genus of the CHROMATIACEAE.

ectothrix infection (see RINGWORM)

ectotrophic mycorrhiza (see MYCORRHIZA)

ED_{50} Effective dose (50%)—that dose of a given agent which, when given/applied to each of a number of experimental test animals/systems, affects 50% of the experimental samples under given conditions.

ED pathway (see ENTNER–DOUDOROFF PATHWAY)

EDTA Ethylenediaminetetraacetic acid, $(CH_2.N(CH_2COOH)_2)_2$. A chelating agent which e.g. prevents blood clotting *in vitro* by complexing the plasma calcium ions (i.e. it is an ANTICOAGULANT). EDTA also inhibits COMPLEMENT-fixation.

Edwardsiella A genus of gram-negative, motile bacteria of the ENTEROBACTERIACEAE; strains of the sole species, *E. tarda*, occur e.g. as parasites or pathogens of animals—including man. The organisms give a positive INDOLE TEST and METHYL RED TEST, and a negative VOGES–PROSKAUER TEST and CITRATE TEST. Acid and gas are produced from glucose at 37 °C; lactose is not fermented. Lysine and ornithine decarboxylases are produced. H_2S is formed in TSI medium.

EES Ethylethanesulphonate—see ALKYLATING AGENTS.

efficiency of plating (see EOP)

effused-reflexed (effuso-reflexed) (*mycol.*) Refers to a sheet-like fruiting body which lies flat against the substratum except at the edge(s)—the latter growing away from the substratum.

egg-yolk agar (EYA; lecithovitellin (LV) agar) An agar-based medium which incorporates a saline extract of the yolks of hens' eggs (lecithovitellin). Egg-yolk agar is commonly used to demonstrate the production of bacterial extracellular LECITHINASES: the colonies of lecithinase-producing strains are surrounded by zones of opacity, or opalescence, due to the deposition of insoluble diglycerides as a result of lecithin cleavage. (see also NAGLER'S REACTION)

EGTA Ethyleneglycol-*bis*(β-aminoethylether)-*N,N,N',N'*-tetraacetic acid; the magnesium salt chelates calcium ions—see e.g. COMPLEMENT-FIXATION.

eguttulate (*mycol.*) Refers to a spore which does not contain GUTTULES.

EHC (ethylhydrocuprein) (see OPTOCHIN)

Eijkman test A test used to distinguish between certain strains of *coliform* organism. The modern form of the test determines the ability of an unknown strain to produce gas from lactose at 44 °C ± 0.2 °C; a positive Eijkman test is presumptive evidence of *Escherichia coli*. A variety of media have been used for the test, e.g. lactose-ricinoleate broth, MacConkey broth, lactose-peptone water. The Eijkman and IMVIC TESTS are used in water bacteriology for the detection of faecal coliforms in samples of treated water.

Eimeria A genus of protozoan intracellular parasites of the sub-order EIMERIINA. (For a generalized account of the morphology and life cycle of *Eimeria* and related organisms see COCCIDIA.) *Eimeria* is distinguished from related genera (e.g. ISOSPORA) by the production of oocysts containing four sporocysts, each sporocyst containing two sporozoites. No species is parasitic in man, but many are pathogenic for domestic animals: see COCCIDIOSIS.

Eimeriina (Eimeriorina) (*protozool.*) A sub-order of the EUCOCCIDIA in which *syzygy* does not occur (cf. Adeleina), in which motile zygotes are not formed (cf. HAEMOSPORINA), and in which the sporozoites are typically enclosed within a *sporocyst*; in monoxenous species, resistant oocysts (containing sporocysts or naked sporozoites) function as disseminative units. The sub-order contains a number of pathogens. Genera include EIMERIA, ISOSPORA, TOXOPLASMA and TYZZERIA. N.B. Organisms in this sub-order are commonly referred to as the COCCIDIA (*q.v.*).

El Tor vibrio (see VIBRIO)

elaioplast A lipid-storing PLASTID.

electroendosmosis (electro-osmosis) A phenomenon the effects of which may be observed (for example) during the IMMUNOELECTROPHORESIS of serum in agar gel. During this procedure (which is often carried out at pH 8.6) some of the (negatively) charged proteins (e.g. γ-globulins) migrate towards the cathode (negative pole). This effect results from the ionization of the AGAR; the agar carries a negative charge, and the *fluid* contains an excess of positive ions. The fluid therefore moves towards the cathode—the flow being sufficiently strong to overcome the attraction of the anode for certain proteins and to bear them towards the cathode. Thus, *electroendosmosis* refers to the movement of a charged fluid—relative to a fixed medium carrying the opposite charge—under the influence of an electrical gradient.

electron microscopy A form of MICROSCOPY in which the image is formed by a beam of electrons which has passed through, or has been reflected from, the specimen. The advantages of electron microscopy (compared with light microscopy) include improved *resolving power* and *magnification*; the main disadvantage is that *living* organisms cannot be studied.

The limit of resolution of an electron microscope can be as small as 2–3Å under certain conditions; this compares with 2000–2500Å for the light microscope. For most biological specimens the resolving power is, at best, approximately 10Å. The useful magnification ranges from about 1000x to over 800,000x. In light microscopy the RESOLVING POWER depends partly on the wavelength of the light incident upon the specimen; similarly, in electron microscopy the resolving power is a function of the wavelength of the electron beam. The wavelength of an electron beam is inversely proportional to the square root of the accelerating voltage used in its production (see later).

Except in scanning electron microscopy (see separate heading below) image formation by the electron beam depends on the deflection or scattering of electrons by components of the specimen; only electrons whose paths are *not* significantly deviated by the specimen contribute to the final image-forming beam. Certain specimens, e.g. viruses, can be examined whole; other types of specimen (e.g. mammalian cells, bacteria) must be examined in the form of thin sheets ('sections') of *maximum* thickness 1000Å. Such ultrathin sections and particles of biological material cause minimal deflection or scattering in the electron beam; hence, if examined without further treatment little detail would be seen. Such specimens—or the regions surrounding them—are *stained* with certain heavy metals (or their salts); this enhances the electron-deflecting properties of specific regions of the specimen and/or increases the contrast between specimen and background (see *Preparation of Specimens* below). The image of the specimen may be formed either on a fluorescent screen or on a photographic plate.

The Electron Microscope. The electron beam originates at the *electron gun*: electrons are emitted by a heated metal filament, and are accelerated towards, and through, an aperture in the *anode plate* of the gun; acceleration is a consequence of the high potential difference ('accelerating voltage') maintained between the anode plate (at earth, or zero potential) and the filament (about -20 kV to over -100 kV with respect to the anode). (1 kV $=$ 1000 volts) The various *electromagnetic lenses* each consist basically of a soft-iron reel on which are wound several thousand turns of copper wire; the passage of an electric current through the winding generates an intense magnetic field capable of controlling (focusing etc.) an electron beam passing axially through the lens. The focal length of an electromagnetic lens can be varied by adjusting the current passing through the winding; hence, magnification can be varied continuously (over a certain range) in a way analogous to that of the ZOOM MICROSCOPE. The interior of the instrument must be effectively evacuated in order to prevent e.g. scattering and deflection of electrons by molecules of gas; the air pressure within an electron microscope is therefore reduced to values ranging from 0.5 micrometres of mercury to less than 1 nanometre of mercury, according to the type of instrument. The *screen* consists of a metal plate which has been coated with a layer of fluorescent material.

Preparation of Specimens. Specimens are examined on a particular form of sample holder, the *grid*—a disc of copper, 2–3 mm in diameter, perforated by a regular series of square holes of side approximately 0.1 mm. Before use, the upper surface of the grid is

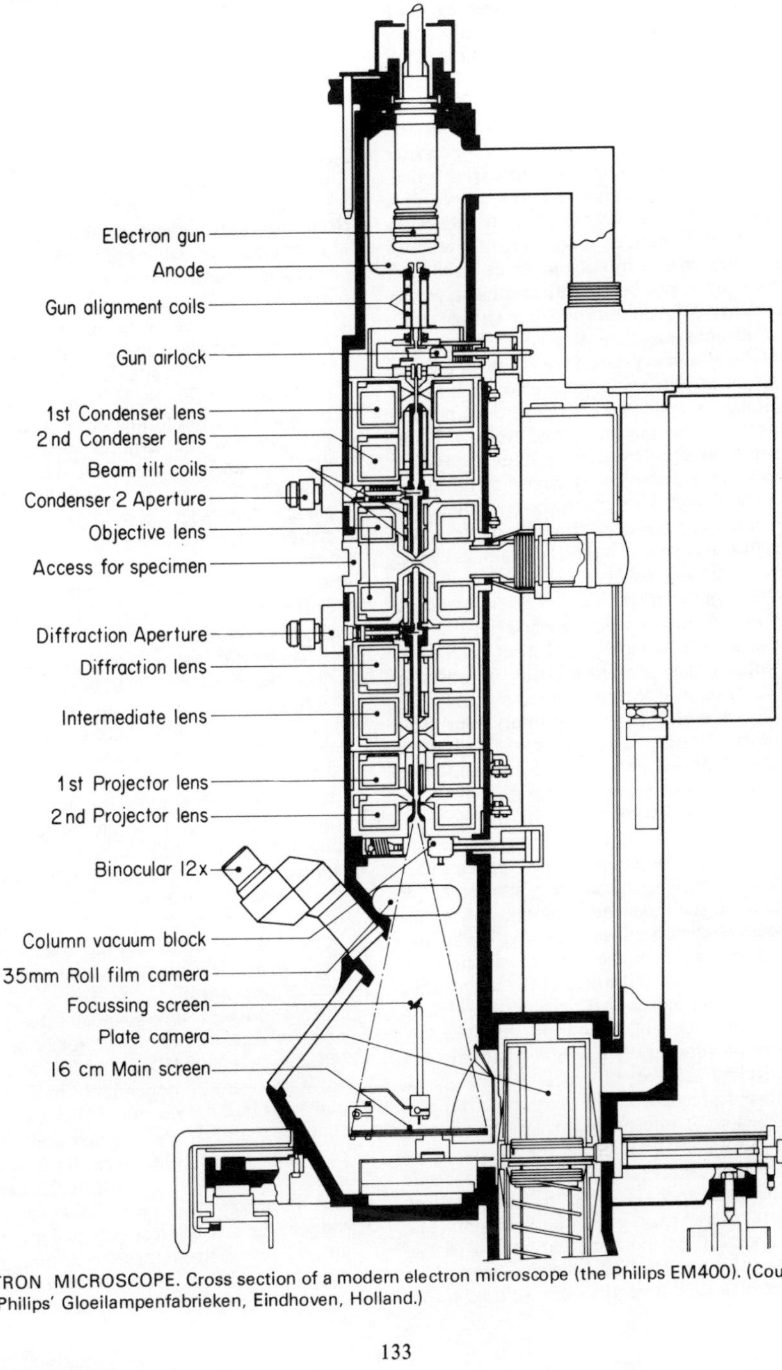

ELECTRON MICROSCOPE. Cross section of a modern electron microscope (the Philips EM400). (Courtesy of N. V. Philips' Gloeilampenfabrieken, Eindhoven, Holland.)

covered with a *support film*—a sheet of plastic, 100–200Å thick—on which the specimen rests; the material of the support film (e.g. nitrocellulose) is chosen such that the film has a low, uniform electron-scattering power.

Apart from staining, viruses and macromolecules require little preparation prior to examination in the electron microscope. Cells, e.g. bacteria, must first undergo FIXATION; frequently-used fixatives include OSMIUM TETROXIDE and/or GLUTARALDEHYDE. Following fixation, the specimen undergoes *dehydration*. The dehydrated specimen is then *embedded*, i.e. impregnated with an organic compound (e.g. a methacrylate) which, when heated, polymerizes and hardens without significant change in volume. Once embedded, the specimen can be cut into sections typically less than 1000Å in thickness (see MICROTOME). The sections are then stained with an agent containing a heavy metal—e.g. uranyl acetate, or a salt of phosphotungstic acid; such agents combine or complex with particular components of the specimen, e.g. proteins, lipids. (Images obtained by electron microscopy thus reflect the distribution of heavy metal atoms and the groups with which the staining agents combine.) The stained section, resting on a support film and grid, can then be examined under the electron microscope.

Other procedures. (a) *Negative staining* (negative-contrast). This method is used for particles such as viruses and macromolecules. Essentially, the specimen is partly embedded in a thin layer of electron-opaque material (e.g. uranyl acetate) on the support film; this material penetrates the interstices of each particle but does not permeate the particle. Thus, electron transmission is greater in those locations occupied by the (relatively) electron-transparent particles. In the image, such particles appear light against a dark background; details of their surfaces correspond to the pattern of stain on the contours and in the crevices of each particle. Thus, e.g. the hollow, cylindrical virion of the tobacco mosaic virus appears as a light rod with a dark, central axial line; the latter indicates the presence of stain in the hollow core of the virion. (b) *Shadow-casting* (shadowing). This method is particularly useful for the study of viruses. Essentially, an unstained preparation of virus (on a support film and grid) is exposed to the *vapour* of a heavy metal; the source of the vapour is so placed that metal is deposited only on *one side* of the specimen—leaving a non-metalized (electron-transparent) 'shadow' on the remainder of the specimen and on that part of the support film shielded from the vapour. The image of a shadowed virus can be used to obtain information on the shape and size of the virion; the size can be calculated from the length of the shadow and the angle at which the metal vapour is incident upon the virus. (c) See FREEZE-ETCHING.

Scanning Electron Microscopy. This form of microscopy is used to examine the *surface* details of a specimen. Essentially, a beam of electrons scans the surface of a specimen which has been coated with a thin layer of gold (or other metal); electrons *reflected* by the specimen are collected and focused, sequentially, to form the image. The resolving power of a scanning electron microscope is at best about 100–200Å.

electron transport chain (respiratory chain; ETC) A sequence of redox agents along which electrons are transferred with concomitant energy conversion. The electrons are donated by certain reduced products of metabolism (e.g. $NADH_2$ and succinate formed in the TRICARBOXYLIC ACID CYCLE); the electrons are transferred—*via* the chain—to molecular oxygen (with the formation of water). (In certain *bacteria* the terminal electron acceptor may be an inorganic ion—see ANAEROBIC RESPIRATION.) Electron transfer along the chain is believed to result in the formation of an energized state or entity (the nature of which is uncertain—see OXIDATIVE PHOSPHORYLATION) which can be used for a number of energy-dependent processes. (see also PHOTOSYNTHESIS)

Eucaryotic system. The ETC occurs in the inner membrane of the MITOCHONDRION and is composed basically of CYTOCHROMES, ubiquinone (see QUINONES), sulphur-containing nonhaem-iron proteins, and two flavoproteins ($NADH_2$-dehydrogenase and succinate dehydrogenase). The precise sequence in which these carriers act is still uncertain—see diagram for a widely-accepted scheme. They appear to be asymmetrically orientated across the mitochondrial inner membrane, i.e. some carriers occur on the matrix side of the membrane, others occur at the outer face. (cf. *chemiosmotic hypothesis* in OXIDATIVE PHOSPHORYLATION)

Some authors have suggested that the ETC may be regarded as three *groups* of carriers (groups I, II and III) linked by carriers with variable, energy-dependent half-reduction potentials (see REDOX POTENTIAL); each *group* contains carriers with similar half-reduction potentials—which differ from those of carriers in other groups. Cytochrome b_T links groups II and III, and cytochrome a_3 links group III and oxygen; the link between groups I and II is unknown. As electrons pass from one group of carriers to the next—or from group III to oxygen—energy conversion occurs (at sites I,

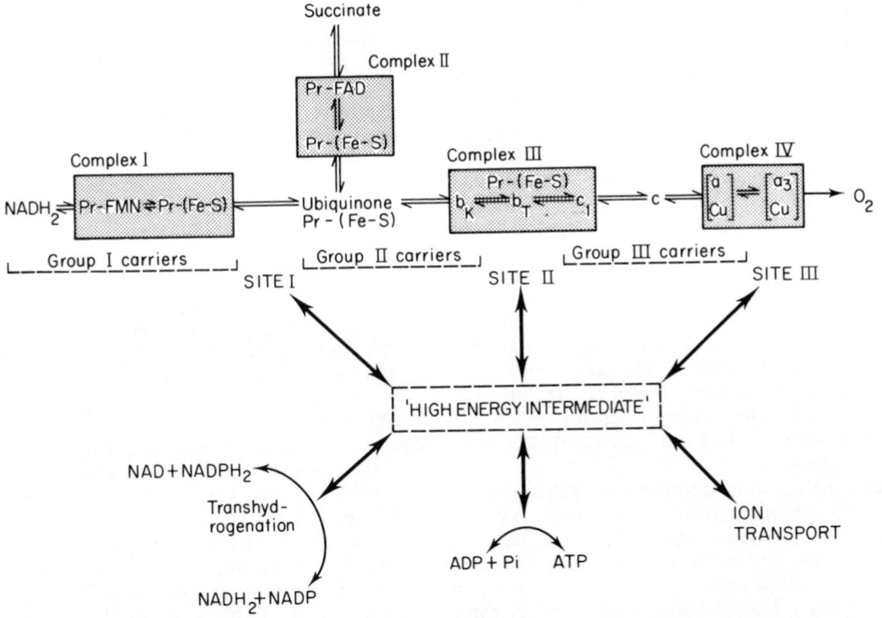

MITOCHONDRIAL ELECTRON TRANSPORT CHAIN—a tentative scheme. ⇌ = electron flow; ⟷ = energy conversion; Pr-FMN = $NADH_2$-dehydrogenase; Pr-FAD = succinate dehydrogenase; Pr-(Fe-S) = nonhaem-iron sulphur proteins; b_K, b_T, c_1 etc. = cytochromes; complexes I, II, III and IV correspond to preparations which can be obtained by treating sonicated mitochondria with a surfactant e.g. deoxycholate.

II and III—see diagram). The relevance of the variable half-reduction potentials to energy transduction is still unclear. (see also RESPIRATORY CONTROL; RESPIRATORY INHIBITORS)

Bacterial systems. The composition of the bacterial ETC appears to exhibit wide variability among different species, and may even vary in the same species grown under different conditions. The chain is located in the CELL MEMBRANE; it may include a range of carriers—e.g. cytochromes of the *a*, *b*, *c*, and/or *d* type, and quinones of the ubiquinone and/or naphthoquinone type. Of those bacteria which have been investigated, many contain a number of membrane-bound dehydrogenases which can apparently donate electrons directly to the chain; two or three terminal cytochrome oxidases are commonly present. In some bacteria (at least), electrons may follow a *branched* pathway between the dehydrogenases and oxidases; for example, in *Azotobacter vinelandii*, the following system has been proposed: electrons from the various dehydrogenases pass to cytochrome b_1 *via* a ubiquinone; from b_1 the flow may follow one of three paths: (a) $b_1 \rightarrow d \rightarrow O_2$; (b) $b_1 \rightarrow c_4 \rightarrow a_1 \rightarrow O_2$; (c) $b_1 \rightarrow c_5 \rightarrow o \rightarrow O_2$. The particular pathway used may depend on the conditions of growth. In anaerobic respiration, part of the aerobic chain may be used—electrons being passed to the terminal acceptor from a member of the chain *via* a *reductase*. A number of obligate anaerobes of the rumen (e.g. *Bacteroides ruminicola* and *B. succinogenes*) have been shown to contain cytochromes; there is evidence to suggest that these organisms possess an ETC in which organic acids (e.g. fumaric acid) act as terminal electron acceptors. (cf. FERMENTATION)

electro-osmosis (see ELECTROENDOSMOSIS)

electrophoresis An analytical procedure by means of which an heterogenous population of charged particles (e.g. protein molecules) may be separated by virtue of their dissimilar migration rates in an electric field; electrophoresis is used e.g. for the separation and quantitative estimation of plasma proteins. In *zone electrophoresis* a small droplet of the sample is applied to one end of a strip of paper or cellulose acetate, or placed in a well cut into a strip of AGAR or starch gel; electrophoresis is carried out in a closed container in which the paper, gel, etc. is first allowed to equilibrate with the vapour of an appropriate buffer solution. An electrical potential difference is then applied lengthwise across the strip—each end of which is in communication with buffer in a separate trough within the container. (For the electrophoresis of plasma proteins the buffer commonly maintains a pH of 8.6.) Following electrophoresis (and after suitable preparation) the paper (or gel) strip may be stained (e.g. with one of the protein stains azocarmine B, amidoschwartz 10B, or nigrosin) when discrete zones, bands, or spots of the various protein components become visible. Quantitative estimations are based on the proportionality between staining density and amount of protein—the stained strip being examined e.g. with a scanning densitometer. (see also ELECTROENDOSMOSIS and IMMUNOELECTROPHORESIS)

Elek plate An *in vitro* method for detecting the formation of toxin by a given strain of *Corynebacterium diphtheriae.** A plate of suitable medium is inoculated with a single streak of the strain under test; the streak is then overlaid, at right angles, with a strip of paper impregnated with diphtheria antitoxin. Following incubation, toxigenicity is indicated by the presence of lines of white precipitate which approximately bisect each of the right angles formed by the paper strip and the line of bacterial growth; these lines result from the combination of toxin with antitoxin following their outward diffusion from the line of growth and the paper strip respectively. Non-specific lines of precipitate may be formed—hence known positive and negative strains are used as controls. *Toxin formation by other organisms may be detected by an analogous procedure.

elementary body (EB) (*virol.*) A synonym of *virion.*

elementary particles (see MITOCHONDRION)

ELISA (*immunol.*) *Enzyme-linked immunosorbent assay*; a technique used for detecting and quantifying specific serum antibodies. Initially the serum is allowed to react with specific antigens which have been adsorbed e.g. to the surface of a plastic tube. Antibodies which combine with the antigens are then detected by treating the test system with a *conjugate*, i.e. anti-immunoglobulin which has been linked with a particular enzyme; the test system is washed free of any uncombined conjugate and examined for the presence of bound enzyme: the system is incubated with an appropriate substrate and the products of enzymic cleavage assayed e.g. by spectrophotometry.

Elphidium A genus of the FORAMINIFERIDA.

Elsinoë A genus of fungi of the class LOCULOASCOMYCETES; species include some important plant pathogens—e.g. *E. veneta*, a pathogen of the raspberry (*Rubus* sp). *Elsinoë* spp form globose asci in uniascal locules which are located at various levels in the pseudoparenchymatous ascostroma; the hyaline or yellowish ascospores each have three transverse septa. The ascostroma is the overwintering stage. The conidial (imperfect) stage of a number of species of *Elsinoë* correspond to members of the *form*-genus *Sphaceloma.*

elute (*immunol. virol.*) Antibodies are said to *elute* from their homologous antigens when they dissociate from them. For example, COLD AGGLUTININS may be eluted from their homologous antigens by incubation for an appropriate period at 37 °C. Similarly, some haemagglutinating viruses *elute* from erythrocytes.

emarginate (*mycol.*) In an agaric: refers to a gill whose lower edge slants sharply upwards at its junction with the stipe. (see also DECURRENT)

EMB agar A medium used e.g. to distinguish between certain members of the Enterobacteriaceae. Constituents of the medium include (w/v) peptone (1%), lactose (1%), eosin Y (0.04%), methylene blue (0.0065%), and agar (1.5%). On EMB agar colonies of different appearance are produced e.g. by *Escherichia coli* and *Klebsiella pneumoniae. E. coli* carries out the MIXED ACID FERMENTATION, and the consequent reduction in pH is reported to result in the precipitation of an eosin-methylene blue complex; *K. pneumoniae* carries out the BUTANEDIOL FERMENTATION which results in a higher final pH in the medium.

Embden–Meyerhof–Parnas pathway (EMP pathway; Embden–Meyerhof pathway; hexose diphosphate pathway; glycolysis) A sequence of reactions in which glucose is broken down to pyruvate; the pathway involves a net production of two molecules each of pyruvate, $NADH_2$, and ATP for each molecule of glucose metabolized (see Appendix I(a)). The EMP pathway occurs in a wide range of microorganisms and plays a central role in both fermentative and respiratory modes of carbohydrate metabolism. In oxidative (respiratory) organisms the pyruvate is converted to acetyl-CoA and enters the TRICARBOXYLIC ACID CYCLE; in fermentative organisms the metabolic fate of pyruvate depends e.g. on the species (see Appendix I(a)). The possession by an organism of the enzyme *fructose diphosphate aldolase* is sometimes taken as evidence for the occurrence of the EMP pathway in that organism. (see also ENTNER–DOUDOROFF PATHWAY; HEXOSE MONOPHOSPHATE PATHWAY)

embryonated egg (embryonating egg) (*virol.*) An egg (usually a hen's or duck's egg) which contains a live embryo. Such eggs are used in virology e.g. for diagnostic purposes (see e.g. POXVIRUS), for culturing certain viruses (e.g. influenza viruses for vaccines), and for preparing TISSUE CULTURES.

Emericella A genus of fungi of the order EUROTIALES; the conidial (imperfect) stages of the organisms correspond to *Aspergillus nidulans* and related *form*-species. In *Emericella* spp the ascocarp (a globose cleistothecium) consists of a number of spherical or ovoid evanescent asci contained within a thin—commonly red or violet—peridium; the cleistothecium is surrounded by a layer of hülle cells. The ascospores are typically some shade of red or violet.

Emerson enhancement effect In plant and algal cells illuminated with light of wavelength 680–700 nm: the increase in photosynthetic activity which occurs when the cells are illuminated with an *additional* beam of light of shorter wavelength. The level of photosynthetic activity with the two beams is greater than the sum of the activities obtained using each beam separately. (see also RED DROP)

emetine An alkaloid, obtained from ipecacuanha root, which is active against *Entamoeba histolytica* (see AMOEBIASIS). Although it is still used for severe infections its toxicity precludes wider application. Emetine acts against trophozoite amoebae but—at permissible dosages—not against cysts. The mode of action of emetine is unknown but may involve inhibition of protein synthesis. Clinically-useful derivatives include emetine bismuth iodide and 2-dehydroemetine.

Emmonsiella capsulata (see HISTOPLASMA CAPSULATUM)

EMP pathway (see EMBDEN–MEYERHOF–PARNAS PATHWAY)

EMS Ethylmethanesulphonate—see ALKYLATING AGENTS.

emulsification (*bacteriol.*) The process of preparing a *suspension* of bacterial cells in distilled water, saline etc; by means of a loop, bacterial growth is transferred to, and mixed with, a drop of water on a slide.

enations (*plant pathol.*) Localized proliferations of leaf tissue caused e.g. by viral infection.

encephalitis Inflammation of the brain.

Encephalitozoon cuniculi Synonym of *Nosema cuniculi*.

end point (*virol.*) (1) (endpoint titration) A method of assaying viral infectivity. *Infectivity* may refer either to (a) the number of infectious virions per unit volume, or (b) the 'strength' of a viral suspension in terms of the extent to which it must be diluted in order that a given volume of the dilution corresponds to the LD_{50} (*q.v.*) or, more generally, the ED_{50} of the virus. The endpoint titration determines infectivity as defined in sense (b). Initially, log (or half-log) dilutions of the sample are prepared. Each dilution is then examined, individually, by inoculating each of a number of similar test units with a fixed volume of that dilution. Test units may be susceptible laboratory animals, or tissue cultures which undergo detectable forms of degeneration when infected with the virus. After incubation, every test unit is examined. Animals (or cultures) inoculated from the lower dilutions tend to react positively in every case; of those inoculated from the higher dilutions, some react positively while others are unaffected. The important dilution is that of which a given volume corresponds to the ED_{50} of the virus; of the animals (or cultures) inoculated from this dilution, 50% react positively while the remainder are unaffected. Such a '50% endpoint' is not always obtained experimentally, but may be calculated from the experimental data—provided that some test units have given negative reactions. The 50% endpoint dilution is often calculated by the Reed & Muench method or by the Kärber method; examples of the use of these methods are given with reference to the specimen data tabulated below. (N.B. The Reed & Muench and Kärber methods can be used only if the dilutions are in a regular logarithmic sequence, and if a similar number of test units are used for each dilution.) The logarithm of the 50% endpoint dilution lies between −5 and −4; in the *Reed & Muench method* the value of this logarithm is obtained by linear interpolation from the appropriate values in column E:

$$\text{Log 50\% endpoint} = -5 + \frac{50 - 33{\cdot}3}{85{\cdot}7 - 33{\cdot}3}$$

$$= -4{\cdot}68$$

Column A	*Column B*	*Column C*	*Column D*	*Column E*
Dilution of virus.	Mortalities in 5 test animals.	Cumulative mortalities.	Cumulative survivors.	(C/C+D)%
10^{-6}	0/5	0	9	0
10^{-5}	2/5	2	4	33.3
10^{-4}	4/5	6	1	85.7
10^{-3}	5/5	11	0	100

In the *Kärber method*, substitutions are made in the equation:

$$\text{Log } 50\% \text{ endpoint} = L_1 - L(S - 0.5)$$

in which L_1 = log of lowest dilution tested; L = log interval between dilutions; S = sum of the proportion of positive reactions at each dilution; 0.5 is a constant. (In the accompanying table S = 0 + 0.4 + 0.8 + 1.0.) Substituting:

$$\text{Log } 50\% \text{ endpoint} = -3 - 1(2.2 - 0.5) = -4.7$$

An *estimate* of the number of infectious virions in the original suspension can be derived from the expression:

$$n = -\log_e (1 - p)$$

in which n = average number of infectious virions *per inoculum*; p = proportion of positive reactions in the test units inoculated from a particular dilution; e = 2.7183; the value of n—calculated for a particular dilution—multiplied by the dilution factor gives an estimate of the viral titre of the original suspension. This type of estimate is less reliable than that made from the results of the *plaque method* (see ENUMERATION OF MICROORGANISMS). (2) Generally, that point in a viral titration (e.g. a titration based on haemagglutination) at which the effects of the virus are minimal.

end-point titration (*immunol.*) A method of assaying the concentration of a particular entity (e.g. antibody or antigen) in a given sample by finding the highest dilution of the sample which gives a reaction regarded as positive under the conditions of the test. In an AGGLUTINATION test, for example, the *end-point* is commonly the highest dilution which exhibits agglutination. (see also TITRE)

Endamoeba A genus of cyst-forming amoebae; species are parasitic in various invertebrates. The organisms differ from ENTAMOEBA in that the nucleus lacks both a karyosome and the peripheral nucleic acid or nucleoprotein.

endemic (of diseases) Refers to any disease which is commonly present in a given geographical region; such diseases generally exhibit a low morbidity.

endergonic reaction Any reaction which is accompanied by an uptake of energy (cf. EXERGONIC REACTION).

endodyogeny A form of reproduction in which two daughter cells form within the parent cell; endodyogeny has been observed e.g. in *Toxoplasma*, certain other coccidians, and in the alga *Chlamydomonas*.

endoenzymes Enzymes which cleave bonds *within* a polymer chain (cf. EXOENZYMES (2)); for examples, see α-AMYLASES, and *restriction endonucleases* under RESTRICTION.

endogenote (see MEROZYGOTE)

endogenous infection Infection by an organism which is part of the common or established body microflora. (see e.g. THRUSH; see also OPPORTUNIST PATHOGENS)

Endogone A genus of fungi of the ZYGOMYCETES.

endolithic (of a lichen) Refers to a lichen whose thallus is embedded in rock; the rock may be penetrated to a depth of up to several millimetres. The lichen is believed to achieve penetration by releasing substance(s) which dissolve the rock. The fruiting bodies of the lichen occur at the rock surface. Examples of lichens which may be endolithic include *Lecanora dispersa* (see LECANORA) and *Verrucaria hochstetteri* (see VERRUCARIA). (cf. EPILITHIC)

endolysin The LYSOZYME coded for by bacteriophage genes. (see also ONE-STEP GROWTH EXPERIMENT)

endomitosis A process in which chromosome replication occurs without nuclear division (cf. MITOSIS).

Endomycetaceae A family of fungi of the ENDOMYCETALES; species occur e.g. in the soil and on decaying wood. The vegetative form of the organisms is a well-developed septate mycelium; asexual reproduction involves the formation of arthrospores and blastospores. Sexual reproduction involves the *fusion of gametangia*; eight (or fewer) ascospores develop in each ascus. Genera include *Eremascus* and *Endomyces*.

Endomycetales An order of FUNGI of the class HEMIASCOMYCETES; constituent species are distinguished from those of other orders in that, during sexual reproduction, the ascus typically develops from a zygote or from a cell derived mitotically from a zygote. (cf. LIPOMYCES) The zygote may be formed by the fusion (plasmogamy and karyogamy) of two cells or by the fusion of two gametangia. (Asci may also develop parthenogenically from single cells.) The order includes both unicellular and mycelial organisms. The constituent families are Ascoideaceae (mycelial organisms which form *multi*spored asci), ENDOMYCETACEAE, SACCHAROMYCETACEAE, and Spermophthoraceae (organisms which form eight or fewer needle- or spindle-shaped ascospores per ascus).

endomycorrhiza (see MYCORRHIZA)

endoparasitic slime moulds (see PLASMODIOPHORALES)

endoperidium In members of the LYCOPERDALES: the inner layer of the peridium. (cf. EXOPERIDIUM)

endophloeodal (endophloedal; endophloeodic; hypophloeodal) (of a lichen) Refers to a lichen whose thallus occurs *within* the bark of a tree; the thallus grows between, but does not penetrate, the cells of the bark. (cf. EPIPHLOEODAL and ENDOLITHIC)

endopolygeny A form of reproduction in which many daughter cells form within the parent cell; endopolygeny occurs e.g. in *Toxoplasma* and in the alga *Chlamydomonas*. (cf. ENDODYOGENY)

endoral membrane (in *Paramecium*) A row of cilia found along the right-hand margin of the buccal overture; it is considered to be equivalent to the undulating membrane in *Tetrahymena*.

endosome (1) A term which may refer either to a PINOSOME or to a PHAGOSOME. (2) An intranuclear, RNA-containing, Feulgen-negative body which is similar in appearance to a NUCLEOLUS but which persists—and divides—during mitosis. Endosomes occur e.g. in the nuclei of euglenoid flagellates.

endospore (bacterial) (see SPORES (bacterial))

endosymbiont (see SYMBIOSIS)

endothrix infection (see RINGWORM)

endotoxins (1) Certain (toxic) substances found within bacterial cells and which are released only on cell lysis. (2) Currently, the term refers mainly to the LIPOPOLYSACCHARIDES of gram-negative bacteria. The toxic character of the lipopolysaccharides is believed to reside in the lipid portion, *Lipid A*; the polysaccharide subunit is believed to contribute to toxicity by promoting the water-solubility of the lipid. (see also SHWARTZMAN REACTION and EXOTOXINS)

endotrophic mycorrhiza (see MYCORRHIZA)

enhancement effect (see EMERSON ENHANCEMENT EFFECT)

enniatins (see DEPSIPEPTIDE ANTIBIOTICS)

enriched medium Any *basal* medium (see e.g. NUTRIENT BROTH) which has been supplemented (enriched) with serum, blood etc.—or any of a range of nutrients or growth factors—in order to be able to support or enhance the growth of particular organisms.

enrichment (*microbiol.*) Any form of culture, in a *liquid* medium, which results in an increase in the numbers of a given type of organism relative to the numbers of other types of organism which may be present in the inoculum; an enrichment medium may contain substance(s) which encourage the growth of the required organism or which inhibit the growth of the other types of organism. For example, if a faecal specimen from a case of typhoid fever is cultured directly on a solid medium, colonies of *Salmonella* may be difficult to detect among a heavy growth of the normal faecal flora; by culturing the faeces in an enrichment medium (e.g. SELENITE BROTH—which suppresses the normal faecal flora), and subculturing to solid media, the detection of *Salmonella* is facilitated.

enrichment medium Any medium used for ENRICHMENT; an example is SELENITE BROTH. (cf. ENRICHED MEDIUM)

ensilage (1) The process of preparing SILAGE. (2) The silage itself.

Entamoeba A genus of amoebae of the order AMOEBIDA; only one *non*-parasitic species (*E. moshkovskii*) has been reported (in sewage). *Entamoeba* spp are distinguished by the presence of nucleic acid, or nucleoprotein, at the inner surface of the nuclear membrane. Most species form cysts—*E. gingivalis* is an exception; some cysts contain CHROMATOID BODIES and may also contain glycogen vacuoles.

E. histolytica. Trophozoites living as non-invasive forms in the gut lumen are generally about 20 microns in diameter; invasive forms, which penetrate and develop within the mucosa (or other tissues)—see AMOEBIASIS—are usually 20–50 microns in diameter. The single nucleus contains a central body, the *karyosome*. The trophozoites stain well e.g. with iron-haematoxylin after fixation in PVA (polyvinyl alcohol). Spherical or ovoid CYSTS (about 10 microns or more in diameter) are formed in the gut lumen; chromatoid bodies are usually seen in young cysts. Mature cysts typically contain *four* nuclei—each containing a central karyosome. In iodine-stained cysts the cytoplasm stains yellow and the glycogen (if any) stains reddish-brown. Media which have been used for the (anaerobic) culture of *E. histolytica* (37 °C) have contained materials such as inspissated horse serum, rice starch and egg albumen—together with a supplement of bacteria or flagellate protozoans; the organism has also been cultured in sterile media.

E. hartmanni. A non-pathogen found in human stools. It resembles *E. histolytica* e.g. in that the mature cyst contains four nuclei, but both the trophozoite (5–10 microns) and the cyst (less than 10 microns) are smaller.

E. coli. A non-pathogen found in human stools. Trophozoites and cysts are similar in size to those of *E. histolytica*, but e.g. the karyosome is usually off-centre, and the mature cyst usually contains *eight nuclei*.

E. gingivalis. A non-pathogen found in the buccal cavity of man and other animals; it is some 10–30 microns in diameter. Other species include *E. bovis* (in cattle) and *E. invadens* (a pathogen of reptiles).

enteritis Inflammation of any part of the intestine—particularly of those tissues which line the small intestine. (cf. GASTRO-ENTERITIS)

Enterobacter A genus of gram-negative, motile bacteria of the ENTEROBACTERIACEAE; strains occur e.g. in the alimentary tract of man and other animals. The organisms give a positive CITRATE TEST and (typically) a positive VOGES-

PROSKAUER TEST; the INDOLE TEST and the METHYL RED TEST are negative. Acid and gas are formed from glucose at 37 °C and lactose is fermented. H_2S is not produced. Strains which form lysine decarboxylase, hydrolyse aesculin, and produce gas from glycerol are placed in the species *E. aerogenes*; strains which give negative results in these tests form the species *E. cloacae*. (*Some* of the organisms previously designated as *Aerobacter aerogenes* are now designated *E. aerogenes*.) Type species: *E. cloacae*; GC% range for the genus: 52–59.

Enterobacteriaceae A family of gram-negative, facultatively anaerobic, asporogenous, peritrichously flagellate or non-motile, rod-shaped bacteria; species are widespread as parasites or pathogens of man, other animals, and plants, and as saprophytes in soil and water. The organisms are chemoorganotrophs which can express a respiratory (oxidative) and/or fermentative type of metabolism according to conditions; growth usually occurs on relatively simple media, e.g. nutrient agar. All species are oxidase-negative and (with the exception of *Shigella dysenteriae* serotype 1) catalase-positive. Most species reduce nitrates to nitrites; at least one species, *Klebsiella pneumoniae*, carries out NITROGEN-FIXATION. Genera and species are distinguished biochemically (see e.g. IMVIC TESTS) and serologically (see e.g. KAUFFMANN-WHITE CLASSIFICATION SCHEME for *Salmonella*); phage TYPING and COLICIN typing are also used. Constituent genera: CITROBACTER, EDWARDSIELLA, ENTEROBACTER, ERWINIA, ESCHERICHIA, HAFNIA, KLEBSIELLA, PROTEUS, SALMONELLA, SERRATIA, SHIGELLA and YERSINIA.

enterobactin (see SIDEROCHROMES)

enterochelin (see SIDEROCHROMES)

enterococci (see STREPTOCOCCUS)

Enterographa A genus of crustose LICHENS in which the phycobiont is *Trentepohlia*. In e.g. *E. crassa* the areolate thallus varies in colour from grey to reddish-brown; a black hypothallus is visible between the areolae. Immersed in the areolae are minute dark apothecioid pseudothecia which vary in shape from circular to thread-like. *E. crassa* occurs on trees in deeply-shaded situations.

enterotoxin Any EXOTOXIN which, when ingested, or produced within the intestine, is absorbed by the gut and affects, directly or indirectly, the functioning of the intestinal mucosa. The CHOLERA enterotoxin is a protein (M.Wt. approximately 84,000) which stimulates adenyl cyclase activity in the mucosa of the small intestine—thus causing raised levels of cyclic AMP (cAMP) and the consequent hypersecretion of water and electrolytes. Antigenically-similar heat-labile enterotoxins produced by enteropathogenic strains of *Escherichia coli* have a similar mode of action; enterotoxin production is determined by a plasmid. The staphylococcal enterotoxins (A, B, C_1, C_2, D and E) are small heat-stable proteins whose mode of action is unknown. (see also FOOD POISONING) Other bacteria, e.g. *Pseudomonas aeruginosa*, have been reported to form enterotoxins.

Enterovioform (*proprietary name*) (see IODINE)

Enterovirus A genus of VIRUSES of the family PICORNAVIRIDAE. Enteroviruses occur typically in the gastrointestinal tract of vertebrates; they cause diseases of the gastrointestinal and respiratory tracts and of the central nervous system—though infection is often asymptomatic. The non-enveloped, icosahedral *Enterovirus* virion—20–30 nm in diameter—contains single-stranded RNA; enteroviruses are ether-resistant and are not inactivated by exposure to pH in the range 3–10. High ionic concentrations (e.g. 1M $MgCl_2$) increase the resistance of enteroviruses to heat inactivation. Tissue cultures used for the cultivation of enteroviruses include human amnion, human embryonic kidney, HeLa, HEp-2, WI-38, African green monkey (*Cercopithecus aethiops*) kidney, and rhesus monkey (*Macaca mulatta*) kidney.

Enteroviruses from *human* sources are divided into three subgroups: polioviruses, coxsackieviruses, and echoviruses.

Polioviruses. The three poliovirus serotypes are designated Brunhilde (type 1), Lansing (type 2), and Leon (type 3)—each of which can cause POLIOMYELITIS in man. Polioviruses remain infective in milk, water etc. for periods of the order of hours/days at room temperatures.

Coxsackieviruses. (The name derives from the town of Coxsackie, in New York State, where these viruses were first isolated.) The coxsackieviruses (about 30 serotypes) are divided into two groups on the basis of the type(s) of lesion they produce in *newborn mice*; group *A* coxsackieviruses cause inflammation and destruction of muscle tissue, while those in group *B* (6 serotypes: B1–B6) give rise to lesions in the central nervous system. In *man*, coxsackievirus infections are often subclinical; diseases of coxsackievirus aetiology include aseptic meningitis (causal agents e.g. serotypes A4, A7, B2, B4), paralytic disease (e.g. A7, B4, B5), respiratory tract disease (e.g. A21, B3, B4, B5), and HERPANGINA (e.g. A2, A5, A6, A8).

Echoviruses. 'Echo' derives from 'enteric cytopathogenic human orphan'—i.e. echoviruses occur in the enteric tract, produce cytopathic effects (CPE) in tissue cultures, and—when originally named—were not clearly associated with any human disease (i.e. they were 'orphans'). In man, the echoviruses (approximately 30 serotypes) can give rise to subclinical infections or to mild or severe diseases of the respiratory tract, intestinal tract, or central nervous system.

Entner–Doudoroff pathway (ED pathway) The major pathway for glucose utilization in e.g. *Zymomonas* spp and some species of *Pseudomonas*; such organisms apparently lack a (complete) EMBDEN–MEYERHOF–PARNAS PATHWAY. The ED pathway is given in Appendix I(c). The glyceraldehyde-3-phosphate may be converted to pyruvate by a series of reactions common to the EMP pathway; the fate of pyruvate depends on the organism and conditions. In (oxidative) *Pseudomonas* spp pyruvate can be metabolized *via* the TRICARBOXYLIC ACID CYCLE; in (fermentative) *Zymomonas* spp pyruvate is decarboxylated to acetaldehyde which in turn is reduced to ethanol. Intermediates of the ED pathway may enter the HEXOSE MONOPHOSPHATE PATHWAY to provide pentoses and reduced NADP for biosynthetic activity.

Entodiniomorphida An order of protozoans of the SPIROTRICHIA.

entomopathogenic Pathogenic for insects.

Entomophthora A genus of fungi of the class ZYGOMYCETES; *Entomophthora* spp are typically parasites or pathogens of insects, but some species are pathogenic for man and other animals and others are saprophytic. The vegetative form of the organisms is a septate or aseptate mycelium which commonly exhibits a tendency to fragment; fragments of mycelium are termed *hyphal bodies*. Asexual reproduction typically involves the formation of conidia—which are projected violently from their branched or unbranched conidiophores. Sexual reproduction may involve the formation of ZYGOSPORES or AZYGOSPORES, according to species.

entomopoxviruses A group of VIRUSES whose hosts are insects—including members of the Coleoptera (beetles), Diptera (flies), Lepidoptera (butterflies, moths), and Orthoptera (cockroaches, grasshoppers, locusts); entomopoxviruses resemble poxviruses (see POXVIRUS) in morphology, structure, and intracytoplasmic replication, and some authors regard them as members of a distinct *Poxvirus* subgenus. Some of the entomopoxviruses are larger than poxviruses—e.g. a virus which infects the cockchafer (*Melolontha melolontha*) is approximately 450 nm, maximum dimension; many of the entomopoxviruses contain a *core* which differs from that typical of poxviruses—e.g. in some the core is unilaterally-concave. Entomopoxviruses cause diseases (known as *spheroidoses*) in the larval and/or other stages of insects; it appears that infection is commonly fatal. Entomopoxvirus-infected cells may contain one or both of two types of inclusion body: the protein *spindle* (in no way relatated to the *mitotic* spindle), and the (virus-containing) *spherule*. (see also INSECT DISEASES)

entry exclusion (see SEX FACTOR)

enumeration of microorganisms *Microorganisms other than viruses*. The total number of (living and dead) cells detectable in a sample is termed the *total cell count*; the number of living cells detectable in a sample is termed the *viable cell count*. A *total cell count* may be determined with electronic aids (see e.g. COULTER COUNTER and NEPHELOMETRY) or by direct (visual) counting (see e.g. COUNTING CHAMBER). Liquid samples containing low concentrations of cells can be membrane-filtered and the (stained) cells counted, on the membrane, under the microscope. A *viable cell count* may be determined by: (a) The examination, in a counting chamber, of a sample stained with a vital stain (see VITAL STAINING). (b) A dilution method—see e.g. MULTIPLE-TUBE METHOD. (c) Colony counts: the number of viable cells in a sample (or aliquot) is assessed from the number of colonies which develop on incubation of a suitable (solid) medium which has been inoculated with the sample (or aliquot); the selectivity of the medium and/or conditions of incubation may significantly affect the number of viable cells which form colonies. (See also FILTRATION, MILES AND MISRA'S METHOD, and ROLL-TUBE TECHNIQUE; POUR PLATES and SPREAD PLATES may also be used for viable counts.) Some types of cell tend to clump or to grow in chains etc.; an accurate viable count of such organisms cannot be evaluated by the above methods, and in these cases counts may be given in terms of the number of *colony-forming units*/volume.

The above methods are generally applicable to most bacteria and spores; some of these methods may be used for certain algae, fungi, and protozoans. The counting of rapidly-moving protozoans, under the microscope, may be facilitated by the use of a viscous suspending medium—such as 2–5% methylcellulose—to reduce the rate of movement of the organisms.

Viruses. Viruses may be counted by electron microscopy; the larger viruses, suitably stained, may be counted under the light microscope. (Such methods are infrequently used; they do not distinguish between infective and non-infective virions.) In the *plaque method* of assay serial dilutions of the viral suspension are initially prepared, and a known volume from each dilution is examined for its content of infective, plaque-forming viruses. (see PLAQUE (1); see also PLAQUE ASSAY) An alternative procedure is the **END POINT** method of assay. Plant-pathogenic viruses are commonly assayed on suitable **LOCAL LESION HOSTS**.

envelope (*bacterial*) (see **CELL ENVELOPE**)

envelope (peplos) (*virol.*) A lipoprotein coat which forms the outermost layer of the virion in certain types of virus—e.g. herpesviruses, togaviruses; the envelope is usually derived from the membrane(s) of the host cell. Virus-coded protein particles which project from the envelope—forming a 'fringe' over the surface of the virion—are termed *peplomers*. In at least some cases the removal of the envelope from a virion (e.g. by treatment with lipid solvents, detergents) results in loss of infectivity. Enveloped viruses generally lose infectivity rapidly at room temperature and at 4 °C; stability may be increased by storage at −20 °C (or better, −70 °C) in e.g. a balanced salt solution supplemented with protein.

enzootic An outbreak of disease, in a given species or type of animal, which is limited to certain districts or particular geographical regions.

enzootic porcine encephalomyelitis (see **TESCHEN DISEASE** and **TALFAN DISEASE**)

enzymes Proteins which function as highly efficient biological *catalysts*. An enzyme increases the *rate* of a (thermodynamically feasible) reaction without altering the equilibrium constant for that reaction; reactions which, in the absence of an enzyme, require extreme conditions of temperature, pH etc. can occur rapidly under physiological conditions in the presence of an appropriate enzyme. Enzymes are generally highly specific in respect of the nature of the reaction catalysed and of the substrate(s) utilized. Many enzymes require a non-protein **COFACTOR** for activity; others may consist of two or more protein subunits (e.g. RNA polymerase—see **RNA**). (see also **ISOENZYME, ENDOENZMES, EXOENZYMES**)

Classification and nomenclature. Originally, enzymes were given arbitrary names (e.g. *pepsin, trypsin*); subsequently attempts were made to indicate the substrate and/or the nature of the reaction catalysed (e.g. *urease* splits urea, *alcohol dehydrogenase* dehydrogenates alcohol). Most of these names are still widely used; however, a systematic scheme for the classification and nomenclature of enzymes has been established by the International Union of Biochemistry Commission on Enzymes. The classification involves the division of all known enzymes into six general classes: 1—**OXIDOREDUCTASE**; 2—**TRANSFERASE**; 3—**HYDROLASE**; 4—**LYASE**; 5—**ISOMERASE**; 6—**LIGASE**. According to the nature of the reaction catalysed, each class is divided into subclasses and further divided into sub-subclasses; within a sub-subclass each individual enzyme is given a specific serial number. Thus any enzyme may be precisely identified by a classification number consisting of four figures; e.g. *pyruvate kinase* has the classification number EC 2.7.1.40, where EC = Enzyme Commission, 2 = **TRANSFERASE**; 7 = phosphate group transferred; 1 = alcohol group accepting the phosphate group; 40 = serial number of pyruvate kinase. Systematic *names* are intended to reflect the nature of the substrate(s), the nature of the reaction catalysed, and any coenzyme involved. Thus e.g. the enzyme which catalyses the oxidative deamination of L-glutamate ('glutamate dehydrogenase') becomes *L-glutamate:NAD oxidoreductase* (*deaminating*); pyruvate kinase becomes *ATP:pyruvate phosphotransferase*.

enzymes I and II (see **TRANSPORT**)

enzymes, induction and repression of (see **OPERON**)

EOP (efficiency of plating) The **PLAQUE TITRE** expressed as a proportion of the total number of viruses present in a given suspension. Under ideal conditions the EOP of some bacteriophages may exceed 90%; the EOP of animal viruses is typically less than 10%.

eosin (eosin Y) Tetrabromofluorescein (a substituted xanthene): a red acidic dye and **FLUOROCHROME**. The dye is very soluble in water and appreciably soluble in ethanol; it is used e.g. as a cytoplasmic stain and as a constituent of **EMB AGAR**. Eosin fluoresces greenish-yellow. (see also **FLUORESCENT DYES** (as antimicrobial agents))

epibasidium (*mycol.*) (1) The **METABASIDIUM** of *Auricularia*, *Septobasidium* and related fungi. (2) In *Tremella* and related fungi each of the sterigmata is sometimes referred to as an epibasidium.

Epichloë A genus of fungi of the order **SPHAERIALES**; species occur e.g. as parasites and pathogens on graminiferous hosts. *E. typhina* (the causal agent of 'choke' of grasses) forms a band of conidiophore-bearing mycelium which encircles the stem of the graminiferous host; this subsequently gives rise to an annular *stroma*—in the superficial layer of which develop a number of perithecia. (The perithecia are embedded in the stroma with their ostiolar pores at the surface.) Each ascus contains eight thread-like ascospores; these subsequently fragment within the ascus.

epidemic In a human population: an outbreak of a given infectious disease in which, for a limited period of time, a high proportion of the population exhibits overt symptoms of the disease; epidemics are characterized by a sudden onset. An analogous situation among plants is referred to as an *epiphytotic*; among animals it is referred to as an *epizootic*. (cf. **ENDEMIC**)

epidemiology The study of the interrelationships which exist between a given pathogenic organism, the environment, and *groups*, or *populations*, of the relevant hosts; the object is to investigate the factors and mechanisms which govern the *spread* of disease within a community or population. Three basic factors are involved. (a) The VIRULENCE of the given pathogen; since the observed virulence of a pathogen is related to host susceptibility (a variable) it is difficult to define in absolute terms. (b) *Herd immunity*. This may be regarded as the *collective* immunity or resistance to a given disease exhibited by a community or population (human or animal) in the setting of its own environment. (Immunity in individual members of a population may arise e.g. through VACCINATION or through previous recovery from the disease.) An important feature of herd immunity is the ratio of *immunes* (individuals resistant to the disease) to *susceptibles* (individuals able to contract the disease); the existence of a high proportion of immunes reduces the probability of contact between an infectious case and a susceptible—thus impeding the spread of disease. The proportion of susceptibles may often be drastically reduced by vaccination, and the risk to remaining susceptibles may be decreased by placing in quarantine any persons or animals who have contracted the particular disease. Under such conditions CARRIERS may be important factors in the persistence of disease within the community. (c) The environment. Environmental conditions (extrinsic factors) play an important part in the containment or spread of disease. Particularly important are those features of society which favour or impede the *transmissibility* of the pathogen, e.g. standards of personal hygiene, overcrowding, the quality of communal WATER SUPPLIES, and the general state of sanitation; when transmission of a disease involves an intermediate VECTOR the prevalence of such vectors and the measures (if any) taken to reduce their numbers are important factors. Since, in the present context, a population and its environment are inseparable, these extrinsic factors determine, in part, the observed herd immunity of a given population. (The totality of factors which determine the transmissibility of disease and the distribution of immunes and susceptibles within a population has been termed the *herd structure* of that population.)

An increase in the number of susceptibles in a population may lead from an ENDEMIC to an EPIDEMIC situation; thus, the incidence of diphtheria among school children may reach epidemic levels if the number of susceptibles in that group is permitted to exceed about 30%. An epidemic may also be initiated when a population immune to a given pathogen is exposed to a variant strain of that pathogen; the effects of ANTIGENIC DRIFTING in influenza viruses is a case in point.

In tracing the epidemiology of a given outbreak of disease it cannot be automatically assumed that the infection arose from a common source. For this reason the causal organisms isolated from each patient may be examined by TYPING procedures in order to detect any variation which may help to establish the source(s) and routes of infection.

Epidermophyton A genus of fungi within the order MONILIALES and included within the category DERMATOPHYTES; the sole species, *E. floccosum*, is parasitic or pathogenic on skin and nails (though not hair), and is a causal agent of athlete's foot. (see also RINGWORM) The vegetative form of the organism is a septate mycelium. Microconidia are not produced; the typical macroconidium is smooth-walled, has up to four septa, and is approximately 10–15 microns in length with one pole rounded and the other square-ended.

epigean (*adjective*) Occuring above ground.

epilithic Growing on the surface of rock or stone (cf. ENDOLITHIC).

epimastigote (crithidial form) A form assumed by certain species of the Trypanosomatidae during a particular stage of development. The epimastigote is elongated and flagellate; the kinetoplast and basal body are situated close to the nucleus, and the flagellum emerges laterally *via* a pellicular depression (flagellar pocket)—becoming free at the anterior end of the cell (see TRYPANOSOMA). An undulating membrane may or may not be observed. Epimastigotes are produced e.g. by species of *Blastocrithidia* and *Trypanosoma*—but not by *Crithidia* spp.

epimerite In members of the Eugregarinida, a differentiated region of a developing trophozoite by which the latter remains attached to the host cell until maturation is complete. (see also PROTOMERITE)

epinasty (*plant pathol.*) The downward curvature of a plant structure due to the faster growth of the dorsal side of the structure relative to the ventral side; e.g. such uneven growth of a petiole (leaf stalk) causes the leaf to point downwards. Epinasty can be induced e.g. by ETHYLENE; it is an early symptom in certain plant diseases.

epineuston (see NEUSTON)

epiphloeodal (epiphloedal) (of a lichen) Refers to a lichen which grows on the surface of the bark of trees (cf. ENDOPHLOEODAL and CORTICOLOUS).

epiphytotic (see EPIDEMIC)

epiplasm (1) (*protozool.*) See PELLICLE. (2)

(*mycol.*) That cytoplasm, of an ASCUS, which is not incorporated into ascospores.

episome A PLASMID which can integrate (reversibly) with the chromosome of its bacterial host; in the integrated state the plasmid behaves as part of the chromosome. Integration appears to require a region of homology between the plasmid and the chromosome; the homologous regions can pair, and a single cross-over between the two circular molecules results in integration. Pairing and crossing-over between two homologous regions also result in excision of the integrated plasmid. Since homology with the bacterial chromosome appears to be essential for integration, a plasmid may behave as an episome in one cell but not in another—i.e. the distinction between a plasmid and an episome is not complete. Integration of a F FACTOR leads to the formation of an Hfr strain—see CONJUGATION. (cf. *prophage* in LYSOGENY)

epitheca (of a diatom) The larger of the two parts of the cell wall (frustule) of a diatom.

epithecium (*mycol.*) In some types of APOTHECIUM: a layer of prosenchymatous tissue formed at the surface of the layer of asci; the epithecium consists of the interwoven, branched tips of the paraphyses. (cf. PSEUDOEPITHECIUM) An analogous structure is found in certain loculoascomycetes.

epitope (*immunol.*) A specific antigenic DETERMINANT.

epixylous Living or growing on wood.

epizootic (see EPIDEMIC)

Epstein–Barr virus (see HERPESVIRUS)

equivalence (*immunol.*) (see OPTIMAL PROPORTIONS)

eradicant (*plant pathol.*) Any chemical agent which *eliminates* a given pathogen (or pathogens) from diseased plants treated with that agent (cf. PROTECTANT).

Eremascus A genus of fungi of the ENDOMYCETACEAE.

ergosome A synonym of polyribosome—see PROTEIN SYNTHESIS.

ergot (see CEREALS, DISEASES OF)

ergot alkaloids A group of compounds obtained from *Claviceps* spp; the group includes ergotamine, ergotoxine and ergometrine. The ergot alkaloids stimulate smooth muscle and have been used e.g. to induce childbirth and to prevent excessive bleeding after childbirth. Ergotamine is used in the treatment of migraine. (see also CEREALS, DISEASES OF)

ergotism A condition of intoxication which follows the ingestion of excessive amounts of ERGOT ALKALOIDS; symptoms may include vomiting, diarrhoea, thirst and convulsions, and gangrenous lesions may subsequently develop at the extremities.

erumpent Bursting or breaking *through*.

Erwinia A genus of gram-negative, chemoorganotrophic, asporogenous bacteria of the family ENTEROBACTERIACEAE; species include plant pathogens and saprophytes. The cells are typically straight rods—commonly 0.5–1 × 1–3 microns; the majority of strains are motile. Some species form pigments. Culture is often carried out on nutrient agar containing glucose, sucrose and/or yeast extract; the optimum growth temperature is generally about 28 °C. The carbon and energy-yielding substrates of aerobic metabolism include acetate, malate and succinate. Acid is commonly produced from glucose, fructose, sucrose, sorbitol and ribose; gas production is uncommon. The type species is *E. amylovora*; GC% range for the genus: 50–58. Species include:

E. amylovora. A motile, non-pigmented organism which does not attack pectins; the causal agent of FIRE BLIGHT of apple and pear. Identification can be made by agglutination with specific antiserum.

E. carotovora. A motile, non-pigmented organism which attacks PECTINS and which may or may not produce gas from glucose; the causal agent of a variety of soft ROT diseases in plants—see e.g. CARROTS, DISEASES OF and POTATO, BACTERIAL DISEASES OF. *E. carotovora* var *carotovora* (formerly *Pectobacterium delphinii*) is capable of growth at 36 °C and produces reducing* end-products from sucrose metabolism. *E. carotovora* var *atroseptica* does not grow at 36 °C and does not produce reducing products from sucrose; the causal agent of *blackleg* of potato. *as determined by Benedict's reagent. (see also PLANT PATHOGENS AND DISEASES)

erysipelas An acute, inflammatory, cutaneous disease characterized by spreading areas of inflamed and thickened skin, and attended by varying degrees of systemic involvement. The causal organism, *Streptococcus pyogenes*, gains access to the body *via* wounds, abrasions etc.; septicaemia may develop if the infection is not treated. (cf. ERYSIPELOID; see also SWINE ERYSIPELAS)

erysipeloid An infectious, cutaneous disease in which the erythematous, oedematous lesions commonly occur on the fingers and hands. The causal organism, *Erysipelothrix rhusiopathiae* (= *E. insidiosa*) gains access to the tissues *via* wounds, abrasions etc. Occasionally, systemic manifestations may occur. Erysipeloid is contracted mainly by persons who handle fish products and meats. (cf. ERYSIPELAS)

Erysipelothrix A genus of gram-positive, asporogenous, non-motile bacteria of uncertain taxonomic affinity. The sole species, *E. rhusiopathiae* (*E. insidiosa*) occurs as a parasite of man and other animals, including fish and shellfish, and is the causal agent of ERYSIPELOID and SWINE ERYSIPELAS. The organism may occur as short, straight or curved, pleomorphic rods (e.g. in smooth colonies) or as filaments, or chains of cells (e.g. in rough colonies). *E. rhusiopathiae* is aerobic or facultatively anaerobic; growth is satisfactory on media enriched with e.g. serum, and is said to be improved by 5–10% CO_2. Greening (α-haemolysis) may surround colonies on blood agar. *E. rhusiopathiae* is catalase-negative, indole-negative and urease-negative; H_2S is produced. Stab-inoculated gelatin may give rise to the 'test-tube brush' form of growth, but gelatin is not liquefied. *E. rhusiopathiae* is generally sensitive to penicillins and tetracyclines.

Erysiphaceae (the 'powdery mildews') A family of fungi of the order ERYSIPHALES; all species are obligate parasites and pathogens of higher plants. The organisms give rise to a hyaline (non-pigmented) septate mycelium which develops superficially on the host plant. Conidia are usually formed in abundance; the conidiophores are typically erect, unbranched hyphae on which the conidia develop basipetally. Dark-coloured cleistothecia may subsequently develop among the mycelium; the plasmogamy which precedes ascocarp formation is believed to involve gametangial contact or somatogamy. Some species are homothallic, others are heterothallic. The asci contain unicellular, hyaline ascospores. (cf. PERISPORIACEAE) The genera include ERYSIPHE, PODOSPHAERA, and SPHAEROTHECA.

Erysiphales An order of filamentous (mycelial) fungi of the class PYRENOMYCETES, subdivision ASCOMYCOTINA; the organisms are haustorium-forming parasites and pathogens of higher plants. In members of the Erysiphales the ascocarp is a dark-coloured cleistothecium which typically contains an hymenial layer of asci; cleistothecia develop among the mycelium on the surface of the host plant. The order may be divided into two families: the ERYSIPHACEAE ('powdery mildews') and the PERISPORIACEAE.

Erysiphe A genus of fungi of the family ERYSIPHACEAE; species are obligate parasites and pathogens on members of the Gramineae and on certain other plants. (see also MILDEW and CEREALS, DISEASES OF) The organisms form patches or regions of confluent septate mycelium on the surface of the host plant; most species form globose haustoria, but those of *E. graminis* are digitate (i.e. each bears a number of finger-like projections). Conidia are typically formed in abundance; those of *E. polygoni* are ovoid and highly vacuolated—while those of *E. graminis* are elongated and highly vacuolated. The conidiophores are simple, erect, hypha-like structures which in many species—including *E. graminis*—give rise to basipetally-formed chains of conidia; the conidiophores of *E. graminis* are distinctive in that in each the basal cell is slightly swollen. Conidia germinate by means of a germ tube which subsequently gives rise to an appressorium; following penetration of the plant host appressoria give rise to haustoria. In the sexual process plasmogamy appears to involve either gametangial contact or somatogamy—the precise mode being unknown. The dark-brown or black cleistothecia—each containing more than one ascus—develop among the mycelium on the surface of the host organism; each cleistothecium bears a number of hypha-like appendages.

erythritol $CH_2OH(CHOH)_2CH_2OH$; the alcohol corresponding to erythrose.

erythrocytes Red blood cells.

erythromycin A MACROLIDE ANTIBIOTIC produced by *Streptomyces erythreus*.

eschar (*med.*) Necrosed tissue which separates (sloughs) from the underlying living tissue. An *eschar* forms the primary lesion in SCRUB TYPHUS, and is formed in other diseases.

Escherichia A genus of gram-negative, chemoorganotrophic bacteria of the family ENTEROBACTERIACEAE. The sole species, *E. coli*, occurs in the intestinal tract of animals (including man) and is common in soil and water; *E. coli* can cause urinary-tract and other infections, and certain serotypes cause enteric diseases—see DYSENTERY (bacterial). The cells are typically straight, round-ended rods—commonly 0.5–1 × 1–4 microns—which stain well with aniline dyes; cells usually occur singly or in pairs, but chains of cells may be formed by mutant strains. The CELL WALL contains PEPTIDOGLYCAN and LIPOPOLYSACCHARIDES. *Most* strains are motile; flagellation is *peritrichous*. Some strains form a CAPSULE, or microcapsule. FIMBRIAE may or may not be present. Spores are not formed. Growth occurs aerobically or anaerobically on basal media (e.g. NUTRIENT AGAR) and e.g. on MACCONKEY'S AGAR—but not on deoxycholate-citrate agar; some strains exhibit haemolysis on blood agar. EMB AGAR has been used to distinguish between *E. coli* and certain related organisms. The optimal temperature and pH for growth are commonly 37 °C and 7 respectively. On nutrient agar (24 hours incubation at 37 °C) *E. coli* typically forms smooth, colourless, slightly opaque, round, entire, low-convex colonies of 1–3 mm diameter; the growth is butyrous and easily emulsifiable. SMOOTH-ROUGH VARIATION occurs. Reproduction occurs by binary fission; the generation time—see GROWTH (bacterial)—is about 20 minutes under optimum conditions. Under aerobic conditions *E. coli* carries out a respiratory (oxidative) metabolism. (see also NITRATE RESPIRATION) Anaerobically, *E. coli* carries out a MIXED ACID FERMENTATION—producing equal amounts of hydrogen and CO_2; thus, e.g. acid and gas are formed from glucose at 37 °C. 'Faecal' type strains of *E. coli* give a positive EIJKMAN TEST. Other typical test reactions: INDOLE TEST positive, METHYL RED TEST positive, VOGES–PROSKAUER TEST negative, CITRATE TEST negative; biochemically aberrant strains occur (see later). *E. coli* does not produce urease, hydrolyse gelatin, or form H_2S in TSI medium, and fails to grow on potassium cyanide media.

The numerous strains of *E. coli* are commonly distinguished serologically on the basis of their O ANTIGENS, K ANTIGENS and H ANTIGENS. (Certain serotypes are frequently isolated from cases of infantile diarrhoea; these—which include 026:K60:H11, 055:K59:H6, 055:K59:H7 and 0111:K58:H2—are referred to as *enteropathogenic* strains.) *E. coli* participates in bacterial CONJUGATION and other forms of genetic transfer, and is susceptible to a number of bacteriophages; bacteriophage TYPING has been used to study strains isolated from urinary-tract infections. COLICINS are formed by some strains.

E. coli is killed by PASTEURIZATION and by many common DISINFECTANTS (when these are used at appropriate concentrations). Growth is inhibited by SODIUM AZIDE (0.025% w/v) and by certain dyes, e.g. malachite green. Diseases caused by enteropathogenic strains have been treated with TETRACYCLINES, KANAMYCIN, or NEOMYCIN (to which some strains are resistant) or with certain other antibiotics; urinary-tract infections have been treated with NALIDIXIC ACID or nitrofurantoin (see NITROFURANS) etc. Resistance to certain antibiotics may be due to an R FACTOR.

The GC% range for *E. coli* strains is 50–51.

Atypical strains. These include: (a) the anaerogenic ALKALESCENS–DISPAR GROUP. (b) A strain capable of NITROGEN-FIXATION has been obtained experimentally by a conjugal cross with *Klebsiella pneumoniae*.

esculin (see AESCULIN)

espundia In man, a *mucocutaneous* LEISHMANIASIS found in South America. The causal agent, *L. brasiliensis*, invades *via* wounds or is transmitted by the bites of sandflies (genus *Phlebotomus*). Ulcerative lesions are formed in the mouth, nasal cavities, on other mucous membranes, and elsewhere. For chemotherapeutic agents see *antimony* under HEAVY METALS. *Laboratory diagnosis*: direct observation of *L. brasiliensis* in material from lesions; a skin test (the *Montenegro test*) may be used.

established cell line Any CELL LINE which appears to be capable of unlimited *in vitro* propagation.

ethambutol (myambutol) Dextro-(2,2′-(ethylenediimino)-di-1-butanol); a synthetic drug used for the treatment of TUBERCULOSIS. Ethambutol is inhibitory to dividing cells of many strains of *Mycobacterium tuberculosis* but its mode of action is unknown. Resistance to ethambutol develops quite readily; it is generally used in combination with other drugs e.g. ISONIAZID.

ethanol (as an antimicrobial agent) (see ALCOHOLS)

ethanol fermentation (see ALCOHOLIC FERMENTATION)

ethidium bromide 2,7-Diamino-10-ethyl-9-phenylphenanthridinium bromide; a trypanocidal agent. (see also KINETOPLAST)

ethirimol (*plant pathol.*) 5-*n*-Butyl-2-ethylamino-4-hydroxy-6-methyl pyrimidine. An agricultural *systemic* antifungal agent which has been used against powdery MILDEWS of CEREALS. Resistant strains of the fungus (*Erisyphe graminis*) readily emerge. (see also DIMETHIRIMOL)

ethylene (*plant pathol.*) Ethylene is produced by a range of plant tissues and is believed to play a part in normal growth regulation in plants; however, in abnormally high concentrations it can cause e.g. EPINASTY, premature fruit ripening, leaf abscission or other effects. Ethylene may be responsible for the symptoms of certain plant diseases; thus, e.g. premature ripening of fruit occurs when banana plants are infected with *Pseudomonas solanacearum*—a pathogen known to produce ethylene *in vitro*. Some plants have been shown to increase ethylene production as a response to infection; thus, e.g. when *Physalis floridana* is infected with POTATO VIRUS Y the plant produces abnormally large amounts of ethylene, and the observable symptoms are epinasty, leaf abscission etc. (see also PLANT PATHOGENS AND DISEASES)

ethylene oxide (as a DISINFECTANT or STERILANT) Ethylene oxide is a colourless, highly-reactive, water-soluble cyclic ether with a boiling point of 10.8 °C. The gas forms highly-explosive mixtures with air in widely varying proportions; thus, for use, it is diluted with carbon dioxide, nitrogen, or fluorohydrocarbons. (*Carboxide* (trade name) contains 10% ethylene oxide, 90% carbon dioxide.) Under acid conditions ethylene oxide is hydrolysed to ethylene glycol. Ethylene oxide is an ALKYLATING AGENT; it brings about the substitution of cellular proteins etc. with hydroxyethyl groups. Under appropriate conditions it is active against all types of microorganisms, including spores and viruses; however, its action is slow, and exposure for at least several hours is usually necessary—exact exposure times depending on conditions. Ethylene oxide is more effective in the presence of water vapour than under dry conditions; however, some disagreement appears to exist regarding optimum humidity. The gas has good powers of penetration, but is absorbed by a number of materials—including rubber, hessian and certain plastics; such absorption may significantly reduce its effective concentration. Ethylene oxide is used for the STERILIZATION of medical equipment, bed linen, and certain heat-labile materials; it is also used in the food and pharmaceutical industries. It may be employed in an apparatus resembling an AUTOCLAVE; prior evacuation of the apparatus increases the rate of penetration of the gas. The sterilizing activity of ethylene oxide at a particular location may be monitored by means of a *Royce indicator sachet*. This consists of a small plastic sachet containing an acid solution of magnesium chloride; a colour-change occurs when the sachet is exposed to ethylene oxide for time/concentration combinations which effect sterility. In microbiology, ethylene oxide has been used for the sterilization of liquid media; a 1% aqueous solution has been used for this purpose. Ethylene oxide blisters the skin and is said to be carcinogenic.

ethylethanesulphonate (EES) (see ALKYLATING AGENTS)

ethylmethanesulphonate (see ALKYLATING AGENTS)

etiology American spelling of AETIOLOGY.

Eubacteriales ('the eubacteria') A vaguely-defined order of bacteria which is no longer recognized.

eucarpic *Eucarpic* fungi are those in which only *part* of the thallus is involved in the formation of the reproductive structures. (cf. HOLOCARPIC)

eucaryotic organisms (eukaryotic organisms) (see PROCARYOTIC ORGANISMS)

euchrysine Synonym of ACRIDINE ORANGE.

Eucoccidia An order of protozoans of the subclass COCCIDIA; species are distinguished from those of the PROTOCOCCIDIA by the occurrence of SCHIZOGONY. The eucoccidians are monoxenous or heteroxenous parasites of vertebrates and invertebrates. The suborders: ADELEINA, EIMERIINA and HAEMOSPORINA.

eu-form rusts (see MACROCYCLIC RUSTS)

Euglena A genus of algae of the division EUGLENOPHYTA; a few species are marine organisms but most occur in freshwater environments—particularly in waters rich in organic matter. *Euglena* is a fusiform, unicellular, flagellate organism of a size which varies with species; *E. gracilis* is approximately 50 × 10 microns. The anterior of the cell is invaginated such that a flask-shaped *reservoir* (or 'gullet') communicates with the exterior *via* a narrow *canal*. A single long locomotory FLAGELLUM arises from one of two basal bodies (situated near the posterior wall of the reservoir) and passes to the exterior *via* the canal; a short, non-emergent flagellum arises from the other basal body. The longer flagellum has a lateral swelling in the region within the reservoir; this swelling lies opposite an EYESPOT situated in the cytoplasm adjacent to the reservoir. A contractile vacuole discharges its contents into the reservoir. Beneath the cell membrane a flexible, proteinaceous *pellicle* encompasses the cytoplasm—except in the region of the reservoir. The cell contains chloroplasts (type depending on species), a single—often conspicuous—nucleus, and granules of PARAMYLUM. Apart from typical flagellate motility, *Euglena* spp exhibit *euglenoid movement*—a motion which involves progressive waves of swelling and constriction of the cell body. Reproduction occurs by longitudinal binary fission; sexual reproduction is unknown.

Euglenophyta (euglenids, euglenoids) A division of the ALGAE; species of the Euglenophyta include freshwater, brackish, and marine organisms. The euglenids are unicellular organisms, though palmelloid phases occur in at least some species. All species are motile or (at least) have a motile phase; basically the organisms are biflagellate—though in some species only one flagellum may emerge from the cell even when two basal bodies are present. The flagella of biflagellate cells may be equal or unequal in length. (More than two flagella are reported to occur in some parasitic euglenids.) All species contain CHLOROPHYLLS *a* and *b*, β-carotene, and a range of xanthophylls (see CAROTENOIDS); intracellular storage products include PARAMYLUM. Sexual processes are unknown within the group. Genera include *Astasia*, EUGLENA, *Phacus*, *Trachelomonas*.

Euglypha A genus of protozoans: see ARCELLINIDA.

eugonic (*adjective*) Refers to a strain of organism which, on culture, produces growth which is more luxuriant than that produced by other (*dysgonic*) strains of the same species. The *growth* (as well as the organism) may also be described as eugonic or dysgonic. These terms are most commonly applied to species of *Mycobacterium*.

Eugregarinida An order of protozoans (subclass GREGARINIA) in which SCHIZOGONY does not occur (cf. SCHIZOGREGARINIA). Developing eugregarine trophozoites usually remain temporarily attached to the host cells from which they have emerged; attachment is made *via* the EPIMERITE until maturity is reached. The genera include MONOCYSTIS.

eukaryotic organisms (see PROCARYOTIC ORGANISMS)

eumycetes (see EUMYCOTA)

Eumycota (Mycobionta; eumycetes; the 'true' fungi) One of the two divisions which constitute the FUNGI; members of the Eumycota do not form a plasmodium or a pseudoplasmodium (cf. division Myxomycota—see SLIME MOULDS). The subdivisions are MASTIGOMYCOTINA, ZYGOMYCOTINA, ASCOMYCOTINA, BASIDIOMYCOTINA, and DEUTEROMYCOTINA.

euploid (1) Refers to any (eucaryotic) nucleus or cell in which the genetic complement consists of the basic haploid number of chromosomes or any *exact* multiple of this number—e.g. DIPLOID, triploid etc. (cf. ANEUPLOID) (2) May refer to the possession of a normal chromosome complement—i.e. a complement which is characteristic of the species. (cf. HETEROPLOID)

Euplotes A genus of ciliate protozoans of the order Hypotrichida, subclass SPIROTRICHIA. Species occur in a variety of freshwater and brackish environments, and some (e.g. *E. moebiusi*) are quite common in British activated-sludge SEWAGE TREATMENT plants. Individuals are roughly ovoid, 50–150 microns in length, with the cytostome located ventrally near the anterior of the cell; the flattened ventral surface carries a number of *cirri*—other forms of somatic ciliation being absent. The distal part of the AZM forms a conspicuous fringe. The macronucleus is band-like and curved, the micronucleus small and spherical. A contractile vacuole is present. Asexual reproduction occurs by binary fission, and CONJUGATION takes place between individuals of appropriate mating types (see SYNGEN). Food consists of bacteria, algae, and small protozoans. (see also STYLONYCHIA)

Eurotiales An order of filamentous (mycelial) fungi of the class PLECTOMYCETES, sub-division ASCOMYCOTINA; species are widespread in nature—occuring e.g. in the soil and on decomposing plant and animal matter. The majority of species are saprophytic organisms; they include fungi which are commonly better known as *form*-species of the form-genera *Aspergillus* and *Penicillium*. Parasitic and pathogenic species include organisms whose conidial (imperfect) stages are classified among the dermatophytes. Ascocarps may develop following gametangial contact (in which antheridia and ascogonia may become mutually intertwined) or through parthenogenic development of ascogonia. According to species the ascocarp peridium may be virtually non-existent (as e.g. in *Byssochlamys*), may consist of an insubstantial tangle of loose hyphae (as e.g. in *Gymnoascus*), or may have a definite structure (as e.g. in *Emericella* and *Eurotium*); in some species (e.g. *Gymnoascus*, *Myxotrichum*) the fruiting body may bear a number of peripheral curved and/or branched hyphal appendages. Some species (e.g. *Monascus*, *Onygena*) form stalked (stipitate) cleistothecia. The genera include *Arthroderma*, BYSSOCHLAMYS, EMERICELLA, EUROTIUM, *Gymnoascus*, *Monascus*, *Myxotrichum*, NANNIZZIA, *Onygena*, *Sartorya*, and TALAROMYCES.

Eurotium A genus of fungi of the order EUROTIALES; the conidial (imperfect) stages of the organisms correspond to *Aspergillus glaucus* and related form-species. In *Eurotium* spp the ascocarp (a globose cleistothecium) consists of a number of spherical or ovoid evanescent asci contained within a smooth-surfaced—commonly yellow—peridium which is one cell thick; during ascocarp development the antheridium and ascogonium become intertwined—but whether or not the antheridium donates nuclei appears to be unknown. The asci disintegrate prior to the breakdown of the cleistothecial peridium.

eury- A prefix which signifies *wide* or *extensive*. The prefix is used e.g. to indicate the capacity of an organism to tolerate a wide range of degrees of a given influence—e.g. *euryoxic* (able to tolerate a wide range of oxygen tensions), *euryhaline* (able to tolerate a wide range of salt concentrations. (cf. STENO-)

eutrophic (of lakes etc.) Rich in those nutrients which support the growth of aerobic photosynthetic organisms (cf. OLIGOTROPHIC).

eutrophication The enrichment of natural waters (lakes, rivers etc.) with inorganic materials—in particular, compounds of nitrogen and phosphorus—such that they support excessive growth of plants and algae. Eutrophication may occur, for example, as a result of the self-purification process which often takes place in a polluted river or stream; the process of self-purification involves microbial mineralization of organic debris.

evanescent (*mycol.*) (*adjective*) Having a transient existence.

Evernia A genus of fruticose LICHENS in which the phycobiont is *Trebouxia*. The thallus is strap-shaped and dorsiventral. Species include e.g.:

E. prunastri (oakmoss). The thallus is branched, 2–6 centimetres in length, greenish-grey on the upper surface and pale to white on the lower surface; soredia are usually present. *E. prunastri* occurs on trees and fences and occasionally on rocks; in sandy areas it may be unattached. *E. prunastri* is used commercially in the perfume industry.

evernic acid A depside produced by certain lichens, e.g. *Evernia prunastri*; it has antibacterial properties.

exciple A synonym of EXCIPULUM.

excipulum (1) (*mycol.*) The sterile supportive tissue of an APOTHECIUM. (2) (*lichenol.*) The sterile supportive tissue of an apothecium *or* the peridium (wall) of a PERITHECIUM.

excision repair (of DNA) (see DARK REPAIR)

exclusion (genetics) (1) See SEX FACTOR. (2) See PRE-ZYGOTIC EXCLUSION.

exergonic reaction Any reaction which is accompanied by a liberation of energy (cf. ENDERGONIC REACTION).

exflagellation (in *Plasmodium*) The process of development and maturation of the male gametes (microgametes).

Exidia A genus of fungi of the order TREMELLALES.

Exobasidiales An order of fungi of the sub-class HOLOBASIDIOMYCETIDAE; constituent species are leaf endoparasites which do not form definite fruiting bodies but which form hymenia on the surface of parasitized plants (cf. BRACHYBASIDIALES). One species, *Exobasidium vexans*, is the causal agent of BLISTER BLIGHT of tea.

Exobasidium A genus of fungi of the EXOBASIDIALES.

exobiology Extra-terrestrial biology.

exocellular (*adjective*) Refers e.g. to enzymes or structures which are external to the CELL MEMBRANE but which are nevertheless part of (attached to) the cell. (see also EXTRACELLULAR)

exoenzymes (1) Enzymes which occur either attached to the outer surface of the CELL MEMBRANE or in the PERIPLASMIC SPACE; the term may also be applied to enzymes which are released into the medium—i.e. *extracellular* enzymes. (2) Enzymes which attack polymers by sequentially removing units from *one end* of a polymer chain. Thus, e.g. *exonucleases* progressively remove the terminal nucleotides of a polynucleotide chain. (see also β-AMYLASES; cf. ENDOENZYMES)

exogenote (see MEROZYGOTE)

exoperidium In members of the LYCOPERDALES: the outer layer of the peridium; the exoperidium may be a composite structure—e.g. the exoperidium in *Geastrum* is triple-layered.

exosporium (see SPORES (bacterial))

exotoxins Protein or proteinaceous toxins which are produced by certain species of microorganism and which—in at least some cases—appear to be secreted by *living* cells; in some instances exotoxin production is determined by a prophage or plasmid. (cf. ENDOTOXINS) Exotoxins include: (a) Diphtheria toxin: a heat-labile, iron-free protein (M.Wt. approximately 62,000) which consists of a single polypeptide and whose synthesis is coded for by a prophage; it appears to be released by growing, lysogenic cells as well as by cells lysed by the phage. Diphtheria toxin appears to inhibit protein synthesis in eucaryotic cells. (b) Neurotoxins produced by *Clostridium botulinum* and *C. tetani*; botulinum toxin blocks neural transmission by inhibiting the synthesis or release of acetylcholine at pre-synaptic sites. (c) ENTEROTOXINS. (d) HYALURONIDASE. (e) LEUCOCIDINS. (f) STREPTOLYSIN-O. (g) Many other extracellular microbial products.

exponential phase (see GROWTH (bacterial))

extracellular (*adjective*) (1) (of enzymes etc.) Refers to substances which are external, but not contiguous, to the cell(s) that produced them. (2) Some authors use the term to refer to substances or structures (e.g. the bacterial CAPSULE) which are external to the CELL WALL—whether or not they are contiguous to it. (cf. EXOCELLULAR)

extract broth (see NUTRIENT BROTH)

EYA (see EGG-YOLK AGAR)

eye, common microflora of The surface of the eye and the conjunctival membranes normally bear only a sparse microflora owing to the constant flushing action of the eye secretions; these secretions contain LYSOZYME which is inhibitory to gram-positive bacteria. The most abundant organisms normally present are probably diphtheroids.

eyespot (stigma) A group of closely-packed CAROTENOID-containing globules which, according to species, occurs within or outside the CHLOROPLAST envelope in certain algae. Eyespots which occur within the chloroplast membrane are characteristic e.g. of the Chlorophyta, while eyespots external to the chloroplast are found e.g. among the Euglenophyta; among the dinoflagellates the eyespot may occur within or outside the chloroplast—according to species. Certain species of dinoflagellate (e.g. *Nematodinium* spp) contain a large, complex eyespot which incorporates a lens (composed of a stack of vesicles) and a cup-shaped structure composed of a row of pigment granules; such an eyespot is termed an *ocellus*. The eyespot appears to be involved in the detection of light; however, although a number of theories have been advanced, the function of the eyespot cannot be stated definitively.

F

f. (see FORMA SPECIALIS)

F_1 (first filial generation) The first generation of progeny derived from two given parent organisms. The second and third generations are designated F_2 and F_3 respectively.

F^+ donor (see CONJUGATION (bacterial))

F′ donor (see CONJUGATION (bacterial))

F-duction (see CONJUGATION (bacterial))

F factor A SEX FACTOR which contains only those genes concerned with its own replication and conjugal transfer (see *bacterial* CONJUGATION); it is a circular DNA duplex of molecular weight of the order of 5×10^7. The F factor can behave as an EPISOME: in an Hfr strain it is integrated with the bacterial chromosome; when dissociating from the chromosome an F factor may carry with it one or more chromosomal genes—and is then referred to as an *F′ factor* (*F-prime factor*) (see CONJUGATION). The extrachromosomal F factor normally replicates in synchrony with the bacterial chromosome; replication of the *extrachromosomal* F factor can be inhibited e.g. by acridines—so that the F factor is effectively diluted out of a growing bacterial population. The ability of an F factor to initiate conjugation may be repressed if it shares a cell with e.g. an *fi^+* R FACTOR.

F′ factor (see CONJUGATION (bacterial))

F-like pilus (see SEX PILI)

F-prime donor (see CONJUGATION (bacterial))

F-prime factor (see CONJUGATION (bacterial))

F^- recipient (see CONJUGATION (bacterial))

f.sp. (see FORMA SPECIALIS)

F value The number of minutes required to heat-inactivate (kill) an *entire* population of cells or spores, in aqueous suspension, at a temperature of 121 °C. (see also DECIMAL REDUCTION TIME)

Fab fragment (antigen-binding fragment) Either of two identical fragments which are produced when one molecule of IgG (or a structurally-similar IMMUNOGLOBULIN) is cleaved by PAPAIN; the remainder of the molecule is termed the *Fc fragment*. (For the general structure of immunoglobulin molecules see IGG.) Each Fab fragment (M.Wt. about 45,000) consists of a light chain linked, *via* an S-S bond, to the N-terminal portion of a heavy chain; a Fab fragment is thus one of the two limbs of the Y-shaped immunoglobulin molecule. Each Fab fragment of an IgG *antibody* contains a single COMBINING SITE and behaves as a non-precipitating univalent antibody. By treating a Fab fragment with MERCAPTOETHANOL, and lowering the pH, the light chain can be separated from its segment of heavy chain; the latter is referred to as the *Fd fragment*. The Fc fragment (crystallizable fragment—M.Wt. about 50,000)—the *stem* of the Y-shaped immunoglobulin molecule—consists of the C-terminal sections of the two heavy chains linked by one or more disulphide bonds; it is the site of COMPLEMENT-FIXATION in complement-fixing antibodies.

Pepsin cleavage of the immunoglobulin molecule produces an $F(ab')_2$ fragment—the remainder of the molecule being degraded. The $F(ab')_2$ fragment contains both Fab fragments and includes the flexible, proline-rich *hinge region*—i.e. the junction of the Fc fragment and Fab fragments in the intact molecule. The $F(ab')_2$ fragment of an *antibody* behaves as a bivalent, precipitating, *non*-complement-fixing antibody.

$F(ab')_2$ fragment (see FAB FRAGMENT)

facilitated diffusion (see TRANSPORT)

facultative (*microbiol.*) Literally: optional; refers to the ability of an organism to adopt an alternative life style or mode of nutrition etc. Thus, an ANAEROBE may be a facultative AEROBE, or a saprophyte may be a facultative parasite. In general usage, the word 'facultative' is commonly followed by the mode which is *not* normally adopted by the organism; thus, a facultative aerobe refers to an organism which is normally associated with anaerobic conditions but which can grow under aerobic conditions.

FAD Flavin adenine dinucleotide (see RIBOFLAVIN; see also Appendix V(c)).

false tinder fungus The basidiomycete *Fomes igniarius*.

falx A specialized pellicular region in Opalinids.

FAME Fatty acid methyl ester. Species of *Thiobacillus* have been placed into categories (Types I, II, III) on the basis of their FAME profiles, i.e. their content (range, amount) of fatty acid methyl esters.

family A taxonomic group: see TAXONOMY and NOMENCLATURE.

farcy (see GLANDERS)

farmer's lung An (immunologically) hypersensitive condition in which individuals sensitized to the spores of certain bacteria (e.g. *Micropolyspora faeni*) develop severe pulmonary dysfunction on inhalation of such spores; the condition is an example of an Arthus-type reaction.

fasciation (*plant pathol.*) A type of abnormal plant growth which resembles a flattened bundle of coalesced shoots. Fasciation is sometimes observed in diseases of microbial origin. (see also ENATIONS and GALLS)

favus (see RINGWORM)

Fc fragment (Fc piece) (see FAB FRAGMENT)
Fd fragment (Fd piece) (see FAB FRAGMENT)
fentin Triphenyltin; two derivatives, *fentin acetate* ($(C_6H_5)_3SnO.CO.CH_3$) and *fentin hydroxide* ($(C_6H_5)_3SnOH$) are used (sometimes in conjunction with MANEB) as antifungal agents, e.g. for the control of *late blight* of potato (*Phytophthora infestans*).
ferbam Ferric dimethyldithiocarbamate. (see DIMETHYLDITHIOCARBAMIC ACID)
ferment (1) (*noun*) An archaic name for an ENZYME. (2) (*verb*) To carry out FERMENTATION.
fermentation A mode of energy-yielding metabolism that involves a sequence of oxidation-reduction reactions in which both the substrate (primary electron donor) and the terminal electron acceptor(s) are *organic* compounds. (cf. RESPIRATION, ANAEROBIC RESPIRATION) Thus, e.g. in certain species of *Lactobacillus* glucose is fermented in a process known as the HOMOLACTIC FERMENTATION: each molecule of glucose (the primary electron donor) is converted to two molecules of pyruvate *via* the EMBDEN–MEYERHOF–PARNAS PATHWAY (see Appendix I(a))—during which two molecules of NAD are reduced to $NADH_2$; the latter are reoxidized by the reduction of pyruvate (the terminal electron acceptor) to lactate (the fermentation product). The terminal electron acceptor is frequently an intermediate formed during the metabolism of the substrate (e.g. pyruvate in the above example) but may be derived from an exogenous source (e.g. in the STICKLAND REACTION—in which one amino acid is the electron donor and a second is the electron acceptor). Energy is obtained during fermentation largely by SUBSTRATE LEVEL PHOSPHORYLATION; in addition there is evidence that, in certain bacteria (e.g. species of *Bacteroides* and *Propionibacterium*), phosphorylation of ADP may be coupled to an electron transport system involving cytochrome(s) and/or quinone(s) and using e.g. fumarate as terminal electron acceptor—a process which may be analogous to OXIDATIVE PHOSPHORYLATION associated with respiration.

Fermentation occurs in a wide range of bacteria and fungi (particularly yeasts) and in certain protozoans (see e.g. RUMEN); in such organisms fermentation may be either an obligate or a facultative mode of metabolism. The nature of the substrates used and products formed may vary widely among different organisms and is often an important taxonomic characteristic. For examples of fermentation pathways see ACETONE-BUTANOL FERMENTATION, ALCOHOLIC FERMENTATION, BUTANEDIOL FERMENTATION, BUTYRIC ACID FERMENTATION, HETEROLACTIC FERMENTATION, HOMOLACTIC FERMENTATION, MIXED ACID FERMENTATION, PROPIONIC ACID FERMENTATION, STICKLAND REACTION.

Fernández-Morán particles Stalked particles observed when the inner membrane of a MITOCHONDRION is examined by certain techniques—e.g. the examination of negatively-stained preparations by electron microscopy.
Fernandez reaction (see LEPROMIN TEST)
ferredoxins A class of iron-sulphur proteins found in a wide range of microorganisms; they have low redox potentials and function as electron carriers. Ferredoxins contain equimolar amounts of iron and labile sulphide (which can be released as hydrogen sulphide on acidification); clostridial ferredoxin, for example, contains four iron and four sulphur atoms, while ferredoxins from higher plants and algae contain only two of each. Ferredoxins are involved in a range of reactions e.g. NITROGEN FIXATION, PHOTOSYNTHESIS, evolution/utilization of hydrogen (see e.g. BUTYRIC ACID FERMENTATION). (see also FLAVODOXINS)
ferrichrome (see SIDEROCHROMES)
ferrimycin (see SIDEROMYCINS)
ferrioxamines (see SIDEROCHROMES)
ferritin An electron-dense iron-containing protein used in the IMMUNOFERRITIN TECHNIQUE. M.Wt. 750,000.
ferruginous Rust-coloured.
Feulgen reaction A reaction by which DNA (*q.v.*) may be specifically stained *in situ*. Mild acid hydrolysis (with e.g. 1N HCl) results in the removal of the purine bases from DNA and consequent release of the aldehyde groups of deoxyribose; the aldehyde groups react with SCHIFF'S REAGENT to give a purple or magenta coloration. (Among sugars, the ability to restore colour to Schiff's reagent is specific to 2-deoxysugars since, in these, a high proportion of the aldehyde form is normally present.) The Feulgen reaction can be used quantitatively.
FH_4 Abbreviation for tetrahydrofolic acid—see FOLIC ACID.
***fi^+*, *fi^-* R factors** (see R FACTOR)
fibrinolysin (plasmin) A proteolytic enzyme capable of dissolving (or preventing the formation of) a fibrin clot. Certain bacteria, e.g. species of *Staphylococcus* and *Streptococcus*, produce extracellular factors (*kinases*) which are able to promote the reaction plasminogen → plasmin (fibrinolysin); these kinases, commonly (though erroneously) called *fibrinolysins*, have been termed *staphylokinase* and STREPTOKINASE respectively. (see also COAGULASE TEST)
fibroma A benign tumour of connective tissues.
field diaphragm (in MICROSCOPY) (1) (also field iris) The adjustable diaphragm which controls the amount of light leaving the lamp or illuminator. (2) (also field stop) A metal annulus attached to the inside of an eyepiece at the focal plane of the eyepiece lens.
fièvre boutonneuse (Marseilles fever) (see BOUTONNEUSE)

Fiji disease (of sugarcane) (see SUGARCANE FIJI DISEASE VIRUS)

filament (1) (*bacteriol.*) Any *elongated* bacterial cell—e.g. one in which the length is approximately 10 or more times greater than the width. (2) (*bacteriol.*) A *chain* of cells—which may or may not be sheathed. (3) In blue-green algae: a *sheathed* TRICHOME; some authors use the term to refer to a trichome regardless of whether or not it is sheathed.

Fildes' enrichment agar (Fildes' digest agar) A medium used for the isolation of *Haemophilus* spp. It is prepared by the addition of 2–5% (v/v) FILDES' PEPTIC DIGEST OF BLOOD* to molten NUTRIENT AGAR which has been cooled to 56 °C. *Before use, the chloroform preservative should be driven off by heating to 56 °C for 30 minutes.

Fildes' peptic digest of blood A preparation which, when gelled with NUTRIENT AGAR, forms a useful medium for the isolation of *Haemophilus* spp; it contains the X FACTOR and the V FACTOR. *Preparation.* Defibrinated sheep blood is mixed with 3 times its volume of physiological saline (0.87% NaCl); for each 100 ml of this preparation are added 3 ml conc HCl and 0.5 gm granular pepsin. The whole is incubated at 56 °C for approximately 4 hours with occasional shaking. The pH is then brought to 7 by adding an appropriate volume of strong sodium hydroxide solution, and 0.25% (v/v) chloroform is added as preservative. The digest is stored at 4 °C.

filiform Thread-like.

filipin (see POLYENE ANTIBIOTICS)

film A synonym of SMEAR.

Filobasidiella neoformans (see CRYPTOCOCCUS)

filopodium A slender, branched or unbranched *pseudopodium*: see MOTILITY.

filterable viruses An obsolete term previously applied to infectious agents which were able to pass through the early microbiological filters, e.g. the *Berkefeld candle*. (see FILTRATION) Such agents were subsequently identified as VIRUSES.

filtration (microbiological aspects) Filtration can be used to separate microorganisms from a liquid (or gas) in which they are dispersed; it is a particularly useful method for sterilizing heat-labile liquids and solutions which cannot be autoclaved. *Sterilization by filtration.* A liquid known to be free of viruses may be sterilized by passing it through a filter capable of retaining all non-viral organisms (i.e. bacteria, fungi, etc.). Many different types of filtering apparatus have been devised; in the majority of these the liquid is drawn through the filter by reducing the pressure in the receiving vessel. The *candle filter* consists essentially of a thick-walled, test-tube-shaped structure made either of unglazed porcelain (e.g. the *Chamberland candle*) or wholly or partly of DIATOMACEOUS EARTH (e.g. the *Berkefeld candle*). When assembled, the cavity of the candle filter communicates with the receiving vessel, and the liquid sample is contained in a jacket which surrounds the exterior of the candle. Such filters may be used repeatedly—but should be cleaned and sterilized after each use. Other types of porcelain filter include the *Selas* and *Doulton* filters; the *Mandler* filter incorporates diatomaceous earth. The *Seitz* filter consists of a flat pad of asbestos (or asbestos and cellulose) suitably mounted between the sample container and the receiving flask; the filter pad is used once and discarded. In *sintered glass* filters the filter proper consists of a layer of minute glass particles which have been fused into a porous mass. Such filters may be re-used after cleaning and sterilization; they may be sterilized in a hot-air oven (160–170 °C). *Membrane filters* (e.g. the *Gradocol* filter) are thin films of cellulose nitrate (*collodion*), cellulose acetate, or other material; they are manufactured in a range of accurately-defined pore sizes. Membranes with the smallest pores are capable of retaining all known viruses and may therefore be used to sterilize a liquid regardless of the nature of the contaminating organisms; prefiltration with e.g. a Seitz filter may be necessary to avoid blocking the fine pores of the membrane with coarse particulate matter. (Filtration of such small particles is often referred to as *ultrafiltration.*) In the *Hemming filter* filtration is effected by centrifugal force. Two *bijou* bottles are clamped together (mouth-to-mouth) with a filter pad between them; one bijou contains the sample while the other (sterile) bijou acts as a receiver. The whole is so arranged in a centrifuge that, on centrifugation, liquid passes through the filter into the receiver.

Although filtration depends largely on a mechanical sieving action, suspended particles (and organisms) may be adsorbed to the filter as a result of electrostatic or other forces; some filters, e.g. the Seitz filter, are highly adsorptive, while others, e.g. the modern porcelains and membrane filters, are relatively non-adsorptive. Conversely, leaching of the material of the filter may occur; thus, e.g. the Seitz filter tends to enrich the filtrate with magnesium ions.

The uses of filtration include: (a) The preparation of cell-free culture filtrates e.g. for the recovery of extracellular enzymes, antibiotics etc. (b) The enumeration of small numbers of bacteria in a large volume of liquid—e.g. in water bacteriology; a measured volume of the sample is passed through a membrane filter (pore size about 0.45 microns) and the membrane is subsequently incubated (face upward) on a pad of absorbent material (e.g.

Whatman No 17 absorbent pad) which has been saturated with a suitable liquid culture medium. The number of bacteria per unit volume is obtained from the number of colonies which subsequently develop on the membrane. Membranes overprinted with a grid facilitate colony counting. (c) Estimation of the size of virus particles. The viral suspension is filtered through a series of membranes of progressively smaller pore size; the size of the virion corresponds *approximately* to the pore size of the first membrane to effect significant retention of the particles.

fimbriae (common pili) (*sing. fimbria*) In bacteria: an heterogenous class of PILI whose formation is controlled by *chromosomal* genes (cf. SEX PILI). On the basis of certain characteristics—e.g. diameter—several types of fimbriae can be distinguished; these are designated type I, type II etc. Each cell may have up to several hundred fimbriae; fimbriae may be distributed over the entire surface of a cell (e.g. as in *Escherichia coli*) or may occur as a polar tuft (e.g. in *Pseudomonas aeruginosa*). The function of fimbriae is unknown. Fimbriate cells tend to adhere to one another and to form a pellicle at the surface of (undisturbed) liquid culture media; they also adhere to erythrocytes and cause haemagglutination. (Fimbriae may play a role in the pathogenicity of dysentery-producing species of *Shigella* by promoting adherence of the bacteria to the intestinal mucosa.) The pilin of type I fimbriae has a molecular weight of *circa* 17,000, and contains a high proportion of amino acids with non-polar side chains; it has been suggested that this property may be responsible for the tendency of fimbriate cells to adhere to hydrophobic surfaces.

fimbriate (*adjective*) (1) Having FIMBRIAE. (2) Having a fringed edge.

fire blight (of apples, pears) A disease characterized by browning and necrosis of blossoms, fruit, leaves and twigs. The causal agent, *Erwinia amylovora*, enters the plant e.g. *via* the blossom, stomata and/or wounds. Dissemination of the disease is aided by rain and by insects. Fire blight is a NOTIFIABLE DISEASE.

fission (1) (*microbiol.*) A type of asexual reproduction in which a cell divides to form two or more daughter cells which are (typically) of equal or similar size. (see also BINARY FISSION; cf. BUDDING) (2) (*mycol.*) In certain yeasts (e.g. *Schizosaccharomyces*): a form of cell division which involves septum-formation (rather than budding).

Fistulina A genus of fungi of the family Fistulinaceae, order APHYLLOPHORALES. *F. hepatica* (the 'beef-steak fungus') forms (edible) fruiting bodies in which the hymenophore consists of a layer of closely-packed but separable tubules arranged parallel to one another; the annual, sessile or laterally stipitate pileus may be flat and tongue-like, or hoof-like, and it develops with the (ventral) spore-bearing surface orientated horizontally. The context of the pileus is monomitic; it is red and fleshy and exudes a deep red liquid when cut. The upper surface of the pileus is dark reddish-brown, and the undersurface is pale yellow when unbroken. Cystidia are not formed in the hymenium; the spores are smooth-walled, spherical to ovoid—each approximately 5×4 microns—and are pinkish-brown *en masse*.

FITC Fluorescein isothiocyanate. A fluorescent substance used for labelling proteins (antibodies etc.) with which it readily combines under appropriate conditions of pH. (see also IMMUNOFLUORESCENCE) Fluorescence is greenish-yellow.

fixation (of biological material) The process of killing cells while preserving their structure and internal organization in as life-like a condition as possible. Fixation involves the inactivation of enzyme systems—particularly those responsible for *autolysis*; ideally it hardens cellular structures and renders cell components compatible with particular stains. Most (if not all) methods of fixation create ARTEFACTS. (a) *Heat fixation.* This crude method is used almost exclusively for the fixation of bacterial smears—see FLAMING. A heat-fixed smear is suitable for certain identification procedures (e.g. the GRAM STAIN) but is unsuitable if fine cellular details are to be examined. (It should be noted that heat fixation—as commonly practiced—cannot be relied upon to kill all species of bacteria.) (b) *Chemical fixation.* Chemical *fixatives* permeate cells and stabilize their components by binding to them or denaturing them. Most fixatives react with proteins or lipids; some react with both. Some fixatives precipitate proteins and thereby introduce considerable disturbance in cellular organization; they include ethanol, mercuric chloride and picric acid. Others do not cause precipitation and tend to preserve proteins *in situ*; they include FORMALDEHYDE, GLUTARALDEHYDE, potassium dichromate and OSMIUM TETROXIDE. Osmium tetroxide and potassium dichromate are also important *lipid* fixatives.

Specimens for light microscopy are often fixed in BOUIN'S FLUID, CARNOY'S FLUID, ethanol, formalin, methanol, POLYVINYL ALCOHOL FIXATIVE, SCHAUDINN'S FLUID or ZENKER'S FLUID.

The fixation of specimens for electron microscopy is more critical since structural disturbances must be minimal. For such work glutaraldehyde and osmium tetroxide are commonly used—usually in an alkaline phosphate buffer or an acetate-veronal buffer; in a frequently-used schedule, protein fixation is first accomplished with glutaraldehyde, and lipid fixation achieved in a subsequent *post-fixation* with osmium tetroxide.

Specimens for AUTORADIOGRAPHY should be fixed in solutions which do not contain heavy metals since these may block or divert radiation from point sources in the specimen.

fixative (*microbiol.*) A substance used for FIXATION.

fixed virus (of the *rabies virus*) A strain of the rabies virus which has undergone serial passage through rabbits or e.g. chick embryos. (cf. STREET VIRUS; see also FLURY VIRUS; SEMPLE VACCINE)

flabelliform (flabellate) Fan-shaped.

flacherie A disease of the silkworm (*Bombyx mori*). The causal agent appears to be a virus.

flagella Plural of FLAGELLUM.

flagellins Soluble, globular proteins which constitute the subunits of *bacterial* flagella (see FLAGELLUM). The composition of flagellin varies with species; there is usually a low content of aromatic amino acids and a relatively high content of glutamate and aspartate. Cysteine is usually absent. The flagellins of certain species of *Salmonella* contain ε-*N*-methyllysine. All flagellins have similar physical properties; in many species the molecular weight of flagellin is approximately 40,000. The depolymerization of flagella into soluble subunits may be achieved by treatment with detergents or by reduction of the pH. Spontaneous re-assembly of the subunits may occur under certain conditions—e.g. the re-establishment of neutral pH. The presence of a *primer* (e.g. small fragment of flagellum) seems to be necessary for the polymerization of many flagellins. X-ray diffraction studies have revealed similarities between flagellins and *myosin* (a mammalian muscle protein). (see also H ANTIGENS)

flagellum (bacterial) A thread-like appendage which occurs singly, in groups or tufts, or as a covering layer on the cells of certain bacteria; the *arrangement* of flagella on a cell is a genetically stable characteristic and is an important taxonomic feature. (see AMPHITRICHOUS, LOPHOTRICHOUS, and PERITRICHOUS; see also SPIROCHAETALES) Flagella are responsible for MOTILITY in the *majority* of motile species. Bacterial flagella are commonly some 3–20 microns in length and have a uniform diameter of 120–250Å; they can be detected by *dark-field* microscopy but cannot be resolved by bright-field microscopy—unless previously subjected to special STAINING techniques. (see LEIFSON'S FLAGELLA STAIN) Flagella constitute important antigenic sites of the cell. (see H ANTIGENS and KAUFFMANN-WHITE CLASSIFICATION SCHEME)

The bacterial flagellum consists of several (often 3) longitudinally-orientated protein strands—each strand being composed of a chain of globular FLAGELLIN subunits; these strands are closely associated—like the strands of a rope—either in a spiral or parallel arrangement or in some intermediate form. Labelling experiments, which indicate *distal* flagellar growth, infer that the flagellum may be hollow and that the flagellin subunits synthesized within the cell may pass through the core of the flagellum prior to incorporation at the distal end. The proximal end of the flagellum terminates in a structure known as the BASAL BODY which appears to be closely associated with the cell envelope.

In vivo, the form of the flagellum may be a loose helix—the pitch of the helix (i.e. the 'wavelength' of the flagellum as seen on a 2-dimensional slide preparation) being characteristic for a given species. Some organisms exhibit *biplicity*, i.e. individual cells of such species may display two types of flagellum—each type being characterized by a given pitch. Flagella may be removed, mechanically, without loss of cell viability; they commonly regenerate with great rapidity. Enzymic removal of the CELL WALL of a motile bacterium produces a flagellated but *non-motile* protoplast. The mechanism of flagellar movement has yet to be elucidated; the movement itself is believed to be rotatory in nature—the rotations originating from the basal body of the flagellum. ATPase activity has not, thus far, been detected in bacterial flagella. The rate of locomotion in peritrichously flagellated cells is generally lower than that in polarly flagellated cells; the latter are said to attain speeds of 50 microns/second or more.

flagellum (eucaryotic) A thread-like appendage which occurs on the cells of certain algae, fungi, and protozoa; *flagellate* cells (i.e. cells which have one or more flagella) usually exhibit MOTILITY. Eucaryotic flagella from widely differing sources are remarkably similar in structure. In general they may reach 40 or more microns in length and are some 0.2–0.3 microns in diameter. Each flagellum consists of an outer membranous *sheath* (a continuation of the cell membrane) which encloses a matrix containing the *axoneme*. This is essentially a system of parallel, longitudinally-arranged microtubules—each about 25 nm in diameter—which extend the length of the flagellum. The tubules are arranged in a characteristic pattern: two closely-adjacent tubules, which form the centre of the axoneme, are surrounded by a circle of nine uniformly-arranged compound tubules (doublets)—each doublet consisting of a pair of tubules. One tubule of each doublet bears a regular series of projections (*arms*) along the greater part of its length; these arms are formed at right angles to the axis of the tubule—some being orientated tangentially, others radially, with respect to the circular cross-section of the flagellum. The proximal part of the axoneme, i.e. that part within the cell body, terminates in the BASAL BODY. The microtubules are composed of a globular protein, TUBULIN; the arms which project from the peripheral doublets are believed to consist of a different (higher M.Wt.) protein. The surface (sheath) of the flagellum may be smooth or it may bear fine filamentous processes (*flimmer*) along much or all of its length; when such processes are relatively short and rigid they may be referred to as mastigonemes (mastigonemata). (In algology and mycology, smooth flagella are referred to as the *whiplash* (or *peitschengeissel*) type while those bearing flimmer are said to be of the *tinsel* (or *flimmergeissel*) type.)

Flagella often move in an undulatory manner—the undulations commonly originating at the proximal end; movement may occur in one or more planes. The beating of a *smooth* flagellum generates forces which act on the surrounding fluid in the direction of the propagated waves; hence, the cell moves in a direction opposite to that of the flagellar waves. (Thus, flagellar waves are propagated from tip to base in those organisms—e.g. *Trypanosoma* spp—which swim with the flagellum pointing in the direction of motion.) Flagella which bear mastigonemes generate forces which act on the surrounding medium in a direction opposite to that of the propagated waves. Little appears to be known of the mechanism which governs flagellar (and ciliar) movement. Viscous drag may be important in the synchrony often observed in the beating of closely-adjacent flagella (or cilia). Flagellar (and ciliar) movements have been studied with the aid of the stroboscope and high-speed cinephotography. (see also CILIUM (1) and HAPTONEMA)

flaming (1) The brief exposure of an object (or part of an object) to the hottest part of a flame in order to *sterilize* the surface thus treated; flaming is thus an example of dry-heat STERILIZATION. In microbiology flaming is part of the ASEPTIC TECHNIQUE; objects flamed include the LOOP and the STRAIGHT WIRE (exposure time approximately 3–5 seconds) and the rims of bottles etc. from which sterile (non-inflammable!) substances are to be poured. (2) Flaming is also used for the FIXATION of SMEARS of bacteria prior to staining; such *heat-fixed* smears are suitable e.g. for the GRAM STAIN, but are not suitable when the object of staining is to observe the structural details of the cells. (Smears are flamed by passing the slide, smear-to-flame, *rapidly* through the flame several times.)

flask-fungi Those fungi which form perithecia—see PERITHECIUM and SPHAERIALES.

flat-field objective lens An objective lens (of a microscope) which can provide an image in which all parts of the field are simultaneously in focus. (see also PLANACHROMAT)

flat-sours (see FOOD, MICROBIAL SPOILAGE OF)

flavacol 3-Hydroxy-2,5-diisobutylpyrazine; a compound produced by *Aspergillus flavus*.

Flavobacterium A genus of gram-negative, motile or non-motile, pigmented, asporogenous bacilli or coccobacilli found in soil and in marine and fresh water. Taxonomic affinity: uncertain.

flavodoxins A class of flavoproteins which have low redox potentials and which can replace FERREDOXINS as electron carriers in many reactions; flavodoxins contain 1 FMN per molecule, and do not contain metal or labile sulphide. In many organisms (e.g. *Clostridium* spp, certain algae) flavodoxins are synthesized in place of ferredoxins in response to iron deficiency; they are less efficient than ferredoxins as electron carriers. (The algal flavodoxin is also known as *phytoflavin*.) An unusual flavodoxin—*azotoflavin*—functions as an electron carrier in NITROGEN FIXATION in *Azotobacter vinelandii*.

flavoprotein (see RIBOFLAVIN)

Flavovirus A genus of mosquito-borne and tick-borne ARBOVIRUSES of the family TOGAVIRIDAE; members of the genus exhibit cross-reactivity in haemagglutination-inhibition tests, and are distinguished from viruses of the genus ALPHAVIRUS in that e.g. the latter are insensitive to trypsin. The genus includes the causal agents of DENGUE, Japanese B encephalitis, and YELLOW FEVER.

flax, diseases of These include (a) *Rust*. A disease caused by the autoecious rust *Melampsora lini*. Initial signs of disease are the orange *sori* of the *aecial* or *uredial* stages of *M. lini* which appear on the leaves and stems; these are later replaced by dark-brown *telia*. (b) *Wilt*. A soil- or seed-borne disease in which chlorosis and retarded growth (or death) follow vascular penetration by *Fusarium oxysporum* f.sp. *lini*.

flax retting (see RETTING)

Flexibacter A genus of bacteria of the CYTOPHAGACEAE.

flexuous hyphae (receptive hyphae) Hyphae which project from a *pycnium* and which function as female structures—see RUSTS (1) and SPERMATIZATION.

flimmer (flimmergeissel) (see *eucaryotic* FLAGELLUM)

flocculation (*immunol.*) (1) In certain

PRECIPITIN TESTS, the development of precipitate in the form of flaky or fluffy masses; flocculation occurs e.g. during the titration of diphtheria toxoid against horse-derived antitoxin. (2) More loosely, precipitation in general—or even H-agglutination (the agglutination of bacteria by means of antiflagellar antibodies).

flood plate A plate prepared by covering the entire surface of the medium with liquid inoculum (e.g. diluted broth culture) and withdrawing excess inoculum (with a sterile pasteur pipette) prior to incubation; a LAWN PLATE may be prepared in this way.

floridean starch (see STARCH)

floridoside 2-Glycerol-α-D-galactopyranoside; a storage compound formed by members of the Rhodophyta.

flotation (as a method for concentrating protozoan cysts) Flotation methods involve the dispersion of a specimen (e.g. faeces) in a solution of such specific gravity (S.G.) that any cysts present in the specimen float to the surface of the solution. (a) *Zinc sulphate flotation.* A solution (S.G. 1.18) is made by adding distilled water to 331 gm of zinc sulphate ($ZnSO_4.7H_2O$) to give a final volume of 1 litre. A small sample of faeces (or other specimen) is added to some 20 ml of the solution in a universal bottle. The whole is well mixed and allowed to stand for 20 minutes. With a flamed and cooled loop, a loopful of liquid from the *surface* of the solution is transferred to a drop of iodine solution on a slide; a coverslip is applied and the preparation examined under the microscope. (b) *Salt flotation.* The use of a salt solution (sodium chloride, 70%–100% saturated) is commonly used for the concentration of coccidian oocysts from faecal samples. The homogenized sample is sieved (to remove the grosser débris) and subjected to several cycles of centrifugation in water (1,500 r.p.m./3 minutes)—the supernatant being discarded each time. The sediment is then dispersed in the salt solution and centrifuged at 1000 r.p.m. for 5 minutes. The *surface* layer of supernatant is transferred (by pipette) to ten times its volume of water; this is centrifuged and the sediment collected for study. Prolonged contact with saturated salt solution causes plasmolysis; however, the oocysts generally remain viable, and they regain their original appearance on transfer to water. (see also CYST and FORMOL-ETHER METHOD)

fluctuation test A test devised by Luria and Delbrück (1943) for investigating the mode in which bacterial populations respond to changes in the environment. At that time two hypotheses co-existed: (a) Genetic changes occur *adaptively*, i.e. as a result of environmental influences on the cells; those cells which adapt to the environment survive and proliferate. (b) Genetic changes occur *spontaneously*, i.e. independently of the environment; those cells which have become particularly well suited to certain environmental conditions proliferate (preferentially) if such conditions arise.

If bacteria *adapted* to a new environment, a similar proportion of adapted cells should arise in each of a number of independent cultures. If, however, mutants arose *spontaneously* and randomly, the number of mutant cells should vary considerably from culture to culture.

In the original test, a measurement was made of the statistical *variance* among the numbers of bacteriophage-resistant cells in each of ten separate cultures of phage-sensitive bacteria; the variance (3498) could not be attributed to sampling error since a variance of only 27 was obtained for the numbers of resistant cells in each of ten samples taken from a single (bulk) culture. (Phage-resistant cells were counted by plating an aliquot of each culture/sample, subjecting the plate to virulent phage, and incubating; resistant cells formed colonies.) The wide fluctuation (high variance) in mutant numbers among the separate cultures suggested that mutant clones had been initiated in some cultures much earlier than in others—i.e. resistant cells had appeared at different times; this indicated that mutant (resistant) cells had arisen *before* the cells had been exposed to phage—thus lending support to hypothesis (b). (see also NEWCOMBE EXPERIMENT and REPLICA PLATING)

fluid mosaic model (see CELL MEMBRANE (bacterial))

fluorescein A yellow dye and FLUOROCHROME which is used e.g. for labelling antibodies etc. in techniques such as IMMUNOFLUORESCENCE; fluorescein, a substituted xanthene—chemically related to EOSIN, is commonly used in the form of its reactive derivative, FITC (*q.v.*). The dye fluoresces greenish-yellow. The solubility of fluorescein in water is very low; in ethanol the solubility is approximately 2% w/v at room temperatures.

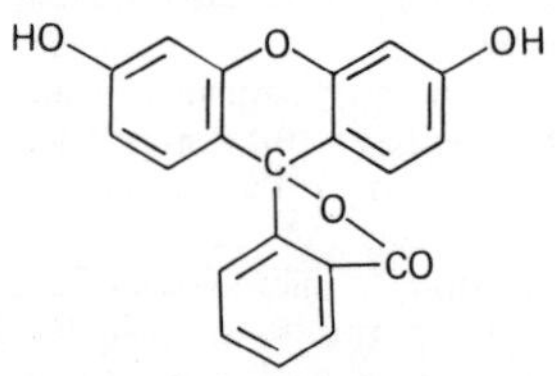

Fluorescein

fluorescence The emission of light by certain substances (FLUOROCHROMES) on absorption of an exciting radiation; the wavelength of the emitted light is longer than that of the absorbed radiation, and on cessation of irradiation emission of light ceases within a small fraction of a second. (cf. PHOSPHORESCENCE) Fluorochromes emit light of characteristic wavelengths.

fluorescence microscopy (see MICROSCOPY)

fluorescent antibody technique (see IMMUNOFLUORESCENCE and SANDWICH TECHNIQUE)

fluorescent dyes (as antimicrobial agents) On staining with certain fluorescent dyes (e.g. EOSIN, ROSE BENGAL, ACRIDINE ORANGE) some microorganisms (particularly bacteria) may be killed if they are subsequently exposed to strong light. This effect, known as the *photodynamic effect*, is due largely to the formation of SINGLET OXYGEN. (see also DYES (as antimicrobial agents))

fluorine and compounds (as antimicrobial agents) Fluorine, a powerful oxidizing agent, is too reactive for use as a general antimicrobial agent. However, sodium and ammonium fluorides and silicofluorides have been used as antifungal wood preservatives; their efficacy is said to be greater than that of zinc or copper compounds. The antimetabolite 5-fluorocytosine has been used systemically to treat cryptococcosis and certain other fungal infections. 5-Fluorouracil is mutagenic for RNA viruses. (see also CHLORINE, BROMINE and IODINE)

fluorochrome ('fluor') Any substance which exhibits FLUORESCENCE. Fluorochromes such as *fluorescein** are used e.g. for labelling antibodies in IMMUNOFLUORESCENCE techniques. *see also FITC.

Flury virus An avianized strain of the *rabies virus* (see also RHABDOVIRUS). The LEP (*q.v.*) and HEP Flury viruses are used for anti-RABIES vaccination in domestic animals. (see also FIXED VIRUS)

fly agaric A common name for the basidiomycete *Amanita muscaria*; *A. muscaria* produces MUSCARINE which is toxic for flies.

FMN Flavin mononucleotide (see RIBOFLAVIN and Appendix V(c)).

foamy agents Viruses of subgenus *D* of the genus LEUKOVIRUS. Foamy agents have been isolated from e.g. cats, cattle, and primates, but have not been shown to be oncogenic or to be associated with any other form of disease. In tissue cultures the foamy agents give rise to cytopathic effects (CPE) which typically include the formation of syncytia and the development, within infected cells, of the appearance of a 'foamy' (highly vacuolated) cytoplasm.

Foettingeria A protozoan genus of the APOSTOMATIDA.

folic acid (pteroylglutamic acid) A water-soluble, photolabile VITAMIN consisting of *pteroic acid* linked to glutamic acid. (Pteroic acid = 2-amino-4-hydroxy-6-methylpteridine linked at the 6-position to the amino group of *p*-AMINOBENZOIC ACID.) *Polyglutamate* forms of folic acid also occur (e.g. *pteroyltriglutamic acid*); certain of these may be specifically required for certain reactions.

The coenzyme form of the vitamin is 5,6,7,8-tetrahydrofolic acid (THF)—formed when folic acid is reduced by dihydrofolate reductase and $NADPH_2$; 7,8-dihydrofolic acid is an intermediate. THF functions as a carrier of one-carbon units (e.g. formyl, methyl, hydroxymethyl groups); such groups may be attached either at the N^5 or N^{10} positions, or may bridge these two positions. THF is involved e.g. in glycine-serine interconversion (see Appendix IV(c)), in the biosynthesis of methionine (see Appendix IV(d)) and of purines (see Appendix V(a)), and in the catabolism of histidine.

Certain microorganisms require an exogenous source of folic acid for growth; these organisms include e.g. *Streptococcus faecalis* and *Lactobacillus casei* (both used in the BIOASSAY of folic acid), and the protozoa *Tetrahymena* spp, *Crithidia fasciculata*, and *Leishmania tarentolae*. *N^5*-

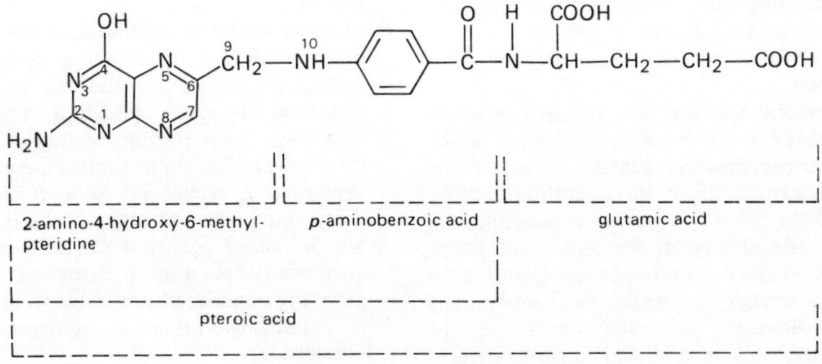

Folic acid (pteroylglutamic acid)

formyltetrahydrofolate—also known as *citrovorum factor*, *folinic acid*, or *leucovorin*—was discovered as a growth requirement for *Leuconostoc citrovorum* (= *L. cremoris*). Most fungi can synthesize folic acid, although a few require the precursor *p*-aminobenzoic acid. (see also BIOPTERIN, SULPHONAMIDES, TRIMETHOPRIM, PYRIMETHAMINE)

foliicolous Growing on leaves.

folinic acid (1) N^5-formyltetrahydrofolate; see FOLIC ACID. (2) Occasionally used (incorrectly) to refer to other derivatives of folic acid, e.g. tetrahydrofolate.

foliose (of a lichen) Refers to a lichen whose thallus is leaf-like (i.e. flattened and DORSIVENTRAL). The thallus may be lobed (laciniate) and part or most of the lower surface is more or less firmly attached to the substratum (e.g. by means of *rhizinae*); alternatively, the thallus may be disc-like and attached to the substratum by a holdfast in the centre of the lower surface (an *umbilicate* thallus). Foliose lichens include e.g. members of the genera LOBARIA, PARMELIA, PELTIGERA, and XANTHORIA. (cf. CRUSTOSE (2) and FRUTICOSE)

folpet *N*-(trichloromethylthio)phthalimide. An agricultural antifungal agent closely related to CAPTAN—with which it shares a similar antifungal spectrum.

Fomes A genus of lignicolous fungi of the family POLYPORACEAE. *Fomes* spp form perennial fruiting bodies in which the hyphae of the light brown context have clamp connections and exhibit trimitic organization; the hymenophore is typically stratified. Some species which were formerly included within the genus *Fomes* have been transferred to other genera; thus, e.g. *Fomes annosus* is now classified as *Heterobasidion annosum* (see HETEROBASIDION). (*Fomes* is often used as a *form genus*; when used as such it includes those polypores which form hard—corky or woody—perennial fruiting bodies.)

fomites Objects and materials which have been associated with infected persons or animals and which potentially harbour PATHOGENIC microorganisms. (see also VECTORS)

food, microbial sources of (see MICROORGANISMS AS FOOD)

food, microbial spoilage of *Spoilage* refers to any change in the condition of food in which the latter becomes less palatable, or even toxic; these changes may be accompanied by alterations in taste, smell, appearance or texture. (see also FOOD POISONING and FOOD, PRESERVATION OF) Organisms responsible for spoilage include a variety of bacteria and fungi—although particular foods may be susceptible only to a limited range of species. The involvement of a given species is determined e.g. by the types of nutrient present in the food, its water content and pH, the presence of additives or preservatives, and the conditions of storage—particularly temperature and accessibility of air.
(a) *Dairy products*. BUTTER may be attacked by the psychrophilic bacterium *Pseudomonas fragi* and by species of fungi, e.g. *Alternaria*, *Aspergillus*, *Cladosporium*—the fungi producing dark, patchy regions of surface growth. Surface spoilage of CHEESE may be due to species of *Aspergillus*, *Cladosporium* or *Penicillium*; *Geotrichum* may cause spoilage of high-moisture cheeses, e.g. cottage cheese. During the ripening process the formation of undesirable holes (*eyes*) may be due to the presence of coliform bacteria or species of *Bacillus*. (b) *Meat*. Refrigerated fresh meat is susceptible to surface attack by *Pseudomonas* spp and e.g. *Lactobacillus* spp—*Pseudomonas* being one of the commonest of the odour and slime-producing bacteria. Bacterial spoilage is encouraged if the cut surfaces are moist and the humidity high. Spoilage is discouraged by acidity; for this reason, animals should not be unduly excited prior to slaughter since, by retaining high levels of muscle glycogen, suitable levels of lactic acid will subsequently be present in the meat. (c) *Canned foods*. Spoilage may be due to thermoduric and/or aciduric organisms such as the ascomycete *Byssochlamys fulva* which may cause deterioration in canned fruits. *Bacillus* and *Clostridium* spp (spore-formers) are particularly important as indicators of inadequate CANNING. Some species, e.g. *Bacillus polymyxa*, ferment carbohydrates with the production of acid and gas—the latter causing the can to swell. *B. stearothermophilus* produces acid only, and gives rise to a non-swollen can whose contents have been soured; such cans are referred to as *flat sours*. Spoilage may result from the use of contaminated cooling water which, during cooling, may be drawn into the can through an imperfect seal. (d) For spoilage of beers, see BREWING; see also CIDER SICKNESS.

food, preservation of Methods of food preservation include the traditional ones of pickling, salting and smoking, and the more recent processes of PASTEURIZATION, refrigeration, CANNING, dehydration (including FREEZE-DRYING) and the use of synthetic preservatives. Preservation techniques seek to prevent or delay microbial and other forms of spoilage, and to guard against FOOD POISONING. (see also FOOD, MICROBIAL SPOILAGE OF) It is generally considered preferable to use physical means of preservation (e.g. refrigeration) when possible.
(a) *Heating*. Pasteurization is used for the treat-

ment of dairy products, e.g. milk, and for beers (see BREWING) and wines. Foods at or near neutral pH (e.g. some vegetables, mushrooms) are canned at sterilizing temperatures (upwards of 115 °C)—which ensures the total destruction of the spores of *Clostridium botulinum.* Lower temperatures can be used for the treatment of certain *acidic* foods (mainly fruits)—approximately pH4 or below—since organisms (including spores) are generally less resistant to heat under acid conditions. (b) *Low temperature preservation.* (i) *Above 0 °C.* The object is to reduce the metabolic activities of contaminating organisms and/or the activity of endogenous food enzymes; this *delays* spoilage. Organisms which cause spoilage between 0 °C and 10 °C include species of *Pseudomonas* (meat, fish), *Xanthomonas* and *Erwinia* (certain vegetables) and species of the fungi *Penicillium* and *Cladosporium* (dairy products, fruits). *Clostridium botulinum* (types E and F) is reported to be able to grow and produce toxin at 3.3 °C. (ii) *Temperatures below 0 °C.* Freezing effectively reduces the food's available water, thus making the environment less hospitable for contaminating organisms. Nevertheless, some spoilage organisms (e.g. *Pseudomonas*, *Cladosporium*, *Penicillium*) continue to metabolize at temperatures several degrees below 0 °C; care must therefore be taken to exclude or reduce contamination before freezing. (c) *Dehydration.* If water is progressively withdrawn from a foodstuff, a point is reached at which the WATER ACTIVITY of the food is such that it will not support the growth of contaminating organisms. For the majority of *bacteria* an a_w greater than about 0.95 is required for active metabolism, although some, e.g. the *halophilic* bacteria, tolerate or prefer lower levels. Fungi grow in even drier conditions—at values of a_w down to about 0.65 or lower. In foods, the available water may be reduced by heat-drying (e.g. dried milk), by freeze-drying, or by the addition of salt (*salting*) or sugar syrups. *Smoking* (e.g. fish products) combats spoilage by (i) dehydration, and (ii) impregnating the foodstuff with antimicrobial substances, e.g. phenolic compounds. (d) *Acidification* (*pickling*)—in which a foodstuff is permeated by an organic acid, usually acetic acid (as VINEGAR). (see also SAUERKRAUT) (e) *Preservatives.* Preservatives may be added directly or indirectly to a foodstuff. The use of *added preservatives* in particular foods (in controlled quantities) is subject to governmental consent in various countries; preservatives used in one country may be banned in another. A preservative should be innocuous to the consumer; it should not be an ALLERGEN; it should not be inactivated by constituents of the food or by the products of microbial contaminants; it should be microbicidal rather than microbistatic towards the range of possible contaminants. Few, if any, preservatives fulfil all of these requirements. BENZOIC ACID is a permitted preservative in the U.K., U.S.A., Canada and many other countries; it is widely used in fruit juices, cordials, sauces and other products, and is both harmless and tasteless at permitted concentrations. NISIN is a permitted preservative in the U.K. and some other countries, though not in the U.S.A.; it is particularly effective against heat-resistant organisms and has been used in certain types of canned foods as well as in cheese and *clotted* cream. Sodium and potassium NITRITES are permitted additives in the U.K., U.S.A. and elsewhere; they are used e.g. in certain meat products in which one of their functions is to maintain an attractive coloration. SORBIC ACID is fungistatic and is also active against certain bacteria; its use is permitted in the U.K., U.S.A. and many other countries. Sorbic acid is used, e.g. in cheeses, flour confectionery and marzipan. ANTIBIOTICS, particularly those which have a medical use, have limited application in food preservation; TETRACYCLINES are sometimes used to extend the storage life of refrigerated poultry. Other preservatives include DIETHYLPYROCARBONATE, parabenzoate, fatty acids (particularly *propionic acid* and its salts) and SULPHUR DIOXIDE (in fruit juices, cordials etc.).

food intoxications (see FOOD POISONING (bacterial) and FOOD POISONING (fungal))

food poisoning (bacterial) Bacterial food poisoning can be considered in two categories: (a) *Food-borne intoxications*—which result from the ingestion of food containing specific bacterial EXOTOXINS, or from the elaboration of such toxins within the intestine. Food intoxications include: (i) *Staphylococcal food poisoning*—an acute, non-febrile condition caused by the ENTEROTOXINS of certain strains of *Staphylococcus.* The incubation period (1–7 hours) is followed by nausea, vomiting, diarrhoea and prostration; recovery is usually rapid. Toxinogenic, coagulase-positive staphylococci are commonly involved, but toxinogenic coagulase-negative strains have been described. Food in which effective concentrations of toxin have been produced does not usually exhibit any detectable change in condition. Food may be examined for toxins by preparing a saline extract, concentrating by dialysis, purifying, and using a GEL DIFFUSION technique with known antitoxins. (ii) *Botulism*—a severe, often fatal, intoxication due to the toxin(s) of *Clostridium botulinum* (see CLOSTRIDIUM). (Very infrequently, botulism has resulted from wound contamination.) Symptoms commonly appear in 3–36 hours; they may include weakness, nausea, vomiting, blurred vision, difficulty in swallowing, dysphonia, and paralysis of the respiratory system. The toxins block neural transmission by inhibiting the synthesis or release of acetylcholine at presynaptic sites; toxin reaches these sites *via* the blood following absorption through the intestine. Therapeutic measures include the administration of polyvalent antitoxin. In *cattle* the disease known as *lamziekte* is usually associated with *C. botulinum* type *D. Limberneck* of fowl may involve *C. botulinum* type *A*. (iii) Infantile gastroenteritis/diarrhoea, travellers' diarrhoea etc. may be due to enteropathogenic strains of *Escherichia coli*; such strains produce cholera-like ENTEROTOXINS.

(b) *Food-borne infections*—acute, enteric diseases in which the causal organism appears to give rise to an active infection of the intestinal tract. Although the following are traditionally regarded as infections, toxin-production has been reported in each case; the significance of these toxins has yet to be evaluated. (i) *Clostridial food poisoning* (*perfringens food poisoning*)—a self-limiting condition characterized by abdominal pain and diarrhoea; symptoms commonly begin some 12 hours after the ingestion of food which is heavily contaminated with strains of *Clostridium perfringens* type *A*. (ii) *Salmonella food poisoning* typically involves gastroenteritis with vomiting, abdominal pain, diarrhoea, headache and fever; symptoms commonly develop 6–24 hours after the ingestion of contaminated food. Causal agents include *Salmonella typhimurium*. (iii) Other food-borne pathogens include strains of *Vibrio parahaemolyticus* (mainly in sea foods).

food poisoning (fungal) Instances of food poisoning which follow the ingestion of fungi are probably exclusively food *intoxications*, i.e. conditions which result from the intake of fungal *toxins*—see also FOOD POISONING (bacterial). Fungal food poisoning occurs in man and other animals. Fungal toxins include AFLATOXINS, AMATOXINS, PHALLOTOXINS, RUBRATOXINS and MUSCARINE. (see also ERGOTISM)

food vacuoles (gastrioles) (*protozool.*) Closed intracellular sacs, bounded by a unit-type membrane, which contain particles of food that have been ingested by the protozoan. Food vacuoles are formed e.g. by ciliates and by amoebae.

*Food vacuoles in ciliates**. Food particles are ingested at the CYTOSTOME. The particles initially become enclosed within a flask-like invagination of the cytostomal membrane; subsequently the invaginated membrane becomes detached and passes into the cytoplasm as a closed food vacuole. In at least some ciliates vacuole formation appears to be stimulated by the presence of particulate matter. Soon after its formation the vacuole decreases in volume and its contents become acidic. Digestion is apparently brought about by a variety of enzymes which are emptied into the vacuole by lysosomes; these enzymes may include acid phosphatases, nucleases, amylases, and esterases. Digested food appears to enter the cytoplasm by pinocytosis; in many cases, small pinocytotic vesicles have been observed to develop as invaginations of the vacuolar membrane into the cytoplasm. Undigested food is voided when the vacuolar membrane coalesces with that of the cytopyge. More than one food vacuole may be present in a given cell at a given time. Under certain conditions, e.g. starvation, a cell may digest its own internal structures (*autophagy*). *The mechanism of food vacuole formation in *suctorians* is unknown.

foot (*mycol.*) In fungi of the class LABOULBENIOMYCETES: a specialized region of the thallus by means of which the fungus attaches to its host.

foot and mouth disease (aphthous fever) An acute, highly infectious disease which affects mainly cloven-hoofed animals—e.g. cattle, pigs, sheep; horses are immune to the disease, and man is affected only rarely. The causal agent is a picornavirus (see RHINOVIRUS). Infection commonly occurs *via* the oral or nasal route, and the incubation period—usually a few days—may be as long as 10 days in the pig. The symptoms include fever and the formation of *vesicles* (fluid-filled, blister-like lesions) on the mucous membranes of the mouth. (The copious saliva contains viruses derived from the ruptured vesicles; viruses are also found in the blood, milk, urine, and faeces.) Lesions also appear on the teats, in the rumen, and on the feet—in particular where the horny material and skin are contiguous. In adult animals the disease is not usually fatal; however, it causes serious economic losses in terms of milk and meat production. Foot and mouth disease is a NOTIFIABLE DISEASE in the United Kingdom—where the policy is one of strict quarantine and slaughter of diseased and suspect animals. (An important consideration is the existence of a carrier state; pharyngeal carriage of virus may persist in cattle, goats and sheep for several years following the initial infection.) *Disinfection*. The most effective disinfectants are those which subject the virus particles to extremes of pH (see RHINOVIRUS); sodium hypochlorite (bleach)—although effective—tends to be rapidly inactivated by organic matter, and phenolic disinfectants are ineffective. *Laboratory diagnosis* may involve e.g. culturing the virus (from a pharyngeal swab etc.) and/or the performance of a complement fixation test.

In *man*, the disease involves fever, malaise, and the formation of vesicles in the mouth and on the lips, hands and feet; infection may result from wound contamination or the ingestion of contaminated dairy products. Complete recovery commonly occurs within weeks. (see also SWINE VESICULAR DISEASE, VESICULAR STOMATITIS)

foot cell (*mycol.*) (1) A hyphal cell which differentiates (branches) to give rise to a CONIDIOPHORE. (2) A foot-shaped terminal cell in a multiseptate macroconidium of *Fusarium* spp.

Foraminiferida An order of free-living, testate, amoeboid, marine protozoans within the superclass SARCODINA; most species are benthic, but some (e.g. *Globigerina bulloides*) are pelagic (see ZOOPLANKTON). Foraminiferans are omnivorous—feeding e.g. on small protozoans and microalgae. Typically, the organisms are multinucleate, but uninucleate forms or stages occur. Some species form a single-chambered test, but the tests of most species are multiloculate (i.e. multichambered). In multiloculate species, new chambers are sequentially added to the first-formed chamber (*proloculus*) as the organism grows; the succession of chambers may be linear (e.g. in *Nodosaria*), spiral (e.g. in *Elphidium*), or in some other arrangement. Each chamber communicates with the adjacent (distal and/or proximal) chamber *via* an opening (*foramen*—pl. *foramina*) so that all the chambers are in communication. The test is composed basically of secreted organic materials which may incorporate calcareous or other matter, or to which may become attached small grains of sand, diatom frustules, etc. The finely granular *reticulopodium* (see MOTILITY) extends from the distal opening (*aperture*) of the most recently formed chamber and may also emerge from small perforations in the test wall; in some cases a thin cytoplasmic layer may form external to the test. In size, foraminiferans range from less than 0.5 millimetres to 5 millimetres or more. At least some species exhibit an ALTERNATION OF GENERATIONS (unique among protozoans). In the life cycle of *Elphidium*, meiosis occurs in the mature, multiloculate, diploid organism (*agamont*) with the subsequent liberation of numbers of haploid cells (young *gamonts*)—each of which grows to form a mature haploid organism (*gamont*) with multiloculate test. The gamont eventually releases numerous biflagellate isogametes; after fertilization, each zygote (young *agamont*) develops to form an agamont. (The haploid or diploid condition of an organism may be reflected in the size or morphology of the test; thus, e.g. in *Elphidium crispum*, the proloculus in the agamont test is smaller than that in the gamont test.)

Fossil foraminiferans occur in oceanic deposits and in certain sedimentary rocks. Identification of such fossils in rock drillings assists in determining the nature of the substratum and is a valuable aid in oil prospecting.

forespore (see SPORES (bacterial))

form-genus, form-family etc. A *form*-genus or *form*-family etc. is a *non-phylogenetic* group of organisms classified together on the basis that all have in common one or more major characteristics (e.g. a particular morphological feature) which can be ascertained with relative ease. (see e.g. DEUTEROMYCOTINA, NOMENCLATURE, POLYPORUS, PORIA, SACCARDIAN SYSTEM)

forma specialis (f.sp.—often written f.; pl. *formae speciales*, f.spp.) (*plant pathol.*) A taxon which embraces those individuals in a given species which may be regarded as a subgroup—mainly or solely on the basis of physiological characteristics, e.g. pathogenicity for specific plant host(s). Example: *Puccinia graminis* f.sp. *tritici* (also written *Puccinia graminis tritici*, *P. graminis* f. *tritici*, or *P. graminis* var. *tritici*) refers to that sub-species (*forma specialis*) of *P. graminis* which is pathogenic for wheat and *some* other graminiferous plants. Just as a species may be divided into *formae speciales*, a *forma specialis* may be divided into a number of *physiologic*(*al*) *races*—each race being specifically pathogenic for a particular variety of the given host plant(s). A given race may be referred to by a number, letter etc.

formaldehyde (HCHO) (as an antimicrobial agent) Formaldehyde is a colourless, pungent-smelling gas* which is very soluble in water and in alcohol; it is an effective disinfectant either in the gaseous state or in solution. Gaseous formaldehyde is active against bacteria, fungi, spores, and many types of virus; a relative humidity in excess of 50% is required for optimum activity. Formaldehyde has been used as a *fumigant*—although it has poor powers of penetration; at (gas) concentrations of about 3 mg/litre formaldehyde tends to polymerize and form a white deposit on exposed surfaces. Organisms within dried organic matter (sputum, blood etc.) may be largely protected from the action of formaldehyde. A 40% aqueous solution of formaldehyde (*formalin*) is used (in various dilutions) for e.g. the FIXATION and preservation of biological specimens, for soil disinfection, and as a general disinfectant; alcoholic solutions of formaldehyde may be used for the sterilization of surgical instruments.

Formaldehyde behaves as an ALKYLATING AGENT—substituting proteins and nucleic acids with hydroxymethyl groups; it is more effective against single-stranded than double-stranded nucleic acids. The antimicrobial action of formaldehyde is to some extent reversible; thus cells may be revived if excess formaldehyde is removed and suitable nutrients are present. (Certain bacteria can metabolize small amounts of formaldehyde.) Since formaldehyde can inactivate certain types of virus without causing excessive loss of specific antigenicity, it has been used for the preparation of inactivated viral vaccines (see also TOXOID). *Formaldehyde may be prepared by heating tablets of the formaldehyde polymer *paraformaldehyde*—a white solid, $HO(CH_2O)_nH$. It may also be prepared by treating formalin (cautiously) with potassium permanganate—heat produced by the ensuing oxidation driving off the remaining formaldehyde. (see also GLUTARALDEHYDE and HEXAMETHYLENETETRAMINE)

formalin (see FORMALDEHYDE)

formate hydrogen lyase In certain bacteria: an enzyme system which splits formic acid into carbon dioxide and hydrogen (see e.g. MIXED ACID FERMENTATION, BUTANEDIOL FERMENTATION). The system includes a *formate dehydrogenase* (which eliminates CO_2 from formic acid) and a *hydrogenase* (which catalyses the transfer of electrons to hydrogen ions to form H_2). Electrons derived from the first reaction appear to be transferred to hydrogenase *via* one or more electron carriers; in *Escherichia coli* a low-potential *c*-type cytochrome (c_{552}) has been implicated in this role.

formol-ether method (for concentrating protozoan cysts) The sample (e.g. 1 or 2 grams of faeces) is emulsified in physiological saline and filtered through a double layer of wet gauze; the filtrate is centrifuged (1,500 r.p.m./3 minutes) and the supernatant discarded. More saline is added, the sample again centrifuged, and the supernatant discarded. To the sediment is added 10% formalin (10 ml) and diethyl ether (3 ml) and the whole is thoroughly mixed and centrifuged (1,500 r.p.m./2 minutes); cysts, if present, are found in the sediment. Compared with the FLOTATION method, the formol-ether method causes less distortion in the cysts—which, as a consequence, are more suited to microscopical examination and identification. The method is unsuitable for the concentration of coccidian cysts if the latter are subsequently to be sporulated.

Forshay test (*immunol.*) A skin test, similar in form to the TUBERCULIN TEST, used to detect DELAYED HYPERSENSITIVITY in respect of infection by *Francisella tularensis*. (see also TULARAEMIA) The antigen, injected intradermally, consists of a suspension, or protein-extract, of killed *F. tularensis*.

Forssman antigen (see HETEROPHILE ANTIGENS)

forward mutation A MUTATION which results in the formation of a phenotype differing from that of the WILD TYPE STAIN. (cf. BACK MUTATION)

foulbrood The name of each of two distinct diseases of the honey bee (*Apis mellifera*). (a) *American foulbrood* is caused by *Bacillus larvae*; ingestion of this organism by larvae is followed by fatal infection and subsequent putrefaction of the larvae. (b) *European foulbrood* is caused by *Streptococcus pluton*; secondary invaders include *Streptococcus faecalis*.

foveate Pitted.

fowl cholera An acute, infectious, septicaemic disease of domestic poultry characterized by fever and severe, bloody diarrhoea; haemorrhages occur in the internal organs and mucous membranes, and death usually follows within days of the onset of symptoms. The causal agent is *Pasteurella multocida*.

fowl pest (1) FOWL PLAGUE. (2) *Newcastle disease*. An acute, infectious disease of fowl; the causal agent, Newcastle disease virus (NDV), is a paramyxovirus. Infection occurs by droplet inhalation or by the ingestion of contaminated food or water. The incubation period is commonly 4–12 days. Symptoms include laboured breathing, a nasal discharge, drowsiness and paralysis; mortality rates may be very high, but the disease also occurs in a mild form. *Laboratory diagnosis* may include culture and characterization of NDV and/or serological tests. In man, NDV causes a form of conjunctivitis.

fowl plague (see also FOWL PEST) A disease of fowl in which the causal agent, an orthomyxovirus, invades *via* the respiratory tract. Symptoms may include those of a respiratory tract infection and/or encephalitis. Fowl plague may be attended by high mortality rates; it is a NOTIFIABLE DISEASE in the United Kingdom. Vaccines against fowl plague are available.

foxfire BIOLUMINESCENCE in decaying wood.

Fragilaria A genus of pennate diatoms of the BACILLARIOPHYTA.

frambesia (see YAWS)

frameshift mutation (see PHASE-SHIFT MUTATION)

framycetin A broad-spectrum AMINOGLYCOSIDE ANTIBIOTIC produced by a strain of *Streptomyces lavendulae*; it may be identical to one of the components of NEOMYCIN.

Francisella A genus of gram-negative, aerobic, chemoorganotrophic bacteria of uncertain taxonomic affinity; they have been isolated from e.g. wild rodents, water, and clinical specimens. The cells are non-motile coccobacilli, or pleomorphic bacilli, about 0.2 x 0.2–0.6 microns; the smaller forms pass through Berkefeld-type filters. The cells generally stain poorly; bipolar staining may be exhibited. Growth occurs only on enriched media; optimum temperature: 37 °C. The cells are soluble in sodium ricinoleate solution.

F. tularensis (formerly *Pasteurella tularensis*). The causal agent of TULARAEMIA. *F. tularensis* is capable of anaerobic growth (on suitably buffered media) but grows best aerobically; the organism requires cysteine or cystine, and grows well on a correctly-formulated agar medium containing cysteine, glucose, ferrous iron and blood. It may be identified serologically by agglutination with specific antiserum, or by IMMUNOFLUORESCENCE. *F. tularensis* is generally sensitive to streptomycin and tetracyclines.

Frankiaceae A family of the ACTINOMYCETALES; species of the sole genus, *Frankia*, occur as soil saprophytes and as endosymbionts in nitrogen-fixing ROOT NODULES in a range of non-leguminous plants.

free cell formation (in *ascomycetes*) (see ASCUS)

free-living Living independently of other organisms, i.e. not a parasite, symbiont or commensal.

freeze-drying (lyophilization) A process used e.g. for the long-term preservation of cultures of certain types of microorganism and certain other types of material (including foodstuffs). When a bacterial culture is to be freeze-dried the cells are suspended in an appropriate medium and rapidly frozen—e.g. by means of a freezing mixture consisting of solid carbon dioxide ('dry ice') and ethanol; suitable media (e.g. solutions of glucose, gelatin, milk, serum, or other protein-rich substances) permit amorphous (non-crystalline) solidification to occur during the freezing process. The cells are dehydrated under vacuum—water passing directly from the solid to the vapour phase. Freeze-dried materials are susceptible to atmospheric oxidation and are therefore stored *in vacuo* or under an inert gas (e.g. nitrogen); when required for use such materials are *reconstituted* with distilled water.

freeze-etching A technique used to examine the topography of a surface exposed by fracturing or cutting a deep-frozen cell. (The fact that viable organisms can be recovered after the initial freezing process suggests that minimal damage is sustained by cells undergoing such treatment; thus, artefacts are probably reduced to a minimum.) Essentially, freeze-etching involves: (a) Rapid freezing of the specimen to about −150 °C. (b) Fracturing or sectioning (slicing) the specimen; fracturing (as opposed to sectioning) has the advantage that part of the cleavage may follow the boundaries of internal structures. (c) Etching; this involves sublimation of water (and other volatiles) from the frozen surface by raising the temperature of the specimen to about −100 °C in a vacuum. By this means, non-volatile materials and structures are thrown into relief. The vapour of carbon and/or a heavy metal (often platinum) is then allowed to condense on the deep-frozen surface; provided that this process is carried out within seconds the specimen undergoes minimal damage, and a thin carbon/metal replica of the surface topography is formed. The replica is cleaned, rinsed in distilled water, mounted, and examined by electron microscopy.

freezing (as a method of preservation) (*synonym:* cryopreservation) Sub-zero (centigrade) temperatures are routinely used for the PRESERVATION of certain foodstuffs (see FOOD, PRESERVATION OF) and for the maintenance of populations of viable microorganisms. The long-term preservation of viable microbial populations by freezing or FREEZE-DRYING is generally preferred, where practicable, to the method of repeated subculture. (see also STABILATE) Freezing may be used for the preservation of many species of bacteria, fungi and protozoa, a range of viruses, and for tissue cultures.

In the absence of special precautions, the subjection of cellular organisms (e.g. bacteria) to sub-zero temperatures commonly results in a decrease in the viable count. (Repeated cycles of freezing and thawing may in fact be used specifically to decimate a cellular population.) Cellular damage is commonly attributed to one (or both) of two main causes. Firstly, the formation of intracellular ice crystals which are believed to damage the cell mechanically. Secondly, the development of intracellular microenvironments consisting of concentrated solutions of salts. These solutions develop as the temperature continues to fall below 0 °C—just as the concentration of a pure salt solution increases as the *eutectic* temperature (about −20 °C for NaCl) is approached. The continued presence of these solutions at temperatures found in certain types of commercial freezing apparatus (e.g. −10 °C) precludes the use of such apparatus for effective microbial preservation. These concentrated solutions are believed to bring about irreversible damage to certain cellular components—e.g. enzymes.

In order to avoid, as far as possible, the risk to cell viability, *rapid freezing* has been used with the object of encouraging non-crystalline (amorphous) solidification, i.e. *vitrification*. In this method the sample reaches its final temperature within several seconds. Vitreous solidification may be further encouraged by the addition of certain agents (*cryoprotectants*) to the liquid in which the cells/organisms are suspended; these agents include glycerol, ethylene glycol and DMSO (DIMETHYL SULPHOXIDE). Glycerol and ethylene glycol appear to be more widely used for bacterial preservation while DMSO has been extensively used for the cryopreservation of protozoa and the cells of tissue cultures.

Some workers prefer to use a *slow freezing* technique when using cryoprotectants which dehydrate cells, or which permeate cells, prior to freezing. For example, in the cryopreservation of tissue culture cells, the cells may be left initially to equilibrate for 1 hour at 4 °C in a freezing medium consisting of growth medium, serum, and 7.5–15% (v/v) DMSO; this permits penetration of the DMSO. The cells, in sealed glass ampoules, are then transferred to the 'freezing chamber' of a liquid nitrogen apparatus and, after 20–30 minutes, are placed in liquid nitrogen (−196 °C).

The thawing of deep-frozen cells or organisms is carried out rapidly; ampoules in liquid nitrogen may be allowed to equilibrate in solid CO_2 (−78 °C) before being transferred to a 37 °C water bath.

(When working with a deep-freeze apparatus the use of a plastic face-shield and rubber gloves are wise precautions.)

Freund's adjuvant (1) *Freund's complete adjuvant*. A water-in-oil emulsion which contains dead mycobacteria dispersed in the oily phase; mineral oil is commonly used. (2) *Freund's incomplete adjuvant* differs from the above by the absence of mycobacteria. In both types of ADJUVANT the antigen is dispersed in the water phase.

Friedländer's bacillus A species of *Klebsiella* isolated by Friedländer (1883) from the lungs of *post mortem* cases of pneumonia.

front focal plane (of a convex lens) The plane, perpendicular to the optical axis, which passes through the focal point of the lens on the side nearest the source of light. (cf. BACK FOCAL PLANE)

Frontonia A genus of protozoans of the PENICULINA.

fructans Polymers of fructose. (a) *Inulin-type fructans*. These linear polymers of β-(2 → 1)-linked fructofuranose residues are synthesized by certain plants as storage polysaccharides, e.g. dahlia tubers contain large amounts of inulin. (b) *Levan-type fructans*. These are linear polymers of fructofuranose residues in which the β-(2 → 6)-linkage predominates. Levans are widespread as storage polysaccharides in the plant kingdom, especially in grasses. Certain bacteria (e.g. some species of *Bacillus*, *Streptococcus*) form high molecular weight levans as extracellular slime. Bacterial levan, structurally similar to the grass levans, is known as *phlein-like* levan (after *Phleum*, timothy grass). The chains of both types of fructan terminate in a non-reducing sucrose residue, and each is apparently synthesized by the addition of fructose residues to a sucrose molecule. Fructans can be hydrolysed by a wide range of microorganisms.

fructification (*mycol.*) (see FRUITING BODY)

fructosan Synonym of FRUCTAN.

fructose diphosphate aldolase (see EMBDEN–MEYERHOF–PARNAS PATHWAY)

fruit (*mycol.*) (*verb*) To produce a FRUITING BODY.

fruiting body (1) (*mycol.*) (fructification; fruit body) A specialized fungal structure which bears sexually- or asexually-derived spores; examples include the ACERVULUS, the various types of ASCOCARP, and the CONIDIOPHORE, COREMIUM, and PYCNIDIUM. The fruiting bodies of some basidiomycetes (e.g. species of *Heterobasidion*, *Poria*) are perennial. (2) (*bacteriol.*) (see MYXOBACTERALES)

frustule The silicaceous CELL WALL of a diatom. (see also EPITHECA and HYPOTHECA)

fruticose (of a lichen) Refers to a lichen whose thallus is thread-like (terete) or strap-like (more or less flattened) and either erect and shrubby or pendulous (hanging from branches of trees etc.); the thallus may be attached to the substratum e.g. by a holdfast or may be unattached. The mechanical strength of the thallus is derived either from a dense axial strand (e.g. in *Usnea* species) or from hyphae of the cortex; thalli of the latter type may or may not be hollow. Fruticose lichens include e.g. members of the genera ALECTORIA, EVERNIA, RAMALINA, THAMNOLIA, and USNEA. (cf. CRUSTOSE (2) and FOLIOSE)

FTA-ABS (fluorescent treponemal antibody-absorbed serum) A sensitive serological diagnostic test for SYPHILIS. Initially the patient's serum is absorbed to remove non-specific antibodies*—which could otherwise lead to false-positive results. The absorbed serum is applied to a slide on which there is a film of killed, fixed cells of *Treponema pallidum*. The slide is then briefly incubated to enable any specific antibodies (in the serum) to combine with the surface antigens of *T. pallidum*. The slide is rinsed free of serum and briefly re-incubated with the appropriate area flooded with a fluorescent *conjugate* (anti-human globulin conjugated with a fluorescent dye—in buffered aqueous solution). The slide is rinsed free of uncombined conjugate and examined by fluorescence MICROSCOPY; any serum antibodies which have combined with *T. pallidum* combine also with the fluorescent conjugate—their presence being indicated by the appearance of fluorescent treponemes. Adequate controls must be used. *See REITER ANTIGENS.

fuberidazole 2-(2′-furyl)-benzimidazole. An agricultural *systemic* antifungal agent. It is particularly effective for the prevention of *Fusarium* infections of seeds. (see also BENZIMIDAZOLE DERIVATIVES (as antifungal agents))

fuchsin (basic fuchsin) A reddish-purple dye which commonly consists of a mixture of basic dyes of the triaminotriphenylmethane group, the *rosanilins*; the main constituent of the commercial product appears to be *pararosanilin*, while rosanilin and magenta-2 may also be present in varying proportions. In microbiology fuchsin is often used in the form of CARBOLFUCHSIN. *Acid-fuchsin* is derived from basic fuchsin by sulphonation; it is used e.g. in ANDRADE'S INDICATOR. (see also TRIPHENYLMETHANE DYES)

fucidin Sodium fusidate; see FUSIDIC ACID.

fucoidin (fucoidan) A complex sulphated polysaccharide which occurs in members of the Phaeophyta (brown algae); it contains a high proportion of L-fucose (= 6-deoxy-L-galactose) residues. Fucoidin occurs in the algal cell wall and in the intercellular matrix. Algae which are normally totally submerged contain less fucoidin than do those which undergo periodic exposure to the atmosphere; the hygroscopic nature of fucoidin may serve to prevent desiccation of the alga on such exposure.

fucoxanthin (see CAROTENOIDS)

Fucus A genus of algae of the division PHAEOPHYTA; species occur mainly in marine intertidal zones, although some are found in brackish waters. The vegetative organism has a leathery, flattened *blade* which is generally dichotomously branched and has a pronounced central midrib; the blade varies in width and may have entire or serrated edges—according to species. The midrib of the blade is continuous with a *stipe*, and the organism is anchored to the substratum by means of a branched *holdfast*. In some species (e.g. *F. vesiculosus*: bladderwrack) gas-filled vesicles (PNEUMATOCYSTS) give buoyancy to the thallus. Sexual reproduction is oogamous.

Fuligo A genus of the ACELLULAR SLIME MOULDS.

fumagillin An ANTIBIOTIC produced by *Aspergillus fumigatus*; it has a complex structure which includes two epoxide rings and is an ester of decatetraenedioic acid. Fumagillin has no action against bacteria or fungi but is a potent amoebicide; it has been used in the treatment of amoebiasis. It also inhibits the development of certain bacteriophages, and has shown some antiviral activity in tissue culture.

fumigation The exposure of enclosed spaces to the action of gaseous or vapour-phase DISINFECTANTS or STERILANTS. Fumigation may be used to disinfect e.g. rooms vacated by persons suffering from infectious diseases, or glasshouses which have contained infected plants. Fumigating agents (*fumigants*) include ETHYLENE OXIDE, FORMALDEHYDE, SULPHUR DIOXIDE, DAZOMET.

functional immunity (see PROTECTIVE IMMUNITY)

fungi (*sing.* fungus) A group of diverse and widespread unicellular and multicellular (including coenocytic) *eucaryotic* microorganisms; the fungi do not contain chlorophyll (cf. ALGAE) and most species form a rigid polysaccharide cell wall (cf. PROTOZOA). (The distinction between fungi and other eucaryotic microorganisms is not absolute owing to the existence of intermediate forms—e.g. achlorophyllous strains of algae.) Taxonomically, the fungi are divided into two divisions: EUMYCOTA (the 'true fungi') and Myxomycota (see SLIME MOULDS).

General characteristics of fungi. In most fungi the vegetative (somatic) form of the organism (the *thallus*) consists of hyphae—see HYPHA. (see also CELL WALL (fungal); CELL MEMBRANE (fungal); MYCELIUM) In certain fungi (see e.g. YEASTS) the vegetative thallus is predominantly unicellular—though in certain of these species the unicellular form can give rise to a pseudomycelium or a true mycelium. Some fungi (e.g. certain species of *Mucor*) which normally form a mycelial thallus can, under appropriate environmental conditions, give rise to unicellular forms. (see also DIMORPHIC FUNGI) The majority of fungi are non-motile throughout their life cycles; however, motile (flagellated) reproductive/disseminative forms are produced by many of the LOWER FUNGI. (Phylogenetically, motility is regarded as a 'primitive' feature among the fungi.)

Occurrence. According to species, fungi occur in soil, in water, on plant debris, and as symbionts, parasites, and pathogens of animals (including man), plants, and other microorganisms. Saprophytic species are important e.g. in the decomposition of plant litter (and thus in the cycles of matter). For examples of symbiotic associations involving fungi see MYCORRHIZA, LICHENS, MYCETOCYTES. (see also AMBROSIA FUNGI; FUNGUS GARDENS) Human and animal diseases caused by fungi include e.g. ASPERGILLOSIS; CANDIDIASIS; COCCIDIOIDOMYCOSIS; CRYPTOCOCCOSIS; HISTOPLASMOSIS; PARACOCCIDIOIDOMYCOSIS; PITYRIASIS VERSICOLOR; RINGWORM; WHITE PIEDRA. (see also FOOD POISONING (fungal); PLANT PATHOGENS AND DISEASES; TIMBER, DISEASES OF; TURKEY X DISEASE)

Nutrition, metabolism, conditions of growth.

All known fungi are chemoheterotrophs, and nutrients enter the fungal cell/hypha in solution. Some fungi (e.g. the common moulds *Aspergillus* and *Penicillium*) can use any of a wide range of compounds as sources of carbon and energy—while others (e.g. certain obligate pathogens) have highly specific nutritional requirements; a number of fungi require specific growth factors—e.g. particular VITAMINS (see e.g. BIOTIN, THIAMINE). Carbon sources used by fungi include e.g. certain ALIPHATIC HYDROCARBONS, CELLULOSE, LIGNINS, PECTINS, and a wide range of soluble sugars; fungi which utilize polymers (e.g. the wood-rotting fungi) produce extracellular enzymes which degrade such polymers to soluble products. In most fungi metabolism is respiratory (see RESPIRATION) while other species (e.g. many yeasts) are facultatively fermentative (see FERMENTATION). A particular species grows only within a certain range of pH, temperature, osmolarity, and WATER ACTIVITY. Although a number of fungi are facultative anaerobes no *obligately* anaerobic species are known. (see also BIOLUMINESCENCE)

Growth. (see GROWTH (fungal); see also IDIOPHASE; TROPHOPHASE; TROPISM) In many fungi nuclear division has been found to differ in detail from the classical pattern—see MITOSIS. For some fungi *light* is an important environmental factor which influences the development of reproductive structures, spores etc.—see PHOTOINDUCTION AND PHOTOINHIBITION.

Reproduction. A variety of modes of reproduction occur among the fungi; most species are EUCARPIC—holocarpic species occuring chiefly among the lower fungi. In many fungi reproduction involves the formation of specialized structures—see FRUITING BODY. *Asexual reproduction* occurs in most species of fungi though it appears not to occur e.g. in members of the class Laboulbeniomycetes or in species of the order Pezizales; according to species asexual reproduction may involve e.g. the formation of CONIDIA or SPORANGIOSPORES and/or certain other type(s) of spore/propagule (see e.g. BLASTOSPORE, CHLAMYDOSPORES, GEMMA, OIDIUM). *Sexual reproduction* occurs in a wide range of fungi—though not e.g. in many members of the Deuteromycotina. Sexual reproduction may involve e.g. isogamy (e.g. in certain members of the Chytridiales), anisogamy (e.g. in *Allomyces* and in certain other members of the Blastocladiales), oogamy (e.g. in members of the order Monoblepharidales and the class Oomycetes), or the fusion of isogametangia (e.g. in the majority of zygomycetes). In many basidiomycetes, and in some ascomycetes, the fruiting bodies which bear the sexually-derived spores are of macroscopic dimensions—e.g. the mushrooms, puffballs, morels, cup fungi etc. (see also COMPATIBILITY (in fungal sexuality) and HORMONAL CONTROL (of sexual processes in fungi))

Genetic aspects. Changes in the genetic constitution of a fungus may occur e.g. during sexual processes (which involve MEIOSIS and KARYOGAMY) or as a result of MUTATION(s) (see also DIKARYOTIZATION and PARASEXUALITY IN FUNGI).

Methods of studying fungi. Unicellular fungi (e.g. yeasts) are commonly manipulated by means of a LOOP or a STRAIGHT WIRE (see also ASEPTIC TECHNIQUE and MICROMANIPULATION); macroscopic fungal tissues may be handled with a scalpel or a hand microtome, mounted needles, forceps etc. Certain features of fungal cells/hyphae can be determined by light MICROSCOPY—specimens often being prepared as a WET MOUNT. (see also FIXATION and STAINING) Media used for the CULTURE of fungi include e.g. SABOURAUD'S DEXTROSE AGAR, CORNMEAL AGAR, and NUTRIENT AGAR.

Identification of fungi. The identification of fungi is based largely on morphological and/or structural features. Some fungi are sufficiently distinctive in appearance (and/or habitat) to permit identification to genus (or even species) level; however, in most cases identification requires the use of microscopical and (less frequently) cultural techniques. Features used in identification include e.g.: the morphology of the vegetative thallus (e.g. unicellular or hyphal); the nature of the hyphae (e.g. septate or aseptate, branched or unbranched etc.); the nature of any asexual reproductive structures present (e.g. conidia, sporangiospores, chlamydospores, acervuli, pycnidia etc.); the nature of any sexual reproductive structures present (e.g. asci or basidia, cleistothecia, perithecia etc.); habitat.

Destruction/inactivation and parasitization of fungi. Fungi may be killed/inactivated e.g. by STERILIZATION techniques, by ULTRASONICATION, and by a range of FUNGICIDES AND FUNGISTATS. Some fungi are parasitized by certain viruses and by certain other fungi; thus, e.g. *Boletus parasiticus* is parasitic on *Scleroderma*, and *Piptocephalis* spp parasitize a range of fungi—chiefly other members of the class Zygomycetes. (see also MUSHROOM, DISEASES OF)

Importance to man. Fungi are important agents in the spoilage of a wide range of materials—see e.g. FOOD, MICROBIAL SPOILAGE OF; GLASS (microbial spoilage of);

PAINTS (microbial spoilage of); PAPER (microbial spoilage of); RUBBER (microbial spoilage of); TIMBER, DISEASES OF. Fungi are exploited by man as food (see e.g. MICROORGANISMS AS FOOD), in various fermentation processes (see e.g. BAKING, BREWING, CIDER-MAKING, WINE-MAKING), and in the manufacture of ANTIBIOTICS, certain kinds of CHEESE, certain organic acids (e.g. citric acid) etc. (see also MICROORGANISMS, EXPLOITATION OF)

Fungi Imperfecti (deuteromycetes) Fungi of the DEUTEROMYCOTINA.

fungicides and fungistats Chemical agents which (respectively) kill, or inhibit the growth and reproduction of, fungi. Some examples are given below.

Antifungals of medical and general interest. BISPHENOLS, CYCLOHEXIMIDE (*Actidione*), GRISEOFULVIN, 8-HYDROXYQUINOLINE, POLYENE ANTIBIOTICS, POLYOXINS, SALICYLIC ACID AND DERIVATIVES, TOLNAFTATE, TRIPHENYLMETHANE DYES (see also *nifuroxime* under NITROFURANS and *nitrophenols* under PHENOLS)

Antifungals produced by higher plants. See: PHYTOALEXINS and PHYTONCIDES.

Antifungals of agricultural interest. (a) SYSTEMIC ANTIFUNGAL AGENTS: BENZIMIDAZOLE DERIVATIVES (e.g. BENOMYL), ETHIRIMOL, DIMETHIRIMOL. (b) BORDEAUX MIXTURE, BURGUNDY MIXTURE, CAPTAN, CHESHUNT COMPOUND, CHLORONITROBENZENES, DINITROPHENOLS (e.g. DINOCAP), DITHIOCARBAMATE DERIVATIVES, DODINE, FOLPET, GLYODIN, QUINONES, SULPHUR. (see also *calomel*, *copper*, *nickel*, *tin* and *zinc* under HEAVY METALS; see, additionally, ERADICANT and PROTECTANT)

fungicidin (nystatin) (see POLYENE ANTIBIOTICS)

fungicole An organism which grows on fungi.

fungimycin (perimycin) (see POLYENE ANTIBIOTICS)

fungus (see FUNGI)

fungus gardens Fungal growth which is cultivated, on plant material, by parasol ants (sub-family Myrmicinae, tribe Attini); in some (unknown) way the fungi are induced to form swellings (*bromatia*) which are used as food by the ants. Fungus cultivation (for food) is also practiced by certain termites.

funicle (funiculus) (*mycol.*) (see CYATHUS)

funicular cord (*mycol.*) (see CYATHUS)

funoran A mucilaginous, sulphated galactan produced by the alga *Gloiopeltis furcata*. Chemically, funoran is similar to AGAR in that it consists mainly of D-galactose and 3,6-anhydro-L-galactose residues; L-galactose residues are also present. In Japan, funoran is used as an adhesive and sizing agent.

furacin (nitrofurazone) (see NITROFURANS)

furadantin (nitrofurantoin) (see NITROFURANS)

furazolidone (furoxone) (see NITROFURANS)

furcellaran (furcelleran; Danish agar) A complex, sulphated D-galactan produced e.g. by strains of the red alga *Furcellaria*. It has been used as a gelling agent in the food industry.

Furcellaria A genus of red algae (RHODOPHYTA).

furfuraceous Scurfy; covered with small scales.

furoxone (furazolidone) (see NITROFURANS)

fusaric acid 2-Carboxyl-5-*n*-butyl-pyridine, a phytotoxin produced by *Fusarium* spp. Fusaric acid alters the permeability of plant cells, and is thus (at least partly) responsible for the wilting of plants infected with *Fusarium* spp.

Fusarium A genus of fungi within the order MONILIALES; species include a number of plant pathogens—e.g. *F.oxysporum* (the causal agent of e.g. Panama wilt disease of banana, wilt of flax, and a yellows disease of celery), and *F. coeruleum* (which causes a dry rot in stored potato tubers). The vegetative form of the organisms is a septate mycelium. According to species, reproduction may involve the formation of microconidia, chlamydospores, and/or macroconidia; typically, macroconidia are produced from single or grouped phialids on conidiophores which are clustered on a sporodochium. The hyaline, typically crescent-shaped macroconidium may contain one, two, or many transverse septa, or may be aseptate—according to species and to the state of maturity of the macroconidium; a characteristic feature of the macroconidium of *Fusarium* is the *foot cell*—a foot-shaped cell, forming one pole of the spore, in which a 'heel' is typically present. (The opposite pole of the spore tapers uniformly.) Certain species of *Fusarium* are known to have sexual stages in ascomycete genera (e.g. *Nectria*, *Gibberella*).

fusel oil (see SPIRITS)

fusidic acid An ANTIBIOTIC produced by *Fusidium coccineum*; fusidic acid is a steroid structurally related to *cephalosporin P*—with which it exhibits cross-resistance. The sodium salt is marketed as *fucidin*. Fusidic acid is active mainly against gram-positive bacteria—particularly *Staphylococcus* spp; it is inactive against coliforms and fungi. Fusidic acid inhibits PROTEIN SYNTHESIS in cell-free systems of both procaryotes and eucaryotes by acting on *factor G*; it apparently prevents the dissociation of factor G and GDP from the ribosome following translocation. Some instances of plasmid-mediated resistance have been reported.

Fusidium A genus of fungi within the order MONILIALES. The vegetative form of the organisms is a septate mycelium; elongated, hyaline, non-septate conidia are formed from conidiophores which closely resemble the vegetative hyphae. The sexual stage of at least one species is believed to be a discomycete. (see also FUSIDIC ACID)

fusiform (*adjective*) Spindle-shaped—i.e. tapered at both ends.

Fusiformis fusiformis (see FUSOBACTERIUM)

Fusobacterium A genus of gram-negative, anaerobic, asporogenous, non-motile or peritrichously flagellate bacilli of the family BACTEROIDACEAE; species occur in the mouth, intestinal and respiratory tracts, and elsewhere, in man and other animals. Some species appear to be opportunist pathogens—often being isolated from purulent and necrotic lesions.

Details of culture and cell morphology are similar to those of BACTEROIDES; the cells may or may not be fusiform, and may or may not exhibit pleomorphism. Butyric acid is among the main products of sugar or peptone fermentation; isobutyric and isovaleric acids are not formed.

F. nucleatum (formerly *Fusiformis fusiformis*, *Sphaerophorus fusiformis*, *Fusobacterium polymorphum*). The type species; the cells are non-motile fusiform rods or filaments. The organisms produce indole from tryptophan and form propionate from threonine; negligible quantities of gas are produced from glucose. At least some strains form large amounts of H_2S.

F. necrophorum (formerly *Sphaerophorus necrophorus*). The (non-motile) cells are typically round-ended rods or filaments; biochemical reactions are similar to those of *F. nucleatum* except e.g. that (a) copious amounts of gas are formed from glucose, and (b) propionate is formed from lactate as well as from threonine.

G

Gaeumannomyces A genus of fungi of the order SPHAERIALES, class PYRENOMYCETES. *G. graminis* (formerly known as *Ophiobolus graminis*) causes take-all disease of certain cereals (see CEREALS, DISEASES OF); the organism forms a thick-walled (true) perithecium within which the asci occur free.

galactans Polymers of galactose, e.g. GALACTOCAROLOSE. (see also AGAR, CARRAGEENAN, FUNORAN, FURCELLARAN)

galactocarolose An extracellular product of *Penicillium charlesii* consisting of (1 → 5)-linked D-galactofuranosyl residues.

Gallionella A genus of gram-negative, aerobic bacteria of uncertain taxonomic affinity; the organisms occur in chalybeate waters and in certain soils. Each reniform, bacilliform or coccoid cell gives rise to a long 'stalk' which contains ferric hydroxide; post-division cells are motile by polar or subpolar flagella. Metabolism appears to be chemolithotrophic; energy is thought to be obtained by the oxidation of ferrous compounds. Type species: *Gallionella ferruginea*.

galls (cecidia) (*plant pathol.*) Abnormal plant structures formed in response to parasitic attack by certain insects or microorganisms. Galls may develop either by localized cell proliferation or by increase in cell size. In at least some cases, gall-formation may involve a plant hormone imbalance—see AUXINS, CYTOKININS. Gall-inducing organisms often show considerable host specificity; the host plant is rarely killed.

Fungus-induced galls (*mycocecidia*). (a) *Synchytrium endobioticum* causes *wart disease* of the POTATO—in which warty outgrowths form on the tubers. (b) WITCHES' BROOM due to *Taphrina* spp. (c) *Club root* of CRUCIFERS. (d) *Cedar apples*—see RUSTS (1). Galls are also incited by *Albugo candida* on certain crucifers, and by *Ustilago maydis* on maize (*Zea mays*).

Bacterium-induced galls include CROWN GALL, some instances of FASCIATION (*Corynebacterium fascians*), and 'olive knot'—a gall of the olive tree due to *Pseudomonas savastanoi*. (see also ROOT NODULES)

Virus-induced galls are produced e.g. by the SUGARCANE FIJI DISEASE VIRUS and the WOUND TUMOUR VIRUS; ENATIONS are characteristic symptoms of a number of plant diseases of viral origin.

galvanotaxis The locomotion of an organism as a result of its presence within an electrical potential gradient. If an organism moves with its normal mode of locomotion (i.e. anterior end directed forwards) towards the negative pole (cathode) it is said to be negatively galvanotactic; *Paramecium* is negatively galvanotactic—though, with a higher potential difference, this organism moves towards the positive pole (anode) with its (normally) posterior end pointing in the direction of motion. Some protozoans are *positively* galvanotactic—i.e. they move towards the positive pole with their anterior end pointing in the direction of motion.

gametangial contact (*mycol.*) A process in which one or more male nuclei pass from a male gametangium to a female gametangium while the protoplasts of the two gametangia are in communication; according to species, the nucleus or nuclei are transferred *via* a trichogyne or *via* a pore or fertilization tube which forms between the gametangia. In some species karyogamy occurs immediately after the entry of the male nucleus/nuclei; karyogamy is delayed in those ascomycetes in which gametangial contact occurs.

gametangial copulation (*mycol.*) (1) The occurrence of plasmogamy and karyogamy in a process which may involve either the complete fusion of two gametangia ('gametangial fusion') or the passage of the *entire* contents (i.e. the protoplast) from one gametangium to the other. In gametangial fusion the adjacent walls of contiguous gametangia break down, and a single diploid cell results from the fusion of the two protoplasts; gametangial fusion occurs e.g. in the zygomycetes. In protoplast transfer the two gametangia come into contact and the protoplast passes through a pore which develops in the adjacent walls of the juxtaposed gametangia; this process is characteristic of certain chytrids. (2) The meaning of the expression gametangial copulation is sometimes extended to include GAMETANGIAL CONTACT.

gametangial fusion (see GAMETANGIAL COPULATION)

gametangium A structure which gives rise to gametes or which, in its entirety, functions as a gamete.

gametes (1) Haploid reproductive cells or nuclei; male and female gametes are commonly differentiated in size, structure etc. During fertilization the fusion of a male and a female gamete produces a *zygote*. (2) Undifferentiated somatic cells which function as reproductive cells. (3) Diploid cells which develop parthenogenically in APOMIXIS.

gametic meiosis MEIOSIS preceding gamete formation.

gametocyte Any cell which, on undergoing meiosis, gives rise to gametes. (cf. GAMONT)

gametophyte (see ALTERNATION OF GENERATIONS)

gametothallus (*mycol.*) A haploid, gamete-forming thallus (cf. SPOROTHALLUS). In e.g. *Allomyces* spp the gametothallus forms male and female gametes which fuse in pairs and give rise to sporothalli. (see also ALTERNATION OF GENERATIONS)

***gamma* globulins** A subclass of the serum globulins distinguished by the lowest *electrophoretic mobility* towards the anode under neutral or alkaline conditions. (see ELECTROPHORESIS) Many of the IMMUNOGLOBULINS are found within this fraction. *NB.* When serum is subjected to electrophoresis in agar gels the *gamma* globulins may move towards the *cathode* (i.e. the *negative* terminal) owing to the effects of *electroendosmosis*. (see IMMUNOELECTROPHORESIS)

***gamma* rays** Electromagnetic radiation of short wavelength and great penetrating power. *Gamma* rays are used e.g. for STERILIZATION.

gamone (see HORMONAL CONTROL (of sexual processes in fungi))

gamont (1) Synonym of GAMETOCYTE. (2) A mature individual in the haploid phase of an ALTERNATION OF GENERATIONS. (see also FORAMINIFERIDA)

Ganodermataceae A family of fungi of the order APHYLLOPHORALES. Constituent species form highly characteristic spores—in each of which a number of small, needle-like processes emanate from the inner, brown layer of the spore wall and penetrate deeply into the outer hyaline layer of the spore wall; the ovoid spores are typically flattened at one pole. The annual or perennial fruiting body typically consists of a flattened, corky, semicircular pileus which may be non-stipitate or may bear a short lateral stipe; the context of the pileus is trimitic, and the underside of the pileus bears a porous hymenophore. The organisms occur on felled timber and one species, *Ganoderma applanatum*, causes heart rots in a number of different species of tree—gaining entry *via* wounds.

gas bladders (see PNEUMATOCYSTS)

gas gangrene (myonecrosis) A condition which may develop (in man and other animals) when certain species of bacteria are present in anaerobic wounds or necrotic tissues. These organisms, mainly *Clostridium perfringens* type *A*, *C. septicum* and *C. novyi*, produce a variety of extracellular enzymes (see CLOSTRIDIUM) which promote the destruction of muscle and connective tissue; the incubation period may be 6 hours to 6 days. Pockets of gas (hydrogen, carbon dioxide) form within tissues surrounding the lesion, and the pressure thus exerted may disrupt blood supplies to nearby tissues—thus promoting the spread of necrosis. The tissues become oedematous and assume an unhealthy-looking discoloration. Diagnosis is mainly clinical, but direct microscopical examination of wound material may be useful. Remedial measures often involve immediate surgery; antitoxins are administered, and suitable antibiotics are given locally and parenterally. (see also BLACKLEG)

gas vacuoles Membrane-limited, gas-filled vacuoles which occur, commonly in groups, in the cells of a number of blue-green algae (including species of *Anabaena*, *Gloeotrichia*, *Microcystis*, *Oscillatoria*, and *Trichodesmium*) and in certain bacteria (including species of *Halobacterium*, *Pelodictyon*, and *Rhodopseudomonas*). The vacuoles of blue-green algae are cylindrical with conical ends and are approximately 300 × 70 nm; the vacuole membrane (some 3 nm thick) is rigid and appears to consist mainly (or exclusively) of protein. Vacuoles commonly occur in groups of a hundred or more. (Some authors describe each vacuole as a *gas vesicle**—reserving the name *gas vacuole* for a group of contiguous vesicles.) The vacuolar membrane appears to be freely permeable to gases; the gases in the vacuoles are thus (presumably) in equilibrium with the gases dissolved in the cell's cytoplasm. On a sudden increase in external pressure (of several atmospheres) gas vacuoles collapse; they are apparently unaffected by reduced pressure. Vacuolated species lose buoyancy when their vacuoles are destroyed by pressure. *also used as a synonym of PNEUMATOCYST.

gas vesicles Synoym of (1) PNEUMATOCYSTS, and of (2) GAS VACUOLES.

Gasteromycetes A class of fungi of the subdivision BASIDIOMYCOTINA; constituent species are distinguished from those of other classes e.g. in that they form well-developed, typically *angiocarpous* fruiting bodies in which the basidiospores are statismospores. The gasteromycetes exhibit an extensive range of types of fruiting body. Only about half of the gasteromycetes form an hymenium; in the remaining species the fertile cells are scattered, either singly or in groups, throughout the spore-bearing tissue (*gleba*). The various types of fruiting body have distinct modes of development; *Secotium* and other *secotioid gasteromycetes* (all essentially angiocarpous species and all forming statismospores) have been recently transferred to the order AGARICALES on the basis that they have fundamental similarities with members of this order in the modes of development of the fruiting bodies. The basidiospores of the gasteromycetes are formed on sterigmata or develop directly on the basidium; when sterigmata are formed each basidiospore is usually orientated symmetrically on its sterigma. The gasteromycetes are typically saprophytic fungi which occur on soil and on dead wood or dung; some species form mycorrhizal associations. The orders of gasteromycetes include e.g. LYCOPERDALES (which includes the 'puffballs' and 'earthstars'), NIDULARIALES (which includes the 'bird's-nest fungi'), PHALLALES (which includes the 'stinkhorns'), and SCLERODERMATALES (which includes the 'earthballs').

gastriole Synonym of FOOD VACUOLE.

gastro-enteritis Inflammation of the tissues of the stomach and intestine (usually the small intestine) (cf. ENTERITIS).

gastrointestinal tract, common microflora of Many of the microorganisms which are ingested with food are inhibited or inactivated by the low pH which normally occurs in the human stomach. Further along the alimentary canal the range and numbers of microorganisms increase with the progressive rise in pH. In the large intestine the predominating organisms are generally species of *Bacteroides*, *Clostridium*, *Escherichia*, *Lactobacillus*, and *Streptococcus*; in addition there may be species of *Alcaligenes*, *Proteus*, and *Pseudomonas*. The non-bacterial flora may include yeasts (e.g. *Candida*) and viruses (e.g. certain echo-viruses have been isolated from the faeces of apparently healthy individuals).

The nature of the intestinal microflora may be influenced by diet. In *breast-fed* infants the flora may consist almost exclusively of *Bifidobacterium bifidum*, while infants given cows' milk develop a mixed flora which commonly includes e.g. *Lactobacillus acidophilus*, coliforms, and *Clostridium* spp. The intestinal flora may undergo a radical (but temporary) change following the administration of antibiotics. (see also MOUTH, COMMON MICROFLORA OF, and RUMEN)

GC types (see BASE RATIO)

GC values (GC ratios; %GC values) (of microbial DNA) The ratio:

$$\frac{G + C}{A + T + G + C}$$

in which G, C, A, and T represent the amounts of guanine, cytosine, adenine, and thymine, respectively, in a sample of DNA. (cf. BASE RATIO) GC values are usually expressed as percentages (%GC). Organisms known to be genotypically or phenotypically similar are commonly found to have similar values of %GC—although a similarity in base composition does not necessarily signify a close taxonomic relationship. The GC value is a fundamental characteristic of an organism and is used as a criterion in microbial TAXONOMY; it is particularly well suited for use in numerical (Adansonian) taxonomy. The range of %GC values found among microorganisms is far greater than that which occurs among the higher organisms. In bacteria and fungi the approximate range is 25–75%, in algae 35–70%, and in protozoa 20–70%; in comparison: vertebrates 35–45%. (The possession of a PLASMID by a bacterium may alter its observed GC value.)

While the GC value indicates the base composition of an organism's DNA, information of more taxonomic significance may be gained by comparison of the corresponding base *sequences* in organisms—since base sequences constitute the genes. Base sequences may be compared e.g. by DNA hybridization experiments; thus, if the DNA from each of a pair of organisms can be extensively hybridized, those organisms are considered to be closely related. (see also DNA BASE COMPOSITION, DETERMINATION OF)

Geastrum A genus of fungi of the LYCOPERDALES.

gel diffusion (*immunol.*) A procedure in which antibodies (usually in antiserum) and antigens diffuse independently through a gel medium (usually an AGAR gel)—forming a precipitate within the gel where homologous antigens and antibodies meet in OPTIMAL PROPORTIONS. Usually, lines or bands of precipitate are obtained within hours, or a day or two, of the commencement of the experiment. The number and position of such lines or bands depend on the number and nature of the homologous pairs of antigens and antibodies within the diffusing system. The precipitate may or may not be visible to the naked eye; the gel may be washed free of uncombined antigen and antibody and the precipitate stained *in situ* with protein stains e.g. *azocarmine*, *thiazine red*.

When antigen and antibody are *both* actively diffusing through the medium, the procedure is referred to as a *double-diffusion* technique (e.g. the OUCHTERLONY TEST). In the *Oudin* procedure (a *single-diffusion* technique) the antibody is evenly distributed throughout a column of agar gel (or similar medium) in a test-tube, and an aqueous solution of the antigen is layered on top. A band of precipitate forms where the antigen, diffusing downward into the agar column, meets homologous antibody in optimal proportions. (see also OAKLEY-FULTHORPE PROCEDURE)

gelatin The product obtained on boiling COLLAGEN. Gelatin is soluble in water at temperatures above about 40 °C; strong solutions form gels below about 28–35 °C. The gel can be liquefied by any of a range of nonspecific, extracellular proteolytic enzymes (produced by a number of bacteria and certain fungi) and by specific *collagenases* produced by *Clostridium* spp. Many gelatin-splitting enzymes require, or are stimulated by, calcium ions. In many organisms, gelatinase activity is higher at sub-optimal growth temperatures.

Gelidium A genus of red algae (RHODOPHYTA).

Gemella A genus of bacteria of the STREPTOCOCCACEAE.

gemma (*mycol.*) A thick-walled, irregularly-shaped or spheroidal, asexually-derived spore produced from a portion of a somatic hypha; gemmae may be formed in terminal or intercalary positions—either singly or in chains. On germination each gemma gives rise to a germ tube.

gene A sequence of nucleotides (in DNA or, in certain viruses, RNA) which specifies a particular polypeptide chain (see PROTEIN SYNTHESIS), rRNA, or tRNA (see RNA). Certain other nucleotide sequences, which do not specify products, are sometimes referred to as genes—e.g. the 'operator gene' (see *operator* in OPERON). In BACTERIOPHAGE ϕX174 evidence has been obtained that the nucleotide sequences of different genes may overlap. (see also ALLELE, CISTRON, CYTOPLASMIC GENE, and GENETIC CODE)

gene conversion (see RECOMBINATION)

gene-for-gene relationship (*plant pathol.*) (see RESISTANCE (of plants to disease))

genera Plural of *genus*. (see TAXONOMY)

generalised transduction (see TRANSDUCTION)

generation time (see GROWTH (bacterial))

generative hyphae (see HYPHA)

generic name (see SPECIFIC EPITHET)

genetic code The code in which information for the synthesis of proteins is contained in the sequence of nucleotides in DNA (or, in certain viruses, in RNA)—see DNA and PROTEIN SYNTHESIS. During the synthesis of proteins the coded information in DNA is initially transmitted to *messenger* RNA (mRNA), i.e. the sequence of deoxyribonucleotides is *transcribed* into a complementary sequence of ribonucleotides. (see RNA for details of mRNA synthesis) Each amino acid in a polypeptide is coded for by an appropriate sequence of three nucleotide bases in the mRNA; this triplet of bases is known as a *codon**. The nucleotide sequence of mRNA is thus made up of a continuous series of non-overlapping triplets (codons) which, collectively, specify the sequence of amino acids in a given polypeptide. A codon is not recognized *directly* by its corresponding amino acid; the amino acid must first be linked to an adaptor molecule, *transfer* RNA (tRNA), which contains a base triplet complementary to the codon—the *anticodon*. Thus, e.g. the codon adenine-cytosine-guanine ($5' \rightarrow 3'$) is recognized by the anticodon uracil-guanine-cytosine ($3' \rightarrow 5'$); codons and anticodons are usually designated by the initials of their constituent nucleotides written in the $5' \rightarrow 3'$ direction—ACG and CGU respectively in the example quoted above. Polypeptide synthesis begins at a specific *initiator codon* on the mRNA (see PROTEIN SYNTHESIS) and continues sequentially—thus ensuring that the triplets are read in phase by the translating mechanism. (cf. PHASE-SHIFT MUTATION)

There are 64 possible triplet combinations obtainable from the four bases adenine (A), guanine (G), cytosine (C), and uracil (U). 61 of these code for 20 amino acids; thus, many amino acids are coded for by more than one codon (a phenomenon known as *degeneracy*)—e.g. lysine has two codons, valine four, and serine six. The remaining three codons—UAA, UAG and UGA—are referred to as *nonsense codons*; they specify polypeptide chain *termination*. (see also AMBER CODON, OCHRE CODON) The three bases of a codon do not contribute equally to the specificity of that codon; the second base contributes maximally and the third (3′ end) base contributes minimally. Thus, certain codons which specify the same amino acid differ from each other only in the third base—e.g. UCU, UCC, UCA and UCG all code for serine. (see also WOBBLE HYPOTHESIS) There is also some relationship between codons which specify related amino acids—e.g. all codons in which U is the second base code for hydrophobic amino acids.

The genetic code appears to be universal, i.e. a particular codon generally specifies the same amino acid in all organisms.

*'Codon' also refers to the corresponding triplet of bases in the DNA.

genetic drift (see DRIFT)

genitourinary tract, common microflora of In the human male the distal portion of the urethra usually contains gram-positive cocci and diptheroids and, occasionally, gram-negative coccobacilli (*Acinetobacter* sp). The female urethra may be sterile or may contain gram-positive cocci. From the external genitalia of both sexes may be isolated *Mycobacterium smegmatis*—a saprophytic acid-fast organism. (see also VAGINA, COMMON MICROFLORA OF)

genome (1) The *genes* which, *in toto*, specify all the expressed and potentially expressible features associated with a given organism; *genome* does not connote the allelic nature of its component genes—cf. GENOTYPE. Thus, a given organism or cell has but one genome—regardless of whether the organism or cell is heterozygous or homozygous, or whether it is haploid, diploid, or polyploid. (2) *Genome* may be used to refer to the actual chromosome (in bacteria) or haploid set of chromosomes (in eucaryotes); in this sense, a diploid cell has two genomes, a triploid cell three etc. (3) *Genome* is sometimes used—very loosely—as a synonym of genotype.

genophore A term which may refer to (a) a *procaryotic* CHROMOSOME, (b) a viral genome, or (c) a nucleic acid structure—functionally analogous to a chromosome—which occurs in certain eucaryotic organelles, e.g. the CHLOROPLAST or MITOCHONDRION.

genotype Of an individual organism: the genetic information contained in the entire complement of *alleles—or* in certain, named, alleles. (cf. PHENOTYPE) (*Genotype* is occasionally used—very loosely—as a synonym of GENOME, sense (1).)

gentamicin (gentamycin) An AMINOGLYCOSIDE ANTIBIOTIC produced by *Micromonospora* spp; it is particularly useful against *Pseudomonas aeruginosa* infections. Plasmid-borne resistance to gentamicin (and other antibiotics) occurs in some stains of *Pseudomonas aeruginosa.*

gentian violet An impure dye consisting mainly of CRYSTAL VIOLET. (see also DYES)

genus A taxonomic group: see TAXONOMY. (see also BINOMIAL and NOMENCLATURE)

geotropism A *gravity*-dependent TROPISM.

GERL A region of the endoplasmic reticulum which is associated with the GOLGI APPARATUS and which appears to be the source of LYSOSOMES.

germ sporangium (*mycol.*) A sporangium produced at the distal end of a germ tube formed by a germinating ZYGOSPORE.

germ tube (*mycol.*) A short, hypha-like process which develops on the germination of many types of fungal spore; a germ tube commonly gives rise to a hypha.

German measles (see RUBELLA)

germicide Used (loosely) to mean either (1) a microbicidal DISINFECTANT or (2) a STERILANT.

germinant (see SPORES (bacterial))

germination (of bacterial spores) (see SPORES (bacterial))

germination by repetition (*mycol.*) On the germination of a spore: the formation of a secondary spore—as opposed to a germ tube.

giant cells Large *multinucleate* cells found in the tissues e.g. in certain pathological conditions—e.g. TUBERCULOSIS, CHICKENPOX, MEASLES. Some giant cells are believed to be formed from aggregates of macrophages.

Giardia A genus of parasitic flagellate protozoans of the order DIPLOMONADIDA. Morphologically similar strains occur in the intestinal tracts of a wide range of vertebrates; these strains are often named according to the host in which they are found, e.g. *G. canis* (dog)—but such speciation may or may not be justified. Host-host transmission involves CYST formation. *Giardia lamblia* (synonyms: *G. intestinalis*; *Lamblia intestinalis*). *G. lamblia* is quite common in the small intestine of man; it occurs in other primates and in the pig. Morphologically, the organism resembles one half of a longitudinally-bisected pear—the flat (actually concave) side being ventral; it is commonly 10–20 microns by 5–10 microns. The two nuclei occur symmetrically in the broad anterior region. Between, and slightly anterior to, the nuclei are the origins of the eight flagella—three pairs of which leave the body laterally, while the distal ends of the fourth pair become free at the extreme posterior end of the body. The anterior ventral surface of the body forms a sucker by means of which the parasite adheres to the intestinal wall of its host. One or two deeply-staining rod or comma-shaped bodies (*median bodies*—function unknown) are found—more or less at right angles to the long axis of the body—near the centre of the organism. Reproduction of the trophozoite occurs by binary fission. The trophozoites stain well with iron-haematoxylin and other dyes. No satisfactory method of *in vitro* culture of *G. lamblia* appears to have been devised.

The cysts of *G. lamblia* are ovoid, usually about 12–15 × 8–10 microns; when stained (with iodine etc.) visible details include the nuclei and portions of the median bodies and flagella (cf. CHILOMASTIX). *Mature* cysts generally contain four nuclei—which are commonly clustered at one pole of the cyst.

G. lamblia may cause acute or chronic diarrhoea (giardiasis), particularly in children, and heavy infestations are believed to inhibit fat absorption in the intestine; tissue invasion does not occur. Cysts are more likely to be observed in solid rather than diarrhoeic stools. Giardiasis has been treated with drugs such as chloroquine, diodoquin, and METRONIDAZOLE.

Gibberella A genus of fungi of the order SPHAERIALES, class PYRENOMYCETES; species occur e.g. as parasites and pathogens of certain plants—see e.g. BAKANAE DISEASE OF RICE. *Gibberella* spp form blue- or violet-coloured perithecia at the surface of the substratum; ascospores are generally ovoid or fusiform. Many species have conidial (imperfect) stages which are classified in the form-genus *Fusarium*—e.g. the imperfect stage of *G. fujikuroi* is *Fusarium moniliforme.* (see also GIBBERELLINS)

gibberellins A group of complex plant hormones which are involved in the regulation of stem elongation, seed germination etc. The gibberellins share a common tetracyclic structure but vary in the position and nature of their substituents; they are referred to as gibberellin A_1...gibberellin A_n. ('Gibberellic acid' is gibberellin A_3.) Gibberellins were discovered in the fungus *Gibberella fujikuroi* (*Fusarium* *moniliforme*)—the causal agent of *bakanae* disease of rice; relatively few gibberellins are common to both *G. fujikuroi* and higher plants. Gibberellin-like substances have been found in bacteria (e.g. *Arthrobacter* and *Azotobacter*) and in algae. The stunting of plants infected with certain viruses may be reversed by the application of gibberellins; it has been suggested that such viruses in some way affect the functioning of the endogenous plant gibberellins. (see also AUXINS and CYTOKININS)

Giemsa's stain One of a number of ROMANOWSKY STAINS used e.g. for the detection and/or identification of parasitic protozoans (such as *Plasmodium* spp, trypanosomes) in blood films (blood smears). Giemsa's stain exists in many modified forms; constituents of the stain commonly include the dyes azure *A* and azure *B*. Solutions of the stain are usually prepared in a mixture of methanol and glycerol. (see also STAINING)

Gigartina A genus of marine algae of the division RHODOPHYTA; species occur in intertidal and subtidal zones. The numerous species vary widely in form: the dark, purplish-red, filamentous thallus may be terete or flattened (ribbon-like), and may be more or less branched; the surface of the thallus frequently bears numerous papillate outgrowths. The thallus arises from a disc-shaped holdfast. The size of the thallus ranges from a few centimetres to several metres—depending on species. *Gigartina* spp are sometimes used as a source of CARRAGEENAN.

gills (lamellae) (*mycol.*) Blade-like sheets of tissue which hang vertically from the underside of the pileus in the fruiting bodies of many fungi of the order Agaricales; most or all of the surface of each gill consists of fertile tissue (the *hymenium*)—i.e. tissue which subsequently gives rise to basidiospores. In many agarics the gills are arranged radially around a central stipe; in some other agarics—e.g. *Pleurotus*—they radiate, fan-wise, from a lateral stipe. Gills may exhibit branching and fusion, and their colour usually reflects the colour of the spores they bear. (see also SPORE PRINT) The shape of the gills at their junctions with the stipe is a taxonomically-important feature (see e.g. DECURRENT and EMARGINATE). The inner, sterile tissue of a gill (the *trama*) consists of an arrangement of hyphae/cells which is characteristic for a particular group of agarics; in certain genera (e.g. *Lactarius*, *Russula*) the trama contains SPHAEROCYSTS.

In most agarics (including e.g. *Agaricus bisporus*) the gills are *aequihymeniiferous* (= aequihymenial); in such gills the development and maturation of the basidia occur at approximately the same time at all regions of the gill. Each gill is wedge-shaped in transverse cross section—i.e. it tapers from the attached to the free end.

Inaequihymeniiferous gills are typical of *Coprinus* spp; in these fungi the basidia at the lower (free) end of the gill reach maturity first—the remaining basidia maturing progressively from the free end of the gill. In *Coprinus* spp each gill is rectangular in transverse cross section—i.e. the opposite faces of a gill are parallel. (see also BASIDIOSPORES, DISCHARGE OF)

Giménez stain (for rickettsiae) (see RICKETTSIA)

gin (see SPIRITS)

glabrous (*adjective*) Hairless, smooth.

glanders Primarily a disease of the horse, mule and donkey; some other animals—including man—are susceptible. The causal agent, *Pseudomonas mallei*, appears to enter the body *via* the mouth or nose or *via* lesions in the skin. The disease may be acute or chronic and, in either case, may be fatal if untreated. In the horse, symptoms include fever, prostration, a highly-infectious nasal discharge, and ulceration of the nasal mucosa; nodular lesions are formed in the trachea, lungs, spleen, liver and elsewhere. Ulcerative skin lesions often involve the regional lymph nodes—which may become necrotic and suppurative. In *farcy*, ulcerative lesions occur in cutaneous and subcutaneous tissues; the regional lymph nodes and lymph ducts become swollen and hard—the so-called *farcy buds* and *farcy pipes* respectively. Farcy may represent a cutaneous extention of glanders or may be a localized, cutaneous disease. *Laboratory diagnosis* may involve microscopic or cultural examination of discharges or material from lesions. For chronic cases a SKIN TEST, the *mallein test*, may be useful; the reagent may be used subcutaneously or may be applied to one of the horse's conjunctivae. In man, the disease affects mainly the respiratory system and/or cutaneous tissues—but the viscera, muscles or bones may also be affected.

glandular fever (see INFECTIOUS MONONUCLEOSIS)

glass (microbial spoilage of) In humid, tropical regions, glass surfaces (e.g. microscope lenses) may be etched by the extracellular metabolic products of certain fungi which grow on substrates adjacent to the glass (e.g. lens sealing compounds, paint components) and which extend onto the glass surface. Such fungi include *Aspergillus glaucus* and *A. fumigatus* and species of *Cladosporium* and *Penicillium*. Damage may be avoided by frequent cleaning, storage under dry conditions, and/or the use of fungicides. Certain lichens can grow on—and etch the surface of—stained glass windows etc.

gleba (*mycol.*) In the fructifications of members of the Gasteromycetes and the Tuberales: the sporebearing tissue which is enclosed by the wall (peridium) of the fructification.

gliding bacteria Those bacteria which exhibit *gliding movements* (see MOTILITY) during at least some stage of their development cycle. There are two orders of gliding bacteria, the CYTOPHAGALES and the MYXOBACTERALES, together with a number of taxonomically unplaced genera. All species are gram-negative.

gliding movement (see MOTILITY)

Globigerina A genus of pelagic foraminiferans which are common constituents of marine ZOOPLANKTON. The organisms (about 300 microns to 1–2 millimetres) form (calcareous) tests of spirally-arranged spherical chambers; numerous fine, calcareous spicules emanate from the test. (see also FORAMINIFERIDA)

globose Spherical.

Gloeocapsa A genus of unicellular colonial BLUE-GREEN ALGAE; *Gloeocapsa* spp are capable of NITROGEN-FIXATION.

Gloiopeltis furcata A marine red alga.

Glossina (tse-tse fly) A genus of blood-sucking flies of the sub-order Cyclorrhapha (order Diptera). Some 20 species are known; they occur almost exclusively in the African continent in latitudes south of the Sahara Desert. *Glossina* spp are important vectors in trypanosome diseases of man and animals; see e.g. SLEEPING SICKNESS and NAGANA.

glucans Polymers of glucose; glucans include e.g. CELLULOSE, GLYCOGEN, LAMINARIN, LUTEOSE, PARAMYLUM, and STARCH. A highly-branched glucan—containing β-(1 → 6) and β-(1 → 3) linkages—occurs as the fibrillar component in the cell walls of certain yeasts; glucans also occur as amorphous (matrix) components in the cell walls of other fungi (see also CELL WALL (fungal)).

glucitol (see SORBITOL)

gluconeogenesis (glucogenesis) The biosynthesis of glucose from non-carbohydrate substrates (e.g. acetate). Initially, phosphoenolpyruvate is synthesized from pyruvate (by phosphoenolpyruvate synthetase in a reaction requiring ATP) or from malate or oxaloacetate (see Appendix II(b)). (TRICARBOXYLIC ACID CYCLE intermediates used to form phosphoenolpyruvate must be replenished—from acetate—*via* the glyoxylate cycle.) From phosphoenolpyruvate, fructose-1,6-diphosphate is formed by a reversal of reactions of the EMBDEN–MEYERHOF–PARNAS PATHWAY (Appendix I(a)). Fructose-1,6-diphosphate is hydrolysed to fructose-6-phosphate by fructose diphosphatase; fructose-6-phosphate is converted to glucose-6-phosphate by glucose phosphate isomerase.

Gluconobacter Synonym of *Acetomonas* (*q.v.*).

glucose effect (glucose repression) (see CATABOLITE REPRESSION)

glutamate (biosynthesis of) (see Appendix IV(a))

glutamine (biosynthesis of) (see Appendix IV(a))

glutaraldehyde ($CHO.CH_2.CH_2.CHO$) (as an antimicrobial agent) Under appropriate conditions a 2% aqueous solution of glutaraldehyde is an effective STERILANT; its mode of action resembles that of FORMALDEHYDE—*viz*. the substitution of e.g. amino groups in proteins and nucleic acids. Since glutaraldehyde possesses two aldehyde groups it can form cross-links between two amino groups within a molecule—or two adjacent molecules—of protein and/or nucleic acid. This property, which enhances antimicrobial activity, makes glutaraldehyde a useful *fixative*—particularly in ELECTRON MICROSCOPY. (see also β-PROPIOLACTONE, ETHYLENE OXIDE, and FIXATION)

glutarimide antibiotics A group of antibiotics which have, as a common feature, the structural components β-(2-hydroxyethyl)glutarimide and a cyclic or non-cyclic ketone. They include CYCLOHEXIMIDE and *inactone*.

glyc- Prefix denoting any sugar or sugar derivative. Examples: (a) Glycoprotein (sugar + protein). (b) Glycans (polysaccharides). (see also AGLYCON)

glycerol (as an antimicrobial agent) (see ALCOHOLS)

glycerol, microbial production of Glycerol is frequently a minor product of ALCOHOLIC FERMENTATION e.g. in *Saccharomyces* spp. However, it can become a major product under certain conditions—e.g. in the presence of alkali or bisulphite. Under *alkaline* conditions, dihydroxyacetone phosphate (an intermediate in the EMBDEN–MEYERHOF–PARNAS PATHWAY—see Appendix I(a)) is reduced to glycerol-3-phosphate by $NADH_2$ formed in the EMP pathway; thus, $NADH_2$ is not available for the reduction of acetaldehyde. (Acetaldehyde does not accumulate; it is converted to acetic acid and ethanol.) *Bisulphite* forms an addition compound with acetaldehyde and thus prevents its reduction to ethanol; as acetaldehyde is unavailable as an electron acceptor, $NADH_2$ (from the EMP pathway) donates electrons to dihydroxyacetone phosphate to form glycerol-3-phosphate. The glycerol-3-phosphate formed in each of these processes is dephosphorylated to form glycerol. This effect of bisulphite on the alcoholic fermentation of yeast has been used in the commercial production of glycerol, although alternative sources are now used.

Glycerol is also produced by certain bacteria; for example, certain *Bacillus* spp ferment glucose with the formation of 2,3-butanediol and glycerol.

glycine (biosynthesis of) (see e.g. Appendix IV(c))

glycobiarsol (see *Bismuth* under HEAVY METALS)

glycogen A highly-branched glucan containing α-(1 → 4) and α-(1 → 6) linkages; glycogen granules occur as food reserves in many bacteria, fungi (especially yeasts), and protozoa. Glycogen stains reddish-brown with iodine and may be (non-specifically) stained e.g. with Best's carmine stain and by the PAS REACTION.

Glycogen synthesis and breakdown. In bacteria, the initial step in glycogen biosynthesis is the formation of ADP-glucose from glucose-1-phosphate (formed from glucose-6-phosphate by the action of phosphoglucomutase) and ATP in a reaction catalysed by ADP-glucose pyrophosphorylase. In eucaryotes an analogous reaction leads to the formation of UDP-glucose. The subsequent transfer of glucose residues from ADP(UDP)-glucose to glycogen is mediated by glycogen synthetase. Glycogen is degraded to glucose-1-phosphate by glycogen phosphorylase in a reaction requiring inorganic phosphate.

glycolate pathway (in algae) (see PHOTORESPIRATION)

glycols (as antimicrobial agents) (see ALCOHOLS)

glycolysis (1) A synonym of EMBDEN–MEYERHOF–PARNAS PATHWAY. (2) The pathway in which *lactate* is the product of glucose metabolism *via* the EMP pathway.

glycopeptide A synonym of PEPTIDOGLYCAN.

glycosaminopeptide A term sometimes used synonymously with PEPTIDOGLYCAN.

glyodin 2-heptadecyl-2-imidazoline acetate; an agricultural antifungal agent used for the control of apple scab (*Venturia inaequalis*) and fungal diseases of cherries and ornamental plants.

glyoxylate cycle (see TRICARBOXYLIC ACID CYCLE)

glyoxysomes MICROBODIES which contain enzymes of the *glyoxylate cycle*.

gnotobiotic (*adjective*) Refers to a microbiologically-monitored environment—i.e. an environment in which the identity of any microorganisms which may be present is known. (see also SPECIFIC PATHOGEN FREE)

gold compounds (as antimicrobial agents) (see HEAVY METALS)

golden algae (golden-brown algae) (see CHRYSOPHYTA)

Golgi apparatus (Golgi body; dictyosome; parabasal body) An organelle—at least one of which occurs in most eucaryotic cells. The Golgi apparatus consists typically of a stack of flattened, membranous saccules, but a more diffuse arrangement of the saccules occurs in certain cells—e.g. those with a specialized secretory role; the saccules are associated with a high level of enzymic activity. The Golgi apparatus may have many functions; these appear to include (a) the formation of vesicles containing secretory products, enzymes, etc, and (b) the synthesis of complex polysaccharides—e.g. cell wall components.

gonal nucleus (see CONJUGATION (*ciliate protozool.*))

Gonapodya A genus of fungi of the MONOBLEPHARIDALES.

gonidium A cell which is involved in asexual reproduction. For example, in bacteria of the LEUCOTRICHACEAE the *gonidium* is a single cell which is released from the end of a vegetative filament and which functions as a disseminative unit; in VOLVOX the cell from which a new colony develops (asexually) is called a gonidium. (The term was once used to refer to each of the algal cells in a *lichen*—in the mistaken belief that such cells represented the reproductive cells of the organism.)

gonococcus (see NEISSERIA)

gonorrhoea An acute or chronic, usually sexually-transmitted contagious disease which, in nature, affects only man; the causal agent is *Neisseria gonorrhoeae.* The incubation period is commonly 2–8 days. In typical, uncomplicated gonorrhoea, infection is confined to the urino-genital tract; *N. gonorrhoeae* occurs intracellularly and extracellularly. In the male there is a sudden onset with painful urination and a yellowish mucopurulent urethral discharge—in which gonococci occur free or within pus cells. In the female the disease may be mild or asymptomatic; infection may involve the urethra, cervix and associated regions, including Skene's glands, and there may be a urethral and/or vaginal discharge. In either sex the disease may subsequently lead to permanent sterility; joints, heart valves etc. may also be affected. Eye infections (*gonococcal ophthalmia*) may follow gonococcal contamination of the conjunctivae. In some places, the eyes of newly-born infants are treated with aqueous silver nitrate (1% w/v)—the Credé procedure—as a prophylactic measure against gonococcal *ophthalmia neonatorum*—which may follow transmission of the gonococcus by an infected mother. Chemotherapeutic agents used for the treatment of gonorrhoea include PENICILLINS, TETRACYCLINES and erythromycin. *Laboratory diagnosis*: microscopical/cultural examination of discharges, rectal swabs etc.; a CFT is sometimes used. In *chronic* cases, repeated attempts may be required in order to isolate the pathogen.

Gonyaulax A genus of algae of the DINOPHYTA.

Gradocol filter (see FILTRATION)

Grahamella A genus of gram-negative bacteria of the family BARTONELLACEAE. The organisms are non-motile, rod-shaped cells which are intraerythrocytic parasites or pathogens in certain mammals, though not in man; a cell wall is present. Type species: *G. talpae*.

gram stain An important bacteriological staining procedure discovered empirically in the 1880s by the Danish scientist, Christian Gram. When bacteria are stained with certain basic dyes some species (*gram-negative species*) can be easily decolorized with organic solvents such as ethanol or acetone; others (*gram-positive species*) resist decolorization. The gram stain is useful as an initial step in the identification of bacteria; the ability of bacteria either to retain or lose the stain reflects fundamental differences between gram-positive and gram-negative species and is an important taxonomic feature. *The gram stain.* A heat-fixed SMEAR of bacteria (on a SLIDE) is stained with one or other of certain basic dyes of the triphenylmethane series, commonly CRYSTAL VIOLET; the stain is allowed to act for a few minutes and excess stain is rinsed off with tap water. All bacteria take up the stain. The smear is then flooded with LUGOL'S IODINE (the mordant) for several minutes and subsequently rinsed in tap water. Decolorization is then attempted by treating the smear with ethanol, acetone, or iodine-acetone; this is the critical step—the solvent being allowed to act only for as long as the stain runs *freely* from the smear. Decolorization is carried out by running the solvent (from a dropper) over the surface of the tilted slide. When the stain no longer runs freely from the smear (approximately 1 to 3 seconds) the slide is *immediately* rinsed under a running tap. The smear is now *counterstained*; this is carried out since any gram-negative bacteria present in the smear will now be colourless and hence not clearly visible under the microscope. Counterstaining is carried out by flooding the smear with a dye (such as dilute CARBOLFUCHSIN or safranin) so as to stain any gram-negative bacteria with a colour (red) which contrasts with that (violet) of any gram-positive cells which may be present. After counterstaining, the slide is briefly rinsed in tap water and blotted (or allowed to dry in an incubator); it is then ready for examination under the microscope.

Originally it was thought that the ability of a given bacterium to retain the stain depended on some chemical component in the cell wall. However, disrupted cells of gram-positive species do not retain the stain—i.e. the integrity of the cell wall is an essential requirement for gram-positivity. Currently, it is generally believed that a lower permeability of the cell wall is responsible for the retention of the dye-iodine complex by gram-positive species; there is some evidence to support this—e.g. the PEPTIDOGLYCAN of gram-positive cell walls is much more extensively cross-linked than that of gram-negative species.

Some bacterial species often give a *gram-variable* reaction, i.e. they are sometimes gram-positive, sometimes gram-negative; this may reflect minor variations in the staining technique used. Other species, e.g. *Bacillus* spp, are strongly gram-positive as young cells but tend to become gram-negative in ageing cultures; this may reflect degenerative changes in the nature of the cell wall. (*Mammalian* cells are uniformly gram-negative.)

gramicidins A group of linear polypeptide ANTIBIOTICS which have formyl groups at their N-terminals and ethanolamine residues at the C-terminals. (Gramicidin S is misnamed, and belongs to the TYROCIDINS.) Gramicidins act against gram-positive bacteria, being bacteriostatic at low concentrations and bactericidal at higher concentrations. Gram-negative species are resistant; their cell walls apparently adsorb the antibiotic, and thus may protect gram-positive organisms from its action. Gramicidins act by increasing the permeability of the CELL MEMBRANE to ions. (see also TYROTHRICIN)

granulocyte (see POLYMORPHONUCLEAR LEUCOCYTE)

granulose (1) (*noun*) An intracellular glucan found in species of *Clostridium*; granulose resembles amylopectin (see STARCH). (2) (*adjective*) Having a granular appearance.

granuloses (*entomopathol.*) Viral diseases of Lepidopterous larvae in which the causal virions occur embedded in intracellular crystalline protein inclusion bodies (termed *granules*, or *capsules*); each capsule usually contains a single virion (cf. POLYHEDROSES). (The capsules—which are ellipsoidal, or rod-shaped with rounded ends—are just visible by light microscopy.) The rod-shaped granulosis virions contain DNA.

Granville wilt (of tobacco) (see TOBACCO, DISEASES OF)

Graphis A genus of crustose LICHENS in which the phycobiont is *Trentepohlia*. The thallus is light or dark grey; apothecia are lirelliform with black DISCS. *Graphis* spp occur mainly on trees.

gratuitous inducer (see OPERON)

green algae (see CHLOROPHYTA)

green-ear (of millet) (see PHYLLODY)

green monkey fever In man, a disease caused by the MARBURG AGENT. In 1967 an outbreak occurred in Marburg (Germany) among laboratory personnel working with the tissues of the African green monkey; there were a number of fatalities. The disease was recorded in England in 1976—where it is now a NOTIFIABLE DISEASE. Symptoms include fever and haemorrhages from the mucous membranes; the liver, kidneys and central nervous system may be affected.

green oak Oak wood which has been stained green as a result of infection by the fungus *Chlorosplenium aeruginosum*. It is valued as a medium for inlay work etc.

green sulphur (see SULPHUR)

green sulphur bacteria (Chlorobacteriaceae) (see suborder CHLOROBIINEAE)

gregarin A *gregarine* trophozoite.

gregarine movement (see MOTILITY)

gregarines Protozoans of the sub-class GREGARINIA.

Gregarinia A subclass of protozoans of the TELOSPOREA. The *gregarines* are parasites of invertebrates. SCHIZOGONY occurs only in the smaller of the two orders, SCHIZOGREGARINIDA; in both orders (EUGREGARINIDA and Schizogregarinida) the *mature* trophozoites usually occur extracellularly. SYZYGY occurs during the sexual phase.

grex A synonym of *pseudoplasmodium*—see CELLULAR SLIME MOULDS.

grid The specimen holder in ELECTRON MICROSCOPY.

Griffith's tube An apparatus used for grinding tissues. It consists of a stout, large-diameter glass tube which is closed at the bottom; the lower, *inner* part of the tube has a roughened surface—against which the tissue is ground by means of a close-fitting pestle-like glass rod. Before use, the apparatus is sterilized in an autoclave or hot-air oven. (Tissue-grinding may be necessary e.g. in order to examine a biopsy for the presence of *Mycobacterium tuberculosis*; the ground tissue is used to inoculate a suitable medium.)

Grifola A genus of fungi of the family POLYPORACEAE.

grisein An antibiotic believed to be identical with one of the components of *albomycin*—see SIDEROMYCINS.

griseofulvin (7-chloro-4,6-dimethoxycoumaran-3-one-2-spiro-1′-(2′-methoxy-6′-methylcyclohex-2′-en-4′-one)) An ANTIBIOTIC produced by a number of *Penicillium* spp e.g. *P. griseofulvum* and *P. patulum*. It is inhibitory for a range of fungi but is not active against fungi with cellulose walls, bacteria, or yeasts. Griseofulvin causes the tips of growing hyphae to become characteristically curled—hence the early name of *curling factor*. The mode of action of griseofulvin is not clear; it was at first thought to interfere with CHITIN biosynthesis but, as some chitin-containing fungi are resistant, this has been disputed. It is now believed that griseofulvin interferes with DNA replication; in relatively high concentrations it inhibits mitosis in animal cells. Griseofulvin is widely used in the treatment of DERMATOPHYTE infections of the skin, hair and nails; it accumulates in the keratin of these tissues. Griseofulvin is not effective against systemic mycoses. Resistance to the antibiotic can be obtained *in vitro* but has not been observed clinically. Griseofulvin is taken up by the roots of higher plants and has limited use as a systemic antifungal agent in plants. It has low phytotoxicity and is effective against certain infections e.g. those involving *Botrytis* spp, *Alternaria solani*. (see also FUNGICIDES)

group phase antigens (of *Salmonella*) (see PHASE VARIATION, IN SALMONELLA)

group translocation (see TRANSPORT)

group treponemal antigens (see REITER ANTIGENS)

grouping antiserum An antiserum which can agglutinate all the members of a *group* of SEROTYPES within a particular bacterial genus, i.e. serotypes which have in common one or more antigens which are not found in serotypes outside the group. For example, an antiserum used to identify the serotypes of *Salmonella* group *D* would contain antibodies to the major somatic antigen (O9) found exclusively in members of that group; this antiserum would fail to agglutinate the serotypes of other groups—which do not possess the O9 antigen.

growth (bacterial) In a single living cell: a *coordinated* increase in the mass of cell constituents; in a bacterial population: a similar, coordinated increase in biomass, or an increase in cell numbers, in the population. (An increase in cell mass does not necessarily signify growth; it may be due e.g. to the synthesis of a storage compound.) The growth of an individual cell normally culminates in (asexual) reproduction—which commonly occurs by BINARY FISSION. (see also COLONY)

Measurement of growth. The growth of mycelial bacteria can be quantified e.g. by measuring increases in the dry weight of cells, or by measuring the amount of a radioactive metabolite taken up by the cells in a given time; the growth of other types of bacteria is often measured by monitoring cell numbers—see ENUMERATION OF MICROORGANISMS.

The growth of bacterial populations. When cells divide by *binary* fission, a single cell will give rise to 2,4,8,16,32. . . . cells after 1,2,3,4,5. . . . *generations* (rounds of division). Hence, the number of cells (N) derived from a single cell after *n* generations will be given by

$$N = 2^n$$

Similarly, after *n* generations, the number of cells derived from an initial cell population of N_0 will be given by

$$N = 2^n N_0$$

The growth of a bacterial population can be

expressed in the form of a *growth curve*: a graph in which the increase in cell numbers, or in biomass, is plotted against time. For bacteria which undergo *binary* fission, cell numbers are conveniently plotted on a $\log_2$ scale; this gives, directly, the number of bacterial generations per unit time, and indicates clearly any changes in the growth rate.

The growth curve obtained by monitoring a BATCH CULTURE commonly depicts four main phases of growth; these phases occur sequentially following the inoculation of a medium. (a) *The lag phase.* During this phase the rate of increase in cell numbers initially remains static but subsequently rises to a value dictated by environmental conditions. The extent of the lag phase is influenced by the cultural history of the cells of the inoculum and by the chemical and physical nature of the growth medium. Thus, e.g. if slowly-dividing cells are placed in a medium which supports a higher growth rate, a lag phase occurs during which the organisms become adapted to the new environment; adaptation involves e.g. an increase in the number of ribosomes per cell—thus permitting more rapid protein synthesis and an increase in growth rate. During such adaptation growth is said to be *unbalanced*; in *unbalanced growth* there is a change in the rate of synthesis of some cellular components (e.g. rRNA, protein) relative to that of other components. A lag phase may also occur e.g. when rapidly-dividing cells are transferred to a medium containing a different source of carbon and/or nitrogen; under such conditions appropriate amounts of suitable enzyme(s) may have to be synthesized before the growth rate begins to rise. (see also DIAUXIE) A long lag phase may occur e.g. when only a small proportion of the cells of the inoculum are capable of using the source of carbon and/or nitrogen. (When actively-dividing cells are transferred to a medium similar to that in which they were growing a lag phase is not observed.) (b) *The exponential* (*logarithmic*) *phase.* From the lag phase the cells enter the exponential phase—in which, for a given organism, the growth rate is both constant and maximal for the particular medium and growth conditions. Under these conditions, cell numbers (or biomass) increase exponentially with time, and there is a proportional increase in all cell constituents; this type of growth is referred to as *balanced growth.* For cells undergoing binary fission at a constant rate, the number of doublings (of biomass or cell population) per unit time is given by the exponential *growth rate constant*, *k*:

$$k = \frac{\log_2 N_t - \log_2 N_0}{t}$$

where N_0 = population at time zero, N_t = population at time *t*. *k* is usually expressed as the number of doublings per *hour.* The reciprocal of *k* ($1/k$) gives the *mean generation time* (= *mean doubling time*) i.e. the time required for the cell population (or biomass) to double. (Variation may occur among the doubling times of individual cells within a population.) For some cells (e.g. those of *Bacillus* spp, *Escherichia coli*) the *minimum* generation times (doubling times) are of the order of 15–20 minutes; for others (e.g. *Mycobacterium* spp) they may be many hours. (c) *The stationary phase.* In batch cultures the rate of growth eventually declines (due e.g. to the exhaustion of nutrient(s) or to the accumulation of toxic metabolic wastes) and finally becomes zero (*stationary phase*). (Cells can be maintained in the exponential phase by the use of CONTINUOUS-FLOW CULTURES.) (d) *The death phase.* In this phase the number of viable cells (maximal in the stationary phase) declines.

The growth of individual cells. CELL WALL growth has been studied e.g. by time lapse experiments using fluorescent antibodies homologous to cell wall components. In certain gram-positive cocci new material appears to be added to the cell wall at restricted sites (polarly and equatorially in *Streptococcus pyogenes*); in *Escherichia coli* and other gram-negative bacteria new material appears to be incorporated diffusely (throughout the wall). The CHROMOSOME is attached to the CELL MEMBRANE or to a MESOSOME; following replication (see DNA) the daughter chromosomes appear to be drawn into their respective cells as new material is incorporated in the membrane of the parent cell during cellular growth and elongation. SEPTUM-formation begins after the completion of DNA replication. (Mutant cells which are defective in septation may give rise e.g. to filaments or to MINICELLS.) Wall growth, septation, and cell separation are believed to involve the action of AUTOLYSINS.

Current ideas on the relationships between overall growth, chromosome replication, and cell division are embodied in the *Helmstetter–Cooper model.* (This model is based largely on work carried out on SYNCHRONOUS CULTURES of *Escherichia coli.*) The cell division cycle is divided into a linear sequence of three main periods. During the initial period, *I*, there occur certain (unknown) biochemical events which necessarily precede, and culminate in, the initiation of DNA replication. The length of the *I* period (which corresponds to the generation time) decreases with increased growth rate; in *Escherichia coli* it is a minimum of about 20 minutes. The next (*C*) period extends from the initiation to the ter-

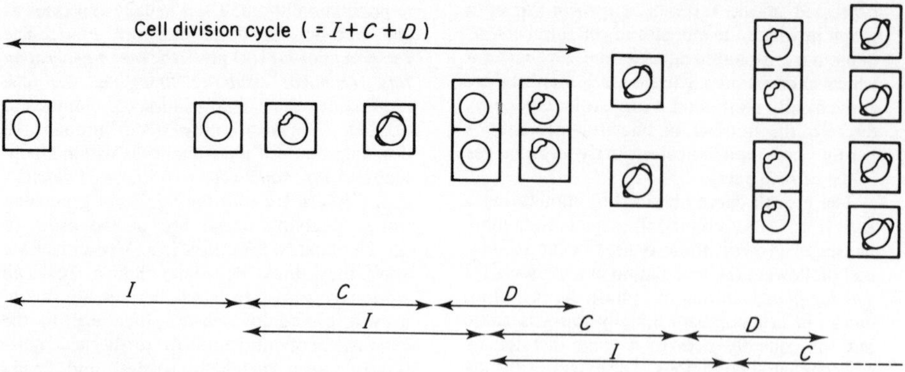

GROWTH (bacterial). Figure 1. *Helmstetter–Cooper Model*—slowly-dividing cells undergoing balanced growth. The *I*, *C* and *D* periods (see text) of different cell generations occur in a series of overlapping sequences. A cell and its progeny (squares and oblongs) are shown at various stages in the cell division cycle. Chromosomes at corresponding stages of replication are shown within the cells.

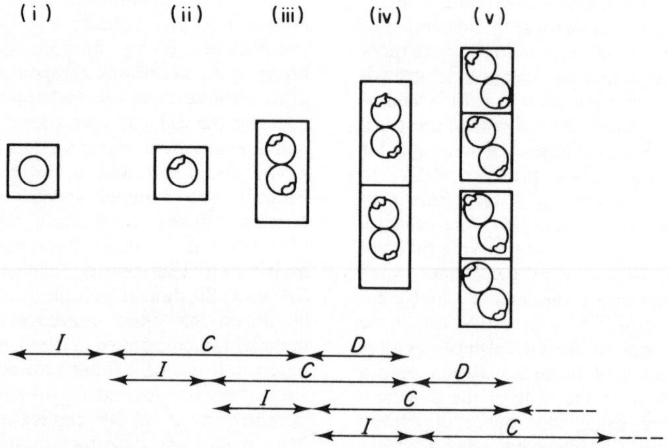

GROWTH (bacterial). Figure 2. *Helmstetter–Cooper Model*—rapidly-dividing cells undergoing balanced growth. With *I* shorter than *C* (see text) DNA replication becomes dichotomous at iii.

mination of DNA replication; in *E. coli* the *C* period is normally about 40 minutes. In the final (*D*) period, septum-formation occurs; cell separation takes place at the end of the *D* period. The *D* period occupies a minimum of about 20 minutes in *E. coli*—being longer when the growth rate is low. The cell division cycle (*I* + *C* + *D*) is repeated at a constant rate during steady-state growth; at the end of each *I* period another *I* period begins. (In *Bacillus subtilis* the *D* period is reported to be about 140 minutes over a range of growth rates; thus, single cells are formed during slow growth (*I* very long) while chains of cells are formed during rapid growth.)

Cellular events in *slowly-dividing* cells are depicted in figure 1. (In both figures 1 and 2 the sequence begins with a single cell which contains one chromosome; the subsequent events are those which would occur if the cell were to be placed in an environment in which balanced growth could commence immediately.) In figure 1, *I* is somewhat longer than *C*; DNA replication for a given round of division begins before the previous cell division has been completed.

Cellular events in *rapidly-dividing* cells are depicted in figure 2. *I* is shorter than *C*; the replication of DNA is *dichotomous*—i.e. more than one pair of replication forks (see entry

DNA) occur in the bi-directionally replicating DNA. If the *D* period is of sufficient length, new sites of septation may begin to develop before the completion of those begun earlier. One of the consequences of rapid growth is the occurrence of a high chromosome/cell ratio. (Since, in general, the rate of DNA synthesis is constant, increases in cellular DNA depend on the frequency at which DNA replication is *initiated*.)

growth (fungal) In a single living cell: a *coordinated* increase in the mass of cell constituents; in a population of single cells (e.g. yeasts) or in fungal mycelium: a similar increase in biomass or (of single cells) an increase in cell numbers. According to species and conditions, growth may involve, or lead to: (a) An increase in cell size or (in filamentous fungi) an extension in hyphal length and/or hyphal branching. (b) Vegetative differentiation—e.g. a phase change in DIMORPHIC FUNGI. (c) Asexual reproduction. (d) Sexual reproduction. (see also COLONY)

Modes and phases of growth. In hyphae and in some elongated yeasts (e.g. *Pichia*) and fission yeasts (e.g. *Schizosaccharomyces*) growth is predominantly *apical*—i.e. 'tip growth' in which CELL WALL synthesis and cellular extension occur mainly at the end(s) of a hypha or at the pole(s) of a yeast cell. Cell wall extension is believed to involve the coordinated action of autolytic and synthetic enzymes. In the *yeast-phase* cells of *Mucor rouxii* autoradiographic studies (following the uptake of tritiated *N*-acetyl-D-glucosamine) indicate that new cell wall materials are incorporated diffusely—i.e. throughout the existing wall. In *populations* of unicellular fungi increases in cell numbers commonly occur by BUDDING or by BINARY FISSION.

As in *bacterial* GROWTH (*q.v.*), batch cultures of unicellular fungi exhibit lag, exponential, stationary, and death phases.

The formation of SECONDARY METABOLITES, in batch cultures, appears to be maximal when growth rates are restricted during unbalanced growth—or in the complete absence of growth. (see also IDIOPHASE and TROPHOPHASE)

growth rate constant (see GROWTH (bacterial))

GTP Guanosine-5′-triphosphate. (see ATP)

guanidine ($(H_2N)_2C{=}NH$) (as an antiviral agent) Salts of guanidine (e.g. guanidine hydrochloride: $(H_2N)_2C{=}NH_2^+Cl^-$) selectively inhibit a number of picornaviruses (including polioviruses) in tissue culture systems—apparently by inhibiting viral RNA biosynthesis. Essential cell processes do not appear to be significantly affected. Guanidine salts are not used as chemotherapeutic agents owing to the rapid development of resistant mutants. (Guanidine-dependent mutants have also been isolated.) (see also HBB under BENZIMIDAZOLE DERIVATIVES and ANTIVIRAL AGENTS)

guanosine-5′-monophosphate (biosynthesis of) (see Appendix V(a))

Guarnieri bodies Eosinophilic bodies, surrounded by a clear non-staining region, which are characteristically found in the cytoplasm of cells infected with *vaccinia* or *variola* virus; one or more such bodies may be present per infected cell. Guarnieri bodies contain aggregates of virus particles.

gullet (1) In some flagellates (e.g. *Euglena*) a vestibule or vestibular depression formed by an invagination of the body; the gullet is not necessarily involved in feeding. (2) When used in ciliate protozoology gullet may refer to the buccal cavity, to the combined buccal cavity and VESTIBULUM, or even to the CYTOPHARYNX.

guttulate (*mycol.*) Refers to a spore which contains one or more GUTTULES. (cf. EGUTTULATE)

guttule (*mycol.*) A globule or droplet (of oil?) within a spore.

Gymnoascus A genus of fungi of the order EUROTIALES.

gymnocarpic development (gymnocarpous development) (*mycol.*) In a fruiting body: a form of growth in which the spores are exposed to the environment from the time of their initial differentiation on the sporophore. Fungi whose fruiting bodies develop gymnocarpically include e.g. the basidiomycete *Cantharellus cibarius*. (cf. ANGIOCARPIC DEVELOPMENT)

Gymnodinioides A protozoan genus of the APOSTOMATIDA.

Gymnodinium A genus of algae of the DINOPHYTA.

Gymnosporangium A genus of fungi of the order UREDINALES. (see also RUSTS (1)) *Gymnosporangium* spp are mainly heteroecious rusts (one species is autoecious); typically, the primary host (on which occurs the telial stage) is a species of *Juniperus*, and the alternate host (on which occurs the aecial stage) is a member of the Rosaceae. (Thus, e.g. the heteroecious rust *G. clavariaeforme* is parasitic on the juniper (*Juniperus*) and on the hawthorn (*Crataegus*).) Only two species of *Gymnosporangium* are known to have a uredial stage; the remaining species exhibit stages O, I, III, and IV (i.e. they are *demicyclic* rusts). *Gymnosporangium* spp often develop their telial and aecial stages in GALLS which arise on the host organism in response to the presence of the rust; thus, e.g. *G. juniperi-virginianae*, which parasitizes the juniper and apple, produces telia which develop within (and subsequently project from) spherical galls—'cedar apples'—which form on the juniper.

Gymnostomatida An order of protozoans

(subclass HOLOTRICHIA, subphylum CILOPHORA). *Gymnostomes* are generally considered to be the most primitive ciliates; they occur in fresh, brackish, and marine waters and in various types of soil. Individual cells are commonly large, and the (typically apical—i.e. polar) cytostome opens directly to the environment—i.e. it does not occur at the base of a vestibulum or buccal cavity. The cytopharyngeal region contains a microtubular endoskeletal structure. The suborders: RHABDOPHORINA and CYRTOPHORINA.

Gyrodinium A genus of algae of the DINOPHYTA.

gyrose (*adjective*) (*lichenol.*) Refers to an apothecium in which there is a maze-like pattern of fertile ridges and sterile furrows. Gyrose apothecia occur e.g. in *Umbilicaria* spp.

H

H antigens (*microbiol.*) Bacterial *flagellar* ANTIGENS. (see FLAGELLUM (bacterial) and FLAGELLINS) H antigens are important in the serological classification of certain genera e.g. *Salmonella* (see KAUFFMANN-WHITE CLASSIFICATION and PHASE VARIATION, IN SALMONELLA). AGGLUTINATION of bacteria with (homologous) H antiserum occurs as a floccular aggregate—in contrast to agglutination with O antiserum which is commonly granular. Flagellar antigens are heat-labile and are destroyed by alcohol but not by dilute formalin.

***h* mutants** (of bacteriophage) (see HOST-RANGE MUTANTS)

HAA (see HEPATITIS B ANTIGEN)

hadacidin ($CHO.NOH.CH_2.COOH$) An analogue of L-aspartic acid which specifically inhibits *adenylo-succinate synthetase*—and thus inhibits the synthesis of adenine nucleotides (see Appendix V(a)); it has no significant effect on other reactions involving aspartate. Hadacidin has antimicrobial and anti-tumour activity. (see also AZASERINE)

haem (American: *heme*) (1) A generic term for any iron-PORPHYRIN (including iron-chlorins); thus e.g. iron-mesoporphyrin = mesohaem. The four pyrrole nitrogen atoms each contribute an electron-pair to form co-ordinate bonds with the iron atom—which may be in the ferrous or ferric state. The 5th and 6th co-ordination positions of the iron may be filled by electron-pairs donated e.g. from water, chloride, or basic amino acid residues of proteins—see e.g. CYTOCHROMES. (2) *Haem* may refer specifically to iron-protoporphyrin IX (= protohaem IX).

haemadsorption The adherence of erythrocytes to cells which are infected with certain viruses, e.g. MYXOVIRUSES; thus, the active infection of a TISSUE CULTURE by such viruses may be detected by its ability to adsorb the erythrocytes of certain species under appropriate conditions. Such a technique is particularly useful for myxoviruses which do not regularly produce observable *cytopathic effects* in the cells of a tissue culture. (see also HAEMADSORPTION-INHIBITION TEST)

haemadsorption-inhibition test (*virol.*) A test used for detecting serum antibodies homologous to HAEMADSORPTION-promoting viruses. The serum is allowed to react with a suspension of such a virus—which is then used to inoculate a tissue culture. Following incubation, the tissue culture is tested for haemadsorption; the *absence* of haemadsorption indicates specific neutralization of the virus, i.e. a positive test.

haemagglutination The AGGLUTINATION of erythrocytes. Haemagglutination may be brought about by: (a) The interaction between erythrocytes and antibodies homologous to the surface antigens of those erythrocytes (HAEMAGGLUTININS). (b) Certain viruses. Some viruses (e.g. orthomyxoviruses, paramyxoviruses) agglutinate the erythrocytes of particular species; the virions attach to specific sites on the red cell membrane and thus act as links between cells. Following haemagglutination some viruses (e.g. orthomyxoviruses) elute spontaneously (see NEURAMINIDASE). (c) Passive haemagglutination: see PASSIVE AGGLUTINATION. (d) The action of non-specific substances such as LECTINS. (e) Infection by *Plasmodium falciparum* (see MALARIA). (f) Fimbriate bacteria.

haemagglutination-inhibition test (1) (HI or HAI test) A serological test for detecting serum antibodies homologous to haemagglutinating viruses; an HI test is used e.g. in the diagnosis of RUBELLA. Essentially, the test determines the ability of the patient's serum to inhibit or abolish the haemagglutinating properties of a given virus. A suspension of the virus is allowed to react with the patient's serum and is subsequently examined for its ability to agglutinate erythrocytes; the absence of agglutination (a *positive* test) indicates that the viruses have reacted with homologous neutralizing antibodies. Known positive and negative sera must be used as controls since it is important to exclude the effects of *non-specific* inhibitors of haemagglutination which are normally present in some sera. (2) The test may be used e.g. to titrate a *soluble* antigen with its homologous antibody; such a test involves *passive haemagglutination* (see PASSIVE AGGLUTINATION). Serial dilutions of antigen are each allowed to react with a fixed amount of antibody; to each dilution is then added a fixed amount of antigen-coated erythrocytes (see BOYDEN PROCEDURE). After incubation, a given dilution will exhibit haemagglutination only if the reaction mixture added to it contains free (uncombined) antibody—thus giving a measure of the antigen concentration in the original sample.

haemagglutinin Any agent which can bring about HAEMAGGLUTINATION. Haemagglutinins include: (a) Antibodies homologous to the surface antigens of erythrocytes; such antibodies agglutinate erythrocytes by forming bridges or links between them. (b) LECTINS. (c) The haemagglutinating regions of the envelope in certain types of virus—see e.g. ORTHOMYXOVIRUS.

Haemamoeba A sub-genus of PLASMODIUM.

haematin (as a bacterial growth factor) (see X FACTOR)

haemin (as a bacterial growth factor) (see X FACTOR)

haemolysin (1) An antibody homologous to the surface antigens of erythrocytes (see also IMMUNE HAEMOLYSIS). (2) Any agent of microbial origin which causes HAEMOLYSIS—e.g. STREPTOLYSIN-O, PHALLIN, α-HAEMOLYSIN, β-HAEMOLYSIN etc.

α-haemolysin (of staphylococci) A protein EXOTOXIN which is produced mainly by COAGULASE-positive staphylococci. It lyses the erythrocytes of the sheep, dog, and pig—but is most active against those of the rabbit. The α-haemolysin can be lethal for man and other mammals, and is a leucocidal agent—being also known as the *Neisser–Wechsberg* LEUCOCIDIN; in some species it has dermonecrotic effects when injected subcutaneously, and it is also able to aggregate and lyse the blood platelets of certain species. (see also α-HAEMOLYSIS (by staphylococci))

β-haemolysin (of staphylococci) A protein EXOTOXIN which is produced mainly by those COAGULASE-positive strains of staphylococci isolated from animals other than man; the toxin functions as a sphingomyelinase. β-Haemolysin causes HOT-COLD LYSIS of ox and human erythrocytes and (more effectively) of sheep erythrocytes. It has been reported that β-haemolysin can be lethal for the rabbit and certain other mammals. It does not function as a LEUCOCIDIN. (cf. α-HAEMOLYSIN and δ-HAEMOLYSIN)

δ-haemolysin (of staphylococci) A protein EXOTOXIN believed to be produced mainly by COAGULASE-positive staphylococci. It is active against the erythrocytes of many species—including those of man, the rabbit, sheep and horse; haemolytic activity is said to be inhibited by certain factors present in normal serum. The δ-haemolysin is a LEUCOCIDIN and has a lytic action on the protoplasts and sphaeroplasts of certain bacterial species. Staphylococci which do not produce the δ-haemolysin may become active producers when lysogenized by a particular temperate bacteriophage.

haemolysis The lysis of red blood cells (erythrocytes). Among the various ways in which haemolysis may occur are the following: (a) The action of microbial or other types of HAEMOLYSIN—which may or may not be COMPLEMENT-dependent; thus, e.g. on a suitable blood agar medium, a zone of haemolysis develops around each colony of those species of bacteria which produce extracellular haemolysin(s). (see also α-HAEMOLYSIS, β-haemolysis under STREPTOCOCCUS, γ-HAEMOLYSIS, HOT-COLD LYSIS, and IMMUNE LYSIS) (b) Osmotic lysis; this occurs when red blood cells are placed in a suitably hypotonic medium. (c) Viral haemolysis. Certain viruses (e.g. the mumps virus) are able to lyse the erythrocytes of certain species. (d) Passive haemolysis. This may occur as a result of a complement-fixing union between antibody and *cell-bound* antigen, i.e. antigen which has been adsorbed or attached to the surface of the erythrocyte. (e) REACTIVE LYSIS.

α-haemolysis (by staphylococci) A wide zone of clear haemolysis (on e.g. rabbit or sheep-blood agar) which surrounds the colonies of those strains of *Staphylococcus* which produce α-HAEMOLYSIN. Strains which form only the δ-HAEMOLYSIN may give rise to a similar—though narrower—zone of haemolysis. (see also HOT-COLD LYSIS)

α-haemolysis (by streptococci) (see STREPTOCOCCUS)

β-haemolysis (by streptococci) (see STREPTOCOCCUS)

γ-haemolysis (*bacteriol.*) Refers to the *absence* of haemolysis.

haemolytic immune body (HIB) Antibody to the surface antigens of erythrocytes (red blood cells); see also AMBOCEPTOR.

haemolytic system (in the COMPLEMENT-FIXATION TEST) A suspension of *sensitized* erythrocytes, i.e. red blood cells coated with antibody homologous to their surface antigens. The system exhibits no HAEMOLYSIS in the absence of COMPLEMENT; in the presence of excess complement total haemolysis occurs. For quantities of complement between these extremes, the percentage of cells which lyse is related to the amount of complement added to the system. Thus, the haemolytic system can indicate the amount of complement (if any) remaining free (unbound) in a given dilution of anti-serum at the end of the first stage in a complement fixation test. This gives an assessment of the extent to which antibody–antigen combination has occurred in that dilution, and thus, indirectly, an assessment of the quantity of antibody which has combined with the fixed quantity of antigen.

Haemophilus A genus of gram-negative, facultatively anaerobic, asporogenous, parasitic or pathogenic bacteria of uncertain taxonomic affiliation; species occur e.g. in the respiratory tract of man and other animals. The organisms are pleomorphic coccobacilli, bacilli or filaments—bacillary forms often being about 0.4 × 1–2 microns; some strains are capsulated. For growth, some species require the X FACTOR, some the V FACTOR, and some require both. Incubation under 5% CO_2 is essential for some species, and enhances the growth and haemolytic activity of others. Media used for the primary isolation of *Haemophilus* spp include FILDES' ENRICHMENT AGAR, LEVINTHAL'S AGAR and CHOCOLATE AGAR. (The growth of most species is reported to be inhibited by fresh blood.) 24-Hour colonies may be less than 0.5 mm to almost 1 mm in diameter—according to media and growth conditions. (see also *satellite* colonies under SYNTROPHISM) *Haemophilus* spp are generally sensitive e.g. to tetracyclines and sulphonamides; most or all species are killed by exposure to 60 °C for 30 minutes. Type species: *H. influenzae*; GC% for the genus: 38–42. Species include:

H. influenzae. A non-haemolytic species, common in the human nasopharynx, which is often a secondary invader in cases of INFLUENZA; it is the cause of one form of meningitis in children. *H. influenzae* requires the X and V factors; it grows on chocolate agar with or without 5% CO_2.

H. parainfluenzae. A non-haemolytic species similar to *H. influenzae* but requiring only the V factor.

H. haemolyticus. A haemolytic species which requires both X and V factors; growth occurs on fresh blood agar but is richer on chocolate agar. (Wide zones of haemolysis may be formed—similar to those produced by β-haemolytic streptococci.) It is parasitic or pathogenic in the upper respiratory tract of man.

H. parahaemolyticus. Similar to *H. haemolyticus* but requires only the V factor.

H. suis (*H. influenzae-suis*). A non-haemolytic species which requires both X and V factors; it is similar to *H. influenzae* but differs e.g. in that the indole test is consistently negative. *H. suis* is a secondary invader in SWINE INFLUENZA. (*H. parasuis* is similar but requires the V factor only.)

H. ducreyi. The causal agent of CHANCROID in man; *H. ducreyi* requires the X factor only and produces slight or delayed haemolysis.

Haemoproteidae A family of protozoans of the sub-order HAEMOSPORINA.

Haemosporina A sub-order of protozoans of the EUCOCCIDIA in which SYZYGY does not occur and in which a motile zygote (the ookinete) is formed. Members are heteroxenous, the sexual phase being completed within an insect and the asexual phase taking place in a vertebrate. Three families are distinguished: (a) Plasmodiidae, consisting of one genus, PLASMODIUM. The invertebrate host (vector) is a mosquito, and schizogony occurs in the erythrocytes (and other cells) in the vertebrate. A pigment (haemozoin) is formed in intraerythrocytic forms. (b) Haemoproteidae, in which the insect vectors are not mosquitoes, and in which schizogony does not occur in erythrocytes; haemozoin is formed in the gametocytes, which develop in erythrocytes. Haemoproteids are weakly or non-pathogenic; they occur in mammals and birds. (c) Leucocytozoidae, in which the vectors are not mosquitoes, and in which schizogony does not occur in the erythrocytes; haemozoin is not produced. Species occur mainly in birds; few are pathogenic. Despite these differences, the life cycles of all members of the sub-order are basically similar.

haemozoin (see PLASMODIUM)

Hafnia A genus of gram-negative, motile bacteria of the ENTEROBACTERIACEAE; strains occur e.g. in the alimentary tract of animals, including man. The organisms give a positive CITRATE TEST and grow on deoxycholate-citrate agar and on media containing potassium cyanide; at 37 °C the METHYL RED TEST is typically negative and the VOGES–PROSKAUER TEST may be positive or negative. Lactose is not fermented. One species: *H. alvei*.

HAI test (see HAEMAGGLUTINATION-INHIBITION TEST)

halazone (see CHLORINE)

Halimeda A genus of green algae (CHLOROPHYTA).

halo blight (of beans) (see BEANS, BACTERIAL DISEASES OF)

Halobacteriaceae A family of asporogenous, aerobic bacteria* which grow only in media containing at least 15–20% sodium chloride; species occur in the Great Salt Lake (Utah) and the Dead Sea, and in salted fish etc. A high concentration of NaCl maintains the integrity of the CELL WALL in these organisms, and is (apparently) necessary for normal metabolic activity. The cell wall—which lacks peptidoglycan, and which is composed mainly of lipoprotein—disintegrates in unfavourable ionic conditions. The organisms are chemoorganotrophs; metabolism is respiratory (oxidative) (see however PURPLE MEMBRANE), and reproduction occurs by binary fission. The cells contain CAROTENOID pigments of some shade of red. The genera are:

Halobacterium. Polarly-flagellate, or non-motile, bacilli approximately 0.5–1 × 1–4 microns; GAS VACUOLES are common. *H. halobium* is a motile, oxidase-positive organism containing a PURPLE MEMBRANE.

Halococcus. Non-motile cocci, diameter 1±0.5 microns. High concentrations of NaCl are required for *growth*, but the cell walls are said to be more stable than those of

Halobacterium in low ionic concentrations. *H. morrhuae* (formerly *Sarcina morrhuae*) is oxidase-positive; cells occur in regular or irregular groups. *Probably gram-negative.

halogens (and their compounds) (as antimicrobial agents) (See: FLUORINE, CHLORINE, BROMINE and IODINE.)

halophilic Refers to those organisms (*halophiles*) whose requirement for salt (sodium chloride)—as an environmental factor—exceeds that of other (non-halophilic) organisms; extreme halophiles (e.g. members of the HALOBACTERIACEAE) require for growth salt concentrations of at least 15–20%.

hamycin (see POLYENE ANTIBIOTICS)

hanging drop A procedure used for detecting the presence of *motile* microorganisms in a fluid medium, e.g. a liquid CULTURE (see MOTILITY). The method is shown diagrammatically (in cross-section):

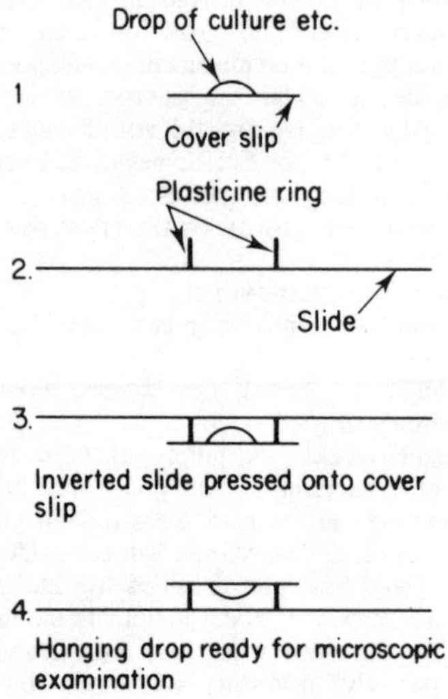

HANGING DROP

Hanks' BSS (Hanks' balanced salt solution) One of a number of balanced salt solutions, i.e. solutions used in TISSUE CULTURE to provide a normal ionic, pH and osmotic environment for cell development. (see also EARLE'S BSS) Without supplement, such solutions are often used for washing cells; when supplemented with nutrients etc. they are used as growth or maintenance media. Hanks' BSS, supplemented with protein, is used as a viral TRANSPORT MEDIUM. Hanks' BSS contains (in grams/100 ml distilled water): NaCl (0.8), KCl (0.04), $CaCl_2$ (0.014), $MgSO_4.7H_2O$ (0.01), $MgCl_2.6H_2O$ (0.01), $Na_2HPO_4.H_2O$ (0.006), KH_2PO_4 (0.006), $NaHCO_3$ (0.035), glucose (0.1) and phenol red (0.002). The pH of the medium is 7–7.2. Antibiotics are sometimes added (as antibacterial agents).

Hansen's bacillus *Mycobacterium leprae.*

Hansen's disease (see LEPROSY)

Hansenula A genus of fungi within the family SACCHAROMYCETACEAE; species have been isolated from e.g. tree exudates, soil, and sewage. The vegetative form of the organisms may be a single spheroidal or elongated cell, a pseudomycelium, or a true mycelium—according to species and conditions; asexual reproduction occurs by budding. The asci each contain 1–4 smooth-surfaced ascospores which may be spherical, hemispherical, Saturn-shaped, or shaped like a bowler hat. Homothallic and heterothallic species have been reported. Most species exhibit fermentative as well as oxidative metabolism; non-fermentative species include *H. wingei. Hansenula* spp assimilate nitrate.

haploid Refers to any (eucaryotic) nucleus or cell which has a PLOIDY of *one.* (cf. DIPLOID, POLYPLOID)

haploidization (see PARASEXUALITY IN FUNGI)

haplont (1) (*noun*) An organism in whose life cycle only the zygote is diploid. (2) (*noun*) The haploid form, thallus etc. of an organism—e.g. a *gametophyte* (see ALTERNATION OF GENERATIONS). (3) (*adjective*) Refers to the haploid phase of an organism.

haplophase In the life cycles of those organisms which reproduce sexually: the haploid phase which begins at MEIOSIS and ends during fertilization. In many of the higher fungi haplophase and DIPLOPHASE are separated by the DIKARYOPHASE.

Haplosporea A poorly-understood class of protozoans of the SPOROZOA; members of the class are parasitic in fish and invertebrates. Haplosporeans form spores. Sexual reproduction is unknown.

hapten (*immunol.*) A substance which elicits antibody-formation *only when combined with other molecules or particles* (*carriers*); if administered on its own it fails to stimulate antibody production. The *unbound* hapten is able to combine with antibodies raised against the hapten-carrier complex; such combination may or may not result in the precipitation of a hapten-antibody complex. A given substance may act as a hapten in one species but may act as a complete antigen in other species; for example, the pneumococcal capsular polysaccharides are haptenic in the rabbit but antigenic in man. A hapten may be artificially bound to a protein in order to increase the immunological *specificity* of the latter. *Auto-coupling haptens* bind spontaneously with tissue carriers *in vivo* and thus become antigenic.

hapteron (*mycol.*) (see CYATHUS)

haptonema (*pl. haptonemata*) A thread-like organelle found on members of the Haptophyta, e.g. *Chrysochromulina.* In fine structure the haptonema bears some resemblance to the *eucaryotic* FLAGELLUM; however (in contrast) the haptonema contains fewer microtubules, and the sheath comprises several concentric membranes—the inner membrane enclosing the microtubules. The haptonema appears to be used to attach the organism to solid objects in the environment.

Haptophyta A division of the ALGAE. Species of the Haptophyta are typically biflagellate (isokont), unicellular organisms which possess a HAPTONEMA; they are predominantly marine algae. Many species have a mineralized CELL WALL. The organisms contain CHLOROPHYLLS *a* and *c*, *β*-carotene, and fucoxanthin (see CAROTENOIDS); intracellular storage products include CHRYSOLAMINARIN. Genera include *Apistonema*, *Coccolithus*, and *Chrysochromulina.*

Hartig net (see MYCORRHIZA)

haustorium (*mycol.*) A specialized hyphal structure formed by certain plant-parasitic fungi—e.g. *Peronospora*, *Erysiphe*—by means of which the fungus obtains nutrients from its host; an haustorium is formed *within* a host cell by the development of a small hyphal branch which penetrates the host cell wall—though in at least some cases the haustorium remains outside the cell membrane of the host cell. According to the species of fungus, an haustorium may be club-shaped, spherical, or elongated—the latter being branched or unbranched. (Haustoria are also formed by the mycobionts in certain lichens.)

HB-Ag, HBAg (see HEPATITIS B ANTIGEN)

HBB (see BENZIMIDAZOLE DERIVATIVES)

HD_{50}(CH_{50}; $C'H_{50}$) 50% haemolytic dose; the quantity of COMPLEMENT which lyses 50% of a standardized suspension of *sensitized* erythrocytes. In COMPLEMENT-FIXATION TESTS the HD_{50} may be used as the unit of complement in place of the MHD (*q.v.*); this gives a more accurate end-point (since the graph '%lysis of cells *versus* quantity of complement' is sigmoidal).

heavy chain A polypeptide component of the IMMUNOGLOBULIN molecule; two heavy chains occur in each molecule of IgA, IgE and IgG, and ten in the molecule of IgM. (The basic structure of the immunoglobulin molecule is described under IGG.) Each immunoglobulin class has a class-specific heavy chain; these chains differ e.g. in their antigenic determinants, and are designated *alpha*, *gamma*, *delta*, *epsilon* and *mu*—corresponding to IgA, IgG, IgD, IgE and IgM respectively. Heavy chain molecular weights are: IgA: 55,000, IgG: 50,000, IgD: ?65,000, IgE: 65,000, IgM: 70,000. Immunoglobulins may be divided into *sub*-classes on the basis of minor differences in their heavy chains. (see also LIGHT CHAIN and FAB FRAGMENT)

heavy metals (and their compounds) (as antimicrobial agents) Although *trace* amounts of certain heavy metal ions are essential for the growth of most microorganisms, higher concentrations may exhibit antimicrobial activity. They are believed to inactivate enzymes by binding to certain groups in proteins (particularly sulphhydryl groups); in sufficiently high concentrations such ions *precipitate* proteins. The reaction between some metals (e.g. mercury) and sulphhydryl groups may be reversed by the addition of substances such as cysteine, glutathione, hydrogen sulphide etc.; such metals act as *microbistatic* agents. Some metals in elemental form exhibit an OLIGODYNAMIC EFFECT; others are active either as inorganic salts or as components of organic complexes. Occasionally the antimicrobial activity of one metal may be antagonized by the presence of another; thus, the inhibition of *Lactobacillus arabinosus* by zinc may be reversed by the addition of manganese or strontium. Certain metal ions may contribute to, or be essential for, the activity of other antimicrobial agents—see *copper* (below).

Antimony. Compounds of antimony have long been used as general therapeutic agents—though their toxic effects were frequently greater than the benefits they bestowed. Antimonials are currently used to treat leishmaniasis and related diseases—e.g. espundia and kala-azar. It is believed that *trivalent* antimony is necessary for antimicrobial activity, and that pentavalent compounds owe their activity to *in vivo* conversion to the trivalent state. Antimonials have been shown to combine with essential sulphhydryl groups in trypanosomal enzymes involved in glucose metabolism; such inhibition may be reversed by thiol compounds, e.g. cysteine. Trivalent antimonials include (a) Potassium antimonyl tartrate (*tartar emetic*)—formerly used for the treatment of trypanosomiasis, and (b) *stibophen* (sodium antimony *bis*(pyrocatechol-3,5-disulphonate)). Pentavalent antimonials are often unstable; *stibanilic acid* (*p*-aminobenzenestibonic acid) and its sodium salt (*stibamine*) are constituents of a number of widely-used drugs—e.g. ethyl stibamine (*neostibosan*) which is used for the treatment of espundia and kala-azar.

Bismuth. The most important of the antimicrobial bismuth compounds is probably *glycobiarsol* (bismuth *N*-glycolyl-*p*-arsanilate) which is amoebicidal and is used (with other agents) for the treatment of amoebic dysentery.

Copper. Trace amounts of copper are required by a variety of microorganisms; thus, copper-containing proteins include e.g. cytochrome aa_3 and tyrosinase. Copper exhibits the oligodynamic effect; its organic compounds (e.g. copper naphthenate) and inorganic salts are strongly bacteriostatic and fungistatic and are used in industry and in agriculture as antifungal agents. (see also BORDEAUX MIXTURE, BURGUNDY MIXTURE and CHESHUNT COMPOUND) *Copper sulphate* has been used to combat algal BLOOMS and as an agricultural antifungal agent. Copper ions are apparently necessary for the antifungal action of 8-HYDROXYQUINOLINE; copper-8-hydroxyquinoline is used in agriculture and for the rot-proofing of materials. (Media prepared from some brands or batches of peptone have been found to be inhibitory to microbial growth owing to the high copper content of the peptone; similarly, media prepared with water that has been distilled in a copper still may be inhibitory.)

Gold. Gold salts were formerly used for the treatment of a variety of diseases (e.g. tuberculosis, syphilis) though without significant results; they are no longer used for the treatment of diseases of microbial aetiology.

Mercury. Many mercury compounds are effective antibacterial and antifungal agents; their action is generally microbistatic and reversible (see above). Some mercury compounds inhibit certain viruses; all are virtually without effect against bacterial spores—though they may delay germination. *Inorganic mercurials* are toxic, irritant to tissues, and are readily inactivated by organic matter; blood or serum may reverse the antimicrobial activity of these compounds. *Mercuric chloride* (corrosive sublimate) is used for the disinfection of inanimate objects (when other methods are unsuitable) and as a preservative for timber, paper etc. *Mercurous chloride* (calomel) has restricted use as a fungicide in horticulture, e.g. it is used against *club root*. *Yellow mercuric oxide* has been used in an ointment for the treatment of eye infections. *Aminomercuric chloride* (ammoniated mercury) has been used in ophthalmic preparations and for the treatment of superficial mycoses. The *organic mercurials* are generally less toxic, less irritant, and more effective antimicrobial agents—though none are sporicidal. *Mercurochrome* (*merbromin*; the di-sodium salt of 2′,7′-dibromo-4′-hydroxymercurifluorescein) was the first of the organic mercurial antiseptics but is among the least effective; it is readily inactivated by blood etc. and is no longer widely used. *Thiomersal* (thimerosal, merthiolate)—sodium ethylmercurithiosalicylate—has low toxicity and is an effective antibacterial agent—see THIOMERSAL. *Nitromersal* (metaphen)—4-nitro-3-hydroxymercuri-*o*-cresol anhydride—has properties similar to those of thiomersal but is slightly less effective. *Phenylmercuric-nitrate, borate* and *acetate* have similar activities; the acetate is the most soluble. The nitrate has been used as an antiseptic, as a preservative for pharmaceutical preparations, and for the treatment of superficial mycoses. The acetate and nitrate are used as preservatives for paper, leather etc. and for the treatment of seeds.

Nickel. Certain inorganic nickel salts have been used as agricultural antifungal agents, e.g. as eradicants in the control of certain wheat rusts. Their use is limited by their phytotoxicity.

Silver. The antibacterial action of silver is believed to be due to the combination of silver *ions* with microbial proteins. Silver ions are *slowly* released from such protein complexes—thus accounting for the prolonged bacteriostatic activity of silver and its salts. Silver and its *insoluble* salts liberate insufficient ions for *bactericidal* action, but they are bacteriostatic at extremely low concentrations. *Katadyn* (spongy silver containing traces of palladium, gold etc.) is highly bacteriostatic and has been used (deposited on sand or other filtering medium) for the disinfection of water; its activity is reduced by organic and certain inorganic materials, and it is inactive against bacterial spores, most fungi, and protozoa. *Colloidal* silver preparations are used in dentistry and medicine e.g. for treating mucous membrane infections and burns. *Silver proteinates* (e.g. *protargol*, *argyrol*) are less irritant than most silver salts, but are less effective as antiseptics; their activity depends on the amount of silver ions *liberated* rather than on the proportion of silver in the preparation. The silver proteinates are bacteriostatic; they possess astringent properties and are sometimes used for treating infections of the eye, urinogenital tract etc. Of the *soluble* silver salts only the nitrate is widely used; this has astringent though irritant and corrosive properties. It has been used as an external antiseptic e.g. for the treatment of burns, and in the *Credé procedure* (see GONORRHOEA). Silver nitrate may be bactericidal or bacteriostatic. *Ammoniacal silver nitrate* is used in dentistry for the disinfection of tooth cavities.

Tin. Some organotin derivatives have been used as disinfectants in hospitals and laundries—though they are mainly used in agriculture and industry e.g. for disease control and as preservatives. (see also FENTIN)

Zinc. Trace amounts of zinc are required by many microorganisms; it is an essential constituent of certain enzymes (e.g. alcohol dehydrogenase) and zinc ions function as activators for a number of enzymes. However, in effective concentrations, zinc and zinc compounds are useful antifungal agents. Thus, zinc oxide and zinc undecylenate are used topically for the treatment of certain superficial mycoses (e.g. athlete's foot) and zinc is a constituent of a number of agricultural antifungal agents—e.g. ZINEB, ZIRAM.

HeLa An ESTABLISHED CELL LINE which was started, in 1952, from a human cervical carcinoma; the cells are epithelioid. HeLa cell cultures support the replication of a wide range of viruses—including polioviruses (types 1, 2 and 3), poxviruses (e.g. variola and vaccinia), adenoviruses, mumpsvirus, influenza *B* virus.

Helber chamber (see COUNTING CHAMBER)

helicosporae (see SACCARDIAN SYSTEM)

heliozoans Protozoans of the subclass Heliozoia (class Actinopodea, superclass SARCODINA). Heliozoans are free-living organisms which occur mainly in freshwater habitats and which feed chiefly on flagellates, ciliates, and microalgae; most species are free-floating organisms while others (e.g. *Clathrulina*) are commonly attached to the substratum by a stalk. The heliozoans are spherical cells; the (many) pseudopodia which radiate from each cell are typically AXOPODIA, though filopodia are formed by some species. The organisms are typically naked, but some species bear silicaceous scales or spines. According to species, one or more nuclei, together with e.g. contractile vacuole(s) and food vacuoles, are found in the endoplasm; the ectoplasm (outer zone of cytoplasm) is often highly vacuolated. A central capsule is *not* present (cf. RADIOLARIANS). Reproduction normally occurs by binary fission; a quasi-sexual process, AUTOGAMY, has been recorded in some species. In the species of one order (Centrohelida) the axopodia radiate from a central body, the *centroplast*; this organelle appears to play some part in the development of the mitotic spindle. Centrohelids typically possess a silicaceous test or skeleton of scales or spines. In members of the order Actinophryida, both centroplast and skeletal structures are absent. Members of the remaining order, Desmothoracida, have reticulate skeletal structures composed of silica-impregnated organic material; they do not contain centroplasts. Common heliozoans include: (a) *Actinophrys sol* (the 'sun animalcule'). A small organism (approximate diameter: 50 microns) with a single, central nucleus; a member of the Actinophryida. (b) *Actinosphaerium* spp. Large, multinucleate organisms (200–800 microns or more) in which the highly vacuolated ectoplasm is sharply contrasted with the granular endoplasm; a member of the Actinophryida. (c) *Raphidiophrys pallida*. A small, uninucleate centrohelid (about 50 microns in diameter) in which numerous silicaceous spicules are embedded in an outer gelatinous layer—particularly at the bases of the axopodia.

Helminthosporium A genus of fungi within the order MONILIALES; species include a number of plant pathogens—e.g. *H. oryzae*, the causal agent of *brown spot* disease of rice. The vegetative form of the organisms is a septate mycelium. Each (unbranched) hypha-like conidiophore bears—terminally and laterally—a number of dark-coloured, elongated or ovoid conidia—each conidium containing two or more transverse septa; conidia are not formed in chains. Several species of *Helminthosporium* are known to have ascigerous sexual stages. (see also VICTORIN)

Helmstetter–Cooper Model (see GROWTH (bacterial))

Helotiales An order of fungi within the class DISCOMYCETES; the order includes saprophytes and parasites or pathogens of higher plants—e.g. *Monilinia*. Constituent species form inoperculate, unitunicate asci in sessile or stipitate ascocarps; the wall of the ascus is *not* greatly thickened at the apex (i.e. in the region of the apical pore), and in most species the ascospores are *not* filiform (thread-like)—cf. OSTROPALES. The ascocarp does not develop within a fungal stroma.

helotism (*mycol.*) Literally, *serfdom. Helotism* has been used to refer to the (supposedly) subservient role of the alga in a lichen.

Helvella A genus of fungi of the order PEZIZALES. The fruiting body of *Helvella* spp (an APOTHECIUM) consists of a stipitate cup-shaped or saddle-shaped pileus; the smooth or warty, quadrinucleate and guttulate ascospores are commonly hyaline or pale brown.

heme (see HAEM)

Hemiascomycetes A class of fungi of the subdivision ASCOMYCOTINA; constituent species are distinguished from those of other classes in that (a) they form completely naked asci—i.e. the asci are not formed on or in an organized fruiting body (ASCOCARP), and (b) the asci do not develop from a system of *ascogenous hyphae* (see ASCUS). The class Hemiascomycetes includes the majority of yeasts (e.g. *Saccharomyces* spp) and a number of plant-pathogenic fungi. The constituent orders are ENDOMYCETALES, Protomycetales, and TAPHRINALES.

Hemibasidiomycetes A former name for the class TELIOMYCETES.

hemicellulose An heterogenous group of polysaccharides found in the amorphous matrix of the cell walls of plants and certain algae. Hemicelluloses include: (a) the *cellulosans* (hemicellulose A), e.g. XYLANS, arabinogalactans, and (b) the *polyuronides* (hemicellulose B), polymers of xylose and glucuronic acid, arabinose and galacturonic acid. (cf. CELLULOSE)

Hemispeira A genus of protozoans of the order THIGMOTRICHIDA.

Hemitrichia A genus of the ACELLULAR SLIME MOULDS.

Hemming's filter (see FILTRATION)

HEP High egg passage—refers to the number of times that a given strain of virus has been transferred from one chick embryo to another. This procedure (*serial passage*) is employed to reduce the virulence of the virus for a non-avian host. The HEP strain of the FLURY VIRUS has undergone over 180 transfers.

HEp-2 An ESTABLISHED CELL LINE which was started from a human laryngeal carcinoma; the cells are epithelioid. HEp-2 cell cultures support the replication of a wide range of viruses—including herpes simplex, respiratory syncytial virus, poliovirus type 3, influenza *B* virus, and the measles and mumps viruses.

hepatitis (viral hepatitis) In man, either of two distinct viral diseases which are characterized by inflammation of the liver and which frequently involve jaundice. (*Non*-viral icteric diseases—e.g. Weil's disease—are not generally referred to as hepatitis.) In both diseases the early symptoms often include malaise, loss of appetite, and nausea; subsequent clinical and biochemical patterns are also similar. Infection by hepatitis viruses may be inapparent (a *carrier* state is recognized), or may result in mild and non-icteric hepatitis, or severe jaundice and death. (a) *Infectious hepatitis* (*hepatitis A; epidemic hepatitis*). The causal agent (hepatitis virus *A*) can be transmitted orally and parenterally; the incubation period is 2–6 weeks. Infectious hepatitis often occurs as a mild disease—the mortality rates generally being much lower than those in serum hepatitis. (b) *Serum hepatitis* (*hepatitis B*; formerly: *homologous serum jaundice*). Transmission of the causal agent (HEPATITIS VIRUS B) occurs parenterally and probably occurs orally and venereally. The incubation period is 2–6 months. In serum hepatitis the blood contains HEPATITIS B ANTIGEN during the late incubation period and in the acute phase of disease; it is generally not detectable during the convalescent stage, but in some patients it persists for months or years.

Treatment of hepatitis. An antiserum against HBAg has been used to provide PASSIVE IMMUNITY in persons known to have become infected with hepatitis virus B. Current research aims at developing vaccines consisting of purified suspensions of the 22 nm spherical particles of HBAg (see HEPATITIS B ANTIGEN)—or of polypeptides derived from these particles.

Laboratory diagnosis. For *serum hepatitis* this includes the demonstration of HBAg in serum; methods used include immunoelectrophoresis, radioimmunoassay, and a complement fixation test. Biochemical tests include an assay of serum alanine aminotransferase—the level of which becomes raised in cases of serum hepatitis and infectious hepatitis.

hepatitis B antigen (synonyms: Australia antigen; HBAg; HB-Ag; serum hepatitis antigen, SH antigen; hepatitis-associated antigen, HAA) An ANTIGEN detectable in the blood of patients suffering from *serum* HEPATITIS and in the sera of *carriers* of the disease. The presence of the antigen (in serum) correlates with the presence of small particles of DNA-free lipoprotein; most of these particles are spherical (approx. 22 nm in diameter) while others are tubular or filamentous. Sera containing these particles also contain spherical structures of about 45 nm diameter (Dane particles). The Dane particles appear to contain an outer and an inner protein shell—the latter enclosing double-stranded DNA; they appear to be the causal agents of serum hepatitis. Antibodies homologous to the antigen of the 22 nm lipoprotein particles react with the *surface* antigen of the Dane particles; the latter antigen has been designated HB_sAg. (see also HEPATITIS VIRUS B)

hepatitis virus B The causal agent of *serum* HEPATITIS; it appears to correspond to the *Dane particle*—see HEPATITIS B ANTIGEN. The agent is resistant to many of the chemical and physical agents which are generally inhibitory to viruses; it has been inactivated by e.g. β-PROPIOLACTONE, formalin, and hypochlorite, and has been reported to be inactivated, in serum, by exposure to 98 °C for 1 minute.

HEPES *N*-2-Hydroxyethylpiperazine-*N′*-2-ethanesulphonic acid; a pH buffer used e.g. in TISSUE CULTURE. The pK_a value of the buffer, at 25 °C, is approximately 7.5.

herbivorous (of protozoans) Refers to protozoans which feed on bacteria and/or algae.

herd immunity and **herd structure** (see EPIDEMIOLOGY)

Herellea vaginicola (see ACINETOBACTER)

Hericium A genus of fungi of the family HYDNACEAE.

hermaphroditism The condition in which an individual organism possesses both male and female sexual organs; hermaphroditic fungi may be homothallic or heterothallic—see HOMOTHALLISM and HETEROTHALLISM.

herpangina An acute, infectious pharyngeal disease which affects mainly children and adolescents; the causal agent is any of a number of strains of group *A* coxsackieviruses (see ENTEROVIRUS). Following an incubation period of approximately one week there is a sudden onset with fever, sore throat, stiff neck, headache, vomiting (or nausea), and the formation of vesicles (one to several millimetres in diameter) in the pharyngeal region; the vesicles subsequently ulcerate. The disease is usually self-limiting.

herpes-zoster (see SHINGLES)

Herpesvirus A genus of VIRUSES whose hosts include man, other primates, amphibians, birds, and other vertebrates; in man, herpesviruses are associated with a variety of localized and systemic diseases—infection characteristically remaining latent for long periods between recurrent outbreaks of clinical disease. The complex, spheroidal, enveloped *Herpesvirus* virion is approximately 150 nm in diameter; enclosed by the lipid-glycoprotein envelope is the icosahedral capsid (100 nm in diameter, 162 capsomeres) which contains the linear, double-stranded DNA genome. Assembly of the *Herpesvirus* capsid occurs within the *nucleus* of the host cell; envelopment of the capsid occurs as the latter buds through the host's nuclear membrane during the maturation process. There appears to be no antigen common to all herpesviruses, but groups within the genus can be distinguished by specific capsid antigens. In clinical specimens herpesviruses tend to be unstable at room temperatures; they are inactivated by lipid solvents (e.g. ether) and by lipases. Herpesviruses can be cultivated in certain types of tissue culture—e.g. human embryonic kidney, human amnion, HeLa, HEp-2, chick embryo, African green monkey kidney, WI-38; cytopathic effects (CPE) typically include the formation of syncytia (multinucleate 'giant cells'). The characteristic inclusion body of herpesviruses is an eosinophilic, Feulgen-negative intranuclear body (*Lipschütz body*) formed late in infection. *Diseases of Herpesvirus aetiology*. The *Herpes simplex* viruses can cause e.g. gingivostomatitis (inflammation of the gums and mouth), COLD SORE, herpetic whitlow, and encephalitis, and can give rise to lesions on the eye, genitals, and anal region; herpes simplex type 2 *may* be associated with cervical carcinoma. (see also CYTARABINE and IUDR) The *varicella-zoster virus* is the causal agent of CHICKENPOX and SHINGLES. *Cytomegaloviruses* can cause diseases of e.g. the salivary glands, and may cause degeneration of the central nervous system in the newborn. *Epstein–Barr virus* (EB virus) is the causal agent of INFECTIOUS MONONUCLEOSIS, and *may* be associated with Burkitt's lymphoma and nasopharyngeal carcinoma. *B virus* occurs naturally in the rhesus and other monkeys; this virus can cause encephalitis in man. Herpesviruses which infect other animals include the *Lucké virus* (the causal agent of renal carcinoma in the frog—*Rana* sp), and *Marek's disease virus* (an oncogenic agent in birds).

Herpetomonas A genus of flagellate protozoans of the sub-order TRYPANOSOMATINA; species are gut parasites of invertebrates (chiefly insects).

Herpetosoma A sub-genus of TRYPANOSOMA within the group STERCORARIA; most species are parasitic in rodents. The organisms are some 30 microns in length; a subterminal kinetoplast lies between the nucleus and posterior extremity of the cell, and there is a free flagellum. Species include *T.(Herpetosoma)lewisi*.

Herxheimer-type reaction Adverse effects (e.g. exacerbation of lesions) which may be caused by carrying out chemotherapy on patients suffering from certain chronic diseases, e.g. brucellosis, syphilis. The reaction is thought to be either one of HYPERSENSITIVITY in respect of antigens released by the organisms, or a response to the release of endotoxins—or perhaps a combination of both.

heteroauxin 3-indole-acetic acid (see AUXINS).

Heterobasidion A genus of lignicolous fungi of the family POLYPORACEAE. The sole species, *H. annosum* (formerly classified as *Fomes annosus*) forms perennial bracket-shaped (sometimes resupinate) fruiting bodies which are characteristically corky or woody; the bracket-shaped fructifications are typically rusty-brown on the upper surface and pale-coloured on the lower, porous surface. The context of the fruiting body consists of hyphae which exhibit dimitic organization and which do not have clamp connections; in bracket-shaped fructifications of several years' standing the hymenophore consists of vertically-stacked ('stratified') layers of vertically-orientated tubules. The non-pigmented basidiospores are spheroidal, approximately 5 × 4 microns. *H. annosum* is a pathogen of certain trees—particularly conifers, but also e.g. the red oak (*Quercus borealis*). (see also TIMBER, DISEASES OF)

heterocaryosis (see HETEROKARYOSIS)

heterocysts Cells which occur in the trichomes of *some* filamentous blue-green algae, e.g. *Anabaena*, *Nostoc* and *Tolypothrix*; they may be important sites of NITROGEN FIXATION. Heterocysts may occupy terminal or intercalary positions in the trichome. They are produced by the differentiation of vegetative cells during active growth; their formation is inhibited by the presence of nitrogenous compounds. A heterocyst is usually rounded or cylindrical and is commonly larger than the adjacent vegetative cell(s); external to the cell wall there is a conspicuous multilayered *envelope*. The heterocyst communicates with adjacent cell(s) by a pore channel; in the region of the pore channel the envelope is considerably thickened, and this (polar) part of the heterocyst is referred to as the *polar thickening* or *polar nodule*. The cytoplasm is agranular (unlike vegetative cells) and contains a system of membranes; it usually contains less phycocyanin than is found in the vegetative cells.

heterocytotropic antibody A CYTOPHILIC ANTIBODY which is able to attach both to cells of the species in which it was formed and to certain cells of other species. (cf. HOMOCYTOTROPIC ANTIBODY)

heteroecism The obligate use of at least two distinct host species by certain parasitic fungi for the completion of their life cycles.

heterofermentation Fermentation in which more than one major product is formed; the term is frequently used as a synonym of HETEROLACTIC FERMENTATION.

heterogenote (see MEROZYGOTE)

heteroimmunization (see IMMUNIZATION)

heterokaryon (heterocaryon) In fungi, a hyphal cell, mycelium, organism, or spore which contains genetically different nuclei (cf. HOMOKARYON).

heterokaryosis (heterocaryosis) In fungi: the condition in which genetically different nuclei exist in the same hyphal cell, mycelium, organism, or spore. Heterokaryosis may arise e.g.: (a) by a mutation in a binucleate or multinucleate homokaryotic cell or mycelium; (b) by the entry of a genetically different nucleus into a HOMOKARYON—e.g following hyphal fusion. Heterokaryosis plays an important role e.g. in PARASEXUALITY IN FUNGI, and is a characteristic feature in the life cycles of basidiomycetes.

heterokont Refers to a pair of flagella of *dissimilar* length (cf. ISOKONT).

Heterokontae Former name for XANTHOPHYTA.

heterolactic fermentation A type of fermentation in which a range of products—including lactic acid, acetic acid and/or ethanol, and carbon dioxide—results from the fermentation of glucose (see Appendix III(a) (ii)). (cf. HOMOLACTIC FERMENTATION) A key enzyme in this pathway is *phosphoketolase* which cleaves xylulose-5-phosphate to acetyl phosphate and glyceraldehyde-3-phosphate (cf. HMP pathway—Appendix I(b)). (If—instead of glucose—a *pentose* is the substrate, acetate may be formed instead of ethanol since the reducing power generated by the oxidative decarboxylation of glucose is absent.) Heterolactic fermenters include *Leuconostoc* spp and certain species of *Lactobacillus* (e.g. *L. brevis* and *L. fructivorans*). *Leuconostoc* spp form D(−)lactate, while *L. brevis* and *L. fructivorans* produce DL-lactate.

heterologous (*immunol.*) (1) Of an antibody (or antigen): one which is not HOMOLOGOUS to a given antigen (or antibody). An heterologous antibody, for example, may be a CROSS-REACTING ANTIBODY or may not combine at all with a given antigen. (2) May refer to the source of e.g. a serum: derived from a species other than that being referred to. (cf. HOMOLOGOUS)

heteromerous (1) (of a lichen) Refers to a lichen in which the phycobiont occurs in a definite layer within the thallus (cf. HOMOIOMEROUS). (2) (*mycol.*) In an agaric: refers to a pileus and stipe that consist of tissue which is clearly differentiated into two types of cell—SPHAEROCYSTS and hyphae; such differentiation is characteristic of members of the Russulaceae (cf. HOMOIOMEROUS).

heteromixis A term which refers, collectively, to DIMIXIS, DIAPHOROMIXIS, and SECONDARY HOMOTHALLISM.

heteromorphic alternation of generations An ALTERNATION OF GENERATIONS in which the sporophyte and gametophyte are morphologically dissimilar.

heterophile antigens (heterogenetic antigens) Immunologically-related antigens found in *unrelated* species. The heat-stable glycolipid *Forssman antigen* occurs e.g. in *Streptococcus pneumoniae*, on the erythrocytes of the horse, sheep and dog, and in some tissues of the guinea-pig. It is absent in the rabbit.

heteroploid Refers to any (eucaryotic) nucleus or cell in which the genetic complement differs from that characteristic of cells of the species. An heteroploid cell may be EUPLOID (e.g. a polyploid cell derived from a normal, diploid parent cell) or it may be ANEUPLOID.

heterothallism (*mycol.*) In sexual reproduction: the system in which syngamy occurs only between genetically-*different* gametes and in which the gametes derive from different thalli. (cf. HOMOTHALLISM)

In *morphological* heterothallism syngamy occurs between two gametes which derive, respectively, from male and female thalli (see DIMIXIS). In all other forms of heterothallism (collectively: *physiological* heterothallism) compatibility between gametes is determined by the presence of particular *mating type alleles* at certain loci in the chromosomes of the gametes; gametes which have identical alleles at a given mating type locus are *incompatible*. (Thus, syngamy can occur only between gametes which have different alleles at mating type loci.) In contrast to morphological heterothallism, in physiological heterothallism morphological differences are not generally exhibited between the thalli, or between the male and female gametes and gametangia, of the various mating types.

In *bipolar heterothallism* compatibility is

determined at *one* mating type locus. In some (*dimictic*) bipolar heterothallic organisms there are but two mating types—i.e. the mating type locus can be occupied by one of only two alleles; if these alleles be designated *A*, *a* then mating can occur if one individual (mating type) possesses the *A* allele and the other individual the *a* allele. In other bipolar heterothallic organisms there are many mating types—i.e. the mating type alleles correspond to an *allelomorphic series* (see ALLELE).

In *tetrapolar heterothallism* compatibility is determined at *two* mating type loci; if these loci be designated *A*,*B* (with alleles *a*,*b* respectively) four mating types can be distinguished: *AB*, *ab*, *Ab*, *aB*. Mating can occur only between individuals whose nuclei contain *complementary* alleles at *both* loci; thus, compatible mating types are *AB*:*ab* and *Ab*:*aB*. In some tetrapolar heterothallic organisms there are more than four mating types—the mating type alleles at one (or each) locus corresponding to an allelomorphic series.

Heterothallism provides a constraint on inbreeding (self-fertilization) and, in consequence, increases the potential for genetic reassortment among the individuals of a species. (see also COMPATIBILITY)

Heterotrichida An order of protozoans of the sub-class SPIROTRICHIA.

heterotroph (organotroph) Any organism which requires a range of exogenous organic compounds for growth and reproduction (cf. AUTOTROPH). An heterotroph which obtains energy from light (see PHOTOSYNTHESIS) is said to be a *photoheterotroph* or *photoorganotroph*; one which lives chemotrophically (see CHEMOTROPH) is referred to as a *chemoheterotroph* or *chemoorganotroph*. Photoheterotrophs include bacteria of the family Rhodospirillaceae; the chemoheterotrophs include an extensive range of microorganisms.

heteroxenous (*adjective*) Refers to a parasitic organism which completes its life cycle in or on two or more host species. (see also HETEROECISM)

heterozygote A cell which is HETEROZYGOUS in respect of a *specified* gene, or genes.

heterozygous (1) Of eucaryotic genes: refers to the existence of one or more pairs of different alleles at *specified* loci on homologous chromosomes; thus, a cell (or nucleus) is said to be heterozygous for a particular gene or genes. (2) Of bacterial genes: refers to the existence of one or more pairs of different alleles at *specified* loci in a MEROZYGOTE.

hexachlorophene (hexachlorophane) (see BISPHENOLS)

hexamer (*virol.*) In an icosahedral capsid: a group of *six* capsomeres in a particular geometrical arrangement (cf. PENTAMER).

hexamethylenetetramine (hexamine; methenamine) A condensation product of ammonia and formaldehyde. It has no antimicrobial activity but slowly liberates FORMALDEHYDE (a powerful disinfectant) under acid conditions. It was widely used as a urinary antiseptic; it is not active against urea-splitting organisms since these raise the pH of urine by producing ammonia. Hexamethylenetetramine is generally used as the salt of MANDELIC ACID (methenamine mandelate) or hippuric acid; methenamine mandelate is active against *Escherichia coli*, *Staphylococcus aureus*, and certain strains of streptococci, and may still be used in cases where resistance precludes the use of antibiotics.

Hexamita A genus of protozoans of the order DIPLOMONADIDA.

hexon (*virol.*) In an icosahedral capsid: a capsomere to which *six* other capsomeres are adjacent (cf. PENTON).

hexose diphosphate pathway (see EMBDEN–MEYERHOF–PARNAS PATHWAY)

hexose monophosphate pathway (hexose monophosphate shunt; pentose phosphate pathway/cycle; Warburg–Dickens pathway; phosphogluconate pathway) A metabolic pathway which involves the oxidative decarboxylation of glucose-6-phosphate to ribulose-5-phosphate, followed by a series of reversible, non-oxidative sugar interconversions (see Appendix I(b)). Two molecules of NADP are reduced for each molecule of hexose decarboxylated, and an important function of this pathway is the provision of $NADPH_2$ for biosynthesis e.g. of fatty acids. The pathway also provides intermediates for a number of biosynthetic pathways—e.g. pentoses for nucleotide biosynthesis, erythrose-4-phosphate for aromatic amino acid synthesis. The HMP pathway may operate as a shunt of the EMBDEN–MEYERHOF–PARNAS PATHWAY (EMP pathway)—common intermediates being glucose-6-phosphate, fructose-6-phosphate, and glyceraldehyde-3-phosphate. Bacteria which possess the HMP pathway can synthesize hexoses from pentoses *via* this pathway, and hence can use certain pentoses (*via* fructose-6-phosphate and the EMP pathway) as energy sources; the pentose must first be converted to an intermediate of the pathway—e.g. arabinose → ribulose → ribulose-5-phosphate; xylose → xylulose → xylulose-5-phosphate.

Hfr donor (see *bacterial* CONJUGATION)

HFT (high frequency transfer) (see SEX FACTOR)

HFT lysates (see TRANSDUCTION)

HI test (see HAEMAGGLUTINATION-INHIBITION TEST)

HIB (*immunol.*) (see HAEMOLYTIC IMMUNE BODY)

hibitane Proprietary name for CHLORHEXIDINE.

high mutability gene (see MUTATOR GENE)

higher fungi Fungi of the sub-divisions ASCOMYCOTINA, BASIDIOMYCOTINA, and DEUTEROMYCOTINA.

Hikojima serotype (see VIBRIO)

Hill reaction In PHOTOSYNTHESIS: the light-dependent evolution of oxygen (with concomitant reduction of suitable electron acceptors) by isolated CHLOROPLASTS in the absence of carbon dioxide.

hilum (*mycol.*) That part of a spore which is in contact with the sporophore prior to spore discharge.

hinge region (of the immunoglobulin molecule) (see FAB FRAGMENT)

hippurate hydrolysis The ability to hydrolyse sodium hippurate ($C_6H_5.CONH.CH_2.CO_2Na$) is characteristic of certain bacteria e.g. some streptococci. Hippurate is cleaved by *hippuricase* to benzoate and glycine. In one form of test for hippuricase activity, benzoate is detected by the formation, and *persistence*, of a precipitate on addition of a *known volume** of acid ferric chloride solution to the medium. *Determined by gradually adding the reagent, with shaking, to an *uninoculated* (control) tube of medium of the same volume; the required volume is that which just dissolves the precipitate (of hippurate) initially formed in the control tube.

hircinol A phenanthrene derivative closely related to ORCHINOL which it resembles in activity. Hircinol is produced by the orchid *Loroglossum hircinum* in response to infection by certain fungi, e.g. *Rhizoctonia* spp.

histamine A physiologically active amine formed by the decarboxylation of histidine. It causes rapid contraction in many smooth muscles, and is an effective vasodilator. Histamine is present in *mast cells*—from which it is released (together with other substances) during the immunological reactions which lead to ANAPHYLACTIC SHOCK and other manifestations of IMMEDIATE HYPERSENSITIVITY.

histidine (biosynthesis of) (see Appendix IV(g))

Histomonas meleagridis The only species of the protozoan genus *Histomonas* (order RHIZOMASTIGIDA) and the cause of BLACKHEAD in turkeys. The organisms are uninucleate; they reproduce by binary fission. Cysts are not formed. Intracellular (amoeboid) cells of *H. meleagridis* are oval or spherical, 10–20 microns in diameter; within the caecal lumen of the turkey the cells may be both amoeboid and flagellate. (Each flagellate cell commonly has one flagellum, but up to four flagella may be present.)

histomoniasis (in turkeys) (see BLACKHEAD)

histones Basic proteins (rich in arginine and lysine) which occur in close association with the nuclear DNA of many *eucaryotic* cells—including eucaryotic microorganisms; histones have not been found in procaryotes or in certain eucaryotes, e.g. dinoflagellates*. Histones are believed to be involved in the control of DNA transcription. *Very low concentrations of histones (by weight, less than 0.5% of the DNA) have been reported in *Gyrodinium cohni*.

Histoplasma capsulatum A dimorphic fungus which is the causal agent of HISTOPLASMOSIS; *H. capsulatum* occurs intracellularly as yeast-like ovoid cells, each approximately 2–3 × 3–4 microns, which reproduce by budding. In culture, at room temperature or e.g. 30 °C, the organism grows as a mycelium; the mycelial stage reproduces by means of conidia (= microconidia), each 2–5 microns in diameter, and macroconidia, each approximately 8–15 microns in diameter. The macroconidium typically bears a surface covering of short, finger-like projections which are believed to be globules of a mucopolysaccharide. The ascigerous stage of *H. capsulatum* is *Emmonsiella capsulata*. *E. capsulata*, which is heterothallic, produces eight-spored asci in spherical or spheroid cleistothecia.

histoplasmin A culture filtrate of *Histoplasma capsulatum* used in the HISTOPLASMIN TEST.

histoplasmin test A skin test, similar in form to the TUBERCULIN TEST, used in the diagnosis of HISTOPLASMOSIS; the test antigen is HISTOPLASMIN. Skin sensitivity usually develops within 1 to 2 months of the onset of disease and may persist for many years. Cross-reactions occur with other fungal antigens, e.g. coccidioidin.

histoplasmosis (Darling's disease) A disease of man and other animals in which the causal agent is the dimorphic, intracellular fungus *Histoplasma capsulatum*. *H. capsulatum* occurs in soil containing the droppings of birds, bats and other animals; infection commonly occurs *via* the respiratory route. The pathogen generally affects the reticuloendothelial system. In man, *primary histoplasmosis* is a mild, self-limiting disease of the lungs; lesions in the lungs and regional lymph nodes may calcify on healing. *Progressive disseminated histoplasmosis* is a severe disease in which ulcerative lesions may occur in the nose, mouth and gastrointestinal tract; enlargement of the liver and/or spleen is common. Amphotericin B has been used for treatment. In other animals the symptoms are similar to those in man. *Laboratory diagnosis* includes microscopical and cultural examination of e.g. sputum, lymph node aspirates; serological tests include the HISTOPLASMIN TEST and a CFT.

HMP pathway (see HEXOSE MONOPHOSPHATE PATHWAY)

hog cholera (see SWINE FEVER)

holdfast An avascular, root-like organ by means of which certain algae (e.g. *Laminaria* spp) remain attached to the substratum; it is not an absorptive organ. (cf. RHIZOID)

Holobasidiomycetidae A sub-class of fungi of the class HYMENOMYCETES; constituent species give rise to *holobasidia*—i.e. aseptate basidia—which are usually present in an hymenium on a well-developed fruiting body. Members of the Holobasidiomycetidae are initially divided into (a) species which are internal parasites in the leaves of vascular plants *and* which do not form a definite fruiting body external to the host plant (orders BRACHYBASIDIALES and EXOBASIDIALES), and (b) all other species (orders AGARICALES, APHYLLOPHORALES, DACRYMYCETALES, and TULASNELLALES).

holobasidium (*mycol.*) An *aseptate* BASIDIUM.

holocarpic *Holocarpic* fungi are those in which the *entire thallus* takes on a reproductive function.

holoenzyme (1) (see COFACTOR) (2) See *RNA polymerase* in entry RNA.

Holophrya A genus of protozoans of the suborder RHABDOPHORINA.

holophytic nutrition *Plant*-type nutrition—in which nutrients are formed by endogenous photosynthesis.

holotrich A ciliate of the subclass Holotrichia.

Holotrichia A sub-class of protozoans of the sub-phylum CILIOPHORA. In *holotrichs* the somatic ciliature is typically simple and uniform—but is arranged spirally in the Apostomatida, and is absent in mature organisms of the Chonotrichida. Specialized oral ciliature is absent in the majority of orders and, when present, is of a simple nature. The following orders are recognized: APOSTOMATIDA, ASTOMATIDA, CHONOTRICHIDA, GYMNOSTOMATIDA, HYMENOSTOMATIDA, THIGMOTRICHIDA, and TRICHOSTOMATIDA.

holozoic nutrition *Animal*-type nutrition—involving the *ingestion* of organisms, or components of organisms.

homocaryon (see HOMOKARYON)

homocytotropic antibody (1) A CYTOPHILIC ANTIBODY which attaches only to cells of the species in which it was formed (cf. HETEROCYTOTROPIC ANTIBODY). (2) Synonymous with the meaning given under sense (1) of REAGINIC ANTIBODY.

homodiaphoromixis (see SECONDARY HOMOTHALLISM)

homodimixis (see SECONDARY HOMOTHALLISM)

homoeomerous A synonym of HOMOIOMEROUS.

homofermentation Fermentation in which only one major product is formed; the term is frequently used as a synonym of HOMOLACTIC FERMENTATION.

homogeneous immersion (see IMMERSION OIL)

homogenote (see MEROZYGOTE)

homoheteromixis A synonym of SECONDARY HOMOTHALLISM.

homoiomerous (homoeomerous) (1) (of a lichen) Refers to a lichen in which the phycobiont does not occur in a distinct layer but is more or less evenly distributed throughout the thallus. Homoiomerous lichens include members of the genera *Collema*, *Lempholemma*, and *Leptogium* (cf. HETEROMEROUS). (2) (*mycol.*) In an agaric: refers to a pileus and stipe which consist of hyphae only (cf. HETEROMEROUS).

homokaryon (homocaryon) In fungi: a hyphal cell, mycelium, organism, or spore which contains genetically identical nuclei (cf. HETEROKARYON).

homolactic fermentation A type of fermentation in which lactic acid is the main product of the fermentation of glucose (see Appendix III(a)(i)); small amounts of other compounds may also be formed. (cf. HETEROLACTIC FERMENTATION) Homolactic fermentation is carried out by certain species of *Lactobacillus* and by some members of the family Streptococcaceae. The optical activity of the lactate produced is characteristic of the organism producing it: e.g. *Streptococcus* spp produce L(+)lactate; *Lactobacillus casei*, *Pediococcus* spp produce predominantly L(+)lactate; *L. bulgaricus*, *L. delbrueckii*, *L. lactis*, *L. leichmannii* produce D(−)lactate; *L. helveticus* and *L. acidophilus* produce DL-lactate. Lactate dehydrogenases from these organisms are specific for either one or the other of the two optical isomers of lactic acid. Species which produce racemic mixtures of lactate appear to possess either a lactate racemase or lactate dehydrogenases with different specificities.

homologous (genetics) Of two chromosomes, chromatids, nucleic acids: containing identical sequences of genes—though not necessarily identical alleles.

homologous (*immunol.*) (1) The antibody elicited by a given antigen is said to be *homologous* to that antigen; similarly, the antigen is said to be homologous to the antibody. (2) May refer to the source of e.g. a serum: derived from the same species as that being referred to. (cf. HETEROLOGOUS)

homologue (of chromosomes) Each of a pair of (usually) morphologically similar CHROMOSOMES in a eucaryotic DIPLOID nucleus; in each *homologous pair* one chromosome derives from the male and one from the female of the cell's original progenitors. In general, both chromosomes of an homologous pair carry identical sequences of genes—though not necessarily identical ALLELES.

homomixis A synonym of HOMOTHALLISM.

homothallism (homomixis) (*mycol.*) The system in which those nuclei (or gametes) which fuse during sexual reproduction are genetically-similar (though not necessarily homozygous) and in which the nuclei are usually derived from the same thallus. *Homothallic* fungi are thus self-fertile. (cf. HETEROTHALLISM; see also SECONDARY HOMOTHALLISM)

homothetogenic (*adjective*) Refers to the mode of cell division typical of ciliate protozoans. In such division, which is transverse, or perkinetal, the daughter cells (PROTER and OPISTHE) are not mirror images of one another—although they may be similar or identical. (Thus, when *Tetrahymena* divides, the anterior end of the opisthe is adjacent to the plane of division.) (cf. SYMMETROGENIC)

homozygote A cell which is HOMOZYGOUS in respect of a *specified* gene, or genes.

homozygous (1) Of eucaryotic genes: refers to the existence of one or more pairs of identical alleles at *specified* loci on homologous chromosomes; thus, a cell (or nucleus) is said to be homozygous for a particular gene or genes. (2) Of bacterial genes: refers to the existence of one or more pairs of identical alleles at *specified* loci in a MEROZYGOTE.

honey fungus A common name for the basidiomycete *Armillaria mellea.*

HOQNO (see HYDROXYQUINOLINE-N-OXIDES)

horizontal resistance (*plant pathol.*) If a cultivar exhibits similar levels of resistance to each of the races of a given pathogen it is said to display horizontal resistance with respect to that pathogen. Horizontal resistance, unlike VERTICAL RESISTANCE, tends to be a permanent (though not necessarily a complete) form of resistance which does not readily break down as a result of genetic changes in the pathogen population; it is commonly inherited polygenically.

horizontal transmission (of a parasite) Transmission from one individual to another, contemporary individual. (cf. VERTICAL TRANSMISSION)

hormocyst (*lichenol.*) A type of propagule which is known only in species of the gelatinous lichen *Lempholemma*; a hormocyst consists of filament(s) of the phycobiont *Nostoc* enclosed in a gelatinous sheath with which hyphae of the mycobiont are—or may be—associated. Hormocysts are formed in specialized, swollen regions of the thallus called *hormocystangia.*

Hormodendrum resinae A synonym of *Cladosporium resinae.*

hormogonium (1) In filamentous blue-green algae: a portion of a trichome or filament which has separated from the parent organism (see also NECRIDIUM); a hormogonium is also formed by the germination of an *akinete.* (2) (*bacteriol.*) A fragment of a filament.

hormonal control (of sexual processes in fungi) A fungal hormone is a compound which is produced by a given fungus (or part of a fungus) and which, in low concentrations, elicits a specific response from another individual of that species or of a closely-related species (or from another part of the same individual); hormones appear to coordinate (spatially and temporally) the sequence of events which precedes karyogamy in a wide range of fungi—although no system is understood in detail. Some examples:

In *Allomyces* spp the female gametangia and female gametes release into the (liquid) medium a compound called *sirenin* (a sesquiterpene diol) which acts as an attractant to the male gametes; the latter thus move towards the female gametes by positive chemotopotaxis (see TAXIS). Sirenin is inactivated by the male gametes, but the mechanism of the chemotactic response is unknown.

In *Achlya* spp the formation of antheridia and oogonia is governed by diffusible hormones. The female vegetative mycelium produces a steroid hormone (*antheridiol*—also called 'hormone A') which diffuses into the medium and stimulates the formation of antheridial hyphae in the male mycelium; the latter take up and inactivate the hormone and grow towards its source, and they produce a second hormone ('hormone B') which is structurally related to antheridiol. Hormone B stimulates the female mycelium to produce oogonia; antheridiol production appears to increase as the oogonia develop. Subsequently the oogonium and antheridium come into contact and the gametangia are delimited; the male nucleus then migrates to the female cell and karyogamy occurs. The mechanism of the hormonal action is unknown.

In certain zygomycetes (e.g. *Blakeslea trispora*, *Mucor mucedo*) the heterothallic strains each comprise two mating types which are designated (+) and (–); in both mating types the differentiation of the (iso)gametes is governed by the same hormone—a *trisporic acid* (formerly known as 'gamone'). The hormone is synthesized (from β-carotene) in a reaction sequence which can be completed only when two organisms of opposite mating type grow in close proximity; each mating type produces a specific trisporic acid precursor or *prohormone* (prohormones P^+ and P^-—formerly plus-progamone and minus-progamone)—each prohormone being convertible to trisporic acid only by a strain of the opposite mating type. Thus, P^+ is converted to trisporic acid by a (–) strain, while P^- is converted to trisporic acid by a (+) strain. In each mating type the trisporic acid initiates gametangium formation and derepresses those enzymes involved in prohormone synthesis—leading to increased levels of the latter. The zygophore initials grow towards the source of prohormone of opposite mating type until the gametangia make contact. (A similar hormonal system appears to occur in homothallic strains.)

In the ascomycete *Ascobolus stercorarius* the exposure of an oidium to a mycelium of appropriate mating type results in the initiation of ascogonial initials in the latter; such initiation is apparently due to the diffusion of a substance from the oidium to the mycelium. (It appears that the oidium develops male function only after it has been exposed to mycelium of the appropriate mating type.) The trichogyne of the ascogonium grows chemotropically towards the oidium and eventually fuses with it.

host-controlled modification If a bacteriophage is cultured on a given bacterial strain, the DNA of the phage progeny will carry the MODIFICATION pattern characteristic of that strain. If these phages are used to infect bacteria of a *different* strain, a low efficiency of plating is generally obtained since, in most of the cells, infection is followed by RESTRICTION of the phage DNA. However, in a few cells restriction fails to occur and the phage DNA acquires the modification pattern characteristic of the new strain. Thus the modified DNA can replicate—new strands being modified prior to separation from their template strands; progeny so formed will plate with high efficiency on cultures of this strain. If such phage progeny are used to re-infect bacteria of the original strain a low efficiency of plating will be obtained since the pattern of phage DNA modification is that characteristic of the second strain.

host-dependent mutant (see CONDITIONAL LETHAL MUTANT)

host-range mutant (*h* mutant) (of bacteriophage) A mutant bacteriophage which can infect a bacterial strain which had previously undergone a mutation rendering it resistant to the non-mutant phage. For example: a strain of *Escherichia coli* may undergo a mutation which causes a structural alteration of the phage receptor sites on the bacterial cell wall—resulting in resistance to phage T2; a T2 phage may then undergo a mutation which allows it to adsorb to the mutant *E. coli* cell wall. (This host-range mutant is designated T2*h*.) In at least some cases the ability of T2*h* to infect the wild type strain of *E. coli* is unimpaired.

host specificity Refers to the host range of a given parasite, i.e. the number of host species or strains (one, several, many) susceptible to the parasite. Some parasites (and pathogens) exhibit a high degree of specificity; thus, e.g. man is the only natural host of *Neisseria gonorrhoeae*, and even higher specificities are found among the plant-pathogenic rust fungi (see RUSTS (1)) and certain bacteriophages.

hot-air oven An apparatus used for STERILIZATION.

hot-cold lysis A mode in which staphylococcal β-HAEMOLYSIN lyses susceptible erythrocytes, e.g. sheep RBCs. When sheep-blood-agar is inoculated with a β-haemolytic staphylococcus and incubated at 37 °C, the colonies produced are each surrounded by a *darkened* zone. If the plate is then incubated overnight at 4 °C, or room temperature, the darkened zones become lighter and more transparent, i.e. haemolysis occurs. The mechanism of this *hot-cold lysis* has yet to be elucidated. Sensitive erythrocytes may suffer lysis without the period of 'cold' incubation when a strain of *Streptococcus agalactiae* is grown closely adjacent to a β-haemolytic *Staphylococcus*; this is the basis of the CAMP TEST.

HPr (see TRANSPORT)

HSA Human serum albumin.

Huffia A sub-genus of PLASMODIUM.

Hugh and Leifson's test (see OXIDATION-FERMENTATION TEST)

hülle cells (*mycol.*) Thick-walled cells which form a layer around the ascocarp in certain ascomycetes (e.g. *Emericella*); the cells are borne, in profusion, directly on branched hyphae. The function of hülle cells is unknown.

humoral immunity (1) ANTIBODY-dependent IMMUNITY. (2) More generally, immunity derived from factors in the body fluids; besides antibodies these factors include COMPLEMENT, INTERFERONS, LYSOZYME, PROPERDIN.

hyaline (*adjective*) Transparent or translucent.

Hyalococcus A genus of algae of the division CHLOROPHYTA. Species occur e.g. as the phycobiont in certain LICHENS.

hyaloplasm Synonym of CYTOSOL.

hyaluronic acid A linear polysaccharide which is composed of the repeating disaccharide unit: D-glucuronic acid-β-(1 → 3)-*N*-acetyl-D-glucosamine—the disaccharide units being linked β-(1 → 4). Hyaluronic acid occurs e.g. as an intercellular constituent of animal tissue, as a component of synovial fluid, and in the capsular envelope of certain strains of group A streptococci. Hyaluronic acid forms highly viscous solutions in water and readily forms gels.

hyaluronidase (hyaluronate lyase) An enzyme which depolymerizes HYALURONIC ACID by splitting the β-(1 → 4) linkages between the glucosamine and glucuronic acid residues. An extracellular hyaluronidase is produced by certain gram-positive pathogenic bacteria—e.g. *Clostridium* spp (the μ toxin of *Clostridium perfringens*), *Staphylococcus* spp; such an hyaluronidase (formerly known as *spreading factor*) is believed to promote the invasiveness of the pathogen—due to its ability to break down hyaluronic acid in the tissues of the host.

hybrid Any (diploid) organism which results from a cross between two genetically dissimilar parent organisms; such an organism is HETEROZYGOUS for one or more pairs of alleles.

Hydnaceae A family of fungi of the order APHYLLOPHORALES. Constituent species form fruiting bodies which—according to species—may be e.g. non-stipitate, or stipitate and mushroom-shaped, and in which the hymenophore is in the form of a layer of pendulous, tooth-like processes ('teeth' or 'spines'); the context of the pileus is non-xanthochroic and contains thin-walled generative hyphae which exhibit clamp connections. The spores are smooth-walled, hyaline or pigmented, and are neither amyloid nor cyanophilous. A number of species are edible; these include e.g. *Hericium coralloides* and *Hydnum repandum*.

Hydnum repandum forms a mushroom-shaped fruiting body in which the thick, fleshy pileus is approximately 5–10 centimetres in diameter and is fawn to pinkish-buff on the upper surface; the spore-bearing surface (the underside of the pileus) consists of a layer of unequally-sized teeth which are of a colour similar to that of the cap—though paler. The stipe is short and thick and is frequently eccentric in relation to the pileus; the teeth are often decurrent. The non-pigmented spores are spheroidal, approximately 6 × 8 microns.

hydrocarbons (microbial utilization of) For examples see ALIPHATIC HYDROCARBONS (microbial utilization of) and METHANE (microbial utilization of).

hydrogen bacteria (see KNALLGAS BACTERIA)

hydrogen peroxide (as an antimicrobial agent) (see OXIDIZING AGENTS)

hydrogen sulphide production H_2S is produced during the anaerobic metabolism of sulphur-containing amino acids and by DISSIMILATORY SULPHATE REDUCTION. The production of H_2S can be detected by its ability to form black sulphides with ferrous, ferric or lead salts—see e.g. TRIPLE-SUGAR-IRON AGAR. (see also SULPHUR CYCLE)

hydrogenase An enzyme which catalyses the reaction $2H^+ + 2e^- \rightleftharpoons H_2$. Hydrogenases occur in a wide range of bacteria and algae—see e.g. FORMATE HYDROGEN LYASE, KNALLGAS BACTERIA, and PHOTOREDUCTION.

Hydrogenomonas An obsolete bacterial genus; organisms originally placed in this genus have now been allocated to one or other of the genera *Alcaligenes* and *Pseudomonas*. Thus, e.g. *H. facilis* = *Pseudomonas facilis*, *H. ruhlandii* = *Pseudomonas ruhlandii*, *H. eutropha* = *Alcaligenes eutropha*. (see also KNALLGAS BACTERIA)

hydrolase A category of ENZYMES (EC class 3) which encompasses those enzymes which catalyse hydrolysis—e.g. of ester bonds, glycosidic bonds, peptide bonds.

hydrophobia (see RABIES)

hydroxamates (in iron transport) (see SIDEROCHROMES)

hydroxylamine (as a MUTAGEN) Hydroxylamine (NH_2OH) appears to react exclusively with *cytosine* residues in DNA—modifying this base so that, during subsequent replication of the DNA, it pairs with adenine rather than guanine; thus hydroxylamine causes only G-C to A-T transitions. When used to treat bacteriophage or isolated DNA, hydroxylamine is a highly specific mutagen; in bacteria, however, any mutagenic effects may fail to be expressed owing to the toxic or lethal effects of the agent.

8-hydroxyquinoline (8-quinolinol; oxine) (as an antimicrobial agent) A chelating agent which, in the presence of *copper* ions, is an effective *fungicide*. Copper-oxine antifungal sprays have been used in agriculture, and 8-hydroxyquinoline derivatives have been used in the treatment of *Dutch elm disease*. Copper-oxine is also used as a rot-proofing agent, e.g. for textiles. Some halogen derivatives of 8-hydroxyquinoline are used for the treatment of intestinal *amoebiasis*: 5-chloro-7-iodo-8-hydroxyquinoline (*Enterovioform*; *iodochlorohydroxyquin*; *Vioform*); 5,7-diiodo-8-hydroxyquinoline (*diiodohydroxyquin*); 7-iodo-8-hydroxyquinoline-5-sulphonate (*chiniofon*). The mode of action of these compounds is not clear; it has been suggested that *chiniofon* chelates ferrous ions necessary for the growth of amoebae. (The iodine in the molecule is now believed to play a secondary role.)

hydroxyquinoline-*N*-oxides (HOQNO) 2-Alkyl-4-hydroxyquinoline-*N*-oxides function as RESPIRATORY INHIBITORS in mitochondria and in certain bacteria; some bacteria appear to be impermeable to HOQNOs. The site of inhibition is between *b*- and *c*-cytochromes in the ELECTRON TRANSPORT CHAIN.

Hygrocybe A genus of fungi of the HYGROPHORACEAE.

hygrophilous Preferring a moist environment.

Hygrophoraceae A family of fungi of the order AGARICALES; constituent species are distinguished from those of other families e.g. in that they form thick—but sharp-edged—*waxy* gills (which are triangular in cross section) and typically give rise to elongated basidia. The organisms occur e.g. in pastures and meadows, and many species are brightly coloured. The genera include e.g. *Hygrocybe* and *Hygrophorus*.

hymenium (*mycol.*) (1) In ascomycetes: a layer of *asci* (see ASCUS) on or within a fruiting body; each individual ascus in an hymenium is more or less parallel to the surrounding asci and approximately perpendicular to the plane of the hymenium at that location. The hymenium may contain structures such as paraphyses and pseudoparaphyses—according to species. (2) In basidiomycetes: a layer of basidia (see BASIDIUM) on or within a fruiting body; hymenia occur e.g. on the gill surfaces of agarics, and lining the tubules of polypores. (see also CYSTIDIUM)

Hymenomycetes A class of fungi of the sub-division BASIDIOMYCOTINA; the typical fruiting body of the hymenomycetes is a well-developed, macroscopic, gymnocarpous or semiangiocarpous structure in which the basidia—which bear ballistospores*—are organized into an hymenium. Hymenomycetes which form phragmobasidia are placed in the sub-class PHRAGMOBASIDIOMYCETIDAE; those which form holobasidia are placed in the sub-class HOLOBASIDIOMYCETIDAE. *See however *secotioid gasteromycetes* under AGARICALES.

hymenophore (*mycol.*) That part of a fruiting body on which the hymenium is *directly* borne—e.g. the walls of the tubules in a polypore, or the tramae of the gills in an agaric.

Hymenostomatida An order of protozoans of the sub-class HOLOTRICHIA, sub-phylum CILIOPHORA. The *hymenostomes* typically have uniform body ciliature and a ventral buccal cavity; the latter is invested with specialized oral ciliature consisting basically of one undulating membrane and an AZM of three membranelles. Three sub-orders are recognized: PENICULINA, PLEURONEMATINA, and TETRAHYMENINA.

hyperauxiny (see AUXINS)

hyperimmune (*immunol.*) Refers to the condition of a subject (often an animal) whose serum contains a high titre of a specific antibody following an intensive course of immunization.

Hypermastigida An order of parasitic protozoans of the class ZOOMASTIGOPHOREA; species inhabit the gut of termites and other insects. Each (uninucleate) organism bears numerous flagella which arise chiefly from the anterior half of the cell. Sexual reproduction occurs in at least some species. The sub-orders are Lophomonadina and Trichonymphina—the latter including TRICHONYMPHA.

hyperplasia Abnormal proliferation of tissue cells resulting in the formation of a TUMOUR (in animals) or a GALL (in plants); hence, the adjective *hyperplastic*.

hypersensitivity (allergy) (*immunol.*) Refers to the condition of a PRIMED individual who tends to give an exaggerated immunological response on further exposure to the relevant antigen; an hypersensitivity reaction may cause varying degrees of damage to the subject's tissues and may be fatal. (Hypersensitivity is manifested e.g. in the rejection, by a tuberculous animal, of a subcutaneous inoculum of tubercle bacilli (see KOCH'S PHENOMENON); such a rejection indicates a developing resistance to tubercle bacilli.) Hypersensitivity reactions are of two main types: (a) ANTIBODY-mediated IMMEDIATE HYPERSENSITIVITY, and (b) CELL-MEDIATED IMMUNITY.

hypersensitivity (*plant pathol.*) A mechanism which may enable a higher plant to resist certain diseases. Hypersensitivity involves the *rapid* death of plant cells at the site(s) of infection; this results in the formation of typical small necrotic lesions to which the infection is confined. Cell death is frequently associated with e.g. PHYTOALEXIN production or accumulation of phenolic compounds. Where an infection involves an *obligate* parasite the death of the host cells may in itself be sufficient to prevent the spread of the parasite to uninfected tissues. Hypersensitivity can also be induced by *facultative* parasites; in such cases associated production of toxins by the plant may be important in restricting the infection. Hypersensitivity may occur in response to infection by fungi, bacteria or viruses. (see also RESISTANCE (of plants to disease))

hypha (*pl.* hyphae) In many (mycelial) fungi, and in some bacteria (see ACTINOMYCETALES): one of a number of unbranched or branched filaments which, collectively, constitute the vegetative form of an organism (and which, in some species, form the sterile portion of a fruiting body); a mass of vegetative hyphae is referred to as a MYCELIUM.

The *fungal* hypha is a tubular structure whose shape is maintained by a rigid CELL WALL; the wall is external to a CELL MEMBRANE which encloses the cytoplasm, nuclei, mitochondria, and other cell components typical of eucaryotic organization. The diameter of a fungal hypha, which depends e.g. on species, varies from one micron or less to values of macroscopic dimensions. According e.g. to species, a hypha may or may not be *septate*—see SEPTUM (*mycol.*). Hyphae grow mainly by apical extension—see GROWTH (fungal).

In basidiomycetes the *primary* hyphae are uninucleate, haploid hyphae which develop following the germination of a uninucleate, haploid basidiospore; *secondary* hyphae are dikaryotic hyphae which develop following DIKARYOTIZATION.

The sterile part of a *basidiocarp* may consist of one, two, or three distinct types of hypha—according to species; if the hyphal system consists of one, two, or three types of hypha it is referred to as *monomitic*, *dimitic*, or *trimitic*, respectively. The types of hypha are: (a) *Generative hyphae*. These are typically thin-walled, branched hyphae which frequently exhibit clamp connections; while invariably thin-walled at and near the region of growth, the hyphae may subsequently become thick-walled. Generative hyphae, which contain relatively dense cytoplasm, give rise to the basidia and to the other two types of hypha. (b) *Skeletal hyphae*. These are thick-walled, unbranched hyphae which do not exhibit clamp connections and in which the lumen is narrow. (c) *Binding* (*ligative*) *hyphae*. These are thick-walled hyphae which are usually profusely branched; they appear to bind together other hyphae within the basidiocarp.

hyphal bodies (*mycol.*) Hyphal fragments which function as propagules or gametes; they are formed by certain zygomycetes—see e.g. ENTOMOPHTHORA.

Hyphales A synonym of MONILIALES.

Hyphochytridiomycetes A class of FUNGI of the sub-division MASTIGOMYCOTINA; constituent species are distinguished from those of other classes in that they give rise to zoospores each of which possesses a single anteriorly-directed tinsel flagellum. The hyphochytridiomycetes are aquatic saprophytes and parasites which, in general, resemble members of the CHYTRIDIALES in morphology and life cycle—though sexual processes appear not to occur. Genera include *Rhizidiomyces*.

Hyphomycetes A class of fungi within the sub-division DEUTEROMYCOTINA; constituent species are distinguished from those of the class COELOMYCETES in that their conidia develop superficially—i.e. on conidiophores at the surface of the substratum—rather than within cavities. The fructifications of the hyphomycetes include individual and grouped conidiophores which, in some species, may be borne on stromata. The sole order: MONILIALES.

hyphopodium In fungi of the family PERISPORIACEAE: a short, one- or two-celled branch of a somatic hypha; some types of hyphopodium function as organs of attachment—the distal cell acting as an appressorium and giving rise to an haustorium.

hypobiosis A synonym of DORMANCY (sense 1).

hypochlorites (as antimicrobial agents) (see CHLORINE)

hypogean (*adjective*) Occuring under the ground.

hyponeuston (see NEUSTON)

hypophloeodal (hypophloedal; hypophloeodic) (of a lichen) A synonym of ENDOPHLOEODAL.

hyposensitization (see DESENSITIZATION)

hypothallus (*lichenol.*) In certain LICHENS: a layer of non-lichenized (i.e. alga-free) hyphae which extends from the periphery of the lichen thallus—and which also occurs between the areolae of certain areolate crustose species. The hypothallus is generally different in colour from the thallus. (see e.g. OCHROLECHIA, PANNARIA, RHIZOCARPON.

hypotheca (of a diatom) The smaller of the two parts of the cell wall (frustule) of a diatom.

hypothecium (see APOTHECIUM)

Hypotrichida An order of protozoans of the sub-class SPIROTRICHIA.

Hypoxylon A genus of fungi of the order SPHAERIALES.

I

I-like pilus (see SEX PILI)

IAA 3-Indole-acetic acid (see AUXINS).

IBT ISATIN-β-THIOSEMICARBAZONE.

Iceland moss The lichen *Cetraria islandica*.

icosahedron A solid figure contained by 20 plane faces—the form of many types of virus, e.g. enteroviruses.

ICR compounds (see ACRIDINES (as MUTAGENS))

icterus Jaundice.

ID_{50} Infectious dose (50%)—that dose of a given infectious agent which, when given to each of a number of experimental test systems/animals, brings about the infection of 50% of the experimental systems/animals under given conditions.

-id reaction The formation of skin lesions at sites which are remote from a focus of bacterial or fungal infection. Such lesions are commonly sterile and it is generally believed that they represent a local HYPERSENSITIVITY response to the infection. If the infecting organism is the fungus *Trichophyton* (for example) the reaction is termed trichophyt*id*.

idiogram (see KARYOTYPE)

idiophase (of fungal metabolism) During batch culture, that phase in which *secondary* metabolism is dominant over primary (growth-directed) metabolism.

idoxuridine (see 5-IODO-2′-DEOXYURIDINE)

IEP An abbreviation for ISOELECTRIC POINT.

Ig (*immunol.*) Abbreviation for IMMUNOGLOBULIN.

IgA Immunoglobulin A—a class of IMMUNOGLOBULINS. The IgA molecule (see IGG for details of structure) contains alpha HEAVY CHAINS, and includes about 8% carbohydrate; the molecular weight is 160,000 and the sedimentation coefficient (S^{20}_{w}) is about 7. Polymeric forms also occur. In man, *secretory* IgA occurs e.g. in COLOSTRUM, saliva, tears; the molecule incorporates an IgA *dimer* and the so-called *secretory piece*—a protein (M.Wt. about 60,000) of unknown function. Polymers of human IgA bring about COMPLEMENT-FIXATION by the *alternative pathway*. (see also *J chain* under IGM)

IgD Immunoglobulin D—a class of IMMUNOGLOBULINS. The IgD molecule contains δ HEAVY CHAINS and includes about 12–13% carbohydrate; its molecular weight is about 180,000 and its sedimentation coefficient (S^{20}_{w}) about 7. IgD occurs in minute traces in normal human serum but appears not to possess antibody activity.

IgE Immunoglobulin E—a class of IMMUNOGLOBULINS. The IgE molecule contains ε HEAVY CHAINS and includes 12% carbohydrate; its molecular weight is about 190,000 and its sedimentation coefficient (S^{20}_{w}) is 8. IgE is heat-labile (56 °C/3–4 hours); it does not fix complement. (see also REAGINIC ANTIBODIES)

IgG Immunoglobulin G—a class of IMMUNOGLOBULINS. The IgG molecule contains γ HEAVY CHAINS and includes about 3% carbohydrate. IgG has a molecular weight of about 150,000 and a sedimentation coefficient (S^{20}_{w}) of 7. The IgG molecule contains two identical heavy chains which are adjacent for much of their length—being linked by a number of disulphide bonds. The N-terminal ends of the chains are not linked but form the limbs of the Y-shaped molecule; parallel to each limb is attached a LIGHT CHAIN (linked by a disulphide bridge)—the N-terminals of the light chains and heavy chains being adjacent. (The Y-shaped molecule is basically similar in all immunoglobulins, including the monomers of IgM and IgA.) The antibody COMBINING SITES are located on the limbs of the Y-shaped molecule (see also V-REGION); the *stem* of the Y (the *Fc region*) is important in COMPLEMENT-FIXATION. The molecule can be broken into its constituent chains by the action of MERCAPTOETHANOL and the subsequent lowering of pH. For the action of PAPAIN and pepsin on IgG see FAB FRAGMENT.

In man, IgG is the major class of serum immunoglobulins; IgG antibodies are bivalent and, with the exception of IgG subclass 4, fix COMPLEMENT *via* the classical pathway during reactions with homologous antigens. IgG antibodies are present in milk, and are able to pass the placenta in many of the mammalian species studied—including man; they are important as opsonins and as antitoxins both in the plasma and extravascular regions. (see also PROTEIN A)

IgM (macroglobulin) Immunoglobulin M—a class of IMMUNOGLOBULINS. The IgM molecule contains μ HEAVY CHAINS and includes about 12% carbohydrate. IgM (M.Wt. 900,000) has a sedimentation coefficient (S^{20}_{w}) of 19. The molecule consists of five IgG-like sub-units (see IGG) which are arranged radially and linked by disulphide bonds; the combining sites are located circumferentially. Each molecule of IgM contains, near the centre of the complex, a single cysteine rich polypeptide—the *J chain* (M.Wt. 15,000). The J chain (which is also found in *polymers* of IgA) appears to be essential for the *in vivo* formation of IgM—although IgM can be assembled from its subunits *in vitro* in the absence of the J chain.

In man, IgM is a minor class of immunoglobulin. IgM antibodies have a theoretical valency of 10 but often exhibit only 5 effective combining sites; they are COMPLEMENT-FIXING ANTIBODIES. IgM antibodies are unable to pass from the blood into the tissues; they cannot pass the placenta, and are absent (or present in minutes quantities) in milk. IgM antibodies are often the first to be formed following antigenic stimulation. (The *Wassermann* antibody and the RHEUMATOID FACTOR are largely IgM.) IgM antibodies are particularly effective against particulate antigens which display a pattern of repeated antigenic determinants—e.g. bacterial cells.

immediate hypersensitivity (*immunol.*) ANTIBODY-mediated HYPERSENSITIVITY. The three forms of immediate hypersensitivity are: (a) *Anaphylaxis* (Type 1 reaction). This involves the union of antigen with homologous MAST CELL-bound *reaginic* (IgE) antibodies; complement-fixation is not involved. The reaction—which develops in minutes—is associated with mast cell degranulation. Manifestations of anaphylaxis include hay fever and ANAPHYLACTIC SHOCK. (b) TYPE 2 REACTION. (c) *Complex-mediated hypersensitivity* (Type 3 reaction). This involves the formation of antigen–antibody *complexes*; examples of type 3 reactions include the ARTHUS REACTION and SERUM SICKNESS. The reaction, which normally involves complement-fixation *via* the classical pathway, reaches peak intensity in 3–12 hours; immediate hypersensitivity reactions are sometimes referred to as immediate-*type* hypersensitivity reactions to include such phenomena.

immersion oil Oil (of refractive index about 1.52) used with the *oil-immersion* objective lenses of a microscope. (For theory see RESOLVING POWER.) Cedarwood oil is widely used; synthetic oils are also available. A condition of *homogeneous immersion* is obtained when the immersion oil, the glass of the objective lens, and the glass of the cover slip all have the same refractive index. For fluorescence microscopy a non-fluorescing immersion oil is an essential requirement. Immersion oil should be removed from the objective (with a lens tissue) immediately after use. Oil which has dried on the lens may be removed with a tissue moistened with benzene; benzene, xylol, and other oil solvents may damage lens mountings on prolonged contact. (see also MICROSCOPY)

immobilization test (*immunol.*) Any test in which antibody is detected and quantified by its inhibitory action on the movement of motile microorganisms. Antibodies may immobilize a flagellated organism by bringing about the simple adhesion of flagella. A more complex inhibitory action, which is COMPLEMENT-dependent, occurs during the immobilization of *Treponema pallidum* in the presence of homologous antibody. The *Treponema pallidum* Immobilization Test (TPI) is important in the diagnosis of SYPHILIS.

immune (*immunol.*) (of persons, animals) (1) Refers to the condition, which follows initial contact with a given antigen, in which antibodies (or primed lymphocytes) specific for that antigen are present in the body. (2) In a more restricted sense: refers to a condition of relative insusceptibility to a given *disease* following VACCINATION against, or recovery from, that disease, or following the administration of pre-formed specific antibody.

immune adherence A phenomenon in which antigen–antibody complexes adhere firmly to macrophages, polymorphs, primate erythrocytes, or the blood platelets of many non-primate species in the presence of COMPLEMENT. During the early stages of complement fixation, a group of complement components binds to the antigen–antibody complex; following the attachment of component C3, and its enzymic modification to C3b, the antigen–antibody complex adheres strongly (*via* a hydrophobic site on the C3b component) to the surfaces of cells and platelets.

immune complex Synonymous with antigen–antibody complex.

immune cytolysis Lysis of sensitized cells (cells coated with antibodies to their surface antigens) in the presence of COMPLEMENT.

immune haemolysis Lysis of *sensitized* erythrocytes (red blood cells coated with antibodies to their surface antigens) in the presence of COMPLEMENT.

immune lysis Refers to IMMUNE CYTOLYSIS.

immune response The specific response of the body to the introduction of an ANTIGEN or autocoupling HAPTEN. The response may involve (a) ANTIBODY-FORMATION; (b) HYPERSENSITIVITY; (c) IMMUNOLOGICAL TOLERANCE. (see also ANTIGENIC COMPETITION)

immune serum Synonymous with ANTISERUM.

immunity (*immunol.*) The *relative* insusceptibility of a person or animal to active infection by pathogenic microorganisms or to the harmful effects of certain toxins; the term is also used to refer to those conditions in which the body *over-reacts* towards such agents (HYPERSENSITIVITY). Some aspects of immunity are clearly advantageous to the person or animal (see e.g. ANTIBODY) while others may be less advantageous, disadvantageous, or positively lethal (see e.g. ANAPHYLACTIC SHOCK). ACTIVE IM-

MUNITY may be engendered e.g. by VACCINATION. (see also PASSIVE IMMUNITY, NON-SPECIFIC IMMUNITY, and SPECIFIC IMMUNITY)

immunization (1) *Active immunization.* Any procedure in which an ANTIGEN or autocoupling HAPTEN is introduced into the body, either orally or parenterally, with the object of producing a specific IMMUNE RESPONSE. When immunization is carried out *prophylactically* (see PROPHYLAXIS) it is commonly termed VACCINATION. (*Immunization* and *vaccination* are sometimes used synonymously.) (2) *Passive immunization.* Parenteral administration of pre-formed antibody (i.e. antibody taken from an *immune* subject), or *primed* lymphocytes, with the object of producing PASSIVE IMMUNITY in respect of a given antigen, or antigens. (3) *Isoimmunization.* The production of an immune response in respect of antigens derived from the tissues of another individual of the same species. (4) *Heteroimmunization.* The production of an immune response in respect of antigens derived from other species.

immunoassay Any procedure in which serological methods are used to detect and quantify a specific biological substance. An example is the LATEX PARTICLE TEST for pregnancy. During (human) pregnancy a sex hormone (human chorionic gonadotrophin—HCG) appears in the urine. One drop of urine is allowed to react with one drop of an antiserum containing antibodies to HCG. If the urine contains HCG, combination will occur between the HCG and anti-HCG antibodies; the absence of *uncombined* anti-HCG antibodies (indicating a *positive* test) is detected by an absence of agglutination on the addition of one drop of latex-bound HCG.

immunoconglutinins Antibodies homologous to the determinants of certain bound components of COMPLEMENT—particularly C3b. Titres of serum immunoconglutinins are raised in long-term autoallergic conditions and in some diseases of microbial aetiology.

immunocyte adherence technique (*syn*: rosette technique) One of the procedures used e.g. for detecting cells which form specific antibodies. When these cells are incubated with the homologous particulate antigen (or with erythrocytes coated with homologous soluble antigens) the antigenic particles adhere to specific *receptor sites* on the surfaces of the antibody-forming cells—forming rosettes.

immunodiffusion (see GEL DIFFUSION)

immunodominant (*adjective*) Refers to that part of an antigenic DETERMINANT which appears to bind with greatest strength to the antibody. (see also ANTIGENIC COMPETITION)

immunoelectrophoresis A two-stage procedure used for the analysis of materials containing mixtures of distinguishable proteins, e.g. serum. The sample is first subjected to ELECTROPHORESIS in an agar gel. This leads to the formation of discrete zones each containing a protein (or proteins) characterized by a specific electrophoretic mobility. (Some proteins migrate in a direction opposite to that which may be expected—see ELECTROENDOSMOSIS.) The agar strip in which electrophoresis has been carried out bears a rectangular slot, or trough, arranged parallel to the line of electrophoretic migration. This trough is now filled with an antiserum containing antibodies to some, or all, of the proteins in the sample being analysed. The proteins separated by electrophoresis, and their homologous antibodies, now diffuse towards each other through the agar and later give rise to lines or arcs of precipitate.

immunoferritin technique A procedure used to locate specific antigenic sites in sections of tissue. Initially, homologous antibody, labelled with FERRITIN, is allowed to combine with the relevant sites. The preparation then undergoes fixation with osmium tetroxide (or glutaraldehyde) and is further prepared for examination under the *electron microscope*.

immunofluorescence (*immunol.*) Any of a variety of techniques used to detect a specific antigen (or antibody) by means of homologous antibodies (or antigens) which have been conjugated with a *fluorescent* dye. A suitably prepared section of tissue, or a smear, is initially incubated with the *conjugate* (i.e. the dye-linked antibody or antigen); the preparation is subsequently washed free of conjugate and examined by fluorescence MICROSCOPY. Fluorescence is observed only when specific antigen–antibody combination has occurred*. An immunofluorescent technique is used, for example, in the FTA-ABS diagnostic test for syphilis. (see also FITC and SANDWICH TECHNIQUE) *In immunofluorescence techniques control procedures are necessary to preclude the possibility of an interpretation based on the non-specific localization of fluorescent conjugate.

immunogen (1) A substance which, when administered on its own, is capable of eliciting an IMMUNE RESPONSE: in this sense synonymous with ANTIGEN. (cf. HAPTEN) (2) A substance which elicits a *positive*-type immune response, e.g. antibody-production—but which does not give rise to immunological tolerance. (3) A substance capable of eliciting an immune response which provides PROTECTIVE IMMUNITY.

immunogenic Having the ability to function as an IMMUNOGEN.

immunoglobulins A varied class of proteins found in plasma and in other body fluids; all the known *antibodies* are immunoglobulins.

The major portion of the immunoglobulin molecule consists of equal numbers of HEAVY CHAINS and LIGHT CHAINS linked by ionic forces and interchain disulphide bonds; the specific antibody COMBINING SITES are located near the N-terminals of these chains. (For details of basic immunoglobulin structure see entry IGG.) The structure and function of immunoglobulins have been studied with the aid of proteolytic enzymes such as PAPAIN and pepsin, and with MERCAPTOETHANOL. Important products of molecular cleavage are described under FAB FRAGMENT. Five classes of immunoglobulins are distinguished on the basis of (a) the nature of the heavy chain; (b) molecular size; (c) sedimentation coefficient; (d) carbohydrate content; see IGA, IGD, IGE, IGG, and IGM (see also ANTIBODY).

immunological paralysis Commonly synonymous with IMMUNOLOGICAL TOLERANCE.

immunological tolerance Persons or animals which are immunologically tolerant to *a specific antigen* fail to produce antibody or to develop cell-mediated immunity when exposed to that antigen. (Unrelated antigens elicit a normal, positive response in the same subject.) Immunological tolerance may follow if a foetus or neonate is exposed to an antigen, or if an adult is exposed to a very large dose of antigen (high zone tolerance) or to continuous or interrupted contact with the antigen at a very low level (low zone tolerance). Immunological tolerance may persist for months or years.

immunological unresponsiveness (1) Absence of a positive response to a *particular* antigen: see IMMUNOLOGICAL TOLERANCE. (2) An absent or reduced response to antigens *in general* due, for example, to IMMUNOSUPPRESSION.

immunoradiometric assay A highly sensitive serological technique for the assay of specific antigens; the method can be used in place of RADIOIMMUNOASSAY when it is not possible to prepare a radioactive antigen. *Assay of specific antigen.* Initially, radioactive *antibodies* are prepared as follows. The antigen is adsorbed to an inert carrier (e.g. cellulose) and exposed to purified immunoglobulins prepared from a serum containing the antigenic-specific antibody. Those antibodies which do not combine with the adsorbed antigens are removed (by washing); antibodies which have combined with the adsorbed antigens are then made radioactive e.g. by iodination with ^{125}I. The radioactive antibodies are then eluted and used in the assay. In the assay, excess radioactive antibody is allowed to react with the antigen preparation of unknown concentration. The reaction mixture is then exposed to a fresh preparation of cellulose-adsorbed antigens; these antigens adsorb those antibodies which have not combined with the antigens in the specimen. Those antibodies which have combined with antigens in the specimen are removed and their level of radioactivity is used as an indication of the concentration of antigen in the specimen; for this purpose, use is made of a previously-prepared standard curve of radioactivity plotted against antigen concentration.

immunosuppression Suppression or modification of the normal IMMUNE RESPONSE by e.g. (a) The administration of *antimetabolites*—e.g. *methotrexate* (a folic acid antagonist), *6-mercaptopurine.* (b) *X-rays.* The intensity and mode of suppression vary with the effective dose. (c) The administration of ALS.

imperfect (*mycol.*) Refers to the *asexual* stage or condition of fungi.

impetigo An inflammatory disease of the skin characterized by the occurrence of discrete pustules which appear chiefly on the nose and around the mouth; the pustules become crusty and subsequently rupture. The causal agent(s) include *Streptococcus pyogenes* and *Staphylococcus aureus.*

IMVEC tests A group of tests which comprise the IMVIC TESTS and the EIJKMAN TEST.

IMViC tests A group of tests used in the identification of bacteria of the family *Enterobacteriaceae.* The tests are: (a) the INDOLE TEST; (b) the METHYL RED TEST; (c) the VOGES-PROSKAUER TEST, and (d) the CITRATE TEST. (see also EIJKMAN TEST)

Inaba serotype (see VIBRIO)

inactivated vaccine Any VACCINE containing microorganisms which have been so treated that they are no longer capable of multiplying and causing disease. (Some viral pathogens are attenuated prior to inactivation.) Organisms may be inactivated by agents such as β-PROPIOLACTONE, phenol, formalin etc. (see e.g. SALK VACCINE; SEMPLE VACCINE) *Heat-inactivated suspensions of Streptococcus mutans* have been used as prophylactic vaccines against DENTAL CARIES. Agents used for inactivation should not materially affect the PROTECTIVE ANTIGENS of the particular pathogen.

inactivation (of microorganisms) A term used to refer either to (a) the *permanent* loss of viability, or (b) a temporary, *reversible* arrest in the ability to function as a viable organism.

inactivation (of serum) (*immunol.*) Heat treatment (56 °C/30 minutes) which abolishes COMPLEMENT activity in the serum.

inaequihymeniiferous (inaequihymenial) (see GILLS)

inclusion bodies (1) (*virol.*) Discrete assemblies of virions and/or viral components (sometimes with cellular material) which develop within virus-infected cells. *Intracytoplasmic* inclusion bodies include those formed by poxviruses (see e.g. GUARNIERI BODIES) and by rhabdoviruses (see e.g. NEGRI BODIES); *intranuclear* inclusion bodies include e.g. the Lipschütz bodies formed by herpesviruses (see HERPESVIRUS). (see also POLYHEDROSES) (2) (*bacteriol.*) *Inclusion bodies* may refer to any of a range of bodies found within bacterial cells—e.g. granules of storage or metabolic products (e.g. poly-β-hydroxybutyrate, sulphur), GAS VACUOLES, and CARBOXYSOMES. (3) (*bacteriol.*) See ANAPLASMA.

incompatible (noncompatible) (*plant pathol.*) Used to describe a plant-pathogen relationship in which disease does *not* develop.

incomplete antibodies (1) Synonymous with the meaning given under sense (1) of BLOCKING ANTIBODIES. (2) May refer to FAB FRAGMENTS.

incubation The maintenance of a particular ambient temperature* for inoculated media or for other types of material. Incubation is commonly achieved within closed, heat-insulated, thermostatically-controlled chambers (*incubators*) or within vessels suspended in a WATER BATH; usually, any one of a range of temperature settings may be selected. Refrigerated incubators maintain the lower range of temperatures. Temperatures commonly used are those which promote maximum growth rates in particular organisms, or maximum reaction rates in biochemical or immunological tests etc. Specialized incubation permits control of humidity, light, radiation, and other factors in addition to temperature. **Incubation* is sometimes used to refer to the subjection of a specimen/culture etc. to environmental (laboratory) temperatures for a given period of time—an imprecise form of incubation known as 'incubation at room temperature'.

incubation period (of a disease) (*med.*) The period of time between the commencement of active infection and the onset of symptoms.

independent assortment (genetics) During MEIOSIS: the *chance* distribution of *unlinked* genes (see LINKAGE) among progeny cells. Thus, e.g. if the (diploid) parent cell contained the unlinked allele pairs A/a and Z/z, any one of four possible allele combinations may occur, with equal probability, in a given (haploid) progeny cell: A and Z, A and z, a and Z, and a and z.

indicator organisms (in water pollution studies) (a) In quantitative studies on *faecal* pollution (in rivers etc.) counts of the common faecal bacteria *Escherichia coli* and *Streptococcus faecalis* are often used to indicate the degree of pollution; spores of *Clostridium perfringens* have been used to estimate the extent of *previous* faecal contamination. (b) A range of microorganisms act as indicators in the SAPROBITY SYSTEM.

indicator viruses *Haemagglutinating* viruses which have been heated, to destroy RDE, but which have retained the ability to cause HAEMAGGLUTINATION. Such viruses do not spontaneously elute from erythrocytes. (see also NEURAMINIDASES)

indicators (pH) (see PH INDICATORS)

indirect selection (see REPLICA PLATING)

3-indole-acetic acid (see AUXINS)

indole test A test used in the identification of certain bacteria, e.g. members of the Enterobacteriaceae. (see also IMVIC TESTS) The test determines the ability of an organism to produce indole from tryptophan (see Appendix IV(f)). The organism is grown for 48 hours (at 37 °C) in peptone water or tryptone water*; KOVÁCS' REAGENT (0.5 ml per 5 ml culture) is then added and the (stoppered) tube shaken. Indole, if present, dissolves in the reagent which becomes pink, or red, and forms a layer at the surface of the medium. In a simpler form of the test a paper strip, impregnated with oxalic acid, is placed, prior to incubation, between the tube and its stopper or cap; during incubation, the indole (volatile at 37 °C) reacts with the test paper—which becomes pink or red. *Tryptone water contains a peptone rich in tryptophan which is prepared e.g. by the digestion of *casein* with pancreatic enzymes.

induction (of bacteriophage) (see LYSOGENY)

induction (of enzymes) (see OPERON)

infantile paralysis (see POLIOMYELITIS)

infection The condition in which pathogenic microorganisms have become established in the tissues of a host; although such a condition does not necessarily constitute, or lead to, disease, the terms *infection* and *disease* are often used as though synonymous. ('Infection' is sometimes used to refer to the mere presence of a pathogen in or on tissues.)

infection court (*plant pathol.*) That region of the host in which a pathogen passes the incubation stage of a disease.

infection peg (see APPRESSORIUM)

infectious centre (in bacteriophage infection) (see ONE-STEP GROWTH EXPERIMENT)

infectious disease Any disease which can be transmitted from one person, animal or plant to another.

infectious hepatitis (see HEPATITIS)

infectious mononucleosis (glandular fever) An acute, infectious disease which primarily affects the lymphoid tissues. Typical symptoms include fever, sore throat, enlarged and tender lymph nodes, an enlarged spleen, and the presence of abnormal leucocytes in the blood; in the majority of cases, certain heterophile antibodies become detectable in the blood during the course of the disease. Hepatitis is a common complication. The causal agent, the Epstein–Barr virus (EB virus, EBV), appears to enter the body *via* the nose or mouth.

infectious nucleic acids (of VIRUSES) Nucleic acid, extracted from the virions of *certain* types of virus (e.g. papovaviruses, picornaviruses, togaviruses), which can infect a cell and give rise to viral progeny. Nucleic acids obtained e.g. from virions which incorporate a *transcriptase* are not infectious; those obtained from e.g. polioviruses are infectious for a range of cell types wider than that susceptible to poliovirus virions. (see also TRANSFECTION)

infectivity (of diseases) The ease with which an infectious disease may be contracted under given conditions.

infectivity (of pathogenic microorganisms) The ability of a pathogen to become established on or within the tissues of a host.

infectivity, viral (see END POINT (*virol.*))

influenza In man, an acute, highly communicable disease which tends to occur in epidemic or pandemic form. The causal agent, a *myxovirus*—see INFLUENZA VIRUSES, invades *via* the mouth or nose. Following a 1–3 day incubation period, there is a sudden onset with malaise, headache, fever, and marked prostration associated with varying degrees of coryza; viraemia and systemic involvement are rare. Influenza is usually a self-limiting disease, but secondary bacterial infections (e.g. of the respiratory tract, nasal sinuses) may occur. Anti-influenza activity is exhibited e.g. by certain derivatives of *amantadine* under experimental conditions.

influenza viruses The three members of the genus ORTHOMYXOVIRUS (influenza virus types *A*, *B*, and *C*) which are distinguished on the basis of the nucleocapsid and envelope protein antigens; subtypes (strains) of a given type are distinguished on the basis of antigenic differences in the haemagglutination (H) and neuraminidase (N) peplomers (see ORTHOMYXOVIRUS). The type *A* subtypes include causal agents of diseases in man (INFLUENZA), swine (swine influenza), and avians (FOWL PLAGUE). The types *B* and *C* influenza viruses have been isolated only from cases of influenza in man; little appears to be known about type *C*. Type *A* viruses have been responsible for most, or all, of the influenza pandemics of the present century. The current system of nomenclature of the influenza viruses includes: the virus *type*—i.e. antigenic type *A*, *B*, or *C*; the animal from which the virus was first isolated—omitted if isolated from man; the place of initial isolation; the year of isolation; the particular haemagglutination (H) and neuraminidase (N) antigens. Thus e.g. strain

A/Singapore/1/57(H2N2)

was involved in the pandemic of 1957*, while strain

A/Hong Kong/1/68(H3N2)

was responsible for the 1968 pandemic. Influenza viruses (types *A* and *B*) can be cultivated in e.g. embryonated hens' eggs or chick embryo tissue cultures. Tests involving haemagglutination are carried out at 4 °C (to inhibit elution by neuraminidase) with erythrocytes from e.g. man, fowl, or the guinea-pig. Influenza viruses are inactivated by lipid solvents and detergents, and are labile at room temperatures and 4 °C; they can be stored at −70 °C. *this strain was formerly known as strain A2.

informosome In eucaryotic cells: a structure consisting of newly-synthesized mRNA complexed with (non-ribosomal) protein—believed to be the form in which mRNA is transported from the nucleus to the cytoplasm prior to translation. The function of the associated protein is unknown; it may protect the mRNA from enzymatic degradation, or may serve some regulatory function.

infraciliature (*ciliate protozool.*) The subpellicular system which consists mainly of the ciliar *basal bodies* (*q.v.*) and their associated *kinetodesmal fibrils*.

infundibulum The buccal cavity of *peritrich* ciliates.

infusion agar Infusion broth (see NUTRIENT BROTH) gelled with 1.5%–2.0% AGAR.

infusion broth (see NUTRIENT BROTH)

infusoria An archaic term which embraced the wide range of microscopic organisms—particularly motile protozoans—found in aqueous infusions of vegetable matter.

INH 1-Isonicotinic acid hydrazide—ISONIAZID.

initial(s) Any part (of an organism) which subsequently differentiates to form a characteristic structural feature.

ink cap fungi A common name for species of *Coprinus* in which the gills (and sometimes the pileus) undergo 'auto-digestion' to form a black spore-containing liquid. (see also BASIDIOSPORES, DISCHARGE OF)

innate (*mycol.*) (*adjective*) Refers to any fungal structure which is embedded in a substratum or in the fungal thallus such that its outer surface is more or less level with the surface of the substratum or thallus.

inner veil (*mycol.*) Synonym of PARTIAL VEIL.

inoculation (1) (*microbiol.*) The deposition of material (the INOCULUM) on or into a MEDIUM, living system (e.g. animal, TISSUE CULTURE) etc. Inoculation may be carried out e.g. with the object of initiating a CULTURE—in which case the medium is subsequently incubated (see INCUBATION) for an appropriate period of time. Inoculation is carried out with an ASEPTIC TECHNIQUE. (2) (*immunol.*) The oral or parenteral introduction of an antigen, or antigens, into the body of a person or animal. (see also IMMUNIZATION and VACCINATION)

inoculative infection (*parasitol.*) A mode of infection in which pathogenic microorganisms are introduced into the body of the host from the ANTERIOR STATION of an arthropod vector. (cf. CONTAMINATIVE INFECTION)

inoculum (1) (*microbiol.*) The material used to inoculate (see INOCULATION) a MEDIUM, living system (e.g. animal, TISSUE CULTURE) etc.; an inoculum typically comprises or contains viable microorganisms. (2) (*plant pathol.*) Pathogenic microorganisms which have penetrated and become established within a plant and which may give rise to disease.

inoperculate (*mycol.*) Refers to those structures (e.g. some types of ascus) in which an OPERCULUM is not present.

inosine-5′-monophosphate (IMP) (in purine nucleotide biosynthesis) (see Appendix V(a))

inositol Hexahydroxycyclohexane. Theoretically there are nine possible stereochemical forms of this compound; of these, *meso*-inositol (= *myo*-inositol), scyllitol, *d*-inositol and *l*-inositol occur naturally, and only *meso*-inositol is common. *Meso*-inositol occurs widely in the phospholipids of microbial and higher systems; it is required, in trace amounts, as a growth factor by a number of microorganisms—including many yeasts e.g. *Kloeckera* spp, *Schizosaccharomyces* spp. Inositol is used as a substrate in biochemical characterization tests for bacteria; it is attacked by e.g. *Klebsiella pneumoniae.* (see also CYCLITOL ANTIBIOTICS)

insect diseases (a) Viral diseases include *spheroidoses* (see ENTOMOPOXVIRUSES), GRANULOSES, and POLYHEDROSES. (see also IRIDESCENT VIRUSES) (b) Fungal pathogens include species of *Aspergillus*, *Beauveria*, *Entomophthora*, and *Septobasidium*; the host range of pathogenic fungi is generally wide. Infection often results from direct fungal penetration of the insect cuticle; some entomopathogenic fungi form toxins. (c) Protozoan pathogens include *Malpighamoeba* (a pathogen of the honey-bee, *Apis mellifera*) and species of the flagellates *Crithidia*, *Herpetomonas*, and *Leptomonas* (pathogenic in a range of dipterous and hemipterous hosts). Many parasites and pathogens of insects occur among the Sporozoa and the Cnidospora, e.g. *Nosema bombycis* (see PÉBRINE). (d) Bacterial pathogens include *Bacillus popilliae* and *B. lentimorbus* (agents of milky disease in the Japanese beetle) and *B. thuringiensis* (pathogenic in many insect species). (see also BIOLOGICAL CONTROL; FOULBROOD; SILKWORM DISEASES)

insertion (addition) (genetics) A type of MUTATION in which a nucleotide, or two or more contiguous nucleotides, are added to DNA (or, in certain viruses, RNA). If the number of nucleotides gained is not 3 or a multiple of 3 (i.e. not one or more complete codons) the insertion will be a PHASE-SHIFT MUTATION. (cf. DELETION)

inspissation Literally, the process of thickening. In microbiology the term is applied to a process in which certain heat-coagulable substances (e.g. serum, homogenized hens' eggs etc.)—or media containing such substances—are subjected to temperatures in the region of 80 °C for one hour on each of three successive days; in addition to coagulation such substances also undergo TYNDALLIZATION.

instructive theories of antibody-formation Theories which hold that the ability to produce specific antibodies develops in the antibody-producing cells only *after* their initial contact with antigen, i.e. it is assumed that pre-existing specificity does *not* occur. (cf. CLONAL SELECTION THEORY)

interference (genetics) The phenomenon in which one recombinational event increases or decreases the chances of another occuring in the closely-adjacent region. For example: in *chiasma interference* the occurrence of one chiasma appears to prevent the occurrence of a second, adjacent chiasma (*positive interference*); the effect of chiasma interference tends to decrease with increasing distance between the markers under consideration. In genes which are closely linked, a high localized *negative interference* is sometimes observed—i.e. the occurrence of one recombinational event appears to *promote* recombination in the adjacent region; this may be due to the correction of adjacent sequences in hybrid DNA during gene conversion (see RECOMBINATION).

interference (*virol.*) The phenomenon in which the replication of a particular ('challenge') virus, in given cells or tissues, is partially or completely inhibited as a result of interaction(s) between those cells or tissues and another ('interfering') virus. In *homologous interference* replication of the challenge virus is inhibited as a result of prior infection by a similar or identical type of virus; in *heterologous interference* the challenge and interfering viruses are of different types. Interference may also be due e.g. to the destruction of virus-attachment sites by the interfering virus.

interference factors Protein factors which inhibit

the initiation factor F_3 (IF-3) in PROTEIN SYNTHESIS; interference factors are believed to be involved in the control of translation.

interferons Proteins which are produced by animal cells and which prevent the replication of any of a wide range of viruses in those cells. (Interferon molecules released from one cell may adsorb to other cells and inhibit viral replication in them.) Interferons are synthesized in response to exposure of cells to any of a variety of animate or inanimate inducers—e.g. (a) DNA or RNA viruses—including certain inactivated virions; (b) complex polysaccharides, endotoxins, double-stranded RNA—e.g. poly(I:C)*; (c) certain bacteria—e.g. species of *Brucella* and *Rickettsia*. Interferon production is reported to commence within minutes of the initial stimulation; each species of animal gives rise to its own particular type of interferon—which may or may not be effective in other species. The mechanism of virus inhibition by interferons is unknown; it appears that, on exposure to an interferon, a cell synthesizes a protein (the *translation inhibitory protein*—TIP, or *antiviral protein*) which may block the translation of viral mRNA.

Interferons have been assayed by means of a *plaque-reduction method*, i.e. the determination of the extent to which a given dilution of interferon inhibits plaque-formation in a tissue culture challenged with virus. *A double-stranded polymer which consists of one strand each of polyinosinic acid and polycytidylic acid.

interkinesis (see MEIOSIS)

intermediate donor (F′ donor) (see CONJUGATION (bacterial))

interphase (see MITOSIS)

interrupted mating A technique, devised by Wollman and Jacob (1955), for studying chromosome transfer during bacterial CONJUGATION. An Hfr population is mixed with an F^- population at time zero. At regular intervals of time a sample is withdrawn from the mating mixture and subjected to vigorous agitation to separate the mating pairs; each sample is diluted to reduce the probability of further pairing. The ex-conjugants are then plated on a medium which inhibits the growth of Hfr cells but which promotes the growth of certain recombinant recipients, i.e. those recipients which contain a particular donor gene, or genes. Samples withdrawn from the mating mixture within (about) 5 minutes of time zero fail to give rise to recombinants. Samples withdrawn after increasing periods of time give rise to recombinants which include increasingly greater proportions of the donor gene sequence. For a particular Hfr strain, a given donor gene appears in a recombinant only after a characteristic interval of time; this time depends on the distance which separates the gene from the *origin*—i.e. the leading end of the donor strand—and on the efficiency with which the gene integrates, by recombination, with the recipient's chromosome. The greater the distance between the origin and a given gene the longer the time which must elapse before that gene appears in a recombinant; however, genes remote from the origin are transferred with less efficiency owing to the (normal) supervention of donor strand breakage. The time required for transmission of the entire donor strand has been measured by using several different Hfr strains—the origin in each strain being at a different locus. The entire donor strand of *Escherichia coli* is transferred in 90 minutes at 37 °C; thus, the position of any gene on the *E. coli* chromosome may be referred to by the time taken (in minutes) for the transfer of that gene relative to a given origin. For example, on a standardized, arbitrarily-selected origin—the *thr* (threonine) locus—the *lac* genes occur at approximately 10 minutes.

Intoshellina A genus of protozoans of the order ASTOMATIDA.

***intra vitam* staining** (intravital staining) (see VITAL STAINING)

inulins (see FRUCTANS)

invasiveness (of pathogenic microorganisms) The ability of a pathogen to *spread* through a host's tissues; the degree of invasiveness of a pathogen reflects its relative insusceptibility to the defence mechanisms of the host. (see also AGGRESSINS)

inversion (genetics) (see CHROMOSOME ABERRATION)

invertase (sucrase) An enzyme which splits sucrose into glucose and fructose.

involution forms (of bacteria) Morphologically or otherwise aberrant cells often seen in old cultures or among organisms cultured under harsh or hostile conditions—e.g. in the presence of sub-lethal concentrations of antibacterial substances. Such forms are usually regarded as degenerate—degeneracy commonly being ascribed to the action of autolysins, failure of cell wall synthesis, or failure of cell division without cessation of macromolecule synthesis. Involution forms are frequently produced by *Clostridium septicum* and are often seen in old cultures of *Yersinia pestis*. (see also L FORMS)

Iodamoeba bütschlii A species of amoeba (order AMOEBIDA) common in the human intestine; it is rarely, if ever, pathogenic. The trophozoite is about 8–15 microns in diameter; the (single) nucleus—which contains no peripheral nucleoprotein (cf. ENTAMOEBA)—has a large, commonly off-centre, *karyosome*. The spherical or oval CYST (usually 10–15 microns in diameter) contains a single nucleus (with a large karyosome) and a large, persistent glycogen vacuole.

iodine (as a stain) (see LUGOL'S IODINE and CYSTS)

iodine and compounds (as antimicrobial agents) Iodine is a broad-spectrum bactericidal agent*; it is effective against e.g. the tubercle bacillus but possesses only poor sporicidal activity. Iodine is a weak oxidizing agent; its antimicrobial activity appears to be due to the combination of molecular iodine with cellular proteins (e.g. tyrosine is irreversibly iodinated). Iodine is readily soluble in organic solvents; *tincture of iodine* (used as an ANTISEPTIC) contains 2% (or more) iodine in a water-ethanol solution of potassium iodide. The low solubility of iodine in water can be increased by the presence of iodide, the tri-iodide (I_3^-) being formed. The tri-iodide itself is not antimicrobial, but it readily decomposes releasing free iodine. For skin antisepsis aqueous solutions are generally preferred since they are less irritant than alcoholic solutions; alcoholic solutions have been used for the disinfection of surgical appliances, thermometers etc. The antimicrobial activity of iodine solutions is maximal under acid conditions—pH6 or below.

The staining, corrosive and irritant properties of iodine can be overcome by complexing the element with certain high molecular weight surface-active compounds, e.g. polyvinyl pyrrolidone, certain quaternary ammonium compounds; these complexes are referred to as *iodophores* (or *iodophors*). Iodophores possess all the antimicrobial properties of iodine together with the detergent qualities of the surfactant; however, only some 70–80% of the bound iodine can be released. As with solutions of iodine, iodophores exhibit maximum activity under acid conditions. Iodophores have been used for the disinfection of drinking water, swimming-bath water etc.; polyvinylpyrrolidone-iodine (*povidone-iodine*) has been used against skin and mucous membrane infections involving species of *Streptococcus*, *Staphylococcus* and *Candida*.

Iodoform (CHI_3), once widely used as an antiseptic, is now rarely employed; its antimicrobial activity is poor compared with that of iodine. *Iodonium* compounds (R_2IX) contain trivalent iodine; such compounds (e.g. diphenyliodonium chloride) are strongly basic and possess antimicrobial properties, but their activity against spores is doubtful. The iodine derivatives of certain quinoline compounds are widely used as antiprotozoal agents; these include 5-chloro-7-iodo-8-HYDROXYQUINOLINE (*Enterovioform*) and 5,7-diiodo-8-hydroxyquinoline (*Diodoquin*)—both of which are amoebicidal.

5-IODO-2′-DEOXYURIDINE has been found to inhibit vaccinia virus and herpesviruses *in vitro* (i.e. in tissue culture) and has been used in human medicine. (see also CHLORINE, FLUORINE and BROMINE)

*Under certain conditions iodine abolishes the infectivity of *tobacco mosaic virus*.

5-iodo-2′-deoxyuridine (IUdR; Idoxuridine) An analogue of thymidine (5-methyl-2′-deoxyuridine) which acts as an antiviral agent. IUdR becomes incorporated into viral DNA, causing errors in viral transcription. Selective toxicity tends to be poor; IUdR is highly toxic to mammalian cells and is rarely used systemically. As an antiviral agent, IUdR may be used topically e.g. in the treatment of certain herpesvirus infections.

iodoform (see IODINE)

iodonium compounds (see IODINE)

iodophores (iodophors) (see IODINE)

ionizing radiation (in sterilization) (see STERILIZATION)

ionophores Substances which cause an increase in permeability to specific ions of natural or synthetic lipid (or lipoprotein) membranes. Certain ANTIBIOTICS owe their activity to this property—see e.g. GRAMICIDINS and DEPSIPEPTIDE ANTIBIOTICS.

ipomeamarone 2-Methyl-2-(4-methyl-2-oxopentyl)-5-(3-furyl)-tetrahydrofuran; a PHYTOALEXIN produced by the sweet potato (*Ipomea batatas*) following infection by certain fungi.

IPTG Isopropyl-β-D-thiogalactoside—see OPERON.

Irgasan (*proprietary name*) (see SALICYLIC ACID AND DERIVATIVES)

Iridaea A genus of red algae (RHODOPHYTA).

iridescence The production of coloured light when white light is reflected from both surfaces of a thin transparent or translucent layer or film; the effect is due to interference between light rays of certain wavelengths.

iridescent viruses Those VIRUSES which, when present in masses, exhibit the property of IRIDESCENCE; the iridescent viruses (genus *Iridovirus*) are parasitic in a wide range of insects—including species of the Coleoptera (beetles), Diptera (flies) and Lepidoptera (butterflies, moths). Iridescence is associated with pellets of purified virus, with certain virus-infected insects (living or dead), and with parts of infected insects. The iridoviruses are complex, non-enveloped icosahedral virions of diameter approximately 130–180 nm; they contain linear double-stranded DNA. Iridoviruses replicate within the *cytoplasm* of the host cell. Iridoviruses include the *Tipula* iridescent virus (natural host the cranefly,

Tipula) and the *Sericesthis* iridescent virus (isolated from a larva of the beetle *Sericesthis pruinosa*); both of these viruses have a wide host range and are capable of infecting insects of several different orders. Other iridescent viruses have been reported to cause disease in a number of mosquito species (particularly *Aedes* spp). Infections by iridescent viruses appear to be commonly fatal. (see also INSECT DISEASES)

Iridovirus (see IRIDESCENT VIRUSES)

Irish moss The red alga *Chondrus crispus*. (see also CARRAGEENAN)

ironophores (see SIDEROCHROMES)

isatin-β-thiosemicarbazones (IBT) (as antiviral agents) (Thiosemicarbazone: $R_2C{=}N.NH.CS.NH_2$) Isatin-β-thiosemicarbazone and some of its derivatives (e.g. *methisazone*: *N*-methyl-isatin-β-thiosemicarbazone) inhibit the replication of poxviruses and certain adenoviruses; they appear to inhibit the production of those viral proteins which are normally synthesized late in the replicative cycle. The exact mode of action is unknown. Methisazone has been used prophylactically in cases of exposure to smallpox infection and (prophylactically) against alastrim. Other derivatives of isatin-β-thiosemicarbazone inhibit certain enteroviruses. Some workers have reported that *in vitro* and *in vivo* resistance to isatin-β-thiosemicarbazones develops in certain strains of poxviruses. (see also ANTIVIRAL AGENTS)

isidium (*lichenol.*) A small protuberance (0.01–0.3 × 0.5–3.0 millimetres) which arises from the surface of the thallus in certain lichens; isidia may be e.g. spherical, flattened, cylindrical, branched ('coralloid'), according to species—see e.g. COLLEMA, UMBILICARIA, USNEA. An isidium contains algal cells and medullary tissue and is typically covered by a cortex. Isidia are believed to function (at least in some cases) as vegetative propagules.

isoelectric point (IEP; p*I*) The value of pH at which there is a net charge of zero on the molecules of substances (e.g. amino acids, proteins) which can be either positively or negatively charged according to the pH of the medium. The electrophoretic mobility of a protein molecule at its IEP is zero. (see also ZETA POTENTIAL)

isoenzyme (isozyme; multiple enzyme) An enzyme which occurs in more than one form in a given species—i.e. isoenzymes differ from each other in structure although they catalyse the same reaction(s).

isogamy Fertilization in which the gametes are similar in appearance and behaviour.

isoimmunization (see IMMUNIZATION)

isokont Refers to a pair of flagella of *similar* length (cf. HETEROKONT).

isolation (*microbiol.*) Any procedure in which a given species of organism, present in a particular sample or specimen, is obtained in pure culture. If the specimen contains a mixed flora (i.e. if several different types of organism are present) the isolation of a given species of bacterium or fungus may initially involve an ENRICHMENT technique; a SPORE-forming species of bacterium may be isolated e.g. by initially subjecting the specimen to a temperature at which all non-sporing organisms are inactivated. The isolation of viruses may involve e.g. one or more stages of FILTRATION (to remove non-viral organisms) and the use of the filtrate to inoculate a tissue culture which is sensitive to the particular virus sought; viral contaminants (i.e. unwanted viruses present in the filtrate) may be inactivated by physical, chemical, or serological agents to which the required virus is insensitive. Many protozoans and microalgae may be isolated by MICROMANIPULATION techniques—using a fine glass pipette and a low-power microscope; bacterial and/or fungal contaminants may be greatly reduced or eliminated by passing the required organism through a series of sterile solutions which contain antibiotics or other selective antimicrobial agents.

isoleucine (biosynthesis of) (see Appendix IV(d))

isomerase A category of ENZYMES (EC class 5) which encompasses those enzymes which catalyse isomerizations (i.e. intramolecular rearrangements). Examples include *racemases* (e.g. L-amino acid ⇌ D-amino acid) and *epimerases* (e.g. D-ribulose-5-phosphate ⇌ D-xylulose-5-phosphate).

isomorphic alternation of generations An ALTERNATION OF GENERATIONS in which the sporophyte and gametophyte are morphologically similar.

isoniazid (INH) A synthetic ANTIBIOTIC, 1-isonicotinic acid hydrazide. INH is active against *Mycobacterium tuberculosis* and, to a lesser extent, other *Mycobacterium* species; it is a bactericidal agent. Many theories have been put forward to account for the action of isoniazid but none are completely satisfactory.

isopycnic centrifugation (see CENTRIFUGATION)

Isospora A genus of protozoans of the suborder EIMERIINA; the organisms are intracellular parasites or pathogens of certain animals, including man. (For a generalized account of the morphology and life cycle of *Isospora* and related coccidians see COCCIDIA.) *Isospora* spp are distinguished from those of related genera (e.g. EIMERIA) by the production of an oocyst containing two sporocysts which (each) contain four sporozoites. At least two species may infect man, *I. belli* and *I. hominis*; infections are believed to be commonly asymptomatic, but *I. belli* (at least) may be responsible for a mild, self-limiting intestinal illness. *I. bigemina* may infect the small intestine of the cat and dog; symptoms, if present, may include bloody diarrhoea, dehydration and emaciation.

isozyme (see ISOENZYME)
IUdR (see 5-IODO-2′-DEOXYURIDINE)
iwatake The edible lichen *Umbilicaria esculenta*.

J

J chain (see IGM)

Japanese river fever (see SCRUB TYPHUS)

jelly fungi Those basidiomycetes which form gelatinous fruiting bodies—e.g. *Auricularia auricula*, *Calocera viscosa*, *Dacrymyces deliquescens*, *Exidia glandulosa*, and *Tremella mesenterica*.

jew's ear fungus (Judas' ear fungus) The basidiomycete *Auricularia auricula*.

Johne's bacillus *Mycobacterium paratuberculosis*.

Johne's disease (chronic bacillary diarrhoea) A chronic inflammatory infection of the intestines of cattle and certain other animals, e.g. goats. The causal organism, *Mycobacterium paratuberculosis*, enters the body on contaminated grass or feedingstuffs. Symptoms include diarrhoea, weakness and wasting.

johnin A substance prepared from the culture filtrate of *Mycobacterium paratuberculosis* (*M. johnei*). It is used (in a SKIN TEST) for the detection of JOHNE'S DISEASE in cattle. (see also DELAYED HYPERSENSITIVITY)

juglone 5-Hydroxy-1,4-naphthoquinone; an antifungal compound produced in the leaves and roots of *Juglans* spp (the walnut). (see also PHYTONCIDES)

jungle yellow fever (sylvan yellow fever) (see YELLOW FEVER)

K

K antigen A somatic antigen external to the cell wall proper or forming part of a bacterial CAPSULE. Important examples: the VI ANTIGENS and the capsular antigens of the *pneumococci*.

'K' virus (see POLYOMAVIRUS)

KAF (conglutinogen activating factor; C3b-inactivator) An enzyme which occurs in normal serum; KAF modifies the bound C3b component of COMPLEMENT—thus abolishing the strong immune adherence properties of the unmodified component. Modified C3b is involved in *conglutination*.

Kahn test A standard test for syphilis (see STS).

kala-azar Human *visceral* LEISHMANIASIS, a disease found in Africa, Asia, South America and the Mediterranean area. The causal agent, *L. donovani*, is transmitted (inoculatively) mainly by sandflies of the genus *Phlebotomus*. In man, the pathogen develops intracellularly (in the AMASTIGOTE form) in e.g. cells of the liver and spleen—which become enlarged. For chemotherapeutic agents see *antimony* under HEAVY METALS. *Laboratory diagnosis*: blood, material from lesions, liver biopsies etc. may be examined microscopically and culturally.

kanamycin A broad-spectrum AMINOGLYCOSIDE ANTIBIOTIC produced by *Streptomyces kanamyceticus*.

***kappa* particles** Bacterium-like particles which occur in the cytoplasm of certain (*killer*) strains of *Paramecium aurelia*; such strains are able to kill other (sensitive) paramecia, i.e. strains which lack *kappa* particles. Two types of *kappa* particle have been distinguished: (a) B-type, spherical *kappa* particles ('brights') which are about one micron in diameter, and (b) N-type (non-bright), rod-shaped *kappa* particles, about 3 × 0.5 microns. The N and B particles appear to be different stages in the life cycle of *kappa*; only the N particles seem to be capable of self-replication. B particles appear to be formed from N particles; only the B particles are able to confer the killer characteristic on the host *Paramecium*. B particles contain one or more refractile inclusion bodies which, under the electron microscope, appear as coiled ribbons. It appears that B particles (or parts of B particles) are released by a killer strain, and that these particles ('paramecin') are ingested by, and kill, organisms of a sensitive strain; isolated refractile inclusion bodies are also lethal for sensitive strains. When killer paramecia undergo binary fission, *kappa* particles are passed to each daughter cell; the intracellular multiplication of *kappa*—on which depends the continuance of the killer strain—requires the presence within the host organism of a certain dominant allele, *K*.

Kappa particles stain well with toluidine blue; they appear to have walls similar to those of bacteria, divide by binary fission, and are sensitive to certain antibiotics. They have not been cultured *in vitro*. Other particles (e.g. LAMBDA PARTICLES, *mu* particles, *pi* particles etc.) have also been found in strains of *P. aurelia*.

Kärber equation (see END POINT (*virol.*))

karyogamy The coalescence of nuclei. Karyogamy follows PLASMOGAMY during fertilization.

karyogram (see KARYOTYPE)

karyokinesis A synonym of MITOSIS.

karyolysis (*histopathol.*) The dissolution of a cell's nucleus with consequent loss of affinity for basic dyes; a sign of cellular abnormality or death. (see also KARYORRHEXIS)

karyonide (see CARYONIDE)

karyorrhexis (*histopathol.*) The fragmentation of a cell's nucleus; a sign of cellular abnormality or death. (see also KARYOLYSIS and PYKNOSIS)

karyosome A term used to refer to a NUCLEOLUS or to a nucleolus-like body.

karyotype (caryotype) The chromosomal constitution of a (eucaryotic) cell in terms of the number, size, and morphology of the chromosomes during *metaphase*. A systematized *diagrammatic* representation of a karyotype is referred to as an *idiogram*; a systematized *photographic* representation may be referred to as an idiogram or a *karyogram*.

katadyn (see *Silver* under HEAVY METALS)

Katodinium A genus of algae of the DINOPHYTA.

Kauffmann–White classification scheme A method of identifying the numerous SEROTYPES of *Salmonella* for epidemiological and other purposes; each serotype is defined in terms of its surface antigens, i.e. somatic and, where applicable, capsular and flagellar antigens. Serotypes thus identified have been placed in one or other of a number of *groups*—the serotypes in each group having in common at least one somatic antigen (the *group antigen*) which is not found in the members of other groups. Thus, the ANTIGENIC FORMULA of *Salmonella typhi*, written 9,12,Vi:d: –, shows that this organism is in the same group (D) as, for example, *Salmonella panama*: 1,9,12:l,v:1,5, the group D antigen (9) being common to both. The formula denotes that *S. typhi* possesses the somatic (O) antigens 9 and 12 and the VI ANTIGEN together with the *phase 1* d flagellar antigen. (see PHASE VARIATION, IN

SALMONELLA) The dash indicates that there are no phase 2 flagellar antigens. Similarly, *S. panama* possesses somatic antigens 1, 9 and 12, and is able to express either the phase 1 flagellar antigens l and v or the phase 2 flagellar antigens 1 and 5. Well over 1000 serotypes of *Salmonella* have been identified.

KCN test (*bacteriol.*) (see CYANIDE TEST)

KDO 2-Keto-3-deoxyoctonate: a constituent of the LIPOPOLYSACCHARIDES of certain gram-negative bacteria (e.g. *Salmonella* spp).

kelp A non-specific term for any large brown seaweed (algae of the division PHAEOPHYTA)—e.g. *Laminaria, Sargassum.*

keratin A highly insoluble protein which occurs in e.g. hair, wool, horn, skin. *Keratinases* are produced by relatively few organisms—including *Streptomyces* spp, *Candida albicans*, dermatophytes.

kieselgühr (see DIATOMACEOUS EARTH)

killer paramecia (see KAPPA PARTICLES and LAMBDA PARTICLES)

kinesis A stimulus-incited locomotion in which the *rate* (but *not* the direction) of motion is governed by the strength of the stimulus. (cf. TAXIS)

kinetid (*protozool.*) In ciliates, a unit of the KINETY; it consists of a CILIUM with its associated BASAL BODY, *kinetodesmal fibril* and microtubules together with the adjacent pellicular alveoli (see PELLICLE). (see also KINETODESMA)

kinetin (see CYTOKININS)

kinetochore (kinetochor) The part of a CHROMATID to which are attached the microtubules of the SPINDLE; the kinetochore is an electron-dense body found in the region of the centromere. (*Kinetochore* is sometimes used as a synonym of CENTROMERE.)

kinetodesma (*syn:* kinetodesmos) In ciliate protozoans, a series of overlapping endoplasmic fibrils which, under the light microscope, may appear as a single fibre running parallel with, and to the right of, the row of basal bodies of a KINETY. By electron microscopy it can be seen that a kinetodesmal fibril arises from a BASAL BODY, passes a little to the right, and then runs anteriorly to terminate in a position where it overlaps the fibril(s) of the more anterior basal bodies. Kinetodesmal fibrils are striated, the cross-striations having a periodicity of approximately 30 nm. The role of the *kinetodesmata* is at present unknown; the consensus opinion appears to be that they have little or no fuction in the coordination of ciliar movement. (see also PELLICLE)

kinetoplast An organelle located within an expanded region of the mitochondrion in members of the order Kinetoplastida (which includes the trypanosomes). The kinetoplast is situated close to the BASAL BODY—with which it is sometimes confused. It consists of a mass of linear and circular DNA which, some believe, contains the genetic determinants which code for certain of the mitochondrial components. During cell division the kinetoplast—and mitochondrion—divide prior to nuclear division. Kinetoplast replication may be inhibited by certain drugs; these include the aromatic diamidines (e.g. *berenil*) and the phenanthridines (e.g. *ethidium*)—some of which are used for the treatment of human and/or animal trypanosomiases. The kinetoplast gives a positive FEULGEN REACTION; with Romanowsky dyes the kinetoplast stains red—as does the nucleus.

Kinetoplastida An order of flagellate protozoans of the class ZOOMASTIGOPHOREA; typically, each organism possesses a KINETOPLAST. Species with two or more flagella (e.g. BODO) are placed in the sub-order Bodonina; some are free-living (occuring e.g. in sewage-polluted waters) while others are parasitic. Species with one flagellum are placed in the sub-order TRYPANOSOMATINA.

kinetosome (*protozool.*) (see BASAL BODY)

kinety (*protozool.*) In ciliates, a longitudinal (meridianal) row of cilia (see CILIUM) with their associated *basal bodies* (*q.v.*) and KINETODESMA. The location of a kinety corresponds to a primary meridian in the SILVER LINE SYSTEM. (see also KINETID)

kinins (see CYTOKININS)

Klebs–Löffler bacillus *Corynebacterium diphtheriae.*

Klebsiella A genus of gram-negative, non-motile bacteria of the ENTEROBACTERIACEAE; strains are widespread in soil and water, and occur as parasites or pathogens of animals, including man. The organisms are capsulated rods which usually form convex, viscid colonies—particularly on media containing fermentable carbohydrates; growth occurs optimally at or about 37 °C. Type species: *K. pneumoniae.* GC% range for the genus: 52–56.

K. pneumoniae. This species includes *some* of the organisms which are sometimes referred to as *Aerobacter aerogenes*, and those organisms sometimes referred to as *K. aerogenes* and *K. edwardsii.* Typically, *K. pneumoniae* produces acid and gas from glucose, ferments lactose, produces UREASES, grows on potassium cyanide media, and gives a positive VOGES-PROSKAUER TEST and a negative METHYL RED TEST. Certain strains carry out nitrogen fixation. Some strains are causal agents of pneumonia.

K. ozaenae. Most strains are MR+ and all are VP−; lactose is fermented by some strains. Variable characteristics include: citrate utilization, urease production, and the formation of lysine decarboxylase. The majority of strains do not utilize malonate.

K. rhinoscleromatis. MR+ and VP−; malonate is utilized by most strains. The characteristics given as variable in *K. ozaenae* are negative in *K. rhinoscleromatis*.

Kloeckera A genus of fungi within the family CRYPTOCOCCACEAE; species have been isolated from e.g. various fruits, tree bark, and soil. The vegetative form of the organisms is an ovoid, lemon-shaped, or elongated cell; some strains form a pseudomycelium. Reproduction occurs by *bipolar* budding. All species ferment glucose and some ferment sucrose; none ferments maltose. Nitrate is not assimilated. Inositol and pantothenic acid are growth requirements for all species. In culture the organisms generally appear cream, yellow, or brown.

Kluyveromyces A genus of fungi of the family SACCHAROMYCETACEAE.

knallgas bacteria ('hydrogen bacteria') A group of bacteria which can obtain energy for growth from the oxidation of molecular hydrogen by the (overall) reaction: $2H_2 + O_2 \rightarrow 2H_2O$ (the *knallgas* or *oxyhydrogen reaction*). These bacteria are facultative autotrophs—autotrophy requiring an atmosphere of hydrogen, oxygen, and carbon dioxide; they can also grow mixotrophically—using the energy derived from hydrogen oxidation for the utilization of organic substrates. (Heterotrophic growth can also occur.) The knallgas bacteria include species of *Alcaligenes* and *Pseudomonas* (formerly HYDROGENOMONAS spp), and *Paracoccus denitrificans* (formerly *Micrococcus denitrificans*). Species of the former two genera have been found to contain two types of HYDROGENASE. One type is membrane-bound and transfers electrons from hydrogen to a cytochrome- and quinone-containing electron transport chain—the terminal electron acceptor generally being oxygen, although some species can also use nitrate (cf. NITRATE RESPIRATION); OXIDATIVE PHOSPHORYLATION is believed to be associated with this electron transport chain. The second type of hydrogenase is soluble and transfers electrons from hydrogen directly to NAD; this type is referred to as *hydrogen dehydrogenase* to distinguish it from the first type.

Koch–Weeks bacillus *Haemophilus aegyptius.*

Koch's blue bodies Macroschizonts of THEILERIA.

Koch's phenomenon (*immunol.*) Koch observed that different responses were given by healthy and *tuberculous* guinea-pigs to a subcutaneous injection of virulent tubercle bacilli. After 10–14 days, healthy animals developed a nodule at the injection site; this nodule subsequently ulcerated and remained in that condition until death ensued within 6 to 8 weeks. Enlargement and subsequent necrosis of the regional lymph nodes were also observed. In *tuberculous* animals a (different) reaction was observed within 2 days of the injection; the area surrounding the injection site became *indurated* (hardened) and *erythematous* (reddened) and later became necrotic. The dead skin finally sloughed and the area quickly healed. Infection of the regional lymph nodes did not occur. Koch later demonstrated that *tuberculous* animals give a similar reaction when injected with killed tubercle bacilli and even with culture filtrates of the tubercle bacillus*. Reactions in the *tuberculous* guinea-pigs were manifestations of DELAYED HYPERSENSITIVITY—a form of CELL-MEDIATED IMMUNITY. *Healthy animals give no visible reaction (or react minimally) in response to such injections.

Koch's postulates Robert Koch (1843–1910) regarded the fulfilment of the following requirements as essential in establishing the causal relationship between a given organism and a given disease. (a) The organism must be present in every case of the disease. (b) It can be isolated in pure culture. (c) When susceptible experimental animals are inoculated with the isolated organism a similar disease must be produced. (d) The (same) organism must be isolable from the diseased animals.

Koehler illumination (see KÖHLER ILLUMINATION)

Köhler illumination (in MICROSCOPY) A mode of illumination in which the entire field of view is subjected to light of uniform intensity—regardless of any irregularities of intensity in the light source. (cf. CRITICAL ILLUMINATION) The procedure for obtaining Köhler illumination with an external light source having a condensing lens and an iris diaphragm external to it: (a) Close the microscope condenser diaphragm. (b) Open the diaphragm of the lamp (the field diaphragm). (c) Position the lamp so that an image of the bulb's filament is formed (by reflection from the *plane* mirror of the microscope) on the underside of the microscope condenser diaphragm. The lamp and microscope should then be kept in the same positions relative to one another. (d) Open the condenser diaphragm and almost close the field diaphragm. With a low-power objective focused on the object (specimen) make a *slight* adjustment to the microscope condenser so that the edge of the field diaphragm becomes sharply in focus in the plane of the specimen. (e) Open the field diaphragm until the small disc of light in the centre of the field has increased in size to cover the entire field. (Should the disc of light not be in the centre of the field it can be centred by adjusting the position of the mirror.)

Step (d) ensures that Köhler illumination has been achieved, i.e. that the image of the lamp filament is in the lower focal plane of the microscope condenser. Under these conditions, *each* point of the filament's image gives rise to a bundle of parallel rays which leave the upper surface of the microscope condenser lens and pass through the specimen.

kojibiose (α-D-glucopyranosyl-(1 → 2)-α-D-glucopyranose) A disaccharide which occurs linked to the C2 position of glycerol in the TEICHOIC ACIDS of group D streptococci; the kojibiose may be esterified with D-alanine.

kojic acid (5-hydroxy-2-(hydroxymethyl)-4-pyrone) A substance synthesized by *Aspergillus* spp (particularly the *flavus-oryzae* group) during the fermentation of glucose, xylose and certain other sugars. It is a chelating agent with weak antibiotic properties; antibiotic activity is enhanced by certain metal ions. Susceptibility to kojic acid is greater among gram-negative bacteria than among gram-positive species. Kojic acid is used in the preparation of insecticides and fungicides. In culture filtrates kojic acid gives a strong blood-red reaction with ferric salts.

Kolpoda (see COLPODA)

Koplik's spots Lesions commonly formed on the mucous membranes of the mouth in patients suffering from MEASLES. They are small red spots with whitish centres.

Koster's stain (for *Brucella*) A procedure for detecting cells of the (gram-negative) genus BRUCELLA among (gram-negative) mammalian tissues. The tissue SMEAR or section is first stained for 1 minute in a solution prepared from two parts saturated aqueous *safranin* and five parts *normal* potassium hydroxide. The slide is then washed in tap water and differentiated in 0.1% sulphuric acid (several changes) for a *total* period of between 10 and 20 *seconds*. After washing in tap water the preparation is counterstained in 1% carbol methylene blue. *Brucella* spp stain orange; tissues stain blue.

koumiss (kumiss) An acidic, mildly alcoholic fermented milk product native to certain regions of the USSR. It is prepared from mares' milk; the organisms used include species of *Streptococcus* and *Lactobacillus* and lactose-fermenting strains of yeast. (cf. YOGHURT)

Kovács' oxidase reagent A 1% aqueous solution of tetramethyl-*p*-phenylenediamine dihydrochloride; a reagent used in the OXIDASE TEST.

Kovács' reagent A reagent used in the INDOLE TEST. It contains *p*-dimethylaminobenzaldehyde (5% w/v) in acid-alcohol (concentrated hydrochloric acid and isoamyl (or amyl) alcohol in the proportions 1:3 by volume).

krankheitsherd A name given by Woronin (1878) to a cluster of enlarged, infected cells in the tissues of crucifers affected by *clubroot* disease.

Krebs cycle (see TRICARBOXYLIC ACID CYCLE)

Kupffer cells (see MACROPHAGES)

Kurthia A genus of gram-positive, obligately aerobic, asporogenous bacteria of uncertain taxonomic affinity; strains occur e.g. in various meat products and in premises associated with meat handling. The organisms occur as chains of rods (or short filaments) or as peritrichously-flagellate bacilli; coccoid forms predominate in ageing cultures. The sole species, *K. zopfii*, is a non-fermentative chemoorganotroph.

kuru A slowly-progressive disease of man observed in New Guinea. The long incubation period (months, years) is followed by a pathological process occupying some 1 to 2 years; the symptoms are those associated with dysfunction of the central nervous system. The nature of the causal agent is unknown.

L

L forms (L-phase variants) Defective bacterial cells of indefinite or spherical shape formed either spontaneously (e.g. by *Streptobacillus moniliformis*) or as a result of any of a variety of stimuli, e.g. temperature shock, osmotic shock, or the presence of antibiotics which inhibit cell wall biosynthesis; in L forms the cell wall is either defective or totally absent. Upon removal of the stimulus L forms may resume cell wall synthesis and revert to the condition of the original strain. Alternatively they may continue to reproduce as L forms; they are then termed *stable L forms*. L forms have been observed in a range of bacteria—including species of *Bacillus*, *Proteus*, *Streptococcus*, and *Vibrio*. L form colonies often resemble those of *Mycoplasma* spp. L forms were named after the Lister Institute of Preventive Medicine (London). (cf. PROTOPLASTS and SPHAEROPLASTS)

L-membrane (see OUTER MEMBRANE)

Laboulbeniomycetes A class of fungi within the sub-division ASCOMYCOTINA; constituent species are obligately parasitic mainly on insects (including members of the Coleoptera, Diptera, and Hemiptera). The minute, non-mycelial but multicellular thallus remains attached to the host by means of a specialized organelle—the foot. Some species are monoecious, others dioecious. The ascocarp is a perithecium, and each elongated ascus typically contains four elongated, uniseptate ascospores. Asexual reproduction appears to be unknown in the members of this class. All species are placed in the single order Laboulbeniales.

Labyrinthulales (Labyrinthulida) (see NET SLIME MOULDS)

***lac* operon** (see OPERON)

Lachnospira A genus of gram-variable, anaerobic, motile (flagellate) bacteria of uncertain taxonomic affiliation; the cells are curved rods. Metabolism is fermentative. The type species, *L. multiparus*, occurs in the RUMEN.

laciniate Lobed.

lacrimoid Tear-shaped.

β-lactamases Enzymes which hydrolyse the β-lactam linkage in PENICILLINS and CEPHALOSPORINS—thus rendering them inactive as antibiotics. (see e.g. PENICILLINASE) β-Lactamases are synthesized by a number of gram-positive and gram-negative bacteria (e.g. many strains of *Staphylococcus* and *Proteus*). In gram-positive bacteria synthesis of the enzyme may be inducible and the enzyme is commonly liberated into the medium; in gram-negative species synthesis is usually constitutive and the enzyme remains within the cell. Genes coding for the synthesis of β-lactamase may be carried by the bacterial chromosome or by an R FACTOR. Organisms which synthesize β-lactamases are generally resistant to the action of one or more of the penicillins or cephalosporins; a given β-lactamase may be more effective against some antibiotics than others (see PENICILLINS). (see also ANTIBIOTICS, RESISTANCE TO)

Lactarius A genus of fungi of the family RUSSULACEAE; constituent species are distinguished from those of the genus *Russula* in that the fruiting body exudes a milk-like fluid ('latex') when cut. *Lactarius* spp are commonly known as 'milk caps'.

lactic acid fermentation (see HOMOLACTIC FERMENTATION; HETEROLACTIC FERMENTATION; see also Appendix III(a))

Lactobacillaceae A family of bacteria of which LACTOBACILLUS is the sole genus.

Lactobacillus A genus of gram-positive, asporogenous, catalase-negative*, anaerobic, microaerophilic or facultatively aerobic bacteria of the family Lactobacillaceae; the organisms are straight or curved bacilli, or coccobacilli, which occur singly or in chains. Most strains are non-motile. The GC% range for the genus is reported to be approximately 35–53. Metabolism is fermentative and all species form lactate as a major end product of the fermentation of glucose. Lactobacilli which carry out HOMOLACTIC FERMENTATION include *L. acidophilus*, *L. bulgaricus*, *L. casei*, *L. delbrueckii* (the type species), and *L. lactis*; those which carry out HETEROLACTIC FERMENTATION include *L. brevis* and *L. fructivorans*. (see also Appendix III(a)) Lactobacilli are nutritionally-demanding—their requirements generally including a wide range of vitamins.

No pathogenic lactobacilli are known. Species occur as saprophytes (see e.g. SILAGE) and e.g. in the human mouth, intestinal tract, and vagina. Lactobacilli are important in the preparation of certain fermented foods, e.g. CHEESE, SAUERKRAUT, and YOGHURT. *See however CATALASE TEST.

Lactobacillus bifidus Synonym of *Bifidobacterium bifidum*.

lactoflavin (see RIBOFLAVIN)

lactophenol cotton blue A combined clearing agent and stain used as a general mounting fluid for the microscopical examination of fungi. Phenol (20 g) is dissolved in distilled water (20 ml), and the solution thoroughly mixed with lactic acid (20 g) and glycerol (30 ml); cotton blue (water-soluble *Anilin blue*) (0.05 g) is dissolved in this mixture.

lactose A disaccharide—D-galactopyranosyl-β-(1 → 4)-D-glucopyranose—which occurs in milk. The ability to metabolize lactose is an important characteristic in the identification of bacteria of the family Enterobacteriaceae; organisms which metabolize exogenous lactose possess both β-galactosidase and galactoside 'permease' (i.e. a transport system for galactosides—see TRANSPORT). (see also *lac* OPERON)

laeoplectic metachrony (see METACHRONAL WAVES)

lag phase (see GROWTH (bacterial))

laking (of erythrocytes) Lysis—with release of haemoglobin.

LAL test (see LIMULUS AMOEBOCYTE LYSATE)

***lambda* particles** Endosymbiotic, rod-shaped, flagellate bacteria (approximately 4 × 1 microns) found in certain strains of *Paramecium aurelia*—on which they confer the *killer* characteristic. (see also KAPPA PARTICLES) Unlike *kappa*, *lambda* particles can be cultured *in vitro*—i.e. in a cell-free medium; particles which have been cultured *in vitro* are able to kill sensitive strains of paramecia. *Lambda* particles stain well with toluidine blue.

Lamblia intestinalis (see GIARDIA)

lamella (*pl. lamellae*) (of basidiomycetes) (see GILLS)

lamina (blade) (*algol.*) The leaf-like part of the thallus of certain ALGAE.

Laminaria A genus of algae of the division PHAEOPHYTA; species occur in subtidal marine environments. The organism shows an heteromorphic ALTERNATION OF GENERATIONS. The *sporophyte* consists of a large flattened *blade* which, depending on species, may be e.g. long and narrow with straight or crinkled edges, or broad and more or less deeply divided to form a strap-like fringe. The blade is connected, *via* a *stipe*, to a branching *holdfast* which attaches the organism to the substratum. The sporophyte exhibits a considerable degree of tissue differentiation. The *gametophyte* is a microscopic, filamentous structure.

laminarin (laminaran) The main food reserve in members of the Phaeophyta (brown algae); laminarins are essentially water-soluble, linear, β-(1 → 3)-linked GLUCANS—though branching may occur with β-(1 → 6) branch points. Some types of laminarin contain 2–3% D-mannitol; these contain approximately 50% of each of two types of chain: the *M-chain*—with terminal D-mannitol residues, and the *G-chain*—with terminal (reducing) D-glucose residues. Laminarin is attacked by certain fungi and by species of *Nocardia*.

lamziekte (see FOOD POISONING (bacterial))

Lancefield's streptococcal grouping test A procedure for identifying streptococci; it involves the extraction and serological examination of cellular components known as C SUBSTANCES. By means of this procedure most strains of *Streptococcus* can be placed into one of the following *Lancefield groups*: *A*, *A-variant*, *B*, *C*, *D*, *E*, *F*, *G*, *H*, *K*, *L*, *M*, *O* and *S*; all the strains of a particular group possess a particular C substance which does not occur in the strains of any other group. The C substance may be extracted by a number of methods, e.g. acid-extraction (hydrochloric acid) or enzymatic extraction. The test is often carried out as a RING TEST in a capillary tube: the extract is layered onto an antiserum to a given C substance and the test repeated with antisera to different C substances until a positive result is obtained. Once a strain has been assigned to a group further tests may be carried out to determine the particular sub-group or *type* to which the strain belongs. (see also M PROTEIN (streptococcal))

lapinized vaccine A vaccine containing live organisms whose virulence (for a given host) has been suitably attenuated by serial passage through rabbits.

Lassa fever In man, an acute, infectious, systemic disease caused by the Lassa fever virus, an *arenavirus*; transmission appears to occur by droplet inhalation and by the contamination of wounds with infected material. Symptoms may include tiredness, headache, aches in the back and limbs, an inflamed and ulcerated throat, fever and convulsions; in fatal cases (50% of known cases) death generally occurs within 10–14 days of the onset of symptoms. Various organs, e.g. the kidneys, may be affected. Convalescent serum has been used therapeutically. *Laboratory diagnosis* (in specially-equipped laboratories): cultural/serological/electron-microscopical examination of specimens which include throat swabs, blood, urine. Lassa fever, first recorded in Nigeria, has occurred in travellers from Africa, and their contacts; it is a NOTIFIABLE DISEASE in the United Kingdom.

late blight, of potato (see POTATO, FUNGAL DISEASES OF)

latent period (in bacteriophage infection) (see ONE-STEP GROWTH EXPERIMENT)

latex A product of the rubber tree, *Hevea brasiliensis*. Natural latex is a colloidal suspension of polyisoprenoid particles in an aqueous phase which contains carbohydrates, proteins and other substances; it is readily attacked by a variety of microorganisms. (see also RUBBER (microbial spoilage of))

latex particle test (latex test) (*immunol.*) A PASSIVE AGGLUTINATION test which involves the use of a *soluble* antigen which has been adsorbed to minute particles of polystyrene latex. In the presence of antibody specific to the latex-bound antigen the latex particles are agglutinated. The test may be used quantitatively.

lattice hypothesis (*immunol.*) An hypothesis advanced by Marrack (1934) to account for the observation that, in a PRECIPITIN TEST, antibody and antigen may combine *in varying proportions*. The hypothesis postulates that each molecule of antibody has a VALENCY of at least two, and that antigens are multivalent; thus, when antigens and antibodies combine in suitable proportions, giant insoluble aggregates (*lattices*) are formed. If either antigen or antibody is present in considerable excess, smaller, *soluble* complexes are formed. Under conditions of antibody excess, for example, a soluble complex may consist of a single molecule of antigen with each of its determinants satisfied by a molecule of antibody; thus, such complexes could not unite to form larger, insoluble aggregates. Apart from the *size* of a complex, precipitability may depend e.g. on its degree of hydrophobicity. The lattice hypothesis has considerable experimental support.

laver bread (laverbread) Certain edible species of the macroscopic red alga *Porphyra*. (see also MICROORGANISMS AS FOOD)

Laverania A sub-genus of PLASMODIUM.

lawn plate A solid medium which bears a superficial, *confluent* growth of microorganisms of one species or strain. In the preparation of a lawn plate the inoculum may be spread by means of a sterile swab; a *liquid* inoculum may be spread by tilting the plate in various directions and withdrawing excess inoculum with a sterile Pasteur pipette.

LD_{50} Lethal dose (50%)—that dose of a given lethal agent which, when given to each of a number of test animals, brings about 50% mortality in that population of test animals under given conditions.

leaf blotch (of barley) (see CEREALS, DISEASES OF)

leaf curl (see PEACH, DISEASES OF)

leaky mutation Any MUTATION which results in the formation of a *partially*-active gene product.

Lecanora A genus of crustose or squamulose LICHENS in which the phycobiont is *Trebouxia*. The apothecia are lecanorine and may be sessile or sunk into the surface of the thallus. Species include e.g.:

L. dispersa. The crustose thallus may be superficial and white to grey, or may be endolithic—when it appears only as a dark stain on rocks, walls etc. The apothecia have yellowish-brown to dark brown discs and are often pruinose when young. *L. dispersa* is highly tolerant of atmospheric pollution and occurs on concrete, asbestos etc.

L. conizaeoides. The crustose thallus is grey-green, granular, sorediate, and frequently areolate. Apothecia are usually present; they have yellowish-brown or greenish-brown discs and sorediate margins. *L. conizaeoides* is pollution-tolerant and occurs on trees and on non-basic rocks, bricks etc.

L. atra. The crustose thallus is thick, grey, and areolate. The apothecia have black discs and may reach 1.5 millimetres in diameter. *L. atra* is widespread on non-basic rocks, walls etc.

L. muralis. The thallus is circular in outline and greenish to yellow-brown; the centre is crustose and areolate, while the radiating squamules at the periphery are paler. The apothecia are frequently crowded together at the centre of the thallus; the discs are yellowish-brown to reddish-brown. *L. muralis* is pollution-tolerant and occurs on rocks, concrete etc.

L. lacustris. The crustose thallus of this aquatic species is smooth and pale fawn to orange-brown. The apothecia have pinkish-brown to orange discs and are immersed in the thallus; they are initially concave, subsequently plane. *L. lacustris* occurs on rocks in streams.

lecanorine (of a lichen apothecium) Refers to an apothecium which has a THALLINE MARGIN; the PROPER MARGIN may be more or less obscured by the thalline margin. Lecanorine apothecia occur e.g. in lichens of the genera *Lecanora*, *Usnea*, and *Xanthoria*. (cf. LECIDEINE)

Lecidea A genus of crustose or squamulose LICHENS in which the phycobiont is a green alga—e.g. *Trebouxia*. The apothecia are lecideine. Species include e.g.:

L. limitata. The crustose thallus is greenish-grey to yellowish-grey, smooth or slightly warty, and sometimes bordered and intersected by a dark hypothallus. The apothecia have dark grey or black discs. *L. limitata* occurs on trees, fences etc.

L. macrocarpa. The crustose thallus is grey, sometimes with rust-coloured or yellowish patches, areolate, and with a black hypothallus. The apothecia have black discs; each is initially flat with a thick margin, subsequently becoming convex with no margin. *L. macrocarpa* occurs on rocks and stones.

L. scalaris. The pale green to pale brown thallus is squamulose—consisting of a mass of tiny, imbricated, shell-like scales (squamules); the outer edges of the squamules are upturned and bear soralia. Ascocarps are frequently absent. *L. scalaris* occurs mainly on trees.

lecideine (of a lichen apothecium) Refers to an apothecium which lacks a THALLINE MARGIN—i.e. an apothecium which has only a PROPER MARGIN. Lecideine apothecia occur e.g. lichens of the genera *Buellia*, *Lecidea*, and *Rhizocarpon*. (cf. LECANORINE)

lecithinases (bacterial) Extracellular, lecithin-splitting enzymes produced by certain bacteria, e.g. *Clostridium perfringens* (the α-toxin), strains of *Cl. oedematiens* (= *Cl. novyi*) and *Cl. bifermentans*, other clostridia, some *Bacillus* spp, and strains of *Staphylococcus aureus*. These enzymes cleave LECITHINS with the production of phosphocholine and insoluble diglycerides; precipitated diglycerides form a zone of opacity around the colonies of lecithinase-producing organisms on EGG-YOLK AGAR.

lecithins Phosphatidylcholines: phospholipids found e.g. in egg-yolk and mammalian cell membranes. Some bacteria produce extracellular LECITHINASES.

lecithovitellin agar (LV agar) (see EGG-YOLK AGAR)

lectins Plant proteins which exhibit antibody-like reactions towards carbohydrate sites on animal cells and materials. HAEMAGGLUTINATION by lectins has been studied in the context of blood grouping. Some lectins behave as MITOGENS—see PHYTOHAEMAGGLUTININ.

leek, diseases of (see ONION, DISEASES OF)

leghaemoglobin A red pigment (resembling mammalian haemoglobin) found in leguminous ROOT NODULES. The genes coding for the protein and haem moieties appear to occur in the plant cells and bacteroids respectively. The function of leghaemoglobin is unknown. It appears to be indirectly involved in NITROGEN FIXATION—possibly acting as an oxygen-carrier, supplying oxygen for respiration and protecting the oxygen-labile nitrogenase.

Leifson's flagella stain (for bacterial flagella) *The stain* is prepared by mixing the following solutions in equal proportions: sodium chloride (1.5% in water); tannic acid (3% in water); pararosaniline acetate (0.9%) and pararosaniline hydrochloride (0.3%) in 95% ethanol. (A suitable grade of basic FUCHSIN may be substituted for the pararosanilines.) The stain remains usable for several weeks at 4 °C.

The procedure. Initially, the bacteria are prepared as a medium-free, slightly cloudy suspension in distilled water. On one side of a flamed and cooled, chemically-clean slide an area (equal to half the slide) is ringed by a wax pencil; one loopful of the bacterial suspension is placed onto the ringed area, spread by tilting the slide, and allowed to dry in air at room temperature. No fixation is carried out. To the prepared film is added 1 ml of stain which is allowed to act for approximately 10 minutes and then rinsed off under the tap. If required, the preparation can be counterstained with e.g. 1% methylene blue. Stained flagella appear to be coated with a colloidal precipitate (a tannic acid-dye complex) which renders them visible under the light microscope.

Leighton tube (*virol.*) A large test-tube in which part of the wall (near the closed end of the tube) forms a flat, rectangular region; the latter region is designed to accommodate a cover slip on which a cell *monolayer* (see TISSUE CULTURE) may be prepared.

Leishman–Donovan bodies (see AMASTIGOTE)

Leishmania A genus of parasitic flagellate protozoans of the sub-order TRYPANOSOMATINA; *Leishmania* spp cause various diseases in man: see ESPUNDIA, KALA-AZAR, and ORIENTAL SORE. In order to complete their life cycles the organisms require two hosts: (a) an invertebrate (sand-flies of the genus *Phlebotomus*) in which occurs the promastigote form, and (b) a vertebrate (e.g. man) in which occurs the amastigote form. *Leishmania* spp can be cultured *in vitro*—e.g. in NNN MEDIUM.

leishmanial form (see AMASTIGOTE)

leishmaniasis Any disease caused by protozoa of the genus *Leishmania*, e.g. KALA-AZAR, ESPUNDIA.

Leishman's stain A preparation used e.g. for staining blood smears. Leishman's stain may be made by mixing equal volumes of a solution of eosin Y (0.1% aqueous) and an aqueous solution of *polychrome* methylene blue (containing 0.33% methylene blue); the precipitate which develops is washed (with distilled water), dried, and dissolved in pure, acetone-free methanol to give a solution of 0.15% w/v.

Lempholemma A genus of gelatinous LICHENS in which the phycobiont is *Nostoc*; the thallus is homoiomerous and lacks a cortex. Certain species form hormocystangia (see HORMOCYST). *Lempholemma* species occur e.g. on mosses growing on the mortar of walls.

lens-shaped A phrase which some authors use to describe the shape of a structure or colony etc. As a descriptive phrase it is of doubtful value since a lens may be double-convex, plano-convex, meniscus etc.

lenticular Shaped like a double-convex lens.

Lentinus A genus of fungi of the family TRICHOLOMATACEAE. Constituent species form fruiting bodies which are tough but pliant and which bear gills whose free edges are serrated at maturity; the organisms form smooth-walled, non-amyloid basidiospores. *L. lepideus* forms a mushroom-shaped fruiting body in which the fawn-coloured pileus bears dark-coloured scales and is supported on a stipe that is central or eccentric; the whitish, decurrent gills bear ellipsoidal, hyaline basidiospores—each approximately 10 × 5 microns. Cystidia are not formed in the hymenium. *L. lepideus* is frequently found on telegraph poles, mine timbers etc. since it is able to tolerate quite high concentrations of CREOSOTE. (see also TIMBER, DISEASES OF and SHII-TAKE)

Lenzites A genus of fungi of the family POLYPORACEAE.

LEP Low egg passage—refers to the number of times that a given strain of virus has been transferred from one chick embryo to another. This procedure (*serial passage*) is employed to reduce the virulence of the virus for a non-avian host. The LEP strain of the FLURY VIRUS has undergone between 40 and 60 transfers. (see also AVIANIZED VACCINE and HEP)

Lepiotaceae A family of fungi of the AGARICALES.

Lepraria A genus of LICHENS in which the thallus consists of loosely-associated fungal hyphae and green algal cells; the thallus forms a soft, powdery crust over the surface of the substratum. Fruiting bodies are unknown. Species include e.g.:

L. incana. The thallus is soft and thick, whitish to greenish-grey; parts of the crust may consist only of the white mycobiont. The phycobiont is *Trebouxia. L. incana* is fairly tolerant of atmospheric pollution; it occurs widely e.g. on walls, tree-trunks etc. in damp situations.

L. candelaris. The thallus is thin and bright yellow. *L. candelaris* occurs in crevices on the trunks of trees—particularly oaks and elms.

lepromin test (*immunol.*) A SKIN TEST used in the diagnosis of LEPROSY. The test antigen is *lepromin*, a substance extracted from the skin lesions of *lepromatous* patients. A positive reaction may be preceded by a non-specific inflammatory response (the *Fernandez* reaction) which occurs 24–48 hours after the injection. On subsidence of this reaction there develops a nodular infiltration which becomes maximal within 3 to 4 weeks; this 'late' or *Mitsuda* reaction is the diagnostically significant part of the response. A positive lepromin test is usually given by patients suffering from *tuberculoid* leprosy, but the test is usually negative in *lepromatous* leprosy.

leprose (of a lichen thallus) Refers to a primitive type of thallus which consists of a layer of algal cells and fungal hyphae in loose association; the thallus may thus resemble a layer of soredia (see SOREDIUM). A leprose thallus occurs e.g. in LEPRARIA.

leprosy (Hansen's disease) A chronic, contagious disease which, in nature, affects only man; laboratory animals—e.g. the mouse—can be infected experimentally. The causal agent, *Mycobacterium leprae*, probably invades *via* skin lesions or through mucous membranes. *Lepromatous* leprosy is characterized by disfiguring granulomata of the face, hands, feet, and other regions of the body. *Tuberculoid* leprosy involves the peripheral nerves and gives rise to areas of skin anaesthesia and discoloration. Chemotherapeutic agents include *dapsone* (see SULPHONES) and *rifampicin* (see RIFAMYCINS). *Laboratory diagnosis* includes microscopical examination of lesion material, biopsies etc.; Ziehl-Neelsen or fluorescence staining may be used. Foot-pad inoculation of mice is an additional diagnostic technique. (see also LEPROMIN TEST)

lepto- Combining form meaning fine, narrow, thin.

Leptogium A genus of gelatinous LICHENS in which the phycobiont is *Nostoc*. Members of the genus are similar to COLLEMA species but differ in that they tend to be less gelatinous and in that they possess a distinctly differentiated cortical layer of cells.

leptomonad form (see PROMASTIGOTE)

Leptomonas A genus of flagellate protozoans of the sub-order TRYPANOSOMATINA; species are gut parasites of invertebrates (chiefly insects).

Leptospira A genus of chemoorganotrophic, strictly aerobic, motile bacteria of the SPIROCHAETALES; some strains are parasitic or pathogenic while others—found typically in fresh or salt waters—appear to be saprophytic. The cells are commonly 5–20 microns in length, 0.1 micron in width; one or both ends of a cell may be bent or hooked. Metabolism is respiratory (oxidative). The organisms can be cultured in cell-free media—e.g. serum-enriched basal media; many (or all) strains have growth requirements which include certain vitamins. Currently, only one species is recognized: *L. interrogans*; constituent strains are divided into complexes, serogroups and serotypes. (see also WEIL'S DISEASE)

leptospirosis Any disease, of man or animals, in which the causal agent is a strain of *Leptospira*. In man, leptospiroses include WEIL'S DISEASE and other diseases which do not involve jaundice.

leptotene stage (see MEIOSIS)

lesion A region of tissue which has been damaged mechanically or altered by any pathological process.

lethal mutation A mutation which kills the organism in which it occurs. (cf. CONDITIONAL LETHAL MUTANT)

lettuce big vein virus The agent of *big vein disease* of lettuce and a pathogen of certain other members of the Compositae. The virus has yet to be characterized. On lettuce (*Lactuca sativa*) symptoms include vein-clearing and vein-banding. The virus is transmitted by the fungus *Olpidium brassicae*.

leucine (biosynthesis of) (see Appendix IV(b))

leuco compound The colourless form of a dye

produced e.g. by reduction—see e.g. METHYLENE BLUE.

leucocidal effect (of streptococci) (see LEUCOTOXIC EFFECT (of streptococci))

leucocidin (leukocidin) An extracellular bacterial product which can kill the leucocytes (white blood cells) of certain species and which may, in addition, exhibit toxic or lytic activity towards other cells, e.g. erythrocytes. Leucocidins are produced e.g. by pathogenic species of *Staphylococcus* and *Streptococcus*. *Staphylococcal leucocidins*. These include the α-HAEMOLYSIN (the Neisser–Wechsberg leucocidin), the δ-HAEMOLYSIN, and the *non*-haemolytic *Panton-Valentine* leucocidin—which is active against human and rabbit polymorphs and macrophages. The Panton-Valentine leucocidin ('staphylococcal leucocidin') comprises the F (fast) and S (slow) components—two protein constituents named in accordance with their respective electrophoretic migration rates. Both constituents are essential for leucocidal action—which involves activity against the cell membrane and (in the presence of calcium ions) cell degranulation. *Streptococcal leucocidins* include e.g. STREPTOLYSIN-S.

leucocytes The blood's white cells; however, the term is often used to refer solely to the POLYMORPHONUCLEAR LEUCOCYTES (*granulocytes*)—thus excluding both the LYMPHOCYTES and the monocytes.

Leucocytozoidae A family of protozoans of the sub-order HAEMOSPORINA.

Leuconostoc A genus of gram-positive bacteria of the STREPTOCOCCACEAE; species occur in a variety of dairy products and in fermenting vegetables (see also SAUERKRAUT). The cells may be coccal or lenticular and typically occur in pairs or short chains. Some strains form a CAPSULE. Growth requirements include a suitable carbohydrate and a range of vitamins, e.g. BIOTIN, THIAMINE, NICOTINIC ACID and FOLIC ACID. Glucose is metabolized by HETEROLACTIC FERMENTATION—D(-)lactate, CO_2, and either ethanol or acetic acid being formed. Growth occurs optimally at about 25 °C. Type species: *L. mesenteroides*; GC% range for the genus: 38–44. Species include:

L. mesenteroides. The cells have a maximum dimension of about 1 micron; a DEXTRAN capsule is often formed.

L. citrovorum (= *L. cremoris*). In the presence of a suitable carbohydrate *L. citrovorum* attacks citrate with the formation of acetoin, diacetyl, acetate and CO_2. Cultures are used in the production of BUTTER.

leucoplast Any colourless PLASTID, e.g. those found in some apochlorotic algae.

leucosin (see CHRYSOLAMINARIN)

Leucothrix A genus of gram-negative, filamentous, chemoorganotrophic bacteria of the LEUCOTRICHACEAE. Species commonly occur as epiphytes on *marine* algae.

leucotoxic effect (of streptococci) The death of mammalian phagocytes following ingestion, by the phagocytes, of the cells of certain strains of *Streptococcus*; the leucotoxic effect may be due to the action of cell-bound STREPTOLYSIN-S. The *leucotoxic effect* has been distinguished from the *leucocidal effect* of cell-free filtrates of such strains; the latter effect is believed to be due to extracellular streptolysin-S and/or streptolysin-O.

Leucotrichaceae A family of freshwater and marine bacteria of the CYTOPHAGALES. The organisms are non-pigmented, sessile (*non*-motile) filaments, but the disseminative units (*gonidia*—see GONIDIUM) exhibit gliding movements; the filaments are usually less than 5 microns wide but are often in excess of 50 microns in length. ROSETTES are commonly formed in culture. The organisms are believed to be strictly aerobic. The genera are LEUCOTHRIX and THIOTHRIX.

leucovorin N^5-formyltetrahydroFOLIC ACID.

leukaemia A malignant disease in which there is an abnormal increase in the number of white blood cells.

Leukovirus A genus of VIRUSES whose hosts include a range of avians and mammals; some members of the genus are the causal agents of leukaemias, sarcomas, and non-neoplastic diseases, while others appear not to cause disease. The *Leukovirus* virion is a spheroidal, enveloped virion (of complex structure) which contains a fragmented single-stranded RNA genome; all leukoviruses contain an RNA-dependent DNA polymerase (REVERSE TRANSCRIPTASE). Envelopment of the virion occurs during the process of budding at the cytoplasmic membrane or intracellular membranes of the host cell. Some authors recognize four subgenera on the basis of morphological, antigenic and other criteria. *Subgenus A* includes the C-TYPE PARTICLES; *subgenus B* consists of a single virus: the mouse mammary tumour virus (see B-TYPE PARTICLES); *subgenus C* includes certain viruses which cause respiratory or demyelinating diseases in sheep (e.g. VISNA); *subgenus D* includes the FOAMY AGENTS.

levans (see FRUCTANS)

Levinthal's agar A medium used for the isolation of *Haemophilus* spp; it contains both the X FACTOR and the V FACTOR. Defibrinated rabbit or horse blood is added to a suitable broth (e.g. brain-heart infusion broth) to give a final concentration of 10% blood; this mixture is heated to 100 °C, stirred, and the whole filtered first through sterile glass wool and then through a membrane filter. The filtrate is then added to an equal volume of sterile, molten agar base, mixed, and dispensed into petri dishes.

LFT (low frequency transfer) (see SEX FACTOR)

LFT lysate (see TRANSDUCTION)

lichen (see LICHENS)

lichenan (lichenin; lichen starch; moss starch) A polysaccharide which occurs e.g. in the lichen *Cetraria islandica* (Iceland moss). Lichenan is a D-glucan which appears to be a linear polymer containing β-(1 → 4) and β-(1 → 3) linkages. An analogue of lichenan, *isolichenan*, also occurs in *C. islandica*; isolichenan contains only alpha linkages.

lichenicolous Growing on lichens.

lichenometry A method for dating rocks, old buildings etc. which is based on a knowledge of the growth rates of crustose LICHENS. It is assumed that the largest crustose thallus present on a given exposed surface started to grow when the surface was first exposed to the environment; the thallus is measured, and its age is estimated from a standard growth curve for the species—thus giving an approximate age for the substratum. Certain crustose lichens have been estimated to be over 1000 years old.

lichens A large group of composite organisms—each of which consists of a fungus (the *mycobiont*) in symbiotic association with an alga (the *phycobiont*). The vegetative body of a lichen (i.e. the *thallus*) is essentially crust-like, leaf-like, or filamentous (erect and shrubby or pendulous)—according to species. Lichens occur in a wide range of terrestrial habitats—e.g. on trees, rocks, and soil—and some species are aquatic.

The *mycobiont* is generally an ascomycete (see ASCOMYCOTINA); in a few lichens it is a basidiomycete. It is believed that the mycobiont does not normally occur independently in nature—although, in some cases, it can be isolated from the lichen and cultured. The mycobiont is generally the dominant partner in a lichen.

The *phycobiont*, which is constant for a given species of lichen, is either a member of the CHLOROPHYTA (e.g. *Trebouxia*, *Trentepohlia*, *Coccomyxa*, *Myrmecia*) or of the BLUE-GREEN ALGAE (e.g. *Nostoc*, *Calothrix*); the lichen *Solorina crocea* (see SOLORINA) contains two algae—*Coccomyxa* and *Nostoc*. (Many lichens which have a green phycobiont have, in addition, a blue-green alga in specialized regions of the thallus—see CEPHALODIUM.) The phycobiont may differ in certain ways from its free-living counterpart—e.g. the cells of the phycobiont may be smaller and lack motile stages.

The lichen thallus differs morphologically both from that of the mycobiont and that of the phycobiont when these are grown separately in culture. In a lichen thallus the hyphae of the mycobiont may be closely appressed to the cells of the phycobiont—or they may form haustoria (see HAUSTORIUM) which penetrate more or less deeply into the algal cells. The form of the lichen thallus may be (according to species) CRUSTOSE, FOLIOSE, or FRUTICOSE, or may have some intermediate form (see e.g. SQUAMULOSE). In certain lichens (e.g. CLADONIA) the thallus has a dual nature: a *primary* (*basal* or *horizontal*) thallus (which may be e.g. crustose or squamulose—and either persistent or evanescent) from which arises a fruticose *secondary* (*vertical*) thallus (see PODETIUM and PSEUDOPODETIUM).

The simplest type of lichen thallus consists of a layer of fungal hyphae and algal cells in loose association with little or no differentiation—see e.g. LEPRARIA. In gelatinous lichens (e.g. *Collema*, *Lempholemma*, *Leptogium*) the hyphae of the mycobiont appear to grow in the gelatinous sheath of the (blue-green) phycobiont and there is relatively little internal differentiation. In such lichens the phycobiont and mycobiont are more or less evenly distributed throughout the thallus; such a thallus is said to be *homoiomerous*. In the majority of crustose lichens the mycobiont is differentiated to form an upper (surface) layer (the upper *cortex*—a layer of compressed, gelatinized hyphae) and the *medulla* (a fibrous tissue of loosely-interwoven hyphae) which is in contact with the substratum; the phycobiont occurs as a layer of cells between the cortex and the medulla. A thallus in which the phycobiont is restricted to a definite layer is said to be *heteromerous* (or *stratified*). In crustose lichens the thallus is attached to the substratum by hyphae of the medulla. Foliose lichens have a similar type of internal organization—although the majority have a lower cortex as well as an upper cortex; in some species the lower cortex bears *rhizinae* (see RHIZINA). In many fruticose lichens the internal organization of the thallus is radial—i.e. the medulla, in the central region, is surrounded by a layer of algal cells and an outer, concentric layer of the fungal cortex; in some species (e.g. *Usnea*) a dense central cord runs along the axis of the thallus, while in other species the thallus may be hollow. One or more somatic structures may occur on or in a given lichen thallus, according to species: see e.g. CEPHALODIUM, CILIUM (2), CYPHELLA, ISIDIUM, and PSEUDOCYPHELLA; in most species reproductive structure(s) are also present—see later.

Lichen physiology. The carbon requirement of a lichen appears to be fulfilled mainly by carbon dioxide fixed photosynthetically by the phycobiont. A product of photosynthesis is transferred from the phycobiont to the mycobiont; blue-green algal phycobionts appear always to transfer glucose to the mycobiont, while green algal phycobionts transfer a sugar alcohol—e.g ribitol is transferred by *Trebouxia*, erythritol by *Trentepohlia*. In each case the compound received by the mycobiont appears always to be converted, by the mycobiont, to e.g. mannitol or arabitol. Lichens which contain *Nostoc* or *Calothrix*—either as phycobiont or in cephalodia—obtain nitrogen by NITROGEN FIXATION; fixed nitrogen has been shown to pass from the blue-green alga to the mycobiont. Other lichens are presumed to absorb nitrogen (as e.g. nitrate) from the environment. Vitamins are required for the growth of (at least) some mycobionts in culture; in nature it is assumed that these are synthesized by the phycobiont. (There is no evidence that nutrients are transferred from the mycobiont to the phycobiont.) The mineral requirements of lichens are not well known; lichens can accumulate high concentrations of certain ions (e.g. lead, zinc) by an unknown mechanism. Water (as liquid or vapour) is readily absorbed by lichens; it occurs intercellularly and cannot be stored—being readily lost to the atmosphere by evaporation. Absorption can occur over the entire surface of a lichen thallus.

Growth. A lichen grows apically or centrifugally—according to whether it is e.g. fruticose or crustose; thus, e.g. crustose lichens, whose thalli are more or less circular in outline, grow by increasing the diameter of the thallus. (Certain lichens—e.g. *Ochrolechia parella*, *Pertusaria pertusa*—form concentric and distinguishable growth zones—see OCHROLECHIA and PERTUSARIA.) Characteristically, lichens grow slowly; thus e.g. the thallus of *Xanthoria parietina* has been reported to increase in radius by about 2.5 millimetres per year, while for *Umbilicaria cylindrica* the corresponding figure is in the range 0.01–0.06 millimetres. Factors which affect the rate of growth include rainfall, light intensity, and temperature. In certain crustose and squamulose lichens (e.g. *Pannaria*, *Pertusaria*, *Rhizocarpon*) a phycobiont-free margin may be formed around the thallus as the hyphae of the mycobiont grow outwards; this margin is known as a *hypothallus* (or 'prothallus'). The phycobiont replicates by cell division or by the formation of aplanospores; the daughter cells subsequently spread into the region of new fungal growth. There is evidence that free-living algae, present on the substratum, can be incorporated into the thallus during growth of the mycobiont. (see also LICHENOMETRY)

Reproduction. The principal vegetative (asexual) disseminative units are *soredia* (see SOREDIUM; see also SORALIUM); in many species *isidia* (see ISIDIUM) appear to have a reproductive function—being easily detached from the thallus. Other modes of asexual reproduction include the dispersal of fragments or lobes of the thallus, and the formation of HORMOCYSTS. The mycobionts of many lichen species can form *ascocarps* which are analogous to those produced by non-lichenized ascomycetes. The commonest type of ascocarp is the apothecium, but perithecia are formed e.g. by *Dermatocarpon* and *Verrucaria*, pseudothecia e.g. by *Opegrapha* and *Enterographa*, and mazaedia e.g. by *Calicium* and *Sphaerophorus*. In contrast to non-lichenized fungi, the lichen ascocarp tends to be long-lived and may continue to produce ascospores over a period of several years; ascospores released by the lichen do not contain the phycobiont, and the mode in which a new lichen thallus develops from a germinating ascospore is generally unknown. In some species the ascospores germinate only in the presence of free-living algae; in others, germination gives rise to a mat of hyphae which subsequently incorporates algal cells that are present in the vicinity. (It appears that the incorporation of algal cells is not a selective process—but that all algal species except the prospective phycobiont are killed by the mycobiont.) In the lichen *Endocarpon*, cells of the phycobiont occur in the hymenial layer of the ascocarp; these cells adhere to the ascospores as they are released. In those lichens whose phycobiont is *Trebouxia* (an alga which is very rare in the free-living state) the phycobiont of a new thallus may possibly be derived from *Trebouxia*-containing soredia present on the substratum.

Habitat. Lichens occur in a wide range of habitats and may be found—according to species—attached to trees, rocks, soil, mosses etc.; some lichens are not attached to a substratum. Saxicolous and corticolous species frequently show a distinct preference for a particular type of rock/tree; other species appear to be relatively indifferent to the nature of the substratum. (see also ENDOLITHIC and ENDOPHLOEODAL) Some species can grow on—and etch the surface of—glass, marble etc. Many lichens can grow in situations where they are exposed to extremes of temperature and desiccation which can be tolerated by few other forms of life—e.g. in desert and polar regions and at high altitudes; in some of these regions terricolous lichens play an important role in the stabilization of the soil surface. *Aquatic* lichens include e.g. *Dermatocarpon fluviatile*, *Lecanora lacustris*, and *Verrucaria aquatilis*. Some lichens occur on rocks by the sea—at or above (according to species) the intertidal zone; such lichens include e.g. *Anaptychia fusca*, *Caloplaca marina*, *Lichina* spp, *Roccella* spp, *Verrucaria maura*. Lichens have been found in some unusual locations—e.g. on the carapaces of giant tortoises and on the backs of (living) flightless weevils of the genus *Gymnopholus*.

Classification of lichens. Lichens are classified on the basis of the nature and origin of their fruiting bodies and on the morphology of their thalli; no taxonomic significance is attached to the phycobiont. (The name of a lichen refers to the *mycobiont* and thus is also used to refer to the isolated mycobiont in culture.) It is currently accepted, in principle, that the lichens should be integrated in a mycological taxonomic scheme; however, although several schemes have been proposed none has yet gained general acceptance.

Lichen substances (*'lichen acids'*). Many lichens produce a range of extracellular compounds which may be deposited on the medullary hyphae or on the surface of the cortex—frequently giving the surface a pruinose appearance; these compounds include e.g. weak phenolic acids, aliphatic acids, depsides, anthraquinones. The function of these compounds is unknown; it has been suggested that they may provide a means for the removal of excess carbon fixed by the phycobiont. Some of the compounds (e.g. USNIC ACID) exhibit antibiotic activity against certain fungi and gram-positive bacteria. The reactions of lichen substances with certain reagents provide the basis of the *colour tests* widely used in the identification of lichens; the reagents are: saturated aqueous calcium hypochlorite ('C'), 10–25% aqueous potassium hydroxide ('K'), and 1–5% ethanolic *p*-phenylenediamine ('P' or 'PD'). The reagent is applied (e.g. with the tip of a glass rod) to a fragment of the lichen thallus (cortex or medulla) and any colour reaction is noted.

Lichens and pollution. Many lichens are highly sensitive to atmospheric pollution; they are killed by low concentrations of sulphur dioxide, fluoride, fertilizer dust etc.—probably as a result of their ability to accumulate high concentrations of the pollutant. The pollutant primarily responsible for lichen death in industrial regions appears to be sulphur dioxide; sulphur dioxide reacts with the chlorophyll of the phycobiont—apparently converting it to phaeophytin. The response of a lichen to pollution may be affected by moisture and temperature. Corticolous communities of lichens are more susceptible to pollution than are saxicolous communities. Some lichens (e.g. *Lecanora conizaeoides*, *L. dispersa*, *L. muralis*, *Lepraria incana*, *Stereocaulon pileatum*) are more or less resistant to pollution; the reason for their resistance is unknown. *Lecanora conizaeoides* appears actually to favour moderately polluted environments—being common in cities and absent from areas where the air is pure; it has been suggested that this species has a requirement for sulphur.

The capacity of lichens to accumulate ions has resulted in their accumulation of radioactive isotopes (e.g. strontium-90, caesium-137) in arctic regions which have been subjected to the fall-out from nuclear tests; such lichens form the base of the food chain: lichen → reindeer (caribou etc.) → man.

The uses of lichens. Some lichens are used as food in certain countries—e.g. *Cetraria islandica* in Iceland, *Umbilicaria esculenta* (as 'iwatake') in Japan; species of *Cladonia* and *Cetraria* form a major part of the diet of reindeer, caribou, and other deer during winter in arctic regions. A number of lichens have been used as sources of dyestuffs—e.g. *Ochrolechia tartarea* (the purple dye *cudbear*), *Parmelia* spp (a brown dye), *Roccella* spp (LITMUS and the purple dye *orchil*). Some lichens, e.g. *Evernia prunastri*, are used as a raw material in the perfume industry. Lichens were formerly widely used in folk medicine; currently, the lichen substance *usnic acid* is a constituent of certain preparations used for the topical treatment of wounds etc.

Lichen genera include e.g.: ALECTORIA, ANAPTYCHIA, BAEOMYCES, BUELLIA, CALICIUM, CALOPLACA, CANDELARIELLA, CETRARIA, CLADONIA, COLLEMA, DERMATOCARPON, ENTEROGRAPHA, EVERNIA, GRAPHIS, LECANORA, LECIDEA, LEMPHOLEMMA, LEPRARIA, LEPTOGIUM, LICHINA, LOBARIA, NEPHROMA, OCHROLECHIA, OPEGRAPHA, PANNARIA, PARMELIA, PELTIGERA, PERTUSARIA, PHYSCIA, PILOPHORUS, PSEUDOCYPHELLARIA, RAMALINA, RHIZOCARPON, ROCCELLA, SOLORINA, SPHAEROPHORUS, STEREOCAULON, STICTA, THAMNOLIA, UMBILICARIA, USNEA, VERRUCARIA, XANTHORIA.

Lichina A genus of small, fruticose LICHENS in which the phycobiont is the blue-green alga *Calothrix*. The thallus is gelatinous, brown to black, erect, branching, terete (in *L. confinis*) or flattened (in *L. pygmaea*)—forming tufts or mats (respectively) on rocks on the seashore. *L. confinis* may reach 0.5 centimetre in height and occurs at, or just above, high tide mark; *L. pygmaea* may reach 1 centimetre or more and occurs in the upper regions of the intertidal zone. Apothecia may occur immersed in swollen ends of branches.

ligase (synthetase) A category of ENZYMES (EC class 6) which encompasses those enzymes which catalyse reactions in which a bond is formed between two substrate molecules using energy obtained from the cleavage of a pyrophosphate bond (e.g. in ATP). Subclasses are recognized on the basis of whether the bond formed is carbon-oxygen, carbon-sulphur, carbon-nitrogen, or carbon-carbon. Examples of the first subclass are aminoacyl-tRNA synthetases (= amino acid-RNA ligases).

ligative hyphae (binding hyphae) (see HYPHA)

light (as an environmental factor: microbial aspects) (see e.g. PHOTOINDUCTION; PHOTOSYNTHESIS; PHOTOTAXIS; PURPLE MEMBRANE; (photo)TROPISM)

light chain (*immunol.*) The smaller of the two types of polypeptide component which form the basic IMMUNOGLOBULIN molecule; IGG and structurally-similar immunoglobulins contain two light chains and two HEAVY CHAINS per molecule. (The basic structure of the immunoglobulin molecule is described under IGG.) Light chains are of two types, *kappa* and *lambda,* which differ structurally and antigenically; both types occur in all classes of immunoglobulin, but a given molecule contains only one type of light chain—either *kappa* or *lambda*. Each light chain in IgG has a molecular weight of about 22,000. (see also FAB FRAGMENT)

light reaction (in photosynthesis) (see PHOTOSYNTHESIS)

light trap (1) A region of focused light in a liquid medium containing phototactic organisms (see PHOTOTAXIS). If the light-dark boundary offers a suitable intensity gradient, *positively* photophobotactic organisms which have entered the trap (as a result of random motion) are unable to leave it, and thus tend to accumulate in the illuminated zone. (2) Sometimes used to refer to the reaction centre of a photosynthetic pigment system—see PHOTOSYNTHESIS.

lignicolous (*adjective*) (of fungi etc.) Using wood as a source of nutrients; living in or on wood.

lignins A class of complex polymers composed of phenyl-propanoid subunits—e.g. 3-methoxy-4-hydroxyphenyl prop-1-ene-3-ol (*coniferyl alcohol*). Lignins vary in the nature and linkage of their subunits. They occur in the woody material of the higher (vascular) plants—in close association with CELLULOSE and HEMICELLULOSES in the plant cell walls. (Wood contains approximately 20–35% by weight of lignin.)

Lignins are extremely resistant to chemical and enzymatic degradation; lignin breakdown tends to occur at a late stage in the decay of plant material. Lignins are attacked almost exclusively by fungi—especially by members of the Basidiomycotina (e.g. polypores and agarics), but also by certain members of the Ascomycotina. These fungi secrete enzymes which split the polymer into its constituent subunits; the subunits are then susceptible to further degradation by a number of microorganisms (e.g. *Pseudomonas* spp). Degradation of lignin may also be involved in certain tree diseases, e.g. in *white rot* infections lignin and carbohydrate are broken down simultaneously, causing thinning of the cell walls. (see also TIMBER, DISEASES OF)

Limacella A genus of fungi of the AMANITACEAE.

***Limax*-type movement** A form of movement similar to that of the common garden slug (*Limax*).

limberneck (see FOOD POISONING (bacterial))

lime sulphur (see SULPHUR)

limit dextrin (see DEXTRINS)

limit of resolution (see RESOLVING POWER)

***Limulus* amoebocyte lysate** (LAL) A lysate prepared from the blood cells (amoebocytes) of the horseshoe crab, *Limulus polyphemus*; LAL forms a gel when incubated at 37 °C with endotoxins, i.e. LIPOPOLYSACCHARIDES, or with gram-negative bacteria. The LAL test is used for the detection and quantitative estimation of LPS, and has been used e.g. for the examination of fluids (saline etc.) intended for parenteral administration; the test can detect quantities of LPS of the order of 10^{-9} gm/ml.

lincomycin An ANTIBIOTIC produced by *Streptomyces lincolnensis*. It consists of a substituted pyrrolidine nucleus linked to methyl-α-thiolincosamine (*lincosamine* = 6-amino-6,8-dideoxyoctose). Lincomycin is active mainly against gram-positive bacteria, e.g. *Clostridium* spp, *Corynebacterium diphtheriae*, many Lancefield group A streptococci; many strains of *Staphylococcus aureus* are sensitive. Clinically, lincomycin is sometimes used as a substitute for penicillin—against penicillin-resistant organisms or in cases of penicillin allergy. Lincomycin inhibits PROTEIN SYNTHESIS but the precise mode of action is unknown; in gram-positive bacteria it binds to the 50S subunit of 70S RIBOSOMES where it may (a) prevent binding of aminoacyl-tRNA to the ribosome, (b) prevent peptide-bond formation, or (c) function in some as yet undetermined way. Resistance to lincomycin develops readily; there is some degree of cross-resistance with the MACROLIDE ANTIBIOTICS. Antagonism may also occur: strains normally sensitive to lincomycin may be resistant to it in the presence of erythromycin. *Clindamycin* (7-chloro-7-deoxy-lincomycin) is rather more active than lincomycin.

linkage (genetics) The association of genes on a common chromosome (or extrachromosomal genetic structure); such genes tend to be inherited *en bloc*. During reproduction, the genes on a particular chromosome—i.e. *linked genes*—do not exhibit the characteristics of independent assortment; such genes are said to constitute a *linkage group.* The alleles in a linkage group may be separated—e.g. by CROSSING-OVER during MEIOSIS. The distance between two linked genes can be estimated from their RECOMBINATION FREQUENCY.

linkage group (see LINKAGE)

linked gene (see LINKAGE)

lipoic acid (α-lipoic acid; thioctic acid) A coenzyme which is involved in the oxidative decarboxylation of keto-acids—e.g. pyruvate and α-ketoglutarate are decarboxylated to acetyl-CoA and succinyl-CoA respectively. (see Appendix II(a)) Lipoic acid may be involved in the reduction of thiosulphate by *Thiobacillus denitrificans.* The *reduced* form of the coenzyme, dihydrolipoic acid, is 6,8-dithiooctanoic acid—$CH_2SH.CH_2.CHSH.(CH_2)_4.COOH$; in lipoic acid the two -SH groups form a disulphide bond (-S-S-). Lipoic acid is required as a growth factor e.g. by *Tetrahymena*; it can replace acetate as a growth factor for *Lactobacillus casei.* (see also THIAMINE)

Lipomyces A genus of unicellular fungi within the family SACCHAROMYCETACEAE; species occur in the soil. The vegetative form of the organisms is a spherical or ovoid cell which reproduces asexually by budding; individual cells are enclosed by a mucoid capsule which may contain starch. The cells of *Lipomyces* spp characteristically accumulate lipids and each cell may contain a large lipid globule. Asci—each containing up to 20 ascospores—are typically produced from so-called *active buds*; an active bud resembles a vegetative bud but its contents give rise to ascospores—either following conjugation (coalescence) with e.g. another active bud (derived from the same cell or from a different cell) or in the absence of conjugation. *Lipomyces* spp are non-fermentative organisms; they do not assimilate nitrate.

lipopolysaccharides (LPS) Components of the OUTER MEMBRANE of gram-negative bacteria. LPS are important antigenic factors, and are ENDOTOXINS.

General structure. An LPS molecule consists of a heteropolysaccharide chain covalently linked to a glycolipid (*lipid A*). *Part* of the heteropolysaccharide chain (known as the *core polysaccharide*) has a structure which appears to be similar, or identical, in closely-related strains of bacteria; the *remainder* of the chain (the *O-specific chain*) is highly variable in composition (see later). The overall structure of LPS can be represented thus:

Lipid A—core polysaccharide—O-specific chain
(⟵heteropolysaccharide chain⟶)

Core polysaccharide. This has been extensively studied in *Salmonella*; in all strains thus far examined, the core components include glucose, galactose, *N*-acetyl-D-glucosamine, 2-keto-3-deoxyoctonate (KDO) and L-glycero-D-mannoheptose ('heptose'). Phosphate and ethanolamine residues may also be found in the core—ethanolamine may be linked to sugar residues *via* phosphate or pyrophosphate. The core polysaccharide is linked to lipid A *via* a KDO residue, and to the O-specific chain *via* a glucose residue. There may be phosphodiester cross-links between the core polysaccharides of adjacent LPS molecules.

O-specific chain. This is a chain of identical oligosaccharide subunits; depending on serotype, each subunit may contain either 3, 4 or 5 monosaccharide residues and may be either linear or branched. The number of *subunits* in each chain may vary—even in the same organism. In *Salmonella* the constituent monosaccharides may include hexoses, pentoses, 6-deoxyhexoses and 2,6-dideoxyhexoses. For example, each oligosaccharide in the O-specific chains of *Salmonella typhimurium* contains galactose, mannose, rhamnose and abequose. The nature of the O-specific chain, in terms of its constituent sugars and structure, is important immunologically since these factors determine the identity of cell-surface antigens. In many serotypes of *Salmonella* the antigenic specificity of the O-specific chain is partly determined by the presence of one or other of the dideoxyhexoses COLITOSE, ABEQUOSE, PARATOSE and TYVELOSE; the nature and configuration of the other saccharide components are also determining factors. In general, the antigenic characteristics of the O-specific chain may be affected by: (a) the presence or absence of certain sugars; (b) variation in the positions of glycosidic linkages (e.g. 1 → 4 or 1 → 6); (c) alternative anomeric configuration (i.e. α or β linkage); (d) the substitution of sugar(s) e.g. by acetyl group(s). Minor structural variations may occur even within a single serotype (see O ANTIGEN).

Lipid A. This consists of a disaccharide (two glucosamine residues generally linked β-(1 → 6) fully substituted with long-chain (e.g. 12–16C) fatty acids, β-hydroxymyristic acid (3-hydroxytetradecanoic acid), phosphate, and the core polysaccharide. Adjacent disaccharides are apparently linked *in vivo* by pyrophosphate bridges. Lipid A appears to be the component responsible for the endotoxic activity of LPS.

Treatment of whole cells with trichloroacetic acid (TCA) extracts an LPS-protein complex (the *Boivin* antigen) while phenol-water extracts a protein-free LPS. Some of the LPS can be removed from cells by treatment with chelating agents such as EDTA. Thus it appears that LPS is bound, within the outer membrane, both by hydrophobic bonds and by ionic bonds *via* divalent and/or polyvalent cations.

Biosynthesis of LPS. The relevant enzymes are located in the CELL MEMBRANE. Details of lipid A biosynthesis are not known. The core polysaccharide is built up by the sequential addition of sugar residues to lipid A; the first of these, a KDO residue, is added from a CMP precursor, while the residues of glucose, galactose and *N*-acetylglucosamine are added from UDP precursors. Precursors have not yet been identified for the heptose residues or for the remaining KDO residues. The oligosaccharide subunits of the O-specific chain are built up on BACTOPRENOL carrier molecules by the sequential addition of sugars from nucleotide precursors. The subunits are then polymerized to form an O-specific chain. This is subsequently transferred to the core polysaccharide of an incomplete LPS molecule, the bactoprenol pyrophosphate being concomitantly released. The completed LPS is then transferred to the outer membrane. (see also SMOOTH-ROUGH VARIATION)

liposomes Artificial phospholipid vesicles (sacs) used as a model system for the study of CELL MEMBRANES.

Lipschütz body (see HERPESVIRUS)

lirella (lirelliform ascocarp) (*lichenol.*) An ascocarp which is *slit-like*—i.e. elongated and narrow—and which may or may not be branched. Lirellae are characteristic of certain lichen genera—e.g. GRAPHIS and OPEGRAPHA.

lissamine rhodamine (*immunol.*) (*synonyms*: sulphorhodamine B; acid rhodamine B) A red FLUOROCHROME used for labelling proteins—particularly those involved in IMMUNOFLUORESCENCE (e.g. antibodies). The dye is initially treated with phosphorus pentachloride and acetone to form the highly reactive sulphonyl chloride derivative. Protein-dye conjugation is effected by a sulphamido condensation. The dye fluoresces orange-red.

Listeria A genus of gram-positive, asporogenous, microaerophilic (facultatively aerobic) bacteria of uncertain taxonomic affinity; species of *Listeria* are parasitic in a wide range of animals. The organisms are pleomorphic bacilli or coccobacilli, about 0.5 × 1–2 microns, which may occur singly, in groups ('Chinese letter' arrangement), or in palisade formation (two, three or more bacilli in contact, with their long axes parallel); filaments occur in rough colonies. Cells cultured at 25 °C are peritrichously flagellate, while those grown at 37 °C are polarly flagellate or (infrequently) non-motile; motile cells exhibit a characteristic 'tumbling' motion. *Listeria* spp form small colonies which appear blue-grey or blue-green in obliquely-transmitted light; some strains form a soluble haemolysin. *Listeria* spp hydrolyse aesculin, produce catalase, and are indole-negative and urease-negative. *L. monocytogenes*, the type species, is the causal agent of e.g. one form of meningitis in children; the organism is commonly sensitive to ampicillin and tetracyclines.

lithotroph (see AUTOTROPH)

litmus An amphoteric dye, obtained from certain species of the lichen *Roccella*, which consists of a complex mixture of compounds. Litmus is widely used as a pH indicator: pH 4.5 (red) to pH 8.3 (blue).

litmus milk A bacteriological medium which consists of skim-milk (whole milk less cream) to which has been added a volume of litmus solution sufficient to impart a blue-purple coloration. Prior to use the medium may be either steamed (20/30 minutes on each of 3 successive days) or autoclaved (115 °C/10 minutes).

A given strain of bacterium growing in litmus milk may give one of the following reactions: (a) No visible change in the medium. (b) Acid production from lactose (indicated by the pH indicator). (c) Alkali production—commonly due to hydrolysis of *casein.* (d) Reduction (decolorization) of the litmus. (e) The production of an *acid clot*; the clot, which is soluble in alkali, forms at about pH5—at or about the ISOELECTRIC POINTS of the various types of casein. The production of acid *and* gas may give rise to a STORMY CLOT. (f) The formation of a clot at or about pH7. Such clots are formed by certain bacteria (including species of *Bacillus*) which produce rennin-like extracellular enzymes; reduction of the litmus may precede or accompany clot formation, and peptonization (digestion of the clot) may subsequently occur. (In the dairy industry such clotting is referred to as 'sweet curdling'.)

littoral region (see BENTHIC ZONE)

liver of sulphur (see SULPHUR)

Lobaria A genus of foliose LICHENS in which the phycobiont is either *Myrmecia* or *Nostoc*; species containing *Myrmecia* may also have internal cephalodia containing *Nostoc*. The lobed thallus is more or less extensively tomentose on the undersurface. Species occur mainly on old trees—particularly in ancient woodland—but may occasionally occur on rocks, mosses, or heather stems; members of this genus are highly sensitive to atmospheric pollution. Species include e.g.:

L. pulmonaria (tree lungwort). Phycobiont: *Myrmecia*. The thallus is large (up to 20 centimetres) with truncated lobes. The upper surface is bright green when wet, pale greenish-brown when dry, and has a reticulated system of ridges and depressions; soredia or isidia may occur along the ridges. The lower surface of the thallus is brown; the region corresponding to the ridges above is tomentose, while the convex regions corresponding to the depressions are naked. Apothecia are uncommon; when present they are small with reddish-brown discs and occur along the margin of the thallus.

L. scrobiculata. Phycobiont: *Nostoc*. The lobes are broad and rounded and the thallus has ridges and depressions similar to those of *L. pulmonaria*. The upper surface is bluish-green when wet, greyish-yellow when dry; bluish-grey soredia occur along the ridges and margins. The lower surface is light brown—the regions corresponding to the ridges above being tomentose. Ascocarps are uncommon; when present they are small with dark red discs and thick margins.

lobopodium A broad and blunt *pseudopodium*; see MOTILITY.

local lesion (*plant pathol.*) On various plant tissues (leaves, stem etc.): a discrete region of e.g. chlorosis or necrosis due to viral infection. (see also RING SPOT and VIRUS DISEASES OF PLANTS)

local lesion host (*plant pathol.*) A plant which develops LOCAL LESIONS when infected by a given virus. Such hosts are used e.g. for assaying viral infectivity: the number of infective virus particles in a preparation may be assessed from the number of lesions produced when the leaves of the plant are treated with a known volume of the preparation.

lockjaw (see TETANUS)

locule (*mycol.*) A cavity or chamber. (see also ASCOSTROMA)

Loculoascomycetes A class of filamentous (mycelial) fungi within the sub-division ASCOMYCOTINA; all species exhibit two main characteristics: (a) their fruiting bodies are ascostromata (see ASCOSTROMA), and (b) they form *bitunicate* asci. The loculoascomycetes include saprophytes and parasites of higher plants, insects, and other fungi; constituent genera include ELSINOË, *Mycosphaerella*, *Myriangium*, and VENTURIA.

locus (genetics) (*pl.* loci) The location of a given gene, operon, mutation etc. on a chromosome.

Loeffler's methylene blue (see METHYLENE BLUE)

Loeffler's serum A solid medium (usually prepared as a *slope*) used e.g. for the culture of *Corynebacterium diphtheriae*. The medium contains sterile serum and nutrient broth (3:1 by volume) and glucose (0.5% w/v); this mixture may be coagulated and tyndallized by heating to 80 °C on each of three successive days—the temperature being raised *slowly* during the initial period of heating.

log dilutions (see SERIAL DILUTIONS)

logarithmic phase (see GROWTH (bacterial))

lomasomes Membranous structures which have been observed between the cell wall and cell membrane in fungal hyphae and in certain algae; lomasomes may appear as tubular or vesicular convolutions of membrane or as sac-like structures which enclose particulate matter. It has been suggested that lomasomes may be involved in cell wall synthesis.

loop An instrument widely used in bacteriology and mycology for the manipulation of small quantities of material (solid or liquid)—e.g. during INOCULATION. The rod-shaped metal handle holds a short length of wire which is fashioned into a loop at the free end; the wire, ideally platinum, is commonly nickel-steel, and is usually 3–10 cms in length. Before *each* use the terminal loop and adjacent wire* is sterilized by heating it to redness in a flame (see FLAMING); the loop is allowed to cool prior to use. In use, the sterile loop is brought into contact with a colony (or other material) such that a small quantity of material adheres to the loop; this material can then be used e.g. to inoculate a MEDIUM. If a loop is dipped into a broth CULTURE and withdrawn, the loop retains a small circular film of liquid (containing microorganisms) which may be used as an INOCULUM; the volume of liquid removed may be varied by adjusting the size of the loop prior to use. After each use the loop should be flamed to avoid contamination of the bench etc.
*Long-wired loops are used e.g. for inoculating media in bottles; in use, no part of the handle should enter the bottle. When sterilizing, the wire must be effectively flamed along its entire length. (see also STRAIGHT WIRE; STREAKING)

lophotrichous (*adjective*) Refers to the arrangement of the flagella on a bacterium: lophotrichous flagella occur as a tuft, or group, either at one end or both ends of a cell.

lorica (1) (*protozool.*) (*syn*: test) The shell—i.e. outer rigid wall—found in certain species of protozoan. (2) (*algol.*) (*syn:* tunica) A synonym of THECA (sense 2).

louping-ill An acute or subacute, commonly fatal, infectious disease which affects mainly sheep and cattle; the causal agent, a *togavirus*, is transmitted by the bites of ticks (*Ixodes ricinus*, *Ixodes reduvius*). Viraemia and fever are shortly followed by signs of neural incoordination e.g. leaping ('louping') movements, and by paralysis and death. In man, the virus causes a mild, febrile illness.

Löwenstein-Jensen medium A complex medium used for the isolation of *Mycobacterium* spp from e.g. sputum. The medium is prepared by the inspissation of a mixture consisting of homogenized hens' eggs and a solution containing asparagine, potassium dihydrogen phosphate, magnesium sulphate, magnesium citrate, glycerol, potato starch, and malachite green. The medium is dispensed (as slopes) in screw-cap bijou or universal bottles.

lower fungi (the 'phycomycetes') Fungi *other*

than the ascomycetes, basidiomycetes, and deuteromycetes. The lower fungi are generally considered to be 'primitive' on the evolutionary scale; they usually form non-septate mycelium and typically form sporangiospores rather than conidia.

LPS (see LIPOPOLYSACCHARIDES)

Lucibacterium A genus of bacteria of the VIBRIONACEAE.

luciferase, luciferin (see BIOLUMINESCENCE)

Lucké virus (see HERPESVIRUS)

Lugol's iodine An aqueous solution which contains iodine (5%) and potassium iodide (10%); for use (e.g. in the GRAM STAIN) this solution is commonly diluted with four times its volume of distilled water.

lumber, diseases of (see TIMBER, DISEASES OF)

luminescence (see BIOLUMINESCENCE)

lumper (see TAXONOMY)

lumpy jaw A chronic disease of cattle and other animals in which dense nodular lesions are formed in the bones of the jaw and in other tissues; the lesions may cause appreciable swelling of the jaw, and often develop into pus-discharging abscesses. The causal agent is *Actinomyces bovis*, and infection is probably *endogenous*. The pathogen may be isolated from the small granular particles found in the pus.

lumpy skin disease (LSD) An infectious viral disease of cattle which occurs in certain parts of Africa. The causal agent is the LSD virus—a POXVIRUS; insects are presumed to be mechanical vectors of the disease. Following an incubation period of 3–12 days the infected animal exhibits a crop of cutaneous nodules, an obvious generalized lymphadenitis, and dysgalactia; the limb(s) may become oedematous. Although the mortality rate is usually less than 10%, LSD is economically important since it results in general unthriftiness. A similar, though less severe, disease (pseudolumpyskin disease) is caused by a *Herpesvirus*.

lutein (see CAROTENOIDS)

luteose A β-(1 → 6)-linked GLUCAN synthesized by *Penicillium luteum*.

LV agar Lecithovitellin agar; see EGG-YOLK AGAR.

lyase A category of ENZYMES (EC class 4) which encompasses those enzymes which catalyse the non-hydrolytic removal of a group from a substrate (often resulting in the formation of a double bond), or the reverse reaction. Subclasses are recognized on the basis of whether a carbon-carbon, carbon-oxygen, or carbon-nitrogen bond is split. The category includes decarboxylases, aldolases, and dehydratases.

Lycogala A genus of the ACELLULAR SLIME MOULDS.

lycopene (see CAROTENOIDS)

Lycoperdales An order of fungi of the class GASTEROMYCETES. Constituent species form fruiting bodies in which the highly-convoluted tramal tissue—bearing a continuous hymenium—is enclosed by a peridium that is essentially bi-layered; throughout the hymenium the basidia reach maturity at about the same time, and the (typically) light-coloured, globose basidiospores are dispersed—together with the powdery remains of the spore-bearing tissue—either *via* a peridial pore or following the rupture of the peridium. In some species the fruiting bodies are sessile; in other species the sporophore is borne on a *pseudostem*. Among the 'earthstars' (e.g. *Geastrum*)—in which the fruiting body is globose, or onion-shaped—the outer layer of the peridium (the *exoperidium*) ruptures and peels back to form a stellate fringe and to expose the inner layer of the peridium (*endoperidium*) in which an apical pore develops to permit spore release. The 'puffballs' include species of *Calvatia* (in which spores are released on peridial rupture) and *Lycoperdon* (in which the peridium develops a pore); the larger, globose puffballs may attain a diameter greater than one metre.

Lycoperdon A genus of fungi of the LYCOPERDALES.

lycoperdon nut The subterranean fruiting body of certain ascomycetes—e.g. *Elaphomyces*; it may reach several centimetres in diameter.

lymphadenitis (*med./vet.*) Inflammation of lymphatic *node*(*s*). (cf. LYMPHANGITIS)

lymphangitis (*med/vet.*) Inflammation of the lymphatic *vessels*. (cf. LYMPHADENITIS)

lymphocyte A type of cell found in blood, tissues and lymph; lymphocytes are essential factors in CELL-MEDIATED IMMUNITY and HUMORAL IMMUNITY. They are roughly spherical (some 8–15 microns in diameter) and may be actively motile; the rounded, non-lobed nucleus forms the major part of the cell.

B lymphocytes. In mammals, the precursors of the B lymphocytes—the undifferentiated *stem cells*—arise in the bone marrow and in the foetal liver; the stem cells subsequently become *immunologically competent*, i.e. able to produce specific IMMUNOGLOBULINS. On antigenic stimulation, competent B lymphocytes undergo TRANSFORMATION—becoming either memory cells or PLASMA CELLS which form antibodies homologous to the given antigen. (see ANTIBODY-FORMATION) (The antibody-forming response of B lymphocytes to certain—*T-dependent*—antigens requires the co-operation of *T* lymphocytes; antigens to which B lymphocytes can respond unaided are termed *T-independent* antigens.) B lymphocytes form a minority of the population of *circulating* lymphocytes; their surfaces carry specific immunoglobulin antigen-binding *receptor sites* which can be demonstrated by immunofluorescence techniques. B lymphocytes are more sensitive than T lymphocytes to X-ray inactivation, and less sensitive to ALS.

T lymphocytes develop from bone marrow stem cells which mature in the *thymus* gland—each cell becoming immunologically competent, i.e. antigen-specific, and able to participate in the reactions of CELL-MEDIATED IMMUNITY. On initial antigenic stimulation, T lymphocytes form *memory cells*; subsequent antigenic stimulation promotes TRANSFORMATION and LYMPHOKINE production.

Morphologically, the B and T lymphocytes are identical but they differ e.g. in surface antigens; T lymphocytes are more numerous in blood. Separation of B and T lymphocytes may be achieved by allowing the suspension to flow through a column of glass beads which have been coated with antiglobulin; the B cells are readily adsorbed onto the coated glass.

lymphocytic choriomeningitis virus (see ARENAVIRUS)

lymphocytosis A significant increase in the number of LYMPHOCYTES in the blood or tissue fluids.

lymphokines A varied group of biologically-active extracellular products formed by activated T-lymphocytes and possibly by other types of cell (e.g. macrophages); lymphokines appear to be involved in the reactions of CELL-MEDIATED IMMUNITY. Many lymphokines are proteins—e.g. the *chemotactic factor for macrophages*, the (antigen-independent) *mitogenic factor*, and the *migration-inhibitory factors*.

lymphoma A tumour of *lymphoid* tissues.

lymphon (*immunol.*) The entire *immune system* of an individual—including e.g. lymphoid tissues, LYMPHOCYTES, MACROPHAGES and the COMPLEMENT system.

lymphopoiesis The formation of LYMPHOCYTES.

Lyngbya A genus of filamentous BLUE-GREEN ALGAE; species occur in freshwater and marine environments. The trichome typically occurs within a clearly-defined sheath; HETEROCYSTS and AKINETES are not formed. Reproduction occurs by the formation of hormogonia.

lyophilization (see FREEZE-DRYING)

lysate The products of cell lysis.

lysine (biosynthesis of) (see AMINOADIPIC ACID PATHWAY and DIAMINOPIMELIC ACID PATHWAY)

lysins Antibodies or other entities which, under appropriate conditions, are capable of causing the lysis (rupture) of cells (see e.g. HAEMOLYSIN; see also BACTERIOLYSIS).

lysis from without (see BACTERIOPHAGE)

lysis inhibition (in bacteriophage infection) A delay in lysis which occurs, in some bacteria, as a result of superinfection (further infection) by another phage of the same strain during the *latent period* (see ONE-STEP GROWTH EXPERIMENT). The occurrence of lysis inhibition is usually indicated by the PLAQUE morphology; such plaques typically have an indistinct periphery—cf. RAPID LYSIS MUTANTS.

lysogen A lysogenized cell, or a population of lysogenized cells. (see also LYSOGENY)

lysogenic bacteria (see LYSOGENY)

lysogenic conversion (see BACTERIOPHAGE CONVERSION)

lysogenic immunity (prophage immunity) (see LYSOGENY)

lysogeny The non-disruptive infection of a bacterium* by a (*temperate*) BACTERIOPHAGE—a condition in which the phage genome becomes integrated with the host's chromosome; during subsequent cell divisions the integrated phage genome (*prophage*) is replicated as part of the bacterial chromosome. Bacteria which carry a prophage are said to be *lysogenic*. (see also LYSOGEN) Several *different* prophages may be carried by the chromosome of a given cell; several identical prophages may be carried by a given cell owing to the presence of several chromosomes within that cell—see GROWTH (bacterial). Lysogenic bacteria may exhibit characteristics not displayed by the corresponding non-lysogenic cells—see BACTERIOPHAGE CONVERSION.

The infection of a population of susceptible bacteria with temperate phage does not necessarily result in the lysogenization of every infected cell; under favourable growth conditions many, or most, of the infected cells may be lysed. Lysogenization is generally encouraged by sub-optimal growth conditions and/or the presence of certain inhibitors of protein synthesis—e.g. chloramphenicol. Whether or not lysogeny occurs is partly determined by environmental conditions such as temperature and ion concentration; conditions which determine the lysogenic or lytic pathway of infection may influence the synthesis or activity of certain phage-coded proteins involved in prophage establishment and maintenance (see later).

The mechanism of lysogeny. The account below refers specifically to lysogeny involving *Escherichia coli* and BACTERIOPHAGE LAMBDA. (i) *Establishment and control of lysogeny.* The intracellular, circularized *lambda* genome consists essentially of two OPERONS—the so-called left and right operons—and a complex immunity region whose control mechanisms are incompletely understood. The genes of the *left* operon, reading from the promoter (P_L), include *N*, *cIII*, *red*, *xis* and *int*; genes *xis* and *int* are concerned with the integration and excision of the phage genome. The genes of the *right* operon, reading from the promoter (P_R), include *cro* (also called *tof* or *fed*), *cII*, *O*, *P*, *Q*, *S*, *R*, and genes (*A*, *W*, *B*. . . *J*) concerned with the synthesis of phage structural components. Following infection of a susceptible cell, the host RNA polymerases commence to transcribe the left and right operons from P_L and P_R respectively; genes *N* and *cro* are transcribed immediately. The product of gene *N* is required for the transcription of *cIII* and subsequent genes of the left operon, and for the transcription of *cII* and subsequent genes of the right operon. The products of genes *cII* and *cIII* jointly permit transcription of gene *cI* from promoter P_{re} in the immunity region of the genome; *cI* codes for the *repressor protein*—which, initially, is synthesized in large quantities. Repressor can bind to the operator regions of the left and right operons—thus inhibiting further transcription of these operons; in this way the repressor can inhibit vegetative (lytic) development of the phage while, with the involvement of the *int* gene product, integration of the phage genome with the bacterial chromosome can be effected. However, during the initial period of phage protein synthesis the phage-cell system is committed neither to lysogeny nor to lytic development—since both operons are transcribed. Whether lysogeny or lysis supervenes may depend on the relative dominance of two phage-coded products: the repressor and the product of gene *cro*; the functions of these products are mutually antagonistic—e.g. each inhibits synthesis of the other—and the occurrence of lysogeny or lysis may depend on the prevailing activity of repressor or *cro* product respectively. In the *lysogenic* pathway the left and right operons are repressed; transcription of *cI* from the promoter P_{re} thus ceases—since *cII* and *cIII* products (derived from the right and left operons respectively) are no longer produced. However, repressor protein existing within the cell stimulates transcription of *cI* from an alternative promoter, P_{rm} —i.e. repressor induces its own synthesis from P_{rm}; this ensures continuing synthesis of repressor throughout lysogeny. (Certain mutant strains of phage *lambda* produce a temperature-sensitive repressor protein which, at elevated temperatures, becomes inactivated; transcription from P_{rm} thus ceases and, in the absence of intracellular repressor, the prophage detaches from the chromosome and enters the lytic cycle—i.e. becomes de-repressed.) (ii) *Integration and excision of the phage genome.* A single cross-over between the circularized phage genome and the bacterial chromosome causes the former to become linearly inserted between the *gal* and *bio* genes of the chromosome; integration requires the product of the phage *int* gene. During insertion of the circularized genome the latter breaks at the *att* (attachment) site, and the linear sequence of genes of the inserted genome (prophage) is *J*. . .*A*. . .*R*. . .*N*. . .*int*; this differs from the gene sequence of the pre-circularized genome—i.e. that of the phage virion—(*A*. . .*J*. . .*int*. . .*N*. . .*R*) since *att* lies between *J* and *int*. Excision (detachment) of the prophage occurs by a process which is the reverse of integration; a single cross-over is involved—as are the phage genes *int* and *xis*. Excision occurs when the prophage is de-repressed—e.g. during *induction* (see below)—and is normally followed by lytic development of the phage and cell lysis.

Induction. Free (infective) phage usually occur in cultures of lysogenic bacteria; they derive from the spontaneous lysis of a small number of lysogenic cells (about 1 per 10^2–10^5 cells per generation)—owing to the change prophage → vegetative (lytic) phage within those cells. The change prophage → vegetative phage may be brought about in the majority of cells in a lysogenic population by treatment with e.g. ultraviolet radiation, X-radiation, and certain chemical agents which inhibit DNA synthesis—e.g. MITOMYCIN C, NITROGEN MUSTARDS; such a transition, which results in lysis of the majority of cells, is termed *induction*. The mechanisms of induction are incompletely understood; it has been suggested that ultraviolet radiation causes the intracellular accumulation of an adenine derivative (a normal intermediate in adenine metabolism) which, in some way, inhibits repressor-operator interaction—thus derepressing the prophage. (see also ZYGOTIC INDUCTION)

Superinfection immunity. A lysogenic cell is immune to lysogenic or lytic infection by phages whose operators respond to the repressor formed by the prophage; such *superinfection immunity* (= *prophage immunity*, *lysogenic immunity*) results from operator-repressor interaction on the genome of the superinfecting phage.

Virulent mutants of lambda. Mutations which can cause a temperate phage to become virulent include those which affect the *cII* and *cIII* genes (no repressor formed), the *cI* gene (inactive or poorly active repressor), and/or the phage operator regions (failure of repressor to bind); the latter type of mutant could infect and lyse a lysogenic cell since superinfection immunity would not be operative. Virulent mutants form *clear* PLAQUES. (see also PSEUDO-LYSOGENY) *See also CYANOPHAGE.

lysol (see PHENOLS)

lysosomes Closed, intracellular, membranous sacs which contain hydrolytic enzymes; they occur widely in animal cells, and similar structures may be present in plant cells. Lysosomes are believed to originate in the Golgi apparatus; they are often one-half to several microns in size, though the smallest—'Golgi vesicles'—are of the order of 50 nanometres. Lysosomes are involved in intracellular digestion; collectively, lysosomal enzymes can degrade all classes of biological macromolecules. (These enzymes—which have a pH optimum in the acid range—include phosphatase, nuclease, β-glucuronidase, lipase and hyaluronidase.) Lysosomes are functional e.g. in AUTOPHAGY and in the digestion of material in FOOD VACUOLES. A *primary lysosome* is one which has yet to coalesce with an autophagic or food vacuole; following fusion, the whole is termed a *secondary lysosome*. (When digestion is complete, the vacuole with its contents is termed a *residual body*.) Lysosomes may be disrupted or damaged by certain agents, e.g. streptolysin-O and phalloidin.

lysozyme (*N*-acetylmuramidase) An enzyme which hydrolyses the *N*-acetylmuramyl-β-(1 → 4)-*N*-acetylglucosamine linkages in PEPTIDOGLYCAN; lysozyme thus functions as a bactericidal agent towards those species in which the CELL WALL peptidoglycan is exposed to enzymic action. (*Micrococcus luteus*—formerly *M. lysodeikticus*—is particularly susceptible.) Lysozyme has a molecular weight of approximately 14,500 and an isoelectric point in the region of pH 11. It is found in egg-white, milk, tears, and in many types of animal cell, e.g. polymorphs, macrophages.

lytic cycle (in the infection of bacteria by bacteriophage) The sequence of events which begins with the adsorption of a phage to a susceptible bacterial cell and terminates with cell lysis and release of the phage progeny—see BACTERIOPHAGE.

lytic infection (of bacteria by bacteriophage) The infection and subsequent lysis of susceptible bacteria by BACTERIOPHAGE. (cf. LYSOGENY)

M

M antigen (1) The M PROTEIN of streptococci. (2) One of the major cell-surface antigens of *Brucella*. (3) A cell-surface component found in certain bacteria of the Enterobacteriaceae, e.g. *Escherichia coli*.

M bands (M fibres) (of *Stentor*) (see MYONEMES)

M protein (of *Escherichia coli*) A protein involved in the galactoside TRANSPORT system; the M protein is specified by the *y* gene of the *lac* operon.

M protein (M antigen) (of *Streptococcus*) An antigen found e.g. in certain (M^+) strains of *Lancefield group A* streptococci; the presence of the M protein is associated with virulence. In group *A* strains the M protein occurs mainly as a cell wall component; it occurs in a number of different antigenic forms—thus providing a basis for the serological TYPING of M^+ strains. (see also LANCEFIELD'S STREPTOCOCCAL GROUPING TEST) On appropriate media M^+ strains often give rise to *matt* or *mucoid* colonies. Under certain conditions of culture some M^+ strains produce an extracellular *streptococcal proteinase* which breaks down their own M protein; strains cultured under such conditions may therefore be unsuitable for M protein typing. Under different cultural conditions (e.g. a lower incubation temperature) the proteinase may be produced in an inactive form.

MacConkey's agar A solid medium used e.g. for the primary isolation of *Salmonella* or *Shigella* from faeces. The medium includes: peptone (2%); lactose (1%); sodium taurocholate (or other suitable bile salt) (0.5%) to inhibit the growth of *non*-enteric organisms; sodium chloride (0.5%); NEUTRAL RED as a pH indicator; agar (1.5% w/v). The pH of the uninoculated medium should be about 7.4. On MacConkey's agar *Escherichia*, *Klebsiella*, and other lactose-fermenting organisms produce pink or red colonies; *Salmonella*, *Shigella*, *Proteus*, and other non-lactose-fermenting species (NLFs) produce colourless colonies.

MacConkey's broth A liquid medium whose constituents are identical to those of MACCONKEY'S AGAR—*except* that no *agar* is included. (Some workers use BROMCRESOL PURPLE, instead of neutral red, as the pH indicator.) The medium has been used e.g. in the EIJKMAN TEST.

macrocyclic rusts (1) (*eu-form* rusts) Those rusts—see RUSTS (1)—which, during their life cycles, produce all the forms of rust spore (i.e. pycniospores, aeciospores, uredospores, teliospores, and basidiospores). (2) Those rusts which produce more than one type of *dikaryotic* spore—i.e. aeciospores and/or uredospores in addition to teliospores.

macrogamete (a) A large GAMETE. (b) A *female* gamete.

macroglobulin Any globulin of high (but unspecified) molecular weight; macroglobulin may refer specifically to IgM (*q.v.*).

macrolide antibiotics A large group of ANTIBIOTICS—few of which find wide clinical usage; the group includes *erythromycin*, *oleandomycin*, and *tylosin*. The macrolides contain a large *lactone ring* substituted with various functional groups and linked glycosidically to one or more sugars—including an aminosugar and, frequently, an unusual deoxy sugar; thus, erythromycin consists of a 14-membered ring linked to the aminosugar *desosamine* and the deoxysugar *cladinose*. (Macrolides which contain conjugated polyene systems form a separate group: see POLYENE ANTIBIOTICS.) The macrolides are active mainly against gram-positive bacteria—including *Clostridium* spp, Lancefield group *A* streptococci and *Streptococcus pneumoniae*; certain gram-negative organisms are susceptible, e.g. *Haemophilus*, *Neisseria*, some species of *Bacteroides*. The macrolides inhibit PROTEIN SYNTHESIS by binding to the 50S subunits of 70S RIBOSOMES; the molecular basis of inhibition is unknown—but it is thought that the translocation step may be affected. Several instances of ribosome-based resistance to erythromycin have been reported; this may involve the modification of a protein component within the 50S subunit. Some types of resistance appear to be plasmid-mediated.

macronucleus (*ciliate protozool.*) The larger of the two types of nucleus (see also MICRONUCLEUS) found in ciliates. Many ciliates contain a single macronucleus, but more than one may be present; according to species, the macronucleus may be ovoid, elongated, moniliform, C-shaped or complex etc. Macronuclei are usually highly *polyploid* and are not involved in genetic recombination during sexual reproductive processes. They contain nucleoli, and are concerned only with the somatic functions of the cell. During sexual processes the macronucleus normally disintegrates and is regenerated from the micronuclei; exceptionally, one or more fragments of the macronucleus may give rise to a new macronucleus.

macrophages (mononuclear phagocytes) Large, actively phagocytic cells found e.g. in the spleen, liver and lymph nodes and in the blood;

macrophages are important factors in NON-SPECIFIC IMMUNITY, ANTIBODY-FORMATION and CELL-MEDIATED IMMUNITY. Macrophage (extracellular) products include certain components of COMPLEMENT (e.g. C4) and INTERFERON.

Macrophage precursors arise in the bone marrow and localize in various body tissues following a brief period in the bloodstream; macrophages include the blood *monocytes* and the sedentary *Kupffer cells* of the liver.

Macrophages are found in acute and chronic lesions; their ability to break down ingested particles is facilitated by an abundance of intracellular LYSOSOMES. PHAGOCYTOSIS is stimulated by e.g. certain LYMPHOKINES, bacterial ENDOTOXINS and double-stranded RNA.

Macrophages may be maintained for some time in culture. Viable macrophages adhere to glass and plastics and accumulate dyes such as trypan blue. (see also GIANT CELLS)

macroscopic (*adjective*) Of a size which is visible to the *unaided* eye.

macrostome (see TETRAHYMENA)

macrotetralides A group of ANTIBIOTICS which act in a manner similar to that of the DEPSIPEPTIDE ANTIBIOTICS. The group includes the so-called *actins* (e.g. *nonactin, monactin*)—cyclic structures which comprise four similar tetrahydrofuran-containing units linked *via* ester bonds. Open-chain macrotetralides include *monensin* and *nigericin.*

macula (macule) (*med.*) On tissues (particularly skin): a discrete, unelevated region of discoloration.

Madura foot A chronic disease of the foot in which granulomata and pus-discharging lesions form in subcutaneous tissues and bones; it occurs in tropical and sub-tropical regions where footwear is not worn. The causal agent, which commonly invades *via* wounds etc., may be any of a variety of bacteria (e.g. *Nocardia, Streptomyces*) or fungi (e.g. *Allescheria boydii, Aspergillus* spp, *Madurella* spp); fungal infections are termed *maduromycoses. Laboratory diagnosis*: microscopical/cultural examination of pus etc.

maduromycoses (see MADURA FOOT)

magnification The extent to which the image of an object is larger than the object itself. With a compound (light) microscope the magnification is calculated as the product of the individual magnifications of the objective and eyepiece lenses; such a figure gives the *angular magnification*, i.e. it indicates the ratio:

$$\frac{\text{angle subtended by magnified image at eye}}{\text{angle subtended by object at unaided eye}}$$

The amount of magnification is determined by the focal lengths of the objective and eyepiece lenses; the RESOLVING POWER of the optical system involves other factors. Magnification and resolving power must therefore be clearly distinguished; they are not directly related. Thus, it is possible to have a number of objective lenses of the same magnification but of different resolving powers; further, the resolving power (but not the magnification) of a given objective may be influenced by sub-optimal conditions of illumination. An increase in magnification without an accompanying increase in resolution is said to be *empty magnification*.

The maximum *useful* magnification obtained with a given objective lens is approximately 1000 times its *numerical aperture* (N.A.). With good-quality light microscopes the maximum magnification obtainable is usually about 1200×. The ultraviolet microscope gives somewhat higher magnifications, while the electron microscope can give values greater than 800,000×.

maintenance medium (see TISSUE CULTURE)

mal de caderas A fatal wasting disease of the horse and mule and of certain other animals. The causal agent, *Trypanosoma equinum*, is transmitted by blood-sucking horse-flies of the family *Tabanidae.* Mal de caderas occurs in South America. In some parts of South America *mal de caderas* refers to a fatal paralytic disease of cattle in which the causal agent is the RABIES virus, and the vector is the blood-sucking vampire bat.

mal du coit (see DOURINE)

malachite green A light-green basic dye of the triphenylmethane group. Malachite green has been used e.g. as a spore stain (hot, 5% aqueous) and as a selective bacteriostatic agent in certain media. (see also TRIPHENYLMETHANE DYES (as antimicrobial agents) and ALBERT'S STAIN)

malaria An acute (often fatal), or chronic (recurrent) infectious disease of man and other animals; malaria is endemic in parts of Africa, Asia and South America. The causal agent is a species of PLASMODIUM. In *man*, *P. vivax*, *P. falciparum*, or *P. malariae* (rarely *P. ovale*) is usually transmitted by the bite of an infected (female) mosquito (*Anopheles*); the type of malaria produced by each species of *Plasmodium* differs e.g. in severity and periodicity of symptoms. A mosquito becomes infected by ingesting blood containing the *gametocytes* of *Plasmodium*; the parasite reproduces (sexually) within the mosquito, and eventually forms *sporozoites* which appear in the insect's salivary glands. The mosquito can now transmit *Plasmodium* inoculatively. (For details of the life cycle in the mammalian host see: PLASMODIUM.) In man, the incubation period is usually 7–14 days, but may be 1 month or longer. Typical symptoms include periodic chills, fever, and intense sweating; enlargement of the spleen subsequently occurs. In *quartan malaria* (*P. malariae*) 3 days elapse between paroxysms; in *benign tertian malaria* (*P. vivax*) the corresponding period is 2 days. *Malignant tertian malaria* (*P. falciparum*) exhibits the severest symptoms; erythrocytes infected with *P. falciparum* tend to agglutinate and block capillaries—leading e.g. to *cerebral malaria* with attendant coma and death. Untreated *malariae* or *vivax* malaria may recur—but may become self-limiting within 2–3 years. Antimalarial drugs include QUININE and AMINOQUINOLINES. Genetically-based resistance to *falciparum* malaria occurs in persons heterozygous for the gene of sickle-cell anaemia; such persons produce a proportion of abnormal haemoglobin which inhibits disease development. In the control of malaria an important factor is the eradication of the mosquito vector. *Laboratory diagnosis*: stained blood smears are examined for infected erythrocytes; Giemsa's stain is often used. *Malaria in animals. P. knowlesi* produces a fatal disease in rhesus monkeys, and *P. gallinaceum* is but one species pathogenic for fowl, including chickens. New antimalarial drugs are often tested against *P. gallinaceum* infections in chickens.

malarial pigment (see PLASMODIUM)

Malassezia (see PITYROSPORUM)

malefic mutation A mutation which is disadvantageous to the organism in which it occurs.

mallein test (see GLANDERS)

Mallomonas A genus of algae of the CHRYSOPHYTA.

malo-lactic fermentation A type of fermentation in which L-malic acid is converted to lactic acid and carbon dioxide. The malo-lactic fermentation is carried out by lactic acid bacteria (including species of *Lactobacillus*, *Leuconostoc*, and *Pediococcus*); it can occur in e.g. wines, cider, etc.—in which it has the effect of reducing acidity.

Malta fever (see BRUCELLOSIS)

Maltaner antigen A cardiolipin-lecithin antigen used in syphilis serology (see STS).

maltose A disaccharide: D-glucopyranosyl-α-(1 → 4)-D-glucopyranose; it is formed e.g. on hydrolysis of GLYCOGEN or STARCH. Maltose is metabolized by a wide range of fungi and bacteria. (see also AMYLASES)

mancozeb An agricultural antifungal agent consisting of a complex of zinc and MANEB; it is used mainly against *late blight* of POTATO (*Phytophthora infestans*).

mandelic acid ($C_6H_5.CHOH.COOH$) Formerly widely used as a urinary antiseptic; it is excreted unchanged in the urine. Mandelic acid may be used in combination with HEXAMETHYLENETETRAMINE to provide the low pH necessary for the activity of the latter.

Mandler filter (see FILTRATION)

maneb (dithane M-22) (*plant pathol.*) Manganese ethylenebisDITHIOCARBAMATE, an important agricultural antifungal agent with a spectrum of activity similar to that of ZINEB; maneb gives good control of infections involving *Alternaria solani* and *Phytophthora infestans*.

mannans Polymers of mannose. β-(1 → 4)-linked mannans occur in certain plants (e.g. ivory nuts, palm seed) and in some algae—in which they may replace CELLULOSE as the main structural component of the CELL WALL. A highly-branched mannan which contains α-(1 → 2), α-(1 → 3), and α-(1 → 6) linkages, and which may be extensively phosphorylated, can be extracted with hot dilute alkali from the CELL WALLS of certain yeasts.

mannitol The polyhydric alcohol corresponding to mannose, i.e. a product of the reduction of mannose or fructose. Mannitol occurs widely in plants; it is found in many algae—particularly in species of the Phaeophyta and Rhodophyta. It occurs as a constituent of some LAMINARINS. Mannitol is metabolized by a number of bacteria including many strains of *Escherichia coli* and *Salmonella* spp and some strains of *Bacillus*, *Mycobacterium* and *Streptococcus*.

Mantoux test (see TUBERCULIN TEST)

map unit (genetics) A unit used for indicating the distance between two given markers on a chromosome. The number of map units between two given markers is equal to the (statistically corrected) RECOMBINATION FREQUENCY of the markers expressed as a percentage; for example, a corrected RF of 10% gives a distance of 10 map units.

Marburg agent (Marburg virus) A virus detected (in 1967) in the tissues of the African green monkey (*Cercopithecus aethiops*). The Marburg agent resembles a rhabdovirus in morphology, is ether-sensitive, and is believed to contain RNA. (see GREEN MONKEY FEVER)

Marek's disease virus (see HERPESVIRUS)

marfanil (sulphamylon) *p*-Aminomethylbenzene sulphonamide, $NH_2CH_2.C_6H_4.SO_2NH_2$. This differs from other SULPHONAMIDES in that it is not a derivative of *sulphanilamide*, i.e. a *p*-aminomethyl group ($-CH_2NH_2$) replaces the *p*-amino group; it also differs in that it is not inactivated by *p*-AMINOBENZOIC ACID, pus, etc. Formerly used topically to treat wounds, marfanil is now used in the treatment of burns. It is active against *Pseudomonas aeruginosa*.

marker (genetics) A chromosomal locus which is associated with a particular phenotypic characteristic, e.g. the production of an enzyme or toxin.

mast cells Cells which contain granules of

HISTAMINE, SEROTONIN and heparin. The *tissue mast cell* is present in heart, lungs, liver, spleen etc., and is particularly numerous in connective tissues and around blood vessels. The *mast leucocyte* (or *basophil*) is less numerous; in human blood it forms about 1% of the total white cell population. During ANAPHYLACTIC SHOCK mast cells undergo *degranulation* i.e. they release histamine and other substances as a result of the combination of antigen with cytophilic antibodies. Degranulation also results from the action of ANAPHYLATOXINS.

Mastigomycotina A sub-division of FUNGI of the division EUMYCOTA; constituent species are distinguished from those of other sub-divisions in that they typically give rise to motile (flagellated) gametes/spores. (Zoospores are not formed by certain of the more advanced species of the class Oomycetes.) The classes are CHYTRIDIOMYCETES, HYPHOCHYTRIDIOMYCETES, and OOMYCETES.

mastigoneme (see FLAGELLUM (eucaryotic))

Mastigophora A superclass of protozoans (subphylum SARCOMASTIGOPHORA); locomotion in constituent species is effected chiefly or exclusively by means of *flagella* (pseudopodia are formed e.g. by *Histomonas*). Cell division is typically SYMMETROGENIC. The classes are PHYTOMASTIGOPHOREA and ZOOMASTIGOPHOREA.

mastigosome (*protozool.*) (see BASAL BODY)

mate killer The term for certain strains of *Paramecium* which, during or after CONJUGATION, cause the death of the conjugal partner; the mechanism of this phenomenon is unknown.

maternal inheritance A form of CYTOPLASMIC INHERITANCE in which certain features may be transmitted to sexually-derived progeny only from the *female* parent cell—see e.g. POKY MUTANT.

mating type (1) (in ciliates) (see SYNGEN) (2) (in fungi) (see HETEROTHALLISM)

Maurer's clefts (see SCHÜFFNER'S DOTS)

mazaedium (*mycol., lichenol.*) A type of ascocarp within which the asci disintegrate to release a powdery mass of ascospores. Mazaedia occur in certain lichens (e.g. members of the genera *Calicium* and *Sphaerophorus*) and in certain fungal genera of the Plectomycetes (e.g. *Onygena* spp).

MBSA (*immunol.*) Methylated bovine serum albumin.

McCartney bottle (universal bottle) A cylindrical (shoulderless) screw-cap glass jar with a capacity of approximately 25 millilitres.

McIntosh and Fildes' anaerobic jar (see ANAEROBIC JAR)

mean generation time (see GROWTH (bacterial))

measles (morbilli; rubeola) An acute, systemic, highly communicable disease of man (mainly children) and other primates; the causal agent (see PARAMYXOVIRUS) invades *via* the oral or nasal route. The incubation period (about 12 days) is followed by coryza, cough and fever and the development of KOPLIK SPOTS. Subsequently a rash appears on the head and face and then on the limbs and trunk. Viruses are excreted in the urine and in the secretions of the eye, nose and mouth during the prodromal period—and for several days following the appearance of the rash. Sequelae may include e.g. encephalitis, pneumonia. Antimeasles vaccines are available; long-term (usually life-long) immunity follows recovery from the disease. *Laboratory diagnosis* may include microscopical examination of smears of the nasal mucosa for GIANT CELLS and inclusion bodies and/or characterization of the infecting virus.

mechanical transmission (of pathogens) (see VECTORS and VIRUS DISEASES OF PLANTS)

medallion (see CLAMP CONNECTIONS)

median lethal dose (see LD_{50})

medical flat A flat-sided (rectangular cross-section) glass screw-cap bottle obtainable in various sizes.

medium (*microbiol.*) In general, any material which supports the growth/replication of microorganisms. As commonly used in microbiology: any liquid or solid *preparation* used for any of a number of purposes—which may include (a) the growth (CULTURE) of microorganisms (including ENRICHMENT and other differential processes); (b) diagnostic (identification) tests—in which the identity of a given organism may be deduced from its growth characteristics in or on particular media; (c) the storage and/or transportation of microorganisms—see TRANSPORT MEDIUM; (d) the growth or maintenance of cells in a TISSUE CULTURE. Before use a medium is normally *sterile*.

Liquid culture media for *bacteria* are often prepared from *basal* media such as NUTRIENT BROTH or PEPTONE WATER—which, without supplement, support the growth of a number of species. (see also ENRICHED MEDIUM) Autotrophic bacteria are commonly cultured in media which consist largely or exclusively of mixtures of inorganic salts. ANAEROBES are usually cultured in reduced media (e.g. BREWER'S THIOGLYCOLLATE MEDIUM, ROBERTSON'S COOKED MEAT MEDIUM) or on reduced or non-reduced media in an ANAEROBIC JAR. Media used for diagnostic tests ('test media') are generally made by the addition of specific substance(s) to basal or enriched media—see e.g. PEPTONE WATER SUGARS. Some media incorporate a PH INDICATOR—the purpose of which is to give information about the metabolic activities of the organism(s) being cultured on such media. Many algae, fungi, and protozoans may be cultured in appropriate liquid media.

Solid media (for bacteria, fungi, and certain algae and protozoans) usually consist of an AGAR or gelatin *gel* which is supplemented with appropriate nutrients etc. prior to solidification.

A *defined* medium is one in which all the constituents (including trace substances) are *quantitatively* known. An *inhibitory* medium is one which contains substance(s) added in order to suppress the growth of certain type(s) of microorganism. *Selective medium* may refer either to (a) an inhibitory medium, or (b) a medium designed to encourage the growth of certain type(s) of microorganism in preference to other types which may be present in the inoculum. (According to some authors only a *solid* medium should be included within the category 'selective media'.)

megacins (see BACTERIOCINS)

Megasphaera A genus of gram-negative, anaerobic, non-motile cocci of the bacterial family Veillonellaceae. The type species, *M. elsdenii* (formerly *Peptostreptococcus elsdenii*), occurs e.g. in the RUMEN.

Megatrypanum A sub-genus of TRYPANOSOMA within the group STERCORARIA; species are parasitic e.g. in rodents and bovines. The organisms are large (length may exceed 100 microns) and have a kinetoplast situated (approximately) mid-way between the nucleus and the posterior extremity of the cell; there is a free flagellum. Species include *T. (Megatrypanum) theileri*.

meiocyte Any cell whose nucleus undergoes MEIOSIS.

meiosis (reduction division) (in diploid cells) The process in which a diploid cell gives rise to haploid cells and in which genetic RECOMBINATION usually occurs; meiosis thus provides a mechanism for the re-distribution of parentally-derived genetic information among progeny cells. Meiosis occurs during the formation of GAMETES from diploid cells, and at the inception of haplophase in those organisms which exhibit an ALTERNATION OF GENERATIONS. (cf. MITOSIS)

The following *generalized* account of meiosis describes those phases which are commonly distinguished; in some organisms certain phases are omitted.

Meiosis involves *two* distinct nuclear divisions. *The first meiotic division. Prophase I.* Individual chromosomes become visible as long, *single* threads (*leptotene* stage). In the next (*zygotene*) stage each chromosome pairs, lengthwise, with its HOMOLOGUE—i.e. the two chromosomes come into close contact to form a so-called *bivalent*; the mechanism of pairing (= *synapsis*) is unknown. (see SYNAPTONEMAL COMPLEX) In the following (*pachytene*) stage the chromosomes of each homologous pair appear to shorten and thicken. In the penultimate (*diplotene*) stage of prophase I, each chromosome of an homologous pair is seen to consist of two (sister) chromatids; thus, at diplotene, each homologous pair occurs as a four-chromatid structure—referred to as a *tetrad*. During the diplotene stage each chromosome begins to draw away from its homologue—except in region(s) where connections have been established between homologous (non-sister) chromatids; such connections (*chiasmata*) are believed to be manifestations of the occurrence of CROSSING-OVER. Although a given cross-over occurs at a particular site, the associated chiasma tends to move slowly towards the ends of the chromatids ('terminalization') as the homologous chromosomes pull apart; consequently, chiasmata cease to indicate the sites of crossing-over. During the final stage in prophase I (*diakinesis*) chromatid shortening and thickening reaches a maximum and each tetrad exhibits a configuration dictated by the number and positions of chiasmata; thus, a tetrad with a single chiasma forms the shape of a cross, while tetrads with two or more chiasmata appear as circles or multiple loops. By the end of diakinesis the nucleolus and nuclear membrane cease to be demonstrable. *Metaphase I.* A SPINDLE is formed; the centromeres of each homologous pair take up positions either side of the equatorial plane of the spindle. *Anaphase I.* The chromosomes separate; of each homologous pair, one chromosome moves towards one pole of the spindle while its homologue moves to the opposite pole. *Telophase I.* Each of the two groups of chromosomes (separated during anaphase) becomes enclosed by a nuclear membrane, and a nucleolus develops in each of the newly-formed (haploid) nuclei. *Interkinesis.* An interval between the first and second meiotic divisions during which the chromosomes (if visible) appear less distinct (i.e. less condensed).

The second meiotic division is analogous to a mitotic division (see MITOSIS); the phases are prophase II, metaphase II, anaphase II, and telophase II. During anaphase II the chromatids of each chromosome move to opposite poles of the spindle; in telophase II each (single) chromatid assumes the role of a chromosome in one of the newly-formed (second generation) haploid nuclei.

If each meiotic division is followed by CYTOKINESIS, the original diploid cell will give rise to four haploid progeny cells—each containing one chromatid from each tetrad.

meiosporangium (*mycol.*) (see SPORANGIUM)

meiospore (see ALTERNATION OF GENERATIONS)

Melampsora A genus of fungi of the order UREDINALES. (see also RUSTS (1)) *Melampsora* spp include both heteroecious and autoecious rusts. Typically, *Melampsora* spp form non-peridiate aecia (*caeomata*) and peridiate uredia; the teliospores are sessile. (see also FLAX, DISEASES OF)

Melanconiales An order of fungi within the class COELOMYCETES; the (asexual) fruiting body formed by members of the Melanconiales is an ACERVULUS (cf. SPHAEROPSIDALES). Species include a number of plant pathogens (see e.g. ANTHRACNOSE). Genera include e.g. COLLETOTRICHUM.

melarsoprol (see ARSENICALS)

melezitose α-D-Glucopyranosyl-(1 → 3)-β-D-fructofuranosyl-(2 → 1)-α-D-glucopyranoside. Melezitose is metabolized e.g. by certain species of *Lactobacillus*.

melibiose D-Galactopyranosyl-α-(1 → 6)-glucopyranose.

Meliola A genus of fungi of the PERISPORIACEAE.

Melosira A genus of centric diatoms (division BACILLARIOPHYTA); species occur in either freshwater or marine environments. The cylindrical cells are typically joined valve to valve to form a long (sheathed) chain.

Meltzer's reagent A stain used in mycology and lichenology for studying e.g. the septation of spores and the nature of the ascus wall and apex. The reagent consists of iodine (0.5 gram), potassium iodide (1–1.5 grams), chloral hydrate (20 grams), and distilled water (20 mls).

membrane (bacterial) (see CELL MEMBRANE and OUTER MEMBRANE)

membrane filtration Any form of microbiological FILTRATION in which a cellulose-ester (or similar) membrane is used.

membranelle (*ciliate protozool.*) A discrete group of *oral* cilia which behave as a unified organelle; membranelles, which are often grouped together to form an AZM (*q.v.*), are primarily involved in the process of feeding, i.e. in the production of water currents directed towards the cytostome. (cf. CIRRUS)

memory cells (*immunol.*) (see TRANSFORMATION, OF LYMPHOCYTES and Y CELLS)

menadione (see QUINONES (in electron transport))

menaquinones (see QUINONES (in electron transport))

Mendel's laws (see SEGREGATION, PRINCIPLE OF and INDEPENDENT ASSORTMENT)

meningitis Inflammation of the membranes which cover the brain and spinal cord. Organisms may gain access to the meninges *via* the lymph or blood vascular systems, through head wounds, or by invasion from the sinuses. Examples of aetiological agents include: *Neisseria meningitidis* ('epidemic meningitis'), *Haemophilus influenzae* (particularly in children), *Streptococcus pneumoniae, Listeria monocytogenes* and *Cryptococcus neoformans*. Meningitis may develop secondarily to tuberculosis (*Mycobacterium tuberculosis*) or syphilis (*Treponema pallidum*). *Aseptic meningitis* is of viral aetiology; members of the *coxsackie* and *echovirus* groups are among the causal agents.

meningococcus (see NEISSERIA)

mercaptoethanol ($HSCH_2CH_2OH$) A reagent used in immunochemistry to reduce the disulphide bonds of IMMUNOGLOBULINS (see also FAB FRAGMENT). Reduction of the disulphide bond (-S-S-) gives two -SH groups.

mercurochrome (see *Mercury* under HEAVY METALS)

mercury (and compounds) (as antimicrobial agents) (see HEAVY METALS)

meri- Combining form signifying 'part'.

meristem Region(s) of an organism at which new, permanent tissue is formed, i.e. the 'growth region(s)' of a plant or multicellular alga.

meristoderm In certain algae (e.g. *Laminaria*): the surface layer of cells which, on division, cause an increase in the thickness of the thallus.

meristogenous development (*mycol.*) The development of certain fungal structures (e.g. pycnidia, stromata) by the proliferation and differentiation of a small number of adjacent cells in a single hypha. (cf. SYMPHOGENOUS DEVELOPMENT)

mero- Combining form signifying 'part'.

merogony Synonym of SCHIZOGONY.

merosporangium (*mycol.*) (see SPORANGIUM)

merozoite (see SCHIZOGONY)

merozygote A bacterium which is partially diploid, partially haploid; a merozygote may be formed in any of the processes (CONJUGATION, TRANSDUCTION, TRANSFORMATION) in which bacterial genes are transferred from one cell (the donor) to another (the recipient). The genetic complement of a merozygote comprises the *endogenote* (i.e. the chromosome of the recipient) and the fragment of genetic material (the *exogenote*) received from the donor; chromosomal and extraneous genes which are allelic constitute the diploid region of the merozygote. If corresponding alleles of the exogenote and endogenote are identical, the merozygote is said to be an homogenote; if a pair of unlike alleles occurs at one or more loci, the merozygote is termed an *heterogenote*. Recombination may occur between the exogenote and endogenote.

merthiolate (thiomersal; thimerosal) (see THIOMERSAL and *mercury* under HEAVY METALS)

Merulius lacrymans (see SERPULA)

mesophile (mesophil) Any organism whose optimum growth temperature is in the range 20–45 °C.

mesoplankton (see PLANKTON)

mesosaprobic zone (see SAPROBITY SYSTEM)

mesosomes (chondrioids) Intracellular, membranous structures which have been observed in many bacteria; they appear to be infoldings of the CELL MEMBRANE. Mesosomes are often more apparent in gram-positive than gram-negative species. In gram-positive species they may take various forms, e.g. parallel or concentric lamellae, or vesicles (perhaps tubules) which may be interconnected; in gram-negative species mesosomes appear to be smaller and less intricate, and are often seen as simple invaginations of the cell membrane. The function of mesosomes is unknown—although various roles have been attributed to them. Thus, e.g. they are commonly associated with the developing septum during cell division, and it has been suggested that they may play some part in septum-formation. Others suggest that mesosomes may be involved in DNA replication, or that they form centres of cell respiration.

metabasidium (*mycol.*) That part of a BASIDIUM from which arise the basidiospore-bearing sterigmata. In e.g *Tremella* and related fungi, the metabasidium is the septate part of the basidium; in *Tulasnella* and related fungi the metabasidium is an aseptate structure; in *Septobasidium* and related fungi the metabasidium is the transversely-septate part of the basidium which arises from the PROBASIDIUM; in members of the Uredinales (rust fungi) the metabasidium is the septate structure (also called the *promycelium*) which arises from the teliospore.

metabolic inhibition test A method of quantifying a cytocidal agent (e.g. a virus or toxin) by monitoring the metabolic activity in a series of tissue cultures—each having been inoculated with one of a range of dilutions of the cytocidal agent. In the *absence* of a cytocidal agent, the growth medium of a tissue culture normally becomes acidic (after a given period of time); this pH change (detected by a pH indicator in the medium) is due e.g. to the cells' metabolism of glucose—from which acidic products are formed. In the *presence* of a cytocidal agent, such a change in the growth medium is delayed or totally inhibited. The titre of a cytocidal agent may be estimated by determining the highest dilution of the agent which brings about such metabolic inhibition in the culture.

Another form of metabolic inhibition test is used to detect serum antibodies homologous to an acid-forming strain of *Mycoplasma*. A sample of serum is added to a suspension of the organism in a suitable growth medium containing an appropriate pH indicator; following incubation, the presence of specific antibodies is indicated by growth-inhibition and the absence of acidic metabolic products.

metaboly A former name for *euglenoid movement*—see EUGLENA.

metachromasy The phenomenon in which certain biological entities (e.g. METACHROMATIC GRANULES) become stained with a colour which is different to that of the dye used to stain them.

metachromatic granules (volutin granules; Babes-Ernst granules) Microbial intracellular granules which, on staining with certain basic dyes (e.g. polychrome methylene blue, toluidine blue), exhibit METACHROMASY. These electron-dense granules are found in yeasts, bacteria (e.g. *Corynebacterium diphtheriae*, *Mycobacterium* spp) and a variety of other microorganisms; they contain POLYPHOSPHATE together with lipid and/or protein. Metachromatic granules are formed typically during the stationary phase of growth, and their formation is influenced by the nature of the growth medium; such granules are believed to function as a reserve of phosphate. In some bacteria, e.g. *C. diphtheriae*, a single, conspicuous metachromatic granule may often be observed at each pole of the cell; such granules are sometimes referred to as *polar bodies*. (see also ALBERT'S STAIN)

metachronal waves The collective movement of coordinated cilia (see CILIUM (1)); metachronal waves may be likened to the effect produced by an intermittent breeze on a field of barley. Although the cilia on a particular ciliate beat with similar or identical *frequencies*, the beating, in general, is not *synchronous*—i.e. all the cilia do not pass simultaneously through the same phase or stage of the beat cycle. In metachronal rhythm, adjacent cilia pass through slightly different stages of the beat cycle at any given instant; thus, when a metachronal wave travels along a row of cilia, the movement of each successive cilium within the metachronal wave is phased slightly ahead (or behind) that of the preceding cilium. (This results in the smooth motion characteristic of ciliates.) In one form of metachronal rhythm (*symplectic metachrony*) those cilia which are passing (at a given instant) through the effective stroke of the beat cycle tend to bunch together—thus forming the crest of a metachronal wave. In symplectic metachrony the crests of the metachronal waves move in the direction of the effective stroke; cilia arranged in the direction of the effective stroke are successively further *behind* in the beat cycle. In *antiplectic metachrony* the metachronal crests move in a direction *opposite* to that of the effective stroke; cilia arranged in the direction of the effective stroke are successively further *ahead* in the beat cycle. Symplectic metachrony is quite rare in ciliates; antiplectic metachrony is more common. Other forms of metachronal rhythm are known. Ciliar movements (i.e. the effective and recovery strokes) may occur in a plane at right angles to the propagated metachronal wave; the effective stroke may be to the left (*laeoplectic metachrony*) or to the right (*dexioplectic metachrony*) of the propagated wave.

The mechanism of ciliar coordination—responsible for metachronal rhythm—is at present unknown; evidence has been presented for and against each of several hypotheses. These include: (a) the *neuroid hypothesis*—which supposes that stimulatory impulses are transmitted sub-pellicularly, from cilium to cilium, along conducting fibres; (b) the *hydro-mechanical linkage* hypothesis—which supposes that ciliary coordination is the result of mutual interference involving hydrodynamic forces; (c) a hypothesis which supposes that ciliary coordination may involve electrical charges at the cell surface. The mechanism of ciliar movement is not necessarily the same in all ciliates.

metacyclic forms (metatrypanosomes) Trypanosomes which are infective for the *vertebrate* host. The metacyclic form is a TRYPOMASTIGOTE which (typically) lacks a *free* flagellum; such forms are produced at the end of the period of cyclical development within an invertebrate vector. (see also TRYPANOSOMA)

metaphase The second stage in MITOSIS and analogous stages in MEIOSIS.

metaphen (nitromersal) (see *Mercury* under HEAVY METALS)

metatrypanosomes A synonym of METACYCLIC FORMS.

methacycline (rondomycin) 6-Demethyl-6-deoxy-5-hydroxy-6-methylenetetracycline (see TETRACYCLINES).

metham sodium (vapam) Sodium *N*-methylDITHIOCARBAMATE; $CH_3.NH.CS.SNa$. Metham sodium is used as a soil fumigant; in the soil it breaks down to form the volatile methylisothiocyanate. (see also DAZOMET and QUINTOZENE)

methane (microbial production of) (see METHANOBACTERIACEAE)

methane (microbial utilization of) A number of bacteria (e.g. members of the METHYLOMONADACEAE) can use methane (CH_4) as the sole source of carbon. Methane metabolism is obligately aerobic—an oxygenase being involved in the initial oxidation of methane. Details of the reactions are unknown. It has been widely accepted that the sequence involved is: methane → methanol → formaldehyde → formic acid → carbon dioxide; however, there is evidence to suggest that the initial product of methane oxidation may be dimethyl ether, and that this is subsequently converted *via* a number of steps to methanol. Formaldehyde is withdrawn from the pathway as a source of cellular carbon. Methane-utilizing organisms appear to use one of two known cyclic pathways for the incorporation of formaldehyde: (a) formaldehyde is condensed with ribose-5-phosphate to form allulose-6-phosphate—which is then converted to fructose-6-phosphate; ribose-5-phosphate is regenerated in a series of reactions analogous to reactions of the Calvin cycle. (b) Formaldehyde is transferred, *via* tetrahydrofolate, to glycine to form serine; part of the serine is assimilated and part is used to regenerate glycine. (see also ALIPHATIC HYDROCARBONS)

methane-producing bacteria The METHANOBACTERIACEAE.

Methanobacteriaceae A family of methane-*producing* bacteria comprising the genera* *Methanobacterium*, *Methanococcus* and *Methanosarcina*; species occur in the anaerobic muds of ponds, marshes etc., in sewage sludge, and in the RUMEN. All species are strict anaerobes and appear to be highly sensitive to oxygen. At least some strains are believed to be facultative autotrophs. A form of energy production common to all species is the anaerobic oxidation of hydrogen gas using carbon dioxide as electron acceptor—the metabolic products being methane (CH_4) and water; some strains can oxidize formate in place of hydrogen. VITAMIN B_{12} is involved in the production of methane from carbon dioxide. Some species produce methane by the anaerobic metabolism of acetate, methanol or carbon monoxide. *These genera contain mixtures of gram-positive and gram-negative (or gram-variable) species; *Methanobacterium* and *Methanococcus* both include motile and non-motile species.

methenamine (see HEXAMETHYLENETETRAMINE)

methicillin (see PENICILLINS)

methionine (biosynthesis of) (see Appendix IV(d))

methisazone *N*-methyl-ISATIN-β-THIOSEMICARBAZONE.

methyl bromide (CH_3Br) (*plant pathol.*) A gas which is used as a soil fumigant; it has herbicidal, insecticidal, and fungicidal activity.

methyl-green pyronin stain (Unna-Pappenheim stain) A stain used for distinguishing DNA (which stains green) from RNA (which stains red) in histological sections etc.; it is useful e.g. for distinguishing LYMPHOCYTES from (ribosome-rich) PLASMA CELLS. (see also ACRIDINE ORANGE)

methyl orange (PH INDICATOR) pH3.0 (red) to pH4.4 (orange-yellow). pK_a 3.6.

methyl red (PH INDICATOR) pH4.4 (red) to pH6.2 (yellow). pK_a 5.1.

methyl red test (MR test) A test used in the identification of certain bacteria, e.g. members of the Enterobacteriaceae; it determines the ability of an organism, growing in a phosphate-buffered glucose-peptone medium, to produce sufficient acid (e.g. by the MIXED ACID FERMENTATION) to reduce the pH of the medium from 7.5 to about 4.4 or below. The inoculated medium is incubated (48 hours/37 °C*) and the pH of the medium is tested by the addition of 2 drops of 0.04% METHYL RED solution. (After reading the test the same culture can be used for the VOGES–PROSKAUER TEST.) **Not* overnight incubation, since early acidification of the medium (giving a false-positive reaction) may be produced by MR-negative organisms, e.g. *Klebsiella pneumoniae*; on further incubation the latter metabolizes the acidic products (with attendant rise in pH) and thus gives the characteristic negative reaction. (see also IMVIC TESTS)

methyl violet A mixture of methylated triamino-TRIPHENYLMETHANE DYES which includes CRYSTAL VIOLET.

methylene blue (Swiss blue) A non-fluorescent, deep-blue, basic dye chemically related to phenothiazine. Methylene blue is employed e.g. as a general stain (see STAINING), as a vital stain (see VITAL STAINING), and as a redox indicator ($E_0 = +11$mV at pH7); it is also used in certain tests—see e.g. METHYLENE BLUE TEST (for milk) and TOXOPLASMA DYE TEST.

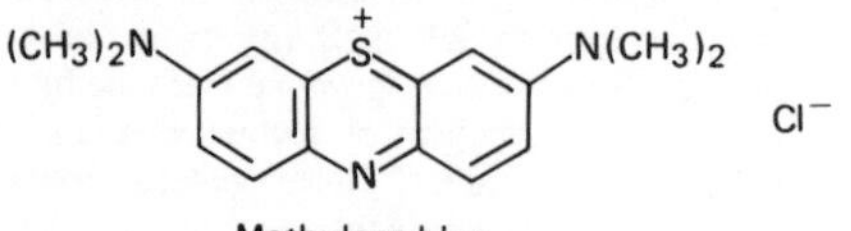

Methylene blue

Polychrome methylene blue is a mixture of dyes which includes methylene blue and certain of its oxidation products, e.g. dyes of the *azure* series—which impart a distinct violet coloration to the solution; polychrome methylene blue forms when an alkaline solution of methylene blue is allowed to stand in contact with the air for long periods of time.

Loeffler's methylene blue is prepared by the addition of 0.01% potassium hydroxide solution (100 ml) to a 1% solution of methylene blue in 95% ethanol (30 ml)—the whole being allowed to mature (i.e. to oxidize slowly) over a period of several months. The matured stain is also referred to as 'polychrome methylene blue'.

Leuco methylene blue is the reduced (colourless) form of methylene blue.

methylene blue test (for milk) A test used to assess the microbiological quality of raw (i.e. untreated) MILK. To a standard volume of milk is added a fixed quantity of METHYLENE BLUE* solution; the whole is incubated at 37 °C and examined at regular intervals for any change in colour. If the milk contains large numbers of organisms which are metabolically active at 37 °C the redox potential of the milk falls rapidly, and the dye will be quickly reduced to the colourless *leuco* form; conversely, decolorization of the dye will be delayed if the milk contains relatively few organisms. The time taken for dye decolorization is thus a *guide* to microbiological quality. *In some countries methylene blue thiocyanate is used. (see also RESAZURIN TEST)

Methylomonadaceae A family of gram-negative, obligately aerobic bacteria which use only certain 'one-carbon compounds'* as the sole source of carbon. The organisms can be cultured on mineral salts agar under a methane-air mixture. The family contains the genera *Methylococcus* and *Methylomonas*.

Methylomonas methanica (formerly *Pseudomonas methanica*). The cells are polarly-flagellated bacilli, about 0.5 × 1–3 microns. Energy and carbon requirements are satisfied only by methane or methanol. *M. methanica* is catalase-positive, oxidase-positive.

Methylomonas methanitrificans (formerly *Pseudomonas methanitrificans*). The cells are motile, pigmented bacilli, about 1 × 2 microns. The organism obtains carbon and energy only from methane, and is capable of nitrogen fixation.

*Generally taken to mean compounds containing no carbon-carbon bonds—e.g. methane, methanol, dimethyl ether. (see also METHANE (microbial utilization of))

methylotroph An organism which can utilize, as the sole source of carbon, certain 'one-carbon compounds'—i.e. compounds which contain no carbon-carbon bonds, e.g. METHANE, methanol, dimethyl ether.

metiram An agricultural antifungal agent; it is a complex of ZINEB and *polyethylene* THIURAM DISULPHIDE. It is used for the control of *late blight* of POTATO.

metronidazole 1-(β-Hydroxyethyl)-2-methyl-5-nitroimidazole. A drug used e.g. for the treatment of *Trichomonas vaginalis* infections and of bacterial infections involving *obligate anaerobes* (e.g. *Bacteroides fragilis*). Metronidazole is also active against *Giardia lamblia* and *Entamoeba histolytica*. Its mode of action is unknown.

metula (primary sterigma) In the conidiophores of certain fungi (e.g. certain *Aspergillus* spp): a short, hypha-like branch from the distal end of which arises one or more phialids.

MHD (1) (*immunol.*) *Minimum haemolytic dose*. (a) The smallest quantity of COMPLEMENT needed to lyse completely a fixed quantity of a standardized suspension of *sensitized* erythrocytes. (b) The smallest quantity of complement-dependent haemolysin needed to lyse completely a fixed quantity of erythrocytes in the presence of excess complement. In (a) and (b) above the MHD may be referred to as *1 unit* of complement or haemolysin respectively. (see also HD_{50}) (2) (*virol.*) *Minimum haemagglutinating dose.* The smallest amount of haemagglutinating virus capable of bringing about maximum agglutination of the erythrocytes in a given volume of a standardized suspension.

MIC (of antibiotics) Minimum inhibitory concentration; the lowest concentration at which a given antibiotic is effective against a given microorganism. The MIC is commonly quoted in micrograms/ml.

microaerophilic (*adjective*) (1) (of an environment) Refers to an environment in which the partial pressure of oxygen is lower than that which occurs under normal atmospheric conditions—but which is not quite an ANAEROBIC environment. (2) (of organisms) Refers to organisms which exhibit a maximum rate of growth in microaerophilic environments.

Microbacterium A genus* of gram-positive, aerobic or facultatively anaerobic, non-motile, asporogenous, pleomorphic rod-shaped bacteria of uncertain taxonomic affinity; species occur e.g. in dairy products. *M. lacticum* and *M. liquefaciens* grow optimally at about 30 °C but survive higher temperatures; *M. lacticum* is said to survive 72 °C/15 minutes and 80–85 °C/10 minutes. All species form L(+)-lactic acid. *Not universally accepted as a valid genus; *Microbacterium* spp have GC% values in the range 36–70.

microbiology A discipline which is concerned with the study of MICROORGANISMS and their interactions with other organisms and with the environment. There is considerable overlap between microbiology and certain other disciplines—e.g. biochemistry, biophysics, genetics, and immunology; microbiology is also directly relevant to certain aspects of e.g. medicine, veterinary science, agricultural science, forestry, the food industry, many manufacturing processes (e.g. brewing, antibiotic production), ecology and pollution studies.

Microbispora A genus of bacteria of the MICROMONOSPORACEAE.

microbodies Particles which occur within the cells of many types of eucaryotic organism; a microbody consists of a collection of functionally-related enzymes contained within a membranous envelope. Microbodies are usually less that 1.5 microns in diameter. (see PEROXISOMES and GLYOXYSOMES)

microcapsule (bacterial) (see CAPSULE)

Micrococcaceae A family of gram-positive, facultatively anaerobic (or strictly aerobic), chemoorganotrophic, motile or non-motile, asporogenous, coccal bacteria; species occur as saprophytes and as parasites and pathogens of animals, including man. All strains produce CATALASE. The genera: MICROCOCCUS, PLANOCOCCUS and STAPHYLOCOCCUS.

Micrococcus A genus of gram-positive, catalase-positive, strictly aerobic cocci of the MICROCOCCACEAE; species occur as saprophytes and as parasites of animals, including man. The cells are generally non-motile, commonly about 1–2 microns in diameter—but larger and smaller strains occur. Metabolism is respiratory (oxidative). All strains tolerate at least 5% salt (NaCl). Type species: *M. luteus*; GC% range for the genus: 66–75.

M. luteus (synonyms: e.g. *M. lysodeikticus*; *Sarcina pelagia*). Cocci occuring singly or in pairs, tetrads, packets or clusters; a yellow pigment is commonly formed. The majority of strains are sensitive to LYSOZYME.

M. denitrificans: see PARACOCCUS.

M. roseus (synonyms: e.g. *M. tetragenus*; *M. roseofulvus*). Red or pink-pigmented cocci occuring in tetrads, packets, clusters etc.; flagellated strains reported.

M. radiodurans. A red-pigmented, radiation-resistant coccus which possesses a cell wall atypical of the genus *Micrococcus*.

microconidium (*mycol.*) In *ascomycetes*, a minute cell or spore capable of acting either as a male gamete (see SPERMATIUM, SPERMATIZATION) or as a propagule—in the latter case germinating to form a vegetative thallus. Some authors use the terms microconidium and spermatium interchangeably.

microcycle sporogenesis (*bacteriol.*) The occurrence of sporulation immediately following outgrowth—see SPORES (bacterial).

microcyclic rusts Those rust fungi—see RUSTS (1)—which form only one type of *dikaryotic* spore (the teliospore) (cf. MACROCYCLIC RUSTS).

microcyst (1) A type of resting cell formed by some species of the MYXOBACTERALES. (2) Coccoid forms, with increased heat resistance, which develop from the vegetative cells of certain members of the Nocardiaceae.

Microcystis A genus of unicellular BLUE-GREEN ALGAE; species occur in freshwater environments and are often present in water BLOOMS. (see also GAS VACUOLE) The cells may occur in an aggregate within a gelatinous matrix.

microflora The totality of microorganisms associated with a given environment or location.

microforge An apparatus used for the preparation of glass microtools (see MICROMANIPULATION). The microforge is essentially a means of controlling the movement of a thin glass rod or tube relative to an electrically-heated filament—the whole being observed under the low power of a microscope.

microfungi A term used to refer to those fungi whose vegetative thalli or fructifications are microscopic or are too small to be immediately obvious to the unaided eye.

microgamete (a) A small GAMETE. (b) A *male* gamete.

micromanipulation (micrurgy) (*microbiol.*) Any procedure involving the manipulation of individual cells (or spores etc.) or their components. Examples include: the isolation and culture of single cells for genetic studies; dissection of the ciliate infraciliature in cytophysiological studies; transplantation of nuclei, mitochondria etc.; studies on bacterial conjugation by the isolation, culture and examination of individual ex-conjugants. Dissection and transplantation procedures are sometimes referred to as *microsurgery*.

Micromanipulation may be carried out with any conventional microscope that is fitted with a good-quality mechanical stage and a high-power dry (i.e. non-immersion) objective; *phase contrast* working is preferable for certain types of operation. A vibration-free workplace is essential. The material under study may be arranged e.g. on the surface of an agar block or as a hanging-drop preparation; the preparation is commonly placed on the microscope stage in a partially-enclosed moist chamber. (The use of a heated objective lens mounting prevents condensation of water vapour on the lower lens.) The actual manipulation is done with *microtools*—which are fashioned in a MICROFORGE; microtools are usually made of glass. They include: (a) *Needles*. These are fine glass filaments which, in the region of the point, may have diameters of the order of 0.2 microns. (b) *Loops* of various sizes. (c) *Pipettes*, which may be less than 1 micron in diameter, are made by drawing out glass tubing which has been heated to plasticity. Such pipettes have been used e.g. for holding cells during microsurgery: the (heat-polished) tip of the pipette is brought into contact with the cell and a gentle suction applied. Each microtool is fixed in a metal holder which is clamped to a *micropositioner*; essentially, this is a means whereby the coarse movements of the hand are scaled down to give the fine movements necessary for the control of the microtools. Various elaborate forms of micropositioner are available; by such means, microtools can make controlled movements, with speed and precision, along all three axes in space. In a simpler method (used for uncomplicated operations) the microtool holder is clamped to the nosepiece of a second microscope; movement of the fine focusing control of this microscope moves the microtool in a vertical plane, while movements of the microtool relative to the specimen in a horizontal plane can be made by moving the *specimen* with the mechanical stage of the first microscope.

micrometre (μm; micron) 10^{-6} metre—i.e. one thousandth of a millimetre.

Micromonospora A genus of gram-positive bacteria of the MICROMONOSPORACEAE; the organisms are mainly aerobic, mesophilic, soil saprophytes. Aerial mycelium is not formed, and spores are produced singly on the substrate mycelium. Many species metabolize cellulose, protein and starch, and most form pigments. *M. purpurea* is a source of GENTAMICIN.

Micromonosporaceae A family of gram-positive or gram-variable bacteria of the ACTINOMYCETALES; the organisms occur in soil, composts, mouldy hay etc. Most species are aerobic and mesophilic; some are thermophilic. All species form non-fragmenting mycelium; *Micromonospora* spp form substrate mycelium only, while all other species form both substrate and aerial mycelium. Spores are formed in ones, twos, or in short chains—either on the hyphal surfaces or at the tips of hyphal branches (*sporophores*); spores are formed either from the aerial or substrate mycelium, or from both, according to species. The cell walls of all species contain *meso*-diaminopimelic acid (cf. STREPTOMYCETACEAE). The genera include ACTINOBIFIDA, *Microbispora*, MICROMONOSPORA and MICROPOLYSPORA.

micron (μ; micrometre, μm) 10^{-6} metre—i.e. one thousandth of a millimetre.

micronucleus (*ciliate protozool.*). One of two types of nucleus (see also MACRONUCLEUS) found in the majority of ciliates; a cell may contain one, several or many micronuclei—according to species. (Micronuclei are absent e.g. in *Stephanopogon* and in some (viable) strains of ciliate species which are normally micronucleate.) Micronuclei are generally less than five microns in size, and are commonly *diploid*; they are involved in genetic recombination and in the regeneration of macronuclei during sexual reproductive processes. They do not contain nucleoli.

microorganisms The term *microorganism* commonly includes the following types of organism: ALGAE, BACTERIA, BLUE-GREEN ALGAE, FUNGI, LICHENS, PROTOZOA, VIROIDS and VIRUSES. (see also PROCARYOTIC AND EUCARYOTIC ORGANISMS)

microorganisms, exploitation of (by man) The use of microorganisms by man may be considered under three main headings: (i) *Microbial products*. These include e.g. AGAR,

ANTIBIOTICS, certain organic acids (e.g. citric acid—produced by *Aspergillus niger*), and litmus (obtained from lichens of the genus *Roccella*). Certain fungal *enzymes* are used in a range of manufacturing processes; such enzymes include proteases (used e.g. for dehairing hides in the manufacture of leather, and for tenderizing meat) and pectinases (used e.g. for wine clarification and in coffee bean fermentation). A fungal enzyme, glucose oxidase, is incorporated in diagnostic test papers used for detecting the presence of glucose in urine. In the past, certain microbial fermentations have been used for the large-scale production of acetone and ethanol—see ACETONE-BUTANOL FERMENTATION and ALCOHOLIC FERMENTATION. (see also BREWING, CIDER, WINE-MAKING, VINEGAR) Ustilagic acid (produced by the smut fungus *Ustilago zeae*) is used as a raw material in the manufacture of certain perfumes. (ii) *Microbial activity*. Microorganisms are involved in e.g. BIOASSAY, BIOLOGICAL CONTROL, flax RETTING, SILAGE production, SEWAGE TREATMENT, and in the manufacture of BUTTER, CHEESE, SAUERKRAUT, YOGHURT. (iii) *Microorganisms*. Certain microorganisms are used as indicator organisms e.g. in oil prospecting (see FORAMINIFERIDA and ALIPHATIC HYDROCARBONS) and in water pollution studies (see e.g. SAPROBITY SYSTEM). (see also MICROORGANISMS AS FOOD and DIATOMACEOUS EARTH)

microorganisms, preservation of A population of viable microorganisms may be preserved by one or other of the following methods: DESICCATION, FREEZE-DRYING, FREEZING, and SUBCULTURE (see also STABILATE); some species do not survive certain of these procedures.

microorganisms as food For *man*, traditional sources of food include: (i) The cultivated mushroom (*Agaricus bisporus*) and other fungi—e.g. the chanterelle (*Cantharellus cibarius*), the morel (*Morchella* spp), the truffle (species of the Tuberales), and yeasts (as yeast extract). (ii) Seaweeds, e.g. the red alga *Porphyra* (the 'nori' of Japan), and the brown alga *Laminaria*. (iii) The blue-green alga *Spirulina*—eaten e.g. in parts of Africa. (see also BLACKFELLOWS' BREAD and SHII-TAKE)

Single-cell protein (*SCP*) refers to protein sources which include bacteria, yeasts, filamentous (mycelial) fungi, and microalgae. The potential use of SCP as food (for man and other animals) is being studied in a number of countries; the desirable/essential features of SCP include: (a) An absence of inherent toxicity. (b) An acceptable content of protein. (c) An acceptable content of *essential* amino acids—e.g. leucine, lysine, methionine and valine (for man). Compared with animal protein, microbial protein (in general) contains smaller proportions of these amino acids—particularly the *sulphur*-containing amino acids, e.g. methionine. (d) An acceptably low content of nucleic acids—excess intake of which may be harmful. The problem of reducing levels of nucleic acids is common to all forms of SCP. (e) Palatability. (f) Ease and cheapness of manufacture. Organisms currently being studied include species of the algae *Chlorella* and *Scenedesmus*, the cyanophyte *Spirulina*, yeasts, and species of *Fusarium*. Some algae are being grown autotrophically, while others are being cultured mixotrophically/heterotrophically—acetic acid, glucose etc. being supplied as carbon sources. Weight for weight, algal protein (from *Scenedesmus*) has been reported to provide about 81% of the biological value of egg protein; the inclusion of *Scenedesmus* in human diet appears to have produced no adverse effects in short-term experiments. The culture of algae in sewage disposal ponds (in the USA) is being studied as a source of cheap protein for animal feeds. *Spirulina* is regarded as a potential poultry feed and as a supplement to human diet. The culture of yeasts on hydrocarbon substrates (paraffins) is currently providing a source of animal feed protein. In the preparation of SCP, simple inorganic compounds (e.g. ammonia, nitrates) may be used as nitrogen sources; vitamins are required in some substrates. By using CONTINUOUS-FLOW CULTURE, secondary metabolism (and thus waste of substrate) can be minimized and a uniform product more easily obtained. The ratio protein/RNA—and the protein content—can, to some extent, be controlled by manipulating growth conditions. (see also AMBROSIA FUNGI, BAKING, BREWING, CHEESE, FUNGUS GARDENS, SILAGE, VINEGAR and YOGHURT)

micropalaeontology The study of fossils of microscopic organisms, e.g. foraminiferans, radiolarians.

Micropolyspora A genus of gram-positive bacteria (family MICROMONOSPORACEAE); species form chains of from 1 to 20 spores on both aerial and substrate mycelia. Most species are pigmented. The spores of *M. faeni* (found in large numbers e.g. in mouldy hay) are believed to be important allergens in *farmer's lung*. *M. faeni* and several other species are thermophilic; the type species, *M. brevicatena*, is mesophilic.

micropore (1) (micropyle; ultracytostome) (in coccidia) A small organelle, visible only under the electron microscope, consisting of an invagination of the outer layer of the pellicle coincident with a discontinuity of the inner layer; micropores (one or more may be present) occur in cells formed at various stages of the life cycle—e.g. sporozoites, merozoites, microgamonts. In some species the micropore appears to function as a site for the ingestion of food. (2) (*mycol.*) (see SEPTUM (*mycol.*))

micropyle (of the oocysts of some coccidian species) A thin, polar region in the oocyst wall. (*Micropyle* was formerly used to refer to the MICROPORE.)

microscopy The use of an optical or electronic instrument (*microscope*) for viewing objects or details too small to be visible (or clearly visible) to the naked eye. In microbiology, microscopy may be used e.g. for: (a) The detection and/or identification of living and/or dead microorganisms. (b) Examination of the gross morphology, fine structure, or composition of living or dead cells, parts of cells, colonies, viruses etc. (c) The enumeration of microorganisms—see e.g. COUNTING CHAMBER. (d) MICROMANIPULATION. (see also PHOTOMICROGRAPHY) The two basic types of microscopy are (i) ELECTRON MICROSCOPY, and (ii) light microscopy—the various forms of which include: (a) *Bright-field microscopy*. The commonest form of light microscopy; unless otherwise specified it is the form generally referred to by the term 'microscopy'. For bright-field microscopy the arrangement of the microscope is as indicated in figure 1. Light from the source is concentrated by the condenser and passes through the plane of the object (specimen) to the objective lens. The magnified image formed by the objective is further magnified by the eyepiece prior to reception by the eye.

The object to be studied is normally examined on a SLIDE which rests on the *stage*. Methods of specimen preparation include the SMEAR, HANGING DROP and WET MOUNT. Tissues (suitably fixed and mounted) are generally examined in the form of thin sheets ('sections')—5–10 microns in thickness—which have been cut on a MICROTOME.

For image formation, light transmitted by the specimen must differ from the background light in amplitude (intensity) and/or in composition (colour); thus, images are seen against a lighter (bright-field) background. Many

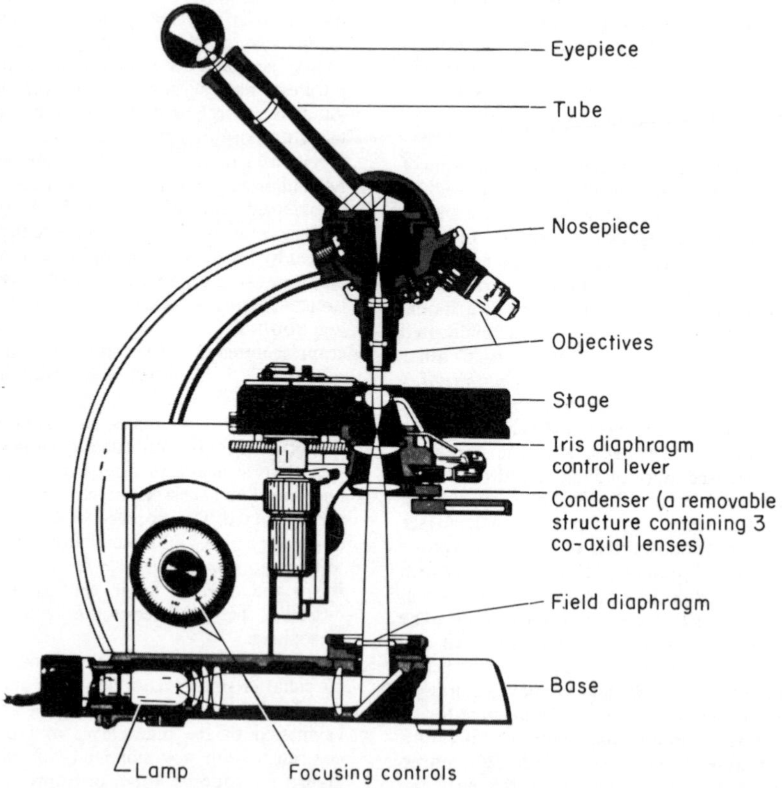

MICROSCOPY. Figure 1. The Zeiss Standard RA microscope. (Courtesy of Carl Zeiss, Oberkochen, W. Germany)

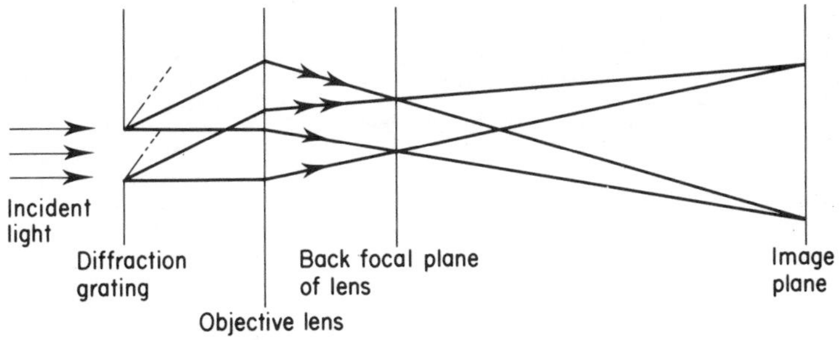

MICROSCOPY. Figure 2. Image-formation in bright-field microscopy (according to the Abbe theory). Zero-order beam: ——► ; first-order beam: ——►► . The terminally-dashed lines represent beams of the second-order. (See text for explanation)

microbiological specimens, e.g. bacteria, are commonly translucent; such specimens can be made light-absorbing by suitable STAINING.

Image formation is often depicted by means of a ray diagram. However, for objects below a certain size image formation can be described satisfactorily only in terms of *wave* optics: light is *diffracted* by the object, and the image is formed by interference between diffracted and non-diffracted light. The formation of an image is shown in figure 2—in which the object is a diffraction grating; biological specimens form images in an analogous way.

From figure 2 it can be seen that, for image formation, the objective lens must collect (at least) the undeviated (zero-order) beam and the first-order diffraction beam. (For a perfect image *all* the diffracted light should be collected—but this is impracticable.) In order to collect the maximum amount of diffracted light it is necessary to use an objective lens of high *numerical aperture*. (see RESOLVING POWER for a discussion of numerical aperture, and for the principle of the *oil-immersion lens*; see also MAGNIFICATION.) The resolving power of an objective lens can be exploited fully only when the object (specimen) is correctly illuminated; illumination should be uniform, and the (divergent) beam from the specimen should fill the *entire* aperture of the objective lens. (see CRITICAL ILLUMINATION and KÖHLER ILLUMINATION) When using dry (i.e. non-immersion) objectives of high numerical aperture, refraction by the cover slip is an important factor in optimal image formation; most objectives are designed for use with cover slips of thickness between 0.17 mm and 0.18 mm. (see also CORRECTION COLLAR)

(b) *Dark-field microscopy*. The dark-field microscope differs from the bright-field instrument only in the design of the sub-stage condenser. The lamp-side of the condenser is fitted with a central opaque disc (the *stop*) which covers much of the area of the condenser lens; thus, the plane of the specimen is illuminated by the apex of a *hollow* cone of rays which, on diverging, are at such an angle that none enters the aperture of the objective lens. Hence, in the absence of a specimen the field of view appears black; only light reflected or scattered by the specimen can enter the objective. Under these conditions the image of a specimen appears as a luminous object against a dark background; such an image is generally poor in detail. The intensity of illumination must be greater in dark-field than in bright-field microscopy since only a small fraction of the incident light enters the objective. Dark-field microscopy is used e.g. for examining very small specimens; thus, e.g. it is possible to see organisms whose dimensions place them beyond the resolving power of the objective under bright-field conditions.

(c) *Phase-contrast microscopy*. A colourless specimen which absorbs a negligible quantity of light (e.g. a non-pigmented, living cell) can be viewed by phase-contrast microscopy provided that it is not optically homogenous with the surrounding medium. Such a specimen produces a first-order diffraction beam which, compared with the zero-order beam, is retarded in phase by some quarter of a wavelength ($\frac{1}{4}\lambda$; 0.5π radians) and is of smaller amplitude (figure 3a); these two beams interfere to produce an image-forming wave of amplitude similar to that of light in the area surrounding the specimen (figure 3b). The light waves which form the image differ slightly, in phase, from those in the areas surrounding the image; however, the eye cannot detect differences in phase, and, since the waves are of similar amplitude (intensity), no image can be seen. In phase-contrast microscopy a slight retardment in phase ($\frac{1}{4}\lambda$) is effected in all the undeviated

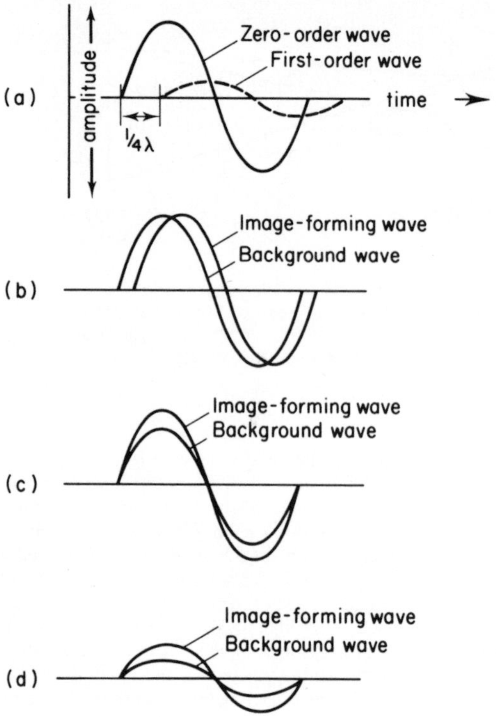

MICROSCOPY. Figure 3. Amplitude and phase relationships in waves of diffracted and non-diffracted light. Only the zero-order and first-order waves are considered. (a) Compared with the zero-order wave, the first-order wave is retarded by $\frac{1}{4}$-wavelength and is of smaller amplitude. (b) The image-forming wave results from additive interference between the zero-order wave and the first-order diffraction wave. The image-forming and background waves are of similar amplitude (intensity) so that no image is seen. (c) The zero-order wave has been retarded by $\frac{1}{4}$-wavelength; it interferes constructively (additively) with the first-order wave to produce an image-forming wave of amplitude greater than that of the background wave. A visible image is thus formed. (d) Contrast between the image and background waves has been increased. (See text)

(undiffracted) light—i.e. in the light waves of the zero-order beam and those of beams in areas surrounding the image (figure 3c); thus, the amplitudes (intensities) of the image-forming waves are greater than those of waves in the area surrounding the image—since constructive (additive) interference occurs between the zero and first-order waves (figure 3c). In this way a visible image is produced. The difference in intensities (contrast) between image-forming and background waves can be increased by reducing the amplitude of the background and zero-order beams (figure 3d).

The condenser of the phase-contrast microscope has an *annular* aperture in its front focal plane (figure 4); this permits a *hollow* cone of rays to be focused (as a small bright ring) on a *phase plate* located in the back focal plane of the objective lens. The phase plate is a glass disc on which has been deposited a ring of material (usually magnesium fluoride) of such thickness that light transmitted *via* the material of the ring is retarded in phase by $\frac{1}{4}\lambda$; in the absence of a specimen all the incident light is focused on, and transmitted *via*, the ring. When a specimen is examined, the zero-order beams—and those of the background light—are transmitted *via* the phase ring; the first and higher-order beams of diffracted light are transmitted by the phase plate *via* points not located in the phase ring (see figure 4). In the image plane, interference occurs between the zero and higher-order beams; this results in the production of image-forming waves of amplitudes greater than those of waves in the background light (figure 3c). The situation shown in figure 3d can be achieved by depositing on the ring a thin layer of (light-absorbing) metal.

The scheme described above is *negative phase-contrast* microscopy: the zero-order beam is *retarded* $\frac{1}{4}\lambda$ with respect to the first-order beam; the specimen appears *lighter* than the background. In *positive-phase contrast* microscopy the zero-order beam is *advanced* ($\frac{1}{4}\lambda$) with respect to the first-order beam; the specimen thus appears *darker* than the background—since *destructive* interference occurs between the zero-order and first-order beams.

(d) *Fluorescence microscopy*. In fluorescence microscopy the specimen is stained or con-

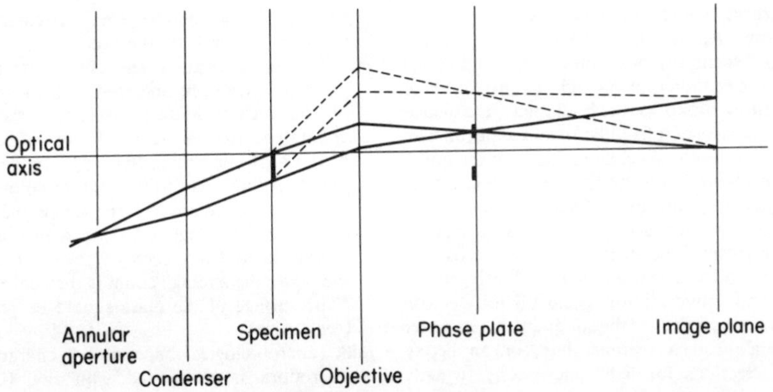

MICROSCOPY. Figure 4. Image-formation in phase-contrast microscopy. Zero-order beam: (————); first-order beam: (- - - - - - - -). Simplified diagrammatic scheme showing spatial relationships between the zero-order and first-order beams. (See text for explanation)

jugated with a suitable FLUOROCHROME ('fluor'); radiation incident upon the specimen is of wavelength (approx.) 300–450 nm ('near' ultraviolet and shortwave visible violet). (Such radiation is of wavelength *longer* than that used for ultraviolet STERILIZATION.) The specimen thus exhibits FLUORESCENCE and is examined by a method similar, in principle, to that used in bright-field microscopy.

In the fluorescence microscope the exciting (ultraviolet/violet) radiation is produced by a mercury arc lamp (or tungsten lamp) and an optical filter. Radiation which passes the filter is concentrated by the condenser and is focused in the plane of the specimen. Light from the *fluorescent* specimen enters the objective lens, and image-formation occurs as in bright-field microscopy; the ultraviolet and violet radiation received by the objective is subsequently absorbed by a BARRIER FILTER. In the absence of a specimen the field of view appears dark (or black). It is clearly necessary that use be made of non-fluorescing lenses, slides, cover slips, immersion oil etc.

In microbiology, fluorescence microscopy is used e.g. to examine/detect objects or details which are difficult or impossible to see by other forms of microscopy. Uses include the FTA-ABS TEST, other IMMUNOFLUORESCENCE techniques, the detection of acid-fast bacteria—see e.g. AURAMINE-RHODAMINE STAIN, and the demonstration of bacterial DNA—see ACRIDINE ORANGE. Fluorescence microscopy is sometimes referred to as 'ultraviolet microscopy'—but see (e).

(e) *Ultraviolet microscopy*. A form of microscopy in which ultraviolet radiation is used for illumination of the specimen *and for image formation*—i.e. a *non*-fluorescent specimen is used (cf. fluorescence microscopy); the image is recorded photographically and/or on a fluorescent screen. Compared with bright-field microscopy, this form of microscopy can resolve finer detail—since the wavelength of the image-forming radiation is shorter than that of visible light (see RESOLVING POWER).

microsource (see BIOLUMINESCENCE)

Microsporida An order of protozoans of the subphylum CNIDOSPORA.

Microsporidea A class of protozoans of the subphylum CNIDOSPORA.

Microsporum A genus of fungi within the order MONILIALES and included within the category DERMATOPHYTES; species occur as soil saprophytes and as parasites and pathogens in the skin, hair, and nails of man and other animals (see RINGWORM). The vegetative form of the organisms is a septate mycelium. In culture, most species produce both microconidia and macroconidia—the latter typically being fusiform and multiseptate with thick, rough-surfaced walls; the conidia are borne singly (i.e. not in chains) on conidiophores which closely resemble somatic hyphae. Macroconidia are commonly 40–120 microns in length, according to species; they are generally produced in abundance—except by *M. audouinii*. A number of species of *Microsporum* have perfect (sexual) stages in the ascomycete genus NANNIZZIA.

microstome (see TETRAHYMENA)

microstrainer (see SEWAGE TREATMENT)

microsurgery (see MICROMANIPULATION)

microtome An instrument used for cutting thin sheets ('sections') of tissues, or of individual cells, for examination by light MICROSCOPY or ELECTRON MICROSCOPY. Essentially, the instrument consists of a fixed *knife* and a mechanism which moves the specimen relative to the knife; after each section is cut the specimen is moved towards the knife by a distance equal to the thickness of the section. To ensure rigidity during cutting, specimens for light microscopy are embedded in paraffin wax (following FIXATION and dehydration); specimens for electron microscopy are embedded in a stronger material—e.g. a methacrylate, or an epoxy resin. Sections for light microscopy (usually 5–10 microns in thickness) are generally cut with a steel knife; the microtome used to prepare sections for electron microscopy (*ultramicrotome*) commonly employs a glass knife, and is more sophisticated in construction. When using the ultramicrotome, sections (maximum thickness about 1000 Å) are cut and selected with the aid of a dissecting microscope.

microtubules Protein tubules—each commonly 20–25 nm in diameter—which occur within all types of eucaryotic microbial cell (and in cells of the metaphyta and metazoa); they occur e.g. in cilia and eucaryotic flagella, in the pellicles and sub-pellicular structures of sporozoans, flagellates, and ciliates, in the axopodia of heliozoans, and in mitotic and meiotic spindles. Microtubules occur singly, in pairs or triplets, or in bundles—according to location; the wall of each tubule is composed of some 12 or 13 adjacent rows of TUBULIN subunits. Microtubules are easily disrupted by extremes of temperature and pH, and by chemical agents which bind tubulin.

micrurgy Working with, or manipulating, minute objects (see MICROMANIPULATION).

MIF (*immunol.*) Macrophage migration inhibitory factor—see LYMPHOKINES.

mildew Any of a variety of certain plant diseases in which mycelium of the parasitic fungus is visible at the surface of the affected plant. *Downy mildews* are caused by members of the Peronosporales; *powdery mildews* are caused by members of the Erysiphaceae. (see also PERISPORIACEAE and PLANT PATHOGENS AND DISEASES)

Miles and Misra's method (drop method) A method for determining the *viable cell count* of a bacterial suspension (see ENUMERATION OF MICROORGANISMS). Serial dilutions (often log dilutions) of the suspension are prepared. One drop (of known volume) from each dilution is placed at a separate, recorded position on the surface of an agar plate*; the drops are allowed to dry. The plate is then incubated until visible colonies develop in the (small, circular) areas formerly occupied by the drops.

Provided that a sufficient number of dilutions have been prepared, a drop from one of the higher dilutions will give rise to a countable number (say, 1–20) of discrete colonies. By assuming that each viable cell in the drop gave rise to a separate colony (and knowing the volume of the drop) the viable count can be calculated. (In practice, a count is calculated from each of several drops on the plate; the average count is then calculated.) *The surface of the plate should be pre-dried (see PLATE).

milk (microbiological aspects) A characteristic microflora is associated with raw (i.e. untreated) milk; organisms typically present include species of *Streptococcus* (e.g. *S. lactis*), *Lactobacillus*, *Leuconostoc*, and various yeasts. Additionally, milk may harbour any of a variety of pathogenic microorganisms. Diseases which may be spread by the consumption of raw milk include BRUCELLOSIS, DYSENTERY, Q FEVER, salmonellosis and TUBERCULOSIS. Raw milk may be tested for microbial content (and hence keeping quality) by the METHYLENE BLUE TEST, the RESAZURIN TEST, or by direct plating and counting techniques. Milk for human consumption is commonly subjected to PASTEURIZATION; after such treatment it may again be examined by the methylene blue test and by the PHOSPHATASE TEST. Milk may be subjected to other tests if the donor animal is suspected of suffering from certain diseases, e.g. bovine mastitis (see CAMP TEST) or brucellosis (see MILK RING TEST). (see also COLOSTRUM)

milk caps A common name for the fruiting bodies of species of the basidiomycete *Lactarius*.

milk ring test (*immunol.*) A method for detecting the presence in milk of antibodies to *Brucella abortus*. To the milk sample (in a small test-tube) is added one drop of a dilute suspension of killed, *stained* cells of *Brucella abortus*; the tube is shaken and then incubated for up to one hour. In a *positive* test a coloured band forms at the surface (colour depending on the stain used); in the presence of specific antibodies the bacteria are adsorbed by the globules of fat which rise to the surface.

milky disease (of *Popillia japonica*) (see BIOLOGICAL CONTROL)

millimicron (mμ; nanometre) 10^{-9} metre.

Mima polymorpha (see ACINETOBACTER)

mineralization The microbial breakdown of organic materials into inorganic materials; mineralization is effected mainly by bacteria and fungi. Mineralization in freshwater environments (rivers, lakes, etc.) may lead to EUTROPHICATION; the mineralizing activities of saprophytes are essential in the cycles of matter: see e.g. NITROGEN CYCLE.

minicell A small bacterial cell which lacks a chromosome; minicells are formed e.g. from parent cells in which the septation process is imperfect.

minimal medium (MM) A type of culture medium used e.g. in experimental genetics. The medium lacks certain growth factors—i.e. it does not support the growth of some or all auxotrophic strains of a given organism, but permits the growth of prototrophic strains. (cf. COMPLETE MEDIUM)

minimum haemagglutinating dose (see MHD)

minimum haemolytic dose (see MHD)

minimum inhibitory concentration (of antibiotics) (see MIC)

minimum lethal dose (MLD) (1) The minimum dose of a given lethal agent sufficient to cause 100% mortality in a population of test animals under given conditions. (2) The minimum dose of a given lethal agent sufficient to kill an individual of a given species of specified body weight.

minus and plus strands of DNA and RNA (see PLUS AND MINUS STRANDS OF DNA AND RNA)

minute colonies (due to abortive transduction) (see TRANSDUCTION)

minutes (in relation to bacterial genetic loci) (see INTERRUPTED MATING)

mirror yeasts Those imperfect yeasts which form ballistospores—e.g. *Sporobolomyces*.

mis-sense mutation A mutation in which a codon specifying one amino acid is altered so as to specify a different amino acid. (cf. NONSENSE MUTATION)

mitochondrion (*syn*: chondriosome) A semi-autonomous organelle—one or more (usually many) of which occur in most *eucaryotic* cells; mitochondria are the sites of respiration and certain other cellular processes. Mitochondria vary greatly in shape, size, and number, according to the nature of the cell in which they occur; the number and size of the mitochondria appear to reflect the level of metabolic activity of the cell. Mitochondria may be spherical, ovoid, elongated, calyciform, or irregularly-shaped, and some are branched; in size they range from less than one micron to more than ten microns.

Structure of the mitochondrion. The mitochondrion comprises two closed membranous sacs, one fitting closely within the other, and an amorphous *matrix* enclosed by the inner membrane. The inner membrane is extensively corrugated and forms plate-like, tubular, or finger-like structures (*cristae*) which project into the matrix—often more or less at right angles to the longitudinal axis of the mitochondrion. The inner surface of the inner membrane (including that of the cristae) bears numerous spherical particles—each of which is attached to the membrane by a short stalk; these structures have been referred to as stalked particles, Fernández-Morán particles, elementary particles, and oxysomes. The spherical particles (each about 85 Å in diameter) contain the mitochondrial coupling factor F1 (= F_1 or mitochondrial 'ATPase') which is involved in ATP *synthesis*. (The dephosphorylation of ATP by F1 is believed to be prevented *in vivo* by an 'inhibitor protein'—which, it has been suggested, may occur in the stalk.) Recent FREEZE-ETCHING studies indicate that, *in vivo*, the F1 spherical particles may be in close contact with the inner membrane; this contrasts with (earlier) evidence, obtained by electronmicroscopy of negatively-stained specimens, in which each particle appears to project, on its stalk, into the matrix. The components of the ELECTRON TRANSPORT CHAIN are located in the inner membrane. The mitochondrial matrix contains e.g. proteins, lipids, and small amounts of nucleic acids, and is the location of most of the components of the tricarboxylic acid cycle.

Origin and semi-autonomy of the mitochondrion. Continuity between the outer mitochondrial membrane and the cell membrane, nuclear membrane, or endoplasmic reticulum has occasionally been observed—suggesting the possibility that mitochondria may be derived from these structures; however, it is generally believed that mitochondria arise by the division of existing mitochondria or of *promitochondria*. Promitochondria are regarded as precursors or degenerate forms of mitochondria in that they are structually-aberrant and are deficient e.g. in cytochromes. In the yeast *Saccharomyces cerevisiae* it is believed that promitochondria, present during anaerobic growth, develop to form fully-functional mitochondria under aerobic conditions in the absence of high concentrations of glucose or sucrose; this implies that reversion occurs under anaerobiosis and/or in the presence of high concentrations of glucose and/or sucrose.

Biochemical and genetic studies indicate that at least some of the genetic information required for the synthesis of mitochondrial components occurs in the mitochondrion itself; that nuclear genes play some part in the synthesis and function of the mitochondrion has been inferred e.g. from the existence of *segregational petite* mutants of certain yeasts (see PETITE COLONIES). Little is currently known of the function of mitochondrial genes—or of the way in which dual control of mitochondrial synthesis and function may be exercised by nuclear and mitochondrial genes;

mitochondrial DNA is typically double-stranded and circular, and differs from nuclear DNA e.g. in base composition and buoyant density. The mitochondrion is equipped with all the components necessary for PROTEIN SYNTHESIS, although mitochondrial and cytoplasmic protein synthesis differ in certain respects—the former resembling more closely that which occurs in bacteria. Thus e.g. (a) mitochondrial ribosomes appear to resemble 70S bacterial ribosomes; (b) in mitochondria *N*-formylmethionine is the first amino acid to be incorporated in the polypeptide chain; and (c) mitochondrial protein synthesis is inhibited by antibiotics which inhibit bacterial protein synthesis—e.g. CHLORAMPHENICOL, ERYTHROMYCIN—to which cytoplasmic protein synthesis is insensitive. (These features lend support to a currently popular hypothesis which supposes that mitochondria have their evolutionary origin in endosymbiotic bacteria.) The mitochondrial DNA and RNA polymerases, the *proteins* of mitochondrial ribosomes, and the elongation factors, are all believed to be coded for by nuclear genes and to be synthesized in the cytoplasm; mitochondrial tRNAs and rRNAs are among components believed to be synthesized in the mitochondrion. At least some of the cytochromes and the enzymes of the tricarboxylic acid cycle appear to be coded for by nuclear genes. (see also POKY MUTANT)

mitogens (*immunol.*) Agents (e.g. phytohaemagglutinins) which promote mitosis in lymphocytes.

mitomycins A group of ANTIBIOTICS produced by *Streptomyces* spp; they are toxic for bacteria and mammalian cells and possess antitumour activity. The principal member of the group is *mitomycin C* ('mitomycin'); the molecule includes a quinone moiety and an aziridine (ethyleneimine) group. Treatment of cells with mitomycin causes cessation of DNA replication—followed by degradation of DNA; the synthesis of proteins and RNA continues for a time. The activity of mitomycin depends on its *in vivo* reduction to the highly reactive hydroquinone derivative; reduction involves an $NADPH_2$-dependent enzyme. *Reduced* mitomycin behaves as a bifunctional ALKYLATING AGENT which forms covalent cross-links between complementary strands of DNA; cross-links may involve the guanine residues of DNA. Mitomycins also *induce* many lysogenic bacteriophages.

mitosis (karyokinesis) A sequence of events which results in the division of the (eucaryotic) nucleus into two genetically-identical nuclei during asexual (vegetative) cell division; mitosis is thus a mechanism which permits each of the two progeny cells to receive an identical genetic complement. (cf. MEIOSIS; see also PARASEXUALITY IN FUNGI) The form of mitosis described below is that which occurs typically in the cells of animals, certain lower plants, and some microorganisms. Chronologically, the events are: (a) *Prophase*. During this phase, individual CHROMOSOMES become visible and later become shorter and thicker—apparently by a process of coiling; each chromosome is seen to consist of two longitudinally adjacent CHROMATIDS which are joined at the CENTROMERE. The nuclear membrane and the NUCLEOLUS become indistinct and, by the end of prophase, cease to be demonstrable. (b) *Metaphase*. The SPINDLE is formed between polar CENTRIOLES (two centrioles per pole), and the chromosomes become arranged in a plane perpendicular to the axis of the spindle and midway between its poles (the 'metaphase plate'). (Mitosis can be artificially arrested at metaphase by the use of COLCHICINE.) (c) *Anaphase*. The chromatids of each chromosome separate and move to opposite poles of the spindle. (d) *Telophase*. A nuclear membrane develops around each of the two groups of chromatids to form two new nuclei—in each of which a new nucleolus appears. Telophase is followed by CYTOKINESIS. The chromatids assume the role of chromosomes in the newly-formed cells. The resting (i.e. non-dividing) nucleus is referred to as the *interphase* nucleus; DNA replication occurs during interphase. (see also CELL CYCLE)

In many microorganisms—including algae, fungi and protozoans—the form of mitosis differs fundamentally from that described above. Thus, e.g. the nuclear membrane commonly persists throughout mitosis—two new nuclei being formed by the constriction of an elongated parental nucleus. Usually the spindle is wholly or partly intranuclear, but in some organisms (e.g. *Mucor*) no spindle is formed. According to species, connections between chromosomes and the microtubules of the spindle may or may not be apparent. In many microorganisms centrioles are not present at the poles of the spindle—even when the latter is partly extranuclear; in the yeast *Saccharomyces* (spindle wholly *intra*nuclear) the microtubules of the spindle terminate in localized thickenings of the nuclear membrane—the so-called *centriolar plaques*. In the hypermastigid flagellates (e.g. *Spirotrichonympha*) mitosis involves a persistent nuclear membrane and an extranuclear spindle; the poles of the spindle are occupied by organelles which act, simultaneously, as centrioles and BASAL BODIES. (see also AMITOSIS)

mitosporangium (*mycol.*) (see SPORANGIUM)

mitotic crossing-over (see PARASEXUALITY IN FUNGI)

Mitsuda reaction (see LEPROMIN TEST)

mixed acid fermentation A type of fermentation which is carried out by many members of the Enterobacteriaceae—including *Escherichia coli*, species of *Proteus*, *Salmonella*, *Shigella*, *Yersinia*, and some species of *Erwinia*. The fermentation pathway is given in Appendix III(b); acetic, lactic, succinic, and formic acids are formed in relative proportions which vary with the organism and growth conditions. Those organisms which possess a FORMATE HYDROGEN LYASE enzyme system may decompose formate to form carbon dioxide and hydrogen gas; *E. coli* forms CO_2 and H_2 under acid conditions, but accumulates formate under alkaline conditions. The ratio CO_2:H_2 observed during the fermentation of glucose may be less than 1 owing to the assimilation of some of the CO_2 (e.g. during succinate synthesis) and to the solubility of CO_2 in the medium. Anaerogenic organisms (e.g. *Salmonella typhi*, *Shigella* spp) lack the formate hydrogen lyase system; species which are normally aerogenic may give rise to anaerogenic strains e.g. by mutational loss of a functional formate hydrogen lyase. Organisms which carry out the mixed acid fermentation give a positive METHYL RED TEST. (see also BUTANEDIOL FERMENTATION)

mixed culture (see CULTURE)

mixed vaccine A single vaccine designed to give protection against more than one pathogen; such a vaccine contains protective antigens from each of the organisms concerned. (cf. POLYVALENT VACCINE)

mixoploid Refers to a population of cells in which the chromosome complement per cell varies among the cells of the population.

mixotrophy The simultaneous use, by an organism, of autotrophic and heterotrophic metabolic processes—e.g. the concomitant use of organic compounds as sources of *carbon* and inorganic compounds as sources of *energy*.

Miyagawanella (see CHLAMYDIA)

MK Menaquinone—see QUINONES (in electron transport).

MLD Minimum lethal dose (*q.v.*).

MMTV (see B-TYPE PARTICLES)

Mobilina A sub-order of protozoans of the subclass PERITRICHIA.

modification (of DNA) In bacteria: newly-synthesized DNA is modified by specific enzyme(s) in a manner characteristic of the particular bacterial strain. (see also HOST-CONTROLLED MODIFICATION) Modification generally involves a specific pattern of *methylation* of nucleotide residues in the DNA; the methylating enzymes ('methylases') act on duplex DNA in which one or both strands are unmodified, and appear to require *S*-adenosylmethionine as methyl donor. The sites at which modification occurs are also recognized by corresponding RESTRICTION endonucleases; these endonucleases cannot act on DNA in which these sites have been modified in one or both strands. This suggests that modification serves to protect DNA from the cell's own restriction endonucleases. Since modification does not interfere with normal base-pairing, semi-conservative replication of modified duplex DNA will result in two duplexes—each containing one (modified) parental strand and one unmodified daughter strand. The daughter strand is modified before further replication—the modified parental strand preventing restriction.

The DNA of the T-even BACTERIOPHAGES undergoes a specific pattern of *glucosylation* (presumably mediated by phage-coded enzymes)—some or all of the 5′-hydroxymethylcytosine residues being glucosylated; this may serve to protect the phage DNA from the action of phage-coded nucleases.

Møller's KCN medium (see CYANIDE TEST)

Mollicutes (the mycoplasmas; PPLOs) A class of minute procaryotic organisms which do not form cell walls; some regard the organisms as bacteria while others regard them as procaryotes unrelated to bacteria. Species are saprophytes, parasites, or pathogens of animals; certain plant diseases are caused by organisms resembling mycoplasmas. The organisms can be grown on cell-free media; parasitic growth usually occurs extracellularly. The mode of reproduction has not been established, but budding, fragmentation and/or other processes may be involved. All species are placed in the order Mycoplasmatales—which is divided into the MYCOPLASMATACEAE and ACHOLEPLASMATACEAE.

molybdoferredoxin (see NITROGENASE)

monactin (see MACROTETRALIDES)

-monas A suffix signifying 'a unit'.

Monascus A genus of fungi of the order EUROTIALES.

monensin (see MACROTETRALIDES)

monera The procaryotic protists; see PROTISTA. (see also PROCARYOTIC ORGANISMS AND EUCARYOTIC ORGANISMS)

Monilia A synonym of CANDIDA.

Moniliales (Hyphales) A large order of fungi whose characteristics are those of the class HYPHOMYCETES. Constituent genera include ALTERNARIA, ASPERGILLUS, BEAUVERIA, BOTRYTIS, CEPHALOSPORIUM, CERCOSPORA, CLADOSPORIUM, EPIDERMOPHYTON, FUSARIUM, HELMINTHOSPORIUM, MICROSPORUM, PENICILLIUM, PYRICULARIA, STACHYBOTRYS, *Thielaviopsis*, TRICHOPHYTON, VERTICILLIUM.

moniliform (*adjective*) Resembling a string of beads.

Monilinia A genus of fungi of the order HELOTIALES.

Monoblepharidales An order of FUNGI within the class CHYTRIDIOMYCETES; the vegetative form of the organisms is a well-developed, branched mycelium, and sexual reproduction is *oogamous* (cf. CHYTRIDIALES and BLASTOCLADIALES). Most (or all) species are aquatic saprophytes. Organisms within this order are characterized by hyphae which contain a highly vacuolated ('foamy') protoplasm; in at least some species the hyphal wall contains microfibrils of chitin. Asexual reproduction involves the formation of elongate, typically cylindrical sporangia in which develop large numbers of zoospores; on germination, each zoospore produces a germ tube and, subsequently, a new vegetative thallus. Sexual reproduction occurs by a process which is unique among the fungi: motile male gametes are formed in, and liberated from, an antheridium which is formed on an expanded, terminal portion of the thallus; the expanded region (oogonium) contains a single non-motile female gamete (the oosphere). (In some species the relative positions of the antheridia and oogonia are reversed, and the oogonium may form more than one gamete.) In most species fertilization occurs while the female gamete is still within the oogonium. After fertilization the oosphere develops into a thick-walled resting structure, the oospore. The oospore germinates to form a germ tube and, subsequently, a new vegetative thallus. The genera are *Monoblepharis*, *Monoblepharella*, and *Gonapodya*.

monocentric (*mycol.*) Refers to a type of thallus which gives rise to a single reproductive centre.

Monocystis A protozoan genus of the EUGREGARINIDA; many species are parasitic in the seminal vesicles of the earthworm (*Lumbricus*). *Life cycle*. Sporozoites (see later) enter the seminal vesicles (*via* the gut?), pass into the sperm morulae, and grow (*trophozoite* stage). The fully-grown *acephaline* trophozoites (*gametocytes*)—which are free within the vesicles—associate in pairs (syzygy) and encyst (*gametocyst*). Within the cyst each gametocyte divides into gametes; later, these fuse in pairs and the zygotes encyst (*sporocysts*, *pseudonavicellae*). Each zygote divides to form eight sporozoites. The ways in which sporocysts are released from the gametocyst, and in which infection occurs are apparently not known.

monocytes Amoeboid, phagocytic leucocytes (white blood cells)—approximately 10–20 microns in diameter—which are present in normal human blood at a concentration of 200–800/mm^3. Each monocyte contains a single, large, spherical or indented nucleus; the cytoplasm is commonly agranular. Monocytes are derived from the bone marrow. (see also MACROPHAGES)

monoecious fungi Fungi in which the male and female reproductive structures occur in the same individual.

monogenic resistance (*plant pathol.*) See: RESISTANCE (of plants to disease).

monolayer culture (see TISSUE CULTURE)

monomitic (see HYPHA)

monomorphic trypanosomes Those trypanosomes (e.g. *T. equinum*) which exhibit more or less constant morphology—cf. POLYMORPHIC TRYPANOSOMES.

mononuclear phagocytes (see MACROPHAGES)

mononucleosis The presence of an increased number of mononuclear leucocytes in the blood.

monophasic serotypes, of *Salmonella* (see PHASE VARIATION, IN SALMONELLA)

monoplanetic (see DIPLANETISM)

monopodial Refers to the form or mode of development of a branching structure (e.g. a conidiophore): formed or developing like a tree—i.e. with a main, central trunk or stem from which develop lateral branches. (cf. SYMPODIAL)

monosome (see PROTEIN SYNTHESIS)

monosomic (see CHROMOSOME ABERRATION)

monospecific antiserum An antiserum capable of reacting *specifically* with a *particular* antigen or antigenic determinant. A GROUPING ANTISERUM, containing antibodies against a single major group antigen, is one example of a monospecific antiserum. (see also POLYVALENT ANTISERUM)

monostromatic (*algol.*) Refers to a parenchymatous thallus which is one cell thick (cf. POLYSTROMATIC).

monothetic classification Any form of classification in which a small number of carefully selected criteria are used as the basis on which organisms are arranged into groups. (see also TAXONOMY) Since the choice of criteria may include an element of subjectivity, a given organism may be classified differently by different taxonomists. Monothetic classifications include the various forms of *dichotomous key* used as guides for the classification of unidentified organisms. (cf. POLYTHETIC CLASSIFICATION and see also NUMERICAL TAXONOMY)

monotrichous Refers to the possession of *one* FLAGELLUM by each cell or zoospore (cf. AMPHITRICHOUS and PERITRICHOUS).

monoxenous (*adjective*) Refers to a parasite which completes its life cycle on or in a single host species. (see also AUTOECIOUS)

Monsur agar A medium for the primary isolation of *Vibrio cholerae*. The medium contains pancreatic digest of casein (1%), sodium chloride (1%), sodium taurocholate (0.5%), sodium carbonate (0.1%), gelatin (3%), agar (1.5%), and potassium tellurite (final concentration 1 in 200,000).

Monsur transport medium A transport medium used for samples suspected of containing *Vibrio cholerae*. The base consists of pancreatic digest of casein (1%), sodium chloride (1%), sodium taurocholate (0.5%), and sodium carbonate (0.1%); potassium tellurite (final concentration 1 in 100,000) is added before inoculation.

Morax-Axenfeld bacillus *Moraxella lacunata*.

Moraxella A genus of gram-negative, aerobic, chemoorganotrophic, asporogenous, aflagellate bacilli or coccobacilli of the family NEISSERIACEAE; species are parasitic on the mucous surfaces of warm-blooded animals, including man. The cells, which are 1.5–3.0 × 1–1.5 microns, occur in pairs or short chains. The organisms do not utilize carbohydrates; they are oxidase-positive and usually catalase-positive. All species are sensitive to penicillin. *M. lacunata* (the type species) may cause conjunctival inflammation in man.

morbidity (*med.*) The proportion of a population which has contracted a given disease at a given time, or during a specified period of time.

morbilli Synonym of MEASLES.

Morchella A genus of fungi of the order PEZIZALES; the fruiting bodies of *Morchella* spp ('morels') are among the most highly-prized of the edible fungi. The fruiting body (an APOTHECIUM) consists of a hollow, sponge-like, ovoid or elongated pileus which is continuous with a stout, hollow stipe—the entire fruiting body being some 5–12 centimetres high, and some 5 centimetres at the broadest part; the external surface of the pileus is conspicuously ridged and pitted—the hymenium-lined pits being separated by the sterile ridges. The pileus is yellowish or tan and is generally described as waxy and brittle. The cylindrical, operculate asci each contain eight multinucleate ascospores; the large, hyaline, oval ascospores are generally eguttulate—though in fresh specimens a group of small guttules may be seen at each pole of the spore. The fruiting bodies of *Morchella* species develop during the spring.

morels The edible fruiting bodies of species of the ascomycete genus *Morchella*.

Moro test (see TUBERCULIN TEST)

morphological heterothallism (see DIMIXIS)

morphological unit (*virol.*) (see CAPSOMERE)

Mortierella A genus of fungi of the ZYGOMYCETES.

mosaic (*plant pathol.*) A form of leaf variegation found in certain systemic VIRUS DISEASES OF PLANTS. Two or more colours, or shades of colour, develop—the boundaries between regions of different colour being relatively sharp and clear (cf. MOTTLE). (Leaf patterns which mimic virus-induced mosaic may be due e.g. to inherited abnormalities or to toxins secreted by leaf-feeding arthropods.) Leaves which become infected subsequent to a certain stage in their development fail to exhibit mosaic.

moss starch (see LICHENAN)

motile (of microorganisms). Having the capacity for independent locomotion. (see also MOTILITY)

motile colonies (*bacteriol.*) The *colonies* of certain bacteria—in particular those of *Bacillus circulans*—have the ability to move across the surface of the culture plate; the course of such movement is often marked by lines of bacterial growth which arise from cells which have been left behind by the migrating colony.

motility Motile organisms (i.e. those which are capable of independent locomotion) and organisms with motile stages occur among the algae, bacteria, fungi, and protozoans; motility enables an organism e.g. to respond to certain environmental stimuli (see e.g. TAXIS). Locomotion may be achieved in any of a variety of ways—according to species: (a) *Amoeboid movement* occurs e.g. in the sarcodine protozoans—*Amoeba*, *Arcella*, *Difflugia* etc. In *Amoeba* the cell body initially becomes extruded in the direction of motion; cytoplasm then flows forward into the developing extrusion (the *pseudopodium*) which expands to accommodate the remainder of the cell's contents. Such motion is believed to involve reversible sol-gel interconversions. Among the sarcodines pseudopodial form and function vary widely; forms include the broad, blunt *lobopodium* (of e.g. *Amoeba*), the slender (branched or unbranched) *filopodium* (of e.g. *Euglypha*), and the network of thin, branching filaments, the *reticulopodium* (of e.g. *Nodosaria* and other foraminiferans). Amoeboid movement is also exhibited by the myxamoebae of the cellular slime moulds. Amoeboid movement is slow compared with the movements of e.g. ciliates or flagellates; its mechanism is not fully understood. In some organisms (e.g. *Histomonas*, *Naegleria*) amoeboid and flagellar movement may occur in the same individual. (b) *Gregarine movement*. This slow gliding motion is exhibited by many members of the Sporozoa; these protozoans lack obvious locomotive organelles. One hypothesis supposes that such movement results from wave-like contractions of the pellicle. (c) *Euglenoid movement*. (see EUGLENA) (d) A *gliding movement* (mechanism unknown) is exhibited by the GLIDING BACTERIA—e.g. *Beggiatoa*, *Leucothrix*; such movements occur when the organisms are in contact with a solid surface. A similar type of movement is found in certain blue-green algae—e.g. *Oscillatoria.* (e) The characteristic movements of the *spirochaetes* include bending, coiling, and rotation about the longitudinal axis—a corkscrew-like movement; these movements are believed to involve the axial fibrils of the organisms. (f) The motion of *diatoms* is accompanied by cytoplasmic streaming and by the production of a mucilaginous secretion; contact between the cell and the (solid) substratum appears to be essential for motility, and the (unknown) locomotive forces are believed to be transmitted *via* the raphe system of the cell. (g) In members of the Ciliophora and the Opalinata locomotion is effected by coordinated beating of cilia (see CILIUM (1); see also METACHRONAL WAVES). Individual cilia often beat fifteen or more times per second, and the locomotion of ciliates may be quite rapid. (h) *Flagellum-mediated locomotion* occurs among protozoans (superclass Mastigophora), bacteria (many taxa), fungi (limited to certain lower fungi—e.g. the zoospores of chytrids), and in many groups of algae. Flagellar movements (see FLAGELLUM) generate forces which act against the ambient fluid medium; the reaction to these forces gives rise to cellular motility.

Direct (microscopical) determinations of cell motility are commonly achieved without difficulty; however, with some organisms—particularly bacteria—BROWNIAN MOVEMENT may be mistaken for true motility. (see also HANGING DROP) Motility in certain bacteria (e.g. *Proteus* spp) may be determined macroscopically: see SWARMING. Motile bacteria may be separated from non-motile species e.g. by CRAIGIE'S TUBE METHOD.

mottle (*plant pathol.*) A form of leaf variegation found in certain systemic VIRUS DISEASES OF PLANTS. Two or more colours, or shades of colour, develop—the boundaries between regions of different colour being indistinct. Leaves may exhibit symptoms intermediate between mottling and MOSAIC.

mould (American: mold) A general term used to refer to any fungus which normally forms visible *mycelial* growth. (see MYCELIUM)

mountant (mounting medium) (*microbiol.*) Any substance used to impregnate and/or embed (or immerse) a specimen or smear prior to examination under the microscope. For temporary preparations, e.g. a WET MOUNT, saline (or glycerol etc.) is often used as the mountant. For permanent preparations the mountants most commonly used include CANADA BALSAM and DPX; tissue sections must be dehydrated and cleared (see CLEARING) before being mounted in these substances. For fungal material LACTOPHENOL COTTON BLUE is a useful mountant (and stain) for temporary or permanent preparations.

mouse mammary tumour virus (see B-TYPE PARTICLES)

mouth, common microflora of The microflora commonly present in the human mouth includes *alpha-* and non-haemolytic species of *Streptococcus* (e.g. *S. salivarius*, *S. sanguis*), fusiform bacilli, *Veillonella* spp, non-pathogenic species of *Staphylococcus*, and yeasts (e.g. *Candida*). Microaerophilic and anaerobic crevices and microenvironments (e.g. regions between the teeth, cavities in teeth etc.) support a characteristic flora which includes spirochaetes such as non-pathogenic *Treponema* spp and *Borrelia* spp. (see also DENTAL CARIES)

MPN (most probable number) (see MULTIPLE-TUBE METHOD)

MR test (see METHYL RED TEST)

mRNA Messenger-RNA—see RNA and PROTEIN SYNTHESIS.

MS-1 (see STRINGENT CONTROL)

mucigel A mucilaginous layer on the rhizoplane of certain plants; it frequently contains large populations of soil bacteria.

mucocyst (*protozool.*) (see PELLICLE)

mucopeptide A term sometimes used synonymously with PEPTIDOGLYCAN.

mucopolysaccharide A polysaccharide containing a high proportion of *hexosamine* residues, e.g. HYALURONIC ACID.

mucoprotein A complex which contains both protein and MUCOPOLYSACCHARIDE components.

Mucor A genus of fungi within the class ZYGOMYCETES. *Mucor* spp occur typically as saprophytes on soil and dung—though some species can behave as pathogens in man and other animals; the vegetative form of the organisms is a branched coenocytic mycelium. Asexual reproduction involves the formation of spherical, columellate (see COLUMELLA), thin-walled, dehiscent sporangia on branched or unbranched sporangiophores; sporangia, which contain numerous aplanospores, develop terminally on the sporangiophores. Sexual reproduction involves the fusion of gametangia with the subsequent formation of ZYGOSPORES; most species are reported to be heterothallic.

multicellular (see CELLULAR)

multicistronic mRNA (see POLYCISTRONIC MRNA)

multicomponent viruses (see COVIRUS)

multiline cultivar (*plant pathol.*) A variety of cultivated plant the individual members of which exhibit different degrees of resistance to each of a number of races of a given pathogen. Such a cultivar inhibits the spread of a *par-*

ticular pathogen and prevents total crop loss during an epiphytotic.

multiple drug resistance (see ANTIBIOTICS, RESISTANCE TO and R FACTOR)

multiple-tube method (most probable number (MPN) method) A method for estimating the concentration of viable organisms (usually bacteria) in a suspension. Five tubes, containing a suitable liquid growth medium, are each inoculated with a fixed volume of the suspension. Another five tubes, containing the same medium, are each inoculated with a smaller fixed volume of the suspension, and further sets of tubes are similarly inoculated with progressively smaller volumes of the suspension. Following incubation, each tube is examined for the presence or absence of growth. Growth usually occurs in those tubes inoculated with the larger volumes of suspension—since these tubes are more likely to have received at least one viable organism in the inoculum. (It is assumed that growth will occur in any tube which receives at least one viable organism in the inoculum.) The concentration of viable cells in the original suspension may be calculated from the pattern of positive (growth) and negative (no growth) tubes—using statistical (probability) tables. The use of a selective medium may enable the concentration of a particular type of organism to be assayed. The multiple-tube method is characterized by a large sampling error; it is used e.g. in the bacteriological examination of natural and treated waters. (see also ENUMERATION OF MICROORGANISMS)

multiplicity of infection Refers to the ratio of virus particles to susceptible cells in a given system; thus, e.g. a high multiplicity refers to a relatively large number of virus particles per susceptible cell.

multisite mutation A MUTATION which involves the alteration of two or more contiguous nucleotides. Often used as synonymous with DELETION mutation.

mumps (infectious or epidemic parotitis) An acute, infectious disease of man (mainly children) characterized by inflammation and swelling of the salivary glands, particularly the parotid. The causal agent, the mumps virus—a paramyxovirus, invades *via* the oral or nasal route; the incubation period appears to be 1–3 weeks. Salivary gland enlargement may persist for about a week; possible sequelae include orchitis, mastitis, meningoencephalitis. Recovery confers long-term or permanent immunity to the disease. *Laboratory diagnosis*: culture/characterization of viruses from the saliva, and/or serological tests.

muramic acid An amino sugar which occurs uniquely (as the *N*-acetyl derivative) in PEPTIDOGLYCAN.

murein A synonym of PEPTIDOGLYCAN.

muriform (of a spore) Refers to a spore which has both transverse and longitudinal septa.

murine An adjective which denotes an association with members of the family Muridae (mice and rats).

murine typhus (see TYPHUS FEVERS)

murogenous development (of conidia) (see CONIDIA)

muscarine A quaternary ammonium compound, 2-methyl-3-hydroxy-5-trimethylammonium-methyltetrahydrofuran. Muscarine is produced by *Amanita muscaria* (the *fly agaric*) and by certain other fungi, notably *Inocybe* spp. The compound is toxic for flies, and may (infrequently) cause a fatal intoxication in humans. (see also MYCOTOXINS)

muscicolous Growing on mosses.

mushroom (1) Any of the edible species of the (basidiomycetous) genus *Agaricus*. These include the *field mushroom* (*A. campestris*), the *cultivated mushroom* (*A. bisporus*; *A. campestris bisporus*; *A. hortensis*), the *horse mushroom* (*A. arvensis*), the *woodland mushroom* (*A. silvicola*) and other species such as *A. rodmani*. (2) Loosely, any umbrella-shaped fruiting body. (see also TOADSTOOL)

mushroom, diseases of Diseases of the cultivated mushroom (*Agaricus bisporus*) include: (a) *Dry bubble* (*brown spot*) caused by *Verticillium malthousei*. (b) *Wet bubble* (*bubbles*) caused by *Mycogone perniciosa*. (c) *Cobweb* (*mildew*) caused by *Dactylium dendroides*. (d) *Hard-gill*, *watery stipe*, and *stunting* have been attributed to infection by certain viruses.

must The ingredients (e.g. fruit pulp, juice) which are fermented in WINE-MAKING.

mustard gas (see SULPHUR MUSTARDS)

mutagen (mutagenic agent) Any agent (chemical or physical) which promotes the occurrence of MUTATIONS; in general, a given mutagen induces a particular type of mutation. Chemical mutagens include: (a) base analogues, e.g. 5-BROMOURACIL, 2-aminopurine; (b) agents which react directly with the bases in nucleic acids, e.g. ALKYLATING AGENTS, HYDROXYLAMINE, NITROUS ACID; (c) see ACRIDINES (as mutagens). Physical mutagenic agents include: (a) ionizing radiations, e.g. X-rays, gamma rays; (b) non-ionizing radiations, e.g. ULTRAVIOLET RADIATION.

mutagenesis The generation of one or more MUTATIONS. Mutagenesis can be a spontaneous phenomenon or it can be induced by a MUTAGEN. Spontaneous mutations occur at a rate specific for a given gene (see MUTATION RATE). Most spontaneous mutations are believed to occur through errors during DNA replication and repair; however, some may be due to factors such as ULTRAVIOLET RADIATION in sunlight, or the mutagenic activity of certain metabolic intermediates (e.g. peroxides). (see also MUTATOR GENE)

mutant (1) (*noun*) Any organism which differs from the WILD TYPE STRAIN as the result of one or more MUTATIONS. (2) (*adjective*) Refers to any organism, population of organisms, chromosome, gene etc. which, as a consequence of one or more mutations, differs from the wild type.

mutation A stable, heritable change in the nucleotide sequence of a genetic nucleic acid (DNA or, in RNA viruses, RNA) which may occur spontaneously, or through the activity of a MUTAGEN—see MUTAGENESIS; the term *excludes* changes in genetic content due to RECOMBINATION or to the incorporation of foreign genetic material, e.g. an episome or prophage. A mutation often results in a change in the phenotype of the organism (*mutant*) in which it occurs, due to the alteration of the information content of the genetic material—see GENETIC CODE. (see also FORWARD MUTATION) However, some mutations do not alter the phenotype of the cell—e.g. a mutation may not significantly affect the activity of the product of the gene in which it occurs; in a *diploid* cell, a mutant gene may be *recessive* to its wild-type allele. Certain mutations may be lethal to the cell in which they occur; others may be lethal only under certain conditions—see CONDITIONAL LETHAL MUTANT. A mutation, or the effects of a mutation, may be reversed by a second mutation—see BACK MUTATION and SUPPRESSOR MUTATION. (cf. PHENOTYPIC SUPPRESSION) A wide range of types of mutation are known: see e.g. DELETION; INSERTION; LEAKY MUTATION; MIS-SENSE MUTATION; NONSENSE MUTATION; PHASE-SHIFT MUTATION; PLEIOTROPIC MUTATION; POINT MUTATION; POLAR MUTATION; TRANSITION; TRANSVERSION.

mutation frequency The proportion of mutants, of a given type, present at a given time in a growing population of cells; the mutation frequency depends on the MUTATION RATE for the given type of mutation and on the degree of selectivity of the environment for the corresponding mutant—i.e. it indicates the cumulative total of mutants.

mutation rate (*bacteriol.*) The probability that any given cell in a growing population will undergo, spontaneously, a specific type of mutation during its division cycle. The concept of mutation rate can be expressed:

$$a = \frac{m}{d}$$

where a = mutation rate, m = the number of specific mutations within the population, and d = the number of individual cell division cycles within the population. Since each cell division increases the population by 1, d is $(n - n_0)$ where n is the final number of cells, and n_0 is the initial number of cells. However, when—as is usual—growth occurs *non*-synchronously, the determination of n by viable count involves plating cells which are in various stages of the division cycle; thus, chromosome replication will have occurred in some cells which have still to divide—and which, since they form single colonies, are counted as single cells. (see Helmstetter–Cooper model under GROWTH (bacterial)) Hence, the value of $(n - n_0)$ indicates only the number of completed cell divisions, and does not exactly reflect the number of newly-formed or partly-formed chromosomes—i.e. the number of genomes in which mutation could occur; accordingly, a statistical correction is applied, and the expression for mutation rate becomes:

$$a = \frac{m}{(n - n_0)/0.69} \quad \text{or} \quad a = \frac{m \times 0.69}{(n - n_0)}$$

Example: determination of the rate of mutation to prototrophy in a population of auxotrophs. An auxotrophic population is plated on solid MINIMAL MEDIUM enriched with limiting amounts of essential growth factor(s). After each cell has divided several times (forming a total of n cells) growth of all auxotrophic cells ceases, and only prototrophic mutants continue to grow and form colonies—each colony being scored as 1 mutation; from the number of mutations scored is subtracted the number of prototrophic mutants present in the original inoculum—determined by plating, on (unenriched) minimal agar, an aliquot of the culture from which the inoculum was taken. In order to determine n, all (prototrophic) colonies are first removed from the plate; the remaining cells are washed from the plate and their number is determined by a viable count on COMPLETE MEDIUM. n_0 is determined by a viable count, on complete medium, of an aliquot of the original culture equal in volume to that of the inoculum.

The value obtained for a given mutation rate varies with the method used in its determination—owing to errors and assumptions inherent in the methods. Some examples of mutation rates: (a) *Escherichia coli*: resistance to bacteriophage T1 = 3×10^{-8}. (b) *E. coli*: inability to ferment galactose = 1×10^{-10}. (c) *Staphylococcus aureus*: resistance to penicillin = 1×10^{-7}. A value such as 3×10^{-8} means that, on average, one mutant is formed for every 3×10^{8} cell division cycles within the population. For two independent mutations, having rates, say, of 10^{-6} and 10^{-8}, the mutation rate for a double mutant (i.e. a cell having both mutations) would be 10^{-14}.

mutational equilibrium In a bacterial culture growing under *non*-selective conditions: the state in which the ratio of spontaneously-arising mutants to parental (non-mutant) cells has become, and remains, constant. Such a state requires that the number of mutants be limited to a constant proportion of the total cell population by appropriate rates of back mutation.

mutator gene (high mutability gene) A gene within which certain mutation(s) cause an increase in the spontaneous mutation rate in other genes. A mutator gene is thought to code for some factor involved in DNA replication and/or repair; e.g. a mutation in a gene coding for a DNA polymerase may result in an enzyme which permits a higher than normal number of errors during replication/repair. A particular mutator may favour a particular type of mutation (e.g. G-C to T-A transversion).

Mutation(s) in this gene may cause a *decrease* in the spontaneous mutation rate in other genes; the gene is then referred to as an *antimutator gene*. The product of the antimutator gene may be a DNA polymerase which permits a *lower* than normal number of errors during replication/repair.

Mutinus A genus of fungi of the order PHALLALES.

mutualism A stable condition in which two organisms of different species live in close physical association—each organism deriving some benefit from the association. (see also COMMENSALISM and SYMBIOSIS)

Mycelia Sterilia Fungi of the class AGONOMYCETES.

mycelium Collectively, a group or mass of discrete *hyphae* (see HYPHA): the form of the vegetative organism in many types of fungi and in certain types of bacteria (see ACTINOMYCETALES). Some fungi form structures which consist of an organized arrangement of hyphae—see e.g. PLECTENCHYMA. (see also SPROUT MYCELIUM)

mycetocytes Cells, found in certain insects, which carry intracellular bacterial or fungal *symbionts*. (If the symbiont is a bacterium the term *bacteriocyte* may be used; however, *mycetocyte* is commonly used whether the endosymbiont is a bacterium or a fungus.) Mycetocytes may be irregularly distributed in certain tissues, e.g. the gut lining, or may be aggregated into specialized organs (*mycetomes*) which are usually associated with the gut. In (at least) some cases the microflora of the mycetomes supplies essential nutrients to the insect host. (see also SYMBIOSIS)

mycetome (see MYCETOCYTES)

Mycetozoa (see SLIME MOULDS)

Mycobacteriaceae A family of gram-positive, aerobic, non-motile, ACID FAST bacteria of the ACTINOMYCETALES; the organisms do not form mycelium but may form branching rods or fragile filaments as well as straight or curved rods and coccoid forms. All species have a Type IV cell wall (see ACTINOMYCETALES). The family includes both saprophytes and pathogens—all of which are placed within the sole genus MYCOBACTERIUM.

Mycobacterium A genus of gram-positive, aerobic, non-motile, non-sporing, ACID FAST bacteria of the family MYCOBACTERIACEAE; species occur e.g. in soil and water and in the tissues of a variety of animals. The organisms are straight or curved rods, approximately 0.2–0.8 × 1–10 microns, but may occur as branched rods or as fragile filaments; cells may stain uniformly or exhibit banding or beading. The cell wall, which contains MYCOLIC ACIDS, is classified as Type IV (see ACTINOMYCETALES). (see also CORD FACTOR and PEPTIDOGLYCAN) Some strains are pigmented: see PHOTOCHROMOGENS and SCOTOCHROMOGENS. Some species grow rapidly (macroscopic growth within 1 week under optimal conditions); such species include saprophytes, parasites, and weakly-invasive, low-grade pathogens. Other species grow slowly (macroscopic growth only after approximately 1–3 weeks under optimal conditions); these species include important pathogens e.g. *M. tuberculosis*. Most ATYPICAL MYCOBACTERIA are slow-growing species.

The type species is *M. tuberculosis*; the GC% range for the genus is approximately 60–70.

Tests used to distinguish between *Mycobacterium* spp include: (a) NIACIN TEST. (b) T2H TEST. (c) Nitrate reduction test. (d) Arylsulphatase test—in which an organism is tested for its ability to cleave tripotassium PHENOLPHTHALEIN disulphate; the phenolphthalein formed (in a positive test) is detected by the addition of alkali. (e) TWEEN 80 HYDROLYSIS. (f) The persistence of CATALASE activity following incubation for 20 minutes at 68 °C in neutral phosphate buffer.

M. tuberculosis (*M. tuberculosis* var *hominis*). A causal agent of TUBERCULOSIS (*q.v.* for isolation and culture); a slow-growing species which generally forms rough, raised, pale buff colonies frequently containing cord-like strands consisting of masses of aligned rods. Growth is stimulated by glycerol. *M. tuberculosis* gives a positive niacin test, reduces nitrate, and is resistant to T2H at test concentrations. Optimum temperature and pH range: 37 °C and approximately 6.5–7 respectively.

For drugs used against *M. tuberculosis* see TUBERCULOSIS. Mutant, ISONIAZID-*resistant* strains of *M. tuberculosis* are catalase-negative, and their virulence for the guinea-pig—but not for man—is greatly reduced. Disinfectants effective agaihst *M. tuberculosis* (and *M. bovis*) include sodium hypochlorite, ethanol (70%), phenol (5%), glutaraldehyde (2% aqueous), and formalin; *M. tuberculosis* is killed when effectively subjected to 65 °C for 30 minutes.

M. bovis (*M. tuberculosis* var *bovis*). A causal agent of TUBERCULOSIS; its pathogenicity for animals other than man is generally greater than that of *M. tuberculosis*. *M. bovis* is a slow-growing species which is culturally and antigenically similar to *M. tuberculosis*; growth is stimulated by pyruvate but not by glycerol. It is reported to be microaerophilic on primary isolation. *M. bovis* is now infrequently the cause of human tuberculosis in the United Kingdom and USA. It is used to prepare BCG vaccine (*q.v.*).

M. avium (*M. tuberculosis* var *avium*). The cause of tuberculosis in fowl; it has only rarely been regarded as an agent of human tuberculosis and is not pathogenic for the guinea-pig.

M. smegmatis. A rapidly-growing, non-chromogenic, non-pathogenic species found in smegma.

M. intracellulare and *M. kansasii*. Slow-growing ATYPICAL MYCOBACTERIA; *M. kansasii* is photochromogenic, *M. intracellulare* non-chromogenic.

M. xenopi. A slow-growing, niacin-negative, 'atypical' species; it is able to grow at 45 °C. Ageing cultures develop a yellow pigment.

M. paratuberculosis (*M. johne*). The causal agent of JOHNE'S DISEASE. *In vitro* cultivation requires the use of media containing either dead mycobacteria or MYCOBACTINS.

M. leprae. The causal agent of LEPROSY in man; *M. leprae* is strongly acid fast. It cannot be cultured *in vitro* but can be grown in the footpads of mice. (see also LEPROMIN TEST)

M. phlei (Timothy-grass bacillus). A rapidly-growing, scotochromogenic, saprophytic species capable of growth at 52 °C; it occurs in soil and grasses.

mycobactins A group of complex, lipophilic compounds produced by species of MYCOBACTERIUM (excluding *M. paratuberculosis*) in response to growth-limiting concentrations of iron. Mycobactins chelate metals—being strongly selective for *ferric* iron. They are thought to occur in the cell envelope and function in iron transport; iron is released following enzymic reduction to the ferrous form. (see also SIDEROCHROMES)

mycobiont The fungal partner in a LICHEN.

Mycobionta (see EUMYCOTA)

mycocecidia GALLS induced by fungi.

mycolic acids α-Substituted, β-hydroxylated fatty acids (R′CHOH.CHR″.COOH) found in the cell walls of *Mycobacterium* spp, *Nocardia* spp and *Corynebacterium* spp. Mycobacterial mycolic acids are within the range C_{60}–C_{90}, nocardomycolic acids C_{40}–C_{60}, and corynemycolic acids C_{28}–C_{40}. (see also WAX D)

mycology The study of FUNGI.

mycophenolic acid An ANTIBIOTIC, produced e.g. by *Penicillium stoliniferum*, which has antitumour and antimicrobial activity. Mycophenolic acid inhibits the synthesis of xanthosine monophosphate from inosine monophosphate—hence blocking guanosine monophosphate synthesis (see Appendix V(a)).

Mycoplasma A genus of gram-negative procaryotic organisms of the family MYCOPLASMATACEAE, class MOLLICUTES; species are parasitic or pathogenic in a range of animals, and *Mycoplasma*-like organisms are the causal agents of a number of plant diseases. The cells, which lack cell walls, are highly pleomorphic; spherical or ellipsoid forms vary from about 150 to 300 nm in diameter, and filaments may be many microns in length. Some strains exhibit a gliding-type motility. Colonies on agar media vary from about 10 to 500 microns in diameter and typically have a 'fried egg' appearance—i.e. a central, granular region, embedded in the agar, surrounded by non-granular surface growth. All species are chemoorganotrophs with fermentative and/or oxidative-type metabolism; growth usually occurs under aerobic or anaerobic conditions. *In vitro* cultivation requires the use of complex media—containing factors such as serum and yeast extract; all species require sterols (e.g. cholesterol) as well as other growth factors. (The sterol content of the CELL MEMBRANE is associated with the susceptibility of *Mycoplasma* spp to POLYENE ANTIBIOTICS.) Optimum growth temperature: 37 °C; optimum pH: 7.0. *Mycoplasma* spp are sensitive to surfactants (e.g. digitonin) and to certain antibiotics, e.g. tetracyclines, erythromycin. The type species is *M. mycoides*; GC% range for the genus: 23–40.

M. mycoides. Certain strains cause a pleuropneumonia in cattle. All strains are erythromycin-sensitive.

M. pneumoniae (Eaton's agent). A slow-growing species which is the causal agent of an atypical pneumonia in man. On primary isolation the organism forms colonies which are not of the characteristic 'fried egg' type: the colonies, which are entirely granular, are up to 100–150 microns in diameter. Growth first appears after 5–6 days incubation. A zone of clear haemolysis surrounds colonies formed on guinea-pig blood agar. (A suspension of *M. pneumoniae* can agglutinate guinea-pig erythrocytes—an important diagnostic feature.) The organism is highly sensitive to erythromycin.

M. laidlawii: see **ACHOLEPLASMATACEAE**.

T-strains of *Mycoplasma* typically form small colonies (less than about 40 microns in diameter) and hydrolyse urea; they occur in the urinary tract and elsewhere in man and other animals.

mycoplasmas Members of the class **MOLLICUTES**.

Mycoplasmataceae A family of the Mycoplasmatales, class **MOLLICUTES**; species of **MYCOPLASMA**, the sole genus, differ from members of the **ACHOLEPLASMATACEAE** in that they have a sterol growth-requirement.

mycorrhiza (*pl. mycorrhizae* or *mycorrhizas*) A stable, symbiotic association (see **SYMBIOSIS**) between a fungus and the root of a plant; the term is also used to refer to the root-fungus structure itself. Mycorrhizae occur in a very wide range of vascular plants—including trees, shrubs, herbaceous plants, and ferns; similar associations occur among the bryophytes. The fungus may be a basidiomycete, an ascomycete, or a deuteromycete; it is always associated with the primary cortex of the root. Mycorrhiza formation is encouraged by low levels of plant nutrients (e.g. phosphorus and nitrogen). Two major types of mycorrhiza are recognized.

Ectomycorrhiza (ectotrophic mycorrhiza). In this type of mycorrhiza the fungal hyphae occur outside the root and between the cortical cells of the root; they do not normally penetrate the cortical cells. Ectomycorrhizae are particularly common among forest trees; the majority of fungi involved in ectomycorrhizae are basidiomycetes (e.g. agarics) but a few are ascomycetes (e.g. *Tuber* spp). A given tree may associate with more than one type of fungus. In a typical ectomycorrhiza the fungus forms a sheath of tissue (the *mantle*) over the surface of the root; hyphae from the mantle penetrate the soil surrounding the root and the root itself—penetrating between the cortical cells of the root to form a network of hyphae (the *Hartig net*) which enmeshes the individual cortical cells. A root involved in an ectomycorrhizal association undergoes a number of morphological changes; thus e.g. it lacks root hairs and a root cap; it is thicker than an uninfected root and may be a different colour; it may branch extensively and characteristically—e.g. pinnately (in *Fagus* spp) or dichotomously (in *Pinus* spp). In certain cases an ectomycorrhiza may develop in the form of *nodules* (= tubercles)—a nodule consisting of a rounded, dense mass of mycorrhizal roots.

Ectomycorrhizal fungi cannot utilize complex carbohydrates (e.g. cellulose) and obtain sugars (e.g. glucose, fructose, sucrose) from the plant root; these sugars may be stored in the mantle in the form of e.g. mannitol or glycogen. The plant appears to benefit from the association in a number of ways: (a) nutrients (e.g. nitrogen, phosphorus, potassium) are taken up more efficiently by an infected plant than by an uninfected plant; (b) the association may help to protect the plant from infection by certain pathogens—e.g. the fungus may produce antibiotic(s), or the mantle may function as a mechanical barrier; (c) many mycorrhizal fungi produce a number of plant hormones—including **AUXINS**, **GIBBERELLINS**, **CYTOKININS**—but although these are assumed to influence the development of the plant, their precise effect is unknown. In certain cases (e.g. certain species of pine (*Pinus*)) a mycorrhizal association appears to be essential for the normal development of a plant.

Endomycorrhiza (endotrophic mycorrhiza). In this type of mycorrhiza the fungus develops *within* the cells of the root cortex; a fungal sheath external to the root is not usually present, although hyphae may penetrate the soil around the root. Endomycorrhizae may be divided into those in which the fungus has septate hyphae, and those in which the fungus has aseptate hyphae. The latter type is extremely common among angiosperms, many gymnosperms, pteridophytes and bryophytes; the fungus involved is frequently a species of *Endogone*. This type of mycorrhiza is commonly of the *vesicular-arbuscular* type (see **ARBUSCULE**). Fungi with septate hyphae form endomycorrhizae with a smaller range of plants—including e.g. members of the Orchidaceae and Ericaceae. In a typical endomycorrhiza the fungus hyphae penetrate the cortical cells of the root and undergo a period of intracellular development; subsequently, the hyphae are digested by the root cell—leaving a knot of undigested hyphal wall material in the cell. As the root grows, the fungus invades new cells behind the growing region of the root; thus a balance is set up between fungal invasion of the plant and plant digestion of the fungus.

In orchids the mycorrhizal association appears to be essential—at least for seed germination and for the growth of the seedling; orchid seeds are very small and have little or no food reserves, and it appears that the nutrients required for germination and early growth must be supplied (under natural conditions) by an invading fungus. Saprophytic species of orchid may continue to depend on the mycorrhizal fungus for their supply of carbon; these plants are associated with fungi such as *Fomes*, *Armillaria* etc. which are capable of degrading complex compounds (e.g. cellulose, lignin) to simple compounds (e.g. sugars) which become available to the orchid. Green orchids are generally associated with fungi of the form-genus *Rhizoctonia*.

Plants associated with endomycorrhizae generally appear to benefit from the association; much of the observed improvement in growth and development has been attributed to an increased uptake of phosphate by the mycorrhizal root.

mycosamine (see POLYENE ANTIBIOTICS)

mycosin A synonym of CHITOSAN.

mycosis Any disease in which the causal agent is a fungus.

Mycosphaerella A genus of fungi of the class LOCULOASCOMYCETES.

mycotoxicosis Any disease in which the aetiological agent is a MYCOTOXIN. (see also FOOD POISONING (fungal))

mycotoxins Toxic substances produced by fungi. Those toxic for man and animals include: AFLATOXINS, AMATOXINS, ERGOT ALKALOIDS, MUSCARINE, PHALLOTOXINS and RUBRATOXINS; phytotoxins include: ALTERNARIC ACID, FUSARIC ACID and VICTORIN.

mycotrophic (*of plants*) Refers to any plant in which the root system participates in a mycorrhizal association—see MYCORRHIZA.

myonemes (*ciliate protozool.*) Bundles of intracellular protein microfilaments observed in a number of ciliates; they occur, e.g. in the zooid and stalk of *Vorticella*, and—below the level of the infraciliature—in the spirotrichs *Spirostomum* and *Stentor*. (In *Stentor* the myonemes are also known as *M bands* or *M fibres*.) Myonemes are believed to be contractile—and may thus account for contractility in organisms which contain them; contraction in such organisms appears to be promoted by increased environmental and/or intracellular levels of calcium ions.

Myriangium A genus of fungi of the class LOCULOASCOMYCETES.

Myrmecia A genus of algae of the division CHLOROPHYTA. Species occur e.g. as the phycobiont in certain LICHENS—e.g. species of DERMATOCARPON and LOBARIA.

myxamoeba A non-flagellated amoeboid cell which occurs in the life cycles of the ACELLULAR SLIME MOULDS, the CELLULAR SLIME MOULDS, and in members of the PLASMODIOPHORALES.

myxo- A combining form derived from the Greek word for *mucus* or *slime*.

myxobacter Any species of the MYXOBACTERALES.

Myxobacterales An order of GLIDING BACTERIA consisting of those species which form *fruiting bodies* (see below); myxobacters occur in soil, decaying plant material, and on dung. The vegetative cells, which are embedded in slime, are slender (often fusiform) gram-negative rods, generally 0.4–1.5 × 2–15 microns—depending on species and growth conditions; they reproduce by transverse binary fission. The cells often contain CAROTENOID pigments—glycoside carotenoids are said to be common. The myxobacters are obligate aerobic chemoorganotrophs. They can *lyse* many species of bacteria, yeasts, and filamentous fungi—these organisms commonly forming the primary food source. Most species grow on solid media containing live or dead bacteria; some can use cellulose as the sole carbon source. The colony ('swarm' or 'pseudoplasmodium') is usually flat and wrinkled.

The *fruiting body* is a resting stage; its formation is incited e.g. by starvation. Each species forms fruiting bodies which are characteristic in size, shape and composition; they are often brightly coloured. A fruiting body consists of an orderly aggregation of *resting cells*. Some resting cells appear to be slightly modified vegetative cells (*myxospores*); another type (the *microcyst*) is a myxospore enclosed by a shell of hardened slime. The simplest form of fruiting body consists of a mass of resting cells in a droplet of slime. In another type, a spherical or spheroid mass of resting cells occurs at the top of a *stalk*—the latter consisting of hardened slime. In the fruiting bodies of some species, numbers of resting cells are enclosed within hard, sac-like structures termed *cysts* or *sporangia*; sporangia may occur singly or in clusters, and may be either attached directly to the substratum or elevated on the tip(s) of simple or branched stalks. The genera include CHONDROMYCES and MYXOCOCCUS.

Myxobionta (see SLIME MOULDS)

Myxobolus A genus of protozoans of the subphylum CNIDOSPORA.

Myxococcus A genus of gram-negative bacteria of the MYXOBACTERALES; vegetative cells are about 0.5 × 2–10 microns, and often contain orange or yellow carotenoid pigments. Most species form stalkless fruiting bodies, and sporangia are not formed; the *microcysts* are coccal or coccoid. No species is cellulolytic.

myxomatosis An acute, infectious disease of the *rabbit*; the causal agent is a *poxvirus*. Infection may occur by wound contamination or by the bite of an infected arthropod vector (e.g. the rabbit flea)—the insect acting as a *mechanical* vector. Symptoms include the development of tumours on the face and in the genito-anal tissues. The disease is rapidly fatal to susceptible animals. Myxomatosis was introduced into Australia in the early 1950s in order to control the rabbit population—an example of

BIOLOGICAL CONTROL; initially the disease had a mortality rate of about 99%—but this subsequently declined.

Myxomycetales (see ACELLULAR SLIME MOULDS)

Myxomycetes (see ACELLULAR SLIME MOULDS)

Myxomycota (see SLIME MOULDS)

Myxophyceae (see BLUE-GREEN ALGAE)

myxospore A resting cell in the fruiting body of members of the MYXOBACTERALES.

Myxosporidea A class of protozoans of the subphylum CNIDOSPORA.

Myxotrichum A genus of fungi of the EUROTIALES.

myxotrophy A term sometimes used (incorrectly) as a synonym of MIXOTROPHY.

myxoviruses A group (not a taxon) of enveloped, single-stranded RNA viruses which are generally associated with diseases of the respiratory tract in man and other vertebrates; such diseases include FOWL PEST, FOWL PLAGUE, and INFLUENZA. Individual myxoviruses are currently referable to one or the other of the genera ORTHOMYXOVIRUS and PARAMYXOVIRUS.

myxoxanthophyll (see CAROTENOIDS)

N

N.A. (numerical aperture) (see RESOLVING POWER)

nabam (dithane D-14) Disodium ethylenebisDITHIOCARBAMATE. An agricultural antifungal agent which is no longer widely used—except in conjunction with zinc sulphate and lime, when ZINEB is formed. (see also DITHANE)

NAD Nicotinamide adenine dinucleotide—formerly known as diphosphopyridine nucleotide (DPN). NAD is a coenzyme and functions as a hydrogen carrier in a wide range of hydrogen transfer reactions; the oxidized form may be written as NAD or NAD^+, and the reduced form as $NADH_2$ or $NADH+H^+$. (see also NADP; NICOTINIC ACID; Appendix V(c))

NADP Nicotinamide adenine dinucleotide phosphate—formerly known as triphosphopyridine nucleotide (TPN). NADP functions as a coenzyme in a manner analogous to that of NAD (*q.v.*). (see also NICOTINIC ACID; Appendix V(c))

Naegleria A genus of amoebae of the order AMOEBIDA. Most species appear to be free-living, but some (e.g. *N. aerobia*) can give rise to an acute and fatal meningoencephalitis in man and other animals.

N. aerobia is approximately 15 microns in diameter when rounded; the single nucleus contains a large, central nucleolus. The boundary between endoplasm and ectoplasm is clearly defined. A contractile vacuole is present. The life cycle includes a biflagellate stage, and spherical or ovoid CYSTS are formed. In man, infection probably occurs *via* the nasal mucosa; the organism has been isolated from the cerebrospinal fluid and has been detected in sections of the brain.

nagana The name of any of several types of trypanosomiasis which affect domestic animals in various parts of Africa; the vector is often a species of *Glossina* (the tse-tse fly). *T. brucei* causes a fatal disease of equines and of certain other animals, and a non-fatal disease of cattle. *T. congolense* causes progressive anaemia and wasting in cattle and other animals. *T. vivax* causes a disease (also called *souma*) similar to that produced by *T. congolense*.

Nagler's reaction A test used to detect *Clostridium perfringens*—strains of which produce a diffusible LECITHINASE (the α-toxin)*. When cultured on EGG-YOLK AGAR such strains give rise to growth which is surrounded by a zone of opalescence. In the Nagler reaction, *one* half of a plate of EYA is spread with *anti-α*-toxin; the plate is then inoculated with the unknown strain—the inoculum forming a line which passes into both halves of the plate. Following incubation, strains of *Cl. perfringens* produce opalescence only in that section of the plate which is free from antitoxin—although growth occurs in both halves of the plate. *An antigenically-similar toxin is produced by *Clostridium bifermentans*.

nalidixic acid A synthetic ANTIBIOTIC consisting of a substituted 1,8-naphthyridine. It is active mainly against gram-negative bacteria, e.g. many strains of *Proteus* and *E. coli*; *Pseudomonas aeruginosa* and gram-positive bacteria are generally resistant. Nalidixic acid specifically inhibits the replication of bacterial DNA but appears to have no effect on eucaryotic nuclear DNA synthesis. The mechanism of action is unknown. Nalidixic acid is used mainly in the treatment of urinary tract infections. Resistance to the antibiotic develops readily.

naming of microorganisms (see NOMENCLATURE)

NANA *N*-Acetylneuraminic acid (see NEURAMINIC ACIDS).

Nannizzia A genus of heterothallic fungi of the order EUROTIALES; species occur in the soil, and the imperfect stages of *Nannizzia* spp correspond to dermatophytes of the *form*-genus MICROSPORUM. The ascocarp is a cleistothecium whose peridium consists of a loose network of differentiated hyphae; the individual cells of the peridial hyphae are rough-surfaced, and each cell is slightly swollen at both poles. Typically, the free ends of the peridial hyphae terminate in coiled appendages.

Nannomonas A sub-genus of TRYPANOSOMA within the group SALIVARIA; the type-species, *T.* (*Nannomonas*) *congolense*, is parasitic or pathogenic e.g. in cattle, sheep, and horses. The organisms are less than 20 microns in length; the subterminal kinetoplast typically appears close to the pellicle, and there is no *free* flagellum.

nannoplankton (see PLANKTON)

nanometre (nm) 10^{-9} metre; one millimicron.

naphthoquinones (see QUINONES (in electron transport))

naramycin B (see CYCLOHEXIMIDE)

nasse (*mycol.*) A structure which occurs in the apical region in most *bitunicate* asci; a nasse may consist e.g. of four vertically orientated rods, arranged in a group, or a system of branched or reticulate rods—according to species. The nasse stains with iodine or e.g. cotton blue.

Nassula A genus of freshwater protozoans of the suborder CYRTOPHORINA. The cells are ovoid or elongated, about 200 microns in length; the alveolate pellicle is completely covered with cilia. The cytostome is lateral or ventral, and there is a conspicuous pharyngeal basket. Food consists of filamentous algae, e.g. the blue-green alga *Phormidium*.

natural antibodies Those IMMUNOGLOBULINS, normally present in the plasma, which behave as antibodies towards antigens with which an individual has apparently never been in immunological contact.

Navicula A genus of motile, pennate, spindle- or boat-shaped diatoms (division BACILLARIOPHYTA) which occur in freshwater and marine environments and on damp soil.

NCIB The National Collection of Industrial Bacteria maintained at the Torry Research Station, Aberdeen, Scotland, United Kingdom.

NCTC (National Collection of Type Cultures) A collection of well-characterized bacteria of medical importance kept at the Central Public Health Laboratory, Colindale, London, U.K.

NCVPOO (see VIRUSES—*translation of the viral genome*)

necridium In the trichome or filament of a blue-green alga: any dead intercalary cell which promotes the fragmentation of the trichome or filament. (see also HORMOGONIUM)

nectin (see CELL MEMBRANE (bacterial))

Nectria A genus of fungi of the order SPHAERIALES, class PYRENOMYCETES; species occur e.g. as parasites and pathogens of trees. *N. cinnabarina* (the coral spot fungus) occurs on the exposed roots, trunk, and branches of a number of trees—particularly maple (*Acer* spp). Asexual reproduction involves the formation of simple or sparsely-branched conidiophores which develop on superficial *sporodochia* (see SPORODOCHIUM and STROMA); the elongated, unicellular, bright orange-pink conidia develop terminally on the conidiophores. These reproductive structures are commonly from one to several millimetres in size. The fungus overwinters in the perithecial stage: superficial, dark-red perithecia first develop at the margin of the sporodochium (i.e. at the sporodochium-host junction) and may subsequently form over its entire surface—supplanting the conidiophores. Each ascus contains eight two-celled banana-shaped ascospores.

N. galligena differs from *N. cinnabarina* in that e.g. the conidia are large and multiseptate, and the perithecia are formed singly without the development of a stroma.

A number of species of *Nectria* have conidial (imperfect) stages which are classified in the form-genus *Fusarium*; other species have conidial stages which appear to correspond to other form genera.

negative phase (*immunol.*) Reduction in the *in vivo* titre of a particular antibody which immediately follows the administration of homologous antigen. The fall in titre is due to the combination of antigen with circulating antibody.

negative staining A procedure used in light microscopy (see STAINING) and in ELECTRON MICROSCOPY.

Negri bodies Acidophilic, intracytoplasmic inclusion bodies which develop in cells of the central nervous system in cases of RABIES. Negri bodies may be present only in the late stages of the disease; they are not formed as a result of (experimental) inoculation with attenuated ('fixed') rabies virus. Negri bodies may contain virions or viral components.

Neisser–Wechsberg leucocidin The α-HAEMOLYSIN of staphylococci. (see also LEUCOCIDIN)

Neisseria A genus of gram-negative, aerobic, chemoorganotrophic, asporogenous, non-motile, oxidase-positive and catalase-positive bacteria of the family NEISSERIACEAE; species are obligately parasitic or pathogenic in mammals. The cells are coccoid, 0.6–1 micron in diameter; typically they occur in pairs, with adjacent sides flattened or concave, but may be found singly or in tetrads. *Neisseria* spp are highly susceptible to desiccation and to exposure to direct sunlight; most species are susceptible to penicillin and tetracyclines. Type species: *N. gonorrhoeae*; GC% range for the genus: 47–52.

N. gonorrhoeae (the *gonococcus*). The causal agent of GONORRHOEA and ophthalmia neonatorum in man; not pathogenic for other animals. Media for isolation and culture include CHOCOLATE AGAR and THAYER-MARTIN SELECTIVE MEDIUM; optimum growth temperature 35–36 °C, optimum pH 7.2–7.5. Growth is stimulated by 3–10% CO_2. Colonial details vary with strain. The cells stain well with aniline dyes or methylene blue. *N. gonorrhoeae* produces acid from glucose but not from maltose; it does not produce H_2S. Antigenically the species is diverse; a number of heat-stable antigens are common to *N. gonorrhoeae* and *N. meningitidis*. *N. gonorrhoeae* is very sensitive to moist heat (55 °C/5 minutes is lethal) and to the common disinfectants (1% phenol is commonly lethal within 2–5 minutes).

N. meningitidis (the *meningococcus*). A causal agent of MENINGITIS in man. Morphology, culture, sensitivities and staining reactions are similar to those of *N. gonorrhoeae*. Acid is produced from glucose and from maltose; H_2S is not produced.

The other species (*N. flavescens*, *N. mucosa*, *N. sicca* and *N. subflava*) produce H_2S; *N. flavescens* forms a yellow pigment. (*N. catarrhalis* has been placed in the genus *Branhamella*—*B. catarrhalis*.)

Neisseriaceae A family of gram-negative coccoid and rod-shaped bacteria which includes the genera ACINETOBACTER, MORAXELLA and NEISSERIA.

nekton A term used to refer, collectively, to those organisms (e.g. fish) which are actively motile in a body of water (e.g. lake, sea) and which are thus distinguished from those organisms which float or drift passively in the water (i.e. PLANKTON).

nematodesma (*pl.* nematodesmata) (see TRICHITES)

neoantigen (*immunol.*) An ANTIGEN which has gained new antigenic specificity, i.e. acquired new antigenic DETERMINANTS, by any form of modification or by the addition of a HAPTEN.

neomycin A broad-spectrum composite AMINOGLYCOSIDE ANTIBIOTIC produced by *Streptomyces fradiae*; *Pseudomonas aeruginosa* and the obligate anaerobes are among those species which are resistant to neomycin.

neoplasm (*med.*) The result of abnormal and excessive proliferation of the cells of a tissue; if the progeny cells remain localized (at least initially) the resulting mass is termed a TUMOUR.

neostibosan (see *Antimony* under HEAVY METALS)

neoteny The persistence, in a mature organism, of features which are characteristic of the immature organism.

nephelometry A method for determining the concentration of a suspension of cells (often in terms of cell mass) by measuring the amount of light scattering which occurs when a beam of light is passed through the suspension. Light scattering is due to: (a) Intracellular refraction; this depends on a difference between the refractive index of the cell interior and that of the suspending medium. (b) Diffraction. (c) Reflection from the cell surface.

A nephelometer consists essentially of a light source, a cuvette (receptacle) for the sample, and several photosensitive cells which receive the incident and scattered light; the degree of excitation of the photosensitive cells is measured electronically. The relationship between the intensities of the incident and scattered light and the concentration of cells is given by:

$$\log_{10}(I_0/I_{scat}) = -xcl$$

where I_0 = intensity of incident light; I_{scat} = intensity of scattered light; x = a constant which depends on the type of cell being examined, and on the characteristics of the suspending medium (e.g. refractive index); c = concentration of cell suspension; l = length of light path through suspension. The graph $\log(I_0/I_{scat})$ *versus* c is thus a straight line of negative slope. The amount of light scattering increases with increase in cell size and with a decrease in the wavelength of the incident light; monochromatic light (green or orange) is used. The intensity of the scattered light varies with its angular displacement from the axis of the incident light—exhibiting several peaks of intensity in a wide arc on either side of the axis. (These peaks are due to interference phenomena.) The scattered light which is measured is that which emerges from the suspension within certain angular limits of the axis of incident light. The sensitivity of nephelometric determinations decreases under conditions of high cell density—owing to the effects of *secondary scattering*, i.e. the redirection of a deflected ray by a second (or further) cell. (see also ENUMERATION OF MICROORGANISMS)

Nephroma A genus of foliose LICHENS in which the phycobiont is a blue-green alga (*Nostoc*) or a green alga (*Coccomyxa*)—according to species. The thallus is dark-green to brown and may be more or less lobed, according to species; in some species (e.g. *N. laevigatum*) the thallus is dark-coloured when wet, pale when dry. Structures which are present in certain *Nephroma* species include *Nostoc*-containing cephalodia (e.g. in *N. arcticum*), tubercles in the lower surface (e.g. in *N. resupinatum*), soredia on the upper surface (e.g. in *N. parile*). When present, apothecia occur on the *lower* surface of the ends of the thallus lobes—the ends being turned back so that the discs face upwards. (In e.g. *N. laevigatum* the discs are reddish-brown.) Some species rarely form apothecia. *Nephroma* species occur mainly on mossy trees and rocks.

net slime moulds (Labyrinthulales; Labyrinthulida) A group of poorly-understood eucaryotic organisms which are classified in both mycological and protozoological taxonomic schemes; the naked, typically spindle-shaped cells form characteristic reticulate aggregations in which individual cells are joined together by slime filaments. The net slime moulds appear to occur mainly in aquatic environments; one species, *Labyrinthula macrocystis*, is reported to be a pathogen of eel grass (*Zostera marina*).

Neuberg's fermentations Fermentations obtained on the addition of alkali or bisulphite to yeast cultures carrying out ALCOHOLIC

FERMENTATION—see GLYCEROL, MICROBIAL PRODUCTION OF for the mechanism.

neuraminic acids 9-Carbon, 3-deoxy-5-amino sugars; *N*-acylneuraminic acids are referred to as *sialic acids*. *N*-Acetylneuraminic acid (NANA) is derived from *N*-acetylmannosamine and pyruvic acid; it occurs as the terminal residue in mucoprotein molecules in the erythrocyte membrane—the so-called *receptor sites* of viral HAEMAGGLUTINATION. Sialic acids also occur e.g. in various mucous secretions, serum, urine, brain glycolipids (gangliosides), milk (*N*-acetylneuraminyllactose). Polymers of NANA are produced by strains of *Escherichia coli* (see COLOMINIC ACID) and by *Neisseria meningitidis* as the group B and C surface antigens.

neuraminidases (sialidases) Enzymes which hydrolyse the ketosidic linkages between NEURAMINIC ACIDS and other sugars in oligosaccharides, glycoproteins etc. Extracellular neuraminidases are produced by certain bacteria—e.g. *Vibrio cholerae*, *Streptococcus pneumoniae*. A standard substrate for the assay of neuraminidase is *N*-acetylneuraminyllactose (found in milk) in which the *N*-acetylneuraminyl-2→3-D-galactosyl bond is attacked; neuraminidases are generally inhibited by EDTA. The neuraminidases vary in specificity, though most will not hydrolyse COLOMINIC ACID.

In certain viruses (see e.g. ORTHOMYXOVIRUS) a neuraminidase, known as *receptor destroying enzyme* (RDE), occurs as one of the two types of peplomer—the other type of peplomer being a haemagglutinin. HAEMAGGLUTINATION by such viruses results from the attachment of the haemagglutinins to mucoprotein receptor sites on the membranes of erythrocytes. After a period of time (which depends on e.g. temperature) the membrane receptor sites are destroyed by the viral neuraminidases so that the virions spontaneously elute (i.e. dissociate from the erythrocytes) and agglutination is reversed. Erythrocytes which have undergone this process cannot again be agglutinated by the same strain of virus; they are said to be *stabilized erythrocytes*. (Sometimes, erythrocytes stabilized by one strain of virus will be agglutinated by another strain.)

Neurospora A genus of fungi of the order SPHAERIALES, class PYRENOMYCETES; the organisms occur e.g. on decomposing and/or burnt vegetation. *Neurospora* spp form branched, multinucleate mycelium. In two species (*N. crassa*, *N. sitophila*) asexual reproduction involves the formation of conidia or microconidia (= spermatia) which are borne in chains on branched sporophores; on germination both conidia and microconidia give rise to somatic mycelium. (Other species have no conidial stages.) All species reproduce sexually—some being homothallic, others heterothallic; the ascocarp is a PERITHECIUM, and the ascospores have a striate (ridged) surface which is characteristic of the genus.

N. sitophila (the red bread mould) is an hermaphroditic, heterothallic species which forms dark-coloured pyriform perithecia that contain eight-spored asci. Ascocarp development initially involves the formation of a female gametangium (ascogonium) which bears a number of long hyphal processes (trichogynes); a male gametangium is not formed, and plasmogamy may occur in any of a number of ways—e.g. by spermatization (in which the microconidia function as male gametes) or by conidiation (in which the conidia act as male gametes).

N. crassa is similar to *N. sitophila* in that it is an hermaphroditic, heterothallic species which forms eight-spored asci; the remaining species (e.g. *N. tetrasperma*—which forms four-spored asci) are either homothallic or secondarily homothallic.

neuston The community of organisms (mainly algae) whose habitat is the surface film of a body of water. Some (the *epineuston*) rest on the surface and project into the atmosphere; others (the *hyponeuston*) are associated with the *underside* of the surface film and project into the water below.

neutral red (PH INDICATOR) pH 6.8 (red) to pH 8.0 (yellow).

neutralization test Any test in which antibody neutralizes (prevents, inhibits) the expression of a biological function by a microorganism or toxin. Thus, e.g.: (a) bacteriophage-neutralizing antibody may be detected by its ability to inhibit plaque-formation by the homologous phage; (b) a microorganism may be identified, quantified etc. by its failure to exhibit certain features (e.g. growth, ability to infect cells) in the presence of a known, specific antiserum; (c) a toxin may be identified, quantified etc. by its failure to produce toxic effects (in an experimental animal) following admixture with specific antitoxin.

Newcastle disease (of fowl) (see FOWL PEST)

Newcastle disease virus (see PARAMYXOVIRUS)

Newcombe experiment An experiment by which Newcombe (in 1949) demonstrated that the response of bacterial populations to an altered environment may involve the selection of spontaneously-arising mutants.

A number of nutrient agar plates are each spread with the same quantity of inoculum from a single culture of phage-sensitive bacteria. The plates are incubated for a few hours during which time each viable cell gives rise to a CLONE. The plates are then divided into two groups. In group A the bacterial growth on each plate is re-distributed (with a sterile glass spreader) so that the cells of each clone are dispersed over the agar surface. Group B plates are left undisturbed. All plates are then sprayed with a suspension of virulent phage, and re-incubated. During this second period of incubation, all the sensitive bacteria are lysed; thus, colonies formed on any particular plate must have developed from cells which were, or had become, resistant to phage. In practice, the number of colonies formed on group A plates greatly exceeds the number formed on group B plates.

If phage-resistant bacteria developed *adaptively*, resistant cells would be formed *only after the plates had been exposed to phage*. The re-distribution of growth in group A plates (*before* phage inoculation) would therefore not influence the subsequent development of resistant cells, and similar numbers of resistant cells, and hence colonies, should appear on all the plates. If, however, phage-resistant cells arose *spontaneously* during the initial incubation (i.e. *before phage inoculation*) each resistant cell (mutant) would form a discrete clone of resistant cells; subsequent re-distribution of growth on the group A plates would disperse the resistant cells over the agar surface. Thus, following phage inoculation and re-incubation, *each resistant cell* on a group A plate would form a *single colony*. On the group B plates, however, *each undisturbed clone* of resistant cells would form a *single colony*. Hence, far fewer colonies would be formed on the group B plates. (see also FLUCTUATION TEST)

NGU (see NON-GONOCOCCAL URETHRITIS)

niacin Nicotinic acid; some authors use the term to include both NICOTINIC ACID and nicotinamide.

niacin test A diagnostic test used to detect those species of *Mycobacterium* which produce—and liberate into an agar medium—an easily detectable amount of *niacin* (i.e. NICOTINIC ACID). An aqueous extract of the agar medium is obtained by allowing a small volume of distilled water to remain in contact with the surface of the medium—near a region of growth—for 15 minutes; the extract is then placed on a white tile or in a small test-tube. To 0.5 ml of the extract is added, firstly, 0.5 ml aniline (4% in ethanol), and subsequently 0.5 ml cyanogen bromide (10% aqueous). An immediate yellow coloration indicates the presence of niacin. (After the test, cyanogen bromide must be hydrolysed with excess sodium hydroxide prior to autoclaving the apparatus.)

Nichols' treponeme (a) Pathogenic variety: strains of *Treponema pallidum*. (b) Non-pathogenic, cultivable variety: strains of *Treponema refringens*.

nickel salts (as antimicrobial agents) (see HEAVY METALS)

nicotinic acid (niacin) Pyridine-3-carboxylic acid; the amide (*nicotinamide*) is a component of two important coenzymes: *nicotinamide adenine dinucleotide* (NAD) and *nicotinamide adenine dinucleotide phosphate* (NADP)—both of which are involved in hydrogen transfer reactions. (see also Appendix V(c)) Nicotinic acid is synthesized by most bacteria from aspartic acid and a 3-carbon unit derived from glycerol; *Xanthomonas* spp and (at least) most fungi appear to synthesize it from tryptophan. A range of microorganisms require for growth an exogenous source of nicotinic acid (or nicotinamide); these include *Lactobacillus* spp, *Trichophyton equinum*, some yeasts, *Tetrahymena pyriformis*, and a number of parasitic protozoans (e.g. coccidia). *Lactobacillus* spp may be used in the BIOASSAY of nicotinic acid. (Certain *Haemophilus* spp require the preformed nucleotide coenzymes as growth factors: see V FACTOR.) (see also NIACIN TEST)

Nidulariales An order of fungi of the class GASTEROMYCETES. Constituent species form fruiting bodies (approximately 1 centimetre, maximum dimension) in which the gleba is differentiated into one or more basidiospore-containing *peridioles* (glebal masses) which are forcibly projected from the fruiting body by a species-dependent mechanism; an hymenium is not formed, and the basidia occur—singly or in groups—scattered among the gleba or glebal masses. The fruiting bodies of members of the Nidulariales develop e.g. on soil, on dead wood, and on dung; they may be e.g. spheroidal (e.g. *Sphaerobolus*—see also BASIDIOSPORES, DISCHARGE OF) or funnel-shaped (e.g. CYATHUS).

***nif* genes** (see NITROGEN FIXATION)

nifuroxime (see NITROFURANS)

nigeran A linear GLUCAN which contains alternating α-(1 → 3) and α-(1 → 4) linkages. It is synthesized by *Aspergillus niger*.

nigericin (see MACROTETRALIDES)

nigrosin A black composite dye which includes one or other of the *indulins*; it is used e.g. for negative STAINING in light microscopy.

nisin (as a food preservative) A polypeptide ANTIBIOTIC which occurs naturally in certain cheeses. In the UK nisin is a permitted preservative for cheeses and canned foods, but in the latter it is not acceptable as an alternative to adequate heat treatment (see CANNING). Nisin is not a permitted preservative in the USA. (see also FOOD, PRESERVATION OF)

Nitella A genus of algae of the division CHAROPHYTA.

nitratase Nitrate reductase—see NITRATE RESPIRATION and ASSIMILATORY NITRATE REDUCTION.

nitrate broth Nutrient broth containing potassium nitrate (final concentration 0.1–0.2% w/v).

nitrate reduction (see ASSIMILATORY NITRATE REDUCTION and NITRATE RESPIRATION)

nitrate reduction test (nitrate test) A test used to determine whether or not a given bacterial strain can reduce nitrate*. The organism is cultured in NITRATE BROTH for 2–5 days, and the medium then examined for the presence of *nitrite*; the presence of nitrite indicates a positive reaction. To the culture is added 0.5 ml 'reagent A' (0.8% w/v sulphanilic acid in 5N acetic acid) followed by 0.5 ml 'reagent B' (0.6% w/v *N,N*-dimethyl-α-naphthylamine in 5N acetic acid); if present, nitrite reacts with sulphanilic acid to form a diazonium salt which, with the naphthylamine, forms a soluble red azo dye. The absence of coloration could mean that either (i) nitrate has not been reduced, or (ii) the *nitrite* has been reduced. The subsequent addition of a trace of zinc dust (which reduces any nitrate present) resolves this problem: the development of a red colour indicates possibility (i) (i.e. the test is *negative*), while failure to develop colour indicates (ii). *The ability to reduce nitrate may depend on the conditions of growth; thus, e.g. certain bacteria carry out NITRATE RESPIRATION under anaerobic/microaerophilic conditions. ASSIMILATORY NITRATE REDUCTION generally occurs under aerobic conditions in the absence of sources of reduced nitrogen. *Mycobacterium scrofulaceum* has been reported to reduce nitrate only in the presence of lactic acid.

nitrate respiration (dissimilatory nitrate reduction) Under anaerobic or microaerophilic conditions, some bacteria can use nitrate as a terminal electron acceptor for respiratory metabolism. (see ANAEROBIC RESPIRATION) Thus certain bacteria capable only of oxidative metabolism (e.g. *Pseudomonas aeruginosa*, *Paracoccus denitrificans*) can grow under anaerobic conditions *only* in the presence of nitrate. Nitrate is reduced to nitrite by a *nitrate reductase* in a reaction coupled with an ELECTRON TRANSPORT CHAIN (which may or may not be similar to that operative under aerobic conditions). Nitrate reductase receives electrons from a *b*- or a *c*-type cytochrome (depending on species); the electrons may derive from organic substrates or, in certain cases, from inorganic sources e.g. hydrogen (*Paracoccus denitrificans*) or thiosulphate (*Thiobacillus denitrificans*). Organisms which reduce nitrate only as far as nitrite can employ nitrate respiration only to a limited extent since nitrite is toxic. Other organisms can reduce nitrite by a *nitrite reductase*; in *Pseudomonas aeruginosa*, *Paracoccus denitrificans* and *Alcaligenes faecalis* this enzyme is also called CYTOCHROME d_1c. Nitric oxide is formed as an intermediate and may be released in conditions of high nitrite concentration; it is normally reduced to nitrous oxide and/or gaseous nitrogen. (When the products of nitrate respiration are *gaseous* the process is called DENITRIFICATION.) In e.g. certain *Bacillus* spp the end product of nitrate respiration is ammonia. During nitrate respiration energy is obtained by OXIDATIVE PHOSPHORYLATION; the energy yield is generally less than that obtained during aerobic respiration.

Escherichia coli (and some other facultative anaerobes), when grown anaerobically in the presence of nitrate, reduces nitrate—*via* nitrite—to ammonia. In these organisms, nitrite inhibits aconitase and fumarase—hence, under these conditions the tricarboxylic acid cycle cannot function, and normal respiratory metabolism cannot occur. In *E. coli* nitrate reductase appears to be closely associated with the FORMATE HYDROGEN LYASE system—nitrate acting as an electron acceptor in the oxidation of formate. (cf. ASSIMILATORY NITRATE REDUCTION)

nitrification The process in which ammonia is oxidized to nitrite, and nitrite to nitrate, by the strictly aerobic, chemolithotrophic bacteria of the family NITROBACTERACEAE*. The process occurs in two stages: (a) *ammonia to nitrite* ('nitrosofication'). This stage is carried out e.g. by species of *Nitrosomonas*. Initially, ammonia is oxidized to hydroxylamine. Hydroxylamine is then oxidized to nitrite (*via* unknown intermediates)—electrons being transferred to oxygen *via* an ELECTRON TRANSPORT CHAIN; OXIDATIVE PHOSPHORYLATION accompanies this electron transport. (One of the oxygen atoms in nitrite is derived from molecular oxygen.) (b) *Nitrite to nitrate*. This stage is carried out e.g. by species of *Nitrobacter*. The oxidation (in which the oxygen is derived from water) provides energy (as ATP) but cannot directly supply electrons for the reduction of pyridine nucleotides; the latter are believed to be reduced by an ATP-dependent reversal of the electron transport chain.

Nitrification occurs mainly in aerated, neutral to slightly alkaline soils. The process can be agriculturally disadvantageous since nitrates are highly soluble and are easily leached from the soil by rain; attempts have been made to counteract the effects of nitrification by using an inhibitor of ammonia oxidation (e.g. 2-chloro-6-(trichloromethyl)pyridine: 'N-Serve') in conjunction with nitrogenous fertilizers. (see also SEWAGE TREATMENT and NITROGEN CYCLE) *Many heterotrophic microorganisms can carry out nitrification under certain conditions; however, they do not gain energy from the process—which appears to be of little ecological significance.

nitrifying bacteria (see NITROBACTERACEAE)

nitrites (as food additives and preservatives) Nitrites (of sodium and potassium) are used e.g. in bacon and canned meats; they inhibit bacterial growth—particularly under acid conditions—and maintain an attractive coloration in various types of processed meat. The permitted concentrations of nitrites are limited since they are physiologically active substances. (see also FOOD, PRESERVATION OF)

Nitrobacter A genus of gram-negative, rod-shaped pyriform or coccoid, non-sporing bacteria of the NITROBACTERACEAE; strains occur in the soil and in aquatic environments. Cells of the only species, *N. winogradskyi*, are 1–2 microns in length and less than 1 micron wide; they are commonly non-motile. Peripherally arranged cytomembranes form a cap at one pole of the organism. Division occurs by budding. *Obligate* chemolithotrophy obtains in some strains; these fix carbon dioxide, and obtain energy by oxidizing nitrite. Other strains are facultative chemoorganotrophs. Storage compounds include poly-β-hydroxybutyrate. Growth occurs over a wide range of temperatures under alkaline conditions (pH 7.5–8.0). GC ratio: approximately 61%.

Nitrobacteraceae (nitrifying bacteria) A family of gram-negative, obligately aerobic, chemolithotrophic* bacteria; these organisms, which occur in fresh and marine waters and in the soil, obtain energy either by the oxidation of ammonia to nitrite or by the oxidation of nitrite to nitrate. According to species, the cells are rod-shaped, coccoid, spiral, ovoid or lobate, and flagella (when present) may be peritrichous or subpolar. Spores are not formed. Most species divide by binary fission; budding occurs in *Nitrobacter winogradskyi*. Species which oxidize ammonia to nitrite have been placed in the following genera: NITROSOMONAS, *Nitrosococcus*, *Nitrosolobus* and *Nitrosospira*; these organisms are referred to, by some authors, as the *nitrosofying bacteria*. Species which oxidize nitrite to nitrate have been placed in the following genera: NITROBACTER, *Nitrococcus* and *Nitrospina*. *Strains of *Nitrobacter winogradskyi* are facultative chemoorganotrophs. (see also NITRIFICATION)

Nitrococcus A genus of gram-negative cocci of the NITROBACTERACEAE; the cells occur singly or in pairs, each cell being greater than 1.5 microns in diameter. Each cell carries one flagellum or two flagella with basal bodies close together. A random system of tubular cytomembranes occurs in the cytoplasm.

nitrofurans A group of synthetic ANTIBIOTICS; they are derivatives of 2-substituted 5-nitrofurans. The nitrofurans are active against a wide range of gram-negative and gram-positive bacteria, certain fungi, and some protozoa; their mode of action is unknown but it has been suggested that they may function as electron acceptors and thus interfere e.g. with microbial respiratory systems. They are bacteriostatic at low concentrations but may be bactericidal at higher concentrations. *Nitrofurazone* (*furacin*; 5-nitro-2-furaldehyde semicarbazone). A broad-spectrum agent used topically to treat skin infections, bacterial infections of wounds, burns etc., and infections of the mucous membranes. *Nitrofurantoin* (*furadantin*; *N*-(5-nitro-2-furfurylidine)-1-aminohydantoin). A broad-spectrum antibacterial agent widely used in the treatment of urinary-tract infections; following oral administration it is excreted in the urine. Nitrofurantoin is generally effective against coliforms, enterococci and some strains of *Staphylococcus*; *Klebsiella* and *Proteus* are sometimes resistant, and *Pseudomonas aeruginosa* is commonly resistant. *Furazolidone* (*furoxone*) is active against *Salmonella* and *Shigella* and has also been used e.g. for the treatment of giardiasis (*Giardia lamblia*) and vaginitis (*Trichomonas vaginalis*). *Nifuroxime* is an important antifungal agent which is also active against many bacteria and some protozoa. It is used for the topical treatment of *Candida* infections of mucous membranes, and has been used (in conjunction with furazolidone) to combat *Trichomonas* infections of the vagina.

nitrofurantoin (furadantin) (see NITROFURANS)

nitrofurazone (furacin) (see NITROFURANS)

nitrogen cycle A series of interrelated processes in which nitrogen and its compounds are interconverted by living organisms—see diagram.

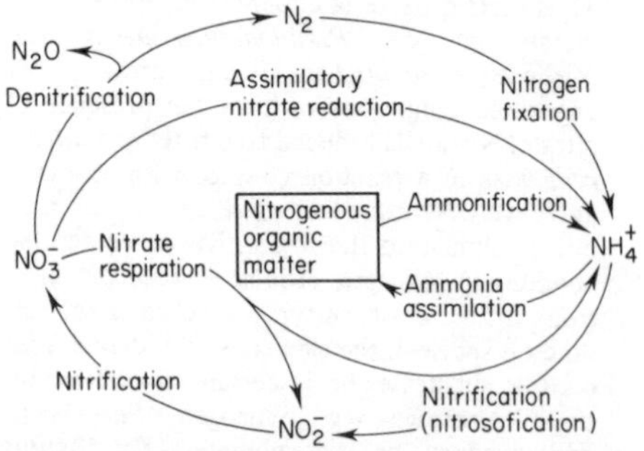

THE NITROGEN CYCLE—microbiological aspects. (See separate entries for individual processes)

nitrogen fixation (*biological*) The reduction of gaseous nitrogen ('dinitrogen') to ammonia.

Nitrogen fixation can be carried out only by certain procaryotes; these include the heterocystous (see HETEROCYST) blue-green algae *Anabaena*, *Calothrix*, *Nostoc*, *Stigonema*, and *Tolypothrix*, the non-heterocystous blue-green algae *Gloeocapsa*, *Plectonema*, and *Trichodesmium*, and the bacteria *Bacillus macerans*, *B. polymyxa*, *Clostridium pasteurianum*, *C. butyricum*, *Desulphovibrio* spp, *Klebsiella pneumoniae*, *Rhizobium* spp, and members of the Azotobacteraceae and Rhodospirillaceae.

Mechanism. This appears to be basically similar in all nitrogen-fixing organisms. The reaction requires NITROGENASE, ATP (as Mg-ATP), and a strong reducing agent; it is believed to involve the sequential transfer of three pairs of electrons to (nitrogenase-bound) nitrogen. Probable (enzyme-bound) intermediates are diimide and hydrazine. Some 4–5 molecules of ATP are hydrolysed for each pair of electrons transferred. The electrons are donated to nitrogenase by a FERREDOXIN or a FLAVODOXIN—which in turn receives electrons from any of a variety of substrates; the latter may include e.g. formate (in *B. polymyxa*, *C. pasteurianum*, *K. pneumoniae*), hydrogen, $NADH_2$ or $NADPH_2$ (in *Azotobacter*, blue-green algae), and the artificial reductant dithionite. In *C. pasteurianum* reducing power and ATP can be obtained from pyruvate: reducing power is provided by the oxidative decarboxylation of pyruvate to acetyl-CoA; acetyl-CoA is converted to acetylphosphate—from which phosphate can be transferred to ADP in a substrate-level phosphorylation. A similar system has been found in *B. polymyxa*, *K. pneumoniae*, *Chromatium* spp, and some blue-green algae.

Assimilation of fixed nitrogen. The following pathway appears to be important in many organisms:

$$NH_4^+ + \text{glutamate} \xrightarrow{\text{glutamine synthetase}} \text{glutamine}$$

$$\text{glutamine} + \alpha\text{-ketoglutarate} \xrightarrow[\text{NADH}_2/\text{NADPH}_2]{\text{glutamate synthetase}} 2\ \text{glutamate}$$

$$\text{glutamate} + \alpha\text{-ketoacid} \xrightarrow{\text{aminotransferase}} \alpha\text{-ketoglutarate} + \text{amino acid}$$

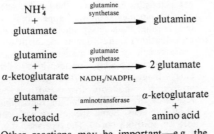

Other reactions may be important—e.g. the formation of alanine from pyruvate, and of citrulline from ornithine.

Genetic aspects. In bacteria the nitrogen-fixing (*nif*) genes appear to constitute an operon which, in *K. pneumoniae*, occurs near the *his* operon; the *nif* genes can be transferred by TRANSDUCTION or during conjugation (see CONJUGATION (bacterial)). (Nitrogen-fixing strains of *Escherichia coli* have been obtained following conjugation between *E. coli* and *K. pneumoniae*.)

Non-symbiotic nitrogen fixation. In certain environments (e.g. rice paddies) most of the nitrogen available to plants is thought to be provided by nitrogen-fixing blue-green algae. Free-living nitrogen-fixing bacteria are thought to contribute minimally to soil nitrogen.

Symbiotic nitrogen fixation. Examples include (a) symbioses between bacteria (*Rhizobium*, *Frankia*) and plants—see ROOT NODULES. In the case of *Rhizobium*, at least, the *nif* genes appear to be located in the endosymbiont. (b) *Klebsiella* spp have been reported to fix nitrogen in the intestines of man and other animals—although the nutritional significance of this to the animal is unknown. (c) Nitrogen is fixed by blue-green algae in a number of LICHENS (see also CEPHALODIUM); at least some of the fixed nitrogen is transferred to the mycobiont. (d) *Anabaena azollae* fixes nitrogen in leaf cavities of the floating fern *Azolla*; this fern is sometimes used in rice paddies etc. as a green manure. (e) *Nostoc punctiliforme* fixes nitrogen in nodules within the stem at the leaf bases of *Gunnera* (a rhubarb-like plant); *N. punctiliforme* is thought to be capable of supplying the entire nitrogen requirement of the plant. (see also NITROGEN CYCLE)

nitrogen mustards Highly reactive, mutagenic, *bifunctional* ALKYLATING AGENTS—the majority of which have the generalized formula: $ClCH_2.CH_2.NR.CH_2.CH_2Cl$. Nitrogen mustards are toxic, cause blisters on contact with the skin, and bring about IMMUNOSUPPRESSION. (see also SULPHUR MUSTARDS)

nitrogenase (dinitrogenase) The enzyme which catalyses biological NITROGEN FIXATION. It consists of two protein components—neither of which can function alone; one (*molybdoferredoxin*, *azofermo*) contains iron, labile sulphide and molybdenum, while the other (*azoferredoxin*, *azofer*) contains iron and labile sulphide only. In the presence of Mg-ATP, nitrogenase can catalyse the reduction of e.g. nitrogen, acetylene, cyanide, azide and nitrous oxide. (The reduction of acetylene to ethylene—detectable by gas chromatography—is utilized in nitrogenase assay.) Azoferredoxin is highly oxygen-labile, and aerobic organisms which contain nitrogenase appear to have evolved various mechanisms to prevent its inactivation; thus, e.g. the presence of an 'oxygen-scavenging' respiratory system in *Azotobacter chroococcum* has been suggested. In heterocystous blue-green algae, nitrogenase may be confined to the strongly reducing environment of the HETEROCYST;

non-heterocystous species may fix nitrogen only under microaerophilic conditions. LEGHAEMOGLOBIN may serve to protect nitrogenase in leguminous ROOT NODULES. The biosynthesis of nitrogenase is inhibited by ammonia and possibly by glutamine synthetase (an enzyme involved in ammonia assimilation).

nitromersal (see *Mercury* under HEAVY METALS)

Nitrosococcus A genus of gram-negative cocci or coccoid bacteria of the NITROBACTERACEAE; the cells, which are about 1.5 microns in diameter, are each bisected by a central band of flattened, lamellate cytomembranes. *N. nitrosus* appears to be non-motile; strains of *N. oceanus* may be motile. All strains are obligate chemolithotrophs. Growth occurs over a wide range of temperatures. GC% ratio: approximately 51.

nitrosofying bacteria (see NITROBACTERACEAE)

Nitrosolobus A genus of gram-negative, obligately chemolithotrophic bacteria of the NITROBACTERACEAE. Cells are irregularly-shaped and lobate, about 1–1.5 microns across; the peripheral part of the cytoplasm appears to be divided into a number of membrane-limited vesicle-like regions. Cell division is reported to occur 'by constriction'. GC ratio: approx. 54–55%.

Nitrosomonas A genus of gram-negative, obligately aerobic, non-sporing bacteria (family NITROBACTERACEAE) found in the soil and in freshwater and marine habitats. All strains are obligate chemolithotrophs—obtaining carbon from carbon dioxide, and energy from the oxidation of ammonia. The cells are round-ended rods, 1–2 microns in length and somewhat less than 1 micron wide; cytomembranes occur peripherally. Cells may be non-motile or may have one or two subterminal flagella. Growth occurs in the pH range 6–8 over a wide range of temperatures. GC% approx. 48–51.

Nitrosospira A genus of gram-negative, motile or non-motile, obligately chemolithotrophic bacteria of the NITROBACTERACEAE; individual cells are spiral-shaped, and cytomembranes are absent.

Nitrospina A genus of gram-negative, non-motile bacteria of the NITROBACTERACEAE; the rod-shaped cells are less than 0.5 microns wide and greater than 2.5 microns in length. Cytomembranes are reduced or absent.

nitrous acid (as a MUTAGEN) Nitrous acid (HNO_2) causes the oxidative deamination of guanine, cytosine, and adenine—forming xanthine, uracil, and hypoxanthine, respectively. During the replication of DNA which has been treated with nitrous acid, *xanthine* pairs with cytosine which, in subsequent replication, pairs with guanine—hence no mutation results from the deamination of guanine. *Uracil* pairs with adenine; thus, deamination of cytosine results in a G-C to A-T transition. *Hypoxanthine* pairs with cytosine; thus, deamination of adenine results in an A-T to G-C transition.

Nitzschia A genus of pennate diatoms of the BACILLARIOPHYTA.

NLF(*bacteriol.*) A non-lactose-fermenting organism.

nm (nanometre) 10^{-9} metre; one millimicron.

NNN medium A blood-agar medium used for the culture of certain species of *Trypanosoma* and *Leishmania*. It is prepared by the addition of 1 ml of (whole) vertebrate blood to 5 mls of molten agar (1.5% w/v agar in physiological saline) at 50 °C; after mixing, the molten blood-agar is allowed to set in a universal bottle placed in a sloping position. After inoculation, incubation may be carried out for periods ranging from 3 days to several weeks at temperatures within the range 28–36 °C. Organisms which grow on the medium may be obtained, for examination, by (aseptically) removing a loopful of fluid from the base of the agar slope.

Culture of a pathogenic flagellate on a blood-agar medium may result in the development of only one morphological type; thus, e.g. when *Leishmania donovani* is cultured (at 25–30 °C) only the *promastigote* (leptomonad) form is produced. However, cultures of *Trypanosoma cruzi* (28 °C) may contain a range of types—including a small proportion of the infective (metacyclic) forms.

noble rot (*pourriture noble*) A rot of over-ripe grapes caused by *Botrytis cinerea*. The infected grapes lose water—their sugars etc. becoming more concentrated. Such grapes are used for making high-quality sweet wines. (see also WINE-MAKING)

Nocardia A genus of gram-positive, aerobic bacteria of the NOCARDIACEAE; the genus characteristics are those of the family. According to species, colonial appearance and texture range from glossy and butyrous to matt and leathery; the former description is characteristic of those species which tend to fragment more readily. Species which undergo early fragmentation often give rise to *microcysts*—coccoid forms with a heat resistance greater than that of the ordinary vegetative cells. Some species are pathogenic (see e.g. MADURA FOOT). The type species is *N. farcinica* and the GC% range for the genus is 61–72. The antibiotic RISTOCETIN is derived from certain *Nocardia* spp.

Nocardiaceae A family of gram-positive, aerobic bacteria of the ACTINOMYCETALES; the organisms may form scant or well-developed mycelium—according to species—but mycelial fragmentation eventually occurs with the production of cocci, bacilli or branched, filamentous forms. All species have a Type IV cell wall (see ACTINOMYCETALES); some species are ACID FAST or weakly acid fast. A few species are pigmented. The genera are NOCARDIA and *Pseudonocardia.*

Noctiluca A genus of algae of the DINOPHYTA.

Nodosaria A genus of the FORAMINIFERIDA.

nomen rejiciendum In taxonomy, any name which has been formally described as unacceptable.

nomenclature (*microbiol.*) The *naming* of organisms: one of the functions of TAXONOMY. The practice of nomenclature is governed by an array of Codes, Rules, and priorities laid down by *ad hoc* Committees—though such guidelines are not always universally adhered to; quite frequently different authors use different names when referring to the same taxon.

With some exceptions (see later) each species of microorganism is referred to by a unique BINOMIAL. (The names of species and genera are usually printed in italics; in typewritten and handwritten manuscripts such names are usually underlined.) Where ambiguity cannot arise the generic name may be abbreviated; thus, e.g. *Staphylococcus aureus* may be abbreviated to *Staph. aureus* or *S. aureus.* The name of a species is sometimes followed by an *authority citation*; this gives the name(s) of the author(s) who first proposed the name in a 'valid publication' and often gives the date of that publication—e.g. *Bacillus cereus* Frankland and Frankland 1887. (A *valid* publication includes a description of the organism and a correctly formulated Latin or latinized name; such a publication must be in printed form and must also be made available to the public.) Sometimes a *double citation* is used—e.g. *Escherichia coli* (Migula) Castellani and Chalmers 1919; the name(s) in parenthesis refer to the author(s) who first described the organism and referred to it by its *original* name, while the name(s) which follow indicate the author(s) responsible for the organism's *present* name. (Authors' names may appear in abbreviated form.)

Sub-species, i.e. physiologically or otherwise distinguishable organisms within a given species, may be referred to e.g. by *trinomials*—such as *Puccinia graminis tritici*: see e.g. FORMA SPECIALIS. SEROTYPES may be referred to by their antigenic formulae (see e.g. KAUFFMANN–WHITE CLASSIFICATION SCHEME) or may be named individually (see e.g. *poliovirus* under ENTEROVIRUS).

While an organism may bear only one valid name, certain fungi (members of the DEUTEROMYCOTINA) may also be referred to by a second binomial. These fungi typically occur in the imperfect (asexual) state and are classified, and named, on the basis of their (asexual) reproductive structures and spores; as fungal taxonomy is based preferentially on the characteristics of the *sexual* stage (which reflect natural affinities), classification based on asexual characteristics is indicated by referring to such taxa as *form*-families, *form*-genera, *form*-species etc. When the sexual stage of a given form-species is discovered the organism is usually referable to a *genus* within the ascomycetes or (less frequently) the basidiomycetes; such an organism thus bears the name of the form-species (used to refer only to the imperfect—asexual—stage of the organism) and the name of a *species* within an ascomycete or basidiomycete genus. The latter name is the valid name of the organism and may be used to refer to the sexual and asexual stages of the fungus. The name of the form-species may be written, in parenthesis, after that of the perfect stage—e.g. *Leptosphaeria avenaria* (*Septoria avenae*).

Each taxon above the rank of genus is referred to by a uninomial, i.e. a name consisting of one term only. The rank of a particular named taxon is often indicated by a standardized suffix; some of these suffixes are given in the accompanying table.

SUFFIXES USED IN MICROBIAL NOMENCLATURE

Taxon	*Algae*	*Bacteria*	*Fungi*	*Protozoa*
Division (Phylum)	-phyta*			
Subdivision	-phytina*		-mycotina*	
Class	-phyceae*		-mycetes*	
Subclass	-phycidae*		-mycetidae*	
Order	-ales	-ales	-ales	
Suborder	-ineae	-ineae	-ineae	
Superfamily				-oidea*
Family	-aceae	-aceae	-aceae	-idae
Subfamily	-oideae	-oideae	-oideae	-inae
Tribe	-eae	-eae	-eae	-ini*
Subtribe	-inae	-inae	-inae	
Genus				

* Though not mandatory, these suffixes are recommended in the relevant Code.

Viruses often bear *trivial* names, but many have been assigned to viral genera—see 'taxonomy' under VIRUSES.

nonactin (see MACROTETRALIDES)

non-agglutinating antibodies (see BLOCKING ANTIBODIES)

non-cellular (see CELLULAR)

noncytopathogenic (*virol.*) Refers to those viruses which do not cause cytopathic effects (CPE) in tissue cultures.

non-disjunction (genetics) Any abnormal distribution of chromosomes, chromatids, or parts of chromatids to the poles of the spindle during mitosis or meiosis. (see also PARASEXUALITY IN FUNGI)

non-gonococcal urethritis (NGU) URETHRITIS of unspecified, but non-gonococcal, causation; NGU may refer e.g. to urethritis due to infection by *Chlamydia trachomatis*.

Nonidet (*proprietary name*) A non-ionic detergent used e.g. in structural and serological analyses of certain types of virus.

non-mendelian inheritance (see CYTOPLASMIC INHERITANCE)

non-persistent viruses (*plant pathol.*) (see STYLET-BORNE VIRUSES)

non-reciprocal recombination (see RECOMBINATION)

nonsense codon (see GENETIC CODE)

nonsense mutation A mutation in which a codon specifying an amino acid is altered to a polypeptide chain-terminating codon (i.e. a nonsense codon)—see GENETIC CODE and PROTEIN SYNTHESIS. (cf. MIS-SENSE MUTATION)

non-specific immunity (*immunol.*) The constitutive or acquired resistance of the body to foreign materials in general (living and non-living)—including those with which it has had no previous contact. Factors which may be involved in non-specific immunity include e.g. (a) PHAGOCYTOSIS. (b) The presence of LYSOZYME in secretions (e.g. tears). (c) The activities of INTERFERONS and PROPERDIN. (cf. SPECIFIC IMMUNITY)

non-specific urethritis (NSU) URETHRITIS of unspecified, but non-gonococcal, causation; NSU may refer e.g. to urethritis due to infection by *Chlamydia trachomatis*.

nonsulphur purple bacteria (Athiorhodaceae) (see RHODOSPIRILLACEAE)

nori (see MICROORGANISMS AS FOOD)

North American blastomycosis (see BLASTOMYCOSIS)

Nosema A genus of parasitic protozoans (class Microsporidea, sub-phylum CNIDOSPORA). Many species are parasitic in insects; thus, e.g. *N. apis* is the causal agent of an intestinal disease of the honey-bee (*nosema disease*), and *N. bombycis* causes PÉBRINE in the silkworm. *N. cuniculi* (= *Encephalitozoon cuniculi*) is a common parasite of rodents; it was recorded as the causal agent in a single case of human encephalitis—being isolated from the cerebrospinal fluid and urine of the patient. The spores of *N. cuniculi* are ovoid, approximately 2.5 × 1.5 microns.

nosocomial infection An infection acquired while in hospital.

Nostoc A genus of filamentous BLUE-GREEN ALGAE. Species occur in freshwater and marine environments and in symbiotic association with other organisms; *Nostoc* spp occur in certain LICHENS (see e.g. PELTIGERA), and *N. punctiliforme* forms a symbiotic relationship with *Gunnera* (a rhubarb-like plant). The filament resembles a string of beads and is commonly blue-green or olive-green in colour; *Nostoc* spp form AKINETES and HETEROCYSTS and carry out NITROGEN FIXATION.

notatin A glucose oxidase which is produced, intracellularly, by *Penicillium notatum* and other fungi. The enzyme is a flavoprotein which oxidizes glucose to gluconic acid (*via* gluconolactone) with accompanying uptake of oxygen and formation of hydrogen peroxide. Notatin has been used e.g. in analytical processes for the detection of glucose in urine etc.; it can also show antimicrobial activity due to the production of hydrogen peroxide.

notifiable diseases (of animals) Those diseases the occurrence (or suspected occurrence) of which must, by law, be notified to the police or to the appropriate government authority. In Great Britain, the notifiable diseases (in 1976) *included*: ANTHRAX, cattle plague (RINDERPEST), DOURINE, FOOT AND MOUTH DISEASE, FOWL PEST, GLANDERS and farcy, bovine pleuropneumonia, RABIES, SHEEP POX (variola ovina), sheep scab, SWINE FEVER, SWINE VESICULAR DISEASE, TESCHEN DISEASE of pigs, bovine TUBERCULOSIS of the udder—and tubercular emaciation and cough accompanied by other, definite symptoms of tuberculosis.

notifiable diseases (human) Those diseases the occurrence of which must, by law, be reported to the appropriate government authority. Notifiable diseases in England and Wales *include*: acute POLIOMYELITIS, ANTHRAX, CHOLERA, DIPHTHERIA, DYSENTERY, FOOD POISONING, GREEN MONKEY FEVER, LASSA FEVER, LEPROSY, LEPTOSPIROSIS, MALARIA, MEASLES, ophthalmia neonatorum (see GONORRHOEA), PARATYPHOID FEVER, PLAGUE, RABIES, RELAPSING FEVER, SCARLET FEVER, SMALLPOX, TETANUS, TUBERCULOSIS, TYPHOID FEVER, TYPHUS FEVERS, WHOOPING COUGH, and YELLOW FEVER.

notifiable diseases (of plants) Those diseases the occurrence (or suspected occurrence) of which must, by law, be notified to the appropriate government authority. In England and Wales the notifiable diseases are FIRE BLIGHT, PLUM POX, *progressive wilt* of the hop (*Humulus lupulus*), *red core* of the STRAWBERRY, and *wart disease* of the POTATO.

novobiocin (albamycin) An ANTIBIOTIC produced by *Streptomyces spheroides*, *Streptomyces niveus* and certain other species. It is a substituted coumarin, 4,7-dihydroxy-3-amino-8-methylcoumarin, linked to a substituted *p*-hydroxybenzoic acid and to a substituted hexose, *noviose*. Novobiocin is active mainly against gram-positive bacteria (particularly staphylococci and certain streptococci) and against some gram-negative species (e.g. *Neisseria* spp, *Haemophilus influenzae*). Novobiocin inhibits DNA polymerase and, subsequently, cell wall and protein synthesis. Resistance to the antibiotic develops readily, and it has been used jointly with tetracycline.

Novyella A sub-genus of PLASMODIUM.

Nowakowskiella A genus of eucarpic, polycentric fungi within the order CHYTRIDIALES; *Nowakowskiella* spp are typically aquatic saprophytes. On germination, a zoospore gives rise to a RHIZOMYCELIUM bearing a number of branched rhizoids which penetrate the substratum. *Operculate* sporangia (see OPERCULUM) develop on different parts of the rhizomycelium; the vegetative cycle is completed when zoospores are liberated from the sporangia. Thick-walled resting spores are produced—but it is not known whether or not sexual processes occur.

NSU (see NON-SPECIFIC URETHRITIS)

nuclear area (see NUCLEUS)

nucleic acids NUCLEOTIDE polymers—see DNA and RNA.

nucleocapsid (*virol.*) (1) (nucleoprotein) A helical structure composed of nucleic acid (the viral genome) and protein (the CAPSID). In some viruses—e.g. the rhabdoviruses—the nucleocapsid is contained within a lipoprotein outer membrane; in e.g. the tobacco mosaic virus the nucleocapsid is naked, i.e. non-enveloped. (cf. CORE) (2) *Nucleocapsid* is also used to refer to any complete non-enveloped (or unenveloped) virion.

nucleoid body (see NUCLEUS)

nucleolar organizer That region of a eucaryotic chromosome which is associated with the formation and function of the NUCLEOLUS; the region contains the genes for ribosomal RNA (see RNA). Normally, at least one nucleolar organizer is present per haploid set of chromosomes.

nucleolus An RNA-rich, Feulgen-negative, intranuclear body which is not bounded by a limiting membrane; the nucleolus is the site of rRNA synthesis in eucaryotes (see entry RNA). Nucleoli can be observed in interphase nuclei but they do not persist during mitosis; typically they become undemonstrable in prophase (cf. ENDOSOME). Nucleoli appear to absent e.g. from the micronuclei of ciliate protozoans.

nucleoside A class of compound in which a purine or pyrimidine base is linked to a pentose sugar—usually ribose (in *ribonucleosides*) or 2-deoxyribose (in *deoxyribonucleosides*). The ribonucleoside corresponding to each of the bases adenine, guanine, cytosine, uracil, thymine, and hypoxanthine, is called (respectively) adenosine, guanosine, cytidine, uridine, thymidine*, and inosine. For the corresponding deoxyribonucleosides, these names are given the prefix *deoxy*—e.g. deoxyadenosine. *('Thymidine' is sometimes used to refer to the *deoxy*ribonucleoside, since thymine occurs principally in DNA.)

nucleoside antibiotic An ANTIBIOTIC whose structure includes a NUCLEOSIDE; examples include: decoyinine, the POLYOXINS, PSICOFURANINE, and PUROMYCIN.

nucleotide A phosphorylated NUCLEOSIDE—the phosphate group(s) occurring on the sugar residue. (For the structure of some important nucleotides see Appendix V(c); see also DNA and RNA. For the biosynthesis of ribonucleotides see Appendixes V(a) and V(b).) In *Escherichia coli* deoxyribonucleotides are generally synthesized by the reduction of ribonucleoside-5′-diphosphates* (in a reaction involving THIOREDOXIN); in *Lactobacillus leichmannii* ribonucleoside-5′-*tri*phosphates are reduced (in a reaction involving vitamin B_{12}). Antibiotics which interfere with nucleotide biosynthesis include AZASERINE, 6-DIAZO-5-OXO-L-NORLEUCINE, HADACIDIN, MYCOPHENOLIC ACID, and PSICOFURANINE. *(*Primed* numbers, e.g. 5′, refer to positions on the *sugar* residue.)

nucleus (1) (eucaryotic) An intracellular, membrane-limited body which contains e.g. the CHROMOSOMES and one or more nucleoli and/or endosomes (see NUCLEOLUS and ENDOSOME); the number of chromosomes per nucleus depends e.g. on species. (see also PLOIDY) The nuclear membrane appears to be a *double* unit-type membrane which is perforated by a number of pores; the space between the two component membranes is termed the *perinuclear space*.

A cell may contain more than one nucleus (see also COENOCYTE and SYNCYTIUM); ciliate protozoans typically have two *kinds* of nuclei: see MACRONUCLEUS and MICRONUCLEUS. *Fungal* nuclei are generally smaller than those of most other eucaryotes. (see also MITOSIS and MEIOSIS)

(2) (procaryotic) The 'nucleus' of a procaryotic cell (synonyms: *chromatinic body*, *nuclear area*, *nucleoid body*) corresponds to

the chromosome(s); procaryotic CHROMOSOMES appear to be in direct contact with the cytoplasm, i.e. they are not enclosed within a nuclear membrane, and they are not associated with nucleoli or endosomes. The number of chromosomes per bacterial cell appears to be influenced by the growth rate—see Helmstetter–Cooper Model in GROWTH (bacterial).

numerical aperture (N.A.) (see RESOLVING POWER)

numerical taxonomy The modern form of ADANSONIAN TAXONOMY; in numerical taxonomy an organism is assigned to a particular group solely on the basis of observable similarities between that organism and those of the group—i.e. it is a purely *phenetic* system of classification. (This is in contrast to classical (*phylogenetic* or *phyletic*) taxonomy in which the prime aim is to group organisms according to their *evolutional* affinities.) (see also TAXONOMY)

In numerical taxonomy each organism (OTU—operational taxonomic unit) among those to be classified is initially defined by a number of parameters which may include—for example—morphological, biochemical, and/or physiological characteristics; the same parameters are used for the definition of each of the organisms being classified. Any number of characteristics, above a certain minimum, may be used. Each characteristic is expressed in a manner such as (+), (–) or 0,1; thus, e.g. the characteristic 'growth at 45 °C' could be expressed as (+) (growth occurs at 45 °C) or (–) (growth does not occur at 45 °C). *Each* OTU is thus defined by a series of pluses and minuses—each plus and minus indicating the nature of the organism in respect of a particular characteristic; this is the form in which the data are fed to a computer. (The use of a computer is not strictly essential, but the calculations involved would otherwise take an inordinate length of time.)

Within the computer the first step in classification is the comparison of each OTU with every other OTU—likenesses being expressed as 'percentage similarities'; for example, two given OTUs which correspond in any 60 out of 80 characteristics used would have a mutual similarity of 75%. It follows that equal significance is accorded to each of the characteristics used in the comparison—a central feature of numerical taxonomy. The result of these initial comparisons is a number of values—each of which is the percentage similarity between two given OTUs. The computer is programmed to search these values for pairs, or groups, of OTUs which exhibit a mutual similarity of 100%—i.e. pairs or groups of 'identical organisms'; such a pair or group is referred to as a '100% cluster'. On the computer 'printout' such clusters—if present—may appear thus:

```
100% 5-10-11-19-
     43-47-
     29-31-33-
```

Each number (5,10 etc.) is the designation of one of the OUTs being classified; the printout indicates that three clusters are formed at the 100% level of similarity. Although strains within a given cluster are mutually 100% similar, no information is given on the degree of similarity which exists between the OTUs of one cluster and those of another. The remaining OTUs are progressively added at successive, lower levels of similarity (e.g. 97%, 95%, 80%, 75%, 50%, 40%) to one or other of the initially-formed clusters. Additionally, two or more clusters may coalesce at given levels of similarity; thus, e.g. (in the above example) the clusters 43-47- and 29-31-33- may form a single cluster at, say, 75%—meaning that all five OTUs have a mutual similarity of 75%. The mode in which OTUs are added to clusters, and in which clusters coalesce, may be regulated according to the purpose of the taxonomist.

The term *phenon* has been proposed to refer to any group of organisms established by the process of numerical taxonomy; thus an 80% phenon (or an 80-phenon) would correspond to a cluster at the 80% similarity level etc. However, the phenon is not necessarily capable of direct translation into terms of taxonomy since a significant range of organisms may have been omitted from the survey used for its definition.

To date, numerical taxonomy appears to have been used mainly for studying intra-genus relationships in bacteria.

nutrient agar NUTRIENT BROTH gelled with 1.5%–2.0% AGAR: a general-purpose solid medium which may be made selective for, or inhibitory to, particular organisms by the addition of appropriate substances.

nutrient broth A *basal* MEDIUM, i.e. one which supports the growth of a range of nutritionally undemanding chemoorganotrophs; such a medium may be supplemented with particular nutrients and/or growth factors (to support the growth of nutritionally fastidious organisms) and/or with selectively inhibitory substances (to suppress the growth of unwanted organisms). *Nutrient broth* may refer either to extract broth or to infusion broth. *Extract broth* (the type most commonly used) is an aqueous solution of beef extract (0.5%–1.0%), peptone (0.5%–1.0%), and sodium chloride (0.5%). The medium is sterilized by autoclaving; its final pH should be approximately 7.4. Powdered (dehydrated) extract broth is available commercially; it is reconstituted with distilled (or tap) water, dispensed into tubes etc., and sterilized by autoclaving. *Infusion broth* is prepared by infusing overnight, at 4 °C, 50 gm of minced lean beef in 100 ml of distilled water; peptone (1.0%) and sodium chloride (0.5%) are added to the mixture—which is boiled, filtered, and sterilized by autoclaving. The final pH should be about 7.5.

nystatin (fungicidin) (see POLYENE ANTIBIOTICS)

O

O/129 The vibriostatic agent 2,4-diamino-6,7-diisopropyl pteridine. (see AEROMONAS and VIBRIO)

O antigens (Boivin antigens) The LIPOPOLYSACCHARIDE-protein somatic antigens of gram-negative bacteria—in particular those of the ENTEROBACTERIACEAE. Antigenic specificity is determined by a region of the polysaccharide known as the *O-specific chain* (see LIPOPOLYSACCHARIDES). O antigens are important in the serological characterization of *Salmonella*, *Shigella* and other bacteria. (see KAUFFMANN–WHITE CLASSIFICATION SCHEME) O antigen may be subject to modification or variation—see e.g. SMOOTH–ROUGH VARIATION. In some strains of *Salmonella* a certain mutational event inhibits the synthesis of normal O-specific chains and gives rise exclusively to *TI side chains* which contain residues of ribose and galactose. (see also ENDOTOXINS)

O-specific chain (see LIPOPOLYSACCHARIDES and O ANTIGENS)

Oakley–Fulthorpe procedure (*immunol.*) A PRECIPITIN TEST of the *double-diffusion* type. The antibody is incorporated in a layer of agar at the bottom of a narrow test-tube. Above this is a layer of plain agar and an aqueous solution of antigen is placed on top. Antigen and antibody diffuse into the plain agar in which the band of precipitate forms.

oakmoss A common name for the lichen *Evernia prunastri*—see EVERNIA.

obligate (*microbiol.*) (*adjective*) Refers to any state or condition which is an essential attribute of a given organism. Thus, e.g. an *obligate anaerobe* is an organism which grows only under anaerobic conditions; an *obligate parasite* is an organism which, in nature, can grow only as a parasite. (Some obligate parasites can be grown in artificial media; some authors do not regard these organisms as *obligate* parasites.) (cf. FACULTATIVE)

ocellus (*pl.* ocelli) A large complex EYESPOT which occurs in certain species of dinoflagellate.

ochre codon The codon UAA (in mRNA) which codes for polypeptide chain termination—see PROTEIN SYNTHESIS and GENETIC CODE. (cf. AMBER CODON, UMBER CODON)

ochre mutation A NONSENSE MUTATION which gives rise to the OCHRE CODON.

ochre suppressor (see SUPPRESSOR MUTATION)

Ochrolechia A genus of crustose LICHENS in which the phycobiont is *Trebouxia*. The apothecia are lecanorine. Species include e.g.:
O. parella. The thallus is grey, granular, circular in outline, with concentric light and dark zones—each light and dark zone representing one year's growth; the periphery of the thallus is surrounded by a white hypothallus. The apothecia have pinkish discs and are pruinose. *O. parella* occurs on rocks and walls and occasionally on trees.
O. tartarea (the 'cudbear lichen'). The thallus is grey to brownish-grey and coarsely granular. The apothecia are large (3–5 millimetres in diameter); each has a light brown to pinkish-orange disc and a thick thalline margin. *O. tartarea* occurs on trees, rocks, mosses, other lichens, etc. It was formerly used commercially as a source of the purple dye *cudbear*.

Ochromonas A genus of algae of the CHRYSOPHYTA.

ocular The eyepiece of a microscope.

Odontostomatida An order of protozoans of the subclass SPIROTRICHIA.

O-F test (see OXIDATION-FERMENTATION TEST)

Ogawa serotype (see VIBRIO)

oidium (*pl.* oidia) (*mycol.*) A thin-walled cell formed by the fragmentation of a hypha; an oidium may function as a conidium or as a male gamete.

oil-immersion objective (see RESOLVING POWER)

Okazaki fragments (see DNA)

old tuberculin (OT) A filtrate of a heated broth culture of *Mycobacterium tuberculosis*.

oleandomycin A MACROLIDE ANTIBIOTIC produced by *Streptomyces antibioticus*; it is closely related to *erythromycin*.

oligodynamic effect Generally refers to the bacteriostatic action exerted by certain metals in elemental form, e.g. silver, copper. (Other microorganisms, particularly certain algae, may also be inhibited.) The oligodynamic effect may be studied by placing small metal discs on a freshly-inoculated lawn plate and measuring the zones of non-growth surrounding each disc following appropriate incubation. (see also HEAVY METALS)

oligomycin An antibiotic which binds to a component of the mitochondrial inner membrane and causes inhibition of OXIDATIVE PHOSPHORYLATION but not of other energy-linked processes (ion transfer etc.)—unless these are ATP-linked (see ELECTRON TRANSPORT CHAIN). Electron transport is (indirectly) inhibited due to RESPIRATORY CONTROL.

oligosaprobic zone (see SAPROBITY SYSTEM)

Oligotrichida An order of protozoans of the subclass SPIROTRICHIA.

oligotrophic (of lakes etc.) Poor in those nutrients which support the growth of aerobic photosynthetic organisms (cf. EUTROPHIC).

olive knot (see GALLS)

Olpidium A genus of endobiotic, holocarpic fungi of the order CHYTRIDIALES; *Olpidium* spp are parasitic on algae and other aquatic organisms and on a range of terrestrial plants. The asexual cycle is initiated when a zoospore settles on a susceptible host and encysts; the contents of the cyst enter the host cell and, following growth and repeated nuclear division, eventually form a sporangium which contains a large number of zoospores. The zoospores subsequently escape *via* one or more pores. The sexual cycle is initiated when two gametes (morphologically indistinguishable from zoospores) fuse and subsequently encyst on the surface of a susceptible cell; the contents of the cyst pass into the cell and develop into a thick-walled resting sporangium. Karyogamy (and probably meiosis) occurs prior to germination of the resting sporangia—each of which produces a number of zoospores.

omasum (see RUMEN)

***omega*-oxidation** (of hydrocarbons) (see ALIPHATIC HYDROCARBONS (microbial utilization of))

omnivorous (of protozoans) Refers to protozoans which feed on other protozoans and on bacteria and/or algae.

oncogene hypothesis An hypothesis which postulates that most or all vertebrate cells contain, as a normal part of the genome, genetic information which specifies a C-type RNA tumour virus (see LEUKOVIRUS); genes which specify the virus, and which are normally in a repressed state, are termed *virogenes*. Certain of the virogenes are termed *oncogene(s)*; if expressed, the oncogene(s) specify changes which lead to cell TRANSFORMATION. The hypothesis postulates that, in the presence of oncogenic agents, some or all of the virogenes become derepressed; according to which of the virogenes are derepressed, the effects of the oncogenic agent may include virion synthesis and/or cell transformation. The oncogene hypothesis is consistent with a number of observations which include: (a) the 'spontaneous' intracellular appearance of C-type RNA tumour viruses following the repeated subculturing of certain cells which gave no indication of previous virus infection; (b) the induction of viruses (by oncogenic agents) in cells which, previously, gave no indication of containing viruses or their antigens; (c) the demonstration, by hybridization studies, of the occurrence of C-type viral genetic information in cells which are not synthesizing C-type virus. These observations are also compatible with the PROVIRUS HYPOTHESIS.

oncogenic (*adjective*) Capable of giving rise to a TUMOUR or bringing about cell TRANSFORMATION.

oncornaviruses (thylaxoviruses) Oncogenic RNA viruses—see LEUKOVIRUS.

one-step growth experiment A procedure first used by Ellis and Delbrück (1939) for the quantitative study of BACTERIOPHAGE multiplication. A dilute suspension of phage is mixed with a culture of sensitive bacteria such that the *bacteria* are in considerable excess; under these conditions a given bacterium is unlikely to be infected by more than one phage. Several minutes are allowed for phage adsorption, after which any unadsorbed phages are neutralized by the addition of anti-phage antiserum. The mixture is then diluted with warm broth (to 'dilute out' the antiserum) and incubated at the growth temperature of the bacterium. Every few minutes (for 30 minutes) a sample is withdrawn from the mixture and used to inoculate a LAWN PLATE of sensitive bacteria. The plates are incubated (usually 6–18 hours) and a count subsequently made of the number of PLAQUES on each plate. Samples withdrawn during the initial stage of the incubation (the *latent period*) give rise to a fairly uniform number of plaques. However, samples taken in the minutes immediately following the latent period produce plaque counts which indicate a sudden, sharp increase in the number of plaque-forming particles in the mixture; samples taken subsequently produce plaque counts which reach a plateau. These observations are interpreted as follows:

During the latent period none of the infected cells has lysed so that, when plated, each cell—the potential source of many phage particles—forms a single *infectious centre* which gives rise to a *single* plaque; the number of plaques per plate thus indicates the number of infected cells per sample. (An infectious centre is thus analogous to a PLAQUE-FORMING UNIT.) The latent period, i.e. the time between infection and the onset of lysis, may differ for different phage-bacterium systems; for the system phage T2/*Escherichia coli* B the latent period is 20–25 minutes. At the end of the latent period lysis begins to occur among the infected cells. Every infected cell does not lyse at the same instant; thus, samples withdrawn immediately after the latent period contain a mixture of infectious centres (infected, unlysed cells) and the phage progeny from lysed cells. Since each cell, on lysis, releases many phage particles, and since each phage is potentially capable of forming a plaque, a considerable increase in the plaque count occurs. Samples withdrawn later contain even greater numbers of phages since more cells will by then have lysed; when all the initially infected cells have lysed the mixture will contain a peak number of phage particles (plaque count plateau). From the plaque counts of the latent period and those obtained from the plateau, the *burst size* can be estimated, i.e. the average number of phages released per infected cell.

If samples of phage-infected bacteria are lysed prematurely (e.g. disrupted by chloroform) at different times during the latent period, it is possible to distinguish the so-called *eclipse period*. The eclipse period begins at the same time as the latent period but is of shorter duration. In the eclipse period no infectious phage particles can be obtained from experimentally-disrupted infected cells since, during this period, phage components are being synthesized and assembled.

An analogous procedure may be used with viruses other than phages.

onion, diseases of These include (a) *White rot*. Causal agent: *Sclerotium cepivorum*. The leaves become yellow and necrotic, while roots and scale bases decay and become covered with white fluffy mycelium. Black *sclerotia* develop on or in necrotic tissues. The disease is encouraged by low soil temperatures. Leeks and garlic may also be attacked. (b) *Neck rot* (*grey mould neck rot*). Causal agent: *Botrytis allii*. Infection *via* neck tissues commonly occurs during harvest—particularly under wet conditions. Softening of bulb tissues is followed by the development of a thick grey mycelial mat. Shallots may also be attacked. (c) *Smudge*. Causal agent: *Colletotrichum circinans*. Mainly the white varieties of onion are attacked. Irregular black patches appear on the bulb surface; these patches consist of fungal *stromata* which may, under favourable (wet) conditions, give rise to masses of conidiophores interspersed with long dark setae. Leeks may also be attacked.

ONPG test A test which detects β-D-galactosidase activity in bacteria. Organisms are grown in buffered peptone water (pH 7.5) containing ONPG (*o*-nitrophenyl-β-D-galactopyranoside). β-Galactosidase hydrolyses ONPG to galactose and *o*-nitrophenol—the latter being yellow at pH 7.5. Organisms which give a positive ONPG test (i.e. produce *o*-nitrophenol from ONPG) do not *necessarily* split lactose since, although they contain β-galactosidase, they may not synthesize lactose 'permease'. (ONPG can enter cells without a specific 'permease'.)

ontogeny The mode of development of an individual—from embryonic form to maturity. The supposition that 'ontogeny recapitulates PHYLOGENY' was made by the 19th Century biologist Haeckel.

onychomycosis Any fungal infection of the nail(s); the causal agent may be a species of *Candida* (usually *C. albicans*) or a dermatophyte (e.g. *Trichophyton mentagrophytes*)—infection by a dermatophyte being termed *tinea unguium*.

Onygena A genus of fungi of the order EUROTIALES.

oocyst (1) (*mycol.*) An OOGONIUM. (2) (*protozool.*) A cyst in which *sporozoites* are formed; oocysts are formed e.g. by *Plasmodium* spp and by the coccidia. (3) A fertilized female gamete which has encysted.

oogamy A form of fertilization which may involve (a) a motile male gamete and a relatively large non-motile female gamete, or (b) GAMETANGIAL COPULATION or GAMETANGIAL CONTACT in which the gametangia are morphologically differentiated.

oogonium (*oocyst*) (*mycol.*) A female structure within which develop one or more gametes (*oospheres*); on fertilization, the oospheres differentiate to form thick-walled resting spores (*oospores*).

ookinete A motile zygote formed by members of the HAEMOSPORINA. (see also PLASMODIUM)

Oomycetes A class of FUNGI of the sub-division MASTIGOMYCOTINA; constituent species are distinguished from those of other classes in that they typically give rise to zoospores each of which possesses one anteriorly directed tinsel flagellum and one posteriorly directed whiplash flagellum. (Certain advanced oomycetes—see PERONOSPORA—do not form zoospores.) The oomycetes include aquatic, terrestrial, saprophytic and parasitic species—the latter including a number of important pathogens of cultivated crops. According to species the form of the vegetative organism ranges from a simple, chytrid-like, holocarpic unicellular thallus to a well-developed, extensively-branched mycelium. The oomycete cell/hyphal wall appears to lack chitin—see CELL WALL (fungal). Sexual reproduction is oogamous. Oomycete genera include *Achlya*, ALBUGO, *Aphanomyces*, *Dictyuchus*, PERONOSPORA, PHYTOPHTHORA, *Plasmopara*, PYTHIUM, and SAPROLEGNIA.

oospheres (*mycol.*) (see OOGONIUM)

oospores (*mycol.*) (see OOGONIUM)

opal codon A synonym of UMBER CODON.

Opalinata A superclass of ciliated protozoans within the subphylum SARCOMASTIGOPHORA. The *opalinids* are parasites of the amphibian intestinal tract. Individual cells are usually ovoid, pyriform or elongated, and are generally extremely flat and leaf-like; according to species, the cells vary from about sixty to six hundred microns in length. A cytostome is lacking in all species. Most opalinids contain many nuclei—but, in contrast to the CILIOPHORA, only one type of nucleus is present. Asexual reproduction occurs by binary fission—which may be transverse or longitudinal. In *Opalina* spp (rectal parasites of the frog) a phase of rapid asexual multiplication and encystment coincides with the breeding season of its amphibian host; it is believed that this phase is triggered by the host's hormonal activity. Cysts are the infective stages; they are voided by the frog and subsequently ingested by tadpoles—in which excystment is followed by a number of longitudinal fissions and the production of gametes. After syngamy, the encysted zygote is voided by the tadpole and ingested by another—in which the zygote may undergo fission and gamete-formation or develop to form the mature, multinucleate trophic stage. Other opalinid genera include *Cepedea* and *Zelleriella*.

Opegrapha A genus of crustose LICHENS in which the phycobiont is *Trentepohlia*; the apothecioid pseudothecia are black and lirelliform (see LIRELLA). *Opegrapha* spp occur on rocks and on trees.

open culture Open-system culture: CONTINUOUS-FLOW CULTURE.

operator (operator gene) (see OPERON)

Opercularia A genus of ciliate protozoans of the suborder Sessilina, subclass PERITRICHIA. *Opercularia* is found in various freshwater habitats, and some species (e.g. *O. coarctata*) are common in British SEWAGE TREATMENT plants—both biological filter and activated-sludge types. In general appearance the organism is not unlike *Vorticella* (*q.v.*); however, *Opercularia* is usually smaller and is a *colonial* organism—one zooid being carried on each branch of the (*non*-contractile) stalk.

operculum (*mycol.*) A hinged lid or easily-detachable portion of the wall in any sac-like spore-bearing structure; opercula occur e.g. in certain types of ASCUS and in the sporangia of certain chytrids. In operculate members of the order Pezizales the operculum may be apical or (less commonly) subapical.

operon A group or cluster of STRUCTURAL GENES whose coordinated expression is controlled by a *regulator gene*. (Opinions differ as to whether or not the regulator gene should be regarded as part of the operon.) Enzyme induction and repression can be explained in terms of the operon concept:

Enzyme induction. In bacteria, the enzymes of certain metabolic pathways are synthesized only in the presence of the initial substrate of the pathway, or (in some cases) certain analogues of that substrate; this phenomenon occurs most commonly among enzymes involved in catabolic reactions—e.g. the breakdown of sugars. An example of an inducible enzyme system is the lactose (*lac*) operon of *Escherichia coli*. This operon contains genes specifying *β-galactosidase* (which e.g. hydrolyses lactose to glucose and galactose), *β-galactoside permease* (a β-galactoside transport protein), and *thiogalactoside transacetylase* (an enzyme of unknown significance). The (coordinated) transcription of these genes requires the presence of certain galactosides (*inducers*)—which may or may not be substrates for the enzymes induced. (Thus, e.g. the inducer isopropyl-β-D-thiogalactoside, IPTG, is not a substrate for β-galactosidase, while the enzyme can hydrolyse the non-inducer phenyl-β-D-galactoside; inducers which are not substrates of the induced enzymes are termed *gratuitous inducers*.) *In vivo*, the inducer of the *lac* operon appears to be *allolactose* (rather than lactose itself) which is formed by the action of β-galactosidase on lactose. In the absence of an inducer, the proteins specified by the *lac* operon are produced only in very small amounts. The presence of an inducer does not *necessarily* lead to transcription—see CATABOLITE REPRESSION.

The contiguous structural genes of the *lac* operon are designated *z* (β-galactosidase), *y* (galactoside permease), and *a* (transacetylase):

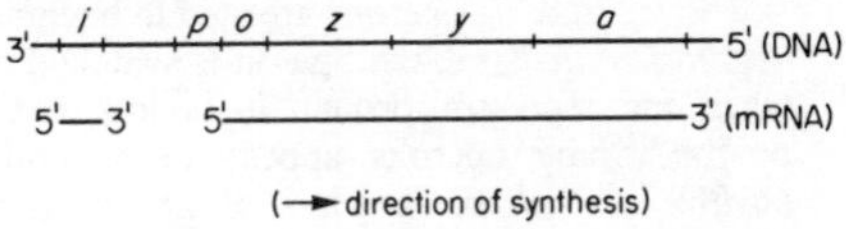

p is the *promoter*—believed to be the site of binding of RNA polymerase (see RNA). *i* is the *regulator gene*; it has its own promoter and is transcribed constitutively—independently of the remainder of the operon. The (protein) product of the *i* gene is termed the *repressor*. In the absence of an inducer, the repressor binds to the *operator* (*o*); this inhibits transcription of the structural genes. (The binding of the repressor to the operator appears not to inhibit the binding of the RNA polymerase to the promoter.) If present, an inducer binds to the repressor and—possibly by causing a conformational change—decreases its affinity for the operator; the repressor thus releases the operator, and transcription of the structural genes can proceed. The product of transcription of the operon is a POLYCISTRONIC MRNA. Removal of the inducer results in a rapid cessation of protein synthesis—thus the mRNA must be short-lived; the mRNA appears to be degraded enzymically soon after its synthesis.

The wild-type *i* gene (which produces a functional repressor) is designated i^+. Mutations in the *i* gene may cause the production of an inactive repressor (e.g. one which does not recognize the operator), or the failure to produce a repressor; such mutants (i^-) transcribe the structural genes *constitutively*. In another type of mutant (i^s) the repressor protein has the normal affinity for the operator but has negligible affinity for the inducer; the repressor (called a *super-repressor*) thus remains permanently bound to the operator, and the operon is *non-inducible.* Certain mutations in the operator reduce its affinity for the repressor; such mutants transcribe the *lac* genes constitutively and are designated o^c.

Enzyme repression. The synthesis of enzymes involved in certain metabolic (usually anabolic) pathways may be repressed ('turned off') in the presence of the end-product (or a derivative of it) of that pathway. A regulator gene codes for *apo-repressor* protein which does not itself recognize the operator; repression is brought about by the interaction of the metabolic end-product (the *co-repressor*) with the apo-repressor to produce a functional repressor which can recognize and bind to the operator. An example is the histidine (*his*) operon of *Salmonella typhimurium*: all the enzymes involved in the biosynthesis of histidine are coordinately repressed in the presence of histidine; the co-repressor is not histidine itself, but histidyl-tRNA.

The *lac* and *his* operons are said to be under *negative control*, i.e. the operon is switched *off* when the repressor protein is bound to the operon. Some operons appear to be under *positive control*: the product of the regulator gene is an *activator* protein which is necessary for transcription of the operon. (cf. CATABOLITE REPRESSION) Such operons may be inducible or repressible. The arabinose (*ara*) operon of *E. coli* appears to be under both positive and negative control; this operon codes for enzymes involved in the conversion of L-arabinose to D-xylulose-5-phosphate. A regulator gene, *C*, codes for a repressor protein, P1, which can prevent transcription (negative control) by binding to the operator. In the presence of arabinose (the inducer), P1 is converted to P2; P2 is an *activator* protein which promotes transcription of the *ara* operon (positive control) by binding to the *initiator* (a region between the operator and structural genes). Thus, the expression of the *ara* operon depends on the relative concentrations of P1 and P2. (P1 is also known as the *apo-activator*, arabinose as the *co-activator*; P2 is sometimes called the *inducer protein.*)

Genes coding for proteins with related functions are not always organized into clusters. For example, of the genes coding for arginine biosynthesis in *E. coli*, four occur in a cluster while the remaining five are dispersed around the chromosome. However, all the arginine cistrons are controlled by a single regulator gene. Each unlinked gene, and the gene cluster, appears to have an operator which can bind the product of the regulator gene; this product appears to be an apo-repressor which is activated by arginine. Repressor molecules can bind the operator associated with each locus and thus achieve some measure of coordinated control. (see also LYSOGENY)

In higher organisms, related genes tend not to occur in clusters; the mechanism of coordinated regulation in eucaryotes is unknown.

Ophiobolus A genus of fungi of the class LOCULOASCOMYCETES; the organisms occur on the tissues of dicotyledonous plants—particularly members of the Compositae. (The organism referred to in many texts as *Ophiobolus graminis* (the causal agent of 'take-all' of cereals) has been re-classified as a species of the genus GAEUMANNOMYCES.)

ophiobolus patch disease (see CEREALS, DISEASES OF)

Ophiostoma A synonym of CERATOCYSTIS.

ophthalmia neonatorum (see GONORRHOEA)

opisthe The *posterior* cell of the two cells formed during binary fission in a ciliate protozoan.

opportunist pathogens Organisms which exist as part of the normal body microflora but which may become pathogenic under certain conditions—e.g. when the normal antimicrobial defence mechanisms of the host have been impaired. (see also ENDOGENOUS INFECTION) Thus, e.g *Bacteroides* spp—common inhabitants of the large intestine in man—may give rise to peritonitis etc. following bowel surgery. (see also LUMPY JAW)

opsonic index An *in vitro* assessment of the relative concentration of OPSONINS in a sample of blood. The opsonic index is the ratio of phagocytic activity in the patient's blood to that in the blood of a normal (non-immune) subject. It is determined by initially incubating both (citrated) blood samples for 15 minutes with a standard number of the relevant antigen particles (usually microbial cells). Blood smears are then prepared from each sample, and the average number of particles engulfed by the phagocytes in each sample is determined; the two averages are expressed as a ratio (the opsonic index).

opsonin An antibody, or other component of serum, which, when combined with a *particulate* antigen*, increases the susceptibility of the latter to PHAGOCYTOSIS. *e.g. a microbial cell.

opsonization The process by which a *particulate* antigen (e.g. a microbial cell) becomes more susceptible to PHAGOCYTOSIS by combination with an OPSONIN.

optimal proportions (optimal ratio) (*immunol.*) In a titration involving antibodies and *soluble* antigens, the ratio in which antibody and antigen are present when precipitation occurs at a maximum rate. For a given antigen/antibody system two different values may be obtained for the optimal ratio—each value corresponding to a particular experimental procedure. In the DEAN AND WEBB TITRATION the *constant-antibody optimal ratio* is obtained. In this titration an optimally reacting mixture often contains antibody and antigen at *equivalence*, i.e. following antigen–antibody combination no free antigen or antibody remains. The RAMON TITRATION gives the *constant-antigen optimal ratio.* In this titration an optimally-reacting mixture frequently contains an excess of free antibody following antigen–antibody combination.

optochin Ethylhydrocuprein: a reagent which strongly inhibits the growth of *Streptococcus pneumoniae* (the pneumococcus); viridans streptococci, which may form colonies similar to those of pneumococci, are resistant to ten or more times the concentration of optochin required to inhibit the pneumococcus. Optochin may be used to detect the presence of pneumococci among a mixed flora which may include viridans or similar streptococci. An inoculum from such a sample is used to streak the surface of a solid medium, and a paper disc, impregnated with optochin, is placed on the surface of the medium prior to incubation. A noticeable decrease in the number of colonies in the vicinity of the disc may indicate the presence of pneumococci. (see also BILE SOLUBILITY)

orange peel fungus The ascomycete *Aleuria aurantia* whose fruiting body (an apothecium) resembles orange peel.

Orbivirus A genus of ARBOVIRUSES recognized by some authors. The *Orbivirus* virion is a non-enveloped icosahedral virus, 50–60 nm in diameter, which contains a *fragmented* linear double-stranded RNA genome—i.e. a genome consisting of 10 or more pieces; an RNA polymerase forms part of the virion. Replication occurs within the *cytoplasm* of the host cell. Orbiviruses include the causal agents of e.g. African horse sickness and BLUETONGUE. (cf. REOVIRUS)

orchil (*orseille*) A purple dye obtained from species of the fruticose lichen ROCCELLA.

orchinol 2,4-dimethoxy-6-hydroxy-9,10-dihydrophenanthrene; a wide-spectrum antifungal agent produced by the tubers (and to a lesser degree by the roots and stems) of certain orchid species in response to fungal infection. Orchinol is formed e.g. by *Orchis militaris* infected with the mycorrhizal fungus *Rhizoctonia repens*; other parasitic fungi, and certain soil bacteria, can stimulate orchinol production.

orcinol 3,5-Dihydroxytoluene—a compound obtained from a number of species of lichen.

order A taxonomic group: see TAXONOMY and NOMENCLATURE.

ORF (see CONTAGIOUS PUSTULAR DERMATITIS)

organelle Any structure which forms part of a microorganism and which performs a specialized function—e.g. a FLAGELLUM, or the *undulating membrane* of a ciliate.

organotroph (see HETEROTROPH)

oriental sore (Delhi boil) Human *cutaneous* LEISHMANIASIS, a disease found in Asia and in the mediterranean area. The causal agent, *L. tropica*, is transmitted chiefly by sandflies of the genus *Phlebotomus*. In 'classical' oriental sore one or more persistent, dry, ulcerative skin lesions are formed; secondary infections may occur. Amphotericin B has been used therapeutically. *Laboratory diagnosis*: microscopical demonstration of *L. tropica* in smears of lesion material. (see also ESPUNDIA and KALA-AZAR)

origin (in DNA) (1) See DNA. (2) See *bacterial* CONJUGATION.

ornithine An amino acid formed e.g. as an intermediate in the biosynthesis of arginine—see Appendix IV(a).

ornithosis (see PSITTACOSIS)

orotate An intermediate formed in the biosynthesis of pyrimidine nucleotides—see Appendix V(b).

oroya fever In man, an acute disease characterized by fever and severe anaemia; the causal agent, *Bartonella bacilliformis*, is transmitted by the bites of sandflies of the genus *Phlebotomus*.

orseille (see ORCHIL)

Orthomyxovirus A genus of VIRUSES whose hosts include man, avians, swine, and other vertebrates; orthomyxoviruses are characteristically associated with diseases of the respiratory tract—e.g. INFLUENZA, swine influenza, and FOWL PLAGUE. The *Orthomyxovirus* virion (80–120 nm in diameter) is an enveloped (ether-sensitive) single-stranded RNA virus which contains a fragmented helical nucleocapsid and an RNA-dependent RNA polymerase; the glycoprotein peplomers of the envelope are of two types: one type is associated with HAEMAGGLUTINATION (and HAEMADSORPTION), the other type with NEURAMINIDASE activity. *Filamentous* forms of the virion (several hundred nanometres in length) are common. Orthomyxoviruses are replicated at least partly in the cytoplasm of the host cell, and mature by budding through the cytoplasmic membrane; viral replication is inhibited by certain agents—e.g. ACTINOMYCIN D. The genus *Orthomyxovirus* currently consists of the types *A*, *B*, and *C* INFLUENZA VIRUSES. (cf. PARAMYXOVIRUS)

Oscillatoria A genus of motile, filamentous BLUE-GREEN ALGAE; species occur in freshwater and marine environments. In many species of *Oscillatoria* the trichome appears to be unsheathed, although a sheath is present in *O. princeps*; AKINETES and HETEROCYSTS are not formed. Reproduction occurs by the formation of hormogonia. (see also GAS VACUOLE)

osmic acid Incorrect name for OSMIUM TETROXIDE.

osmiophilic Refers to structures which become electron-dense when fixed with osmium tetroxide.

osmium tetroxide ('osmic acid') (OsO_4) A volatile, moderately water-soluble compound used as a fixative—particularly in ELECTRON MICROSCOPY. (see FIXATION) The material to be fixed is either immersed in a buffered solution (0.5%–2%) or exposed to the vapour of the solution. Osmium tetroxide has been used for the fixation of bacteria, protozoans, algae, viruses and tissue sections; it is believed to react with proteins and with double bonds in lipids. (Both the vapour and solution of osmium tetroxide are extremely harmful—particularly to the eyes and respiratory tract.)

osmophilic Refers to those organisms which grow best (or only) in or on media of relatively high osmotic pressure.

osmotic shock Any disturbance/disruption in a cell, subcellular organelle etc., which occurs when the cell, organelle etc, is transferred to a significantly hypertonic or hypotonic medium. Osmotic shock has been used e.g. as a means by which certain cellular components (e.g. BINDING PROTEINS) may be released from gram-negative bacteria. The cells are initially suspended in a buffered, hypertonic medium containing EDTA; subsequently they are exposed to an ice-cold hypotonic medium—the change from hypertonic to hypotonic medium being made rapidly. On centrifugation the liberated components remain in the supernatant. The function of the EDTA appears to involve the disruption of LIPOPOLYSACCHARIDES in the cell wall.

ostiole (*mycol.*) (1) (of a perithecium or pseudoperithecium) The neck-like or pore-like opening to the exterior through which the asci or ascospores are discharged at maturity. (see also PERIPHYSES) (2) The aperture of a pycnidium or other fructification.

Ostropales An order of fungi within the class DISCOMYCETES. Constituent species form inoperculate, unitunicate asci which contain filiform (thread-like) multiseptate ascospores; the wall of the ascus is greatly thickened at the apex (i.e. in the region of the apical pore)—at least in young (immature) asci. (cf. HELOTIALES) The ascocarp does not develop in a fungal stroma.

otitis Inflammation of the ear. *Otitis media* (inflammation of the *middle* ear) may occur e.g. as a bacterial complication of a viral upper respiratory-tract infection; in children under five years of age the commonest causal agent is *Haemophilus influenzae*, while in older children and adults it is often *Streptococcus pneumoniae* or *Streptococcus pyogenes*.

OTU Operational taxonomic unit; see NUMERICAL TAXONOMY.

Ouchterlony test (*immunol.*) A qualitative analytical PRECIPITIN TEST involving GEL DIFFUSION of the *double-diffusion*, double-dimension* type. A pattern of wells is cut into a

OUCHTERLONY TEST.

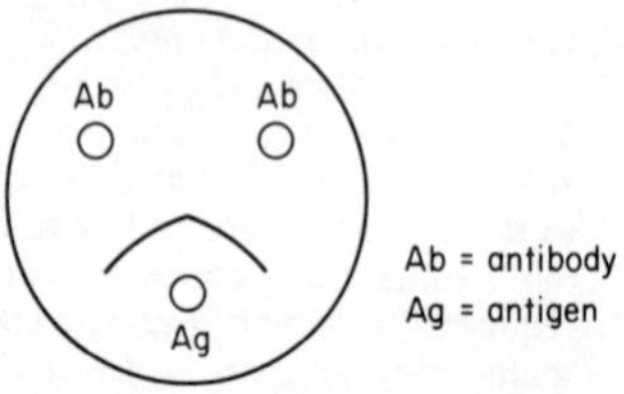

A. *Reaction of Identity.* With wells located symmetrically, the lines of precipitate are symmetrical and do not overlap. This is evidence that the antibodies in each of the top two wells are serologically identical.

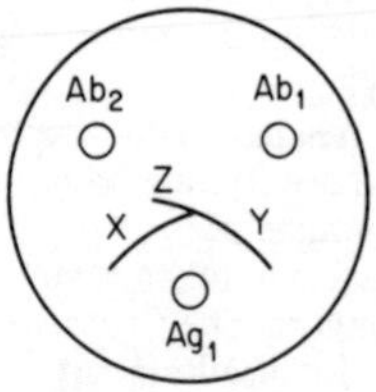

B. *Reaction of Partial Identity.* Y is a region where Ag_1 and its *homologous* antibody Ab_1 meet in optimal proportions. At X, Ag_1 forms a precipitate with the *cross-reacting* antibody Ab_2; Ab_2 has less affinity for Ag_1 so that some Ag_1 is not precipitated at X, but diffuses further to form a precipitate with Ab_1 at Z. (Z is a *spur.*) The spur points towards the antibody of lower specificity, i.e. the cross-reacting antibody.

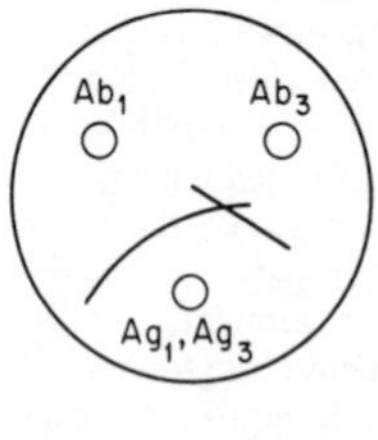

C. *Reaction of Non-Identity.* Two independent precipitating systems are present.

sheet of agar gel or similar substance. (The number and arrangement of the wells depend on the number and nature of the samples to be examined, and on the object of the examination.) In the test, each well is filled with an antigen, antibody, or control preparation. (The 'antibody' usually consists of a sample of *antiserum*.) The whole is then covered—to prevent excessive drying-out—and examined for lines of precipitation after a period of hours or days.

The diagram shows an arrangement of wells, and some possbile results, for the serological comparison of two samples of antiserum. Identical antisera give the *reaction of identity* shown at *A*. Cross-reacting antibodies, which have less affinity than the homologous antibody for the particular antigen, give the *reaction of partial identity* shown at *B*; the lower the affinity of the cross-reacting antibody the weaker will be the line of precipitation at *X*, and the stronger will be the *spur*, *Z*. When the antisera contain unrelated antibodies (diagram: *C*) (and the antigen well contains antigens homologous to each of the unrelated antibodies) the two precipitating systems act independently and give a *reaction of non-identity*.

The Ouchterlony test has certain advantages over techniques such as the *Oudin* procedure and the OAKLEY-FULTHORPE PROCEDURE; it is simpler to perform, and from its results comparison can be made between two or more precipitating systems.

*(Diffusion of the *double-dimension* type refers to radial diffusion in one plane; thus, material diffusing from a given well is able to meet and react with material diffusing towards it from two or more different directions. This contrasts with the Oudin procedure, for example, in which diffusion takes place in one direction only, i.e. it is diffusion of the single-dimension type.)

Oudemansiella A genus of fungi of the family TRICHOLOMATACEAE. Constituent species form fruiting bodies in which the stipe is typically central, and the broad, non-serrated gills are well spaced; the spores are non-amyloid. The organisms grow on wood or in soil containing woody remains.

Oudin procedure (*immunol.*) (see GEL DIFFUSION)

outer membrane (*bacteriol.*) (1) (L membrane) The layer of LIPOPOLYSACCHARIDE, protein, and lipoprotein external to the PEPTIDOGLYCAN in the CELL WALL of gram-negative bacteria. (2) See SPIROCHAETALES.

outer veil (*mycol.*) A synonym of UNIVERSAL VEIL.

overoxidation (see ACETOBACTER)

Oxford strain (of *Staphylococcus*) (NCTC strain 6571) A strain of *S. aureus* which is sensitive to a wide range of antimicrobial agents and which (for this reason) is often used as a control in antibiotic *sensitivity tests*. (see ANTIBIOTICS, RESISTANCE TO)

oxidase An OXIDOREDUCTASE which catalyses a reaction in which electrons removed from a substrate are donated *directly* to molecular oxygen; e.g. cytochrome oxidase accepts electrons from the penultimate carrier of the ELECTRON TRANSPORT CHAIN and donates them to oxygen with the formation of water.

oxidase reagent (see KOVÁCS' OXIDASE REAGENT)

oxidase test A test used to detect the presence, in bacteria, of cytochrome *c* oxidase. Oxidase-*positive* organisms (which contain cyt. *c* oxidase) include species of *Neisseria*, *Photobacterium*, *Pseudomonas* and *Vibrio*. *The test.* A small area of filter paper is moistened with a few drops of KOVÁCS' OXIDASE REAGENT*; using a *platinum* loop, or a glass spatula, a small amount of bacterial growth is smeared onto the moist filter paper. With oxidase-positive species a violet coloration develops immediately or within 10 seconds. The reagent donates electrons to cyt. *c*, becoming oxidized to a violet-coloured compound; cyt. *c* passes the electrons to oxygen *via* cyt. *c* oxidase. *Other reagents are sometimes used.

oxidation-fermentation test (O-F test; Hugh and Leifson's test) (*bacteriol.*) A test used to indicate whether a given carbohydrate (usually glucose) is utilized oxidatively or fermentatively by a given strain of bacterium. The peptone-agar medium includes sodium chloride, dipotassium hydrogen phosphate, and the given carbohydrate (1% w/v), and incorporates a pH indicator (usually BROMTHYMOL BLUE); the (unused) medium is green (if bromthymol blue is used) and has a pH of 7.1. In the test, each of two tubes of the above medium is stab-inoculated with the test organism; in *one* of the tubes the medium is immediately covered by a layer of sterile liquid paraffin to a depth of approximately one centimetre. Both tubes are then incubated at a temperature appropriate to the test organism and subsequently examined for evidence of carbohydrate utilization (acid production: bromthymol blue becomes yellow) in each tube. Yellowing of the indicator in *both* tubes indicates that the test organism can attack the carbohydrate fermentatively; an acid reaction only in the medium not covered by paraffin indicates that the carbohydrate is attacked oxidatively. No reaction in either tube indicates that the particular carbohydrate is not attacked by the test organism.

oxidation-reduction potential (see REDOX POTENTIAL)

oxidative (*adjective*) (of an organism) Refers to an organism which has a respiratory mode of metabolism—see RESPIRATION.

oxidative phosphorylation The reaction (ADP + Pi ⇌ ATP + H_2O) mediated by the high-energy intermediate (HEI) generated by the ELECTRON TRANSPORT CHAIN. The nature of the HEI, and of the mechanism by which its energy is used for ATP formation, is the subject of three main hypotheses:
The chemical hypothesis. The HEI is assumed to be a discrete chemical compound. It is postulated that, at the coupling sites (sites I, II and III) a reduced redox component (C) of the electron transport chain complexes with a reactant (I) forming C~I (~ denotes a bond which has a high energy of hydrolysis). C~I reacts with another entity, X, and X~I is formed; X~I is said to function as the HEI. It is postulated that, in oxidative phosphorylation, X~I reacts with inorganic phosphate to form X~P—the phosphate group (P) subsequently being transferred to ADP with the formation of ATP. The identities of C, X, and I are unknown.
The chemiosmotic hypothesis. This supposes that the electron transport chain consists of alternate *hydrogen*-carriers (QUINONES, flavoproteins) and *electron*-carriers (CYTOCHROMES, iron-sulphur proteins); these carriers are thought to be spatially orientated within the mitochondrial inner membrane* such that, basically, the *electron*-carriers are situated at the outer face of the membrane, and the *H*-carriers at the inner face. Essentially, of the hydrogen *atoms* carried by a reduced H-carrier, the *electrons* are transferred to an electron-carrier situated at the opposite face of the membrane, and the protons are concomitantly released at that (outer) face. Electrons subsequently pass from the electron-carrier to the next H-carrier in the chain; the consequent (negative) charge on this H-carrier is balanced by an uptake of protons from the mitochondrial matrix. This cycle of events is repeated as electrons pass along the electron transport chain; the result is a transfer of protons from one side of the membrane to the other. Since the membrane is impermeable to protons (in the absence of specific transport systems), pH and electrical gradients are set up across the membrane—the mitochondrial matrix becoming alkaline and electrically negative with respect to the exterior. The force due to the tendency of protons to re-enter the matrix is referred to as the *proton motive force* (pmf) and may be regarded as the HEI. An ATPase is said to be activated (directly or indirectly) by this gradient—ATP being formed during the inward translocation of protons. (see also TRANSPORT and PURPLE MEMBRANE)
The conformational hypothesis. The energy generated by the electron transport chain is conserved in the form of a *strained conformational state* of membrane components. In the presence of ADP and Pi, ATP is synthesized with concomitant relief of the conformational strain.
(cf. SUBSTRATE LEVEL PHOSPHORYLATION)
*These hypotheses generally refer to mitochondrial systems; however, a similar situation is thought to obtain in bacterial systems and in photophosphorylation during PHOTOSYNTHESIS.

oxidizing agents (as antimicrobial agents) Strong or moderately strong oxidizing agents may behave as effective antimicrobial agents when present in adequate concentrations under appropriate conditions; however, their use as DISINFECTANTS is often limited by the high degree of inactivation which occurs in the presence of excess organic matter. Antimicrobial action involves oxidation of various groups in proteins and lipids, e.g. sulphhydryl and hydroxy groups, double bonds etc. *Hydrogen peroxide* (H_2O_2) is rapidly broken down to water and oxygen in the presence of certain metal ions and organic matter—particularly when the latter incorporates CATALASE. Dilute solutions of hydrogen peroxide are used for the treatment of wounds; here, the action may be largely mechanical—tissue catalase causing a rapid evolution of oxygen which dislodges contaminating foreign matter thus facilitating its removal. However, the aerobic environment thus created within the wound may help to prevent the development of pathogenic ANAEROBES, and the nascent oxygen may exert some positive antimicrobial activity. *Sodium perborate* ($NaBO_3$) in aqueous solution behaves as a mixture of borax and hydrogen peroxide. Prepared as a paste with water or glycerol this substance was widely used for the treatment of oral infections involving anaerobes, e.g. Vincent's angina. *Potassium permanganate* ($KMnO_4$) has been used for the treatment of certain superficial mycoses; however, it stains tissues and fabrics and may be irritant at high concentrations. Algae are highly susceptible. *Ozone* (O_3) readily oxidizes most organic matter and is active against a wide range of microorganisms—including fungi and viruses. It is more active at lower temperatures, and requires 60–90% relative humidity for optimal activity. Ozone is sometimes used for the disinfection of water supplies and for the preservation of certain foods; in the latter function it is active only against surface contaminants, and is not suitable for use with fats, meat etc. which are likely to be oxidized. Ozone has the advantage that it leaves no residue. (see also HALOGENS)

oxidoreductase A category of ENZYMES (EC

class 1) which encompasses all those enzymes which catalyse oxidation-reduction reactions. This category includes enzymes of the following types: DEHYDROGENASE, OXIDASE, OXYGENASE, PEROXIDASE.

oxine 8-HYDROXYQUINOLINE (*q.v.*).

oxygenase An OXIDOREDUCTASE which catalyses the incorporation of one or both atoms of molecular oxygen into a substrate molecule. When only one oxygen atom is incorporated, the other is reduced to water; the enzyme catalysing this reaction is called a *monooxygenase* (or a *mixed function oxidase*). When both oxygen atoms are incorporated, the enzyme is called a *dioxygenase*. Oxygenases frequently catalyse the first step in the bacterial degradation of aromatic and paraffin hydrocarbons.

oxyhydrogen reaction (see KNALLGAS BACTERIA)

Oxyrrhis A genus of algae of the DINOPHYTA.

oxysomes (see MITOCHONDRION)

oxytetracycline (terramycin) 5-Oxytetracycline. An ANTIBIOTIC obtained from *Streptomyces rimosus*. (see TETRACYCLINES)

oyster fungus (oyster-cap fungus) The edible basidiomycete *Pleurotus ostreatus* which grows on wood and produces oyster-shaped fructifications. The oyster fungus can be pathogenic for certain trees, especially beech (*Fagus*).

ozone (as an antimicrobial agent) (see OXIDIZING AGENTS)

P

P-450 (see CYTOCHROMES)
P site (of a RIBOSOME) (see PROTEIN SYNTHESIS)
PAB (PABA) (see *p*-AMINOBENZOIC ACID)
pachytene stage (see MEIOSIS)
packet (of cells) (*bacteriol.*) A 3-dimensional group of cells formed when parent cell(s) divide successively along three mutually perpendicular axes and the progeny cells fail to separate. Packets are formed e.g. by *Micrococcus roseus*, *Methanosarcina*, and *Sarcina*—the packets commonly consisting of eight cells.
paints (microbial spoilage of) The susceptibility of paint to attack by microorganisms depends on e.g. the chemical nature of the paint and its solvent, the nature of the surface covered, and the conditions to which the paint film is exposed. The final, hardened paint film may be attacked by fungi (e.g. species of *Pullularia*, *Cladosporium*, *Aspergillus*, *Penicillium*) which commonly obtain nutrients from the material beneath the film; thus the hardness, thickness and integrity of the film are important factors influencing spoilage. Some paints contain substances (such as LATEX and CELLULOSE) which can be used by microorganisms as nutrients, and/or antimicrobial substances (e.g. zinc oxide). PRESERVATIVES which have been added to paints include *Shirlan* (salicylanilide) and halogenated phenols. ALGICIDES are incorporated into marine paints.
palisade (*bacteriol.*) A term used to describe the arrangement of cells seen in the stained smears of certain bacteria: two, three or more bacilli in contact, with their long axes parallel.
palmelloid phase (palmella phase) Of certain motile algae (e.g. CHLAMYDOMONAS): a phase in which the characteristic structure is an aggregate of non-motile (non-flagellate) cells embedded in a gelatinous matrix.
paludrine (paludrin; proguanil) (see CHLORGUANIDE)
pandemic An outbreak of disease which affects large numbers of people in a major geographical region, or which has reached epidemic proportions, simultaneously, in many different parts of the world.
Pannaria A genus of LICHENS in which the phycobiont is *Nostoc*; the thallus is generally squamulose and possesses a dark hypothallus. Apothecia are lecanorine. Species include e.g.: *P. rubiginosa*. The thallus is circular in outline, flat, and composed of long radiating squamules with shorter, imbricated squamules towards the centre; the upper surface is bluish-grey to brown, lighter at the margins, while the hypothallus is dark blue-grey. Apothecia are frequently numerous on the upper surface in the centre of the thallus; each has a reddish-brown disc and a notched thalline margin. *P. rubiginosa* occurs mainly on mossy trees.
pansporoblast (see CNIDOSPORA)
pantetheine (see PANTOTHENIC ACID)
Panton–Valentine leucocidin A staphylococcal LEUCOCIDIN.
pantothenic acid A water-soluble VITAMIN which is a component of COENZYME A; it consists of *pantoic acid*($CH_2OH.C(CH_3)_2.CHOH.COOH$) in amide linkage with a residue of β-alanine. Linked to 2-mercaptoethylamine (= cysteamine) pantothenic acid forms *pantetheine* (= pantotheine). Coenzyme A consists of pantetheine-4′-phosphate linked to adenosine-3′,5′-diphosphate—see Appendix V(c). Pantetheine-4′-phosphate also forms the prosthetic group of the *acyl-carrier protein* (ACP) involved in fatty acid biosynthesis.

Exogenous pantothenic acid is required for growth by some microorganisms—including many yeasts (e.g. *Saccharomyces* spp, *Candida* spp), the filamentous fungus *Polyporus texanus*, certain bacteria (e.g. *Acetobacter* spp, *Lactobacillus* spp), and certain protozoa (e.g. *Tetrahymena*, *Crithidia*, *Leishmania*). Precursor(s) of pantothenic acid (e.g. β-alanine, pantoic acid) may satisfy the requirement for pantothenic acid.
papain A proteinase (proteolytic enzyme) derived from the unripe fruit of the papaya (pawpaw) tree (*Carica papaya*); the enzyme consists of a single polypeptide chain of M.Wt. 21,000. Papain has been used e.g. in the analysis of IMMUNOGLOBULINS. (see also IGG)
paper (microbial spoilage of) *Pulpwood* (timber selected for pulping) may be attacked e.g. by species of the cellulolytic fungi *Stereum* (particularly *S. sanguinolentum*) and *Poria*; such damage may result in the subsequent production of inferior (weaker) paper. Wood *pulp* may be attacked e.g. by species of *Cladosporium*, *Fusarium*, *Penicillium*, *Phoma*, *Pullularia* and *Stereum*—which produce rots and/or stains. Pulp is used as a watery suspension which may be rich in microbial nutrients (e.g. hemicelluloses, sugars, proteins)—particularly when it is prepared by purely mechanical methods. The suspension is susceptible to slime-forming organisms—e.g. species of *Alcaligenes*, coliforms, and other bacteria, and fungi (e.g. *Fusarium*, *Geotrichum*); *Serratia* forms red slime. Slime may cause weaknesses, holes, and/or discolorations in the paper; slime-forming organisms may be controlled e.g. by

chlorine or organic mercurials. Finished paper has a low water content, and is subject only to *fungal* spoilage—e.g. by *Chaetomium globosum*, *Stachybotrys atra* and other cellulolytic fungi; *Aspergillus* and *Penicillium* are common causes of discoloration. Preservatives, e.g. zinc naphthenate, phenol derivatives, mercury compounds, may be incorporated in paper.

Papillomavirus A genus of viruses of the family PAPOVAVIRIDAE; the icosahedral virion, 50–55 nm in diameter, contains a circular double-stranded DNA genome. Constituent species include the human papilloma virus (= wart virus), the bovine papilloma virus, and the Shope rabbit papilloma virus.

Papovaviridae A family of VIRUSES—most (or all) of which, under appropriate conditions, can cause tumours in vertebrate hosts. The virion consists of a non-enveloped icosahedral capsid (40–55 nm in diameter, 72 capsomeres) which contains a circular, double-stranded DNA genome; no carbohydrate or lipid components have been detected. Papovaviruses are assembled in the *nucleus* of the host cell. The family consists of two genera: PAPILLOMAVIRUS and POLYOMAVIRUS; papillomaviruses are typically difficult to cultivate *in vitro* (i.e. in tissue culture), while polyomaviruses are readily cultivable.

PAPS 3′-Phosphoadenosine-5′-phosphosulphate (also called adenosine-3′-phosphate-5′-sulphatophosphate or 3′-phosphoadenylylsulphate). (see also ASSIMILATORY SULPHATE REDUCTION and Appendix V(c))

papule (*med.*) On tissues (particularly skin): a discrete, elevated, solid, erythematous lesion.

parabasal body (*protozool.*) The GOLGI APPARATUS; in some of the earlier texts the term *parabasal body* was apparently used to refer to the KINETOPLAST.

parabenzoates (parabens) (as food preservatives) Esters of *p*-hydroxyBENZOIC ACID; they are mildly bacteriostatic and fungistatic—generally being more active under acid conditions. Parabens are stable and non-toxic but poorly soluble in water. (see also FOOD, PRESERVATION OF)

paracentric (of a chromosome) Refers to an intrachromosomal structural alteration (see CHROMOSOME ABERRATION) which does *not* include the centromere; e.g. in a paracentric inversion, the inverted segment does not contain the centromere. (cf. PERICENTRIC)

paracoagulation The clumping which occurs on the addition of plasma, or soluble complexes of fibrin monomers*, to a suspension of bacterial cells which have been coated with protamine sulphate; the protamine sulphate thus simulates bound COAGULASE. *A fibrin monomer is a fibrinogen molecule which, by the action of thrombin or a similar agent, has been structurally modified—certain polypeptide chains having been excised.

Paracoccidioides A genus of fungi of uncertain taxonomic affinity. *P. brasiliensis* (the causal agent of PARACOCCIDIOIDOMYCOSIS) is a dimorphic fungus which occurs within tissues in a yeast-like (unicellular) form; the individual, spherical or spheroid cells are commonly several microns to thirty microns in diameter. In the unicellular form the organism reproduces by budding. Buds may form at any position on the cell surface; each bud is attached by a narrow neck to the parent cell such that the latter always appears to be spherical. A given cell may have one or a few buds, or may have almost a confluent layer of buds. The yeast-like form of the organism can be cultured on blood agar at 37 °C. When cultured at 30 °C, or at room temperature, on Sabouraud's agar the organism gives rise to a branching, septate mycelium; the mycelial form may give rise to chlamydospores, and some strains have been reported to form aleuriospores. The cell wall of the unicellular form differs in composition from the hyphal wall of the mycelial form—see CELL WALL (fungal). Sexual reproduction is unknown.

paracoccidioidomycosis (South American blastomycosis) A chronic, commonly fatal mycosis which occurs principally in Brazil; the disease is characterized by lesions in the lungs and viscera and ulcerative granulomata in the tissues of the mouth, nose, and skin. The causal agent is *Paracoccidioides brasiliensis*. (see also PARACOCCIDIOIDES; cf. BLASTOMYCOSIS)

Paracoccus A genus of gram-negative, aerobic, asporogenous, non-motile bacteria of uncertain taxonomic affinity; species occur e.g. in the soil. The cells are coccoid, approximately 0.5–1 micron in diameter; they occur singly or in pairs or groups. The organisms are chemoorganotrophs (*P. denitrificans* is facultatively chemolithotrophic); metabolism is oxidative (respiratory). (see also NITRATE RESPIRATION) *P. denitrificans* (formerly *Micrococcus denitrificans*) is oxidase-positive and catalase-positive.

paracolon A term formerly much used to refer to those enteric bacteria which *do not* metabolize lactose within 24/48 hours at 37 °C. (cf. COLIFORM)

paraffins (microbial utilization of) (see ALIPHATIC HYDROCARBONS (microbial utilization of))

paraformaldehyde (see FORMALDEHYDE)

parainfluenza viruses (see PARAMYXOVIRUS)

paramecin (see KAPPA PARTICLES)

Paramecium (the 'slipper animalcule') A genus of freshwater protozoans (suborder PENICULINA). The organisms are ovoid or elongated with uniform somatic ciliature. Length varies with species and strain: *P. caudatum* is 180–300 microns; *P. aurelia* 120–200 microns; *P. bursaria* 90–150 microns; *P. trichum* 50–120 microns, and *P. multimicronucleatum* 200–300 microns. The ventrally located vestibulum leads to the buccal cavity which contains a single ENDORAL MEMBRANE, two peniculi (see PENICULUS) and the QUADRULUS. Most species have one micronucleus and one macronucleus; *P. aurelia* has an additional micronucleus, and *P. multimicronucleatum* has at least four micronuclei. The presence of two contractile vacuoles is typical, and most species have large numbers of trichocysts; paramecia are negatively galvanotactic.

Asexual reproduction occurs by binary fission, and the sexual processes are CONJUGATION and AUTOGAMY; mating types are arranged into SYNGENS.

Food consists of bacteria and small flagellates etc. *P. bursaria* may derive nutrients from its numerous algal endosymbionts (zoochlorellae). Certain species of *Paramecium* are the natural food of *Didinium*.

Paramecium has been used extensively as a test organism in various types of research, e.g. cytology, genetics. (see also KAPPA PARTICLES)

paramural body (*microbiol.*) Any of a variety of types of membranous structure which occur close to the inner surface of the cell wall, e.g LOMASOMES.

paramylon Synonym of PARAMYLUM.

paramylum (paramylon) A storage polysaccharide characteristic of members of the Euglenophyta; it is an insoluble polymer consisting of β-(1 → 3)-linked D-glucose residues. Paramylum granules occur in various forms, e.g. discs, rings, rods—a given form generally being characteristic of a particular species. The granules are highly refractile bodies; they give no colour reaction with iodine.

Paramyxovirus A genus of VIRUSES whose hosts include man, avians and other vertebrates; paramyxoviruses include the causal agents of e.g. FOWL PEST and MUMPS. The *Paramyxovirus* virion (commonly 120–350 nm in diameter) is an enveloped (ether-sensitive) single-stranded RNA virus which contains a non-fragmented helical nucleocapsid and associated RNA-dependent RNA polymerase; the peplomers of the envelope are of two types: one type is associated with HAEMAGGLUTINATION (and HAEMADSORPTION) and the other type with NEURAMINIDASE activity. Under certain conditions, paramyxoviruses are able to lyse the cells (including erythrocytes) of some vertebrates. The paramyxoviruses are labile at room temperatures and at 4 °C—infectivity generally diminishing by 90% or more within hours. Paramyxoviruses are assembled within the cytoplasm of the host cell and mature by budding through either the cytoplasmic membrane or the intracytoplasmic membranes; viral replication is not inhibited by actinomycin D (cf. ORTHOMYXOVIRUS). Paramyxoviruses can be cultivated in certain types of tissue culture (e.g. chick embryo, HeLa, hamster kidney) in which they promote cell fusion with the formation of syncytia. Tests involving haemagglutination are carried out at 4 °C (to inhibit elution by neuraminidase) with erythrocytes from e.g. man, fowl, or the guinea-pig. The currently accepted members of the genus include the mumps virus, the Newcastle disease virus (causal agent of fowl pest), and the parainfluenza viruses types 1–4 (causal agents of e.g. bronchopneumonia and minor upper respiratory tract diseases in man and other vertebrates).

Tentatively assigned to this genus are the causal agents of e.g. DISTEMPER, MEASLES, and RINDERPEST, and the RESPIRATORY SYNCYTIAL VIRUS; these viruses closely resemble the accepted members of the genus in structure, size and general properties—though none of them exhibits neuraminidase activity. The *measles* virus promotes haemagglutination (e.g. with monkey erythrocytes, 37 °C) and exhibits the haemolytic and syncytia-forming properties of the paramyxoviruses; measles virus can be cultivated in e.g. monkey kidney, chick embryo, HeLa and HEp-2 tissue cultures. The *rinderpest* virus (which closely resembles the measles virus) can be cultivated in bovine kidney cells.

paraphyses (*mycol.*) (*sing.* paraphysis) Sterile, unbranched or branched filaments which occur among the asci of the hymenium in the fructifications of certain ascomycetes; the (basally-attached) paraphyses form part of the hymenium, and may or may not project above the level of the asci. The function of the paraphyses is unknown. (Some authors use this term to refer to the BASIDIOLES in the hymenia of the basidiomycetes; see also CYSTIDIUM.)

paraplectenchyma (see PLECTENCHYMA)

parasexuality in fungi Certain fungi can form recombinant nuclei by a process in which meiosis and fertilization are not involved. This *parasexual cycle* involves: (a) The formation of heterokaryotic hyphal cells or mycelium. (see HETEROKARYOSIS) (b) The (infrequent) fusion of two haploid nuclei (in the heterokaryon) to form a single diploid nucleus. (c) Mitotic crossing-over: the (infrequent) exchange of chromatid segments between chromosomes of an homologous pair during *mitotic* propagation of the diploid nucleus. (d) Haploidization involving mitotic NON-DISJUNCTION: the formation of haploid recombinant nuclei from those (diploid) nuclei in which mitotic crossing over has occurred; haploidization is believed to occur by the step-wise loss of one member of each pair of homologous chromosomes during successive mitoses.

For the imperfect fungi parasexuality brings the advantages of genetic recombination; it is probably of little significance, genetically, in fungi which have a sexual stage.

parasite Any organism which lives on (*ectoparasite*) or within (*endoparasite*) the tissues of another living organism (the *host*) from which it derives nutrients. The presence of a parasite may or may not significantly harm the host; a parasite which harms or kills the host is generally referred to as a *pathogen.* (see also COMMENSALISM and SYMBIOSIS)

parasitism The mode of life of any PARASITE; an organism may be a FACULTATIVE parasite or an OBLIGATE parasite. Many parasites can parasitize any of a variety of hosts while others require a particular species or strain of a given organism. When compared with its free-living (non-parasitic) relatives the parasite often exhibits certain aspects of degeneracy—e.g. a reduction in, or loss of, structures, organelles, or metabolic mechanisms involving the ingestion and digestion of food, respiration, osmoregulation, or excretion. Adaptation to the parasitic mode of life may be regarded as a form of specialization; some parasites are so well integrated into the economy of their hosts that they can be cultivated *in vitro* only with difficulty or not all. (cf. SYMBIOSIS)

parasitology The study of certain PARASITES—their life cycles, ecology, transmission etc.; those parasites studied in parasitology include some which are classified as microorganisms (e.g. amoebae, trypanosomes, malarial parasites) as well as various arthropods, helminths etc. Pathogenic bacteria, fungi, and viruses are not included within the scope of parasitology—even though they live parasitically.

parasol mushroom A common name for the fruiting body of the basidiomycete *Lepiota procera.*

parasomal sacs In many ciliates (e.g. *Paramecium*): small, flask-like invaginations of the surface membrane near the ciliary bases; function: unknown.

paratope (*immunol.*) The *combining site* of an ANTIBODY. (see also EPITOPE)

paratose (3,6-dideoxy-D-glucose) A sugar which occurs e.g. in the O-specific chains of LIPOPOLYSACCHARIDES in certain serotypes of *Salmonella.* Paratose, first isolated from *S. paratyphi,* contributes to the specificity of somatic (O) antigen 2 in the group *A* salmonellae.

paratyphoid fever (paratyphoid) In man, enteric disease caused by certain salmonellae other than *S. typhi*—commonly strains of *S. paratyphi* or *S. typhimurium.* Infection generally results from consumption of contaminated food or water; the incubation period is usually 1–7 days. Symptoms are generally less severe than those of TYPHOID FEVER; gastroenteritis is common and there is frequently an early bacteraemia—though skin lesions rarely develop.

Parazoa The *sponges* (Porifera)—multicellular organisms which exhibit a degree of organization lower than that found in other multicellular organisms (Metazoa). (see also PROTOZOA)

parcentric objectives (see PARFOCAL OBJECTIVES)

parenteral (of the administration of antibiotics, vaccines etc.) Administered by any route *except via the alimentary canal.*

parenthesome (septal pore cap) (see SEPTUM (*mycol.*))

parfocal objectives (in MICROSCOPY) A set of objective lenses which, when used in a given microscope, permit the observer to change from one objective lens to another (by using the revolving nosepiece) without the necessity of adjusting the focusing controls. If the objective lenses are *parcentric* each objective provides a field of view whose central point is identical to that observed through the other objectives.

parietin (*lichenol.*) An orange anthraquinone pigment which occurs in certain lichens—e.g. species of *Caloplaca* and *Xanthoria.* Parietin gives a purple colour reaction with potassium hydroxide.

Park nucleotide UDP-*N*-acetylmuramyl pentapeptide: an intermediate in the biosynthesis of PEPTIDOGLYCAN.

Parmelia A genus of foliose LICHENS in which the phycobiont is *Trebouxia.* Apothecia—when present—are lecanorine. Species include e.g.:

P. saxatilis. The thallus consists of deeply-cut lobes which form a flat rosette; the upper surface is grey (or greenish-grey in shaded positions) while the lower surface is dark brown to black in the centre and lighter brown towards the periphery. On the upper surface pseudocyphellae form a reticulated pattern of white lines; isidia may also be present. Rhizinae occur on the lower surface—extending to the margin. Apothecia are uncommon; when present the discs are dark brown. *P. saxatilis* occurs on rocks, walls, trees etc.

P. glabratula. The thallus forms a flat rosette—light greenish-brown above and black below; the radiating lobes are glossy at the periphery. Isidia may occur towards the centre of the thallus. *P. glabratula* occurs on trees and fences.

paromomycin A broad-spectrum

AMINOGLYCOSIDE ANTIBIOTIC produced by *Streptomyces rimosus*. It has been used for the treatment of localized intestinal infections.

parthenogenesis (1) *Haploid* parthenogenesis: the development of a female gamete or gametangium without the involvement of a male gamete or gametangium; development may or may not lead to the formation of a mature (adult) individual. (2) The development of a *diploid* egg/gamete in APOMIXIS.

partial veil (inner veil) (*mycol.*) In the immature fruiting bodies of certain agarics, e.g. *Agaricus* spp: a membranous tissue which extends from the margin of the pileus to the stipe; when intact, the partial veil seals off the gill cavity. During the development of the fruiting body the partial veil tears—thus leaving a ring of tissue (the *annulus*) around the stipe and (often) a ragged ring of tissue (the *cortina*) attached to the periphery of the pileus. (see also UNIVERSAL VEIL)

partition The region between adherent THYLAKOIDS.

partridge wood A white *pocket rot* of oak (*Quercus*) produced by the basidiomycete *Stereum frustulatum*.

parvobacteria A poorly-defined term sometimes used to refer to some or all of the small gram-negative bacilli, e.g. *Brucella*, *Pasteurella*.

Parvovirus (*syn*: picodnavirus) A genus of VIRUSES whose hosts include a range of vertebrates. The non-enveloped, spheroid or icosahedral *Parvovirus* virion is 20–25 nm in diameter and has a genome of linear, single-stranded DNA; parvoviruses are highly resistant to physical and chemical agents—being able to withstand e.g. exposure to 60 °C for at least 30 minutes. Subgenus *A* consists of those parvoviruses which are capable of autonomous replication within appropriate host cells; subgenus *B* consists of those parvoviruses which can replicate in cells only when the latter are infected by adenoviruses—the so-called *adenovirus-associated viruses* (AAVs). Parvoviruses are assembled in the *nucleus* of the host cell.

PAS (*para*-aminosalicylic acid) A therapeutic agent which has bacteriostatic activity against certain species of *Mycobacterium*—e.g. *M. tuberculosis*. Within the bacterium, PAS is incorporated into FOLIC ACID in place of the normal metabolite, *para*-aminobenzoic acid (PABA); the analogue of folic acid thus formed exhibits impaired cofactor activity. (PAS is competitively antagonized by PABA.) PAS has been used, in conjunction with e.g. ISONIAZID or STREPTOMYCIN, in the treatment of TUBERCULOSIS; the use of several antimycobacterial agents together delays the development of resistant strains of *M. tuberculosis*.

PAS reaction (periodic acid-Schiff reaction) A procedure which can be used to demonstrate the presence of certain carbohydrates—e.g. CELLULOSE, CHITOSAN, GLYCOGEN, STARCH. Adjacent hydroxyl groups in the carbohydrate are oxidized to aldehyde groups on treatment with periodic acid (see diagram); on subsequent treatment with SCHIFF'S REAGENT the aldehyde groups give a magenta coloration. (Tetraacetic acid or chromic acid may be used in place of periodic acid.) A positive PAS reaction is also given by sugars containing an unsubstituted amino group adjacent to an hydroxyl group (as in CHITOSAN). The PAS reaction has been used e.g. for the microscopic detection of certain pathogenic fungi (e.g. *Blastomyces dermatitidis*, *Histoplasma capsulata*) in tissue sections.

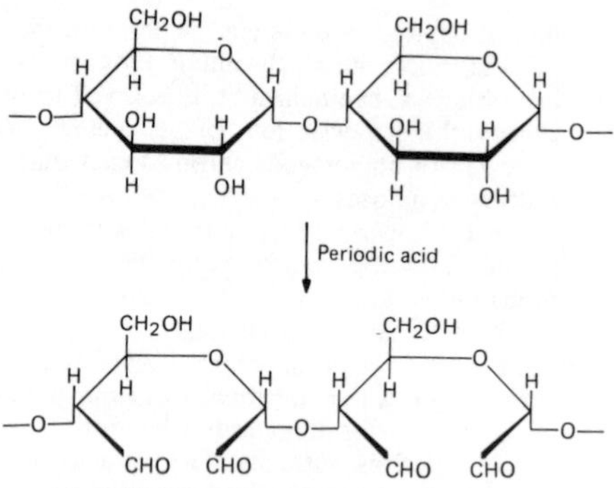

Periodic acid oxidation of an *alpha*–(1→4) glucan

passage, passaging (see SERIAL PASSAGE)

passive agglutination (*immunol.*) A procedure in which the combination of antibody with a *soluble* antigen is made readily detectable by the prior adsorption of the *antigen* to erythrocytes or to minute particles of organic or inorganic materials (e.g. *latex* or *bentonite* respectively). The union of antibody with particle-bound antigen brings about the agglutination of such particles—the antibodies acting as links or bridges between them. If the antigen has been adsorbed to *erythrocytes* the reaction with homologous antibody brings about *passive haemagglutination*. Passive agglutination tests are highly sensitive methods for the detection and quantitation of antibodies to soluble antigens; such tests give positive reactions with sera containing amounts of antibody far below the operational limit of the PRECIPITIN TEST.

passive haemolysis (see HAEMOLYSIS)

passive immunity (adoptive immunity) Short-term IMMUNITY brought about by the transfer of pre-formed antibody (or specifically sensitized lymphocytes) from an immune subject to a non-immune subject; the non-immune subject therefore becomes immune without necessarily having had contact with the appropriate antigen. (cf. ACTIVE IMMUNITY)

passive immunization (see IMMUNIZATION)

Pasteur effect In those organisms (e.g. *Saccharomyces*) capable of both fermentative and respiratory modes of metabolism: the inhibition of glucose utilization in anaerobically-grown cells on exposure of those cells to oxygen. (This effect reflects the increased energy yield obtained by the respiratory metabolism of glucose as compared with that obtained by the fermentation of glucose.) The Pasteur effect appears to result from the operation of a complex series of control mechanisms—which may differ, in detail, in different microorganisms.

pasteur pipette An open-ended glass tube (commonly about 5 mm internal diameter) one end of which is drawn out to form a tube of much smaller diameter (1 mm or less); the other end is fitted with a rubber teat. The pasteur pipette is generally used for non-quantitative transfer of liquids; however, it may be 'calibrated' by counting the number of drops delivered per millilitre—e.g. for a rate of 50 drops/ml, each drop will be 0.02 ml. Liquids being transferred by pasteur pipette may be protected from microbial contamination (from the rubber teat) by 'plugging' the pipette: a small plug of cotton wool is inserted into the wider end of the glass tube before sterilization of the pipette prior to use. (In microbiological work graduated (analytical) pipettes are also plugged for similar reasons.)

Pasteurella A genus of gram-negative, facultatively anaerobic, chemoorganotrophic, asporogenous bacilli or coccobacilli of uncertain taxonomic affiliation; species include pathogens of domestic and other animals. The cells are non-motile and are about 0.5 × 1–1.5 microns; bipolar staining is not uncommon. The majority of strains give positive oxidase and catalase tests, and a range of sugars is fermented with the production of acid but no gas. Most strains (other than those of *P. haemolytica*) fail to grow on MacConkey's agar. Type species: *P. multocida*; GC% for the genus: approximately 36–43.

P. multocida (*P. septica*). A non-haemolytic species which is typically oxidase-positive, indole-positive, and urease-negative, and which fails to ferment lactose; the causal agent of FOWL CHOLERA, of haemorrhagic septicaemia in ruminants, and (in conjunction with *P. haemolytica*) of *shipping fever* in cattle.

P. haemolytica. A haemolytic, indole-negative, urease-negative species; the causal agent e.g. of *shipping fever* (either alone or with *P. multocida*).

P. pneumotropica. A non-haemolytic, indole-positive, urease-positive species sometimes associated with respiratory-tract infections in laboratory rodents.

P. ureae. A haemolytic, indole-negative, urease-positive species.

P. pestis and *P. pseudotuberculosis: see* YERSINIA; *P. tularensis*: see FRANCISELLA.

pasteurization (of milk) A form of heat treatment which is lethal for the causal agents of a number of MILK-transmissible diseases (e.g. salmonelloses, tuberculosis)—as well as for a proportion of the normal milk microflora. In the older method (LTH: low temperature holding method) milk is held at 63 °C for 30 minutes. In the modern method (HTST: high temperature, short time) the milk is held at 72 °C for 15 seconds*. Pasteurization also inactivates certain bacterial enzymes (notably lipases) which may cause deterioration in milk; the inactivation of these enzymes and the reduction in milk microflora enhance the keeping qualities of the milk. Pasteurization is also used e.g. for wine and beer. *Both the LTH and HTST methods are reported to be occasionally inadequate for the inactivation of *Coxiella burnetii*, the agent of Q FEVER.

pasteurized Refers to any substance which has undergone PASTEURIZATION.

patent period (*med.* and *vet.*) That period during which parasitic organisms (or their cysts) can be demonstrated in the blood, tissues or faeces of the host.

pathogen An organism which is capable of causing disease in an animal, plant or microorganism.

pathogenesis The process or mechanism of disease development.

pathogenic Able to cause disease.

pathognomonic (of symptoms) Characteristic of, or specific to, a given disease.

Paxillaceae A family of fungi of the AGARICALES.

PBS Phosphate-buffered saline.

PCNB (see QUINTOZENE)

pea, diseases of These include: (a) *Root rot.* A soil-borne disease caused by the oomycete *Aphanomyces euteiches.* Following fungal penetration, the external tissues of the root and stem base soften; subsequently the root system may decay. (b) *Leaf spot* (*pod spot*). A soil- and seed-borne disease caused by *Ascochyta pisi.* Small brown lesions with dark edges are formed on leaves and pods, and dark *pycnidia* may be observed.

peach, diseases of These include (a) CROWN GALL. (b). *Leaf curl.* Causal agent: *Taphrina deformans.* Infected leaves become discolored and distorted, and a white or silvery film develops on their upper surfaces. Fruit, and the infected leaves, fall prematurely. The almond may also be attacked. (c) PLUM POX.

pear, diseases of (see e.g. FIRE BLIGHT)

pébrine A fatal disease of the silkworm (*Bombyx mori*) caused by the protozoan *Nosema bom-*

bycis. Infection may occur e.g. by ingestion of spores; these germinate in the gut, and vegetative cells of *N. bombycis* penetrate the gut wall and enter the haemocoel—becoming widely distributed in the tissues. Dark spots characteristically appear on infected individuals.

pectinate (*adjective*) Comb-like.

pectinella Either of the two bands of cilia which encircle *Didinium*.

pectins A group of complex polysaccharides which consist basically of α-(1 → 4)-linked D-galacturonic acid residues (*pectic acid*) in which a high proportion of the (C-6) carboxyl groups are methylated. Residues of L-rhamnose may occur at intervals within the chain, and side chains of neutral sugars (e.g. arabinose, galactose) may be present. Structural details vary with source. Pectins form gels under certain conditions of e.g. ionic concentration.

Pectins occur in plants as an intercellular cement (middle lamellae), or in the primary cell wall in association with e.g. HEMICELLULOSES; many fruits have a high pectin content. Pectins also occur in the cell wall and/or slime of certain algae.

Pectins can be attacked by a wide range of microorganisms, e.g. species of *Erwinia* and *Xanthomonas* (plant pathogens), and the ascomycete *Byssochlamys fulva* (which causes spoilage of canned fruits); pectinolytic organisms also occur in the RUMEN. (see also RETTING) At least two types of enzyme are involved in pectin degradation; *pectin methyl esterases*, which remove the methyl groups, and *polygalacturonases*, which break down the galacturonan chain. The specificity and mode of action of these enzymes appears to vary considerably.

Pectobacterium delphinii (see *E. carotovora* under ERWINIA)

Pediastrum A genus of coenobial algae of the division CHLOROPHYTA.

pedicel In the fruiting bodies of certain species of the MYXOBACTERALES, one of a number of individual stalks arising from a common stalk—each individual stalk bearing a single sporangium.

Pediococcus A genus of gram-positive bacteria of the STREPTOCOCCACEAE; species occur in fermented beverages and in fermenting vegetable material. The cells are coccoid, up to 1 micron in diameter, and occur in pairs or tetrads. A number of vitamins and amino acids are essential for growth—which occurs optimally under reduced oxygen tension. Glucose is attacked *homofermentatively* with the formation of lactic acid but without gas production. Type species: *P. cerevisiae*; GC% range for the genus: 34–44.

peitschengeissel The smooth ('whiplash') type of *eucaryotic* FLAGELLUM.

pelagic zone That region of an aquatic environment (e.g. lake, sea) which comprises the entire body of water but which excludes the mud, sand etc. which forms the bed or bottom of that environment. (cf. BENTHIC ZONE; see also PLANKTON, NEUSTON, NEKTON)

pellet A spherical colony formed by the growth of a filamentous fungus within a liquid medium.

pellicle (1) (*bacteriol.*) A continuous or fragmentary film which sometimes forms at the surface of a liquid culture. The film may consist entirely of cells or may be formed largely of extracellular products of the cultured organism(s); *Acetobacter xylinum* forms a tough cellulose pellicle. (2) (*mycol.*) Infrequently used to refer to any superficial membranous structure which may be easily removed from the tissues beneath it. (3) (*protozool.*) (synonym: *periplast*). A composite membranous structure which forms the limiting envelope in many protozoans. Among the simplest pellicles are those of the Sporozoa; these often consist of two or more layers of unit-type membrane underlain by a system of MICROTUBULES. Those of certain flagellates (e.g. *Trypanosoma* spp) consist of a single cell membrane and associated microtubules; some authors do not regard these structures as pellicles. More complex pellicles occur among the ciliates. The typical ciliate pellicle (found e.g. in *Paramecium*, *Tetrahymena*) consists of three layers of unit-type membrane, the outermost of which invests the whole organism—including the cilia. The two inner layers are sometimes referred to as the *alveolar membranes*. The space between these membranes is partitioned so as to form a number of vesicles, or alveoli; adjacent alveoli may or may not be in communication. This alveolar-type pellicle does not cover the whole organism, and in some ciliates may form only a relatively small part of the cell envelope. Even when such a pellicle comprises the major part of the envelope, it may be interrupted by *mucocysts* (see below) and its inner membranes are penetrated by the ciliar basal bodies which project below the level of the innermost membrane. *Mucocysts* are small invaginations of the outermost membrane which form flask-shaped vessels whose inner, closed ends are located in the cytoplasm. Under certain circumstances the mucocysts are believed to release a mucoid secretion onto the surface of the organism. The innermost of the three ciliate pellicular membranes may bear, on its inner (cytoplasmic) surface, a dense fibrous and/or microtubular layer known as the *epiplasm*. The epiplasm may be the site of connection and/or anchorage of sub-pellicular tubules and/or fibrils. (4) (*algol.*) (see e.g. THECA (sense 1) and EUGLENA)

Pelodictyon A genus of non-motile,

photosynthetic bacteria of the CHLOROBIACEAE. The cells are rod-like or coccoid and contain GAS VACUOLES; they may form chains which, following *ternary fission* in certain members of the chain, develop into colonial structures consisting of three-dimensional networks.

Pelomyxa A genus of free-living protozoans within the order AMOEBIDA (superclass SARCODINA); species occur in soil and water. The organisms are naked, multinucleate cells which feed mainly on small algae (e.g. diatoms) and bacteria; they reproduce asexually by PLASMOTOMY. The 'giant amoeba', *P. carolinensis* (also called *Chaos chaos*), may reach 3 to 5 millimetres (maximum dimension), while *P. palustris* is commonly greater than 0.5 mm.

pelotons Coils or balls of fungal hyphae found within the cortical cells of the root in certain endotrophic mycorrhizal associations. (see also MYCORRHIZA)

Peltigera A genus of foliose LICHENS in which the phycobiont is a blue-green alga (*Nostoc*) or a green alga (*Coccomyxa*)—according to species. The thallus tends to be large (1–10 centimetres) and more or less deeply lobed; the lower surface is tomentose and veined. When present, apothecia occur on the upper surface of the ends of the lobes. *Peltigera* spp generally occur on soil. Species include e.g.:

P. aphthosa. Phycobiont: *Coccomyxa*. The thallus is large (up to 10 centimetres) and broadly-lobed—bright green when wet, greyish when dry; the lower surface is brown. The upper surface of the thallus bears scattered, wart-like cephalodia which contain *Nostoc*. *P. aphthosa* occurs on soil among damp rocks—particularly in mountainous regions.

P. canina (the 'dog lichen'). Phycobiont: *Nostoc*. The thallus is broadly-lobed and brownish when wet, grey when dry; the upper surface is downy—particularly towards the margins—while the lower surface is pale and has long whitish rhizinae. The apothecia are round with reddish-brown discs. *P. canina* occurs on mossy walls and rocks and on the ground in woods, moors etc.

P. polydactyla. Phycobiont: *Nostoc*. The lobed thallus is approximately 3–5 centimetres; the upper surface is smooth and glossy, dark grey when wet, while the lower surface exhibits conspicuous brown 'veins'. The apothecia occur at the ends of the lobes and are oblong with reddish-brown discs; the ends of those lobes which bear apothecia are held vertically. *P. polydactyla* occurs among grass in heaths, woodland etc.

penicillin amidase An enzyme produced by certain bacteria (e.g. strains of *Escherichia coli*, *Proteus* spp); penicillin amidase cleaves the side chain of the penicillin G molecule (see PENICILLINS) to form 6-aminopenicillanic acid—a substance with little antibacterial activity. (cf. PENICILLINASE)

penicillinase A β-LACTAMASE which hydrolyses the β-lactam linkage of many of the PENICILLINS—yielding penicilloic acid, a substance which has no antibiotic activity. Penicillinase is produced by certain bacteria—including species of *Bacillus* (e.g. strains of *B. anthracis* and *B. cereus*), *Clostridium*, and *Staphylococcus*. Penicillinase-producing strains of *Staphylococcus* are generally resistant to the action of many of the penicillins (and to certain other antibiotics)—though frequently sensitive to e.g. *methicillin*. Penicillinase production in *Staphylococcus* is determined by a PLASMID; plasmid transmission between cells is effected by TRANSDUCTION. Some bacteria produce penicillinase constitutively; in others the enzyme may be *inducible*. (see also PENICILLIN AMIDASE)

penicillins A group of natural and semi-synthetic ANTIBIOTICS. In general, penicillins are more active against gram-positive than gram-negative bacteria, although certain semi-synthetic derivatives are effective against gram-negative species. Penicillins inhibit the formation of cross-links in the PEPTIDOGLYCAN of *growing* bacteria—possibly by acting as analogues of D-alanyl-D-alanine; the structurally weak peptidoglycan thus formed permits osmotic lysis of the cells. The lower sensitivity of gram-negative bacteria to penicillins may be due, at least in part, to the relative inaccessibility of the peptidoglycan in their CELL WALLS. Certain bacteria produce an enzyme (PENICILLINASE) which can inactivate many penicillins.

Penicillin G (benzylpenicillin) and *penicillin V* (phenoxymethylpenicillin) are obtained e.g. from *Penicillium chrysogenum* and *P. notatum*; valine and cysteine are precursors of the penicillin nucleus. The yield of penicillin G may be increased by the addition of phenylacetic acid to the culture medium. Production of the *semi-synthetic* penicillins involves the enzymatic removal of the benzyl side-chain from penicillin G (see formula; see also PENICILLIN AMIDASE); the resulting 6-aminopenicillanic acid can be acylated to produce a wide range of derivatives which vary in their antibacterial spectrum. Thus e.g. *ampicillin* and *carbenicillin* are more effective against gram-negative bacteria—but less effective against gram-positive bacteria—than are other penicillins; *methicillin* and *cloxacillin* are not readily inactivated by penicillinase and can be used against penicillinase-producing bacteria. Other penicillins include e.g. *penicillin N* (R =

The Penicillins

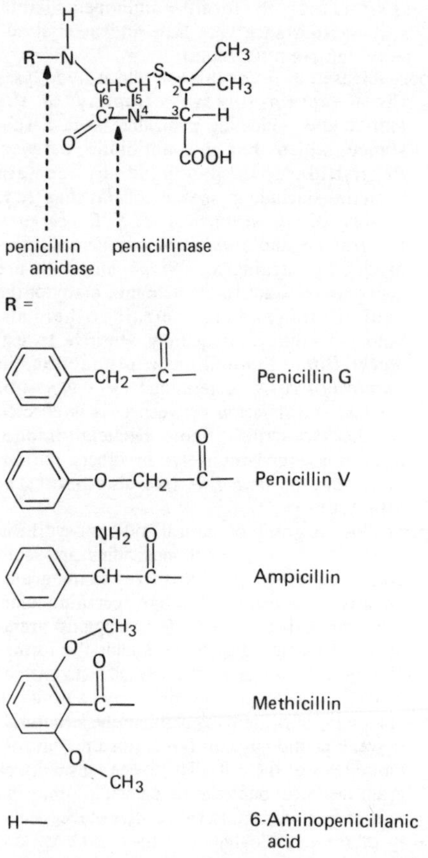

$-CO.(CH_2)_3.CH(NH_2)COOH)$—also called CEPHALOSPORIN N—produced by *Cephalosporium* spp, and *phenethicillin* (phenoxyethylpenicillin, $R = -CO.CH(CH_3).O.C_6H_5$).

Infrequently, the administration of a penicillin to a patient may induce a state of HYPERSENSITIVITY—i.e. the antibiotic may act as an *allergen*. The penicillins commonly exhibit *cross-allergenicity*.

Penicillium A genus of fungi within the order MONILIALES; *Penicillium* spp are typically saprophytic organisms—only a few species (e.g. *P. marneffei*) have been causally or tentatively connected with disease in animals. The vegetative form of the organisms is a septate mycelium. Conidiophores arise directly from the somatic hyphae; the conidiophore (= *penicillus*) consists of an erect hypha which bears either a terminal, hand-like cluster of phialids or a terminal cluster of short, compact or divergent branches (= 'primary sterigmata' or 'metulae')—each branch bearing one or more terminal phialids. The ellipsoidal, non-septate conidia—which are typically some shade of green (often grey-green)—are produced in basipetally-formed chains from the phialids. Some species of *Penicillium* are imperfect stages of TALAROMYCES. (see also GRISEOFULVIN and PENICILLINS)

penicillus (see PENICILLIUM)

Penicillus A genus of green algae (CHLOROPHYTA).

Peniculina A sub-order of protozoans of the order HYMENOSTOMATIDA. The *peniculines* are commonly large, elongated organisms; typically, a vestibulum is present, and the buccal cavity contains a number of specialized membranelles which, in the family Parameciidae, are arranged in the basic tetrahymenal fashion. Certain of the membranelles are termed *peniculi* (see PENICULUS)—from which the sub-order derives its name. Genera include e.g. *Frontonia* and PARAMECIUM.

peniculus (*ciliate protozool.*) A ribbon-like MEMBRANELLE consisting of a number of rows of cilia; two or more peniculi occur in the oral cavity of members of the suborder Peniculina, e.g. *Paramecium*.

pennate diatoms Those diatoms which are bilaterally symmetrical in valve view—e.g. *Navicula* (cf. CENTRIC DIATOMS).

pentamer (*virol.*) In an icosahedral capsid: a group of *five* capsomeres in a particular geometrical arrangement (cf. HEXAMER).

Pentatrichomonas (see TRICHOMONAS)

penton (*virol.*) In an icosahedral capsid: a capsomere to which *five* other capsomeres are adjacent; a penton occurs at each of the twelve vertices of an icosahedral capsid. In *adenoviruses* each penton carries a projecting protein *fibre*. (According to some authors, each penton in an adenovirus consists of the *penton base* (i.e. the capsomere proper) together with the fibre.)

pentose phosphate pathway (pentose phosphate cycle) (see HEXOSE MONOPHOSPHATE PATHWAY)

peplomer (*virol.*) (see ENVELOPE)

peplos (*virol.*) (see ENVELOPE)

pepsin (action on IMMUNOGLOBULINS) (see FAB FRAGMENT)

peptidoglycan (murein; mucopeptide; glycosaminopeptide) The rigid component of the CELL WALL in the majority of bacteria and in blue-green algae. The cell walls of gram-positive bacteria contain approximately 50–80% peptidoglycan; in gram-negative bacteria the proportion is approximately 1–10%. Peptidoglycan also occurs in bacterial spores (see later).

The backbone chains of peptidoglycan consist of alternating residues of *N*-acetyl-D-glucosamine and *N*-acetylmuramic acid linked

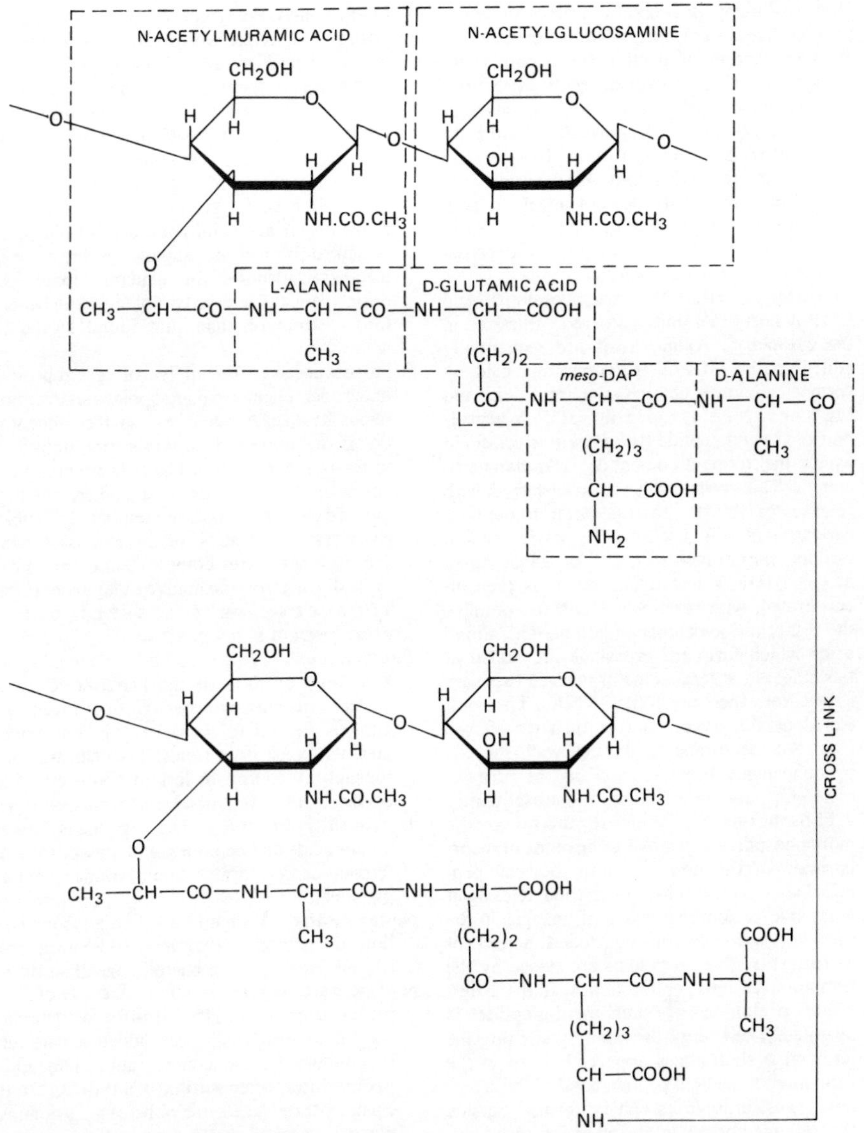

Peptidoglycan: cross-linked subunits in *Escherichia coli.*

β-(1 → 4) throughout. (Minor variations in the backbone chains may be found in certain species—e.g. *N*-glycolylmuramic acid occurs in *Mycobacterium* spp.) A tetrapeptide (L-alanine-D-glutamic acid-*R*-D-alanine) is linked to the carboxyl group of each muramic residue; *R* is species-dependent and is often an L-diamino acid—e.g. L-lysine (as in *Staphylococcus aureus*) or *meso*-diaminopimelic acid (*meso*-DAP) (as in *Escherichia coli*). A given tetrapeptide may or may not be linked to another tetrapeptide on an adjacent polysaccharide chain. When cross-linking does occur it is generally between the D-alanine of one tetrapeptide and the diamino acid of the other; this linkage may either be direct (as in *E. coli*) or may involve a short peptide (as in the pentaglycine cross-link of *S. aureus*). The polysaccharide chains—linked one to another by peptide bridges as described—form a single, giant, hollow molecule which is sometimes referred to as the murein *sacculus*. The extent of cross-linking is generally greater in gram-positive than in gram-negative bacteria.

The structure of peptidoglycan can be investigated e.g. by treatment with LYSOZYME. This enzyme hydrolyses β-(1 → 4) linkages between *N*-acetylmuramic acid and *N*-acetylglucosamine—thus splitting the polysaccharide chains into disaccharide subunits. Such treatment does not affect the peptide links, so that the size of the fragments obtained gives an indication of the extent of cross-linking in the original molecule.

Biosynthesis. UDP-*N*-acetylglucosamine and UDP-*N*-acetylmuramic acid are synthesized in the cytoplasm. Amino acids are sequentially added to the UDP-*N*-acetylmuramic acid by specific enzymes (which require ATP and Mg^{++} or Mn^{++}) to form UDP-*N*-acetylmuramyl *penta*peptide (the *Park nucleotide*) in which the terminal dipeptide is D-alanine-D-alanine. This precursor is then transferred, with release of UMP, to the monophosphate derivative of a lipid-soluble C_{55} polyisoprenoid alcohol (BACTOPRENOL) in the CELL MEMBRANE. UDP-*N*-acetylglucosamine is then incorporated, with release of UDP, to complete the disaccharide-pentapeptide unit. The amino acids which form the cross-link are added at this stage; in at least some organisms these are added from their respective tRNAs. The completed peptidoglycan unit is then transferred from the membrane to the cell wall, the C_{55} pyrophosphate being released in the process. (The C_{55} pyrophosphate is subsequently dephosphorylated, thus regenerating the monophosphate derivative.) The mode of incorporation of the new unit into the wall peptidoglycan is not fully understood; clearly it must involve some breakage of linkages in the sacculus—possibly by controlled action of AUTOLYSINS. The cross-links are closed by the formation of interpeptide bonds, and the terminal D-alanine of each pentapeptide is liberated. The energy provided by splitting the terminal D-ala-D-ala is apparently used in the formation of the final peptide bond of the cross-link. Thus, cross-links can be formed outside the cell membrane in the absence of an additional energy source. The liberated D-alanine re-enters the cell *via* a specific transport system.

In gram-positive bacteria, peptidoglycan is covalently linked to TEICHOIC ACIDS probably *via* phosphodiester links at the 6-position of the muramic acid. In gram-negative bacteria, peptidoglycan is covalently linked to lipoprotein of the outer membrane, e.g. (in *E. coli*) linkages are formed between the lipoprotein and *meso*-DAP residues of the peptidoglycan.

ANTIBIOTICS which inhibit peptidoglycan biosynthesis include: BACITRACIN, CEPHALOSPORINS, CYCLOSERINE, PENICILLINS, PHOSPHONOMYCIN, RISTOCETIN, VANCOMYCIN.

Spore peptidoglycan. Peptidoglycan of the bacterial SPORE *cortex* may differ considerably in structure from that of the corresponding vegetative cell. In *Bacillus subtilis*, for example, the lactyl groups of some muramic acid residues form lactam rings which involve the N on C2, the acetyl group being displaced; the remaining *N*-acetylmuramic acid residues may be linked to a single L-alanine residue or to the usual tetrapeptide. In general, spore peptidoglycan has less cross-linking, and hence a looser structure, than that found in the cell wall.

Peptococcaceae A family of gram-positive, anaerobic, chemoorganotrophic, asporogenous bacteria; species occur e.g. in the alimentary tract of man and animals (including the RUMEN) and in the soil. The cells are non-motile cocci, up to 2.5 microns in diameter, which occur singly, in pairs, in tetrads or cubical PACKETS, in clusters or in chains. Growth requirements are complex; metabolism is typically heterofermentative. The genera: PEPTOCOCCUS, PEPTOSTREPTOCOCCUS, RUMINOCOCCUS and SARCINA.

Peptococcus A genus of gram-positive, anaerobic bacteria of the PEPTOCOCCACEAE; species occur as parasites (and pathogens?) of animals (including man)—e.g. in the alimentary and respiratory tracts. The cells are cocci, typically 1 micron or less in diameter, which occur singly or in pairs, tetrads, clusters or (infrequently) in chains. The organisms ferment amino acids and peptones, and may or may not ferment carbohydrates. Some strains form dark pigments.

peptone water A slightly alkaline aqueous solution of peptone (1%) and sodium chloride (0.5%) used e.g. as a bacteriological MEDIUM.

peptone water sugars ('sugars') A range of liquid media each of which consists of PEPTONE WATER to which has been added a particular carbohydrate (final concentration 1%) and a pH indicator, often ANDRADE'S INDICATOR; the carbohydrate used may be e.g. arabinose, glucose, lactose, maltose, mannitol, mannose, rhamnose, sucrose, or xylose. Such media are used to determine which of a series of carbohydrates are metabolized by a given strain of bacterium; bacterial growth in a given peptone water sugar may give rise to the production of acid or acid and gas—acid production being detected by the pH indicator, and gas production by a DURHAM TUBE.

Some bacteria, e.g. *Bacillus* spp, produce an alkaline reaction in peptone water sugars owing to the formation of alkaline breakdown products from the peptone; since this would mask the effects of carbohydrate metabolism, a medium containing inorganic salts and a carbohydrate may be used for such organisms. Certain bacteria, e.g. many species of *Corynebacterium*, grow in peptone water sugars only when such media are supplemented with serum.

peptones Soluble products of protein hydrolysis; they are not coagulated by heat and are not precipitated by saturation with ammonium sulphate. Peptones can be precipitated by phosphotungstic acid.

peptonization Proteolysis (hydrolysis of proteins) in which soluble products are formed. (see PEPTONES)

Peptostreptococcus A genus of gram-positive anaerobic cocci of the PEPTOCOCCACEAE; cells are typically formed in chains, but the organisms are otherwise very similar to species of PEPTOCOCCUS.

Peptostreptococcus elsdenii Former name for *Megasphaera elsdenii*—see MEGASPHAERA.

perfect (*mycol.*) Refers to the *sexual* stage or condition of fungi.

pericentric (of a chromosome) Refers to an intrachromosomal structural alteration (see CHROMOSOME ABERRATION) which *includes* the centromere; e.g. in a pericentric inversion, the inverted segment contains the centromere. (cf. PARACENTRIC)

peridinin (see CAROTENOIDS)

Peridinium A genus of algae of the DINOPHYTA.

peridiole (peridium) (*mycol.*) (see NIDULARIALES and CYATHUS)

peridium (*mycol.*) The outer enveloping wall or coat(s) forming part of the fruiting body in certain fungi—e.g. the leathery capsule of *Scleroderma*, the multilayered coats of *Geastrum* and other earthstars, and the wall of a CLEISTOTHECIUM, PERITHECIUM etc.

Périgord truffle The fruiting body of the ascomycete *Tuber melanosporum*.

perimycin (fungimycin) (see POLYENE ANTIBIOTICS)

perinuclear space (see NUCLEUS)

periodic acid-Schiff reaction (see PAS REACTION)

periodic selection (in a bacterial culture) Periodic changes in the constitution of a growing bacterial culture due to the upsurge and decline of successive mutant strains—each strain being better suited than its predecessor to the prevailing environment. Each mutant strain, in its turn, may become numerically dominant in the culture.

periphyses (*mycol.*) (*sing.* periphysis) Short, sterile filaments which line the ostiolar canal of a perithecium. (Analogous structures may occur in pseudoperithecia.)

periphyton community (see AUFWUCHS)

periplasmic space (of bacteria) The region between the CELL MEMBRANE and the CELL WALL.

periplast (*protozool.*) Synonym of PELLICLE.

Perisporiaceae ('dark mildews', 'black mildews') A family of fungi within the order ERYSIPHALES; the organisms are similar in many ways to members of the family ERYSIPHACEAE but differ in several respects: (i) they form *dark*-coloured septate mycelium and hyphopodia; (ii) they do not form conidia; (iii) they form *dark*-coloured ascospores each of which consists of two or more cells. The genera include *Meliola*.

peristome (*ciliate protozool.*) A term which may refer either to the buccal cavity or buccal area of a peritrich or spirotrich, or to the convoluted vestibular apparatus of a chonotrich.

perithecioid pseudothecia (see ASCOSTROMA)

perithecium (*mycol.*) A hollow, spheroidal or flask-shaped ASCOCARP which opens to the exterior *via* a pore or a tubular neck (OSTIOLE); asci develop within the perithecium commonly in the form of an hymenial layer or as a basal tuft. Perithecia are the characteristic ascocarps of members of the order SPHAERIALES. Perithecia are formed (by *ascohymenial* development—see ASCOCARP) on the surface of a substratum or fungal stroma or embedded in a fungal stroma—according to species. The wall (peridium) of a perithecium may be membranous or brittle, brightly-coloured or dark-coloured—according to species; in *Podospora* spp the peridium is semi-transparent. In size, perithecia are commonly 100–300 microns—that of e.g. *Chaetomium globosum* being about 200 microns; the perithecia of some species are somewhat larger. At maturity the ascospores or asci (depending on species) are discharged—often forcibly or explosively—*via* the perithecial opening. The nature and arrangement of sterile structures within the perithecium (e.g. paraphyses) are taxonomically important. (cf. *pseudoperithecium* under ASCOSTROMA)

peritonitis Inflammation of the *peritoneum* (the membrane which lines the abdominal cavity and which covers the viscera).

peritrich A ciliate of the subclass Peritrichia.

Peritrichia A subclass of freshwater and marine protozoans of the subphylum CILIOPHORA. *Mature* peritrichs generally lack *somatic* cilia but have well-developed oral ciliature; the latter includes rows of cilia which—looking *into* the buccal cavity—spiral downwards towards the cytostome in an anticlockwise direction. A fringe of (external) cilia is commonly found in the peristomal (peribuccal) region. According to species, individual cells may be shaped like an egg, bell, bottle etc.; some are loricate. In some species the cell (*zooid*) is attached to the substratum by a contractile or non-contractile *stalk*; such sessile forms may occur singly or in colonies. (The zooid is often of the order of

50–100 microns in size, while the stalk may be several hundred microns in length.) Typical ciliate nuclear dimorphism occurs. Asexual reproduction may occur e.g. by budding—motile, ciliated larvae being produced. CONJUGATION has been observed in many species.

All peritrichs are placed in the order Peritrichida; predominantly sessile species are classified in the suborder Sessilina, while mobile, unstalked species are placed in the suborder Mobilina. The genera include CARCHESIUM, OPERCULARIA and VORTICELLA.

peritrichous (*adjective*) (*bacteriol.*) Refers to the arrangement of a cell's flagella in which the flagella are more or less uniformly distributed over the surface of the cell.

perkinetal (*ciliate protozool.*) (*adjective*) In binary fission, refers to the plane of division which cuts *across* the kineties (rows of somatic cilia).

permeability (of bacterial cell membranes) (see TRANSPORT and CELL MEMBRANE)

permease This term was originally applied to a specific inducible protein component of the β-galactoside TRANSPORT system in *Escherichia coli*. The term *permease* may still be used loosely to refer to a whole transport system, or to designate a specific (protein) component within a given transport system. (see also CRYPTIC MUTANT)

permissive condition, host etc. (see CONDITIONAL LETHAL MUTANT)

Peronospora A genus of fungi within the class OOMYCETES; *Peronospora* spp are obligate parasites and pathogens of higher plants and are the causal agents of *downy* MILDEWS. The vegetative form of the organisms is a well-developed, coenocytic, intercellular mycelium from which arise short, club-shaped (or branched, filamentous) haustoria. Asexual reproduction involves the formation of sporangia which, on germination, give rise in the majority of species to germ tubes rather than to zoospores*. (*P. tabacina* is reported to form sporangia which give rise to zoospores.) The sporangia develop on branched sporangiophores which emerge from the stomata of the host plant; the sporangia are borne on the tips of the curved, tapering sporangiophores. Sexual reproduction is oogamous, and thick-walled resistant oospores are formed; on germination an oospore typically produces a germ tube. Species include *P. parasitica* (a pathogen of crucifers), *P. destructor* (parasitic on onions), and *P. pisi* (the causal agent of downy mildew of the pea). *Sporangia which form germ tubes are referred to as *conidia* by some authors—particularly plant pathologists.

perosamine (see POLYENE ANTIBIOTICS)

peroxidase An OXIDOREDUCTASE which catalyses a reaction in which electrons removed from a substrate are donated to hydrogen peroxide. An example of a peroxidase is CATALASE.

peroxisomes MICROBODIES which contain D-amino acid oxidase, α-hydroxy acid oxidase, catalase, and other enzymes; peroxisomes occur e.g. in yeasts and in certain protozoans (e.g. *Tetrahymena*).

persistent viruses (*plant pathol.*) (see CIRCULATIVE VIRUSES)

perthophyte Any organism which grows in the dead tissues of a living plant—e.g. a fungus which attacks the heartwood of living trees.

Pertusaria A genus of crustose LICHENS in which the phycobiont is *Trebouxia*. Ascocarps frequently occur several together in wart-like protuberances (verrucae) on the thallus; they are initially perithecium-like with a pore-like opening, but may subsequently open out to form flat or concave discs. Species include e.g.:

P. pertusa. The thallus is grey-green, smooth or areolate, roughly circular in outline, often with concentric light and dark zones—each light and dark zone representing one year's growth; the periphery of the thallus is surrounded by a white hypothallus. Ascocarps occur in groups in verrucae; they do not open as they mature. *P. pertusa* occurs on trees and occasionally on rocks.

P. amara. The thallus is grey with scattered white, flattened soralia which are 0.5–1.5 millimetres in diameter; it is characterized by a bitter taste due to the presence of the lichen substance picrolichenic acid. Ascocarps are rarely formed; when present they are solitary and the mature discs are open and sorediate. *P. amara* occurs mainly on trees.

pertussis (see WHOOPING COUGH)

petechial (*adjective*) (of haemorrhage(s)) Refers to very small, discrete regions of haemorrhage in the skin or in mucous membranes etc.

petite colonies In certain yeasts (e.g. *Saccharomyces cerevisiae*): abnormally small colonies which occur with an incidence of about 1 per 500 normal colonies; the incidence of such colonies may be greatly increased by the inclusion of certain agents, e.g. acridine dyes, in the growth medium. The cells which form petite colonies are mutants which lack certain components of the respiratory chain, e.g. cytochromes *a* and *b*; such cells, which cannot carry out oxidative phosphorylation, exhibit a reduced growth rate probably because they are restricted to the less-efficient fermentative mode of metabolism.

In *segregational* petite strains the mutant characteristic is determined by mutation(s) in the *chromosomal* DNA; when segregational petite strains are crossed with wild-type strains the progeny consists of segregational petite and wild-type strains in the ratio 1:1.

In *cytoplasmic* petite strains the mutant characteristic is determined by mutation(s) in the *mitochondrial* DNA. When *neutral cytoplasmic* petites are crossed with wild-type strains the zygote and ascospores are of the wild-type. When *suppressive cytoplasmic* petites are crossed with wild-type strains, the zygote, and the majority of (diploid) cells budded from the zygote, are *petite*; if ascospore-formation occurs *immediately* after zygote-formation, all the ascospores produced are usually *petite*. If ascospore-formation is delayed, the (diploid, budded) progeny of the zygote lose the ability to sporulate; only a *rare* budded cell with normal phenotype can produce ascospores—all of which are of the wild-type. (see also POKY MUTANT)

petri dish A round, shallow, flat-bottomed dish with a vertical edge together with a similar, slightly larger structure which forms a loosely-fitting lid. Petri dishes are made of glass or plastic; they are manufactured in various sizes—those having a diameter of about 10 centimetres being the most common. They are widely used in bacteriology and mycology e.g. as receptacles for various types of solid media.

Peziza A genus of fungi of the order PEZIZALES; the organisms form fawn or light brown, sessile, discoid or cup-shaped apothecia—often several centimetres in diameter—which lack a surface covering of hair-like processes. The (operculate) asci each contain a number of uninucleate, hyaline or brown, spherical, ovoid or elongated ascospores; each ascospore typically contains two guttules—some contain one or (infrequently) three. The apical region of the ascus stains strongly with Meltzer's reagent.

Pezizales An order of fungi within the class DISCOMYCETES; most or all species are saprophytic organisms which occur e.g. on wood, soil, and dung. Members of the Pezizales typically form operculate asci (see OPERCULUM); the few inoperculate species discharge their spores *via* a vertical slit which develops at the apex of each ascus (rather than *via* a preformed pore). In the majority of species the asci are arranged, with paraphyses, as a distinct hymenial layer. According to species the ascocarp (APOTHECIUM) may be barely visible (to the unaided eye) or may be up to 10 centimetres or more in diameter; it may be dark or brightly-coloured. The majority of species appear to have no conidial (imperfect) stage. The genera include ALEURIA, HELVELLA, MORCHELLA, PEZIZA, PYRONEMA, *Sarcoscypha*, SCUTELLINIA, and VERPA.

Pfeiffer phenomenon The rapid lysis of *Vibrio cholerae* when incubated with specific antibody and COMPLEMENT.

p.f.u. PLAQUE-FORMING UNIT.

pH indicators (in microbiology) Microbial metabolism in a given medium may bring about an increase or decrease in the pH of that medium due to the accumulation of alkaline or acidic metabolic products; certain tests used for the identification of bacteria (e.g. the METHYL RED TEST) involve the detection of any change in the colour of a pH indicator (incorporated in, or subsequently added to, the medium) following appropriate incubation. Some pH indicators: ANDRADE'S INDICATOR; BROMCRESOL GREEN; BROMCRESOL PURPLE; BROMPHENOL BLUE; BROMTHYMOL BLUE; CHLORPHENOL RED; CONGO RED; CRESOL RED; LITMUS; METHYL ORANGE; METHYL RED; NEUTRAL RED; PHENOL RED; PHENOLPHTHALEIN; THYMOL BLUE; THYMOLPHTHALEIN.

PHA (see PHYTOHAEMAGGLUTININ)

Phacus A genus of ALGAE of the division EUGLENOPHYTA.

Phaeophyta (brown algae) A division of the ALGAE; the brown algae are predominantly marine organisms. The thallus may be filamentous or complex—the latter involving leaf-like, stem-like, and root-like organs—and some species may reach 50 or more metres in length; there are no unicellular or colonial species. All species contain CHLOROPHYLLS *a* and *c*, β-carotene, and several xanthophylls—including fucoxanthin (see CAROTENOIDS); intracellular storage products include LAMINARIN. Sexual processes vary from isogamy to oogamy, according to species; the biflagellate, pyriform gametes each have one anteriorly-directed *tinsel* FLAGELLUM and one posteriorly-directed *whiplash* flagellum. Many species exhibit an isomorphic or heteromorphic ALTERNATION OF GENERATIONS, and life cycles are often complex. Genera include *Cutleria*, FUCUS, LAMINARIA, *Sargassum*, and *Turbinaria*. (see also ALGINIC ACID and FUCOIDIN)

phaeophytin A compound which consists of a CHLOROPHYLL molecule *minus* its magnesium atom.

phage An abbreviation of BACTERIOPHAGE.

phage typing (see TYPING)

phagocyte (*immunol.*) Any of a variety of cells which ingest and (commonly) break down certain categories of particulate matter, e.g. bacteria. The most actively phagocytic cells are the MACROPHAGES; other cells, e.g. the *Kupffer* cells of the liver, and the *neutrophils*, are also phagocytic. (see also PHAGOCYTOSIS)

phagocytosis The process in which particulate matter is ingested by a cell or organism (cf. PINOCYTOSIS). In phagocytosis the cell membrane invaginates to form a pocket containing the particulate matter; subsequently the invaginated membrane forms a closed intracellular sac or *vacuole*. The contents of the vacuole are digested or degraded by enzymes which are introduced when the vacuole coalesces with one or more LYSOSOMES. (see also FOOD VACUOLE) Phagocytosis is carried out e.g. by certain metazoan cells (e.g. macrophages, monocytes) and by many types of protozoan.

Within the animal body phagocytosis is a non-specific activity and is an important factor in NON-SPECIFIC IMMUNITY; the susceptibility of microorganisms to phagocytosis is increased following OPSONIZATION.

phagosome A vacuole formed by PHAGOCYTOSIS.

phagotrophy The ingestion of nutrients in *particulate* form.

phallacidin (see PHALLOTOXINS)

Phallales An order of fungi of the class GASTEROMYCETES. Constituent species form fruiting bodies which are, at least initially, hypogean; the immature fruiting body (the 'egg') is spheroidal and, according to species, varies from less than one centimetre to several centimetres in diameter. The 'egg' consists essentially of a gleba which is enveloped in a layer of gelatinous material (the 'mesoperidium')—the whole being enclosed within a membranous peridium. In some species little further morphological development occurs: the peridium is indehiscent, and the enclosed basidiospores are dispersed on rupture of the peridium by burrowing animals. In other species (the 'stinkhorns') the 'egg' contains an axial, solid or hollow mass of spongy tissue (the 'receptacle') present within, or partly within, the gleba; during the development of the fruiting body the peridium ruptures, and the receptacle elongates to form a vertical, columnar structure (a *pseudostem*) with the hymenium-bearing gleba present as a layer around the apex. Concurrently, autolysis occurs in the hymenial trama, and the fertile tissue subsequently forms a dark, typically evil-smelling liquid which contains the basidiospores; the spores are subsequently dispersed by insects which are attracted to the apex of the fruiting body. In *Mutinus*, *Phallus*, and other stinkhorns the basidiospores are bacilliform, less than five microns maximum dimension.

phallin A haemolytic, cytotoxic, thermolabile substance, of unknown chemical composition, isolated by Kobert (1891) from *Amanita phalloides*. (see also PHALLOTOXINS and FOOD POISONING (fungal))

phallin B (see PHALLOTOXINS)

phallisin (see PHALLOTOXINS)

phalloidin (see PHALLOTOXINS)

phallotoxins A group of toxic cyclic peptides which occur in the fungus *Amanita phalloides* and in some other species of *Amanita*, e.g. *A verna*. Phallotoxins are less toxic than the AMATOXINS which occur in the same fungi; they produce clinical effects within a few hours of ingestion. Symptoms include severe vomiting and diarrhoea; degenerative changes occur in the cells of the liver. The phallotoxins include phallin B, phalloidin, phallacidin and phallisin.

Phallus A genus of fungi of the order PHALLALES.

pharyngeal basket (*ciliate protozool.*) (see TRICHITES)

phase-contrast microscopy (see MICROSCOPY)

phase-shift mutation (frameshift mutation) A type of MUTATION which causes out-of-phase transcription of the base sequence (see GENETIC CODE); such mutations arise from the addition or deletion of nucleotide(s) in numbers other than 3 or multiples of 3. For example:

A B C A B C A B C A B C A B C.....
(wild type)

A C A B C A B C A B C A B C A.....
(mutant—1 nucleotide deleted)

(A, B, and C represent nucleotide bases.) See also ACRIDINES (as mutagens).

phase variation, in *Salmonella* Certain serotypes of *Salmonella* are genetically stable in respect of their *flagellar* (H) antigens; when isolated, such serotypes always have the same flagellar antigens. These serotypes are said to be *monophasic*.

In *diphasic* serotypes the flagellar antigens remain constant for a number of bacterial generations, but, owing to the activation of a certain *genetic* mechanism, a *different* group of flagellar antigens appears spontaneously in later generations of progeny cells. Reversion to the original flagellar antigens occurs in subsequent generations, and so on. Thus, two different antigenic states are possible; these are referred to as *phase 1* and *phase 2*.

Among the salmonellae many serotypes are diphasic. Each of the *phase 1* antigens is specific to a single serotype or occurs in a relatively small number of serotypes; thus, phase 1 antigens were previously referred to as *specific phase* antigens. Each of the *phase 2* antigens may occur in many serotypes; these were originally designated *group phase* antigens.

Bacteria in a given phase may be encouraged to change to the alternative phase; this may be achieved by repeated subculture or by the use of media containing antibodies to the flagellar antigens of the existing phase. Phase variation is a manifestation of the expression of alternative structural genes. The gene(s) corresponding to each phase specify the synthesis of FLAGELLIN of appropriate antigenic structure.

Flagellar (and somatic) antigens form the basis of the KAUFFMANN-WHITE CLASSIFICATION SCHEME for salmonellae.

phaseollin (phaseolin) (see PHYTOALEXINS)

PHB (see POLY-β-HYDROXYBUTYRATE)

phenetic classification (see NUMERICAL TAXONOMY)

phenocopy A cell whose phenotype has become modified as a result of environmental conditions in a way that mimics a phenotypic modification due to a genetic change. For example, F^+ *donors* of *Escherichia coli* (see CONJUGATION) may lose their donor characteristics under certain conditions—e.g. when grown to maximum density in aerated broth; such cells exhibit the characteristics of recipients (i.e. they mimic F^- cells) and are called F^- *phenocopies*. When subcultured on fresh medium, F^- phenocopies revert to the donor state. (see also PSEUDO-LYSOGENY)

phenol (as an antiseptic/disinfectant) (see PHENOLS)

phenol coefficient A number which expresses the antibacterial power of a *phenolic* DISINFECTANT—relative to that of phenol—*under standard test conditions*. It is determined experimentally (by e.g. the RIDEAL-WALKER TEST or the CHICK-MARTIN TEST) using a single bacterial strain as test organism. The use of phenol as a standard has been criticized e.g. on the grounds that the properties of phenol are not typical of those phenolic compounds which are assessed by these methods. (see also PHENOLS)

phenol red (PH INDICATOR) pH 6.8 (yellow) to pH 8.4 (red). pK_a 7.9.

phenolphthalein (PH INDICATOR) pH 8.3 (colourless) to pH 10.0 (red). pK_a 9.6.

phenols (as ANTISEPTICS and DISINFECTANTS) Phenol and its derivatives are microbicidal or microbistatic, according to concentration and temperature; phenolic compounds denature proteins, and their primary mode of action probably involves the disruption of the CELL MEMBRANE. The antimicrobial activity of phenolic compounds is greatly reduced by dilution and by decrease in temperature; in general it is not significantly affected by the presence of organic matter. *Phenol* (carbolic acid, C_6H_5OH) is soluble in water and, at appropriate concentrations and temperatures, may be bactericidal, fungicidal, and virucidal; at room temperatures its sporicidal activity tends to be very slow. Alkyl- and halogen-substituted phenols, and BISPHENOLS, generally exhibit antimicrobial activity greater than that of phenol; they also tend to be appreciably less toxic and caustic. These compounds are generally water-insoluble; some can be solubilized by the addition of alkali, while others form colloidal suspensions or emulsions in the presence of surface-active agents (e.g. SOAPS). *Cresols* (methyl-phenols) are used as disinfectants—e.g. *Lysol* (proprietary name) contains *o*-, *m*-, and *p*-cresols solubilized by a relatively high proportion of soap. (*Excessive* concentrations of soap may completely abolish the antimicrobial activity of cresols.) *Amyl*-m-*cresol* is used e.g. in antiseptic sweets and mouthwashes. *Xylenols* (dimethyl-phenols) are effective microbicides—e.g. *Sudol* (proprietary name) contains xylenols and other alkyl-phenols; Sudol is non-irritant and is used as a skin antiseptic and as a disinfectant. Other alkyl-phenols include THYMOL (5-methyl-2-isopropyl-phenol)—used e.g. as a PRESERVATIVE, and *menthol* (3-methyl-6-isopropyl-phenol) and *carvacrol* (2-methyl-5-isopropyl-phenol)—both of which are used in antiseptic sweets and mouthwashes. *Polyhydric* phenols generally have weak antimicrobial activity; *n-hexyl resorcinol* (4-*n*-hexyl-1,3-dihydroxybenzene) is used as a topical and urinary antiseptic. *Nitrophenols* are often more active than the corresponding phenols; e.g. *picric acid* (2,4,6-trinitrophenol) is antibacterial and antifungal. (see also DINITROPHENOLS (as antifungal agents)) *Halogenation* of phenols generally increases their antimicrobial activity and renders them less toxic; however, their activity is more susceptible to inhibition by organic matter. The activity of halogenated phenols is generally greatest when the halogen is in a position *para* to the hydroxyl group. *Dettol* (proprietary name) contains chloroxylenols and is used e.g. as a skin antiseptic and general disinfectant; dilution of chloroxylenol disinfectants may drastically reduce or abolish their activity against e.g. *Pseudomonas* spp. p-*Chloro*-m-*cresol* has been used as a paint preservative. (*see also* PHENOL COEFFICIENT and CREOSOTE)

phenon A category established by NUMERICAL TAXONOMY.

phenotype Of an individual organism: the totality of observable structural and functional characteristics—*or* particular, named, characteristic(s) and/or properties; the phenotype of an individual is determined *jointly* by its GENOTYPE and its environment.

phenotypic lag (in bacteria) A delay in the phenotypic expression of a mutation; phenotypic lag may be due e.g. to SEGREGATION LAG.

phenotypic mixing In a cell simultaneously infected by different types of virus: (a) incorporation of the genome of one virus in the capsid of another type (*transcapsidation*), or (b) the formation of progeny virus(es) in which the capsid subunits derive from different viruses.

phenotypic suppression The suppression of a mutant phenotype by non-genetic (environmental) factors (cf. SUPPRESSOR MUTATION). Phenotypic suppression may occur at the level of (a) mRNA, or (b) the ribosomes. (a) During mRNA transcription, 5-fluorouracil can replace uracil, and can be mis-read as cytosine during translation; thus, e.g. the nonsense codon UAG may occasionally be mis-read as CAG (a codon which specifies glutamine)—thus simulating a *nonsense* SUPPRESSOR MUTATION. (b) The ribosomes may be altered such that mRNA may be mis-read during translation. For example, when sub-lethal amounts of STREPTOMYCIN bind to the ribosomes, mis-reading of certain codons during translation may lead to suppression of certain mutations.

β-phenoxyethyldimethyldodecylammonium bromide (see QUATERNARY AMMONIUM COMPOUNDS)

phenylalanine (biosynthesis of) (see Appendix IV(f))

phialid (phialide) (*mycol.*) In some species of fungi: an elongated, flask-shaped or bottle-shaped structure within which—or at the extremity of which—CONIDIA are formed; the conidia may be immediately deciduous or may remain attached to one another so as to form a chain. One or more phialids may form the terminal portion of a conidiophore (and of each of its branches, if any), and in some species the *conidial head* bears a confluent surface layer of phialids. Phialids are sometimes referred to as sterigmata—see e.g. ASPERGILLUS; see also PENICILLIUM. (cf. METULA and STERIGMA)

phisohex (pHisoHex) (proprietary name) An antibacterial lotion which contains hexachlorophene—see BISPHENOLS.

phlogistic (*med.*) Refers to any agent which produces inflammation.

phloretin (see PHLORIDZIN)

phloridzin A phenolic glycoside found in apple leaves. The aglycone portion of phloridzin, *phloretin* (β-(4-hydroxyphenyl)-propionophlorophenone), can be liberated by β-glycosidases and oxidized by polyphenol oxidase to yield polymers toxic to e.g. *Venturia inaequalis*. (see also PHYTONCIDES)

phobotaxis (phobotactic movement) (see TAXIS)

Pholiota A genus of fungi of the family Strophariaceae, order AGARICALES.

Phoma A genus of fungi of the order SPHAEROPSIDALES; species include a number of plant pathogens—e.g. the causal agent of *blackleg* in crucifers. *Phoma* spp form hyaline, ovoid, aseptate conidia; the spores develop singly on conidiophores which are borne in spherical or spheroid brown, thin-walled, papillate, uniloculate and ostiolate pycnidia that are immersed in the substratum.

Phomopsis A genus of fungi of the SPHAEROPSIDALES.

-phore A suffix signifying 'carrier of' or 'bearer of'. Examples: *sporophore*, *ionophore*.

Phormidium A genus of filamentous BLUE-GREEN ALGAE.

phoront (see APOSTOMATIDA)

phosphatase test (for milk) (Aschaffenburg-Mullen phosphatase test) A test used to detect the presence of alkaline phosphatase in *pasteurized* milk. Alkaline phosphatase, an enzyme normally present in raw (i.e. untreated) milk, is inactivated by PASTEURIZATION; hence, the test examines the efficiency of the pasteurization process. In the test, milk, containing alkaline-buffered *p*-nitrophenyl phosphate, is incubated at 37 °C for 2 hours; the milk is then examined (e.g. by colorimetry) for the presence or intensity of yellow coloration due to *p*-nitrophenol. A false positive result can be obtained if, subsequent to pasteurization, milk is contaminated with a phosphatase-producing organism. (see also MILK)

phosphatases (*bacteriol.*) Phosphatases are enzymes which hydrolyse esters of phosphoric acid; they are produced by a range of bacteria—including species of *Lactobacillus*, *Proteus*, *Pseudomonas*, *Salmonella*, and *Staphylococcus*.

Staphylococcal phosphatases. An *acid* phosphatase—i.e. one which exhibits maximum activity under acid conditions—is produced by many COAGULASE-positive strains and (generally in smaller amounts) by relatively fewer coagulase-negative strains; coagulase-negative strains which initially appear to be phosphatase-negative often produce the enzyme on prolonged incubation. It appears that acid phosphatases may occur extracellularly or may be loosely or firmly bound to the cell surface—depending e.g. on the pH of the medium. *Alkaline* phosphatase has been reported to be produced by some strains of *Staph. aureus*. (In general, it appears that the pathogenicity of staphylococci correlates less with phosphatase production than with coagulase production.)

Phosphatase production may be demonstrated by growing the test strain on a solid medium containing the sodium salt of phenolphthalein diphosphate; after incubation the plate is exposed to gaseous ammonia. Colonies of phosphatase-producing strains (and probably the surrounding medium) become deep pink owing to the effect of the higher pH on the liberated PHENOLPHTHALEIN.

phosphoenolpyruvate-hexose phosphotransferase transport system (see TRANSPORT)

phosphogluconate pathway (see HEXOSE MONOPHOSPHATE PATHWAY)

phosphoketolase pathway In certain organisms: a pathway, which involves the enzyme *phosphoketolase*, in which hexoses and/or pentoses are degraded—e.g. the pathways of

HETEROLACTIC FERMENTATION and of the fermentation of glucose by *Bifidobacterium* spp (see Appendix III(a) (ii) and (iii)).

phosphonomycin (*cis*-1,2-epoxypropyl-1-phosphonate) An antibiotic which blocks PEPTIDOGLYCAN synthesis by inhibiting the formation of UDP-*N*-acetylmuramic acid from UDP-*N*-acetylglucosamine. Phosphonomycin is a structural analogue of phosphoenolpyruvate and irreversibly binds the active site of the enzyme *pyruvate-UDP-N-acetylglucosamine transferase.*

phosphorescence A phenomenon, analogous to FLUORESCENCE, in which the emission of light persists for some time after the cessation of irradiation. The term is sometimes (incorrectly) applied to BIOLUMINESCENCE.

phosphoroclastic split A reaction (analogous to hydrolysis) in which a molecule is cleaved with the addition of the components of phosphate. The term is usually applied to the reaction:

$$\begin{matrix} \text{pyruvate} & & \text{acetyl phosphate} \\ + & \rightarrow & + \\ \text{phosphate} & & \text{formate} \end{matrix}$$

which occurs e.g. during MIXED ACID FERMENTATION; however, the initial product of pyruvate cleavage is acetyl-CoA—acetyl phosphate being formed subsequently by the action of a phosphate acetyltransferase (phosphotransacetylase) on acetyl-CoA.

photic zone That zone in a body of water (lake etc.) which receives light of sufficient intensity to enable aerobic photosynthesis to occur. *Bacterial* PHOTOSYNTHESIS can occur below the photic zone.

photoautotroph (see AUTOTROPH)

Photobacterium A genus of gram-negative, oxidase-positive, luminescent* marine bacteria of the VIBRIONACEAE; the cells are short rods or coccobacilli, and are either polarly flagellate or non-motile. The two species (*P. phosphoreum* and *P. mandapamensis*) are sensitive to O/129; they grow on common nutrient media but require at least 0.5% NaCl (optimum 3% NaCl). Optimum growth temperature 20–30 °C. *see BIOLUMINESCENCE.

photochromogen Any strain of *Mycobacterium* which develops a definite pigment *only* when the culture is exposed to light under appropriate conditions. (cf. SCOTOCHROMOGEN) For pigment production it is usually necessary to illuminate *young* cultures. *M. kansasii* is one example of a photochromogenic species.

photodynamic effect (see FLUORESCENT DYES)

photoheterotroph (see HETEROTROPH)

photoinduction and photoinhibition (in fungi) In many fungi light encourages, or is required for, the development of fruiting bodies or for sporulation (*photoinduction*). Light may be required only for the completion of a particular phase in the development of the fruiting body; thus, e.g. the initiation of fruiting in *Pyronema omphalodes* and the normal development of the pileus in *Lentinus lepideus* are both light-dependent processes. In certain species of *Coprinus* the fruiting bodies develop more rapidly in the presence of light—i.e. light encourages fruiting but is not an essential developmental factor. In e.g. *Trichoderma viride* conidium formation is photoinducible, while in some fungi, e.g. *Choanephora cucurbitarum*, the formation of conidia is stimulated by the exposure of the organism to a particular sequence of light and dark periods; thus, in *C. cucurbitarum* conidium formation is promoted by exposure of the organism to a period of darkness which is *preceded* by exposure to light. Photoinduction in relation to the development of new tissues is referred to as *photomorphogenesis*. The synthesis of certain compounds is photoinducible—e.g. the synthesis of carotenoids in *Phycomyces blakesleeanus* is increased in the presence of light, while some fungi which normally synthesize negligible quantities of carotenoids (e.g. *Fusarium* spp, *Verticillium* spp) may be induced to form significant amounts of these compounds in the presence of light.

For some organisms (e.g. *Alternaria* spp, *Penicillium* spp) growth and/or sporulation may be *inhibited* by light of certain intensities (*photoinhibition*). Light is reported to inhibit the development of the stipe and pileus in *Agaricus compestris*.

The wavelengths of light which are effective in photoinduction/photoinhibition commonly lie within the blue-violet region of the spectrum. The mechanism(s) of these phenomena are unknown.

photokinesis In certain motile microorganisms: a change in the *rate* of locomotion brought about by an increase or decrease in light intensity. In *positive* photokinesis locomotion becomes faster under higher intensities of light; in *negative* photokinesis light antagonizes locomotion. (cf. PHOTOTAXIS)

photolithotroph (see AUTOTROPH)

photomicrography Any process in which a photograph is taken of the magnified image of a microscopic object; photomicrography commonly involves the attachment of a specialized camera to the eyepiece of a microscope. When photographing *fluorescent* objects (see fluorescence MICROSCOPY) the eyepiece of the microscope must be fitted with a BARRIER FILTER; this not only protects the eyes (during focusing) but prevents fogging of the film by ultraviolet radiation.

photomorphogenesis (see PHOTOINDUCTION)

photoorganotroph (see HETEROTROPH)

photophobotaxis (see PHOTOTAXIS)

photophosphorylation (see PHOTOSYNTHESIS and PURPLE MEMBRANE)

photoreactivation (photorestoration) In many microorganisms: a mechanism whereby the effects of ULTRAVIOLET RADIATION on DNA may be reversed by exposure to radiation of wavelengths in the range 320–500 nm. An enzyme (photolyase) binds specifically to pyrimidine dimers—cleaving the dimers in the presence of suitable radiation. (see also DARK REPAIR)

photoreduction (*algol.*) In certain algae (e.g. species of *Chlorella*, *Chlamydomonas*, *Scenedesmus*): a light-dependent process in which *hydrogen* functions as an electron donor for carbon dioxide reduction; HYDROGENASE is required, and oxygen is not evolved. Photoreduction occurs only after anaerobic incubation of the algae with hydrogen, followed by illumination with low light intensities. (At higher light intensities normal photosynthesis occurs.) Carbon dioxide reduction proceeds *via* the Calvin cycle—as in PHOTOSYNTHESIS; the light-dependent reactions are not understood, but appear to involve both photosystems I and II.

photorespiration (in algae) Light-dependent oxygen uptake and carbon dioxide production (distinct from mitochondrial respiration); photorespiration is stimulated by high oxygen concentrations, low carbon dioxide concentrations and high light intensities. The process is believed to involve the formation and metabolism of glycolate ($CH_2OH.COOH$) —the 'glycolate pathway'. A possible reaction sequence is: ribulose-1,5-diphosphate* $\xrightarrow{O_2}$ phosphoglycolate → glycolate $\xrightarrow{O_2}$ glyoxylate → glycine → serine → hydroxypyruvate → glyceric acid → 3-phosphoglyceric acid*. Carbon dioxide is produced during the formation of serine from two molecules of glycine. In certain algae, glycolate is excreted into the medium.

The physiological significance of photorespiration is not clear; suggested functions include: (a) synthesis of serine and glycine, (b) elimination of excess carbon and reducing power formed during PHOTOSYNTHESIS. *Intermediates in the Calvin cycle.

photosynthate Any product of PHOTOSYNTHESIS, e.g. starch.

photosynthesis In certain (photosynthetic) organisms: the process in which radiant energy is absorbed by CHLOROPHYLLS and is subsequently converted to chemical energy. (cf. PURPLE MEMBRANE) *Visible* light is the energy source for photosynthesis in plants, algae, blue-green algae, and the photosynthetic bacteria (RHODOSPIRILLALES); the photosynthetic bacteria can also use near infra-red light. Photosynthesis involves (a) the *light reaction*—in which radiant energy is harnessed by the cell, and (b) the (light-independent) *dark reaction*—in which the harnessed energy is used in metabolic processes.

(1) *Photosynthesis in green plants, algae, and blue-green algae.* During photosynthesis in these organisms, carbon dioxide is reduced and oxygen is evolved. The *light reaction* takes place in the grana or THYLAKOIDS of CHLOROPLASTS (in green plants and algae) or in the thylakoid membranes of the blue-green algae. Two functionally-distinct systems are involved: *photosystems I* and *II*; each has a light-harvesting *pigment system* containing chlorophylls and ACCESSORY PIGMENTS, but pigment system I absorbs light of longer wavelength than that absorbed by pigment system II. Only pigment system I can absorb red light (wavelength 680–700 nm). In each pigment system, light energy absorbed by the accessory pigments is transferred (by intermolecular resonance) to chlorophyll *a* and thence to the *reaction centre*. In pigment system I the reaction centre appears to be a specialized form of chlorophyll *a* known as P-700 (due to its absorption maximum); one or two molecules of P-700 occur for every 300 molecules of chlorophyll *a*. The nature of the reaction centre in pigment system II is not known. In both pigment systems, energy transferred from the accessory pigments to the reaction centre causes the expulsion of electrons from the latter; these electrons are transported *via* a system of carriers—see diagram (page 307). In *non-cyclic electron flow*, electrons from pigment system II pass *via* Q to pigment system I and thence—*via* Z—to NADP; the oxidation of water to oxygen provides electrons which replace those lost by pigment system II. In *cyclic electron flow* only pigment system I is involved—see diagram. During electron transfer along these redox potential gradients energy conversion occurs with the formation of ATP (*photophosphorylation*) and with the reduction of NADP. (NB. Both ATP and $NADPH_2$ are formed during *non-cyclic* electron flow, while ATP only is produced during *cyclic* electron flow.) The sites and mechanism of photophosphorylation are unknown; the hypotheses put forward to explain OXIDATIVE PHOSPHORYLATION are also applicable to photophosphorylation.

The *dark reaction* takes place in the chloroplast stroma or in the cytoplasm of blue-green algae; it involves carbon dioxide fixation. In plants, and in those algae which have been investigated, the main pathway appears to be the *Calvin cycle*—see diagram (page 308). Six molecules of 3-phosphoglyceric acid are

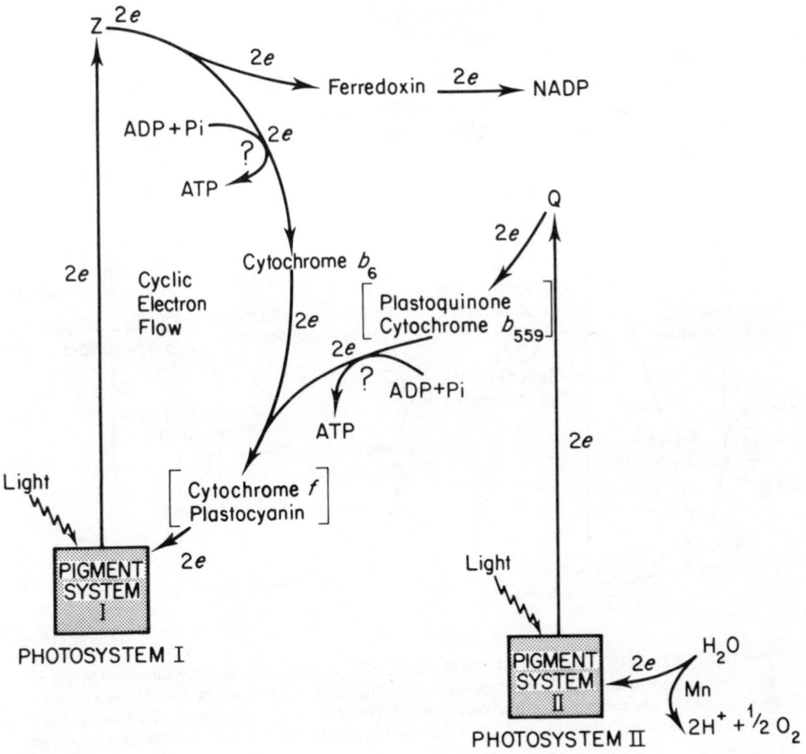

PHOTOSYNTHESIS. *A widely-accepted scheme for the light reaction in plants.* (In algae the system is similar—though minor differences may occur e.g. in the nature of the electron carriers.) Q, Z, are unidentified electron acceptors. Square brackets indicate uncertain order—e.g. cytochrome *f* and plastocyanin may act in parallel or in series. *e* denotes an electron.

formed for every three molecules of carbon dioxide fixed; of these, five are required to regenerate the carbon dioxide acceptor (i.e. ribulose-1,5-diphosphate) and one can be used for other metabolic processes. Other intermediates may be drawn off—e.g. hexose phosphate for the synthesis of storage polysaccharides. The carboxylation reaction itself is poorly understood; it involves *carboxydismutase* (= ribulosediphosphate carboxylase) which is believed to be present in all lithotrophs. The fixation of one molecule of carbon dioxide in the Calvin cycle requires three molecules of ATP and two of $NADPH_2$; these are supplied by the light reaction described above.

(2) *Bacterial photosynthesis* occurs only under strict anaerobic conditions; oxygen is not produced since bacteria cannot use water as an electron donor.

In photosynthetic bacteria only *one* type of reaction centre is believed to exist—apparently consisting of a specialized form of bacteriochlorophyll referred to as P840 in *Chlorobium*, P870 in *Rhodopseudomonas spheroides*, or P890 in *Rhodospirillum rubrum*. (Numbers refer to absorption maxima; P870 is sometimes used as a general term to refer to any bacterial reaction centre.) Electrons expelled from the excited reaction centre are transferred through a series of carriers—the nature of which depends on species; generally present are CYTOCHROME(S) (*c*-type in members of the Chromatiaceae and Chlorobiaceae, *b*- and *c*-types in members of the Rhodospirillaceae), QUINONE(S), and FERREDOXIN(S). The pathways of electron transfer are not known in detail, but two main types are believed to occur: (a) *non-cyclic*—in which electrons from the reaction centre pass, *via* ferredoxin, to NADP; electrons lost by the reaction centre are replaced with electrons from various donors (see below) *via* a chain of quinone(s) and/or cytochrome(s). (b)

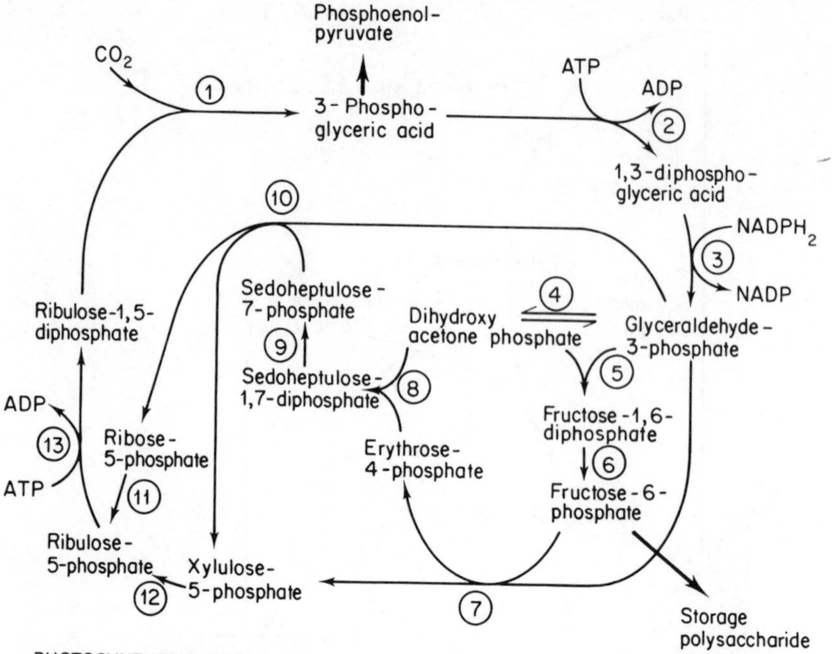

PHOTOSYNTHESIS. *The Calvin cycle.* Enzymes (circled): 1 carboxydismutase; 2 3-phosphoglycerate kinase; 3 triose phosphate dehydrogenase; 4 phosphotriose isomerase; 5 aldolase; 6 phosphatase; 7 transketolase; 8 aldolase; 9 phosphatase; 10 transketolase; 11 phosphoribose isomerase; 12 phosphoketopentose epimerase; 13 phosphoribulose kinase.

Cyclic—in which electrons from the reduced ferredoxin pass to e.g. a quinone, and thence back to the reaction centre. Photophosphorylation appears to be associated only with the *cyclic* pathway. The source of reducing power for non-cyclic electron flow depends on species and on growth conditions. Members of the Chlorobiaceae and Chromatiaceae can use inorganic electron donors (e.g. hydrogen sulphide, sulphur); from these compounds electrons are passed (*via* the non-cyclic pathway) to NADP. Many photosynthetic bacteria can also use organic compounds as sources of reducing power and/or carbon. Under lithotrophic conditions (at least) carbon dioxide fixation appears to occur mainly by the *Calvin cycle* (see above). In *Chlorobium thiosulphatophilum* and *Rhodospirillum rubrum* an additional pathway—the *reductive carboxylic acid cycle*—may be important. This is essentially a reversed, extended form of the TRICARBOXYLIC ACID CYCLE, in which four molecules of carbon dioxide are fixed and one molecule of oxaloacetate is synthesized for each turn of the cycle; reduced ferredoxin is required for two of the carboxylations.

Apart from the photochemical event, most of the reactions which take place in photosynthetic bacteria have been found in other species of bacteria; thus, e.g. the Calvin cycle occurs in chemolithotrophic bacteria.

photosystems I and II (see PHOTOSYNTHESIS)

phototaxis Light-induced TAXIS. In *photophobotaxis* (see TAXIS) an organism responds to a *temporal* change in light intensity. This response is said to be *positive* when triggered by a *decrease* in light intensity (as occurs e.g. when an organism crosses a light-dark boundary, or when there is a generalized diminution of light intensity in the environment); a *negative* photophobotactic response is one triggered by an *increase* in light intensity. Organisms which respond in a positive photophobotactic manner tend to *accumulate* in a LIGHT TRAP, while organisms which respond in a negative photophobotactic manner tend to be *excluded* from a light trap.

In *phototopotaxis* an organism moves towards (positive phototopotaxis) or away from (negative phototopotaxis) the light source.

In general, a positive phototactic response gives way to a negative response when the light intensity increases above a certain threshold value. (see also PHOTOKINESIS)

phototopotaxis (see PHOTOTAXIS)

phototrophs Organisms whose sole or principal *primary* source of energy is *light*. Such organisms may be photoAUTOTROPHS or photoHETEROTROPHS. Some phototrophs are facultative CHEMOTROPHS.

phototropism A *light*-motivated TROPISM.

Phragmobasidiomycetidae A sub-class of fungi of the class HYMENOMYCETES; constituent species form *phragmobasidia*—i.e. partly or completely septate basidia—which are organized into an hymenium in fruiting bodies which may be dry, fleshy, gelatinous, or waxy. The sub-class includes the orders AURICULARIALES, Septobasidiales, and TREMELLALES.

phragmobasidium (*mycol.*) A *septate* BASIDIUM.

phragmosporae (see SACCARDIAN SYSTEM)

phthisis A synonym of TUBERCULOSIS.

phycobiliproteins (biliproteins) Water-soluble pigments found in the Cyanophyta and Rhodophyta (see PHYCOBILISOMES) and in the Cryptophyta (apparently between the thylakoids). The molecule consists basically of a protein linked covalently to a *linear tetrapyrrole* (*phycobilin*); the latter resembles protoPORPHYRIN IX minus the methene bridge between rings I and II. Phycobiliproteins include the (red) phycoerythrins and the (blue) phycocyanins—which may occur together in various proportions. Phycobiliproteins are believed to transfer light energy to CHLOROPHYLLS during PHOTOSYNTHESIS.

phycobilisomes In members of the Cyanophyta and Rhodophyta: granules found on the surfaces of the THYLAKOIDS; they are thought to be aggregates of PHYCOBILIPROTEINS.

phycobiont The algal partner in a LICHEN.

phycocyanins (see PHYCOBILIPROTEINS)

phycoerythrins (see PHYCOBILIPROTEINS)

phycology The study of ALGAE.

Phycomyces A genus of fungi of the ZYGOMYCETES.

phycomycetes A term used to refer to the LOWER FUNGI.

phycovirus A synonym of CYANOPHAGE.

phyletic classification (see TAXONOMY)

phyllocladium (*lichenol.*) A scale-like, finger-like, or branched ('coralloid') projection—a number of which arise from the thallus in species of STEREOCAULON.

phyllody (*plant pathol.*) Of a plant: a malformation in which normal floral components are replaced by leaf-like structures. Phyllody is symptomatic of certain fungal diseases, e.g. *green-ear* of millet (*Pennisetum typhoides*) caused by the downy mildew fungus *Sclerospora graminicola*.

Phylloporus A genus of fungi of the BOLETACEAE.

phylloquinone (see QUINONES (in electron transport))

phyllosphere (phylloplane) An ecological niche comprising the surfaces of leaves. (see also RHIZOSPHERE)

phylogenetic classification (see TAXONOMY)

phylogeny The range of developmental stages in the evolution of an organism. (see also ONTOGENY)

phylum A taxonomic group: see TAXONOMY and NOMENCLATURE. (see also DIVISION)

Physarum A genus of the ACELLULAR SLIME MOULDS.

Physcia A genus of foliose LICHENS in which the phycobiont is *Trebouxia*. The thallus generally consists of small, narrow, branching lobes which are radially orientated to form a rosette which is more or less flat; the lower surface bears rhizinae by means of which the thallus attaches to the substratum. Apothecia are lecanorine and usually have dark brown to black discs. Species include e.g.:

P. adscendens. The thallus is light grey and the narrow lobes tend to ascend; short brown or black cilia occur along the margins. Pseudocyphellae appear as small pale dots in the upper surface. The lobe ends may form erect hoods beneath which soralia are formed. *P. adscendens* occurs on trees, rocks, walls etc.

P. caesia. The thallus is flat and blue-grey; the upper surface contains pseudocyphellae and bears globose laminal soralia. *P. caesia* occurs on rocks, walls etc.—particularly where these are enriched with nitrogen (e.g. by bird droppings).

P. pulverulenta. The thallus is grey-green to green-brown, and pruinose—particularly towards the ends of the lobes. Soralia are not formed. *P. pulverulenta* occurs mainly on trees.

physiological heterothallism (see DIMIXIS)

physiologic(al) races (*plant pathol.*) (see FORMA SPECIALIS)

phytoalexins (stress metabolites) (*plant pathol.*) Fungitoxic substances produced by higher plants in response to fungal infection—or to certain other stimuli, e.g. physical damage, exposure to certain chemicals. Under pre-stimulus conditions, phytoalexins may be either absent from plant tissues or present in extremely low concentrations. The production of phytoalexins appears to be a *local* effect—their translocation has yet to be conclusively demonstrated. The ability to produce phytoalexin(s) may enable a plant to resist disease development following infection by a fungal pathogen. Pathogens which successfully invade such a plant may (a) fail to induce phytoalexin production; (b) be insensitive to the phytoalexin(s) produced; or (c) be able to degrade the phytoalexins as they are formed. A phytoalexin may be specific to the plant which produces it. For example, (a) *Pisatin*, 3-hydroxy-7-methoxy-4′,5′-methylenedioxy chromanocoumaran, is produced by the PEA (*Pisum sativum*) in response to infection by certain fungi, e.g. *Ascochyta pisi* (a pathogen), *Colletotrichum lindemuthianum* (a non-pathogen); it is also produced on treatment with salts of copper or mercury. (b) *Phaseollin*, 7-hydroxy-3′,4′-dimethylchromanocoumarin, is produced by the french bean (*Phaseolus vulgaris*). (see also ORCHINOL, HIRCINOL, IPOMEAMARONE) Other phytoalexin-like compounds (e.g. CHLOROGENIC ACID, CAFFEIC ACID) may be produced by a number of different plants. In some cases the production of phytoalexins appears to be associated with an HYPERSENSITIVITY reaction in the plant. (see also PHYTONCIDES)

phytoflavin (see FLAVODOXINS)

phytohaemagglutinin (PHA) A generic name for those LECTINS which induce mitosis and transformation in lymphocytes.

Phytomastigophorea A class of flagellate organisms within the super-class MASTIGOPHORA. Members of the class typically possess photosynthetic pigments, and are commonly classified in both botanical and zoological taxonomic schemes; in *botanical* taxonomic schemes these organisms are classified in the divisions DINOPHYTA and EUGLENOPHYTA.

Phytomonas (*bacteriol.*) Obsolete name for XANTHOMONAS.

Phytomonas (*protozool.*) A genus of flagellate protozoans of the sub-order TRYPANOSOMATINA. *Phytomonas* spp parasitize a number of plants (e.g. certain succulents); the vectors are sap-sucking insects of the order Hemiptera. Amastigote and promastigote forms both occur in the insect vector and in the plant.

phytoncide (*plant pathol.*) A term generally used to denote an antimicrobial substance produced by a higher plant. Such substances occur normally in the tissues of certain plants (cf. PHYTOALEXINS) and may play an important role in the resistance of the plant to disease: see RESISTANCE (of plants to disease). Examples of phytoncides include AVENACIN, JUGLONE, PHLORIDZIN, PINOSYLVIN, TANNINS, THUJAPLICINS, WYERONE.

Phytophthora A genus of fungi within the class OOMYCETES; *Phytophthora* spp include a number of important pathogens of cultivated plants—e.g. *P. infestans*, the causal agent of *late blight* of potato (see POTATO, FUNGAL DISEASES OF), and *P. fragariae*, the causal agent of *red core* of strawberry (see STRAWBERRY, DISEASES OF). The generalized life cycle of *P. infestans* is as follows. The organism overwinters, as mycelium, in an infected potato tuber and, in the spring, grows systemically into the growing shoot which develops from the tuber; ovoid, deciduous sporangia subsequently develop on the branched sporangiophores which emerge from the stomata on the lower surface of the leaves. The sporangia are dispersed by the wind. On reaching a susceptible host plant (and providing that conditions are suitable) a sporangium germinates to form either a germ tube or zoospores (depending on temperature); zoospores subsequently give rise to germ tubes. A germ tube may enter the plant *via* a stoma or directly (by means of an appressorium). Within the new host an extensively-branched mycelium develops intercellularly, and a number of elongated haustoria are formed; subsequently, sporangiophores emerge from the stomata. Sexual reproduction occurs between strains of appropriate mating type; during gametangial development the oogonial initial penetrates and grows through the antheridium such that, when fully developed, the expanded, globose oogonium carries the antheridium as an encircling collar around its base. The oospore, within which fertilization is reported to occur, develops from the oogonium; the oospore is a thick-walled, resistant structure which, on germination, gives rise to a germ tube that usually terminates in a zoosporangium. (see also PYTHIUM)

phytoplankton (see PLANKTON)

phytotoxin Any substance which is toxic for plants—e.g. ALTERNARIC ACID, FUSARIC ACID, α-PICOLINIC ACID.

pI (see ISOELECTRIC POINT)

Pi Abbreviation for inorganic orthophosphate (PO_4^{3-}).

Pichia A genus of fungi of the family SACCHAROMYCETACEAE.

picodnaviruses (see PARVOVIRUS)

α-picolinic acid Pyridine-2-carboxylic acid; a phytotoxin produced by *Pyricularia oryzae*, the causative organism of blast disease of rice. (see also PIRICULARIN)

Picornaviridae A family of VIRUSES which occur typically in the gastrointestinal and respiratory tracts of man and other vertebrates. Picornaviruses are non-enveloped, ether-resistant, icosahedral viruses, 20–40 nm in diameter, which contain linear single-stranded RNA; they replicate within the *cytoplasm* of the host cell. Picornaviruses give rise to a variety of localized and systemic diseases which include mild infections of the gastrointestinal and respiratory tracts, diseases of the central nervous system (e.g. POLIOMYELITIS, TESCHEN DISEASE, aseptic MENINGITIS), and diseases which involve the formation of lesions on cutaneous and mucous surfaces (e.g. FOOT AND MOUTH DISEASE, SWINE VESICULAR DISEASE). Many picornaviruses can infect subclinically, and the carrier state is recognized in at least some of the diseases referred to above. The family Picornaviridae comprises the genera ENTEROVIRUS, RHINOVIRUS, and CALICIVIRUS.

picric acid 2,4,6-trinitrophenol (see PHENOLS).

piericidin A An antibiotic, produced by *Streptoverticillium mobaraense* (formerly *Streptomyces mobaraensis*), which acts as a RESPIRATORY INHIBITOR in both mitochondrial and bacterial ELECTRON TRANSPORT CHAINS. It binds (non-covalently) to a site between $NADH_2$-dehydrogenase and the quinone. *Intact* cells may be impermeable to piericidin A.

pileus (*mycol.*) Commonly: the fleshy structure which supports spore-bearing tissue in the fruiting bodies of certain basidiomycetes and ascomycetes; examples include the *cap* of the common (edible) mushroom, and the ridged and pitted cylindrical structures of the morels (*Morchella* spp). Some authors use the term to refer also to the corky or leathery fruiting bodies of certain wood-rotting fungi.

pili (fimbriae*) (*sing.* pilus) Filamentous appendages which project from the cell surface of certain gram-negative bacteria—e.g. members of the Enterobacteriaceae and species of *Pseudomonas*. Two principal classes of pili occur: FIMBRIAE (*common* pili) and SEX PILI (also called *sex fimbriae*); pili of both classes may occur on the same cell—on which may also occur flagella. Pili are believed to consist of one or two strands which are wound helically to form a tube or tube-like structure; each strand is composed of a chain of protein (*pilin*) subunits. In diameter, pili are reported to range from 3 nm to 25 nm—according to type; the diameter of the axial hole (or groove) is usually stated to be about 2 nm. Length may vary according to type, although some sex pili appear to be longer than fimbriae; both types are generally shorter than flagella (see *bacterial* FLAGELLUM). Pili can be removed from cells by mechanical agitation—e.g. in a Waring blender; bacteria so treated readily develop new pili. (Pili appear to be assembled from a 'pool' of preformed pilin.) *The terms *pili* and *fimbriae* are commonly used interchangeably; however, *fimbriae* may be used to refer specifically to *common* pili, while *pili* may refer specifically to *sex* pili.

pilin (see PILI; FIMBRIAE; SEX PILI)

Pilobolus A genus of coprophilous fungi within the class ZYGOMYCETES; species occur on the dung of herbivores. *Pilobolus* spp are remarkable mainly in respect of the method in which the sporangium is discharged from the (positively-phototropic) sporangiophore; during discharge, the substantial (non-dehiscent) sporangium is propelled on a jet of liquid which escapes, explosively, from a subsporangial vesicle at the distal end of the sporangiophore.

Pilophorus A genus of LICHENS in which there is a granular *primary* (basal) thallus and erect *pseudopodetia* (see PSEUDOPODETIUM) which may bear terminal, convex, black apothecia; the phycobiont is *Trebouxia*. Cephalodia are generally present. Those species which rarely form apothecia produce soredia.

pilus (see PILI)

pimaricin (tennecetin) (see POLYENE ANTIBIOTICS)

pine-oil disinfectants DISINFECTANTS which contain volatile oils obtained by steam-distillation of pine wood; the most active antimicrobial agent among these oils is α-terpineol. Pine-oil disinfectants are active mainly against gram-negative bacteria; they have little or no activity against gram-positive species. Pine-oils are insoluble in water but form emulsions in the presence of e.g. soaps. (see also PHENOLS)

pinocytosis The ingestion, by a cell, of minute droplets of the ambient *liquid*. Pinocytosis involves *vacuole* formation—which occurs in a manner similar to that in PHAGOCYTOSIS; pinocytotic vesicles (vacuoles) likewise coalesce with LYSOSOMES. Pinocytosis occurs in certain metazoan cells and in a number of microbial cells.

pinosome A pinocytotic vesicle—see PINOCYTOSIS.

pinosylvin 3,5-dihydroxy-*trans*-stilbene; an antifungal substance produced by many species of pine (*Pinus*). Pinosylvin appears to be at least partly responsible for the relative resistance of pine heartwood to fungal attack. (see also TIMBER, DISEASES OF)

pinta (mal-del-pinto; spotted sickness) A chronic, infectious human disease which occurs in central and South America. The causal organism is *Treponema carateum*. Pinta gives rise to spots or patches of abnormally-pigmented or depigmented skin; the disease rarely affects other tissues.

Piptocephalis A genus of fungi within the class ZYGOMYCETES; *Piptocephalis* spp are obligately parasitic on other fungi—mainly other zygomycetes (e.g. *Mucor*). The vegetative hyphae develop extracellularly on the host organism—from which nutrients are abstracted by means of haustoria. Asexual reproduction involves the formation of cylindrical sporangia (*merosporangia*) each of which contains a single row of spores; the (deciduous) merosporangia develop as terminal clusters on long sporangiophores which are dichotomously branched at their distal ends. Sexual reproduction involves gametangial fusion and zygospore formation; most species are reported to be homothallic.

Piptoporus A genus of fungi of the family POLYPORACEAE. Constituent species form *non*-stipitate fruiting bodies in which the whitish or pale brown, corky context consists of hyphae which exhibit dimitic organization; the material of the context does not darken on application of potassium hydroxide solution (10% aqueous). The typically dimidiate, ungulate, or reniform fruiting bodies are usually of the bracket type, and the upper surface of the mature basidiocarp is commonly invested with a thin, paper-like skin; the hymenophore is porous, and the outer surfaces of the dissepiments are sterile. The fructifications are typically annual; in perennial fructifications the hymenophore does not consist of vertically stacked layers of tubules.

P. betulinus (often referred to as *Polyporus betulinus*—see POLYPORUS) is saprophytic and parasitic on the birch (*Betula*); the upper surface of the fruiting body is pale brown and the lower, porous surface is white. The non-pigmented, cylindrical basidiospores are approximately 5 × 2 microns.

piricularin (pyricularin) A phytotoxin produced by *Pyricularia oryzae*, the causal agent of *blast disease* of rice; piricularin is also toxic to *P. oryzae* itself. The fungus also produces a copper-containing 'piricularin-binding-protein' which, by complexing with piricularin, renders it non-toxic to the fungus without affecting its toxicity to the rice plant. CHLOROGENIC ACID can complex with and inactivate piricularin.

Piroplasmea A class of protozoans of the subphylum SPOROZOA; the class contains a single order: Piroplasmida. The organisms parasitize the red or white blood cells of vertebrates; tick vectors are the only known means of transmission. *Piroplasms* are small, pyriform, spherical or elongated, uninucleate cells which reproduce by binary fission or by schizogony; some authors believe that sexual processes occur in certain species, but definitive evidence appears to be lacking. Spores are not formed and pigments are not elaborated from haemoglobin. Some species are important pathogens of animals; man is rarely parasitized. Genera include e.g. BABESIA and THEILERIA. (see also ANAPLASMA)

piroplasms Members of the PIROPLASMEA.

pisatin (see PHYTOALEXINS)

pityriasis versicolor (*med.*) A chronic disease of the skin characterized by discrete regions of superficial brown or fawn discoloration which coincide with areas of epidermal scaling; the disease persists unless treated. The causal agent is the imperfect yeast *Pityrosporum orbiculare* (= *Malassezia furfur*).

Pityrosporum (*syn*: *Malassezia*) A genus of fungi within the family CRYPTOCOCCACEAE; species have been isolated from e.g. the human scalp and skin—*P. orbiculare* (= *Malassezia furfur*) being the causal agent of PITYRIASIS VERSICOLOR. The vegetative form of the organisms is a spheroid or elongated cell—though highly elongated cells/short, curved hyphae may develop in the epidermis; the size of an individual cell generally lies within the range 2–7 microns. Reproduction occurs by *monopolar* budding—i.e. repeated budding from one region of the cell. *Pityrosporum* spp are non-fermentative organisms.

placodioid (placoid) (of a lichen thallus) Refers to a crustose lichen thallus which is generally circular in outline and which is lobed at the periphery. A placodioid thallus occurs e.g. in *Buellia canescens*.

plague An acute, infectious disease which affects man and other animals, principally rodents. Plague occurs sporadically; in earlier centuries* there were devastating epidemics and pandemics involving enormous loss of life. The causal agent is *Yersinia pestis*; reservoirs of infection are found in populations of plague-resistant rats and other animals (e.g. some species of mice). An epidemic of human plague is commonly preceded by an epizootic of plague among susceptible animals in the rat population. *Y. pestis* is transmitted from rat to rat, and from rat to man, by the bites of rat fleas—or, from man to man, by the human flea, *Pulex irritans*; exceptionally, infection may be contaminative. The incubation period apparently lies within the range 1 to 8 days. In man, *bubonic plague* (the commonest form) is characterized by infected, swollen, necrotic (and often suppurative) lymph nodes (*buboes*); haemorrhagic lesions occur in the skin. In *septicaemic plague* death occurs swiftly—without the appearance of buboes. *Pneumonic plague* involves a massive infection of the lungs and is highly communicable by droplet inhalation. Mortality rates are usually very high and may approach 100% in cases of pneumonic and septicaemic plague. *Laboratory diagnostic procedures* include the demonstration of specific agglutinins in the patient's serum, and the microscopical and cultural examination of sputum (pneumonic plague) and of blood and aspirates from buboes. *Y. pestis* has been reported to occur in the throats of symptomless human carriers. *e.g. the Black Death (14th Century) is said to have caused over 25 million deaths.

planachromat (*plan*-objective) (microscopy) An ACHROMATIC OBJECTIVE which provides an image in which all parts of the field of view are simultaneously in focus. (When using a *non*-plan objective the peripheral region of the field tends to be out of focus when the central region is in focus. During visual observation the peripheral regions of a field can be focused by a small adjustment of the fine-focusing control of the microscope—although the central region of the field then becomes out of focus. Such an objective is unsuitable for good-quality PHOTOMICROGRAPHY.) *Planapochromats* are apochromatic objectives analogously corrected. (see also MICROSCOPY)

plankton A term which refers, collectively, to the microorganisms (chiefly microalgae and protozoans), small metazoans (e.g. copepods and other small crustaceans), and the embryonic and/or larval forms of other metazoans, which drift, passively, in the pelagic zone of lakes, seas, and other bodies of water. The plant or plant-like members of the plankton are referred to, collectively, as *phytoplankton*, while the animal or animal-like members are called *zooplankton*. Plankton—particularly phytoplankton—is important as the base of the aquatic food chain; it is consumed by a range of invertebrate animals and by certain fish and whales. Microorganisms often found in *marine* plankton include dinoflagellates (e.g. *Ceratium*, *Gonyaulax*, *Peridinium*), diatoms (e.g. *Chaetoceros*), foraminiferans (e.g. *Globigerina*), and a range of radiolarians (e.g. *Hexacontium*). Microorganisms often found in *freshwater* plankton include blue-green algae (e.g. *Anabaena*, *Oscillatoria*), diatoms (e.g. *Asterionella*, *Tabellaria*), desmids (e.g. *Arthrodesmus*, *Staurastrum*), other algae and heliozoans. Various schemes have been proposed for the classification of planktonic organisms according to size. Thus, e.g. the category *nannoplankton* may include (depending on author) organisms which are less than 20 microns, less than 50 microns, 5–50 microns, or 5–60 microns; similarly, *mesoplankton* may include organisms which are 0.5–1 millimetre, 1–5 mm, 0.2–2 mm, or greater than 1 mm.

Planococcus A genus of gram-positive, flagellated, strictly aerobic cocci of the MICROCOCCACEAE; species occur as marine saprophytes. The organisms are chemoorganotrophs with a respiratory metabolism.

planogamete A *motile* gamete.

planospore A *motile* spore.

plant pathogens and diseases (a) *Plant-pathogenic microorganisms*. (i) *Fungi*. Examples include species of the lower fungi: *Synchytrium*, *Pythium*, *Phytophthora*, *Albugo*, and *Plasmodiophora*; the ascomycetes: *Ceratocystis*, *Erysiphe*, *Gibberella*, and *Venturia*; basidiomycetes: members of the Uredinales and Ustilaginales; the *fungi imperfecti*: *Alternaria*, *Botrytis*, *Fusarium*, *Helminthosporium*, and *Rhizoctonia*. (ii) *Viruses*. These are mainly of the RNA type; some (e.g. *cowpea mosaic virus*) are COVIRUSES. (see also SATELLITE VIRUS and VIRUS DISEASES OF PLANTS) (iii) *Bacteria*. These include species of *Erwinia*, *Pseudomonas*, and *Xanthomonas*; *Streptomyces scabies* and *Corynebacterium fascians* are examples of gram-positive pathogens. A number of YELLOWS diseases have recently been attributed to *Mycoplasma*-like organisms. (iv) *Algae*. *Cephaleuros* causes RED RUST disease of citrus, coffee, tea, and other crops. (v) *Protozoa*. The flagellate *Phytomonas* causes disease in certain succulent plants.

(b) *Plant diseases*. (i) *Host-parasite relationships*. Some microorganisms are *facultative* plant parasites while others (e.g. rusts, viruses) are *obligate* parasites; as a rule, host-parasite specificity is greater in the latter category. Some pathogens exhibit HETEROECISM—see e.g. RUSTS (1). (ii) *Modes of infection*. A pathogen may be soil-borne or seed-borne, it may be transmitted from a diseased to a healthy plant by wind or rain, or by a *vector*—sometimes a specific vector: see VIRUS DISEASES OF PLANTS. Fungal infection may also occur e.g. by *rhizomorph* extension from an adjacent focus of disease—see e.g. *Armillaria mellea* in TIMBER, DISEASES OF. A pathogen may enter the plant *via* natural pores (e.g. stomata, lenticels) or *via* wounds; fungi may enter by direct penetration—see APPRESSORIUM. Little is known of the manner in which bacteria enter a host plant; they appear to be unable to breach an intact plant surface. (iii) *Types of disease*. Disease may be systemic or localized, and may kill or damage the plant—the degree of damage often being influenced e.g. by environmental conditions. According to the symptoms displayed, a disease may be designated e.g. a MILDEW, ROT, RUST, SCAB, SMUT, or WILT. (see also GALLS) Certain symptoms are peculiar to virus diseases. (iv) *Pathogenesis*. As a disease develops, a number of physiological changes may be observed within the plant—e.g. an increased rate of respiration. In some diseases, the cause-effect relationships underlying the symptoms are understood—e.g. wilting due to vascular blockage, tissue death (necrosis) due to viral replication, and disintegration of host tissues by PECTINolytic bacteria. In some diseases, symptoms appear to result from an imbalance in certain plant growth-regulating substances due to the presence of the pathogen—see AUXINS, CYTOKININS, ETHYLENE and GIBBERELLINS. Symptoms may also be due to the presence of *phytotoxins*—e.g. ALTERNARIC ACID, FUSARIC ACID, VICTORIN. (v) *Resistance to disease*. See RESISTANCE (of plants to disease). (vi) *Diagnosis of disease*. The identity of a pathogen may be deduced from the symptoms it produces in a given plant (see e.g. DIFFERENTIAL HOST), or it may require identification by e.g. microscopical and/or cultural examination; in some bacterial and viral diseases serological tests have been used. (vii) *Control of disease.* Whenever possible disease-resistant varieties are cultivated. Fungal diseases may be controlled by the application of appropriate FUNGICIDES AND FUNGISTATS; however, some diseases, e.g. *white rot* of onion (caused by *Sclerotium cepivorum*) may be quite intractable owing the persistence of resistant *sclerotia* in the soil. In diseases caused by heteroecious rusts some benefit may be derived from the elimination of the alternate host. Volunteer plants may form a reservoir of disease in a given location. See also entries for diseases of APPLE, BEANS, BEET, CARROTS, CELERY, CEREALS, CLOVER, CRUCIFERS, FLAX, ONION, PEA, PEACH, POTATO, RICE, STRAWBERRY, TOBACCO and TOMATO.

plaque (1) A discrete, macroscopic or microscopic, usually circular region in a cell monolayer (see TISSUE CULTURE) or in a film or layer of bacterial growth (see LAWN PLATE, PLAQUE ASSAY) in which some or all of the cells have been lysed as the result of viral activity. (Similar regions in a film of bacterial growth may be due e.g. to the lytic activity of the parasitic bacterium *Bdellovibrio*.) It is generally assumed that each isolated plaque is formed as a result of the development of a single virion: when the first-infected cell lyses the progeny of this virus diffuse outwards—progressively infecting and lysing other cells until a visibly depopulated area (the plaque) is produced. Plaque characteristics—e.g. size, shape, nature of the plaque margin—can be useful e.g. in helping to establish the identity of the infecting virus, or in determining whether or not a cell population is homogenous in its susceptibility to the virus; thus e.g. clear plaques (total cell lysis) on a lawn of susceptible bacteria are typically produced by *virulent* strains of BACTERIOPHAGE, while turbid plaques are characteristic of *temperate* phages—which lyse only some of the cells in the plaque area. (2) (*verb*) To inoculate a lawn plate, or a cell monolayer, with a viral suspension in order e.g. to determine the nature of the plaque produced. (3) (dental) A thin film (up to about 0.5 mm thick) closely adherent to the surfaces of teeth; plaque consists of a mixed microbial flora embedded in a matrix composed largely of extracellular bacterial polysaccharides and salivary factors. Plaque is a predisposing factor in DENTAL CARIES and periodontal disease. Organisms commonly found in dental plaque include various streptococci which produce dextran or dextran-like polymers (e.g. *Strep. mutans*, *Strep. sanguis*) and species of *Lactobacillus*, *Bacteroides*, and *Corynebacterium*. (4) (*vet.*) A type of skin lesion which occurs in DOURINE.

plaque assay (of bacteriophage) A technique used for counting the number of PLAQUE-FORMING UNITS in a given suspension. One method is as follows: log dilutions of the suspension are initially prepared. From a given dilution a measured volume is removed and added to a known volume of molten (46 °C) semi-solid agar containing a known concentration of sensitive bacteria; the whole is well mixed and poured onto a plate of solid nutrient agar so as to form an upper layer of 1–2 mm thickness. An identical procedure is carried out with each of the log dilutions of the bacteriophage suspension. When set, the plates are incubated at the growth temperature of the bacterium and subsequently examined for PLAQUES. Each plaque is assumed to have been caused by a single phage so that, knowing the volumes and dilutions involved, the PLAQUE TITRE of the original phage suspension can be calculated; the plaque count is taken from those plates which exhibit an easily-countable number of well-separated plaques.

plaque-forming unit A virion which has the ability to infect and lyse particular cells and to give rise to a PLAQUE under suitable conditions. (see also EOP, PLAQUE TITRE, and ONE-STEP GROWTH EXPERIMENT)

plaque-reduction method (in the assay of interferons) (see INTERFERONS)

plaque titre The number of virus particles (per unit volume of a given suspension) which are able to form a PLAQUE under appropriate conditions. (see also EOP)

plaque-type mutants (of bacteriophage) Mutants of a given phage which form plaques that differ, morphologically or otherwise, from the plaques formed by the wild type phage on cultures of a particular strain of sensitive bacteria.

plasma (1) (*in vivo*) The fluid (non-particulate) fraction of blood. (2) (*in vitro*) Fluid obtained (by centrifugation etc.) from blood which has been pretreated with an ANTICOAGULANT; plasma differs from SERUM e.g. in that it contains the clotting factors (e.g. fibrinogen).

plasma cells Antibody-forming cells which develop from B LYMPHOCYTES—see ANTIBODY-FORMATION. Typically, plasma cells are ovoid, about 15 × 10 microns, with a rounded, off-centre nucleus; when stained, the nucleus has a 'clock-face' or 'cartwheel' appearance due to the arrangement of chromatin. The cytoplasm is non-granular and contains a well-developed ribosome-rich endoplasmic reticulum. The METHYL-GREEN PYRONIN STAIN is useful for distinguishing between plasma cells and lymphocytes; in the latter the cytoplasm is commonly far less basophilic. Plasma cells are commonly found in the spleen, in other lymphoid tissues, and particularly in lymph nodes which drain an infected site or antigenically-stimulated region.

plasma membrane (see CELL MEMBRANE)

plasmagene In eucaryotic cells: a (postulated) cytoplasmic particle which carries self-replicating extrachromosomal gene(s).

plasmalemmasomes (plasmalemmosomes) (1) (*bacteriol.*) Synonym of MESOSOMES. (2) (*mycol.*) Synonym of LOMASOMES.

plasmid (*bacteriol.*) An extrachromosomal genetic structure which can replicate independently within a bacterial cell and which is normally dispensible to the cell—although its presence may be advantageous to the cell in certain circumstances; those plasmids which have been investigated appear to be circular double-stranded DNA molecules of molecular weight approximately 10^6–10^8. Genes specifying a wide range of functions may be carried by plasmids. Such functions include: antibiotic resistance (see R FACTOR); colicin synthesis (see COLICIN FACTOR); toxin formation (e.g. enterotoxin formation in *Escherichia coli*); formation of certain metabolic enzymes (such as those which degrade e.g. camphor and naphthalene in *Pseudomonas putida*, and enzymes for lactose fermentation*); RESTRICTION endonucleases*. Some (*cryptic* or *silent*) plasmids appear not to affect the phenotype of the host cell.

Many plasmids can bring about their own transfer from one cell to another by CONJUGATION; such *transmissible* plasmids are referred to as SEX FACTORS. *Non-transmissible* plasmids may be *co-transferred* during conjugation induced by a superinfecting sex factor, or they may be transferred by TRANSDUCTION. (In *Staphylococcus* conjugation is unknown; plasmid transfer occurs by transduction.)

Two models for plasmid replication have been proposed; there is evidence to support each model, and a given model may describe events under a particular set of conditions. In the CAIRNS MODEL both strands of the circular duplex are replicated *without initial breakage*—the strands separating as replication proceeds (see DNA). The second model (the ROLLING CIRCLE MODEL) may describe events which occur during plasmid transfer in CONJUGATION. Plasmid replication appears to involve a specific membrane attachment site; on cell division, this site may be concerned in the distribution of the plasmids to daughter cells. When plasmid replication is synchronized with that of the chromosome (i.e. the chromosome: plasmid ratio remains constant, e.g. 1:1) replication is said to be under *stringent control*. Under *relaxed control* the replication of plasmids greatly exceeds that of the chromosome—sometimes resulting in 20 or more plasmids per cell. Under conditions of *stringent* control, two identical plasmids, or closely-similar plasmids, cannot normally coexist stably in the same cell; such plasmids are said to be *incompatible*, and the phenomenon is referred to as *superinfection immunity*. (Incompatible plasmids are segregated into separate cells on cell division, and/or they may undergo recombination.) Plasmids which can stably coexist in a cell are said to be *compatible*. A number of classes of plasmid (*incompatibility classes* or, sometimes, compatibility classes) have been distinguished on the basis of mutual incompatibility. It has been suggested that incompatibility may result from competition between plasmids for a particular site on the bacterial cell membrane—attachment to which is necessary for replication. (The determination of incompatibility can be complicated by *exclusion* (see SEX FACTOR) or RESTRICTION.)

Occasionally, the replication of a plasmid may lag behind that of the chromosome; in such a case only one daughter cell receives a plasmid on cell division. This effect may be achieved experimentally e.g. by treating certain plasmid-containing cells with ACRIDINE ORANGE at concentrations insufficient to inhibit chromosome replication; such treatment appears to inhibit the replication of certain plasmids, and is known as *curing* or *elimination*.

In some cases, certain plasmids may become incorporated in the bacterial chromosome (see EPISOME); the integrated plasmid replicates as part of the chromosome—i.e. it ceases to be autonomous. RECOMBINATION can occur between a plasmid and a homologous region of the chromosome, and may also occur between two plasmids occupying the same cell. *Such plasmids can lead to erroneous identification of bacteria when tests such as lactose-fermenting ability and phage TYPING are used.

plasmin (see FIBRINOLYSIN)

Plasmodiidae A family of protozoans of the HAEMOSPORINA.

Plasmodiophorales (endoparasitic slime moulds) An order of eucaryotic organisms usually classified among the fungi (see also SLIME MOULDS); they are obligate intracellular parasites in a range of higher plants (e.g. cabbage and potato), certain algae (e.g. *Vaucheria*), and some aquatic fungi (e.g. *Achlya*, *Saprolegnia*). Some species are important plant pathogens—e.g. *Plasmodiophora brassicae* (the causal agent of *clubroot* of cabbage—see CRUCIFERS, FUNGAL DISEASES OF), and *Spongospora subterranea* (the causal agent of powdery scab of potato).

The somatic form of a member of the Plasmodiophorales is an (intracellular) *plasmodium*—i.e. a multinucleate mass of cytoplasm; plasmodia are believed to be formed during two separate phases in the life cycle. At the end of one phase the plasmodium appears to give rise to uninucleate, biflagellate (heterokont) zoospores (haploid gametes?); each zoospore (or fused pair of zoospores?) invades a susceptible host cell and initiates another plasmodium. This second plasmodium, if diploid, is the stage during which meiosis occurs; the plasmodium gives rise to a number of resting spores—each of which subsequently germinates to form a biflagellate zoospore. Each zoospore subsequently appears to develop into a non-flagellate amoeboid cell (*myxamoeba*) which, on invading a fresh host cell, initiates a plasmodium. The organisms do not form hyphae or fruiting bodies.

plasmodium (*mycol.*) (see ACELLULAR SLIME MOULDS and PLASMODIOPHORALES)

Plasmodium A protozoan genus of the HAEMOSPORINA. Certain species cause MALARIA in man; others parasitize lower animals. The life cycle is basically similar in all species of *Plasmodium*. Plasmodial *sporozoites* (see later) in the salivary glands of the (female) mosquito enter the vertebrate host when the insect takes a blood meal. The subsequent course of infection differs somewhat in mammals and non-mammals.

In *mammals*, sporozoites enter the parenchymatous cells of the liver, grow, and undergo SCHIZOGONY; this is *pre-erythrocytic schizogony* or (primary) *exoerythrocytic schizogony*. The *merozoites* enter the *erythrocytes* of the mammal. (Until recently it was believed that the merozoites of most *Plasmodium* species could enter fresh liver cells and undergo (secondary exoerythrocytic) schizogony; this seemed to explain the periodic relapses which characterize most types of mammalian malaria. It now seems probable that the merozoites of all mammalian strains of *Plasmodium* infect *only* erythrocytes, and that relapses are due to the persistence, in liver cells, of slowly-developing or dormant sporozoites or schizonts derived from the initial infection.) The intraerythrocytic merozoite (*trophozoite*) initially assumes a characteristic *ring form* ('signet-ring stage'): in stained preparations this appears as a ring of cytoplasm with a central vacuole and a peripheral nucleus; two nuclei are sometimes present. The trophozoite grows and undergoes multiple nuclear division, becoming a *schizont*—from which between 6 and 24 merozoites are subsequently liberated into the blood stream (*erythrocytic schizogony*). (SCHÜFFNER'S DOTS etc. may be seen in suitably stained preparations of parasitized erythrocytes. Granules of pigment—malarial pigment or *haemozoin*—may be observed in trophozoites and other intraerythrocytic stages; haemozoin is derived from the host cell's haemoglobin.) The merozoites derived from erythrocytic schizogony infect fresh red cells and schizogony again occurs. At a later stage of infection some intraerythrocytic merozoites develop into uninucleate male *microgametocytes* while others become uninucleate female *macrogametocytes*; these remain quiescent until subsequently ingested by the mosquito vector. Within the mosquito's gut the gametocytes emerge from their erythrocytes—the macrogametocyte undergoing little further differentiation in becoming a mature gamete. The microgametocyte undergoes *exflagellation*: three nuclear divisions occur, eight peripheral flagella are formed, and the microgametocyte breaks up to form eight uninucleate, uniflagellate male gametes. After fertilization, the non-flagellated but motile zygote (the ookinete) passes through the insect's gut wall and encysts on the outer side of the wall. SPOROGONY occurs in the encysted zygote (oocyst), large numbers of *sporozoites* subsequently being released; these are uninucleate, haploid, fusiform cells of length about 6–18 microns. Sporozoites pass through the haemocoel and enter the insect's salivary glands.

A number of sub-genera of *Plasmodium* are defined on the basis of (a) vertebrate host, and (b) physical characteristics of the erythrocytic schizont and gametocyte. Sub-genera include: (i) *Plasmodium*. Species (e.g. *P. malariae*, *P. vivax* and *P. ovale*) occur in primates and are transmitted by *Anopheles* mosquitoes; erythrocytic schizonts large, gametocytes spherical. (Species are also written *Plasmodium* (*Plasmodium*) *vivax* etc. or *Plasmodium* (*P.*) *vivax* etc.) (ii) *Laverania*. Parasitic in primates and transmitted by *Anopheles*. Schizonts large, gametocytes elongated, crescentic. Species include *Plasmodium* (*L.*) *falciparum*. (iii) *Vinckeria*. Parasitic in *non*-primate mammals, transmitted by *Anopheles*. Schizonts small, gametocytes spherical. (iv) *Haemamoeba*. Parasitic in birds, *naturally* transmitted by *Mansonia crassipes*. (*Experimental* vectors include *Aedes aegypti*, *Anopheles* spp and *Culex* spp.) Schizonts large, gametocytes spherical. Other (avian) sub-genera are *Giovannolaia*, *Novyella* and *Huffia*.

plasmogamy The coalescence of the cytoplasm of two or more cells or hyphae; in *fertilization* plasmogamy is followed by KARYOGAMY. Plasmogamy is also used e.g. to refer to an early phase in GAMETANGIAL CONTACT in which communication is established between the gametangial protoplasts; during gametangial contact the complete intermingling of cytoplasm appears not to occur.

Plasmopara A genus of fungi of the OOMYCETES.

plasmotomy In certain multinucleate protozoans (e.g. *Pelomyxa*): an asexual reproductive process in which fission results in the formation of two or more multinucleate daughter cells—nuclear division and fission of the parent cell apparently being unlinked.

plastids A class of membrane-bound organelles found within the cells of higher plants and algae (though not in *blue-green* algae or other procaryotes); plastids contain pigments and/or certain products of the cell (e.g. starch, lipid). CHLOROPLASTS are familiar examples. (see also AMYLOPLAST, CHROMOPLAST, ELAIOPLAST and LEUCOPLAST)

plastocyanin A copper-containing protein (2Cu/molecule) which occurs bound to the thylakoid membranes in algae and higher plants. It is involved in electron transport associated with PHOTOSYNTHESIS.

plastoglobuli Osmiophilic lipid granules in the *stroma* of a CHLOROPLAST.

plastome The genetic complement of a PLASTID.

plastoquinones (see QUINONES (in electron transport))

plate (1) An unused (sterile) solid MEDIUM—usually an AGAR-based medium—in a PETRI DISH. (Gelatin, or gelatin-agar, is used in certain cases.) A plate is prepared by pouring the molten medium into a petri dish (to a depth of 3–5 mm) and allowing it to set. An ASEPTIC TECHNIQUE must be used. Freshly-poured plates which are to be used for STREAKING, or for any other process designed to give individual (discrete) colonies, should be dried prior to use; drying prevents unwanted spreading of the inoculum in the surface film of moisture. The plates are dried by leaving them for about 20 minutes in a hot-air incubator (at 37 °C) until the *surface* moisture has evaporated; to facilitate evaporation, the petri dish is incompletely closed. (2) A *plate culture*: see CULTURE. (3) (*verb*) To inoculate a plate—see PLATING.

platelets (thrombocytes) Small anucleate bodies (about 2.5 microns in diameter) which occur in the blood of mammals. They assist in the mechanism of blood clotting—though clotting can occur in their absence. Platelets contain HISTAMINE and SEROTONIN. They are formed from giant cells (megakaryocytes) which occur in the red bone marrow.

plating The distribution of an INOCULUM on the surface of a solid MEDIUM in a PETRI DISH. Plating may be carried out e.g. by STREAKING. (see also LAWN PLATE) Plating should always be carried out with an ASEPTIC TECHNIQUE.

plectenchyma (*mycol.*) A tissue which consists of fungal hyphae or cells. In one form (*prosenchyma*, *prosoplectenchyma*) the constituent hyphal elements are distinguishable; prosenchymatous tissue may be found e.g. in *stromata* (see STROMA). In another form (*pseudoparenchyma*, *paraplectenchyma*) the fungal cells are rounded, and the tissue resembles the parenchyma of higher plants; pseudoparenchymatous tissue may occur e.g. in stromata and in *sclerotia* (see SCLEROTIUM).

Plectomycetes A class of filamentous (mycelial) FUNGI within the sub-division ASCOMYCOTINA. Constituent species are distinguished from those of other classes on the basis of a group of criteria taken *collectively*; the plectomycetes typically (a) produce asci at different levels within the fruiting body—i.e. they do not form a hymenial layer within the fruiting body; (b) produce *closed* ascocarps (cleistothecia) which contain no paraphyses; (c) produce evanescent asci which usually contain eight ascospores per ascus. All species are placed within the single order EUROTIALES.

Plectonema A genus of filamentous BLUE-GREEN ALGAE; *Plectonema* spp are capable of NITROGEN FIXATION.

pleiotropic mutation A single mutation which causes changes in two or more phenotypic characteristics.

pleomorphism (1) (*bacteriol.*) The variation, in size and form, between individual cells in a CLONE, or (as generally understood) between individual cells in a pure culture. (2) (*mycol.*) The loss of ability to form conidia and/or pigment which occurs in laboratory-propagated strains of some fungi, e.g. certain *dermatophytes*.

Plesiomonas (C_{27} organisms) A genus of gram-negative, oxidase-positive bacteria of the VIBRIONACEAE; the rod-shaped cells commonly have two or more polar flagella. Strains of the sole species, *P. shigelloides* (formerly *Aeromonas shigelloides*), do not liquefy gelatin (cf. *Aeromonas*, *Vibrio*), do not produce gas from glucose, and do not grow in media containing 7.5% NaCl. Acid is produced from inositol, and lysine is decarboxylated; most strains are sensitive to O/129. Optimum growth temperature: 30 °C.

Pleurococcus A genus of green algae (CHLOROPHYTA).

pleuronematic Refers to a *eucaryotic* FLAGELLUM which bears fine hairs.

Pleuronematina A suborder of freshwater and marine ciliate protozoans of the order HYMENOSTOMATIDA. The *pleuronematines* are characterized by a well-developed undulating membrane which protrudes from the side of the organism; the AZM is poorly developed, and there is no vestibulum. The pellicle is thinly populated with cilia, but there is often a single, conspicuous cilium at the posterior end of the organism. Genera include *Cyclidium* and *Pleuronema*.

Pleurostomata A tribe of protozoans of the sub-order RHABDOPHORINA.

Pleurotus A genus of fungi of the family TRICHOLOMATACEAE. Constituent species form fleshy, non-stipitate fruiting bodies in which the gills are sharp-edged and the basidiospores are non-amyloid; the organisms occur as saprophytes and parasites on the wood of various types of deciduous tree—particularly beech (*Fagus*). The fruiting bodies of *P. ostreatus* (the oyster fungus) are edible.

plicate Folded.

Plistophora A genus of protozoans of the sub-phylum CNIDOSPORA.

ploidy Of a (eucaryotic) nucleus or cell: the number of sets into which the *entire* chromosomal complement of the nucleus can be *exactly* fitted. (see also HAPLOID, DIPLOID, POLYPLOID; cf. ANEUPLOIDY)

plugging (of pipettes) (see PASTEUR PIPETTE)

plum pox (Sharka disease) An aphid-borne disease, of viral aetiology, which affects many species of *Prunus* (e.g. plum, peach, apricot, blackthorn). Symptoms include mottling of leaves and may include various types of lesion (rings, bands etc.) on the fruit; the fruit often drops prematurely. Plum pox is a NOTIFIABLE DISEASE in England and Wales.

plus and minus strands of DNA and RNA (*virol.*) A system of nomenclature used to distinguish between complementary strands of nucleic acid. In the BALTIMORE CLASSIFICATION of viruses, a given strand of nucleic acid in the viral genome is designated 'plus' or 'minus' according to whether its base sequence is equivalent or complementary, respectively, to the virus-specified mRNA; in all cases virus-specified mRNA is regarded as a *plus* strand. In an alternative system of nomenclature, the genome of an RNA virus is designated '+' (*plus*)—regardless of whether it is equivalent or complementary to mRNA.

pmf Proton motive force—see OXIDATIVE PHOSPHORYLATION (*chemiosmotic hypothesis*).

pneumatocysts (*synonyms*: air bladders/vesicles; bladders; gas bladders/vesicles) Gas-filled regions normally present within the tissues of many algae, e.g. species of *Fucus*, *Laminaria*, *Macrocystis* and *Sargassum*; pneumatocysts, which are readily observable as distinct swellings, are a source of buoyancy. Analysis of the gas in bladders of *Fucus vesiculosus* indicates that it consists of air; the oxygen content of the bladder gas increases by day and decreases by night—suggesting the influence of photosynthetic activity. Studies on *F. vesiculosus* and other brown algae show that new bladders are formed annually. (cf. GAS VACUOLE)

pneumococcus *Streptococcus pneumoniae.*

pneumococcus capsule swelling reaction (see QUELLUNG PHENOMENON)

pneumonia Inflammation of the lungs; pneumonia is commonly caused by microorganisms—but may be due e.g. to chemical irritants. Causal agents include: *Streptococcus pneumoniae*, *Haemophilus influenzae*, *Klebsiella pneumoniae*, and certain viruses (see also PRIMARY ATYPICAL PNEUMONIA).

Among animals, similar diseases may be produced by bacteria or viruses. *Shipping fever*, which may occur in cattle under stress conditions, may be caused by *Pasteurella haemolytica* or by paramyxoviruses.

pock (1) (*med.*) A cutaneous pustule; pocks are formed e.g. in SMALLPOX. (2) (*virol.*) A lesion produced on the chorioallantoic membrane during the cultivation or assay of certain viruses, e.g. vaccinia virus.

pocket rot (of timber) (see TIMBER, DISEASES OF)

podetium (*lichenol.*) A vertical (fruticose) secondary thallus which develops from the sterile portion of the ascocarp initials in the primary thallus (i.e. develops from that part of the ascocarp initials which occurs below the ascogonial complex); podetia may or may not bear ascocarps. Podetia are formed e.g. by members of the genera BAEOMYCES and CLADONIA; they may be branched or unbranched and are generally hollow. In e.g. *Cladonia* spp the apices of the podetia may be pointed or may each terminate in a cup-shaped or funnel-shaped structure (a SCYPHUS). (cf. PSEUDOPODETIUM)

Podophrya A genus of protozoans of the sub-class SUCTORIA.

Podosphaera A genus of fungi of the family ERYSIPHACEAE; the organisms form cleistothecia which bear appendages that are highly branched at their distal ends—each cleistothecium containing a *single* ascus.

Podospora A genus of fungi of the order SPHAERIALES, class PYRENOMYCETES; species occur on dung and in the soil. *Podospora* spp form non-evanescent asci in perithecia which develop at the surface of the substratum; the dark-coloured ascospores each bear one or more gelatinous appendages. In *P. anserina* the ascus contains four binucleate ascospores; each spore contains nuclei of compatible mating type—i.e. the organism exhibits *secondary* homothallism. Ascocarp development initially involves plasmogamy by spermatization.

point mutation A type of MUTATION in which a

single base in DNA (RNA in some viruses) is replaced by another; a point mutation may be either a TRANSITION or a TRANSVERSION. Due to the degeneracy of the GENETIC CODE, a point mutation may not affect the product of the gene in which it occurs; the mutated codon may specify the same amino acid as did the original codon. However, the mutated codon may specify an alternative amino acid (a *mis-sense* mutation), or it may be one of the three chain-terminating codons (a *nonsense* mutation). MUTAGENS which induce point mutations include 5-BROMOURACIL, HYDROXYLAMINE, and NITROUS ACID.

poky mutant A mutant strain of *Neurospora* in which the rate of growth is lower than that in wild-type strains; the reduced growth rate is associated with the absence or deficiency of certain components of the respiratory chain. The *poky* gene (= *mi-1* or *maternal inheritance-1* gene) appears to occur in the *mitochondrial* DNA, and is transmitted to sexually-derived progeny only when the female (ascogonial) parent exhibits the *poky* phenotype; such MATERNAL INHERITANCE is presumably a consequence of the absence of mitochondria in the (male) spermatium (microconidium). (see also PETITE COLONIES)

polar bodies Intracellular granules or deposits found at one or both poles of a cell—e.g. the METACHROMATIC GRANULES in *Corynebacterium diphtheriae* or deposits of poly-β-hydroxybutyrate in *Beijerinckia*.

polar capsule (cnidosporean) (see CNIDOSPORA)

polar filament (cnidosporean) (see CNIDOSPORA)

polar flagellation (of bacteria) Refers to the arrangement of a cell's FLAGELLUM or flagella; in polarly flagellated cells one or more flagella occur at one pole or at both poles of the cell.

polar mutation In an OPERON: a MUTATION which affects not only the gene in which it occurs, but also all gene(s) in that operon on the side of the mutation furthest from the operator (*operator-distal* genes); genes between the operator and the mutation (*operator-proximal*) are not affected by the mutation. Thus, e.g. in the *lac* OPERON, a NONSENSE MUTATION in the *z* gene may prevent the synthesis of β-galactosidase, and also reduces the amounts of permease and transacetylase synthesized; similarly, a nonsense mutation in the *y* gene may eliminate permease synthesis, does not affect galactosidase synthesis, but causes lower amounts of transacetylase to be synthesized. The mutation is said to be *strongly* polar when the expression of operator-distal genes is strongly inhibited; weakly-polar mutations cause only slight inhibition. The strength of polarity depends on the distance between the mutation and the end of the gene in which it occurs (i.e. between the mutation and the next site of initiation of translation)—being greatest when this distance is greatest.

The mechanism by which such mutations exhibit polarity is uncertain. Translation (see PROTEIN SYNTHESIS) begins at the operator-end of the POLYCISTRONIC MRNA and proceeds in the 5′ to 3′ direction, concomitantly with transcription (see RNA). When the translating mechanism reaches the nonsense mutation, polypeptide chain synthesis is terminated and the ribosomes and incompleted polypeptide chain are released. Normally, ribosomes are thought to protect the mRNA from degradation by nucleases. It has been suggested that mRNA transcribed beyond the mutation is not translated and hence is exposed to degradation by nucleases; thus, as the initiator region for the next gene is transcribed, it may be degraded before ribosomes can bind and initiate translation. (No exonuclease is known which degrades RNA in the 5′ to 3′ direction; degradation may be achieved by the concerted action of an endonuclease together with an exonuclease which removes nucleotides from the 3′ end.) Since some translation of the distal genes may occur, it is assumed that some polypeptides are initiated before the nucleases can act.

PHASE-SHIFT MUTATIONS may also be regarded as polar mutations.

polarilocular (of a lichen spore) Refers to a two-celled spore in which the two cells—one at each pole of the spore—are separated by a thick median wall which is perforated by a central tubular canal. Polarilocular spores are formed e.g. by species of CALOPLACA and XANTHORIA.

polaroplast (see CNIDOSPORA)

poliomyelitis (polio; infantile paralysis) In man, an acute, infectious disease which is more common among young adults; the causal agent, a *picornavirus*, may enter the body *via* the mouth or nose. The incubation period may be 3–35 days, but is commonly 1–2 weeks. Most cases are mild respiratory or gastrointestinal infections, but in some the disease develops with headache, stiffness of the neck muscles, and varying degrees of paralysis; characteristic lesions occur in the spinal cord and brain. *Bulbar* poliomyelitis is frequently fatal since the effects often include paralysis of the respiratory musculature. *Laboratory diagnosis*: pharyngeal or rectal swabs may be examined culturally for poliovirus; serological tests may be carried out. Prophylactic vaccines include the SABIN VACCINE and SALK VACCINE.

poliovirus (see ENTEROVIRUS)

polyacetylenes (polyynes) (in fungi) Linear compounds which contain both double and triple bonds and which are produced by certain of the higher fungi—especially basidiomycetes. The polyacetylene molecule usually terminates in an oxygen-containing group such as a carboxyl group or an alcohol group; many are highly unstable. Some of the polyacetylenes exhibit antibiotic activity against both bacteria and fungi.

polycentric (*mycol.*) Refers to a type of thallus which gives rise to a number of reproductive centres that are located at different parts of the thallus. (cf. MONOCENTRIC; see also HOLOCARPIC and EUCARPIC)

polychrome methylene blue (see METHYLENE BLUE)

polycistronic mRNA (polygenic messenger; multicistronic mRNA) In bacteria: an mRNA molecule which codes for the synthesis of several (normally) functionally-related proteins; it is formed by the transcription of an OPERON, and must contain nucleotide sequences which specify the initiation and termination of each polypeptide (see PROTEIN SYNTHESIS). Ribosomes bind to the free 5′-end of mRNA immediately after initiation of transcription, and translation closely follows transcription; thus, cistrons near the operator are translated before those further from it. (see also POLAR MUTATION)

polyene antibiotics A functionally-allied group of MACROLIDE ANTIBIOTICS; they are synthesized by members of the Streptomycetaceae. Polyenes contain a macrocyclic lactone comprising a lipophilic region of conjugated double bonds and a hydrophilic region which includes hydroxyl groups; the former region may contain between 4 and 7 conjugated double bonds—generally in the all-*trans* configuration. The lactone may be substituted by amino sugars, aliphatic side-chains etc. The polyenes are poorly soluble in water but are soluble in organic solvents; solutions exposed to light may be unstable owing to photo-oxidation of the double bonds. Polyenes interact with *sterols* in the CELL MEMBRANES of sensitive cells causing leakage of small molecules. They are active against the majority of fungi (particularly those with yeast-like forms), protozoa, and algae, but are without effect on bacteria and blue-green algae—whose membranes do not contain sterols (see, however, MYCOPLASMA). Polyene antibiotics may be microbistatic or microbicidal according to concentration. Examples: *Nystatin* (*fungicidin*) is usually classed as a *tetra*ene although it has a total of *six* double bonds: four conjugated double bonds separated from two others by two methylene groups. It is substituted with *mycosamine* (3-amino-3,6-dideoxy-D-mannose). Nystatin was named after the New York State Health Department where it was discovered; it is produced by *Streptomyces noursei*. It has been used in the treatment of superficial mycoses, e.g. CANDIDIASIS. *Pimaricin* (*tennecetin*) is a tetraene which contains mycosamine; it is produced by *S. natalensis*. *Filipin* is produced by *S. filipinensis* and is a methyl-substituted *penta*ene. The *hepta*enes include *amphotericin B, candicidin B, hamycin, perimycin* and *trichomycin*. *Amphotericin B* is produced by *S. nodosus*; it contains mycosamine. It is active against a wide range of fungi and against certain pathogenic protozoans; it has been used to combat systemic mycoses. *Candicidin B* is produced by *S. griseus*; its spectrum of antifungal activity is similar to that of amphotericin B. It is particularly effective against *Candida albicans*. *Hamycin*, produced by *S. pimprina*, is active against a limited range of fungi, but is particularly effective against *C. albicans*. *Perimycin* (*fungimycin*), produced by a strain of *S. coelicolor*, contains the sugar *perosamine* (4-amino-4,6-dideoxy-D-mannose). *Trichomycin* is produced by *Streptoverticillium hachijoense*; it contains mycosamine and has been used against infections by *C. albicans* and *Trichomonas vaginalis*.

polygenic messenger (see POLYCISTRONIC MRNA)

polygenic resistance (*plant pathol.*) See: RESISTANCE (of plants to disease).

polyhedra (*entomopathol.*) (see POLYHEDROSES)

polyhedral bodies (see CARBOXYSOMES)

polyhedroses (*entomopathol.*) INSECT DISEASES of viral aetiology in which the causal virions occur embedded in intracellular polyhedral crystalline protein inclusion bodies—the *polyhedra*; the polyhedra are visible by light microscopy. Each polyhedron contains many virions (cf. GRANULOSES). The polyhedrosis viruses are primarily pathogens of the larvae of e.g. Diptera (flies), Lepidoptera (butterflies, moths), and Hymenoptera (sawflies). *Nuclear polyhedrosis viruses* (NPVs) replicate only in the nucleus of the host cell; the virions are rod-shaped and contain DNA. *Cytoplasmic polyhedrosis viruses* (CPVs) replicate only in the cytoplasm of the host cell; the virions are icosahedral and contain RNA. NPVs have been used for the BIOLOGICAL CONTROL of the European pine sawfly (*Neodiprion sertifer*) and the European spruce sawfly (*Gilpinia hercyniae*); infection of larvae occurs by ingestion of polyhedra and results in a high mortality rate.

poly-β-hydroxybutyrate (PHB) A linear polymer of D(−)-3-hydroxybutyrate (β-hydroxybutyrate): $H\text{-}(O.CH(CH_3).CH_2.CO)_n\text{-}OH$. In a number of procaryotic microorganisms PHB occurs in refractile, membrane-bound, intracellular granules which vary widely in size;

the intracellular granules can be stained with the lipophilic dye Sudan Black. (Sudan Black is not taken up by purified PHB—suggesting that the staining of PHB granules may be due to the presence of lipid associated with the granules.) In many organisms PHB is accumulated when the growth medium is rich in carbon compounds but poor in nitrogenous compounds, and is broken down when these conditions are reversed; PHB thus functions as a reserve of carbon and/or energy. The pathways of synthesis and breakdown of PHB appear to differ in detail in different organisms. The substrate for PHB synthesis is acetyl-CoA; PHB synthesis occurs *via* acetoacetyl-CoA and β-hydroxybutyryl-CoA—the latter being converted to PHB by the action of *PHB synthetase* (which is closely associated with the PHB granule). PHB is broken down, *via* free β-hydroxybutyrate and acetoacetate, to acetyl-CoA. Certain organisms (e.g. some species of *Pseudomonas*) can degrade exogenous PHB. Bacteria which synthesize PHB include species of *Azotobacter*, *Bacillus*, *Beggiatoa*, *Beijerinckia*, *Paracoccus*, *Pseudomonas*, *Rhizobium*, *Sphaerotilus*, and members of the Chromatiaceae and Rhodospirillaceae. (see also GLYCOGEN and POLYPHOSPHATE)

poly(I:C) (see INTERFERONS)

polykaryocytosis The formation of GIANT CELLS.

polymetaphosphate (see POLYPHOSPHATE)

polymorphic trypanosomes (*syn*: pleomorphic trypanosomes) Those trypanosomes (see TRYPANOSOMA) whose normal life cycle includes several morphologically-distinct forms; examples: *T. brucei*, *T. gambiense*. (cf. MONOMORPHIC TRYPANOSOMES)

polymorphonuclear leucocyte ('polymorph'; granulocyte) A white blood cell which has a lobed nucleus and granular cytoplasm. Cells of this type include the *neutrophils*, *basophils* and *eosinophils*—so named because of the affinity of their cytoplasmic granules for particular dyes. Cell sizes range from about 10 microns to 14 microns in diameter.

polymyxins A group of polypeptide ANTIBIOTICS produced by *Bacillus polymyxa*. Polymyxins are active against most gram-negative bacteria (including *Pseudomonas* spp)—exceptions including *Neisseria* spp, most *Proteus* spp, and *Vibrio cholerae* biotype *eltor* (which differs in this respect from *Vibrio cholerae* biotype *cholerae*); polymyxins are bactericidal to growing and to non-growing cells. Most gram-positive bacteria, and most fungi, are resistant to polymyxins, although *Candida tropicalis* is sensitive.

The polymyxin molecule consists of a cyclic polypeptide linked, *via* α,γ-diaminobutyric acid (DAB), to a short linear peptide which terminates in a branched-chain fatty acid residue. The various polymyxins differ in their amino acid content, but always include several residues of DAB; at least some of the amino groups on the DAB residues must remain unsubstituted for the molecule to exhibit antibacterial activity. Polymyxin *E* (*colistin*) and polymyxin *B* each consist of two components which differ only in their fatty acid residue; thus, B_1 and E_1 each contain 6-methyloctanoic acid, while B_2 and E_2 each contain 6-methylheptanoic acid.

Polymyxins bind strongly to the CELL MEMBRANE—apparently *via* phosphate groups—causing an increase in the permeability of the membrane to small molecules. It is believed that the fatty acid portion of the antibiotic penetrates the lipophilic region of the membrane while the peptide portion interacts with the polar region. (cf. TYROCIDINS)

Polyomavirus A genus of VIRUSES of the family PAPOVAVIRIDAE; the constituent species, which are typically oncogenic under appropriate conditions, include polyoma virus and 'K' virus (both mouse viruses) and SV40 (simian virus 40)—a virus isolable from the tissues of the African green monkey. The icosahedral virion, 40–45 nm in diameter, contains a circular double-stranded DNA genome. Some species (e.g. polyoma virus—but not SV40) can agglutinate the erythrocytes of the guinea-pig (and of certain other species). Polyomaviruses are generally readily cultivable *in vitro*; thus, e.g. SV40 can be cultivated in tissue cultures prepared from e.g. African green monkey kidney or rabbit kidney.

polyoxins A group of ANTIBIOTICS produced by a strain of *Streptomyces cacaoi*; they are derivatives of pyrimidine nucleosides. The polyoxins are active specifically against certain fungi. Polyoxin D, for example, appears to inhibit CHITIN biosynthesis—possibly by acting as an analogue of UDP-*N*-acetylglucosamine. Polyoxins have been used for the control of certain plant pathogens.

polyphosphate A phosphate polymer (polymetaphosphate—see formula) which occurs in many microorganisms (see e.g. METACHROMATIC GRANULES). Polyphosphate

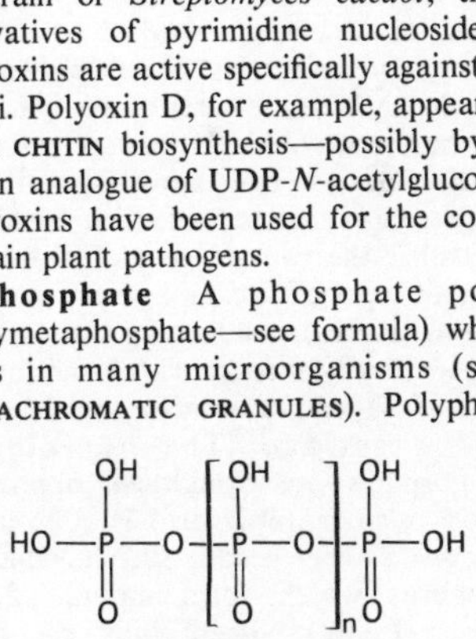

Polyphosphate

is believed to function as a reserve of phosphate; it is synthesized by a reaction—catalysed by polyphosphate kinase—in which the terminal phosphate group of ATP is transferred to pyrophosphate or polyphosphate. Several enzymes (including polyphosphate kinase) are known to degrade polyphosphate, although it is unknown which of these functions under physiological conditions.

polyphyletic (*adjective*) Of, or containing, members of a number of phyla.

polyploid Refers to any (eucaryotic) nucleus or cell which has a PLOIDY of greater than two. (cf. HAPLOID, DIPLOID; see also EUPLOID and ANEUPLOID)

Polyporaceae A family of fungi of the order APHYLLOPHORALES; the organisms are typically saprophytic and lignicolous. Constituent species form annual or perennial fruiting bodies which vary (with species) from crust-like (resupinate) to pileate—the latter being either sessile or stipitate; the context may be monomitic, dimitic, or trimitic, and may be corky, leathery, or woody. In no species is the context xanthochroic. The hymenophore is *typically* tubular (porous), and the outer, exposed surfaces of the dissepiments are sterile—i.e. the hymenium is confined to the walls of the tubules. According to species the basidia may form two or four spores; basidiospores are typically spheroidal, ovoid, or cylindrical, and are usually smooth-walled and hyaline. The genera include CORIOLUS, DAEDALEA, FOMES, HETEROBASIDION, *Lenzites*, PIPTOPORUS, POLYPORUS, PORIA, and *Trametes*.

polypore (*mycol.*) Any species of the order APHYLLOPHORALES in which the fruiting body contains a porous hymenophore.

Polyporus A genus of lignicolous fungi of the family POLYPORACEAE. Constituent species form stipitate fruiting bodies (stipe central, lateral, or eccentric) in which the fleshy or leathery context is white or pale brown (at least when immature); the pores of the hymenophore are non-hexagonal in cross-section. The basidiospores are smooth-walled and elongated. *P. squamosus* (the 'saddle back fungus') is the causal agent of heart rot in a range of types of deciduous tree—particularly the elm (*Ulmus*); the fruiting body is laterally or eccentrically stipitate with a light brown upper surface and a whitish or cream lower, porous surface. The non-pigmented basidiospores are cylindrical, approximately 12 × 5 microns. (*Polyporus* is often used as a *form genus*; when used as such it includes those polypores which form annual, fleshy to leathery fruiting bodies. Thus, e.g. the form species *Polyporus betulinus* (the 'razor-strop fungus') is referable to the species *Piptoporus betulinus*.)

polyribosome (see PROTEIN SYNTHESIS)

polysaprobic zone (see SAPROBITY SYSTEM)

polysome (see PROTEIN SYNTHESIS)

Polysphondylium A genus of the CELLULAR SLIME MOULDS.

Polyspira A protozoan genus of the APOSTOMATIDA.

Polystictus versicolor (see CORIOLUS)

polystromatic (*algol.*) Refers to a parenchymatous thallus which is several/many cells thick.

polythetic classification Any form of classification in which a *large number* of criteria form the basis on which organisms are arranged into groups—e.g. NUMERICAL TAXONOMY. (cf. MONOTHETIC CLASSIFICATION)

polyvalent antiserum In general, any antiserum which contains antibodies to a number of different antigens.

polyvalent vaccine A single vaccine containing the protective antigens from each of a number of different strains of *a given species* of pathogen. (cf. MIXED VACCINE)

polyvinyl alcohol fixative A fixative used e.g. in protozoological work. It consists of SCHAUDINN'S FLUID prepared with five times the normal concentration of glacial acetic acid and supplemented with polyvinyl alcohol—$(CH_2.CHOH)_n$—(5% w/v) and glycerol (1.5 ml per 100 ml fluid). PVA fixative is particularly useful in that the specimen can remain in the solution, without damage, for prolonged periods of time. (see also FIXATION)

polyynes (see POLYACETYLENES)

porfiromycin The *N*-methyl derivative of mitomycin C—see MITOMYCINS.

Poria A genus of lignicolous fungi of the family POLYPORACEAE. Constituent species form resupinate fruiting bodies which are variable in appearance—the white or cream porous hymenophore occurring in patches, or confluently, on the white, superficial mycelial growth; the thickness of the fruiting body may range from a few millimetres to over one centimetre. The white or pale-coloured context contains hyphae which exhibit monomitic organization and which do not form clamp connections; in perennial fruiting bodies the tubules are vertically-stacked ('stratified'). Cystidia are not formed. The smooth-walled, non-pigmented basidiospores are elliptical. *P. vaillantii* is a common cause of decay in structural timbers—e.g. in mines and in wet situations in residential buildings; it does not spread to dry timbers—cf. *Serpula lacrymans*. The spores are approximately 6 × 4 microns. (*Poria* is often used as a *form-genus*; when used as such it includes those polypores which form annual or perennial *resupinate* fruiting bodies.)

porospores (*mycol.*) (see CONIDIA)

Porphyra A genus of marine algae of the division

RHODOPHYTA; species occur in supratidal, intertidal, and subtidal regions. The thallus is parenchymatous and varies in form according to species; it may be one or two cells thick, membranous or ribbon-like, and may range in colour from shades of reddish-purple to olive-green. In some species the thallus may reach 50–60 cms in length. The individual cells of the thallus are surrounded by thick gelatinous material; each cell contains a single nucleus and one or two (depending on species) stellate chloroplasts. The thallus is attached to the substratum—or sometimes to other algae—by means of a rhizoid-like holdfast. Several species of *Porphyra* are used as food in various countries—e.g. as *nori* in Japan, and as *laver bread* or *purple laver* in the British Isles.

porphyrans Complex, mucilaginous polysaccharides produced by *Porphyra* spp (species of red algae). Porphyrans contain D- and L-galactose, 3,6-anhydro-L-galactose, sulphate and methyl residues. (see also AGAR and CARRAGEENAN)

Porphyridium A genus of algae of the division RHODOPHYTA; species occur in marine, brackish, and freshwater environments and on moist soil, damp stones, etc. When viewed in mass, the spherical, unicellular (non-flagellate) organisms vary in colour from blood-red (*P. cruentum*) to blue-green (*P. aerugineum*)—depending on the relative proportions of phycoerythrin and phycocyanin which they contain. (see also PHYCOBILIPROTEINS and PHYCOBILISOMES) The uninucleate cell contains a single stellate chloroplast which has a central pyrenoid; there is no cell wall—each cell being surrounded by a gelatinous capsule of polysaccharide.

porphyrins Derivatives of *porphin*—a tetrapyrrole ring in which the pyrroles are linked by their α-carbon atoms *via* methene (–CH=) bridges. In porphyrins, the pyrrole β-carbon atoms are variously substituted. Isomers of a given porphyrin are usually denoted by Roman numerals, e.g. *aetioporphyrin* I (= *etioporphyrin* I) is 1,3,5,7-tetramethyl-2,4,6,8-tetraethylporphin, and *aetioporphyrin* III is 1,3,5,8-tetramethyl-2,4,6,7-tetraethylporphin. Other porphyrins: *mesoporphyrins* (4 methyl, 2 ethyl and 2 propionic acid substituents) and *protoporphyrins* (methyl, vinyl and propionic acid substituents); *protoporphyrin* IX is 1,3,5,8-tetramethyl-2,4-divinylporphin-6,7-dipropionic acid. Porphyrins readily form chelates with a variety of metals; *iron protoporphyrin* IX* (*haem*)—in which the iron atom is coordinated to the four pyrrole nitrogen atoms—is a component of e.g. haemoglobin, CATALASE and certain CYTOCHROMES. Porphyrins are generally identified by their absorption spectra. *Also known as *protohaem.*

port (see WINE-MAKING)

posterior station (*parasitol.*) The posterior part of the alimentary canal of an arthropod vector. Pathogens which become infective in the hind gut of the vector (e.g. *T. cruzi* in CHARGAS' DISEASE) cannot be transmitted by the *bite* of the vector—infection of the vertebrate host occurring by the contamination of wounds etc. by the *faeces* of the vector. (cf. ANTERIOR STATION)

post-meiotic segregation (see RECOMBINATION)

postreplication repair (see RECOMBINATION-REPAIR)

postzone (see PROZONE)

post-zygotic exclusion (see PRE-ZYGOTIC EXCLUSION)

potassium permanganate (as an antimicrobial agent) (see OXIDIZING AGENTS)

potato, bacterial diseases of These include: (a) *Common scab*; causal agent: *Streptomyces scabies.* Dark, corky scabs of variable size develop on the surface of the tubers. Infection may occur *via* wounds. (b) *Blackleg.* A soft rot characterized by yellowing of the foliage and the development of slimy lesions within the tubers. The causal agent, *Erwinia carotovora*, commonly gains entry *via* wounds.

potato, fungal diseases of These include: (a) *Early blight*; causal agent: *Alternaria solani.* The disease characteristically appears early in plant development. Small, dark spots appear on the leaves; later, concentric rings of necrotic tissue give the lesions a characteristic 'target' appearance. The tubers may develop a superficial brown, dry, corky rot. (b) *Late blight*; causal agent: *Phytophthora infestans.* Purplish-black patches form on stems and on the upper surfaces of leaves, with corresponding patches of a white 'mildew' of sporangia and sporangiophores on the undersides of the leaves. The tubers develop purplish-black patches on their surfaces, and subsequently succumb to a dry, brown rot; tubers may be infected during storage. Disease development requires a high relative humidity (90% or above). Secondary infection by soft rot bacteria may occur. Late blight was responsible for the Irish potato famine in the mid-19th century. (c) *Wart disease* (*black-wart disease*); causal agent: *Synchytrium endobioticum.* A soil-borne disease characterized by cauliflower-like outgrowths (GALLS) from the tubers and the lower portion of the stem; entire tubers may be converted to a warty mass. The warts may continue development in harvested tubers during storage. Wart disease is a NOTIFIABLE DISEASE.

potato slope (see SLOPE)

potato virus Y An RNA virus which can infect a range of species of the Solanaceae—e.g. potato (*Solanum* sp) and tobacco (*Nicotiana tabacum*); the virion, approximately 730 × 10 nm, may be stylet-borne by aphids (particularly *Myzus* spp) or transmitted mechanically. On potato symptoms of infection may include leaf-drop, mosaic, veinal necrosis, or discrete necrotic leaf lesions. (see also ETHYLENE and VIRUS DISEASES OF PLANTS)

pour plate (*bacteriol.*) (1) A method of culture in which the inoculum is dispersed uniformly in molten agar or other medium (usually at 45–48 °C) in a petri dish; the medium is then allowed to set and is subsequently incubated. The pour plate method is used e.g. for counting those cells in the inoculum which are capable of growing on and within the medium—each colony which develops being assumed to have arisen from a single cell. (see also ENUMERATION OF MICROORGANISMS) (The pour plate method may be unsuitable for counting obligate aerobes owing to limitation of oxygen within the medium—an environment which may be (or become) microaerophilic or anaerobic.) Certain reducing agents—e.g. sodium thioglycollate—may be incorporated in the medium so that, even under aerobic incubation, obligate anaerobes can grow *within* the medium. (2) A culture which has been prepared by the pour plate method.

povidone-iodine (see IODINE)

powdery mildews (see MILDEW)

Poxvirus A genus of VIRUSES whose hosts include mammals and birds. (see also ENTOMOPOXVIRUSES) Poxviruses can cause localized and systemic human diseases which typically involve the formation of vesicular lesions in the superficial tissues of the body; certain poxviruses give rise to tumours. The complex, ellipsoidal or 'brick-shaped' *Poxvirus* virion is approximately 200–300 nm, maximum dimension. The virion contains a bilaterally-concave discoid structure (the *core*) which incorporates the genome (linear, double-stranded DNA); either side of the core, coincident with the concavity, is an ellipsoidal structure known as the *lateral body*, and the whole (core and associated lateral bodies) is enclosed by a lipoprotein outer membrane. The outer membrane is synthesized *de novo* in the cytoplasm of the host cell; *Poxvirus* virions are assembled in *viroplasmic matrices*—distinct 'virus factories' located within the cytoplasm. Several enzymes—including a DNA-dependent RNA polymerase—are associated with the core of the virion. The poxviruses of vertebrates share a common antigen, and sub-genera are distinguished on the basis of minor antigens; members of a given subgenus may be serologically indistinguishable. Poxviruses in clinical specimens tend to be stable (i.e. they remain infective) at room temperatures; they are significantly resistant to desiccation, and most are not inactivated by ether. They are inactivated e.g. by chloroform and by exposure to 60 °C for approximately 20 minutes. Poxviruses can be cultivated in certain types of tissue culture—e.g. human embryonic kidney, monkey kidney, HeLa, HEp-2, and on the chorioallantoic membrane (CAM) of a hen's egg; the characteristic inclusion body of the poxviruses is an eosinophilic intracytoplasmic body. (see also GUARNIERI BODIES)

The *vaccinia* subgenus includes the following viruses: variola, vaccinia, cowpox, ectromelia (mousepox), and monkeypox. The *variola major virus* is the causal agent of SMALLPOX. (see also ISATIN-β-THIOSEMICARBAZONE) The *natural* hosts of this virus are man and monkey; the virus can cause lesions in the skin of a rabbit but cannot be propagated from one rabbit to another. When cultivated on the CAM the variola major virus gives rise to *small* circular lesions (pocks). *Vaccinia* virus (serologically similar or identical to the variola major virus) is generally non-pathogenic for man, and strains of the virus are used in smallpox vaccines. The host range of vaccinia is much wider than that of variola major—and includes the rabbit, sheep, and calf; vaccinia virus can be propagated from one rabbit to another (cf. variola major). When cultivated on the CAM, vaccinia virus gives rise to *large* pocks which are often haemorrhagic (cf. variola major). (Pock size is one important criterion used to distinguish between variola major and vaccinia viruses.) The *cowpox virus* (causal agent of COWPOX) differs antigenically from vaccinia and the other members of the subgenus; in tissue cultures it gives rise to *strongly* eosinophilic cytoplasmic inclusion bodies, and produces haemorrhagic ulcers on the CAM.

The *orf* subgenus includes the *orf virus* and e.g. the virus of pseudocowpox. The orf virus is the causal agent of CONTAGIOUS PUSTULAR DERMATITIS.

The *sheep pox* subgenus includes the *sheep pox virus* (see SHEEP POX) and the *goatpox* and LUMPY SKIN DISEASE viruses. The *sheep pox virus* is smaller than other poxviruses (being approximately 200 nm, maximum dimension); it is inactivated by ether and chloroform and by 2% phenol.

The *myxoma-fibroma* subgenus includes a group of oncogenic viruses—e.g. the causal agent of MYXOMATOSIS.

PPD (purified protein derivative) Specific tuberculo-proteins prepared from the filtrate of an (unheated) culture of *Mycobacterium tuberculosis* by precipitation with e.g. ammonium sulphate. PPD is used in TUBERCULIN TESTS on humans and animals.

ppGpp (see STRINGENT CONTROL (in the synthesis of bacterial RNA))

PPi Abbreviation for inorganic pyrophosphate ($P_2O_7^{4-}$).

PPLOs Pleuropneumonia-like organisms (see MOLLICUTES).

Prasinophyta A division of the ALGAE.

Prausnitz-Küstner antibodies Reaginic (IgE) antibodies.

precipitation (*immunol.*) The formation of a precipitate when *soluble* antigens and antibodies react in *suitable* proportions. The precipitate consists of an antigen–antibody complex. (see also OPTIMAL PROPORTIONS; ANTIGEN EXCESS; GEL DIFFUSION)

precipitin Any antibody which gives rise to a precipitate following interaction with its homologous, *soluble* antigen. *Precipitin* sometimes refers to the precipitate itself.

precipitin test Any serological test in which the interaction of antibodies with *soluble* antigens is detected by the formation of a precipitate.

pregnancy (serological test for) (see IMMUNOASSAY)

Preisz-Nocard bacillus *Corynebacterium pseudotuberculosis* (formerly *Corynebacterium ovis*).

premunition (*immunol.*) The maintenance of a state of immunity, in respect of a given pathogen, due to the persistence of a latent infection with that pathogen.

prepatent period (*med.* and *vet.*) The period between infection and the appearance of parasitic organisms (or their cysts) in the blood, tissues, or faeces.

pre-reduced medium (see ANAEROBE and REDOX POTENTIAL)

preservation In the context of microbiology: the use of physical and/or chemical means which either kill or retard/inhibit the growth of those microorganisms which cause spoilage of any of a variety of products or substances. Physical means of preservation (e.g. heat STERILIZATION, FREEZING) either kill contaminating organisms or render the products or substances unsuitable for the (rapid) growth of those organisms likely to be present as contaminants. Chemical PRESERVATIVES may be microbistatic or microbicidal. (see also FOOD, PRESERVATION OF; TIMBER, DISEASES OF, and entries for PAINTS and PAPER)

For the 'preservation'—i.e. maintenance—of a population of viable microorganisms (for reference purposes etc.) FREEZING and FREEZE-DRYING are commonly used. (see also DESICCATION)

preservatives Chemicals used for the PRESERVATION of e.g. PAINTS, PAPER, pharmaceutical products and timber (see also FOOD, PRESERVATION OF). Preservatives used in medical, pharmaceutical and laboratory materials include FORMALDEHYDE, PHENOLS, SODIUM AZIDE, THYMOL and THIOMERSAL; those used in ophthalmic preparations include benzalkonium chloride (see QUATERNARY AMMONIUM COMPOUNDS).

prespore (see SPORES bacterial))

pretrichocysts (see TRICHOCYSTS)

pre-zygotic exclusion During processes such as bacterial CONJUGATION and TRANSDUCTION a given donor gene may not appear in the recombinant—either because the gene never entered the recipient (*pre-zygotic exclusion*) or because, having entered the recipient, the gene was excluded during recombination (*post-zygotic exclusion*).

primaquine An antimalarial AMINOQUINOLINE.

primary atypical pneumonia PNEUMONIA caused by *Mycoplasma pneumoniae.* (see also COLD AGGLUTININS and STREPTOCOCCUS M.G.)

primary constriction (see CENTROMERE)

primary culture (1) (*microbiol.*) (a) The process of preparing a CULTURE by incubating a medium that has been inoculated with (or from) a specimen. (cf. SUBCULTURE) (b) A culture prepared as in (a). (2) (tissue culture) A culture prepared from cells or tissues taken *directly* from an animal or plant, i.e. not by subculture from an existing tissue culture. (see also CELL LINE)

primary host In the life cycle of *heteroecious* RUSTS: the (plant) host in which the *telial* stage occurs.

primary hyphae (in basidiomycetes) (see HYPHA)

primary type zoospores (in oomycetes) (see DIPLANETISM)

primed (*immunol.*) (of persons, animals) Refers to those individuals who have experienced the intitial immunological contact with *a particular antigen.* The relevant LYMPHOCYTES of a primed subject are referred to as *primed lymphocytes* or *memory cells.*

primordium (*microbiol.*) Synonym of INITIAL(s).

probasidium (*mycol.*) (1) In members of the Uredinales (rust fungi) and Ustilaginales (smut fungi): the contents of a teliospore—*or* the teliospore itself. (2) (promycelium) In smut fungi: that structure which bears the basidiospores. (3) In e.g. *Septobasidium* and related fungi: a thick-walled cell from which arises the septate, basidiospore-bearing part of the basidium (the *metabasidium*); the probasidium and metabasidium constitute the basidium. (4) That part or stage of a developing basidium in which karyogamy takes place.

procaryotic organisms and eucaryotic organisms

PROCARYOTIC AND EUCARYOTIC ORGANISMS: some distinguishing features

CHARACTERISTICS	EUCARYOTES	PROCARYOTES
Organization of nuclear material	Chromosomes enclosed within a membranous sac. (see NUCLEUS)	Chromosome(s) *not* enclosed within a membrane. (see NUCLEUS)
Histones	Present in most species.	Absent in all species.
Reproduction	Mitosis and sexual processes common.	Mitosis does not occur; in bacteria CONJUGATION is probably the closest approximation to a sexual process.
Genetic recombination	Occurs during sexual processes. (see also PARASEXUALITY IN FUNGI)	In bacteria: may occur as a result of CONJUGATION, TRANSDUCTION or TRANSFORMATION.
Cell wall	When present, contains e.g. CELLULOSE, CHITIN, CHITOSAN.	When present, usually contains PEPTIDOGLYCAN. (cf. HALOBACTERIACEAE, MYCOPLASMA)
Cell membrane	Contains sterols.	Sterols absent in most species. (Present in species of MYCOPLASMA)
Mitochondria	Generally present.	Absent.
Chloroplasts	Present in photosynthetic species.	Absent. CHROMATOPHORES or CHLOROBIUM VESICLES present in photosynthetic bacteria; naked THYLAKOIDS present in blue-green algae.
Ribosomes	80S RIBOSOMES in cytoplasm; 70S ribosomes in chloroplasts and mitochondria.	70S ribosomes.
Storage compounds	May include e.g. STARCH, LAMINARIN, PARAMYLUM.	May include e.g. POLY-β-HYDROXYBUTYRATE; cyanophycean starch in blue-green algae.
Flagella	When present: structurally complex—see FLAGELLUM (eucaryotic).	When present: structurally simple—see FLAGELLUM (bacterial).
Nitrogen fixation	Unknown.	Occurs in certain bacteria and blue-green algae. (see NITROGEN FIXATION)

procaryotic organisms and eucaryotic organisms (prokaryotic and eukaryotic organisms) *Cellular* microorganisms (i.e. non-viral species) may be divided into the *procaryotes* (BACTERIA, BLUE-GREEN ALGAE—see also PROCHLOROPHYTA) and *eucaryotes* (ALGAE, FUNGI, PROTOZOA). Procaryotic and eucaryotic cells differ in a number of ways—see accompanying table.

Prochloron didemni (see PROCHLOROPHYTA)

Prochlorophyta A recently-proposed division of algae erected to accommodate certain organisms which have characteristics intermediate between those of BLUE-GREEN ALGAE and eucaryotic ALGAE; these organisms occur associated with ascidians (sea squirts)—e.g. *Didemnum*—which inhabit tropical coastal waters in the Pacific area and adjacent regions. The organisms are procaryotic and unicellular, green, spheroidal to ovoid, 6–20 microns in size; they contain chlorophylls *a* and *b* and carry out photosynthesis with the evolution of oxygen and the fixation of carbon dioxide. Within the cells the thylakoids are reported to occur in stacks of two or more; phycobilisomes are reported to be absent. One member of the Prochlorophyta, *Prochloron didemni*, was formerly classified as a species of the blue-green algal genus *Synechocystis* (*S. didemni*).

prodigiosin Any of a number of red, non-diffusible pigments produced by certain strains of *Serratia marcescens* and by some species of *Streptomyces*. The prodigiosins are tripyrrole derivatives; they are soluble e.g. in ethanol.

prodromal (*med.*) Refers to the earliest symptoms of a disease; prodromal symptoms are not necessarily those which characterize the disease. The prodromal *period* is the period between the onset of symptoms and the appearance of fever or rash.

proenzyme (see ZYMOGEN)

proflavine 3,6-Diaminoacridine (see ACRIDINE DYES)

progametangium (*mycol.*) In the zygomycetes: an enlarged terminal portion of a hyphal branch; the progametangium differentiates into a proximal SUSPENSOR and a distal *gametangium* prior to gametangial fusion and ZYGOSPORE formation.

progamone (see HORMONAL CONTROL (of sexual processes in fungi))

proguanil (paludrine; paludrin) (see CHLORGUANIDE)
prohormone (see HORMONAL CONTROL (of sexual processes in fungi))
prokaryotic organisms (see PROCARYOTIC ORGANISMS)
proline (biosynthesis of) (see Appendix IV(a))
proloculus (see FORAMINIFERIDA)
promastigote (leptomonad form) A form assumed by many species of the Trypanosomatidae during a particular stage of development. The promastigote is elongated and flagellate; the kinetoplast and basal body are situated at the *anterior* end of the cell—from which end the flagellum emerges *via* a pellicular depression (*flagellar pocket*) (see TRYPANOSOMA). Promastigotes are produced e.g. by *Leishmania* spp in an invertebrate host, and by *Leptomonas* spp.
promitochondria (see MITOCHONDRION)
promotor (promoter) (see RNA and OPERON)
promycelium (see BASIDIUM)
prontosil (see SULPHONAMIDES)
pronucleus (1) The haploid nucleus of a gamete. (2) One of the haploid nuclei formed during AUTOGAMY or during (ciliate) CONJUGATION.
propagative viruses (*plant pathol.*) Those CIRCULATIVE VIRUSES which multiply in the insect vector. It is believed by some authors that (at least) some 'propagative viruses' may, in fact, by *Mycoplasma*-like organisms.
propagule Any disseminative unit of an organism, e.g. a spore, a mycelial fragment.
proper margin (*lichenol.*) A rim at the margin of an APOTHECIUM formed by the projection of the upper part of the excipulum; a proper margin does not contain algal cells and is frequently the same colour as the disc. (cf. THALLINE MARGIN; see also LECIDEINE and LECANORINE)
properdin The *properdin system* consists of a group of proteins which occur in normal serum; these proteins are: properdin, Factor B and Factor D. (COMPLEMENT component C3 may be referred to as Factor A.) In the presence of certain substances, and magnesium ions, the properdin system is able to bring about *in vitro* or *in vivo* COMPLEMENT-FIXATION *via* the *alternative pathway*; these substances include polysaccharides (e.g. ZYMOSAN, inulin), ENDOTOXINS, and aggregates of human IgA. Factor B (and thus the entire properdin system) is heat-labile.
prophage (see LYSOGENY)
prophage immunity (see LYSOGENY)
prophase The first stage in MITOSIS and analogous stages in MEIOSIS.
prophylaxis Measures taken to *prevent* the occurrence of a disease—e.g. DISINFECTION, IMMUNIZATION. (see also PROTECTANT)
β-propiolactone (BPL) A microbicidal, water-miscible, non-inflammable liquid (boiling point 155 °C) which acts as an ALKYLATING AGENT—substituting cellular proteins etc. with propionic acid residues. BPL is bactericidal, fungicidal, virucidal, and sporicidal (in the vapour phase or in solution) and has been used e.g. for the sterilization of blood plasma and surgical instruments and as an inactivating agent in the preparation of certain vaccines. The vapour may be used to sterilize surface areas within enclosed spaces (e.g. operating theatres); however, it requires a high relative humidity for optimum activity and has low powers of penetration. BPL is readily hydrolysed (to β-hydroxypropionic acid) at temperatures above 25 °C; under acid conditions it forms an open-chain ester-linked polymer. BPL causes blistering of the skin, irritation of the eyes, and has been reported to be carcinogenic.
Propionibacterium A genus of gram-positive, asporogenous, anaerobic bacteria which have been classified (along with e.g. *Corynebacterium*) as relatives of the actinomycetes; *Propionibacterium* spp ferment glucose (and lactic acid) with the formation of propionic, acetic and other acids. (see PROPIONIC ACID FERMENTATION; see also Appendix III(e)) The organisms are highly pleomorphic—individual cells may be cocci or branched or unbranched rods, the latter often being coryneform; in stained preparations groups of cells often give the impression of 'Chinese letters'. All species are non-motile. *P. freudenreichii* subsp. *shermanii* (formerly *P. shermanii*) and *P. jensenii* (formerly *P. peterssonii*) are important in the manufacture of certain types of CHEESE. *Propionibacterium* spp are found in dental plaque and in the alimentary tract of man and other animals.
propionic acid fermentation A fermentation—carried out by species of *Propionibacterium*—in which the major products from glucose (or lactate) are propionic and acetic acids (see Appendix III(e)). When lactate is the substrate, lactate dehydrogenase and fumarate reductase apparently form components of an energy-yielding electron transport system which appears also to involve a *b*-type cytochrome and a quinone.
proplastid An immature, colourless PLASTID which may subsequently develop into a LEUCOPLAST or CHROMOPLAST.
Prorocentrum A genus of algae of the DINOPHYTA.
Prorodon A protozoan genus of the RHABDOPHORINA.
prosenchyma (prosoplectenchyma) (see PLECTENCHYMA)
prosoplectenchyma (see PLECTENCHYMA)

prostheca (*pl.* prosthecae) A cell wall-limited appendage forming a narrow extension of a procaryotic cell, e.g. the stalk of *Caulobacter spp.*

prosthetic group (of an enzyme) (see COFACTOR)

Prostomata A tribe of protozoans of the suborder RHABDOPHORINA.

protargol Any of a range of silver proteinates. These preparations are used e.g. to stain protozoan flagella (and other structures) and as veterinary antiseptics for application to the eyes and mucous membranes. (see also *silver* under HEAVY METALS)

protectant (*plant pathol.*) Any chemical agent which *prevents* the occurrence of a given disease (or diseases) among plants subjected to that agent; also called a *prophylactic*. (cf. ERADICANT)

protective antigens Those antigens which, if isolated from a pathogen, could (on their own) elicit an immune response giving protection against infection by the intact (viable) organism.

protective immunity (functional immunity) IMMUNITY to a pathogen or other harmful agent (e.g. a toxin) *with which the immune subject may come into contact*; the object of VACCINATION is to engender protective immunity.

protein A A cell-wall protein found in certain strains of *Staphylococcus aureus* (e.g. NCTC 8530). Protein A combines with the Fc portion of most types* of human IgG; it combines minimally, or not at all, with human IgM. Suitably processed cells of *S. aureus* are used to absorb IgG from patients' sera and thus permit the subsequent detection of IgM antibodies—which may be diagnostically significant. (cf. A-PROTEIN) *the exception being subclass IgG3.

protein synthesis The biosynthesis of proteins is directed by the nucleotide sequence of DNA (or, in some viruses, RNA)—see GENETIC CODE. Initially, *messenger RNA* (mRNA) is transcribed from DNA—see RNA for details of transcription; mRNA functions as a template for the assembly of polypeptide chains—a process known as *translation*. Since amino acids cannot directly recognize their specific codons (see GENETIC CODE) on the mRNA, each must first be linked to a specific *transfer* RNA (tRNA) which contains the correct anticodon. A tRNA is charged with its specific amino acid by a two-step reaction: (a) *activation*, in which the amino acid reacts with ATP in the presence of a specific *aminoacyl-tRNA synthetase* to form an aminoacyl-AMP-synthetase complex—pyrophosphate being released; (b) *transfer*, in which the aminoacyl group is transferred to the 3′ position of the terminal adenosine in the appropriate tRNA (see RNA)—AMP and the synthetase being released. The binding of aminoacyl-tRNAs (aa-tRNAs) to the mRNA and subsequent peptide bond formation are mediated by the RIBOSOMES. A ribosome binds to a specific initiation site on the mRNA; it moves along the mRNA and thus permits the sequential reading of the codons. Each ribosome has two sites for tRNA binding—each site apparently being shared by the two ribosomal subunits; these sites are the *A site* (acceptor site, entry site) and the *P site* (donor site). The orientation of the ribosome on the mRNA is such that each codon in turn is exposed to the A site; incoming aa-tRNAs (excluding the first) enter the A site.

Initiation of polypeptide chain synthesis. In bacteria, the first (i.e. *N*-terminal) amino acid to be incorporated appears always to be *N*-formylmethionine—in which the amino group is blocked by formylation. Two tRNAs are specific for methionine (Met): $tRNA_f^{Met}$ and $tRNA_m^{Met}$. When charged with methionine (Met-tRNA), Met-$tRNA_f$ is formylated by a reaction requiring N^{10}-formyltetrahydrofolic acid, and the resulting fMet-$tRNA_f$ is involved in the initiation of polypeptide synthesis; Met-$tRNA_m$ cannot be formylated and is involved in the incorporation of methionine during chain elongation. (In the eucaryotic cytoplasm—cf. MITOCHONDRION—the situation appears to be similar, although Met-$tRNA_f$ is *not* formylated, i.e. methionine appears to be the first amino acid incorporated.) *In vivo*, the initiator codon in mRNA appears usually to be AUG, although GUG can also be used *in vitro*. In addition to their function as initiator codons, AUG and GUG code for methionine and valine respectively. It is not known how the initiator codons are recognized from other AUG (or GUG) codons in the mRNA (including those out of phase); it has been suggested that a sequence of nucleotides preceding the initiator codon is recognized, or that non-initiating AUG (GUG) codons are masked by folding of the mRNA molecule. The bacterial 30S ribosomal subunit binds to mRNA at the initiation site, and fMet-$tRNA_f$ subsequently binds to the mRNA-30S complex at the initiator codon; unlike other aa-tRNAs, fMet-$tRNA_f$ binds directly to the *P* site. A number of protein *initiation factors* are required for the formation of this initiation complex. *Factor F_3* (IF-3, B) causes dissociation of the 70S ribosomes into their 30S and 50S subunits, and also stimulates the binding of the 30S subunit to the correct initiator site(s) of the mRNA. (see also INTERFERENCE FACTORS) *Factor F_2* (IF-2, C) is involved in the binding of fMet-$tRNA_f$ to the 30S-mRNA complex in a reaction requiring GTP. F_2 appears to complex with GTP prior to binding fMet-$tRNA_f$; the F_2-GTP-(fMet-$tRNA_f$) complex then binds to the 30S-mRNA complex. An additional factor, F_1 (IF-1, A), appears to be required, although its function is unknown; F_1 appears to stimulate the activities of F_2 and F_3 and may be involved in the stabilization of the initiation complex. The final stage of initiation involves the binding of the 50S ribosomal subunit with concomitant release of initiation factors and the hydrolysis of GTP to GDP and phosphate. Once chain initiation is complete and elongation is under way, the formyl group at the N-terminal end may be removed by a *deformylase*; sometimes the methionine residue is also removed.

Polypeptide chain elongation. Elongation can begin only after the association of the 50S subunit with the initiation complex. The aa-tRNA specified by the codon next to AUG enters the A site on the ribosome. Peptide bond formation occurs between the α-amino group of the incoming amino acid and the carboxyl group of *N*-formylmethionine; this leaves uncharged $tRNA_f^{Met}$ in the P site, and the dipeptidyl-tRNA (*N*-formylmethionyl-aminoacyl-$tRNA^{aa}$) in the A site. The ribosome now moves relative to the mRNA-dipeptidyl-tRNA by three nucleotides (i.e. one codon) in the $5' \rightarrow 3'$ direction (*translocation*); the $tRNA_f^{Met}$ is ejected, the dipeptidyl-tRNA enters the P site, while the vacated A site and the next codon become juxtaposed. This cycle of events is repeated at each codon, the growing peptide chain always being transferred to the α-amino group of the incoming aa-tRNA in the A site. Elongation requires the participation of certain protein *elongation factors* (transfer factors) and GTP. *Factor T* (EF-T) is involved in the binding of the aa-tRNA to the vacant A site; this factor has two components: Tu (EF-Tu, S_3) and Ts (EF-Ts, S_1). Tu-Ts binds with GTP, Ts being displaced and Tu-GTP being formed. Tu-GTP complexes with aa-tRNA (*not* fMet-$tRNA_f$) to form (aa-tRNA)-GTP-Tu; this complex binds to the A site of the ribosome. Subsequently GTP is hydrolysed and Tu-GDP and phosphate are released. Ts displaces GDP to reform Tu-Ts, which can then react with GTP to repeat the cycle. Peptide bond formation does not require protein factors or GTP; it is achieved by the enzyme *peptidyl transferase*, a component of the 50S ribosomal subunit. The mechanism of translocation is not understood; it requires *factor G* (EF-G, S_2) and GTP hydrolysis. Factor G may also be involved in the ejection of the uncharged tRNA from the donor site.

Polypeptide chain termination. This occurs when a specific termination codon (UAA, UAG, or UGA) is reached; in *Escherichia coli*, UAA (ochre codon) is apparently used for normal peptide chain termination, although UAG (amber codon) and UGA (umber codon) can also be recognized. Termination codons are recognized in some (unknown) way by protein *release factors* (R factors, termination factors); *factor R_1* (RF-1) recognizes codons UAA and UAG, while *factor R_2* (RF-2) recognizes UAA and UGA. The binding of an R factor is followed by the hydrolysis of the ester bond between tRNA and the polypeptide chain, thus releasing the latter. A third protein, the S protein (RF-3), appears to stimulate the activities of R_1 and R_2. The mechanisms for the release of the last tRNA, and of the ribosome from mRNA, are unknown.

As each ribosome passes along the mRNA, another ribosome may bind to the vacated initiation site and initiate the translation of another polypeptide chain. Thus the mRNA may carry a number of ribosomes situated at intervals along its length—a structure known as a *polyribosome* (*polysome*, *ergosome*); a single ribosome is sometimes referred to as a *monosome*. (see also POLYCISTRONIC MRNA)

ANTIBIOTICS which interfer with protein synthesis include: AMINOGLYCOSIDE ANTIBIOTICS, CHLORAMPHENICOL, CYCLOHEXIMIDE, FUSIDIC ACID, LINCOMYCIN, MACROLIDES, PUROMYCIN, SPARSOMYCIN, and TETRACYCLINES.

proteoses Soluble products of protein hydrolysis; they are not coagulated by heat but may be precipitated in saturated aqueous ammonium sulphate.

proter The *anterior* cell of the two cells formed during binary fission in a ciliate protozoan. (cf. OPISTHE)

Proteus A genus of gram-negative, motile bacteria of the ENTEROBACTERIACEAE; strains occur in soil and sewage, and as parasites and pathogens of animals, including man. The organisms are generally rod-shaped, about 0.5 × 2 microns, but irregularly-shaped cells (including filaments) may be produced. Many strains exhibit SWARMING (see also DIENES PHENOMENON). Growth occurs on basal media (and on media containing potassium cyanide) and is generally prolific between 20 °C and 40 °C. *Proteus* spp produce acid from glucose—but may or may not produce gas; lactose is not fermented. The METHYL RED TEST is typically positive but the VOGES–PROSKAUER TEST varies according to strain; with the exception of *P. inconstans* all species produce UREASES. *Proteus* spp deaminate phenylalanine (to phenylpyruvic acid). Many strains—particularly of *P. vulgaris*—produce β-LACTAMASES. Type species: *P. vulgaris*; GC% range for the genus: 38–50.

P. vulgaris. An indole-positive species which produces H_2S in TSI medium; it does not produce ornithine decarboxylase. Swarming strains are common.

P. mirabilis. Indole may or may not be produced; H_2S is produced in TSI medium and ornithine decarboxylase is produced. Swarming strains are common.

P. morganii. An indole-positive, H_2S-negative species which produces ornithine decarboxylase; gelatin is not liquefied.

P. rettgeri. An indole-positive, H_2S-negative species which does not produce ornithine decarboxylase and does not liquefy gelatin.

P. inconstans (formerly *Providencia inconstans, Providencia alcalifaciens,* etc.). A urease-negative, indole-positive species which does not produce H_2S.

prothallus (*lichenol.*) (1) A term which is used to refer to a lichen thallus in the initial stages of its development. (2) *Prothallus* is sometimes used as a synonym of HYPOTHALLUS.

Protista (the 'protists') A taxon (kingdom) which includes the algae, fungi, and protozoa (all of which constitute the eucaryotic protists) and the bacteria, blue-green algae, and members of the Prochlorophyta (which constitute the procaryotic protists); the latter are sometimes referred to as the *monera*. The kingdom Protista formerly included only the eucaryotic protists.

protoaecia (see *stage I* under RUSTS (1))

Protococcidia An order of protozoans of the sub-class COCCIDIA; species are distinguished from those of the EUCOCCIDIA by the absence of schizogony. The protococcidians are parasites of marine invertebrates.

protohaem (see PORPHYRINS)

protokaryote An earlier term for *procaryote*.

protomerite In members of the Eugregarinida, the middle region of a *developing* trophozoite; it is separated (by cross walls) from the EPIMERITE and the DEUTOMERITE. (see also CEPHALINE)

Protomycetales An order of fungi of the class HEMIASCOMYCETES.

proton motive force (see OXIDATIVE PHOSPHORYLATION and TRANSPORT)

protoperithecium (*mycol.*) Perithecial initials prior to fertilization.

protoplast membrane (see CELL MEMBRANE)

protoplasts (1) Spherical or near-spherical osmotically-sensitive structures formed when cells are suspended in an isotonic medium and their cell walls are *completely* removed e.g. by enzymic action; thus, e.g. a bacterial protoplast consists of an intact cell membrane and the cytoplasm, nucleoid body (or bodies) etc. contained within the membrane. Bacterial protoplasts are able to metabolize under appropriate cultural conditions; they are resistant to infection by bacteriophage—but protoplasts prepared from previously infected cells are able to support phage replication. (cf. SPHAEROPLAST; see also L FORMS) (2) *Protoplast* may refer to the cell membrane and its contents within an *intact* cell.

protoporphyrin IX (see PORPHYRINS)

Prototheca A chloroplast-free, achlorophyllous strain of CHLORELLA.

prototroph A microorganism whose nutritional requirements do not exceed those of the corresponding wild type strain. (cf. AUXOTROPH)

protozoa A phylum of diverse, *eucaryotic*, typically unicellular microorganisms; they are classified within the animal kingdom —although certain species (e.g. the SLIME MOULDS and members of the PHYTOMASTIGOPHOREA) are also classified in *botanical* taxonomic schemes.

General characteristics, structure, size. Considerable variation (e.g. in form, structure, and life style) occurs among species of each of the sub-phyla CILIOPHORA, CNIDOSPORA, SARCOMASTIGOPHORA, and SPOROZOA. In general, cellular components include one or more nuclei (which may be of different types—see e.g. CILIOPHORA), mitochondria, Golgi apparatus, and a more or less differentiated limiting membrane or wall—the PELLICLE. Many species have external and/or internal skeletal structures which may consist largely of calcareous, silicaceous or other durable material—see e.g. ARCELLINIDA, CILIOPHORA, FORAMINIFERIDA, RADIOLARIANS; see also AXOSTYLE, TEST, TRICHITES. Specialized structures found in some species include AXOPODIA, the AZM (of ciliates), CONTRACTILE VACUOLES, CYTOSTOME, *eucaryotic*-type flagella (see FLAGELLUM) or cilia (see CILIUM), and TRICHOCYSTS. CYSTS and/or spores are formed by a number of species; such structures may have a disseminative and/or reproductive function. According to species, an organism may be motile (see MOTILITY) or non-motile; some species are sessile. Very few protozoans form aggregates or colonies; colonial species include certain peritrichs—e.g. CARCHESIUM. In size, protozoans range from several microns (e.g. the amastigote form of *Leishmania*) through millimetres (e.g. PELOMYXA, SPIROSTOMUM, many radiolarians) to centimetres (e.g. many foraminiferans—particularly fossil species).

Occurrence. Free-living protozoans typically occur in aquatic environments—e.g. puddles, freshwater ponds, lakes and rivers, and in estuaries, seas and oceans; freshwater species include many ciliates and flagellates and the majority of HELIOZOANS, while marine species include foraminiferans and radiolarians. (see also PLANKTON) Additionally, many species are found e.g. in muds and moist soils, and some can survive (in encysted form) when available water is minimal. Parasitic and pathogenic species generally occur in the blood and/or alimentary or reproductive tracts of their hosts.

Some protozoans are involved in symbiotic relationships with bacteria (see e.g. KAPPA PARTICLES), algae (see e.g. RADIOLARIANS), insects (see e.g. TRICHONYMPHA), and other organisms. (see also RUMEN) Fossil protozoans (e.g. foraminiferans, radiolarians) occur in the beds of seas and oceans and in sedimentary rocks.

Growth conditions. Probably most species are aerobic, but some free-living and parasitic protozoans survive or grow microaerophilically or anaerobically—some growing only under such conditions. Protozoans which tolerate or require low oxygen tensions, or anaerobiosis, include flagellates (e.g. *Bodo* spp, *Trichonympha* spp), sarcodines (e.g. *Entamoeba histolytica*, *Pelomyxa palustris*), and ciliates (e.g. *Spirostomum*, RUMEN ciliates). According to species, growth temperatures range from a few degrees above freezing-point to over 50 °C (for organisms found in hot springs). In general, protozoans appear to exhibit some tolerance of fluctuations in pH; in some species high acidity appears to encourage encystment. Light is essential for phototrophic species.

Nutrition. The phytoflagellates (Phytomastigophorea) are typically photolithotrophic; other protozoans are chemoorganotrophic. Among the chemoorganotrophs, feeding may occur saprozoically and/or by pinocytosis (e.g. in TRYPANOSOMA) and/or holozoically (e.g. in many amoebae, ciliates, and others). (see also FOOD VACUOLES) According to species, and environment, the diet of free-living protozoans may include bacteria and/or algae (herbivorous species), protozoans (carnivorous species), or both (omnivorous species). In some holozoic feeders there is a specialized region of food intake, the CYTOSTOME. Some species exhibit growth requirements which may include amino acid(s) and/or vitamins—see e.g. BIOPTERIN, FOLIC ACID, NICOTINIC ACID, PANTOTHENIC ACID, PYRIDOXINE, and THIAMINE.

Reproduction, genetic transfer. Asexual reproduction (only)—involving e.g. binary and/or multiple fission—appears to occur in a number of species (e.g. *Amoeba proteus* and many other sarcodines, euglenoid flagellates, *Trypanosoma* spp, and the ciliate *Stephanopogon*). Fission, SCHIZOGONY, and budding are all forms of asexual reproduction which may occur in those species which also exhibit sexual phenomena. Sexual reproduction occurs in sporozoans (e.g. *Monocystis*, *Plasmodium*, *Eimeria* and other coccidians), flagellates (e.g. *Trichonympha*), and other organisms (e.g. the opalinids and foraminiferans). The processes of CONJUGATION and/or AUTOGAMY occur e.g. in many ciliates and in some HELIOZOANS; these processes bring about genetic transfer and/or genetic re-arrangement but are not directly concerned with cell multiplication. According to species, a protozoan may be haploid, diploid, or polyploid; thus, e.g. sporozoans are typically haploid organisms while ciliates are typically diploid or polyploid. An alternation of haploid and diploid generations (a feature unique among protozoans) has been recorded in a number of species of foraminiferans.

Preservation of protozoans. Populations of viable protozoans are commonly preserved either by FREEZING (suitable e.g. for *Plasmodium*, *Tetrahymena*, *Trypanosoma*) or by repeated sub-culturing; species which form resistant CYSTS (e.g. certain amoebae, the ciliate *Didinium*) may be encouraged to encyst (and thus survive for protracted periods) by e.g. slow desiccation or by starvation.

Destruction/inactivation of protozoans. Protozoans may be destroyed/inactivated e.g. by STERILIZATION techniques and, for those species which do not form resistant cysts or spores, by desiccation; see also ANTIPROTOZOAL AGENTS.

Protozoans as parasites/pathogens. Protozoans parasitize a range of organisms which include plants (see PHYTOMONAS), annelid worms (e.g. MONOCYSTIS, *Triactinomyxon*), insects (e.g. NOSEMA), and man and/or other animals (e.g. BABESIA, BALANTIDIUM COLI, EIMERIA, ENTAMOEBA, GIARDIA, HISTOMONAS MELEAGRIDIS, NAEGLERIA, PLASMODIUM, TOXOPLASMA, and TRYPANOSOMA).

Exploitation of protozoans by man. Certain species are used in BIOASSAY, while others are important as indicator organisms in the SAPROBITY SYSTEM. Some species are important in SEWAGE TREATMENT systems.

Evolutionary relationships between protozoans and the regularly-multicellular animals (the metazoa) are not clear. Sponges (*Porifera*) appear to have developed from zooflagellates of the order Choanoflagellida (class ZOOMASTIGOPHOREA); however, the sponges—referred to as *parazoa*—lack the degree of organization typical of metazoans, and are believed to have developed separately from other multicellular animals.

Providencia An obsolete bacterial genus; the three species are now regarded as strains of *Proteus inconstans*.

provirus hypothesis An hypothesis concerned with the interactions between eucaryotic cells and oncogenic RNA viruses (e.g. Rous sarcoma virus). The hypothesis proposes that, following infection of a cell, the viral RNA is transcribed by a REVERSE TRANSCRIPTASE with the subsequent formation of a *provirus*—i.e. a double-stranded DNA equivalent of the viral genome; the provirus becomes integrated with the host's genome and is replicated and transferred to progeny cells at mitosis. Each cell which contains a provirus thus contains the information necessary for the synthesis of new virions and for maintenance of the transformed state; a transformed cell may continue to synthesize virions—in contrast to cells transformed by DNA viruses. This hypothesis accounts for certain features of the development of oncogenic RNA viruses—e.g. the inhibition of viral growth (and cell transformation) by the presence—at the time of infection—of inhibitors of DNA synthesis, and the inhibition of virion formation by ACTINOMYCIN D. Evidence for the existence of a provirus has been obtained e.g. by hybridization studies. (cf. ONCOGENE HYPOTHESIS)

prozone (*immunol.*) (1) In a titration involving antibodies and *particulate* antigens, an absence of visible agglutination in tubes containing the higher concentrations of antibody. A prozone may be due to: (a) The existence of ANTIBODY EXCESS conditions. (b) The presence of BLOCKING ANTIBODIES. (c) The presence of non-specific inhibitory factors. (2) In a titration involving antibodies and *soluble* antigens, the absence of visible precipitation due to ANTIGEN EXCESS conditions; analogously, a *postzone* may be due to antibody excess conditions.

pruinose (*adjective*) (of surfaces) Having a powdery appearance.

Psalliota campestris A former name of *Agaricus campestris*.

Pseudocoprinus A genus of fungi of the COPRINACEAE.

pseudocyphella (*lichenol.*) In certain lichens: a pore or depression in the cortex of the thallus. Pseudocyphellae occur in a number of lichen genera (e.g. species of *Cetraria*, *Physcia*, *Pseudocyphellaria*) and are more or less constant in size for a given species; they may occur (as warts or shallow depressions) in the upper or lower surfaces of the thallus, and are believed to facilitate gaseous exchange in the thallus. (cf. CYPHELLA)

Pseudocyphellaria A genus of foliose LICHENS in which the phycobiont is *Nostoc* or a green alga—according to species; the lower surface of the thallus is tomentose and contains pseudocyphellae (see PSEUDOCYPHELLA). Species include e.g.:

P. crocata. The thallus is greenish-brown to brown; the upper surface bears bright yellow soralia, while on the lower surface the pseudocyphellae expose the yellow medullary tissue. *P. crocata* occurs on mossy rocks and trees.

pseudocyst (of *Toxoplasma*) (see TACHYZOITES)

pseudoepithecium (*mycol.*) In some types of APOTHECIUM (at the surface of the layer of asci): a layer which consists of the tips of the paraphyses immersed in an amorphous matrix. (cf. EPITHECIUM)

pseudolumpyskin disease (see LUMPY SKIN DISEASE)

pseudo-lysogeny Simulated LYSOGENY. One form of pseudo-lysogeny may be exhibited when a population of susceptible bacteria is exposed to virulent phage at a *low* multiplicity of infection; progeny phage liberated in the first round of lysis may be *unable to adsorb* to other cells in the population if, during the initial period of growth, such cells had acquired a *phenotypic resistance* to the phage. (In at least one case of pseudo-lysogeny, phenotypic resistance is believed to be the result of a reaction between the cell wall phage receptor sites and an enzyme released from the lysed cells.) The cells which are (temporarily) resistant to phage (i.e. *phenocopies*) become susceptible on subculture. In another form of pseudo-lysogeny a phage genome is carried *within* each of a number of cells of a susceptible bacterial population, and each genome is passed to one of the daughter cells during successive cell division cycles; such cells are said to be in the *carrier state*. After several successive transits from cell to daughter cell a phage genome enters the lytic cycle and lyses the host cell; other (previously-uninfected) cells within the population then become infected and enter the carrier state. In this form of pseudo-lysogeny the lytic activity of the phage is temporarily repressed—presumably as a result of an unstable immunity response engendered in the host cells. Unlike truly lysogenized cells, a bacterial population which exhibits pseudo-lysogeny can be rendered phage-free by prolonged incubation with neutralizing antiserum specific to the given phage.

Pseudomonadaceae A family of gram-negative, aerobic*, chemoorganotrophic or facultatively chemolithotrophic bacteria; the genera are ACETOMONAS (*Gluconobacter*), PSEUDOMONAS, XANTHOMONAS and ZOOGLOEA. Most species conform to the broad generic definition of *Pseudomonas*; the organisms are placed in different genera on the basis of ecological and other factors. *See however NITRATE RESPIRATION.

Pseudomonadales A vaguely-defined order of bacteria which is no longer recognized.

Pseudomonas A genus of gram-negative, chemoorganotrophic or (in some species) facultatively chemolithotrophic bacteria of the PSEUDOMONADACEAE; the organisms are widespread in soil and water and are important participants in MINERALIZATION processes. All species use molecular oxygen as the terminal electron acceptor in an obligately oxidative (respiratory) metabolism; some species are capable of NITRATE RESPIRATION. Typically, the cells are polarly flagellate bacilli, about 0.5–1 × 1–4 microns. Growth generally occurs on simple, unenriched media, e.g. nutrient agar; any of a wide range of substrates may be utilized, e.g. camphor, phenols, starch. (see also ENTNER–DOUDOROFF PATHWAY) Poly-β-hydroxybutyrate is a storage product in some species. A number of species form diffusible fluorescent pigments. All species are catalase-positive and most are oxidase-positive; none is sensitive to O/129. *P. aeruginosa* is susceptible to a range of temperate and virulent bacteriophages; lysogenic strains are believed to be extremely common. (Lysogeny appears to occur very infrequently in other species though these may be susceptible to a variety of virulent phages.) Strains of *P. aeruginosa* produce a range of BACTERIOCINS which have been used in a system of typing. (Bacteriocin production seems to be less common among other species.) Genetic transfer between strains of *Pseudomonas* can occur by CONJUGATION or by TRANSDUCTION. Type species: *P. aeruginosa*; GC% range for the genus: approximately 58–69. Species include:

P. aeruginosa (formerly *P. pyocyanea*). An important opportunist pathogen in man—isolated e.g. from urinary tract infections and burns; also pathogenic in certain plants. Growth on nutrient agar is usually accompanied by the production of a blue-green or yellow-green water-soluble phenazine pigment, *pyocyanin* (which becomes red when acidified) and/or one or more diffusible fluorescent pigments; pyocyanin production is peculiar to *P. aeruginosa*. Optimum growth temperature: 37 °C. The organism is capable of DENITRIFICATION. It is often resistant to a range of antibiotics; resistance to one or more antibiotics may be conferred by an R FACTOR. Antibiotics which have been used to treat *P. aeruginosa* infections include POLYMYXIN B and GENTAMICIN. (see also MARFANIL) Some disinfectants, e.g. QUATERNARY AMMONIUM COMPOUNDS, are not effective against *P. aeruginosa*.

P. mallei. A non-motile species, the causal agent of GLANDERS.

P. solanacearum. A species pathogenic in a wide range of plants—including banana, tomato and TOBACCO; infected fruits may exhibit premature ripening (see ETHYLENE; see also SCOPOLETIN).

P. syringae. An oxidase-negative 'umbrella' species which currently includes a large number of named species of *Pseudomonas*—including *P. phaseolicola* and *P. pisi*. Optimum growth occurs at about 25 °C.

P. fragi. A psychrophilic species which may cause spoilage of refrigerated food, e.g. butter, eggs, meat.

P. facilis (formerly *Hydrogenomonas facilis*). A non-pigmented facultative chemolithotroph.

P. fluorescens. A psychrophilic species which usually grows at temperatures below 4 °C. The organism does not grow at 41 °C (unlike *P. aeruginosa* and certain other species); most strains produce diffusible fluorescent pigments.

P. methanica, *P. methanitrificans*. See METHYLOMONADACEAE.

pseudomycelium A synonym of SPROUT MYCELIUM.

pseudonavicellae The sporocysts of *Monocystis*.

pseudoparaphyses (*mycol.*) (*sing.* pseudoparaphysis) Sterile filaments which occur in the perithecia of certain ascomycetes, e.g. *Nectria*, and in the pseudothecia of certain loculoascomycetes. Pseudoparaphyses originate in the upper part of the perithecial or pseudothecial cavity and grow *downwards*—often becoming attached to the basal part of the centrum; their function is unknown.

pseudoparenchyma (paraplectenchyma) (see PLECTENCHYMA)

pseudoperithecium (see ASCOSTROMA)

pseudoplasmodium (1) (*mycol.*) In the CELLULAR SLIME MOULDS: an aggregation of cells formed prior to the production of a fruiting body. (2) (*mycol.*) In the NET SLIME MOULDS: a network of cells linked by slime filaments. (3) (swarm) (*bacteriol.*) In the MYXOBACTERALES: a number of individual cells embedded in a slime matrix.

pseudopodetium (*lichenol.*) A vertical (fruticose) secondary thallus which develops by differentiation of part of the primary thallus—and is thus a purely thalline (somatic) structure (cf. PODETIUM); the ascocarp initials develop later on the branched or unbranched pseudopodetia. Pseudopodetia occur e.g. in members of the genera PILOPHORUS and STEREOCAULON.

pseudopodium (see MOTILITY)

pseudoseptum (see SEPTUM)

pseudostem (pseudostipe) (*mycol.*) In certain gasteromycetes (e.g. *Phallus* spp): the columnar, supportive part of the fruiting body; a pseudostem differs from a STIPE e.g. in the orientation of the constituent hyphae.

pseudothecium (see ASCOSTROMA)

pseudovacuole A synonym of GAS VACUOLE.

pseudo-wild phenotype (see SUPPRESSOR MUTATION)

psicofuranine (9-β-D-psicofuranosyladenine) A nucleoside ANTIBIOTIC produced by a strain of *Streptomyces hygroscopicus*. Psicofuranine has antitumour and antimicrobial activity. It inhibits the final step in GMP biosynthesis, i.e. it inhibits XMP aminase (= GMP synthetase)—see Appendix V(a). *Decoyinine* is a closely related antibiotic which has a similar mode of action.

psittacosis *Psittacosis* is used by different authors in different ways: (1) An infectious disease of the parrot and other psittacine and non-psittacine birds, and of man and other non-avians; the causal agent is *Chlamydia psittaci*. In man, infection commonly results from the inhalation of aerosols of infected bird faeces; the disease ranges from a mild influenza-like condition to a severe, sometimes fatal infection which (typically) involves fever, intense headache, and an atypical pneumonia. In the parrot, infection is often latent; in cases of overt disease the bird may exhibit weakness, diarrhoea, and difficulty in breathing, and lesions may develop in the liver and other organs. *Laboratory diagnosis* involves cultural examination (of e.g. blood, sputum) and/or serological tests. (The term *ornithosis* is sometimes used as a synonym of psittacosis.) (2) *C. psittaci* infection in man and psittacine birds only—the term *ornithosis* being reserved for the infection in non-psittacine birds. (3) *C. psittaci* infection in man only—*ornithosis* being reserved for the disease in psittacine and non-psittacine birds.

psoralen A compound which, in the presence of light, can form cross-links between pyrimidine residues in DNA.

psychrophile (psychrophil) According to many authors: any organism which has an optimum growth temperature below 20 °C. Some authors define a psychrophile as an organism capable of producing a visible colony after 1 week's incubation at 0 °C. Other definitions have also been proposed. (cf. MESOPHILE and THERMOPHILE)

pteroylglutamic acid (see FOLIC ACID)

pubescent Downy.

Puccinia A genus of fungi of the order UREDINALES (see also RUSTS (1)); species are parasitic on a wide range of plant hosts—including monocotyledonous and dicotyledonous species. (see also CEREALS, DISEASES OF) *Puccinia* spp are macrocyclic, heteroecious rusts; they form cupulate aecia, non-peridiate uredia, and stalked (pedicellate) two-celled teliospores. The life cycle of *Puccinia graminis tritici* is described under RUSTS (1).

puerperal fever (puerperal sepsis; childbed fever) An acute febrile condition, following childbirth, caused by infection of the uterus and/or adjacent regions by e.g. streptococci of Lancefield groups *A* or *C*, anaerobic streptococci, *Clostridium* spp. Complications may occur and the mortality rate is high in untreated cases.

puffball The fruiting body of certain members of the basidiomycete order Lycoperdales.

puffing (see ASCOSPORES, DISCHARGE OF)

pullulan An extracellular polysaccharide produced by *Pullularia pullulans*. It is a polymer of D-glucopyranose in which α-(1 → 4) and α-(1 → 6) linkages are found. (see also STARCH)

pulque A Mexican alcoholic beverage made by fermenting the juices of certain succulents (*Agave* spp). Fermentation is effected by *Saccharomyces carbajali* (believed to be identical with *Saccharomyces cerevisiae*)—a yeast present in the plant's natural microflora. Species of *Lactobacillus* and *Leuconostoc* contribute acidity and viscosity (respectively) to the drink. *Tequila* is made by the distillation of pulque. (see also WINE-MAKING and SPIRITS)

pulverulent Powdery.

pulvinate (*adjective*; of structures) Cushion-like; forming a swelling.

punctiform Dot-like.

P.U.O. (*med.*) Pyrexia of unknown origin.

purine nucleotides (biosynthesis of) (see Appendix V(a); see also NUCLEOTIDE)

purity plate A plate which has been inoculated from a given culture and subsequently incubated. The examination of such a plate is carried out to determine whether or not the inoculum came from a pure or contaminated culture—the presence of colonies of two or more types being indicative of a contaminated culture.

puromycin 6-dimethyl-3′-deoxy-3′-*p*-methoxyphenylalanylamino adenosine: an ANTIBIOTIC obtained from cultures of *Streptomyces alboniger* or prepared synthetically. It inhibits PROTEIN SYNTHESIS by functioning as an analogue of the 3′-terminal portion of an aminoacyl-tRNA. When a tRNA-peptide is bound to the *donor* site of a RIBOSOME, and puromycin is bound to the corresponding acceptor site, a peptide bond is formed between the amino group of the puromycin and the C-terminal of the peptide. Since puromycin binds only weakly to the ribosome, the peptidyl-puromycin entity immediately separates—thus causing premature chain termination. Puromycin is active against certain bacteria and protozoa; it has been used to treat amoebiasis and trypanosomiasis, and in studies on protein synthesis.

purple membrane In *Halobacterium halobium*: regions of the CELL MEMBRANE which contain *bacteriorhodopsin* (a retinal-containing, rhodopsin-like protein) as the sole protein component; such regions are formed under microaerophilic conditions in the presence of light, and may occupy up to 50% of the membrane area. In the presence of light, the bacteriorhodopsin oscillates between two conformational states with concomitant outward pumping of protons across the membrane. The resulting proton gradient can be used to drive ATP synthesis—see *chemiosmotic hypothesis* under OXIDATIVE PHOSPHORYLATION. This light-dependent ATP synthesis (*photophosphorylation*) provides an alternative to oxidative phosphorylation under microaerophilic conditions. (see also HALOBACTERIACEAE)

purple nonsulphur bacteria Bacteria of the family RHODOSPIRILLACEAE.

purple sulphur bacteria Bacteria of the family CHROMATIACEAE.

purse (in bird's-nest fungi) (see CYATHUS)

pusule (see CONTRACTILE VACUOLES)

putrefaction The microbial breakdown of protein, under *anaerobic* conditions, with the formation of evil-smelling products. Such products include e.g. amines formed by the decarboxylation of certain amino acids (e.g. cadaverine from lysine, putrescine from ornithine) and hydrogen sulphide (from sulphur-containing amino acids). Putrefaction is brought about typically by proteolytic *Clostridium* spp—e.g. *C. bifermentans*, *C. histolytica*, *C. sporogenes*; *Proteus* spp may be involved.

PVA (see POLYVINYL ALCOHOL FIXATIVE)

pyaemia The condition in which *pyogenic* bacteria are disseminated *via* the blood stream—a particular form of SEPTICAEMIA.

pycnidiospores Conidia formed within a PYCNIDIUM.

pycnidium (*mycol.*) A hollow, typically spherical, flattened, or flask-shaped structure within which conidia (= pycnidiospores) are produced and which, at maturity, is—or becomes—open to the environment (e.g. *via* one or more pores); pycnidia, which may superficially resemble perithecia or cleistothecia, are formed by certain members of the order SPHAEROPSIDALES. According to species, a pycnidium may be thin-walled or thick-walled, unilocular or multilocular, black, brown, or brightly coloured (e.g. orange, yellow), glabrous or setose; it may develop superficially or may be immersed (or partly immersed) within the substratum or a fungal stroma. In size, pycnidia are generally 100–500 microns.

pycniospores (spermatia) Mononucleate haploid cells which are formed within a *pycnium* and which function as male elements—see *stage O* under RUSTS (1).

pycnium (spermogonium; spermagonium) A structure within which the rust fungi (order UREDINALES) form their male and female reproductive structures—see RUSTS (1). (The terms pycnium and PYCNIDIUM are frequently confused.)

pyknosis (*histopathol.*) The shrinkage of a cell's nucleus resulting in the formation of a visibly smaller, densely-staining body; a sign of cellular abnormality or death. (see also KARYORRHEXIS and KARYOLYSIS)

pyo- A combining form meaning pus; e.g. *pyogenic* (pus-forming).

pyocins (see BACTERIOCINS)

pyocyanin (see PSEUDOMONAS)

pyrenoid A dense, proteinaceous body found within the CHLOROPLAST envelope in most types of algae; in at least some species it appears to be the site of polysaccharide synthesis and storage.

Pyrenomycetes A class of filamentous (mycelial) fungi within the sub-division ASCOMYCOTINA; the class has been defined (loosely) in widely-differing ways by various authors. In several taxonomic schemes one of the main characteristics of the ascigerous stage in a *typical* pyrenomycete is the formation of an hymenial layer of (unitunicate) asci within a PERITHECIUM or a CLEISTOTHECIUM; atypical species (e.g. organisms which exhibit certain plectomycete characteristic(s)) are included by some authors in the belief that such organisms have certain affinities with the more typical members of the Pyrenomycetes. The class Pyrenomycetes has been divided into orders and families in a number of ways; a given (named) taxon may be used by different authors to refer to different groups of organisms. Constituent orders include e.g. ERYSIPHALES and SPHAERIALES.

Pyricularia A genus of fungi within the order MONILIALES; species include *P. oryzae*, the causal agent of *blast disease* of rice (see RICE, DISEASES OF).

P. oryzae. The vegetative form of the organism is a branched, septate mycelium. Conidia are borne directly on unbranched (or rarely branched) conidiophores; the elongated, pyriform, typically biseptate, hyaline or pale olive conidia are attached to the conidiophore, at their rounded ends, *via* small projections on the conidiophore. The sexual stage of *P. oryzae* is reported to be a loculoascomycete.

pyridoxine (pyridoxin; vitamin B_6; pyridoxol) A water-soluble, photolabile VITAMIN: 2-methyl-3-hydroxy-4,5-di(hydroxymethyl)pyrimidine.

Two derivatives—*pyridoxal* and *pyridoxamine* (CHO and CH_2NH_2 at the 4-position respectively)—also occur; any of these forms may be referred to as vitamin B_6. The *coenzyme* form of B_6 is *pyridoxal-5-phosphate* (*codecarboxylase*); pyridoxamine-5-phosphate is involved in some reactions. The coenzyme form is important in amino acid metabolism e.g. reactions involving transamination, decarboxylation, racemization (interconversion of D-

and L-amino acids), the interconversion of serine and glycine, the formation of pyruvate from serine etc. Certain microorganisms require an exogenous source of pyridoxine; the derivatives are not necessarily equally effective in this respect. Dependent organisms include species of *Lactobacillus* and *Clostridium*, some yeasts, and the protozoans *Tetrahymena*, *Leishmania tarentolae* and *Crithidium fasciculata*.

pyriform (*adjective*) Pear-shaped.

pyrimethamine 2,4-diamino-5-(*p*-chlorophenyl)-6-ethylpyrimidine. A drug which inhibits dihydrofolate reductase—thus inhibiting FOLIC ACID metabolism. Its main use is in the suppression of *falciparum* MALARIA; some strains of *Plasmodium vivax* are sensitive. Many plasmodia readily develop resistance to it. Pyrimethamine has also been used in combination with other drugs (e.g. SULPHONAMIDES) in the treatment of toxoplasmosis and other diseases. (see also TRIMETHOPRIM)

pyrimidine nucleotides (biosynthesis of) (see Appendix V(b); see also NUCLEOTIDE)

pyrogen (*med.*) Any fever-inducing substance.

Pyronema A genus of fungi of the order PEZIZALES; species occur primarily on burnt or steam-sterilized soil. The fruiting body (apothecium)—which may be white, pale orange, or pale pink—is commonly 1–2 millimetres in diameter; the apothecium develops on a subiculum. The uninucleate, eguttulate, hyaline ascospores are borne in operculate asci.

Pyrrhophyta A synonym of DINOPHYTA.

Pythium A genus of fungi within the class OOMYCETES; *Pythium* spp include aquatic and terrestrial organisms, saprophytes and facultative parasites. Certain species (e.g. *P. debaryanum*, *P. aphanidermatum*) are common causal agents of DAMPING-OFF in the seedlings of various plants. The vegetative form of the organisms is a well-developed, coenocytic, branched mycelium; sporangia are borne on sporangiophores which typically differ little from somatic hyphae. The generalized life cycle of *P. debaryanum* (= *P. de Baryanum*) is as follows. A reniform zoospore (flagella arising from the concave surface) encysts and subsequently gives rise to a germ tube which penetrates the host plant; the fungus forms both intercellular and intracellular hyphae. Haustoria are not formed. Spherical or ovoid sporangia develop, in terminal or intercalary regions of the hyphae, within the tissues of the host plant; the sporangia are not deciduous—i.e. they germinate *in situ* (cf. e.g. *Albugo*). The germination of a sporangium involves the extrusion, from the sporangium, of a thin tube which terminates in a thin-walled, bubble-like vesicle; the multinucleate protoplasm within the sporangium passes, *via* the tube, to the vesicle where differentiation into zoospores takes place. Subsequently the vesicle bursts—releasing the zoospores. (The entire process of germination occurs within 15–30 minutes.) Zoospores escape from the host plant presumably as a result of tissue disintegration brought about by pectinolytic and other enzymes secreted by the pathogen. Sexual reproduction is initiated when a spherical oogonium and a smaller, elongate or club-shaped antheridium develop on the somatic hyphae; male and female gametangia frequently develop on the same hypha, and gametangia may develop in terminal or intercalary positions. (Most species of *Pythium* appear to be homothallic, but at least one species has been reported to be heterothallic.) A fertilization tube forms between antheridium and oogonium—gametangia normally developing in close proximity to one another; a functional male nucleus passes, *via* the tube, to the oogonium where fertilization and zygote-formation occur. The oosphere develops into a thick-walled oospore which, subsequently, germinates to form either a germ tube or a number of zoospores. (It is generally assumed that meiosis occurs within the oospore prior to germination; however, some reports claim that—in *P. debaryanum*, and also in certain species of *Phytophthora*—meiosis occurs in the gametangia.)

Q

Q fever (*query* fever) In man, an acute disease in which—following an incubation period of 2–3 weeks—there is a sudden onset with headache, malaise, fever, muscular pain, and (frequently) respiratory symptoms; there is no rash. The causal agent is *Coxiella burnetii.* Although complications may occur the disease is rarely fatal. Infection is believed to occur mainly by inhalation of contaminated dust, but may also follow the ingestion of infected milk; tick-borne infection in man is said to be rare. Tetracycline therapy has been used. Reservoirs of infection are found in populations of cattle, sheep, bandicoots etc. and in Ixodid and Argosid ticks. *Laboratory diagnosis.* Blood or sputum may be cultured (e.g. in the yolk sac, or guinea-pig) and/or serological tests carried out. The Weil–Felix test is not useful.

Quadrula A genus of testate amoebae which inhabit fresh water and certain mosses.

quadrulus (*ciliate protozool.*) A four-rowed ribbon of cilia which runs, in a spiral, down the gullet of *Paramecium.*

quantasomes (see CHLOROPLAST)

quarantine (1) The isolation of persons (or animals) who are suffering from an infectious disease—the object being to prevent transmission of the disease to others. (2) The isolation of persons or animals e.g. prior to entering a country, for the purpose of determining whether or not they are suffering from a particular infectious disease; the period of isolation should be equal to, or longer than, the incubation period of the suspected disease. The term may have derived from the Italian *quarantina* (= 40 days)—a period of quarantine formerly used for immigrants.

quaternary ammonium compounds (quats) (as antimicrobial agents) A group of cationic detergents which are widely used as ANTISEPTICS and DISINFECTANTS. Quats may be regarded as derivatives of ammonium chloride or ammonium bromide in which one of the hydrogen atoms has been replaced by a long hydrocarbon chain (e.g. C-8 to C-18) while the remaining three hydrogen atoms have been replaced by short-chain alkyl groups, phenyl groups, etc. Quats are bacteriostatic at low concentrations, bactericidal at higher concentrations; gram-positive bacteria are generally more susceptible than gram-negative species. (Quats are not active against e.g. *Mycobacterium tuberculosis*, *Pseudomonas* spp, or bacterial spores.) Quats apparently act by disrupting the bacterial CELL MEMBRANE with resulting increase in permeability of the membrane; at higher concentrations quats enter the cells and bring about denaturation of proteins and nucleic acids. Effective contact between quats and bacterial cells is facilitated by the negative charge normally present at the cell surface. The antibacterial action of quats may be inhibited by low pH, by the presence of certain organic compounds (e.g. phospholipids), soaps, *anionic* detergents, and by certain metal ions—e.g. Ca^{++}, Mg^{++}—which may compete for the anionic sites on the cell surface; quats are strongly adsorbed by many substances, e.g. agar, glass.

Quats are used as skin antiseptics e.g. for preoperative cleansing, surgical dressings, etc.; they are also used for the disinfection of food utensils, dairy equipment, etc.—being relatively non-toxic at the concentrations used. Examples of quats include: cetyltrimethylammonium bromide (proprietary names *CTAB, Cetrimide, Cetavlon*); alkyldimethylbenzylammonium chloride (proprietary names *Benzalkonium chloride*, *Zephiran*); β-phenoxyethyldimethyldodecylammonium bromide (proprietary name *Domiphen bromide*).

quats (see QUATERNARY AMMONIUM COMPOUNDS)

quellkörper (*mycol.*) *Within* the cleistothecium in certain ascomycetes: a mass of gelatinous cells contiguous to the apical part of the peridium; the quellkörper is involved in the rupture of the cleistothecial wall when the ascocarp reaches maturity.

quellung phenomenon (pneumococcus capsule swelling reaction) Apparent swelling and darkening of the capsule of *Streptococcus pneumoniae* on the addition of antibody specific to the capsular polysaccharide.

quinine An alkaloid obtained from the bark of the *Cinchona* tree; it is used in the treatment of MALARIA. Quinine is active against the *intraerythrocytic* stage of the malarial parasite (*Plasmodium* spp); it binds to the plasmodial DNA and is believed to inhibit nucleic acid biosynthesis in the parasite. Quinine is usually used in the form of its sulphate or hydrochloride. (see also AMINOQUINOLINES)

quinones (as antifungal agents) Certain quinones are used as agricultural antifungal agents; these include BENQUINOX, CHLORANIL, DICHLONE and DITHIANON. (see also JUGLONE)

quinones (in electron transport) (a) *Ubiquinones* (coenzymes *Q*). Derivatives of 2,3-dimethoxy-5-methyl-1,4-benzoquinone in which the 6-position carries a polyisoprene chain of

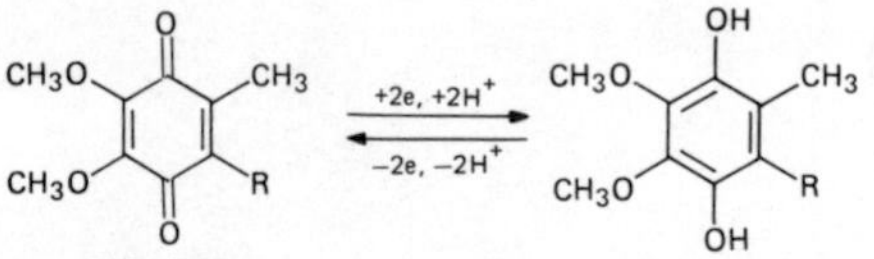

Ubiquinone

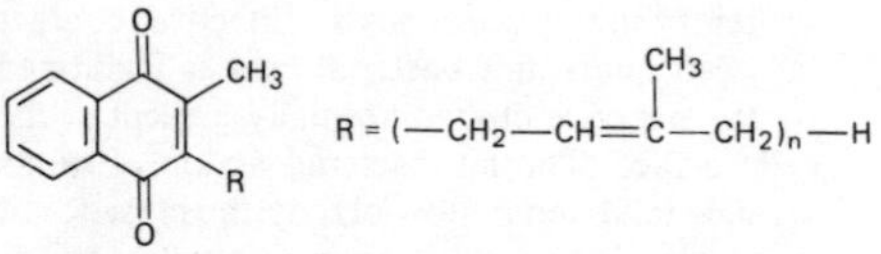

Vitamin K_2
(Menaquinone)

variable length. Ubiquinones are named according to the number (n) of isoprene units in the side chain—ubiquinone-n, UQ-n or ubiquinone$_n$. (Sometimes the name incorporates the number of *carbon atoms* in the side chain; thus, e.g. ubiquinone-50 (UQ-50) is the same compound as ubiquinone-10, coenzyme Q-10, UQ_{10} etc.) Ubiquinones can be reversibly reduced (see diagram) and function as redox agents in eucaryotic and many procaryotic ELECTRON TRANSPORT CHAINS. (b) *Naphthoquinones*. These include the vitamins *K*—derivatives of 2-methyl-1,4-naphthoquinone (menadione, or vitamin K_3) in which the 3-position is substituted. Vitamin K_1 (*phylloquinone*) has a mono-unsaturated phytyl group at the 3-position. Vitamins K_2 (*menaquinones*) have polyisoprene side chains at the 3-position. The vitamins K_2 are named according to the number of isoprene units in the side chain, e.g. menaquinone-6, or MK-6. (Sometimes the number of *carbon atoms* in the side chain is given, thus: vitamin $K_{2(30)}$.) Some bacteria (e.g. *Haemophilus parainfluenzae*) contain derivatives in which the 2-methyl group of vitamin K_2 is replaced by a hydrogen atom (2-demethylvitamins K_2). Naphthoquinones behave as redox agents e.g. in certain bacterial electron transport systems. (c) *Plastoquinones*. 2,3-Dimethylbenzoquinones involved in PHOTOSYNTHESIS.

quinsy Acute tonsillitis with concomitant peritonsillar abscess. Quinsy is always of bacterial causation; a common causal agent is *Streptococcus pyogenes*.

quintozene (PCNB) Pentachloronitrobenzene, an agricultural antifungal agent used mainly for the treatment of soil; it is effective against a number of soil and seed-borne diseases, e.g. DAMPING-OFF, various rots of bulbs. Quintozene is insoluble in water and is used as a dust; it is very stable and has low volatility—hence it is very persistent in the soil. (see also TECNAZENE and DICLORAN)

R

R antigen (see SMOOTH-ROUGH VARIATION (in bacteria) and STREPTOCOCCUS)

R factor (resistance factor; drug resistance factor) (*bacteriol.*) A PLASMID, the intracellular presence of which confers on a cell the property of resistance to certain ANTIBIOTICS (and/or to certain heavy metals—e.g. mercury). R factors occur in a wide range of bacteria—including members of the Enterobacteriaceae and species of *Vibrio*, *Pseudomonas*, and *Staphylococcus*; antibiotics to which resistance may be gained include AMINOGLYCOSIDE ANTIBIOTICS, CHLORAMPHENICOL, ERYTHROMYCIN, PENICILLINS, SULPHONAMIDES, TETRACYCLINES. Resistance may be due to the formation of an enzyme which inactivates the antibiotic (e.g. penicillinases, chloramphenicol acetylase, streptomycin adenylate synthetase) or to the alteration of permeability of the organism to the antibiotic (as in tetracycline resistance).

Some R factors are *non-transmissible* (e.g. the penicillinase plasmids of *Staphylococcus aureus*), but the majority are *transmissible* (see SEX FACTOR). The portion of a transmissible R factor which contains genes involved in CONJUGATION and gene transfer is referred to as the *resistance transfer factor* (RTF); genes coding for antibiotic resistance are called *resistance determinants*. The two parts of the plasmid (i.e. the RTF and resistance determinants) can separate and behave as autonomous plasmids. A number of resistance determinants may become linked to one another and to an RTF, so that resistance to many antibiotics may be carried on a single transmissible plasmid (*multiple drug resistance*). Recombination between an R factor and the bacterial chromosome is rare, probably due to the absence of homology between them.

The RTF is normally *self-repressed*. Certain repressed R factors can inhibit the expression of an F FACTOR present in the same cell; such R factors are called *fi*$^+$ (for fertility inhibition) and are thought to be related to the F factor. R factors which do not inhibit the expression of F are termed *fi*$^-$. (see also ANTIBIOTICS, RESISTANCE TO)

R factors (release factors) (see PROTEIN SYNTHESIS)

R mutant (see SMOOTH-ROUGH VARIATION (in bacteria))

rabies (hydrophobia) An acute, usually fatal disease of man, canines, felines, bats and other animals. The causal agent, the *rabies virus* (a RHABDOVIRUS), is commonly transmitted (in saliva) by the bite of a rabid animal. In *man*, the incubation period is often 2–8 weeks, but periods varying between 6 days and 1 year have been reported; short incubation periods are associated with bites on the face and neck, while longer periods follow bites on the legs and feet. Viral invasion takes a neural route to the central nervous system. Symptoms include irritability, spasm of the pharyngeal muscles on swallowing, and involuntary contractions of other muscles; convulsions and paralysis precede death. In the *dog*, rabies may begin with furtiveness and may be followed by irregular behaviour, violence, paralysis and death; violence and aggression are not invariably displayed. *Laboratory diagnosis* usually centres on the (suspect) rabid animal which, if captured, should (preferably) be allowed to succumb naturally to the disease; in this way, NEGRI BODIES (which are pathognomonic of rabies) are permitted to develop in the brain of the animal. Specific fluorescent staining techniques may be used on neural tissues. Vaccines are available (for man and domestic animals) for use under certain conditions—e.g. following unequivocal human exposure to rabies virus. Anti-rabies vaccines include (a) the SEMPLE VACCINE; (b) a strain of rabies virus grown in embryonated duck eggs and inactivated with β-propiolactone; (c) the live, attenuated LEP and HEP strains of the FLURY VIRUS (used for domestic animals). (Experimental vaccines include a non-attenuated strain of rabies virus (grown in WI-38 tissue culture) which is inactivated by β-propiolactone.) A rabies antiserum may be injected into the wound and/or may be used systemically to provide passive immunity. Rabies is a NOTIFIABLE DISEASE in the United Kingdom.

radioautography (see AUTORADIOGRAPHY)

radioimmunoassay A highly sensitive serological technique used for assaying specific antibodies or antigens. *Antibody assay*. Essentially, antibodies are allowed to react with an excess of their homologous, radioactively-labelled* antigens, and the immune complex is separated from any uncombined antigen molecules; the amount of labelled antigen in the immune complex is then determined by measurement of the radioactivity level of the immune complex. The amount of antibody is determined from a previously-prepared standard curve (radioactivity plotted against antibody concentration). *Antigen assay*. The proportion of a *fixed* amount of radioactively-labelled antigen which combines with a *fixed* amount of homologous antibody is governed by the amount of *un*labelled antigen which forms part of the reaction mixture; the greater the amount of unlabelled antigen the smaller will be the proportion of labelled antigen found in the immune complex—and the lower the level of radioactivity of the immune complex. Thus, it is possible to construct a standard curve from which can be obtained the amount of unlabelled antigen which corresponds to a given level of radioactivity in the immune complex. In assays for antigen, the preparation containing an unknown concentration of the antigen is used as the unlabelled antigen, and its concentration is obtained from the standard curve. *Often by iodination with ^{125}I. (cf. IMMUNORADIOMETRIC ASSAY)

radiolarians Protozoans of the subclass Radiolaria (class Actinopodea, superclass SARCODINA). Radiolarians are free-living organisms which occur almost exclusively in marine (salt water) habitats; some species are benthic, but most are pelagic (see ZOOPLANKTON). The organisms feed on small protozoans and microalgae. The typical radiolarian is a spherical cell; other types of cell are commonly obvious variations of the basic spherical form. The (many) pseudopodia which radiate from each cell are typically filopodia (see MOTILITY), though AXOPODIA are formed by some species. The cells of some species are naked, but most have silicaceous skeletal structures which may consist of e.g. an array of radiating spines, or a delicate internal latticework. A feature of the radiolarian cell (absent in HELIOZOANS) is the *central capsule*; this is a membranous, spherical capsule—concentric with the cell—which separates the ectoplasm from the endoplasm. The capsule is perforated by one, several, or many pores—according to species. One or more nuclei (depending on species) are located in the endoplasm. The peripheral region of the ectoplasm (the *calymma*) is highly vacuolated and is believed to contribute to cellular buoyancy. The ectoplasm may contain endosymbiotic algae (*zooxanthellae*). In size, radiolarian cells range from about 100 microns to several millimetres; certain species form aggregates or colonies which may be larger than 10 centimetres. Reproduction occurs by fission; small, flagellated cells—possibly equivalent to gametes—may indicate the existence of a sexual stage.

Fossil radiolarians are found in ocean depositis ('radiolarian ooze') and in certain sedimentary rocks. (see also FORAMINIFERIDA and DINOFLAGELLATES)

radiomimetic (*adjective*) Refers to those compounds whose effects (e.g. mutagenic) mimic those of ionizing radiations. Radiomimetic substances include NITROGEN MUSTARDS and other ALKYLATING AGENTS.

raffinose α-D-Galactopyranosyl-(1 → 6)-α-D-glucopyranosyl-(1 → 2)-β-D-fructofuranoside; a trisaccharide common in higher plants. Raffinose is metabolized by certain bacteria (e.g. *Klebsiella* spp) and is used as a substrate in biochemical indentification tests. Raffinose fermentation is also used as a criterion for distinguishing between top and bottom yeasts in BREWING.

Ramalina A genus of fruticose LICHENS in which the phycobiont is *Trebouxia*. The thallus is strap-shaped (but not dorsiventral) and erect or pendulous; apothecia are lecanorine. Species include e.g.:

R. farinacea. The thallus is greenish-grey, narrow (3 millimetres or less) and branching, and forms erect or pendulous tufts; whitish soralia occur along the margins. Apothecia are rarely formed. *R. farinacea* occurs mainly on trees and occasionally on rocks.

R. fastigiata. The thallus is greenish-grey with short, slightly swollen branches which form erect tufts 2–5 centimetres in height; the surface is pitted or wrinkled. The apothecia are borne on the ends of the branches; the discs are pinkish-buff. *R. fastigiata* occurs on trees in exposed situations.

R. fraxinea. The thallus is greenish-grey, broad and flat with few branches, and pendulous; the surface is wrinkled and pseudocyphellae are usually present on both sides. The apothecia are marginal or laminal and are borne on short stalks; the discs are pinkish-brown to brown. *R. fraxinea* occurs on trees.

Ramon titration (*immunol.*) A serological titration in which a constant volume of a given antigen preparation is added to each of a number of serial dilutions of antiserum. The end-point is indicated by the tube in which precipitation first occurs. The Ramon titration is often used for diphtheria toxoid-antitoxin titrations. (see also OPTIMAL PROPORTIONS)

raphe A slit or pore in the cell wall of a *diatom*. The raphe system is believed to be associated with MOTILITY. (see also BACILLARIOPHYTA)

Raphidiophrys pallida (see HELIOZOANS)

rapid lysis mutants Bacterial mutants which do not exhibit LYSIS INHIBITION; rapid lysis (*r*) mutants form PLAQUES which are larger and which have a more clearly-defined periphery than those formed by wild type (r^+) strains.

rat-bite fever (sodoku) A human disease which may be contracted from the bite of a rat or (apparently) from the consumption of contaminated milk; the causal agent may be either *Streptobacillus moniliformis* or *Spirillum minor* (= *S. minus*)—both of which are commonly present in the rat's mouth. The disease, which may be fatal, involves recurrent fever and skin lesions; arthritic symptoms characterize the streptobacillary disease.

rate zonal centrifugation (see CENTRIFUGATION)

razor-strop fungus A common name for the basidiomycete *Piptoporus betulinus*.

RBC Red blood cell.

RCF Relative centrifugal field: see CENTRIFUGATION.

RDE (see NEURAMINIDASE)

reaction of identity (*immunol.*) (see OUCHTERLONY TEST)

reaction of non-identity (*immunol.*) (see OUCHTERLONY TEST)

reaction of partial identity (*immunol.*) (see OUCHTERLONY TEST)

reactive lysis (*immunol.*) The lysis of an *un*sensitized cell (e.g. an erythrocyte—*reactive haemolysis*) as a result of the attachment, to its surface, of the COMPLEMENT complex C56—which has been formed elsewhere (see COMPLEMENT-FIXATION); lysis follows the formation of the complex C56789.

reagin (see REAGINIC ANTIBODY)

reaginic antibody (reagin; homocytotropic antibody) (1) An antibody which is elicited by an ALLERGEN and which is able to bind to the surface of MAST CELLS; such antibodies are principally of the IgE type. The combination of cell-bound reaginic antibodies with homologous antigens causes *degranulation* of the mast cells; this occurs in ANAPHYLACTIC SHOCK. (2) The term *reagin* is also used to refer to the complement-fixing antibody of the WASSERMANN REACTION.

receptacle (*mycol.*) In fungi of the order PHALLALES: that structure, present in the immature fruiting body, which develops to form the *pseudostem*.

receptive hyphae (see FLEXUOUS HYPHAE)

receptor destroying enzyme (see NEURAMINIDASE)

receptor sites (of erythrocytes in viral HAEMAGGLUTINATION) (see NEURAMINIC ACIDS and NEURAMINIDASES)

recessive gene (see DOMINANT GENE)

recipient (*bacteriol.*) Any cell which receives genetic information from another (*donor*) cell—see CONJUGATION; TRANSDUCTION; TRANSFORMATION.

reciprocal recombination (see RECOMBINATION)

recombinant (genetics) (1) (*noun*) Any organism whose genotype has arisen as a result of RECOMBINATION. (2) (*adjective*) Refers to any chromatid, nucleic acid etc. which has arisen as a result of RECOMBINATION.

recombination (genetics) (a) *In eucaryotes.* Recombination may refer to: (i) The exchange and/or transfer of genetic information between homologous CHROMATIDS—resulting in the formation of allele combination(s) not found in either parental chromatid; recombination occurs during MEIOSIS and, occasionally, during mitosis—see PARASEXUALITY IN FUNGI. (ii) The re-assortment of unlinked alleles during meiosis.

Reciprocal recombination. This occurs as a result of CROSSING-OVER in which a *symmetrical* exchange of genetic material takes place—i.e. genes lost by one chromatid are gained by the other, and *vice versa.* The molecular mechanism of recombination is unknown; models which have been proposed are generally modifications of the *breakage and reunion model.* The original model proposed that, during crossing-over, two homologous chromatids break at identical sites; the broken ends of each chromatid were supposed to combine with the corresponding ends of the homologous chromatid. (cf. COPY-CHOICE MODEL) Currently, the DNA duplex of each chromatid is believed to undergo two *single-stranded* nicks at different locations; crossing-over is believed to involve base-pairing between complementary strands of DNA from each of the homologous chromatids. This process results in the formation of short regions (one region per chromatid) of *hybrid* DNA—i.e. DNA which contains one strand from each parental duplex. One current model of meiotic recombination is described in the diagram (page 342). Alleles *within* the region of hybrid DNA segregate during the first mitotic division after meiosis (*post-meiotic segregation*)—see also below.

Non-reciprocal recombination (*gene conversion*). This involves the elimination of one allele and its replacement by another. Gene conversion is believed to be a consequence of incomplete complementarity between the two strands in a region of hybrid DNA (see above and diagram). Nucleotides which cannot form normal base-pairs give rise to distortions in the duplex; such distortions may be recognized by DNA repair enzymes—with resultant excision of one (either) strand of the hybrid DNA. The gap left by the excision could be filled by DNA synthesis—using as template the remaining strand of the hybrid region. For example, the hybrid region may contain two alleles of a given gene: Z and z; following conversion ('correction') these may become ZZ or zz. Ratios of e.g. 3Z:1z or 1Z:3z can be obtained on segregation if *both* hybrid regions (see diagram) are corrected; if only one of the hybrid regions is corrected then ratios of 5Z:3z or 3Z:5z will be obtained at *post-meiotic segregation*.

(b) In *procaryotes.* In bacteria, recombination generally involves interaction—presumably crossing-over—between an intact genome and a genome fragment, and the result is a single

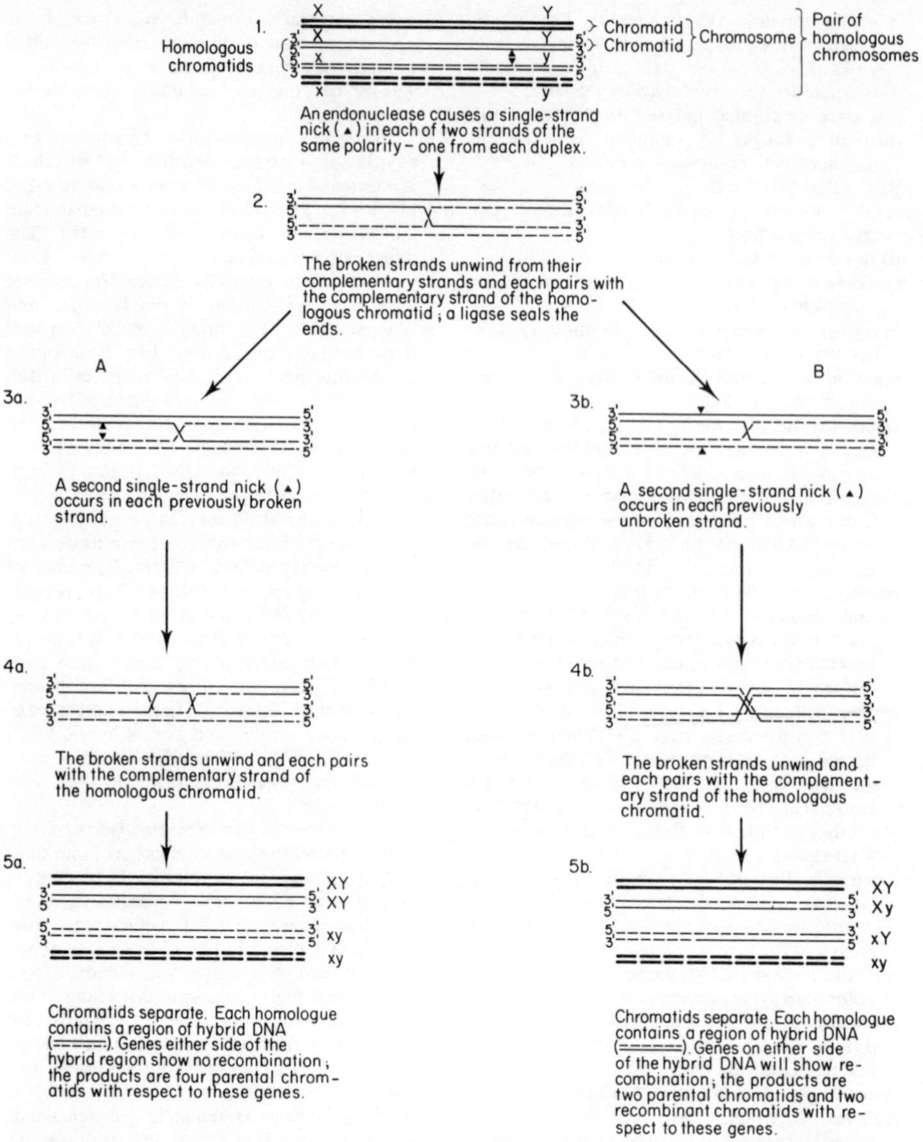

RECOMBINATION. *A model to show a possible mechanism for recombination in eucaryotic cells.* (Each chromatid is assumed to contain a single DNA duplex. X, x, Y, and y are genetic markers)

recombinant cell; opportunities for recombination occur in CONJUGATION, TRANSDUCTION and TRANSFORMATION. Some authors use the term *recombination* to refer to the insertion of a PLASMID or phage genome (see LYSOGENY) into a bacterial chromosome.

recombination frequency (genetics) The average frequency with which two given alleles are separated by RECOMBINATION; it is the number of recombinant progeny (with respect to the alleles considered) expressed as a percentage of the total progeny. A recombination frequency (RF) of the order of 50% indicates that the alleles segregate independently among the progeny—i.e. they occur on different chromosomes; frequencies of a lower order indicate that the alleles occur on the same chromosome. (see also INTERFERENCE)

recombination-repair (postreplication repair) A mechanism for the repair of DNA (*q.v.*) which has been damaged e.g. by ULTRAVIOLET RADIATION. Recombination-repair effectively 'dilutes out' the damaged regions of DNA. During replication of the damaged strand (A), large gaps are left in the daughter strand (B) opposite the sites of damage. It has been proposed that these gaps are filled as a result of recombination between strand B and an undamaged, homologous region of another strand (C) (e.g. the parent strand which is complementary to the damaged template A). The resulting gap in strand C can then be filled by a repair polymerase—using its complementary daughter strand (D) as template.

Rectus Flexibilis In streptomycetes, refers to the *form* of a spore chain, or chains: straight, flexuous, or fascicled (bundles of straight or flexuous chains).

red algae (see RHODOPHYTA)

red drop (in green plant photosynthesis) An abrupt decrease in the rate of PHOTOSYNTHESIS as the wavelength of incident monochromatic light increases beyond 680 nm; the rate is virtually zero at about 700 nm. Light of these wavelengths cannot be absorbed by pigment system II.

red rust (of tea) A disease of the tea plant (*Camellia sinensis*—also known as *Thea sinensis*) which occurs in India and Sri Lanka; the causal agent is a species of *Cephaleuros*—an alga of the division CHLOROPHYTA. Symptoms—which occur on stem and leaves—include the development of elevated regions which subsequently become red; affected plants lose leaves and develop a sickly appearance. The disease may be controlled by the use of copper sprays. (see also BLISTER BLIGHT)

red tide A particular type of water BLOOM in which regions of the sea (or an estuary) become red owing to the presence of immense numbers of certain types of plankton, e.g. the dinoflagellate *Gonyaulax*. A red tide typically causes high mortality rates among the fish and shellfish in the affected waters—due to the production of toxins by the dinoflagellates; these toxins (in sufficient quantity) can cause sickness or death in man.

red-water fever (cattle tick fever) An acute, sometimes fatal disease of cattle and other animals; the causal agent(s) are species of BABESIA. The disease is transmitted by blood-sucking *ixodid* ticks (e.g. species of *Boophilus*, *Rhipicephalus*) either transovarially (in one-host ticks) or by the various stages of the tick when these transfer from one host animal to another. The incubation period is about 4–14 days. *Babesia* multiplies within, and destroys, the host's erythrocytes. Symptoms include an initial high fever, haemoglobin in the urine (hence 'red-water fever'), diarrhoea and anaemia. Immunity, following recovery, is based largely on PREMUNITION. *Treatment.* Useful drugs include trypan blue (against *B. bigemina*) and acriflavine. *Control.* Animals are kept tick-free by the use of acaricides. The disease occurs in Europe, Central and South America, Africa and Australia.

redox potential (oxidation-reduction potential; E_h or E) A measure of the tendency of a given oxidation-reduction system (i.e. *redox system*) to donate electrons (i.e. to behave as a reducing agent) or to accept electrons (i.e. to behave as an oxidizing agent). The E_h of a redox system may be determined by measuring the electrical potential difference—in volts (V) or millivolts (mV)—between the given system and a standard redox system (the hydrogen electrode) whose potential is, by convention, arbitrarily taken to be zero volts.

Redox systems. One example is the Fe^{2+}/Fe^{3+} system (or 'iron electrode'). If a solution contains *ferric* (but no ferrous) ions, the system is totally oxidized and has no reducing capacity—but maximum oxidizing capacity; such a solution exhibits the highest E_h of the Fe^{2+}/Fe^{3+} redox system: approximately +0.9 V. A solution containing only *ferrous* ions is totally reduced and has no oxidizing capacity—but maximum reducing capacity; the corresponding E_h is the lowest possible for this system: approximately +0.6 V. A solution containing ferric *and* ferrous ions can accept electrons—some (or all) of the Fe^{3+} ions being *reduced* to the Fe^{2+} state and the E_h of the system falling to a less positive value; alternatively, the system can donate electrons—some (or all) of the Fe^{2+} ions being *oxidized* to the Fe^{3+} state, and the E_h of the system becoming more positive. A solution containing ferric and ferrous ions exhibits an E_h whose value (between +0.6 V and +0.9 V) is dependent on the ratio of *activities* (i.e. effective concentrations) of the two species of ion—i.e. on the ratio Fe^{2+}:Fe^{3+}. The ratio reduced:oxidized species in a given redox system can be controlled by electrical means, or by another, dominant, redox system (a *poising* system) which can maintain a given electrical potential in the presence of the first redox system; that is, the ratio oxidized: reduced species adjusts so as to conform to a suitable, externally-imposed potential.

Standard redox potential (*half-reduction potential*; *mid-point potential*; E_0—sometimes E^0 or E_m). In any redox system: when the ratio of the activities of the reduced and oxidized species is 1:1 the E_h is referred to as the *standard redox potential*, E_0. In many redox systems the value of E_0 varies with the pH; E_0 determined at pH 7 is usually written E_0'. (Certain cytochromes appear to exhibit energy-dependent variability in their half-reduction potentials—see ELECTRON TRANSPORT CHAIN.)

E_0 values can be used to predict possible interactions between redox systems. Thus, e.g. a given system may oxidize a second of lower (less positive, or more negative) E_0. The E_0 can also be used to calculate the E_h under various conditions:

$$E_h = E_0 - \frac{RT}{nF} \log_e \frac{\text{activity of reduced species}}{\text{activity of oxidized species}}$$

(where R is the gas constant, 8.314 joules $mole^{-1}$ degree abs^{-1}; T is the absolute temperature; F is the Faraday, 96,490 coulombs $gram\text{-}equivalent^{-1}$; n is the number of electrons transferred to/from each atom/molecule/ion of the system—e.g. 1 in the Fe^{2+}/Fe^{3+} system.)

Redox couple refers either to (i) a *half-cell* (i.e. *redox system* as used above), or (ii) a pair of linked redox systems capable of mutual oxidation-reduction.

Measurement of E_h (a) *Potentiometric*. The potential difference between a given system and a hydrogen electrode (or other standard electrode) is measured by a zero-current method. The hydrogen electrode consists of a suitably-prepared platinum plate immersed in 1.228 N hydrochloric acid (activity of $H^+ = 1$ gram ion/litre); when a stream of hydrogen gas (at 760 mm Hg pressure) impinges on the plate, the following equilibrium is set up:

$$H \rightleftharpoons H^+ + \text{electron}$$

and the potential difference between plate and acid is taken, arbitrarily, to be zero volts. (b) *Redox indicators*. Certain dyes (e.g. METHYLENE BLUE, litmus, resazurin) are coloured when oxidized, colourless when reduced; such dyes become colourless in solutions which are at or below characteristic values of E_h—e.g. in the region −20 to −40 millivolts for methylene blue at pH 7.

E_h measurements of microbiological media, cultures etc. Since a medium or culture may contain a large number of different redox systems which may or may not reach a state of true equilibrium, the E_h *measured* does not necessarily indicate the redox state of individual component systems—which may nevertheless be of importance to the organism being studied. However, E_h measurements may be useful; thus, e.g. certain ANAEROBES grow only on media at or below certain values of E_h, and some organisms are characterized by their ability to reduce, to a particular E_h, media in which they are grown. Media for the culture of anaerobes are often *poised* at a certain E_h; poising is analogous to buffering in the context of pH—i.e. while the E_h produced depends on the ratio reduced:oxidized poising agent, the stability of E_h is governed by the absolute amount of poising agent in the medium. Poising agents include e.g. ascorbic acid, cysteine hydrochloride, and thioglycollate.

reduction division A synonym of MEIOSIS.

reductive carboxylic acid cycle (see PHOTOSYNTHESIS)

reductive pentose cycle (see CALVIN CYCLE)

Reed and Muench viral titration (see END POINT (*virol.*))

regulator gene (see OPERON)

reindeer moss A common name for the lichen *Cladonia rangiferina*—see CLADONIA.

Reiter antigens Antigens found in the Reiter treponeme and in many other non-pathogenic and pathogenic treponemes, including *Treponema pallidum*; they are called *group treponemal antigens* since they are common to a range of treponemes. In the FTA-ABS test for SYPHILIS, antibodies to group antigens are removed (absorbed) from the patient's serum by incubating it with extracts of Reiter treponemes; such absorption increases the specificity of the test, since those antibodies which remain in the absorbed serum are specific to the pathogenic treponemes—*T. pallidum*, *T. carateum* and *T. pertenue*. (Serological tests—including the FTA-ABS and TPI—cannot distinguish between the three species of pathogenic treponemes, i.e. a test positive for one species will be positive for the other two.)

Reiter treponeme A strain of *Treponema phagedenis*.

relapsing fever A disease (of man) characterized by recurrent fever; the causal agent, a species of *Borrelia*, is transmitted by ticks (*Ornithodoros* spp) or (contaminatively) by lice (*Pediculus* sp). In non-fatal cases the fever becomes progressively less severe, and finally ceases. Treatment with benzylPENICILLIN is commonly effective. In epidemics mortality rates may be high; at *post mortem* lesions may be seen in the liver, spleen and intestinal tract. *Laboratory diagnosis*: blood or urine, taken during febrile periods, is examined for spirochaetes; serology is of little value.

relaxed control (of bacterial rRNA synthesis) (see STRINGENT CONTROL)

relaxed control (of plasmid replication) (see PLASMID)

reniform (*adjective*) Kidney-shaped.

Reovirus ('respiratory enteric orphan virus') A genus of VIRUSES whose hosts include a range of *vertebrates*—e.g. man, other primates, avians, and canines; reoviruses occur in the enteric and respiratory tracts but are not commonly associated with disease—i.e. as agents of disease they are 'orphans'. The *Reovirus* is a non-enveloped icosahedral virion, approximately 80 nm in diameter, which contains a second (inner) capsid and a genome consisting of linear double-stranded RNA; the genome occurs, within the double capsid, in the form of ten separate RNA fragments. An RNA polymerase forms part of the virion; reoviruses are referable to class III of the BALTIMORE CLASSIFICATION. Replication occurs within the *cytoplasm* of the host cell. Certain plant-pathogenic viruses—e.g. the WOUND TUMOUR VIRUS—appear to be related to the reoviruses. (cf. ORBIVIRUS)

replica plating A technique, devised by Lederberg and Lederberg (1952), by means of which various types of mutant can be isolated from a population of bacteria grown under *non-selective* conditions. Replica plating has been used e.g. to demonstrate the presence of streptomycin-resistant mutants in cultures of streptomycin-sensitive bacteria which had never been exposed to streptomycin; such experiments (which involved *indirect selection*) showed that variant cells arose *spontaneously* in bacterial populations—in the absence of selective agents or environments—and thus reinforced the concept of mutational (rather than adaptive) variation in bacteria.

For the detection of e.g. phage-resistant mutants in a population of phage-sensitive bacteria, a *master plate* is prepared by inoculating the entire surface of a non-selective agar medium with a culture of sensitive cells. Following incubation, the surface of the master plate is pressed lightly onto a disc of sterile velvet which is attached to the top of a cylindrical wooden block. The disc, to which has adhered growth from the master plate, is then used to inoculate one or more plates of medium (the *replica plates*) which have been made selective by treatment with a suspension of phage to which only mutant cells are resistant; the replica plates are then incubated, and the mutant (phage-resistant) cells form colonies. A record is made of the orientation of the disc with respect to the master plate (during contact) and of the disc with respect to the replica plate (during inoculation); thus, the inoculum for a particular colony (of phage-resistant cells) which develops on a replica plate can be traced back to a particular small area of growth on the master plate. The region of agar which contains the mutant cells can then be cut out and a suspension made of the bacteria on its surface. The suspension, which contains phage-resistant mutant cells, is diluted and used to inoculate a second master plate. Replica plating is repeated (with a second, third....master plate) until isolated colonies of phage-resistant cells are obtained on a master plate. By such means it is possible to obtain a pure culture of phage-resistant cells; at no time had such cells—or their progenitors—been exposed to the selective agent (phage).

replicative form (RF) A double-stranded nucleic acid formed from a single-stranded viral genome during the process of replication—see e.g. BACTERIOPHAGE ϕX 174.

replicon (genetics) A nucleic acid molecule which possesses an *origin* (see DNA) and which is therefore capable of initiating its own replication. The DNA of a eucaryotic chromosome consists of many replicons, while the DNA of a bacterial chromosome, PLASMID etc. corresponds to a single replicon.

repression (of enzymes) (see OPERON)

repressor (see OPERON and LYSOGENY)

resazurin test A test used to examine the microbiological quality of MILK; it resembles the METHYLENE BLUE TEST both in principle and method. In this test the indicator is *resazurin*, a member of the quinone-imine group. Resazurin is blue when fully oxidized; when the REDOX POTENTIAL is lowered sufficiently the compound becomes irreversibly reduced to the pink *resorufin*. On further reduction the colourless *hydroresorufin* is produced.

residual body (see LYSOSOME)

resistance (of microorganisms to antibiotics) (see ANTIBIOTICS, RESISTANCE TO)

resistance (of plants to disease) Most plants are susceptible only to a relatively small number of pathogens, i.e. they are naturally resistant to the vast majority of potentially pathogenic microorganisms. When a species is susceptible to a given pathogen different varieties of that species may exhibit different degrees of resistance to that pathogen; varieties which are highly resistant (to one or more pathogens)—and which combine such resistance with other desirable features (e.g. high yields etc.)—are selected by plant-breeders for cultivation.

Features of constitutive disease-resistance in plants include: (a) The presence of physical barriers to infection—e.g. a waxy cuticle. (b) The formation of antimicrobial exudates which may e.g. inhibit spore germination on plant surfaces. (c) The presence of antimicrobial substances within plant tissues (see PHYTONCIDES). (d) The existence of physical and/or chemical conditions within plant tissues which are hostile to microbial incursion—e.g. the absence of specific microbial growth factors, suitable substrate levels or a suitable pH. Features of induced disease-resistance, i.e. resistance which develops following infection, may include: (a) The formation, by the plant, of cork or gum barriers which seal off the invaded tissue(s)—see also TYLOSES. (b) HYPERSENSITIVITY reactions. (c) The production of PHYTOALEXINS. (d) The activation of polyphenol oxidases—which catalyse the oxidation of phenols to toxic quinones and higher polymers.

In some host-pathogen confrontations (e.g. that between flax and *Melampsora lini*) a so-called *gene-for-gene* relationship has been observed: for each host gene concerned with resistance there is a corresponding gene, in the pathogen, concerned with virulence. A host, resistant to a given strain of pathogen, may become susceptible to that strain following a mutational or recombinational event in the pathogen. Host resistance may be described as *monogenic*, *oligogenic* or *polygenic*—depending on whether there is one, several or many host genes demonstrably concerned with resistance to a given pathogen. In terms of host protection, monogenic resistance is not necessarily inferior to oligo- or polygenic resistance. (see also PLANT PATHOGENS AND DISEASES and VERTICAL RESISTANCE)

resistance transfer factor (RTF) (1) That part of a *transmissible* R FACTOR containing the genes involved in CONJUGATION and gene transfer; thus, the RTF and the *resistance determinants* constitute the R factor. (2) RTF is sometimes used as a synonym of R factor.

resolving power (limit of resolution) (of a light microscope) Resolving power refers to the fineness of detail observable in the image of the specimen; the smaller the resolving power the finer the detail observable. (cf. MAGNIFICATION) The resolving power of a microscope depends largely on the characteristics of its objective lens—although optimal illumination of the specimen is necessary in order to exploit the maximum potential of the lens. The factors which govern resolving power are related by the expression*:

$$d_{min} \propto \frac{\lambda}{\mu \sin \theta}$$

in which d_{min} is the resolving power, i.e. the smallest distance separating two resolvable details of the specimen; λ is the wavelength of light used to form the image; μ is the refractive index of the medium between objective lens and specimen; θ is *half* the apical angle of the widest cone of light which can enter the objective lens from the specimen plane.

The quantity $\mu \sin \theta$ is termed the *numerical aperture* (N.A.) of the objective lens; from the above expression, the higher the N.A. of an objective lens the finer the detail which can be resolved. The highest resolution (and magnification) obtainable with an ordinary light microscope requires the use of a high-power *oil-immersion* objective lens. When such a lens is in its working position a drop of IMMERSION OIL forms a liquid bridge between lens and smear or lens and cover-slip—a distance of approximately 0.1 mm; the replacement of air ($\mu = 1$) by immersion oil (μ = about 1.5) enables the lens to receive a larger cone of light from the specimen and thus to exploit its maximum numerical aperture. (With *water-immersion* objectives the space between lens and specimen is filled with distilled water, $\mu = 1.3$.)

The limit of resolution obtainable with the best light microscope is about 2500Å; with *ultraviolet* microscopy the resolving power is about half this figure. With *electron microscopy* the resolving power can be as small as 2–3Å. (see also MICROSCOPY)

*This expression is derived by considering the use of an obliquely-illuminated diffraction grating as the specimen. The slits of the grating can be discerned (resolved) when both the zero and first order diffraction images are received by the objective lens.

respiration A mode of energy-yielding metabolism in which *oxygen* is the terminal electron acceptor for substrate oxidation; in certain bacteria *inorganic compounds* (e.g. nitrate, sulphate) act as terminal electron acceptors for respiratory metabolism under anaerobic conditions—a process known as ANAEROBIC RESPIRATION. (cf. FERMENTATION) The TRICARBOXYLIC ACID CYCLE plays a central role in the respiratory metabolism of the majority of microorganisms capable of respiration; electrons removed from substrates during e.g. the TCA cycle are transferred to oxygen (or to another inorganic electron acceptor) *via* a sequence of oxidation-reduction reactions involving a system of electron carriers—the ELECTRON TRANSPORT CHAIN. ATP is obtained during respiration largely by OXIDATIVE PHOSPHORYLATION associated with this electron transfer. (Some SUBSTRATE LEVEL PHOSPHORYLATION may also occur e.g. during the EMBDEN–MEYERHOF–PARNAS PATHWAY.) (cf. FERMENTATION)

Some of the microorganisms which carry out respiration under aerobic conditions can switch to a fermentative mode of metabolism under anaerobic conditions.

respiratory chain (see ELECTRON TRANSPORT CHAIN)

respiratory control The condition in which the rate of respiration (i.e. electron flow along the ELECTRON TRANSPORT CHAIN) is controlled by OXIDATIVE PHOSPHORYLATION (i.e. the reaction ADP + Pi $\rightleftharpoons$ ATP + H_2O). Electron flow is increased in the presence of excess ADP and Pi (the *active state* or *state 3*) but decreased (or inhibited) by lack of ADP or Pi or by the presence of excess ATP (the *controlled state* or *state 4*). UNCOUPLING AGENTS remove this control—allowing electron transport to proceed in the absence of oxidative phosphorylation. OLIGOMYCIN inhibits electron transport by preventing ATP formation.

respiratory inhibitors Substances which inhibit the flow of electrons along the ELECTRON TRANSPORT CHAIN—e.g. AMYTAL, ANTIMYCIN A, CARBON MONOXIDE, CYANIDE, HYDROXYQUINOLINE-N-OXIDES, OLIGOMYCIN, PIERICIDIN A, ROTENONE. (cf. UNCOUPLING AGENTS)

respiratory syncytial virus (RSV) A virus which is generally classified in the genus PARAMYXOVIRUS on the basis of the structure, size and general properties of the virion, the types of disease caused, and the ability to promote cell fusion in tissue culture; in man, RSV typically gives rise to localized diseases of the respiratory tract—and is an important pathogen of the lower respiratory tract in infants and young children. RSV does not promote haemagglutination or haemolysis, and does not exhibit neuraminidase activity; the virus does not cross-react serologically with any member (or tentative member) of the genus *Paramyxovirus*.

respiratory tract, common microflora of The human nasal passages usually harbour *Corynebacterium* spp (diphtheroids) and *Staphylococcus albus* (= *Staph. epidermidis*); *Staph. aureus* is commonly present in a proportion of individuals. In the nasopharynx the predominating organisms are usually viridans streptococci. Additionally, certain OPPORTUNIST PATHOGENS may be isolated from the nasopharyngeal region of a significant proportion of healthy individuals; these organisms include *Haemophilus influenzae*, *Streptococcus pneumoniae* and species of *Neisseria*. The lower respiratory tract, i.e. the trachea, bronchi and lungs, normally contains a less varied and less numerous flora; the lung tissues are generally believed to be almost free of microorganisms in normal circumstances. (see also MOUTH, COMMON MICROFLORA OF)

resting spore (winter spore) (*mycol.*) Any thick-walled spore (particularly one derived by a sexual process) which germinates only after an extended period of dormancy. (see also SUMMER SPORES)

restricted transduction (see TRANSDUCTION)

restriction (of DNA) A process in which specific endonucleases cleave double-stranded DNA in which *both* strands lack appropriate MODIFICATION. Each restriction endonuclease recognizes a specific nucleotide sequence (the target sequence) in the DNA. Those target sequences which have been identified exhibit a two-fold rotational symmetry, e.g. an endonuclease obtained from *Haemophilus influenzae* recognizes the sequence:

```
                      ↓
5′..........G-T-Py-Pu-A-C..........3′
3′..........C-A-Pu-Py-T-G..........5′
                      ↑
```

(Pu = purine; Py = pyrimidine.) The endonuclease cleaves both strands at a point indicated by the arrows, resulting in large fragments of DNA which subsequently can be degraded by other nucleases. Restriction is assumed to protect the cell from foreign DNA.

restrictive condition, host etc. (see CONDITIONAL LETHAL MUTANT)

resupinate (*mycol.*) (*adjective*) Of a fruiting body: resembling a crust or scab on the surface of the substratum, the spore-bearing surface being outermost. Resupinate fructifications are formed e.g. by *Poria*.

reticulopodium A reticulate *pseudopodium*; see MOTILITY.

Retinaculum Apertum (see SPIRA)

Retortamonadida An order of parasitic protozoans of the class ZOOMASTIGOPHOREA. Depending on species, each organism may have 2, 3, or 4 anteriorly-arising flagella; one flagellum is directed posteriorly along the ventral surface of the cell. The cells have no axostyle or undulating membrane, and do not exhibit bilateral symmetry. (see CHILOMASTIX)

retting An industrial process in which the stems of flax, hemp or jute plants are attacked by PECTINolytic microorganisms—thus releasing the fibres of CELLULOSE (or cellulose + LIGNIN) for use in the manufacture of linen, sacks etc. Certain species of *Clostridium* are used in anaerobic retting; in aerobic ('dew') retting certain fungi are utilized, e.g. species of *Cladosporium*, *Mucor* and *Rhizopus*.

reverse mutation (see BACK MUTATION)

reverse transcriptase RNA-dependent DNA polymerase—an enzyme found in the LEUKOVIRUSES. The enzyme synthesizes DNA on an *RNA* template in a manner analogous to that of bacterial DNA polymerases (see DNA)—i.e. the enzyme adds nucleotides to the 3′-OH ends of short RNA primers; the product appears to be an RNA-DNA hybrid. A ribonuclease (RNase H), which is closely associated with the reverse transcriptase, is believed to remove the RNA strand from the hybrid product. Subsequently, double-stranded DNA is formed.

The formation of a DNA intermediate appears to be necessary for the replication of leukoviruses; in the PROVIRUS HYPOTHESIS the biological role of the reverse transcriptase is postulated to be the formation of a DNA *provirus*. The enzyme can be inhibited by certain RIFAMYCINS.

reversed anaphylaxis (see TYPE 2 REACTION)

revertant A mutant organism which has undergone a second mutation which partially or totally reverses the effects of the first. (see BACK MUTATION and SUPPRESSOR MUTATION)

Rhabdophorina The larger of the two suborders of the GYMNOSTOMATIDA (subphylum CILIOPHORA). The cytostome is either apical (tribe Prostomata) or lateral (tribe Pleurostomata), and in many species the oral region has an armament of toxic TRICHOCYSTS. The rhabdophorine gymnostomes are typically carnivorous. Genera include *Actinobolina*, *Coleps*, *Didinium*, *Dileptus*, *Prorodon*, *Stephanopogon* and *Trachelophyllum*.

Rhabdovirus A genus of VIRUSES whose hosts include vertebrates, invertebrates, and plants; rhabdoviruses include the causal agents of e.g. RABIES, VESICULAR STOMATITIS, and lettuce necrotic yellows. The *Rhabdovirus* virion consists of a helical NUCLEOCAPSID (containing single-stranded RNA) enclosed within a lipoprotein (ether-sensitive) ENVELOPE; an RNA-dependent RNA polymerase is associated with the nucleocapsid. Animal rhabdoviruses are bullet-shaped—i.e. cylindrical, with one planar and one hemispherical end, while some plant rhabdoviruses (e.g. the eggplant mottled dwarf virus) are hemispherical at both ends; in size, the rhabdoviruses are generally 170–250 × 60–80 nm. The (glycoprotein) peplomers of the rabies virus promote haemagglutination (e.g. with goose erythrocytes at 4 °C). Most rhabdoviruses replicate entirely within the cytoplasm of the host cell and mature by budding through the cytoplasmic membrane.

rhamnose (6-deoxy-L-mannose) A sugar which occurs, for example, in many of the streptococcal C SUBSTANCES and in some LIPOPOLYSACCHARIDES (e.g. the O-specific chains of certain *Salmonella* serotypes).

rhapidosomes Rod-shaped or tubular structures which occur intracellularly in e.g. *Pseudomonas* spp; the structures may be released during growth. Rhapidosomes (in different organisms) have been variously regarded as bacteriocins, defective phage particles, and particles of mesosomal membrane.

rheumatoid factor An antibody (generally IgM) commonly present in the sera of patients with rheumatoid arthritis; it combines with slightly denatured human IgG and with *sensitized* sheep erythrocytes. The rheumatoid factor may act as an autoantibody but its role (if any) in the aetiology of rheumatoid arthritis is unknown. (see also ROSE-WAALER TEST)

Rhinovirus A genus of VIRUSES of the family PICORNAVIRIDAE; in size and morphology the rhinoviruses resemble the enteroviruses (*q.v.*) but they differ from them e.g. in that they are intolerant of low pH—being inactivated at pH 3–5. Rhinoviruses occur in the upper respiratory tract in man and other animals; approximately 90 serotypes have been isolated from man—in whom rhinoviruses are responsible for a proportion of cases of the COMMON COLD. The causal agent of FOOT AND MOUTH DISEASE is placed in this genus by some authors; others place it in an *ad hoc* genus: *Aphthovirus*. The virion of the foot and mouth disease virus (FMDV) is 25 nm in diameter; it is labile below pH 6 and above pH 10 under most conditions. The virion is appreciably resistant to desiccation, and its general stability is enhanced by increased ionic concentration and by low temperatures (e.g. 4 °C). Certain strains of FMDV have been reported to survive exposure to temperature/time combinations used in PASTEURIZATION.

Rhizidiomyces A genus of fungi of the class HYPHOCHYTRIDIOMYCETES.

rhizina (*pl.* rhizinae; rhizine, *pl.* rhizines) (*lichenol.*) A root-like structure consisting of a bundle of hyphae—a number of which arise from the lower cortex of the thallus in certain lichens; rhizinae may be unbranched or branched—branching being either 'squarrose' (in which short branchlets emerge from the rhizina at right angles) or dichotomous. Rhizinae serve to anchor the thallus to the substratum; it is uncertain to what extent—if any—the rhizinae function in the uptake of nutrients by the lichen. Rhizinae occur e.g. in *Anaptychia fusca*, *Parmelia saxatilis*, *Peltigera canina*, and *Xanthoria* spp.

Rhizobiaceae A family of gram-negative bacteria; constituent genera: *Agrobacterium* and *Rhizobium*.

Rhizobium A genus of gram-negative, aerobic, non-sporing rod-shaped or pleomorphic bacteria of the family Rhizobiaceae. Cells have either sparse peritrichous flagella or a single lateral or polar flagellum. *Rhizobium* spp are chemoorganotrophs; they are common inhabitants of the soil and are found as nitrogen-fixing endosymbionts in the ROOT NODULES of leguminous plants—in which they occur as *bacteroids*. (The so-called 'ineffective' strains cannot fix nitrogen, though they may incite nodule-formation.) It has recently been shown that at least some free-living rhizobia are also capable of NITROGEN FIXATION under certain conditions. Growth on nutrient media is usually non-pigmented; highly mucoid colonies are commonly formed on carbohydrate-containing media. Optimum temperature: about 25 °C; pH tolerance about 5–8. Some strains form BACTERIOCINS.

Rhizocarpon A genus of crustose LICHENS in

which the phycobiont is *Trebouxia*; the apothecia are lecideine. Species occur on rocks, stones, walls etc., and include e.g.:
R. geographicum. The thallus is yellow (greenish in shaded positions), minutely areolate, with a black hypothallus evident between the areolae and around the periphery of the thallus. The apothecia are small and black and are sunk into the surface of the thallus. *R. geographicum* forms map-like mosaics on silicaceous rocks—particularly in upland areas.

Rhizochrysis A genus of unicellular, wall-less, amoeboid algae of the division CHRYSOPHYTA.

Rhizoctonia A genus of fungi of the class AGONOMYCETES; species include pathogens of crucifers, legumes, and other plants. (see also DAMPING-OFF) Some species (e.g. *R. solani*) have a basidiomycetous sexual stage; an ascomycetous stage has been reported for at least one species.

rhizoid A unicellular or multicellular *avascular* root-like structure which forms part of the thallus in certain algae; a rhizoid anchors the organism to the substratum and acts as an absorptive organ/organelle. (cf. HOLDFAST)

Rhizomastigida An order of protozoans of the class ZOOMASTIGOPHOREA; flagella (one or more) and pseudopodia may occur together in a given cell, or may be present at different periods in the life cycle. Most species are non-parasitic, but HISTOMONAS MELEAGRIDIS is an important pathogen of the turkey.

rhizomorphs (*mycol.*) Macroscopic strands of tissue formed by certain higher fungi—e.g. *Armillaria mellea*; in some cases a rhizomorph appears to consist of a bundle of parallel, compacted hyphae, while in other types of rhizomorph individual constituent hyphae are not evident. The development of rhizomorphs often enables a fungus to spread into the surrounding environment; thus, e.g. the (bioluminescent) rhizomorphs ('shoe-strings') of *A. mellea* enables the fungus to spread from an infected to an uninfected tree *via* the soil. Rhizomorphs are typically resistant, enduring structures.

rhizomycelium A primitive form of mycelium formed by certain chytrids (see NOWAKOWSKIELLA); it consists of a system of branched, reticulate filaments of indefinite width—some regions of the system being nucleate, the remainder being anucleate. (The term is also used to refer to a system of branched processes formed by members of the class Laboulbeniomycetes.)

Rhizophydium (*Rhizophidium*) A genus of epibiotic, eucarpic fungi of the order CHYTRIDIALES; *Rhizophydium* spp are typically aquatic saprophytes or parasites—*R. couchii* is reported to be parasitic on *Spirogyra*. In *R. couchii* the asexual cycle is initiated when a zoospore settles on a susceptible host cell and penetrates the latter with a system of branching rhizoids; that part of the fungus which remains external to the host cell wall develops, by growth and repeated division, to form a sporangium which contains a number of zoospores. The zoospores subsequently escape from the sporangium *via* one or more pores which develop in the sporangial wall. In sexual reproduction infection of the host cell occurs as outlined above, but the external portion of the fungus grows and develops into an oogonium; a second zoospore, which becomes established close to the first, functions as a male gametangium (antheridium). Plasmogamy occurs when the male protoplast passes into the oogonium. Subsequently the oogonium becomes a thick-walled resting spore—within which karyogamy and meiosis are presumed to occur. On germination a resting spore liberates a number of zoospores.

rhizoplane The root surface; part of the RHIZOSPHERE.

Rhizopus A genus of fungi within the class ZYGOMYCETES. *Rhizopus* spp occur typically as saprophytes on soil, fruit etc. (see also BREAD MOULDS)—though some species can behave as pathogens in man and other animals and in certain plants; the vegetative form of the organism is a branched, coenocytic mycelium which attaches to the substratum by means of a system of branching rhizoids. Asexual reproduction involves the formation of spherical, columellate (see COLUMELLA), thin-walled, dehiscent sporangia; sporangia, which contain numerous aplanospores, develop terminally on the sporangiophores. Sexual reproduction involves the fusion of gametangia and the subsequent formation of ZYGOSPORES; most species are reported to be heterothallic. *R. stolonifer* (= *R. nigricans*) colonizes a substrate by producing *stolons*—surface hyphae analogous to the stolons of higher plants; a stolon grows for some distance and then gives rise to a further group of rhizoids and another centre of vegetative growth and asexual reproduction.

rhizosphere An ecological niche which comprises the surfaces of plant roots (the *rhizoplane*) and that region of the surrounding soil in which the microbial population is affected by the presence of the roots; the rhizosphere generally extends a few millimetres or (at most) centimetres from the rhizoplane. From the roots, a number of substances pass into rhizosphere soil; these substances, many of which are microbial nutrients, include carbohydrates, vitamins, and a variety of amino acids. The rhizosphere is usually richer than the surrounding soil in *bacterial* numbers; often a 10 to 50-fold increase may be demonstrated. Rhizosphere bacteria are predominantly gram-negative rods (e.g. *Pseudomonas* spp); gram-positive species appear to be less numerous in the rhizosphere than in the surrounding soil. There appears to be little or no evidence of plant-bacterium species specificity. Numerically, *fungi* may be present in similar or in slightly greater abundance in the rhizosphere (as compared with the surrounding soil). However, there may be qualitative (species) differences between rhizosphere and non-rhizosphere fungi, and there is evidence which indicates that, in some cases at least, the rhizospheres of certain plants encourage, or select, particular fungi. The extent to which the plant benefits from the rhizosphere microflora is often not clear; it seems possible that, by filling the ecological niche, these organisms play some part in protecting the plant from the incursions of soil-borne pathogens. Additionally, the plant may derive some benefit from the mineralizing activities of these organisms. (see also ectotrophic MYCORRHIZA)

rhodamine O (brilliant pink B) A red basic dye and FLUOROCHROME; rhodamine O—a substituted xanthene—has been used e.g. for the fluorescent staining of acid-fast organisms—see e.g. AURAMINE-RHODAMINE STAIN. The dye fluoresces red.

rhodanese (thiosulphate:cyanide sulphurtransferase) An enzyme present in a number of bacteria (e.g. *Bacillus subtilis*, *Desulphotomaculum nigrificans*, *Thiobacillus* spp, some photosynthetic bacteria) and in plant and animal tissues. Rhodanese catalyses the reaction of cyanide with thiosulphate ($S_2O_3^{2-}$), forming thiocyanate (SCN^-) and sulphite. Its metabolic function is unknown.

Rhodomicrobium A genus of bacteria of the RHODOSPIRILLACEAE.

Rhodophyta (red algae) A division of the ALGAE. The red algae are predominantly marine organisms—only a few species occur in freshwater or terrestrial environments. Most species are filamentous, some are parenchymatous, and a few are unicellular; sizes range from microns to metres. Typically, marine species are some shade of red while freshwater species are blue-green or grey-green. There are no flagellate species, and no flagellated gametes are produced. All species contain CHLOROPHYLL *a* and some contain chlorophyll *d*; all contain a number of xanthophylls and β-carotene (see CAROTENOIDS) and phycocyanins and/or phycoerythrins (see PHYCOBILIPROTEINS). Intracellular storage products include *floridean* STARCH and FLORIDOSIDE. Sexual reproduction occurs in most species and oogamy is common; life cycles are often complex. Sexual reproduction appears not to occur in e.g. *Porphyridium*. Certain red algae are used as sources of e.g. AGAR and CARRAGEENAN. Genera of red algae include: *Bangia*, *Bostrychia*, *Ceramium*, CHONDRUS, *Furcellaria*, *Gelidium*, GIGARTINA, *Iridaea*, PORPHYRA, and PORPHYRIDIUM.

Rhodopseudomonas A genus of gram-negative, non-sporing, polarly flagellate, photosynthetic bacteria of the RHODOSPIRILLACEAE; species occur e.g. in mud and stagnant waters. The cells are rod-shaped, coccoid or spherical (according to species)—elongated forms being usually 1–5 microns in length, and cocci up to 4 microns in diameter. Cell division occurs by binary fission or budding. Storage products include poly-β-hydroxybutyrate. Some species require BIOTIN.

Rhodospirillaceae (Athiorhodaceae; purple non-sulphur bacteria) A family of motile bacteria of the suborder RHODOSPIRILLINEAE; species are widespread in soil and water. Typically, they are anaerobic photoorganotrophs with alternative modes of metabolism; some species can grow aerobically in the light or dark. No species can metabolize elemental sulphur, but *low* (sub-inhibitory) concentrations of sulphide can be oxidized to sulphate without the accumulation of sulphur as an intermediate product. The genera are: *Rhodomicrobium* (species may grow as filaments), RHODOPSEUDOMONAS and RHODOSPIRILLUM.

Rhodospirillales An order comprising the *photosynthetic* bacteria*. All species are gram-negative, non-sporing organisms; the morphological range includes cocci, rods, and spiral forms. Cell division occurs, in most species, by binary fission; budding occurs in some species. All species carry out PHOTOSYNTHESIS anaerobically, without the evolution of oxygen, and contain one or more *bacterio*CHLOROPHYLLS as the sole chlorophyllous pigment(s); a range of CAROTENOIDS is found among the various species. NITROGEN FIXATION occurs in at least some species. Two suborders are distinguished: RHODOSPIRILLINEAE and CHLOROBIINEAE. *Organisms described as 'photosynthetic flexibacteria'—which contain bacteriochlorophylls *a* and *c*—have recently been reported. Currently, they are not formally included in the Rhodospirillales.

Rhodospirillineae (purple* photosynthetic bacteria) A suborder of the RHODOSPIRILLALES; species are characterized by the presence of *bacterio*CHLOROPHYLL *a* or *b* (cf. suborder CHLOROBIINEAE). Two families are distinguished: RHODOSPIRILLACEAE and CHROMATIACEAE. *The colour of a cell suspension (not always purple) depends e.g. on the type of CAROTENOID in the cells.

Rhodospirillum A genus of gram-negative, nonsporing, polarly flagellate, photosynthetic bacteria of the RHODOSPIRILLACEAE; species occur e.g. in mud and stagnant waters. The cells are spiral-shaped, 3–9 microns in length and commonly less than 1 micron in width; division occurs by binary fission. Although characteristically anaerobic organisms, *R. rubrum* can grow aerobically and other species can grow microaerophilically. NITROGEN FIXATION has been demonstrated in *R. rubrum*. Storage products include poly-β-hydroxybutyrate.

rhodospirillum haem protein (see RHP)

Rhodotorula A genus of imperfect yeasts within the family CRYPTOCOCCACEAE; species have been isolated from e.g. bark-beetles, tree exudates, various types of plant, soil, and water. The vegetative form of the organisms is a single spheroid, oval or elongated cell and/or pseudomycelium; reproduction occurs by budding. The organisms typically form red, orange, or yellow mucoid colonies. *Rhodotorula* spp are non-fermentative; they do not assimilate inositol.

rhoptries In certain parasitic protozoans, e.g. many members of the *Coccidia*: paired organelles located at the anterior end of the cell. The available evidence suggests that, at least in some cases, these organelles may secrete enzymes which—perhaps in conjunction with the *conoid*—may assist the protozoan parasite to penetrate the cells of its host.

RHP (*Rhodospirillum* haem protein; cytochromoid *c*) Original term for CYTOCHROMES *c′* and *cc′* found in purple photosynthetic bacteria and *Pseudomonas denitrificans*; its function is unknown.

Rhynchodina A sub-order of protozoans of the order THIGMOTRICHIDA.

riboflavin (riboflavine; vitamin B_2; lactoflavin) A water-soluble heat-stable VITAMIN: 6,7-dimethyl-9-(1′-D-ribityl)-isoalloxazine. The important coenzyme forms are: riboflavin-5′-phosphate (=*flavin mononucleotide* or FMN) and *flavin adenine dinucleotide* (FAD, composed of FMN and adenosine-5′-phosphate)—see Appendix V(c). These form the reactive components of *flavoproteins*—acting as hydrogen carriers in a wide range of oxidation/reduction reactions. (The isoalloxazine ring is hydrogenated/dehydrogenated at the N^1 and N^{10} positions.) Some reduced flavoproteins may be re-oxidized directly by molecular oxygen, hydrogen peroxide being a product; others are re-oxidized *via* the electron transport chain. Riboflavin is highly photolabile and may act as a photoreceptor in the phototropic reactions of certain fungi. FMN is involved in bacterial BIOLUMINESCENCE.

The majority of microorganisms appear to be capable of synthesizing riboflavin; *Lactobacillus* spp and the protozoans *Crithidia fasciculata* and *Tetrahymena* spp are among those organisms which require riboflavin as a growth factor. Under certain cultural conditions, some organisms (e.g. *Eremothecium ashbyii*, *Ashbya gossypii*, certain yeasts, *Clostridium* spp) secrete large quantities of riboflavin into the medium, and are important commercial sources of this vitamin.

ribonucleic acid (see RNA)

ribosomes Intracellular structures on which PROTEIN SYNTHESIS occurs. A ribosome is of the order of 20 nm in diameter and consists of RNA and protein. The RNA-protein complex appears to be maintained e.g. by hydrogen bonds and hydrophobic interactions; magnesium ions appear to have an important role in maintaining structural integrity. A ribosome contains one each of a number of proteins; these may include enzymes, structural proteins, and factors required for protein synthesis. The function of the ribosomal RNA is unknown; there is some evidence to suggest that it has a role in the binding of tRNA.

Different types of ribosome are characterized by their *sedimentation coefficients* (see SVEDBERG UNIT). *Bacterial* ribosomes sediment at approximately 70S; each consists of one 30S and one 50S subunit. The 30S subunit contains 16S rRNA (see RNA) and approximately 20 different proteins, while the 50S subunit contains 23S and 5S rRNA and 30–40 proteins. Ribosomes of the *eucaryotic cytoplasm* sediment at approximately 80S; each consists of one 40S and one 60S subunit. The 40S subunit contains 18S rRNA, while the 60S subunit contains 28S and 5S rRNA; details of the proteins are unknown. Ribosomes of the MITOCHONDRION and CHLOROPLAST appear to bear some resemblance to those of procaryotes.

ribovirus Any virus whose genome consists of RNA.

rice, diseases of These include: (a) *Blast disease*; causal agent: *Pyricularia oryzae*. Dark necrotic lesions are formed on the leaves and stem—the latter characteristically breaking in the neck region. The fungus may invade by means of appressoria; seed-borne infection is also believed to occur. The disease is favoured by high levels of soil nitrogen. (see also PIRICULARIN and α-PICOLINIC ACID) (b) BAKANAE DISEASE OF RICE. (see also RICE DWARF VIRUS)

rice dwarf virus (rice stunt disease virus) A double-stranded RNA virus which affects some members of the Gramineae; the virion is polyhedral, approximately 70 nm in diameter. On rice (*Oryza sativa*) symptoms include yellowish spots or streaks on new leaves, stunting, and the formation of many small tillers which (collectively) exhibit a *rosette* appearance. Inclusion bodies are usually present. The virus is transmitted *via* (and replicates within) leafhoppers—in which transovarial passage has been demonstrated.

rice stunt disease virus (see RICE DWARF VIRUS)

rice-water stools (see CHOLERA)

ricinoleic acid 11-Hydroxyheptadec-8-ene-1-carboxylic acid; it is obtained from castor oil.

Rickettsia A genus of gram-negative bacteria of the family RICKETTSIACEAE; species are obligate intracellular parasites or pathogens in vertebrates (including man) and arthropods (fleas, mites, ticks etc.). The cells are non-motile bacilli, 0.3–0.6 × 0.8–2 microns, which stain well by the Giemsa method or with the Giménez stain (carbol basic fuchsin in phosphate buffer (pH 7.45)—with malachite green or fast green as counterstain). The cell walls of the rickettsiae resemble those of other bacteria. No species has been cultivated extracellularly, and the nutritional requirements of *Rickettsia* spp are largely unknown; in the laboratory, rickettsiae may be grown in chick embryo cells, in the yolk sacs of hens' eggs, or in certain tissue cultures (e.g. HeLa, HEp-2, Detroit-6). Rickettsiae grow within the cell cytoplasm and (in certain cases) within the nucleus; depending on species the optimum growth temperature is 32–35 °C. Reproduction occurs by transverse binary fission. An oxidative (respiratory) metabolism appears to be operative, and glutamate is believed to be the main energy-yielding substrate; an electron transport chain involving cytochromes appears to be present. The organisms are sensitive to disinfectants such as sodium hypochlorite and phenol, and are killed within 30 minutes by temperatures above 56 °C; TETRACYCLINES have been used as chemotherapeutic agents in diseases of rickettsial aetiology. Many species share antigens with certain strains of *Proteus* (see WEIL-FELIX TEST). Type species: *R. prowazekii*; GC% range for species examined: 30–32.5.

R. prowazekii. The causative agent of certain TYPHUS FEVERS in man; the organism is pathogenic for the guinea-pig and (much less so) for the mouse. Reproduction occurs within the host cell's *cytoplasm.*

R. mooseri (= *R. typhi*). The causal agent of *murine* typhus; the organism is almost identical to *R. prowazekii* but can be distinguished e.g. by serological methods.

R. rickettsii. The causal agent of ROCKY MOUNTAIN SPOTTED FEVER. This species, and others of the 'spotted fever group'—e.g. *R. conorii* (BOUTONNEUSE) and *R. akari* (rickettsialpox)—develop within the cytoplasm *and* the nucleus of the host cell.

R. tsutsugamushi (*R. orientalis*). The causal agent of SCRUB TYPHUS.

Rickettsiaceae A family of gram-negative bacteria of the order RICKETTSIALES; the organisms are intracellular (and in some cases extracellular) parasites or pathogens of vertebrates and arthropods. (The cells parasitized do not include erythrocytes.) Constituent genera include COXIELLA, RICKETTSIA and ROCHALIMAEA (which, together, form the tribe Rickettsieae).

rickettsiae Members of the family RICKETTSIACEAE.

Rickettsiales An order of bacteria comprising the families RICKETTSIACEAE, BARTONELLACEAE and ANAPLASMATACEAE.

Rideal–Walker test A test used to determine the PHENOL COEFFICIENT of a (phenolic) disinfectant (see PHENOLS); the test involves the determination of that dilution of the test disinfectant which gives the same *rate* of kill (of a test bacterium) as does a standard dilution of phenol. Initially, serial dilutions (in water) of the disinfectant under test are prepared and allowed to equilibrate at 17–18 °C. Each dilution is inoculated with a fixed volume of a 24-hour broth culture of *Salmonella typhi.* A (standardized) loopful of liquid is subsequently removed from each dilution at 2½, 5, 7½, and 10 minutes after inoculation, each loopful being transferred to a separate, fixed volume of sterile broth; these broths are then incubated at 37 °C for 2–3 days. A similar procedure is carried out with phenol. After 2–3 days the tubes of broth are examined for the presence or absence of bacterial growth. The broths inoculated from one particular dilution of the disinfectant will show growth in tubes inoculated at 2½ and 5 minutes, but no growth in tubes inoculated at 7½ and 10 minutes; a similar result will be given by one particular dilution of phenol (commonly a dilution within the range 1/90 to 1/110 phenol w/v). The phenol coefficient (*Rideal–Walker coefficient*) may be obtained from the ratio of the dilution factors of these two dilutions:

$$\frac{\text{dilution factor of disinfectant}}{\text{dilution factor of phenol}}$$

Since in this test the activity of the disinfectant is evaluated in water, the results obtained are not necessarily relevant to the practical application of the disinfectant (cf. CHICK-MARTIN TEST).

Rieckenberg reaction An IMMUNE ADHERENCE phenomenon, first noted by Rieckenberg in 1917, in which blood platelets adhere to trypanosomes when the latter are incubated with homologous antiserum, COMPLEMENT, and the platelets of non-primate species.

rifampicin (see RIFAMYCINS)

rifamycins A group of ANTIBIOTICS produced by *Streptomyces mediterranei*; a rifamycin molecule is a highly substituted macrocyclic ring. *Rifamide* and *rifampicin* are clinically useful semi-synthetic derivatives of natural rifamycins. Rifampicin is highly active against gram-positive bacteria (including *Mycobacterium tuberculosis*); a few gram-negative bacteria (e.g. *Neisseria, Haemophilus*) are inhibited but others are much less sensitive—apparently due to lower permeability to the antibiotic. Rifampicin specifically inhibits bacterial DNA-dependent RNA polymerase; it has no action on the equivalent eucaryotic enzyme. (*Mitochondrial* RNA polymerase appears to be sensitive, but mitochondria are impermeable to rifampicin.) Rifampicin binds to the *beta* subunit of the RNA polymerase (see RNA) and inhibits initiation of transcription; binding of DNA by the enzyme is not inhibited. Once the initiation of transcription is complete, rifampicin does not inhibit chain elongation. *Resistance* of bacteria to rifampicin develops readily; certain mutations modify the *beta* subunit of the RNA polymerase so that binding of the antibiotic is prevented.

Certain rifamycins are also active against certain viruses. In e.g. poxviruses, rifampicin inhibits viral *maturation*; the REVERSE TRANSCRIPTASE of leukoviruses is inhibited by certain rifamycins.

rinderpest (cattle plague; bovine typhus) An acute, infectious disease which affects mainly cattle; the disease occurs in parts of Africa and Asia, but rarely in the United Kingdom—where it is a NOTIFIABLE DISEASE. The causal agent (see PARAMYXOVIRUS) is antigenically similar to the measles virus. Infection presumably occurs *via* the mouth or nasal passages; the incubation period is 4–15 days. Symptoms include fever, thirst, loss of appetite, a nasal and oral discharge, and ragged, ulcerative lesions on the gums; often there is (bloody) diarrhoea. Mortality rates may be extremely high.

ring slide (see SLIDE)

ring spot (*plant pathol.*) One form of LOCAL LESION which consists of single or concentric rings of chlorosis or necrosis—the regions between the concentric rings being green; the centre of the lesion may be chlorotic or necrotic.

ring test (*immunol.*) A qualitative PRECIPITIN TEST, carried out in a narrow tube, in which a solution containing antigen is layered over one containing antibody. A positive reaction is indicated by a band of white precipitate which forms at the interface of the two solutions.

Ringer's solution A solution used in microbiology e.g. as a general diluent. Composition (grams/100 ml distilled water): NaCl (0.9), KCl (0.042), $CaCl_2$ (0.048), $NaHCO_3$ (0.02).

ringworm (*tinea*; dermatomycosis; dermatophytosis) Any mycosis of the skin, hair, or nails—in man or other animals—in which the causal agent is a DERMATOPHYTE. Infection occurs by contact with infected individuals (including animals) or fomites; dermatophytoses can occur in all types of domestic animal and in at least some wild animals.

Ringworm in man. On the skin and scalp the lesions are often roughly circular with a raised border—but may coalesce to form confluent areas of dry, scaling skin which, in severe cases, may ulcerate. When hair is involved the causal organism penetrates the hair follicle and grows downward inside the hair shaft. Some dermatophytes (e.g. *Trichophyton schoenleinii, T. tonsurans*, and *T. violaceum*) cause infections in which the fungal hyphae and conidia occur largely or wholely *inside* the hair shaft; such infections are referred to as *endothrix* infections. Other dermatophytes (e.g. *Trichophyton mentagrophytes*, most or all species of *Microsporum*) grow within the hair shaft (as above) but also form a sheath of arthrospores on the surface of the hair shaft; such *ectothrix* infections are divided into the *small-spored* type (arthrospores 2–3 microns) and the *large-spored* type (arthrospores typically 4–8 microns). Dermatophytoses include *tinea pedis* (athlete's foot)—caused e.g. by *Trichophyton mentagrophytes* or *Epidermophyton floccosum*; *tinea capitis* (scalp ringworm) caused e.g. by *Microsporum audouinii* or *M. canis*; *Tinea corporis* (body ringworm) caused e.g. by species of *Trichophyton* or *Microsporum*; *tinea unguium* (ringworm of the nails) caused e.g. by *T. mentagrophytes*. *Favus* is a severe infection in which yellow, cup-shaped, honeycomb-like crusts (*scutula*) develop from infected hair follicles; the causal agent is commonly *T. schoenleinii*. *Laboratory diagnosis* includes microscopic and cultural examination of epidermal scales and hairs. Hairs infected by certain dermatophytes (e.g. *Microsporum audouinii, M. canis, T. schoenleinii*) may fluoresce under the WOOD'S LAMP, while those infected by most species of *Trichophyton* or by certain species of *Microsporum* (e.g. *M. gypseum*) do not fluoresce. In man, dermatophytoses are treated with GRISEOFULVIN or with certain POLYENE ANTIBIOTICS—e.g. Amphotericin B, Nystatin.

ristocetin A glycopeptide ANTIBIOTIC produced by certain actinomycetes, e.g. *Nocardia* sp. It is bactericidal for many gram-positive pathogens; gram-negative bacteria appear to be resistant. Ristocetin prevents PEPTIDOGLYCAN synthesis by inhibiting the transfer of disaccharide-pentapeptide subunits from bactoprenol to the growing cell wall.

RNA (ribonucleic acid) A linear polymer of riboNUCLEOTIDES in which the ribose residues are linked by 3′,5′-phosphodiester bridges (cf. DNA). The nitrogenous bases attached to each ribose residue may be adenine, guanine, uracil, or cytosine, or a modified form of one of these bases. Three main types of RNA—all of which are involved in PROTEIN SYNTHESIS—occur in both procaryotic and eucaryotic cells:

(a) *Messenger RNA (mRNA).* mRNA is single-stranded and contains no modified bases. It is synthesized on a DNA template*—a process known as *transcription*; the bases of ribonucleoside-5′-triphosphates pair with bases on the DNA template prior to bond formation by a *DNA-dependent RNA polymerase* (RNA polymerase) and concomitant release of pyrophosphate. (Uracil pairs with adenine, guanine with cytosine etc.—see DNA.) *In vivo*, only *one* strand of the DNA duplex functions as template (asymmetric transcription). It has been proposed that the binding of RNA polymerase to DNA causes localized unwinding of the DNA duplex; this allows base-pairing between the nucleotides and the template strand. The RNA polymerase appears to move along the DNA, linking ribonucleotides in the 5′ → 3′ direction (*antiparallel* with the template), so that a transient RNA-DNA hybrid duplex is formed in the region of the enzyme-DNA complex; as the enzyme proceeds, the RNA peels off from the template and the DNA duplex is restored. In *Escherichia coli*, RNA polymerase (the *holoenzyme*) consists of a *core enzyme* (consisting of two *alpha* chains, a *beta* chain, and a *beta-prime* chain) and a (protein) *sigma factor*. The core enzyme has a high affinity for DNA and, in the absence of *sigma* factor, binds to and initiates transcription at random points on either DNA strand. The *sigma* factor appears to be involved in the recognition of specific initiation sites (*promoters*) on the DNA template. Once initiation is complete the *sigma* factor is released—elongation being achieved by the core enzyme only. Elongation continues until a specific termination signal is reached in the template; at this point, in the presence of a protein factor *rho* (at least in *Escherichia coli*), the RNA chain is terminated and released from the DNA. In the absence of *rho* elongation continues through the termination signal. In bacteria, mRNA is short-lived; in eucaryotes, the life of mRNA depends e.g. on the nature of the cell and of the mRNA. (see also INFORMOSOME, POLYCISTRONIC MRNA, and OPERON)

(b) *Ribosomal RNA (rRNA).* rRNA may constitute up to 90% of the total RNA of a cell. It is single-stranded, but helical regions are formed by base-pairing between complementary regions within the strand; a high proportion of nucleotides are methylated. rRNA occurs in the RIBOSOMES; three main types are characterized by their sedimentation coefficients—28S, 18S, and 5S in eucaryotes, 23S, 16S, and 5S in bacteria. rRNA is coded for by several (in bacteria) or many (in eucaryotes) copies of each gene. In *eucaryotes*, rRNA synthesis occurs in the NUCLEOLUS. Initially, a large precursor form of rRNA is synthesized by a specific RNA polymerase; this precursor complexes with protein and undergoes *maturation* in the nucleolus. Maturation involves methylation of specific regions of the precursor molecule and subsequent degradation by endonucleases and exonucleases to yield one each of the 28S and 18S rRNA molecules. (The methyl groups appear to have a protective role; only unmethylated regions are excised by the nucleases.) The 5S rRNA appears to be synthesized separately. In *bacteria*, it is uncertain whether the (23S, 16S, and 5S) rRNAs are transcribed independently or transcribed as a unit and cleaved immediately following transcription; the same RNA polymerase appears to be involved in the transcription of mRNA, rRNA, and tRNA.

(c) *Transfer RNA (tRNA; soluble RNA, sRNA).* tRNA is involved in the assembly of amino acids at the mRNA during PROTEIN SYNTHESIS; for each amino acid there is one or more corresponding tRNAs which can bind it specifically. (tRNAs are also involved in the formation of cross-links during PEPTIDOGLYCAN synthesis.) tRNA is single-stranded, about 75–85 nucleotides in length (sedimentation coefficient about 4S); it contains a high proportion of unusual nucleosides, e.g. pseudouridine, 4-thiouridine, thymidine, various methylated nucleosides. The nucleotide sequence has been elucidated for many tRNAs. A secondary structure which may be common to all is the *clover-leaf* structure: short sequences of base-pairing between complementary regions of the chain fold the molecule into three main *loops*; the two ends of the chain are connected by a region of base-pairing such that the 3′-OH terminus extends beyond the 5′-phosphate terminus by 3 or 4 nucleotides. The 3′-OH end always terminates with the sequence pCpCpA (p = phosphodiester bridge; C = cytidine; A = adenosine) and is the terminus at which binds the specific amino acid.

The *anticodon* triplet (which recognizes the correct codon on mRNA) occurs in the centre of the middle loop. A fourth loop (the 'extra arm') may also be present; this may vary in size from one tRNA to another. The tRNA 'clover-leaf' appears to be folded into a specific tertiary structure which is essential for biological activity. *Transcription* of genes coding for tRNA yields a precursor tRNA which is longer than tRNA by 20–30 nucleotides, and which does not contain modified bases. The precursor undergoes cleavage, followed by specific modification (e.g. methylation of bases, formation of pseudo-uridine, etc.); the mechanism and function of this modification is unknown.

ANTIBIOTICS which inhibit RNA synthesis include ACTINOMYCINS and RIFAMYCINS. *See however RNA viruses under entry VIRUSES.

RNA-dependent DNA polymerase (see REVERSE TRANSCRIPTASE)

Robertson's cooked meat medium (cooked meat medium) A medium used e.g. for the culture of ANAEROBES. The medium (pH 7.4) generally contains minced beef heart (30–50% v/v), beef extract (1% w/v), peptone (1% w/v), and sodium chloride (0.5% w/v); it is sterilized by autoclaving (121 °C/15–30 minutes) and may be stored in tightly-closed screw-cap jars.

Roccella A genus of fruticose LICHENS in which the phycobiont is *Trentepohlia*; the thallus consists of terete to more or less flattened, strap-like branches which form erect or pendulous tufts. The thallus is light grey to bluish-grey or pinkish-grey and may range in size from 3 to 30 centimetres—depending on species; bluish-white soralia may occur on the branches. *Roccella* occurs mainly on rocks by the sea; species are used commercially as sources of litmus and orcinol, and have been used extensively as sources of the purple dye *orchil* (= *orseille*).

Rochalimaea A genus of gram-negative bacteria of the family RICKETTSIACEAE. The organisms are closely allied to members of the genus *Rickettsia* but differ (a) in that they can be grown in cell-free systems, and (b) in that they are normally extracellular parasites of their *arthropod* hosts. *R. quintana* (formerly *Rickettsia quintana*) is the causal agent of TRENCH FEVER.

rock tripe A common name for species of the lichen genus UMBILICARIA.

Rocky Mountain spotted fever A tick-borne, human rickettsial disease which occurs in parts of North America; the causal agent is *Rickettsia rickettsii*. Infection (by the bite of an infected tick) is followed by an incubation period of 2–14 days; symptoms include headache, joint and muscular pains, and a sustained fever. A macular rash develops on the limbs and trunk (in that order) and the lesions may become haemorrhagic. Mortality in untreated cases may be 5–90%. Reservoirs of infection occur in dogs, wild rabbits etc. and (mainly) in *Dermacentor* ticks (which may acquire infection congenitally). *Laboratory diagnosis* may include a CFT and/or the WEIL-FELIX TEST.

roestelioid (cornute) (*mycol.*) Refers to an *aecium* which, before peridial rupture, projects in a finger-like fashion through the epidermis of the host; at maturity the exposed portion of the peridium ruptures by a number of longitudinal slits—the torn strips of peridium curling back in a rosette-like pattern. Roestelioid aecia are formed e.g. by species of *Gymnosporangium*.

roguing (*plant pathol.*) A form of disease control involving the removal of diseased plants from a crop.

roll-tube technique A technique used e.g. for counting the number of viable bacteria in a given sample. A measured volume of the sample (or of a known dilution) is added to, and mixed with, a suitable molten agar medium (45–48 °C) contained in a large test-tube or a small, cylindrical bottle. The tube (or bottle) is then inclined at a slight angle to the horizontal and rotated about its long axis so that the agar solidifies as a thin layer around the inner surface of the container; rotation is continued until all the agar has set. The roll-tube is then incubated, and a count is made of the colonies which develop on and in the agar. The technique is sometimes used for the culture/enumeration of ANAEROBES; for this purpose the tube or bottle is flooded with a constant stream of oxygen-free nitrogen (or other gas).

rolling circle model (of DNA replication) A model which accounts for the asymmetric replication of circular double-stranded DNA—e.g. during certain instances of PLASMID and BACTERIOPHAGE replication. The model proposes that, initially, a breakage occurs in *one* strand of the circular duplex and the 5′-end of this strand begins to unwind. DNA synthesis (see DNA) then proceeds by elongation of the 3′-hydroxyl end of the broken strand, using the unbroken circular strand as template; as replication proceeds, the 5′-end of the broken strand is progressively displaced from the template strand. DNA synthesis may occur on the unwinding strand by the synthesis of short fragments in the 3′ → 5′ direction (see DNA)—yielding a double-stranded product. When a complete genome length has been synthesized, the displaced strand (with or without a complementary strand) may be removed by endonuclease action to yield a (double- or single-stranded) progeny genome. Alternatively, synthesis may continue to produce a (single- or double-stranded) *concatenate* i.e. a continuous sequence of covalently-linked genomes; the concatenate may be split subsequently to form a number of progeny genomes. (cf. CAIRNS MODEL)

It has been suggested that the rolling circle model may account for DNA transfer during bacterial CONJUGATION; this hypothesis is supported by the finding that, following transfer of the whole donor genome, a second copy of the donor genome begins to enter the recipient—i.e. a concatenate appears to be formed.

Romanowsky stains (Romanovsky stains; Romanowski stains) A range of composite dyes (e.g. GIEMSA'S STAIN, *Wright's stain* for parasites) which are based on EOSIN, METHYLENE BLUE, and dyes of the *azure* series. Romanowsky stains are used e.g. for the detection and identification of blood parasites (such as *Plasmodium* spp and trypanosomes) in (air-dried) blood smears; the dyes are commonly prepared as solutions in methanol or a mixture of methanol and glycerol. (In use, the dyes are often diluted, either before or after application to the slide, in buffered distilled water pH 7.2.) *Examples of staining reactions*: erythrocytes and SCHÜFFNER'S DOTS: pink; protozoan nuclei and KINETOPLASTS: red; protozoan cytoplasm: blue/purple.

root nodules (of leguminous* plants) Tumour-like swellings of the root tissues of clover, pea and other legumes; they contain endosymbiotic nitrogen-fixing bacteria which enable the plant to thrive in nitrogen-deficient soils. The roots of leguminous plants appear to secrete substances which encourage the multiplication of nodule-forming bacteria: *Rhizobium* spp; there is a degree of plant-bacterium specificity. Tryptophan, one of the extracellular plant products, undergoes bacterial conversion to IAA—an AUXIN which, possibly in association with an extracellular bacterial product (or products), may bring about deformation in the tips of the root hairs. The production of an extracellular bacterial polysaccharide stimulates the plant to synthesize *polygalacturonase*—an enzyme which, by acting on the wall of the root hair, may facilitate bacterial penetration. Bacterial penetration is followed by the development of an 'infection thread' i.e. a thin column of bacterial growth which gradually extends (within the root hair) to the root cortical cells. Except at the advancing tip, the infection thread is bounded by a membrane which is of plant origin. At first the bacteria multiply; later, the 'bacteroids' (morphologically-differentiated cells of *Rhizobium*) cease multiplying and enlarge. In size, root nodules are generally of the order of millimetres. (see also SYMBIOSIS, NITROGEN FIXATION) *Nitrogen-fixing root nodules occur also in a range of non-leguminous plants, e.g. the alder (*Alnus*); the corresponding actinomycete-like bacterial endosymbionts have been placed in the family Frankiaceae.

ropiness The (spoiled) condition of e.g. milk, beer, bread, in which long strands of certain polymers (DEXTRANS, LEVANS etc.) have been synthesized by microbial contaminants—e.g. species of *Bacillus* and *Streptococcus*. (see also FOOD, MICROBIAL SPOILAGE OF)

rosaniline (rosanilin) Triaminotriphenylmethane—see TRIPHENYLMETHANE DYES. (see also FUCHSIN)

rose bengal A deep-pink fluorescent dye which is chemically related to EOSIN; it is the sodium salt of tetraiodotetrachlorofluorescein. Rose bengal inhibits the spread of some types of rapidly-growing fungi, and has antibacterial properties; for this reason it is sometimes used in media employed for the isolation of certain soil fungi. (see also FLUORESCENT DYES (as antimicrobial agents))

Rose–Waaler test A serological test used to detect and quantify the RHEUMATOID FACTOR by its ability to agglutinate sheep erythrocytes which have been *sensitized** to a sub-agglutinable degree; unsensitized cells are used as a control. *See SENSITIZATION (1).

rosette (1) (*bacteriol.*) In certain gliding bacteria (e.g. *Leucothrix*): a radially-arranged cluster of filaments. In *Caulobacter*: a radially-arranged cluster of cells—the (adhesive) tips of the stalks forming the centre of the rosette. (2) (*ciliate protozool.*) An organelle (?) found near the cytostome in many species of the Apostomatida; its function is apparently unknown. (3) (*plant pathol.*) In a plant: an abnormality in which the leaves form a radial cluster on the stem—a symptom e.g. of certain VIRUS DISEASES OF PLANTS. Rosettes may arise e.g. by stunting in which internodal distances are reduced.

rosette technique (see IMMUNOCYTE ADHERENCE TECHNIQUE)

rot Any of a variety of unrelated plant diseases characterized by primary decay and disintegration of host tissue. A number of *soft* rots are due to PECTINolytic species of *Erwinia*. Species of *Alternaria* may cause *corky* rots—e.g. in the POTATO. See also ONION, PEA, STRAWBERRY, and TIMBER, DISEASES OF.

rotenone A mitochondrial RESPIRATORY INHIBITOR; many bacterial systems are insensitive to rotenone. Rotenone binds (non-covalently) to a site between $NADH_2$-dehydrogenase and ubiquinone (see ELECTRON TRANSPORT CHAIN).

Rothia A genus of the ACTINOMYCETACEAE.

rough-smooth variation (in bacteria) (see SMOOTH-ROUGH VARIATION (in bacteria))

Rous sarcoma virus An oncogenic RNA virus belonging to subgenus *A* of the genus LEUKOVIRUS. RSV was first isolated from a naturally-occuring sarcoma of the chicken; the virus can bring about cell TRANSFORMATION in both avian and mammalian cells, and some strains can promote tumour formation in mammals. RSV belongs to the class of C-TYPE PARTICLES; it can be cultured in chick embryo tissue cultures (or in embryonated eggs) and can be assayed by counting the number of 'microtumours' (colonies of transformed cells) in the monolayer.

routine test dilution (RTD) (1) (*virol.*) The highest dilution of a virus suspension (i.e. the lowest concentration of viruses) which can bring about confluent lysis of a cell monolayer. (2) (*bacteriol.*) The highest dilution of a bacteriophage suspension capable of producing confluent lysis in a LAWN PLATE prepared with a sensitive bacterial strain.

Roux bottle A flat-bottomed rectangular glass container which opens to the exterior, at the side, through a short, straight, wide neck.

Royce indicator sachet (see ETHYLENE OXIDE)

rRNA Ribosomal-RNA—see RNA.

RSV (*virol.*) Abbreviation for (i) respiratory syncytial virus, or (ii) Rous sarcoma virus.

RTD (see ROUTINE TEST DILUTION)

RTF (see RESISTANCE TRANSFER FACTOR)

rubber (microbial spoilage of) Both raw and vulcanized rubbers are susceptible to microbial attack (see also LATEX). Natural rubber is susceptible to staining as a result of microbial colonization e.g. by species of the fungi *Aspergillus*, *Cladosporium* and *Penicillium*, and the bacteria *Bacillus* and *Serratia*. The hydrocarbon (polyisoprenoid) portion of rubber is itself susceptible to breakdown by actinomycetes—e.g. *Actinomyces* spp—and probably by species of *Aspergillus* and *Penicillium*. Vulcanized rubbers are also subject to microbial surface staining and to body decay; the latter may be due e.g. to species of *Fusarium* or *Streptomyces*—or to *Thiobacillus thiooxidans*. Rubber is particularly vulnerable when moist or when contiguous with the soil. Certain chemicals (accelerators) which are added to rubber during vulcanization have antimicrobial properties; however, additional preservatives may be incorporated.

rubella (German measles) An acute, systemic, infectious disease of man; children and young adults are those most commonly affected. The causal agent, rubella virus (*q.v.*), invades *via* the mouth or nose; the incubation period is commonly 2–3 weeks. Early symptoms may include mild fever, malaise, and enlargement of the lymph nodes of the head and neck. Following a stage of viraemia, a rash of discrete lesions appears first on the face and then spreads to the rest of the body. The disease may have teratogenic effects if contracted during the earlier months of pregnancy. Prophylactic vaccines are available; long-term (or life) immunity follows recovery. *Laboratory diagnosis* may involve any of several serological tests, e.g. the HAEMAGGLUTINATION-INHIBITION TEST. Rubella viruses can be isolated from blood and/or nasopharyngeal swabs immediately following the onset of the rash and at other stages in the disease.

rubella virus The causal agent of RUBELLA. The rubella virus consists of an enveloped icosahedral capsid which contains single-stranded RNA; the overall diameter of the virion has been reported, by various authors, to be within the range 50–85 nm. Rubella viruses are inactivated by ether and by other lipid solvents, and they are labile at room temperatures and at 4 °C; they can be stored for periods of 1–2 months at −20°C (1 year or more at −70 °C) in e.g. a protein-supplemented balanced salt solution. Rubella viruses promote haemagglutination under appropriate conditions of pH and temperature; erythrocytes used in haemagglutination-inhibition tests include those of the goose, pigeon, and newly-hatched chick. Rubella viruses are generally considered to be members of the TOGAVIRIDAE.

rubeola Synonym of MEASLES; less commonly, a synonym of RUBELLA.

rubratoxins Polycyclic compounds, produced by *Penicillium rubrum*, which are toxic for certain animals. (see also MYCOTOXINS)

rugose (*adjective*) Wrinkled.

rum (see SPIRITS)

rumen One of the four compartments which form the stomach of a *ruminant*; the remaining compartments are termed the *reticulum*, *omasum*, and *abomasum* (or 'true stomach'). (*Ruminants* include cattle, deer, giraffes, goats and sheep.) The rumen (capacity approximately 100 litres in the cow) contains a large number of microorganisms (mainly bacteria and protozoa) which play an essential role in ruminant nutrition; these organisms break down plant material ingested by the host animal and provide the animal with protein, vitamins, and assimilable carbon and energy-yielding substrates—see below. (The rumen itself does not secrete enzymes.) The rumen contents are anaerobic (the E'_h is reported to be about −350mV); the pH varies with diet but is generally in the region of 6–6.5. The rumen temperature is about 39–40 °C in the cow—slightly higher than the body temperature (38 °C) due to the (exothermic) microbial fermentation.

The types and numbers of microorganisms present in the rumen of a given species depend on the nature of the animal's diet and on the period of time since the last intake of food; the rumen contents are reported to contain, on average, 10^{10} cells/millilitre. The bacteria may include: *Bacteroides amylophilus*, *B. ruminicola*, *B. succinogenes*, *Butyrivibrio fibrisolvens*, *Lachnospira multiparus*, *Lactobacillus ruminis*, *L. vitulinus*, *Megasphaera elsdenii* (formerly *Peptostreptococcus elsdenii*), *Methanobacterium ruminantium*, *Ruminococcus albus*, *R. flavefaciens*, *Selenomonas ruminantium*, *Succinivibrio dextrinosolvens*, *Veillonella alcalescens*. Most protozoans found in the rumen are ciliates—chiefly members of the HOLOTRICHIA (e.g. *Isotricha* spp) and of the SPIROTRICHIA (e.g. species of *Entodinium* and *Diplodinium*). The (anaerobic) rumen protozoa, which have a fermentative metabolism, ingest particulate plant material and rumen bacteria. Protozoans cannot survive in the rumen if the pH of the rumen contents falls to about 5; they appear not to be essential for ruminant nutrition, although when present they are believed to play an important role.

Most of the plant material ingested by the ruminant is resistant to enzymes produced by the animal and must be degraded in the rumen by microbial enzymes. Plant *carbohydrates* (e.g. CELLULOSE, HEMICELLULOSES, PECTINS, STARCH) are degraded by certain bacteria—e.g. cellulose by *Bacteroides succinogenes* and *Butyrivibrio fibrisolvens*, pectins by *Lachnospira multiparus*. Monosaccharides are metabolized to pyruvate—e.g. glucose via the EMBDEN–MEYERHOF–PARNAS PATHWAY, xylose (and other pentoses) *via* the HEXOSE MONOPHOSPHATE PATHWAY; pyruvate is metabolized to a variety of products. The end-products of fermentation formed by individual species may be further metabolized by other species—e.g. lactate, a common fermentation product, is metabolized by *Megasphaera elsdenii* and *Veillonella alcalescens*. Hydrogen and carbon dioxide (derived e.g. from formate by the FORMATE HYDROGEN LYASE system) are converted to methane by *Methanobacterium ruminantium* (see METHANOBACTERIACEAE). The principal products which ultimately result from the various metabolic pathways are fatty acids (acetic, propionic and butyric acids), carbon dioxide and methane; the host animal absorbs the fatty acids from the rumen and from the omasum and abomasum and eliminates the gases by eructation. (The ruminant uses fatty acids, rather than glucose, as primary sources of energy and carbon.) The acidity of the fermentation products is counteracted by the buffering action of the ruminant's saliva—which is produced in copious amounts and contains sodium bicarbonate and sodium hydrogen phosphate.

Dietary *proteins* are degraded, by microbial enzymes, to polypeptides and amino acids—most of which are eventually incorporated into microbial cells. The bacteria can also utilize simple nitrogenous compounds as nitrogen sources for protein synthesis: e.g. *urea* is split by microbial ureases to form ammonia and carbon dioxide—ammonia being the preferred nitrogen source for many rumen bacteria. Urea formed during *ruminant* metabolism passes to the rumen (*via* the rumen wall, and in the saliva) and is used as a nitrogen source by the rumen microflora; thus, dietary nitrogen is conserved, and ruminants can survive on diets high in plant fibre but low in nitrogen. (Similarly, feedstuffs low in protein can be supplemented with urea.) As the microorganisms pass—with the digested material—from the rumen to the omasum, abomasum, and small intestine, they are digested by the ruminant's enzymes; thus, *microbial* protein is the principal source of protein for the ruminant. Vitamins synthesized by the bacteria are also used by the host animal (see e.g. VITAMIN B_{12}).

In weaned and adult ruminant animals, ingested food passes *via* the oesophagus to the rumen where it undergoes microbial fermentation. After some time, a bolus of partially-digested material (a *cud*) enters the reticulum from the rumen and is regurgitated to the mouth. The cud is thoroughly chewed and swallowed, and another cud is regurgitated, etc. The swallowed cuds pass back to the rumen where they may undergo further digestion. Finely-divided material becomes separated from coarse material and passes from the reticulum to the omasum, abomasum, and small intestine. (In the *suckling* animal, the rumen and reticulum are not fully developed, and ingested food passes from the oesophagus *via* the oesophageal groove to the omasum and abomasum—thus by-passing the rumen.)

A (sometimes fatal) condition known as *acidosis* sometimes develops when ruminants are fed excessive amounts of readily-fermentable carbohydrate—e.g. starch, sugar (found e.g. in grain and beet, respectively). Such carbohydrates lead to a proliferation of acid-producing bacteria which cause a fall in pH and consequent loss of protozoans and many species of bacteria; as a result, acidophilic *Lactobacillus* spp predominate and cause a further fall in pH.

Ruminococcus A genus of gram-positive, anaerobic bacteria of the PEPTOCOCCACEAE; species occur e.g. in the RUMEN of cattle and sheep. The cells are coccal or coccoid, typically 1 micron or less in diameter, and occur singly or in pairs or chains. Metabolism is typically heterofermentative with acetate and formate among the main products; lactate may be produced in small amounts. Ammonia appears to be essential as a source of nitrogen. CELLOBIOSE and (commonly) CELLULOSE are fermented, and other sugars may or may not be attacked. Growth requirements may include e.g. B vitamins. Culture media typically include rumen fluid and cellobiose; growth occurs (under strict anaerobic conditions) e.g. at 37 °C. The two species of *Ruminococcus* are *R. albus* and *R. flavefaciens*.

Russula A genus of fungi of the family RUSSULACEAE; constituent species are distinguished from those of the genus LACTARIUS in that they form fruiting bodies which do not contain 'latex'. (see also SICKENER)

Russulaceae A family of fungi of the order AGARICALES; constituent species are distinguished from those of other families in that the tissues of the pileus and stipe contain aggregates of SPHAEROCYSTS, and in that the basidiospores bear amyloid ornamentation. The family contains two genera: LACTARIUS and RUSSULA.

rusts (1) (rust fungi) Fungi of the order UREDINALES; all rusts are obligate plant parasites, and some species cause economically-important diseases in certain crops—see e.g. RUSTS (2). The rust fungi generally exhibit a high degree of host specificity, and strains parasitic on a particular host species may be divided into a number of *physiological races* (see FORMA SPECIALIS).

Morphology and life cycles of the rust fungi. The vegetative (somatic) form of the organisms is a hyaline, mononucleate, haploid mycelium (which occurs during the early part of the life cycle) and a hyaline dikaryotic mycelium (which occurs during the remainder of the life cycle); the mycelium, which occurs intercellularly within the plant host, rarely bears clamp connections. Haustoria are commonly formed.

The life cycles of the rusts involve a series of distinct spore stages which occur in a definite sequence; some species of rust fungi exhibit all of the maximum of five stages while other species exhibit four or fewer stages. (see also MACROCYCLIC RUSTS and MICROCYCLIC RUSTS) In *autoecious rusts* the entire life cycle occurs in one host plant; in *heteroecious rusts* certain spore stage(s) occur in one host species and the remaining spore stage(s) occur in a different host species. (see also PRIMARY HOST) Many rusts are known to be heterothallic.

The life cycle of the rusts is often exemplified by that of *Puccinia graminis tritici*—a macrocyclic heteroecious strain which causes *black stem rust* of wheat and which exhibits all the various rust spore stages; the primary and alternate hosts of *Puccinia graminis tritici* are wheat and the barberry (*Berberis*) respectively. The life cycle:

Stage 0 (*the pycnial stage*). In the spring the overwintering spores (teliospores)—present on and around cereal-growing land—germinate and give rise to (haploid) basidiospores (see BASIDIUM for details); the basidiospores are dispersed by wind. When a basidiospore germinates on a barberry leaf, the germ tube penetrates the tissues of the leaf and gives rise to a mononucleate, haploid, intercellular mycelium. (At this stage the rust is unable to infect wheat.) Within the leaf tissue the fungus develops haustoria and, several days after infection, forms a number of erumpent flask-shaped *pycnia* which appear as groups of small, raised, yellowish lesions on the upper surface of the leaf. Each pycnium (= spermogonium, spermagonium)—which is a wholly fungal structure—is lined with a layer of thread-like processes from the termini of which are produced minute, mononucleate, haploid cells (*pycniospores*, *pycnospores*, *spermatia*) which are exuded in a drop of sweet, sticky 'nectar' from the ostiole (opening) of the pycnium; extending from the ostiole of each pycnium is a tuft of stiff, sterile, orange-coloured filaments and a number of thin-walled *flexuous hyphae*. The pycniospores are male reproductive cells, and the flexuous hyphae (= receptive hyphae) are the female structures; however, since *Puccinia graminis tritici* is heterothallic plasmogamy does not occur between the pycniospores and the flexuous hyphae of a given pycnium. (Plasmogamy fails to occur since both male and female elements in a given pycnium are of the same mating type—having arisen from mycelium derived from a basidiospore of one particular mating type.) Plasmogamy can occur only when the pycniospores of one mating type are carried, by insects, to the flexuous hyphae of the opposite mating type—i.e. from a (+) pycnium to a (–) pycnium or *vice versa*. When a pycniospore of appropriate mating type fuses with a flexuous hypha the male nucleus travels down the hypha and undergoes repeated division.

Stage I (*the aecial stage*). Within the barberry leaf the fungal mycelium differentiates in a number of regions near the lower epidermis of the leaf (opposite the pycnia); these spheroidal, plectenchymatous, differentiated regions are the *aecial initials* or *protoaecia* (= protoaecidia). The protoaecia (whose mating type is identical to that of the pycnia) do not develop further unless they receive nuclei of the appropriate mating type; such nuclei may be provided by pycniospores (*via* the flexuous hyphae) or by the fusion of hyphae of compatible strains ((+) and (–)) which occupy adjacent regions of the leaf. On dikaryotization the aecial initials develop into *aecia* (= *aecidia*) (*sing.* aecium and aecidium, respectively). Each aecium consists of a basal layer of sporogenous cells which give rise to a mass of parallel, basipetally-formed chains of dikaryotic spores (*aeciospores*, *aecidiospores*) arranged at right angles to the host's epidermis; the whole spheroidal mass of spores is contained within a thin wall of fungal origin, the *peridium*. The mature aecium subsequently breaks through the lower epidermis of the leaf, and the exposed peridium ruptures to form a cup-like structure (the *aecial cup* or *cluster cup*) from which the orange-coloured aeciospores are discharged. Aeciospores are liberated from the barberry in late spring or early summer.

Stage II (*the uredial stage*). Aeciospores cannot re-infect the barberry; they infect suitable strains of wheat—within the leaves of which they give rise to a dikaryotic mycelium. Subsequently, regions of spore-bearing fungal tissue appear in the sub-epidermal tissue of the host; these regions, *uredia* (= *uredinia*, *uredosori*) each bear a layer of single-celled, binucleate, rust-coloured spores, *uredospores* (= *urediospores*)—each cell being borne on a thin stalk (*pedicel*). The uredium is not bounded by a peridium. Subsequently the spore mass ruptures the epidermis of the host and the uredospores are dispersed by the wind. Uredospores cannot infect the barberry but they can infect fresh (i.e. uninfected) wheat and may re-infect the original host plant; several cycles of infection and uredospore-formation may occur during the summer. Uredospores are thus the principal dispersal agents of this rust.

Stage III (*the telial stage*). In the late summer/early autumn the fungus ceases to form uredia and commences to form *telia* (= *teleutosori*) (sing. *telium* and *teleutosorus*, respectively). A telium resembles a uredium in that it is an erumpent mass of stalkborne spores unbounded by a peridium; however, each *teliospore* (= *teleutospore*) is a dark-coloured two-celled structure (each cell binucleate) with a wall much thicker than that of a uredospore. By the time the spore has reached maturity karyogamy has taken place in both cells. Teliospores are resting spores—being unable to germinate immediately after their formation. The telial stage is regarded as the sexual or perfect stage of the rust fungi; rusts are classified mainly on the basis of the characteristics of the telial stage.

Stage IV (*the basidial or promycelial stage*). In spring the teliospores germinate to form *promycelia* (= basidia) which bear basidiospores (= sporidia); one basidium can be formed from each cell of the teliospore. Of the four basidiospores formed by each basidium, two are of the (+) mating type and two are of the (–) mating type.

Characteristics of other rust fungi. In some rusts the *pycnia* may be e.g. globose, or indeterminate in shape, and may be located subcuticularly or intraepidermally within the host plant. In relatively few species the *aecia* are *caeomatoid* (i.e. they lack a peridium); in peridiate species the peridium may be e.g. CUPULATE (characteristic of *Puccinia* and many other rusts) or ROESTELIOID (typical of *Gymnosporangium*). In e.g. *Melampsora* the *uredium* is peridiate—but this is atypical of the rusts as a group. *Teliospores* may be sessile (e.g. *Melampsora*) or may be pedicellate—i.e. may develop on a stalk (e.g. *Puccinia*); they may be one-celled (e.g. *Uromyces*), two-celled (e.g. *Puccinia*), or multicelled (e.g. *Xenodochus*).

(2) Plant diseases caused by fungi of the order UREDINALES; such diseases are referred to as 'rusts' as a consequence of the rust-coloured spores formed by many of the causal agents on the surfaces of infected plants. (see also CEREALS, DISEASES OF and FLAX, DISEASES OF) *White rusts* ('white blister') are plant diseases which are generally caused by species of the oomycete *Albugo*—e.g. white rust of mustard caused by *A. candida*; white rust of chrysanthemums is caused by a species of *Puccinia*. (see also RED RUST)

rye, diseases of (see CEREALS, DISEASES OF)

S

s Symbol for *sedimentation coefficient*; see SVEDBERG UNIT.

S Symbol for SVEDBERG UNIT.

***S* phase** (of the cell cycle) (see CELL CYCLE)

S → R (see SMOOTH-ROUGH VARIATION (in bacteria))

Sabin–Feldman dye test (see TOXOPLASMA DYE TEST)

Sabin vaccine An *oral* anti-POLIOMYELITIS vaccine containing live (attenuated) strains of the three types of poliovirus. (Each strain may be administered separately.) The aim of the vaccine is to encourage an active viral infection of the alimentary tract and thus to provide prolonged antigenic stimulation. (see also SALK VACCINE)

Sabouraud's dextrose agar A medium used for the isolation of a number of medically-important fungi. The medium consists of a 2% agar gel which contains 4% glucose (dextrose) and 1% peptone; the pH is adjusted to 5.6. One or more antibiotics may be added—e.g. CYCLOHEXIMIDE (*Actidione*) and CHLORAMPHENICOL may be included in media used for the isolation of DERMATOPHYTES.

Saccardian system A system of classification, introduced by Saccardo (1899), in which the *form*-genera of the *fungi imperfecti* are grouped into *sections* on the basis of the characteristics of their conidia. These sections are: *Amerosporae*. Conidia rounded or slightly elongate; aseptate. *Didymosporae*. Conidia ovoid or slightly elongate; uniseptate. *Phragmosporae*. Conidia elongate, having two or more transverse septa. *Dictyosporae*. Conidia ovoid or elongate, having both transverse and longitudinal septa. *Scolecosporae*. Conidia filamentous, septate or non-septate. *Helicosporae*. Conidia helical, septate or aseptate. *Staurosporae*. Conidia star-shaped, septate or aseptate. Sub-sections may be formed from the above on the basis of conidial colour; thus, e.g. conidia ovoid, uniseptate, colourless: *Hyalodidymae*. The grouping of a number of form-genera into a given section is not intended to signify that any real affinity exists between members of those form-genera; the system is merely one of convenience which helps in identification.

Saccharomyces A genus of unicellular fungi within the family SACCHAROMYCETACEAE; species occur e.g. on certain fruits. *Saccharomyces* spp exhibit both oxidative and fermentative modes of metabolism, and selected strains of some species are used e.g. in BAKING, BREWING, and WINE-MAKING. (see also ALCOHOLIC FERMENTATION) The vegetative form of the organisms is a single spheroidal, ovoid, or elongated cell; such cells may give rise to a sprout mycelium but true hyphae are not formed. Asexual reproduction occurs solely by budding. Sexual reproduction involves the fusion (plasmogamy and karyogamy) of haploid somatic cells (derived from ascospores) or the fusion of ascospores; the resulting diploid somatic cells constitute the predominant vegetative phase in most species of *Saccharomyces*. An ascus, which typically contains 1–4 spheroidal ascospores, develops—following meiosis—from a diploid somatic cell.

Most species of *Saccharomyces* ferment a range of sugars which may include one or more of the following: galactose, glucose, maltose, melibiose, raffinose, sucrose; soluble starch is fermented only by one species (*S. diastaticus*). *Saccharomyces* spp do not assimilate nitrate. Most species are inhibited by CYCLOHEXIMIDE. Species include:

S. cerevisiae. The vegetative cells are commonly 4–6 × 5–10 microns (range approximately 3–10 × 4–20 microns). Each cell contains a small nucleus and a large membrane-limited vacuole which may occupy much of the cell. (see also MITOCHONDRION) The CELL WALL consists largely of glucans and mannans and includes lipid and protein constituents—the protein fraction including certain enzymes (e.g. sucrase); chitin is a minor component and appears to be concentrated in the regions of the bud SCARS. Mitotic division involves the formation of an intranuclear splindle—see MITOSIS. Many strains are reported to be heterothallic; the sexual cycle involves the fusion of (haploid) somatic cells, or ascospores, of appropriate mating type. (see also COMPATIBILITY (in fungal sexuality)) Asci—which contain 1–4 ascospores—develop directly from diploid somatic cells following meiosis; in general, ascus formation is encouraged by media which are deficient in nitrogen and which contain acetate as the main carbon source. *S. cerevisiae* ferments galactose, glucose, maltose, and sucrose (galactose and maltose are sometimes fermented slowly); cellobiose, lactose, and melibiose are among sugars not attacked. All strains are reported to be sensitive to cycloheximide.

S. aceti. A species which ferments glucose and which, under aerobic conditions, oxidizes ethanol to acetic acid.

S. carbajali. A species used in the preparation

of PULQUE; it appears to be identical with *S. cerevisiae.*

S. carlsbergensis. A species used in brewing; some authors consider this organism to be a strain of *S. uvarum.*

S. uvarum. A species used e.g. in CIDER-making. The organisms ferment galactose, glucose, maltose, melibiose, raffinose, and sucrose (galactose, maltose, and melibiose are sometimes fermented slowly); lactose is among sugars which are not fermented. All strains are reported to be sensitive to cycloheximide.

Saccharomycetaceae A family of fungi of the order ENDOMYCETALES, class HEMIASCOMYCETES; species occur in a variety of habitats—e.g. on fruits, in tree exudates, and in soil and sewage. The family includes many YEASTS. The organisms are predominantly unicellular and the cells are typically spherical, spheroidal, ovoid, or elongated; the CELL WALL commonly contains chitin. Some species may form a sprout mycelium (pseudomycelium) and/or a true mycelium though such vegetative forms are rarely well-developed. Asexual reproduction occurs by BUDDING (e.g. in *Saccharomyces*), by bud-fission (e.g. in *Saccharomycodes*), or by fission (in *Schizosaccharomyces*). For the greater part of the life cycle the organisms may be diploid (as e.g. in *Saccharomycodes ludwigii*) or haploid (as e.g. in *Schizosaccharomyces octosporus*); in e.g. *Saccharomyces cerevisiae* both haploid and diploid vegetative phases are propagated by budding. Sexual reproduction typically involves the fusion of somatic cells or the fusion of ascospores—species being either homothallic or heterothallic; an ascus develops, meiotically, from a zygote or from a cell derived mitotically from the zygote. (cf. LIPOMYCES) Typically, each ascus contains 1–8 ascospores (the asci of *Kluyveromyces polysporus* each contain hundreds of ascospores); according to species the shape of the ascospores may be e.g. spherical, ovoid, hemispherical, reniform, Saturn-shaped, or shaped like a bowler hat. (Species which form needle-shaped or spindle-shaped ascospores are placed in the family Spermophthoraceae.) Members of the Saccharomycetaceae typically exhibit both oxidative (respiratory) and fermentative modes of metabolism; some species (e.g. *Hansenula wingeii*) are non-fermentative. Criteria used to distinguish the constituent genera include the mode of asexual reproduction, the ability to assimilate nitrate, and the shape and surface appearance of the ascospores. The genera include *Debaromyces*, HANSENULA, *Kluyveromyces*, LIPOMYCES, *Pichia*, SACCHAROMYCES, SACCHAROMYCODES, *Saccharomycopsis*, SCHIZOSACCHAROMYCES, and *Wingea.*

Saccharomycodes A genus of unicellular fungi within the family SACCHAROMYCETACEAE; strains of the sole species, *S. ludwigii*, have been isolated from wine and from grape must. The vegetative form of the organisms is an ovoid or elongated cell which is commonly about 5 × 10–20 microns. Asexual reproduction occurs by bipolar *bud-fission*—in which a septum forms across the constricted region of a broad-based bud which develops at one or both poles of the cell. Sexual reproduction is initiated when a (diploid) somatic cell undergoes meiosis and gives rise, directly, to an ascus that commonly contains four smooth, spherical ascospores; within the ascus the ascospores fuse in pairs, and each fused pair gives rise to a sprout mycelium from which the vegetative cells are budded. Homothallic and heterothallic strains have been reported. *S. ludwigii* ferments glucose, raffinose, and sucrose, and fails to ferment galactose, lactose, and maltose.

Saccharomycopsis A genus of fungi of the family SACCHAROMYCETACEAE.

saccharose Synonym of SUCROSE.

***Saccharum* virus 2** (see SUGARCANE FIJI DISEASE VIRUS)

sacculus A term sometimes used to denote the single giant hollow molecule of PEPTIDOGLYCAN which functions as the rigid component of the cell wall in the majority of bacteria and in blue-green algae.

saddle-back fungus The basidiomycete *Polyporus squamosus.* (see also DRYAD'S SADDLE)

saddle fungus A common name for the (saddle-shaped) fruiting bodies of *Helvella* spp.

safranin (*syn*: safranin O) A red, basic, water-soluble and ethanol-soluble dye used in microbiology as a general counterstain and in KOSTER'S STAIN.

saké Japanese rice wine. Initially, soluble sugars are obtained from a mash of steamed rice by the action of *Aspergillus oryzae*; these sugars are then fermented by pure cultures of *Saccharomyces saké*, a yeast which possesses high alcohol tolerance.

salicin *O*-Hydroxymethyl-phenyl-β-D-glucopyranoside. Salicin is used as a substrate in biochemical characterization tests for bacteria; it is attacked e.g. by many species of *Streptococcus* (including *Strep. pyogenes*), *Klebsiella pneumoniae*, and *Serratia marcescens.*

salicylic acid and derivatives (as antimicrobial agents) Salicylic acid (*o*-hydroxybenzoic acid) and many of its derivatives are fungistatic but have little activity against bacteria. Salicylanilide (*Shirlan*) is used as a PRESERVATIVE for e.g. textiles, paints, leather goods. Halogenated salicylanilides (e.g. *Irgasan*, 3,5,3′,4′-tetrachlorosalicylanilide) are active against fungi and bacteria and are used e.g. as skin antiseptics in germicidal soaps. Salicylic acid, salicylanilide, and halogenated derivatives of these compounds and of salicylaldehyde may be used in the treatment of cutaneous mycoses. (see also PAS)

saline agglutination (see AUTO-AGGLUTINATION)

salivaria One of two groups of sub-genera of trypanosomes which are parasitic in *mammals*. (see TRYPANOSOMA) The *salivarian* trypanosomes are those in which forms infective for the vertebrate host typically develop in the ANTERIOR STATION of the invertebrate host (vector) and which are thus involved in *inoculative infection* (cf. STERCORARIA). Salivarian sub-genera: DUTTONELLA, NANNOMONAS, *Pycnomonas*, and TRYPANOZOON.

Salk vaccine An anti-POLIOMYELITIS vaccine containing formalin-inactivated strains of the three types of poliovirus. It is administered subcutaneously or intramuscularly. (see also SABIN VACCINE)

Salmonella A genus of gram-negative, chemoorganotrophic, asporogenous bacteria of the ENTEROBACTERIACEAE; species occur as parasites and pathogens of animals, including man. The cells are rods, 0.5–1 × 1–3 microns; *most* strains are motile—but these may form non-motile variants. Growth occurs on basal media (e.g. NUTRIENT AGAR) and on MACCONKEY'S AGAR and deoxycholate-citrate agar; enrichment media include SELENITE BROTH, tetrathionate broth, and strontium chloride enrichment medium. *Typical* reactions (to which there are important exceptions) include: a negative INDOLE TEST and VOGES-PROSKAUER TEST and a positive CITRATE TEST and METHYL RED TEST; typically, H_2S is produced in TSI medium, lysine and ornithine decarboxylases are produced, and ureases are not formed. Gas is produced from glucose (37 °C) by most strains; lactose is not fermented by the majority of strains. *Salmonella* spp are killed by PASTEURIZATION, by routine chlorination in the treatment of WATER SUPPLIES, and by appropriate concentrations of many common DISINFECTANTS. Antibiotics used in the treatment of salmonellosis include ampicillin (see PENICILLINS), CHLORAMPHENICOL, and certain CEPHALOSPORINS. Resistance to a particular antibiotic may be due to an R FACTOR. The type species is *S. cholerae-suis*; GC% for the genus: 50–53.

Classification of the salmonellae. Four unnamed subgenera are distinguished on the basis of biochemical reactions. *Subgenus I* comprises the majority of strains of the salmonellae, and includes *S. typhi*, *S. paratyphi* A, B and C, *S. typhimurium* and *S. cholerae-suis*. *Subgenus III* comprises those organisms previously referred to as the Arizona group (or as species of the genus *Arizona*); typically, these organisms carry out delayed fermentation of lactose.

The numerous strains of *Salmonella* are more familiarly referred to as named serotypes, each serotype being indentified by its antigenic formula (see KAUFFMANN–WHITE CLASSIFICATION SCHEME, O ANTIGENS and H ANTIGENS). The 1000 or more serotypes (which comprise the four subgenera) are arranged into a number of *groups*, each group being referred to by a letter or number. The serotypes which belong to a particular group each have a particular O antigen (the *group antigen*) which does not occur in any other group; thus, the group antigen of group A is O antigen 2, that of group B is O antigen 4 etc. (The *somatic* antigens are affected during SMOOTH-ROUGH VARIATION. See also: PHASE VARIATION, IN SALMONELLA. The antigenic pattern (and name) of a given serotype may be affected by BACTERIOPHAGE CONVERSION.)

S. typhi (occasionally referred to as *S. typhosa*). The causal agent of TYPHOID. *S. typhi* is a motile organism which gives a negative CITRATE TEST and which carries out *anaerogenic* fermentation of glucose and other sugars (e.g. mannitol). H_2S production in TSI medium may be very weak. The antigenic formula is 9,12,Vi:d:– (some strains 9,12:d:–); a group D *Salmonella*—9 being the group antigen. (N.B. Other salmonellae may give rise to anaerogenic variants.)

S. typhimurium. An aerogenic, group B serotype (antigenic formula 1,4,5,12:i:1,2 *or* 1,4,12:i:1,2) which is a common cause of gastro-enteritis.

S. paratyphi A. An aerogenic, citrate-negative, group A serotype (antigenic formula 1,2,12:a:–) which typically fails to produce H_2S in TSI medium.

S. paratyphi B (= *S. schottmuelleri*). An aerogenic, citrate-positive, group B serotype which produces H_2S; antigenic formula 1,4,5,12:b:1,2 *or* 1,4,12:b:1,2.

S. paratyphi C (= *S. hirschfeldii*). An aerogenic, citrate-positive, group C serotype (antigenic formula 6,7,Vi:c:1,5 *or* 6,7:c:1,5) which produces H_2S.

S. cholerae-suis. An aerogenic, citrate-positive, group C serotype (antigenic formula 6,7:c:1,5) which does not produce H_2S.

S. gallinarum. The causal agent of enteric disease in fowl. A *non*-motile group D serotype (antigenic formula 1,9,12:–:–); gas-production (from glucose) and H_2S-formation vary according to strain.

salmonellosis Any disease (of man or animals) in which the causal agent is a species of *Salmonella*. In man, salmonelloses include TYPHOID FEVER and PARATYPHOID FEVER. (see also **DYSENTERY** (bacterial) and **FOOD POISONING** (bacterial))

saltant An organism which exhibits **SALTATION**.

saltation (dissociation) (*mycol.*) A phenotypic change in a thallus or part of a thallus; saltation may be due e.g. to a mutation or to the segregation of nuclei in a heterokaryon.

salting (of food) (see **FOOD, PRESERVATION OF**)

salvarsan (see **ARSENICALS**)

sandwich technique An *indirect* **IMMUNOFLUORESCENCE** technique used e.g. to demonstrate the presence of specific antibodies at the surface of antibody-forming cells. A suitably-prepared section of tissue, or a smear, is incubated with unlabelled homologous antigen. The preparation is then washed, to remove any free (uncombined) antigen, and exposed to fluorescein-conjugated antibody of the same specificity as that being sought on the cells. Thus, the antigen becomes sandwiched between the internal (cellular) antibody and the external fluorescent antibody.

sap-stain (of timber) (see **BLUE-STAIN**)

sap transmission (of plant-pathogenic viruses) (see **VIRUS DISEASES OF PLANTS**)

saprobity system (saprobiensystem) The system by which a body of polluted water (river, stream etc.) that is undergoing **SELF PURIFICATION** may be divided into a number of ecologically-distinct zones—each zone being characterized e.g. by its level of pollution, content of dissolved oxygen, and types of microorganism present (indicator organisms). The zones are: (a) *Polysaprobic zone*. The zone most heavily polluted with organic matter (sewage etc.), i.e. the zone nearest the source of pollution. Oxygen consumption is high, and there is a vigorous production of ammonia and hydrogen sulphide; the concentration of dissolved oxygen is minimal. Organisms found in this zone may include the flagellates *Bodo* and *Euglena*, the tetrahymenid ciliate *Glaucoma scintillans*, the blue-green alga *Oscillatoria*, and the bacteria *Beggiatoa* and *Thiothrix*. (b) *α-Mesosaprobic zone*. Mineralization and oxygen consumption continue vigorously but the content of dissolved oxygen is higher than that in the polysaprobic zone; there is some oxidation of ammonia. Hydrogen sulphide is not produced. Organisms characteristic of this zone may include species of *Oscillatoria*, *Carchesium*, *Stentor* and *Vorticella*. (c) *β-Mesosaprobic zone*. Oxygen consumption is low since mineralization is almost complete; dissolved oxygen content is considerably higher than that in the α-mesosaprobic zone. Levels of free ammonia are low. Organisms characteristic of this zone may include the blue-green alga *Phormidium*, the spirotrich *Halteria grandinella* and the peritrich *Vorticella striata*. (d) *Oligosaprobic zone*. Mineralization is complete and dissolved oxygen abundant. This zone is populated predominantly by aerobic photosynthetic organisms.

Saprodinium A genus of protozoans of the order Odontostomatida, subclass **SPIROTRICHIA**.

Saprolegnia A genus of fungi within the class **OOMYCETES**; *Saprolegnia* spp are typically saprophytic organisms which occur in freshwater environments and in moist soils—though a few species are parasitic on fish. The vegetative form of the organisms is a branched, coenocytic mycelium which attaches to the substratum by means of rhizoidal extensions of the hyphae. Asexual reproduction involves the formation of long, cylindrical, tapering, non-deciduous zoosporangia which develop terminally on the hyphae; zoospores are liberated *via* a pore in the distal end of the zoosporangium. *Saprolegnia* is *diplanetic*—see **DIPLANETISM**. An alternative mode of asexual reproduction involves the formation of *gemmae*—see **GEMMA**. Sexual reproduction is oogamous and involves the formation of spherical oogonia—each containing one or more oospheres—and elongate antheridia; gametangia usually develop terminally on the hyphae. Fertilization takes place when the male nucleus/nuclei enter the oogonium *via* a fertilization tube which links the gametangia. Each fertilized oosphere forms a thick-walled oospore. On germination, an oospore gives rise to a short hypha which characteristically develops a terminal sporangium. *Saprolegnia* spp appear to be commonly hermaphroditic and homothallic.

sapropel The anaerobic, putrefying matter—rich in hydrogen sulphide—which forms the mud of stagnant waters.

saprophyte Any 'plant-like' organism (e.g. bacterium, fungus) whose nutrients are obtained from dead and decaying plant or animal matter in the form of organic compounds *in solution*. Soil saprophytes are important as agents of decay and **MINERALIZATION** and are thus essential links in the cycles of matter. Some saprophytes are facultative parasites.

The term *saprophyte* is sometimes extended to include *saprozoic* organisms—see **SAPROZOIC NUTRITION**.

saprophytic nutrition The type of nutrition characteristic of a **SAPROPHYTE**.

saprozoic nutrition In 'animal-like' organisms (e.g. protozoans): a mode of nutrition in which nutrients are obtained from dead and decaying plant or animal matter in the form of organic compounds *in solution*. (see also **SAPROPHYTE**)

saramycetin (X5079C) A polypeptide **ANTIBIOTIC** produced by *Streptomyces saraceticus*.

Saramycetin, which has a high sulphur content, is active against certain fungi but has no antibacterial activity. It has low toxicity and has had some success in clinical trials against blastomycosis, histoplasmosis and aspergillosis.

Sarcina A genus of gram-positive, anaerobic coccoid bacteria of the PEPTOCOCCACEAE; species occur e.g. in the alimentary tract of animals (including man) and in the soil. The cells are approximately 2 microns or more in diameter and occur in PACKETS of 8 or more cells; they are not pigmented. Metabolism is fermentative. Type species: *S. ventriculi*; GC% range for the genus: 28–31.

S. ventriculi. Many strains form a cellulose capsule.

S. morrhuae: see HALOBACTERIACEAE.

sarcina sickness (of beer) (see BREWING)

Sarcodina A superclass of protozoans of the subphylum SARCOMASTIGOPHORA; locomotion in the sarcodines is typically effected by *pseudopodia* (see MOTILITY)—flagella occuring only in the developmental stages of certain species. Sarcodine morphology ranges from the simple, naked cell (e.g. *Amoeba* spp) to relatively complex forms having internal or external skeletal structures (e.g. foraminiferans, RADIOLARIANS, certain HELIOZOANS); however, the cortical region of the cell fails to exhibit the degree of differentiation common in many other groups of protozoans. The sarcodines are typically free-living inhabitants of the soil and/or fresh and salt waters; a few species (e.g. *Entamoeba histolytica*, *Naegleria* spp) are pathogens. According to species the organisms may be carnivorous, herbivorous or omnivorous. Sexual and quasi-sexual processes (e.g. gametogamy, AUTOGAMY) occur in a number of species, and an alternation of haploid and diploid generations occurs in some species of foraminiferans.

The (many) constituent orders of the Sarcodina include AMOEBIDA, ARCELLINIDA, and FORAMINIFERIDA. Several taxa are commonly found both in protozoological *and* mycological taxonomic schemes; these include the orders Acrasida (CELLULAR SLIME MOULDS), Labyrinthulida (NET SLIME MOULDS), and Plasmodiophorida (endoparasitic slime moulds, e.g. *Plasmodiophora brassicae*).

sarcoma A malignant tumour of *connective tissues*. (cf. CARCINOMA; see also TUMOUR)

Sarcomastigophora A sub-phylum of PROTOZOA; locomotion in constituent species is effected by flagella or pseudopodia, or both. The superclasses are MASTIGOPHORA, SARCODINA, and OPALINATA. (The locomotory appendages of the opalinids are regarded by some as cilia, and by others as 'short flagella'.)

Sarcoscypha A genus of epixylous fungi within the order PEZIZALES.

Sargassum A genus of algae of the PHAEOPHYTA.

Sartorya A genus of fungi of the order EUROTIALES; the conidial (imperfect) stage of *S. fumigata* is the *form*-species *Aspergillus fumigatus*.

satellite colonies (see SYNTROPHISM)

satellite virus (SV) (of TOBACCO NECROSIS VIRUS) A polyhedral single-stranded RNA virus approximately 17 nm in diameter; serologically-distinct strains of the virus have been reported. Satellite virus is able to replicate, in a plant host, only when that host is infected with TOBACCO NECROSIS VIRUS; the replication of SV appears to depend on the presence of enzyme(s) coded for by TNV. TNV is said to function as the 'activator' of SV; different strains of TNV vary in their capacities to activate SV. TNV is able to replicate within a plant in the absence of SV; the lesions produced on the plant in a *combined* infection commonly differ, in number and nature, from those produced by TNV alone. SV and TNV are serologically unrelated; SV (with TNV) can be transmitted mechanically or transmission may occur *via Olpidium brassicae*. (see also COVIRUS)

sauerkraut Cabbage preserved by the acid fermentation products of *Lactobacillus brevis* and *Leuconostoc mesenteroides* (both heterofermentative) and *Lactobacillus plantarum* (homofermentative); these bacteria are naturally present on the inner leaves of the cabbage. Cabbage is salted to extract the natural juices; these are fermented to produce lactic and acetic acids.

saxicolous Growing on (and/or in) rock (or rock-like substrata—e.g. walls). (cf. ENDOLITHIC and EPILITHIC)

scab (*plant pathol.*) Any of a variety of unrelated plant diseases characterized by the formation of discrete, scab-like surface lesions—see e.g. *common scab* of POTATO, and APPLE *scab*.

scabrous (scabrid) With a rough surface.

scales (*mycol.*) (see UNIVERSAL VEIL)

scanning electron microscopy (see ELECTRON MICROSCOPY)

scar (*mycol.*) In yeasts: a differentiated region of the cell wall marking the location at which cell separation occurred during budding or fission. In the multipolar budding yeast *Saccharomyces cerevisiae* two types of scar can be distinguished: the *bud scar* on the parent cell, and the *birth scar* on a daughter cell; each type of scar appears as a raised ring on the cell wall. When cells of *S. cerevisiae* are stained with the dye *primulin* and examined by fluorescence microscopy the *bud scar* appears as a ring of bright fluorescence on the non-fluorescent (or weakly-fluorescent) cell wall; it has been suggested that the ability of the bud scar to take up primulin may reflect a local change in the microfibrillar structure of the cell wall. One, several, or many bud scars may occur on a given wall, and in *S. cerevisiae* the scar rings appear never to overlap. In *S. cerevisiae* the region of the bud scar seems to contain chitin in concentrations greater than those found elsewhere in the cell wall.

scarlatina Synonym of SCARLET FEVER.

scarlet fever (scarlatina) An acute, infectious disease, primarily of children, characterized by sore throat, fever, enlargement of cervical lymphoid tissues, rash, and desquamation of the skin. The causal organism is a toxin-producing strain of *Streptococcus pyogenes*; infection commonly occurs *via* the mouth or nose. Various PENICILLINS have been used for treatment. Susceptibility to scarlet fever may be gauged by the DICK TEST.

Scenedesmus A genus of non-motile green algae of the division CHLOROPHYTA; species occur—often abundantly—in a wide range of freshwater environments. The elongate or fusiform cells occur as *coenobia* which contain four, eight, or occasionally sixteen laterally-adjacent cells; the two terminal cells frequently differ from the others e.g. in that they may bear spine-like processes at each corner of the free side of the cell. The uninucleate cells each contain a single chloroplast which contains one pyrenoid. Reproduction occurs by autospore (autocolony) formation: daughter coenobia are formed within *each* cell of a parent coenobium—the daughter coenobia being released on rupture of the parental cell wall.

Schardinger dextrins *Cyclic* DEXTRINS formed from STARCH or GLYCOGEN by the action of an extracellular enzyme (α-(1 → 4)-glucan 4-glycosyltransferase, cyclizing) produced by *Bacillus macerans*; they contain 6, 7, or 8 α-(1 → 4)-linked D-glucose residues and are known, respectively, as α-, β-, and γ-dextrins.

Schaudinn's fluid A fixative used in protozoological work: mercuric chloride (100 ml saturated, aqueous) and 95% ethanol (50 ml) supplemented (immediately before use) with glacial acetic acid (1.5 ml).

scheduled diseases (*vet.*) Diseases scheduled under the Diseases of Animals Act: NOTIFIABLE DISEASES.

Schick test (*immunol.*) An *in vivo* test used to determine whether or not an individual is susceptible to DIPHTHERIA. The test procedure involves an intradermal injection of the toxin of *Corynebacterium diphtheriae* into one forearm, and (as a control) the injection of a similar amount of heat-inactivated toxin into the other forearm. Individuals whose blood and tissues contain little or no antitoxin, and who are consequently *susceptible* to disease, give a *positive* Schick test; this consists of a localized inflammatory response at the site of the injection of potent toxin. The reaction develops in 2 to 4 days after the injection. Individuals whose blood or tissues contain a certain minimum level of antitoxin (and who, in consequence, are resistant to the disease) give a *negative* Schick test; in such cases the injected toxin is neutralized and causes little or no effect.

Schiff's reagent Leucofuchsin—i.e. basic FUCHSIN which has been decolourized by treatment with sulphurous acid. Schiff's reagent is used e.g. in the FEULGEN REACTION and the PAS REACTION.

schizogony A form of asexual reproduction characteristic of certain groups of sporozoan protozoa. Coincident with cell growth nuclear division occurs several or numerous times—the original cell, now multinucleate, being termed a *schizont*. The schizont subsequently divides (by a process analogous to budding) to form a number of uninucleate cells termed *merozoites*. (cf. SPOROGONY)

Schizogregarinida An order of protozoans of the sub-class GREGARINIA. *Schizogregarines* are distinguished from members of the EUGREGARINIDA by the occurrence of SCHIZOGONY; they are parasites of arthropods and other invertebrates. Genera include e.g. *Selenidium* and *Spirocystis*.

Schizomycetes 'Fission fungi'—the BACTERIA.

schizont (see SCHIZOGONY)

Schizophyceae (see BLUE-GREEN ALGAE)

Schizophyllum commune A species of fungus of doubtful taxonomic affinity; it is currently classified in the family Schizophyllaceae (order APHYLLOPHORALES) and was formerly classified in the order Agaricales. The organism grows saprophytically on various types of wood. *S. commune* forms a non-stipitate, typically fan-shaped, bracket-type fruiting body of diameter approximately 4 centimetres; the upper surface of the fruiting body is grey-brown and finely hairy. The underside of the fruiting body bears a number of thick, gill-like ridges which radiate out from the point at which the fruiting body is attached to the substratum; each ridge is deeply indented, longitudinally, and the hymenium lines the surfaces of the indented region. The hyaline, elongate basidiospores are approximately 6 × 3 microns.

Schizosaccharomyces A genus of fungi within the family SACCHAROMYCETACEAE; species have been isolated from e.g. molasses and grape must. The vegetative form of the organisms is a single spheroidal or cylindrical cell—or a true mycelium which exhibits a tendency to fragment. Asexual reproduction occurs by *fission*—i.e. it involves septum formation (as opposed to budding). Sexual reproduction typically involves the fusion of (haploid) somatic cells, and the development of a 4-, 6-, or 8-spored ascus direct from the zygote. All species ferment glucose; none ferments galactose or lactose. *Schizosaccharomyces* spp do not assimilate nitrate.

Schizotrypanum A sub-genus of TRYPANOSOMA within the group STERCORARIA; species are intracellular parasites e.g. in man, other primates, and rodents. The organisms are small (length about 20 microns); the (relatively large) kinetoplast is situated close to the pointed posterior extremity of the cell, and there is a free flagellum. The type species is *T. (Schizotrypanum) cruzi*.

schlepper (*immunol.*) Any substance which, by forming a complex with a non-antigenic or weakly-antigenic substance, is able to promote the antigenicity of the latter.

schlieren system An optical system for measuring a refractive index *gradient*.

schmutzdecke The microbial film or layer which develops in a slow sand filter (see WATER SUPPLIES); the layer contains algae, bacteria and protozoa.

Schüffner's dots Numerous fine dots often seen in suitably stained erythrocytes infected with certain stages of the malaria parasites *Plasmodium vivax* or *P. ovale*; in Giemsa-stained preparations the dots appear pink. The nature of these dots is unknown. Similar dots which may be seen in erythrocytes infected with certain stages of *P. malariae* are termed *Ziemann's dots*. *Maurer's clefts* (or dots) are stainable cytoplasmic inclusions occasionally seen in erythrocytes infected with certain stages of *P. falciparum*.

scintillon (see BIOLUMINESCENCE)

Scleroderma A genus of fungi of the order SCLERODERMATALES.

Sclerodermatales An order of fungi of the class GASTEROMYCETES. Constituent species form globose or reniform fruiting bodies (several centimetres in diameter) in which the basidia are scattered throughout the gleba (i.e. an hymenium is not formed) and in which the gleba (with the basidiospores) forms a powdery, typically dark-coloured mass at maturity; when mature, the thick, tough peridium becomes dry and ruptures. In most or all species the basidiospores are sessile (i.e. sterigmata are not formed). The genera include e.g. *Scleroderma*.

Sclerospora A genus of fungi of the class OOMYCETES.

sclerotium (*mycol.*) The resting form of certain fungi; sclerotia are hard, resistant, typically plectenchymatous bodies which, under favourable conditions, may give rise e.g. to somatic mycelium or to sexual or asexual fruiting bodies. (see also *sclerotium* under ACELLULAR SLIME MOULDS) The sclerotium of *Claviceps purpurea* (see *ergot* under CEREALS, DISEASES OF) is a black, elongated body about 1 centimetre in length; it is the source of ergotamine and other drugs. (see also BLACKFELLOWS' BREAD)

Sclerotium A genus of fungi of the class AGONOMYCETES; species include a number of plant pathogens—e.g. the causal agent of *white rot* of onion (see ONION, DISEASES OF). Some species have an ascomycetous sexual stage while others have a basidiomycetous sexual stage.

scolecosporae (see SACCARDIAN SYSTEM)

scopoletin 6-methoxy-7-hydroxycoumarin; scopoletin occurs in a number of higher plants—often as *scopolin* (scopoletin-7-glucoside). It accumulates in the tissues of certain plants infected with certain pathogens, e.g. potato tubers infected with *Phytophthora infestans*, tobacco plants infected with *Pseudomonas solanacearum*. Scopoletin inhibits IAA-oxidase and may be a factor in the pathogenesis of certain diseases. (see also AUXINS)

scopula (*ciliate protozool.*) In the larvae of sessile peritrichs and suctorians: a specialized group of cilia involved in stalk formation.

scotochromogen Any strain of *Mycobacterium* which develops a definite pigment (yellow or orange carotenoid pigments) when incubated either in the light or in the dark. (cf. PHOTOCHROMOGEN) The buff or 'yellowish' coloration exhibited by *M. tuberculosis* does not constitute pigmentation in the present context. Scotochromogens occur among slowly-growing mycobacteria e.g. *M. scrofulaceum*, and among the rapidly-growing species, e.g. *M. phlei*.

scours, scouring (see DYSENTERY (bacterial))

SCP Single-cell protein. (see MICROORGANISMS AS FOOD)

scrapie A disease which affects mainly sheep; following a long incubation period (months, years) the animals appear to develop an intense itch which is accompanied by increasing emaciation and weakness until death ensues. The causal agent is unknown.

screening test (1) Any test, often a shortened (or qualitative) form of a more comprehensive (or quantitative) test, used for the detection of a given agent (e.g. a particular type of microorganism) in each of a relatively large number of specimens. In such tests a certain number of false-positive results may be tolerated; specimens which give a positive screening test may then be examined by more detailed or quantitative tests. (2) Any test the results of which enable a given (unknown) species to be assigned to one of a possible range of taxa.

scrofula Tuberculosis of the cervical (neck) lymph nodes.

scrub typhus (Tsutsugamushi disease; Japanese River Fever) An acute, infectious, systemic, human disease which occurs mainly in South-East Asia, Japan, and Northern Australia. The causal agent, *Rickettsia tsutsugamushi* (*R. orientalis*), occurs in wild rodents and in blood-sucking Trombiculid mites which are parasitic on the rodents; man becomes infected through the bite of an infected mite. The incubation period is 1–3 weeks. The initial symptom is frequently the primary lesion (*eschar*) which develops at the location of the bite; subsequent symptoms are similar to those of *epidemic typhus* (see TYPHUS FEVERS). Tetracyclines, chloramphenicol, or PABA have been used therapeutically. *Laboratory diagnosis* may include the WEIL-FELIX TEST.

Scutellinia A genus of fungi of the order PEZIZALES; the organisms form apothecia which bear brown or black, pointed, hair-like processes (*setae*). *S. scutellata* is common on rotting wood; the organism forms brick-red apothecia which are usually several millimetres in diameter.

scutulum (see RINGWORM)

scyllitol (see INOSITOL)

scyphus (*pl. scyphi*) (*lichenol.*) A cup-shaped or funnel-shaped structure which forms the terminal part of a PODETIUM in certain species of CLADONIA. The *margin* of the scyphus may bear stalked or sessile apothecia.

sea lettuce The alga *Ulva lactuca*.

seasoning (of timber) (see TIMBER, DISEASES OF)

secondary homothallism (homoheteromixis) (in fungi) Some fungi which are basically heterothallic can give rise to individuals which are self-fertile; such individuals are said to exhibit *secondary homothallism*. Secondary homothallism may arise e.g. when nuclei of *compatible* mating type (see HETEROTHALLISM) are incorporated in a given spore; on germination the spore gives rise to a self-fertile thallus. (In some species—e.g. *Neurospora tetrasperma*—this is a regular occurrence.)

In one form of secondary homothallism (*homodimixis*) compatibility between nuclei is of the 1-locus 2-allele type; in another form of secondary homothallism (*homodiaphoromixis*) compatibility between nuclei is *diaphoromictic* (see DIAPHOROMIXIS).

secondary hyphae (in basidiomycetes) (see HYPHA)

secondary metabolites Metabolic products which appear to play no part in cell growth and which are formed maximally under conditions of restricted growth or in the absence of growth; typically, a given secondary metabolite is produced only by a limited number of organisms, i.e. one or a few species. Examples of microbial secondary metabolites include PENICILLINS, polyenes, gibberellic acid, etc.

secondary response (*immunol.*) The response of an individual to the second or subsequent contact with a specific antigen—see ANTIBODY-FORMATION and CELL-MEDIATED IMMUNITY.

secondary type zoospores (in oomycetes) (see DIPLANETISM)

secotioid gasteromycetes (see AGARICALES)

Secotium A genus of fungi—see AGARICALES.

sectoring (in a colony) The development of a sector or wedge-shaped region of growth which differs (in appearance) from the rest of the colony.

sedimentation coefficient (s) (see SVEDBERG UNIT)

segregation (genetics) The distribution, to progeny cells, of homologous chromosomes or chromatids during MEIOSIS (or following *mitotic* crossing-over—see PARASEXUALITY IN FUNGI). The meaning of segregation is sometimes extended to include the distribution of chromatids to daughter cells during MITOSIS. (cf. *Post-meiotic segregation* under RECOMBINATION)

segregation, principle of (genetics) In diploid (or polyploid) organisms or cells: the alleles of allelic pairs retain their individual identities (whether or not they are expressed) under diploid (or polyploid) conditions and are separated (i.e. *they segregate*) during MEIOSIS.

segregation lag (in bacteria) If a mutation occurs in one chromosome of a multinucleate bacterium, a delay of one or more cell divisions may elapse before the mutation is expressed in the phenotype; this delay may represent the time required for segregation of the mutant chromosome to a cell which contains no wild type (non-mutant) chromosomes (*segregation lag*). (see also Helmstetter-Cooper model in GROWTH (bacterial))

Seitz filter (see FILTRATION)

Selas filter (see FILTRATION)

selective medium (see MEDIUM)

Selenidium A genus of protozoans of the order SCHIZOGREGARINIDA.

selenite broth (selenite F broth) An ENRICHMENT medium used e.g. for the isolation (from faeces etc.) of species of *Salmonella*—including *S. typhi*; the medium is not suitable for the isolation of *Salmonella cholerae-suis*. Selenite inhibits many of the common enteric bacteria. The medium consists of an aqueous solution of peptone (0.5%), mannitol (0.4%), disodium hydrogen phosphate (1.0%), sodium hydrogen selenite ($NaHSeO_3$) (0.4%); the pH of the broth should be about 7.0. The broth should not be autoclaved but may be steamed for 30 minutes. Incubation of the inoculated broth (at 37 °C) should not exceed 12–18 hours.

Selenomonas A genus of gram-negative, anaerobic, motile (laterally-flagellated), kidney- or crescent-shaped bacteria of uncertain taxonomic affiliation; *Selenomonas* spp ferment glucose with the formation of acetic and propionic acids, while some strains can also ferment lactic acid. *S. ruminantium* occurs in the RUMEN of many ruminant species.

self purification (in polluted waters) The *natural* process in which organic material (faeces etc.) is degraded (see MINERALIZATION) and the resulting simple substances (nitrates etc.) are used as nutrients by photosynthetic organisms. Self purification can occur only if the polluting load is not excessive. (see also SAPROBITY SYSTEM)

selfing (*ciliate protozool.*) The occurrence of AUTOGAMY (instead of conjugation) in ciliates that have paired.

SEM Scanning electron microscope—see ELECTRON MICROSCOPY.

semi-conservative replication (of DNA) (see DNA)

Semple vaccine An anti-RABIES vaccine consisting of a phenol-inactivated preparation of *rabbit-fixed* rabies virus, i.e. rabies virus passaged in rabbit brain.

Sendai virus A strain of parainfluenza virus *1*—a member of the genus PARAMYXOVIRUS.

sensitivity agar (see ANTIBIOTICS, RESISTANCE TO)

sensitization (*immunol.*) (1) (of cells) The union of antibodies with their homologous cell-surface antigens; such *sensitized* cells are lysed in the presence of a lytic form of complement. (2) (of an individual) The *initial* exposure of an individual to specific antigen or allergen such that an HYPERSENSITIVITY reaction occurs when the subject is again exposed to that antigen or allergen. (3) (*serol.*) The coating (sensitization) of erythrocytes with soluble antigens (see BOYDEN PROCEDURE) such that they can be used in passive haemagglutination tests (see PASSIVE AGGLUTINATION).

Sephadex (see DEXTRANS)

septal pore cap (parenthesome) (see SEPTUM (*mycol.*))

septate (*mycol.*) (of hyphae, spores etc.) Containing one or more *septa* (see SEPTUM).

septicaemia (blood poisoning) The condition, attended by severe symptoms, in which the blood contains large numbers of bacteria—which may or may not be actively dividing; from the blood the organisms may localize—forming abscesses etc. in various parts of the body. (see also PYAEMIA; cf. BACTERAEMIA) (The term may also be used to refer to the presence of certain fungi—e.g. *Candida albicans*—in the bloodstream.)

Septobasidiales An order of fungi of the subclass PHRAGMOBASIDIOMYCETIDAE; constituent species occur e.g. as parasites of scale insects.

Septoria A genus of fungi of the order SPHAEROPSIDALES; species include a number of plant pathogens—e.g. *S. apiicola*, the causal agent of *late blight* (*leaf spot*) of celery. *Septoria* spp form hyaline, filiform (thread-like), multiseptate conidia in ostiolate pycnidia that are immersed in the substratum; at least some species have sexual stages in the loculoascomycetes.

septum (cross wall) (bacterial) The partition, formed during cell division, which divides the parent cell into two daughter cells. In gram-positive bacteria septum-formation appears to occur mainly or exclusively by centripetal growth of the CELL WALL at a pre-determined site; the completed septum has the thickness of two cell walls and forms one of the polar wall sections of each of the daughter cells. This type of septum-formation is characterized by the absence of any major reduction in cell width (diameter) at the site of septation. Septa, analogous to those described above, have been observed in some gram-negative cells; in others, cell division appears to involve the *constriction* of the cell wall—a process in which the width of the cell becomes progressively smaller at the site where cell separation is to occur.

Separation of daughter cells is believed to involve the activity of AUTOLYSINS. (see also GROWTH (bacterial))

septum (*pl. septa*) (*mycol.*) One of a number of internal, transverse cross walls which occur at intervals of length within each hypha in certain types of (septate) fungi; septa also delimit individual cells within the multicellular spores of certain fungi (e.g. *Fusarium* spp), and are produced during division by the fission yeasts (e.g. *Schizosaccharomyces* spp). Septate hyphae are characteristic of the higher fungi but are also formed by certain species of the lower fungi; some lower fungi form septa in ageing hyphae and/or may form septa which delimit reproductive structures (e.g. sporangia).

A simple type of septum is typical of the ascomycetes and *fungi imperfecti* (and occurs in some basidiomycetes); this form of septum is a plate-like structure which commonly contains a single, central pore of diameter approximately 0.4–1.0 micron. Such pores permit the passage of cytoplasm and nuclei from one hyphal cell to the next. (*Micropores*—pores of diameter approximately 50 nanometres—have been observed in the septa of *Geotrichum candidum*.) The *dolipore* septum occurs in many basidiomycetes; in this type of septum the septal material immediately surrounding the central pore is considerably thickened—such that the pore forms a small tunnel (diameter approximately 0.1 micron) which passes through the expanded central part of the septal plate. A perforated cup-like or cap-like structure (the *septal pore cap, pore cap*, or *parenthesome*) occurs on each side of the septum such that, in cross-section, these structures appear to form brackets (parenthesis) around the septum. Parenthesomes are believed to be formed from the endoplasmic reticulum.

Septa generally appear to be formed by the inward growth of annular rims which develop on the inner surface of the hyphal wall—cessation of growth near the hyphal axis accounting for the existence of a central pore, where present; at least some types of septum appear to have a multilayered structure. It is often assumed that septa have a chemical composition similar to that of the hyphal wall. (Certain chytrids form *pseudosepta*—i.e. structures analogous to true septa but consisting of material unlike that of the hyphal wall.)

serial dilutions A set of dilutions (of a given sample) in which the dilution factor increases in a regular fashion, e.g. 1/10, 1/20, 1/30 or 1/2, 1/3, 1/4 etc. *Doubling dilutions*: a particular form of serial dilution in which the dilution factor progressively doubles, e.g. 1/2, 1/4, 1/8, 1/16 etc. *Log dilutions*: 1/10, 1/100, 1/1000 etc.

serial passage (1) Any procedure in which a given pathogen (usually a virus) is transferred from one to another of a succession of animals, tissue cultures, or synthetic growth media—growth (or replication) taking place in each animal, tissue culture or other medium. Serial passage is often used to *attenuate* a particular pathogen (i.e. to reduce or abolish its pathogenicity for a given host); certain VACCINES (e.g. the SABIN VACCINE) are prepared from attenuated pathogens. For example, the rabies virus may be attenuated by adapting it to chick embryo tissues—the virus being transferred from one to another of a succession of hens' eggs; when replication has occurred in a given egg the virus is transferred to the next, and so on. (Such a virus is said to have been *avianized.*) Adaptation to chick embryos reduces the pathogenicity of the rabies virus for man and other mammals. (The LEP strain (*q.v.*) of rabies virus has been used for anti-rabies vaccination in dogs, and the HEP strain has been used in man.) The essential features of successful passaging are that the pathogen should (a) become non-pathogenic for a given host, or hosts, and (b) retain its specific antigenicity in order to stimulate the formation of protective antibodies when used in a vaccine. (2) In TISSUE CULTURE, *passage* refers to the transfer of an inoculum of cells (from an existing culture) to fresh growth medium in another culture vessel, i.e. subculture.

***Sericesthis* iridescent virus** (see IRIDESCENT VIRUSES)

series A taxonomic group (see TAXONOMY) sometimes used in mycology—particularly in the classification of ascomycetes and basidiomycetes. In the taxonomic hierarchy a series (which may be divided into sub-series) falls between the taxa subclass and order.

serine (biosynthesis of) (see Appendix IV(c))

serology The *in vitro* study of ANTIGENS and ANTIBODIES, and their interactions.

serotonin 5-Hydroxytryptamine. A physiologically active base derived (indirectly) from the decarboxylation of tryptophan. It occurs in *mast cells* and in the blood platelets of many species. Like HISTAMINE, its effects include rapid contraction in certain smooth muscles, but its contribution to anaphylactic reactions is probably significant only in the rabbit.

serotype One of a number of antigenically-distinguishable members of a single bacterial species; serologically, bacterial strains may exhibit differences which are not apparent from the results of biochemical (metabolic) tests. (see also SALMONELLA)

Serpula A genus of fungi of the family CONIOPHORACEAE; species include *S. lacrymans* (formerly *Merulius lacrymans*)—the causal agent of DRY ROT. In *S. lacrymans* the fruiting body is typically a thick layer of soft, leathery tissue that is initially pale grey but which becomes rust-red when the spores develop; the fruiting body commonly exhibits a grey or whitish border of mycelial growth. The hymenial surface may be wrinkled or shallowly pitted, or may develop (geotropically) in the form of a layer of irregularly-shaped, pendulous, tooth-like processes (the so-called 'stalactite' form of fruiting body). The spores are ellipsoidal, approximately 10 × 5 microns.

Serratia A genus of gram-negative, motile bacteria of the ENTEROBACTERIACEAE; strains occur e.g. in soil and water, and there is evidence that some are pathogenic for man. The organisms give a positive CITRATE TEST and (typically) a negative METHYL RED TEST and a positive VOGES-PROSKAUER TEST; acid and gas may or may not be produced from glucose, and lactose is fermented slowly or not at all. *Some* strains produce a non-diffusible reddish pigment, PRODIGIOSIN; pigment production is generally greater at 25 °C than at 37 °C. All strains are included in the species *S. marcescens*.

serum The fluid fraction of *coagulated* blood; serum differs from PLASMA e.g. in that it does not contain fibrinogen (a precursor in the clotting mechanism). Normal serum contains e.g. various nutrients, electrolytes, albumins, IMMUNOGLOBULINS, and waste products.

serum hepatitis (see HEPATITIS)

serum sickness A form of type 3 reaction (see IM-

MEDIATE HYPERSENSITIVITY) which occurs some days after the *initial* administration of a *large* amount of antigen—e.g. following *passive* immunization. In such cases antigen is still circulating when antibody first appears in the plasma, and *soluble* antigen–antibody complexes are formed as a result of the antigen-excess situation. A generalized inflammatory reaction occurs, lesions being formed e.g. in the joints and the glomeruli of the kidneys. (see also ARTHUS REACTION)

sessile (1) (*bacteriol., protozool.*) Attached to the substratum. (2) (*mycol.*) Of a fruiting body: attached *directly* to the substratum, thallus, or fungal stroma—i.e. without a stalk. Of a spore: attached *directly* to the sporophore; thus, e.g. the (sessile) basidiospores of the smut fungi arise directly from the basidia (i.e. there are no sterigmata).

Sessilina A suborder of protozoans of the PERITRICHIA.

seta (*pl. setae*) (*mycol.*) A microscopic, bristle-shaped structure; setae often form part of a reproductive structure—e.g. they may occur among the conidiophores in an acervulus or sporodochium, arise from the surface of a pycnidium, or form a fringe around the apex of a pseudoperithecium (e.g. that of *Venturia inaequalis*).

setose (*mycol.*) Bearing *setae*—see SETA.

sewage fungus A slimy, filamentous deposit found on rocks etc. in sewage-polluted waters; the main component is the bacterium *Sphaerotilus natans*.

sewage treatment Sewage consists of domestic and/or industrial effluents, and contains substances in suspension, in colloidal form and in solution. Sewage pollution in a river etc. is inimical to the natural flora and fauna and leads to a decrease in the amenity value of the locality; it also encourages the spread of water-borne diseases e.g. typhoid, cholera. Wastes containing much organic material are particularly harmful because they exert a high *biochemical oxygen demand* (BOD); BOD refers to the requirement for molecular oxygen which accompanies the microbial oxidation of biodegradable substances in the sewage. (High-BOD effluents tend to de-oxygenate waters into which they are discharged; the latter may become anaerobic, or near-anaerobic, with consequent asphyxiation of fish and other organisms dependent on adequate dissolved oxygen. The *BOD test* measures the oxygen (mg) consumed per litre of sewage—or a known dilution of sewage—during 5 days at 20 °C.)

Sewage is thus treated to (a) reduce its BOD, and (b) to inactivate its complement of pathogenic microorganisms; however, even treated sewage exerts *some* oxygen demand on the receiving waters, and it frequently contains pathogenic bacteria and viruses. *Primary sewage treatment* may consist merely of passing the raw sewage through a screen of metal bars in order to separate and dispose of the grosser débris; additionally, the screened sewage may be passed through a *comminutor*, i.e. a rotary cutting and shredding device. A more satisfactory form of primary treatment consists in allowing the screened and comminuted sewage to remain for a time in a sedimentation tank—where some of the finer particles in suspension settle out and can be removed as a sludge. *Secondary sewage treatment* aims at reducing the BOD of the sewage to acceptable levels by microbial oxidation of the dissolved organic materials. There are two main methods:

(a) The *activated sludge* process. Effluent from the primary settling tank is piped to a vessel containing *activated sludge*—a mass consisting mainly of saprophytic bacteria and protozoa; the whole is vigorously agitated and aerated for several hours and is then transferred to a sludge settling tank—from which the final effluent is subsequently discharged. During this process, much of the organic matter in the sewage is assimilated or oxidized by the sludge microflora, and the BOD of the sewage becomes considerably reduced. The mass of activated sludge increases—due to microbial multiplication—and a proportion must be periodically removed and disposed of (see below)—while some is returned to the reaction vessel as the inoculum for a fresh volume of sewage. Activated sludge commonly contains e.g. the bacteria *Sphaerotilus natans* and *Zoogloea ramigera*, and species of the protozoa *Opercularia* and *Vorticella*. (b) The *biological filter* consists of a bed of broken rock (usually contained within a circular brick wall) over which the sewage is sprayed—often from a rotary sprinkler. The pieces of rock bear surface films of a characteristic microflora—e.g. *Zoogloea ramigera* and species of *Carchesium*, *Opercularia* and *Vorticella*; these organisms reduce the sewage BOD by assimilating or oxidizing organic materials in the slowly-percolating sewage, and many of the sewage bacteria are ingested by protozoa of the filter microflora. (The system is not a *filter* as such; its purpose is to permit intimate, aerobic contact between the sewage and the microflora of the rock film.) The effluent usually contains particles of microbial film which have been washed from the rocks; these particles are allowed to settle in a *humus tank* and the resulting supernatant discharged as the final effluent.

Tertiary sewage treatment aims at reducing still further the BOD of the final effluent; such treatment may be required e.g. when the effluent is to be discharged into a river in which the dilution factor is relatively low. The process generally involves reducing the amount of suspended matter in the effluent from the sludge settling tank or the humus tank. The *microstrainer* is a hollow cylinder of fine-mesh stainless-steel fabric which rotates on a horizontal axis; one end of the cylinder is closed. The effluent to be treated is pumped into the cylinder and the strained effluent passes out through the mesh; material retained on the inner surface of the cylinder is constantly removed by jets of water sprayed in through the top of the cylinder. Other forms of tertiary treatment include slow sand filters and land irrigation.

The object of secondary and tertiary sewage treatment is usually regarded as the elimination of *carbon* compounds; this process is clearly important since it is mainly by the breakdown and elimination of organic carbon that the BOD of the effluent can be reduced to acceptable levels. It has been suggested that a form of tertiary treatment be used to eliminate the effluent *nitrogen* (present mainly as NH_3 or NH_4^+); residual nitrogen may (a) encourage the formation of algal blooms in the receiving waters, and (b) tend to increase the BOD of the effluent—molecular oxygen being required for the oxidation of NH_3 to NO_3^-. It has been proposed that the elimination of effluent nitrogen be achieved by microbial NITRIFICATION followed by DENITRIFICATION (direct conversion of NH_3 to N_2 appears not to occur in nature).

Methods used to dispose of the sludge derived from sewage treatment include *anaerobic digestion*—which reduces the mass of sludge and gives a less-offensive material which can be de-watered in sludge-drying beds; gases (e.g. methane) produced during this process may be used to drive the motors, generators etc. of the sewage treatment plant.

sex factor (*bacteriol.*) (1) A *transmissible* PLASMID, i.e. a plasmid which achieves its own transfer by promoting CONJUGATION; such a plasmid confers on the cell in which it occurs the properties of a conjugal *donor*. Originally, two classes of sex factor were distinguished on the basis of the morphology, phage specificity, and antigenicity of the SEX PILI which they specify; sex factors which specify sex pili similar to those specified by the F FACTOR (or Col I factor—see COLICIN FACTOR) are said to be *F-like* (or *I-like*). Additional classes are now distinguished.

In most sex factors the genes which specify the ability to conjugate (constituting a 'transfer operon') are normally *repressed* (see OPERON); such factors are thus normally transferred at very low frequencies. (Other genetic determinants carried by the plasmid—e.g. genes for antibiotic resistance—may continue to be expressed.) The F factor is exceptional in that it is normally *de-repressed* and is thus transferred at high frequency. However, if an F factor shares a cell with a related, self-repressed sex factor, the repressor produced by the latter can often repress the expression of the former—see e.g. *fi*$^+$ R factors under R FACTOR. When a cell is shared by two unrelated sex factors, each appears to be expressed independently of the other. When cells which contain a repressed sex factor are mixed with a recipient population, initial transfer of the sex factor occurs infrequently (*low frequency transfer*, LFT). However, in the newly-infected recipient cells, the sex factor becomes *de-repressed* and can be transferred very efficiently throughout the remainder of the population—resulting in an 'epidemic spread' of the sex factor. Such a population is called an HFT (*high frequency transfer*) population. High frequency transfer may be maintained for several generations before low frequency transfer (LFT) is resumed. The reason for this phenomenon is unknown.

The presence of one sex factor in a cell frequently appears to prevent the entry of a second, identical or related sex factor; thus, e.g. an F$^+$ cell cannot normally receive a second F factor. This phenomenon is known as *exclusion* (*entry exclusion*, *surface exclusion*); the mechanism is unknown. (cf. *superinfection immunity* under PLASMID) Exclusion can be overcome under certain conditions—e.g. F$^+$ bacteria which behave as F$^-$ recipients (*F$^-$ phenocopies*) under conditions of starvation can accept a second F factor.

Some sex factors can promote—in addition to their own transfer—the transfer of bacterial genes during conjugation. In the case of the F factor, frequency of chromosome transfer is greatly increased by *integration* of F with the chromosome (see CONJUGATION). It is unknown whether or not integration is a prerequisite for chromosome transfer by other sex factors; transfer is known to occur in situations in which integration has not been demonstrated.

(2) The term *sex factor* may be applied only to those genes which specify conjugation—i.e. the transfer factor or transfer operon.

(3) *Sex factor* is frequently used as a synonym of F FACTOR.

sex factor affinity locus (*sfa* locus) (see *bacterial* CONJUGATION)

sex pili (sex fimbriae) (*sing.* sex pilus) A class of PILI whose formation is controlled by genes on a SEX FACTOR (cf. FIMBRIAE). Sex pili are much less numerous than fimbriae—only one or two (per cell) commonly being present; it has been suggested that one sex pilus may be formed for each sex factor present in the cell. Sex pili differ from fimbriae in several ways: certain phages (see ANDROPHAGES) adsorb to sex pili, but not to fimbriae; sex pili and fimbriae are distinct antigenically; a sex pilus commonly possesses a terminal (distal) swelling or 'knob'—which consists either of fragments of cell wall or of disorganized pilin. Sex pili specified by the F FACTOR are called *F-pili*; the protein subunit (*F-pilin*) has a molecular weight of about 12,000. Sex pili specified by COLICIN FACTOR I are called *I-pili*. These two types of sex pili are distinguishable antigenically, morphologically, and by their adsorption of different types of phage. Sex pili specified by other plasmids may resemble either the F-pilus (*F-like pili*) or the I-pilus (*I-like pili*).

Sex pili appear to be essential for bacterial CONJUGATION—both to maintain cell contact and, possibly, to function as an organelle through which DNA is transferred from donor to recipient. However, sex pili have not yet been detected on the conjugal donors of certain species of *Pseudomonas*.

sexduction (see *bacterial* CONJUGATION)

sexual dimorphism (in fungi) The condition in which the male and female sexual structures are present on separate thalli. In some organisms, e.g. *Blastocladiella variabilis*, the male and female thalli appear to be genetically identical. In other organisms there are genetic differences between the male and female thalli; such organisms are *morphologically* heterothallic (see HETEROTHALLISM).

sexuality in bacteria A sexual process similar to that in higher organisms has not been found among the bacteria; thus, no observation has been made of complete nuclear and cytoplasmic coalescence between two bacterial cells. However, zygotes, or partial zygotes (see MEROZYGOTE), are formed during CONJUGATION, and nuclear material is transferred between cells during TRANSFORMATION and TRANSDUCTION.

***sfa* locus** (see *bacterial* CONJUGATION)

SH antigen (see HEPATITIS B ANTIGEN)

shadowing (shadow-casting) (see ELECTRON MICROSCOPY)

shake culture A method formerly used for the culture and isolation of ANAEROBES. The inoculum is dispersed, by shaking, in a molten agar medium (about 48 °C) contained in a long glass test-tube; the medium may incorporate a reducing agent. After the medium has set the tube is incubated aerobically for an appropriate period of time, cut open, and growth from individual colonies in the lower (anaerobic) part of the agar cylinder removed for subculture. A *Veillon tube* may be used instead of a test-tube. This may be prepared from a 25 cm length of wide glass tubing (1 cm internal diameter) which has the lower end sealed with a rubber bung and the upper end plugged with cotton wool. Following incubation the bung and plug are removed and the agar cylinder is forced out into a sterile container. (see also ANAEROBIC JAR and POUR PLATE)

Sharka disease (see PLUM POX)

sheath (bacterial) In some species of bacteria: a tubular structure formed around a filament (chain of cells) or around a bundle of filaments. Sheathed bacteria include *Sphaerotilus natans* and *Thioploca*. (*Sheath* is also used to refer to the cell wall of members of the SPIROCHAETALES.)

sheep pox (variola ovina) An acute, infectious viral disease to which sheep of all ages are susceptible; no other type of animal appears to be susceptible under natural conditions. The causal agent, the *sheep pox* virus—a POXVIRUS, probably enters the body *via* the mouth or nose or *via* abrasions etc.; following an incubation period of 4–7 days the infected animal becomes febrile, and vesicular lesions subsequently develop—mainly on the face, lips and nostrils, but sometimes—in addition—in the digestive and/or respiratory tracts. During the course of the disease (up to about 1 month) viruses are shed in the secretions of the lesions and in the milk. Mortality rates may rise to about 50% among sheep and may reach 100% in lambs. Sheep pox is a NOTIFIABLE DISEASE in Great Britain.

shelf fungi (see BRACKET FUNGI)

sherry (see WINE-MAKING)

shield cells (*algol.*) (see Charophyta under *sexual reproduction* in ALGAE)

shift up, shift down Refers to those conditions under which the growth rate (of micro-organisms) is increased (shift up) or decreased (shift down).

Shiga-Kruse bacillus *Shigella dysenteriae* type 1.

Shigella A genus of gram-negative, non-motile bacteria of the ENTEROBACTERIACEAE: causative agents of gastro-enteritis and bacterial DYSENTERY. The cells are straight rods, 0.5–1 × 1–3 microns. Growth occurs on basal media (e.g. NUTRIENT AGAR) and on MACCONKEY'S AGAR—but may or may not occur on more inhibitory media. With few exceptions the organisms produce acid only (no gas) from glucose and from other fermentable sugars. The CITRATE TEST and the VOGES-PROSKAUER TEST are negative and the METHYL RED TEST is positive; growth does not occur on potassium cyanide media, and H_2S is not produced on TSI medium. Lysine decarboxylase is not produced. Species are distinguished and subdivided biochemically, serologically, and by

COLICIN TYPING. *Shigella* dysentery has been treated with ampicillin, tetracyclines, kanamycin, colistin and by certain other antibiotics. (Resistance to particular antibiotics may be determined by an R FACTOR.) The type species is *S. dysenteriae*.

S. dysenteriae comprises at least 10 (sero)types—collectively referred to as 'subgroup A'. These strains do not ferment mannitol, sucrose or lactose. Serotype 1 does not produce catalase but the remaining serotypes do. Serotype 1 produces an exotoxin.

S. flexneri (formerly *S. paradysenteriae*). Most of the strains (collectively known as subgroup B) ferment mannitol; sucrose and lactose are not fermented. The *Newcastle* strain of serotype 6 does not ferment mannitol, and produces gas during glucose fermentation; the *Manchester* strain ferments mannitol and produces gas.

S. boydii (subgroup C) comprises at least 15 serotypes—all of which ferment mannitol.

S. sonnei (subgroup D) consists of a single mannitol-fermenting serotype which *slowly* ferments *lactose* and sucrose, and produces ornithine decarboxylase.

shii-take The edible basidiomycete *Lentinus edodes* (= *Cortinellus edodes*) which is cultivated and eaten in Japan. *L. edodes* is grown on wood.

shikimate (3,4,5-trihydroxybenzoate) An intermediate formed in the biosynthesis of aromatic amino acids—see Appendix IV(f).

shingles (herpes-zoster) An acute disease (of man) characterized by an eruption of varicella-like vesicles along the paths of certain peripheral nerves—e.g. the intercostal nerves; the causal agent, a herpesvirus, is identical to that which causes varicella. The lesions are commonly unilateral (found only on one side of the body) and usually appear on the trunk; they may persist for two or more weeks. Shingles may represent a resurgence of virus which has remained latent since an earlier attack of varicella. The disease may develop spontaneously or may appear to be provoked e.g. by chemotherapy or by certain unrelated diseases. (see also CHICKENPOX)

shipping fever (of cattle) (see PNEUMONIA)

Shirlan (see SALICYLIC ACID AND DERIVATIVES)

shock movements (of bacteria) (see TAXIS)

shocking exposure (*immunol.*) (1) Any administration of antigen which brings about ANAPHYLACTIC SHOCK. (2) More loosely, any administration of antigen which causes one or other of the immunological HYPERSENSITIVITY reactions.

shoe-strings The black RHIZOMORPHS of *Armillaria mellea*. (see TIMBER, DISEASES OF)

Shope rabbit papilloma virus (see PAPILLOMAVIRUS)

Shwartzman reaction Symptoms which occur, in man and animals, within hours of a second administration of ENDOTOXIN given some 24 hours after the first. If the first dose is given intradermally and the second intravenously, the result is a localized skin lesion involving vascular necrosis, petechial haemorrhages, and infiltration by polymorphs at the (first) injection site. If both administrations are given intravenously the symptoms may include damage to the kidneys and other organs and may result in death. The Shwartzman reaction appears not to fit into any of the established categories of immunological reaction.

sialic acids (see NEURAMINIC ACIDS)

sickener, the A common name for the poisonous fruiting body of the basidiomycete *Russula emetica*. *R. emetica* is commonly found under or near the conifer *Pinus pinaster*.

sideramines (see SIDEROCHROMES)

Siderocapsaceae A poorly-defined family of gram-negative coccal, coccoid and rod-shaped bacteria which deposit iron or manganese oxides in gelatinous CAPSULES or (in species which lack capsules) void such oxides into the medium; species occur in chalybeate waters. Genera include *Siderocapsa*.

siderochromes (ironophores) Compounds which are produced by a wide range of micro-organisms and which are believed to be involved in the uptake of iron by those organisms. Siderochrome synthesis can be induced by growth on media deficient in iron; when the compounds are secreted into the medium they chelate ferric iron, and the ferric-siderochrome complex is actively transported into the cells. (The growth of some organisms, which do not synthesize siderochromes, is dependent on the provision of these compounds in the growth medium.)

Siderochromes are of two main types—one or both of which may be formed by a given organism. (a) *Sideramines*. These compounds are derivatives of *hydroxamic acid* (R.CO.NHOH) which bind strongly to ferric iron. *Ferrichrome*, a sideramine formed by many organisms—e.g. species of *Aspergillus*, *Neurospora*, *Ustilago*, consists of a cyclic hexapeptide composed of a glycine tripeptide and a tripeptide of δ-*N*-acetyl-L-δ-*N*-hydroxyornithine. Other sideramines include the MYCOBACTINS, the *terregens factor* (required for the growth of e.g. *Arthrobacter* spp), *coprogen* (produced by a range of fungi), and the *ferrioxamines* (produced by species of *Streptomyces*). The intracellular release of iron from the ferric-sideramine complex is believed to involve the enzymic reduction of the ferric iron; ferrous iron binds less firmly to the hydroxamate than does ferric iron. (see also SIDEROMYCINS) (b) Derivatives of *catechol* (*o*-

dihydroxybenzene). For example, certain members of the Enterobacteriaceae (e.g. *Escherichia coli*, *Salmonella* spp) produce *enterochelin* (= *enterobactin*)—a cyclic trimer of 2,3-dihydroxy-*N*-benzoyl-L-serine. The ferric iron bound by catechol derivatives cannot be reduced; it is believed to be released intracellularly by the action of an esterase on the ligand—2,3-dihydroxy-*N*-benzoyl-L-serine being excreted.

(Some authors use the term *siderochrome* as a synonym of *sideramine*.)

sideromycins A class of ANTIBIOTICS which are synthesized by certain actinomycetes; sideromycins contain chelated iron and are structurally related to the *sideramines* (see SIDEROCHROMES). The antibiotic action of sideromycins can be antagonized by sideramines. Sideromycins include e.g. *albomycin* (a composite antibiotic) and *ferrimycin*—analogues of ferrichrome and ferrioxamine respectively.

***sigma* virus** A bullet-shaped RHABDOVIRUS which infects the fruit-fly (*Drosophila*). Following a brief exposure to carbon dioxide (at certain concentrations) infected flies are paralysed and die within a few hours; non-infected flies recover completely.

silage An animal feed prepared by the controlled fermentation of certain green crops, e.g. grasses, maize etc. (Young grasses with a high carbohydrate content are ideal.) The vegetation is packed tightly into a closed container (a *silo*) and molasses may be added to encourage fermentation. Fermentation is carried out by the microflora of the vegetation; the material becomes anaerobic and the pH falls as fermentation proceeds. Initially, the active metabolizers are species of the Enterobacteriaceae, but later the predominant organisms are lactobacilli and streptococci —which form lactic, acetic, butyric and propionic acids; the acidity inhibits a range of putrefactive organisms (thus preserving the vegetable protein) and imparts a flavour which is relished by cattle and sheep. Fermentation is complete in about a month, and the product may be stored with little depreciation.

silicalemma (see CELL WALL (algal))

silkworm diseases Diseases of the silkworm (*Bombyx mori*) include: (a) PÉBRINE. (b) *Muscardine*; causal fungus: *Beauveria bassiana*. (c) FLACHERIE. (d) Hyperplasia in *B. mori* has been associated with the Tipula iridescent virus. (e) *Cytoplasmic* POLYHEDROSES.

silver (and compounds) (as antimicrobial agents) (see HEAVY METALS)

silver leaf (*plant pathol.*) A disease of certain trees, e.g. plum. The causal agent, *Stereum purpureum*, enters the tree *via* wounds and grows within the branches or trunk. The characteristic 'silvering' of leaves is believed to be due to the production of fungal toxins. *S. purpureum* also grows saprophytically causing discoloration of wood but little decay.

silver line system (*protozool.*) A regular pattern of lines observed on or within the cortical layer of silver-stained *ciliates*; these lines of deposited silver were first described by Klein in the 1920s. While some of the lines appear to coincide with cell-surface detail others appear to correspond with the locations of certain sub-pellicular structures. Thus, there is evidence that the main (most densely stained) lines in *Tetrahymena* correspond to the rows of basal bodies which mark the locations of the kineties; these (longitudinal) lines have been termed *primary meridia*. Other more weakly-staining lines—observed between the primaries—have been termed *secondary meridia*. Cross-striations have also been observed. Some workers believe that lines of deposited silver are formed at the junctions of the pellicular alveoli. (see also PELLICLE)

single burst experiment A method for studying the results of phage-mediated lysis of a *single* bacterial cell. The single burst experiment can be used e.g. to determine the BURST SIZE of an individual cell under particular conditions. Essentially, a bacterial culture is infected with phage and subsequently diluted such that each of a large number of aliquots has a low probability of containing a phage-infected cell. Following incubation and lysis, the counts of plaque-forming units in the range of aliquots can be statistically analysed and used to assess individual burst sizes. The single burst experiment is also useful for investigating recombination in bacteriophage: a genetic analysis is made of the phage progeny derived from a single bacterium which has been infected with two genetically-different phages.

single-cell protein (see MICROORGANISMS AS FOOD)

single diffusion (*immunol.*) (see GEL DIFFUSION)

single-dimension (in gel diffusion) (see OUCHTERLONY TEST; see also GEL DIFFUSION)

single-site mutation An alternative name for POINT MUTATION.

singlet oxygen An electronically-excited, uncharged, hyper-reactive form of oxygen. Singlet oxygen is formed e.g. when oxygen accepts energy from photoexcited FLUORESCENT DYES; it is a major cause of the cell damage associated with the *photodynamic effect*. Singlet oxygen may be produced during the spontaneous (non-enzymic) dismutation of SUPEROXIDE ANIONS, and has been reported to be formed during ULTRASONICATION. Singlet oxygen is effectively quenched by CAROTENOID pigments.

sinuate (*mycol.*) In an agaric: refers to a gill

whose lower edge is shallowly indented near the stipe.

siphonaceous In certain algae: refers to a multinucleate (i.e. coenocytic), aseptate thallus. (Septa divide the reproductive structures from the remainder of the thallus, and may be formed on wounding.) Typically, a siphonaceous thallus contains a peripheral layer of protoplasm (containing chloroplasts, nuclei etc.) surrounding a central vacuole. Siphonaceous algae include *Botrydium*, *Bryopsis*, *Codium*, *Halimeda*, *Vaucheria*.

sirenin (see HORMONAL CONTROL (of sexual processes in fungi))

skeletal hyphae (see HYPHA)

skin, common microflora of The organisms most commonly found on human skin include *Staphylococcus albus* (and sometimes *Staph. aureus*), *Sarcina* spp, diphtheroids, and yeasts—including *Candida albicans*; organisms which may also be present include various coliforms and species of *Streptococcus*, *Proteus*, and *Pseudomonas*.

skin tests (*med.*) Tests in which specific antigenic substances are introduced into the skin to determine whether or not they can elicit one or other of the characteristic DELAYED HYPERSENSITIVITY reactions. A positive reaction may indicate previous or current involvement with such antigens. Skin tests may be used to assist in medical diagnoses—see e.g. the TUBERCULIN TEST.

slant (see CULTURE)

sleeping sickness A chronic disease (of man) in which the causal agent may be either *Trypanosoma gambiense* or *T. rhodesiense*. Infection is *inoculative*, i.e. may result from the bite of an infected insect vector. Recurrent fever and general weakness are followed, after several months or a year, by the characteristic symptoms of malaise and tiredness—which reflect the activity of the pathogen in the central nervous system. The disease produced by *T. rhodesiense* (chief vector: *Glossina morsitans*) takes a more rapid course than that due to *T. gambiense* (chief vector: *G. palpalis*). Sleeping sickness is commonly fatal if untreated; chemotherapeutic agents include SURAMIN and *melarsoprol* (see ARSENICALS). *Laboratory diagnosis*. Trypanosomes may be observed in wet mounts or stained preparations of peripheral blood; in later stages of the disease the cerebrospinal fluid may be examined. Sleeping sickness occurs in tropical Africa; an unrelated, non-trypanosomal disease is also called 'sleeping sickness'.

slide (1) A piece of flat, transparent, colourless glass, commonly about 75 × 25 × 1 mm, on which a specimen is placed, or a SMEAR made, for microscopical examination. A *cavity slide* is thicker than that described above, and has a shallow, circular depression of approximately 10 mm diameter in the centre of one face. A *ring slide* (used e.g. for slide agglutination tests) is larger and thicker than either of the above slides; on one face it bears a number of fixed, raised rings of ceramic (or other material) of between 1 and 3 centimetres in diameter. (2) The completed preparation (specimen or smear) made on a slide. A suitably prepared specimen (e.g. a fixed, dehydrated and cleared section of tissue) may be made into a permanent preparation by mounting it (e.g. in CANADA BALSAM) and overlaying it with a small square of thin glass (approximately 23 × 23 × 0.17 mm) known as a *cover-slip* or *cover-glass*. (Some cover-slips are circular or oblong.) When dry, the balsam cements the cover-slip to the slide.

slide agglutination test An AGGLUTINATION test in which small amounts of antibody and antigen preparations are mixed and allowed to interact on the surface of a slide. Such a test is used only when agglutination takes place rapidly (of the order of minutes). Low power microscopic examination of the slide may be used to detect agglutination. Slide agglutination tests are often used as screening tests.

slime layer (bacterial) (see CAPSULE)

slime moulds (Mycetozoa*, Myxobionta, Myxomycota) An heterogenous group of eucaryotic organisms some of which are classified in both mycological and protozoological taxonomic schemes. The group includes the ACELLULAR SLIME MOULDS (= Myxomycetales, Myxomycetes, or 'true' slime moulds), the CELLULAR SLIME MOULDS (= Acrasiales, Acrasida, Acrasina, Acrasiomycetes), the NET SLIME MOULDS (= Labyrinthulales), and the endoparasitic slime moulds (members of the PLASMODIOPHORALES). By 'slime moulds' some authors refer specifically to the acellular and cellular slime moulds collectively. *Mycetozoa is used by some authors to refer collectively to the acellular and cellular slime moulds and to a recently-discovered group of organisms, the *protostelids*.

slipper animalcule Common name for *Paramecium*.

slope A solid (usually agar-based) medium which has been allowed to set in a diagonally-arranged test-tube or bottle.

A *potato slope* is a wedge-shaped piece of potato which has been placed in a test-tube or bottle and autoclaved; potato slopes are used as growth or diagnostic media for certain microorganisms.

sloppy agar *Semi-solid* AGAR.

SM medium A medium used for the culture of CELLULAR SLIME MOULDS. The medium contains (per 100 ml) glucose (1 gram), peptone

(1 gram), yeast extract (0.1 gram), and magnesium sulphate (0.1 gram)—all in phosphate-buffered 2% agar (medium pH: 6.4). The medium is inoculated with suitable bacteria (e.g. *Escherichia coli*) and with spores/myxamoebae; incubation is carried out at approximately 22 °C.

smallpox (variola major) In man, an acute, highly communicable disease caused by the *variola major virus*, a poxvirus; infection probably occurs mainly *via* the respiratory route. (Fomites, e.g. clothing or scabs from patients, remain infectious for extended periods of time.) The incubation period is commonly 10–15 days. Typically, symptoms begin suddenly with fever, headache, nausea and prostration. Within days the cutaneous lesions appear; these develop from macules through vesicles to pustules—which subsequently form scabs. Characteristically, lesions are concentrated on the face and limbs, and are relatively sparse on the trunk. The mortality rate may be 10–50%—higher with confluent or haemorrhagic smallpox. *Variola sine eruptione* is a mild form of smallpox in which cutaneous lesions do not develop; it may occur e.g. in individuals who have recovered from smallpox or who have been vaccinated against it. *Alastrim* (variola minor, amaas) is a form of smallpox in which cutaneous and other symptoms develop but in which the mortality rate is well below 1%; causal agent: the *variola minor virus*. *Laboratory diagnosis* may include: (a) Microscopical/cultural examination of lesion material. (see also GUARNIERI BODIES) (b) The use of fluorescein-conjugated anti-poxvirus antibody (for fluorescence microscopy). (c) Serological tests: antibodies are detectable about 1 week after the onset of symptoms.

smear Any material (blood, wound exudate etc.) which is prepared for microscopical examination as a thin film on a SLIDE; such a thin film is often dried and stained prior to examination under the microscope.

smoking (of food) (see FOOD, PRESERVATION OF)

smooth-rough variation (S → R variation) (in bacteria) A change which may occur in the cell-surface composition in certain species of bacterium; when such a change occurs the organism usually gives rise to colonies which differ, in appearance, from those of the original strain. Variations in colony form were first recorded for species of the Enterobacteriaceae; these organisms typically give *smooth* (glossy, shiny) colonies and, following S → R variation, *rough* (dull, matt) colonies. The expression 'smooth-rough variation' (S → R variation) is now used to refer to any of a number of different types of cell-surface variation which occur in bacteria. The commonest mechanism of smooth-rough variation during culture probably involves a process of genetic selection. In general, a change from the 'smooth' to the 'rough' condition may involve some or all of the following: (a) the formation of colonies which have an altered appearance; (b) the loss of specific smooth-strain surface antigens; (c) a reduction in, or loss of, virulence (in pathogenic species); (d) an alteration in the degree of susceptibility to certain bacteriophages; (e) an increase in spontaneous saline-agglutinability—i.e. rough strains exhibit a greater tendency to agglutinate spontaneously in saline. Many types of S → R variation are reversible; a change from the S to the R state, or *vice versa*, may occur spontaneously on subculture, or may be induced e.g. by varying the components of the growth medium.

In members of the Enterobacteriaceae, *smooth* (S) strains are those in which the cell wall LIPOPOLYSACCHARIDES are entire, i.e. they include the complete *O-specific chain*. (see also O ANTIGEN) *Rough* (R) strains are mutant strains in which the LPS components lack O-specific chains. In rough strains the cell-surface antigen consists of core polysaccharide + lipid A; this antigen (the *R antigen**) may be very similar, or identical, among organisms of a given genus, or of related genera. In some strains not only is the O-specific chain missing, but the LPS also lacks the distal portion of the core polysaccharide; such strains are referred to as *deep roughs*. In *semi-rough* mutants (SR mutants) only the initial oligosaccharide subunit of the O-specific chain occurs in the LPS; these strains appear to lack the ability to polymerize the (normally repeating) subunits of the O-specific chain. Smooth strains are agglutinated by homologous anti-*O antigen* antibodies; rough strains usually fail to agglutinate with any anti-*O antigen* antisera. (For this reason, *smooth* strains are used in agglutination tests such as the WIDAL REACTION.) Smooth-rough variation may proceed independently of any flagellar variation (e.g. PHASE VARIATION, IN SALMONELLA).

Analogous changes in cell-surface components occur in bacteria other than members of the Enterobacteriaceae—including gram-positive species. In pathogenic species the change S → R is, by definition, associated with a reduction in, or loss of, virulence; however, owing to the wide diversity in the composition of cell-surface components, S (virulent) and R (avirulent, or less virulent) strains do not always correspond to 'smooth' and 'rough' colonies respectively. For example, virulent strains of *Streptococcus* may form *mucoid* colonies, while less virulent, or avirulent, strains may form *glossy* colonies; the change mucoid → glossy in this case would be an instance of S → R variation. In e.g. *Strep.*

pneumoniae, S → R variation is associated with the loss of the CAPSULE. *A different, *protein* antigen—also referred to as the 'R antigen'—occurs in the cell wall in certain strains of STREPTOCOCCUS.

smooth strains (of bacteria) (see SMOOTH-ROUGH VARIATION (in bacteria))

smudge (see ONION, DISEASES OF)

smuts (1) (smut fungi) Fungi of the order USTILAGINALES; all smuts are plant parasites and some species cause economically-important diseases in certain crops—see e.g. SMUTS (2) and CEREALS, DISEASES OF.

Morphology and life cycles of smut fungi. The main vegetative (somatic) form of the organisms is a hyaline *dikaryotic* mycelium which may occur systemically or locally in the plant host; the hyphae are commonly intercellular (although *Ustilago maydis* forms intracellular hyphae) and clamp connections are formed by some species. Haustoria may or may not be formed, according to species. Some smuts form conidia, but the majority of species form only two types of spore: *teliospores* (also called brand spores, chlamydospores, pseudospores, resting spores, smut spores, teleutospores) and *basidiospores*; some species do not form basidiospores—see below. Teliospores are the disseminative units and resting stages of the smut fungi, and those of certain species have been reported to remain viable for more than ten years. Teliospores arise from the dikaryotic hyphae; each hyphal cell rounds off and develops a thick wall so that a large number of (typically dark-coloured) teliospores are formed. In some species the teliospores are formed in a loose mass—i.e. the spores are separate or are more or less easily separable; in other species the teliospores may develop in the form of *spore balls*: definite structures which each consist of a characteristically-arranged group of teliospores or of teliospores and sterile material—such arrangements often being genus-specific. *Smut sori* (i.e. discrete masses of teliospores or of spore balls) may occur on leaves or stems or in fruits etc. according to the particular species of smut fungus and host plant. Immature teliospores are dikaryotic but at maturity each spore contains a single diploid nucleus. On germination, each teliospore typically forms a basidium (= *promycelium*) from which a number of (haploid) basidiospores (also referred to as 'sporidia') are budded off—for details see BASIDIUM. (In some species the basidiospores can, *in vitro*, give rise to a yeast-like phase characterized by the repeated budding of basidiospores and their daughter cells; the budded cells are referred to as 'sporidia' or 'sprout cells'.)

Germ tubes which are formed by the uninucleate, haploid basidiospores are usually unable to bring about infection in fresh host plants; infection is commonly or exclusively effected only by *dikaryotic* hyphae. DIKARYOTIZATION may take place in any of a variety of ways—e.g. by the fusion of basidiospores of appropriate mating types, or by the fusion of germ tubes formed by two basidiospores of appropriate mating types. (Many smut fungi are known to be heterothallic.) Some smut fungi (e.g. *Ustilago nuda*, *U. tritici*) do not form basidiospores; in such species each promycelium gives rise to uninucleate, haploid hyphae among which dikaryotization occurs by somatogamy.

(2) Plant diseases which are caused by fungi of the order Ustilaginales and which typically involve the formation of masses of dark-coloured *teliospores*—see SMUTS (1)—on or within the tissues of the plant host. In *covered smuts* of cereals the smut sori remain intact within the glumes of the host plant; thus, the spores are normally not released until the ear is threshed. In *loose smuts* the spore masses are exposed and readily dispersed by the wind etc. (see also CEREALS, DISEASES OF)

soaps (as antimicrobial agents) In general, soaps (sodium salts of long-chain fatty acids) have little antimicrobial activity, although their surface active properties can be exploited to reduce the skin microflora. Some soaps are given positive antimicrobial properties by the incorporation of certain ANTISEPTICS—see e.g. BISPHENOLS; *carbolic* soap contains PHENOL, and some antiseptic soaps contain TCC (3,4,4′-trichlorocarbanilide). Critical concentrations of soap are required to solubilize some disinfectants—see e.g. *lysol* under PHENOLS. Some antimicrobial agents (e.g. QUATERNARY AMMONIUM COMPOUNDS) may be inactivated by soaps.

sodium azide (NaN_3) An antimicrobial agent used e.g. as a preservative (0.1% final concentration) in laboratory reagents and similar products. It is also used in certain selective media (0.025% w/v NaN_3) which are employed in the isolation of *faecal* streptococci from samples of sewage-polluted water; at this concentration coliforms and many other gram-negative bacteria are inhibited. (see also AZIDE)

sodium perborate (as an antimicrobial agent) (see OXIDIZING AGENTS)

sodoku (see RAT-BITE FEVER)

soft rot (1) (of timber) (see TIMBER, DISEASES OF) (2) (of vegetables) (see ROT)

Solorina A genus of foliose LICHENS which contain the green alga *Coccomyxa and* the blue-green alga *Nostoc*. Apothecia occur in the upper surface of the thallus. Species include e.g.: *S. crocea*. The thallus is lobed—greenish-brown or grey above, deep orange-red below.

Nostoc occurs within the thallus as a layer below the green algal layer. Apothecia are flat or slightly convex; the disc is reddish-brown. *S. crocea* occurs in damp soil at high altitudes.
S. saccata. The thallus is lobed and the upper surface is bright green when wet, greyish-green when dry; the lower surface is white to pale brown and tomentose. *Nostoc* occurs within the thallus in discrete colonies. Apothecia are sunk deeply into the upper surface of the thallus; the disc is brown to black. *S. saccata* occurs mainly on soil among limestone rocks.

solvent fermentation (see ACETONE-BUTANOL FERMENTATION)

somatic (1) In general: refers to assimilatory, non-reproductive structures, functions etc. (2) Refers to the 'body' or main part of a cell. (see also SOMATIC ANTIGEN)

somatic antigen (of bacteria) Any ANTIGEN which forms part of the main body of a cell, usually at the cell surface, and thus distinguished from antigens which occur on the flagella or capsule. (see also O ANTIGENS)

somatogamy (*mycol.*) A form of plasmogamy which involves the fusion of somatic (vegetative) cells or hyphae.

sonication (as an antimicrobial procedure) The process in which audible sound waves—produced by a SONICATOR—are used for the destruction of microorganisms in a liquid medium. More effective antimicrobial activity can be obtained by ULTRASONICATION.

sonicator (sonifier etc.) An instrument which provides a source of sound energy for SONICATION or ULTRASONICATION. Sonicators commonly operate on the principle of *magnetostriction*, i.e. the resonant deformation exhibited by certain types of metal in the presence of an alternating magnetic field; the resonating metal is coupled, mechanically, to a metal *probe* (a cylindrical metal rod) which dips into the liquid and acts as the transmitter of sound energy. The sound waves commonly used have frequencies within the approximate range 10 kc/s to 25 kc/s, and amplitudes of about 10–50 microns. (1 kc/s (kilocycle/second) = 1 kHz (kiloHertz) = 1000 cycles/second.)

soralium (*lichenol.*) A delimited mass of soredia (see SOREDIUM) which occurs on the thallus in certain lichens. The site and form of the soralia are characteristic for a given species; thus e.g. soralia may be described as *marginal* (i.e. occurring at the margins of the thallus), *capitate* (i.e. occurring at the tips of lobes or branches of the thallus), *laminal* (occurring on the upper or lower surface of the thallus), etc.

sorbic acid ($CH_3.CH{=}CH{-}CH{=}CH.COOH$) (as a food preservative) Sorbic acid is inhibitory to fungi and to certain catalase-positive bacteria—particularly under acid conditions (pH 5 or below); it is used as a preservative in pickles, marzipan, cheeses and other products, and for the impregnation of cheese wrappers. Sorbic acid is a permitted preservative in the U.K., U.S.A. and other countries. (see also FOOD, PRESERVATION OF)

sorbitol (glucitol) The polyhydric alcohol corresponding to glucose—i.e. sorbitol is formed on reduction of the aldehyde group (C1) of glucose; it occurs, with DULCITOL, in certain algae: e.g. *Iridaea*, *Bostrychia*. Sorbitol is used as a substrate in biochemical characterization tests for bacteria; it is attacked by many species—including strains of Lancefield group *C* streptococci, *Escherichia coli*, *Klebsiella pneumoniae*, and many strains of salmonellae.

sorbose A 2-ketohexose related to SORBITOL.

Sordaria A genus of fungi of the order SPHAERIALES, class PYRENOMYCETES; species occur on dung and in the soil. *Sordaria* spp form (non-evanescent) asci in dark-coloured perithecia which develop at the surface of the substratum; the neck of the perithecium is positively phototropic. The ascospores are brown or black. Many species of *Sordaria* do not form conidia; thus, e.g. *S. fimicola* (which is homothallic) reproduces solely by ascospore formation.

soredium (*lichenol.*) A vegetative propagule, 25–100 microns in diameter, which consists of a few algal cells enmeshed in a few fungal hyphae; there is no cortex. Soredia originate in the medulla of the lichen thallus and are pushed up through pores or cracks in the cortex; this may result in the formation of a cluster of soredia (a SORALIUM) on the surface of the thallus.

Soret bands In spectroscopy: absorption bands in the region 400 nm—characteristic of conjugated tetrapyrroles.

sorocarp (see CELLULAR SLIME MOULDS)

sorus (*pl.* sori) (*mycol.*) A mass of spores or sporangia.

souma (see NAGANA)

South American blastomycosis (see PARACOCCIDIOIDOMYCOSIS)

sp An unspecified *species*. Example of use: *Bacillus* sp (a species of *Bacillus*). spp indicates several or a number of unspecified species or *may* indicate all the species of a particular genus.

sparsomycin An ANTIBIOTIC which inhibits PROTEIN SYNTHESIS in both procaryotic and eucaryotic cells; it binds to the larger ribosomal subunit and inhibits the peptidyl transferase reaction, i.e. it inhibits peptide bond formation.

spasmoneme The MYONEME within the stalk of a contractile peritrich (such as *Vorticella*).

spawn In MUSHROOM culture: certain materials (e.g. manure) containing a strong growth of mycelium; it is used as an inoculum when initiating a fresh culture.

specialized transduction (see TRANSDUCTION)

species (*microbiol.*) (*singular and plural*: species) One of the smallest (i.e. least inclusive) of taxonomic groupings (see TAXONOMY): a species is a category of individuals which display a high degree of mutual similarity—the characteristics which circumscribe the category being defined by a consensus of informed opinion. Species differentiation in many microorganisms (e.g. the fungi) is frequently based on morphological or other differences; in bacteria, small differences in metabolic characteristics may be regarded as sufficient grounds for the creation of separate species. Organisms within a given species may exhibit character variation due to the effects of MUTATION, PLASMID transfer, and/or RECOMBINATION. (see also BINOMIAL and NOMENCLATURE)

specific epithet That part of a BINOMIAL which distinguishes a given species from others in the same genus; thus, e.g. in *Streptococcus equi*, '*equi*' is the specific epithet. '*Streptococcus*' is the *generic name*.

specific immunity The IMMUNITY, specific to a given antigen, which is produced when that antigen elicits an IMMUNE RESPONSE.

specific pathogen free (SPF) Denotes the condition of an animal which was born (or removed from the uterus) and reared under conditions in which specific pathogens have been rigorously excluded. (see also GNOTOBIOTIC)

specific phase antigens (of *Salmonella*) (see PHASE VARIATION, IN SALMONELLA)

specificity (of an antibody) Refers to the range of antigens with which a given antibody may combine; thus, in addition to the homologous antigen, the antibody may function as a CROSS-REACTING ANTIBODY to any of a range of stereochemically-similar antigens.

spectinomycin An antibiotic produced by *Streptomyces spectabilis*. Although often classified with the AMINOGLYCOSIDE ANTIBIOTICS it contains no amino-sugar and has fewer basic groups than are found in the other members of this class. It inhibits PROTEIN SYNTHESIS at the RIBOSOMES as do the aminoglycosides; however, its action is reversible. Spectinomycin does not cause misreading of the genetic message—cf. STREPTOMYCIN. One-step high-resistance mutations occur.

spermagonium (spermogonium; spermatogonium) In certain ascomycetes and basidiomycetes: a structure within which male reproductive cells (*spermatia*) are formed. (In the rust fungi—see RUSTS (1)—spermagonia are referred to as *pycnia*.)

spermatiophore (*mycol.*) A modified or undifferentiated hypha which bears a SPERMATIUM.

spermatium (*mycol.*) In certain ascomycetes and basidiomycetes: a non-motile male reproductive cell; spermatia formed by members of the Uredinales (rust fungi) are referred to as *pycniospores*. (see also MICROCONIDIUM)

spermatization (*mycol.*) In certain ascomycetes and basidiomycetes: the union of a SPERMATIUM, MICROCONIDIUM, or pycniospore (see RUSTS (1)) with a female reproductive structure (e.g. a trichogyne or a flexuous hypha).

spermogonium (see SPERMAGONIUM)

Spermophthoraceae A family of fungi of the ENDOMYCETALES.

SPF Specific pathogen free (*q.v.*).

Sphacelotheca A genus of fungi of the order USTILAGINALES.

Sphaeriales An order of filamentous (mycelial) fungi within the class PYRENOMYCETES, subdivision ASCOMYCOTINA. In species of this order the ascocarp is a PERITHECIUM or (infrequently) CLEISTOTHECIUM which, in the majority of species, contains an hymenial layer of (unitunicate) asci and paraphyses; the order *excludes* (i) organisms which form *cleistothecia* on or among mycelium external to the substratum, and (ii) organisms in which the cleistothecium contains a quellkörper. During sexual reproduction plasmogamy may occur in any of a variety of ways (depending on species)—e.g. gametangial contact, somatogamy, spermatization, conidiation. According to species the ascocarp may be spherical, hemispherical, pyriform, or elongate, and the peridium may be fleshy, membranous, or carbonaceous—either brightly coloured or dark*; a perithecium may develop on the surface of the substratum (as e.g. in *Chaetomium*), on the surface of a fungal *stroma* (as e.g. in *Nectria*), or within (i.e. embedded in) a fungal stroma (as e.g. in *Claviceps*, *Daldinia*, *Xylaria*). According to species, ascospores may be unicellular or septate, and may be spherical, ovoid, reniform, or filamentous—colourless or coloured. Constituent genera include CERATOCYSTIS, CHAETOMIUM, CLAVICEPS, *Cordyceps*, DALDINIA, EPICHLOË, GIBBERELLA, *Hypoxylon*, NECTRIA, NEUROSPORA, PODOSPORA, SORDARIA, and XYLARIA. *In some taxonomic schemes the order Sphaeriales is restricted to those species which form an hymenial layer of unitunicate asci in a dark, membranous or carbonaceous perithecium.

Sphaerobolus A genus of fungi of the order NIDULARIALES.

sphaerocysts (*mycol.*) Spherical cells—clusters of which form part of the sterile tissue of the fruiting body (pileus, gill tramae, stipe) in members of the RUSSULACEAE.

sphaeromastigote A form assumed by certain members of the family Trypanosomatidae during a particular stage of development; the sphaeromastigote is a rounded cell with a free (external) flagellum. (cf. AMASTIGOTE)

Sphaerophorus A genus of fruticose LICHENS in which the phycobiont is a green alga. The grey to greyish-brown thallus is erect and branched and may form tufts up to 5 centimetres in height (*S. globosus*); the branches may be terete or flattened, according to species. Globose *mazaedia* may be formed at the ends of branches. *Sphaerophorus* species occur e.g. on rocks, walls, mossy tree-trunks.

Sphaerophorus necrophorus (see FUSOBACTERIUM)

Sphaerophrya A genus of protozoans of the SUCTORIA.

sphaeroplasts Spherical or near-spherical structures formed from bacteria, yeasts, and other cells, by weakening or *partially* removing the rigid component of the cell wall. According to some authors, resistance to lysis in hypotonic media is a necessary attribute of a sphaeroplast; other authors regard the sphaeroplast as an osmotically-sensitive structure. Thus, opinions appear to differ with regard to the degree to which the wall of a cell must be weakened or removed in order that a sphaeroplast be formed. (cf. PROTOPLASTS and L FORMS)

Sphaeropsidales An order of fungi within the class COELOMYCETES; the (asexual) fruiting body formed by members of the Sphaeropsidales is either a PYCNIDIUM or a fungal STROMA—cf. MELANCONIALES. Genera include ASCOCHYTA, *Chaetomella,* PHOMA, *Phomopsis,* SEPTORIA, and ZYTHIA.

Sphaerotheca A genus of fungi of the family ERYSIPHACEAE; the organisms form cleistothecia which bear unbranched and/or sparingly-branched appendages and which each contain a *single* ascus.

Sphaerotilus A genus of gram-negative, rod-shaped, obligately aerobic, non-sporing bacteria of uncertain taxonomic affiliation; the sole species, *S. natans*, occurs in flowing, organically-polluted fresh water (e.g. sewage-polluted streams) and in activated sludge (see SEWAGE TREATMENT). (see also SEWAGE FUNGUS) Individual cells are about 1–2 × 3–10 microns, and each carries a bundle of subpolar or polar flagella; cells may occur singly, but are often in unbranched, *sheathed* filaments—the sheaths being impregnated or coated with iron oxide in chalybeate waters. The filaments may be attached to the substratum. Reproduction occurs by binary fission. The organisms are chemoorganotrophs, though inorganic electron donors may be used by some strains; poly-β-hydroxybutyrate is a storage product. GC%: about 70.

spheroidoses Diseases caused by ENTOMOPOXVIRUSES.

spherule (*entomopathol.*) (see ENTOMOPOXVIRUSES)

spindle An intracellular organelle present only during MITOSIS and MEIOSIS; it is involved in the alignment of chromatids/chromosomes and in their distribution to progeny cells. Essentially, the spindle consists of a number of MICROTUBULES arranged longitudinally between two 'poles'; each pole is the site of a pair of closely-adjacent CENTRIOLES. The so-called *continuous* microtubules of the spindle extend from one pole to the other, while the *kinetochore* microtubules extend from the KINETOCHORE of each chromatid to one or other of the poles. It has been suggested that the movement (separation) of the chromatids/chromosomes may result from an interaction between the continuous and kinetochore microtubules.

The spindle described above occurs typically in cells of the metazoa and metaphyta—and in certain algae and some other microorganisms. In many algae, protozoans and fungi the spindle is wholly, or mainly, *intranuclear*—the termini of the microtubules often being located on the nuclear membrane; centrioles are commonly absent. In such organisms nuclear division typically involves elongation and constriction of the nucleus—the nuclear membrane persisting throughout.

Spira In streptomycetes, refers to the form of a spore chain, or chains: (a) a helix, of coil diameter less than 5 microns, either compressed or extended along the longitudinal axis, or (b) hook-shaped, or in the form of an extended helix of relatively few turns, and of coil diameter greater than 5 microns. Chains of type (b) are also referred to as the *retinaculum apertum* type. (see also RECTUS FLEXIBILIS)

Spirillaceae A family of *helical* bacteria in which each cell forms an incomplete turn of a helix, or a number of turns, depending on species; the cells are *rigid* (cf. SPIROCHAETALES) and are polarly flagellate. Species occur in fresh and marine waters; some species are parasitic or pathogenic. The genera are CAMPYLOBACTER and SPIRILLUM.

spirilloxanthins (see CAROTENOIDS)

spirillum (*pl.* spirilla) Any *rigid* helical bacterium.

Spirillum A genus of gram-negative, aerobic or microaerophilic, chemoorganotrophic bacteria of the SPIRILLACEAE; species typically occur in fresh and marine waters. The cells are 0.3–1.7 × 2–60 microns; flagellation is lophotrichous, and the organisms exhibit a corkscrew-like motion. Metabolism is obligately oxidative (respiratory); a few species (e.g. *S.*

itersonii) are capable of NITRATE RESPIRATION. All species are catalase-positive, oxidase-positive. Some species produce a water-soluble yellow-green fluorescent pigment. The largest species, *S. volutans*, is the type species; it is obligately microaerophilic. An atypical species, *S. minor* (= *S. minus*), a parasite of rodents, is one cause of RAT-BITE FEVER; the cells are 5 microns (or less) in length.

spirits Alcoholic beverages prepared by the distillation of fermented liquors. The characteristic flavours and qualities of the various types of spirits are determined by the nature of the fermented raw materials, the conditions of distillation, and the effects of ageing. The fermentation stage resembles that in BREWING and WINE-MAKING; endogenous or inoculated strains of the yeast *Saccharomyces* are commonly used. In the manufacture of spirits, use is made of yeast strains which effect maximum *attenuation*, i.e. they convert the maximum amount of carbohydrate to alcohol; these strains are more alcohol-tolerant than the common brewers' yeasts. Furthermore, some strains of distillers' yeasts produce a range of higher alcohols—collectively referred to as *fusel oil*—which contribute to the aroma and flavour of products such as whiskey. In the manufacture of *malt whiskey* the *mash* consists of *malted* barley (see BREWING); *grain whiskey* mashes consist predominantly of *un*malted grain (often barley or maize) with a proportion of malted barley. *Brandy* (*cognac*) is prepared by distillation of grape wine. *Gin* is manufactured by distilling fermented rye or maize worts and redistilling the product in a still which contains herbs and juniper berries ('botanicals'). *Rum* is prepared by distilling fermented molasses. Spirits are not subject to microbial spoilage owing to their high alcohol content. (see also PULQUE)

Spirochaeta A genus of free-living, chemoorganotrophic, obligately or facultatively anaerobic bacteria of the SPIROCHAETALES; species occur e.g. in activated sludge, sewage-polluted waters, and in fresh and salt waters containing H_2S. The cells are 5–500 microns in length; their width is less than 1 micron.

Spirochaetales An order of *flexible*, *helical*, asporogenous bacteria (cf. SPIRILLACEAE). According to species, the (motile) cells range from 3 microns to 500 microns in length, and from less than 0.1 to 3 microns in width; those large enough to be visible by bright-field microscopy are gram-negative. Spirochaetes are chemoorganotrophs; metabolism may be fermentative and/or respiratory (oxidative). Species may be facultatively or obligately anaerobic, or aerobic. The spirochaetes divide by transverse binary fission. The order includes free-living, parasitic and pathogenic species. The sole family, Spirochaetaceae, comprises the genera BORRELIA, CRISTISPIRA, LEPTOSPIRA, SPIROCHAETA and TREPONEMA.

Depending on size, spirochaetes may be seen (a) by bright-field microscopy after STAINING with aniline dyes or silver-deposition techniques, (b) by dark-field microscopy, and/or (c) by fluorescence microscopy following treatment with fluorescein-conjugated antibody preparations. Fine structure is determined by electron microscopy.

Morphology. Basically, the spirochaete is structurally similar to a rod-shaped bacterium in that a membrane-limited protoplast is enclosed in an *outer membrane* or *sheath* (both terms referring to the CELL WALL). One or more *axial fibrils* arise (from plate-like or hook-like structures in the cytoplasm) at *each* pole of the cell. In *Leptospira* a single fibril emanates from each pole and passes (between protoplast membrane and cell wall) to the opposite pole of the cell; these fibrils constitute the straight *axistyle* around which the protoplast is coiled. In other genera each axial fibril passes spirally around the protoplast (between membrane and wall) and terminates at or near the opposite pole of the cell; collectively, the axial fibrils form what appears (by electron microscopy) to be a single, composite, helical structure, the *axial filament*. (In *Cristispira* the axial filament forms a prominent longitudinal ridge, the *crista*.) The number of fibrils which form the axial filament varies with genus. Thus, in *Treponema*, the *total* number of fibrils (i.e. fibrils from both poles of the cell) may be 2, 4 or 6; in *Borrelia* it is usually 10–20, and in *Cristispira* several hundreds.

The characteristic MOTILITY of spirochaetes is believed to be associated with the presumed contractility of the axial fibrils. Recently, the axial fibrils of certain spirochaetes have been shown to be similar, chemically, to bacterial flagella (see FLAGELLUM); accordingly, some authors refer to axial fibrils as flagella. The cell wall contains PEPTIDOGLYCAN but is sufficiently pliable to permit the cell to flex freely.

spirochaetes Members of the SPIROCHAETALES.

Spirochona gemmipara A freshwater protozoan of the order CHONOTRICHIDA. The organism, which is usually about 100 microns in length, is most commonly found on the gill plates of the crustacean *Gammarus pulex*.

Spirocystis A genus of protozoans of the order SCHIZOGREGARINIDA.

Spirogyra A genus of algae of the division CHLOROPHYTA; species are common in freshwater environments and are generally free-floating—though some species occur attached to stones etc. The organism consists of an unbranched filament composed of a chain of cylindrical cells (end to end) which is coated with a layer of mucilage; each cell contains a layer of cytoplasm which lines the cell wall and which surrounds a large central vacuole. The single nucleus occurs in the centre of the vacuole and is connected to the peripheral cytoplasm by cytoplasmic threads. One or more (depending on species) ribbon-like, spirally-arranged chloroplasts are embedded in the peripheral layer of cytoplasm—the helix extending throughout the length of the cell; each chloroplast contains a number of pyrenoids. *Spirogyra* reproduces vegetatively by fragmentation of the filaments. Sexual reproduction occurs by *conjugation*—the protoplasts functioning as gametes. Two filaments become juxtaposed and communication is established—by means of a tubular connection—between pairs of cells (one cell from each filament). (This arrangement is known as *scalariform* conjugation.) In a given pair of cells the cell sap is eliminated from the central vacuoles of one or both protoplasts in a process involving contractile vacuoles; subsequently, a (contracted) protoplast passes from one cell to the other, *via* the conjugation tube, and syngamy occurs. The zygote develops a thick wall and undergoes a period of dormancy. Meiosis occurs prior to germination of the zygote; three of the four products of meiosis abort—so that only one (haploid) vegetative filament develops on germination. Conjugation can also occur between adjacent cells of a single filament (*lateral* conjugation).

Spirostomum A genus of ciliate protozoans of the order Heterotrichida, subclass SPIROTRICHIA. *Spirostomum* is found in fresh and brackish waters—particularly those in which the oxygen content is low or submaximal. The organisms are elongated and cylindrical, up to 3 millimetres in length, with uniform, longitudinally-arranged ciliature. Longitudinal myonemes allow the body to contract to a fraction of its normal length. The cytostome is lateral; food consists mainly of bacteria and algae. A large, posterior contractile vacuole communicates with a canal which runs almost the length of the body. In some species (e.g. *S. ambiguum*) the macronucleus is moniliform; in others (e.g. *S. teres*) it is ovoid.

spirotrich A ciliate of the subclass Spirotrichia.

Spirotrichia A subclass of freshwater and marine protozoans of the CILIOPHORA. *Spirotrichs* have a well-developed and conspicuous oral ciliature; this includes a number of membranelles which, looking into the buccal cavity, spiral downwards towards the cytostome in a clockwise direction. The somatic ciliature is typically sparse or absent, but is uniform in some species of the order Heterotrichida. The subclass includes both free-living and parasitic forms, and contains some of the largest protozoans known. The six poorly-defined orders of the subclass include: (a) Heterotrichida. In *heterotrichs* the somatic ciliature is either absent or uniform, and the body is often large; the order includes pigmented and loricate species. Genera include SPIROSTOMUM and STENTOR. (b) Tintinnida. The *tintinnids* are loricate and motile species which are typically marine. (c) Hypotrichida. *Hypotrichs* are characterized by the presence of *cirri* (see CIRRUS); the body is flattened dorso-ventrally. Genera include EUPLOTES. Other orders of the Spirotrichia: Oligotrichida (e.g. *Halteria*), Entodiniomorphida (e.g. *Diplodinium*), Odontostomatida (originally Ctenostomatida) (e.g. *Saprodinium*).

Spirulina A genus of filamentous BLUE-GREEN ALGAE.

spitzenkörper (apical granule) (*mycol.*) In many types of septate fungi: a small, densely-staining body which occurs close to the cytoplasmic membrane in the tips of *actively* growing hyphae; its function is unknown.

splenic fever (see ANTHRAX)

splitter (see TAXONOMY)

spoilage (of food, paint, rubber etc.) See individual entries, e.g. FOOD, MICROBIAL SPOILAGE OF.

Spongospora subterranea (see PLASMODIOPHORALES)

spontaneous generation (see ABIOGENESIS)

sporabola (*mycol.*) The trajectory of a forcibly-discharged basidiospore.

sporadin A *gametocyte* formed by a member of the GREGARINIA—either from a sporozoite (Eugregarinida) or from a merozoite (Schizogregarinida).

sporangiole (see SPORANGIOLUM)

sporangiolum (sporangiole) (*mycol.*) A type of sporangium which contains a single spore or a small number of spores; in such structures a columella is usually absent. Some fungi form sporangia *and* sporangiola. In some zygomycetes (e.g. *Chaetocladium*) the sporangiolum contains a single spore whose spore wall is distinct from the sporangial wall; other zygomycetes, e.g. *Cunninghamella*, form spores which do not exhibit clearly distinct spore and sporangial walls and which are regarded as true CONIDIA. Many mycologists regard the sporangiolum as the evolutionary intermediate between the sporangium and the conidium.

sporangiophore (*mycol.*) A modified or undifferentiated hypha which bears a SPORANGIUM or a SPORANGIOLUM; sporangiophores may be simple or branched structures, and may be any of a variety of forms and sizes (according to species)—those of some species being several hundred microns in length. The sporangiophores of certain species (e.g. *Pilobolus*) are positively phototropic.

sporangiospores (*mycol.*) Thin-walled, motile or non-motile, uninucleate or multinucleate, asexually or sexually derived spores formed within a SPORANGIUM or a SPORANGIOLUM; the formation of sporangiospores is characteristic of the *lower* fungi. (cf. CONIDIA)

sporangium (*bacteriol.*) (1) The cell in which an endospore is formed. (2) That part of a cell which subsequently develops into an endospore. (3) In the MYXOBACTERALES: a structure containing resting cells. (4) In the Actinoplanaceae: a structure containing spores.

sporangium (*mycol.*) A sac-like structure within which numbers of motile or non-motile, sexually or asexually derived spores (sporangiospores) are formed*; sporangia are the characteristic reproductive structures of the *lower* fungi. The sporangia of the various species differ widely in size and shape; they are commonly globose or elongated and may be produced e.g. singly and terminally on a sporangiophore, in clusters on a branched sporangiophore, or (e.g. in *Albugo*) in basipetally-formed chains. *Merosporangia*—cylindrical sacs, each containing a small number of uniseriate spores—are formed by certain zygomycetes, e.g. *Piptocephalis*, *Syncephalastrum*. In many fungi (e.g. *Mucor*) the sporangium contains a columella. In some species (e.g. *Dictyuchus*) the entire sporangium is deciduous at maturity, while in others (e.g. *Saprolegnia*) it remains attached to the sporangiophore; in either case the spores are generally released *via* pore(s) in the wall of the sporangium or by the dissolution of the sporangial wall. In *Pilobolus* (a zygomycete) the entire sporangium is forcibly discharged from the sporangiophore.

Allomyces spp (which exhibit an alternation of generations) form two types of sporangium on the sporothallus: a thick-walled resistant *meiosporangium* (in which are formed motile, haploid spores), and a thin-walled *mitosporangium* (in which are formed motile, diploid spores).

*In some fungi (e.g. *Phytophthora* spp) the sporangium may give rise either to zoospores or to a germ tube (according to environmental conditions), while in most species of *Peronospora* the deciduous sporangia invariably form germ tubes. (see also SPORANGIOLUM and CONIDIA) In *Pythium* spp the sporangiospores are formed (from the sporangial protoplast) within a vesicle which develops on the sporangium.

spore balls (see SMUTS (1))

spore print (*mycol.*) A spore deposit formed when the pileus (cap) of a mature hymenomycete (e.g. a mushroom or similar fungus) is placed gills-down on a piece of plain paper; a spore print is useful e.g. for determining the colour of the spores.

spores (bacterial) Resistant and/or disseminative forms produced asexually by certain types of bacteria in a process (*sporulation*) which involves differentiation of vegetative cells or structures; spores are characteristically formed in response to particular (commonly adverse) environmental conditions. Under favourable conditions a spore germinates to give rise to a vegetative organism. There are two main types of bacterial spore: (a) The spore typical of actinomycetes (e.g. *Streptomyces* spp). Such spores do not develop *within* hyphae (cf. endospores, below) and are not appreciably more heat-resistant than the corresponding vegetative organisms—though the spores of some species are resistant to desiccation. Spores may be motile or non-motile, according to species; in some species the spores (one, several, many) are formed within closed vessels (*sporangia*) which develop on the vegetative mycelium. (b) The bacterial *endospore*—to which the word 'spore' most commonly refers. An endospore is formed entirely *within* the mother cell or hypha. Endospores are considerably more resistant than are vegetative cells to various inimical agents, e.g. heat, certain types of radiation, certain chemicals; commonly, inactivation (irreversible loss of viability) by means of heat can be achieved only by autoclaving, by boiling for extended periods of time, or by use of the hot-air oven. (see AUTOCLAVE and STERILIZATION) Endospores can remain dormant for extended periods of time: viable endospores (of *Bacillus* spp), estimated to be 500–1000 years old, have been recovered from lake sediments; this suggests that spores are, or can be, in a state of *cryptobiosis*—i.e. a state in which no metabolism is detectable.

Endospore-forming bacteria include species of *Bacillus*, *Clostridium*, *Desulphotomaculum*, and *Sporosarcina*, and species of the (mycelial) actinomycete *Thermoactinomyces*. The spores of all these organisms appear to be basically similar in construction and composition (e.g. all contain dipicolinic acid); the process of sporulation appears to be basically similar in aerobic and anaerobic species. (The following account refers specifically to bacterial *endospores*.)

Sporulation (in *Bacillus* spp). Among the first observable indications of sporulation in a cell is the development of an *axial filament*: two chromosomes which form a longitudinal rod-like mass of DNA. Subsequently the cell membrane invaginates near one end of the cell so as to form two separate protoplasts—each containing a single chromosome; the smaller of the two protoplasts—referred to as the *forespore* or *prespore*—develops to form a single endospore. Initially, the forespore is engulfed by the larger protoplast following invagination of the membrane of the latter; consequently, the forespore is invested with a second membrane (derived from the invaginated region of the larger protoplast). At this stage the forespore—bounded by a double membrane—lies within the cytoplasm of the larger protoplast (the 'mother cell'). The material which forms the spore *cortex* (the rigid layer of the spore wall) is then laid down between the two membranes of the forespore; the rigid material of the cortex is a form of *peptidoglycan* which differs from that found in the bacterial cell wall (see PEPTIDOGLYCAN for details). Subsequently the *spore coat* develops as a lamellate protein structure external to the outer membrane of the forespore; the spore coat protein is rich in cysteine/cystine and lysine residues and is typically hydrophobic. The spores of some species (e.g. *Bacillus cereus*) have a loose outer membrane (the *exosporium*) external to the spore coat; little appears to be known of the composition and origin of this structure. The endospore develops heat resistance only late in sporulation—at about the time of coat formation; the development of thermostability is preceded by the rapid accumulation of calcium ions and the synthesis of DIPICOLINIC ACID. Dipicolinic acid occurs within the protoplast ('core') of the endospore, and its presence has been associated with the thermostability of the mature endospore. In the final event of sporulation the mature endospore is released from the mother cell ('sporangium') following autolysis of the latter.

Dormancy. Freshly-formed spores usually enter a period of *dormancy*— in which little or no metabolic activity, or other changes, can be detected.

Activation is the process which occurs between dormancy and germination; this process is commonly reversible and appears to involve changes in the configurations of the spore macromolecules—presumably early steps in the mobilization of the spore's metabolic potential. Activation of spores may be brought about e.g. by heating them to sublethal temperatures or by subjecting them to low pH or to certain types of chemical; 'ageing' has also been shown to bring about activation.

Germination is the irreversible process in which an activated spore becomes a metabolically-active spore; the process is largely degradative and involves the hydrolysis and depolymerization of certain spore constituents. Germination is characterized by the release, from the spore, of dipicolinic acid, calcium ions, and the breakdown products of cortical peptidoglycan; this is accompanied by a significant decrease in the optical density of the spore, and by a loss of resistance to heat etc. Germination requires, or is promoted by, the presence of particular substances (*germinants*); these may include e.g. certain amino acids (e.g. L-alanine), certain ions (e.g. Mn^{++}), surfactants (e.g. *n*-dodecylamine), glucose, calcium-dipicolinic acid chelate. Germination may also be promoted by subjecting the spore wall to mechanical damage. A number of substances (e.g. D-alanine, sodium bicarbonate) inhibit germination of the spores of some types of bacteria.

Outgrowth is the process in which a vegetative cell develops from a germinated spore. (see also MICROCYCLE SPOROGENESIS)

spores (fungal) (see *aeciospores* under RUSTS (1); ALEURIOSPORE; ASCOSPORES; AZYGOSPORES; BALLISTOSPORES; BASIDIOSPORES; BLASTOSPORE; CHLAMYDOSPORES; CONIDIA; GEMMA; MICROCONIDIUM; OIDIUM; PYCNIDIOSPORES; PYCNIOSPORES; RESTING SPORE; SPORANGIOSPORES; SUMMER SPORES; *teliospores* under RUSTS (1) and SMUTS (1); THALLOSPORE; *uredospores* under RUSTS (1); ZYGOSPORE)

spores (fungal), dormancy in Dormancy is a resting state in which metabolic acitivity is low and there is no synthesis of new cellular material. Dormancy may be *constitutive* or *exogenous*. *Constitutive dormancy* is an inherent property of the spore and is not dependent on external conditions; it may involve: (a) the presence of an endogenous chemical inhibitor in the spore—e.g. CAFFEIC ACID may fulfill this role in the uredospores of *Puccinia graminis tritici*. (The release of such substances by spores may account for the suppression of germination frequently observed in spores *en masse*.) (b) A metabolic block within the spore. (c) The presence of a permeability barrier to nutrients. *Exogenous dormancy* is a dormancy imposed on the spore by environmental conditions—e.g. the absence of water or of a specifically required nutrient, sub-optimal temperature, etc.; exogenous dormancy probably permits germination only under environmental conditions favourable for vegetative growth.

Breakage of dormancy. (The breakage of dormancy results in the germination of the spore.) Exogenous dormancy can be broken simply by eliminating the unfavourable condition(s) which brought it about. Endogenous dormancy can be broken only following *activation* with certain physical or chemical agents (*activators*)—e.g. heat (e.g. in the ascospores of *Neurospora crassa*), cold (e.g. in the teliospores of *Puccinia graminis*), certain chemicals (e.g.

certain organic solvents). Spores which contain endogenous inhibitor(s) can germinate following the removal (by washing) or inactivation (by chemical treatment) of the inhibitor(s); in certain cases mechanical disruption of the spore coat may initiate germination.

sporicide Any chemical agent which irreversibly inactivates spores. (see also STERILIZATION)

sporidium (*mycol.*) (1) A basidiospore formed by members of the order Uredinales (rust fungi). (2) In members of the order Ustilaginales (smut fungi): a yeast-like cell which buds from a basidiospore, or a basidiospore itself.

Sporobolomyces A genus of non-fermentative fungi within the family SPOROBOLOMYCETACEAE; the vegetative form of the organisms is a single ovoid or elongated cell, though some species may form pseudomycelium and/or true septate mycelium. Reproduction occurs by budding and by the formation of reniform or sickle-shaped ballistospores—each of which is orientated *asymmetrically* on its sterigma (cf. BULLERA). In culture the cells of most species appear red or pink.

Sporobolomycetaceae A family of fungi within the class BLASTOMYCETES; species have been isolated from e.g. diseased plants. The family comprises a number of imperfect yeasts and non-budding mycelial organisms; all species form ballistospores. Members of the family Sporobolomycetaceae are believed to be imperfect basidiomycetes; the genera include BULLERA and SPOROBOLOMYCES.

sporocladium (*mycol.*) Part of the (elaborate) sporophore formed by certain zygomycetes (members of the Kickxellaceae).

sporocyst (coccidian) (see COCCIDIA)

sporodochium (*mycol.*) A pad-like or cushion-like fungal stroma which bears a surface covering of CONIDIOPHORES; setae (see SETA) may or may not be present on the sporodochium—according to species. Sporodochia are formed e.g. by *Nectria cinnabarina* and *Fusarium* spp.

sporogony (*protozool.*) In certain sporozoans: the division of the *zygote* into a number of cells (*sporozoites*); sporogony involves meiosis. (see e.g. PLASMODIUM and COCCIDIA; cf. SCHIZOGONY)

Sporolactobacillus A genus of gram-positive, SPORE-forming, chemoorganotrophic, microaerophilic bacteria of the family BACILLACEAE; strains occur e.g. in dairy products. The cells are motile (peritrichous) rods. Lactic acid is produced by the (homofermentative) metabolism of hexoses. Catalase is not produced.

sporont Any cell from which a spore may subsequently be formed.

sporophore (*bacteriol.*) Any structure *on* which spores are formed—e.g. the spore-bearing hyphal branches of certain actinomycetes. (cf. SPORANGIUM)

sporophore (*mycol.*) Any spore-bearing structure—e.g. a conidiophore, a sporangiophore, the pileus of a fruiting body, or an entire, complex fruiting body.

sporophyte (see ALTERNATION OF GENERATIONS)

sporoplasm (1) An intrasporal cell of the CNIDOSPORA. (2) Cytoplasm within a developing ascus or sporangium.

sporopollenins Carotenoid polymers believed to be formed by the oxidative polymerization of carotenes (see CAROTENOIDS). Sporopollenins occur in the zygospore walls of members of the Mucorales (e.g. *Mucor mucedo*), in the walls of certain algae (e.g. species of *Chlorella* and *Scenedesmus*), and in the exine (outer coat) of pollen grains.

Sporosarcina A genus of gram-positive, SPORE-forming, chemoorganotrophic, strictly aerobic bacteria of the BACILLACEAE. The cells are flagellated or non-motile cocci of maximum diameter about 2.5 microns; they occur in tetrads or packets. Metabolism: respiratory.

sporothallus (*mycol.*) A diploid, spore-bearing thallus (cf. GAMETOTHALLUS). In e.g. *Allomyces* spp the sporothallus forms both haploid and diploid spores—the former germinating to form gametothalli while the latter give rise to new sporothalli. (see also ALTERNATION OF GENERATIONS)

Sporozoa A sub-phylum of *parasitic* PROTOZOA in which mature organisms lack cilia and flagella; some species form flagellated gametes. Spores or cysts—which do not contain polar filaments (cf. CNIDOSPORA)—are formed by all groups except the *piroplasms*. In many species the infective stages (sporozoites) are formed within resistant spores or cysts which function as disseminative units; in others (e.g. *Plasmodium* spp) the cysts do not have a disseminative function. Sexual reproduction occurs in many but not all species; SCHIZOGONY is the typical form of asexual reproduction but does not occur in all sporozoans. MOTILITY (involving a poorly understood mechanism) is a feature of some species; pseudopodia are rarely formed. Sporozoans are commonly parasitic in the intestinal tract or blood of vertebrates or invertebrates; some are extracellular, others intracellular parasites. Classification of the sporozoa is based on the nature of the life cycle—which is often complex. Identification is based largely on the morphology of trophozoites, gametes, cysts/spores etc. Three classes are distinguished: TELOSPOREA, PIROPLASMEA, and HAPLOSPOREA.

sporozoite (see SPOROGONY)

sporulation (in bacteria) (see SPORES (bacterial))

sporulation (of coccidian oocysts) (see COCCIDIA)

spotted fevers Diseases of rickettsial aetiology which are transmitted by ticks; examples include ROCKY MOUNTAIN SPOTTED FEVER and BOUTONNEUSE.

spp Plural form of 'sp' (*q.v.*).

spread plate (1) A method of inoculation in which a small volume of liquid inoculum (usually 0.1–0.25 ml) is dispersed, with a sterile glass spreader, over the entire surface of an agar plate. (2) A plate so inoculated. (see also ENUMERATION OF MICROORGANISMS)

spreading factor (see HYALURONIDASE)

sprout mycelium (sprout chain) (*mycol.*) A chain of cells formed by sprouting (*budding*).

spur (*immunol.*) (see OUCHTERLONY TEST)

squamulose (1) Bearing small scales. (2) (of a lichen) Refers to a lichen whose thallus consists of small (of the order of millimetres) scales or lobes (*squamules*)—each of which has one edge free of the substratum; the squamules may be imbricated. The squamulose thallus is regarded as a form intermediate between the CRUSTOSE and FOLIOSE forms of thallus. A squamulose thallus occurs e.g. in *Lecidea scalaris*.

SR mutants (see SMOOTH-ROUGH VARIATION (in bacteria))

stab culture (see CULTURE)

stabilate Any population of microorganisms which has been preserved (e.g. by freeze-drying or low temperature methods) so that the organisms remain viable, with unchanged characteristics, during the period of preservation. (A population of organisms maintained by repeated subculture may be subjected to selective (genetic) changes.)

stabilized erythrocyte (see NEURAMINIDASE)

stachybotryotoxicosis A condition which affects domestic animals (e.g. horses) and man and which results from the ingestion or (in man) the inhalation, or absorption through the skin, of a toxin produced by an organism variously reported as *Stachybotrys alternans* and *Stachybotrys atra*; among horses the toxicosis may be mild or lethal, while in man the effects may include pharyngitis and a decrease in the white cell count.

Stachybotrys A genus of fungi within the order MONILIALES. The vegetative form of the organisms is a septate mycelium; spherical or elongated, dark-coloured conidia are produced in chains from conidiophores which each consist of a straight hypha bearing a terminal cluster of phialids. At least one species is reported to have an ascigerous stage within the order Sphaeriales. (see also STACHYBOTRYOTOXICOSIS)

stages 0, I, II, III and IV (in the life cycles of rust fungi) (see RUSTS (1))

stag's horn fungus The basidiomycete *Lentinus lepideus*. (see TIMBER, DISEASES OF)

staining The treatment of cells or tissues with one or more specific dyes (stains) in order that such cells or tissues (or detail therein) can be detected or differentiated by MICROSCOPY. Staining is usually carried out on dead cells or tissues which have been suitably fixed (see FIXATION)—though certain procedures demand VITAL STAINING. The colour imparted to a particular feature by a given dye does not necessarily correspond to the colour of that dye—see METACHROMASY and METACHROMATIC GRANULES. Heightened contrast between a given stained feature and its surroundings may be achieved by means of a COUNTERSTAIN. In *negative staining* the background (rather than the cell/tissue/feature itself) is stained or made opaque. Negative staining is used e.g. to demonstrate bacterial CAPSULES: the cells are suspended in e.g. India ink (or NIGROSIN) and examined as a SMEAR; negative staining is also used e.g. for the detection of parasitic cysts in samples of faeces (see CYST).

Dyes. The usefulness of a dye often depends on its specificity, i.e. its ability to combine with, adsorb to, or dissolve in cellular components of specific physical or chemical compositions. An *acid dye* is one in which the CHROMOPHORE is an *anion*; such a dye combines with basic (cationic) groups. A *basic dye* is one in which the chromophore is a *cation*; such a dye combines with acidic (anionic) groups. (The pH of a dye solution does not necessarily indicate whether it is an acid or basic dye.) Some dyes are neutral, while others are *amphoteric*—i.e. they behave as acids or bases according to the prevailing pH. The lipophilic or hydrophilic character of a dye has an important bearing on the specificity of that dye; thus, e.g. the ability of a dye to stain a neutral lipid may depend e.g. on the solubility of that dye in the lipid. Dyes which depend on chemical reaction include SCHIFF'S REAGENT (e.g. in the FEULGEN REACTION and the PAS REACTION).

Whether or not a given dye stains a given cellular feature may depend on factors other than the characteristics of the dye; thus, e.g. a dye may be excluded from the cytoplasm by the cell membrane. The ambient pH may also be important; thus, e.g. a protein exposed to a pH on the acid side of its ISOELECTRIC POINT is stained by an acid dye—and on the alkaline side of its IEP by a basic dye. (Hence, it does not necessarily follow that a particular cellular feature is acidic *in vivo* if it is stained *in vitro* by a basic dye—and *vice versa*.)

The uses of dyes. Dyes used for the routine examination and detection of microorganisms include CARBOLFUCHSIN, LACTOPHENOL COTTON BLUE, METHYLENE BLUE, and the ROMANOWSKY STAINS. For dyes used in medical and veterinary diagnostic work see e.g. ALBERT'S STAIN, KOSTER'S STAIN, TOXOPLASMA DYE TEST, and ZIEHL-NEELSEN'S STAIN. The GRAM STAIN is the principle taxonomic stain in bacteriology. Dyes used for the staining of components of particular composition include CONGO RED (cellulose), the METHYL-GREEN PYRONIN STAIN (DNA/RNA), and SUDAN BLACK (lipids).

Certain dyes, silver salts, and silver proteinates are used for the staining of particular morphological features. Thus, e.g. procaryotic and eucaryotic flagella may be stained either by conventional dyes (e.g. carbolfuchsin) or may be subjected to silver deposition; such procedures generally involve the use of a *mordant*—i.e. a substance (e.g. certain tannates) which helps to bind the dye to the target molecule. (see also LEIFSON'S FLAGELLA STAIN and SILVER LINE SYSTEM)

Certain dyes are used as inhibitory agents in some types of microbiological media—e.g. malachite green in Löwenstein–Jensen's medium, thionin in diagnostic media for *Brucella*, ROSE BENGAL in media used for the isolation of certain fungi. (see also DYES (as antimicrobial agents)) Some dyes (e.g. methylene blue, resazurin) are used as redox indicators (see also REDOX POTENTIAL), while a range of dyes are used as PH INDICATORS.

staling (*mycol.*) (Adverse) changes in environmental conditions in the immediate neighbourhood of a growing fungus due to the accumulation of waste metabolic products of that fungus.

stalk (1) (*mycol.*) (see STIPE) (2) (*mycol.*) Part of the fruiting body (*sorocarp*) of the cellular slime moulds. (3) (*bacteriol.*) May refer to a PROSTHECA or to the non-prosthecate appendage of *Gallionella* spp. The stalk may or may not be attached to the substratum. (4) (*bacteriol.*) Part of the fruiting body formed by certain members of the MYXOBACTERALES. (5) (*protozool.*) Part of the vegetative cells of certain organisms, e.g. *Carchesium*, *Vorticella*; some species form contractile stalks.

standard tests for syphilis (see STS)

staphylococcal food poisoning (see FOOD POISONING (bacterial))

staphylococcins (see BACTERIOCINS)

Staphylococcus A genus of gram-positive, facultatively anaerobic, chemoorganotrophic, asporogenous, non-motile, CATALASE-positive bacteria of the MICROCOCCACEAE; species occur e.g. as parasites and pathogens of animals, including man. *Staphylococcus* spp can be responsible for e.g. wound infections, bovine mastitis, boils, and one type of FOOD POISONING. The cells are spherical, commonly about 1 micron in diameter—but larger and smaller strains occur; some strains form orange or yellow CAROTENOID pigments. The CELL WALL components include PEPTIDOGLYCAN and TEICHOIC ACIDS. (see also PROTEIN A) L FORMS occur. Growth occurs on simple media (e.g. NUTRIENT AGAR); tolerance of salt (up to about 15% NaCl) is common. On blood agar some strains cause HAEMOLYSIS (see also α, β, and δ-HAEMOLYSIN, α-HAEMOLYSIS and HOT-COLD LYSIS). Both respiratory (oxidative) and fermentative types of metabolism are exhibited; carbohydrates are fermented anaerogenically. Extracellular products may include: BACTERIOCINS, COAGULASE, DNase, FIBRINOLYSIN, LEUCOCIDINS, PENICILLINASE and PHOSPHATASES; some strains produce ENTEROTOXINS.

Staphylococci are killed by PASTEURIZATION (HTST) and by many common DISINFECTANTS. Many strains are sensitive to PENICILLINS, FUSIDIC ACID, TETRACYCLINES, AMINOGLYCOSIDE ANTIBIOTICS and other ANTIBIOTICS. PLASMID-mediated resistance to particular antibiotics (including penicillins) may be acquired by TRANSDUCTION. (see also ANTIBIOTICS, RESISTANCE TO) The OXFORD STRAIN is used as a control in certain antibiotic-sensitivity tests. Type species: *S. aureus*; GC% range for the genus: approximately 30–39.

Classification. Species are distinguished by biochemical tests and by other criteria (e.g. cell wall structure). Serotyping is based on cell wall antigens; phage TYPING has been used.

S. aureus. Cocci, of 1 micron maximum diameter, which occur in clusters; yellow or orange pigments are usually formed. Coagulase-positive. The cell wall contains *ribitol* teichoic acids. A common pathogen.

S. epidermidis (*S. albus*). The cocci, which may be larger than *S. aureus*, sometimes form yellow (or other) pigments—but are usually non-pigmented. Coagulase-negative. The cell wall contains *glycerol* teichoic acids. *S. epidermidis* may be an OPPORTUNIST PATHOGEN; strains are common on the skin and mucous surfaces of animals, including man.

S. saprophyticus. A coagulase-negative, saprophytic (occasionally pathogenic?) species. The cell wall usually contains ribitol teichoic acids. *S. saprophyticus* differs from the other two species by exhibiting resistance to NOVOBIOCIN (2 micrograms/ml).

staphylokinase (see FIBRINOLYSIN)

starch A composite polysaccharide consisting of *amylose* and *amylopectin* in proportions which depend on the source. Amylose is a linear α-$(1 \rightarrow 4)$-linked D-glucan; amylopectin is a branched α-$(1 \rightarrow 4)$-linked D-glucan with α-$(1 \rightarrow 6)$ branch points. With iodine, starch gives a dark blue coloration—due to the presence of amylose. (Amylopectin gives a red coloration with iodine.) Starch occurs principally as the main reserve polysaccharide in plants and in certain algae (see below).

Starch is attacked by a number of microorganisms: thus, α-AMYLASES are produced by many bacteria and fungi, e.g. *Bacillus subtilis* and *Aspergillus oryzae.* (The α-(1 → 6) branch points of amylopectin are cleaved e.g. by *pullulanase* and by *isoamylase* (found in yeasts).)

Chlorophycean starch. A starch-like compound, consisting of amylose and amylopectin, which is the principal reserve polysaccharide in members of the Chlorophyta; chlorophycean starch differs from the starch of higher plants mainly in that it has a lower iodine retentivity.

Cyanophycean starch. An amylopectin-like polysaccharide found in members of the Cyanophyta.

Floridean starch. An amylopectin-like D-glucan which is the main storage polysaccharide in members of the Rhodophyta (red algae).

Moss starch (lichen starch) (see LICHENAN)

starter In certain industrial processes, e.g. CHEESE-making, a pure (or mixed) culture of microorganisms which is added to the raw material(s) in order to initiate a particular phase of production. (see also *pitching* in BREWING)

stationary phase (see GROWTH (bacterial))

statismospore (*mycol.*) A term used to refer to a spore which is not forcibly discharged from the basidium. (cf. BALLISTOSPORES)

Staurastrum A genus of algae—see DESMIDS.

Staurodesmus A genus of algae—see DESMIDS.

Stauroneis A genus of pennate distoms of the BACILLARIOPHYTA.

staurosporae (see SACCARDIAN SYSTEM)

steam trap (in an AUTOCLAVE) A valve which permits the passage of air and/or water but which blocks the passage of steam. One form of steam trap incorporates a metal bellows which contains a small amount of water; when steam (at the required shut-off temperature) flows round the bellows to the outlet line, the water in the bellows vaporizes—causing the bellows to expand and thus close the valve. Air or water (at sub-shut-off temperatures) fail to cause the bellows to expand to a degree sufficient to close the valve.

steamer A vessel in which objects or materials may be subjected to steam at *atmospheric* pressure i.e. it is a non-pressurized vessel. (cf. AUTOCLAVE)

stellate Star-shaped.

Stemonitis A genus of the ACELLULAR SLIME MOULDS.

steno- A prefix which signifies *narrow* or *limited.* The prefix is used e.g. to indicate the capacity of an organism to tolerate a narrow range of degrees of a given influence—e.g. *stenohaline*: able to tolerate only a narrow range of salt concentrations. (cf. EURY-)

Stentor A genus of freshwater ciliate protozoans of the order Heterotrichida, subclass SPIROTRICHIA. Individual organisms, which are funnel-shaped, may reach 2 millimetres in length. The body, which is uniformly ciliated, and which may exhibit a number of short sensory bristles, is capable of a considerable degree of contraction. The cytostome is found at the broader, flared end, while the narrow end is often attached to the substratum; some species form a mucilaginous lorica. The macronucleus may be moniliform (e.g. *S. coeruleus*), ovoid (e.g. *S. igneus*), or elongated and of uniform width (e.g. *S. roeseli*). A contractile vacuole in the broader part of the organism communicates with a longitudinal canal. Food consists mainly of bacteria, algae and other protozoans. *S. coeruleus* is distinguished by a blue pigment; *S. polymorphus* contains a number of algal endosymbionts (*Chlorella* spp).

Stephanopogon A genus of protozoans of the suborder RHABDOPHORINA. *Stephanopogon* is considered to be one of the most primitive of ciliates; primitive features include an apical cytostome and the possession of a number of *similar* nuclei. Division occurs by multiple fission; sexual processes are unknown.

sterco- A combining form which indicates a connection with faeces, e.g. *stercorous* (resembling faeces).

stercoraria One of two groups of sub-genera of trypanosomes which are parasitic in *mammals.* (see TRYPANOSOMA) The stercorarian trypanosomes are those in which forms infective for the vertebrate host typically develop in the POSTERIOR STATION of the invertebrate host (vector), and which are thus involved in *contaminative infection.* (cf. SALIVARIA) Stercorarian sub-genera: HERPETOSOMA, MEGATRYPANUM, and SCHIZOTRYPANUM.

Stereocaulon A genus of LICHENS in which there is a granular *primary* (basal) thallus, which is frequently evanescent, and erect *pseudopodetia* (see PSEUDOPODETIUM) which are usually solid, often branched, sometimes sorediate, and often bear numerous phyllocladia. The phycobiont is *Trebouxia.* Cephalodia are generally present and contain *Nostoc* or *Stigonema.* Apothecia may occur terminally or laterally on the pseudopodetia; the discs become convex. Species include e.g.:

S. vesuvianum. The primary thallus is evanescent. The pseudopodetia are grey, 1–5 centimetres in height, variable in form, and bear dark-centred, round phyllocladia and brown to black, wart-like cephalodia. *S. vesuvianum* occurs on rocks and walls in hilly regions.

S. pileatum. The primary thallus is persistent and

consists of grey, wart-like granules. The pseudopodetia are short (5 millimetres or less) and generally simple; each may bear either a terminal soralium or, less commonly, a terminal apothecium which has a reddish-brown disc. Brownish, wart-like cephalodia occur on both the primary and secondary thalli. *S. pileatum* occurs on rocks and stones.

Stereum A genus of lignicolous fungi of the family Stereaceae, order APHYLLOPHORALES. Constituent species form leathery or woody, sheet-like or tuberculate fruiting bodies which are typically resupinate or effused-reflexed; the tissue of the fruiting body lacks clamp connections and is usually dimitic, and the surface opposite that of the hymenium is frequently downy or hairy. The spores are smooth, thin-walled, and hyaline.

sterigma (*pl.* sterigmata) (*mycol.*) (1) A short, hypha-like structure which bears a terminal spore, or chain of spores, or which supports a spore-bearing structure; thus, e.g. sterigmata support the peripheral layer of phialids on the conidial head in species of ASPERGILLUS. (2) A synonym of PHIALID. (3) Part of a mature BASIDIUM: a protrusion on which a basidiospore is borne. (Sterigmata are not formed e.g. by the smut fungi.)

sterilant A chemical agent which, under appropriate conditions, is able to *sterilize* objects, materials or environments (see STERILE (1) and STERILIZATION). The term has also been used loosely to refer to DISINFECTANTS. (see also BIOCIDE)

sterile (*adjective*) (1) Refers to the condition of an object, or an environment, which is free of all *living* cells, all *viable* spores (and other resistant and disseminative forms), and all viruses and viroids capable of replication. (An effective STERILIZATION process must therefore irreversibly inactivate *all* organisms.) For most practical purposes the main criterion of sterility is the failure to detect the presence of living cells, spores, viruses etc. in or on a given object, material, or environment; it should be appreciated that a viable organism can be detected *only* if it is provided with conditions suitable for its growth/replication or other form of expression. (2) In general, refers to any tissue which does not form, or is incapable of forming, reproductive structures.

sterility The state or condition of being STERILE.

sterilization Any process by which objects, materials, or environments may be rendered STERILE. (Since sterility is an absolute condition, the expression 'partial sterilization' is meaningless—cf. DISINFECTION.) When contaminating organisms are known to consist entirely of vegetative or other (heat-labile) forms of organism, sterility can usually be achieved by the use of e.g. mild heat or chemical treatment; however, the term *sterilization* commonly refers to those processes which guarantee sterility regardless of the nature of the contaminating flora.

Physical methods of sterilization. Heat is the most reliable (and generally preferred) sterilizing agent (except for heat-labile materials); it is non-selective (all organisms—including spores—being susceptible) and it is able to reach organisms that may be protected from the action of chemical sterilizing agents or radiation. The lethal action of heat is due largely to the denaturation of microbial proteins and nucleic acids; at temperatures used for sterilization, nucleic acids may be deaminated, depurinated, or otherwise degraded. The effects of heat on the lipids of a CELL MEMBRANE may also be important.

Moist heat (steam) is more rapidly lethal than dry heat; it requires lower working temperatures and shorter periods of exposure. The relative ease with which moist heat denatures proteins is believed to be due to facilitated rupture of the protein-stabilizing hydrogen bonds—for example: $\gt C{=}O\cdots HN\lt$ bonds; these bonds are more easily broken when water molecules are available for hydrogen bonding. Moist heat sterilization involves an efficient transfer of heat since steam which condenses on a cooler object delivers up a considerable quantity of (latent) heat. The process is generally carried out in an AUTOCLAVE; temperature/time combinations used for sterilization in an autoclave include e.g. 121 °C (15 lbs/square inch, steam pressure) for 15–20 minutes, 115 °C (10 lbs/square inch) for 35 minutes.

Dry heat sterilization is carried out in a hot-air oven; objects (e.g. *clean* glassware etc.) are exposed for 1–2 hours to a temperature of 160–170 °C. Such conditions denature proteins and cell membranes, desiccate cytoplasm, and oxidize various cellular components. Incineration (combustion) is the extreme form of dry heat sterilization. (see also FLAMING)

ULTRAVIOLET RADIATION (a *non-ionizing* radiation) possesses maximum antimicrobial activity at wavelengths in the region of 260 nm; such radiation, which is readily absorbed by DNA, may give rise to mutagenesis and/or cell death. However, damage brought about by ultraviolet radiation is, to some extent, reversible (see e.g. PHOTOREACTIVATION), and radiation of this wavelength has poor powers of penetration—being readily absorbed by solids and having the ability to penetrate liquids only to a limited extent. Ultraviolet radiation is used mainly for the DISINFECTION of air and of exposed surfaces in enclosed areas.

Ionizing radiations (X-rays, *gamma* rays,

cathode rays etc.) are often used for the sterilization of pre-packed equipment used e.g. in hospitals and microbiological laboratories (e.g. surgical equipment, plastic petri dishes etc.). Compared with ultraviolet radiation, ionizing radiations have higher energy and greater penetrating power and are therefore more efficient antimicrobial agents. Ionizing radiations increase the chemical reactivity of microbial constituents and create charged particles which readily bring about irreversible damage to nucleic acids and other macromolecules. The currently-accepted sterilizing dose of ionizing radiation is 2.5*Mrad.* (One *rad* is the quantity of radiation corresponding to the absorption of 100 ergs of energy per gram of material; 1*Mrad* = 10^6 rads.) Several reports have been made of organisms surviving the standard sterilizing dose.

FILTRATION provides a convenient method for the sterilization of heat-labile liquids. (*Ultrafiltration*, using an appropriate membrane filter, is able to provide a filtrate free of all known viruses.)

ULTRASONICATION is not considered to be a *reliable* method of sterilization.

Chemical methods of sterilization. Among the few chemical agents considered to be reliable STERILANTS are ETHYLENE OXIDE, GLUTARALDEHYDE, and β-PROPIOLACTONE; such agents must be used in effective concentrations and under appropriate conditions.

The kinetics of sterilization and disinfection. In a single-species microbial population, the death rate produced by many types of lethal agent approximately follows 1st order kinetics; such agents include heat, radiation, and a variety of chemical disinfectants and sterilants. However, close approximations to 1st order kinetics are obtained only under certain conditions—e.g. when relatively high concentrations of disinfectants etc. are used, or when high sterilizing temperatures are employed. Under these conditions the sterilization of a microbial population may be summarized by the equation:

$$\ln\left(\frac{N_0}{N}\right) = kt$$

where N_0 is the initial number of viable organisms, N is the number of survivors at time t, and k is the *death rate constant.* ($\ln = \log_e$) Rearranged, this gives:

$$\ln N_0 - \ln N = kt$$

Thus, k can be found by plotting lnN against t. For a given organism, different values of k can be obtained for different sterilizing temperatures and for different concentrations of disinfectant. When a microbial population contains cells having different degrees of resistance to heat, chemicals etc. sterilization processes do not follow 1st order kinetics, and the curve obtained by plotting lnN *versus* t is a composite curve. (see also THERMAL DEATH POINT; TYNDALLIZATION; DISINFECTANTS)

sterilizer (1) A STEAMER. (2) An AUTOCLAVE.

Ster-zac (propietary name) A liquid antiseptic SOAP containing *hexachlorophene* (see BISPHENOLS).

stibamine (see *Antimony* under HEAVY METALS)

stibanilic acid (see *Antimony* under HEAVY METALS)

stibophen (see *Antimony* under HEAVY METALS)

stichobasidial (*adjective*) Refers to the *longitudinal* arrangement of meiotic spindles in a developing BASIDIUM (cf. CHIASTOBASIDIAL).

stichonematic Refers to a *eucaryotic* FLAGELLUM which bears a single row of fine hairs.

Stickland reaction In certain *Clostridium* spp: a type of fermentation in which the oxidation of one amino acid (aa_1) is coupled with the reduction of a second (aa_2)—the reaction yielding most of the energy required by the organism. aa_1 is oxidatively deaminated to the ketoacid—1NAD being reduced. The ketoacid is oxidatively decarboxylated to a fatty acid in a reaction requiring coenzyme A and inorganic phosphate; 1NAD is reduced and 1ATP synthesized. The $2NADH_2$ thus formed are reoxidized during the reduction (or reductive deamination) of aa_2. aa_1 may be e.g. alanine, leucine, valine; aa_2 may be e.g. glycine, proline, ornithine. *Example*: alanine (aa_1) is converted to pyruvate, then to acetic acid; glycine (aa_2) is converted to acetic acid.

Sticta A genus of foliose LICHENS in which the phycobiont is *Nostoc* or a green alga—according to species; the lower surface of the thallus is tomentose and contains cyphellae (see CYPHELLA). Species include e.g.:

S. sylvatica. Phycobiont: *Nostoc.* The thallus consists of rounded lobes and is of the order of 3–5 centimetres in diameter. The upper surface is brown, shallowly pitted, and bears black isidia; the lower surface is tomentose, dark brown in the centre and paler towards the margins, and contains cyphellae which are apparent as whitish pits of the order of 0.5 millimetres in diameter. Fruiting bodies are rarely formed. The species has a fishy smell and occurs on mossy trees and rocks.

Stieda body A body which may form part of, or is intimately associated with, the polar portion of the wall of a coccidian sporocyst. In (at least) some species, excystation—i.e. liberation of the enclosed sporozoites—is associated with the breakdown or detachment of the Stieda body.

stigma Synonym of EYESPOT.

stinkhorns Certain species of fungi of the order PHALLALES.

stipe (*mycol.*) The stalk or stem of a fruiting body; the stipe is composed of hyphae which are typically orientated parallel to the axis of the stipe. *Stipitate* fruiting bodies occur among the basidiomycetes (e.g. members of the Agaricales) and the ascomycetes (e.g. species of *Helvella* and *Morchella*). (cf. PSEUDOSTEM)

stipitate (*adjective*) Possessing a STIPE.

stoneworts Certain species of the CHAROPHYTA.

stopping filter (see BARRIER FILTER)

stormy clot (stormy fermentation) A characteristic effect brought about by the culture of certain organisms in milk-based media (e.g. LITMUS MILK). *Clostridium perfringens*, for example, produces an acid clot which is given a turbulent appearance by the formation of copious amounts of gas within the medium.

straight wire An instrument used in bacteriology e.g. to SUBCULTURE from isolated colonies or to prepare stab-cultures. The rod-shaped metal handle holds, at one end, a short, straight piece of wire; the wire, ideally platinum, is usually nicket-steel and is commonly 5–8 centimetres in length. Before *each* use, the wire is sterilized by heating it to redness in a flame and allowing it to cool. In use, the *tip* of the sterile wire is brought into effective contact e.g. with the centre of a selected colony; the small quantity of bacterial growth which adheres to the wire can then be used to inoculate a medium. The wire should be flamed after use to prevent contamination of the bench etc.

strain A cell, or population of cells, which has the general characteristics of a given type of organism (e.g. bacterium, fungus) or of a particular (named) genus, species, or serotype. The term *strain* may also be used to refer to a cell, or population of cells, which exhibits a particular, named characteristic. (see also TYPING)

strawberry, diseases of These include: (a) *Grey mould*; causal agent: *Botrytis cinerea*. The ripening fruit succumbs to a soft rot—its surface characteristically bearing a grey powdery fungal growth. High humidity favours the disease. (b) *Red core*; causal agent: *Phytophthora fragariae*. A soil-borne disease characterized by root rot and wilting; the centres of infected roots become red. Red core is favoured by cold, wet conditions. (see also NOTIFIABLE DISEASES)

streak (*plant pathol.*) Roughly parallel lines of chlorosis or necrosis which appear on the stems or leaves of certain virus-infected plants—particularly monocotyledonous plants.

streaking (*bacteriol.*) A method of INOCULATION of the surface of a *solid* MEDIUM such that, during subsequent INCUBATION, *individual* bacterial colonies (rather than confluent growth) develop on (at least) part of the medium. One method of streaking is as follows. The inoculum, on a sterile LOOP, is drawn back and forth (i.e. *streaked*) across a peripheral region of the PLATE—following the path as indicated at (A) in the diagram. The loop is then *flamed* (i.e. sterilized—see FLAMING), allowed to cool, and streaked across the medium as shown at (B) in the diagram. Streakings (C), (D), and (E) are similarly made, the loop being flamed and cooled between each streaking —and after the last; in this way the inoculum becomes progressively thinned out as it is distributed along the streak lines—such that well-isolated individual cells are deposited on (at least) some regions of the plate. On incubation, each viable cell (which is capable of growing on the particular medium used) gives rise to a COLONY.

STREAKING. The pattern of inoculation used in one method of streaking.

street virus (of the *rabies virus*) Fully-virulent rabies virus—such as is found in the central nervous system of an animal or person suffering from naturally-acquired RABIES. (cf. FIXED VIRUS)

strep throat An infection of the pharyngeal tissues involving β-haemolytic streptococci (*S. pyogenes*).

Streptobacillus A genus of gram-negative, aerobic or facultatively anaerobic, pleomorphic bacteria of uncertain taxonomic affiliation; the cells may occur as filaments, rods—approximately 0.5 × 1–5 microns—or as chains of bacilli. L FORMS develop spontaneously. Serum, blood etc. is required for *in vitro* culture. The sole species, *S. moniliformis*, is one cause of RAT-BITE FEVER.

Streptococcaceae A family of gram-positive, chemoorganotrophic, asporogenous, facultatively anaerobic, typically non-motile bacteria; the spherical or coccoid cells occur singly, in pairs or tetrads, or in chains. The genera: *Aerococcus*, *Gemella*, LEUCONOSTOC, PEDIOCOCCUS and STREPTOCOCCUS.

streptococcal proteinase (see M PROTEIN (of *Streptococcus*))

Streptococcus A genus of gram-positive, chemoorganotrophic, asporogenous, facultatively or obligately anaerobic cocci or coccoid bacteria of the family STREPTOCOCCACEAE; all strains are catalase-negative (see however CATALASE TEST) and oxidase-negative. Species occur e.g. as parasites and pathogens of animals (including man)—particularly in the respiratory and alimentary tracts. The cells are typically less than 1 micron in diameter, and generally occur in pairs or chains; motile strains have been reported. CAPSULE-formation is common in some species. L FORMS occur. Growth is often improved by the use of media enriched with serum or blood (see also TODD-HEWITT BROTH); pathogenic species grow optimally at 37 °C, and enterococci (intestinal strains, group *D* streptococci) grow well at 45 °C. Typically, carbohydrates are metabolized anaerogenically and *homofermentatively*—yielding lactic acid as the main product (cf. LEUCONOSTOC). On blood agar some species give rise to haemolysis. In β-haemolysis the colony is surrounded by a glass-clear zone in which there are few, if any, intact erythrocytes; in α-haemolysis (see also VIRIDANS STREPTOCOCCI) the colony is surrounded by a zone of green or brownish-green discoloration. (N.B. This terminology is the reverse of that used to describe *staphylococcal* haemolysis.) Streptococcal LEUCOCIDINS include STREPTOLYSIN-O and STREPTOLYSIN-S. Some strains produce extracellular nucleases (see STREPTODORNASE) and some produce STREPTOKINASE. BACTERIOCINS are formed by some strains. TRANSDUCTION and TRANSFORMATION occur among the streptococci.

Streptococci are killed by PASTEURIZATION (HTST) and by many common disinfectants; PENICILLINS are commonly effective against β-haemolytic strains but rather less so against others. TETRACYCLINES and other ANTIBIOTICS have been used therapeutically.

Classification of streptococci. The majority of streptococci can be placed into one of a number of *groups* by means of LANCEFIELD'S STREPTOCOCCAL GROUPING TEST—which identifies specific C SUBSTANCES in the CELL WALL. Other cell wall components (e.g. the M PROTEIN) are involved in streptococcal TYPING. The cell wall *R antigen* has also been used for typing; the (protein) R antigen—found e.g. in strains of groups *B* and *C* streptococci—is not a virulence factor (cf. M protein). *Streptococcus* species include:

S. pyogenes (formerly *S. haemolyticus*). β-Haemolytic Lancefield group *A* strains. *S. pyogenes* is the causal agent of e.g. SCARLET FEVER, ERYSIPELAS, and STREP THROAT; certain lysogenic strains produce an erythrogenic toxin (see also DICK TEST). In broth culture the cells commonly occur in long chains—frequently with an hyaluronic acid capsule. Capsulated strains form *mucoid* colonies, while non-capsulated strains form *glossy* colonies. (Colonies of capsulated strains which have become partially dehydrated—due e.g. to conditions of culture—are sometimes referred to as *matt* colonies.) *S. pyogenes* is the type species of the genus.

S. agalactiae. Lancefield group *B* organisms; β-haemolytic, α-haemolytic, and non-haemolytic types occur. Some strains are pigmented. A causal agent of bovine mastitis—see also CAMP TEST.

S. faecalis. Lancefield group *D* non-haemolytic (or β-haemolytic) streptococci which grow at 45 °C and survive 60 °C/30 minutes; the cells typically occur in pairs or short chains. The organisms occur in the lower intestine in man and other animals. A cause of subacute bacterial endocarditis (SBE) in man. The entomopathogenic organism referred to as *Streptococcus apis* is a strain of *S. faecalis*. Enterococci ('faecal streps') can be isolated from sewage-polluted waters by the use of media containing e.g. SODIUM AZIDE.

S. lactis and *S. cremoris*. Lancefield group *N* streptococci commonly present in dairy products, e.g. BUTTER.

S. mutans. A species associated with dental PLAQUE.

S. pneumoniae (the 'pneumococcus'). A causal agent of PNEUMONIA; *S. pneumoniae* does not belong to a Lancefield group. Cells typically occur as *diplococci*. The organism is sensitive to OPTOCHIN; see also BILE SOLUBILITY. On blood agar the typical form is the DRAUGHTSMAN COLONY; α-haemolysis is exhibited. *S. pneumoniae* often produces a polysaccharide capsule; more than 80 different capsular polysaccharides are known, and this variation is used as the basis of pneumococcal typing. The type III pneumococcal polysaccharide consists of repeating subunits of cellobiuronic acid linked by β-(1 → 3) glucosidic bonds.

S. salivarius and *S. sanguis*. Species which produce extracellular DEXTRANS and/or LEVANS.

***Streptococcus* M.G.** Any of the strains of α-haemolytic (viridans) streptococci (e.g. NCTC 8037) which are agglutinated by the sera of patients suffering from *mycoplasmal* pneumonia. Some strains are reported to belong to Lancefield's group *F*.

streptodornase (streptococcal nuclease) Any of the extracellular nucleases produced by *Streptococcus* spp; such enzymes are frequently synthesized by e.g. strains of Lancefield groups *A* and *E*. At least four serologically-distinct nucleases are produced by strains of Lancefield group *A*; all four types—A, B, C, and D—cleave the backbone of DNA, and types B and D are also active against RNA. The activities of streptococcal nucleases are promoted by divalent cations (particularly calcium or magnesium). In contrast to staphylococcal DNases (*q.v.*) the products of streptococcal nuclease activity are 5′-nucleotides.

streptokinase An extracellular enzyme produced by certain streptococci—including many strains of Lancefield groups *A* and *E*. Streptokinases from human group *A* strains generally catalyse the reaction plasminogen → plasmin (fibrinolysin) in *human* blood; the plasmin thus produced is able to lyse a fibrin clot derived from human blood. The catalytic activity of other streptokinases may be specific to the blood (and fibrin) of other animals. Streptokinase is commonly regarded as an AGGRESSIN; it is assumed that the enzyme may assist the invasiveness of pathogenic streptococci by lysing the fibrin barrier which may enclose streptococcal lesions. (see also FIBRINOLYSIN and STREPTODORNASE)

streptolysin-O An exotoxin produced by the majority of group *A* streptococci and by some of groups *C* and *G*. Streptolysin-O is haemolytic in the reduced state but inactive when oxidized; its haemolytic activity is oxygen-labile at room temperatures but may be restored by certain (mild) reducing agents. Haemolysis is inhibited by free cholesterol; streptolysin-O appears to bind at a cholesterol-containing site on the cell membrane. The haemolytic activity of streptolysin-O does not require the presence or participation of *complement*. Streptolysin-O can also bring about the lysis of LYSOSOMES.

In certain diseases (e.g. rheumatic fever) the *in vivo* titre of antibodies to streptolysin-O may rise significantly (see also ANTISTREPTOLYSIN-O TEST). Serologically, streptolysin-O cross-reacts with e.g. the θ-toxin of *Clostridium perfringens*.

streptolysin-S An oxygen-stable haemolysin and anti-leucocyte factor produced by certain streptococci—particularly by many strains of Lancefield's group *A*. The synthesis of streptolysin-S is promoted by serum proteins, by RNA, and by certain other substances; it appears that some streptolysin-S may remain cell-bound while the rest is released into the medium. (see also LEUCOTOXIC EFFECT) The haemolytic activity of streptolysin-S is inhibited by low levels of certain phosphatides—e.g. phosphatidylethanolamine. In constrast to streptolysin-O, streptolysin-S is not immunogenic. (see also STREPTODORNASE and STREPTOKINASE)

Streptomyces A genus of gram-positive, aerobic bacteria of the STREPTOMYCETACEAE; the organisms form non-verticillate chains of spores on the aerial mycelium. The genus contains several hundred species of which *S. albus* is the type species. Taxonomically-important features within the genus include the morphology of the spore chains (see RECTUS FLEXIBILIS and SPIRA), the types of pigments produced, the nature of the spore wall, and the type of ANTIBIOTICS produced (if any). Some species are pathogenic (see BEET, POTATO, and MADURA FOOT). GC% range in those species examined: approximately 69–73.

Streptomycetaceae A family of gram-positive, aerobic bacteria of the ACTINOMYCETALES; the organisms form mycelium which is generally of a stable, non-fragmenting type (cf. NOCARDIACEAE). All species have Type I cell walls—see ACTINOMYCETALES—(cf. MICROMONOSPORACEAE). The streptomycetes form long spore chains—i.e. chains containing more than about 5 spores, and often more than 50 spores. Many species are pigmented. The genera include STREPTOMYCES, *Streptoverticillium* and *Sporichthya*.

streptomycete Any member of the STREPTOMYCETACEAE.

streptomycin An AMINOGLYCOSIDE ANTIBIOTIC produced by *Streptomyces griseus*. The molecule consists of the base *streptidine* (diguanidylinositol) linked glycosidically to *streptobiosamine* (= *streptose*—an aldehyde-containing monosaccharide—linked to *N*-methyl-L-glucosamine). Streptomycin is active against e.g. *Mycobacterium tuberculosis*, *Brucella* spp, *Yersinia pestis*, *Francisella tularensis*, some strains of *Staphylococcus*, many gram-negative bacilli (including some strains of *Pseudomonas aeruginosa*). Bactericidal activity increases with increase in antibiotic concentration; the antibiotic is inactive or weakly active under anaerobic conditions.

Mode of action. Streptomycin appears principally to affect PROTEIN SYNTHESIS; thus e.g. it appears to inhibit polypeptide chain initiation. Streptomycin binds irreversibly to a protein (designated P_{10} or S_{12}) in the 30S subunit of 70S RIBOSOMES. *Sub*-lethal concentrations of streptomycin cause misreading of certain codons in the mRNA—possibly due to conformational changes within the ribosome caused by streptomycin binding.

Resistance to streptomycin. A one-step mutation to high-level streptomycin resistance is not uncommon; a mutation in the *str* gene gives rise to a modified P_{10} protein which may fail to bind streptomycin or which may suffer no loss of function if the antibiotic is bound. (For other mechanisms of resistance, see AMINOGLYCOSIDE ANTIBIOTICS.)

Streptomycin dependence. Certain mutations in

the *str* gene produce mutants for which streptomycin is an essential growth factor. Such mutants have a non-functional ribosomal component; streptomycin appears to bind to and distort the ribosome in such a way that correct translation can occur. Such mutants are said to be streptomycin-dependent; growth cannot occur in the absence of streptomycin. *Conditional* streptomycin dependence is shown by certain auxotrophs which grow in the absence of their required growth factor when supplied with (sub-lethal) concentrations of streptomycin; 'conditional' refers to the fact that streptomycin dependence is exhibited *only* when the growth factor is absent.

streptovaricins A group of ANTIBIOTICS produced by species of *Streptomyces*; they are closely related to the RIFAMYCINS which they resemble both in antimicrobial spectrum and mode of action.

Streptoverticillium A genus of the STREPTOMYCETACEAE.

streptovitacins (see CYCLOHEXIMIDE)

stress metabolites (in plants) (see PHYTOALEXINS)

strict (*microbiol.*) (1) A synonym of OBLIGATE. (2) A term used for emphasis; thus, e.g. an organism which grows only in the *complete* absence of oxygen (and which may also require a pre-reduced medium) is termed a *strict anaerobe*. A *strict autotroph* is an organism which has the ability to grow in the *complete* absence or organic material. The term has a similar meaning when used to qualify an environment or condition; thus, e.g. *strict anaerobiosis* refers to an environment or condition which is characterized by the complete lack of oxygen. (see also FACULTATIVE)

Strigomonas Synonym of CRITHIDIA.

string of pearls (see *B. anthracis* under BACILLUS)

stringent control (in the synthesis of bacterial rRNA) In the wild-type strains of those bacteria examined, deprivation of an essential amino acid results in the cessation of synthesis of protein and rRNA, and a substantial reduction in the rate of overall RNA synthesis. Such a *stringent response* is exhibited by cells which possess the *rel*$^+$ (= RC^{str}) gene; in mutant, relaxed cells—which possess the *rel*$^-$ (= RC^{rel}) gene—deprivation of essential amino acid(s) does not produce the stringent response.

A nucleotide, guanosine-5′-diphosphate-3′-diphosphate* (ppGpp), is believed to be one of the factors whose intracellular concentration regulates the rate of rRNA synthesis. In *stringent* cells the concentration of ppGpp rises whenever protein synthesis is inhibited—e.g. by amino acid starvation, increased temperature, or by the use of antibiotics etc. which block protein synthesis; the manner in which increased levels of ppGpp (reversibly) restrict or block rRNA synthesis is unknown. *Also called MS-1 (magic spot 1)—from its (chromatographic) discovery.

stringent control (of plasmid replication) (see PLASMID)

stroma (1) (*mycol.*) A compact mass of prosenchymatous or pseudoparenchymatous fungal tissue (or of intermingled fungal and non-fungal tissue) on or within which fructifications commonly develop. *Stromata* are formed by many species of fungi, e.g. *Nectria cinnabarina*, *Xylaria* spp. (see also ASCOSTROMA, SCLEROTIUM, and SPORODOCHIUM) (2) (see CHLOROPLAST)

Strophariaceae A family of fungi of the AGARICALES.

structural gene A gene whose product is e.g. an enzyme, structural protein, tRNA, or rRNA—as opposed to a *regulator gene* (see OPERON) whose product regulates the transcription of structural genes.

STS *Standard tests for syphilis*; those serological tests (e.g. the WASSERMANN REACTION, VDRL TEST, and Kahn test) in which CARDIOLIPIN (i.e. a *non*-treponemal antigen) is used as the test antigen. Such tests are also termed standard serologic tests (SST).

Stuart's transport medium A TRANSPORT MEDIUM used e.g. for a range of anaerobic bacteria, delicate organisms such as *Neisseria gonorrhoeae*, and *Trichomonas vaginalis*. Most formulations include sodium thioglycollate (a strong reducing agent), methylene blue (a redox indicator), and agar (0.2–1%—which gives a liquid or semi-solid medium, respectively). The medium is autoclaved (121 °C, 20 minutes) in bijou bottles—which are filled nearly to the brim; the sterilized medium should be colourless. Specimens which are stored/transported in Stuart's medium are commonly kept at room temperature. (Bottles of medium intended for future use should be capped tightly after sterilization.)

stumpy form (of a trypomastigote) A short, broad form of TRYPOMASTIGOTE (see TRYPANOSOMA) which lacks a *free* flagellum and which has an UNDULATING MEMBRANE between the cell body and the entire length of the flagellum.

stylet-borne viruses (non-persistent viruses) (*plant pathol.*) Viruses transmitted by an insect vector either on its stylet or within its labium; stylet-borne viruses generally remain infective only for a matter of hours. (cf. CIRCULATIVE VIRUSES) This type of transmission is common e.g. in aphids; it appears not to be a passive, mechanical process. (see also VIRUS DISEASES OF PLANTS)

Stylonychia A genus of ciliate protozoans of the order Hypotrichida, subclass SPIROTRICHIA.

Morphologically the organisms resemble species of EUPLOTES—differing e.g. in the number and arrangement of the cirri; the latter include three prominent *caudal* cirri (which project from the posterior end of the cell), and a large number of lateral (*marginal*) cirri which form a fringe each side of the organism.

subclinical infection (*med.*) An INFECTION in which symptoms are not apparent.

subculture (1) (*verb*) To prepare a fresh CULTURE from an existing one; a small quantity of growth from the existing culture is used to inoculate a sterile medium which is then incubated. (2) (*noun*) A culture prepared from another by subculturing. (see also PRIMARY CULTURE)

subhymenium (see APOTHECIUM)

subiculum (*mycol.*) A pad or felt of hyphae on or in which fruiting bodies develop.

substrate level phosphorylation The phosphorylation of ADP (or other nucleoside diphosphate) by a process in which the energy required is derived from an energy-yielding *reaction* (cf. OXIDATIVE PHOSPHORYLATION); the substrate of such a reaction must have a free energy of hydrolysis higher than that of ATP. Such substrates include 'high-energy' phosphates (e.g. phosphoenolpyruvate) and acyl-coenzyme A thioesters (e.g. succinyl-CoA). Examples of substrate level phosphorylations: the phosphoglycerate kinase and pyruvate kinase reactions of the EMBDEN–MEYERHOF–PARNAS PATHWAY, and the succinyl-CoA synthetase reaction of the TRICARBOXYLIC ACID CYCLE. (see also FERMENTATION)

substrate mycelium (*bacteriol.*) In certain actinomycetes: that part of the mycelium which remains at the level of the substrate, i.e. in contact with the medium. (cf. AERIAL MYCELIUM)

Succinivibrio A genus of gram-negative, anaerobic, polarly flagellate (monotrichous) bacteria of uncertain taxonomic affiliation; the cells are curved rods with pointed ends. Metabolism is fermentative. The type species, *S. dextrinosolvens*, occurs in the RUMEN of many ruminant species.

sucrose α-D-Glucopyranosyl-β-D-fructofuranoside. Certain bacteria, e.g. *Leuconostoc* spp, *Streptococcus mutans*, split sucrose into its constituent monosaccharides—one of which is then polymerized to form DEXTRAN or *levan* (see FRUCTANS) while the remaining sugar is metabolized. Sucrose is used as a substrate in biochemical characterization tests for bacteria; it is attacked by many species including strains of *Bacteroides*, *Clostridium*, *Klebsiella* and *Lactobacillus*.

Suctoria A subclass of freshwater and marine protozoans of the subphylum CILIOPHORA. *Mature* suctorians are characterized by the complete absence of cilia—although *infraciliature* persists throughout the life cycle. A cytostome is absent; food (small ciliates etc.) is ingested through the tips of *tentacles* which radiate from the body. Individual cells, according to species, may be spherical, cylindrical, cup-shaped etc., and may be from ten to a hundred or more microns in size. A shell, or *lorica*, is formed by some species. The mature organisms are typically sessile, being attached to the substratum by a (non-contractile) *stalk*; non-stalked species occur. Some suctorians form resting cysts. Typical ciliate nuclear dimorphism occurs, the macronucleus often being complex in form.

Asexual reproduction occurs by budding, a ciliated, mouthless larval form being produced. In some species an atypical form of conjugation has been reported. All suctorians are placed in the single order Suctorida. Genera include *Acineta*, *Dendrosoma*, *Podophrya*, *Sphaerophrya* and *Tokophrya*.

suctorian A ciliate of the subclass Suctoria.

Sudan black B A basic, black or blue-black diazo dye commonly used for staining intracellular lipid deposits in microorganisms; the dye is prepared as a 0.5% solution in 70% ethanol. (Sudan black is a constituent of Burdon's stain for bacterial lipids.) (see also POLY-β-HYDROXYBUTYRATE)

sudol (see PHENOLS)

sugarcane Fiji disease virus (*Saccharum* virus 2) A virus which affects a number of species of the Gramineae; the polyhedral virion contains RNA (probably double-stranded) and is approximately 70 nm in diameter. In susceptible strains of sugarcane (*Saccharum* spp) symptoms include stunting and the formation of characteristic elongated swellings or GALLS on the undersurfaces of leaves; inclusion bodies are formed in the cells of infected plants. Severe crop losses may occur. The virus is transmitted *circulatively* by leafhoppers (*Perkinsiella* spp)—viral replication probably occurring within the leafhopper. (see also VIRUS DISEASES OF PLANTS)

sugars (bacteriological media) (see PEPTONE WATER SUGARS and BROTH SUGARS)

Sulfolobus A genus of gram-negative, aerobic, non-motile, non-sporing bacteria; strains occur e.g. in sulphur-containing hot springs. The (lobed) cells are about 1 micron across; the cell wall apparently lacks peptidoglycan. The organism is a facultative lithotroph which obtains energy by oxidizing elemental sulphur, sulphides etc. Optimum growth temperature, pH: about 75 °C and 2–3 respectively.

sulpha drugs (see SULPHONAMIDES)

sulphamylon (see MARFANIL)

sulphanilamides (see SULPHONAMIDES)

sulphate-reducing bacteria The genera *Desulphovibrio* and *Desulphotomaculum*—see DISSIMILATORY SULPHATE REDUCTION.

sulphate reduction (see ASSIMILATORY SULPHATE REDUCTION and DISSIMILATORY SULPHATE REDUCTION)

sulphonamides (as chemotherapeutic agents) The medically-useful (antibacterial) sulphonamides are derivatives of *sulphanilamide* ($NH_2.C_6H_4.SO_2NH_2$, *p*-aminobenzene-sulphonamide); they are bacteriostatic for a wide range of gram-positive and gram-negative bacteria, including e.g. Lancefield group *A* streptococci, staphylococci, and strains of *Shigella*, *Brucella*, and *Neisseria*. Sulphonamides act by competitively inhibiting the incorporation of *p*-aminobenzoic acid (PABA) during the formation of dihydropteroic acid in FOLIC ACID synthesis—the sulphonamide molecule functioning as an analogue of PABA. Those organisms whose growth requires an exogenous source of folic acid are thus insensitive to sulphonamides. However, the majority of bacteria synthesize folic acid and are generally unable to utilize an exogenous supply of the vitamin; such organisms are commonly sensitive to sulphonamides. A lag period of several bacterial generations occurs between exposure of the cells to a sulphonamide and inhibition of bacterial growth; during this time the cells exhaust their preformed 'pool' of folic acid. This delayed effect enables sulphonamides to be used in conjunction with those antibiotics (e.g. PENICILLINS) which are active only against growing bacteria. The inhibitory effect of the sulphonamides may be neutralized by supplying the cells with those metabolites which normally require folic acid for their synthesis (e.g. purines, certain amino acids); such substances may occur e.g. in pus, so that sulphonamides may be ineffective in the treatment of certain suppurative infections. Bacteria readily develop resistance to sulphonamides; thus, e.g. in *Streptococcus pneumoniae* modification of dihydropteroic acid synthetase by a single-step mutation may reduce the affinity of the enzyme for sulphonamides without significantly reducing its affinity for PABA. Resistance may also be plasmid-borne—see R FACTOR. (see also ANTIBIOTICS, RESISTANCE TO)

The various sulphonamides have similar antimicrobial spectra, but vary e.g. in their toxicity to man. The antibacterial activity of *Prontosil*—first used clinically in the 1930s—was due to the *sulphanilamide* formed during its breakdown *in vivo*; sulphanilamide has been largely superceded by other less toxic and more effective derivatives. Most of the derivatives are substituted at the nitrogen of the sulphonamide group ($NH_2.C_6H_4.SO_2NHR$). Substitution at the *p*-amino group results in the loss of antibacterial activity; however, the (inactive) derivative *p-N-succinylsulphathiazole* is used in the treatment of infections of the lower intestine where it is hydrolysed to succinate and the active derivative *sulphathiazole*. Other sulphonamides in current use include *sulphadiazine*, *sulphacetamide*, *sulphadimidine* (sulphamethazine), and *sulphafurazole*. (see also MARFANIL and SULPHONES)

Sulphonamides are now used mainly in the treatment of urinary tract infections, in certain forms of meningitis, and in veterinary practice.

sulphones (R_2SO_2) A group of antibacterial agents effective against a range of gram-positive and gram-negative species; owing to toxicity they are not widely used in medicine. Their mode of action appears to be the same as that of the SULPHONAMIDES. The most important sulphone derivative is 4,4′-diaminodiphenyl sulphone (*dapsone*; DDS); this is an important drug in the treatment of LEPROSY.

sulphorhodamine B (see LISSAMINE RHODAMINE)

sulphur (as an antifungal agent) Elemental sulphur is effective against a number of fungal plant diseases, e.g. apple scab, powdery MILDEWS, black spot of roses; currently, sulphur is used against many powdery mildews, e.g. of the grape. Elemental sulphur may be applied as a fine dust, in colloidal form, or as a WETTABLE POWDER; the degree of fungitoxicity is related to the size of the sulphur particles—preparations containing the smallest particles generally being the most toxic. The mechanism of the antifungal action of sulphur is still unknown.

Sulphur may be used as *polysulphide*—formed by the reaction of certain metal sulphides with sulphur under alkaline conditions; acidification of polysulphide (e.g. by atmospheric CO_2) liberates elemental sulphur. An early preparation, *liver of sulphur* (potassium polysulphide), was made by the fusion of caustic potash and sulphur. *Lime sulphur*, still widely used against apple scab, powdery mildew of vines etc., is prepared by boiling sulphur with lime—the resulting solution containing calcium thiosulphate and calcium polysulphide. Lime sulphur tends to be phytotoxic, and is incompatible with many chemical pesticides owing to its alkalinity. *Green sulphur* contains finely-divided sulphur and certain iron compounds; it is formed as a by-product of a coal-gas purification process in which hydrogen sulphide reacts with ferric oxide.

Elemental sulphur can cause premature defoliation and dropping of fruit in certain

varieties of fruit trees—a phenomenon known as *sulphur shyness*.
(see also DITHIOCARBAMATE DERIVATIVES)

sulphur cycle A series of interrelated processes in which sulphur and its compounds are interconverted by living microorganisms—see diagram.

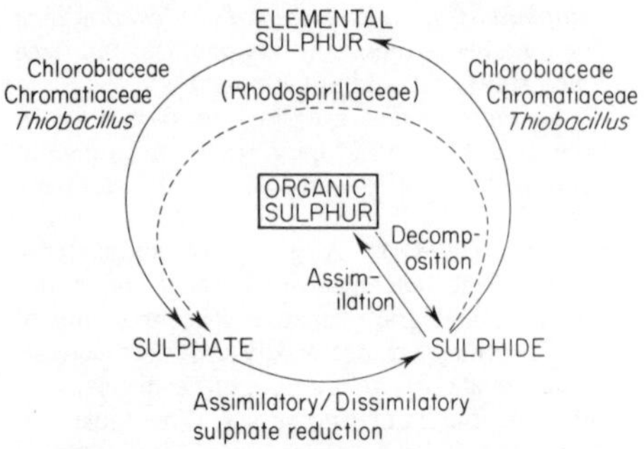

THE SULPHUR CYCLE—microbial aspects. (See also ASSIMILATORY SULPHATE REDUCTION, DISSIMILATORY SULPHATE REDUCTION, HYDROGEN SULPHIDE PRODUCTION, CHLOROBIACEAE, CHROMATIACEAE, RHODOSPIRILLACEAE, THIOBACILLUS.)

sulphur dioxide (as a food preservative) Sulphur dioxide is a permitted preservative in e.g. the UK and USA; it is used e.g. in fruit juices and pulps, jams, beers and wines. The bactericidal and fungicidal properties of sulphur dioxide are believed to depend on its ability to reduce the disulphide bonds of enzymes. In wine-making sulphur dioxide is used to inhibit the growth of contaminating yeasts—the wine-making yeasts being relatively resistant. Bisulphites or acid solutions of sulphites may be used in place of sulphur dioxide. (see also FOOD, PRESERVATION OF)

sulphur fungus The basidiomycete *Polyporus sulphureus*.

sulphur mustards Sulphur-containing analogues of the NITROGEN MUSTARDS—which have similar properties. Example: *mustard gas*, $ClCH_2CH_2SCH_2CH_2Cl$.

sulphur shyness (*plant pathol.*) (see SULPHUR)

summer spores (*mycol.*) Spores which are produced (often in great numbers) during the season of rapid growth and which are able to germinate soon after their formation, e.g. CONIDIA.

sun animalcule *Actinophrys sol*—see HELIOZOANS.

superinfection immunity (1) See PLASMID. (2) See LYSOGENY.

superoxide anion (O_2^-) A hyper-reactive radical formed when an oxygen *molecule* accepts a single electron; such reductions may occur during any of a wide variety of intracellular chemical reactions in procaryotic and eucaryotic organisms. Superoxide anions spontaneously react e.g. with protons and other superoxide anions to form hydrogen peroxide; in a further spontaneous reaction between hydrogen peroxide and superoxide anions the products include hydroxyl ions, hydroxyl free radicals, and (possibly) SINGLET OXYGEN. The intracellular activity of these potentially lethal entities is prevented, in aerobic organisms and aerotolerant ANAEROBES, by enzymes such as SUPEROXIDE DISMUTASE and CATALASE.

superoxide dismutase An enzyme which catalyses the reaction between SUPEROXIDE ANIONS and protons—the products being hydrogen peroxide and (ground-state) oxygen. Superoxide dismutases are metalloenzymes which occur in procaryotic and eucaryotic organisms; those found in eucaryotic cytosol contain copper and zinc and are characteristically cyanide-sensitive, while bacterial dismutases usually contain manganese or iron and are typically insensitive to cyanide. (see also BACTERIOCUPREIN) *Escherichia coli* has been found to contain a manganese-containing cytoplasmic dismutase and an iron-containing periplasmic dismutase. (see also ANAEROBE)

support film (see ELECTRON MICROSCOPY)

suppressor mutation (suppressor) A MUTATION which alleviates the effects of an earlier (*primary*) mutation *at a different locus*; commonly, the wild type phenotype is only partially restored—resulting in a *pseudo-wild* phenotype. (cf. BACK MUTATION) The primary and suppressor mutations may occur in the same gene (*intragenic* suppression) or in different genes which may or may not occur in the same chromosome (*intergenic* suppression).

Intragenic suppression. For example: (a) A (primary) MIS-SENSE MUTATION—causing an inactive gene product—is countered by a second (suppressor) mis-sense mutation (elsewhere in the gene)—allowing the gene product to assume a conformation which permits at least some activity. (b) A primary PHASE-SHIFT MUTATION may be suppressed by a second nucleotide deletion/insertion which restores the correct reading phase; however, the nucleotide sequence between the two mutations will still be read out of phase.

Intergenic suppression. In this mode of suppression the suppressor mutation (intergenic or informational suppressor) occurs in a gene which codes for a specific tRNA; there are various modes of intergenic suppression. (a) During translation, a codon containing a (primary) NONSENSE MUTATION may be recognized by a tRNA whose *anticodon* has been appropriately altered as a result of a *nonsense suppressor*; thus, instead of chain termination, an active or partially active gene product may be formed. Similarly, a codon containing a (primary) mis-sense mutation may be recognized by a tRNA whose anticodon has been altered by a *mis-sense suppressor*; if the aminoacid inserted is more closely related to the correct one, the gene product may be more functional. (b) During translation, a codon in which a mis-sense mutation has occurred may be recognized by a mutant tRNA whose anticodon is unaltered; the mutant tRNA is not recognized by its specific aminoacyl-tRNA synthetase (see PROTEIN SYNTHESIS) but by one which charges it with the 'wrong' amino acid—i.e. an amino acid which does not correspond to the anticodon of the tRNA. In such a case, the amino acid inserted in the polypeptide chain may allow a more functional product than that directed by the mutant codon. (c) An intergenic suppressor has also been found for phase-shift mutations. The mechanism is not known, but may involve the insertion of an extra nucleotide in the anticodon of a tRNA; such a tRNA molecule may recognize a 'codon' of four bases and may thus restore the correct reading phase.

These intergenic suppressors can function only when the tRNA involved is only one of a number of different tRNAs specific for the given amino acid; if an amino acid has only one type of specific tRNA coded for by a single gene, a mutation in that gene will generally be lethal.

Suppressors of the intergenic type will also interfere with the recognition of normal codons, termination codons, etc. and thus tend to be deleterious to the cell. *Ochre suppressors* (which suppress both amber and ochre codons) are necessarily of low efficiency since —because the ochre codon is the normal termination codon (at least in *Escherichia coli*)—a high-efficiency suppressor would be lethal. *Amber suppressors* (which suppress only amber codons) can be moderately efficient.

Indirect suppression. In this form of suppression, the effects of the primary mutation are bypassed. For example, an alternative metabolic pathway may be opened up by the product of the gene containing the suppressor mutation; such a pathway may by-pass that blocked as a result of the primary mutation.

suppressor-sensitive mutation (in bacteriophage) An early term for an AMBER MUTATION—which may be suppressed by an *intergenic* SUPPRESSOR MUTATION in the (permissive) host.

supravital staining (see VITAL STAINING)

suramin A complex derivative of naphthalene sulphonic acid used for the treatment and prophylaxis of trypanosomiasis; it was developed from the trypan dyes. It is used, often in conjunction with tryparsamide, particularly for infections involving *T. rhodesiense*. Suramin does not penetrate the central nervous system and is thus not effective in the later stages of disease. Its mode of action is unknown.

surface exclusion (see SEX FACTOR)

surfactin An extracellular product of *Bacillus subtilis*; surfactin is haemolytic and is also able to lyse the protoplasts of certain bacteria. In common with some other bacterial haemolysins (e.g. STREPTOLYSIN-S) surfactin is not immunogenic. The surfactin molecule consists of an oligopeptide and a lipophilic fatty acid derivative.

surra A serious disease of domestic animals, commonly fatal in *equines*. The causal agent, *Trypanosoma evansi*, is transmitted mechanically by blood-sucking horse-flies of the family *Tabanidae* and by certain other insect vectors. Symptoms include recurrent fevers and wasting. The disease occurs in parts of Africa, Asia and South America.

suspensor (*mycol.*) A clavate or conical structure formed by differentiation within the PROGAMETANGIUM in members of the Zygomycetes. In some species, e.g. *Phycomyces blakesleeanus*, the suspensor gives rise to a number of short filaments, or spines, which subsequently cover the developing ZYGOSPORE.

sutural lines Lines which mark the junction(s) of the spore wall *valves* in species of the CNIDOSPORA.

SV (see SATELLITE VIRUS)

SV40 (simian virus 40) (see POLYOMAVIRUS)

Svedberg unit (S) The unit in which the sedimentation coefficient of a particle (e.g. a macromolecule) is commonly quoted. The sedimentation coefficient (s) is given by:

$$\frac{dx}{dt} \cdot \frac{1}{\omega^2 r}$$

in which $\frac{dx}{dt}$ is the measured sedimentation rate (in the ultracentrifuge), ω is the angular velocity in radians per second, and r (in centimetres) is the distance between the sample and the axis of rotation. When values are substituted in the above expression the result given is in terms of *seconds*; the basic unit (Svedberg unit) is 10^{-13} seconds. Thus, e.g. when IgG is stated to have a sedimentation coefficient of $S^{20}_{W} = 7$, the actual value of the coefficient is 7×10^{-13} seconds; in this example the coefficient has been corrected to that in which the solvent viscosity is that of water at 20 °C. (see also CENTRIFUGATION)

swabs A *swab* usually consists of a thin wooden stick (or a piece of wire) around one end of which has been wound (or otherwise *securely* attached) a small amount of COTTON WOOL* —forming a compact ball of about 5–10 millimetres diameter. A *sterile* swab may be used e.g. for transferring material from the throat, nasal passages, vagina etc. to a culture medium for bacteriological examination—or e.g. for spreading a liquid inoculum during the preparation of a LAWN PLATE. *Or calcium alginate wool (see ALGINIC ACID)—a substance which dissolves in solutions of sodium citrate or sodium hexametaphosphate (e.g. 1% sodium hexametaphosphate in quarter-strength Ringer's solution); in such solutions, alginate swabs release their total complement of organisms.

swan animalcule The ciliate *Lachrymaria olor*.

swan-neck flasks Apparatus used by Pasteur (about 1860) in his successful attempt to refute the doctrine of abiogenesis (spontaneous generation). Each spherical glass flask had a long, slender S-shaped neck which—being bent downwards, then upwards—prevented the access of aerial microorganisms to the pre-sterilized contents of the flask.

swarm (1) (*bacteriol.*) A region of confluent surface growth which has developed as a result of SWARMING. (2) (*mycol.*) A synonym of PSEUDOPLASMODIUM (3).

swarm period (*mycol.*) (see DIPLANETISM)

swarm spores *Motile* spores.

swarming A mode of growth exhibited when certain (motile) bacteria, e.g. *Proteus* spp, are cultured on solid media. Growth occurs as a continuous surface film which radiates from the point of inoculation and which may eventually cover the entire surface of the medium. The film often exhibits an appearance which suggests that growth has occurred in a series of waves—each wave increasing the radius of the surface film. Swarming may be prevented by culturing on stiff AGAR or on certain inhibitory media, e.g. MacConkey's agar; media containing sodium azide or chloral hydrate etc. have also been used.

sweet curdling (of milk) (see LITMUS MILK)

swine erysipelas An acute or chronic disease of (young) pigs. The causal agent, *Erysipelothrix rhusiopathiae*, enters the body *via* the mouth or *via* wounds. In acute erysipelas there is fever, great prostration, erythematous skin lesions and septicaemia; death may occur rapidly without the development of lesions. In a milder acute form red or purple raised rectangular lesions—the 'diamonds'—appear on the skin; the joints may be subsequently affected. In chronic erysipelas there is laboured breathing and progressive unthriftiness; at *post mortem* growths are often found within the heart.

swine fever (hog cholera) An acute, infectious, viral disease peculiar to the pig; the causal agent, a *togavirus*, appears to enter the body *via* the mouth or nose. Following an incubation period of 1–4 days the infected animal exhibits fever, reddish patches on the skin, and commonly a sticky discharge from the eyes; subsequently—owing to central nervous system involvement—the gait is affected. Death usually occurs within two weeks of the onset; mortality rates are typically very high. Control of swine fever may involve prophylactic vaccination and/or slaughter of infected and suspect animals. Swine fever is a NOTIFIABLE DISEASE in the United Kingdom.

swine influenza An acute (often fatal) or chronic respiratory-tract disease of young pigs; characteristic symptoms include coughing, sneezing, and difficulty in breathing. The causal agent is an orthomyxovirus; *Haemophilus suis* (*H. influenzae-suis*) or species of *Brucella*, *Pasteurella* may act as secondary invaders.

swine vesicular disease (SVD) A disease peculiar to the pig; clinically it is indistinguishable from certain other diseases e.g. FOOT AND MOUTH DISEASE. The causal agent is an acid-stable (pH 2–10) picornavirus; infection appears to occur mainly *via* the mouth, and may occur *via* wounds, abrasions etc. Symptoms develop after an incubation period of 2–7 days; viruses are present e.g. in the faeces of infected animals. (The SVD virus remains infective in fomites for periods of the order of days/months; the virus is inactivated by exposure to 60 °C for 30 minutes.) Swine vesicular disease is a NOTIFIABLE DISEASE in the United Kingdom.

symbiont One of the partners in a symbiotic relationship—see SYMBIOSIS.

symbiosis (1) A term which was originally used to refer to any stable condition in which two different organisms (the *symbionts*) live together in close physical association—regardless of the nature of the relationship. (2) Currently, the term is generally taken to refer to any stable condition in which two different organisms live together in close physical association to their *mutual* advantage; in this sense the term is equivalent to *mutualism*. (see also COMMENSALISM) Some authors retain the original meaning of the term when they write of 'mutualistic symbiosis' (mutualism) or 'parasitic symbiosis' (PARASITISM).

The symbiotic relationship is often nutritionally important to one or both symbionts. Thus e.g. certain wood-boring insects—which are incapable of assimilating cellulose—carry within their gut cavities a resident population of the cellulose-hydrolysing protozoan *Trichonympha campanula*; both symbionts benefit nutritionally from the association, and the protozoan also enjoys a stable, protective environment. In this example the protozoan is an *ectosymbiont*, i.e. a symbiont which remains external to the cells and tissues of the other symbiont. Members of the RUMEN microflora are further examples of microbial ectosymbionts. In some symbiotic associations one partner (the *endosymbiont*) penetrates the cells or tissues of the other; thus, e.g. nitrogen-fixing bacteroids of *Rhizobium* spp occur within the ROOT NODULES of leguminous plants. (see also NITROGEN FIXATION) Symbiotic relationships are also found between protozoans (particularly ciliates) and endosymbiotic algae, and between fungi and algae (see LICHENS). (see also MYCETOCYTES and MYCORRHIZA)

In some symbioses certain of the benefits conferred on one or both symbionts may not be nutritional. Thus, e.g. a less obvious role is played by the luminous bacteria found in specialized organs in certain fish (e.g. *Photoblepharon*); it has been suggested that the patches of BIOLUMINESCENCE exhibited (and sometimes voluntarily controlled) by these fish may be signals by means of which members of the species recognize one another.

The symbiotic mode of living may be optional or obligatory for a given symbiont; thus, e.g. for killer strains of *Paramecium aurelia* the endosymbiotic KAPPA PARTICLES are not essential for the continued existence of the protozoan, but for the wood-boring termites (see above) the presence of *T. campanula* is probably indispensible.

The continued association of symbiotic partners, from generation to generation, may involve any of a variety of mechanisms. Thus, e.g. in many alga-protozoan relationships the symbionts divide synchronously or near-synchronously, and in lichens dispersal is often achieved by the formation of propagules which contain both symbionts (e.g. *soredia*). In the event of a change in the environment, the balance which characterizes a symbiotic relationship may be upset—so that one or other symbiont may assume a parasitic role or may become a frank pathogen.

symmetrogenic (*adjective*) Refers to the mode of cell division typical of flagellates. In such division, which is longitudinal, the daughter cells are mirror images of one another. (cf. HOMOTHETOGENIC)

symphogenous development (*mycol.*) The development of certain fungal structures (e.g. pycnidia) by the convergence, interweaving, growth, and differentiation of hyphal branches from a number of different hyphae. (cf. MERISTOGENOUS DEVELOPMENT)

symplectic metachrony (see METACHRONAL WAVES)

sympodial (1) Refers to the form or mode of development of a branching structure (e.g. a conidiophore or a bacterial sporophore): formed with the long axis of the structure consisting of a number of sections—each being a branch of the preceding section (cf. MONOPODIAL). (2) May refer to entities which are formed or united on a common structure or base.

symport (co-transport) Refers to the *linked* TRANSPORT of two solutes across a membrane *in the same direction* (cf. ANTIPORT and UNIPORT).

synapsis (pairing) (of chromosomes) (see MEIOSIS)

synaptonemal complex (synaptinemal complex) A structure which lies between—and parallel with—paired homologous chromosomes during the early stages of MEIOSIS. The complex is visible only by electron microscopy. It consists of three ribbon-like, electron-dense elements (one central and two lateral) which are separated by electron-transparent regions. The significance of the complex is unknown, but it is believed to have a role in the alignment of the homologous chromosomes, and may be involved in CROSSING-OVER.

syncaryon (synkaryon) The (diploid) *nucleus* of a ZYGOTE. (*Syncaryon* is sometimes used as a synonym of zygote.)

synchronous culture A form of microbial culture (or tissue culture) in which, due to appropriate treatment, all the cells in the culture pass through the same stage of the growth-division cycle at approximately the same time. Synchronous culture is used e.g. to determine the stage of the cell cycle at which a given substance is synthesized.

Synchronous cultures of *Tetrahymena pyriformis* have been obtained by *heat shock*—the brief exposure of a non-synchronous culture to an elevated temperature e.g. 40 °C; controlled periods of light and dark have been used to synchronize cultures of algae.

Synchronization of *mammalian* cells (e.g. in tissue cultures) may be achieved e.g. by the *double thymidine blockade*. A non-synchronous culture is exposed to a high concentration of thymidine. Cells passing through the *S* phase (see CELL CYCLE) remain blocked in that phase, while all other cells continue their development to the end of G_1. (Development beyond G_1 does not occur during blockade.) The cells are then exposed to a thymidine-free medium for a period of time equal to that of the *S* phase; during this time all the cells continue their development, and, at the end of the period, no cells are passing through phase *S*. The cells are then re-exposed to thymidine until *all* the cells have reached the end of G_1—i.e. the culture has become synchronous. In a simpler, 'selective' method a tissue culture is gently agitated;

cells undergoing mitosis are more easily detached from the glass, and can be decanted with the growth medium—giving a suspension of approximately synchronous *M* phase cells.

Synchronous cultures of bacteria can be obtained e.g. by adding to the growth medium an inhibitor of protein synthesis—such as CHLORAMPHENICOL. Alternatively, a synchronous culture can be obtained from an asynchronous bacterial population by the use of filtration or centrifugation; in these methods, cells of a similar size or weight—which are assumed to be in the same phase of growth—are separated from the remainder of the population.

After some time synchronous cultures return naturally to an asynchronous condition.

Synchytrium A genus of endobiotic, holocarpic fungi of the order CHYTRIDIALES; *Synchytrium* spp are parasitic on a range of plant hosts, and *S. endobioticum* is the causal agent of *wart disease* of potato (see POTATO, FUNGAL DISEASES OF). The asexual cycle of *Synchytrium* spp resembles that of species of OLPIDIUM—with the main exception that each zoospore may give rise to a group (sorus) of sporangia within the host cell; thick-walled resting sporangia are also formed.

syncytium A term sometimes used as a synonym of COENOCYTE; the term is also used to refer to a multinucleate cell or structure formed by the cytoplasmic *fusion* (but not the nuclear fusion) of a number of individual mononucleate protoplasts, e.g. a GIANT CELL.

syndrome (*med.*) A group of symptoms which, taken collectively, characterize, or are indicative of, a given disease or abnormality.

Synechococcus A genus of unicellular BLUE-GREEN ALGAE.

Synechocystis didemni (see PROCHLOROPHYTA)

synergism (in antibiotic action) When two or more ANTIBIOTICS are acting together: the production of inhibitory effects (on a given organism) which are greater than the additive effects of those antibiotics acting independently. (see e.g. TRIMETHOPRIM)

syngamy The process in which two gametes coalesce to form a ZYGOTE.

syngen (*ciliate protozool.*) (*synonym*: variety) A group or variety of individuals (within a given ciliate species) in which two or more *mating types* can usually be distinguished—CONJUGATION normally taking place only between members of *different* mating types within a given syngen. More than twelve syngens have been found in *Paramecium aurelia*, and six are known in *P. bursaria*. Each syngen is referred to by a number. Similar groupings have been found in e.g. *Euplotes* and *Tetrahymena*.

synnema (see COREMIUM)

synnematin B (see CEPHALOSPORINS)

syntrophism A phenomenon in which the growth, or improved growth, of an organism is dependent on the provision of one or more metabolic factors or nutrients by another organism growing in the vicinity. Thus, e.g. *Haemophilus influenzae* (which requires both the X FACTOR and the V FACTOR for growth) fails to grow on a medium which lacks the V factor; however, if colonies of *Staphylococcus* are grown on the same medium, clusters of *satellite colonies* of *H. influenzae* develop around the colonies of *Staphylococcus*. Growth occurs since V factor, liberated from the growing staphylococci, diffuses into the medium and is utilized by *H. influenzae*.

Some authors regard the terms *cross-feeding* and syntrophism as synonymous; others use the term cross-feeding to refer to the particular type of syntrophism in which *each* organism derives one or more essential factors or nutrients from the other. (see also SYNTROPHISM TESTS)

syntrophism tests Tests in which a syntrophic relationship between two or more organisms of the same species is exploited for the purpose of elucidating a particular biosynthetic pathway which occurs in that species. AUXOTROPHS of the same species which have blocks at different points of the same biosynthetic pathway frequently exhibit a syntrophic relationship. This is because the substrate of a particular blocked reaction tends to accumulate and to diffuse out into the surrounding medium; this substrate may promote the growth of other auxotrophs in the vicinity which are blocked at *earlier* points in the common pathway.

Synura A genus of algae of the CHRYSOPHYTA.

syphilis A chronic, venereal (sexually-transmitted) disease* which, under natural conditions, is peculiar to man; the causal agent is *Treponema pallidum*. Cutaneous symptoms—e.g. the initial ulcerative lesion, the *chancre*—commonly occur early in the course of the disease; later, lesions develop e.g. in the heart and in the tissues of the central nervous system. Diagnosis is based on joint clinical and laboratory evidence. PENICILLINS are commonly used therapeutically. *Laboratory diagnosis*. Material from lesions may be examined by dark-field microscopy and/or by IMMUNOFLUORESCENCE techniques. Serological tests include the VDRL, FTA-ABS and TPI tests (see also WASSERMANN REACTION). *Rare, non-venereal transmission has been recorded. (see also TREPONEMA)

systematics (1) A synonym of TAXONOMY. (2) A term used by some to refer to the range of studies (theoretical and practical) involved in the classification of organisms.

systemic antifungal agents (*plant pathol.*) Antifungal agents which penetrate plant tissues (roots, leaves) and which are then *translocated* within the transport systems of the plant—usually in the xylem vessels (i.e. from the root to the aerial tissues of the plant); the fungal pathogen is thus attacked from within the plant host. *Systemic* antifungal agents are distinct from those agents which, on penetrating the plant, spread *by diffusion* to regions of the plant away from the site of application. Some agents classified as systemic antifungals, e.g. BENOMYL, are apparently converted to fungitoxic derivatives prior to or during translocation. (see also FUNGICIDES AND FUNGISTATS)

systole (vacuolar) (see CONTRACTILE VACUOLES)

syzygy (*protozool.*) A phenomenon found in the *gregarines* (subclass Gregarinia) and in members of the suborder Adeleina (subclass Coccidia): the stage of (sexual) reproduction during which two gametocytes become physically joined to one another *without losing their identity as separate cells*. A pair of gametocytes 'in syzygy' usually encyst and give rise to gametes.

T

T_{90} The period of time required for the occurrence of 90% mortality in a population of microorganisms exposed to a given hostile environment.

t_D (see THERMAL DEATH TIME)

T antigen (streptococcal) One of the main protein antigens in the cell wall of *Streptococcus* spp; variation in specificity in the T antigen among strains of Lancefield group *A* streptococci forms the basis of a system of serological TYPING. The T antigen appears not to be associated with virulence—cf. M PROTEIN.

T-even phages The T2, T4, and T6 BACTERIOPHAGES of *Escherichia coli.*

T2H test A test used to determine whether or not a given strain of *Mycobacterium* is sensitive to thiophene-2-carboxylic acid hydrazine (T2H). The organism is inoculated onto (a) the test medium, containing T2H (10μg/ml), and (b) a T2H-free control medium, and both are incubated; *M. bovis* does not grow on the test medium, but *M. tuberculosis* (and most other mycobacteria) grow on both the test and control media.

T-independent antigens (see ANTIBODY-FORMATION and LYMPHOCYTE)

T lymphocyte (see LYMPHOCYTE)

T-odd phages The T1, T3, T5, and T7 BACTERIOPHAGES of *Escherichia coli.*

T1 side chain (in *Salmonella*) (see O ANTIGEN)

TAB A vaccine used prophylactically against *Salmonella typhi* and the A and B forms of *Salmonella paratyphi.* (see TYPHOID FEVER and PARATYPHOID FEVER) The vaccine contains suitably killed and preserved cells of each of these organisms.

Tabellaria A genus of pennate diatoms of the BACILLARIOPHYTA.

tache noir (see BOUTONNEUSE)

tachyzoites (of *Toxoplasma*) (*synonym*: trophozoites) Rapidly-multiplying vegetative cells which occur, in *acute* TOXOPLASMOSIS, in the cells of various tissues—and in the blood and lymph—of the host. Tachyzoites are uninucleate, cresent-shaped cells about 5–7 microns in length. The term *group* has been proposed for the intracellular progeny of a tachyzoite; such progeny, enclosed by the plasmalemma of the host cell, constitute an arrangement which is commonly called a *pseudocyst*. (see also BRADYZOITES)

taette A Scandinavian product prepared by the fermentation of fresh whole milk by a strain of *Streptococcus lactis*. (cf. YOGHURT)

Takatsi technique A procedure designed for preparing doubling dilutions (see SERIAL DILUTIONS) by means of an instrument consisting of a small spiral wire loop attached to the end of a metal rod (handle). When immersed in a liquid and withdrawn the spiral loop carries with it a fixed volume of the liquid; the spiral loop carrying this liquid can then be immersed in a volume of diluent equal to the volume carried by the loop and, by rotating the handle of the instrument between forefinger and thumb, the liquids can be efficiently mixed to give a 1-in-2 dilution. The spiral loop is then withdrawn and rotated within a fresh volume of diluent, and so on. A modern version of the instrument consists of a small slotted metal block at the end of a metal handle—liquid being transferred within the slots of the block.

take-all (see CEREALS, DISEASES OF)

Talaromyces A genus of fungi of the order EUROTIALES; many of the constituent species have conidial (imperfect) stages which correspond to members of the *form*-genus PENICILLIUM. In *Talaromyces* spp the ascocarp consists of a number of evanescent asci contained within a loosely-constructed hyphal cleistothecium—or simply enclosed by a few wisps of hyphae, according to species; in at least some species the ascocarp develops parthenogenically.

Talfan disease One of several types of viral encephalomyelitis which is peculiar to the pig; Talfan disease was first described in the United Kingdom in 1952. The causal agent of Talfan disease is a *picornavirus* which is closely related to the causal agent of TESCHEN DISEASE; however, the morbidity and mortality associated with Talfan disease are much lower than in Teschen disease, and control measures such as vaccination are generally considered to be unnecessary.

tambour An apparatus sometimes used to study the effects of particular antibiotics on a given test organism. It may consist of a disc of cellophane (grade PT300) secured, by a rubber band, to a plastic or pyrex support ring—forming a shallow 'vessel' some 7–8 cms in diameter and 1–3 cms in depth. *Method of use.* The relevant antibiotics are first allowed to diffuse into a sterile agar plate from strips or discs of antibiotic-impregnated paper placed on the surface of the agar. The strips or discs are then removed, their former positions having been marked on the base of the plate. A sterile (autoclaved) tambour is then placed on the agar so that the (moist) cellophane disc makes perfect contact with the agar surface. The inside of the tambour is flooded with a suspension of the test organism—excess suspension being removed with a sterile pasteur pipette. The whole is then incubated. During incubation, nutrients—and antibiotic(s)—diffuse through the cellophane and permit growth of the test organism on those areas of the disc which are free from inhibitory concentrations of the antibiotic(s). After recording those areas of the disc on which *no* growth has occurred the tambour is removed and re-incubated on a plate of non-inhibitory medium, i.e. a medium containing no antibiotics; the subsequent appearance of growth on these areas indicates that growth-inhibition was due to microbistatic (as opposed to microbicidal) action. The tambour may also be used, for example, to study the effects of antibiotic SYNERGISM.

tanned red cells (see BOYDEN PROCEDURE)

tannic acid (see TANNINS)

tannins An heterogenous group of complex polyhydric phenol compounds which occur in plants. The *hydrolysable tannins* consist of a sugar (often glucose) partially or wholly esterified with polyhydric phenols such as gallic acid (3,4,5-trihydroxybenzoic acid), digallic acid, or ellagic acid; on hydrolysis these tannins yield a sugar and phenolic acids. Tannic acid, a penta-(*m*-digalloyl)-glucose, is one example of a hydrolysable tannin. *Condensed tannins* (non-hydrolysable tannins) are believed to be oligomers of *flavanoid* compounds (i.e. derivatives of flavan); they tend to polymerize, e.g. under acid conditions, to form insoluble reddish compounds called *phlobaphenes*. Tannins occur particularly in woody tissues but are also found in certain leaves, e.g. tea (*Thea*); they may occur in vacuoles, cell walls, or in specialized cells. The presence of tannins in certain plants (e.g. the oak—*Quercus* spp) is believed to be an important factor in the resistance of such plants to fungal attack and decay. Tannins can precipitate and inactivate proteins by forming crosslinks—a property exploited in leather tanning. Tannic acid is used in the BOYDEN PROCEDURE.

Taphrina A genus of fungi of the order TAPHRINALES. *Taphrina* species are parasitic on certain vascular plants (particularly trees)—in which they produce GALLS and other lesions (see e.g. PEACH, DISEASES OF); trees on which *Taphrina* spp are parasitic or pathogenic include e.g. *Alnus* (alder), *Betula* (birch), *Prunus* (cherry, peach, plum), and *Quercus* (oak). Within the host plant the organism forms an intercellular or subcuticular septate mycelium; some species form haustoria. Asci are often formed as a subcuticular layer—each ascus being derived from one of a number of individual cells formed by hyphal fragmentation. (In culture *Taphrina* spp develop only in the yeast phase—i.e. as budding, yeast-like (unicellular) organisms.)

Taphrinales An order of plant-pathogenic fungi of the class HEMIASCOMYCETES; constituent species are distinguished from those of other orders in that, during sexual reproduction, the ascus develops from a binucleate (dikaryotic) cell. Asci may develop from the terminal cells of the (dikaryotic) hyphae or from individual cells formed by the fragmentation of hyphae; in an infected plant host the asci typically form an hymenium-like layer. The order contains one genus: TAPHRINA.

target cell (*immunol.*) Any cell whose surface antigens are the targets for attack by e.g. *primed* lymphocytes or homologous antibodies.

tartar emetic (see *Antimony* under HEAVY METALS)

tartronic semialdehyde pathway (in glyoxylate metabolism) (see TRICARBOXYLIC ACID CYCLE)

***tau* particles** Intracellular particles seen in phage-infected bacteria examined by electron microscopy; they appear to be phage precursors.

taurocholate (see BILE ACIDS and MACCONKEY'S AGAR)

taxa (see TAXON)

taxis (tactic movement) A *directional locomotive response*, to a given stimulus, exhibited by certain (motile) organisms or cells; the *stimulus* may or may not be directional. (see CHEMOTAXIS, PHOTOTAXIS and cf. TROPISM and KINESIS) In *phobotaxis* the organism responds to a change in the concentration or intensity of the stimulus; the response involves an alteration or reversal in the direction of motion—such movement being unrelated to the *direction* (if any) of the stimulus. Thus, e.g. when placed in a nutrient gradient, bacteria may be seen to execute a series of *shock movements*; such movements involve a change in the direction of motion—though not necessarily a *direct* progression towards the more favourable region of the gradient. In order to incite a phobotactic response the stimulus must be greater than a certain minimum (threshold) value. A phobotactic response is said to be *positive* (or *negative*) when it is triggered by a decrease (or increase) in concentration or intensity of the stimulus. In *topotaxis* the response consists of a movement towards (positive) or away from (negative) the stimulus, i.e. the stimulus is directional and the response is related to the direction of the stimulus.

In general, organisms which give a positive tactic response to a given stimulus tend to give a negative response when the strength of the stimulus increases above a certain value.

taxon (*plural*: taxa) Any taxonomic group, e.g.

genus, *family*, *order* etc. (see also NOMENCLATURE)

taxonomy The science of *biological classification*—i.e. the grouping of organisms according to their mutual affinities or similarities. A classified arrangement of organisms permits a logical and informative system of naming (see NOMENCLATURE) and may be used e.g. as a key for the classification of unidentified organisms. Organisms may be classified in a number of ways. Thus, e.g. the classification of higher animals and plants has, as one of its aims, the exposition of evolutionary relationships; such a scheme is referred to as a *phylogenetic* or *phyletic* type of classification. Among microorganisms, however, evolutionary affinities are commonly difficult to discern, and microbial taxonomy currently involves elements of artificiality and subjectivity; accordingly, any particular mode of classifying a given organism, or group of organisms, does not necessarily meet with universal approval. (Attempts to introduce a more objective basis have led to an increased interest in NUMERICAL TAXONOMY.)

For each of the main groups of microorganisms (e.g. the bacteria or fungi) the *taxonomic hierarchy* is a system in which a single, all-inclusive category (e.g. 'bacteria', or 'fungi') is divided and subdivided into progressively smaller and less-inclusive categories; each category contains those organisms which meet certain criteria (i.e. have certain characteristics), and all the categories (*taxa*—sing. *taxon*) are, ideally, mutually exclusive. Thus, e.g. a taxonomic hierarchy may consist of the following main taxa (in descending order of magnitude): DIVISION (*phylum*), *class*, *order*, *family*, *genus*, SPECIES; one or more species constitute a genus, one or more genera constitute a family—and so on. Categories intermediate between those quoted include e.g. *subdivision*, *tribe* (between family and genus), *sub-family, sub-genus*. (see also SERIES) (*Form*-genera, *form*-families etc. are used in the classification of the *fungi imperfecti*—see NOMENCLATURE; see also SACCARDIAN SYSTEM.) The species of certain bacteria and fungi are themselves sub-divided into strains which are distinguishable by serological or other means—see e.g. SEROTYPE and FORMA SPECIALIS; see also TYPING.

The hierarchical system of taxonomy would be optimal if each and every organism could be allotted specifically to one or other of the taxa. However, the characteristics of some organisms are such that they could be placed, with justification, in any of several related species or genera: the 'species end' of the taxonomic tree may exhibit a more or less continuous spectrum of biological types—rather than discrete aggregations of near-identical organisms.

The *kinds of criterion* used for classifying microorganisms include: (for BACTERIA) morphology, staining reactions, spore formation, motility, antigenic structure, biochemical (metabolic) activities, etc.; (for FUNGI) the nature of the thallus, and the form of sexual/asexual reproductive structures, etc.; (for PROTOZOA) the possession or otherwise of flagella or cilia, the modes of locomotion and/or reproduction, the presence or absence (or nature) of specialized intracellular structures or extracellular skeletal structures, etc.; (for ALGAE) the type(s) of pigment, the number and arrangement of flagella on motile cells, the nature of storage carbohydrates etc.; (for VIRUSES) the type of genome (DNA or RNA), the size of the virion, the presence or absence of an envelope, etc.

(Taxonomists are sometimes divided into *lumpers* and *splitters*. In any given system of classification the lumper would tend e.g. to lump together a number of more or less similar species to form a single genus—while the splitter, emphasizing the differences between the species, would establish two or more genera from those same species.)

TB (see TUBERCULOSIS)

Tbilisi (Tb) phage A bacteriophage which (at RTD) lyses certain biotypes of *Brucella abortus*.

TCA cycle (see TRICARBOXYLIC ACID CYCLE)

TCC (3,4,4′-trichlorocarbanilide) (see UREA)

TCD_{50} Tissue culture dose (50%); that quantity of a viral suspension which, when inoculated onto each of a number of tissue cultures, causes degeneration in 50% of the cultures. (see also END POINT (*virol.*))

$TCID_{50}$ Tissue culture infectious dose (50%); an alternative expression for TCD_{50} (*q.v.*).

TCNB (see TECNAZENE)

tea, diseases of Diseases of the tea plant (*Camellia sinensis*—also known as *Thea sinensis*) include BLISTER BLIGHT and RED RUST.

tecnazene (TCNB) 2,3,5,6-Tetrachloronitrobenzene, an agricultural antifungal agent used for the control of dry rot (*Fusarium caeruleum*) of stored potatoes; it also inhibits the sprouting of stored potatoes. (For the treatment of potatoes tecnazene is used as a dust.) TCNB is also used (as a smoke) in greenhouses for the control of *Botrytis* infections. (see also QUINTOZENE)

tegulicolous (of e.g. a lichen) Growing on tiles.

teichoic acids Polymers found in the CELL WALL and/or CELL MEMBRANE of many gram-*positive* bacteria; similar polymers have been isolated from the capsular material of certain of these organisms. Teichoic acids may account for as much as 50% of the dry weight of the cell wall; being negatively charged, these polymers bind cations, and they probably function as a reservoir of those ions (e.g. Mg^{++}) which are necessary for the activity of many of the membrane-bound enzymes. Teichoic acids are believed to be the main serological components of many strains of *Lactobacillus*, and a *glycerol* teichoic acid has been identified as the group-specific antigen of the *Lancefield group D* streptococci. Teichoic acids are highly variable in composition, but are generally characterized by a high proportion of polyhydric alcohol phosphate residues; *wall* teichoic acids are covalently linked to **PEPTIDOGLYCAN**—probably *via* a phosphodiester link to the 6-position of muramic acid.

Glycerol teichoic acids occur in the cell membrane (and possibly in the periplasmic space) as well as in the cell wall. The backbone chain of the polymer is commonly composed of glycerol residues linked (1 → 3) by phosphodiester bridges. The glycerol residues may be further substituted with e.g. D-alanine or a sugar such as *N*-acetylglucosamine, glucose, **KOJIBIOSE** etc. The degree and nature of substitution are highly variable. Alternatively, a *sugar* may be an integral part of the polymer backbone; for example, the chain may consist of a repeating unit of *N-acetylglucosamine-1-phosphate-glycerol phosphate* or *glucosyl-1-glycerol phosphate*. Biosynthesis of glycerol teichoic acids:

(a) α-glycerophosphate + CTP ⟶ CDP-glycerol + PPi (pyrophosphate)

(b) CDP-glycerol + $(\text{glycerophosphate})_n$ ⟶ CMP + $(\text{glycerophosphate})_{n+1}$

Backbone sugars and substituent sugars are added from their corresponding nucleotide derivatives. The enzymes which govern these syntheses are found in the cell membrane; a **BACTOPRENOL**, similar to that which takes part in peptidoglycan synthesis, appears to be involved.

Ribitol teichoic acids have been reported in capsular material but appear to occur mainly in the cell wall; they contain a chain of ribitol residues linked (1 → 5) by phosphodiester bridges, and may include sugar residues as an integral part of the backbone chain. The ribitol residues are usually substituted with D-alanine and may be further substituted with sugars such as *N*-acetylglucosamine, D-glucose etc.; the nature and extent of substitution may be highly variable even among closely related organisms. The biosynthesis of ribitol teichoic acids begins with the reduction of ribulose-5-phosphate (by an $NADPH_2$-dependent enzyme) to give ribitol-5-phosphate. This is then polymerized in a sequence of reactions analogous to those which occur in the synthesis of glycerol teichoic acids. (see also **TEICHURONIC ACID**)

teichuronic acid A cell wall constituent formed by certain species of *Bacillus* when grown in phosphate-deficient media. Teichuronic acid is a polymer consisting of residues of *N*-acetylgalactosamine and glucuronic acid. Teichuronic acid and **TEICHOIC ACIDS** may occur in the same organism.

telemorphosis (*mycol.*) The formation of a branch in one hypha due to the stimulus of the closely-adjacent tip of a second hypha—the tip of the branch and that of the second hypha eventually fusing.

teleutosorus (*pl.* teleutosori) A telium—see **RUSTS** (1).

teleutospore (teliospore) (see **RUSTS** (1) and **SMUTS** (1))

Teliomycetes (Hemibasidiomycetes) A class of fungi of the sub-division **BASIDIOMYCOTINA**; constituent species are distinguished from those of other classes in that they do not form basidiocarps and in that their basidia develop directly following the germination of thick-walled spores (teliospores). All species are plant parasites. The class consists of two orders: the rust fungi or 'rusts' (order **UREDINALES**), and the smut fungi or 'smuts' (order **USTILAGINALES**).

teliospore (teleutospore) (see **RUSTS** (1) and **SMUTS** (1))

telium (see **RUSTS** (1))

telophase The fourth and final stage in **MITOSIS** and analogous stages in **MEIOSIS**.

Telosporea A class of protozoans of the sub-phylum **SPOROZOA**. Sexual reproduction occurs in all members; schizogony does not occur in some. Sub-classes: **GREGARINIA** and **COCCIDIA**.

telotroch (*protozool.*) The motile, *larval* stage of a sessile peritrich (e.g. *Vorticella*). When the *zooid* undergoes binary fission, one daughter cell remains attached to the stalk while the other (the telotroch) develops a ring of cilia (the *trochal band*) and swims away from the parent organism.

temperate (of bacteriophages) (see **BACTERIOPHAGE**)

temperature coefficient (of a given **DISINFECTANT**) A coefficient which indicates the increase in death rate—in a single-species population of microorganisms exposed to that disinfectant—following a rise in temperature of 1 °C (or, more commonly, 10 °C). The temperature coefficient is not necessarily constant over a wide range of temperatures: the coefficient in e.g. the range 20–30 °C may not be the same as that in the range 30–40 °C. (see also **DILUTION COEFFICIENT**)

temperature-sensitive mutant (see CONDITIONAL LETHAL MUTANT)

tennecetin (pimaricin) (see POLYENE ANTIBIOTICS)

tequila (see PULQUE)

teratogenesis (*med.*) The formation of a defective embryo or foetus.

terete (*adjective*) Refers to a stipe, thallus etc. which is circular in transverse section.

terminal redundancy In certain bacteriophages: the presence of an identical nucleotide sequence at each end of the linear genome—see e.g. BACTERIOPHAGE T4.

terminalization (of chiasmata) (see MEIOSIS)

ternary fission The process in which three cells are formed by the division of one cell. (cf. BINARY FISSION) Ternary fission occurs e.g. in PELODICTYON.

terramycin OXYTETRACYCLINE. (see also TETRACYCLINES)

terregens factor (see SIDEROCHROMES)

terricolous Growing on soil, peat, sand, etc.

Teschen disease (enzootic porcine encephalomyelitis) One of several types of viral encephalomyelitis which is peculiar to the pig; Teschen disease was first described in the Teschen (Cieszyn) district of Czechoslovakia. The causal agent of Teschen disease is an acid stable (pH 3–9.5) picornavirus. The disease is generally more common in *young* pigs; symptoms typically include an initial fever followed by an apparent difficulty in locomotion, tremors, convulsions, and increasing evidence of paralysis. Death occurs within 4 days—the mortality rate commonly being greater than 70%. During the course of the disease viruses are excreted e.g. in the faeces. (Subclinical, i.e. inapparent, infection has also been recorded.) According to whether the disease is sporadic or enzootic in a given area, control of Teschen disease has been accomplished by slaughter of infected animals (and their contacts) or by a progamme of vaccination. Teschen disease is a NOTIFIABLE DISEASE in the United Kingdom. (cf. TALFAN DISEASE)

test (*syn*: lorica) The shell of a protozoan.

test cross (genetics) A cross between an organism heterozygous for certain (specified) genes, and an organism known to be homozygous recessive for each of those genes; the phenotypes of the progeny indicate linkage relationships between the specified genes in the heterozygous organism.

Testacida (see ARCELLINIDA)

testate (*protozool.*) Possessing a shell (test).

tetanus (1) (lockjaw) A disease of man, and other animals, in which the symptoms are due to a powerful neurotoxin formed by the causal agent, *Clostridium tetani*, present in an anaerobic wound or other lesion. The incubation period may be days or weeks; short incubation periods correlate with high mortality rates, and conversely. In man, *generalized tetanus* begins with a sustained involuntary contraction of the muscles of the jaw and neck; presently the voluntary muscles of the face, trunk and limbs become painfully contracted, and a contorted facial expression—the *risus sardonicus*—characteristically develops. Death may be due e.g. to asphyxia or exhaustion. In (rare) *local tetanus* spasms occur in muscles near the infected lesion. The diagnosis of tetanus is based mainly on clinical observations. Treatment and short-term prophylaxis involve the administration of antitoxin; vaccines for long-term protection contain tetanus *toxoid*. In the *horse*, tetanus is characterized by rigidity of the voluntary musculature, particularly the jaw muscles, and an extreme state of hypersensitivity to aural and other stimuli; sheep, cattle, dogs etc. are affected less frequently by tetanus. (2) *Tetanus* also refers to the sustained contraction of a muscle brought about, experimentally, e.g. by high-frequency electrical stimuli.

tetracycline (achromycin; tetramycin; polycycline) An ANTIBIOTIC produced semi-synthetically from 7-CHLORTETRACYCLINE. (see TETRACYCLINES)

tetracyclines A group of natural and semi-synthetic ANTIBIOTICS which have in common a modified naphthacene ring (see formula); they include CHLORTETRACYCLINE, DEMETHYLCHLORTETRACYCLINE, DOXYCYCLINE, METHACYCLINE, OXYTETRACYCLINE, and TETRACYCLINE. The tetracyclines are bacteriostatic, and all exhibit a similar broad spectrum of activity. Susceptible organisms include species of *Bacteroides*, *Brucella*, *Chlamydia*, *Corynebacterium*, the rickettsiae, certain spirochaetes, and some anaerobic streptococci; *Proteus* and *Pseudomonas* spp are often resistant, and fungi are not affected. Tetracyclines bind (irreversibly) to the smaller subunits of the RIBOSOMES and (reversibly) to nucleic acids and proteins in the bacterial cell; they appear to inhibit the binding of aminoacyl-tRNAs to the ribosomal acceptor sites—thus inhibiting PROTEIN SYNTHESIS. Tetracycline-sensitive cells appear to accumulate the antibiotic. Resistance to the tetracyclines develops slowly, although plasmid-borne

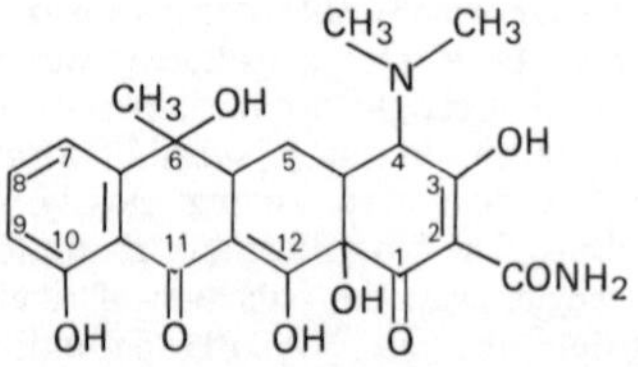

Tetracycline

resistance (see R FACTOR) also occurs; cross-resistance among the tetracyclines is virtually complete.

tetrad (genetics) May refer either to (a) the four chromatids of the bivalent in MEIOSIS, or (b) the four organisms/cells which result from a meiotic division—each organism/cell containing one of the chromatids referred to in (a).

tetrahydrofolic acid (see FOLIC ACID)

Tetrahymena A genus of ciliate protozoans of the suborder TETRAHYMENINA. Species of *Tetrahymena* are found in a variety of freshwater habitats (ponds, lakes etc.)—particularly in those waters which contain decomposing organic matter and which are, as a consequence, rich in the organism's natural food: bacteria; a few species have been reported to live parasitically in metazoans. One species, *T. pyriformis*, has been used extensively as a test organism for studies in genetics, cytology, biochemistry and other fields; one reason for this choice is that the organism can be grown in a sterile, defined medium and obtained in axenic culture. (The ease with which cell division can be synchronized is another important factor.)

T. pyriformis is typically a pear-shaped organism, some 50 microns in length (range about 30–80 microns); variations in body shape are not uncommon. The somatic ciliature consists of some 15 to 26 longitudinal kineties. One macronucleus and one micronucleus are normally present, but amicronucleate strains are known. A contractile vacuole lies near the blunt posterior end. The details and arrangement of pellicular/cortical structures vary among the different SYNGENS. The cytostome is found laterally, near the tapered, anterior end; there is no vestibule, and the buccal cavity contains specialized ciliature consisting of one undulating membrane, on the right, and a group (AZM) of three membranelles on the left—an arrangement referred to, in other organisms, as 'tetrahymenal'. *Microstome* strains are those with small buccal cavities; *macrostomes* have large buccal cavities. The essential nutritional requirements of *T. pyriformis* include the vitamins FOLIC ACID, LIPOIC ACID, NICOTINIC ACID, PANTOTHENIC ACID, PYRIDOXINE, RIBOFLAVIN and THIAMINE. *T. pyriformis* can be kept in a viable condition at −196 °C by using a suitable cryoprotectant, e.g. DMSO. (see also BIOASSAY)

Tetrahymenina A sub-order of protozoans of the order HYMENOSTOMATIDA. Species are often small (less than about 75 microns in length) with uniform somatic ciliature. The specialized oral ciliature consists of an undulating membrane on the right of the buccal cavity and an AZM of three membranelles on the left; a vestibulum is usually absent. Conjugation occurs between organisms of the appropriate mating types—see SYNGEN. Genera include e.g. COLPIDIUM and TETRAHYMENA.

tetrapolar heterothallism (see HETEROTHALLISM)

tetrasporal colony (*algol.*) A type of colonial organization in which a group of cells (not necessarily four) is embedded in a mucilaginous matrix to form a non-motile mass.

thalline margin (*lichenol.*) A rim of thalline tissue (containing algal as well as fungal cells) which surrounds—and sometimes obscures—the PROPER MARGIN of the apothecium in certain lichens. The thalline margin is generally the same colour as the thallus but may differ in colour from the disc. (see also LECIDEINE and LECANORINE)

thallospore (*mycol.*) Any spore formed from part of an *existing* fungal thallus, e.g. a CHLAMYDOSPORE; cf. CONIDIA which are produced as new structures.

thallus (*microbiol.*) (*pl.* thalli) The vegetative (assimilative, somatic) form or part of an alga, fungus, or lichen.

Thamnidium A genus of fungi within the class ZYGOMYCETES; species occur e.g. in the soil and on dung. Asexual reproduction typically involves the formation of both sporangia and sporangiola; in the commonest species, *T. elegans*, the long main axis of the sporangiophore carries a single, terminal, columellate, dehiscent sporangium, while the whorled lateral branches of the sporangiophore carry a number of non-columellate deciduous sporangiola. (If grown in the dark only the sporangiola may be formed.) Sexual reproduction involves gametangial fusion and zygospore formation; at least some species (including *T. elegans*) are reported to be heterothallic.

Thamnolia A genus of fruticose LICHENS in which the phycobiont is *Trebouxia*. The thallus is hollow and may be erect or prostrate; fruiting bodies are unknown. Species include e.g.:

T. vermicularis. The thallus is white to grey, terete, unbranched or sparsely branched, approximately 1–4 centimetres long and 1–2 millimetres wide. *T. vermicularis* occurs among mosses in mountainous regions.

Thayer-Martin selective medium (VCN agar) An enriched growth medium containing the antibiotics VANCOMYCIN, colistin (see POLYMYXINS), and nystatin (see POLYENE ANTIBIOTICS); the range of organisms inhibited by this combination of antibiotics is such that the isolation of *Neisseria gonorrhoeae* or *N. meningitidis* from a mixed flora is facilitated.

theca (1) (*algol.*) (of a dinoflagellate) (*syn*: amphiesma) Essentially, the layer of flattened, membranous vesicles *beneath* the external (limiting) membrane of the organism; the latter membrane appears to be equivalent to the cell membrane (plasmalemma). In some (*unarmoured*) dinoflagellates (e.g. *Gymnodinium*, *Katodinium*, *Oxyrrhis*) the vesicles are either empty or contain plates (one plate per vesicle) which cannot be detected by light microscopy; such a theca is referred to, by some authors, as a *pellicle*. In the *armoured* dinoflagellates (e.g. *Ceratium*, *Prorocentrum*, *Woloszynskia*) each vesicle contains a plate (probably of cellulose) which is thick enough to be visible by light microscopy. The number of thecal plates per organism varies from two principal plates (e.g. in *Prorocentrum*) to a few hundred (e.g. in *Oxyrrhis*); the number and arrangement of plates is of taxonomic importance. In armoured dinoflagellates the edges of adjacent plates (the *suture* region) are commonly bevelled—permitting free movement of the plates relative to one another. Thecal plates often bear characteristic ornamentation on their external surfaces. (2) (*algol.*) (*syn*: lorica) An open or perforated shell-like structure which houses part or all of a cell; such a structure is formed e.g. by *Trachelomonas*. (3) An obsolete term for an *ascus* formed by the mycobiont of a lichen.

thecium A term sometimes used to refer to the HYMENIUM of an ascocarp.

Theileria A genus of protozoans of the class PIROPLASMEA. Several species are important pathogens of domestic animals. Thus, e.g. *T. parva* causes an acute (commonly fatal) tick-borne disease in susceptible cattle and buffalo in Africa; symptoms include fever, enlargement of lymphoid tissues and lung congestion. *Theileria* differs from BABESIA in at least two ways: (a) Development occurs in both red *and* white blood cells of vertebrates, SCHIZOGONY taking place in the white cells. (b) *Transovarial* transmission has not been demonstrated in the acarid vectors. Macroschizonts (10–20 microns)—also called *Koch's blue bodies*—are formed within the lymphocytes of the host but may be seen free (external to ruptured lymphocytes) in stained preparations; the cytoplasm of the schizont stains blue (with Giemsa) and contains a variable number of red-staining nuclei. Merozoites, subsequently released from the schizont, invade the host's erythrocytes. Erythrocytic forms are commonly smaller than those of *Babesia*.

Thelohania A genus of protozoans of the subphylum CNIDOSPORA.

thermal death point The lowest temperature at which a single-species population of microorganisms, in neutral aqueous suspension, can be heat-killed in 10 minutes.

thermal death time (t_D) The time required, at a *given temperature*, for the thermal inactivation (killing) of a single-species population of microorganisms (in aqueous suspension). The value of t_D depends on the size of the microbial population. (see also STERILIZATION)

thermal inactivation point (TIP) (*plant pathol.*) The temperature at which a virus preparation must be held, for 10 minutes, in order to inactivate all the viruses. The TIP of many plant-pathogenic viruses is 50–60 °C; for some (including TMV) it is above 90 °C.

Thermoactinomyces sacchari A mycelial, thermophilic, endospore-forming actinomycete believed to be a causal agent of BAGASSOSIS.

thermoduric (*adjective*) Refers to any organism with the ability to withstand those temperatures which are lethal for most vegetative organisms; for example, some strains of *Microbacterium* survive 70–80 °C for 15 minutes. Among dairy microbiologists the term thermoduric is often used to refer to those organisms which are capable of surviving PASTEURIZATION. (see also THERMOPHILE)

thermophile (thermophil) Any organism having an optimum growth temperature above 45 °C. (cf. PSYCHROPHILE and MESOPHILE) (see also THERMODURIC)

THF Tetrahydrofolic acid (see FOLIC ACID).

thiabendazole 2-(4′-Thiazolyl)-benzimidazole. Originally known as an antihelminthic agent, thiabendazole has been shown to possess antifungal activity and has some use as an antifungal agent in agriculture. It is used mainly to protect against seed-borne smuts and bunts in cereals, and to prevent rots of stored fruits and vegetables. (see also BENZIMIDAZOLE DERIVATIVES (as antifungal agents))

thiacetazone (see UREA AND DERIVATIVES)

thiamine (thiamin; vitamin B_1; aneurin) A water-soluble VITAMIN: 3-(2-methyl-4-amino-5-pyrimidinylmethyl)-5-(β-hydroxyethyl)-4-methylthiazole; this molecule carries a net positive charge and is usually supplied commercially as thiamine chloride hydrochloride (= 'thiamine hydrochloride'). The coenzyme form is thiamine pyrophosphate (TPP, *cocarboxylase*). TPP functions in decarboxylation reactions (e.g. pyruvate → acetaldehyde), in oxidative decarboxylations (e.g. of pyruvate and α-ketoglutarate—in conjunction with LIPOIC ACID—see Appendix II(a)), in α-ketol formation (e.g. acetoin production—see Appendix III(c)), and in transketolation (e.g. the transketolase and phosphoketolase reactions—see Appendix I(b) and Appendix III(a)(ii) and (iii) respectively).

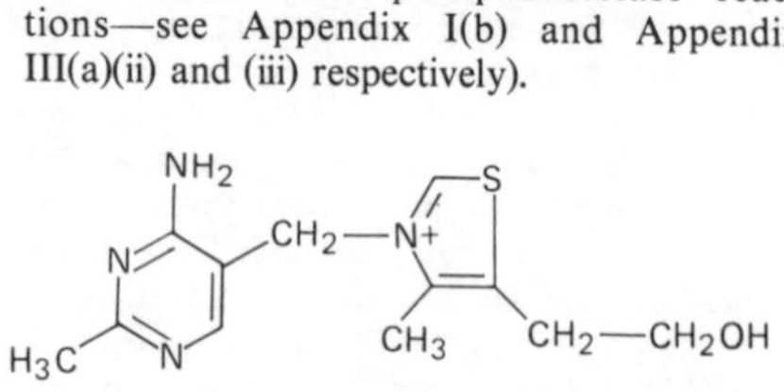

Thiamine

Microorganisms which require exogenous thiamine (or one—or both—of the pyrimidine and thiazole precursors) include: certain algae—e.g. *Euglena*, the fungi *Phycomyces blakesleeanus*, *Phytophthora*, certain species of *Trichophyton*, many yeasts, the bacteria *Lactobacillus* and *Staphylococcus aureus*, and a range of protozoans—e.g. *Acanthamoeba castellanii*, *Crithidia fasciculata*, *Tetrahymena*.

Thielaviopsis A genus of fungi of the class HYPHOMYCETES.

thigmotaxis A *locomotive* movement exhibited by certain organisms as a response to a *tactile* stimulus; the movement may be towards the stimulus (positive thigmotaxis) or away from the stimulus (negative thigmotaxis).

Thigmotrichida An order of protozoans of the subclass HOLOTRICHIA, subphylum CILIOPHORA; the *thigmotrichs*—which, in size, are usually of the order of 50 microns—are generally found as parasites of bivalve molluscs. Typically, the thigmotrich cell bears a tuft of positively thigmotactic cilia—by means of which it maintains contact with the host organism. Members of the suborder Arhynchodina (e.g. *Boveria*, *Hemispeira*) possess both a cytostome and oral ciliature. In *Ancistrocoma* and other genera of the second suborder, Rhynchodina, a sucker replaces the cytostome.

thimerosal (thiomersal; merthiolate) (see THIOMERSAL and *Thiomersal* under HEAVY METALS)

Thiobacillus A genus of gram-negative, non-photosynthetic bacteria which obtain energy by the oxidation of sulphur and/or reduced sulphur compounds. Most species are polarly-flagellated, about 0.5 × 1–2 microns; *T. novellus* and some strains of *T. neapolitanus* are non-motile. *T. thioparus* (the type species), *T. denitrificans*, *T. ferrooxidans*, *T. thiooxidans* and *T. neapolitanus* grow as strict lithotrophs—using carbon dioxide as a carbon source and obtaining energy as described above. (Some authors report that many species of *Thiobacillus* can assimilate organic compounds in the presence of an appropriate inorganic electron donor.) All species, except one, are obligate aerobes; *T. denitrificans* can grow anaerobically when provided with nitrate as electron acceptor. Thiobacilli occur e.g. in sulphur springs, soil, marine mud, mine drainage.

T. thioparus (FAME Type II) grows optimally at neutral pH, 28 °C; electron donors include sulphide and thiosulphate. A sulphur-containing pellicle may be formed in liquid thiosulphate media. GC% 62–68. (During growth in liquid media the pH falls to 4.5 due to the formation of sulphuric acid; growth ceases at this pH. *T. ferrooxidans* and *T. thiooxidans* depress the pH to well below this value.)

T. denitrificans (FAME Type II) grows optimally at neutral pH; electron donors include sulphide, thiosulphate and sulphur. Under anaerobic conditions nitrate is reduced to nitrogen (DENITRIFICATION).

T. ferrooxidans (FAME Type I) grows well at room temperatures in the pH range 3–5; electron donors include thiosulphate, sulphides, sulphur, and *ferrous* compounds.

T. thiooxidans (FAME Type III) grows optimally at pH 1–3.5 over a range of temperatures; electron donors include thiosulphate and (particularly) sulphur. This organism can cause decay in vulcanized rubber.

Thiocapsa (*Thiococcus*) A genus of the CHROMATIACEAE.

Thiococcus (*Thiocapsa*) A genus of the CHROMATIACEAE.

thioctic acid (see LIPOIC ACID)

thiomersal (thimerosal; merthiolate) $C_2H_5.Hg.S.C_6H_4.COONa$. An antibacterial agent often used as a *preservative*, e.g. in biological laboratory reagents. It is effective in low concentrations e.g. 0.01%. (see also *Mercury* under HEAVY METALS)

Thioploca A genus of bacteria of the BEGGIATOACEAE.

thioredoxin A protein containing two thiol groups (cysteine residues) which (at least in *Escherichia coli*) functions as the source of reducing power in the reduction of riboNUCLEOTIDES to deoxyribonucleotides. The cystine residue formed during the reaction is reduced by $NADPH_2$ and a flavoprotein enzyme.

Thiorhodaceae (purple sulphur bacteria) (see CHROMATIACEAE)

thiosemicarbazones For formula see ISATIN-β-THIOSEMICARBAZONES. (see also UREA AND DERIVATIVES)

Thiospira A genus of gram-negative, spiral-shaped, polarly-flagellate bacteria of uncertain taxonomic affinity. The cells, which contain sulphur granules, have not yet been obtained in pure culture.

Thiospirillum A genus of gram-negative, non-sporing, non-motile or motile (lophotrichous) photosynthetic bacteria of the CHROMATIACEAE; species are found e.g. in stagnant waters and sulphide-containing muds. The cells are spiral-shaped, and may be large (e.g. 50 microns, long axis) or small (e.g. 5–10 microns, long axis); they divide by binary fission. Cells contain granules of elemental sulphur; storage products include poly-β-hydroxybutyrate. Some species require vitamin B_{12}.

Thiothrix A genus of gram-negative, filamentous, freshwater and marine bacteria of the LEUCOTRICHACEAE. The organisms form intracellular sulphur granules when grown in the presence of hydrogen sulphide (a feature not exhibited by *Leucothrix*) and appear to be obligate chemolithotrophs.

thiram (TMTD) TetramethylTHIURAM DISULPHIDE; $(CH_3)_2N.CS.S.S.CS.N(CH_3)_2$. Thiram is active against a range of fungi and (mainly) gram-positive bacteria. It is used as an antiseptic, in the topical treatment of dermatophyte infections, and as an agricultural antifungal agent. In the latter context it has a wide range of uses, e.g. as a seed dressing and in the control of apple scab, botrytis fruit rots etc. Being insoluble in water, thiram is used in aqueous suspension. Thiram is formed by the oxidation of DIMETHYLDITHIOCARBAMIC ACID. (see also FUNGICIDES AND FUNGISTATS)

thiuram disulphides $(R_2N.CS.S.S.CS.NR_2)$ These compounds are formed by mild oxidation of DITHIOCARBAMATES; they include some important agricultural antifungal agents: see THIRAM and METIRAM.

Thoma chamber (see COUNTING CHAMBER)

threonine (biosynthesis of) (see Appendix IV(d))

-thrix A suffix signifying a *hair* or *thread.*

thrush In man, infection of the oral mucous membranes by *Candida albicans*; the membranes exhibit whitish patches which sometimes ulcerate, and there may be fever and/or gastrointestinal symptoms. 'Vaginal thrush'—one form of vaginitis—is due to the same pathogen. (see also CANDIDIASIS)

thujaplicins Isopropyltropolones—fungitoxic substances produced by the Western Red Cedar (*Thuja plicata*). Thujaplicins appear to be responsible for the high resistance to fungal attack of timber from *T. plicata.* (see also TIMBER, DISEASES OF)

thylakoids Flattened, membranous vesicles which occur in blue-green algae and in the CHLOROPLASTS of algae and higher plants; the thylakoid membrane contains CHLOROPHYLLS, ACCESSORY PIGMENTS, and electron carriers, and is the site of the *light reaction* in PHOTOSYNTHESIS. (cf. CHROMATOPHORE; see also PHYCOBILIPROTEINS and PHYCOBILISOMES)

Studies on a relatively small number of algae indicate that the arrangement of thylakoids within the chloroplast may have taxonomic significance. Thus, e.g. chloroplasts in the Rhodophyta contain individual (unassociated) thylakoids, while groups of three longitudinally-adjacent (but separate) thylakoids occur in the chloroplasts of the Chrysophyta and Dinophyta; in the Chlorophyta the thylakoids form stacks, or *grana*—similar to those typical of higher plants—each granum consisting of two or more thylakoids.

In blue-green algae (which lack chloroplasts) the thylakoids tend to be more numerous near the periphery of the cell—where they often run parallel to the cell wall; whether or not the thylakoids and cell membrane are continuous appears to be unknown.

thylaxoviruses (oncornaviruses) Oncogenic RNA viruses—see LEUKOVIRUS.

thymidine-5′-monophosphate (biosynthesis of) (see Appendix V(b))

thymol (5-methyl-2-isopropylphenol) An antibacterial *preservative*; it is used e.g. in urine samples pending biochemical tests. (see also PHENOLS)

thymol blue (PH INDICATOR) Acid range: pH 1.2 (red) to pH 2.8 (yellow). Alkaline range: pH 8.0 (yellow) to pH 9.6 (blue). pK_1 1.5; pK_2 8.9.

thymolphthalein (PH INDICATOR) pH 9.3 (colourless) to pH 10.5 (blue). pK_a 9.9.

Tilletia A genus of fungi of the family Tilletiaceae, order USTILAGINALES. *Tilletia* spp are plant parasites and some species are important pathogens—e.g. *T. caries*, the causal agent of bunt ('stinking smut') of wheat (see CEREALS, DISEASES OF); typically, *Tilletia* spp attack the ovaries of the host plant—though leaves may also be parasitized. *Typical life cycle.* On germination, the *teliospore* gives rise to a promycelium which commonly bears a tuft of eight (sometimes more) apical, thread-like, haploid basidiospores ('primary sporidia'); the basidiospores, while still attached to the promycelium, conjugate in pairs (of appropriate mating types) to form a number of H-shaped structures in each of which plasmogamy occurs. Subsequently, each (detached) H-shaped dikaryotic structure—or a mycelium derived from it—gives rise to dikaryotic conidia ('secondary sporidia') which are borne on sterigmata. A fresh host becomes infected by the dikaryotic hyphae which are formed on germination of the conidia. *Spore balls* are not formed.

Tilletiaceae (see USTILAGINALES)

timber, diseases of The main microbial diseases of timber are caused by *fungi*—particularly basidiomycetes; many of these diseases bring about various forms of decay and are known as *rots.* In *brown rots* the CELLULOSE and HEMICELLULOSES of the wood are decomposed while the LIGNIN component remains virtually intact. Brown rots often cause cross-grained cracking and characteristically reduce wood to a soft friable consistency. In *white rots* all components of the wood are decomposed—though not necessarily simultaneously; the wood becomes soft and fibrous. Decay which is restricted to small discrete regions is often referred to as *pocket rot*. Wood is susceptible to fungal attack only in the presence of adequate moisture (usually a minimum of 20% water*) and adequate air; thus, wood which is very dry is resistant to attack (see however DRY ROT) while wood submerged in water is resistant to most basidiomycetes owing to an insufficiency of oxygen. For the latter reason timber is sometimes stored under water (see however *soft rot* below). Certain trees (and their timbers) are resistant to attack by many fungi owing to their content of fungitoxic substances, e.g. TANNINS in oak, PINOSYLVIN in pine, and THUJAPLICINS in Western Red Cedar.

Diseases of standing trees. Most common diseases are due to fungal attack on the (dead) heartwood; the (living) sapwood is infrequently attacked (see however DUTCH ELM DISEASE). The fungi which cause such *heart-rots* normally gain access to the heartwood *via* wounds e.g. damaged roots, broken branches etc. Rotting of the heartwood does not affect the vital processes of the tree, but eventually the tree becomes structurally weakened and easily toppled in high winds. Important tree-rotting pathogens in Britain include *Heterobasidion annosum* (formerly *Fomes annosus*) and *Armillaria mellea*. *H. annosum* often causes a white pocket rot of heartwood—particularly in conifers. *A. mellea*, which causes a white heart-rot, may infect any of a wide variety of trees. The fungus may spread to uninfected parts of the tree, or to adjacent trees, by means of characteristic black rhizomorphs ('shoestrings') which may be found between bark and sapwood or in the surrounding soil. (*A. mellea* may also grow saprophytically e.g. on logs, tree stumps etc.) A brown heart-rot of the oak (*Quercus*) and other trees is caused by the basidiomycete *Polyporus sulphureus* (the 'sulphur fungus'). Another polypore, *Piptoporus betulinus*, is an important pathogen of the birch (*Betula*)—in which it causes a brown cubical heart-rot. Other important tree-rotting fungi include species of *Armillaria*, *Fomes*, *Ganoderma*, *Polyporus*, *Stereum* and *Trametes*.

Diseases of felled timber. In felled timber the *sapwood* is particularly vulnerable to early fungal attack. Logs etc., left open to the weather, may be rotted by any of a wide range of fungi. Thus, e.g. the fallen branches of dead or dying birch trees may rapidly succumb to a brown cubical rot (e.g. *Piptoporus betulinus*) or to a soft, white fibrous rot (e.g. *Fomes igniarius*). Felled timber may also support the *superficial* growth of a number of species of ascomycetes and deuteromycetes; such growth causes little or no damage to the fabric of the timber since such species do not degrade cellulose or lignin. Basidiomycetes of importance in the decay of domestic and industrial timbers include *Serpula lacrymans* (see DRY ROT), *Coniophora cerebella* (see WET ROT) and *Lentinus lepideus* (the 'stag's horn fungus'). *L. lepideus* causes a brown rot in pit props, railway sleepers and other timbers found in damp situations; the fungus is moderately resistant to creosote. In the absence of light (e.g. in mines) normal fructifications are not formed; instead, branching, antler-like growths develop—hence 'stag's horn fungus'. Wood infected with *L. lepideus* has a characteristic balsam-like smell owing to the production, by the fungus, of aromatic compounds (e.g. methyl cinnamate, methyl anisate).

Wood continually in contact with water may suffer a *superficial* rot known as a *soft rot*; causative organisms include members of the Pyrenomycetes—particularly the cellulolytic species *Chaetomium globosum*. Soft rots are important e.g. in the wooden slats of cooling towers.

Preservation of timber. Initially, the timber is *seasoned* (i.e. dried)—a process which (a) renders the wood less susceptible to fungal attack; (b) permits increased penetration by chemical preservatives e.g. CREOSOTE, fluorides, pentachlorophenol, or certain copper compounds. (see also BOUCHERIE PROCESS)

Although, for man, the decay of timber may cause considerable economic loss, in nature, the wood-rotting fungi are essential agents in the re-cycling of carbon and other elements.

*Calculated as follows: weight of water in timber/*dry weight* of that sample of timber; for some wood-rotting fungi the moisture content of wood must be greatly in excess of 20%, e.g. 30–50%.

timber, staining of (by microbial action) A variety of wood-infecting fungi may cause staining—both in standing trees and in felled timber; such fungi may be carried into the timber by wood-boring beetles. The growth of these fungi in or on wood may or may not significantly affect the mechanical strength and/or value of the timber. Staining may be due to the production of a fungal pigment (e.g. see GREEN OAK) or to the visual effects produced by the presence of fungal hyphae among the cells of the wood (see BLUE-STAIN). *Superficial* staining due e.g. to species of *Fusarium* or *Penicillium* is also found. (see also BROWN OAK)

timothy-grass bacillus *Mycobacterium phlei*.

tin (as an antimicrobial agent) (see *Tin* under HEAVY METALS; see also FENTIN)

tincture of iodine (see IODINE)

tinea (see RINGWORM)

tinsel flagellum (see FLAGELLUM (eucaryotic))

Tintinnida An order of protozoans of the subclass SPIROTRICHIA.

TIP (1) See THERMAL INACTIVATION POINT. (2) *Tumour inducing principle*, a substance believed to be responsible for hyperplasia in CROWN GALL formation. (3) Translation inhibitory protein—see INTERFERONS.

***Tipula* iridescent virus** (see IRIDESCENT VIRUSES)

tissue culture (of animal tissues or cells) (1) The maintenance or culture (growth) of isolated tissues. As *commonly* used, tissue culture refers also to the maintenance or culture of populations of individual (single) cells obtained from appropriately disrupted tissues; this procedure (discussed below) is also called *cell culture.* (2) Any population of cells being maintained or cultured.

Tissue (cell) cultures are widely used for the culture of VIRUSES (see VIRUSES, CULTURE OF) and in cancer research, immunology, toxicology, etc.; they are also used to culture certain bacteria, e.g. *Chlamydia, Rickettsia.* Tissue cultures are always prepared and handled with an ASEPTIC TECHNIQUE.

In *monolayer* culture a population of single (disaggregated) cells is grown on a suitable surface where it forms a confluent layer, one cell thick; macroscopically, the monolayer appears as a translucent film. The initial cell suspension may be derived from the tissues of any of a variety of animals—including man and other primates; kidney, neoplastic and embryonic tissues are frequently used.

Culture vessels include medical flats, Roux bottles, test-tubes, LEIGHTON TUBES, etc.; most vessels are of (neutral-pH) glass, though certain plastics (e.g. polystyrene) may be used.

Media. A *growth medium* (e.g. EAGLE'S MEDIUM) permits cellular growth and division; it consists basically of a balanced salt solution (see e.g. HANKS' BSS) supplemented with glucose and an appropriate range of amino acids and vitamins etc. together with serum and/or lactalbumin hydrolysate. (Antibiotics are often added to prevent the development of bacterial contaminants, if present; however, the continual suppression of contaminants by bacteriostatic agents is undesirable, and subcultures on antibiotic-free media should be prepared periodically in order to detect contamination.) Growth media also incorporate a pH indicator (e.g. PHENOL RED) and a buffer system (e.g. bicarbonate, HEPES). A *maintenance medium* is one used to preserve the viability and properties of a population of cells while restricting to a minimum their growth and division; serum-depleted growth media are usually satisfactory maintenance media.

Preparation of a monolayer culture. The selected tissue is cut into small pieces and washed several times with pre-warmed (30–33 °C) phosphate-buffered saline (PBS). The tissue is then digested in PBS containing trypsin (0.05–0.1% w/v)—the whole being maintained at 30–33 °C and stirred mechanically for 20–30 minutes. (Trypsin weakens the intercellular bonds; excessive trypsinization damages cells.) The turbid supernatant (containing damaged cells) is discarded, and the tissue fragments are again digested in (fresh) PBS-trypsin. The supernatant (containing single cells and small aggregates) is centrifuged (800–1000 r.p.m./5 minutes) and the cell deposit resuspended in growth medium. The cell concentration* is determined (see COUNTING CHAMBER) and the suspension diluted (in growth medium) to about 100,000 cells/ml. To the culture vessel is added cell suspension sufficient to form a shallow layer; e.g. a 4-ounce medical flat would require about 15 ml of suspension. During incubation (37 °C for mammalian cells) the cells adhere to the inner surface of the culture vessel and, by growth and division, form a confluent monolayer within several days or a week or more. (A medical flat is incubated on its side. Test-tubes, or bottles of circular cross-section, may be incubated in a near-horizontal position on a roller apparatus; a vessel is rotated about its long axis (at less than 1 r.p.m.) so that the monolayer develops as a continuous band around the vessel's inner surface.) When cultured, animal cells generally become either spindle-shaped (fibroblast-like) or polygonal (epithelioid).

Subculture. The cells are first *stripped* (i.e. detached from the surface on which they have been grown): the monolayer is subjected (for 1 or 2 minutes) to a stripping agent—e.g. PBS containing *versene* (0.02–0.05%) and trypsin (0.1%); after decanting the stripping agent the (detached) cells are resuspended in fresh growth medium. This cell suspension (at a suitable cell concentration) is used to prepare fresh cultures. Cells can be preserved for long periods e.g. by FREEZING.

Transformation. After a certain number of subculturings, cells either die out or develop into an ESTABLISHED CELL LINE. Cells of an established cell line are *heteroploid.* (see TRANSFORMATION (of cells)) The characteristics of cultured cells may be monitored e.g. by immunological means or by examining the KARYOTYPE of the cells; the latter process involves arresting mitosis at metaphase (e.g. with COLCHICINE) and examining the chromosomes.

*The proportion of *living* cells is determined by VITAL STAINING.

tissue cysts (of *Toxoplasma*) (see BRADYZOITES)

titre (American: *titer*) (1) (*virol.*) See END POINT (*virol.*). (2) (*immunol.*) Used in a general, non-quantitative way to refer to the 'concentration' of antibodies or antigens. (3) (*immunol.*) A measure of the concentration of antibodies (or antigens) in a given sample. It is commonly given by the highest dilution of the sample which gives a positive serological reaction (e.g. agglutination, precipitation) with antigen (or antibody) under the conditions of the titration. As an example, the following (*initial*) dilutions of antiserum (containing antibodies) may be prepared in 1 ml volumes:

1/2 1/4 1/8 1/16 1/32 1/64

If now 1 ml volumes of the *antigen* preparation are added to each of the sample dilutions, the *final* dilutions of the sample become:

1/4 1/8 1/16 1/32 1/64 1/128

If, following incubation, the penultimate dilution was the last tube to exhibit a positive reaction, the titre could be given as 1/32 (*initial* antiserum dilution) or 1/64 (*final* antiserum dilution)—or as 32 or 64 respectively. (see also END POINT TITRATION)

Tm (see DNA BASE COMPOSITION, DETERMINATION OF)

TMTD (see THIRAM)

TMV (see TOBACCO MOSAIC VIRUS)

toadstool Any of the poisonous or inedible umbrella-shaped fruiting bodies of the basidiomycetes; even more loosely, any such fruiting body whether poisonous or not. The name may have been derived from the German *todestuhl*—'death's chair'.

tobacco, diseases of These include: (a) *Granville wilt* (*Southern bacterial wilt*). A soil-borne disease (causal agent *Pseudomonas solanacearum*) involving wilting and the yellowing of leaves. (see also SCOPOLETIN) (b) TOBACCO MOSAIC DISEASE. (c) See *tobacco rattle virus* under COVIRUS. (d) *Black root rot*; causal agent: *Thielaviopsis basicola*. (e) *Blue mould*; causal agent: *Peronospora tabacina*. (f) Tobacco club root—caused by the WOUND TUMOUR VIRUS.

tobacco mosaic disease A disease of tobacco (*Nicotiana tabacum*) caused by the (mechanically-transmissible) TOBACCO MOSAIC VIRUS (TMV). The common or field strain of the virus usually produces a systemic disease characterized by the formation of distorted, blistered leaves bearing patches of light and dark green; chlorosis may or may not be evident. Intracellular bodies are generally formed. Other strains of the TMV may cause a non-systemic infection—LOCAL LESIONS being formed. (see also VIRUS DISEASES OF PLANTS)

tobacco mosaic virus (TMV) A mechanically-transmissible, single-stranded RNA virus which is pathogenic for a wide range of plant species (see e.g. TOBACCO MOSAIC DISEASE). The tubular virion (M.Wt. about 40×10^6) consists of RNA (5% by weight) and protein (95% by weight); it is approximately 15 × 300 nm and has a sedimentation coefficient (S^{20}_{w}) of about 200. The RNA—which comprises some 6,300 nucleotides—occurs as a helix of radius 4 nm and pitch approximately 2.5 nm. Surrounding, and intimately associated with, the RNA is an helical array of some 2,100 identical protein coat subunits. Each subunit (a protein of M.Wt. 17,500) consists of a single polypeptide chain—in which there is a high proportion of aspartic acid, glutamic acid, serine, and threonine residues. (see also A-PROTEIN) Reconstitution of TMV from its isolated components (RNA and protein subunits) has been carried out *in vitro*; assembly of the protein subunits commences at the 5′-terminal of the RNA. Preparations of purified TMV can be obtained from infected plant tissues e.g. by precipitation with ammonium sulphate and subsequent differential centrifugation. TMV retains infectivity for over a year in dried leaf tissue. The THERMAL INACTIVATION POINT of the virus is approximately 95 °C. TMV is inactivated by ultrasonication, by pH below 3 or above 8, and by chemical reagents such as formaldehyde, iodine (under certain conditions), or strong solutions of urea. *Local lesion hosts* include *Nicotiana glutinosa* and *Phaseolus vulgaris*. (see also VIRUS DISEASES OF PLANTS)

tobacco necrosis virus (Tulip Augusta disease virus) An RNA virus which may infect a wide range of plant hosts. The virion is polyhedral, approximately 26–30 nm in diameter; a number of serologically-related strains occur. Tobacco necrosis virus is able to cause severe or terminal necroses in relatively few plants; these include young seedlings of tobacco (*Nicotiana tabacum*), the French bean (*Phaseolus vulgaris*), and the tulip (*Tulipa* spp)—Augusta disease of tulips. In tobacco seedlings the symptoms include discrete necrotic leaf lesions. The virus may be transmitted mechanically or transmission may occur *via* the fungal vector *Olpidium brassicae*. (see also SATELLITE VIRUS)

tobacco rattle virus (see COVIRUS)

Todd–Hewitt broth A medium used for the culture of *Streptococcus* spp. It contains an infusion of fat-free beef heart, together with peptone, D-glucose, sodium chloride and a buffer system (sodium carbonate, or bicarbonate, and disodium hydrogen phosphate). The pH of the medium is 7.8.

Togaviridae A family of VIRUSES whose hosts *typically* include arthropods and vertebrates (see also ARBOVIRUSES). The togavirus virion consists of an enveloped icosahedral capsid (overall diameter 40–70 nm) which contains the linear, single-stranded RNA genome; the virion is inactivated by ether and by other lipid solvents. Togaviruses promote the haemagglutination of certain types of erythrocyte (e.g. erythrocytes of the goose, pigeon, newly-hatched chick). Togaviruses replicate within the *cytoplasm* of the host cell; envelopment occurs at the cytoplasmic (cell) membrane in viruses of one genus (ALPHAVIRUS) and within the cytoplasm in viruses of another genus (FLAVOVIRUS). The family includes the causal agents of DENGUE, SWINE FEVER, and YELLOW FEVER. (see also RUBELLA VIRUS)

Tokophrya A genus of protozoans of the subclass SUCTORIA.

tolnaftate 2-Naphthyl-*N*-methyl-*N*-(*m*-tolyl)-thiocarbamate; a compound which is highly active against DERMATOPHYTES but which has little or no activity against most other fungi or bacteria. Tolnaftate is used topically in the treatment of dermatophyte infections.

Tolypella A genus of algae (division CHAROPHYTA).

Tolypothrix A genus of filamentous BLUE-GREEN ALGAE; *Tolypothrix* spp are capable of NITROGEN FIXATION. (see also HETEROCYST)

tomato, diseases of These include: (a) *Early blight* (causal agent: *Alternaria solani*) in which dark lesions form on leaves, stems, and fruits. (Tomatoes may also be attacked by *Phytophthora infestans* (late blight)—cf. POTATO, FUNGAL DISEASES OF). (b) *Leaf mould* (*leaf blotch*); causal agent: *Cladosporium fulvum*. Lesions spread rapidly on the leaves which eventually wither. The disease is favoured by high humidity and is not uncommon in greenhouse plants. (c) Diseases due to certain viruses—e.g. the TOBACCO MOSAIC VIRUS and the TOMATO RING SPOT VIRUS.

tomato ring spot virus A single-stranded RNA virus which can affect a wide range of plant hosts; the virion is spherical, approximately 30 nm in diameter. On the tomato (*Lycopersicum esculentum*) the symptoms include curling and necrosis of the tips of growing shoots, and well delimited necrotic rings and streaks on leaves and stems; symptoms may appear on fruit. X BODIES are formed in certain hosts. Transmission of the virus may be mechanical or (in some cases) seed-borne. (see also VIRUS DISEASES OF PLANTS)

tomentose Downy, woolly.

tomentum (*lichenol.*) A downy or felted mat of hyphae which occurs e.g. on the lower surface of the thallus in certain foliose lichens (e.g. species of *Lobaria* and *Peltigera*).

tomite (1) See APOSTOMATIDA. (2) Any one of the cells formed when *Stephanopogon* undergoes multiple fission.

tomont (see APOSTOMATIDA)

tonoplast The membrane which limits an intracellular vacuole.

tonsillitis Inflammation of the tonsils. One of the commoner causal agents is *Streptococcus pyogenes*.

top necrosis (*plant pathol.*) Death of the terminal bud or of the entire top of a plant.

top yeast (see BREWING)

topical (*med.*) (of treatment etc.) Local, i.e. involving a particular region of the body; not systemic.

topotaxis (topotactic movement) (see TAXIS)

Torulopsis A genus of fungi of the family CRYPTOCOCCACEAE.

torulosis Older name for CRYPTOCOCCOSIS.

torus A thin bacterial CAPSULE impregnated with iron oxide and/or manganese oxide.

total cell count (see ENUMERATION OF MICROORGANISMS)

toxaemia The condition in which toxin(s) are present in the bloodstream. (see also SEPTICAEMIA)

toxicysts (see TRICHOCYSTS)

toxigenic (*adjective*) (of organisms) A synonym of TOXINOGENIC.

toxin (*microbiol.*) Any (organic) microbial substance or product which is harmful or lethal for cells, tissue cultures or organisms. (see also ENDOTOXINS and EXOTOXINS)

toxinogenic (toxigenic; toxicogenic) (of an organism) Able to produce a toxin, or toxins.

toxoid An EXOTOXIN which has been modified (for example, by treatment with formalin) so that its toxicity has been lost while its specific antigenicity is retained. Toxoids are useful for IMMUNIZATION since antibody formed against a toxoid is active against the corresponding exotoxin.

Toxoplasma A genus of protozoans of the suborder EIMERIINA. The sole species, *T. gondii*, is an intracellular parasite in a range of hosts, and is the causal agent of TOXOPLASMOSIS. The uninucleate, crescent-shaped vegetative cells are 5–7 microns in length. (see also TACHYZOITES and BRADYZOITES) *T. gondii* can be grown in tissue cultures.

Asexual development occurs in extra-intestinal tissues (brain, muscle etc.) and is the only form of development observed in non-feline hosts. Typical *coccidian*-type schizogonous and gametogenous development (see COCCIDIA) has been reported to occur in the intestine of the *cat*; oocysts (ovoid, 10–15 microns, of the *Isospora* type) are voided in the faeces. (Certain features of the asexual phase of multiplication—ENDODYOGENY, ENDOPOLYGENY—are not typically coccidian.)

toxoplasma dye test (Sabin–Feldman dye test) A serological test used to detect and quantify serum antibodies to the protozoan pathogen, *Toxoplasma gondii*. (see also TOXOPLASMOSIS) The test is based on the partial or total lysis of laboratory-cultured cells of *T. gondii* in the presence of specific antibodies and a so-called *accessory factor* present in fresh, normal human serum or citrated plasma. Partially or totally lysed cells of *T. gondii* fail to take up alkaline methylene blue—in contrast to undamaged cells which are readily stained.

In the test, the patient's serum (pre-heated to 56 °C for 30 minutes) is incubated with accessory factor and cells of *T. gondii*; uptake, or otherwise, of methylene blue is subsequently determined microscopically. Serum antibodies to *T. gondii* are said to appear within 1–3 weeks of initial infection. Since specific antibodies are commonly found in symptomless persons (sometimes at quite high titres) more importance is attached to a *rising* titre.

toxoplasmosis An acute or chronic disease—of man and other animals—caused by the intracellular pathogen, *Toxoplasma gondii.* The disease may be mild or fatal, but most infections are probably sub-clinical. Transmission is believed to occur by the ingestion of 'tissue cysts' in insufficiently-cooked meats, or by the ingestion of sporulated oocysts. Toxoplasmosis may consist merely of a mild lymphadenitis but, in serious cases, the brain, liver, lungs etc. may be affected. Transplacental transmission occurs; prenatal infection may lead to foetal damage or stillbirth. In the *cat* TOXOPLASMA causes a typical coccidian disease. *Laboratory diagnosis.* (a) Microscopical examination of material from lesions or the deposit from centrifuged cerebrospinal fluid. (b) Lesion material may be inoculated into mice (which are highly susceptible), and smears or sections prepared *post mortem.* (c) See: TOXOPLASMA DYE TEST.

Some infections have been controlled with antimalarial drugs, e.g. PYRIMETHAMINE, in combination with sulphonamides.

TPI (*Treponema pallidum* immobilization test) A serological test used in the diagnosis of SYPHILIS. The test detects antibodies specific to *T. pallidum* (syphilis), *T. pertenue* (YAWS) and to other *pathogenic* treponemes. (The TPI test cannot distinguish between the treponematoses.) Essentially the test depends on the immobilization of living (motile) cells of *T. pallidum* in the presence of specific antibodies and COMPLEMENT. To the serum under test is added (a) a standardized suspension of *T. pallidum*, and (b) a volume of fresh guinea-pig serum as a source of complement. The whole is incubated anaerobically for 18 hours at 37 °C and subsequently examined microscopically. In a positive reaction a specified proportion of treponemes is immobilized. (Immobilization involves death and lysis.) Adequate control tests must be carried out. The test is currently regarded as decisive—being used to resolve the problem of conflicting results from other tests. (see also FTA-ABS)

TPN Triphosphopyridine nucleotide—see NADP.

TPP THIAMINE pyrophosphate.

Trachelomonas A genus of ALGAE of the division EUGLENOPHYTA.

Trachelophyllum A protozoan genus of the RHABDOPHORINA. *T. pusillum* is commonly found in British activated-sludge sewage plants. (see also SEWAGE TREATMENT)

trachoma A disease of the eye which, in nature, is peculiar to man. Infection by the causal agent, *Chlamydia trachomatis*, is *contaminative*. Initially there is inflammation of the inner surfaces of the eyelids (*palpebral* conjunctivae) and of the anterior surface of the eyeball (*bulbar* conjunctiva); *follicles*, which contain accumulations of macrophages, polymorphs and lymphocytes, and which give rise to a discharge, develop within the conjunctival tissues—which subsequently become scarred. A network of blood capillaries (a *pannus*) develops within, and spreads over, the cornea which, as a consequence, becomes more or less opaque—causing partial or total blindness. Subclinically-infected patients constitute a reservoir of infection. Therapeutic agents which have been used include tetracyclines, chloramphenicol and sulphonamides. *Laboratory diagnosis* may include serological examination of discharges etc. for chlamydial antigens and/or culture of the causal agent (in chick embryos, yolk sac etc.). (see also TRIC AGENTS)

trama (*pl.* tramae) (*mycol.*) A sterile tissue which supports an HYMENIUM or which separates adjacent hymenia; for example, in agarics the tramae include the supportive (structural) tissue of a gill (lamella) and the fleshy tissue of the pileus.

Trametes versicolor (see CORIOLUS)

transcapsidation (see PHENOTYPIC MIXING)

transcription *Transcription* most commonly refers to the synthesis of mRNA (see RNA) on a DNA template; the term is also applied to the synthesis of rRNA and tRNA on a DNA template, and to the synthesis of mRNA on the RNA template of a minus-strand RNA virus. (see PLUS AND MINUS STRANDS OF DNA AND RNA; cf. REVERSE TRANSCRIPTASE)

transductant A bacterial cell which has received genetic material from another by TRANSDUCTION.

transduction The transfer of bacterial genes from one bacterium (the *donor*) to another (the *recipient*) by BACTERIOPHAGE. Transduction has been observed in a range of bacteria—including *Escherichia coli* and species of *Bacillus*, *Proteus*, *Pseudomonas*, *Salmonella*, *Shigella*, and *Staphylococcus*.

(a) *Generalized transduction.* In this type of

transduction, any of the donor genes (or a **PLASMID** contained within the donor) may be transduced. Generalized transduction may be demonstrated e.g. with *Salmonella typhimurium* and phage P22. Two genetically different strains of *S. typhimurium* are selected as donor and recipient strains—e.g. the donor may be prototrophic, the recipient auxotrophic. Essentially, a lysate obtained by lytic infection of the donor population is used to infect a recipient population; the mixture is plated on a medium selective for a specific donor characteristic (**MINIMAL MEDIUM** for prototrophy) and examined for colonies following incubation. During phage development in the donor cells, the donor chromosomes are fragmented; during maturation, a small proportion of phage capsids incorporate fragments of donor chromosome *in place of* phage DNA. (The size of the fragment incorporated is limited by the size of the phage head into which it must fit; in the *Salmonella*-P22 system up to 1% of the donor genome may be incorporated.) When the donor cells lyse, and the lysate is mixed with the recipient population, phages which contain bacterial genes will introduce those genes into the recipient cells; provided that the **MULTIPLICITY OF INFECTION** is less than 1 this will not result in lysis since phage DNA is unlikely to enter a cell infected with a donor fragment. The donor fragment may then become aligned with an homologous region of the recipient chromosome and may replace recipient genes by recombination; such a recipient will thus stably inherit those genes (*complete transduction*). However, fragment-chromosome recombination fails to occur in the majority of recipients which receive donor fragments (*abortive transduction*); normally, the fragment cannot replicate but, since it can be transcribed, its genes may be expressed. On division, a cell containing such a fragment can pass the fragment only to one of the daughter cells; however, the other daughter cell may receive sufficient donor gene *products* (e.g. enzymes) to permit expression of the donor phenotype for one or a few cell generations. Abortive transduction is manifested by the formation of *minute* (often microscopic) *colonies* on the selective medium. (Only one cell in each minute colony actually contains a donor fragment.) Since all donor genes appear to have an equal chance of being transduced, and since only a small proportion of a transducing phage population contains donor genes, the transduction of a particular donor gene is a rare event—i.e. generalized transduction occurs only at low frequency. If two or more donor genes are transduced simultaneously (i.e. *cotransduced*) they are assumed to occur on the same fragment of donor DNA— and are thus closely linked; transduction has been used for detailed mapping of short sequences of donor chromosomes.

(b) *Restricted* (*specialized*) *transduction*. This type of transduction occurs only following the induction of a *lysogenic* bacteriophage; only those genes immediately adjacent to one or the other end of the prophage can be transduced. For example, in *Escherichia coli* the lambda (λ) prophage (see **BACTERIOPHAGE LAMBDA**) maps between the *gal* and *bio* regions (specifying galactose utilization and biotin synthesis, respectively). When a λ-lysogenized population of *E. coli* is induced, a small proportion of the resulting virions have genomes which contain *bacterial* genes; this occurs as a result of 'illegitimate' crossing-over between the prophage and bacterial chromosome during excision. (cf. F′ factor formation in **CONJUGATION**) The defective phage thus lacks some genes from one end of its genome and carries, at the opposite end, either the *gal* or the *bio* bacterial genes (depending on the position of the cross-over). In a lysate of λ-lysogenized *gal*$^+$ *E. coli* a small proportion of phage virions contain the *gal*$^+$ bacterial genes and lack certain phage genes which are essential for phage replication; such virions are designated λd*gal* or λ*dg* (*d* signifying 'defective') and occur with a frequency of the order of 1 per 10^6 wild type virions. If such a lysate is used to infect a *gal*$^-$ *E. coli* population, *gal*$^+$ genes will be introduced into a small number of recipient (*gal*$^-$) cells by the λd*gal* virions; such a lysate is therefore called an LFT (low frequency transduction) lysate. Recipients which receive *gal*$^+$ genes (i.e. which become *heterogenotes*—designated *gal*$^-$/λd*gal*$^+$) subsequently exhibit the *gal*$^+$ phenotype; a proportion of these recipients may stably inherit the *gal*$^+$ genes by recombination. If a population of *gal*$^-$ *E. coli* recipients is infected with an LFT lysate at a high multiplicity of infection, any cell which becomes infected by a λd*gal* virion is likely to become simultaneously infected by a wild type (non-defective) virion. Both the defective and normal (wild type) phage genomes can integrate with the recipient chromosome to form a *double lysogen*; when such a cell is induced, the genes of the wild type phage compensate for those missing in the defective phage such that both defective and wild type phages can replicate. Thus, the lysate formed contains approximately equal numbers of λ and λd*gal* virions. If this lysate is subsequently used to infect another population of *gal*$^-$ *E. coli* cells the *gal*$^+$ genes will be transduced with high frequency; such a lysate is termed an HFT (high frequency transduction) lysate.

transfection The infection of *competent* bacterial

cells by *isolated* BACTERIOPHAGE nucleic acid with subsequent production of normal phage progeny. This has been achieved, for example, with certain phages of *Bacillus subtilis*. (see also TRANSFORMATION)

transferase A category of ENZYMES (EC class 2) which encompasses those enzymes which catalyse the transfer of a group containing carbon, nitrogen, phosphorus, and/or sulphur from one molecule to another. Subclasses within this category are recognized and numbered according to the nature of the group transferred; thus e.g. subclass 1—one-carbon groups (e.g. methyl, formyl, hydroxymethyl); 3—acyl groups (e.g. acetyl); 4—glycosyl groups; 6—nitrogenous groups (e.g. amino); 7—phosphate groups; 8—sulphur-containing groups (e.g. coenzyme A). The subclasses are divided into sub-subclasses according to e.g. the specific nature of the groups transferred, the acceptor group, etc.

transformation (bacterial) A mode of genetic transfer in which a DNA fragment derived from one bacterial cell (the *donor*) is taken up by another (the *recipient*) and subsequently undergoes recombination with the recipient's chromosome. Transformation has been observed e.g. in species of *Bacillus*, *Haemophilus*, *Neisseria*, *Rhizobium*, *Staphylococcus*, *Streptococcus*, and *Xanthomonas*.

The uptake of transforming DNA can occur only during a transitory period in the growth cycle of a recipient; cells which can take up DNA are said to be *competent*. The establishment of competence in a recipient cell appears to involve the synthesis of certain extracellular protein(s)—*competence factor(s)* (activator(s)). A competence factor produced by *Streptococcus pneumoniae* can bestow competence on non-competent cells of the same strain; the competence factor of *Bacillus subtilis* appears to be unable to communicate competence to non-competent cells. It has been suggested that a competence factor may be an autolytic enzyme which could create/unmask *receptor sites* for the binding of DNA. In *Strep. pneumoniae* and *Haemophilus influenzae* up to 100% of cells in a population may be competent during the period of maximum competence; in *Bacillus subtilis* the proportion seems not to exceed approximately 10–15%. DNA functional in transformation must be double-stranded and of molecular weight higher than a certain mimimum value—which depends on species; such DNA may be extracted experimentally from donor cells—or it may be derived, under natural conditions, from autolysed cells. Competent cells appear to bind DNA at a number of receptor sites at the cell surface; binding is initially reversible, later irreversible. The mechanism by which DNA penetrates the cell wall and cell membrane is unknown. Within recipient cells of *Strep. pneumoniae* and *B. subtilis* one strand of the double-stranded transforming DNA is degraded; the remaining strand pairs with an homologous region of the recipient's chromosome prior to recombination. (Although binding and penetration of DNA appear not to be specific, recombination can occur only if there are regions of homology between donor and recipient DNA. Thus, most transformations are intraspecific, although interspecific and intergeneric transformations have been reported; it has been suggested that the efficiency of transformation may indicate the closeness of relationship between species.) In *Strep. pneumoniae* and *B. subtilis* a single-strand fragment of the recipient's chromosome is believed to be excised and replaced by the single-stranded transforming DNA; thus, a region of *hybrid* DNA is formed—in which one strand derives from the donor, the other from the recipient. (The nucleotide sequences of the hybrid DNA may not be completely complementary—i.e. it may be *heterozygous*.) During subsequent replication of the recipient's chromosome, the two strands of the heterozygous region are separated into two chromosomes—one resembling the donor and the other the recipient in respect of this region; these chromosomes are segregated on cell division. The transforming DNA is not necessarily incorporated in its entirety: one or both ends may be 'trimmed' by nucleases prior to insertion. In *Haemophilus influenzae* a different mechanism appears to operate: in this organism the transforming DNA remains double-stranded.

Transformants (recipient cells which contain integrated donor genes) can be detected by plating on media selective for certain donor genes.

Transformation does not occur readily in members of the Enterobacteriaceae—apparently owing the inability of DNA to penetrate the cell, and to the presence of nucleases which degrade linear DNA. Low frequencies of transformation can be achieved in *Escherichia coli* e.g. by using sphaeroplasts, by treating cells with calcium chloride (to increase their permeability to DNA), by using strains deficient in certain DNases, or by infecting cells, simultaneously, with transforming DNA and a 'helper' bacteriophage. (see also TRANSFECTION)

transformation (of cells) In TISSUE CULTURE: the conversion of normal cells to cells which exhibit some or all of the properties typical of tumour cells; the meaning of transformation is usually extended to refer to the corresponding process *in vivo**. Transformation may occur 'spontaneously' after the repeated subculturing of a tissue culture, or it may be induced by certain viruses—including DNA viruses (e.g. SV40) and RNA viruses (e.g. Rous sarcoma virus). For a reason as yet unknown, the virus-induced establishment of the transformed state requires at least one round of cell division; some viruses (e.g. SV40) promote the replication of host cell DNA. Transformed cells exhibit some or all of the following characteristics: the ability to grow in soft agar, the development of new surface antigens, the ability to form tumours when injected into animals, an increase in the rate of nutrient transport, and a loss of *contact inhibition*—i.e. inhibition of cell division due to contact with neighbouring cells; the latter change in a monolayer tissue culture permits the development of multilayer colonies from the clones of transformed cells. *The ability of a particular virus to bring about transformation in tissue cultures does not necessarily imply an ability to give rise to tumours in a given species of animal.

transformation, of lymphocytes (blast transformation) (*immunol.*) Morphological and other changes which occur in both B and T LYMPHOCYTES on exposure to antigens to which they are specificially reactive.

B lymphocytes. The *initial* (specific) antigenic stimulation of a clone of competent B lymphocytes is followed by proliferation and the formation of (a) ANTIBODY-producing PLASMA CELLS ('effector cells') and (b) *memory cells*—*primed* cells which, on the second or subsequent antigenic stimulation, are believed to promote *prompt* ANTIBODY-FORMATION (the *secondary response*). *Transformation* refers to the sequence of events in which, following antigenic stimulation, a given cell enlarges and develops a basophilic, ribosome-rich cytoplasm, a prominent nucleolus, and a paler-staining nucleus; such a *blast cell*, which subsequently divides, has a greatly increased rate of macromolecule synthesis. The blast cell is an intermediate form between the B lymphocyte and plasma cell, and (presumably) blast cells are intermediates in the development of memory cells. *Non*-specific transformation and proliferation in B lymphocytes follows exposure to certain LECTINS, e.g. the *pokeweed mitogen*.

T lymphocytes. The *initial* antigenic stimulation of a clone of competent T lymphocytes is followed by proliferation and the formation of *memory cells* (*primed* cells); on the second or subsequent antigenic stimulation, these specifically-reactive cells exhibit the secondary response: (a) they proliferate, and (b) they give rise to LYMPHOKINES. During the secondary response some of the memory cells undergo *transformation*—a process essentially similar to that in B lymphocytes; the blast cells divide to form an expanded clone of memory cells. Whether the lymphokines are released by the blast cell precursors or by the undividing memory cells appears to be unknown. The lymphokine-forming cells are termed 'effector cells' since their products appear to bring about the reactions of CELL-MEDIATED IMMUNITY.

transforming principle Historical name for DNA which brings about bacterial TRANSFORMATION.

transition (genetics) A type of POINT MUTATION in which one purine is replaced by another, or one pyrimidine is replaced by another. (cf. TRANSVERSION)

translation (of mRNA) (see PROTEIN SYNTHESIS)

translocation (genetics) (see CHROMOSOME ABERRATION)

transovarial transmission Transmission of a pathogen from one individual to another *via* the egg of the first. Examples of transovarially transmitted, vector-borne diseases include BLACKHEAD, RED-WATER FEVER, and certain virus diseases of plants. (see also VERTICAL TRANSMISSION)

transport (across bacterial membranes) Only small, non-ionic molecules (especially lipophilic ones) can diffuse passively across the hydrophobic barrier of the CELL MEMBRANE; such molecules include oxygen, carbon dioxide, ethanol, and certain amino acids. Most substances therefore enter or leave the cell as a result of specific *transport mechanisms* which may or may not require metabolic energy, and which are thought to involve specific protein 'carriers' located within the cell membrane.

Facilitated diffusion is a transport mechanism which does not require metabolic energy but which does involve a specific protein carrier. In this mechanism the carrier is assumed to oscillate across the cell membrane (possibly as a result of conformational changes)—combining with the substrate on one side of the membrane and releasing it on the other. Facilitated diffusion does not allow the accumulation of a substrate (by a cell) against a concentration gradient. Such a system is useful to the microorganism in an environment containing high concentrations of metabolites.

Active transport involves the uptake of a substance by a cell *against a concentration gradient*; the process requires metabolic energy and involves a specific carrier system in the membrane. Such transport is necessary where nutrients etc. occur in the environment in low concentrations. Two main types of active transport have been extensively studied:

(a) *Group translocation*. The substrate being

transported (a sugar, or possibly a purine) undergoes a covalent change within the membrane—thus being delivered to the cytoplasm in a modified form. In the *phosphoenolpyruvate-hexose phosphotransferase system* each molecule of substrate is phosphorylated in a reaction which forms an integral part of the translocation mechanism. Initially, a soluble (cytoplasmic) *enzyme I* catalyses the phosphorylation, by phosphoenolpyruvate, of a low M.Wt. heat-stable protein, *HPr*. Subsequent step(s) appear to vary according to species and to the nature of the substrate. In e.g. *Escherichia coli*, a membrane-bound *enzyme II* catalyses the transfer of the phosphate group from phospho-HPr to the substrate. Enzyme II is substrate-specific and may consist of two components—IIa and IIb (IIa being substrate-specific). In e.g. *Staphylococcus aureus*, phosphate is transferred from phospho-HPr to a soluble, substrate-specific protein—*factor III*—and from thence to the substrate *via* a specific membrane-bound enzyme II. (Enzyme IIa of *E. coli* and factor III of *S. aureus* may be functionally analogous.) The phosphorylated derivatives of HPr and factor III are 'high-energy' phosphates (i.e. have a high free energy of hydrolysis). The mechanism of *translocation* (i.e. passage of the substrate across the membrane) is unknown. The PEP-phosphotransferase system is believed to be responsible for the uptake of many sugars by a wide range of bacteria. Evidence has been obtained for a similar system for adenine translocation in *E. coli*.

(b) *Oxidation-linked transport*. (This involves the transport of substances in unchanged form.) The active transport of many substances (including ions, sugars, amino acids) appears to be closely linked to oxidative reactions (i.e. electron transfer) occurring in the membrane. The mechanism of this type of transport is unknown. Oxidation of lactate (to pyruvate) by a flavin-linked, membrane-bound D-lactate dehydrogenase is particularly effective in stimulating transport in *membrane vesicles* prepared from *E. coli*. One hypothesis suggests the existence in the membrane of a protein carrier which has reversibly oxidizable sulphhydryl groups. Only the oxidized form is supposed to have a high affinity for the substrate. Reduction of the carrier *via* D-lactate dehydrogenase results in a conformation change during which the substrate is transported across the membrane. The reduced form of the carrier releases the substrate at the inner surface of the membrane and may then be re-oxidized *via* cytochrome *b*. Little evidence has been obtained for such a system in intact bacteria. An alternative hypothesis suggests that a *proton motive force* (*pmf*) can drive transport. (A *pmf* is said to be generated by the electron transport chain, according to the *chemiosmotic hypothesis*—see OXIDATIVE PHOSPHORYLATION.) Protons extruded from the cell during electron transport are said to re-enter on a carrier which has two sites—one for a proton and a specific site for a substrate 'passenger'. Thus a proton may re-enter the cell with e.g. a sugar or an amino acid (SYMPORT), or may exchange for e.g. an ion within the cell (ANTIPORT). Ions may be transported in response to the membrane potential. ATP hydrolysis can also drive transport; according to this hypothesis a *pmf* is generated by proton extrusion driven by an ATPase present in the cell membrane.

The transport of a particular substance may constitute the rate-limiting step in the metabolism of that substance, and may thus be an important mechanism for control. (see also CRYPTIC MUTANT)

transport medium Any (liquid) medium used for the transportation and/or storage of material from which the isolation of particular organism(s) is subsequently to be attempted. The purpose of such a medium is to maintain the viability and/or infectivity of the organism(s) during the delay between collection and culture of the specimen. An effective transport medium is particularly necessary for anaerobes (which may be killed by oxygen), for delicate organisms (e.g. *Neisseria* spp), and for certain viruses (e.g respiratory syncytial viruses). For anaerobes, and for a number of delicate organisms (including *N. gonorrhoeae*), STUART'S TRANSPORT MEDIUM is usually satisfactory. Other types of transport media have been recommended for particular purposes—e.g. MONSUR TRANSPORT MEDIUM for *Vibrio cholerae*. For many viruses (including the medically-important ones) HANKS' BSS supplemented with protein (e.g. 0.5%–2% bovine albumin) is commonly found to be effective; in such media, viruses can be rapidly frozen and stored at temperatures around −70 °C—usually without too great a loss in infectivity. Some transport media incorporate activated charcoal in order to absorb traces of inhibitory or toxic materials.

transversion (genetics) A type of POINT MUTATION in which a purine is replaced by a pyrimidine, or a pyrimidine by a purine. (cf. TRANSITION)

Trebouxia A genus of unicellular, spherical, non-motile algae of the division CHLOROPHYTA. The cell is of the order of 10 microns in diameter and contains a single large chloroplast. Reproduction generally occurs by the formation of *autospores*. *Trebouxia* species are reported to be the commonest phycobionts found in LICHENS.

tree lungwort A common name for the lichen *Lobaria pulmonaria*—see LOBARIA.

trees, diseases of (see TIMBER, DISEASES OF)

trehalose A non-reducing disaccharide: α-D-glucopyranosyl-α-D-glucopyranoside. Trehalose occurs as a reserve compound in certain fungi (particularly yeasts), algae, and lichens, and is a component of the CORD FACTOR of some strains of *Mycobacterium*. The majority of *α-glucosidases* are unable to cleave trehalose; however, a specific enzyme—*trehalase*—occurs e.g. in many fungi.

Tremella A genus of fungi of the order TREMELLALES.

Tremellales An order of fungi of the sub-class PHRAGMOBASIDIOMYCETIDAE; species occur as saprophytes on a variety of types of wood. Constituent species typically form longitudinally-septate basidia in gymnocarpous fruiting bodies; the basidium (metabasidium) is commonly globose or clavate, and the basidiocarp may be e.g. crust-like, erect and stalked, or sheet-like and convoluted, and is often gelatinous, waxy, or leathery. According to species the basidiocarp may be e.g. orange, fawn, brown, or black. The genera include e.g. *Aporpium*, *Exidia*, and *Tremella*.

trench fever (Wolhynian fever) A louse-borne rickettsial disease of man in which the symptoms include fever, severe muscular pain, and a maculopapular rash; the causal agent is *Rochalimaea quintana*. Mortality rates are low, but complications (e.g. cardiac dysfunction) may add to the seriousness of an attack. XENODIAGNOSIS has been employed.

trench mouth Synonym of VINCENT'S ANGINA.

Trentepohlia A genus of filamentous algae of the division CHLOROPHYTA; species—which are frequently orange-coloured—occur as epiphytes, on stones, and as the phycobiont in a number of LICHENS—e.g. in the genera ENTEROGRAPHA and ROCCELLA.

Treponema A genus of gram-negative, chemoorganotrophic, asporogenous, strictly anaerobic, motile bacteria of the SPIROCHAETALES; species are parasitic or pathogenic—occurring e.g. on the mucous membranes of the oral and genital regions of man and other animals. The cells are 5–20 microns in length, less than 0.5 microns in width; typically, they have pointed ends. Some species, including the major pathogens *T. pallidum*, *T. carateum* and *T. pertenue*, have not been cultivated outside living cells. Species cultivable *in vitro* (i.e in cell-free media) have been classified e.g. by the range of carbohydrates fermented and by the ability to produce H_2S and indole and to hydrolyse aesculin. Species include:

T. pallidum. The type species. The causal agent of SYPHILIS. The cells are commonly 10–15 microns in length and less than 0.2 microns in width. Virulent strains are cultivated intratesticularly in the rabbit. Serologically, *T. pallidum*, *T. pertenue* and *T. carateum* are indistinguishable. (see also NICHOLS TREPONEME)

T. carateum, *T. pertenue*. The causal agents, respectively, of PINTA and YAWS. The cells resemble those of *T. pallidum*.

T. phagedenis. A non-pathogenic species cultivable *in vitro*. The cells are wider than 0.2 microns but otherwise morphologically similar to *T. pallidum*.

treponema pallidum immobilization test (see TPI)

Triactinomyxon A genus of protozoans of the CNIDOSPORA.

tribe A taxonomic group (see TAXONOMY and NOMENCLATURE).

TRIC agents Abbreviation for TRACHOMA and inclusion conjunctivitis agents, i.e. strains of CHLAMYDIA which cause these diseases.

tricarboxylic acid cycle (TCA cycle; Krebs cycle; citric acid cycle) A cyclic sequence of reactions (depicted in Appendix II(a)) which plays a central role in the metabolism of many heterotrophic microorganisms capable of oxidative (respiratory) metabolism. (The TCA cycle is not present in e.g. *Acetomonas suboxydans*.) In eucaryotic microorganisms the reactions of the TCA cycle take place in the MITOCHONDRION—most of the enzymes being located in the matrix; in bacteria most of the enzymes occur in the cytoplasm. (Succinate dehydrogenase is membrane-bound in both mitochondria and bacteria.)

The TCA cycle has both catabolic and anabolic functions—the relative importance of these functions depending on the metabolic state of the cell. During each complete turn of the TCA cycle: one molecule of acetyl-CoA (formed e.g. from pyruvate or by fatty acid degradation) enters the cycle (by condensation with one molecule of oxaloacetate), two molecules of carbon dioxide are eliminated, 3 molecules of NAD(P) and 1 of FAD are reduced, and one molecule of oxaloacetate is regenerated. The reduced coenzymes (NAD(P), FAD) are re-oxidized *via* the ELECTRON TRANSPORT CHAIN (with concomitant release of energy in a form useful to the cell). Thus, *in effect* acetate is completely oxidized to carbon dioxide and water in one turn of the TCA cycle (although the carbon of the carbon dioxide actually derives from oxaloacetate). In theory, a single molecule of oxaloacetate could allow the oxidation of any number of acetate molecules—i.e. oxaloacetate (and other TCA cycle intermediates) have a *catalytic* role in the functioning of the cycle. However, the TCA cycle functions anabolically by providing precursors for a number of biosynthetic pathways (see Appendix II(b)); intermediates withdrawn from the cycle for this purpose cannot be used to regenerate oxaloacetate—thus oxaloacetate must be regenerated in some other way in order to permit the continued operation of the cycle. A number of reactions which achieve this have been found (in various bacteria); these include the direct formation of oxaloacetate by the carboxylation of pyruvate or phosphoenolpyruvate. (Reactions whose function is to replenish intermediates of the cycle are called *anaplerotic* reactions or sequences—see also *glyoxylate cycle* below.)

If an organism uses a di- or tricarboxylic acid (e.g. citrate) as the sole source of carbon, oxaloacetate will be formed by reactions of the cycle, but acetate must be formed if the cycle is to continue. This may be achieved by the decarboxylation of malate (by the 'malic enzyme') to pyruvate which, in turn, is converted to acetyl-CoA by the pyruvate dehydrogenase complex. (Possession of the TCA cycle does not necessarily enable an organism to grow on TCA cycle intermediates since the organism may be impermeable to such compounds.) An organism using *acetate* (or an acetate-generating substrate) as the sole source of carbon cannot generate oxaloacetate by the carboxylation of pyruvate (or phosphoenolpyruvate) since—in aerobic organisms—pyruvate cannot be synthesized directly from acetate. Under these conditions 4-carbon acids are synthesized by the *glyoxylate cycle* (glyoxylate shunt)—a modification of the TCA cycle; two reactions—catalysed by isocitrate lyase and malate synthase—by-pass the decarboxylation steps of the TCA cycle (see Appendix II(b)). Thus, in effect, one molecule of a 4-carbon dicarboxylic acid is synthesized from two molecules of acetate for each turn of the glyoxylate cycle; the TCA and glyoxylate cycles may operate simultaneously—the TCA cycle being predominantly an energy-yielding system, the glyoxylate cycle functioning primarily in biosynthesis. (Pyruvate—which may be required for e.g. GLUCONEOGENESIS—can be formed from acetate by the decarboxylation of oxaloacetate formed by means of the glyoxylate cycle.) If glyoxylate (or a precursor of glyoxylate—e.g. glycolate) is supplied as the sole source of carbon, acetate may be formed *via* 3-phosphoglycerate produced by the tartronic semialdehyde pathway (e.g. in *Pseudomonas* spp) (see Appendix II(b)); acetyl-CoA may then condense with a further molecule of glyoxylate to form malate and hence oxaloacetate.

Certain facultative anaerobes (e.g. *Escherichia coli*) growing under anaerobic conditions, and certain autotrophs, appear to contain all the TCA cycle enzymes with the exception of *α-ketoglutarate dehydrogenase*; in such organisms the pathway ceases to be cyclic and functions, in biosynthesis only, as two linear pathways. (Under aerobic conditions the TCA cycle functions normally in *E. coli*.)

trichites (*ciliate protozool.*) (1) In certain holotrichs (e.g. *Holophrya*): microtubular endoskeletal structures which support the cytostomal region—often forming a 'pharyngeal basket'. They are also referred to as *nematodesmata*. (2) The former name of a particular type of TRICHOCYST.

trichocysts (*protozool.*) Sub-pellicular, capsule-like organelles found in many ciliates (e.g. *Actinobolina*, *Didinium*, *Dileptus*, *Frontonia*, *Paramecium*) and in a number of flagellates. Trichocysts may be few in number or numerous—according to species; they possess the ability to extrude rapidly, into the environment, a thread (filament, shaft) following an appropriate chemical, mechanical, or electrical stimulus. In some species (e.g. *Actinobolina*, *Didinium*) the extremity of the filament appears to bear or contain material which is toxic to other protozoans; trichocysts of this type are termed *toxicysts*. Trichocysts may act as anchoring devices and/or as a form of defence—but their precise role is unknown; toxicysts appear to be offensive weapons which disable a suitable prey. Electron microscope studies of *Frontonia* and *Paramecium* indicate that trichocysts develop from minute membrane-bound vesicles within the cytoplasm, and that the developing structures (*pretrichocysts*) migrate to their final, sub-pellicular positions only shortly before reaching maturity. In *Paramecium aurelia* the mature trichocysts are ovoid, some 5 × 2 microns, orientated approximately at right angles to the pellicle; the pointed, crystalline tip of the filament appears to be the only osmiophilic structure within the undischarged trichocyst. The extruded filament (approximately 15–25 microns) exhibits osmiophilic cross-striations which have a periodicity of about 60 nm.

trichogyne (*mycol.*) In certain ascomycetes: an extension or appendage of the ASCOGONIUM through which the male nuclei pass from the male gametangium; the trichogyne is commonly an elongated structure.

Tricholoma A genus of fungi of the family TRICHOLOMATACEAE. Constituent species form fleshy fruiting bodies in which the stipe is typically central, and the non-serrated gills are typically adnexed to sinuate; the basidiospores are non-amyloid. The organisms grow on soil and humus.

Tricholomataceae A family of fungi of the order AGARICALES. Constituent species form fruiting bodies in which (in stipitate species) the stipe and pileus do not cleanly separate and the gills are typically adnate, adnexed, or decurrent; some species are sessile, and in stipitate species the stipe may be central, eccentric, or lateral. The fruiting body is fleshy, and the spores (which lack a germ pore) may appear white, pale-coloured, or (occasionally) some shade of brown *en masse*. The genera include e.g. ARMILLARIA, *Clitocybe*, *Collybia*, LENTINUS, OUDEMANSIELLA, PLEUROTUS, and TRICHOLOMA.

trichome (1) In blue-green algae: a chain of cells which may or may not include one or more akinetes and/or heterocysts; if the trichome bears a mucilaginous envelope (sheath) the whole is referred to as a *trichome* by some authors, and as a *filament* by other authors. (2) The term is also used, by some authors, to refer to a chain of vegetative bacterial cells.

Trichomonadida An order of parasitic protozoans of the class ZOOMASTIGOPHOREA. Typically, 4, 5, or 6 flagella arise at the anterior end of the cell; one flagellum is directed posteriorly and may form the border of an undulating membrane. Species inhabit the intestinal or reproductive tracts of various animals; they do not form cysts. Reproduction occurs by binary fission; sexual phenomena are unknown. The genera include TRICHOMONAS.

Trichomonas A genus of parasitic protozoans in the order TRICHOMONADIDA; cells are uninucleate, oval or pear-shaped, with a cytostome and endoskeletal structure(s). Of the many species, two are important pathogens: (a) *T. foetus*. (Some authors include this organism in a separate genus: *Tritrichomonas*.) The causative agent of bovine trichomoniasis, a venereal disease of cattle; infection may lead to early abortion. *T. foetus* is a pear-shaped organism, about 20 × 10 microns, with a well-developed *costa* and AXOSTYLE. Four flagella arise from the blunt anterior; three are directed anteriorly while the fourth runs back along the margin of the undulating membrane and extends beyond the posterior limit of the body. The organism has a jerky motion. In cows, infection is commonly self-limiting—the aborted foetus carrying with it all the parasites; following abortion the cow is usually immune to re-infection. (b) *T. vaginalis*. This species may occur, asymptomatically, in the urinary/reproductive tracts of man; no other species is naturally infected. It may give rise to one form of vaginitis; transmission is usually venereal. *T. vaginalis* is pear-shaped, about 15 × 10 microns, with five flagella arising from the blunt anterior; four flagella are directed anteriorly while the short, recurrent fifth flagellum runs back along the margin of an undulating membrane—which extends approximately half-way along the body. A costa and axostyle are present and, as in *T. foetus*, the nucleus is located at the anterior end of the cell. Several drugs have been found useful for the treatment of *T. vaginalis* infections; these include *furoxone* (see NITROFURANS), *trichomycin* (see POLYENE ANTIBIOTICS), and METRONIDAZOLE. *T. foetus* and *T. vaginalis* can be cultured *in vitro* in serum-based media.

T. gallinae produces an inflammatory and ulcerative condition in the upper digestive tract of young birds—particularly pigeons, but occasionally in chickens and turkeys; the organism is similar to *T. vaginalis*, though smaller. Treatment of infected birds with 2-amino-5-nitrothiazole has been found effective.

T. hominis is a common intestinal parasite of man; it resembles the other trichomonads but usually has *five* anteriorly-directed flagella—for which reason some authors place the organism in a separate genus: *Pentatrichomonas*.

Trichomycetes A class of fungi within the subdivision ZYGOMYCOTINA; the vegetative form of the organisms is a simple branched or unbranched hypha or mycelium which attaches to an arthropod host by means of a specialized basal cell.

trichomycin (see POLYENE ANTIBIOTICS)

Trichonympha A genus of cellulose-digesting protozoans (order HYPERMASTIGIDA) which occur in the gut of wood-eating termites. The bell-shaped *T. campanula* (long axis 100–300 microns) bears numerous flagella at the narrow (anterior) end; wood fragments are ingested at the naked posterior end. *T. campanula* supplies its host with products of cellulose digestion and is an indispensible part of the insect's gut microflora.

Trichophyton A genus of fungi within the order MONILIALES and included within the category DERMATOPHYTES; species occur as soil saprophytes and as parasites and pathogens in the skin, hair, and nails of man and other animals (see RINGWORM). The vegetative form of the organisms is a septate mycelium. In species which give rise to endothrix infections (e.g. *T. tonsurans*, *T. violaceum*) the somatic hyphae commonly fragment to form rows of arthrospores within the hair shaft; hairs infected by the endothrix species *T. schoenleinii* exhibit longitudinally-orientated hyphae within the shaft and/or fine tubular canals left by hyphae which have disintegrated. Species which give rise to ectothrix infections (e.g. *T. mentagrophytes*, *T. verrucosum*) produce rows of arthrospores both inside and at the surface of the hair shaft. In culture, most species produce both microconidia and macroconidia—though e.g. *T. schoenleinii* does not produce macroconidia and forms few microconidia. The macroconidia are generally 10–50 microns in length, according to species and strain; they are typically club-shaped and multiseptate with thin, smooth walls. Conidia are borne singly (i.e. not in chains) on conidiophores which closely resemble somatic hyphae. A number of species of *Trichophyton* have perfect stages in the ascomycete genus *Arthroderma*.

Trichosporon A genus of fungi within the family CRYPTOCOCCACEAE; species have been isolated from e.g. insects, wood pulp, and human sputum. (see also WHITE PIEDRA) The vegetative form of the organisms includes budding cells (of various shapes and sizes), pseudomycelium and true mycelium; arthrospores are formed by all species, and many or all form blastospores. According to species fermentative ability is weak or absent; most species fail to assimilate nitrate.

Trichostomatida An order of protozoans of the sub-class HOLOTRICHIA (sub-phylum CILIOPHORA). *Trichostomes* (e.g. *Colpoda*) occur in various freshwater habitats and in the soil; some are parasitic—BALANTIDIUM COLI being the only ciliate which is pathogenic for man. Individual cells may be simple (ovoid, reniform) or complex in shape; cilia may cover the entire body or may be restricted to particular regions of the pellicle. The cytostome may be apical or (in the more advanced species) lateral; it is situated at the base of a VESTIBULUM—no true BUCCAL CAVITY being present. (The vestibular surface may be ciliated, but no membranelles or complex arrangements of cilia are found.)

trimethoprim 5-(3,4,5-Trimethoxybenzyl)-2,4-diaminopyrimidine. A synthetic, broad-spectrum bacteriostatic agent. Trimethoprim inhibits dihydrofolate reductase, thus preventing the reduction of dihydrofolate to tetrahydrofolate in FOLIC ACID metabolism. Thus, trimethoprim and the SULPHONAMIDES act in a similar metabolic region—although at different locations; these drugs are often used in combination (e.g. trimethoprim + sulphamethoxazole = *cotrimoxazole*), the combination being synergistic and often bactericidal—even though each drug is bacteriostatic when acting independently. The affinity of trimethoprim for *bacterial* dihydrofolate reductase is much greater than its affinity for the corresponding mammalian enzyme—hence its selective toxicity is very favourable for clinical use. Trimethoprim is used mainly for treating infections of the upper respiratory tract and the urinary tract; it has also been used against many other diseases, including brucellosis, gonorrhoea and typhoid fever. The antibacterial spectrum includes species of *Klebsiella*, *Haemophilus*, *Staphylococcus* and *Streptococcus*; trimethoprim is not active against *Treponema pallidum*, *Pseudomonas aeruginosa* or *Mycoplasma* spp. Cases of plasmid-borne resistance have been reported. (see also PYRIMETHAMINE)

trimitic (see HYPHA)

triphenylmethane dyes (as antimicrobial agents) This group of basic dyes includes CRYSTAL VIOLET (hexamethyltriaminotriphenylmethane), MALACHITE GREEN (tetramethyldiaminotriphenylmethane), brilliant green (tetraethyldiaminotriphenylmethane), and FUCHSIN. At very low concentrations these dyes are bacteriostatic for many gram-positive bacteria, though relatively inactive against a range of gram-negative species and acid-fast bacteria. Among the gram-positive species, some of the dyes, e.g. brilliant green, have been found to be more inhibitory to aerobic spore-formers than to anaerobic spore-formers; such dyes have therefore found use in a variety of selective media. Some of the dyes have been found to be critically selective; for example, certain concentrations of *basic fuchsin* inhibit the growth of *Brucella suis* without affecting the growth of *Brucella abortus* or *Br. melitensis*. The triphenylmethanes are also fungistatic—e.g. for species of *Candida*, *Torula*, and *Trichophyton*. In general, triphenylmethane dyes are more active under *alkaline* conditions when a higher proportion of molecules are present as *pseudobases*—non-ionized molecules which are relatively more lipophilic and which are able to pass more easily through microbial cell membranes. Within the cell dye *cations* are formed as a new equilibrium is set up, and these cations may then react with nucleophilic groups within the cell interior. (see also DYES)

triple-sugar-iron agar (TSI) A medium used in a SCREENING TEST (sense 2) for members of the Enterobacteriaceae. Constituents of the medium *include* peptone (2%), glucose (0.1%), lactose (1%), sucrose (1%), ferrous sulphate, or ferrous ammonium sulphate (0.02%), sodium thiosulphate (0.02%–0.03%), agar (1.5%–2%) and PHENOL RED (0.0025%). The medium is prepared as a slope with a deep BUTT; the test strain is inoculated onto the surface of the slant *and* stab-inoculated (with a straight wire) deep into the butt. Following appropriate incubation the medium is examined for: (a) Acid/alkaline reaction at the (aerobic) surface of the slant. (b) Acid/alkaline reaction in the (anaerobic) butt. (c) H_2, CO_2 production—detectable as gas pockets, or splitting of the butt. (d) H_2S production. *Abundant* H_2S production is detected by blackening of the medium along the stab line due to the formation of ferrous sulphide. Thus, e.g. *Escherichia coli* (which ferments glucose *and* lactose) produces an acid slant and butt, small amounts of H_2 and CO_2, and no H_2S. *Salmonella* spp (other than *S.*

typhi) and strains of *Edwardsiella*—all of which do not attack lactose or sucrose—produce an acid butt and an alkaline slant; these organisms *ferment* the butt glucose but attack the (aerobic) slant glucose *oxidatively*, producing less acid—which is masked by the alkaline metabolic products. These organisms also form copious H_2S (blackening in butt).

triple vaccine A vaccine containing diphtheria and tetanus TOXOIDS and killed cells of *Bordetella pertussis* (pertussis vaccine). (see also ANTIGENIC COMPETITION and MIXED VACCINE)

TRIS (tris-(hydroxymethyl)-aminomethane) A buffer used in biological work. The pK_a for the reaction:

$(CH_2OH)_3CNH_2 + H^+ \rightleftharpoons (CH_2OH)_3CNH_3^+$ is 8.0.

trisomic (see CHROMOSOME ABERRATION)

trisporic acid (see HORMONAL CONTROL (of sexual processes in fungi))

Tritrichomonas (see TRICHOMONAS)

tRNA Transfer-RNA—see RNA and PROTEIN SYNTHESIS.

trochal band (*ciliate protozool.*) (see TELOTROCH)

trophophase (of fungal metabolism) During batch culture, that phase in which primary (growth-directed) metabolism is dominant over secondary metabolism.

trophozoite A vegetative (feeding) stage in the life cycle of certain protozoans.

tropism A response to a directional stimulus exhibited by bending or growth in an orientation dictated by the stimulus; no *locomotive* movement is involved (cf. TAXIS). For example, the fruiting bodies of *Polyporus* develop horizontally as a response to the force of gravity—a *geotropism*; the tubular pores which open on the underside of the basidiocarp are thus aligned vertically. The sporangiophores of *Pilobolus* exhibit *phototropism*—i.e. they grow in a direction which is related to the direction of maximum light intensity. Phototropism is also exhibited e.g. by the asci in deeply concave apothecia; such asci tend to be positively phototropic (i.e. they tend to become orientated such that the ascospores are discharged in the direction of maximum light intensity). In each of the above examples the tropism results in more effective release and/or dispersal of spores. *Chemotropism* involves growth in a direction dictated by a chemical concentration gradient: see e.g. HORMONAL CONTROL (of sexual processes in fungi).

truffles The edible fruiting bodies of species of fungi of the order Tuberales.

trypan blue An acidic, blue diazo dye. (see also STAINING and VITAL STAINING)

trypan red An acidic, red diazo dye; it was used in the early days of chemotherapy for the treatment of trypanosomiasis. (see also STAINING)

Trypanosoma A genus of parasitic flagellate protozoans of the order KINETOPLASTIDA, suborder TRYPANOSOMATINA; *Trypanosoma* spp cause a range of diseases in man and other animals—principally in parts of Africa and in the Americas.

Morphology. Many species exhibit a life cycle characterized by the successive development of morphologically-distinct forms; a particular form may develop only in certain species of host organism. Two or more of the following forms occur during the life cycle: AMASTIGOTE, EPIMASTIGOTE, PROMASTIGOTE, SPHAEROMASTIGOTE, and TRYPOMASTIGOTE (see diagram); the trypomastigote form is common to all species of *Trypanosoma*. Trypanosomes vary in size (according e.g. to species)—being some 10 microns to more than

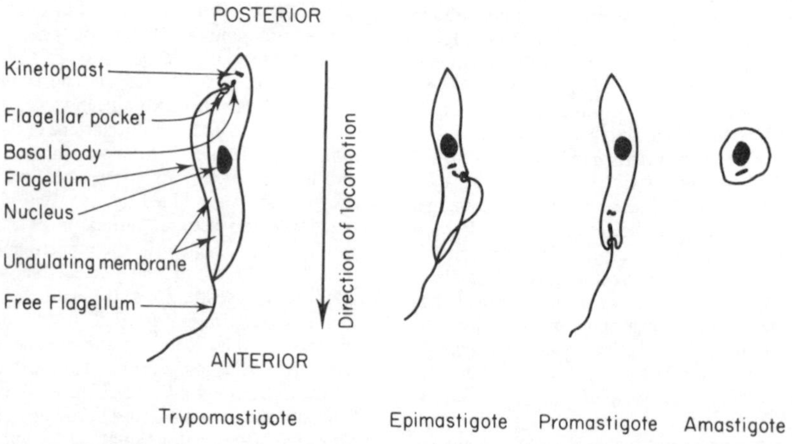

Some forms which occur in the life cycles of species of the Trypanosomatidae.

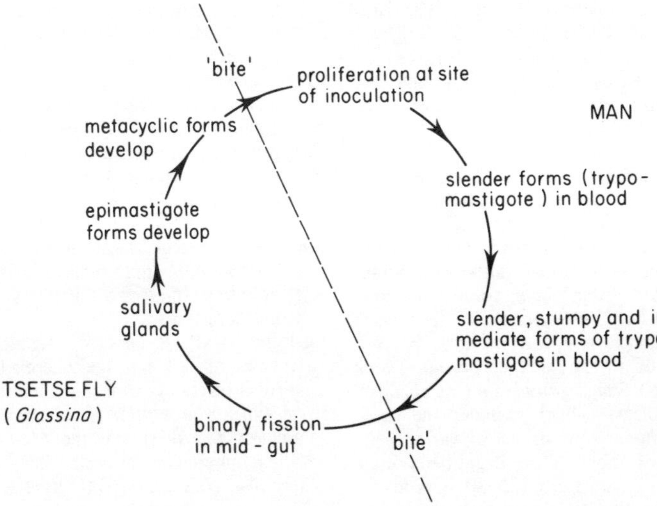

Life cycle of *Trypanosoma* (*Trypanozoon*) *brucei gambiense* (*T. gambiense*).

120 microns in total length (i.e. cell + *free* flagellum); some species exhibit marked pleomorphism.

Nutrition is chemoorganotrophic; intake of nutrients appears to occur saprozoically, but may (also) involve pinocytosis. Many species can be cultured *in vitro* in sterile blood-based media, e.g. NNN MEDIUM.

Reproduction occurs *asexually*: by longitudinal binary fission in the trypomastigote, and e.g. by multiple fission in certain other stages. Fission involves (in order): duplication of the BASAL BODY (a new basal body and flagellum are formed *de novo*), replication of the KINETOPLAST, nuclear division, and cell division.

Parasitism. Trypanosoma spp parasitize a variety of animals (including birds, amphibians, mammals). Species parasitic in *mammals* are classified as members of either of two groups of sub-genera, the SALIVARIA and the STERCORARIA; most are extracellular parasites. Most species require both a vertebrate and an invertebrate host for the completion of their life cycles; the invertebrate hosts (VECTORS), which include insects of the orders Diptera and Hemiptera, are commonly involved in *cyclical transmission*. (see also METACYCLIC FORMS)

Anti-trypanosomal agents include ETHIDIUM BROMIDE and SURAMIN; see also *antimony* under HEAVY METALS.

Trypanosoma spp include:

T. brucei (see TRYPANOZOON and NAGANA)

T. congolense (see NANNOMONAS and NAGANA)

T. cruzi (see SCHIZOTRYPANUM and CHAGAS' DISEASE)

T. equinum (see TRYPANOZOON and MAL DE CADERAS)

T. equiperdum (see TRYPANOZOON and DOURINE)

T. evansi (see TRYPANOZOON and SURRA)

T. gambiense, *T. rhodesiense* (see TRYPANOZOON and SLEEPING SICKNESS)

T. lewisi (see HERPETOSOMA)

T. vivax (see DUTTONELLA and NAGANA)

trypanosomal form (see TRYPOMASTIGOTE)

Trypanosomatidae A family of parasitic flagellate protozoans of the sub-order TRYPANOSOMATINA. Constituent genera: *Blastocrithidia*, CRITHIDIA, *Endotrypanum*, *Herpetomonas*, LEISHMANIA, *Leptomonas*, PHYTOMONAS, and TRYPANOSOMA.

Trypanosomatina A sub-order of flagellate protozoans of the order KINETOPLASTIDA; the uniflagellate, typically slender organisms are parasitic or pathogenic in a range of animals or plants. The genera include CRITHIDIA, *Herpetomonas*, LEISHMANIA, *Leptomonas*, PHYTOMONAS, and TRYPANOSOMA.

trypanosome adhesion test A serological test for detecting the presence of antibodies to trypanosomes. The test is based on the adhesion of erythrocytes (of certain species) to trypanosomes in the presence of specific trypanosomal antibodies and COMPLEMENT.

trypanosomiasis Any of a number of human and animal diseases in which the causal organism is a member of the genus TRYPANOSOMA. Examples include SLEEPING SICKNESS, CHAGAS' DISEASE and DOURINE.

Trypanozoon A sub-genus of TRYPANOSOMA within the group SALIVARIA; the organisms are some 20 to 30 microns in length. The subgenus comprises cyclically-transmitted species (vector: *Glossina*) and mechanically-transmitted species; the latter, although not truly *salivarian*, exhibit evolutionary and other affinities with the former.

Cyclically-transmitted species traditionally include *Trypanosoma brucei*, *Trypanosoma gambiense*, and *Trypanosoma rhodesiense*; *T. brucei* does not parasitize man (see NAGANA), but *T. gambiense* and *T. rhodesiense* are the causal agents of SLEEPING SICKNESS. Some authors regard these three species as constituents of two sub-species of *T. brucei*: *Trypanosoma* (*Trypanozoon*) *brucei brucei** (the agent of nagana), and *Trypanosoma* (*Trypanozoon*) *brucei gambiense* (the agent of sleeping sickness)—which includes the *gambiense* and *rhodesiense* strains or variants. In the vertebrate host these organisms may assume the *slender* form (25–30 microns in length with an undulating membrane and free flagellum), the *stumpy* form (about 20 microns in length, broader than the slender form, with an undulating membrane and—typically—without a *free* flagellum), or some intermediate form.

Of the mechanically-transmitted species, *T. equinum* is believed to be a strain of *T. evansi* which does not possess a kinetoplast. *T. equiperdum* also appears to be related to *T. evansi*. *Abbreviated e.g. as *T.* (*T.*) *brucei brucei*, *T.* (*T.*) *b. brucei*, etc.

tryparsamide (see ARSENICALS)

trypomastigote (trypanosomal form) The form typical of *Trypanosoma* spp in the *vertebrate* host: see diagram under TRYPANOSOMA; the trypomastigote differs from other forms in that the kinetoplast and basal body are located posteriorly. (see also METACYCLIC FORMS)

tryptone water (see INDOLE TEST)

tryptophan (biosynthesis of) (see Appendix IV(f))

tse-tse fly (see GLOSSINA)

TSI TRIPLE-SUGAR-IRON AGAR.

tsutsugamushi disease (see SCRUB TYPHUS)

Tuberales An order of fungi within the class DISCOMYCETES; constituent species typically form closed hypogean fruiting bodies in which the asci are scattered or are arranged in an hymenial layer. The order includes certain edible (and highly-prized) fungi (*Tuber* spp) which are better known as 'truffles'.

tubercle (1) (see TUBERCULOSIS) (2) A deposit formed on the inner surface of an iron or steel water pipe due to bacterial corrosion; it consists mainly of oxides and hydroxides of iron. Bacteria found associated with tuberculation include *Gallionella*, *Crenothrix* and *Leptothrix*. (3) (*lichenol.*) In the thallus of certain lichens: a wart-like protuberance in which there may be a discontinuity in the cortex; a tubercle may be analogous to a CYPHELLA or PSEUDOCYPHELLA. Tubercles occur e.g. in *Nephroma resupinatum*.

tubercle bacillus Refers to any species of *Mycobacterium* which causes TUBERCULOSIS.

tubercular (*adjective*) (1) Resembling the effects, symptoms etc. characteristic of tuberculosis—though not necessarily due to that disease. (2) Any effect or symptom which is an *indirect* result of tuberculosis. (cf. TUBERCULOUS)

tuberculin Protein derived from culture filtrates of *Mycobacterium tuberculosis* (see OLD TUBERCULIN and PPD). PPD is used in the TUBERCULIN TEST.

tuberculin test A test for detecting DELAYED HYPERSENSITIVITY to the tubercle bacillus. The various forms of the test include the *Moro test* (in which TUBERCULIN is smeared over a small area of the skin) and the *Mantoux test* (in which tuberculin is injected intracutaneously). (see also VOLLMER PATCH TEST) A positive test is indicated (within a few days) by a characteristic local reaction (see DELAYED HYPERSENSITIVITY). A positive test is *not* proof of active infection by *Mycobacterium tuberculosis*. (see also SKIN TEST)

tuberculin-type reaction A DELAYED HYPERSENSITIVITY reaction analogous to that seen in the TUBERCULIN TEST but involving a different antigen; examples include reactions seen in the LEPROMIN TEST and the HISTOPLASMIN TEST.

tuberculosis (phthisis, TB, consumption) An infectious disease of man and other animals; it may occur as a chronic, localized, granulomatous disease, or may develop in the disseminated (*miliary*) form. In *man*, the causal agent is usually *Mycobacterium tuberculosis* or *M. bovis*. The routes of infection include the respiratory and alimentary tracts; inhalation of tubercle bacilli may lead to pulmonary tuberculosis—the commonest form of the disease in man. (The pathogen may remain viable for weeks or months in moist or dried sputum.) Tubercle bacilli survive, and multiply, in phagocytic cells; from the initial lesion they may spread, *via* the lymphatic or blood vascular systems, and produce lesions in the regional lymph nodes and elsewhere. The characteristic lesion of tuberculosis is the *tubercle*; this consists of an inner zone—containing tubercle bacilli, epithelioid cells and GIANT CELLS—and a peripheral zone of lymphocytes. In time, the tubercle becomes necrotic and undergoes *caseation*, i.e. becomes converted to a granular, cheese-like mass which often *calcifies*. Alternatively, the tubercle may break down, liberating virulent bacilli. Tuberculous lesions may develop in any of a variety of tissues—e.g. liver, meninges, bones. In untreated cases mortality rates may be very high; drugs which have been used for treatment include ISONIAZID, ethambutol, STREPTOMYCIN, para-aminosalicylic acid (PAS), and the RIFAMYCINS. (see also TUBERCULIN TEST) *Laboratory diagnosis.* Specimens examined microscopically/culturally for tubercle bacilli include sputum, urine sediments, cerebrospinal fluid, and biopsies. Coupled with clinical evidence, an early *presumptive* diagnosis of pulmonary tuberculosis may be made by demonstrating acid-fast bacilli in *sputum*, i.e. material coughed up from the lungs (as opposed to salival secretions); smears of e.g. sputum may be stained by the AURAMINE–RHODAMINE STAIN or ZIEHL-NEELSEN'S STAIN. Culture is often carried out on LÖWENSTEIN–JENSEN MEDIUM or similar media. Mucoid sputum may be digested (liquefied) by an enzymic mucolytic preparation—or by a dilute solution of sodium hydroxide; the latter has the advantage that it also inhibits organisms other than the tubercle bacilli. (Sodium hydroxide may reduce the viability of tubercle bacilli if allowed to act for longer than about 30 minutes.) In handling suspect materials, adequate safety precautions must be taken. The use of a special hood or safety cabinet (which incorporates an air extractor and filter) protects the technician from dangerous aerosols.

Among *animals* tuberculosis occurs more often in cattle and pigs; bovine tuberculosis is a NOTIFIABLE DISEASE. Dogs and cats may be affected, and the disease may also occur in birds. In *cattle*, infection may occur *via* the respiratory or alimentary tracts or *via* wounds; the disease may be contracted from tuberculous attendants. For the milk-consumer, infection of the cow's udder is particularly dangerous since the milk may contain large numbers of tubercle bacilli. In *pigs*, ulcerative lesions may develop from tuberculous lymph nodes or from other infected foci; these lesions may give rise to a thick, purulent discharge. Tuberculosis is uncommon in sheep and horses; it sometimes occurs in goats which have been quartered near tuberculous cattle. (see also ATYPICAL MYCOBACTERIA)

tuberculous (1) Containing one or more *tubercles* (see TUBERCULOSIS). (2) Infected with tubercle bacilli. (cf. TUBERCULAR)

tubulin The protein sub-unit of MICROTUBULES. The polymerization of tubulin is promoted by guanine nucleotides and by certain divalent cations; polymerization is inhibited, and depolymerization promoted, by agents such as COLCHICINE and GRISEOFULVIN—each of which, in appropriate concentrations, binds to tubulin and inhibits SPINDLE-formation in MITOSIS and MEIOSIS.

tularaemia In *man*, tularaemia (O'Hara's disease) is an acute or chronic systemic disease characterized, in the earlier stages, by malaise, fever, and an ulcerative granuloma at the site of infection. The causal agent, *Francisella tularensis*, may penetrate mucous surfaces (e.g. the conjunctivae) or may invade through wounds or insect bites; infection may also occur orally. The pathogen is disseminated *via* the lymph and blood vascular systems; various organs, e.g. spleen, liver, lungs, may be affected—the organism occuring intracellularly. Mortality rates are generally lowest in *ulceroglandular* tularaemia (infection *via* the skin). Streptomycin has been used therapeutically. *Laboratory diagnosis* involves culture and specific staining of material from lesions (see also FORSHAY TEST). Among *animals*, tularaemia is primarily a disease of wild rodents; the rabbit is an important source of human infection. Domestic animals, including sheep, are apparently susceptible to a fatal form of the disease.

Tulasnellales An order of fungi of the sub-class HOLOBASIDIOMYCETIDAE; constituent species are distinguished from those of other orders e.g. in that they form basidia which each give rise to four relatively large, swollen sterigmata. The organisms are typically saprophytic; some species form mycorrhizal associations with certain orchids. The genera include e.g. *Tulasnella*.

tulip Augusta disease virus (see TOBACCO NECROSIS VIRUS)

tulip breaking (*plant pathol.*) (see BREAKING)

tumour (American: tumor) A discrete, non-inflammatory swelling which may occur in any of a variety of tissues and which is due to an abnormal and excessive proliferation of cells at that location. In animals a tumour may be *malignant* or *benign*. A malignant tumour is one which is *invasive*—i.e. extends into surrounding tissues and/or spreads to remote tissues following fragmentation and transport in blood or lymph; a benign tumour lacks the invasive qualities of the malignant tumour. The molecular mechanisms of tumour formation are unknown. Tumour-causing agents (*carcinogens*) include certain types of virus and radiation, and certain chemicals. Among animals tumours are known to be caused by e.g. certain viruses of the genera HERPESVIRUS, PAPILLOMAVIRUS, POXVIRUS, and LEUKOVIRUS (see also BURKITT'S LYMPHOMA). Among *plants* tumours are referred to as *cecidia*, or GALLS.

tunica (*algol.*) A synonym of THECA (sense 2).

tuning fork basidium A type of basidium formed by fungi of the order DACRYMYCETALES.

turbidostat A type of apparatus used for CONTINUOUS-FLOW CULTURE.

Turbinaria A genus of algae of the PHAEOPHYTA.

turkey X disease (see AFLATOXINS)

turnip yellow mosaic virus (TYMV) A single-stranded RNA virus which can affect a number of cruciferous hosts—including Chinese cabbage, cauliflower, swede and turnip; the virion is polyhedral, approximately 28 nm in diameter. On Chinese cabbage (*Brassica chinensis*) disease symptoms commonly include an initial vein-clearing followed by the development of small yellow areas on the leaves; these areas subsequently enlarge to form patches of confluent yellow. (When grown under appropriate conditions Chinese cabbage may be used as a local lesion host.) Inclusion bodies are formed in the cells of infected plants. The virus may be transmitted by a number of biting insects; mechanical transmission also occurs.

Tween (proprietary name) Any of a number of non-ionic polyoxyalkylene derivatives of sorbitan fatty acid esters. Tween 80 (polyoxyethylene derivative of sorbitan monooleate) is used e.g. in a test to distinguish between certain strains of *Mycobacterium*.

Tween 80 hydrolysis A test used to determine whether or not a given strain of *Mycobacterium* can hydrolyse Tween 80 within 5 (or 10) days. *M. kansasii* and *M. phlei* are among species which give a positive test.

tyloses (*plant pathol.*) Outgrowths from xylem parenchyma cells which expand into the lumen of a xylem vessel—thus partially or totally blocking that vessel. Tyloses (sing. *tylose* or *tylosis*) are formed by many plants in response to wounding or infection of the plant's vascular system; they may play a role in resistance to disease by restricting the translocation of the pathogen. Tyloses are frequently associated with vascular wilts due to *Fusarium* spp, *Ceratocystis* spp etc. (see also RESISTANCE (of plants to disease))

tylosin A MACROLIDE ANTIBIOTIC produced by *Streptomyces fradiae*; it is used in veterinary practice.

tylosis (*plant pathol.*) (1) The process of forming TYLOSES. (2) A term sometimes used as a synonym of *tylose*.

TYMV (see TURNIP YELLOW MOSAIC VIRUS)

tyndallization A process sometimes used *with the aim* of sterilizing certain heat-labile materials, i.e. materials which would be adversely affected by autoclaving. The material is heated to 80–100 °C for several minutes on each of three successive days, being incubated at room temperatures, or 37 °C, for the intervening periods. The first period of heating kills all vegetative cells initially present in the material. Incubation during the intervening periods is designed to permit germination of any spores which may be present; vegetative cells which arise from such spores are killed during the subsequent heating. Tyndallization is based on the assumption that materials so treated are able to support the germination of spores, i.e. that they contain all the necessary chemical factors required for the germination of those spores present in the material.

type 1 reaction (*immunol.*) (see IMMEDIATE HYPERSENSITIVITY)

type 2 reaction (*immunol.*) A form of IMMEDIATE HYPERSENSITIVITY in which antibody combines with cell-surface antigen or with cell-bound antigen; the latter reaction has been termed *reversed anaphylaxis*.

type 3 reaction (*immunol.*) (see IMMEDIATE HYPERSENSITIVITY)

type 4 reaction (*immunol.*) CELL-MEDIATED IMMUNITY.

type species (of bacterial genera) (1) The first, or one of the first species assigned to a new genus. (2) An arbitrarily designated species within a given genus. A type species is not necessarily representative (in characteristics) of the relevant genus; however, it is often more fully characterized than other members of the genus. In some genera (e.g. *Corynebacterium*) the type species (*C. diphtheriae*) is a major human or animal pathogen; however, this is not necessarily so: the type species of *Clostridium* is *Cl. butyricum*.

typhoid fever (typhoid) An acute, infectious disease which is peculiar to man; the causal agent, *Salmonella typhi*, invades *via* the oral route. The incubation period is commonly 7–14 days; symptoms include fever, skin lesions, lesions in the intestinal and lymphoid tissues—particularly *Peyer's patches*—and an enlarged spleen. Chemotherapeutic agents used include ampicillin (see PENICILLINS) and CHLORAMPHENICOL. *Laboratory diagnosis. S. typhi* may be isolated by BLOOD CULTURE during the first few weeks of the disease. During the 2nd to 4th week it is sometimes possible to isolate the pathogen from the urine. Positive stool cultures are more likely to be obtained during the 2nd and 3rd weeks of the disease. Serum agglutinins (see WIDAL TEST) reach their highest titres some 3–6 weeks after the onset of disease. In the carrier state, *S. typhi* may localize in the gall bladder.

typhoid Mary A typhoid CARRIER, Mary Mallon, who was responsible for transmitting the disease to almost 30 persons. The name is sometimes used to refer to a potential, suspected, or actual carrier of disease in outbreaks of typhoid and other diseases.

typhus fevers (a) *Epidemic* (*classical* or *louse-borne*) *typhus*. An acute, infectious, systemic disease of man in which the causal agent, *Rickettsia prowazekii*, is transmitted by the body louse or the head louse; infection is contaminative, i.e. it occurs when infected louse faeces are rubbed into the (louse) bite or other lesion. (Infection is also reported to occur by the inhalation of dried, infected louse faeces.) The incubation period is 1–2 weeks. Typical symptoms include shivering, headache, muscular pains, malaise, and a sustained fever which lasts 1–2 weeks; some 5 days after the onset of symptoms a macular rash develops on the trunk and limbs (rarely on the face) and the lesions may later become haemorrhagic. Delirium or stupor may be exhibited. Mortality rates in untreated cases may be 5–70%. Chemotherapeutic agents used include the TETRACYCLINES and CHLORAMPHENICOL. Disease control commonly involves de-lousing measures and the use of prophylactic vaccines. *Laboratory diagnosis* may involve a CFT and/or the WEIL-FELIX TEST. (b) *Brill–Zinsser disease* (*Brill's disease*; *recrudescent typhus*). A mild form of classical typhus which occurs some time after a previous attack; during the supervening period, the pathogen remains latent in the body. (c) *Murine typhus* (*endemic typhus*; *flea-borne typhus*). Similar to (but less severe than) epidemic typhus, murine typhus is primarily a disease of the rat; the causal agent, *R. mooseri* (*R. typhi*), is spread among the rat population by the rat flea. Man becomes infected contaminatively. Treatment, laboratory diagnosis etc. are as for epidemic typhus. (see also SCRUB TYPHUS)

typing (of bacteria) Any of a variety of methods used to distinguish between closely-related strains of bacteria—e.g. members of the same species which exhibit minimal biochemical and biological differences; the division of a pathogenic species into a number of *types* may be useful in elucidating the epidemiology of the corresponding disease.

Typing methods include: (a) *Serological typing*—based on major or minor antigenic differences among the strains of a given organism; such methods include the M PROTEIN and T ANTIGEN typing of Lancefield group *A* streptococci. (b) COLICIN TYPING. (c) *Phage typing*. This method is based on differences in the susceptibility of different strains of an organism to a range of bacteriophages. Initially, a FLOOD PLATE is prepared from a culture of the bacterial strain under test; the plate is dried (see PLATE (1)) and a grid is drawn on the base of the plate. The agar surface corresponding to each square of the grid is inoculated with one drop of a phage suspension (at ROUTINE TEST DILUTION)—each square being inoculated with a different phage; the plate is then incubated. Each of the phages used for typing is lytic for one, or a limited number, of the strains of the species under test; a phage which lyses the particular strain being tested will form a clear macroscopic area amid the opaque layer of surface bacterial growth. The strain can then be defined in terms of the phages to which it is sensitive. (The results of phage typing may be affected if the bacterium under test carries a PLASMID coding for a RESTRICTION endonuclease which destroys phage nucleic acid.) Strains of *Salmonella typhi* which bear the VI ANTIGEN may be typed by a range of phages which are specific for one or more of the Vi strains of *S. typhi*; this is often termed *Vi phage typing*.

tyrocidins A group of *cyclic* decapeptide ANTIBIOTICS (which includes gramicidin S). The ring contains one or more free amino groups, which are essential for activity, and at least one D-amino acid. Tyrocidins are bactericidal, being more effective against gram-positive than gram-negative species; they disorganize the CELL MEMBRANE causing leakage of essential metabolites etc. Since tyrocidins contain both hydrophobic and hydrophilic residues it is thought that they may interact with lipoprotein complexes within the membrane. Tyrocidins are produced by *Bacillus brevis*. (see also TYROTHRICIN; cf. POLYMYXINS)

tyrosine (biosynthesis of) (see Appendix IV (f))

tyrothricin An ANTIBIOTIC which consists of a mixture of GRAMICIDINS (about 20%) and TYROCIDINS. Its activity is similar to that of the gramicidins. Tyrothricin is obtained from *Bacillus brevis*.

tyvelose (3,6-dideoxy-D-mannose) A sugar which occurs e.g. in the O-specific chains of LIPOPOLYSACCHARIDES in certain serotypes of *Salmonella*. Tyvelose, first isolated from *S. typhi*, contributes to the specificity of somatic (O) antigen 9 in the group *D* salmonellae.

Tyzzeria A genus of protozoans of the suborder EIMERIINA; *Tyzzeria* species are distinguished from those of related genera by the production of oocysts containing eight naked sporozoites—i.e. *sporocysts* are not formed. *T. perniciosa* causes an intestinal disease in ducks. (For a generalized account of the morphology and life cycle of this and other coccidian parasites see COCCIDIA.)

U

ubiquinones (see QUINONES (in electron transport))

Udotea A genus of green algae (CHLOROPHYTA).

UDP Uridine-5′-diphosphate. (see NUCLEOTIDE; see also Appendix V(b))

ultracentrifugation (see CENTRIFUGATION)

ultracytostome (of coccidia) (see MICROPORE)

ultrafiltration (see FILTRATION)

ultramicrotome (see MICROTOME)

ultrasonication (as an antimicrobial procedure) The use of sound waves of frequency greater than about 16kc/s* for the destruction of microorganisms in a liquid medium; ultrasonic waves are generated in a SONICATOR. Ultrasonication is used e.g. for the disintegration of cells in studies on microbial enzymes, etc. Sound waves of a wide range of frequencies have been examined for antimicrobial activity; in general, activity increases with frequency but there appears to be no single optimum microbicidal frequency. In general, ultrasonication is more effective against gram-negative than gram-positive cells; bacilli are more susceptible than cocci. The resistance of spores is much greater than that of vegetative cells. Mixed results have been obtained with viruses; virucidal activity may be influenced e.g. by the morphology of the virion. (Ultrasonication is not generally regarded as a reliable *sterilization* process.) The lethal effects of ultrasonication appear to be due to: (a) *Cavitation*—the formation of minute, transient pockets of gas or vapour in those regions of a liquid which correspond (at a given instant) to rarefactions in the transmitted sound waves. The enormous and abrupt changes in pressure which occur in these microenvironments during the formation and collapse of the cavities are believed to account for the disintegration of cells within or adjacent to the cavities. (The effects of *heat*—which is invariably produced by the absorption of ultrasonic energy—should be distinguished from the effects of cavitation.) (b) The formation of hydrogen peroxide and/or singlet oxygen during cavitation in an aerated medium. *kilocycles/second—also written kHz (kiloHertz); 16kc/s = 16kHz = 16,000 cycles/second. 16kc/s is the approximate upper limit of audibility in man.

ultraviolet microscopy (see MICROSCOPY)

ultraviolet radiation (effects on DNA) Ultraviolet radiation (UVR), of wavelengths in the region of 260 nm, is strongly absorbed by the purine and pyrimidine bases in nucleic acids. Such radiation may be MUTAGENic or lethal to organisms exposed to it—depending on organism, intensity of UVR, etc. The main effect is believed to involve the formation of pyrimidine dimers (particularly thymine dimers) in DNA; covalent bonds are formed between the 5,6-positions of two adjacent pyrimidines—a cyclobutane ring being formed between the two residues. A pyrimidine in such a dimer can no longer form hydrogen bonds with the corresponding purine of the base pair (see DNA)—thus localized distortion of the DNA duplex occurs. This results in the inhibition of replication and transcription which—in the absence of effective repair mechanisms—may be lethal to the cell. Several repair mechanisms are known—see DARK REPAIR, PHOTOREACTIVATION, RECOMBINATION-REPAIR; it has been suggested that UVR-induced mutations may arise as errors during the latter process. (see also STERILIZATION)

Ulva A genus of algae of the division CHLOROPHYTA; species occur in marine and brackish environments. The distromatic parenchymatous thallus forms a large sheet which may have undulating margins and may be more or less lobed; the thallus is generally attached to the substratum by a rhizoid-like holdfast and may reach a metre or more in length. The cells of the thallus are uninucleate and each contains one chloroplast in which occur one or more pyrenoids; cells of the holdfast are multinucleate. The life cycle is complex and involves an isomorphic alternation of generations.

umber codon (opal codon) The codon UGA (in mRNA) which codes for polypeptide chain termination—see PROTEIN SYNTHESIS and GENETIC CODE. (cf. AMBER CODON, OCHRE CODON)

Umbilicaria A genus of foliose LICHENS in which the phycobiont is *Trebouxia*. The thallus is discoid and may be entire or lobed; it is attached to the substratum at its centre. Apothecia are lecideine and have smooth or gyrose discs. Species include e.g.:

U. pustulata. The thallus is 3–6 centimetres in diameter and the upper surface has large convexities ('pustules') which correspond with hollows in the lower surface. The upper surface is olive-green when wet, brown when dry, and may bear clusters of black, extensively-branched isidia near the margin; the lower surface is pale to dark brown. *U. pustulata* occurs on rocks and walls in hilly areas.

U. cylindrica. The thallus may be discoid and undivided or deeply divided into lobes. The upper surface is greenish when wet, grey to brown when dry, and may have a mottled appearance; the lower surface is pinkish-brown to pale brown and may bear sparse or dense rhizinae. Cilia are usually present along the margins. Apothecia are generally present and have black gyrose discs. *U. cylindrica* occurs on rocks and walls—particularly in mountainous regions.

umbilicate (1) (*mycol.*) Of a pileus: having a small, localized, central depression. (2) (of a lichen thallus) Refers to a disc-like or plate-like thallus which is attached to the substratum by a central holdfast or *umbilicus*—e.g. in species of UMBILICARIA.

umbilicus (of a lichen) (see UMBILICATE)

umbo (*mycol.*) Of a pileus: having a central convex or conical swelling.

umbonate (*mycol.*) Possessing an UMBO.

UMP Uridine-5′-monophosphate (= uridylic acid).

unbalanced growth (see GROWTH (bacterial))

uncoupling agents Agents which dissipate the high energy intermediate formed during electron flow along the ELECTRON TRANSPORT CHAIN; uncoupling agents release electron transport from RESPIRATORY CONTROL. Thus, e.g. (according to the *chemiosmotic hypothesis*) lipid-soluble weak acids (e.g. 2,4-dinitrophenol) appear to function as proton conductors—dissipating the proton gradient generated by electron transport.

Other agents may *simulate* uncoupling agents under certain conditions—e.g. certain (artificial) electron carriers may allow electrons to by-pass a coupling site by accepting/donating electrons before/after a given coupling site; for example *menadione* can by-pass site I. (Electron flow may occur in the absence of oxidative phosphorylation if the high energy intermediate is used for an alternative energy-requiring process (e.g. the uptake of calcium ions).)

Undaria A genus of algae of the PHAEOPHYTA; species are used, in Japan, as food (*wakame*).

undecaprenol (see BACTOPRENOL)

undulant fever (see BRUCELLOSIS)

undulating membrane (*protozool.*) (1) In some flagellates (e.g. TRICHOMONAS, TRYPANOSOMA): a membrane which occurs between the cell body and the proximal part of the FLAGELLUM. (cf. STUMPY FORM) The undulating membrane appears to consist either of a double layer of cell membrane to which the flagellum has adhered, or a double layer of flagellar membrane which has fused with the cell surface. (2) In some *ciliates*: a row of cilia (see CILIUM) which beat in a co-ordinated manner and which (collectively) give the effect of a membrane.

ungulate (*mycol.*) Refers to the type of fruiting body which is hoof-like in shape.

unialgal Of algal cultures: containing only one species of alga. Such a culture is not necessarily axenic e.g. it may also contain bacteria.

unicellular (see CELLULAR)

uniport Refers to the TRANSPORT of a single solute across a membrane—using a specific carrier mechanism. (cf. ANTIPORT and SYMPORT)

uniseriate (*adjective*) Forming a single row.

unit membrane A conceptual structure proposed in the 1930s by Danielli and Davson for the arrangement of components in biological membranes. Essentially it consists of a bilayer of phospholipid molecules orientated with their fatty acid side-chains at right-angles to the plane of the membrane, and with the hydrophilic portions of the molecules forming the surfaces of the bilayer; a protein layer is supposed to be contiguous with each surface of the lipid bilayer. The term *unit-type membrane* is currently used to refer to any membrane in which a lipid bilayer is believed to be the essential structural component. (See CELL MEMBRANE (bacterial) for some current concepts in membrane structure.)

unitunicate (*mycol.*) (*adjective*) Refers to that type of ASCUS whose wall consists of a single functional layer, i.e. even if the wall possesses a laminar structure the layers do not move relative to one another during spore ejection. (cf. BITUNICATE)

universal bottle (see MCCARTNEY BOTTLE)

universal veil (*mycol.*) In certain agarics, e.g. *Amanita* spp: a membranous tissue which entirely covers the immature (button stage) fruiting body; during the development of the fruiting body the universal veil tears—leaving a cup-shaped fragment (the *volva*) surrounding the base of the stipe, and (sometimes) fragments (*scales*) adhering to the upper surface of the pileus. (see also PARTIAL VEIL)

Unna–Pappenheim stain The METHYL-GREEN PYRONIN STAIN.

UQ Ubiquinone: see QUINONES (in electron transport).

urea and derivatives (as antimicrobial agents) Urea (carbamide: $(NH_2)_2CO$) and certain of its derivatives (e.g. urethanes: $RNH.CO.OC_2H_5$) have mild antibacterial properties. (see however UREASES.) The *carbanilides* are active mainly against gram-positive bacteria; they appear to affect the functioning of the cell membrane. 3,4,4′-*Trichlorocarbanilide* (TCC: $C_6H_3Cl_2.NH.CO.NH.C_6H_4Cl$) is used in antiseptic soaps. *Diphenylthiourea* ($(C_6H_5NH)_2CS$) has some clinical application as an anti-mycobacterial agent. *Thiosemicarbazones* may be regarded as mono-substituted thioureas

($R_2C{=}N.NH.CS.NH_2$); *thiacetazone* (4-acetamidobenzaldehyde thiosemicarbazone) has been used in the treatment of tuberculosis. (see also ISATIN-β-THIOSEMICARBAZONE) Urea itself may be regarded as a CARBAMIC ACID DERIVATIVE.

ureases Enzymes which split urea into carbon dioxide and ammonia. Urease activity in bacteria can be detected by growth on a buffered medium containing urea, glucose, peptone, and PHENOL RED (*Christensen's agar*); when cultured on such a medium urease-producing strains give an alkaline reaction. *Proteus* spp give an alkaline reaction in a strongly buffered medium; a lightly buffered medium (e.g. Christensen's agar) detects (less vigorous) urease activity in a range of bacteria. (see also HEXAMETHYLENETETRAMINE)

Uredinales (rust fungi; 'rusts') An order of fungi of the class TELIOMYCETES; constituent species differ from those of the order USTILAGINALES (smut fungi) e.g. in that the teliospores of the rust fungi arise from the *terminal* cells of the sporogenous hyphae, and in that the basidiospores are formed on sterigmata and are forcibly discharged from the basidium. The rusts are obligate plant parasites which occur on a wide range of plants (including gymnosperms and angiosperms) and some rust fungi cause economically important diseases of certain crops—see e.g. CEREALS, DISEASES OF; certain species of rust fungi have been grown *in vitro*. The rusts are classified mainly on the basis of the *telial* stage; the one hundred or so known genera of rust fungi include GYMNOSPORANGIUM, MELAMPSORA, PUCCINIA, *Uromyces*, and *Xenodochus*. The morphology and life cycles of the rust fungi are described under RUSTS (1).

urediospores Uredospores; see RUSTS (1).

uredium (see RUSTS (1)).

uredosorus (*pl.* uredosori) A uredium—see RUSTS (1).

uredospores (urediospores) (see RUSTS (1))

urethritis Inflammation of the urethra; the causal agent may be any of a variety of organisms —e.g. the gonococcus (see GONORRHOEA). (see also NON-SPECIFIC URETHRITIS and NON-GONOCOCCAL URETHRITIS)

uridine-5′-monophosphate (biosynthesis of) (see Appendix V(b))

Urocystis A genus of fungi of the order USTILAGINALES.

uroid The posterior part of an amoeba; in some amoebae (e.g. *Pelomyxa*, *Trichamoeba*) this region is clearly differentiated and semi-permanent.

Uromyces A genus of fungi of the order UREDINALES.

uronic acids A class of compounds with the general formula $CHO(CHOH)_nCOOH$—i.e. compounds corresponding to aldoses (aldehyde sugars) in which the terminal primary alcohol group is oxidized. For example, the uronic acid corresponding to glucose is *glucuronic acid.*

Usnea A genus of fruticose LICHENS (the 'beard lichens') in which the phycobiont is *Trebouxia.* The thallus is filamentous, radially organized in cross-section, and has a dense axial cord. The apothecia are lecanorine. Species include e.g.:

U. florida. The thallus is grey-green, branching, erect and stiff, and the surface is rough with small papillae; isidia and soredia are absent. Apothecia occur at the ends of branches and are frequently numerous; the discs are fawn-coloured, up to 1 centimetre in diameter, and bear a fringe of hair-like projections. *U. florida* occurs on trees.

U. subfloridana. The thallus is extensively branched, erect or drooping, grey-green with a blackened base; the surface is rough—bearing papillae on which develop clusters of elongated isidia. *U. subfloridana* occurs on trees and occasionally on rocks and walls.

usnic acid An antibiotic produced by certain lichens (e.g. *Usnea* spp, *Cetraria islandica*). Usnic acid is active against many gram-positive bacteria (including *Mycobacterium tuberculosis*) and some fungi. It has been used topically for the treatment of burns, skin diseases etc. and against certain fungal diseases of plants. Although relatively non-toxic to man, usnic acid can cause an allergic skin reaction in sensitive individuals.

ustilagic acids Insoluble extracellular glycolipids produced by *Ustilago* spp cultured in aerated liquid media; ustilagic acids are mixtures of β-D-cellobiosides of ustilic acids (di- and tri-hydroxyhexadecanoic acids)—the cellobiose residues being acylated. Ustilagic acids inhibit gram-positive and gram-negative bacteria but have no clinical use; they are used e.g. in the perfume industry for preparing macrocyclic musks.

Ustilaginaceae (see USTILAGINALES)

Ustilaginales (smut fungi; 'smuts') An order of fungi of the class TELIOMYCETES; constituent species differ from those of the order UREDINALES (rust fungi) e.g. in that the teliospores of the smut fungi commonly arise from the *intercalary* cells of the sporogenous hyphae, and in that the basidiospores are sessile (i.e. are not formed on sterigmata) and are not forcibly discharged from the basidium. The smuts are parasitic on a wide range of angiosperms, and some species cause economically-important diseases of certain crops—see e.g. SMUTS (2) and CEREALS, DISEASES OF; a number of species of smut fungi have been cultured on artificial media. Species which form (tranversely) *septate* basidia from which basidiospores arise *laterally* are placed in the family Ustilaginaceae; species which form basidiospores *apically* on an *aseptate* basidium are placed in the family Tilletiaceae. The genera of smut fungi include *Sphacelotheca*, TILLETIA, *Urocystis*, and USTILAGO; the morphology and life cycles of the smuts are described under SMUTS (1).

Ustilago A genus of fungi of the family Ustilaginaceae, order USTILAGINALES. *Ustilago* spp are plant parasites and some species are important pathogens—e.g. *U. hordei*, *U. kolleri*, *U. nuda* (see CEREALS, DISEASES OF); the smut *sori*—see SMUTS (1)—may develop in a range of locations in the plant host. *Typical life cycle*. On germination the *teliospore* gives rise to a septate promycelium from which bud (laterally and apically) an indefinite number of (haploid) basidiospores ('sporidia'); dikaryotization may occur in any of a variety of ways (e.g. by the fusion of germ tubes from two basidiospores of appropriate mating types) and a fresh host becomes infected by the resulting *dikaryotic* mycelium. In species which do not form basidiospores (e.g. *U. nuda*) the teliospore germinates to form hyphae among which cell fusion (dikaryotization) occurs prior to the infection of a fresh host. *Spore balls* are not formed.

V

V factor A chemical factor which is essential for the growth of certain species of *Haemophilus* (e.g. *H. influenzae*); the V factor requirement is satisfied by nicotinamide adenine dinucleotide (NAD) or by NADP. The V factor may be prepared as the filtrate of a boiled suspension of *Saccharomyces cerevisiae*; it is destroyed at 121 °C/30 minutes and is inactivated e.g. by the culture filtrates of those strains of *Streptococcus* which produce NADase. CHOCOLATE AGAR contains sufficient V factor to permit the growth of NAD(P)-requiring species of *Haemophilus*. (see also X FACTOR and SYNTROPHISM)

V forms (see V-W TRANSITION)

V-region (of an ANTIBODY) The *variable* region: the N-terminal portions of the HEAVY CHAINS and LIGHT CHAINS in which the amino acid constituents exhibit great variability from antibody to antibody. Within the V-region are found the so-called *hypervariable regions*—in which the variability of the amino acids is highest; the antibody COMBINING SITE is associated with these regions. The *C-region* (constant region) of an antibody is that region outside the V-region; the amino acids in the C-region are much less variable. Thus, the heavy (H) and light (L) chains are each divided into variable and constant regions—the basic antibody molecule consisting of C_H, C_L, V_H, and V_L sections. (V_H and V_L both occur on the FAB FRAGMENT. The C_H region consists of C_H1—on the Fab fragment—and C_H2, C_H3, both in the Fc region.)

vaccination IMMUNIZATION carried out prophylactically by the oral or parenteral administration of a VACCINE.

vaccine Any antigenic preparation administered with the object of stimulating the recipient's specific defence mechanisms in respect of given pathogen(s) or toxic agent(s). Some vaccines (e.g. the SABIN VACCINE) are given orally while others (e.g. the SALK VACCINE) are administered parenterally. Vaccines fall into four main categories. (a) INACTIVATED VACCINES e.g. vaccines against *typhoid* and *cholera*. (b) Those comprising suspensions of live (attenuated) pathogens, e.g. vaccines against *yellow fever* and *tuberculosis*. (see BCG) (c) TOXOIDS. (d) Those comprising a solution or suspension of the *antigenic* extracts of specific pathogens, e.g. the vaccine containing polysaccharide capsular material of *Streptococcus pneumoniae*. (see also ANTIGENIC COMPETITION and AUTOGENOUS VACCINE)

vaccinia virus (see POXVIRUS)

vacuoles See separate entries: CONTRACTILE VACUOLE; FOOD VACUOLE; GAS VACUOLE.

vagina, common microflora of The common microflora of the adult (*pre-menopause*) human vagina includes Lancefield group *D* streptococci, coliform bacteria, diphtheroids, staphylococci, lactobacilli, and certain yeasts—e.g. *Candida albicans*; during this time the vagina is an acidic environment. Between birth and puberty, and post-menopause, the vagina is alkaline; under these conditions the predominating organisms include streptococci, diphtheroids, and coliforms. At birth the vagina is normally sterile.

valency (*immunol.*) (1) The number of COMBINING SITES per antibody, or the number of DETERMINANTS per antigen. (2) The number of different antigens in a POLYVALENT VACCINE or of specific antibodies in a POLYVALENT ANTISERUM.

valine (biosynthesis of) (see Appendix IV(b))

valinomycin (see DEPSIPEPTIDE ANTIBIOTICS)

Valonia A genus of algae of the CHLOROPHYTA.

vancomycin A glycopeptide ANTIBIOTIC produced by certain actinomycetes. It inhibits PEPTIDOGLYCAN biosynthesis in a wide range of gram-positive bacteria; gram-negative species appear to be resistant. Vancomycin prevents the transfer of disaccharide-pentapeptide subunits from the bactoprenol to the growing cell wall.

vanillin 3-methoxy-4-hydroxybenzaldehyde. The *vanillin reagent* (5% vanillin in 95% ethanol) is used in the identification of *Clostridium* spp; 10 drops of acidified reagent (50% reagent, 50% conc HCl) are added to 5 ml of the culture, shaken, and left at room temperature for 5–15 minutes. The production of a deep violet colour is a positive reaction.

vapam Synonym of METHAM SODIUM.

var An abbreviation (of 'variant' or 'variety') used in microbial nomenclature to denote a particular atypical—or otherwise distinctive—strain of a given, named species; thus, e.g. *Saccharomyces cerevisiae* var *ellipsoideus*.

variant A strain which differs in some (often minor) way from a particular, named microorganism.

variation (antigenic) (see ANTIGENIC VARIATION)

variation in bacteria Considerable variation, in genotype and/or phenotype, may occur among individuals of a single species. Such variation may arise through: (a) ADAPTATION. (see also FLUCTUATION TEST, NEWCOMBE EXPERIMENT, and SMOOTH-ROUGH VARIATION) (b) *Recombination*—which may occur during CONJUGA-

TION, TRANSDUCTION, and TRANSFORMATION. (c) Neutral mutations (see genetic DRIFT).

varicella see CHICKENPOX)

variegation (*plant pathol.*) Patchy or otherwise irregular colour variation in leaves, petals or other plant components; variegation may be the expression of an inherited characteristic or it may be a disease symptom. (see BREAKING and MOSAIC)

variola Any of certain diseases, caused by poxviruses, in which the symptoms include characteristic skin lesions—e.g. SMALLPOX in man; sheep-pox (variola ovina) is a NOTIFIABLE DISEASE in the UK.

variola virus (see POXVIRUS)

Vaucheria A genus of siphonaceous algae of the division XANTHOPHYTA. Species occur as dark-green mats in pools or running water or on damp soil; the majority of aquatic species occur in fresh water while a few are marine. The thallus of the mature organism is elongate and tubular (tube diameter of the order of 0.05 mm) with infrequent branching; the tubular, cellulosic wall is lined with a thin layer of protoplasm which contains chloroplasts and numerous nuclei. A large central vacuole is continuous throughout the length of the tube. In terrestrial species the thallus may be attached to the substratum by rhizoid-like branches. Asexual reproduction generally involves zoospore formation: the tip of a branch is closed off by a septum to form a zoosporangium—within which the (multinucleate) protoplast develops one pair of flagella for each nucleus it contains. The multinucleate, multiflagellate zoospore thus formed is released and remains motile for a short time; subsequently, the flagella are lost and germination occurs to form a new thallus. Sexual reproduction is oogamous; the zygote develops a thick wall and remains dormant for some time prior to germination—at which time meiosis occurs.

VCN agar (see THAYER-MARTIN SELECTIVE MEDIUM)

V.D. (see VENEREAL DISEASES)

VDRL test (Venereal Diseases Research Laboratory Test) A serological screening test for SYPHILIS. One drop of a suspension of cardiolipin-lecithin-cholesterol particles (the antigen) is added to one drop of the patient's *inactivated* serum on a slide; mixing is conducted under standard conditions for a fixed period of time. Agglutination (aggregation of the antigen particles into clumps) occurs in the presence of antibodies homologous to *Treponema pallidum*—but BIOLOGICAL FALSE POSITIVE REACTIONS can occur. The slide is examined under the microscope (× 100); large clumps are reported as 'reactive', small clumps as 'weakly reactive' and a regular, or finely granular suspension of antigen as 'non-reactive'. Strongly-positive sera may exhibit a PROZONE effect.

vectors (of disease) Organisms which are involved in the *spread* of disease from one individual to another; disease-prevention and disease-control measures are often directed against a specific vector or vectors. Examples of vectors include blood-sucking flies, mosquitoes, ticks, lice etc.; epidemiologically, human CARRIERS of disease may be regarded as vectors.

In *biological* vectors, e.g. *Anopheles* mosquitoes in MALARIA, the pathogen (*Plasmodium* spp in malaria) undergoes, within the vector, one or more specific stages of development in its life cycle. The transmission of a pathogen by a biological vector is referred to as *cyclical transmission*. If the pathogen does not undergo such changes within the vector, the latter is termed a mechanical vector; transmission is referred to as *non-cyclical* or *mechanical*.

Examples of vector-borne diseases include: BOUTONNEUSE; CHAGAS' DISEASE; DENGUE; DUTCH ELM DISEASE; KALA-AZAR; MYXOMATOSIS; OROYA FEVER; RELAPSING FEVER; ROCKY MOUNTAIN SPOTTED FEVER; SCRUB TYPHUS; SLEEPING SICKNESS; SURRA; TRENCH FEVER; TULARAEMIA; TYPHUS FEVERS; YELLOW FEVER. (see also VIRUS DISEASES OF PLANTS)

vegetative (assimilative; somatic; trophic) (of cells etc.) Cells etc. which are engaged in nutrition and growth—i.e. not reproductive or dormant forms.

vegetative bacteriophage The intra-bacterial genome (nucleic acid) of a phage during the *eclipse period* of lytic infection. (see also ONE-STEP GROWTH EXPERIMENT)

Veillon tube (see SHAKE CULTURE)

Veillonella A genus of gram-negative, anaerobic, chemoorganotrophic, non-haemolytic bacteria which are parasitic in the alimentary tract (particularly the mouth) in man and other animals. The cells, which are non-motile cocci, about 0.5 microns or less in diameter, occur singly, in clusters, or as diplococci with adjacent sides flattened. The organisms ferment lactate, pyruvate and other substrates but do not attack sugars. Type species: *V. parvula*.

Veillonellaceae A family of gram-negative, anaerobic cocci which includes the genera VEILLONELLA, *Acidaminococcus*, and MEGASPHAERA.

vein-banding (*plant pathol.*) The occurrence of a narrow band of tissue of altered colouration on either side of the main veins of a leaf—a symptoms of certain VIRUS DISEASES OF PLANTS.

vein-clearing (*plant pathol.*) The clearing (i.e. increased transparency) of the veins in a leaf—an early symptom of certain VIRUS DISEASES OF PLANTS.

venereal diseases (V.D.) Sexually-transmitted diseases—see e.g. CHANCROID, GONORRHOEA, SYPHILIS.

venting The creation of a vertical slit (2–3 mm in depth) in the wall of a plastic petri dish—e.g. by pressing on the edge of the wall with a red-hot STRAIGHT WIRE. Venting is carried out when the material within the petri dish is to be incubated under anaerobic or other non-atmospheric conditions. The slit permits effective communication between the inside of the (closed) petri dish and the ambient gaseous environment—thus preventing the trapping of air within the petri dish.

Venturia A genus of fungi of the class LOCULOASCOMYCETES; *Venturia* spp include a number of plant parasites and pathogens—e.g. *V. inaequalis*, the causal agent of apple scab (see APPLE, DISEASES OF).

V. inaequalis. The host range of *V. inaequalis* includes a number of species of the genus *Malus*. Infection occurs during the spring when an ascospore germinates on the leaf of a susceptible plant and subsequently gives rise to a sub-cuticular stroma. Shortly afterwards the stroma produces numerous short, erumpent conidiophores—each of which bears a single conidium which, in shape, resembles a candle flame. The production of conidia continues until ascocarp formation commences during the late summer or early autumn; the ascocarps–individual pseudoperithecia—develop within the tissues of the dead and dying leaves of the host plant. During ascocarp formation an ascogonium appears within the developing stroma, and a trichogyne subsequently projects through the stromal initial. *V. inaequalis* is an heterothallic organism, and the ascogonium receives male nuclei (*via* the trichogyne) from the antheridium of an adjacent thallus of appropriate mating type; the transfer of nuclei takes place following gametangial contact and the formation of a pore which connects the cytoplasm of the two gametangia. The pseudoperithecium bears a number of setae around the rim of the ostiolar pore; within the ascocarp there is a layer of asci and pseudoparaphyses. The ascospores are yellow or yellowish-green; each has a single septum which divides the spore into two cells which are unequal in size.

vermicule (1) The ookinete of *Plasmodium*. (2) (of *Babesia*) Each of the uninucleate, pyriform or elongated motile cells produced by schizogony in the gut of the insect vector.

Verpa A genus of fungi of the order PEZIZALES; in *Verpa* spp the fruiting body (an APOTHECIUM) consists of a small campanulate (bell-shaped) pileus which develops on a relatively large stipe—the tip of the stipe being attached to the inside of the pileus at its centre. The hymenial layer of asci develops on the outer surface of the pileus.

verruca (see WART)

Verrucaria A genus of crustose LICHENS in which the phycobiont is a green alga; the genus includes aquatic and saxicolous species. The *perithecia* may be superficial or immersed in the thallus. Species include e.g.:

V. maura. The thallus is black, thin and smooth—covering rocks on the seashore in intertidal and supratidal zones.

V. aquatilis is a freshwater species; the brownish-black, thin thallus occurs on rocks in streams.

V. hochstetteri occurs on calcareous rocks, walls etc.; the thallus is white to grey and is generally ENDOLITHIC—only the tips of the (immersed) perithecia being apparent at the rock surface. The perithecia are of the order of 0.5–1 millimetre in diameter; after discharge the perithecia leave pits in the rock surface—each pit being lined by the black excipulum.

verrucose (*adjective*) Refers to a surface which bears wart-like protuberances.

Versene (Versine; Sequestrene) Sodium ethylenediaminetetraacetate. See EDTA.

vertical resistance (*plant pathol.*) If a cultivar is significantly more resistant to some races than to others of a given pathogen it is said to exhibit vertical resistance with respect to that pathogen; when the cultivar exhibits resistance to a given race of the pathogen the resistance is usually complete. Vertical resistance tends to be temporary resistance—i.e. a cultivar resistant to the existing races of a given pathogen will become susceptible when a new race of the pathogen develops. (cf. HORIZONTAL RESISTANCE)

vertical transmission (of a parasite) Transmission from a parent to its offspring during the reproductive process, e.g. transovarial or transplacental transmission. (cf. HORIZONTAL TRANSMISSION)

verticillate (*adjective*) Whorled; thus, e.g. of spore chains, refers to a number of chains arranged radially about a given point of the sporophore.

Verticillium A genus of fungi within the order MONILIALES; species include a number of plant pathogens (see e.g. WILTS). The vegetative form of the organisms is a septate mycelium; small, hyaline, non-septate conidia are formed at the tips of the whorled branches of the conidiophores.

vesicular stomatitis An infectious disease which affects cattle, pigs, and horses. Clinically, vesicular stomatitis is identical to FOOT AND

MOUTH DISEASE; the causal agent, a virus of the genus RHABDOVIRUS, appears to enter the body *via* wounds, abrasions etc. in the skin or in the mucous membranes of the mouth. *Laboratory diagnosis* may involve e.g. culturing the virus and/or the performance of a complement fixation test.

vestibulum (vestibule; pre-oral depression) (*ciliate protozool.*) In some ciliates, a depression in the body surface leading either directly to the CYTOSTOME or to the BUCCAL CAVITY. The vestibular surface may not differ greatly from that of the remainder of the pellicle.

Vi antigens Polysaccharide *microcapsular* antigens found on the cells of certain strains of *Salmonella typhi*, *S. paratyphi C* and *Citrobacter*, and certain atypical strains of *Escherichia coli*. Vi antigens may play some part in the virulence of pathogenic species in relation to certain host animals. The Vi antigen of *S. typhi* has been identified as a polymer of *N*-acetylgalactosaminuronic acid linked α-$(1 \rightarrow 4)$.

Cells bearing Vi antigens are agglutinated by antibodies to the Vi antigen; however, since they are situated at the cell surface, Vi antigens mask the somatic (O) antigens and inhibit agglutination by O-antisera. (The extent of this inhibition is variable.) Vi antigens are destroyed by heating to 100 °C for 30 minutes; cells thus treated may subsequently be agglutinated by homologous O-antiserum.

Vi phage typing (see TYPING)

viable cell count (see ENUMERATION OF MICROORGANISMS)

Vibrio A genus of gram-negative, oxidase-positive, freshwater or marine bacteria of the VIBRIONACEAE; the cells, which are polarly flagellate or (infrequently) non-motile, are curved or straight rods, some 0.5 × 1–5 microns. Most strains are sensitive to O/129, and the majority liquefy gelatin; no strain produces gas from glucose. All vibrios are sensitive to acidity, but can grow in media of pH 9–10. Growth occurs on ordinary nutrient media (e.g. nutrient agar) at temperatures ranging from about 20 °C to 40 °C. Some species form BACTERIOCINS. Type species: *V. cholerae*; GC% range for the genus: approximately 40–50.

V. cholerae (*V. comma*). Certain biotypes are the causal agents of classical or asiatic CHOLERA; pathogenic strains form a NEURAMINIDASE and a potent ENTEROTOXIN. Isolation from clinical specimens (e.g. faeces) may be carried out e.g. on MONSUR AGAR (see also MONSUR TRANSPORT MEDIUM). The cells are polarly flagellate, curved rods about 0.5 × 1–4 microns. All strains are catalase positive, and most are sensitive to O/129 and to NOVOBIOCIN. On nutrient agar colonies are round, entire, glossy, translucent. On blood agar, the colony of *V. cholerae* biotype *cholerae* is surrounded by a narrow zone of partial clearing (*haemodigestion*); the colony of (most strains of) *V. cholerae* biotype *eltor* is surrounded by a zone of complete hydrolysis.

Strains of *V. cholerae* are divided (a) into 6 serological groups (*serogroups*) based on somatic (O) antigens, or (b) into 4 *biotypes* based on a range of properties. The *cholera vibrios* (*V. cholerae* biotype *cholerae* and *V. cholerae* biotype *eltor*) form serogroup 1 (O:1); the *non-cholera vibrios* (*V. cholerae* biotype *proteus* and *V. cholerae* biotype *albensis*) form the remaining 5 serogroups. Serogroup 1 is further divided on the basis of 3 group antigens, A, B and C. *Inaba* serotypes have antigens A and C, *Ogawa* serotypes have antigens A and B, and *Hikojima* serotypes have antigens A, B and C.

V. cholerae biotype *cholerae* strains are sensitive to POLYMYXIN B and to the group IV cholera phages; they give a positive CHOLERA RED TEST.

V. cholerae biotype *eltor* (the El Tor vibrio). Strains are insensitive to polymyxin B and to the group IV cholera phages; they give a positive cholera red test. (The *cholerae* and *eltor* biotypes are susceptible to desiccation, and are killed e.g. in 15 minutes at 60 °C; they are sensitive to tetracyclines and other antibiotics.)

V. cholerae biotype *proteus*. A non-cholera vibrio which may cause gastroenteritis. Strains give a negative cholera red test.

V. cholerae biotype *albensis*. A non-cholera, luminescent vibrio; strains give a negative cholera red test.

Other *Vibrio* spp include:

V. fischeri (formerly *Achromobacter fischeri*). A luminescent marine species. (see also BIOLUMINESCENCE)

V. parahaemolyticus. A halophilic, marine species—similar to *V. cholerae*—which may cause gastroenteritis and other diseases; strains give a negative cholera red test, and are not agglutinated by the serogroup 1 antiserum.

V. fetus: see CAMPYLOBACTER.

vibriocins (see BACTERIOCINS)

Vibrionaceae A family of gram-negative, oxidase-positive, aerobic/facultatively anaerobic*, chemoorganotrophic, marine or freshwater bacteria; the cells are straight or curved rods which are usually polarly flagellate. The genera are AEROMONAS, *Lucibacterium*, PHOTOBACTERIUM, PLESIOMONAS and VIBRIO. *Metabolism may be oxidative or fermentative.

victorin A phytotoxin produced by *Helminthosporium victoriae*, the causal agent of Victoria blight of oats. Victorin does not affect plants resistant to Victoria blight, but is toxic to sensitive plants—apparently causing an increase in cell permeability.

Vincent's angina (trench mouth; Vincent's disease) An ulcerative disease of the mouth and pharynx; *Treponema vincentii* (formerly *Borrelia vincentii*) and *Bacteroides* spp are frequently isolated from the lesions, but the exact identity of the causal agent(s) is still doubtful.

Vinckeia A sub-genus of PLASMODIUM.

vinegar A condiment prepared by the microbial oxidation of ethanol to acetic acid. Ethanol is obtained by the fermentation of malted barley or grape juice. Acetification is carried out by a mixed flora which includes species of ACETOBACTER; the flora may be grown as a surface film on the alcoholic liquor, dispersed in a strongly agitated liquor, or used to coat twigs or shavings of wood over which the liquor slowly trickles. Acetification is an exothermic process, and the system must be cooled to maintain a temperature of about 30–40 °C. Aeration is controlled to prevent microbial oxidation of acetic acid to carbon dioxide. The product is allowed to mature for several weeks, or months, and is filtered, bottled, and pasteurized. During manufacture, spoilage may be caused by the slime-forming species *Acetobacter xylinum*—particularly under sub-optimal temperatures.

Vioform 5-Chloro-7-iodo-8-HYDROXYQUINOLINE.

violacein A violet pigment which occurs in *Chromobacterium* spp; it consists of 2 substituted indole nuclei linked to a pyrrolone residue. Violacein is soluble in ethanol but not in water or chloroform.

viraemia The presence of viruses in the blood stream.

viral hepatitis (see HEPATITIS)

viral infectivity (see END POINT (*virol.*))

viridans streptococci Those strains of *Streptococcus* which give rise to α-haemolysis, i.e. the colony (on blood agar) is surrounded by a zone in which the blood exhibits a greenish discoloration, and in which *some* erythrocytes may have been lysed.

virion (*virol.*) A single, structurally-complete ('mature') virus. (see also VIRUSES)

virogene (see ONCOGENE HYPOTHESIS)

viroids The causal agents of certain diseases—e.g. potato spindle tuber disease; these agents resemble VIRUSES in some ways—but differ from them e.g. in the apparent lack of virus-like structural organization, and in their resistance to a wide range of treatments to which viruses are usually sensitive. The viroid of potato spindle tuber disease appears to be a fragment of single-stranded ribonucleic acid which is closely associated with cell-derived material(s).

viropexis (*virol.*) A term used by some authors to refer to the entry into a cell of one or more virions by the process of PHAGOCYTOSIS.

viroplasmic matrix (see POXVIRUS)

viroplasms Electron-dense bodies which occur within the cytoplasm of cells infected with certain viruses. Viroplasms consist of—or include—accumulations of virus particles or viral components.

viroplasts Synonym of X BODIES.

virulence (of pathogenic microorganisms) The capacity of a pathogen to cause disease—defined broadly in terms of the severity of symptoms in the host. Thus, a highly-virulent strain may cause severe symptoms in a susceptible individual, while a less virulent strain would produce relatively less severe symptoms in the same individual. The observed development of a disease may be regarded as the joint expression of the degree of INFECTIVITY and virulence of the pathogen and the state of immunity (either natural or acquired) of the host in relation to the pathogen.

virulent (of bacteriophages) (see BACTERIOPHAGE)

virus (see VIRUSES)

virus diseases of plants Virus diseases occur in a wide range of plant species. Such diseases may kill the host, but more often result in stunting or general loss of vigour—frequently accompanied by one or more of the following symptoms: CHLOROSIS (see also YELLOWS), ENATIONS (and other GALLS), LOCAL LESIONS, MOSAIC, MOTTLE, RING SPOT, ROSETTE, STREAK, TOP NECROSIS, VEIN-BANDING, and VEIN-CLEARING. Diseased plants often exhibit abnormal *intracellular* bodies; these may include X BODIES, crystalline bodies, and particles which may be abnormal cell organelles or viral protein coats etc.

In a systemic disease the symptoms may be confined to the leaves and stem; plant components fully matured prior to infection may not exhibit symptoms. Non-systemic diseases are generally characterized by local lesions. Within the plant, virus may spread rapidly, *via* the phloem, or slowly, *via* the plasmodesmata. Some virus-infected plants may appear to be free of symptoms—although some stunting may occur; however, disease development may subsequently occur under suitable environmental conditions. Some plants are *normally* virus-infected—e.g. the King Edward variety of potato harbours the potato paracrinkle virus.

Plant-pathogenic viruses. These usually contain RNA (cf. CAULIFLOWER MOSAIC VIRUS). A virus may occur in a variety of modified forms—differing e.g. in the symptom(s) they produce in a given host, and/or in serological and other properties. In general, plant viruses exhibit a low degree of host specificity. A plant-pathogenic virus is often named according to its common host and the main symptom(s) it produces in that host: e.g. *peach rosette mosaic virus*. The identity of a viral pathogen can seldom be deduced from the symptoms it causes; thus, e.g. unrelated viruses may cause similar symptoms in a given plant, and a given virus may cause different symptoms in the same plant under different conditions. Mixed viral infections can also occur. Identification may be assisted e.g. by electron microscopy; see also DIFFERENTIAL HOST and LOCAL LESION HOST. Many viruses are highly immunogenic when injected into animals; the specific anti-virus antibodies thus obtained can be used to identify a virus by any of a variety of serological tests—e.g. precipitation, complement-fixation, gel diffusion.

Transmission of plant viruses. Viruses may be transmitted from plant to plant in a variety of ways—although the mode of transmission of a given virus may be specific and characteristic. Transmission may occur: (a) By insect vectors, such as aphids (e.g. *Myzus persicae*), leaf-hoppers, white-flies. (In some insects transovarial acquisition of virus has been demonstrated.) Insect-transmitted viruses may be STYLET-BORNE and/or CIRCULATIVE. (b) Mechanically (*sap-transmission*): the virus is introduced into abrasions in the leaves or stems of plants; many, but not all, viruses are mechanically transmissible. (Experimentally, this may be simulated by rubbing the leaves of a test plant with a virus preparation—often containing a mild abrasive.) (c) *Via* the soil, fungi, or nematode worms. (d) *Via* the seed or pollen. (e) *Via* parasitic plants, e.g. the dodder (*Cuscuta* spp). (f) By grafting—e.g. infection of the stock by an infected scion.

Control of virus diseases of plants. Viruses in a systemically-infected plant can sometimes be inactivated by subjecting the plant to air temperatures of 30–35 °C (for days/weeks) or to water-immersion at 50 °C for shorter periods. There is no known effective chemical treatment. The control of virus diseases may involve: (a) The use of virus-free seed. (b) Control of vectors. (c) ROGUING. (d) The development of resistant or tolerant cultivars. (e) The development of virus-free apical meristem cultures. (see also BARLEY YELLOW DWARF VIRUS; BEET YELLOWS VIRUS; COVIRUS; LETTUCE BIG VEIN VIRUS; POTATO VIRUS Y; RICE DWARF VIRUS; SATELLITE VIRUS; SUGARCANE FIJI DISEASE VIRUS; THERMAL INACTIVATION POINT; TOBACCO NECROSIS VIRUS; TOMATO RING SPOT VIRUS; TURNIP YELLOW MOSAIC VIRUS; WOUND TUMOUR VIRUS; PLANT PATHOGENS AND DISEASES)

virus pneumonia of calves (see CALF PNEUMONIA)

viruses (archaic: 'filterable viruses') A *virus* is a non-cellular entity which consists minimally of protein and nucleic acid (DNA *or* RNA) and which can replicate only after entry into specific type(s) of living cell; such an agent has no intrinsic metabolism—replication being dependent on the direction of cellular metabolism by the viral genome. Within the host cell viral components are synthesized separately and are assembled, intracellularly, to form mature, infectious viruses (*virions*).

According to type, a virus may infect a particular kind of animal, plant, fungus, blue-green alga (see CYANOPHAGE), bacterium (see BACTERIOPHAGE), or (reportedly) protozoan. Some types of virus invariably cause the death and lysis of the host cell; others normally replicate without causing loss of viability of the host cell—which may continue to grow during viral replication. (see also LYSOGENY) The infection of animal cells by certain types of virus may lead to cell fusion (syncytia formation) or cell TRANSFORMATION (including TUMOUR-formation). The infection of organisms by viruses may or may not give rise to symptoms of disease—depending on e.g. the type of virus involved. Diseases of viral aetiology include: CHICKENPOX, HERPANGINA, INFLUENZA, MEASLES, MUMPS, MYXOMATOSIS, POLIOMYELITIS, RABIES, RINDERPEST, RUBELLA, SMALLPOX, and SWINE FEVER. (see also INSECT DISEASES and VIRUS DISEASES OF PLANTS)

Size of viruses. The smallest viruses (e.g. picornaviruses, phage ϕX174) are approximately 20–25 nm in diameter. The largest non-filamentous viruses (ENTOMOPOXVIRUSES) include some of 450 nm (maximum dimension) while e.g. the filamentous bacteriophage f1 is 800 × 5 nm.

Composition of viruses. Many viruses consist largely, or solely, of nucleic acid (the viral genome) and protein—while others contain, in addition, appreciable amounts of e.g. carbohydrate and/or lipid. The nucleic acid may be single-stranded or double-stranded, linear or circular (covalently closed) DNA or RNA—according to the type of virus. In certain viruses—e.g. orthomyxoviruses, reoviruses—the genome is fragmented, i.e. it occurs within the virion in a number of separate pieces. The virions of some taxa contain certain enzymes—e.g. DNA-dependent RNA polymerase (in poxviruses), RNA-dependent RNA polymerase (in e.g. orthomyxoviruses), RNA-dependent DNA polymerase (in leukoviruses). (see also NEURAMINIDASES)

Structure of viruses. In a mature virus particle (virion) the genome is enclosed by a protein coat (see CAPSID and NUCLEOCAPSID); in some types of virus (e.g. herpesviruses, orthomyxoviruses) the capsid or nucleocapsid is enclosed by a lipid-containing ENVELOPE. (see also BACTERIOPHAGE) (*Populations* of certain types of virus (e.g. polioviruses, the tobacco mosaic virus) can be obtained in crystalline form *in vitro*; some viruses (e.g. adenoviruses) form intracellular crystalline arrays.)

Morphology of viruses. Animal-infecting viruses ('animal viruses') which are enveloped (e.g. orthomyxoviruses, paramyxoviruses, herpesviruses, arenaviruses) are roughly spheroidal or ellipsoidal or, in the case of rhabdoviruses, bullet-shaped or bacilliform; some enveloped viruses (e.g. herpesviruses) contain an icosahedral capsid, while others (e.g. orthomyxoviruses) contain a nucleocapsid. (Animal viruses which contain a nucleocapsid are invariably enveloped.) The non-enveloped animal viruses (e.g. adenoviruses, parvoviruses, orbiviruses) have icosahedral capsids. *Plant*-infecting viruses ('plant viruses') include some with naked icosahedral capsids (e.g. sugarcane Fiji disease virus, tobacco necrosis virus, wound tumour virus) and some which are rod-shaped or filamentous (e.g. the tobacco mosaic virus—a naked nucleocapsid). Enveloped plant viruses include the eggplant mottled dwarf virus—a bacilliform rhabdovirus.

(See also entry BACTERIOPHAGE for general details of phage size, composition, structure and morphology.)

Culture of viruses. (See VIRUSES, CULTURE OF; see also BACTERIOPHAGE.)

Assay of viruses. (See ENUMERATION OF MICROORGANISMS; END POINT (*virol.*); LOCAL LESION HOST; PLAQUE ASSAY.)

Preservation of viruses. Methods commonly used for the preservation and storage of viruses include freezing (at −20 °C or, better, at temperatures in the regions −70 °C or −200 °C) and lyophilization (freeze-drying). Such methods are not equally suitable for all viruses—e.g. lyophilization is reported to cause a marked loss of infectivity in some viruses (e.g. picornaviruses, eggplant mottled dwarf virus) but not in others (e.g. herpesviruses, turnip rosette virus). For many types of virus, a suitable medium for lyophilization or freezing consists of a balanced salt solution supplemented with protein.

Inactivation of viruses. The degree of susceptibility to particular physical and chemical virus-inactivating agents varies widely among the different groups and strains of viruses. Enveloped viruses are generally inactivated by lipid solvents (e.g. ether) and by lipases and detergents; non-enveloped viruses are commonly insensitive to these agents. Although many types of virus are inactivated by extremes of pH enteroviruses remain infective after exposure to pH in the range 3–10. Some viruses (e.g. paramyxoviruses, eggplant mottled dwarf virus) are readily inactivated by desiccation; others are resistant to desiccation—e.g. poxviruses can survive for months in dried scabs etc. Viruses may be inactivated by homologous antisera. (see also ANTIVIRAL AGENTS; STERILIZATION; BACTERIOPHAGE)

The Viral Growth Cycle (the sequence of events which begins with the adsorption of a virus to a susceptible cell, and ends with the release of mature virions from the cell). The growth cycle of a *bacteriophage* is usually completed within minutes (see BACTERIOPHAGE for further details); the growth cycle of an animal virus occupies a number of hours. *Adsorption* (of virus to host cell) involves contact between specific attachment sites on the virion and specific receptor sites on the cell; a cell which lacks the receptor sites appropriate to a particular virus normally cannot be infected by that virus. Typically, animal viruses adsorb optimally in the presence of divalent cations. *Entry* (*penetration*) (of the virion—or part of the virion—into the host cell). Animal viruses appear to enter the cell in one or other (or one or more) of the following ways: PHAGOCYTOSIS by the cell (referred to as 'viropexis' by some authors); direct passage through the cell membrane—reported to occur in some naked icosahedral viruses (e.g. adenoviruses); (in enveloped viruses) fusion of the viral envelope with the cell membrane. Plant viruses presumably enter cells *via* the plasmodesmata. *Uncoating* (the release of the genome from the virion). The mechanisms of uncoating are poorly understood but host cell enzymes commonly appear to be involved for at least part of the process; while some viruses (e.g. poxviruses) are uncoated intracellularly, others (e.g. polioviruses, some enveloped viruses) are uncoated at the cell membrane. *Transcription of the Viral Genome.* (For a generalized account of TRANSCRIPTION and TRANSLATION see entries RNA and PROTEIN SYNTHESIS.) In some viruses (e.g. adenoviruses, herpesviruses) the viral genome is transcribed in the *nucleus* of the host cell, while in others (e.g. poxviruses, many rhabdoviruses) transcription occurs in the *cytoplasm*. A virion-associated DNA-dependent RNA polymerase occurs in poxviruses, and a virion-associated RNA-dependent RNA polymerase occurs e.g. in orthomyxoviruses, paramyxoviruses and rhabdoviruses—RNA viruses whose genomes each consist of a *minus* strand (see PLUS AND MINUS STRANDS OF DNA AND RNA). The DNA-dependent RNA polymerase of the host cell appears to be used by e.g. adenoviruses and herpesviruses. In certain single-stranded RNA viruses (e.g. polioviruses) the nucleic acid of the virion functions as mRNA. Some viruses (e.g.

adenoviruses, poxviruses) exhibit distinct phases of early and late transcription; 'early' genes are transcribed prior to the onset of viral genome replication, while the transcription of 'late' genes must be preceded by replication of the viral genome. The products of 'early' genes may include e.g. enzymes involved in the replication of the viral genome, while 'late' gene products typically include the structural proteins of the virus. The transcription of some viruses (e.g. herpesviruses) appears to give rise to polycistronic transcripts which are subsequently cleaved to form a number of monocistronic mRNA molecules; orthomyxoviruses (and reoviruses) appear to transcribe monocistronic mRNA from each piece of the (fragmented) genome. In many viruses the (monocistronic) mRNA molecules are extended (at the 3′ end) by a segment of polyadenylic acid—such extension being made prior to translation; such 'adenylation' of mRNA has been detected in e.g. adenoviruses, paramyxoviruses, poxviruses, and rhabdoviruses. (In polioviruses—whose genome functions as mRNA—the nucleic acid of the virion is adenylated.) The function of adenylation is unknown. *Translation of the Viral Genome*. Translation occurs in the cytoplasm of the host cell, and all viruses use host cell RIBOSOMES for protein synthesis. Mammalian cells cannot translate polycistronic mRNA which contains multiple initiation sites. While many viruses give rise to monocistronic mRNA molecules (either by direct transcription or by cleavage of polycistronic mRNA) the genome of a poliovirus functions as a single polycistronic mRNA molecule with a single initiation site (see PROTEIN SYNTHESIS); the entire genetic message is translated without interruption, and the resulting polypeptide (referred to as NCVPOO) is cleaved, during and subsequent to translation, by (host-coded?) enzymes. (The products of NCVPOO cleavage include the viral capsid proteins.) Similar cleavage of a giant polypeptide is believed to occur in togaviruses. Following translation, virus-coded proteins may be modified in certain ways—e.g. the glucosylation of peplomer precursors, and the phosphorylation of *Adenovirus* fibre protein. *Replication of the Viral Genome*. (See entry DNA for a generalized account of DNA synthesis.) Genome replication occurs in the host cell nucleus (in e.g. adenoviruses, herpesviruses) or in the cytoplasm (in e.g. picornaviruses, poxviruses, togaviruses); little appears to be known of the mechanisms of genome replication in the animal and plant viruses. For the replication of an RNA genome an RNA-dependent RNA polymerase—not found in any host cell—must be either associated with the virion or coded for by the viral genome; the replication of an RNA genome involves the synthesis of both plus and minus strands of RNA (see also BALTIMORE CLASSIFICATION). In the leukoviruses—whose replication involves the formation of a DNA intermediate—an RNA-dependent DNA polymerase is associated with the virion (see REVERSE TRANSCRIPTASE). Among DNA viruses use may be made of the host's DNA polymerase—though a virus-coded enzyme is used by some of the larger viruses (e.g. herpesviruses). (For details of phage genome replication see BACTERIOPHAGE.) *Assembly of Viruses* (*viral maturation*). Some viruses (e.g. adenoviruses, the capsid of herpesviruses) are assembled in the *nucleus* of the host cell, while others (e.g. paramyxoviruses, picornaviruses, poxviruses, togaviruses) are assembled in the *cytoplasm*; the mechanisms by which animal viruses are assembled are poorly understood. Enveloped viruses typically acquire their lipid membranes from the host cell following the addition of virus-coded proteins to regions of the cytoplasmic membrane (e.g. orthomyxoviruses) or the nuclear membrane (e.g. herpesviruses); the capsid or nucleocapsid is extruded in an outpocketing of the modified region of host membrane—which subsequently detaches from the remainder of the membrane to form the outermost layer of the virion. The latter process is termed *budding*. In certain plant viruses (e.g. the tobacco mosaic virus) virions are assembled spontaneously from their component parts—both *in vivo* and *in vitro*. (see also BACTERIOPHAGE) *Release of viral progeny from the host cell*. In polioviruses (and in virulent BACTERIOPHAGE) progeny virions are released from the host cell following cell lysis; the mechanism of release in polioviruses is unknown. In many types of animal virus the liberation of progeny virions does not require prior lysis of the host cell; thus e.g. enveloped viruses bud from a host cell which is necessarily intact (see budding, above).

Host Cell Damage. Many viruses bring about death of the (intact) host cell by causing irreversible inhibition of the protein and nucleic acid synthesis of the cell; the 'early' gene products of a number of viruses are known to bring about this inhibition. Other cytocidal effects (by animal viruses) may involve e.g. activation of the host's lysosomes, or the accumulation of intracellular virions—with attendant distortion and stress of cellular components.

*Taxonomy of Viruses**. Viruses are classified mainly on the basis of the type of nucleic acid they contain and on their morphology; closely related viruses may differ e.g. in their antigenic characteristics, pathogenicity, etc.

(Taxonomically-important antigenic sites occur on the capsid, nucleocapsid, or envelope of a virion.) A given genus may contain viruses which infect animals and/or plants; thus e.g. the genus RHABDOVIRUS includes the rabies virus, *sigma* virus (which infects the fruit-fly, *Drosophila*), and the eggplant mottled dwarf virus. (see also ADENOVIRUS, ALPHAVIRUS, ARBOVIRUSES, ARENAVIRUS, BARLEY YELLOW DWARF VIRUS, BEET YELLOWS VIRUS, CALICIVIRUS, CAULIFLOWER MOSAIC VIRUS, CORONAVIRUS, ENTEROVIRUS, ENTOMOPOXVIRUSES, FLAVOVIRUS, FOAMY AGENTS, HEPATITIS VIRUS B, HERPESVIRUS, INFLUENZA VIRUSES, IRIDESCENT VIRUSES, LETTUCE BIG VEIN VIRUS, LEUKOVIRUS, MARBURG AGENT, MYXOVIRUSES, ORBIVIRUS, ORTHOMYXOVIRUS, PAPILLOMAVIRUS, PAPOVAVIRIDAE, PARAMYXOVIRUS, PARVOVIRUS, PICORNAVIRIDAE, POLYOMAVIRUS, POTATO VIRUS Y, POXVIRUS, REOVIRUS, RESPIRATORY SYNCYTIAL VIRUS, RHABDOVIRUS, RHINOVIRUS, RICE DWARF VIRUS, ROUS SARCOMA VIRUS, RUBELLA VIRUS, SATELLITE VIRUS, SUGARCANE FIJI DISEASE VIRUS, TOBACCO MOSAIC VIRUS, TOBACCO NECROSIS VIRUS, TOGAVIRIDAE, TOMATO RING SPOT VIRUS, TURNIP YELLOW MOSAIC VIRUS, and WOUND TUMOUR VIRUS; cf. VIROIDS)

*In 1975 the International Committee on the Taxonomy of Viruses recommended certain changes in the taxonomic hierarchy of the viruses and proposed a plethora of new names. In this Dictionary the viral taxa are those in common general usage at the present time.

viruses, culture of Viruses are cultured for research purposes, for the preparation of vaccines, and in medical and veterinary diagnostic work (i.e. isolation and identification of viral pathogens).

Viruses which infect *vertebrates* are generally propagated by incubating a TISSUE CULTURE monolayer (or embryonated egg) which has been inoculated with a viral suspension. Not all types of cultured cell support the replication of a given type of virus, and a particular virus usually replicates in some types of tissue culture more vigorously than in others. Viral replication in a cell monolayer can be detected e.g. by the development of CYTOPATHIC EFFECTS (CPE) or (in some viruses) by HAEMADSORPTION; viruses (such as rubella virus) which fail to cause CPE in certain types of tissue culture may be detected by their ability to cause INTERFERENCE. Viruses can be quantified by culture in cell monolayers—e.g. by counting the number of PLAQUES which develop in a monolayer following inoculation with a particular dilution of a virus preparation and subsequent incubation. (see also METABOLIC INHIBITION TEST) For culture in embryonated eggs, use is commonly made of fertile hens' eggs which have been incubated 5–12 days before inoculation with the virus—the period of incubation depending e.g. on the particular membrane or cavity of the egg which is to be inoculated. Locations preferentially used for the propagation of particular viruses include the yolk sac (egg pre-incubated 5–6 days), chorioallantoic membrane (CAM) (egg pre-incubated 11–12 days), and allantoic cavity (egg pre-incubated 9–11 days); the CAM is commonly used, for example, for the propagation and differentiation of poxviruses—see POXVIRUS. Essentially, an egg is selected, candled (see CANDLING), and inoculated with the appropriate virus *via* a small hole made in the shell; following inoculation the hole is sealed and the egg is incubated (at an appropriate temperature—usually 37 °C—and humidity) for a period (usually 1–7 days) which depends e.g. on the virus being propagated.

Certain *insect*-infecting viruses have been propagated in tissue cultures prepared from the ovarian and other tissues of certain insects.

The culture of *plant* viruses is normally carried out in intact plant hosts; the propagation of plant viruses in plant tissue or cell cultures is an experimental rather than a routine procedure.

visna A chronic, fatal disease of sheep in which the incubation period may be of the order of months or years. The symptoms, e.g. abnormal gait, trembling, paralysis, are associated with a progressive demyelination of the brain and spinal cord. The causal agent is a single-stranded RNA virus.

vital staining The STAINING of *living* cells (or components of living cells) by certain types of dye ('vital stains'); such dyes permit the continuance of viability in the stained cells for periods of time which depend e.g. on the type of dye used. Vital staining has been used e.g. to study differentiation in the pseudoplasmodium of the cellular slime moulds, and is involved in the TOXOPLASMA DYE TEST. Bismark brown, methylene blue, and neutral red are commonly used as vital stains in protozoology, and Janus green is used for staining mitochondria and for following changes in the redox potential in mitochondria; other dyes which, under given conditions, can behave as vital stains include Congo red and trypan blue.

In metazoans, *intra vitam* staining involves the injection or ingestion of a dye which is taken up by *certain* types of cell in the (living) organism; thus, e.g. when trypan blue is injected into a mammal the dye is taken up e.g. by macrophages. (Other types of cell can be stained by trypan blue only when they are dead or damaged; hence, this dye is used to determine the proportion of dead/damaged cells in a cell suspension during the preparation of a TISSUE CULTURE. Thus, trypan blue behaves as a vital stain only in respect of certain types of cell.) The staining of living cells after their removal from the host has been referred to as *supravital staining*.

vitamin (microbiological aspects) A generic term for a group of (unrelated) organic compounds—some or all of which are necessary, in small quantities, for the normal metabolism and growth of microorganisms; they function as coenzymes or as components of coenzymes. Most microorganisms can synthesize the vitamins they require; those which cannot synthesize a particular vitamin must obtain it from the environment (e.g. growth medium). (In some cases a precursor of the vitamin can replace the vitamin itself—e.g. *p*-aminobenzoic acid can sometimes satisfy a requirement for FOLIC ACID.) Microorganisms are important sources of certain vitamins—e.g. vitamin B_{12} apparently can be synthesized only by certain microorganisms. (see also RIBOFLAVIN) Microorganisms which require an exogenous supply of a particular vitamin for growth can be used in the assay of that vitamin—see BIOASSAY.

For details of structure and microbiological function of a range of vitamins, see BIOTIN, BIOPTERIN, FOLIC ACID, LIPOIC ACID, NICOTINIC ACID, PANTOTHENIC ACID, PYRIDOXINE, RIBOFLAVIN, THIAMINE, VITAMIN B_{12}. (see also ASCORBIC ACID and INOSITOL)

vitamin B_1 (see THIAMINE)

vitamin B_2 (see RIBOFLAVIN)

vitamin B_6 (see PYRIDOXINE)

vitamin B_{12} A photolabile, water-soluble VITAMIN. The molecular structure of B_{12} includes a tetrapyrrole ring system (the *corrin* ring) which is highly substituted and which contains a central cobalt atom linked to the four pyrrole nitrogen atoms. (This part of the molecule is called *cobinamide*.) 3-Phosphoribose is linked to a side-chain on one of the pyrrole rings (*cobamide*). Cobamide, variously substituted, forms B_{12} and its analogues, e.g. *cobalamin*, in which 5,6-dimethylbenzimidazole forms a link between the cobalt atom and the (α-)1-position of the ribose residue; the sixth position on the cobalt may be filled by any of a variety of groups, e.g. cyanide, hydroxyl. (The term vitamin B_{12} usually refers to *cyanocobalamin* i.e. α-(5,6-dimethylbenzimidazolyl)cobamide cyanide —but the cyanide group is an artefact of the isolation procedure; cyanocobalamin, as such, probably does not occur *in vivo*.) In analogues of B_{12} the dimethylbenzimidazole group of cobalamin may be replaced by certain purine or benzimidazole derivatives, e.g. α-adenylcobamide cyanide (*pseudovitamin B_{12}*). In the corresponding *cobamide coenzymes* the cyanide group of cyanocobalamin is replaced by 5′-deoxyadenosine linked to the cobalt *via* the 5′-position. These coenzymes are involved e.g. in certain intramolecular rearrangements (glutamate to β-methylaspartate; methylmalonyl-CoA/succinyl-CoA interconversion) and the dehydration of 1,2-diols. Another coenzyme form is *methylcobalamin*, in which a methyl group replaces the cyanide group of cyanocobalamin. Methylcobalamin is involved e.g. in the synthesis of methionine from homocysteine (apparently acting as a methyl carrier) and in the formation of methane by certain bacterial species (the methyl group of methylcobalamin being converted to methane).

Vitamins of the B_{12} group are apparently synthesized only by microorganisms; vitamin B_{12} is obtained commercially from e.g. *Streptomyces* spp as a by-product of antibiotic production. B_{12} is required as a growth factor by many algae and flagellates and by *Lactobacillus* spp; *Euglena gracilis* is often used for BIOASSAY. A wasting disease of ruminants in cobalt-deficient pastures is due to the inability of the rumen microflora to synthesize adequate amounts of B_{12}.

vitamin C (see ASCORBIC ACID)

vitamin H (see BIOTIN)

vitamins K (see QUINONES (in electron transport))

vivotoxin (*plant pathol.*) A substance produced *in a diseased plant*, either by the pathogen or the host, which causes damage to the cells of the plant and which is thus responsible for at least some of the symptoms of the disease.

Voges–Proskauer test (VP test) A test used in the identification of certain bacteria, e.g. members of the Enterobacteriaceae. The test involves the detection of *diacetyl* ($CH_3.CO.CO.CH_3$) or its precursors acetoin and 2,3-butanediol formed e.g. in the BUTANEDIOL FERMENTATION of glucose. A phosphate-buffered glucose-peptone medium is inoculated with the strain of bacterium under test and incubated for 48 hours at 37 °C. In *Barritt's method*, 0.6 ml of an alcoholic solution of α-naphthol and 0.2 ml of 40% potassium hydroxide solution are added to approximately 1 ml of the culture; the tube is shaken vigorously, placed in a sloping position (for maximum exposure of the sample to air), and examined after 15 and 30 minutes. Under these conditions, acetoin and 2,3-butanediol are oxidized to diacetyl; diacetyl reacts with α-naphthol to give a red coloration—a *positive* VP test. *O'Meara's method* is essentially similar except that creatine is used in place of α-naphthol. (see also IMVIC TESTS)

Vollmer patch test A form of TUBERCULIN TEST

in which a patch of tuberculin-impregnated gauze is fastened to the skin with a piece of tape.

volunteer plant A (normally cultivated) plant which occurs as a result of *natural* propagation.

volutin granules (see METACHROMATIC GRANULES)

volva (*mycol.*) (1) A cup-shaped remnant of the UNIVERSAL VEIL which surrounds the base of the stipe in the mature fruiting bodies of certain agarics—e.g. *Amanita* spp. The term *volva* is sometimes used to refer to the (entire) universal veil. (2) A cup-shaped remnant of the peridium which surrounds the base of the stipe in the mature fruiting bodies of certain members of the order Phallales.

Volvox A genus of coenobial freshwater algae of the division CHLOROPHYTA. The individual biflagellate cells resemble CHLAMYDOMONAS; the coenobium consists of a spherical or ovoid gelatinous matrix (up to 0.5 mm in diameter) in which about 500–50,000 cells (depending on species) are arranged in a peripheral layer—flagella projecting outwards. The coenobium has a characteristic rolling motility. In many species the protoplasts of the individual cells of a coenobium are interconnected by fine cytoplasmic threads (*plasmodesmata*).

Asexual reproduction. Certain cells within a colony enlarge to become *gonidia*; each gonidium divides many times to form a hollow sphere of progeny cells orientated such that their flagellate ends face towards the centre of the sphere. Subsequently the sphere of cells everts (*via* a pore) such that the flagellated ends of the cells face outwards—thus forming a daughter coenobium. Daughter coenobia may not be released until the parental coenobium disintegrates.

Sexual reproduction is oogamous—male and female gametes being formed in the same colony or in different colonies—depending on the species. Certain cells may either lose their flagella and swell to become female gametes, or divide many times to form numerous motile male gametes. In dioecious species the male gametes are released and penetrate other colonies containing mature female gametes; pairs of gametes fuse to form zygotes—which subsequently develop thick walls. Zygotes are released by disintegration of the parent colony and undergo a period of dormancy. On germination of the zygote, meiosis occurs and three of the four products of meiosis abort; the remaining cell divides to form a small ('juvenile') colony in a manner which resembles gonidial development (see above). Subsequent daughter colonies, produced asexually, have an increased number of cells—so that eventually full-sized colonies are formed.

Vorticella A genus of ciliate protozoans of the suborder Sessilina, subclass PERITRICHIA. *Vorticella* occurs in fresh, brackish, and marine waters, and some species (e.g. *V. convallaria*) are common in British SEWAGE TREATMENT plants. The *zooid* is commonly bell-shaped or goblet-shaped—the buccal cavity being at the wider end, and the narrow end being attached, by a *stalk*, to the substratum; both the zooid and stalk are contractile. The zooid is usually about 50–150 microns in length; the tubular stalk contains (in its lumen) a contractile filament (the *myoneme* or *spasmoneme*) and is several hundred microns in length—though commonly less than 12 microns in width. The macronucleus is curved, or U-shaped, and the micronucleus is spherical. One or more contractile vacuoles are usually present. Most species feed on bacteria. Motile larvae (see TELOTROCH) are formed asexually; an atypical form of CONJUGATION occurs. *Vorticella* is *solitary*, i.e. only one zooid is carried on each (unbranched) stalk—cf. CARCHESIUM.

VP test (see VOGES–PROSKAUER TEST)

vulvovaginitis Inflammation of the vulva and vagina. Causal agents include *Candida albicans*, herpesviruses, *Trichomonas vaginalis*, and *Neisseria gonorrhoeae*.

VW antigens Protein or lipoprotein antigens found in virulent strains of *Yersinia pestis*. They are believed to be associated with the organism's ability either to resist phagocytosis or to resist digestion within the phagocytes.

V-W transition A *reversible* loss, or re-establishment, of the ability to form the VI ANTIGEN—exhibited, for example, by certain strains of *Citrobacter*. Such a transition may reverse rapidly on subculture. Organisms which, at any given time, are synthesizing the Vi antigen are referred to as *V forms*; cells which are not synthesizing the antigen are known as *W forms*.

W

W forms (see V-W TRANSITION)

wakame Species of the brown alga *Undaria* used, in Japan, as food.

Warburg–Dickens pathway (see HEXOSE MONOPHOSPHATE PATHWAY)

Warburg effect (in PHOTOSYNTHESIS) A decrease in oxygen evolution (and carbon dioxide fixation) in the presence of high concentrations of oxygen; this is due to an increased rate of PHOTORESPIRATION.

wart (verruca) A small benign tumour of the skin; warts usually regress spontaneously—only rarely do they become malignant. In man warts are caused mainly or exclusively by the human papilloma virus.

wart disease (see POTATO, FUNGAL DISEASES OF)

Wassermann reaction (WR) A serological diagnostic test for SYPHILIS. (see also STS) The test detects an entity known as the *Wassermann antibody** which occurs in the sera of syphilitic patients. The Wassermann antibody is detected by means of a cardiolipin-lecithin-cholesterol antigen with which it undergoes a COMPLEMENT-fixing reaction; the Wassermann reaction is thus a form of COMPLEMENT-FIXATION TEST. The test is prone to BIOLOGICAL FALSE POSITIVE REACTIONS. The term *Wassermann reaction* (or WR) may be used loosely to refer to any of the STS (e.g. the VDRL TEST) or to *any* serological diagnostic test for syphilis. *Also known as the Wassermann *reagin*—but not to be confused with IgE antibodies which are also called *reagins*. (see also FTA-ABS and TPI)

water activity (a_w) An expression for the amount of *available* water in a given substrate—e.g. a foodstuff or a medium; water activity may be defined as 1/100th of the relative humidity (R.H.%) of air which is *in equilibrium* with that substrate. Thus, e.g. an R.H. of 95% corresponds to an a_w of 0.95.

Most bacteria fail to grow if the water activity of the medium is below about 0.92; for *Staphylococcus aureus* the figure is about 0.87, while the growth of some halophilic bacteria continues at an a_w of about 0.75. For certain yeasts, e.g. *Saccharomyces cerevisiae*, the minimum a_w compatible with growth is about 0.94—while for others, e.g. *Saccharomyces rouxii*, the minimum value is about 0.65. For certain filamentous fungi, e.g. *Aspergillus* spp, *Penicillium* spp, the minimum a_w compatible with growth falls within the range 0.8–0.9.

water bath A covered metal or glass tank containing water at a thermostatically-controlled temperature; in good-quality water baths a motor-driven propeller ensures rapid mixing of the water and the maintenance of a given temperature in all parts of the tank. In microbiology, water baths are used for the INCUBATION of materials at closely-controlled temperatures; far greater temperature stability is possible in a water bath (e.g. ± 0.2 °C) than in a hot-air incubator.

water bloom (see BLOOM)

water-immersion objective (see RESOLVING POWER)

water supplies (microbiological aspects) Natural waters—particularly *surface* waters (e.g. rivers etc.)—which are used for public water supplies must be adequately treated to prevent the dissemination of water-borne diseases—e.g. typhoid. Bacteriological, chemical and other standards for *raw water* (i.e. water to be treated) are laid down by the World Health Organization and by appropriate governmental departments. The main aim of treatment is the elimination of pathogenic bacteria, protozoans and viruses; processed water entering the distribution system is not necessarily sterile.

Initially, the raw water is seeded with e.g. alum to coagulate the small particles of organic debris; the water is then left in settling tanks (*sedimentation basins*) so that the coagulated particles form a sediment. (A proportion of the microflora becomes trapped within these particles and is thus eliminated.) The water is then passed through a *sand filter*, i.e. a bed of gravel which is overlaid with sand. *Rapid* sand filters function mainly as mechanical sieves—removing about 90% of bacteria in addition to much of the non-coagulated organic matter. In *slow* sand filters the particles of fine sand in the upper layers bear a coating of protozoa and other microorganisms; this biologically-active film reduces the effective pore-size of the sand bed, and reduces the concentration of dissolved organic matter in the (slowly) percolating water. Most bacteria are thus removed mechanically or are ingested by the protozoans of the surface film. (Periodically, sand filters are cleaned by reversing the water flow and blowing air through the sand bed.) The aim of the final stage, *chlorination*, is to kill or inactivate any pathogenic organisms which have passed through the filters. Since the disinfecting power of CHLORINE is reduced by the presence of organic matter, sufficient chlorine must be added to the water to satisfy this *chlorine demand* and to leave some free, available, or *residual chlorine*. *Break-point chlorination* refers to the use of chlorine in quantities greater than that required to satisfy the chlorine demand; usually, the concentration of residual chlorine falls within the range 0.5–2.0 parts per million. (The pH of the water, and the concentration of free ammonia, are but two factors which govern the amount of chlorine required to achieve a given level of residual chlorine.) In some treatment plants both chlorine and ammonia are added to the filtered water; the chloramines thus formed exert prolonged antimicrobial activity. In *superchlorination* the addition of a relatively high concentration of chlorine is followed by a reduction in the level of residual chlorine.

Although some pathogenic viruses are inactivated by efficient chlorination, such treatment may or may not be sufficient to inactivate all water-borne viral pathogens. The cysts of *Entamoeba histolytica* (which may survive chlorination) are commonly greater than 10 microns in diameter and are likely to be retained by an efficient sand filter. Bacteriological tests which assess the quality of fully-treated water aim to detect viable coliform bacteria, particularly *Escherichia coli*; the absence of *E. coli* in water samples argues for the *probable* absence of pathogenic bacteria since, in raw water, the numbers of *E. coli* are normally greatly in excess of those of pathogenic bacteria. (see also SEWAGE TREATMENT)

wax D A component of the cell walls of *Mycobacterium* spp; it consists of an arabinogalactan esterified with MYCOLIC ACIDS and linked (apparently *via* a phosphate group) to the muramic acid of PEPTIDOGLYCAN. Wax D is soluble in ether and chloroform, but not in acetone; it can replace whole mycobacteria in FREUND'S ADJUVANT.

WBC White blood cell.

Weil–Felix test A serological diagnostic test for TYPHUS and certain other rickettsial diseases. Antigens which are present on the causative organisms (species of RICKETTSIA) are shared by certain strains of PROTEUS (strains OX19, OX2 and OXK). Thus, these strains of *Proteus* may be agglutinated by the sera of patients who are suffering from these diseases.

Weil's disease (infectious jaundice) An acute, infectious, systemic disease of man. The causal agent, a strain of *Leptospira*, enters the body *via* wounds or through mucous membranes; pathogenic leptospires are common e.g. in rats' urine. The incubation period is about 5–13 days; a septicaemic phase, with headache, fever and muscular pain, may be followed by jaundice, enlargement of the liver and spleen, renal dysfunction, and haemorrhages. BenzylPENICILLIN and tetracycline have been used therapeutically. *Laboratory diagnosis*: during the first week, blood cultures may yield positive results; subsequently, the urine may be cultured.

wet mount (wet preparation) A (temporary) preparation—for microscopical examination—in which the subject or specimen is present in a liquid medium between a slide and cover slip. In some cases the liquid medium is that in which the specimen normally occurs—as e.g. when a drop of pond water is examined for protozoans. Alternatively, a specimen may be mounted in a solution which functions as a fixative and/or stain; thus, e.g. LACTOPHENOL COTTON BLUE may be used for the examination of fungal mycelium.

wet rot (of timber) A brown rot caused by *Coniophora cerebella* (the *cellar fungus*); only wood having a relatively high moisture content, e.g. 40–60% water, is susceptible. (cf. DRY ROT) Both softwoods and hardwoods may be attacked. Infected timbers usually exhibit a dark, branching surface mycelium; the fruiting bodies, initially pale, become dark green, and the spores may be any of various shades of brown. Control of wet rot may be achieved by effective ventilation and drying of the timbers. *C. cerebella* is generally sensitive to coal tar derivatives, e.g. CREOSOTE. (see also TIMBER, DISEASES OF)

wettable powder (w.p.) A form in which water-insoluble antifungal agents (e.g. CAPTAN) etc. may be prepared for agricultural use. The active ingredient, usually mixed with a filler—e.g. fine clay, DIATOMACEOUS EARTH—is finely ground, and a surface-active 'wetter' is added—usually an alkali salt of an organic acid. The filler permits the active ingredient to be ground into fine particles and prevents caking during storage in an aqueous medium; the wetter allows rapid dispersion of the powder in water.

wheat, diseases of (see CEREALS, DISEASES OF)

whey (see CHEESE)

whiplash flagellum (peitschengeissel) (see *eucaryotic* FLAGELLUM)

whiskey (whisky) (see SPIRITS)

white piedra (*med.*) A chronic disease of the hair in which soft, white or fawn, gelatinous nodules develop on the hairs of the scalp, beard, and/or genital regions; each nodule marks a focus of infection by the causal agent—a species of *Trichosporon*. The hair shaft is subjected to both internal and external infection but the skin is not invaded; hairs break easily in the regions of the nodules.

white rot (of timber) (see TIMBER, DISEASES OF)

white rusts (see RUSTS (2))

white scours (in calves) (see DYSENTERY (bacterial))

W.H.O. plate A rectangular block of plastic (e.g. 18 × 15 × 1.5 cms) one face of which bears a number of rows of circular indentations, or wells. Each well is approximately semicircular in (vertical) cross-section; the volume of each well may be e.g. 1.5 ml. W.H.O. plates are used e.g. in COMPLEMENT FIXATION TESTS and in other quantitative procedures involving the preparation and testing of small volumes of serum etc. in serial dilutions.

whooping cough (pertussis) An acute respiratory-tract disease which occurs mainly in children; the causal agent is *Bordetella pertussis*. Infection commonly occurs by droplet inhalation, and the incubation period is about 10 days. The disease is characterized by paroxysms of coughing—each being followed by an audible inspiratory effort, or *whoop*; whooping begins some 10 days after the onset of the early symptoms—which resemble those of a common cold. TETRACYCLINES have been used therapeutically. *Laboratory diagnosis. B. pertussis* may be isolated e.g. by the COUGH PLATE method.

A milder form of the disease is caused by *B. parapertussis. B. bronchiseptica* causes a respiratory-tract disease in animals (e.g. the dog) and may give rise to whooping cough in man.

Widal reaction (Widal test) An AGGLUTINATION test used for detecting serum antibodies to species of *Salmonella* which cause human enteric fevers, e.g. *S. typhi* and strains of *S. paratyphi*; serial dilutions of serum are tested for the presence of AGGLUTININS to the O and H antigens of these organisms. Suspensions of (killed) bacteria of each type—suitably prepared—are used as the O and H test antigens.

wild type strain Of a given species: a strain whose phenotype is that of the majority of *naturally-occurring* members of its kind; a *non-mutant* strain.

Wilson and Blair's bismuth sulphite medium A medium used for the primary isolation of *Salmonella* spp (e.g. from faeces); the medium inhibits the growth of coliforms. The buffered, alkaline medium contains brilliant green and freshly-precipitated bismuth sulphite. *Salmonella* spp commonly form black colonies on the medium.

wilt disease (*entomopathol.*) Any viral disease of insects—see, e.g. GRANULOSES and POLYHEDROSES.

wilts Plant diseases characterized by a reduction in host tissue turgidity. Common causative agents are species of the fungi *Fusarium* and *Verticillium* and the bacteria *Erwinia* and *Pseudomonas*. The activities of such organisms commonly affect the vascular system of the host plant. (see FLAX, DISEASES OF)

wine-making Wine is produced by the ALCOHOLIC FERMENTATION of grapes. Grapes contain some 10–25% fermentable sugars (mainly glucose and fructose) together with organic acids (mainly tartaric and malic acids) and small amounts of nitrogenous compounds (e.g. amino acids), tannins, vitamins, and minerals. The grapes are crushed, and a calculated amount of sulphur dioxide is added to inhibit undesirable microorganisms (e.g. 'wild yeasts' naturally present on the grape skins). Fermentation is generally initiated by the addition of a starter culture of *Saccharomyces cerevisiae* var *ellipsoideus*; alternatively, the inoculum may derive from the flora on the grape skins or that adhering to the fermentation vessels etc. Fermentation continues for a period of time (usually 4–6 weeks) which depends e.g. on the sugar content of the grapes, and on the temperature (20–35 °C). Particulate matter is allowed to settle out, and the wine is generally allowed to age in oak casks for 1 or more years. Subsequently, the wine may be clarified and/or pasteurized, and is then bottled. *Red wines* are made from black grapes—pigments (e.g. anthocyanins) being leached from the grape skins by the alcohol produced during fermentation. *White wines* may be made from white grapes or from the *juice* of black grapes; oxidative discoloration is inhibited by treatment with sulphur dioxide. *Rosé wines* may be prepared from black grapes—the skins being removed before all the pigment has been extracted—or from less highly-pigmented grapes. In *dry* wines, most or all of the sugar originally present in the grape juice has been fermented to ethanol. In *sweet* wines, fermentation is stopped (e.g. by the addition of sulphur dioxide) before all the sugar has been converted to ethanol; alternatively, grapes with a high sugar content may permit the production of alcohol to concentrations which inhibit the yeast before all the sugar has been fermented. (see also NOBLE ROT) *Sparkling* wines (e.g. Champagne) contain carbon dioxide produced during (residual) fermentation in the bottle. In the production of *fortified* wines, brandy is added to the wine—either during fermentation (e.g. for port) or after fermentation (e.g. for sherry).

Organisms responsible for the spoilage of wines include species of *Acetobacter*, *Leuconostoc*, and *Lactobacillus*. (see also BREWING; CIDER; SPIRITS)

Wingea A genus of fungi of the family SACCHAROMYCETACEAE.

winter dysentery (of cattle) (see DYSENTERY (bacterial))

winter spore (*mycol.*) (see RESTING SPORE)

witches' broom (*plant pathol.*) An abnormal structure which develops mainly in woody plants—often as a result of fungal or bacterial infection. At the site of infection a number of buds develop to form a broom-like cluster of thin twigs which may bear atypical leaves. Examples: *Taphrina betulina* causes witches' broom in the hairy birch (*Betula pubescens*) and *T. turgida* in the silver birch (*B. pendula*); many rusts (e.g. *Puccinia* spp) can cause witches' broom formation. Some forms of witches' broom previously attributed to viral infection are now believed to result from infection by *Mycoplasma*-like organisms. The mechanism of witches' broom formation is not clear. A hormonal imbalance in the plant presumably occurs; many of the causative organisms produce AUXINS *in vitro*. (see also CYTOKININS)

wobble hypothesis An hypothesis, proposed by Crick, which accounts for the observed pattern of degeneracy in the third base of a codon—see GENETIC CODE; the hypothesis thus explains the ability of certain tRNAs to recognize more than one codon when the codons differ only in the *third* base (i.e. the base at the 3′ end). Base-pairing occurs normally between the first and second bases of the codon and their complementary bases (third and second respectively) in the anticodon; the wobble hypothesis proposes that the third base of the codon can undergo unusual base-pairing with the corresponding first (5′ end) base in the anticodon. Thus, when G is the first base of the anticodon it can recognize either C or U as the third base of the codon; similarly, U can recognize either A or G, and I (hypoxanthine) can recognize A, U, or C. When A is the first base of the anticodon it can recognize only U; similarly, C can recognize only G.

Wolhynian fever Synonym of TRENCH FEVER.

Woloszynskia A genus of algae of the DINOPHYTA.

wood, diseases of (see TIMBER, DISEASES OF)

wooden tongue A disease of cattle in which granulomata develop within the tissues of the tongue—the latter becoming stiff or 'wooden'. Causal agent: *Actinobacillus lignierisii.*

Wood's lamp A lamp which emits radiation of wavelength about 360 nm; such radiation causes fluorescence of infected hair and/or scalp in certain dermatophytoses. The Wood's lamp is used in the diagnosis of RINGWORM diseases.

woolsorter's disease Pulmonary ANTHRAX.

working distance (in microscopy) The distance between the specimen (or the upper surface of the coverslip) and the nearest face of the objective lens when the specimen is sharply in focus.

Woronin bodies (*mycol.*) In certain mycelial fungi: small, spheroidal, highly refractile bodies of unknown function which are observed in the cytoplasm—sometimes close to the wall or septa; it has been suggested that they may act as plugs—sealing the septal pore(s) following damage to the hypha.

wort (see BREWING)

wound tumour virus (clover wound tumour virus) A double-stranded RNA virus which is able to cause disease in a wide range of unrelated plants; the virion appears to be icosahedral with a diameter of approximately 60 nm. Disease symptoms may include vein-enlargement, enations, and tumours at wound sites. In *Nicotiana tabacum* (tobacco) the virus causes multiple root tumours—a condition known as *tobacco club root*. Inclusion bodies may be formed. The virus may be transmitted from plant to plant by leafhoppers—in which transovarial passage has been observed; viral multiplication occurs within the leafhopper at sites known as *viroplasms*.

wounds, microbial infection of (*med.*) The nature and extent of wound contamination is influenced e.g. by the nature of the body microflora at the wound site; the nature of the microflora on the object responsible for the wound; the nature of the environment; the nature of any surface covering (e.g. clothing) through which the wound was made. The development of wound infection is influenced by the nature of the contaminating flora; the depth of the wound; the nature of the treatment (if any) given to the wound; immunity (e.g. by vaccination) to contaminating organisms or their toxins.

Common wound contaminants include the pyogenic bacteria *Staphylococcus aureus* and *Pseudomonas aeruginosa*. Anaerobic spore-forming contaminants may cause GAS GANGRENE or TETANUS; the establishment of such organisms within a wound is aided by the presence of facultatively anaerobic bacteria (e.g. coliforms) which help to create an anaerobic microenvironment.

w.p. WETTABLE POWDER.

WR (*immunol.*) (see WASSERMANN REACTION)

Wright's stain (for parasitic protozoans) (see ROMANOWSKY STAINS)

wyerone An acetylenic keto-ester produced by the broad bean (*Vicia faba*). Wyerone is a wide spectrum antifungal agent and may play some part in the resistance of the plant to certain fungal diseases.

X

X bodies (viroplasts) (*plant pathol.*) Spherical or elongated bodies which occur, usually in the cytoplasm, in plant cells infected with certain viruses. Such bodies appear as amorphous structures by light microscopy. X bodies may consist entirely of virus particles or they may include cytoplasmic elements. (see also VIROPLASMS)

X cells (*immunol.*) Immunologically competent LYMPHOCYTES which have not experienced the initial contact with specific antigen. (cf. Y CELLS, Z CELLS)

X factor A chemical factor which is essential for the *aerobic* growth of certain species of HAEMOPHILUS; the X factor requirement is satisfied by a *haemin* (= hemin)—a complex of ferric iron with protoporphyrin IX. (The haemins commonly used are: haemin chloride ('haemin') and haemin hydroxide ('haematin').) The X factor is believed to be required for respiratory activity in the bacterial cell. The X factor may be prepared from the haemoglobin liberated by lysed erythrocytes; CHOCOLATE AGAR contains the X factor and the V FACTOR.

X-rays A type of ionizing radiation; X-rays may be used e.g. for STERILIZATION.

xanthochroic (*mycol.*) Refers to those fruiting bodies in which the context is yellowish-brown but becomes dark brown or olive-brown in potassium hydroxide solution (10% aqueous).

Xanthomonas A genus of gram-negative, obligately aerobic, chemoorganotrophic bacteria of the PSEUDOMONADACEAE; all species are plant pathogens. The cells are rods, about 0.5 × 1–2 microns, which are polarly flagellate; most strains produce a water-insoluble yellow CAROTENOID pigment. The organisms are catalase-positive but only some are oxidase-positive. PECTINolytic activity is common.

X. campestris, the type species, causes disease in a number of CRUCIFERS. A number of organisms (e.g. *X. phaseoli*—see *common blight* of BEANS) are regarded as strains or variants of *X. campestris*. *X. ampelina*, *X. axonopodis* and *X. fragariae* are pathogens of grapevines, certain grasses, and strawberries, respectively.

xanthophylls (see CAROTENOIDS)

Xanthophyta (yellow-green algae) A division of the ALGAE; species of the Xanthophyta are predominantly freshwater organisms though some species are marine or terrestrial. Unicellular, filamentous, siphonaceous, and colonial species occur; vegetative cells (and gametes) which are motile are characterized by two apically-situated flagella of unequal length—one long *tinsel* FLAGELLUM, and one short *whiplash* flagellum. All species contain CHLOROPHYLL *a* (only), β-carotene, and several xanthophylls (see CAROTENOIDS); intracellular storage products include CHRYSOLAMINARIN. Sexual processes occur in a few species—being oogamous in e.g. *Vaucheria* and predominantly isogamous in other species. Genera include the (siphonaceous) organisms BOTRYDIUM and VAUCHERIA.

Xanthoria A genus of foliose LICHENS in which the phycobiont is *Trebouxia*. The thallus contains the orange anthraquinone pigment *parietin* and is attached to the substratum by rhizinae. The apothecia are lecanorine and contain polarilocular ascospores. Species include e.g.:

X. parietina. The thallus is yellow-orange (greenish-yellow in shaded situations) and rosette-shaped with a lobed, crinkled margin; the lower surface is pale or white—particularly towards the margin. Apothecia are numerous and crowded towards the centre of the thallus; the discs are orange, 1–4 millimetres in diameter, and the thalline margins are thin and entire or notched. *X. parietina* is widely distributed and occurs on a variety of substrata—e.g. rocks, asbestos, trees—particularly when these are enriched with nitrogen (e.g. by bird droppings).

xanthosine-5′-monophosphate (XMP) A nucleotide formed e.g. as an intermediate during the biosynthesis of guanosine-5′-monophosphate—see Appendix V(a).

xenodiagnosis A method of detecting the presence of a vector-transmitted pathogen in the blood of a patient. Xenodiagnosis is used e.g. in the examination of persons suspected of having contracted CHAGAS' DISEASE (vector: reduviid bugs); laboratory cultured (trypanosome-free) strains of reduviid bugs are permitted to feed on the suspected case and are subsequently examined for the presence of *T. cruzi*.

Xenodochus A genus of fungi of the order UREDINALES.

xero- Prefix meaning *dry*.

XMP Xanthosine-5′-monophosphate—see Appendix V(a).

xylans Polymers of xylose. Xylans found in plant cell walls (see HEMICELLULOSE) consist of β-(1 → 4)-linked xylopyranosyl residues; the chains may be partially acetylated, and frequently have side chains containing e.g. arabinose, 4-*O*-methyl-D-glucuronic acid. In certain algae (e.g. *Bryopsis*, *Caulerpa*, *Halimeda*, *Udotea*) a linear β-(1 → 3)-linked xylan replaces CELLULOSE as the main structural component of the cell wall.

Xylanases are produced by a range of microbes, e.g. species of *Aspergillus*, *Chaetomium*, *Colletotrichum*, *Streptomyces*, and species of the RUMEN microflora.

Xylaria (*syn*: *Xylosphaera*) A genus of saprophytic fungi of the order SPHAERIALES, class PYRENOMYCETES; species occur e.g. on tree stumps and on felled timber. The somatic mycelium occurs within the substratum, and the organisms form erect, prosenchymatous stromata at the surface; during the asexual reproductive phase the stromata are covered, or partly covered, with conidia. Subsequently, during the sexual phase, perithecia develop within the surface layer of the same stromata. Species include *X. hypoxylon* and *X. polymorpha*.

X. hypoxylon forms clusters of erect branched stromata which, individually, are somewhat flattened; each stroma, some 4–5 centimetres in height, is black and velvety except near the distal ends of the branches—which are antler-like and bear a covering of chalk-white conidia. Perithecia develop in the surface layer of the basal part of the stroma; each perithecium is embedded in the stroma with the ostiolar pore at the surface.

xylenols (as antimicrobial agents) (see PHENOLS)

xylophagous Wood-eating.

xylophilous Having an affinity for wood.

xylose An aldopentose which occurs widely in plants and in algae (see also HEMICELLULOSES). Xylose is metabolized by certain bacteria, including strains of *Lactobacillus* and *Pasteurella*. The sugar is heat-labile: solutions should be sterilized by filtration.

Xylosphaera A synonym of XYLARIA.

Y

Y cells (*immunol.*) Immunologically competent LYMPHOCYTES which have experienced the initial contact with specific antigen—but which are not *effector cells* (see TRANSFORMATION, OF LYMPHOCYTES); Y cells are also called *memory cells* or *primed cells.*

yaws (frambesia) A chronic, infectious disease of man characterized by ulcerative lesions of the skin and (in later stages) lesions in the bones; the causal agent, *Treponema pertenue*, enters the body *via* wounds etc. Yaws occurs in tropical America, Africa and the Far East; in some areas it is endemic. (see also PINTA and SYPHILIS)

yeast (bakers' yeast, brewers' yeast) The common name for the ascomycetous yeast *Saccharomyces cerevisiae*—cf. YEASTS.

yeasts A category of FUNGI defined in terms of morphological and physiological criteria; the yeasts do not constitute a taxon. The 'typical' yeast is a unicellular, saprophytic organism which characteristically ferments a range of carbohydrates and in which asexual reproduction occurs by budding. Although many yeasts conform to this definition, others (regarded as yeasts on the basis of overall affinity) fail to conform in one or more ways. Thus, certain yeasts (*Schizosaccharomyces* spp) reproduce asexually by fission and are capable of forming a pseudomycelium or a true mycelium; some (e.g. *Lipomyces* spp, *Hansenula wingei*) are non-fermentative, while others (e.g. *Candida albicans*) can be pathogenic. Some organisms which are not generally regarded as yeasts (e.g. *Mucor* spp) are capable of growing in yeast-like (unicellular) forms under certain environmental conditions.

Yeasts occur among the ascomycetes (particularly in the family SACCHAROMYCETACEAE) and the basidiomycetes; imperfect yeasts are placed in the class BLASTOMYCETES.

yellow fever (yellow jack) An acute, systemic disease which, in nature, affects man and other primates; the causal agent is a *togavirus*. (a) *Urban yellow fever*. The virus is transmitted from man to man by female mosquitoes of the species *Aedes aegypti*. An incubation period of 2–6 days precedes fever, vomiting, headache and muscular pains; the fever abates but subsequently returns. Degenerative changes occur in the liver, kidneys, spleen, heart and other organs; haemorrhages, jaundice and albuminuria (albumin in the urine) are typical symptoms. (see also COUNCILMAN BODIES) Death generally occurs on the 6th–10th day; some patients recover, and in endemic regions (e.g. Central and South America) mortality rates are of the order of 5–10%. Recovery or vaccination (an attenuated virus vaccine) give long-term immunity. (b) *Jungle or sylvan yellow fever*. Mainly a disease of non-human primates; man is incidental in the natural cycle of disease. The principal vectors are *Haemagogus* spp and *Aedes simpsoni*, but *A. aegypti* may be important in certain areas. Clinically, the disease closely resembles urban yellow fever.

yellow-green algae (see XANTHOPHYTA)

yellows (yellows diseases) (*plant pathol.*) Any of a variety of plant diseases in which the major symptom is a uniform or mottled yellowing of the leaves and/or other plant components. (see also CHLOROSIS) Such diseases may be incited by fungi (see e.g. CELERY, DISEASES OF), by bacteria, or by viruses. Some yellows diseases previously attributed to viruses (e.g. aster yellows and the 'viral' form of celery yellows) are now known to be incited by *Mycoplasma*-like organisms.

Yersinia A genus of gram-negative bacteria of the ENTEROBACTERIACEAE; species occur as parasites or pathogens of man and other animals. The organisms may be cultured on non-enriched media. Typically, lactose is not fermented, but β-galactosidase is produced; acid only is produced from certain other sugars. The METHYL RED TEST is positive. Type species: *Y. pestis*; GC% range for the genus: 46–47.

Y. pestis (formerly *Pasteurella pestis*). Non-motile coccobacilli or bacilli, approximately 2 microns maximum length, which typically stain bipolarly; *Y. pestis* is the causal agent of PLAGUE. Typical reactions: gas not produced from glucose; INDOLE TEST and CITRATE TEST negative; aesculin hydrolysed. Optimum growth temperature about 28 °C. A coagulase, active against rabbit plasma, is produced.

Y. pseudotuberculosis (formerly *Pasteurella pseudotuberculosis*). Coccobacilli or bacilli of maximum length about 6 microns; the cells are motile when grown at 22–30 °C. Coagulase is not produced. Gas is not produced from glucose. Aesculin is hydrolysed.

Y. enterocolitica. Coccobacilli or bacilli which are motile when grown at 22–30 °C. Gas is not produced from glucose. Aesculin is not hydrolysed.

yoghurt (yogurt, yaourt) A fermented milk product. Milk—previously boiled to concentrate the milk solids—is inoculated with cultures of *Lactobacillus bulgaricus* and *Strep-*

tococcus thermophilus; the whole is incubated at about 45 °C until the lactic acid content reaches 0.85–0.9%. In some modern processes, whole milk is supplemented with dry milk solids prior to fermentation. Yoghurt should be stored at about 5 °C.

Z

Z cells (*immunol.*) Effector cells (see TRANSFORMATION, OF LYMPHOCYTES). (cf. X CELLS, Y CELLS)

Zelleriella A genus of protozoans within the superclass OPALINATA.

Zenker's fluid (modified) A fixative used in protozoological and bacteriological work: mercuric chloride (5 g) and potassium dichromate (2.5 g) in distilled water (100 ml) supplemented (immediately before use) by glacial acetic acid (5 ml). (see also FIXATION)

zephiran (see QUATERNARY AMMONIUM COMPOUNDS)

***zeta* potential** (of bacteria) The electrical *potential difference* between the cell surface and surrounding medium. In media of neutral or alkaline pH the majority of bacteria carry a *negative* surface charge owing to the presence of cell surface components whose ISOELECTRIC POINTS are in the acid pH range. The IEP of many bacteria is approximately 3.0; such bacteria tend to agglutinate spontaneously at or near this pH.

zeugite (*mycol.*) Early term for a cell or structure in which KARYOGAMY occurs.

Ziehl–Neelsen's stain A stain used to detect the presence of ACID-FAST ORGANISMS (e.g. *Mycobacterium* spp). A heat-fixed smear of e.g. sputum, on a slide, is flooded with a concentrated solution of carbolfuchsin and heated until steam appears. (The solution should not boil.) The slide is kept hot for about 5 minutes, left to cool, and rinsed in running water. The smear is then passed through several changes of acid-alcohol (e.g. 3% conc. HCl in 90% ethanol), washed in water, and counterstained with an appropriate dye, e.g. 0.5% malachite green. After a final washing the slide is dried in an incubator and examined. Acid-fast organisms stain red, others green. (see also AURAMINE-RHODAMINE STAIN)

Ziemann's dots (see SCHÜFFNER'S DOTS)

zinc (and compounds) (as antimicrobial agents) (see HEAVY METALS)

zineb (dithane Z-78) Zinc ethylenebisDITHIOCARBAMATE: an important agricultural antifungal agent used in the control of a wide range of plant pathogens, e.g. *Botrytis* species, *Phytophthora infestans*, *Cladosporium fulvum* and *Peronospora* spp. It may be used as a wettable powder—or it may be prepared in the field by mixing NABAM, zinc sulphate and lime. (see also DITHANE)

ziram Zinc dimethyldithiocarbamate (see DIMETHYLDITHIOCARBAMIC ACID).

zoites (of *Toxoplasma*) (see BRADYZOITES and TACHYZOITES)

zone electrophoresis (see ELECTROPHORESIS)

zone lines (in decaying wood) Narrow, dark lines which often mark the boundaries between well-rotted and relatively sound timber. They usually consist of regions of dark-coloured mycelium in the rotting wood.

zoochlorellae Green endosymbiotic algae (e.g. *Chlorella*) present in certain protozoans—e.g. species of *Paramecium*, *Stentor*.

Zoogloea A genus of gram-negative, obligately aerobic, chemoorganotrophic bacteria of the PSEUDOMONADACEAE; species occur in natural waters and in certain SEWAGE TREATMENT plants. The cells are approximately 0.8 × 1–3 microns and are polarly flagellate; they form *flocs* (agglomerations of macroscopic dimensions)—particularly when the carbon:nitrogen ratio is relatively high. The organisms are catalase-positive and oxidase-positive; glucose and other sugars are attacked oxidatively. Storage products include poly-β-hydroxybutyrate. Type species: *Z. ramigera*.

zooid (1) A motile spore. (2) The body—as opposed to the stem—of a stalked ciliate.

zoom microscope A compound microscope with which magnification can be varied continuously over a range (or ranges) of values determined by the limits of the instrument; the position of the final image (i.e. that seen by the eye) remains constant. A change in magnification requires no adjustment of the focusing controls. The microscope contains a system of two or more lenses (*zoom lenses**) interposed between the objective and eyepiece—the objective and eyepiece lenses occupying fixed positions relative to one another; variation of magnification is effected by movement of the zoom lenses, along the optical axis, by means of suitable external controls. In *variable-power eyepiece* models the magnification of a given eyepiece may be increased by a factor of two or more. In *variable-power objective* models both the magnification and the effective numerical aperture of the objective may be varied—the latter being varied from sub-maximum to the maximum rated value. Some zoom microscopes incorporate one of the zoom lenses in the objective—the combined zoom lens-objective being movable relative to the eyepiece. *Fixed, *relay lenses* may also form part of the system; these help to achieve a suitable overall length for the optical system.

Zoomastigophorea A class of flagellate protozoans within the superclass MASTIGOPHORA; constituent species lack photosynthetic pigments. The orders are RHIZOMASTIGIDA, KINETOPLASTIDA, RETORTAMONADIDA, DIPLOMONADIDA, TRICHOMONADIDA and HYPERMASTIGIDA.

zoonosis Any disease (e.g. FOOT AND MOUTH DISEASE, TUBERCULOSIS) which can be transmitted, naturally, from animal to man, and *vice-versa*.

zooplankton (see PLANKTON)

zoosporangium Any SPORANGIUM which contains *motile* spores.

zoospores *Motile* (flagellated) spores.

zooxanthellae Yellow or brown endosymbiotic algae found e.g. in radiolarians; they include certain species of the Dinophyta.

Zygomycetes A class of FUNGI within the subdivision ZYGOMYCOTINA; the majority of zygomycetes are terrestrial saprophytes, but the class includes species which are parasites of animals (including insects), plants, and other fungi. The vegetative form of the organisms is a well-developed, branched mycelium which, in most species, is coenocytic (aseptate); some species (e.g. *Mucor* spp) grow in a yeast-like (unicellular) form under appropriate environmental conditions. The hyphal wall is typically chitinous. Asexual reproduction involves the formation of sporangia, sporangiola, or conidia. The primitive zygomycetes are those (such as *Mucor*) which form thin-walled sporangia which contain numerous spores; more advanced species give rise to sporangiola (or sporangia *and* sporangiola—e.g. *Thamnidium*), while the most advanced zygomycetes (e.g. *Cunninghamella*, *Entomophthora*) form conidia. Sexual reproduction typically involves the fusion of two (morphologically-similar) gametangia with the subsequent formation of a thick-walled resting stage, the ZYGOSPORE. (see also AZYGOSPORE) Genera include *Absidia*, *Basidiobolus*, *Chaetocladium*, *Choanephora*, CUNNINGHAMELLA, *Endogone*, ENTOMOPHTHORA, *Mortierella*, MUCOR, *Phycomyces*, PILOBOLUS, PIPTOCEPHALUS, RHIZOPUS, THAMNIDIUM, and *Zygorhynchus*.

Zygomycotina A sub-division of FUNGI of the division EUMYCOTA; constituent species are distinguished from those of other sub-divisions in that they form non-motile (non-flagellate) asexually-derived spores and, typically, reproduce sexually by gametangial copulation with the formation of ZYGOSPORES. The constituent classes are TRICHOMYCETES and ZYGOMYCETES.

zygophore (*mycol.*) A hypha bearing one or more ZYGOSPORES. (see also PROGAMETANGIUM and SUSPENSOR)

Zygorhynchus A genus of fungi of the ZYGOMYCETES.

zygospore (*mycol.*) A thick-walled resting spore formed, subsequent to gametangial fusion, by members of the zygomycetes; a zygospore may be naked or may be covered by a web or mesh of hyphae or other filaments. (see also SUSPENSOR) The nuclear events which occur during and subsequent to zygospore formation appear to differ in different species.

zygote A single, diploid cell formed from two (haploid) parental cells during fertilization; the parental cells undergo cytoplasmic and nuclear coalescence.

zygotene stage (see MEIOSIS)

zygotic induction The change: prophage → vegetative (lytic) phage which occurs when the prophage-bearing chromosome of a *lysogenic conjugal donor* (see bacterial CONJUGATION) enters a non-lysogenic recipient; the change occurs as a result of the absence of repressor protein in the cytoplasm of the recipient (see LYSOGENY). (Zygotic induction may also occur when the recipient contains a prophage whose repressor protein is not recognized by the donor prophage.)

zygotic meiosis MEIOSIS which precedes the formation of haploid vegetative (somatic) cells.

zymogen (proenzyme) An inactive precursor of an enzyme.

zymogram (1) An electrophoretic strip (see ELECTROPHORESIS) on which has been carried out an analysis of the enzymes in a given sample. (2) A table which shows the results of carbohydrate-fermentation tests. (Such tests determine the ability of one or more organisms to ferment each of a range of carbohydrates.)

Zymomonas A genus of gram-negative, motile*, non-sporing bacilli of uncertain taxonomic affinity; the organisms occur in a variety of fermented beverages. *Zymomonas* spp are fermentative, microaerotolerant anaerobes which can form ethanol and carbon dioxide from glucose *via* the ENTNER-DOUDOROFF PATHWAY; pantothenate is a common growth requirement. *Z. anaerobia* causes CIDER SICKNESS. **Z. anaerobia* var *immobilis* is non-motile.

zymosan A polysaccharide found in the cell walls of certain yeasts. In the presence of PROPERDIN zymosan brings about COMPLEMENT-FIXATION by the *alternative* pathway. Certain other polysaccharides (e.g. inulin) have similar activity.

Zythia A genus of fungi within the order SPHAEROPSIDALES; species include *Z. fragariae*—a pathogen of the strawberry both in Europe and in North America. The vegetative form of the organisms is a septate mycelium, and the elongated conidia are formed within light-coloured pycnidia; the sexual stage of *Z. fragariae* is a species of *Gnomonia* (a member of the Sphaeriales).

Appendixes

Abbreviations used in the Appendixes

ADP	Adenosine-5′-diphosphate
AMP	Adenosine-5′-monophosphate
ATP	Adenosine-5′-triphosphate
CoASH	Coenzyme A
e	Electron
FAD	Flavin adenine dinucleotide
GDP	Guanosine-5′-diphosphate
GTP	Guanosine-5′-triphosphate
NAD	Nicotinamide adenine dinucleotide
NADP	Nicotinamide adenine dinucleotide phosphate
(P)	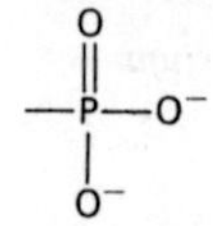
Pi	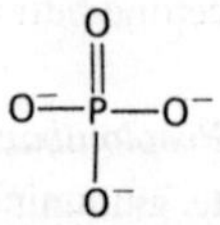 (inorganic phosphate)
PPi	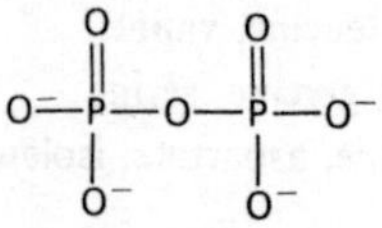 (inorganic pyrophosphate)
TPP	Thiamine pyrophosphate

THE EMBDEN–MEYERHOF–PARNAS PATHWAY

(see entries: EMBDEN–MEYERHOF–PARNAS PATHWAY; GLUCONEOGENESIS; GLYCOGEN)

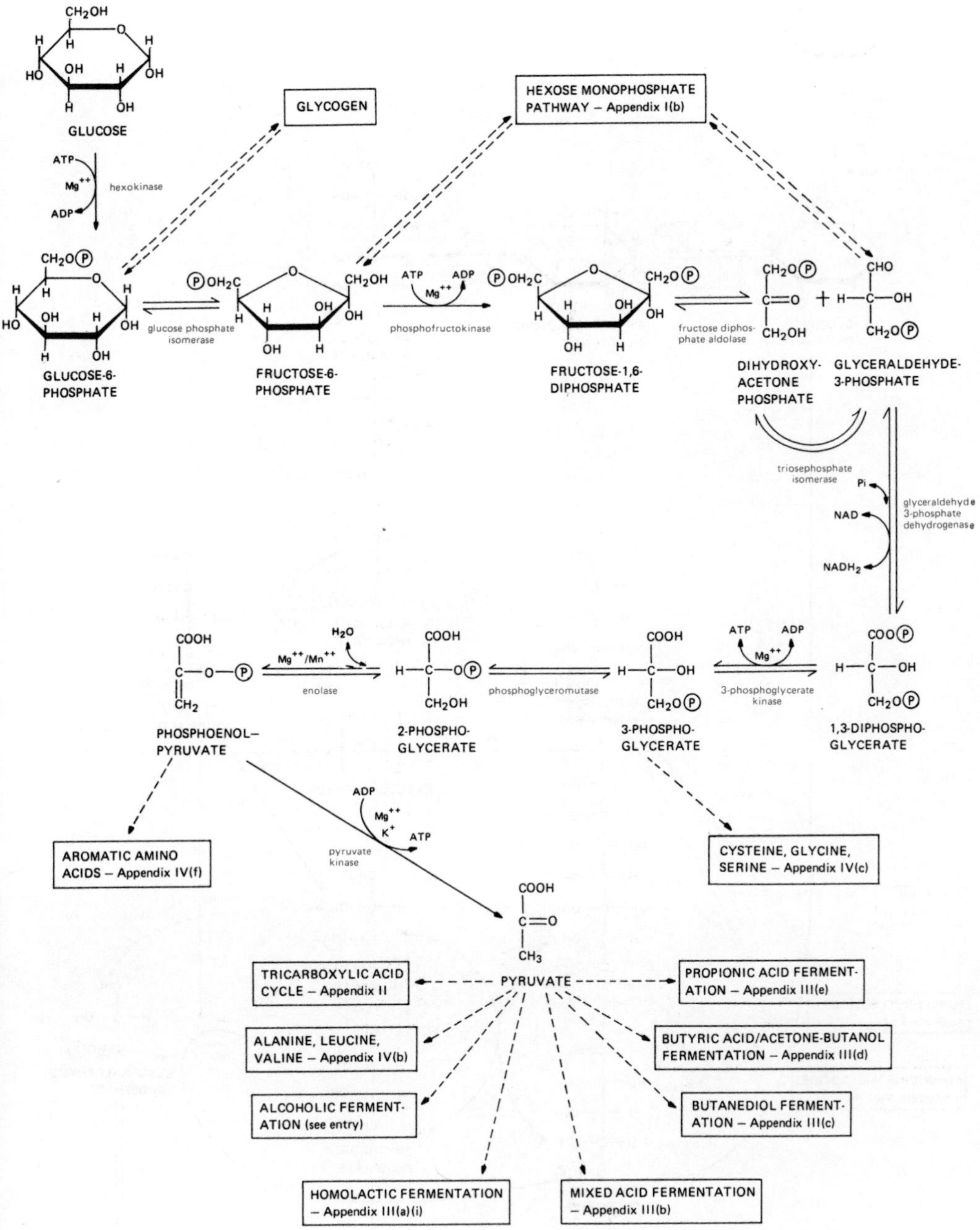

Appendix I(b)

THE HEXOSE MONOPHOSPHATE PATHWAY

(see entry: HEXOSE MONOPHOSPHATE PATHWAY)

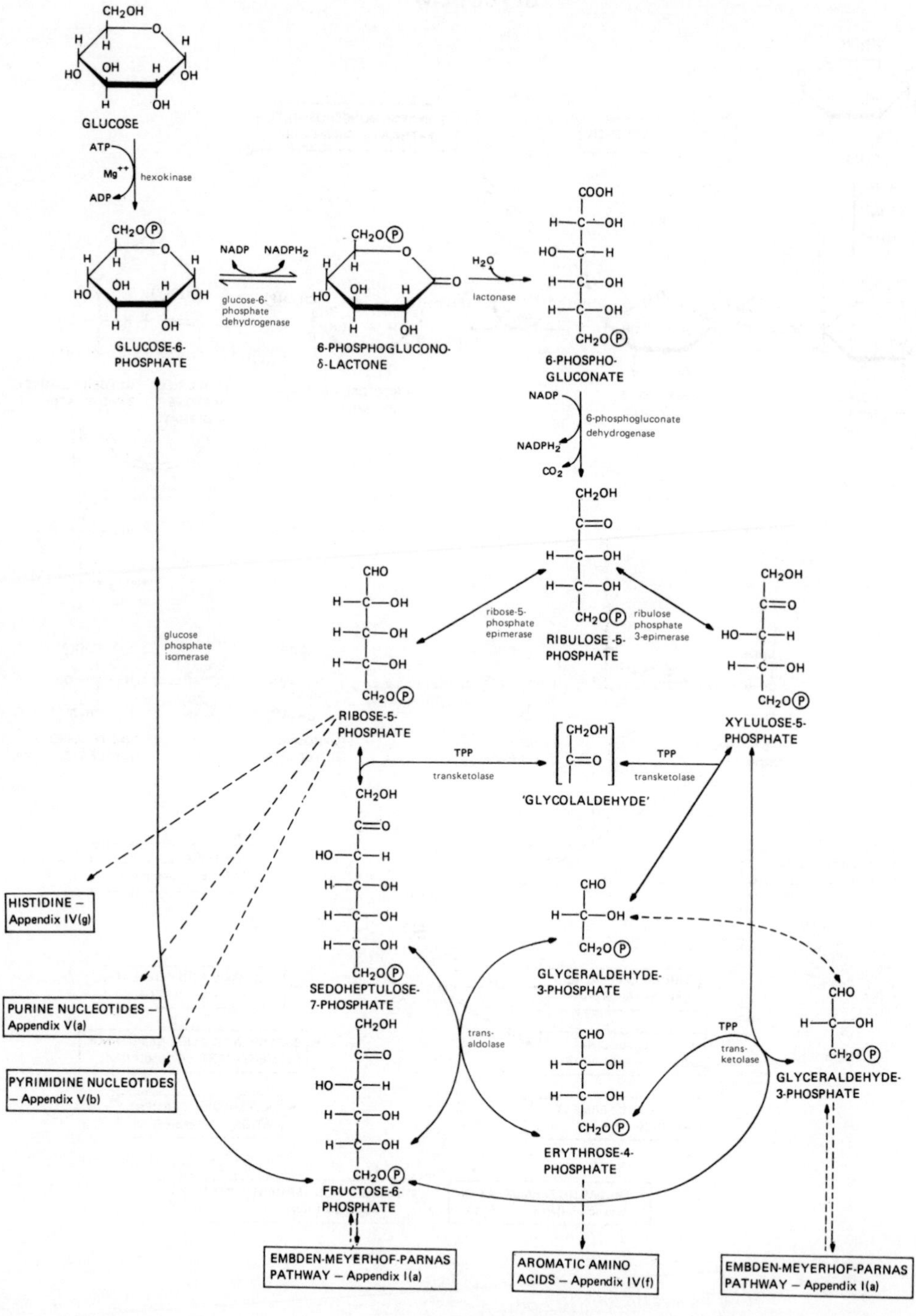

THE ENTNER–DOUDOROFF PATHWAY

(see entry: ENTNER–DOUDOROFF PATHWAY)

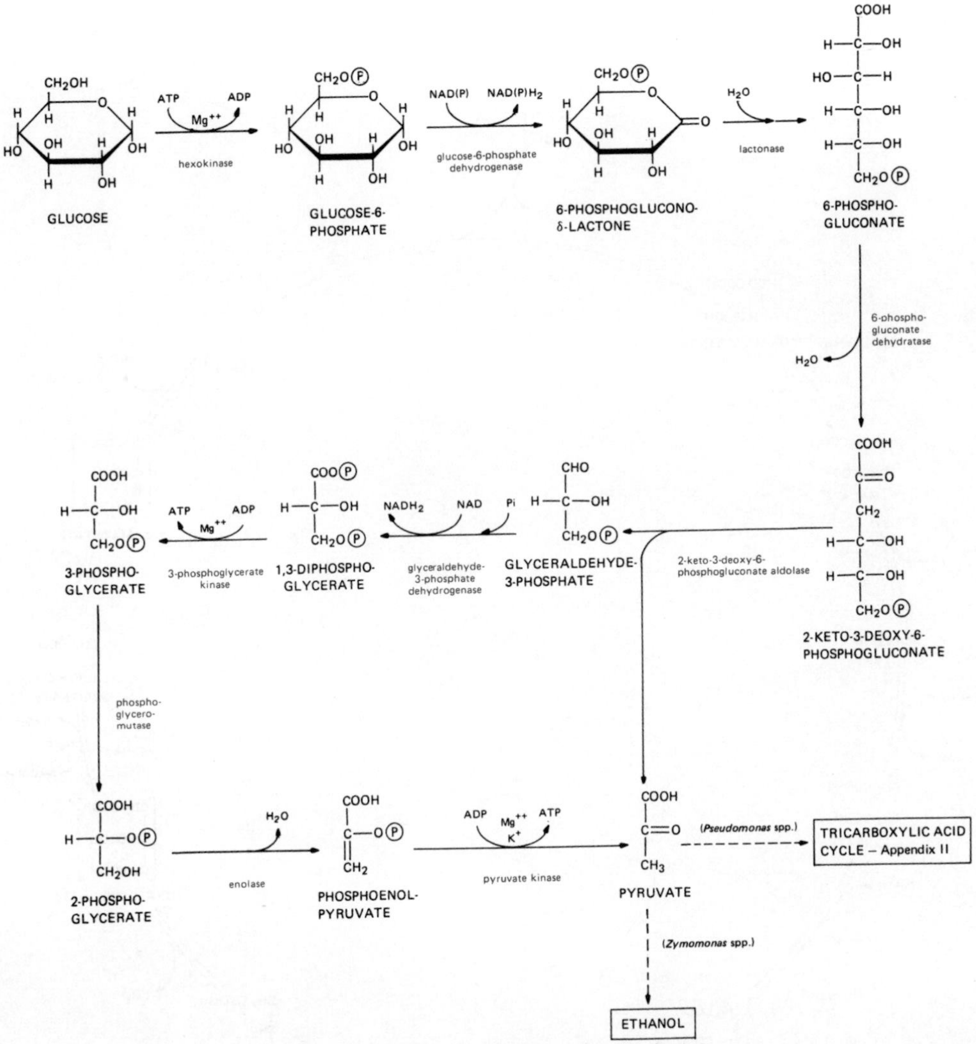

Appendix II(a)

THE TRICARBOXYLIC ACID CYCLE

(see entry: TRICARBOXYLIC ACID CYCLE)

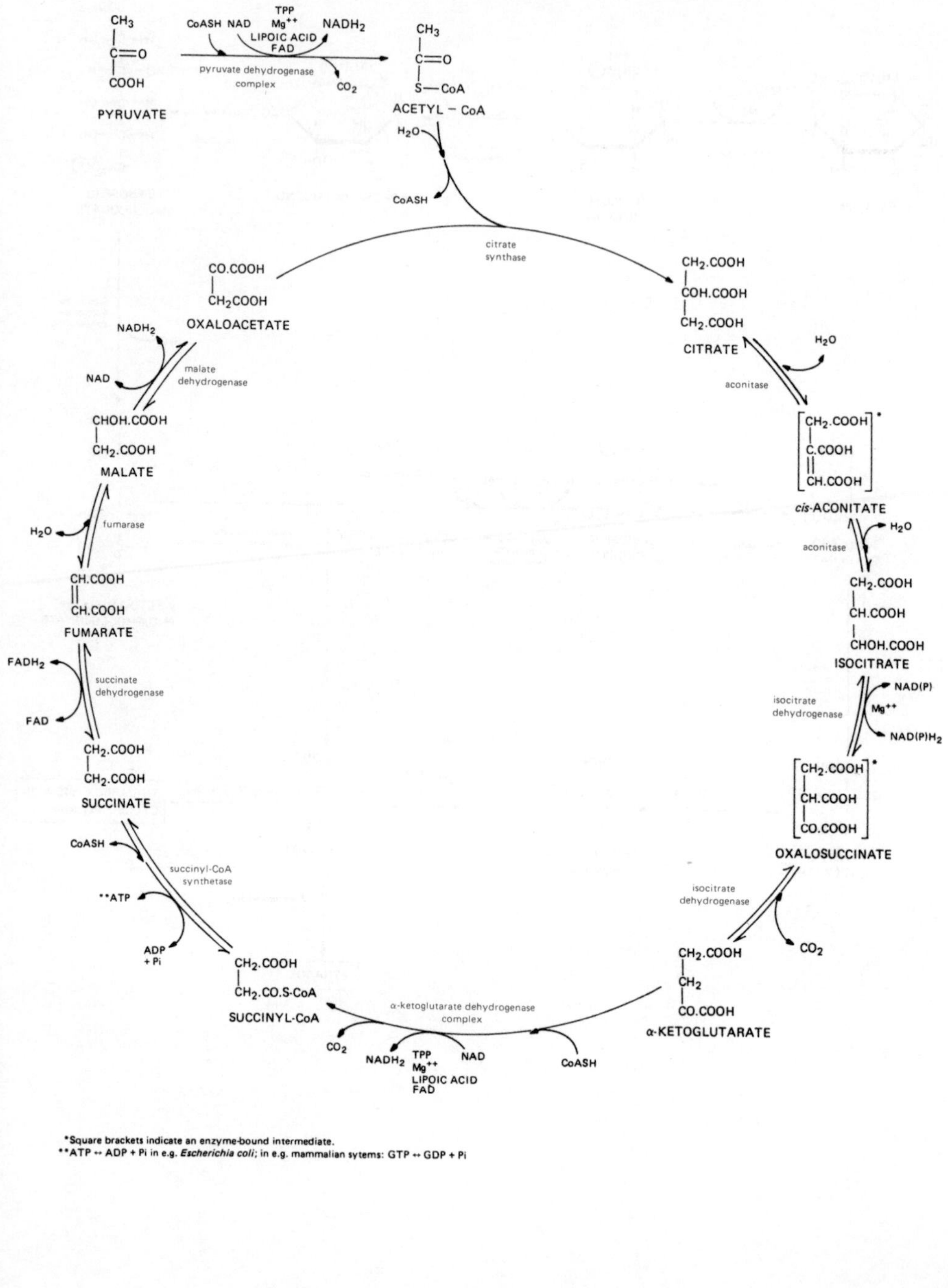

*Square brackets indicate an enzyme-bound intermediate.

**ATP ↔ ADP + Pi in e.g. *Escherichia coli*; in e.g. mammalian sytems: GTP ↔ GDP + Pi

THE TRICARBOXYLIC ACID CYCLE—*showing Anaplerotic Sequences and some Catabolic and Anabolic Interactions*

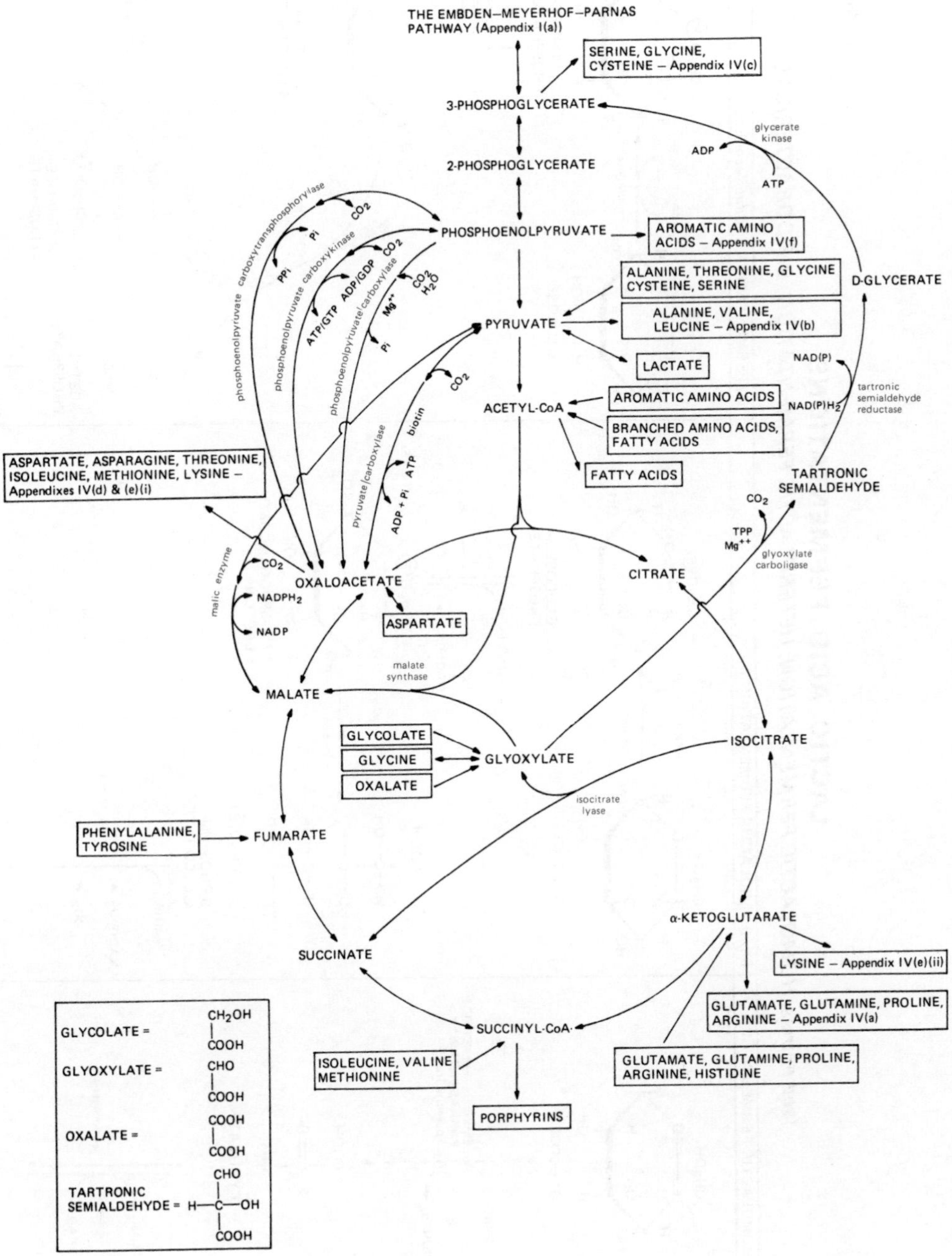

LACTIC ACID FERMENTATIONS

(see entries: HOMOLACTIC FERMENTATION; HETEROLACTIC FERMENTATION; BIFIDOBACTERIUM)

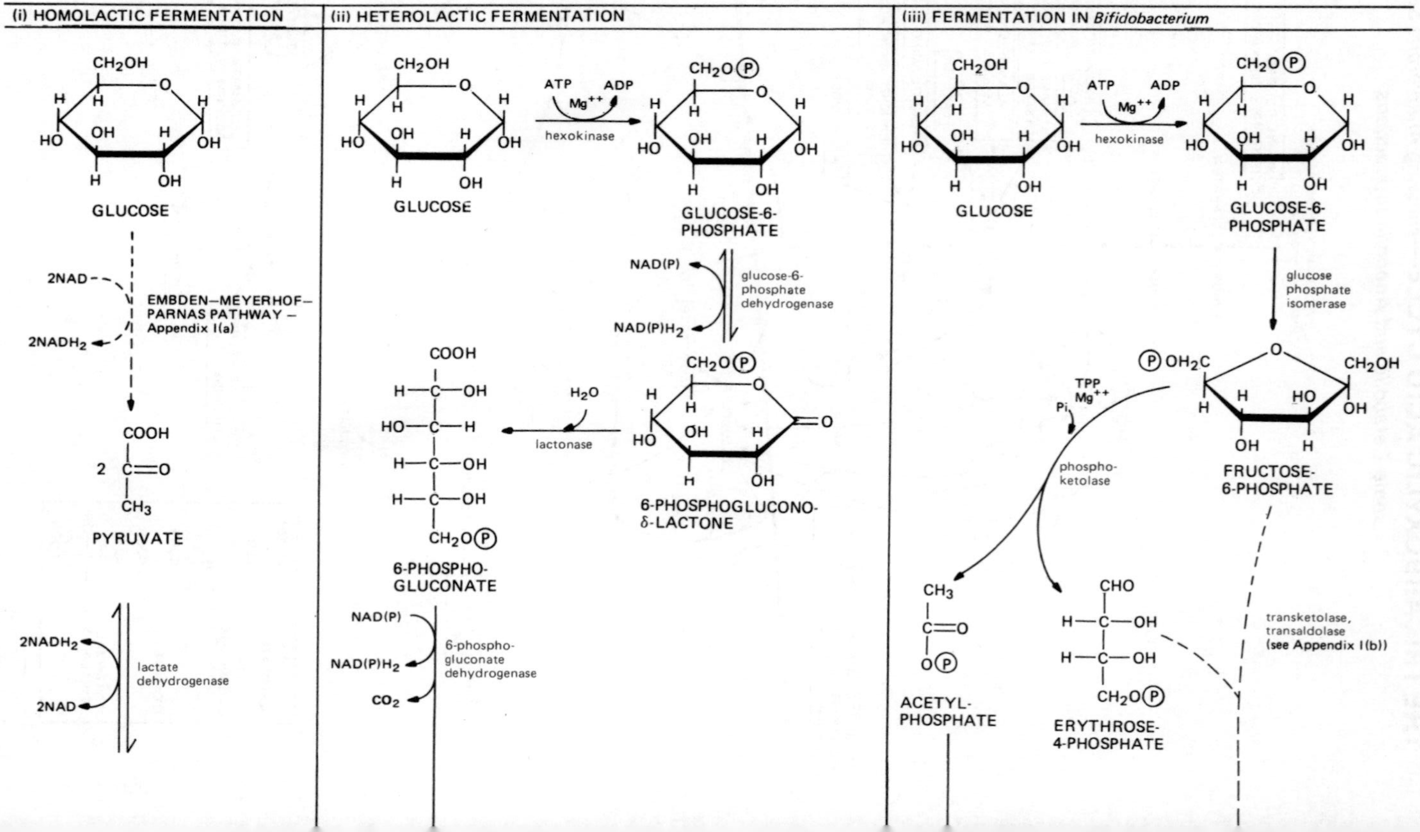

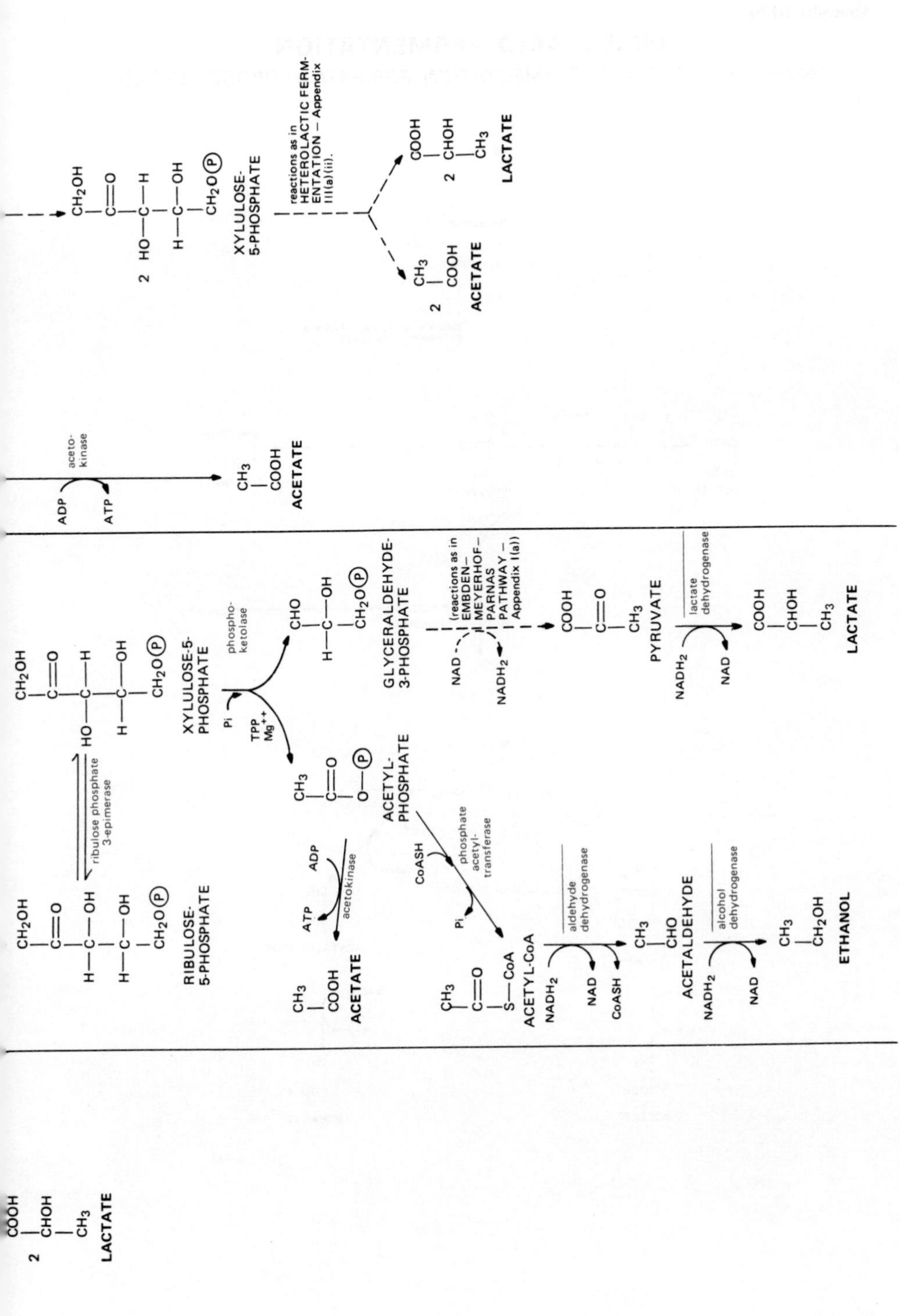
XYLULOSE-5-PHOSPHATE
reactions as in HETEROLACTIC FERMENTATION – Appendix III(a)(ii).
ACETATE
LACTATE
aceto-kinase
ADP
ATP
ACETATE
RIBULOSE-5-PHOSPHATE
ribulose phosphate 3-epimerase
XYLULOSE-5-PHOSPHATE
phospho-ketolase
Pi
TPP
Mg++
GLYCERALDEHYDE-3-PHOSPHATE
(reactions as in EMBDEN–MEYERHOF–PARNAS PATHWAY – Appendix I(a))
NAD
NADH2
PYRUVATE
lactate dehydrogenase
LACTATE
ACETYL-PHOSPHATE
acetokinase
ACETATE
CoASH
phosphate acetyl-transferase
ACETYL-CoA
aldehyde dehydrogenase
ACETALDEHYDE
alcohol dehydrogenase
ETHANOL
LACTATE

Appendix III(b)

MIXED ACID FERMENTATION

(see entries: MIXED ACID FERMENTATION; FORMATE HYDROGEN LYASE)

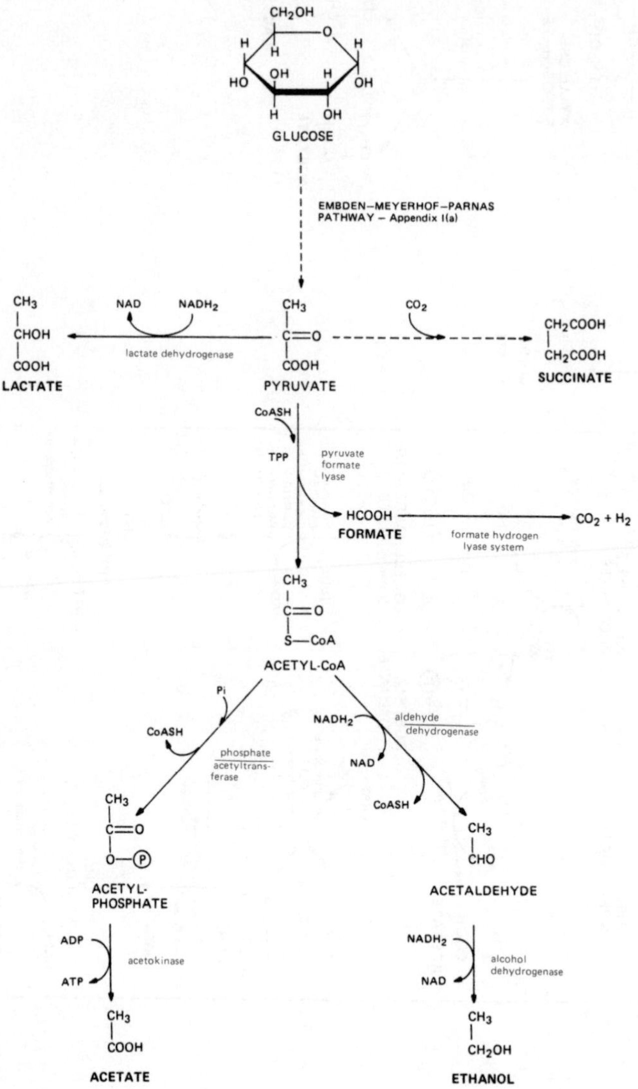

BUTANEDIOL FERMENTATION

(see entries: BUTANEDIOL FERMENTATION; FORMATE HYDROGEN LYASE; VOGES-PROSKAUER TEST)

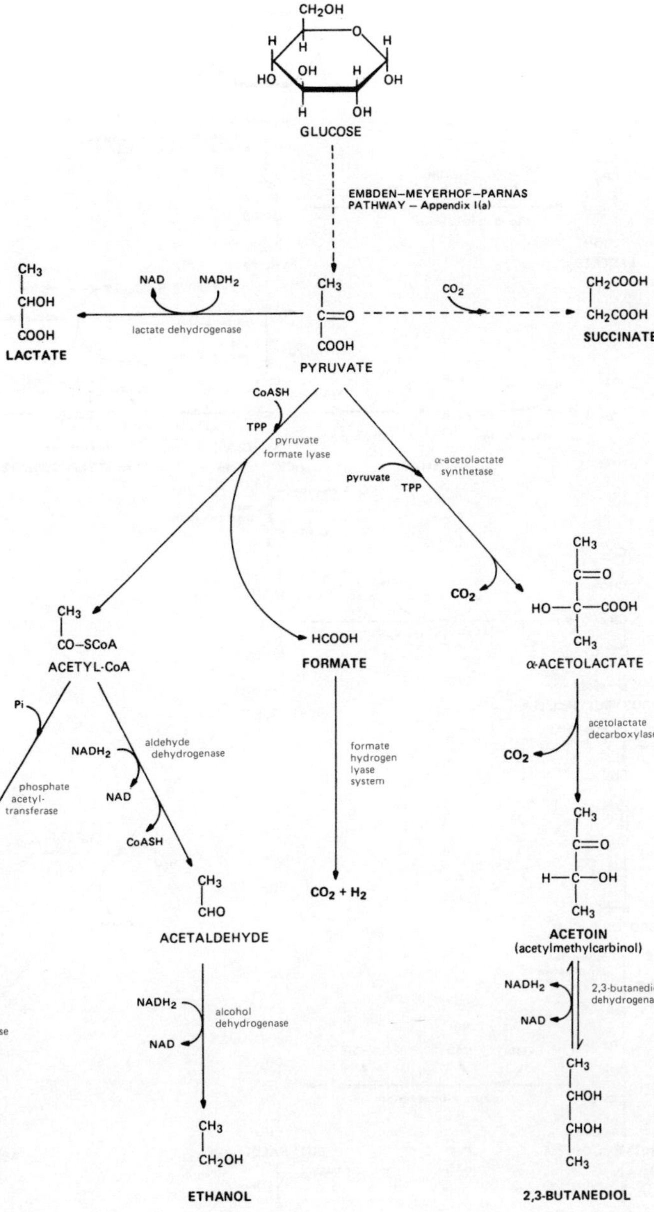

Appendix III(d)

BUTYRIC ACID FERMENTATION AND ACETONE-BUTANOL FERMENTATION

(see entries: BUTYRIC ACID FERMENTATION; ACETONE-BUTANOL FERMENTATION; HYDROGENASE; FERREDOXIN)

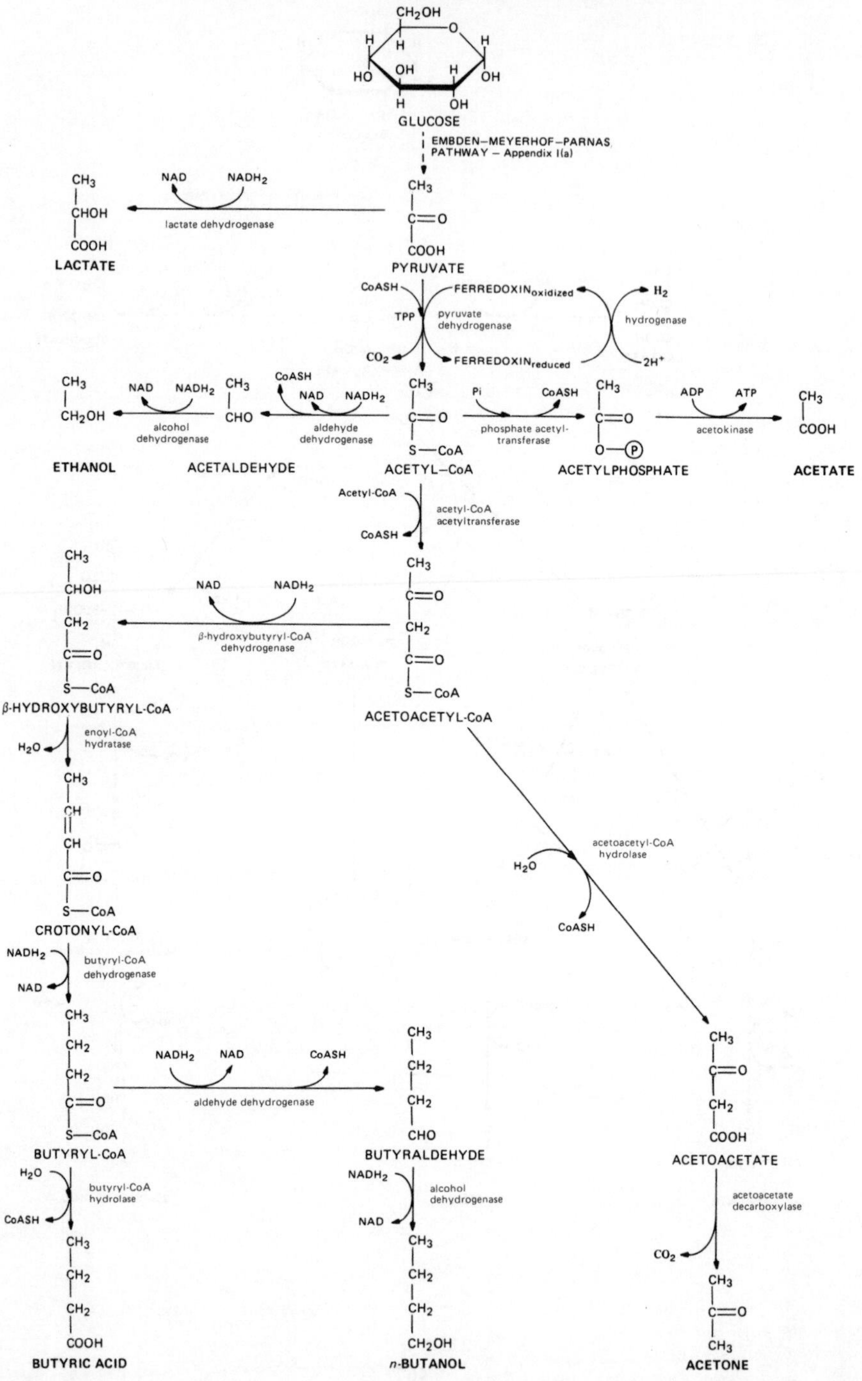

PROPIONIC ACID FERMENTATION
(in *Propionibacterium*)
(see entry: PROPIONIC ACID FERMENTATION)

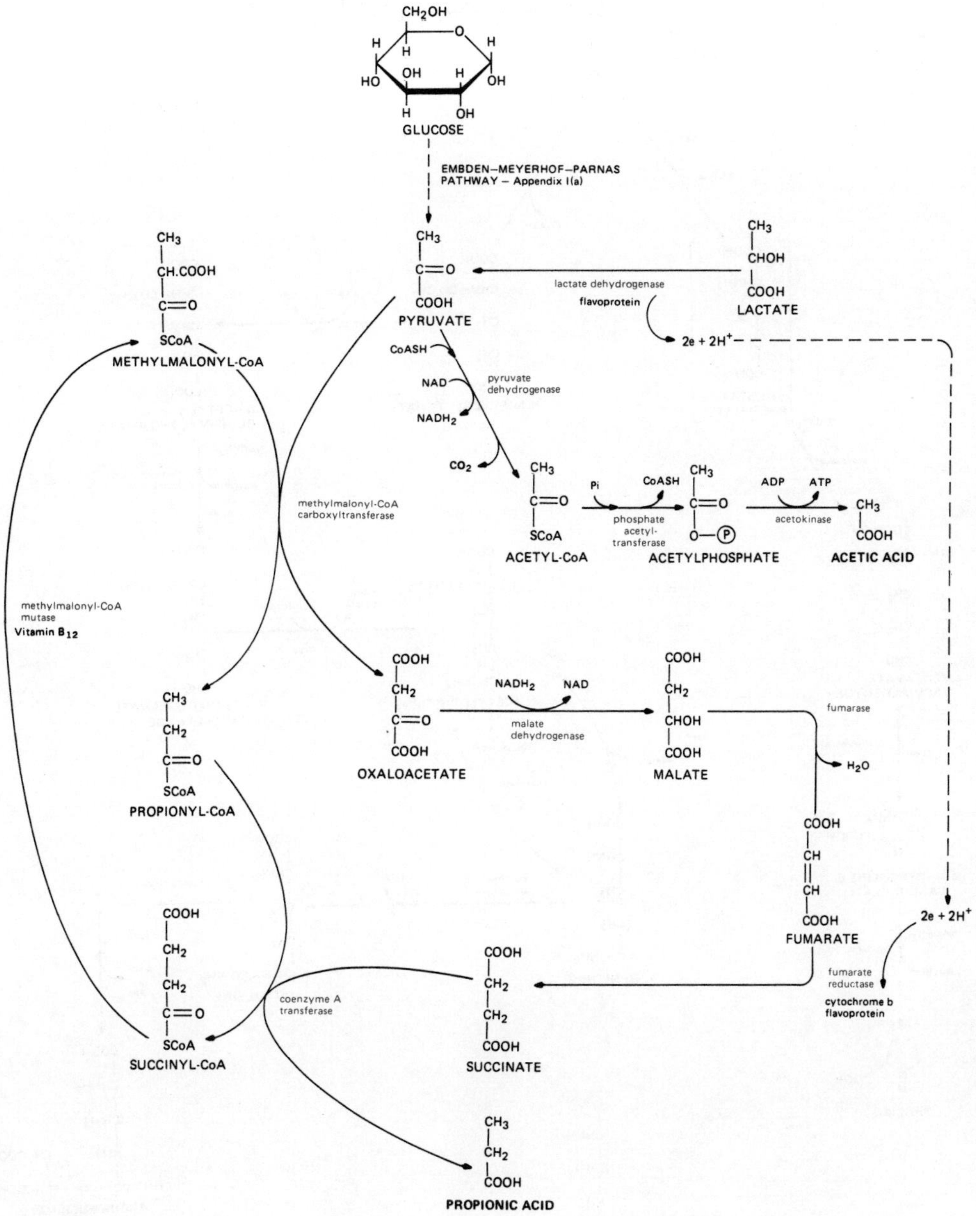

Appendix IV(a)

BIOSYNTHESIS OF ARGININE, GLUTAMATE, GLUTAMINE, PROLINE

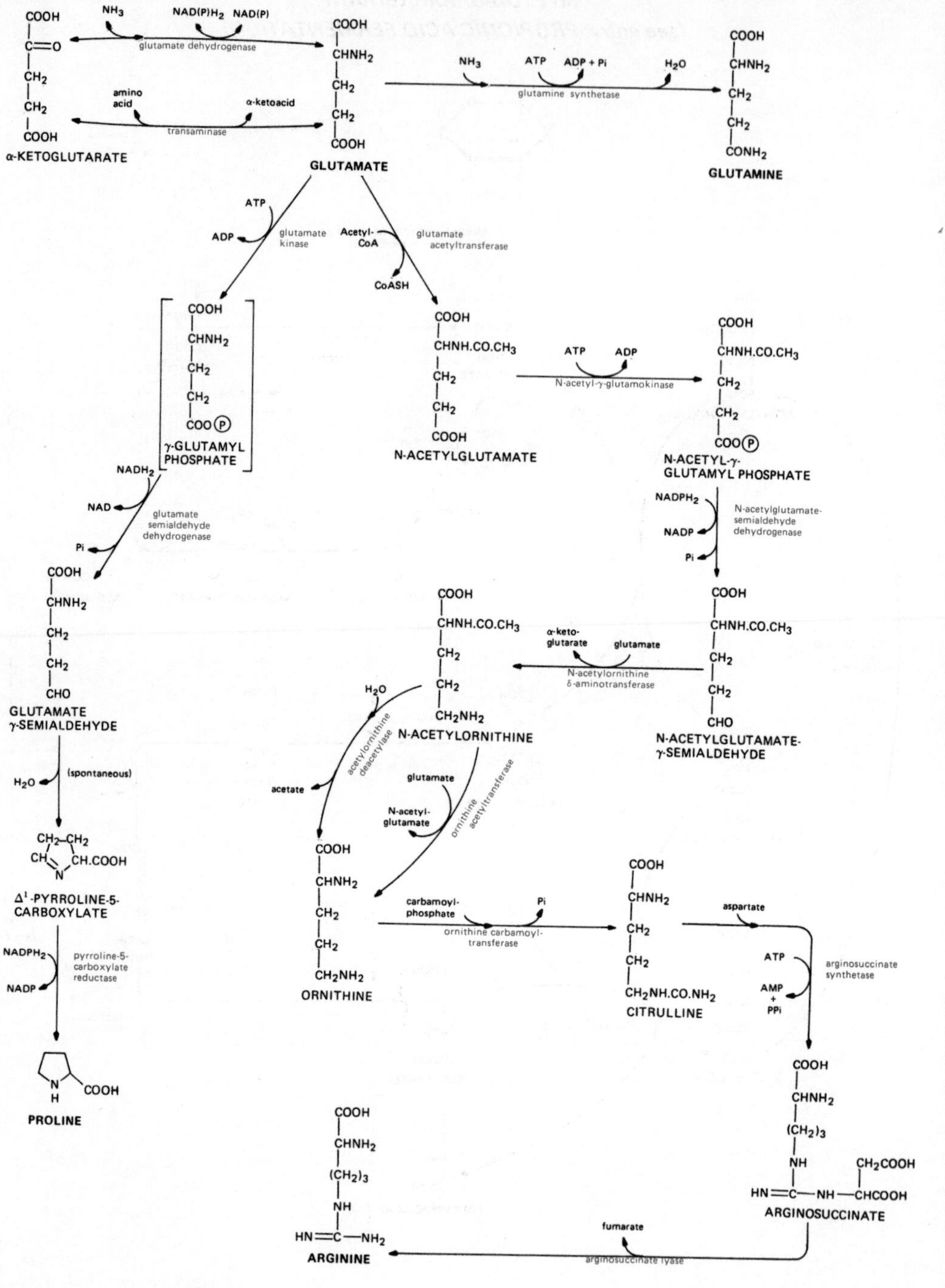

BIOSYNTHESIS OF ALANINE, LEUCINE, VALINE

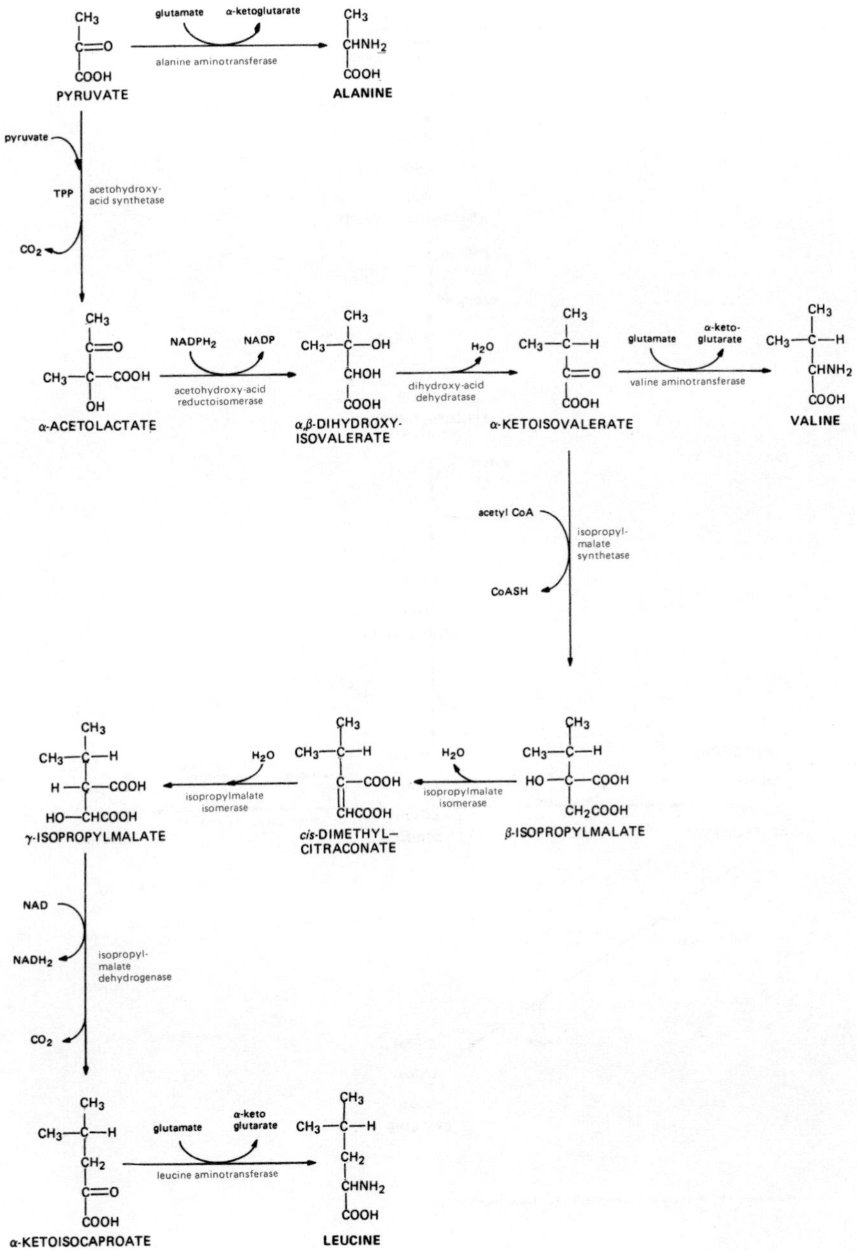

Appendix IV(c)

BIOSYNTHESIS OF CYSTEINE, GLYCINE, SERINE

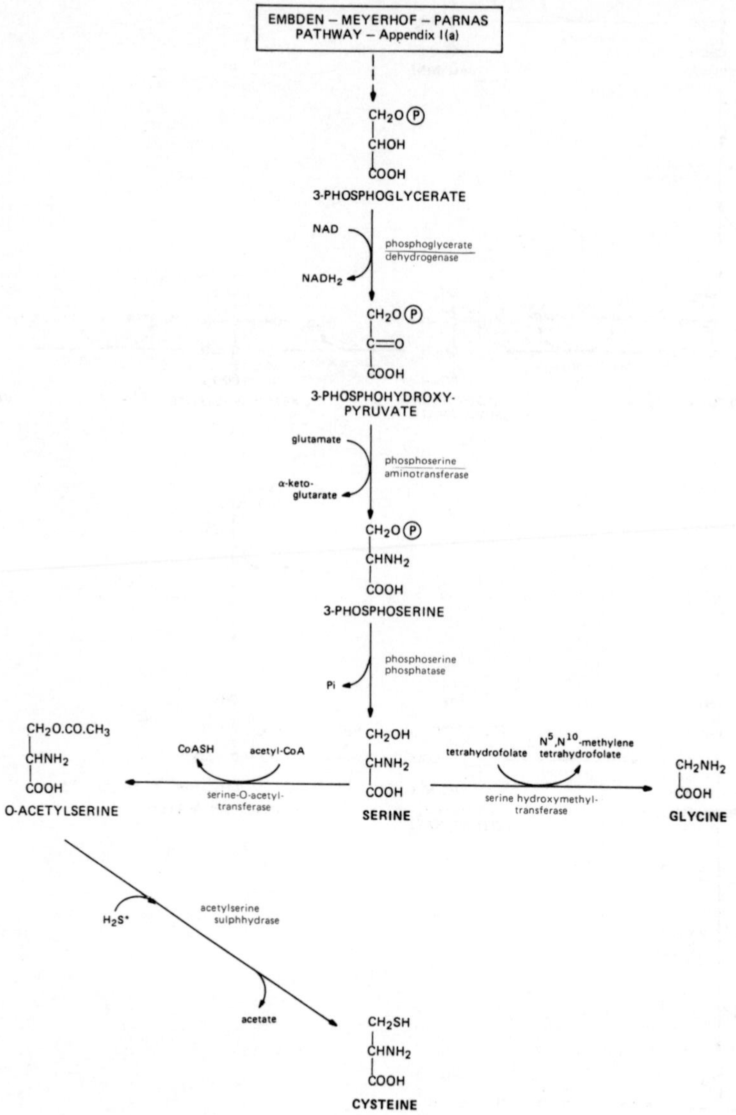

*Formed by ASSIMILATORY SULPHATE REDUCTION (see entry).

BIOSYNTHESIS OF ASPARAGINE, ASPARTATE, ISOLEUCINE, METHIONINE, THREONINE

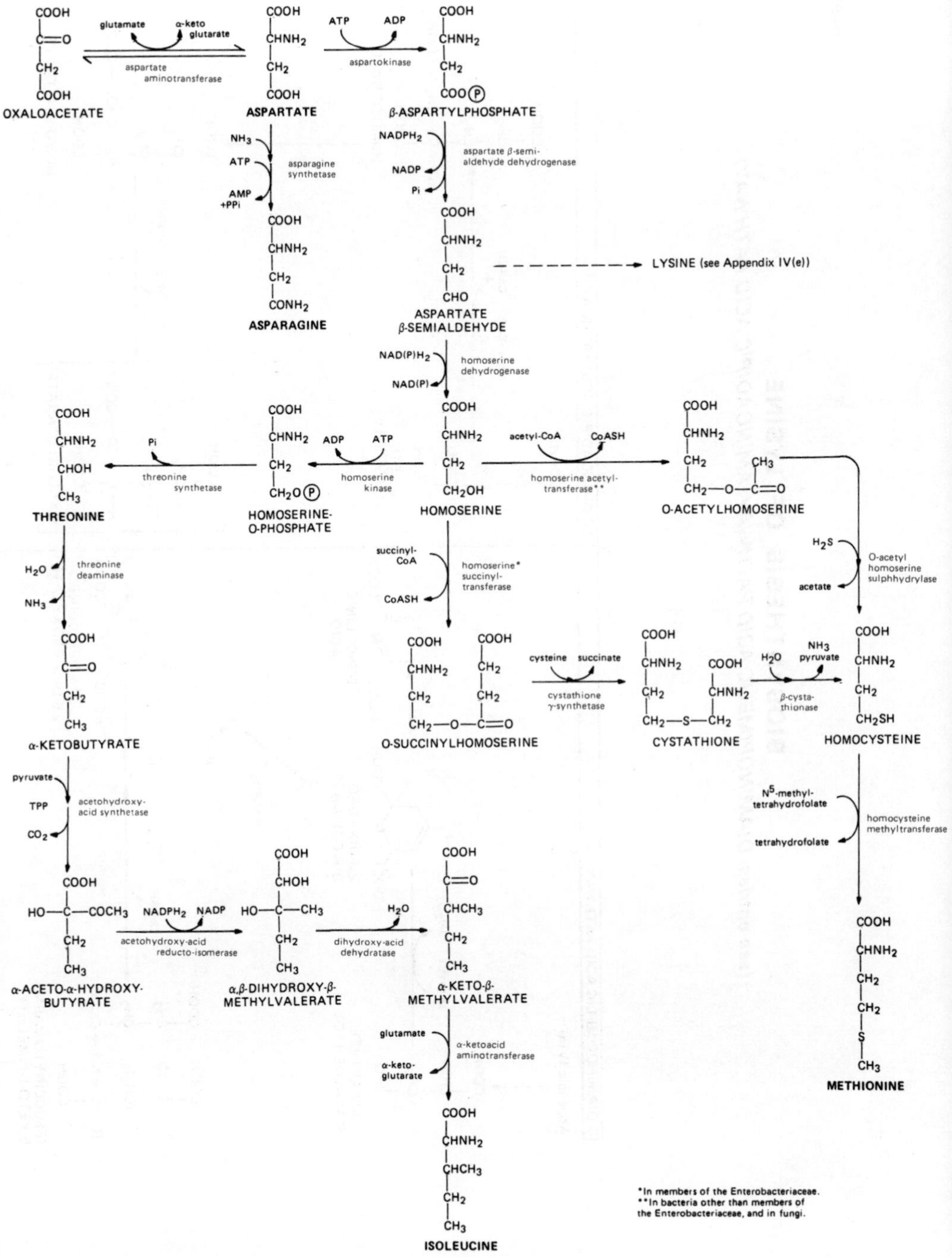

*In members of the Enterobacteriaceae.
**In bacteria other than members of the Enterobacteriaceae, and in fungi.

BIOSYNTHESIS OF LYSINE

(see entries: DIAMINOPIMELIC ACID PATHWAY; AMINOADIPIC ACID PATHWAY)

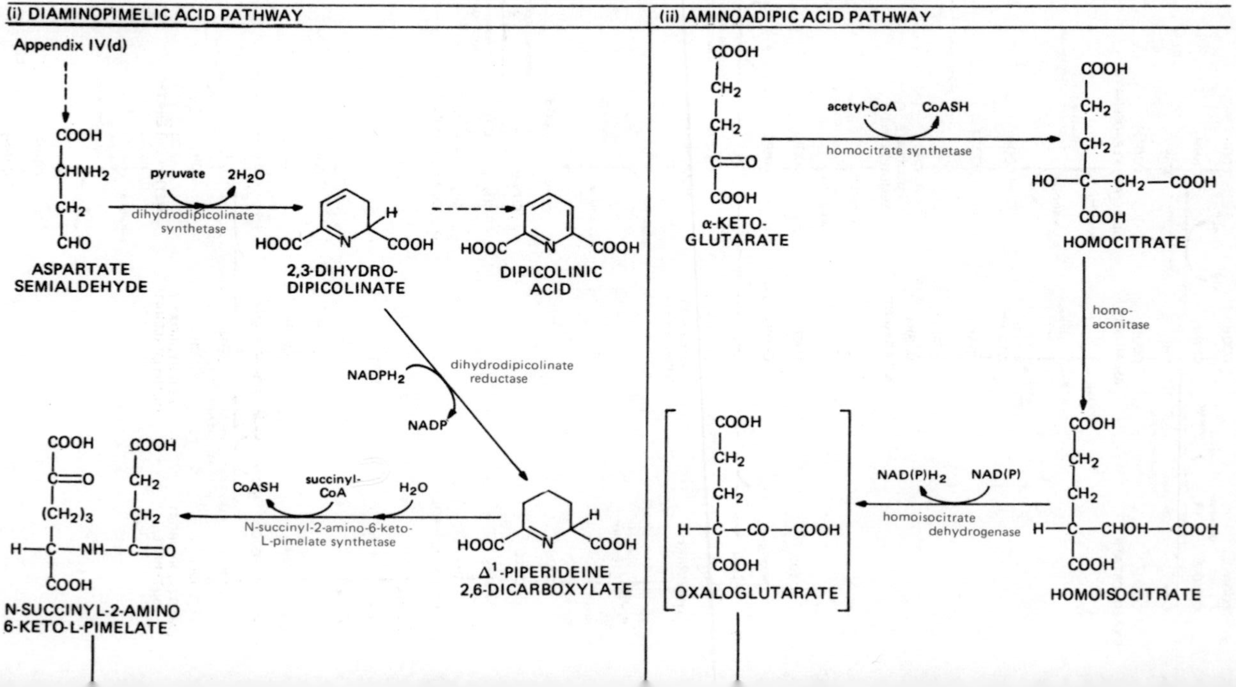

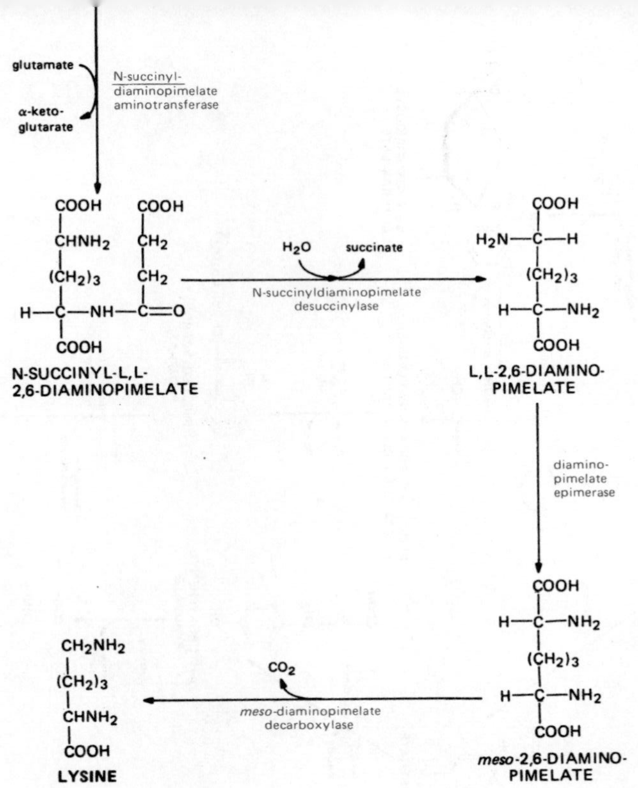
glutamate
α-keto-glutarate
N-succinyl-diaminopimelate aminotransferase
COOH
CHNH2
(CH2)3
H—C—NH—C=O
COOH
COOH
CH2
CH2
N-SUCCINYL-L,L-2,6-DIAMINOPIMELATE
H2O
succinate
N-succinyldiaminopimelate desuccinylase
COOH
H2N—C—H
(CH2)3
H—C—NH2
COOH
L,L-2,6-DIAMINO-PIMELATE
diamino-pimelate epimerase
COOH
H—C—NH2
(CH2)3
H—C—NH2
COOH
meso-2,6-DIAMINO-PIMELATE
CO2
meso-diaminopimelate decarboxylase
CH2NH2
(CH2)3
CHNH2
COOH
LYSINE

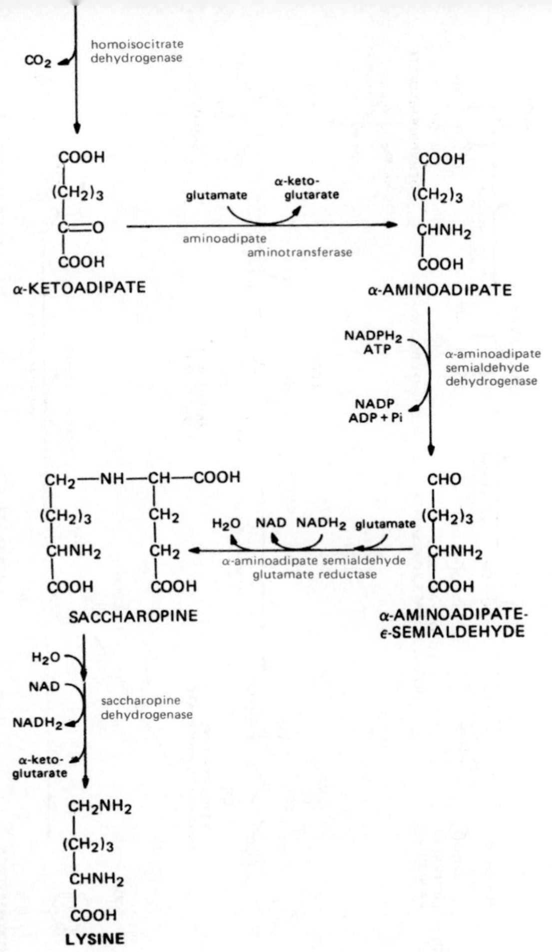
CO2
homoisocitrate dehydrogenase
COOH
(CH2)3
C=O
COOH
α-KETOADIPATE
glutamate
α-keto-glutarate
aminoadipate aminotransferase
COOH
(CH2)3
CHNH2
COOH
α-AMINOADIPATE
NADPH2
ATP
NADP
ADP + Pi
α-aminoadipate semialdehyde dehydrogenase
CHO
(CH2)3
CHNH2
COOH
α-AMINOADIPATE-ϵ-SEMIALDEHYDE
H2O
NAD
NADH2
glutamate
α-aminoadipate semialdehyde glutamate reductase
CH2—NH—CH—COOH
(CH2)3
CHNH2
COOH
CH2
CH2
COOH
SACCHAROPINE
H2O
NAD
NADH2
α-keto-glutarate
saccharopine dehydrogenase
CH2NH2
(CH2)3
CHNH2
COOH
LYSINE

Appendix IV(f)

BIOSYNTHESIS OF AROMATIC AMINO ACIDS

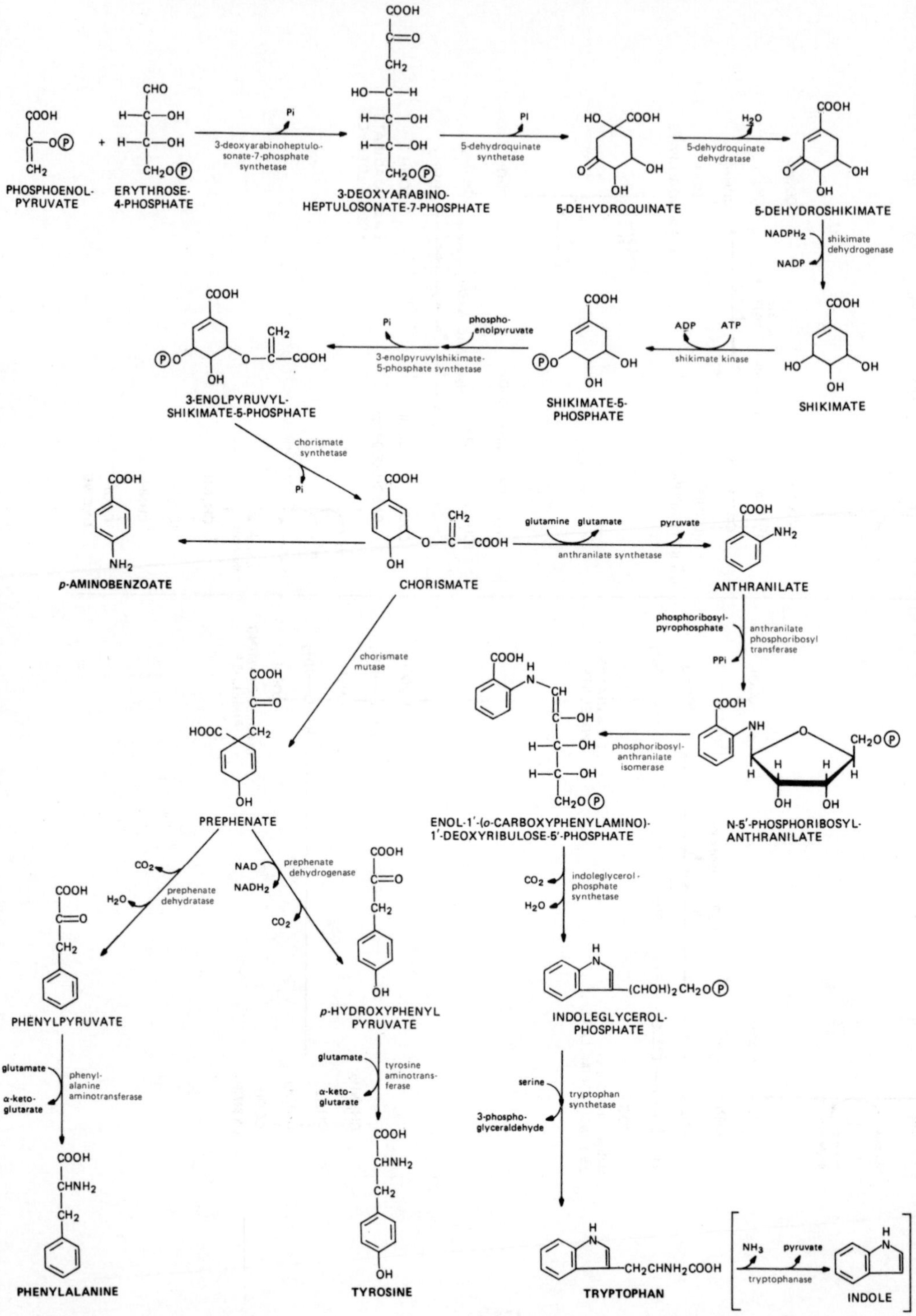

BIOSYNTHESIS OF HISTIDINE

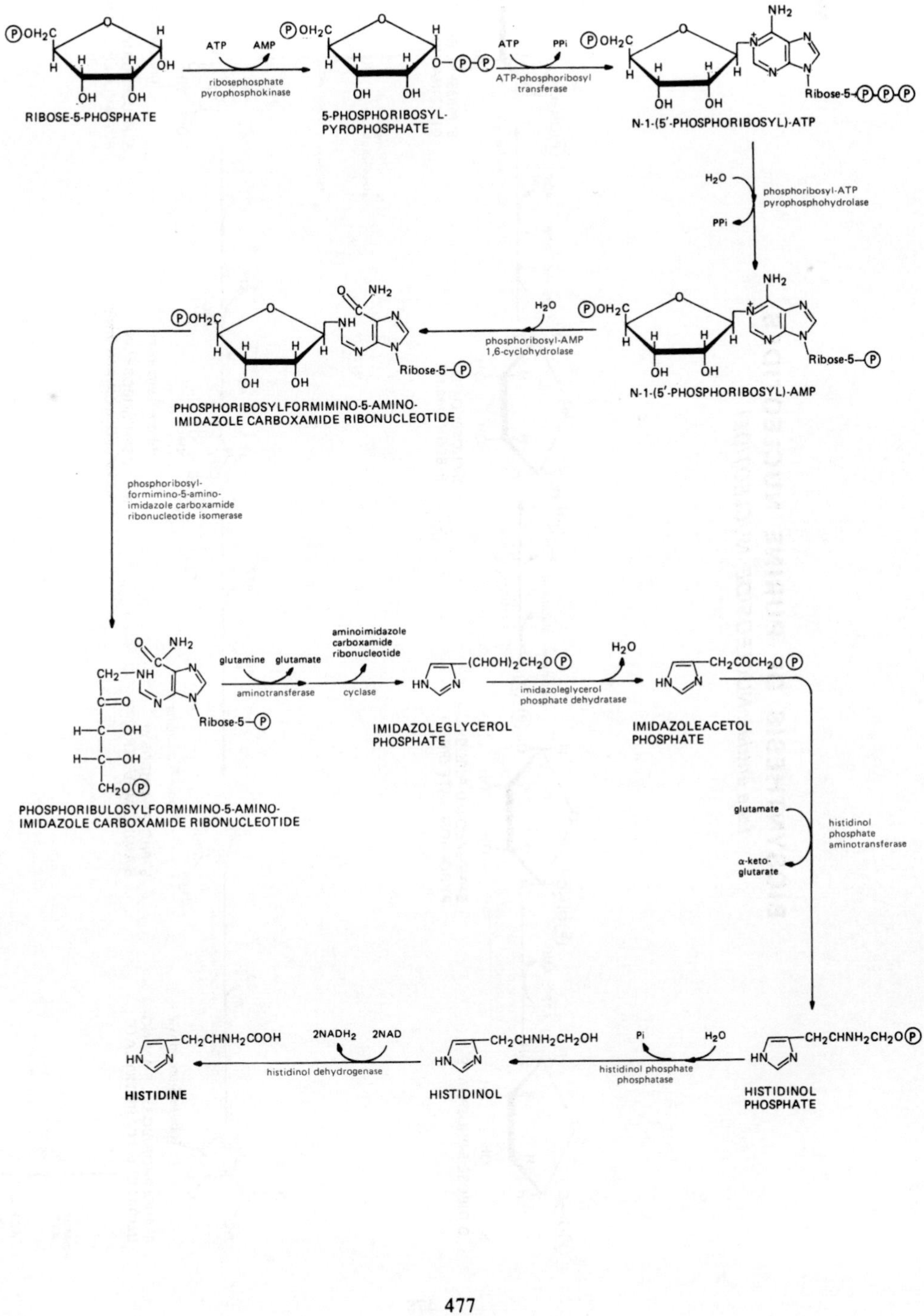

BIOSYNTHESIS OF PURINE NUCLEOTIDES

(see entries: NUCLEOSIDE; NUCLEOTIDE)

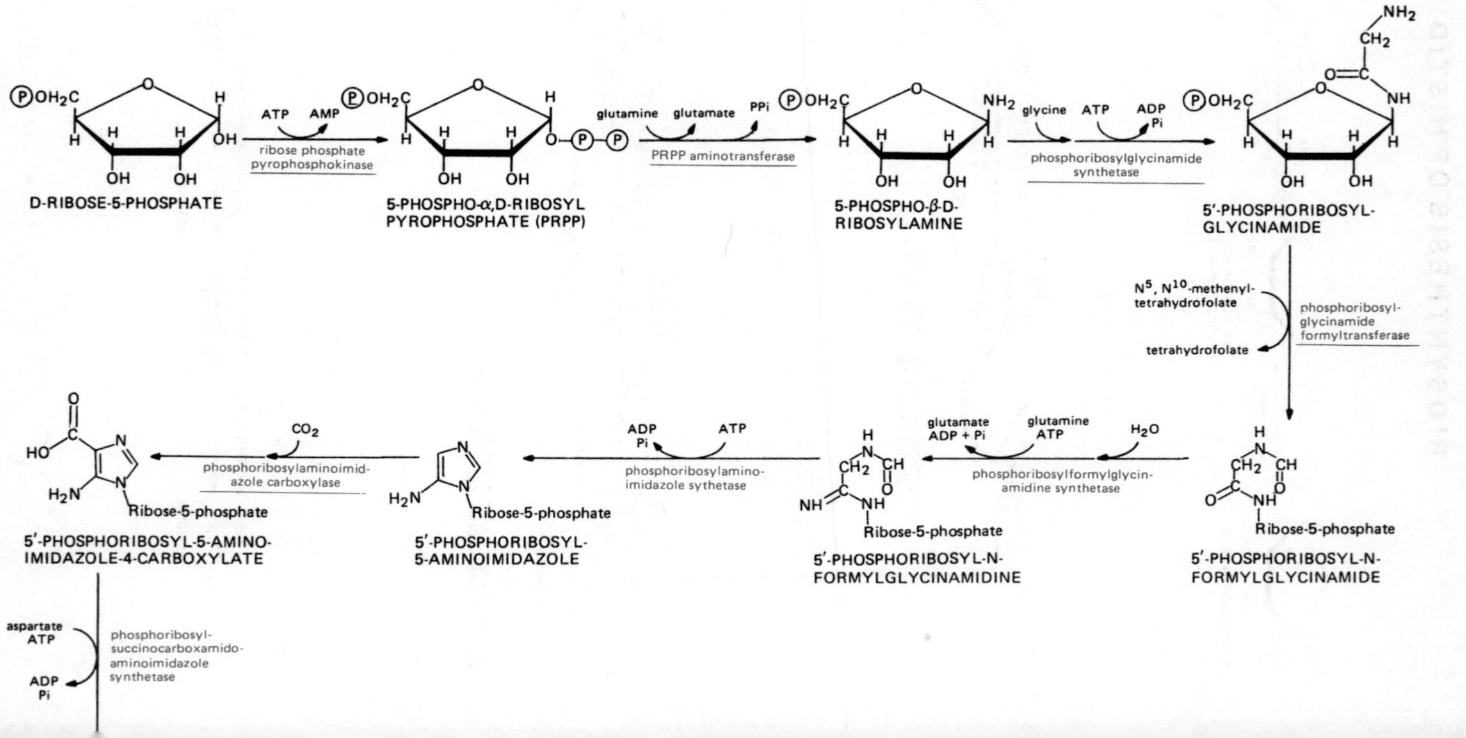

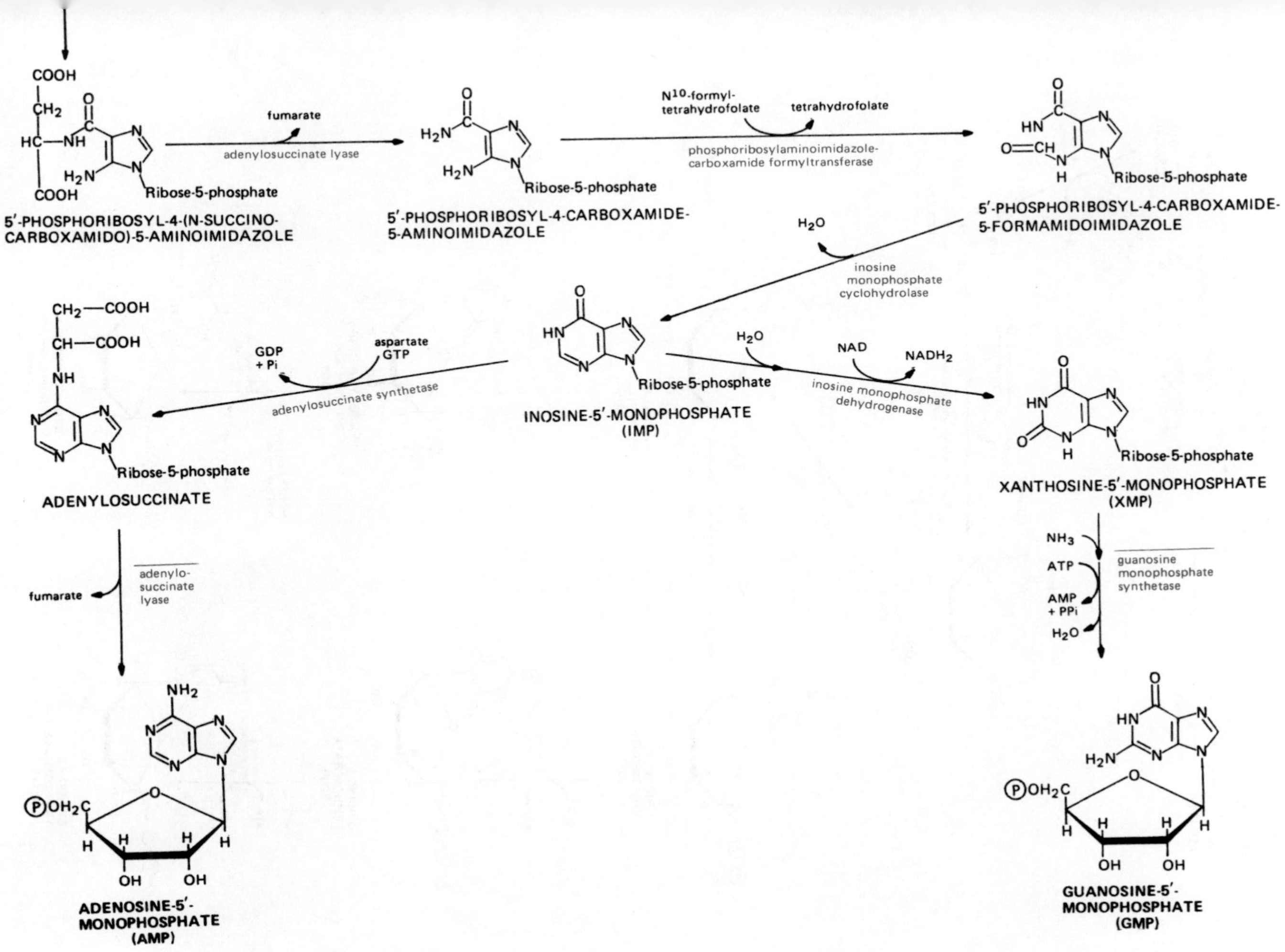
fumarate
adenylosuccinate lyase
N10-formyl-tetrahydrofolate
tetrahydrofolate
phosphoribosylaminoimidazole-carboxamide formyltransferase
Ribose-5-phosphate
5′-PHOSPHORIBOSYL-4-(N-SUCCINO-CARBOXAMIDO)-5-AMINOIMIDAZOLE
5′-PHOSPHORIBOSYL-4-CARBOXAMIDE-5-AMINOIMIDAZOLE
5′-PHOSPHORIBOSYL-4-CARBOXAMIDE-5-FORMAMIDOIMIDAZOLE
H2O
inosine monophosphate cyclohydrolase
GDP + Pi
aspartate GTP
adenylosuccinate synthetase
ADENYLOSUCCINATE
INOSINE-5′-MONOPHOSPHATE (IMP)
NAD
NADH2
inosine monophosphate dehydrogenase
XANTHOSINE-5′-MONOPHOSPHATE (XMP)
adenylo-succinate lyase
NH3
ATP
AMP + PPi
guanosine monophosphate synthetase
ADENOSINE-5′-MONOPHOSPHATE (AMP)
GUANOSINE-5′-MONOPHOSPHATE (GMP)

Appendix V(b)

BIOSYNTHESIS OF PYRIMIDINE NUCLEOTIDES

(see entries: NUCLEOSIDE: NUCLEOTIDE)

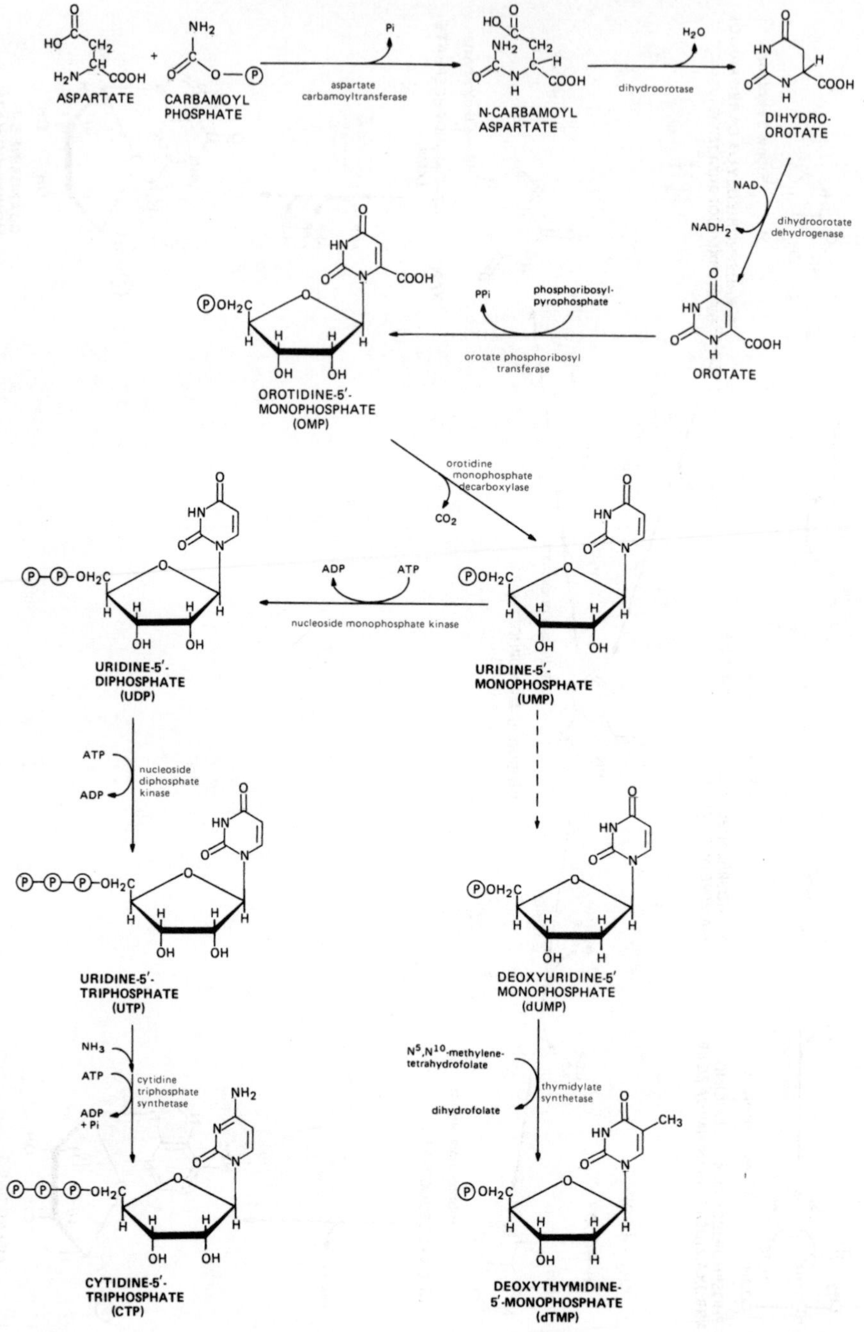

THE STRUCTURES OF SOME NUCLEOTIDES

(see entry: NUCLEOTIDE)

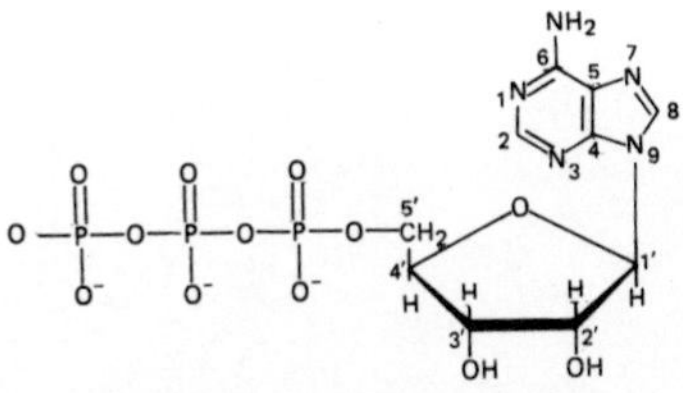

ADENOSINE-5'-TRIPHOSPHATE
(ATP)
(see entry: ATP)

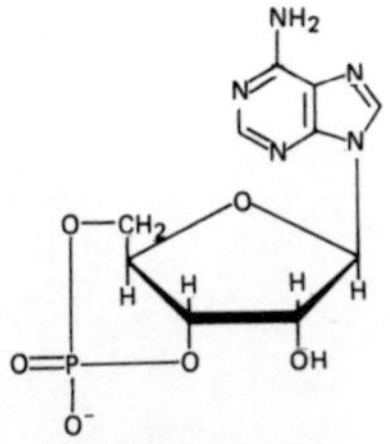

3',5'-CYCLIC ADENOSINE MONOPHOSPHATE
(cAMP)

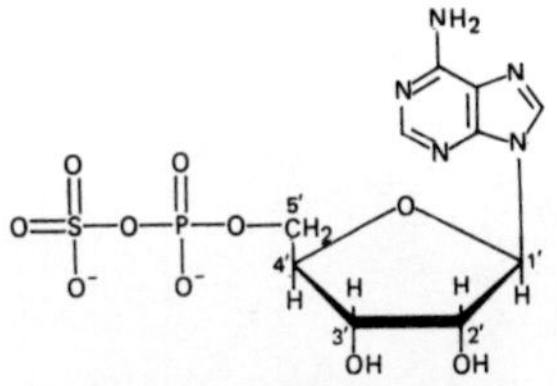

ADENOSINE-5'-PHOSPHOSULPHATE
(APS)
(PAPS = 3'-PHOSPHO-APS)
(see entry: ASSIMILATORY SULPHATE REDUCTION)

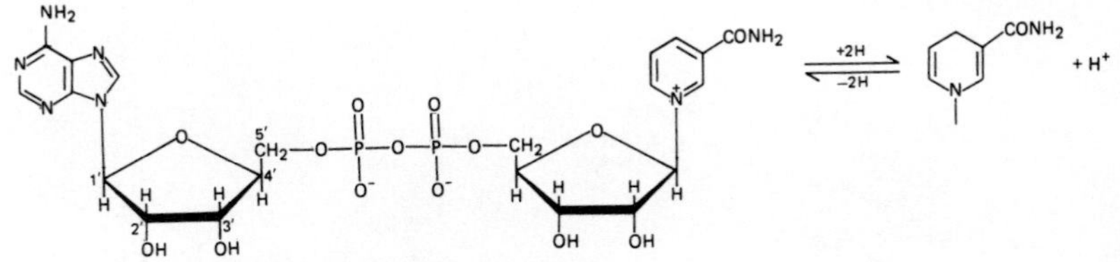

NICOTINAMIDE ADENINE DINUCLEOTIDE (NAD)
(NADP = NAD-2'-PHOSPHATE)
(see entries: NAD; NADP; NICOTINIC ACID)

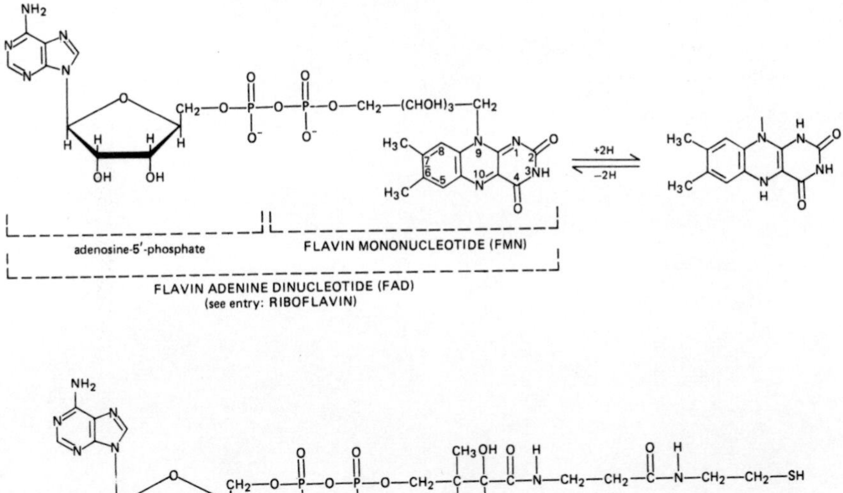

FLAVIN ADENINE DINUCLEOTIDE (FAD)
(see entry: RIBOFLAVIN)

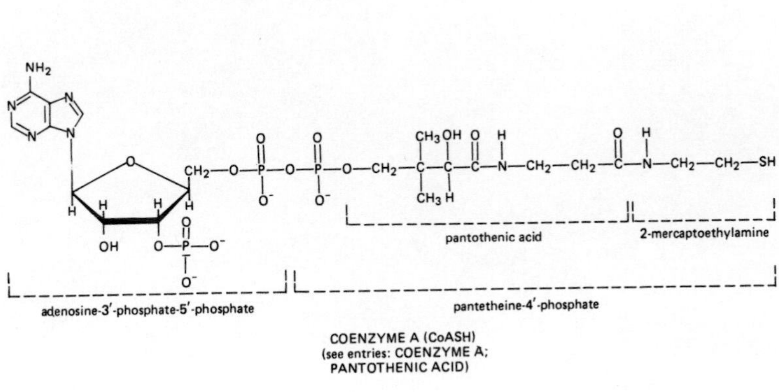

COENZYME A (CoASH)
(see entries: COENZYME A;
PANTOTHENIC ACID)